Lens Design for Imaging

Lens Design for Imaging

Fundamentals of Optical Systems

Volume 1

Herbert Gross

WILEY-VCH

Author

Prof. Dr. Herbert Gross
Luise-Hartmann-Str 5.
73430 Aalen
Germany

Cover Images: © loops7/Getty Images,
Courtesy of Herbert Gross, Aalen,
Germany

Typesetting: Straive, Chennai, India

Printing: CPI Group(UK) Ltd, Croydon CR0 4YY

Library of Congress Card No.: applied for

British Library Cataloguing-in-Publication Data A catalogue record for this book is available from the British Library.

Bibliographic information published by the Deutsche Nationalbibliothek
The Deutsche Nationalbibliothek lists this publication in the Deutsche Nationalbibliografie; detailed bibliographic data are available on the Internet at <http://dnb.d-nb.de>.

Print ISBN: 978-3-527-41456-7
ePDF ISBN: 978-3-527-35498-6
ePub ISBN: 978-3-527-35497-9
oBook ISBN: 978-3-527-35499-3

C9783527414567_290426

To my daughters, Alexandra and Katja, who are always understanding and have endless patience with me and my passion for optics.

Contents

Preface of the Book Series

There already exist many books about aberration theory and optical design, raising the question: why do we need another? I am convinced that a competent working designer needs a deep understanding of aberration theory and huge practical experience for solid and successful work. Therefore, beneath the description of the methods to correct a system, a comprehensive and clear interpretation and understanding of aberrations is necessary. Furthermore, an application and training related section is required to demonstrate the realization and usage of all the theory and methods. From what I know, there are currently two types of books available: the first is mostly written by colleagues in academia; theory and mathematical treatment is in the foreground and the practical applications are less pronounced. On the other side, there are books that deal with the practical realization without a deep theoretical foundation and without a methodology to correct and improve systems successfully. A book that combines both and collects together the more detailed knowledge appears to be missing. This was the reason why I was convinced that one more book would make sense and help the community – and in particular younger colleagues – to learn about this fascinating subject with a considerable scientific systematic background. No book and nobody is perfect, but due to a long time dealing with this topic – from application in industry, as well as from an educational position – I believe I know both worlds and will try to combine them as good as possible. Furthermore, it is well known that optical design is interdisciplinary between mathematics, physic and engineering. Therefore, a lot of knowledge is necessary to work successfully. As one means to make learning easier I always try to visualize and illustrate as many aspects and relations in graphical form as possible. I am sure that this helps a lot, and it is one of the main features of this book series.

Through this series, the considerations are restricted more or less to imaging systems. Illumination systems, as the second large category of optical systems, follow different rules and are not handled here. The overall outline of the five volumes is:

Vol 1: Fundamentals of Optical Systems
Introduction into the basic technical terms of optical systems.

Vol 2: Aberration Theory
Development and practical application of aberration theoretical fundamentals for a better understanding of correction schemes and methods.

Vol 3: Methods of System Correction
Description of the major approaches, rules and strategies to layout, optimize and correct optical imaging systems.

Vol 4: Selected Applications
Here it is demonstrated how existing and well known optical systems are developed on the basis of the knowledge from the first three volumes.

Vol 5: Tutorial and Case Studies
In this last volume, a large variety of practical tasks and exercises are fully worked out to demonstrate to the reader how to practice the theory and to give the opportunity for guided own training.

In particular, one of my goals was to find a style to present the subject that was convenient for self learning. I hope that especially students and young professionals working in small companies without a large experienced team of colleagues will benefit from this attempt.

Aalen, April 2025 *Herbert Gross*

Preface of the First Volume

In this first volume, necessary fundamental knowledge about optical systems is presented. This includes first a consideration of material parameters; in particular, a serious understanding of dispersion is the main goal of this chapter. In the section regarding geometrical optics, the law of refraction, raytrace and the meaning of the paraxial approximation with its simplified calculation options are explained. Furthermore the Lagrange invariant is introduced, which plays an important role in lens design. In the next chapter the function and technology of simple optical components is discussed as well as more sophisticated devices like aspheres, freeform surfaces, gradient index elements and diffractive components. When dealing with geometrical optics, the classical imaging idea is described and the basic principles and terms are introduced. More generalized aspects are presented at the end of this chapter. The basic physics of diffraction theory is presented next in a limited scope focusing on how it is useful for lens design. As special and important representations to assess the quality of systems, including diffraction effects, the point spread function as well as the optical transfer function with corresponding calculation schemes, representation and application are discussed in more detail. Finally some more marginal topics are presented. Gaussian beams are simple representatives of coherent wave optical light fields with important applications for modelling laser light sources. In particular the restrictions and assumptions of this simple model are discussed to get a better understanding of the use and its limitations. It is shown that this approach can be extended to partial coherence and non-paraxial conditions as well. Then the photometry as one important description of the energetic evaluation of optical imaging systems is discussed with simple analytical estimations as well as more rigorous numerical approaches. In particular, the application of the transfer of energy inside optical systems is discussed. In the next chapter, phase space is introduced as a very helpful representation of ray as well as wave optics for a better understanding of their relationship. Here the connection between geometrical and physical modelling is particularly seen to be easily possible through use of the Wigner function and its propagation. Last but not least, a chapter is devoted to digital methods in the computation of quasi-realistic images, the properties of a modern digital image acquisition as well as digital image postprocessing, which nowadays plays an important role in modern systems. Some of the important mathematical concepts used in the book are illustrated and explained in the last chapter of the first book in the form of an appendix. In particular the Fourier transform and its computation properties are presented, together with some more speciailized methods and approaches in Fourier optics and diffraction calculation.

This first volume should give the reader a solid basis of terms, properties and definitions and an understanding of optical systems leaving them well prepared for the more specialized theory of aberrations and methods of system correction and improvement in the following volumes of the book series. In principle, most of the information in the first volume can also be found somewhere else, but in order to unify the notations and collect the necessary basics it seems meaningful to offer this in an introductory volume.

Aalen, April 2025 *Herbert Gross*

Acknowledgements

Of course it is impossible to write and work out everything completely alone. There are many colleagues I have to acknowledge here for their endless help and support. I tried to refer on every source and figure seriously, but this is sometimes complicated. Therefore I beg the pardon of anybody who I haven't dignified correspondingly. I want to express my deepest gratitude to all of them, even if I have forgotten some friendly colleagues.

I want to thank Yi Zhong-Schipp, Siemen Kühl and Patrycja Warkusz for their kind offer to read many chapters. Their hints, comments and proposals helped me a lot. In particular I am deeply indebted to Ziyao Tang for reading several chapters. She dove very deeply into all aspects of correction and gave me comprehensive feedback, which improved the book a lot.

My former colleagues Michael Kempe, Harry Schlemmer and Jürgen Nasse are greatly acknowledged for their permission to use their excellent figures. In addition, I thank Dennis Ochse for preparing two pictures and allowing me to reproduce them in the book.

Aside from those colleagues who actively helped me, I also want to express my gratitude to all my students in the last decade who have inspired me with clever and unexpected questions to think about many problems from a new perspective. In the fruitful and pleasant time I worked with them, I also learned to explain and visualize complicated things more clearly and to accept that many details must be mentioned and added that are not necessarily obvious for a beginner.

Finally, I want thank Martin Preuß from Wiley for his encouraging engagement to make this book series happen. Special gratitude goes to Monica Chandrasekar, who supported me during this time with all my questions and desires to finally solve all issues. I also thank Riya Vats, Farhath Fathima and Kamalini Christilda, as well as the complete team at Wiley, for their support and help.

1
Introduction

Motivation and Outline

This book is the first volume of a series of books about lens design. The overall outline of this series is:

Vol 1: Optical System Properties
 Introduction to the basic technical terms of optical systems.
Vol 2: Aberration Theory
 Development of aberration theoretical fundamentals for a better understanding of correction schemes and methods.
Vol 3: Correction Strategy
 Description of the major approaches, rules and methods to lay out, optimize and correct optical imaging systems.
Vol 4: Selected Applications
 Demonstration of how existing and well-known optical systems are developed from the basis of knowledge from volumes 1–4.
Vol 5: Tutorials and Case Studies
 In this last volume a large variety of practical tasks and examples is fully worked out to demonstrate to the reader how to practice the theory

Throughout this series, considerations are restricted more or less to imaging systems. Illumination systems, as the second large category, follow different rules and are not handled here.

In this first volume, the required basic knowledge for lens design is presented. This includes consideration of material parameters, geometrical optics, the technology of simple optical components as well as consideration of compound systems. Furthermore, the basic physics of diffraction theory is presented in scope: how it is useful for lens design. As special important representations to assess the quality of systems including diffraction effects, the point spread function as well as the optical transfer function are discussed in more detail. Finally, some more marginal topics are presented: Gaussian beams as simple representatives of wave optical light fields with an important relation to the modelling of laser light sources; photometry as one relevant consideration of the energetic evaluation of optical imaging systems; the phase space as a very helpful representation of ray as well as wave optics for better understanding their relationship. Last but not least, a chapter is devoted to digital methods in the computation of quasi-realistic images, the properties of a modern digital image acquisition as well as the digital image postprocessing, which nowadays plays an important role in modern systems.

There are many books about aberration theory and optical design [1–33]. This therefore raises the question: why do we need another? I am convinced that a competent working designer needs a deep understanding of aberration theory and huge, concrete practical experience. Therefore, beneath the description of the methods to correct a system, a comprehensive and clear presentation of aberrations is necessary and an application and training related element is required to demonstrate the realization and application of all the theory and methods. In my understanding, there are currently two types of books available: the first is mostly written by colleagues in academia, theory and mathematical treatment is in the foreground and the practical applications are a less pronounced discussion. On the other hand, there are some books that deal with the practical realization without any theoretical foundation. A book that combines both and collects the knowledge more comprehensively appears to be missing. This is the reason why I am convinced that one more book makes sense and helps the community and, in particular, younger colleagues to learn this fascinating profession with a considerable scientific

(Continued)

Lens Design for Imaging: Fundamentals of Optical Systems, First Edition. Herbert Gross.

Motivation and Outline (Continued)

systematic background. No book and nobody is perfect, but owing to a long history dealing with this topic, from applications in industry as well as from an educational position, I believe I know both worlds and will try to combine them as well as possible. Furthermore, it is well known that optical design is an interdisciplinary field between mathematics, physics and engineering. Therefore, a breadth of knowledge is necessary to work successfully. As one option to make learning easier, I always try to visualize as many aspects and information in graphical form as possible. I am sure that this helps a lot and it is one of the main features of this book series.

Rigorous mathematical strictness is not the goal of this book. With the practical approach it is much more important to gain an understanding to be able to bridge to applications. In Chapter 12, as a mathematical appendix, some more details are explained. It seems more meaningful to avoid this in the running text about optics.

There are several sections that are indicated with an asterisk and a short comment. These are topics that are more specialized and I recommend beginners skip them in the first reading. It should be possible to understand the later chapters without knowing these advanced elements.

One dilemma in this first volume was the fact that many visualizations and explanations require certain knowledge about aberrations. Aberrations are explained in detail in the second volume. Nevertheless, I think it makes sense to use these examples. I have tried to refer to the corresponding later chapters and detailed discussions. But, I know, forward referencing is not a good style. I hope the reader forgives me these inconsistencies in didactic sequence.

Of course, it's impossible to write and work out everything fully alone. There are many colleagues I have to acknowledge here for their endless help and support. I have tried to reference every source and figure accurately, but this is sometimes complicated. Therefore, I beg the pardon of anybody who I haven't dignified correspondingly.

1.1 Modelling and Goal of Lens Design

In general, there are two types of calculation scheme and aim in the simulation of optical systems. If the data and parameters of a system are given, a goal can be to analyse the setup, to calculate the functionality and to evaluate the system performance. This is a direct calculation and is quite straightforward. In this task less creativity is needed.

If a specification is known, which is needed for a special application, the question becomes what kind of system fulfils these requirements – and an inverse problem must be solved. In this case, classical optical design work has to be done to find – through experience or creative thinking – which type of system is able to achieve the request taking certain constraints into account. This second task is much more complicated, often not straightforward and sometimes unsolvable. The corresponding scheme is visualized in Figure 1.1.

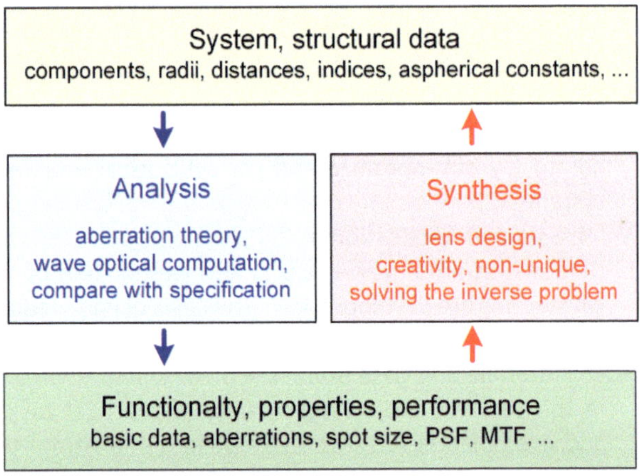

Figure 1.1 Calculation goals for optical system calculations [34].

From the physical point of view, there are several levels of modelling depth used to describe an optical system. On the one hand, a calculation should be fast, effective and accurate. Usually, a more accurate result is needed for deeper physical approaches with fewer approximations. In any case, a decision about a selected model for a specific task must make sure that the investigated effects and results are covered by the approach while, conversely, the computational burden is as small as possible. Figure 1.2 shows certain levels of approximations. Depending on the effects taken into account and the accuracy of the model predictions required, the physical model can be simplified through several approximation steps. Typically, this reduces the computational effort. Furthermore, the simplifications often allow for analytical formulation of solutions, which give much more insight into the dependencies and physical principles.

One of the important questions in simulation or design task reality is: what kind of modelling level is necessary to achieve the necessary accuracy, while keeping the calculation fast and robust.

Optical systems are usually designed, optimized and simulated by raytrace. This kind of approximation level is well established, fast and takes aberrations into account. For the

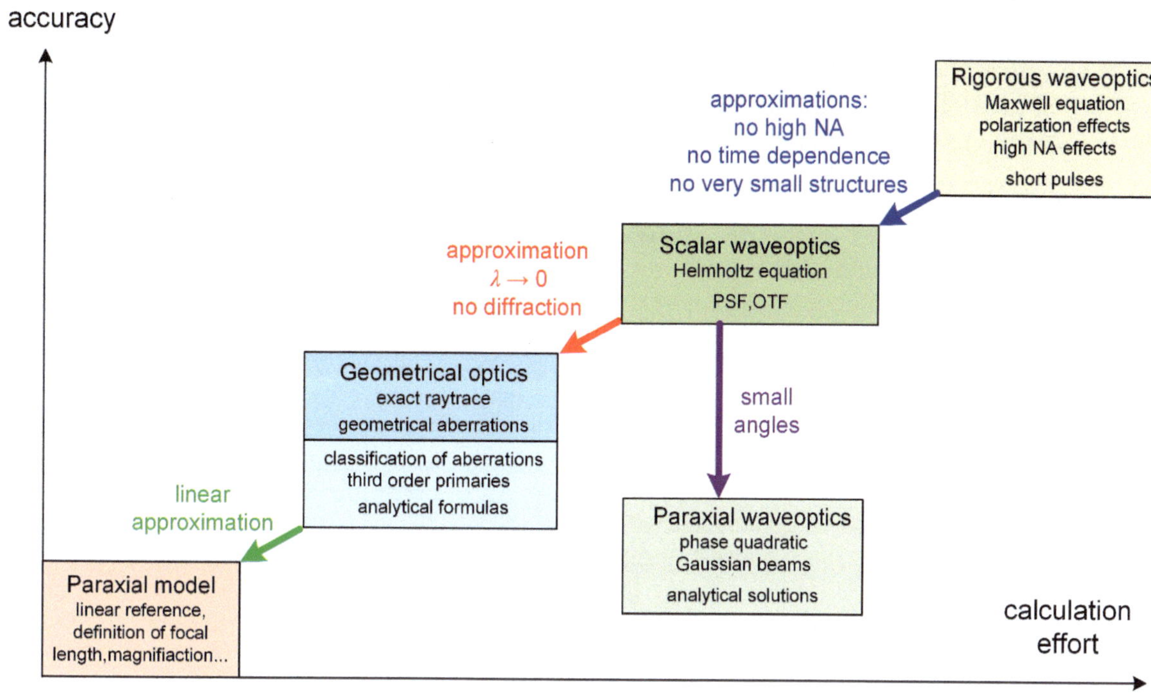

Figure 1.2 Trade-off between calculation effort and accuracy.

majority of systems, this is an acceptable accuracy. In later design phases, if the system is already near the final performance, diffraction related criteria – like the point spread function or optical transfer function – can be included. But the start of calculation with a raw quality is mainly through raytrace. For this purpose, a simple mathematical model is required to describe the essential properties of the system parameters appropriate for this calculation scheme. These include the following.

- Interface surfaces:
  - mathematical modelled surfaces;
  - planes, spheres, aspheres, conics, free shaped surfaces.
- Size of components:
  - thickness and distances along the axis;
  - transversal size, circular diameter, complicated contours.
- Geometry of the setup:
  - special case: rotational symmetry;
  - general case: three-dimensional, tilt angles, offsets and decentrations.
- Materials:
  - refractive indices for all used wavelengths;
  - other properties: absorption, birefringence, non-linear coefficients, index gradients.
- Special surfaces:
  - gratings, diffractive elements;
  - arrays, scattering surfaces.

These data are part of typical system data or the archive files of commercial design software tools. For a concrete reproduction of a calculation, the light data used must additionally be given. These are the wavelengths, the aperture and field size data together with their discrete sampling values and the definition of these properties being used. Furthermore, it can be necessary to fix some more specific cases – like afocal image location, pupil position and telecentricity or vignetting factors – explicitly.

1.2 Optical System Types and Aperture Field Classification

Figures 1.3, 1.7 and 1.10 show some examples as well as the systematic sorting of some special classes of optical systems. These figures should give some idea about the variability and the complexity of real systems. Figure 1.3 is only a small collection of some system types for illustration. Shown are some system types in a diagram indicating the numerical aperture (NA) and the field of view (FoV).

There is a large variety of different optical system types and a long history of developments. To get an overview and a defined structure it makes sense to sort the systems according to their basic properties and characteristic numbers. Since field of view and numerical aperture are very important terms to characterize the imaging capabilities of a system, a diagram with these two properties has been

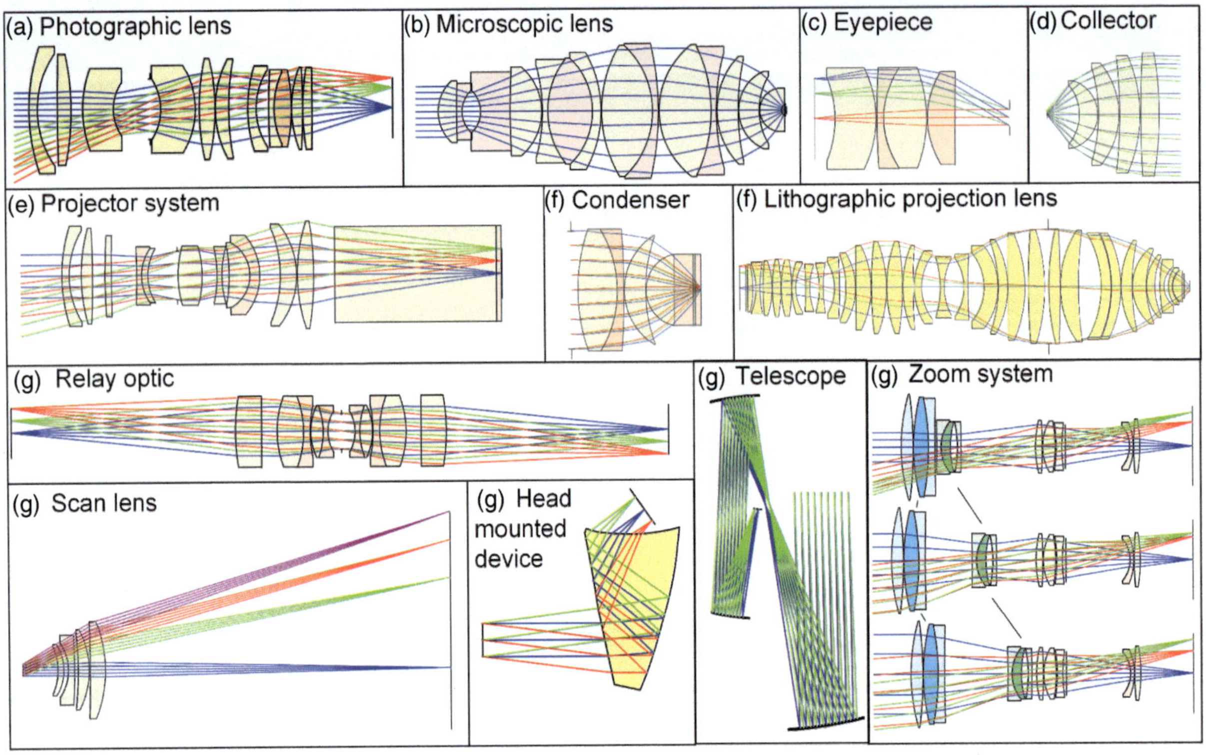

Figure 1.3 Sample systems.

proposed by W. Smith [13]. An example of this diagram is shown in Figure 1.4. For the purpose of comparability, the field angle is used to describe the field of view (FoV) and it is scaled in degree. For telecentric systems, the field is rescaled from the finite field height.

Both properties are essential to describe the imaging options of an optical system. The product of both numbers corresponds to the Lagrange invariant – this is also called etendue or space-bandwidth product. It fixes the complexity of a system and normalizes on the wavelength scales quantitatively the possible information transport capacity. This is discussed and explained in the Chapters 3 and 10 in detail. Therefore, hyperbolic curves describe constant invariants and from practical experience it is known that approximately the same number of lenses is needed for an appropriate correction. Microscopic lenses (large NA, small FoV) and photographic camera lenses (small NA, large FoV) are comparable and roughly contain the same number of lenses. The only exceptions are extremely large apertures (for example in microscopic lenses) or ultra-wide angle fields (fisheye camera lenses); here, the occurrence of many higher orders requires many more lenses.

According to hyperbolic behaviour, systems closer to the origin are easier to correct and simpler in their layout. If a system is placed in the right area, the aperture correction is dominant; spherical aberration correction and marginal ray incidences are most important. In the wide-angle photographic systems, on the other hand, the incidence angles of the chief rays are most prominent and the field-related aberrations are hard to correct. Today's lithographic system has both high field and high NA and they are located in the diagram at an upper right position: the prices are correspondingly high.

It should be mentioned that, in the case of microscopic lenses as well as lithographic projection lenses, field sizes are kept constant due to the application requirements. In particular, for telescopes and EUV lithographic lenses, the field angle is more complicated in reality because the field is arc-shaped and shows certain obscuration effects. Furthermore, the coloured, shaded areas are only raw estimations of the system classes; there is often a huge variety of possible solutions and tradeoffs between field size, aperture and quality. The dots indicate one special realization and are representative, but not unique.

But some aspects are neglected in the classical Smith diagram.

- Spectral range (see Chapter 2): The correction effort for a broadband chromatical correction is higher than for a monochromatic system. Therefore, in reality, one more dimension of complexity is given by $\Delta\lambda$.

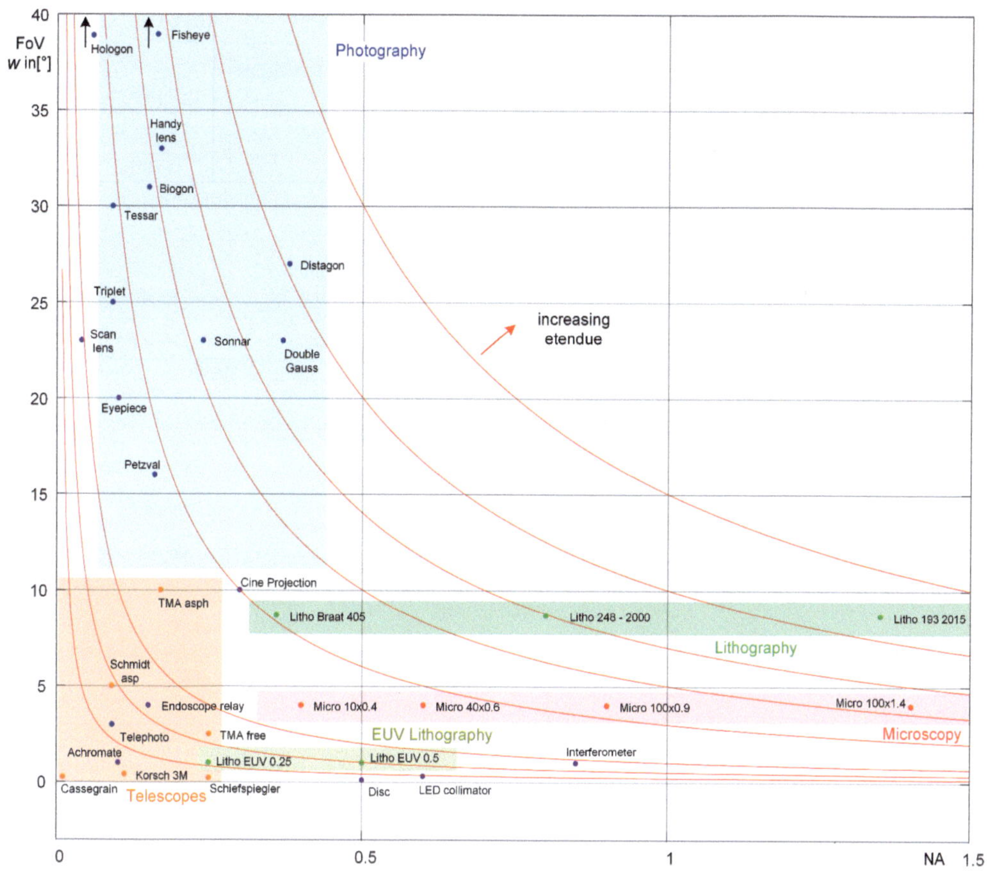

Figure 1.4 Field-aperture diagram according to W. Smith, the red hyperbolic lines indicate a changing etendue value.

- Performance: In principle, the area ranges of the diagram should correspond to the same quality or correction result. But in reality, lithographic lenses, for example, are much better in wavefront correction than usual systems, microscopic lenses are better than photographic lenses and so on. A direct comparison is not trivial.
- The effort of correction also strongly depends on the size, allowed transmission, vignetting and more. Therefore, in the case of an astronomical system with a few mirrors, the comparison of the etendue makes only limited sense; the fabrication complexity is not simply to linear scale with the size. More details are given on this point in Volume 3.
- Dimension: it should furthermore be kept in mind that the systems are working in three dimensions. The simple product of aperture and field is only a one-dimensional simplification and the real energy conservation and information capacity is calculated from the square of this product.
- Definition of FoV and invariant L (see Chapter 3): One additional aspect is the exact definition of the field size. In the original proposal of W. Smith, photographic lenses are mainly considered. These assume the object in infinity and the numerical aperture is measured in the image domain while the field is scaled as an angle in the object domain. This simple definition cannot be taken for finite-finite relay systems or for full afocal systems. Therefore, when comparing systems, this definition should be used carefully.

These facets mean that the diagram is quite useful to classify and compare optical systems, but many special features and a more quantitative evaluation become critical and should be done with care.

Figure 1.5 shows an exemplary diagram, where the one dimensional etendue is applied as the horizontal axis as only one variable and the wavelength range of the application as the vertical axis. Now it is much easier to see that lithographic lenses have a huge $L^{(1D)}$, but only a small spectral request. The same is true for interferometer lenses, simple scan laser scan lenses and more. The problem of not truly comparing the same quality of correction is still present.

To better visualize this spectral third dimension of correction complexity, in Figure 1.6 the diagram is shown

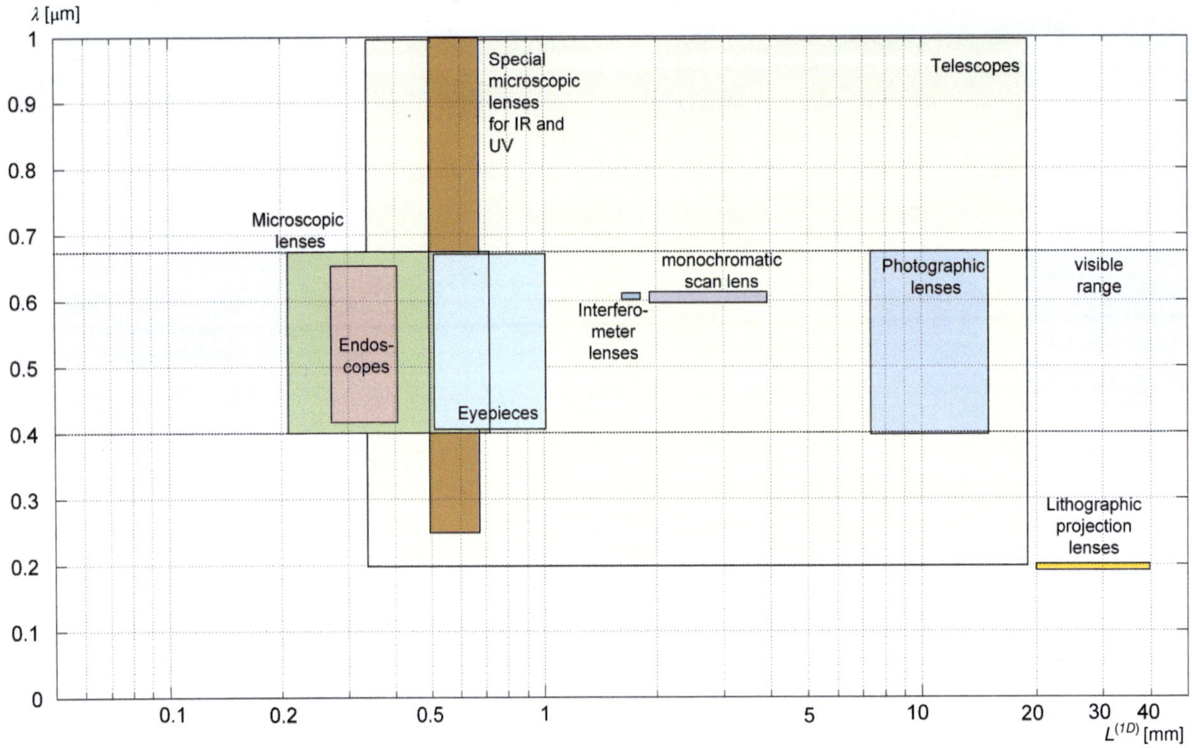

Figure 1.5 Impact of spectral bandwidth correction.

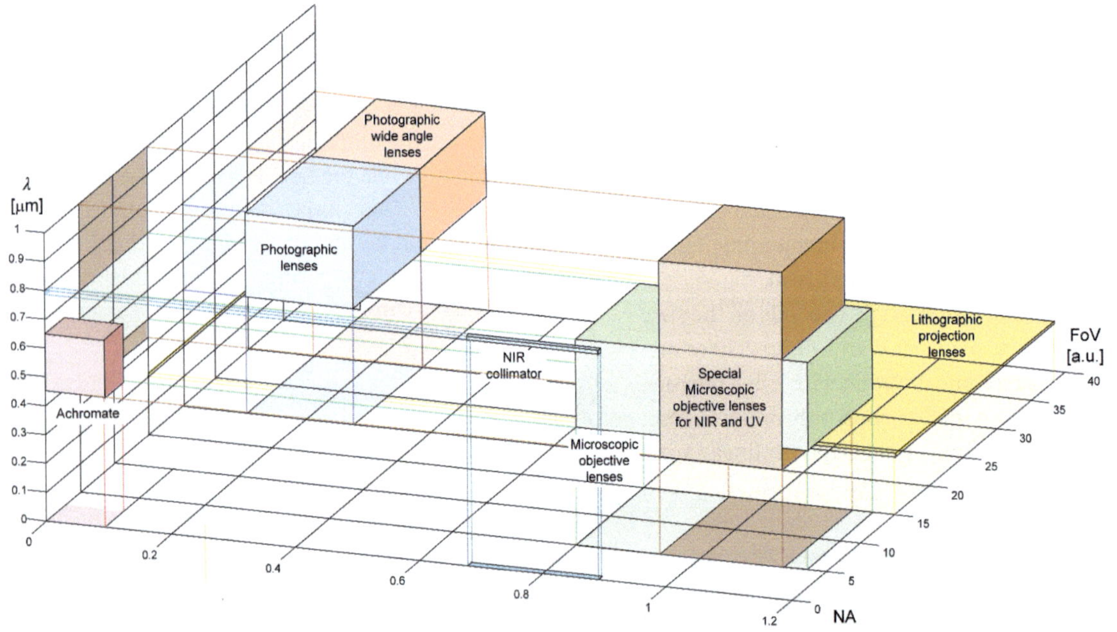

Figure 1.6 Smith's diagram extended by the spectral dimension: the boxes are only raw and schematic.

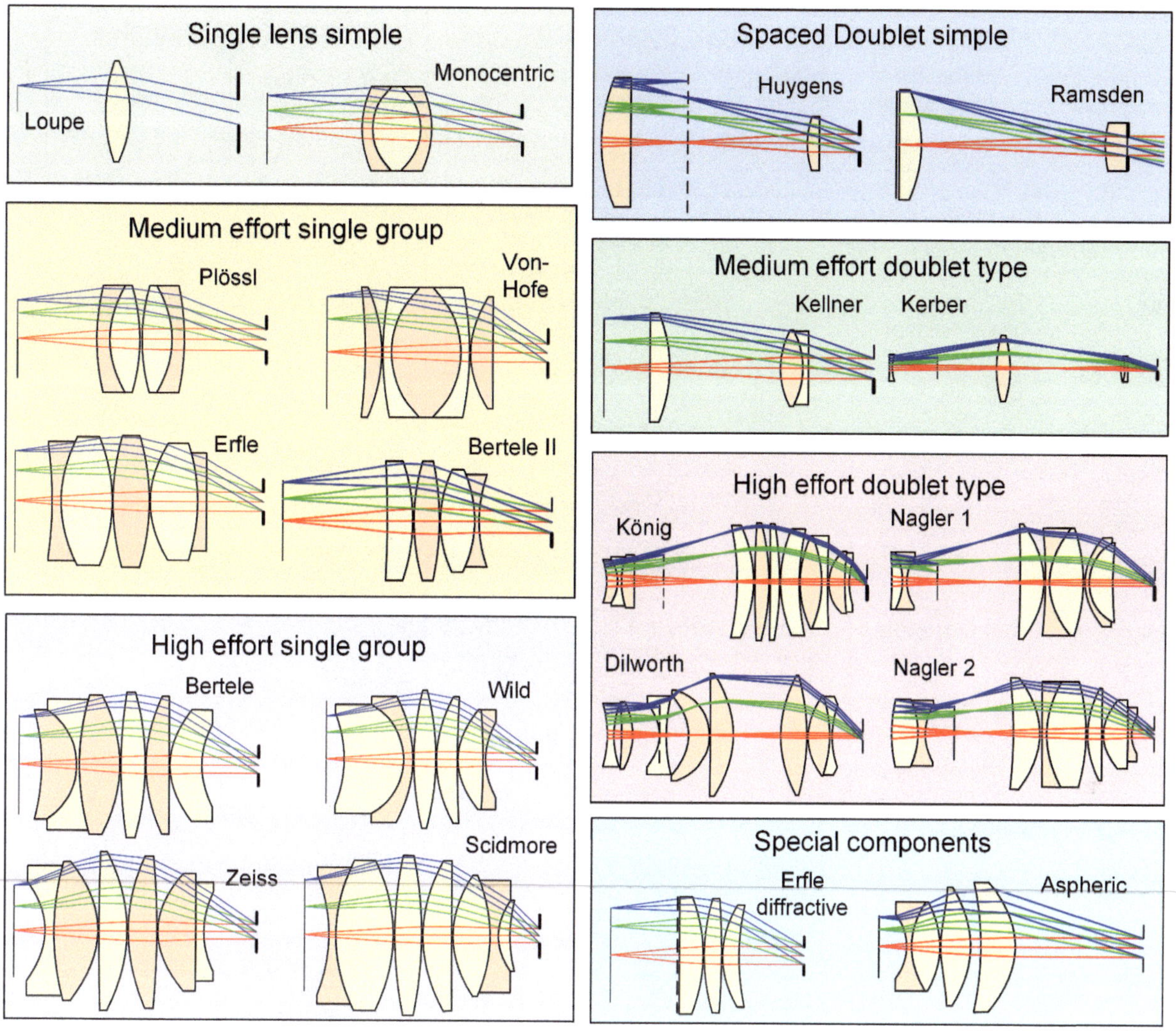

Figure 1.7 Families of eyepiece [35] / with permission of John Wiley & Sons.

in perspective three dimension. For reasons of clarity, only a few system types are included, but it can be seen that, in principle, the volume of the bricks, not just the area in one dimension, is relevant for the correction. This finally shows that the Smith diagram is quite useful for comparisons, but on more detailed examination, some aspects are not considered.

1.2.1 Selected Classes of Systems

In this section, for some examples of system classes the historical development of setups together with the general trend to improve the performance by increasing the number of lenses will be illustrated. These collections and ordering schemes are not fully discussed here; for this, more prior knowledge is required, which cannot be expected in the first chapter of a book. But these figures should give an idea of historical development and the variability of possible solutions. These charts are definitely not complete and different arrangements are possible.

Figure 1.7 shows types of eyepiece with different qualities and complexities. Here, the left column represents the solutions, which come from the one-group approach; the second column on the right side how the results, improve the basic two-lens solution of Huygens with virtual, and Kellner with real, image location.

Figure 1.8 illustrates the families of camera lenses. This very old field has developed over a long period of time. More recent results using aspherical lenses and special new types of glasses, plus the change in detectors from film and photoplates to charge-coupled devices (CCD), deliver different solutions. The main difference is the symmetry of the systems. The larger the field, the more a symmetrical approach helps. Fisheye lenses can be realized only in an asymmetrical setup.

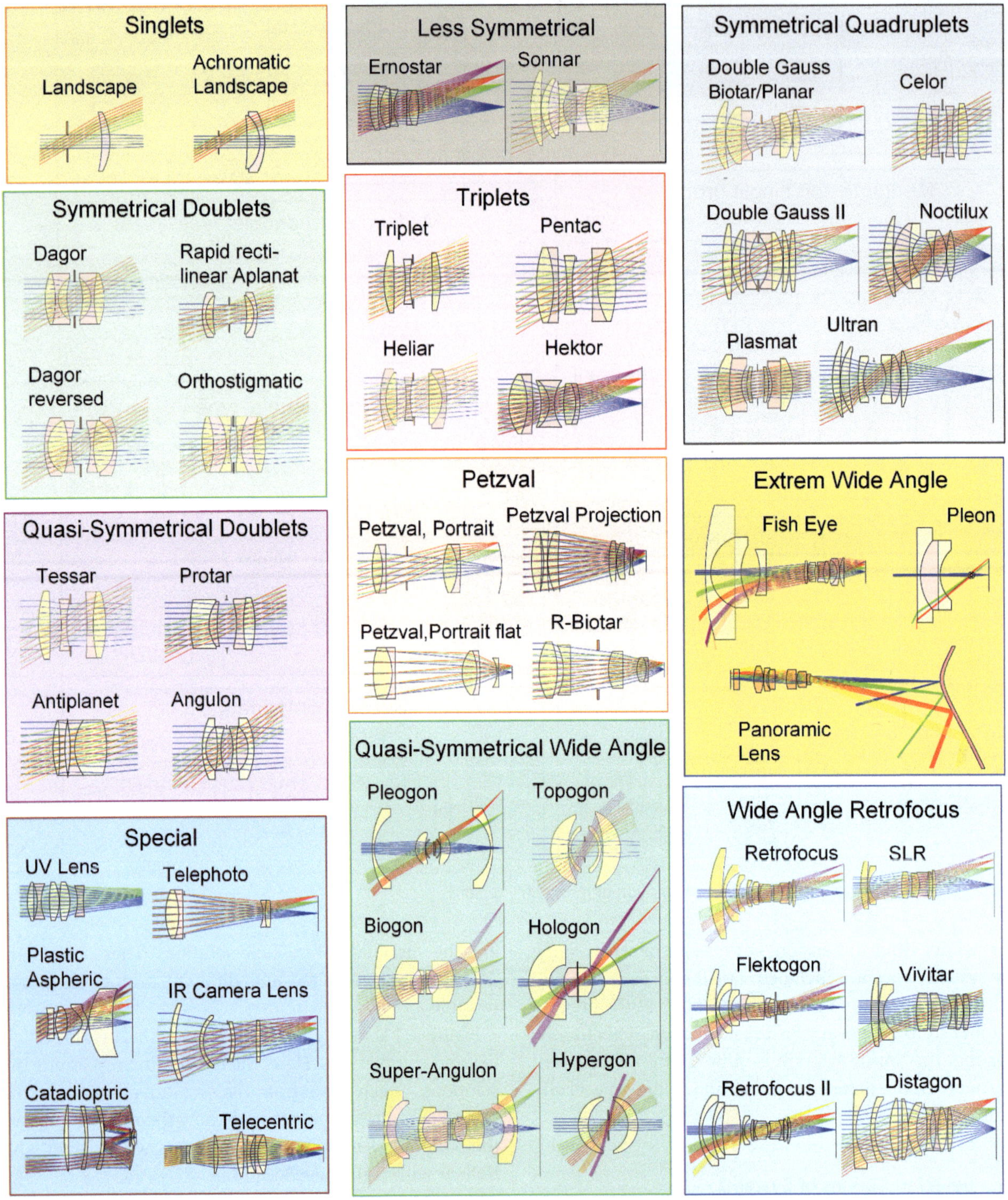

Figure 1.8 Families of photographic lenses.

Figure 1.9 illustrates the development of modern microscope objective lenses for the case of low, medium and high NA with several special aspects. This data basis is extracted from an analysis of 500 patents [36–38]. Most of them come from the four vendors: Olympus, Zeiss, Leica and Nikon. Layouts are typical in every company and correspond to their needs and experiences. The chart should give an idea of the huge variety of possible solutions.

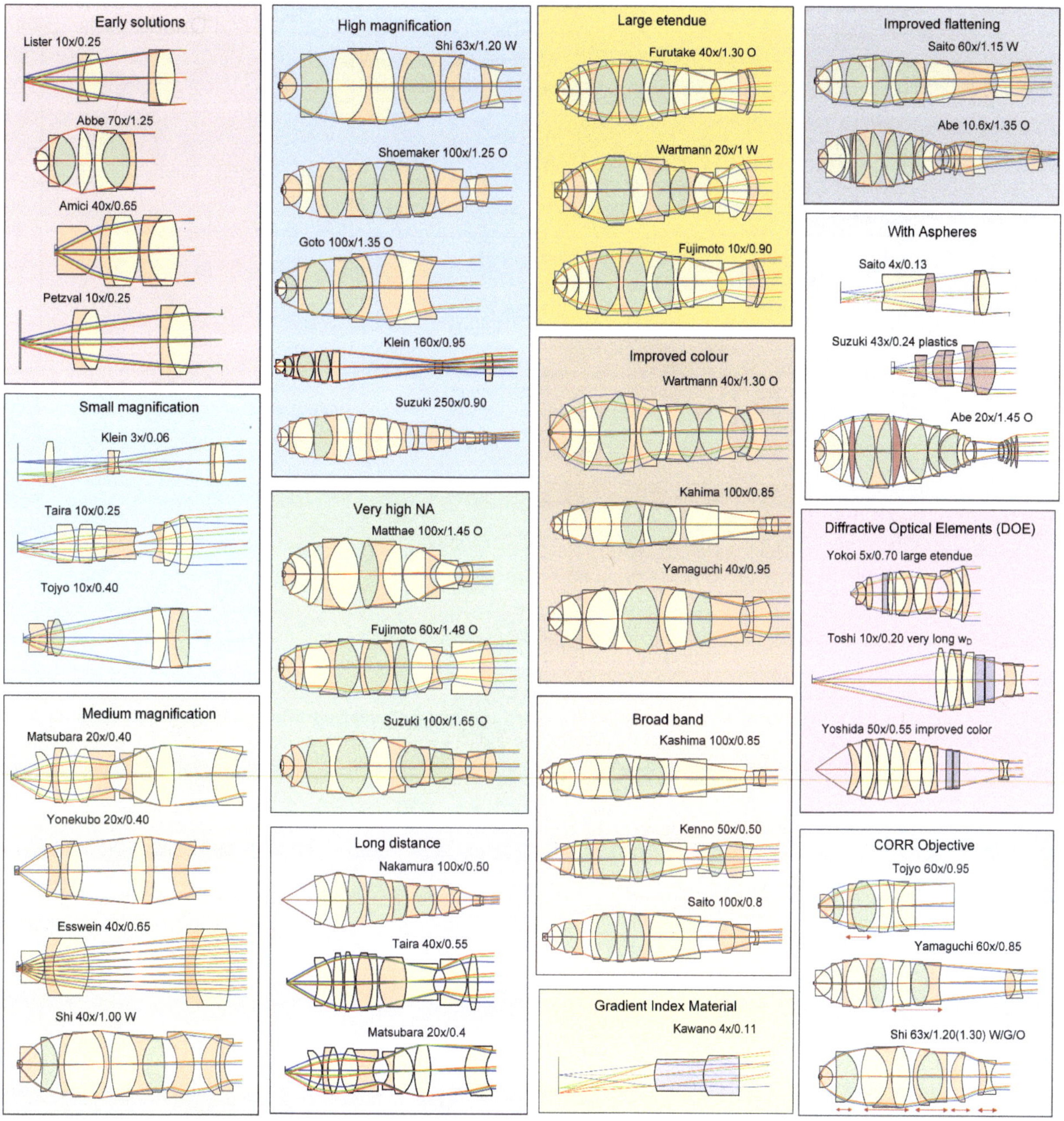

Figure 1.9 Microscopic objective lens types, this overview is far from complete [36–38].

Figure 1.10 illustrates the families of lithographic projection lenses for refractive, catadioptric and fully reflective systems. In the early days, glasses were used, the wavelength was in the visible, no telecentricity was forced, field of view and numerical aperture were moderate in size. Later, the number of waists to correct for field flatness were increased, and then aspheres were used. In the next phase of development, immersion was used to increase the NA and mirrors were used to flatten the field. In the extreme ultraviolet (EUV) mirror solution for lithography, due to the short wavelength of 13.5 nm (soft x-ray), only mirrors can be used. They are used here in nearly normal incidence; therefore, complicated dielectric coatings are necessary to achieve an acceptable throughput.

The overview tables shown here are only examples. They are selective and the list is not complete.

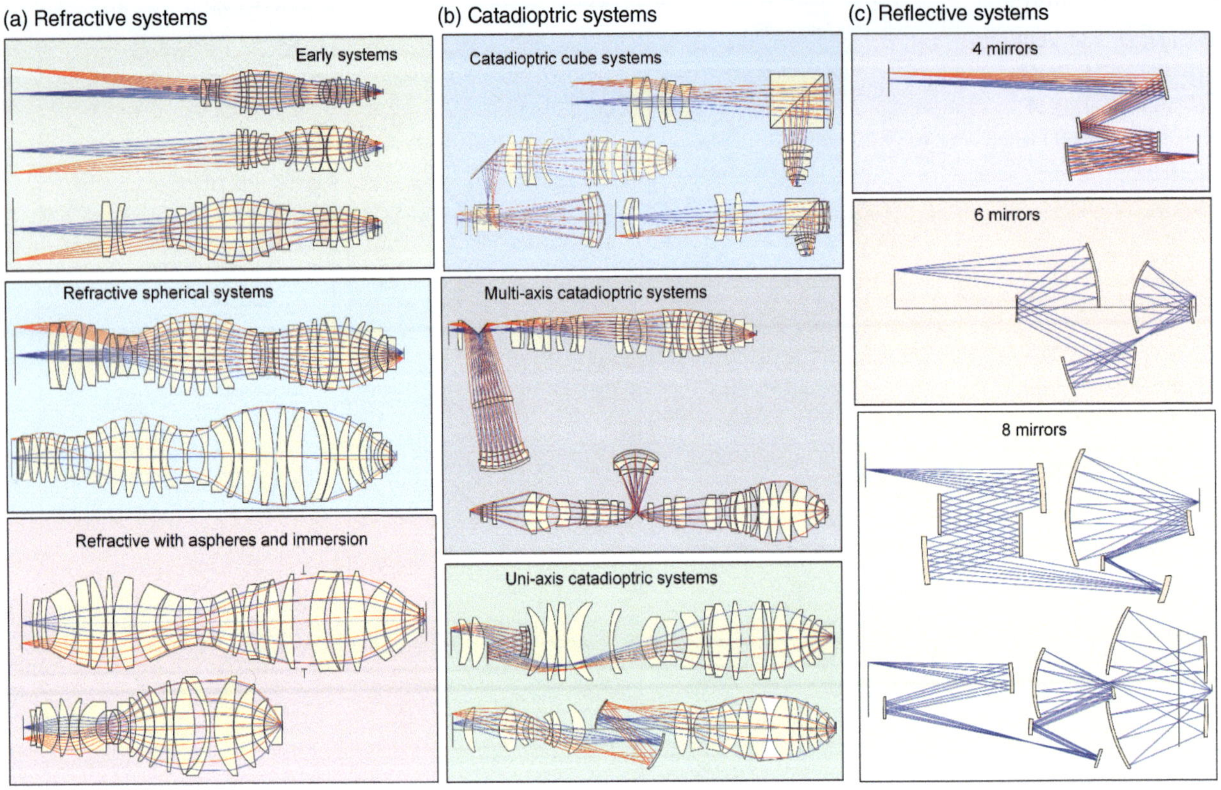

Figure 1.10 Lithographic projection lenses with lenses and mirrors for ultraviolet (UV), deep ultraviolet (DUV) and extreme ultraviolet (EUV) wavelengths, (a) shows purely refractive systems, (b) catadioptric and (c) purely reflective systems.

References

1 Welford, W. (1986). *Aberrations of Optical Systems*. Academic Press.
2 Matsui, Y. and Nariai, K. (1993). *Fundamentals of Practical Aberration Theory*. World Scientific.
3 Slyusarev, G. (1984). *Aberration and Optical Design Theory*. Adam Hilger.
4 Mahajan, V. (1998). *Optical Imaging and Aberrations I, Ray geometrical optics*. SPIE Press.
5 Mahajan, V. (2001). *Optical Imaging and Aberrations II, Wave diffraction optics*. SPIE Press.
6 Sasian, J. (2013). *Aberrations in Optical Imaging Systems*. Cambridge.
7 Haferkorn, H. (1985). *Bewertung optischer Systeme, Deut*. Vlg. der Wiss.
8 Kingslake, R. and Johnson, B. (2010). *Lens Design Fundamentals*. Academic Press.
9 Mouroulis, P. and McDonald, J. (1997). *Geometrical Optics and Optical Design*. Oxford University Press.
10 Laikin, M. (2007). *Lens Design*. Marcel Dekker.
11 Malacara, D. and Malacara, Z. (2013). *Handbook of Lens Design*. Marcel DEkker.
12 Smith, W. (2000). *Modern Optical Engineering*. McGraw Hill.
13 Smith, W. (2005). *Modern Lens Design*. McGraw Hill.
14 Geary, J. (2002). *Lens Design with Practical Examples*. Willmann-Bell.
15 Fischer, R., Tadic-Galeb, B., and Yoder, P. (2008). *Optical System Design*. SPIE Press.
16 Shannon, R. (1997). *The Art and Science of Optical Design*. Cambridge University Press.
17 Smith, G. (1998). *Practical Computer-Aided Lens Design*. Willmann-Bell.
18 Cox, A. (1964). *A System of Optical Design*. Focal Press.
19 Sun, H. (2017). *Lens Design – a practical Guide*. CRC Press.
20 Berek, M. (1970). *Grundlagen der praktischen Optik*. de Gruyter.
21 von Rohr, M. (1904). *Die Theorie der optischen Instrumente*. Springer.
22 Conrady, A.E. (1985). *Applied Optics and Optical Design*, Vol I, 84. Dover.
23 Sasian, J. (2019). *Lens Design*. IOP Publishing.

24 Dilworth, D. (2018). *Lens Design*. IOP Publishing.

25 O'Shea, D. and Bentley, J. (2018). *Designing Optics Using CODEV*. SPIE Press.

26 Velzel, C. (2014). *A Course in Lens Design*. Springer.

27 Kidger, M. (2001). *Fundamental Optical Design*. SPIE Press.

28 Kidger, M. (2004). *Intermediate Optical Design*. SPIE Press.

29 O'Shea, D. (1985). *Elements of Modern Optical Design*. Wiley.

30 Nussbaum, A. (1998). *Optical System Design*. Prentice Hall.

31 Hazra, L. (2022). *Foundations of Optical System Analysis and Design*. CRC Press.

32 O'Shea, D. and Bentley, J. (2024). *Designing Optics Using Zemax OpticStudio*. SPIE Press.

33 Braat, J. and Török, P. (2019). *Imaging Optics*. Cambridge University Press.

34 Richter, W. (2008). *Vorlesung Technische Optik 1*. TU Illmenau.

35 Gross, H., Blechinger, F., and Achtner, B. (2008). *Handbook of Optical Systems, Volume 4: Survey of Optical Instruments*. Verlag Wiley.

36 Zhang, Y. and Gross, H. (2019). Systematic Design of Microscopic lenses I. *Advanced Optical Technologies* 8: 313.

37 Zhang, Y. and Gross, H. (2019). Systematic Design of Microscopic lenses II. *Advanced Optical Technologies* 8: 349.

38 Zhang, Y. and Gross, H. (2019). Systematic Design of Microscopic lenses III. *Advanced Optical Technologies* 8: 385.

2

Optical Materials

2.1 Introduction

Beneath the geometrical shapes of optical components, in the case of refractive components the physical wavelength properties of the materials are quite important in lens design. The most important aspect here is the wavelength dependence of the refractive index. This so-called dispersion is essential for correcting optical systems over a certain spectral range and for the colour quality of images.

In principle, optical materials can be divided into reflective and transmissive media. The transmissive materials can be further subdivided into:

- glasses
- crystals
- plastics
- liquids
- gases
- glues and cements.

Moreover, there are special applications within physical and non-linear optics that require more specialized materials, which are not discussed here.

The most important basic optical parameters for materials are the following.

- Refractive index $n(\lambda)$ and its spectral variation; this dispersion effect is described through different approaches in the literature and this discussion is one of the main points in this chapter.
- Transmission T and absorption α.
- Temperature coefficient of the refractive index dn/dT.
- Non-linear coefficients.
- Autofluorescence.
- Birefringence and index non-isotropy.
- Scattering coefficients.

All these parameters are usually given in data sheets. The refractive indices of solid materials are given relative to air under normal conditions. This means a temperature of $T = 293$ K and a pressure of $p = 1013$ mbar are assumed. Therefore, a refractive index of $n = 1.0$ corresponds to air and not to vacuum. As a consequence, special gases like Helium can have indices $n < 1$.

Other properties of optical materials from a more technical point of view – and in reality addressed as tolerances of non-perfect manufacturing – are as follows.

- The uniformity of the refractive index (striae).
- The anisotropy of the refractive index (induced birefringence which is usually caused by internal stress during the cooling process).
- Bubbles, impurities and contaminations (inclusions and undissolved particles of the melting ingredients).

There are properties of materials used in optical systems that are not directly related to the interaction of the light but have an indirect influence on it. Therefore, they may be important for the design of optical systems from a technical and even a commercial point of view. Amongst these are:

- thermal expansion;
- mechanical properties, such as hardness, density, elasticity;
- chemical properties, such as resistance against acids.

The chemical and mechanical data are important for the manufacturing process and also for further application in connection with mechanical fixtures, as well as coatings, stability and durability under the existing environmental conditions. For the cementing of two glass components, the chemical and particularly the thermal expansion properties are decisive.

Glass is the most important class of materials for optical applications. Glasses have excellent properties in terms of mechanical, thermal and optical parameters. Moreover, they guarantee a wide variability in optical properties even for very different compositions [1–4].

Known suppliers of optical glasses offer about 200 different types. Recently, there has been an attempt to avoid toxic ingredients like lead and cadmium as far as possible, limiting the number to about 120 types of glass. Some important commercial suppliers of optical glasses are the

Lens Design for Imaging: Fundamentals of Optical Systems, First Edition. Herbert Gross.

companies Schott, Hoya, Ohara, Sovirel, Sumita, Hikari, Corning and CDGM.

The glass diagram is a good first description and sorting representation of different glass types. Here the glasses are shown in an n–ν plot. n is the index of refraction and ν the Abbe number, describing the dispersion in the visible wavelength range. The Abbe number is discussed in the next section in more detail. Correspondingly, for a first orientation, a six-digit glass code is assigned containing the first three decimals of the refractive index in the first three digits and the first three valid figures of the Abbe number (rounded) in the digits 4 to 6. For BK7, e.g. we have $n_d = 1.51680$ and $\nu_d = 64.17$, leading to the glass code 517642.

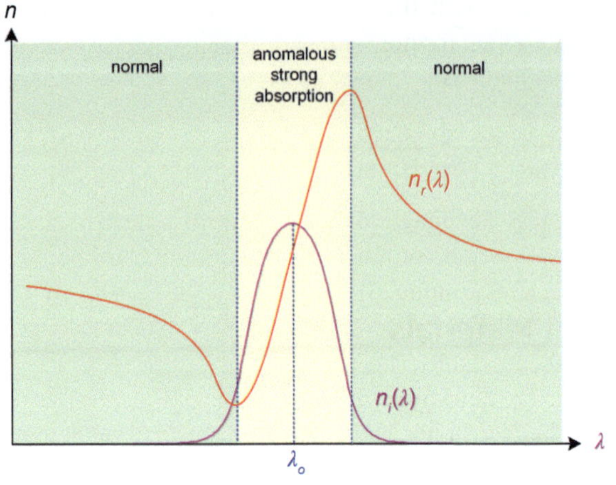

Figure 2.2 Curve of a typical refractive index dispersion with an absorption at the wavelength λ_o. n_r is the real part, n_i the imaginary part.

2.2 Dispersion

2.2.1 Definition

The dependence of the refractive index on the wavelength is called the dispersion and can be formulated as

$$\Delta n = n(\lambda_1) - n(\lambda_2) \tag{2.1}$$

In the case of normal dispersion, the refractive index is higher for shorter wavelengths and the imaginary part of n can be neglected. This can be found in all technical applications of transparent media. So we have

$$\frac{dn}{d\lambda} < 0 \tag{2.2}$$

As an illustration, Figure 2.1 shows this change of index and therefore deflection of rays at a dispersion prism. This means that, in the visible spectral range, blue light is more strongly refracted than red light.

Physically, the refractive index describes the interaction of a light wave with matter. It may happen that the wave is partly absorbed. Then, the refractive index becomes a complex value. In the range of absorption bands of a material there is an anomalous dispersion observed, where the slope of the $n(\lambda)$-curve has a positive sign. Here, there is also a noticeably large imaginary part $n_i(\lambda)$ of the refractive index, which describes the absorption; this is usually a non-attractive aspect for applications. Figure 2.2 shows this behaviour. Anomalous dispersion with $dn/d\lambda > 0$ should not be confused with anomalous partial dispersion, which is often desired for colour correction.

In the classical harmonic oscillator model of the Drude theory for light – matter interaction – the expression for the square of the complex refractive index can be written as [5]

$$\varepsilon = (n_r + i \cdot n_i)^2 = \frac{Ne^2}{2\pi \cdot c\varepsilon_0 m} \cdot \sum_j \frac{f_j \lambda^2 \lambda_j^2}{2\pi \cdot c \cdot \left(\lambda^2 - \lambda_j^2\right) + i\gamma_j \lambda \lambda_j^2} \tag{2.3}$$

where λ_j is the wavelength at the resonant frequency, f_j the oscillator strength of the atomic transition, γ_j the attenuation constant, c the speed of light in vacuum, ε_0 the dielectric constant, e the electron charge, m the mass of the electron and N the density of the absorbing atoms

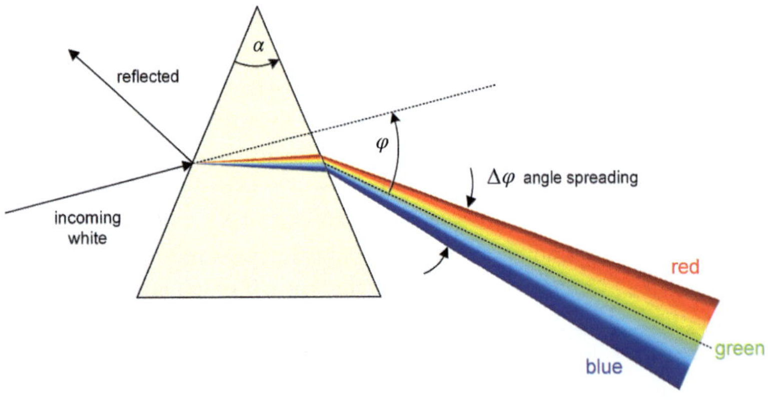

Figure 2.1 Dispersion of glass at a prism, the spreading of the colours in the visible spectrum is seen after the ray refraction through the prism.

or molecules. With the corresponding abbreviation, this can be simplified for non-absorbing conditions far from an absorption band as

$$n^2 = n_o^2 + \sum_j \frac{A_j \lambda^2}{\lambda^2 - \lambda_j^2} \tag{2.4}$$

by neglecting the imaginary term. This formula is the basis for different dispersion formulas originating from the first principles of the atomic description of light–matter interaction. In practice, good transmission of the light in the medium is desirable. Therefore, the materials used should preferably be far away from an absorption line. In this case, normal dispersion always occurs and n_i can be neglected. But, in reality, some absorption lines at λ_j are always present on the longer as well as on the shorter wavelength range, see Figure 2.3. In the application of this dispersion formula in the visible, at least two terms are needed – one on the ultraviolet and one on the infrared side – to describe the behaviour in between accurately. Depending on the distance of the atomic transition resonances to the visible spectrum edges, there are also glasses where three terms are recommended.

Furthermore, there are some hints in the literature that in special cases, in particular for quartz, a system of coupled oscillators exists in the atomic interaction. This means that in the classical Sellmeier formula

$$n^2 = n_o^2 + \sum_j \frac{A_j \lambda^2}{\lambda^2 - \lambda_j^2} \tag{2.5}$$

a term with the modified form with a degenerated resonance wavelength λ_0

$$n^2 = n_o^2 + \frac{A_j \lambda^4}{\left(\lambda^2 - \lambda_o^2\right)^2} + \sum_j \frac{A_j \lambda^2}{\lambda^2 - \lambda_j^2} \tag{2.6}$$

must be added for an accurate description (dispersion formula of [6]). Dispersion of glass is quite important for the management of colour correction in optical systems. Correspondingly a huge amount of literature is found dealing with this subject. Some of it is listed in the references [7–30].

2.2.2 Wavelengths

In optical design usually at least one main wavelength and two secondary colours are used for chromatic characterization in the visible range. The primary colour should be in the centre or midpoint of the spectrum used (in the visible range this is generally green). The two secondary colours are used for evaluation of the behaviour at the red and the blue edges of the spectrum, respectively.

There are two variants in the visible range that are commonly used for the choice of wavelength; they are listed in Table 2.1. The triple F′eC′ is better adapted for the visual sensitivity of the human eye, the choice FdC is more appropriate for CCD based detectors and is preferred for digital image gathering applications.

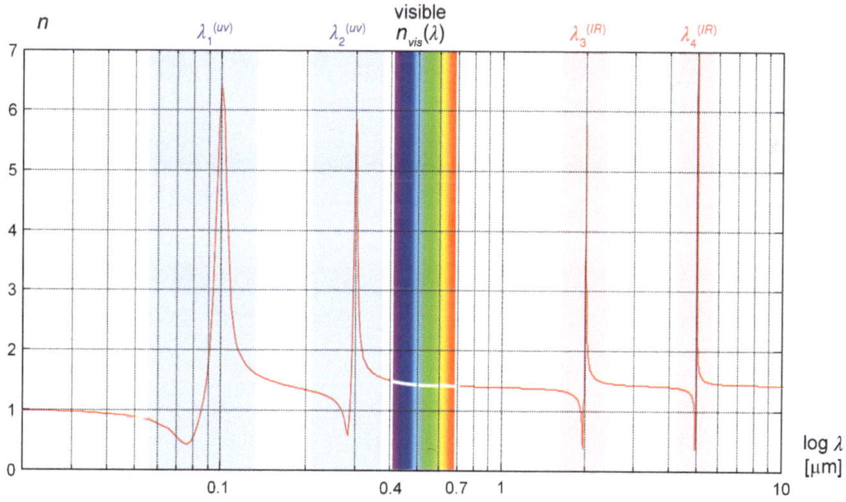

Figure 2.3 Dispersion curve $n(\lambda)$ over a larger range of wavelengths with four absorption bands. The axis of the wavelength is represented logarithmically.

Table 2.1 Common wavelengths for description of dispersion.

Variant	Main Wavelength			1. Secondary colour			2. Secondary colour			Preferred application
I	e	546.07	Green	F′	480.0	Blue	C′	643.8	Red	Microscopy
II	d	587.56	Yellow	F	486.1	Blue	C	656.3	Red	Photography

Table 2.2 Wavelengths of the most important spectral lines.

λ in [nm]	Name	Colour	Element
248.3		UV	Hg
280.4		UV	Hg
296.7278		UV	Hg
312.5663		UV	Hg
334.1478		UV	Hg
365.0146	i	UV	Hg
404.6561	h	Violet	Hg
435.8343	g	Blue	Hg
479.9914	F′	Blue	Cd
486.1327	F	Blue	H
546.0740	e	Green	Hg
587.5618	d	Yellow	He
589.2938	D	Yellow	Na
632.8		Red	HeNe laser
643.8469	C′	Red	Cd
656.2725	C	Red	H
706.5188	r	Red	He
852.11	s	NIR	Cs
1013.98	t	NIR	Hg
1060.0		IR	Nd-glass laser
1529.582		IR	Hg line in the IR
1970.09		IR	Hg line in the IR
2325.42		IR	Hg line in the IR

For chromatic descriptions, the wavelengths listed in Table 2.2 are common as they correspond to easily measurable lines of lasers or special lines found in vapour lamps. The letters in the second column are used for abbreviation, most of them are introduced by Fraunhofer as spectral lines of the spectrum of the sun.

2.2.3 Abbe Number

The easiest method, and historically the first definition of a quantitative description of the dispersion of glass, is the Abbe number. In optical design it is generally used. The Abbe number connects in the visible spectral range the difference between the main wavelength λ_0 in air to the difference between the red and the blue edges of the wavelength interval $\lambda_{1/2}$ under consideration. It is defined as

$$\nu = \frac{n_{\lambda 0} - 1}{n_{\lambda 1} - n_{\lambda 2}} \tag{2.7}$$

In an simple picture, the Abbe number can be interpreted as the inverted mean slope of the dispersion curve in the considered spectral range. The Abbe numbers ν_e and ν_d are used in particular for the visible spectral range and are referenced for these two wavelengths in the middle of the visible. They are defined more concretely as

$$\nu_e = \frac{n_e - 1}{n_{F'} - n_{C'}}, \qquad \nu_d = \frac{n_d - 1}{n_F - n_C} \tag{2.8}$$

It should be mentioned that, in particular for the glass map, the d-related data are used. Optical glasses are usually found in the range $\nu_d = 20$–100. Materials with a low Abbe number show a large dispersion, and vice versa. For large dispersion, the refractive index varies more strongly with the wavelength. The curve of the refractive index plotted against λ has a larger slope for small dispersion. This behaviour is shown in Figure 2.4. The idea of defining the Abbe number in this way can be understood as follows. The motivation of Abbe was to compare the change of the focal length of a lens for the two wavelengths F and C and relate it to the value for the centre wavelength d. The consequence of the glass dispersion on the focal length of a thin lens directly brings us to the definition of the Abbe number. The

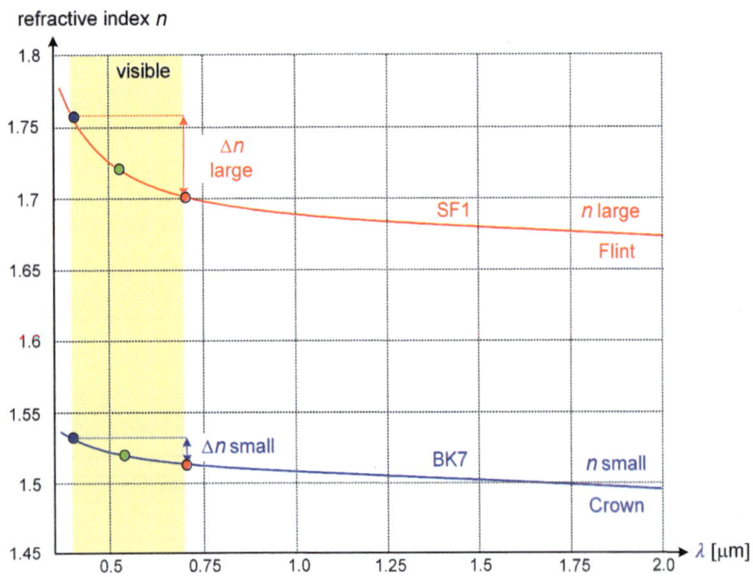

Figure 2.4 Characteristic dispersion for flint and crown glasses SF1 and BK7 respectively.

focal length f of a thin lens for green light is given by ($F = 1/f$ is the focal power, $c_{1/2}$ the surface curvature values)

$$1/f_d = F_d = (n_d - 1)(c_1 - c_2) = (n_d - 1) \cdot \Delta c \quad (2.9)$$

If the outer wavelengths F and C are considered, the difference in focal power is given by

$$\begin{aligned} \Delta F &= F_F - F_C = (n_F - n_C) \cdot \Delta c \\ &= \frac{n_F - n_C}{n_d - 1} \cdot (n_d - 1)\Delta c = \frac{F_d}{\nu_d} \end{aligned} \quad (2.10)$$

or

$$\frac{\Delta F}{F} = -\frac{\Delta f}{f} = \frac{1}{\nu} \quad (2.11)$$

and is therefore directly related to the Abbe number. Since a larger Abbe number means a smaller dispersion, sometimes the inverse value

$$m_d = \frac{1}{\nu_d} = \frac{n_F - n_C}{n_d - 1} \quad (2.12)$$

is used and is called the dispersion number. The general dispersion number for arbitrary wavelengths is then defined as

$$m = \frac{n_{\lambda 1} - n_{\lambda 2}}{n_{\lambda 0} - 1} \quad (2.13)$$

2.2.4 Dispersion of Crown and Flint Glasses

Glasses with low dispersion are called optical crowns. Usually, they also have a smaller refraction index. Glasses with a strong dispersion and a high refractive index are mainly optical flints. The glasses can be classified depending on their refractive indices and Abbe number in the following way [7]:

$$\text{Optical crowns} \quad \begin{array}{lll} \nu_e > 54.7 & if & n_e < 1.6028 \\ \nu_e > 49.7 & if & n_e > 1.6028 \\ & \text{or} & \\ \nu_d > 55 & if & n_d < 1.6 \\ \nu_d > 50 & if & n_d > 1.6 \end{array} \quad (2.14)$$

$$\text{Optical flints} \quad \begin{array}{lll} \nu_e < 54.7 & if & n_e < 1.6028 \\ \nu_e < 49.7 & if & n_e > 1.6028 \\ & \text{or} & \\ \nu_d < 55 & if & n_d < 1.6 \\ \nu_d < 50 & if & n_d > 1.6 \end{array} \quad (2.15)$$

These ranges separate the corresponding glasses with 'K' and 'F' in the name in the glass map shown below in Figure 2.6. But a closer look at the glass map shows (see next section) that there are crown glasses with higher indices than several flint glasses, see Figure 2.6 (b). In principle the chemical ingredients in the SiO_2 matrix are responsible for distinguishing crown and flint. Therefore, there is a discontinuity in the representation and crown/flint selection in the range of $\nu_d = 50\ldots55$ and $n_d = 1.6$.

A typical example is shown in Figure 2.4. Flint SF1 and crown BK7 have very different refractive indices, this is clearly seen in the graph. Furthermore, it can also be observed that the differences of the indices in the visible range between 400 nm and 700 nm are larger for the flint glass, which means it has a stronger dispersion. Figure 2.5 shows as an example a crown and a flint glass with nearly identical refractive indices for green light, where a dispersion discrepancy between these two glass types is observed.

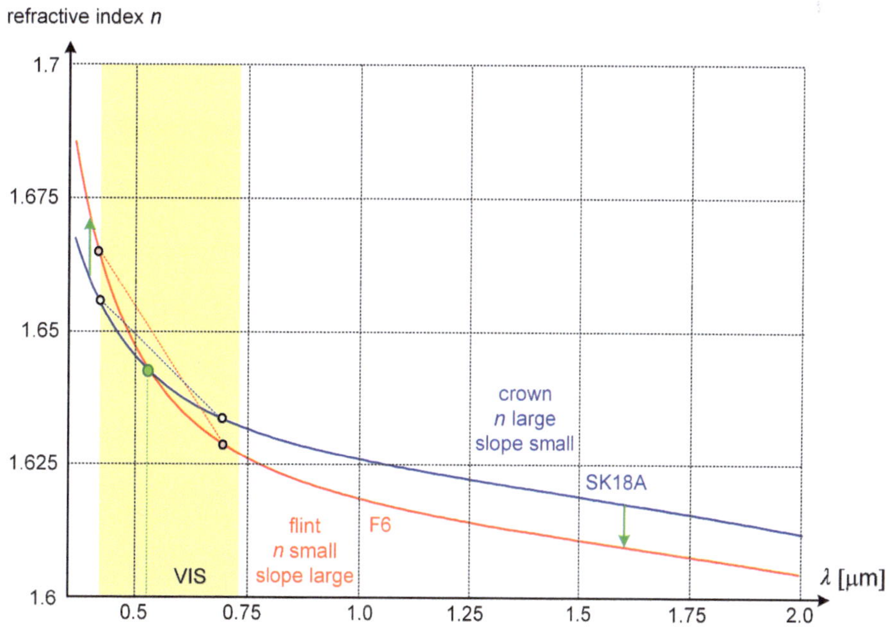

Figure 2.5 Dispersions of a crown and a flint glass with the same refractive index at green.

2.2.5 Glass Diagram

For classification of optical glass materials, in practice glass charts are used that are plotted as a dependence of the refractive index and Abbe number. The ranges in the n–ν plot depend principally only on chemistry and melting processing and can be developed by using corresponding technologies and boundary conditions of melting and cooling. This also explains the characteristic shape of the areas filled with real glasses. There are no conventional glasses that would fill the upper left corner or the lower right one of the chart. Therefore, there are no significant differences between the different suppliers in terms of the filled ranges. This is shown in Figure 2.6 exemplar for the vendor Schott [1]. In part (a) all the glasses are seen as dots, in (b) the flint and crown ranges are indicated. Glass as a material is based on an amorphic and optically isotropic SiO_2, this limits the reasonable range of the refractive index and Abbe number. Both coordinate parameters n and ν are usually given for the reference wavelength $\lambda = 587.56\,\text{nm}$ of the so-called d-line of the Helium spectrum.

The colours in the glass diagram indicate whether the glasses are free of lead and cadmium (green glasses with ‘N’ in the name). The red dots represent glasses that contain these toxic ingredients. The dots with the yellow rings are glasses that can be directly manufactured with precision molding technology.

The n–ν plot of the glass diagram is divided into ranges of glass types, as shown in Figure 2.6. This division approximately shows the manufacturing process or the different substances of the glasses following the characteristic compositions of the melt. But this distinction originates in history. Today, these glass family types are only a rough orientation.

Corresponding to the addition of different ingredients there are also typical differences between glasses within the various glass ranges dependent on their chemical and mechanical properties as well as the ways in which they can be processed.

Since a large ν describes a small dispersion, and vice versa, it is usual to draw in the glass map the corresponding axis in a reverse direction from right to left.

The choice of the available glasses in real glass diagrams is discrete. In practice, however, it can be found that often for the optimization of systems a glass can be substituted without great problem apart from a slight change in the radius and thickness. This is valid, though, only if the relative partial dispersion for the correction of the secondary spectrum does not play a role. Furthermore, the position of the substituted glass must not deviate from the ideal position in the n–ν diagram by more than

- the difference in the refractive indices of $\Delta n = \pm 0.06$
- a ν-difference of $\Delta\nu = \pm 2$ (a few lenses) to $\Delta\nu = \pm 5$ (several lenses).

These numbers are only rules of thumb and can be completely different in the case of sensitive applications.

2.2.6 Interpolation of the Refractive Index

The refractive indices of optical materials are typically measured and listed only for selected wavelengths. In order to determine values of refractive indices for arbitrary wavelengths between these supporting points, dispersion formulas are fitted to the measured data and interpolated at the desired points.

From a physical point of view, the change in the refractive index with the wavelength is caused by characteristic vibrations or rotation levels that are excited on an atomic scale within the structure of the material according to Eq. (2.3). In order to describe the path of the dispersion in a mathematical way, there are two main approaches.

- Modelling the curve $n(\lambda)$ by means of relevant oscillator equation parameters with a sufficient number of terms.
- Modelling a hypothetical behaviour that is mathematically as smooth as possible. These could be, e.g., spline curves, power series expansions or rational approximations.

Both approaches have their advantages and disadvantages. In practice, it can be observed that the first type of approach, following physical first principles, will deliver better results over very large wavelength ranges. Furthermore, potential measurement errors are not a problem as they are not included in the model function, while a purely mathematical approach attempts to describe the interpolation points as closely as possible, and, in the best case, exactly.

The dispersion curves given in the materials catalogues are usually only suitable for interpolation. They do not allow for the determination of refractive indices outside the measured values. This is particularly not recommended for empirical formula approaches that are mathematically motivated. The formulas based on physics do at least allow an extrapolation over small wavelength intervals with an acceptable accuracy.

The interpolation curves that are practically most important for the description of dispersion are listed below. Each parameter has to be determined by fitting a curve for a specific material. Usually, the formulas with a large number of free parameters provide better accuracy. However, in practice, there are simply not enough measurement data to determine all of them exactly. Some historically proposed interpolation formulas are simpler with fewer parameters.

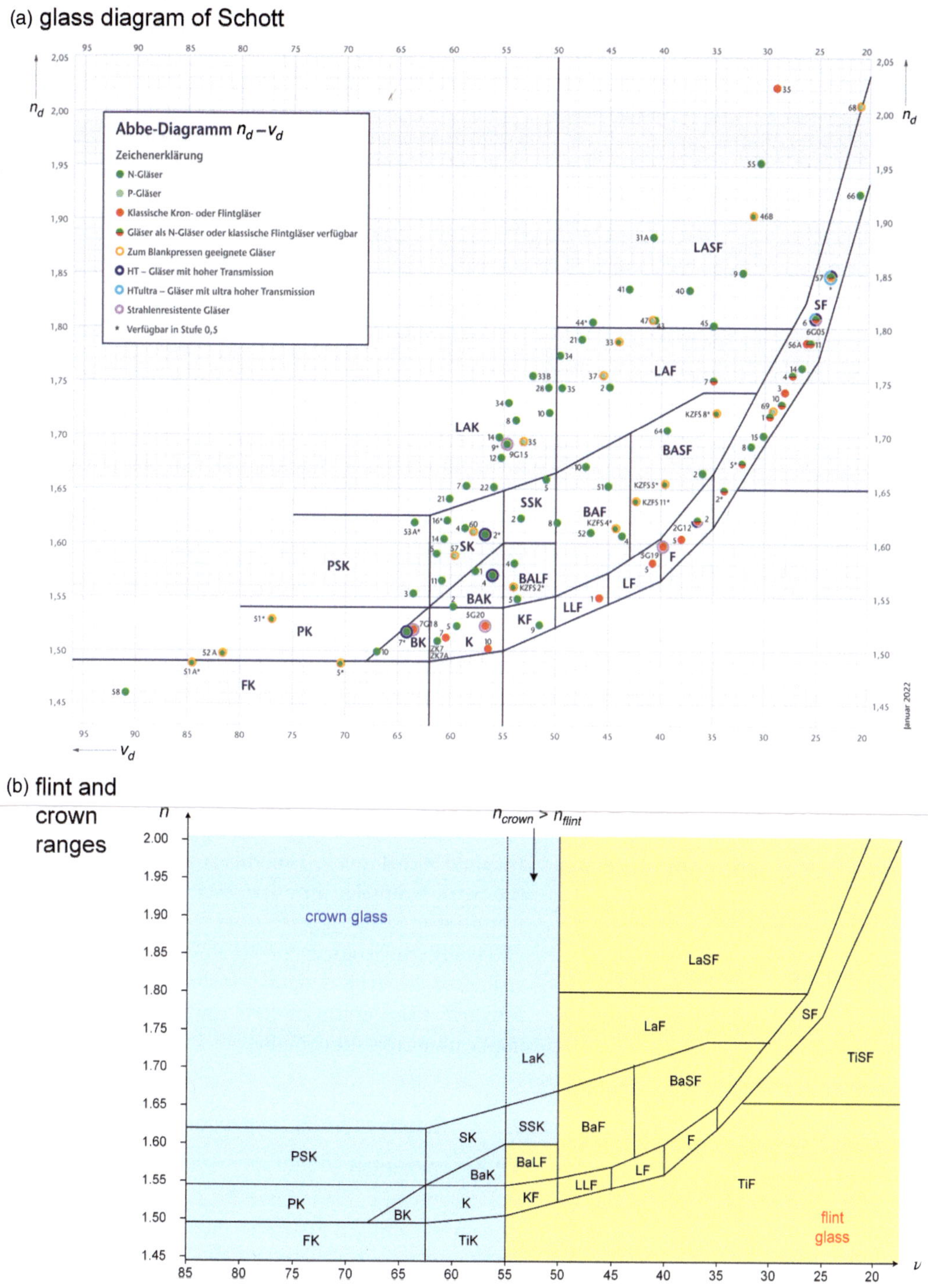

Figure 2.6 Glass diagram of the company Schott in (a) [1, with permission]. It should be noted that usually the Abbe number runs from right to left. Part (b) indicates the ranges of flint and crown glasses.

They are no longer in use today; they stem from an era when the mathematics of fitting and interpolation was done by hand and was cumbersome.

1) Dispersion equation according to Schott (empirical series expansion)

$$n = \sqrt{a_o + a_1\lambda^2 + a_2\lambda^{-2} + a_3\lambda^{-4} + a_4\lambda^{-6} + a_5\lambda^{-8}} \tag{2.16}$$

Because of its even power Taylor-like expansion form, this formula tends to oscillate between the nodes. It is no longer currently recommended by Schott.

2) Dispersion equation according to Sellmeier, which is equivalent to the oscillator model of Eq. (2.3) [23, 24]

$$n = \sqrt{n_o^2 + \sum_j \frac{A_j\lambda^2}{\lambda^2 - \lambda_j^2}} \tag{2.17}$$

This is one of the best approaches and usually used today. In many materials catalogues, only the coefficients are given. If the corresponding glass has only one influencing resonance line in the UV as well as the NIR region near to the visible spectrum, only two terms of the Sellmeier formula give quite good accuracy. If there are two lines on one of the ends, three terms are recommended. Because of the physical basis of the formula, it is also possible to extrapolate indices to a certain extent with acceptable accuracy.

3) Dispersion equation according to Bausch–Lomb. This formula is purely empirical, but delivers excellent results

$$n(\lambda) = \sqrt{A + B\lambda^2 + C\lambda^4 + \frac{D}{\lambda^2} + \frac{E\,\lambda^2}{(\lambda^2 - \lambda_o^2) + \frac{F\lambda^2}{\lambda^2 - \lambda_o^2}}} \tag{2.18}$$

4) Dispersion equation according to Herzberger (mixed form of oscillator model and empirical fit)

$$n(\lambda) = a_o + a_1\lambda^2 + \frac{a_2}{\lambda^2 - \lambda_o^2} + \frac{a_3}{\left(\lambda^2 - \lambda_o^2\right)^2} \quad \text{with}$$

$$\lambda_o = 0.168\ \mu\text{m} \tag{2.19}$$

This dispersion formula is relatively good – especially in the UV range – but not very useful in the IR range.

5) Dispersion equation according to Hartmann I (also more old-fashioned, simple version, empirical)

$$n(\lambda) = a_0 + \frac{a_1}{(a_3 - \lambda)^{1.2}} \tag{2.20}$$

6) Dispersion equation according to Hartmann II (more old-fashioned, simple version, empirical)

$$n(\lambda) = a_0 + \frac{a_1}{a_3 - \lambda} + \frac{a_4}{a_5 - \lambda} \tag{2.21}$$

7) Dispersion equation according to Conrady (more old-fashioned, simple version, empirical)

$$n(\lambda) = a_0 + \frac{a_1}{\lambda} + \frac{a_2}{\lambda^{3.5}} \tag{2.22}$$

8) Dispersion equation according to Kettler–Drude (mixed form of oscillator model and empirical fit)

$$n(\lambda) = \sqrt{A + \frac{B}{\lambda^2 - \lambda_1^2} + \frac{C}{\lambda^2 - \lambda_2^2}} \tag{2.23}$$

9) Dispersion equation according to Cauchy. In principle, this is a simplified form of the old Schott formula

$$n(\lambda) = a_0 + \frac{a_1}{\lambda^2} + \frac{a_2}{\lambda^4} \tag{2.24}$$

The following numerical example and Figure 2.7 shows a typical result for the glass BK7 in the upper row (a). A three-term Sellmeier model dispersion curve is fitted onto 17 measurement data values between $\lambda = 0.365\ldots2.0\ \mu\text{m}$. The overall residuum of the fit is $1.04 \cdot 10^{-6}$ and as is seen in the right-hand diagrams, the worst result for a single measurement point is in the range of $2.5 \cdot 10^{-6}$ for 0.4 μm. In the lower row (b), the same procedure is shown for quartz with 60 measured values of the refractive index and also shorter wavelengths. It is seen that the performance of the fit is worse by a factor of 20 and a systematic error behaviour over the wavelength dependence occurs. The worst-case index is not better than $7 \cdot 10^{-5}$. The reason for this behaviour is the special energy level scheme of quartz mentioned before. It contains a degenerated level and the usual Sellmeier formula describes only single atomic transitions.

Usually an accuracy of 10^{-5} for refractive indices is requested for typical optical system designs, in extreme sensitive cases errors of 10^{-7} can be relevant in critical applications like lithography.

2.2.7 Relative Partial Dispersion

2.2.7.1 Definition

If a dispersion curve is investigated in more detail, it can be seen according to Figure 2.8 that the slope of the curve is significantly changed with wavelength. This means there is a finite curvature in the function and the simple description with the Abbe number alone, more or less characterizing the behaviour by an average slope, must be refined.

The relative partial dispersion gives the dispersion of a glass for two chosen wavelengths λ_1 and λ_2 in relation to the dispersion of the two secondary colours (F, C or F′, C′ in the case of the visible spectrum).

$$P_{\lambda_1\lambda_2} = \frac{n(\lambda_1) - n(\lambda_2)}{n_F - n_C} \tag{2.25}$$

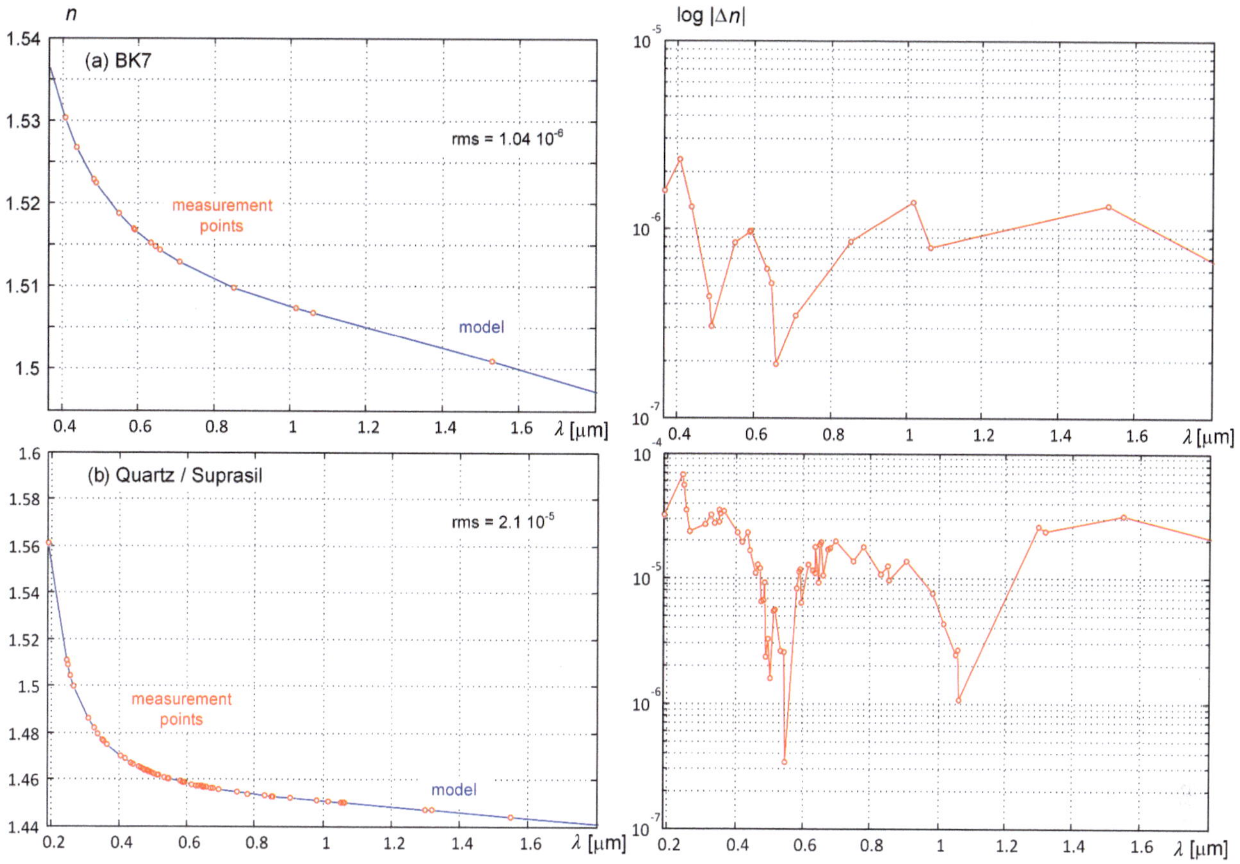

Figure 2.7 Index fit on measured data of (a) BK7 and (b) quartz by a three-term Sellmeier model function. Left-hand sides: measured and fitted dispersion curve $n(\lambda)$; right-hand side: error of the fit as a function of the wavelength.

Figure 2.8 Change of slope in a real dispersion curve; the intervals $\Delta\lambda$ are selected here to have the same lengths.

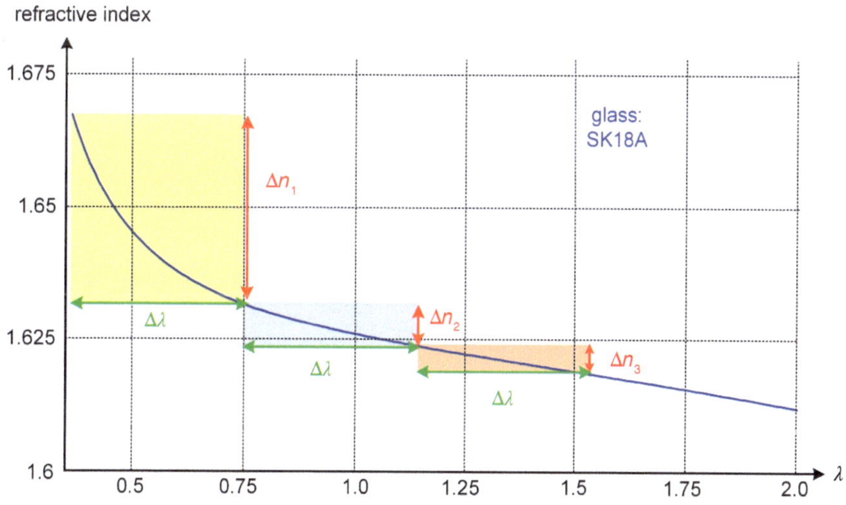

Usually, for comparison, wavelengths in an interval at the red or blue edge of the visible spectral range are chosen. This means that, using this selection, the change in the slope of the dispersion curve between the centre (main dispersion), blue or red end relative to the whole range (determined by λ_F and λ_C) is measured. Hence the partial dispersion is a measure of the curvature of the dispersion curve, similar to the second derivative of the curve $n(\lambda)$. There is not one partial dispersion; at least two values locally at the lower and the higher wavelength range characterize a material. In order to take the different spectral requirements of the optical systems into account, the relative partial dispersions are usually given in the glass catalogues for the following combinations of wavelengths.

$\lambda_1 = 656\,\text{nm},\ \lambda_2 = 1014\,\text{nm}$: $P_{C,t} = \dfrac{n_C - n_t}{n_F - n_C}$ near IR

$\lambda_1 = 656\,\text{nm},\ \lambda_2 = 852\,\text{nm}$: $P_{C,s} = \dfrac{n_C - n_s}{n_F - n_C}$ at the long wavelength end of the visible range

$\lambda_1 = 486\,\text{nm},\ \lambda_2 = 546\,\text{nm}$: $P_{F,e} = \dfrac{n_F - n_e}{n_F - n_C}$ in the visible range near to the centre

$\lambda_1 = 435\,\text{nm},\ \lambda_2 = 486\,\text{nm}$: $P_{g,F} = \dfrac{n_g - n_F}{n_F - n_C}$ in the visible range shifted to blue

$\lambda_1 = 365\,\text{nm},\ \lambda_2 = 435\,\text{nm}$: $P_{i,g} = \dfrac{n_i - n_g}{n_F - n_C}$ at the short wavelength end of the visible range

These five cases are arbitrary; they are illustrated in Figure 2.9. The different secant slopes for the cases should be noted.

For spectral ranges other than the visible, the wavelength selections to define ν and P are arbitrary. For example, in the MWIR between $\lambda = 3\,\mu\text{m}$ and $5\,\mu\text{m}$ the values $\lambda_0 = 4\,\mu\text{m}$, $\lambda_1 = 3\,\mu\text{m}$ and $\lambda_2 = 5\,\mu\text{m}$ are used to formulate the Abbe number.

From the viewpoint of a simple interpretation, the Abbe number indicates the inverse mean slope and the relative partial dispersion of the curvature of the dispersion curve.

This behaviour is sketched in Figure 2.10. In (a) the dispersion curves for artificial glasses with the same centre refractive index and different Abbe numbers are shown, in (b) glasses with the same index and Abbe number but different relative partial dispersion are shown with their dispersion curves. In the latter case, as is seen, the value and the slope for the d-line are the same but the curvatures change.

2.2.7.2 Line of Normal Dispersion

By plotting the relative partial dispersion P for any choice of the wavelengths λ_1 and λ_2, for all glasses vs the Abbe number, points are obtained which lie more or less on a straight line. This behaviour was discovered by Abbe. Those glasses that fulfil this rule with sufficient precision are called normal glasses. Such glasses with a large distance from the normal line are referred to as glasses with anomalous partial dispersion. The linear relation is determined, following convention, by the two glasses K7 and F2. The line equation and its two constants, respectively, depend on the choice of the reference wavelengths λ_1 and λ_2.

The deviation $\Delta P_{\lambda1,\lambda2}$ of the relative partial dispersion from the line of normal dispersion (or iron line) is a measure for the anomalous relative partial dispersion, see also Figure 2.11. This behaviour is now mathematically written as a linear relation with parameters a and b and a deviation ΔP for the description of the anomalous deviation as

$$P_{\lambda1,\lambda2} = a_{\lambda1,\lambda2} \cdot \nu_d + b_{\lambda1,\lambda2} + \Delta P_{\lambda1,\lambda2} \tag{2.26}$$

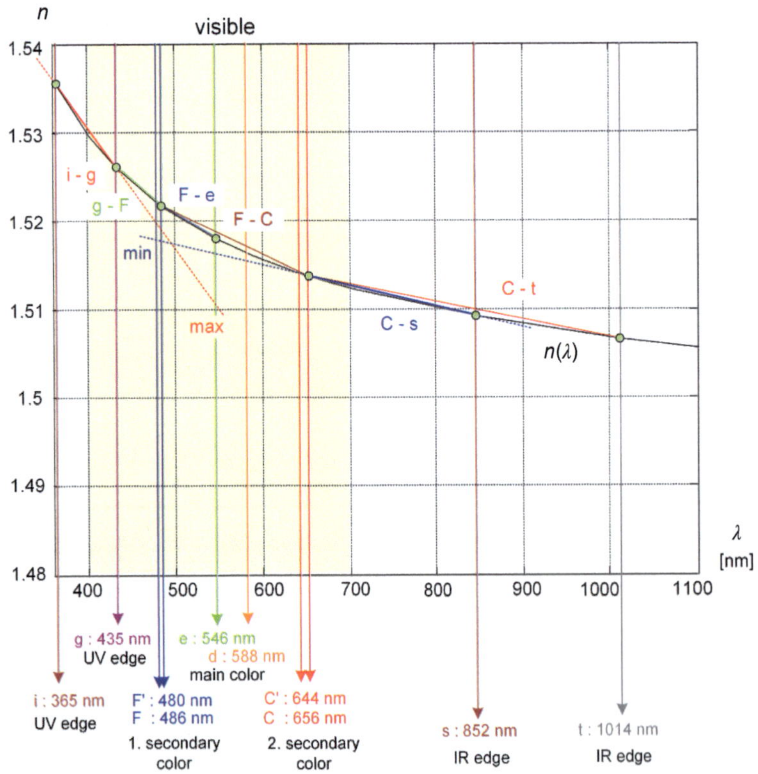

Figure 2.9 Dispersions of the refractive indices as a function of the choice of wavelength. Change of the slope over the whole spectral range and selected partial dispersion definitions by special pairs of the wavelength.

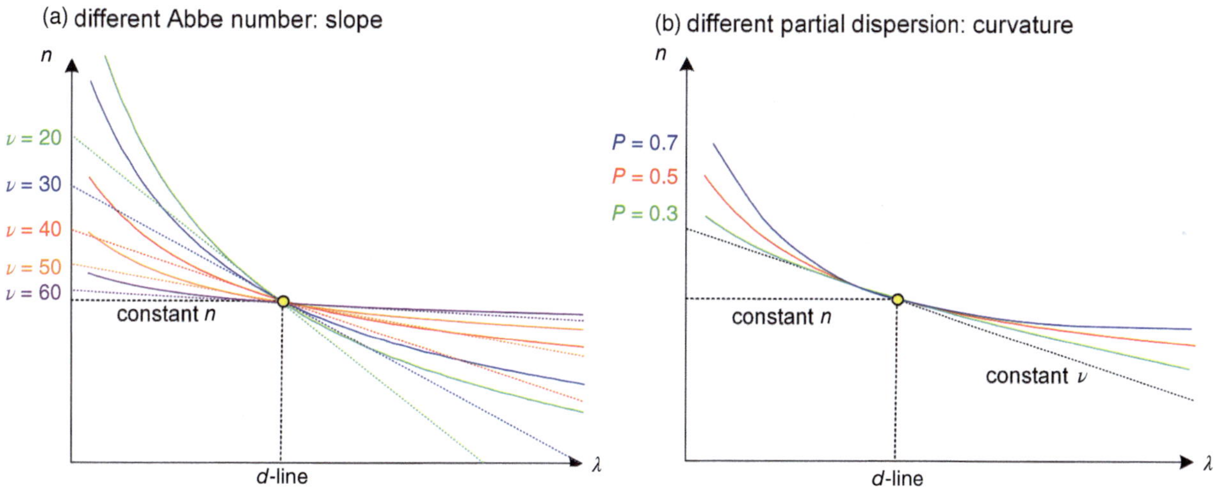

Figure 2.10 Schematic dispersion curves for different (a) Abbe numbers and (b) relative partial dispersion.

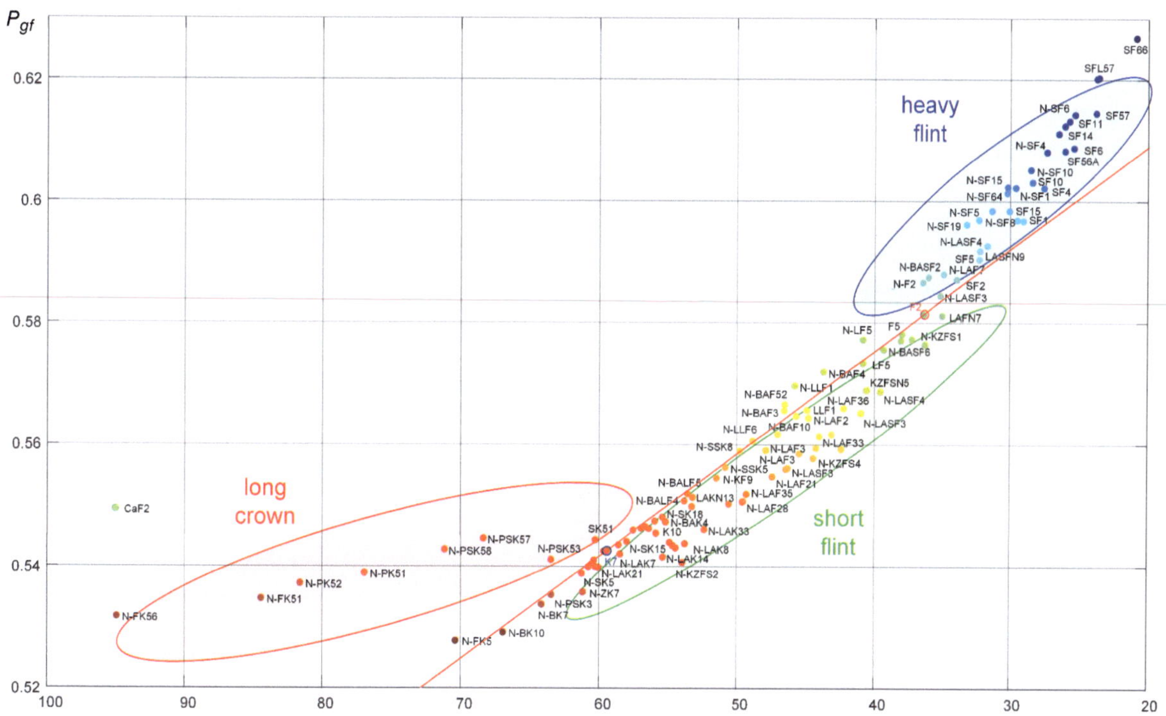

Figure 2.11 ν–P_{gF} diagram with some glasses as well as the normal line (red). The colours of the dots indicate the ν-values.

Describing the behaviour of the dispersion in a similar way by the dispersion number m according to Eq. (2.26), the corresponding linear relation is

$$m_{\lambda 1,\lambda 2} = \alpha_{\lambda 1,\lambda 2} \cdot m_e + \beta_{\lambda 1,\lambda 2} + \sigma_{\lambda 1,\lambda 2} \tag{2.27}$$

In Figure 2.11 the selected wavelengths are g and F, so this describes the relation in the blue range of the visible.

A special material is CaF_2. It is not a glass, but a crystal. It has been used in lens design for colour correction for a long time. Therefore, its large Abbe number $\nu = 95$ and the partial dispersion $P_{gF} = 0.549$ are included in the Figure 2.11 for completeness.

The deviations $\Delta P_{\lambda 1,\lambda 2}$ or $\sigma_{\lambda 1,\lambda 2}$ respectively from the normal line characterize the behaviour of the glass with reference to the normal partial dispersion. These deviations are illustrated in Figure 2.12 by arrows at the P–ν points of the glasses.

In the P–ν plot the normal lines are described, dependent on the choice of the wavelengths, by the following

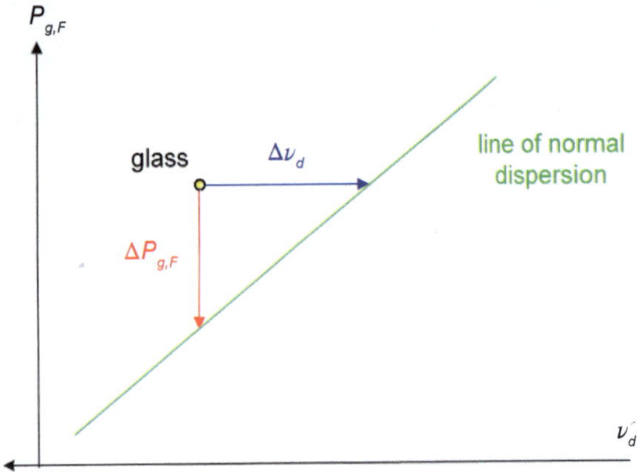

Figure 2.12 Relative partial dispersion as a function of the Abbe number; here only shown for one glass.

equations form NIR over red, blue to the UV range.

$$\begin{aligned}
\tilde{P}_{C,t} &= 0.5450 + 0.004743 \cdot \nu_d \\
\tilde{P}_{C,s} &= 0.4029 + 0.002331 \cdot \nu_d \\
\tilde{P}_{F,e} &= 0.4884 - 0.000526 \cdot \nu_d \\
\tilde{P}_{g,F} &= 0.6438 - 0.001682 \cdot \nu_d \\
\tilde{P}_{i,g} &= 1.7241 - 0.008382 \cdot \nu_d
\end{aligned} \tag{2.28}$$

For the glasses on the normal line, the difference vanishes: $\Delta P_{\lambda 1,\lambda 2} = 0$. Choosing different reference wavelengths, the coefficients immediately change very sensitively. Referring to the refractive indices as e, F' and C' instead of d, F and C, then the fourth of the above equations becomes, for example

$$\tilde{P}_{g,F'} = 0.567959 - 0.0014353 \cdot \nu_e \tag{2.29}$$

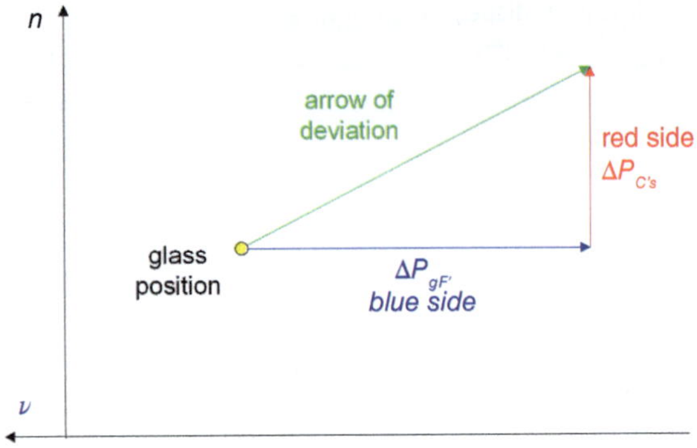

Figure 2.14 Indication of the dispersion behaviour by means of an arrow in the n–ν glass diagram; here only shown for one glass.

Depending on the choice of wavelengths, the normal lines have a different position and slope. This effect is shown in Figure 2.13 for one λ-range in the visible blue (gF) and red (Cs) edge respectively.

Special glasses with a noticeable deviation from the normal line ΔP are necessary in optical design for the correction of the secondary spectrum (apochromatic correction). To indicate the deviation from the normal line in the n–ν glass chart, a vector can be added at the n–ν point of a glass. Its horizontal component gives a deviation ΔP in the blue spectral range and its vertical component gives one in the red range. So, from the corresponding arrows at the glass points, the capability of the glass on the specific colour correction for blue and red can be estimated. This is shown in Figure 2.14.

If the partial dispersion values red vs blue are drawn, the called Herzberger PP-diagram is obtained [29], Figure 2.15 shows this linear correlation for the case P_{Cs} vs P_{gF}:

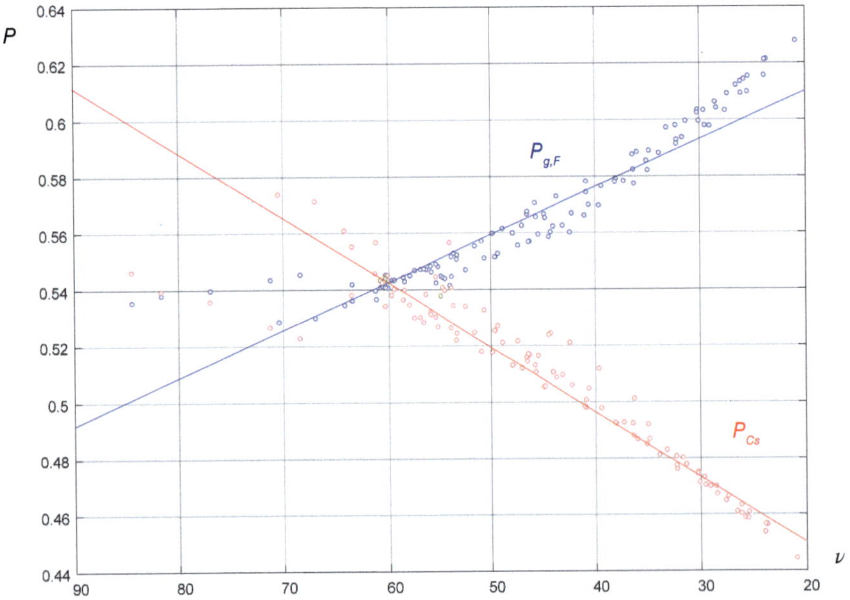

Figure 2.13 Partial dispersion P as a function of the Abbe number ν for two different choices of wavelength pairs [4] / with permission of SPIE.

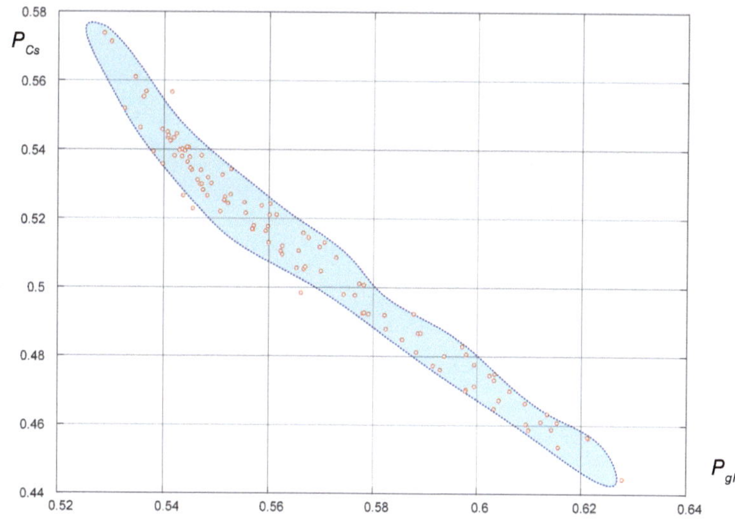

Figure 2.15 Partial dispersion blue vs red side according to Herzberger.

2.2.7.3 Anomalous Partial Dispersion

In order to achieve a colour correction in optical systems over a larger spectral range, materials with a large anomalous relative partial dispersion are necessary. There are three characteristic types: glasses with long-crown characteristics, glasses with short-flint characteristics and the so-called heavy flints. The attributes 'long' and 'short' in the names long-crown and short-flint are derived from evaluation of the length of the spectrum as it becomes visible in a spectrometer. The spectral range with the largest dispersion is always decisive for the length of the spectrum. In the visible spectrum, this is the blue range. So, compared to a normal glass, short-flints show a shortened, and long-crowns an extended, spectrum.

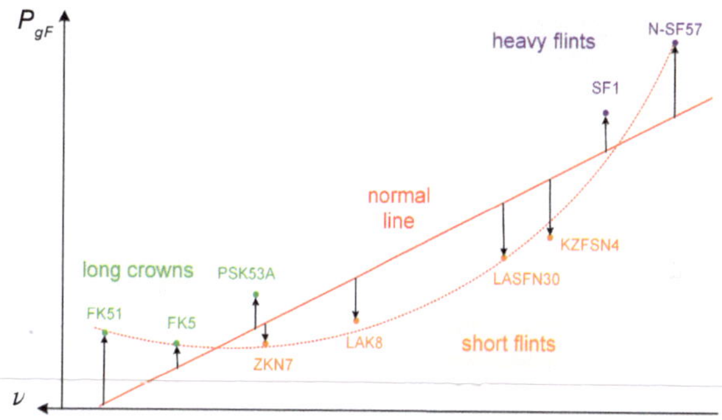

Figure 2.16 Glasses with strong anomalous partial dispersion, vertical distance from the $P_{g,F}$ normal line, with selected concrete glass representatives.

- Long-crowns are optical crowns above the normal line in the $P_{g,F}\nu$-chart. Compared with the normal glasses these so-called long-crowns have a stronger dispersion in the blue wavelength range and a weaker one in the red range. Here, the glass types FK, PK, PSK and the crystal CaF2 are important. The highly refracting dense optical flints also show a long-crown character. In this respect, e.g. N-SF6, SF11, SF14, N-SF15 and N-SF57, should be mentioned.
- Short flints are glasses below the normal line in the $P_{g,F}\nu$-chart. These short-flints show a weaker dispersion in the blue range and a stronger dispersion than normal glasses in the red range. Important representatives of this type are, e.g., SF, TiF, LaF, LaK and LaSF, KZFS types.
- Heavy-flints again lie above the normal line but at small ν-values.

  Typical representatives are SF6, SF57 and more SF-glasses.

Typical representatives of glasses with anomalous partial dispersion are those with large deviations from the normal line. They are necessary for apochromatic corrections and are shown in Figure 2.11 and schematically in Figure 2.16. In this plot, as well as in the former ones, the ν-axis is inverted.

Usually, the curvature of the dispersion curves scales with the slope; this means that the curves are self-similar. Materials with anomalous partial dispersion, on the other hand, have a different curvature than this simple normal behaviour. The long-crowns and the heavy-flints have a larger curvature, the short-flints have a smaller curvature. This behaviour is schematically visualized in Figure 2.17.

A different possibility to visualize the anomalous partial dispersion candidates of the glasses is a classical n–ν glass diagram with the ΔP_{gF} deviation (or other wavelength combination) shown as coloured dots of different sizes

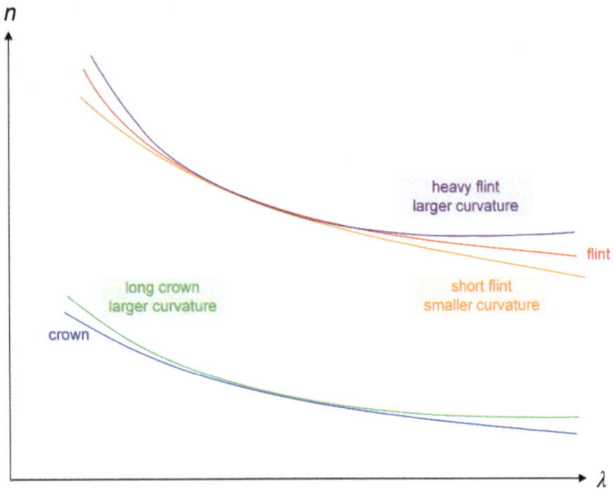

Figure 2.17 Dispersion curves with normal and anomalous partial dispersion ΔP_{gF} for flint and crown glasses.

coding the ΔP-values. Figure 2.18 is an example of such a representation. Small orange bubbles lie near to the normal line, blue and green bubbles are above and red bubbles are below. The larger the bubble, the larger the deviation from the normal line. In this diagram the grouping of the glasses far from the normal line is more pronounced than in the usual P–ν diagram.

2.2.7.4 Hoogland Diagram*

(*) The content of this section is quite specialized and for advanced reading. Knowledge of this section is not necessary to understand the remaining chapters. Beginners are recommended to skip this section.

In the so-called Hoogland diagram, the expression $(P–b)/\nu$ is rearranged from Eq. (2.26) in the form

$$\frac{P_{\lambda 1,\lambda 2} - b_{\lambda 1,\lambda 2}}{\nu_e} = a_{\lambda 1,\lambda 2} + \frac{\Delta P_{\lambda 1,\lambda 2}}{\nu_e} \tag{2.30}$$

and is plotted vs $1/\nu$ [17]. In this representation, all glasses with the same partial dispersion lie on a line through the origin. The normal line is equivalent to a vertical line parallel to the $1/\nu$ axis (see Figure 2.19). So, for an apochromatic correction, the choice of glasses should be on a straight line through the origin and as distant from each other as possible. A large distance between the points on a line with large distance to the normal line leads to small focal powers of the components; they have a quite different ν value. The two glasses at the green dots are an example of a clever glass choice for an apochromatic correction.

A real example of a Hoogland diagram is given in Figure 2.20 for a selected assortment of glasses with the colours e, F′, C′ being chosen for ν and g and F′ for P. Here, one can clearly see the range with the normal glasses lying near to the normal line in a particular distribution. Furthermore, the heavy-flint, long-crown and short-flint glasses with large anomalous dispersion are crowded in the elliptical coloured areas.

2.2.7.5 The Anomalous Dispersion Representation of Münz*

(*) The content of this section is quite specialized and for advanced reading. Knowledge of this section is not necessary to understand the following chapters. Beginners are recommended to skip this section.

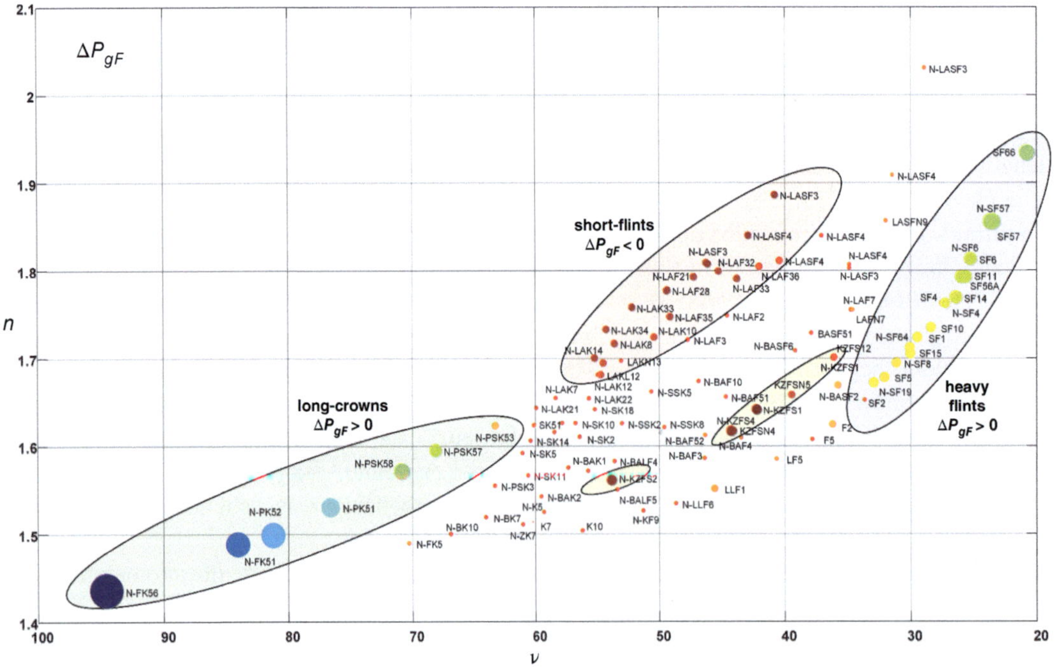

Figure 2.18 n–ν glass map with partial dispersion ΔP_{gF}.

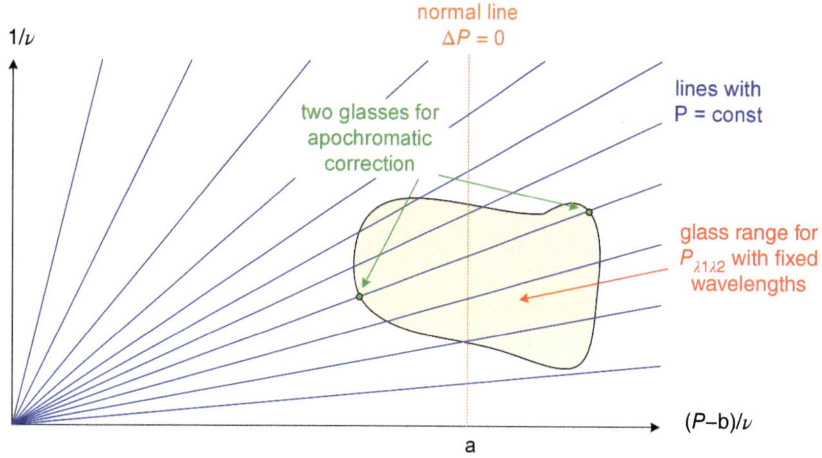

Figure 2.19 Sketch of a Hoogland diagram.

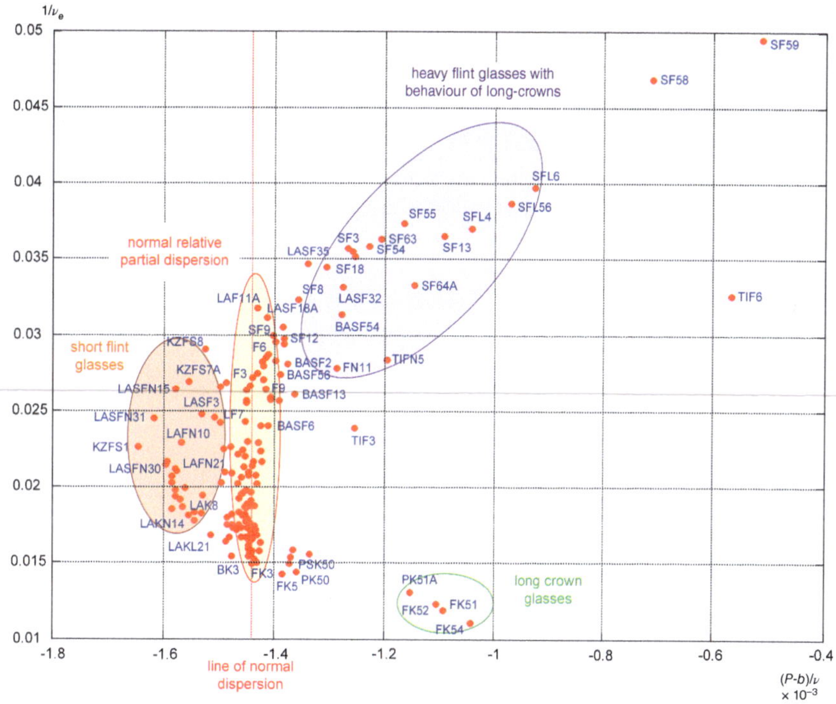

Figure 2.20 Hoogland diagram of Schott glasses for λ_e.

In the literature, there are in addition some further new proposals to describe the non-normal dispersion [26, 30]. The primary dispersion is indicated by the inverse of ν, which is identical to the dispersion number m of Section 2.3.4

$$D = \frac{1}{\nu} = \frac{n_1 - n_2}{n_p - 1} \tag{2.31}$$

with the principal wavelength p. The deviations from the normal line can be characterized by the term

$$D_2 = \frac{n_p - \frac{n_1+n_2}{2}}{n_p - 1} \tag{2.32}$$

In Figure 2.21 a simple interpretation of the term D_2 is sketched. If the secant between the index values of the outer wavelengths in the dispersion curve is evaluated at the midpoint $(\lambda_1 + \lambda_2)/2$ and the index is compared with the value for the principal wavelength λ_p, the vertical difference is proportional to D_2. If the principal wavelength is near to the midpoint, which is mostly a good approximation in reality, the largest vertical extension of the arc-shaped coloured area is a measure for D_2. In the special case that $\lambda_m = \lambda_p = (\lambda_1 + \lambda_2)/2$, by rearranging we can see that the relation

$$D_2 = \frac{P_{2m} - P_{1m}}{2\nu_m} \tag{2.33}$$

connects the D_2 definition with the relative partial dispersions P on the left and on the right side, normalized by ν. In practice, λ_p is mostly very near to the centre λ_m, so this

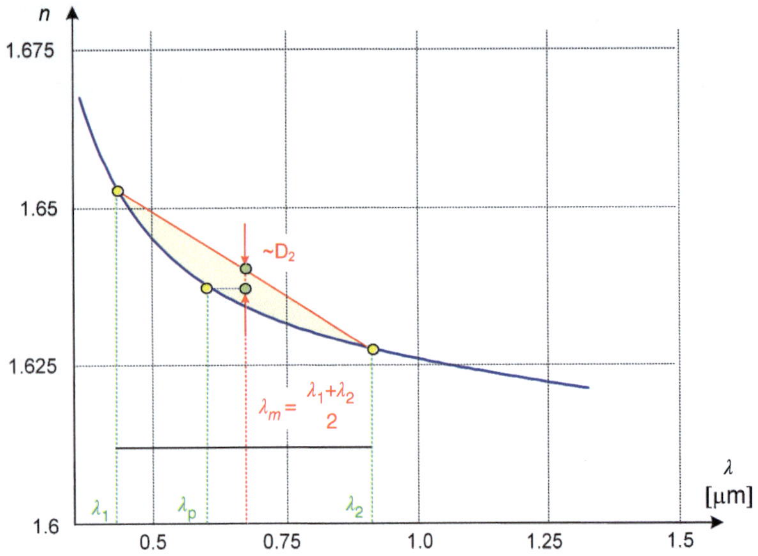

Figure 2.21 Interpretation of the value D_2.

assumption is a good approximation. In the D_2 definition, there is no distinction between the blue and the red side of the spectrum, the partial dispersion is considered symmetrical. This means that the information of a certain asymmetry is lost, but there is only one parameter to describe the partial dispersion and the dependence on the arbitrary choice of the wavelengths for the local slopes is gone.

For normal glasses, D_2 should be proportional to D

$$D_2 = D_{20} + \alpha \cdot D \tag{2.34}$$

Therefore, a plot of the glasses in a diagram D_2 vs D give a good indication of the colour correction opportunities of the glasses, see Figure 2.22 (a). Therefore, for an individual glass with index j a certain small deviation β_j from the normal line is

$$D_{2j} = D_{20} + \alpha \cdot D_j + \beta_j \tag{2.35}$$

and the residual deviation can be written as

$$D_{2effj} = D_{2,j} - \alpha \cdot D_j = D_{20} + \beta_j \tag{2.36}$$

which is called the effective dispersion of higher order and is shown in Figure 2.22 (b). Materials to the right have strong dispersion, materials with considerably higher or lower levels then the zero-line have a strong anomalous dispersion.

If in general M wavelengths are considered and J lenses close together are used to achieve an apochromatic correction, the conditions can be written as a linear system of equations using the Abbe numbers as well as the partial dispersion coefficients P in the form [26]

$$F = \sum_{j=1}^{J} F_j \ , \ \sum_{j=1}^{J} \frac{F_j}{\nu_j} = 0 \ , \ \sum_{j=1}^{J} \frac{P_{3j} \cdot F_j}{\nu_j} = 0, \ \ldots\ldots$$

$$\sum_{j=1}^{J} \frac{P_{mj} \cdot F_j}{\nu_j} = 0 \ , m = 1 \ldots M \tag{2.37}$$

The weighting factors of the component powers F_j can be written as

$$\frac{P_{mj}}{\nu_j} = \frac{n_m - n_j}{n_F - n_C} \cdot \frac{n_F - n_C}{n_{dj} - 1} = \frac{n_m - n_j}{n_{dj} - 1} \tag{2.38}$$

If now the mean values of the refractive index as well as the powers respective to the wavelength range $m = 1, 2 \ldots M$ are considered

$$\overline{F}_j = \frac{1}{M}\sum_{m=1}^{M} F_j(\lambda_m), \quad \overline{n}_j = \frac{1}{M}\sum_{m=1}^{M} n_j(\lambda_m) \tag{2.39}$$

the achromatization condition can be written by summation of the above equations over the wavelengths as

$$F = \sum_{j=1}^{J} \overline{F}_j \ , \ \sum_{j=1}^{J} \frac{n_j(\lambda_m) - \overline{n}_j}{\overline{n}_j - 1} \cdot \overline{F}_j = 0,$$

$$\sum_{j=1}^{J} \delta_j(\lambda_m) \cdot \overline{F}_j = 0, \quad m = 1 \ldots M \tag{2.40}$$

with the abbreviation

$$\delta_j(\lambda_m) = \frac{n_j(\lambda_m) - \overline{n}_j}{\overline{n}_j - 1} \tag{2.41}$$

The δ_j are the normalized index differences and describe the dispersion behaviour in a more comfortable way. A picture of δ vs ν is shown in Figure 2.23 for the wavelengths F, d and C.

The δ-coefficients are highly correlated for different glasses, see Figure 2.24. The residuum of the straight line fit to the measurement points is $4.3 \cdot 10^{-17}$.

The refractive power of a single lens at the wavelength m is now given by

$$\begin{aligned} F_j(\lambda_m) &= \overline{F}_j + \Delta F_j(\lambda_m) = \overline{F}_j + \delta_j(\lambda_m) \cdot \overline{F}_j(\lambda_m) \\ &= \overline{F}_j \cdot \left[1 + \delta_j(\lambda_m)\right] \end{aligned} \tag{2.42}$$

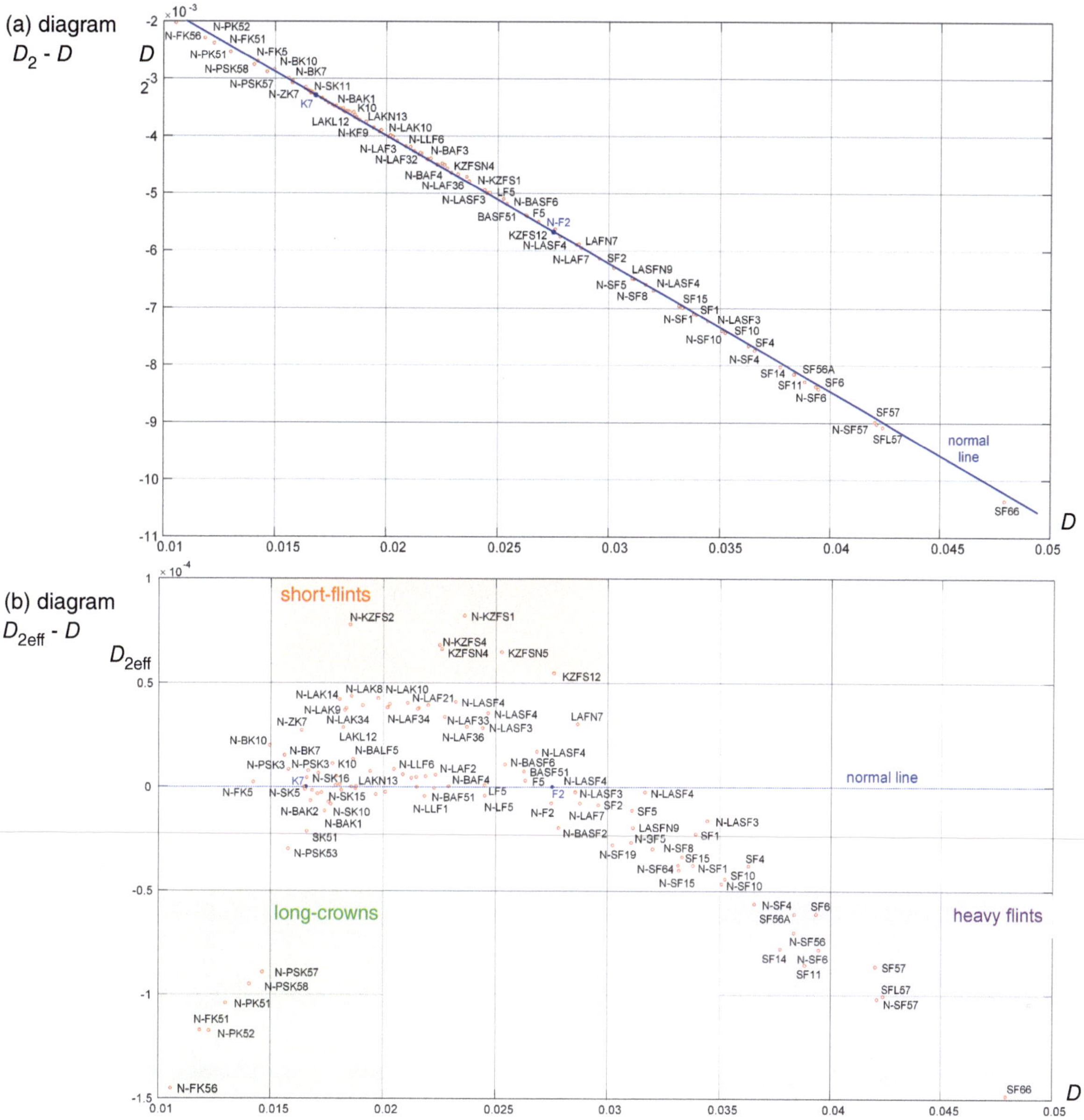

Figure 2.22 Glasses in the (a) D_2 vs D and (b) D_{2eff} vs D representation.

and the paraxial axial colour aberration at wavelength m for a system reads

$$\Delta F(\lambda_m) = \sum_{j=1}^{J} \delta_j(\lambda_m)\overline{F}_j(\lambda_m) \tag{2.43}$$

2.2.8 Generalized Mathematical Descriptions of Dispersion

2.2.8.1 Introduction

There are different options to characterize the dispersive behaviour of optical materials. These are as follows.

- The traditional method proposed by Abbe uses the Abbe number ν and the partial dispersion P. This means that the dispersive behaviour is characterized at concrete wavelengths, so a kind of discretized spectral sampling is used. This approach is mostly used in lens design, although there are many problems in applying this for more broadband chromatical correction or different spectral ranges than the visible, as infrared.

There is one more attempt discussed in the literature by Herzberger. Based on the special dispersion function of Eq. (2.19), an achromatization for four wavelengths is obtained, in principle this is an extension of the ν–P-method with two more wavelengths, but restricted on four discrete spectral sampling points.

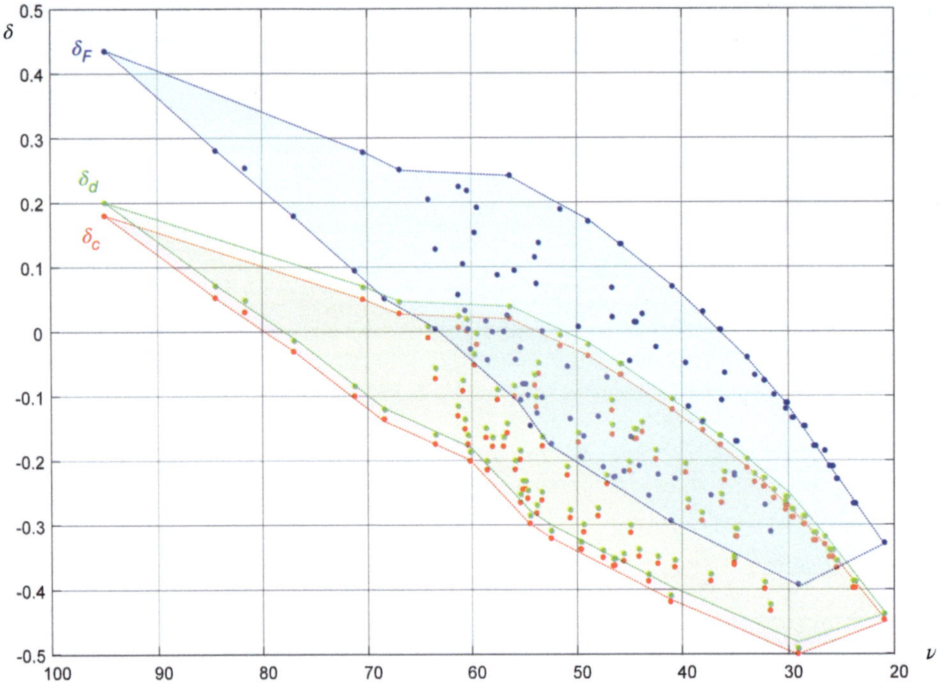

Figure 2.23 Variable δ_C, δ_d and δ_F vs ν.

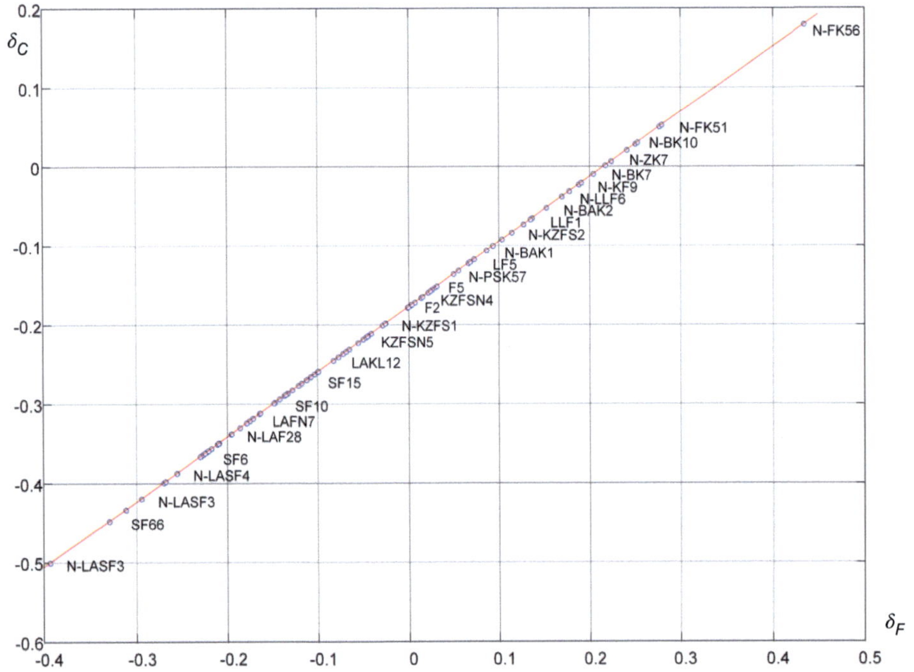

Figure 2.24 Dependence δ_C vs δ_F.

- Another option is to use the derivative $dn/d\lambda$ at the central wavelength together with the second derivative for higher order considerations. This is a kind of local or narrow spectral range approximation in the sense of a Taylor expansion around the centre wavelength. This approach is recommended in [28]. From the mathematical point of view this means, the dispersion curve is fitted locally at the centre.
- A special chromatical coordinate is defined by Buchdahl [16, 18, 19, 22]. From the mathematical point of view, the dispersion curve of a glass is expanded into a series of functions. These functions are all the same for all glasses

and can be interpreted as 'dispersion modes'. This means that the real dispersion curve is approximated over the full wavelength range. The parameter of the shape functions is optimized to provide fast convergence and a small number of terms for the required accuracy. The expansion coefficients describe the dispersion.
- The Sellmeier representation is of this type but uses other mode functions.
- A more general discussion of adapted chromatical coordinates is found in [15, 29].

This means that there are in principle three different types of approaches:

- based on a few discrete wavelengths:
  examples: Abbe, Herzberger;
- based on a Taylor expansion with derivatives locally at the spectral midpoint:
  examples: Rayces, Sasian;
- based on an λ-averaged approximation by appropriate functions:
  examples: Buchdahl, Sellmeier, Sasian.

These three possibilities are sketched in Figure 2.25 for clarification. When deciding on the most comfortable description of the dispersion, the goals in practice are:

- easy and accurate possibilities for fitting and interpolation with a small number of terms, in the best case also with a limited option for extrapolation;
- direct usability and clear interpretation in colour correction for achromates, apochromates etc.

The main approaches mentioned above are discussed in the following sections in more detail.

2.2.8.2 Derivative Based Dispersion Description*

(*) The content of this section is quite specialized and for advanced reading. Knowledge of this section is not necessary to understand the following chapters. Beginners are recommended to skip this section.

If the dispersion curve is considered around the central wavelength λ_o between λ_{max} and λ_{min} with the interval

$$\Delta\lambda = \left(\lambda_{max} - \lambda_{min}\right)/2 \tag{2.44}$$

the dispersion curve can be expressed as a Taylor expansion in the form

$$n(\lambda) = n_0 + \frac{dn}{d\lambda}(\lambda - \lambda_0) + \frac{1}{2}\frac{d^2n}{d\lambda^2}(\lambda - \lambda_0)^2 + \frac{1}{6}\frac{d^3n}{d\lambda^3}(\lambda - \lambda_0)^3 + \dots \tag{2.45}$$

A corresponding definition of a differential Abbe number, describing the slope of the dispersion curve locally at λ_0, can be expressed by [13, 14]

$$\nu_{\lambda 0} = -\frac{1}{2}(n-1)/\left(\frac{dn}{d\lambda}\right) \approx \Delta\lambda \cdot \nu \tag{2.46}$$

and the local partial dispersion correspondingly as

$$P \approx \frac{1}{2} + \Delta\lambda \cdot P_{\lambda 0}, \qquad P_{\lambda 0} = -\frac{1}{4} \cdot \frac{d^2n}{d\lambda^2} / \frac{dn}{d\lambda} \tag{2.47}$$

Similar to Abbe's linear relation between ν and P, here the equation

$$P_{\lambda 0} = A(\lambda) \cdot \nu_{\lambda 0} + B(\lambda) \tag{2.48}$$

can be written. In the limit of a very narrow bandwidth $\Delta\lambda \to 0$ we get the connection to the Abbe parameters a and b of the normal line as

$$A(\lambda) = \frac{a}{\Delta\lambda^2}, \quad B(\lambda) = \frac{2b-1}{2\Delta\lambda} \tag{2.49}$$

If the definition of $\nu_{\lambda 0}$ and $P_{\lambda 0}$ are inserted into the relation (2.45), a differential equation of the form

$$n''(\lambda) + 4B(\lambda) \cdot n'(\lambda) - 2A(\lambda) \cdot [n(\lambda) - 1] = 0 \tag{2.50}$$

is obtained. If this equation is solved by numerical methods, the curves of Figure 2.26 for $A(\lambda)$ and $B(\lambda)$ are found as an approximate averaged functional dependence for all glasses. A represents the slope of the P–ν-curve with wavelength and B the absolute offset. Therefore, this curve summarizes all possible wavelength combinations for the P definitions. There is an interesting anomaly at $\lambda = 1.44$ µm.

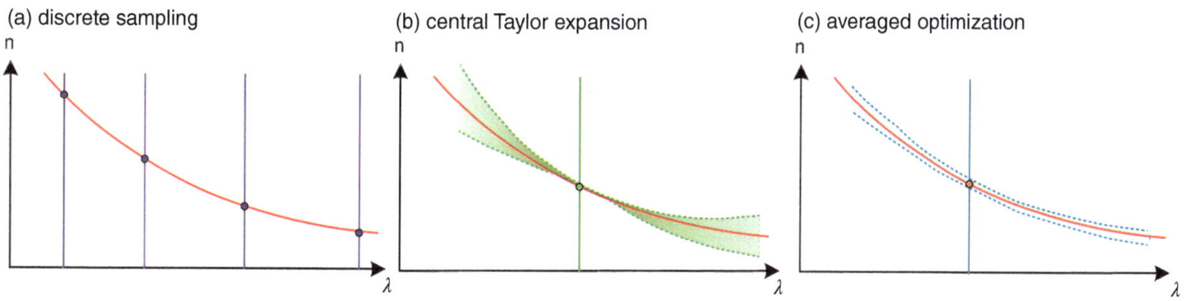

Figure 2.25 Possible approaches for the description of the dispersion curves (a) discrete sampling, (b) Taylor expansion and (c) averaged optimization.

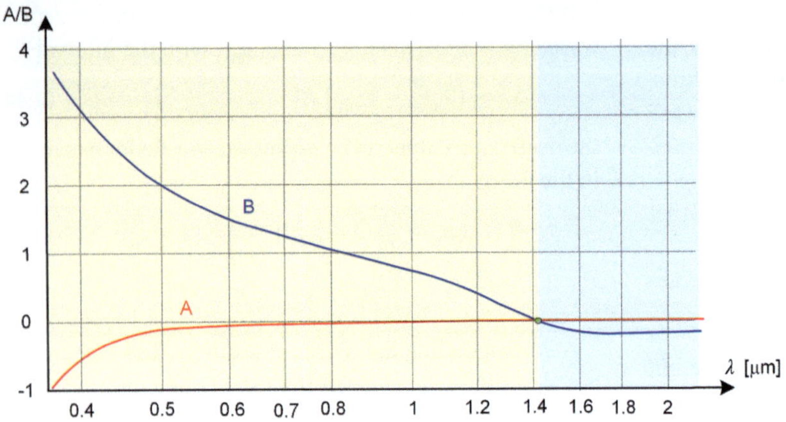

Figure 2.26 Parameters A and B as a function of λ.

Here in the NIR we have $P = 0$ for all glasses. This means, from the viewpoint of application in the colour correction, that all glasses can be used to construct an apochromate at this wavelength.

It should be mentioned that this differential approach with Taylor expansion is more accurate and comfortable the smaller the spectral bandwidth $\Delta\lambda$ of an application is.

2.2.8.3 Buchdahls Chromatical Coordinates*

(*) The content of this section is quite specialized and for advanced reading. Knowledge of this section is not necessary to understand the following chapters. Beginners are recommended to skip this section.

The so-called chromatic coordinate according to Buchdahl is defined as [16, 18, 19, 22]

$$\omega(\lambda) = \frac{\lambda - \lambda_0}{\lambda_{ref} + \alpha \cdot (\lambda - \lambda_0)} \tag{2.51}$$

with the wavelength λ_0 parameter in the mid position of the used spectral range and with $\lambda_{ref} = 1\ \mu\text{m}$. Practical tests give as a result that the constant α is best to be $\alpha = 2.5$, if $\lambda_0 = 574\ \text{nm}$, if the visible range is selected. Other ranges and midpoints must be optimized individually with different α values.

With the help of this coordinate, the refractive index of a glass can be described as a functional expansion after the powers of ω with the coefficients a_j

$$n(\omega) = n_0 + \sum_{j=1} a_j \omega^j = n_0 + a_1 \cdot \omega + a_2 \cdot \omega^2 + a_3 \cdot \omega^3 + \ldots + a_m \cdot \omega^m + \ldots \tag{2.52}$$

Figure 2.27 shows the functions $\omega^j(\lambda)$ over the wavelength in the visible spectral range. The expansion coefficients a_j give a description of the dispersion curve in the j-th order. The odd-even characteristic is clearly seen depending on the power; for a better visualization, here the higher power functions are rescaled. In principle, the Buchdahl representation corresponds to an expansion of the dispersion curve with these mode function powers. The functions are constant for all glasses, the individual dispersion properties of the glass types are coded in the expansion coefficients a_j. The reference wavelength λ_{ref} and the factor α are arbitrary and are selected to guarantee the best convergence of the expansion.

As is demonstrated in the literature, it is mathematically simple to convert the Sellmeier or Schott coefficients to fix the dispersion curve into the a_j parameter analytically or numerical.

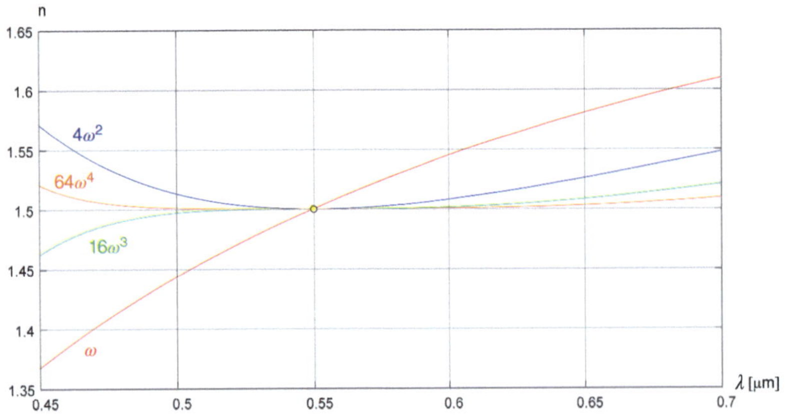

Figure 2.27 Shape functions according to Buchdahls approach for the orders 1–4 as a function of the wavelength in the visible spectrum. For a better representation, the higher powers are rescaled.

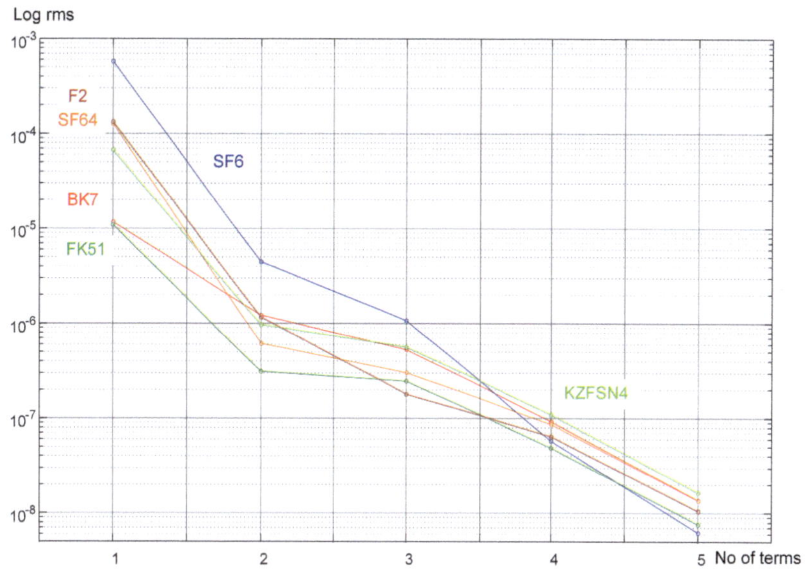

Figure 2.28 Accuracy or residual rms deviation of a Buchdahl expansion for several glasses and different numbers of terms.

The advantage over the classical expansion of $n(\lambda)$ is a significantly better convergence. The promoted fast convergence of the Buchdahl expansion can be seen in Figure 2.28. Here the residual rms error of the refractive index representation is depicted as a function of the number of terms in the sum of Eq. (2.52) for several glasses. It can be seen that for a standard accuracy of 10^{-5} for all glasses this requires only two terms and we have

$$n(\omega) \approx n_0 + a_1 \cdot \omega + a_2 \cdot \omega^2 \tag{2.53}$$

If the transform Eq. (2.51) is resolved for λ, we get the relation

$$\lambda = \lambda_0 + \frac{\omega \cdot \lambda_{ref}}{1 - \alpha \cdot \omega} \tag{2.54}$$

with $\alpha = 2.5$ in a more general formulation. Here the classical description of the dispersion according to Abbe is equivalent to a linear approach. Therefore, the chromatic conditions are already described by the equation of second order with an accuracy higher by one magnitude than by using the simple ν approach. Figure 2.29 shows the chromatic variation of the transverse aberration of the marginal ray for a simple lens in three different ways: the exact calculation; the linear dispersion model using the Abbe number; and the quadratic model according to Buchdahl. The significantly better accuracy of the quadratic approach can be easily seen.

The dispersion function of a glass can be described by a renormalized equation which is similar to the inverse Abbe number definition as

$$D(\lambda) = \frac{n(\lambda) - n_o}{n_0 - 1} = \sum_j \frac{a_j \cdot \omega^j(\lambda)}{n_0 - 1} = \sum_j \eta_j \cdot \omega^j(\lambda) \tag{2.55}$$

The focal power of a thin lens can then be written as

$$\begin{aligned} F(\lambda) &= F(\lambda_0) \cdot [1 + D(\lambda)] \\ &= F(\lambda_0) \cdot [1 + \eta_1 \omega(\lambda) + \eta_2 \omega^2(\lambda) + \ldots] \end{aligned} \tag{2.56}$$

So, a system can be achromatized using the Buchdahl coordinates. Then, using a quadratic approach for the refractive index, together with the focal length condition, an equation for selection of the modified glasses is obtained. In the most simple case, the system of equations is linear. Then pairs of glasses have to be determined whose Buchdahl coefficients obey that relation.

Based on this we can also easily obtain a chromatic correction for more than two wavelengths. For the determination of the solutions, a diagram of the η parameters of the available glasses is needed. This is shown for the Schott glasses in Figure 2.30. The interpretation and practical usage of this kind of diagram is explained in the chapter about colour correction.

As is pointed out in the literature, the concept of Buchdahl can also be used in the infrared; the typical values for the coefficients are slightly different. In particular in the NIR the constant α should be taken as $\alpha = 1.5$ to guarantee a good convergence of the expansion.

2.2.8.4 Adaptive Dispersion Formula of Sasian*

(*) The content of this section is quite specialized and for advanced readingKnowledge of this section is not necessary to understand the following chapters. Beginners are recommended to skip this section.

In [15] it is proposed to define a normalized chromatical coordinate by

$$\Delta\lambda = \frac{\lambda - \lambda_0}{\lambda_{dif}}, \quad \lambda_{dif} = \max(\lambda_{\max} - \lambda_0, \lambda_0 - \lambda_{\min}) \tag{2.57}$$

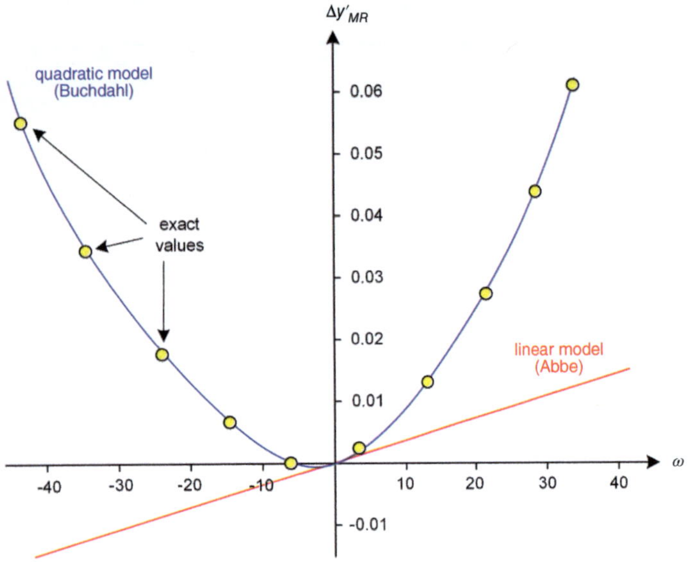

Figure 2.29 Dispersion of the transverse aberration of a marginal ray in a lens. Shown are the exact calculation (yellow dots); linear approximation of Abbe (red line) and Buchdahl second order approximation (blue line).

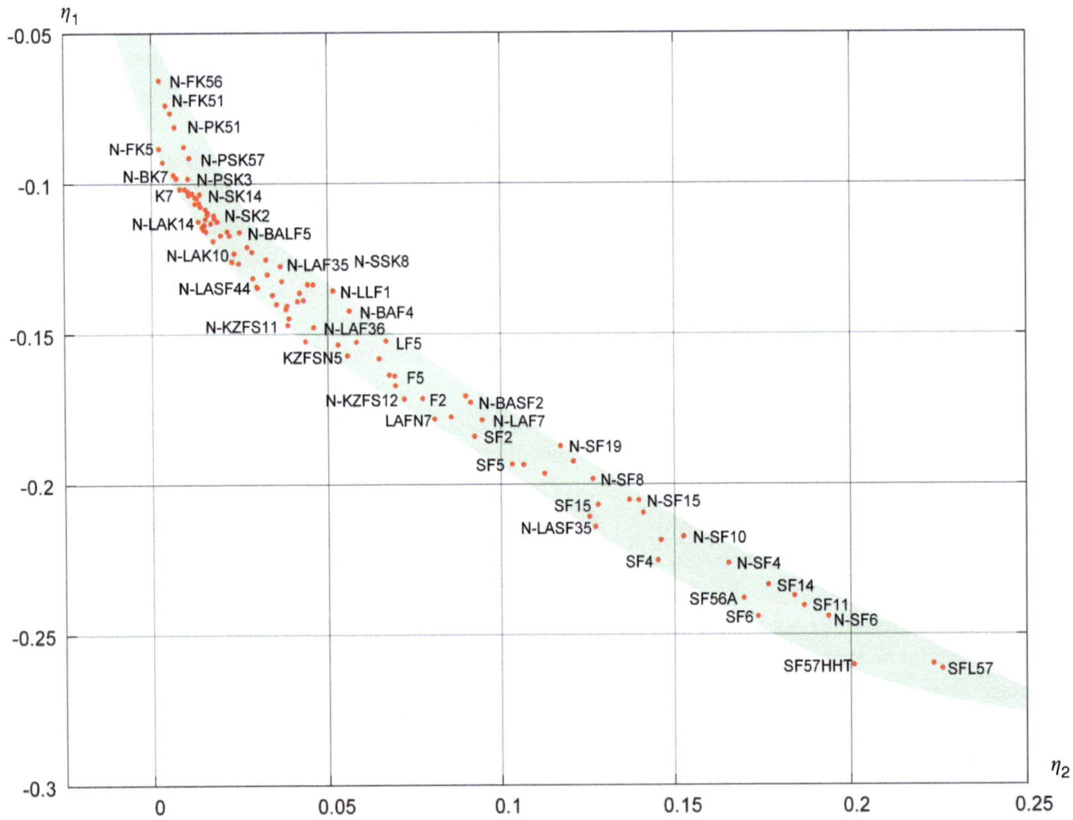

Figure 2.30 Schott glasses modelled by a second order Buchdahl expansion; diagram of the first order coefficient η_1 against the second order η_2. Some of the data points are indicated by the glass names.

where λ_0 is the reference wavelength in or near to the centre of the considered λ range between $\lambda_{min} \dots \lambda_{max}$. Thus, a normalization of the wavelengths with values $|\Delta\lambda| \leq 1$ is obtained and $\Delta\lambda$ has no unit. In the special case that λ_0 is exactly in the centre, the scaled wavelength variable has values in the interval $\Delta\lambda = -1 \dots +1$, This scaled wavelength with $|\Delta\lambda| < 1$ can be used in a comfortable way as a variable for an expansion. Formulated in these normalized coordinates $\Delta\lambda$, the dispersion of a glass can be written with very high accuracy in the form

$$n = n_0 + A_1 \cdot \Delta\lambda + A_2 \cdot \Delta\lambda^2 + A_3 \cdot \Delta\lambda^3 + \dots + A_m \cdot \Delta\lambda^m + \frac{A_{m+1} \cdot \Delta\lambda^{m+1}}{1 + K \cdot \Delta\lambda} \tag{2.58}$$

There are m power terms plus one rational term. Since the complete λ-range under consideration is given by $\Delta\lambda$, the coefficient n_0 gives the average value of $n(\lambda)$, A_1 describes the overall change of n in the interval and is related to the slope, A_2 fixes the quadratic contribution (curvature) of the dispersion curve, A_3 the cubic asymmetrical part and is related to the curvature change. This means the coefficients A_m give the orders of the Taylor expansion of $n(\lambda)$.

Figure 2.31 shows the shape of the lower orders as a function of the normalized chromatical coordinate in (a) and exemplary shapes of the rational term for $K = 0.5$. The rational term contains the highest power as a distorted asymmetric contribution. According to practical experience, this improves the convergence great deal. If the functions are optimized for glasses, the value of K is in the range of $K = 0.5$.

If a system of thin lenses is considered, the change of focal length can be written by wave aberration terms as a sum over the lenses with curvatures c_{j1}, c_{j2}. With

$$\Delta W_{020} = \sum_j \frac{1}{2} y_j^2 (c_{j1} - c_{j2}) \cdot (n_j - n_{j0}) \tag{2.59}$$

and the abbreviation of the pure geometrical influence

$$S_j = \frac{1}{2} y_j^2 \left(c_{j1} - c_{j2}\right) \tag{2.60}$$

used, a chromatical correction requires with growing order:

1) achromatic correction of first order:

$$\sum_j S_j A_{j1} = 0 \tag{2.61}$$

2) correction of secondary spectrum, second order:

$$\sum_j S_j A_{j2} = 0 \tag{2.62}$$

3) correction of tertiary spectrum, third order:

$$\sum_j S_j A_{j3} = 0 \tag{2.63}$$

Therefore, the chromatical correction in higher order over the wavelength range is directly related to the coefficients A_m of the dispersion representation. This means that A_1 corresponds to the Abbe number ν, A_2 to the partial dispersion P.

2.2.9 Miscellaneous

2.2.9.1 Dispersion of Diffractive Surfaces

The usual Abbe number is defined in the visible as

$$\nu_e = \frac{n_e - 1}{n_{F'} - n_{C'}} \tag{2.64}$$

and is used to describe the spectral behaviour of a component in the visible wavelength region. The secondary effects

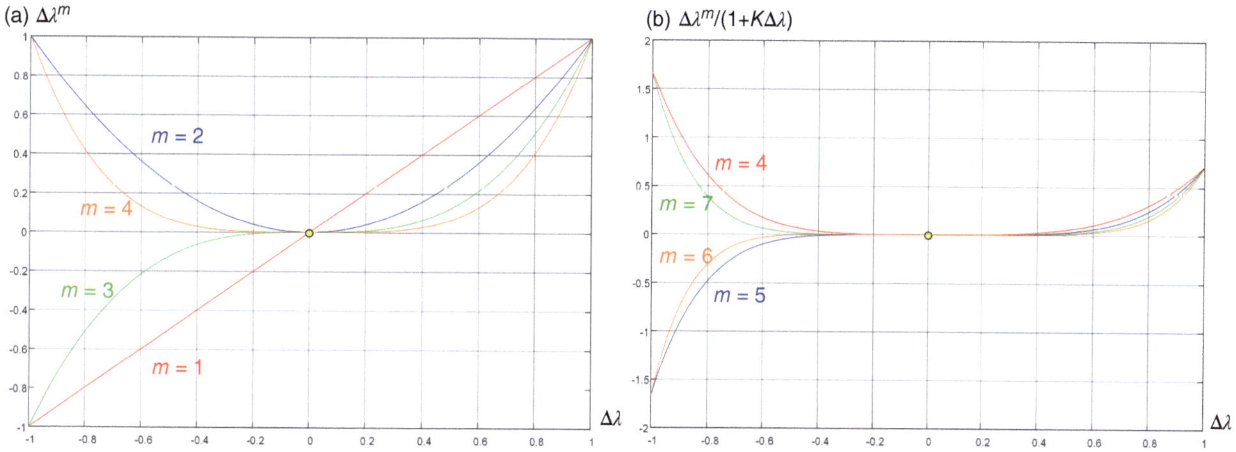

Figure 2.31 Mode functions of Sasian's decomposition with (a) polynomial terms and (b) the rational term with $K = 0.5$.

of the chromatic performance can be included with the help of the relative partial dispersion number

$$P_{g,F'} = \frac{n_g - n_{F'}}{n_{F'} - n_{C'}} \tag{2.65}$$

Now a corresponding equation for the power of a diffractive component is derived here [12]. The angle difference of the grating equation

$$m \cdot \lambda = g \cdot \sin\theta \tag{2.66}$$

gives for the change in intersection length in analogue to the focal length change

$$\frac{\Delta f}{f} = \frac{\Delta(\sin\theta)}{\theta} \approx \frac{\Delta\theta}{\theta} = \frac{m \cdot \Delta\lambda}{g} \cdot \frac{g}{m \cdot \lambda} = \frac{\Delta\lambda}{\lambda} = \frac{1}{\nu} \tag{2.67}$$

Therefore, we get the equivalent Abbe number of a grating

$$\nu_e = \frac{\lambda_e}{\lambda_{F'} - \lambda_{C'}} = -3.330 \tag{2.68}$$

and the partial dispersion

$$P_{g,F'} = \frac{\lambda_g - \lambda_{F'}}{\lambda_{F'} - \lambda_{C'}} = 0.2695 \tag{2.69}$$

These abnormal values for the dispersion and the relative partial dispersion are indicated in Figure 2.32 in the usual P–ν diagram for some typical glass materials by the yellow dot in the lower right corner.

Table 2.3 compares the corresponding values with typical glass materials of crown- and flint-type glasses with diffractive optical elements (DOEs) for different spectral ranges and the choice of the wavelength d, F and C for the visible spectrum.

Table 2.3 Dispersion and relative partial dispersion numbers for glasses and diffractive lenses in comparison for different spectral regions. The Abbe number and the relative partial dispersion are defined according to the wavelengths in the spectral range as indicated. For the visible, ν_d and P_{gF} are used, in the infrared the outer and the centre wavelengths are selected.

λ range	Material	Abbe number	Relative partial dispersion
Visible 486–656 nm	Diffractive	−3.45	0.596
	BK7	64.12	0.693
	F2	36.37	0.706
Mid-IR 3–5 μm	Diffractive	−2.00	0.50
	Germanium	101.5	0.688
	Silicone	250.0	0.670
	ZnSe	179.4	0.625
Far-IR 8 –12 μm	Diffractive	−2.50	0.50
	Germanium	1001.1	0.70
	ZnSe	58.6	0.458

2.2.9.2 Dispersion of Gradient-Index Materials*

(*) The content of this section is quite specialized and for advanced reading. Knowledge of this section is not necessary to understand the following chapters. Beginners are recommended to skip this section.

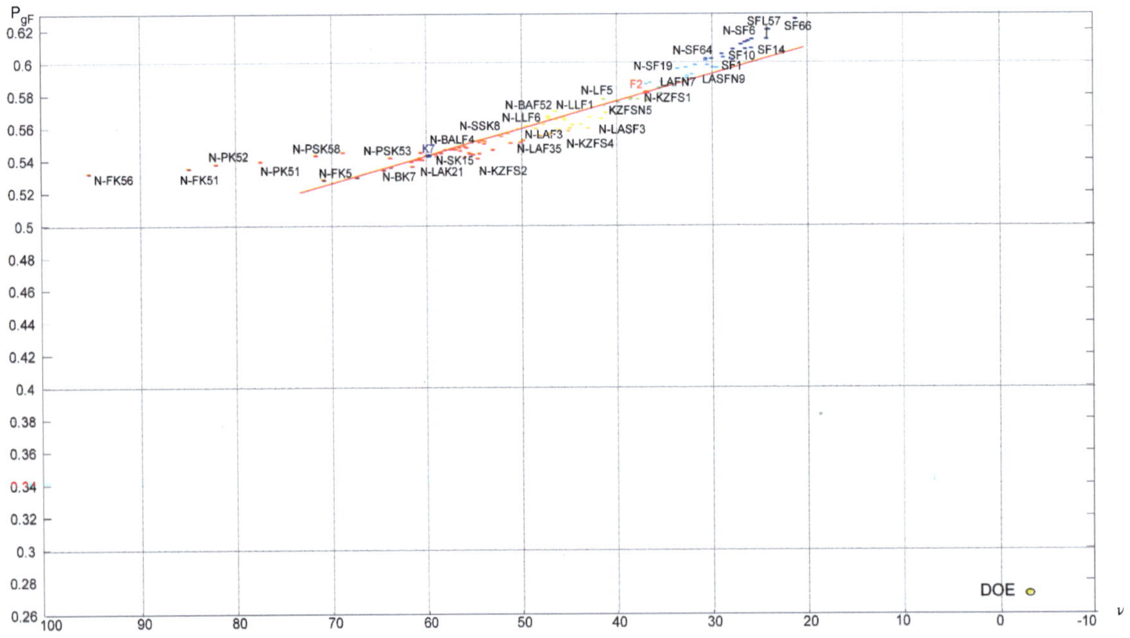

Figure 2.32 Dispersion ν_e and relative partial dispersion $P_{gF'}$ of a diffractive component in comparison with some typical glasses.

Gradient index media have a non-uniform index distribution $n(x,y,z)$ by intention. Usually, the index profile is described by a Taylor expansion for example in the following form

$$n(r,z) = n_{00,\lambda} + N_{10,\lambda}\, r^2 + N_{20,\lambda}\, r^4 + N_{01,\lambda}\, z + N_{02,\lambda}\, z^2 + \ldots \tag{2.70}$$

The dispersion of gradient index media is a special problem. This topic is discussed in the literature in several contributions [31–36]. In principle, all the coefficients of the Taylor expansion of a gradient index (GRIN) profile depend on the wavelength. In reality, the index changes are performed by diffusion processes of materials other than the substrate into the glass. Special ions of the matrix are exchanged and influence the refractive index. The spectral behaviour depends on the choice of the ion-exchange materials that are used for index modification.

For practical reasons, in lens design, the description of the chromatic behaviour of the GRIN lenses is formulated in a similar way as in classical glass dispersion. It is standard to use Buchdahl's chromatic coordinates or to use Sellmeier or Herzberger representations for the various coefficients.

The dispersion of the coefficients in the formula above can then be described; for example, by a classical Abbe number [36]

$$\nu_{00} = \frac{n_{00,d} - 1}{n_{00,F} - n_{00,C}} \tag{2.71}$$

and equivalent expressions for all other coefficients as Abbe numbers and relative partial dispersion coefficients

$$\nu_{jk} = \frac{N_{jk,d}}{N_{jk,F} - N_{jk,C}}, \quad P_{jk} = \frac{N_{jk,d} - N_{jk,C}}{N_{jk,F} - N_{jk,C}} \tag{2.72}$$

It should be noticed here that the '1' is missing in the last formulae in the nominator. These variables are then Taylor expanded as a function of the spatial position

$$\nu_{jk}(r,z) = \sum_{j,k} \nu_{jk} \cdot r^{2j} z^k, \quad P_{jk}(r,z) = \sum_{j,k} P_{jk} \cdot r^{2j} z^k \tag{2.73}$$

Alternatively, the Buchdahl chromatical coordinates ω can be used and the Taylor expansion in r and z is extended in one more dimension as power of ω [31, 32].

2.3 Group Velocity Dispersion and Short Pulses*

(*) The content of this section is quite specialized and for advanced reading. Knowledge of this section is not necessary to understand the following chapters. Beginners are recommended to skip this section.

2.3.1 Introduction

There are several modern applications using quite short laser pulses that are decomposed by a certain wavelength range in coherent superposition. The shape of the pulses and their modification in a beam guiding optical system is important for the applications. If an optical system is used with ultra short pulses, the dispersion of a refractive material changes the pulse shape. Dependent on the order and power dependence, the time of arrival, the width or the symmetry of the pulse is changed. Therefore, in ultra-short pulse optical systems, the dispersion is directly related to the time behaviour and is thus quite important in these types of applications [14, 37–47]. Due to the properties of this application, mostly the circle frequency ω and the propagation constant $k(\omega)$ are used to describe the dispersion effects.

Every spectral component of the pulse propagates in optical media at a different phase velocity due to dispersion in the material $n(\omega)n(\omega)$. Therefore, the pulse phase changes while travelling in a medium.

The propagation of the electromagnetic field in a medium is described by the wave equation. A solution of this equation is a plane wave propagating in the z direction

$$A(z,\omega) = A(0,\omega) \cdot e^{-i \cdot k(\omega) \cdot z} \tag{2.74}$$

Here $k(\omega) = \beta k(\omega)$ is the propagation constant that is governed by the dispersion relation. For the travelling of a wave in the time-spatial domain we can write

$$A(z,t) = A_0 \cdot e^{i \cdot [\omega_0 t - k(\omega_0) z]} = A_0 \cdot e^{i \cdot \omega_0 \cdot \left[t - \frac{z}{c/n_0}\right]} = A_0 \cdot e^{i \cdot \omega_0 \cdot \left[t - \frac{z}{v_p}\right]} \tag{2.75}$$

with the phase velocity v_p. The function $k(\omega)$ can be formulated as a function of the circle frequency ω

$$k(\omega) = \beta(\omega) = k_0 \cdot n(\omega) = \frac{\omega}{c_0} \cdot n(\omega) = \frac{2\pi \cdot n(\omega)}{\lambda_0} \tag{2.76}$$

or the frequency $\nu = \omega/(2\pi)$

$$k(\nu) = \frac{2\pi \cdot n(\nu) \cdot \nu}{c_0} \tag{2.77}$$

The dependence of k as a function of the frequency describes the dispersion, k includes the material properties, which is not the case for ν. A Taylor expansion around the central frequency ν_0 gives the weak dependence of the k-vector on the frequency.

$$k(\nu_0 + \nu) = k(\nu_0) + \nu \cdot \frac{dk}{d\nu} + \frac{\nu^2}{2} \cdot \frac{d^2k}{d\nu^2} \tag{2.78}$$

The wave number k is equivalent to the propagation constant β. In reality, mostly the frequency ω is used instead of ν.

2.3.2 Dispersion Coefficients

To consider higher order dispersion coefficients, the abbreviations

$$\beta_m = \frac{d^m k}{d\omega^m} \tag{2.79}$$

are introduced. With these parameters we obtain

$$k(\omega) = \beta_0(\omega_0) + \beta_1 \cdot (\Delta\omega) + \frac{1}{2}\beta_2 \cdot (\Delta\omega)^2 + \frac{1}{6}\beta_3 \cdot (\Delta\omega)^3 + \ldots \tag{2.80}$$

The sorting and meaning of the various orders are as follows.

- β_0 propagation constant of the centre or carrier frequency, phase velocity of centre frequency, it finally describes a delay of the pulse centre.
- $\beta_1 = \frac{1}{v_g}$ is the inverse group velocity, this causes a delay of the group relative to the phase (PTD).
- $\beta_2 = D$ corresponding to the second order dispersion parameter D, this is identical to the group velocity dispersion (GVD) of first order. The GVD causes a broadening of the pulse; k changes as a function of the frequency deviation $\Delta\omega$ of the individual pulse frequency components.
- $\beta_3 = D_3$ describes to the third order dispersion (TOD) or the second order dispersion of the group velocity. The TOD causes a spectral dependent asymmetry of the pulse.

Dealing with short pulses, the effects of dispersion are often expressed as derivatives with respect to the frequency $\omega = 2\pi v\omega = 2\pi \bullet v$. In optics in general, derivatives with respect to wavelength λ are common. The following formula connects these descriptions and allows switching between the variables and derivatives of the Taylor expansion

$$\frac{d}{d\omega} = -\frac{\lambda^2}{2\pi \cdot c_0} \cdot \frac{d}{d\lambda} \tag{2.81}$$

$$\frac{d^2}{d\omega^2} = \frac{\lambda^2}{(2\pi \cdot c_0)^2} \cdot \left(2\lambda \cdot \frac{d}{d\lambda} + \lambda^2 \frac{d^2}{d\lambda^2}\right) \tag{2.82}$$

$$\frac{d}{dv} = -\frac{\lambda^2}{c_0} \cdot \frac{d}{d\lambda} \tag{2.83}$$

$$\frac{d^2}{dv^2} = \frac{\lambda^2}{c_0^2} \cdot \left(2\lambda \cdot \frac{d}{d\lambda} + \lambda^2 \frac{d^2}{d\lambda^2}\right) \tag{2.84}$$

$$\frac{dk}{dv} = \frac{2\pi n}{c_0} + \frac{2\pi v}{c_0} \cdot \frac{dn}{dv} = \frac{2\pi}{c_0} \cdot \left(n - \lambda \cdot \frac{dn}{d\lambda}\right) \tag{2.85}$$

$$\frac{d^2k}{dv^2} = \frac{4\pi}{c_0} \frac{dn}{dv} + \frac{2\pi v}{c_0} \cdot \frac{d^2n}{dv^2} \tag{2.86}$$

The dispersion coefficients β are then

$$\beta_1 = \frac{1}{v_g} = \frac{n_g}{c_0} = \frac{dk}{d\omega} = \frac{d}{d\omega}\left(\frac{n \cdot \omega}{c_0}\right) = \frac{1}{c_0}\left(n + \omega \cdot \frac{dn}{d\omega}\right) = \frac{1}{c_0} \cdot \left(n - \lambda \cdot \frac{dn}{d\lambda}\right) \tag{2.87}$$

$$\beta_2 = D = \frac{d^2k}{d\omega^2} = \frac{1}{c_0} \cdot \left(2\frac{dn}{d\omega} + \omega \cdot \frac{d^2n}{d\omega^2}\right) = \frac{\lambda^3}{2\pi\, c_0^2} \frac{d^2n}{d\lambda^2} \tag{2.88}$$

$$\beta_3 = D_3 = \frac{d^3k}{d\omega^3} = \frac{1}{c_0}\left(3\frac{d^2n}{d\omega^2} + \omega\frac{d^3n}{d\omega^3}\right) = -\frac{\lambda^4}{4\pi^2\, c_0^3}\left(3\frac{d^2n}{d\lambda^2} + \lambda_0\frac{d^3n}{d\lambda^3}\right) \tag{2.89}$$

According to the equation for β_1 and the fact that for every glass material $dn/d\lambda < 0$, we get the important property that the impact of the dispersion always causes a time delay and never a group behaviour running ahead.

2.3.3 Pulse Changes due to Dispersion

Since the validity of the wave equation can also be expressed as the requirement that for any spectral component within the pulse $\Delta\omega \ll \omega_0$ $\Delta\omega \ll \omega_0$ holds, we can expand the propagation constant about the carrier frequency $\omega = \omega_0 + \Delta\omega$. Now we $\omega = \omega_0 + \Delta\omega$ restrict the consideration to the second order expansion of the wave number as a function of the frequency ω

$$k(\Delta\omega) = k(\omega_0) + \frac{dk}{d\omega}(\Delta\omega) + \frac{1}{2} \cdot \frac{d^2k}{d\omega^2}(\Delta\omega)^2 + \frac{1}{6} \cdot \frac{d^3k}{d\omega^3}(\Delta\omega)^3 \tag{2.90}$$

If the pulse shape is considered, the envelope and the exact high frequency changes must be distinguished. The changes of the envelope due to the global pulse shape according to the coefficients β_1, β_2 and β_3 during a propagation in a homogeneous medium are illustrated in Figure 2.33.

With the abbreviations v_g and D, the propagation constant can be written as

$$k(\omega) = k(\omega_0) + \frac{1}{v_g}\Delta\omega + \frac{1}{2}D \cdot \Delta\omega^2 \tag{2.91}$$

Substituting this relation into the plane wave equation, one obtains

$$A(z, \omega) = A(0, \omega) \cdot e^{-i \cdot \left(k(\omega_0)z + \frac{z}{v_g}\Delta\omega + \frac{z \cdot D}{2}(\Delta\omega)^2\right)} \tag{2.92}$$

The superposition of the plane waves yields the propagated time-dependent field by integration

$$E(z, t) = e^{i \cdot (\omega_0 t - k(\omega_0)z)} \cdot \int_{-\infty}^{\infty} A(0, \Delta\omega) \cdot e^{-i \cdot \left(\Delta\omega \frac{z}{v_g} + \frac{z \cdot D}{2}(\Delta\omega)^2\right)} e^{i \cdot \Delta\omega t} d(\Delta\omega) \tag{2.93}$$

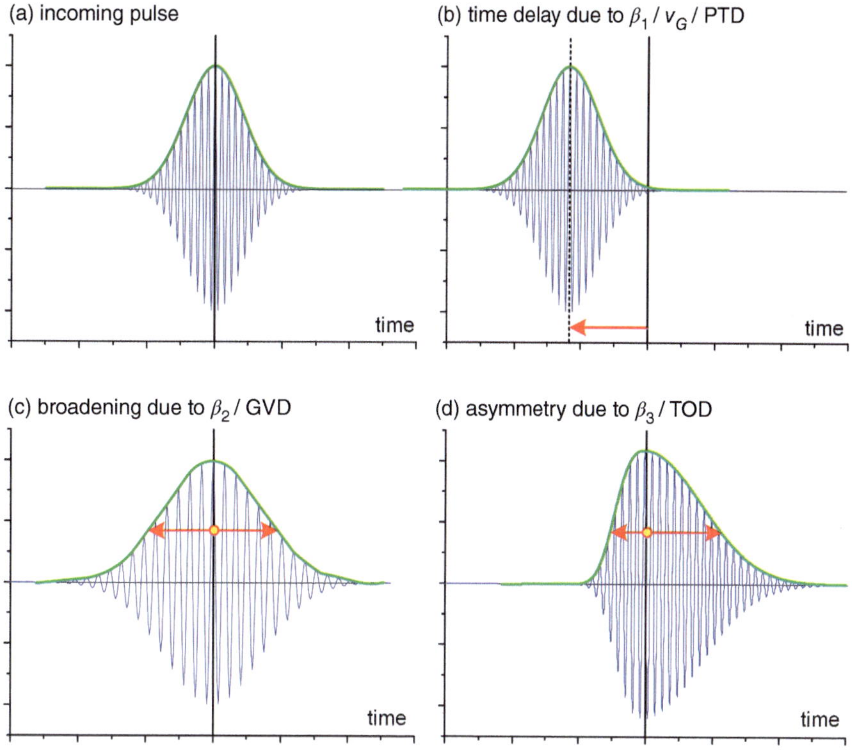

Figure 2.33 Changes of a pulse envelope (in green) in (a) due to dispersion of different orders, (b) describes the delay, (c) the broadening and (d) the asymmetry.

represents the carrier wave travelling with the phase velocity $v_p = c/n_0$, as given in the first factor. Thus, one obtains for evolution of the envelope the remaining integral

$$A(z,t) = \int_{-\infty}^{\infty} \cdot A(0,\Delta\omega) e^{i\cdot\left[\Delta\omega\cdot\left(t-\frac{z}{v_{gr}}\right)-\frac{z\cdot D}{2}(\Delta\omega)^2\right]} d(\Delta\omega) \quad (2.94)$$

$A(z,t)$ is the slowly varying envelope propagating with velocity v_g, which explains its introduction as group velocity. For $D = 0$, the envelope becomes independent of the spatial coordinate, i.e. it retains its shape.

Figure 2.34 visualizes this behaviour. It is seen that due to the broad spectral extend of the pulse in (b), the spectral components on the long and the short spectral side 'see' different propagation constants $k(\omega)$ and therefore the pulse shape is changed. In particular, the non-linearity of the dispersion curve generates different changes on both sides (see orange arrows), so the symmetry of the pulse shape is destroyed.

2.3.4 Pulse Dispersion Data of Glasses

2.3.4.1 Introduction

For optical glasses far from absorption we have the so-called normal dispersion whereby the index of refraction increases with frequency and decreases with wavelength, respectively. As a result, the velocity of light depends on the wavelength and inside the spectral range of a short pulse the speed changes. Therefore, the velocity of the whole pulse is always smaller than the corresponding phase velocity in the material. The second derivative of the index of refraction with respect to frequency, i.e. the group velocity dispersion, is typically positive in the visible wavelength range but can become negative in the near infrared range.

Figure 2.35 shows exemplary the index $n(\lambda)$, the derivative $dn/d\lambda$, the second derivative $d^2n/d\lambda^2$, the normalized group velocity v_g/c and the group dispersion constant (GVD) D for four glass materials as a function of the wavelength. There are two crown-glasses with low dispersion (BK7, FK51) and two flint-glasses with large dispersion (LASF30, SF6) shown. It is seen that the flint-glasses have a quite stronger change and larger values of the derivatives and correspondingly the changes of v_g and D are greater too. It is also seen that for three of the glasses the second derivative as well as D changes to negative values in the near infrared. This means that there are special glass types that for the corresponding wavelength don't disperse the pulse width. This is an attractive result for a corresponding application.

2.3.4.2 Numerical Calculation of Index Derivatives

The simplest way to calculate the derivatives of the index is to use a high precision index calculation scheme using finite differences. The Sellmeier representation usually gives highly accurate index values and is recommended here. For the difference schemes it makes sense to use five

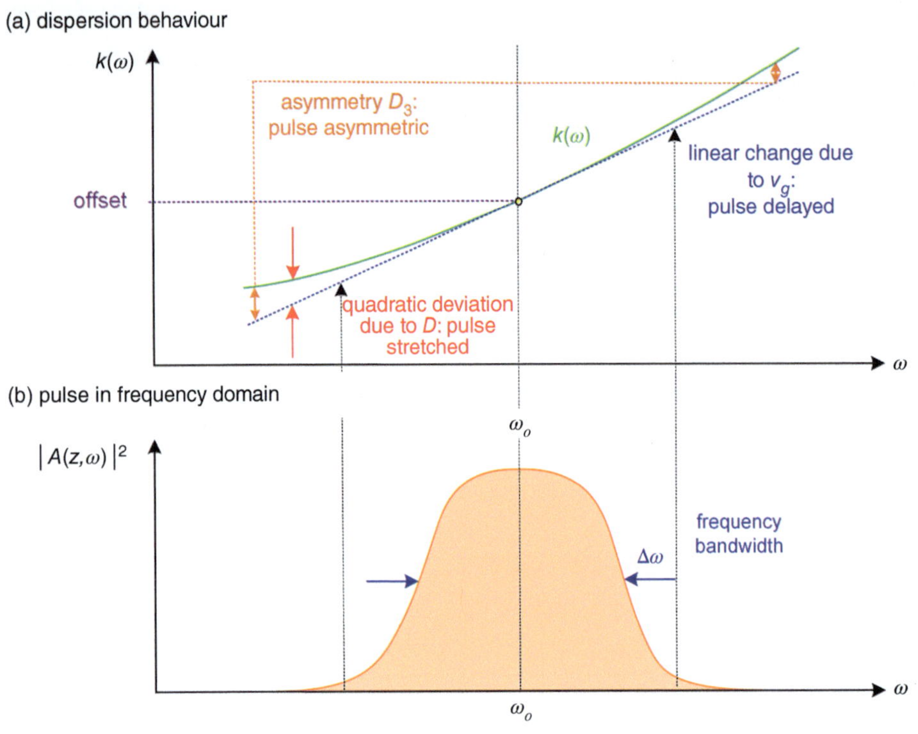

Figure 2.34 Second order propagation constant in (a) and spectral pulse shape (b).

wavelengths around the centre wavelength

$$\lambda_1 = \lambda_0 - 2\Delta\lambda, \quad \lambda_2 = \lambda_0 - \Delta\lambda, \quad \lambda_3 = \lambda_0,$$
$$\lambda_4 = \lambda_0 + \Delta\lambda, \quad \lambda_5 = \lambda_0 + 2\Delta\lambda \tag{2.95}$$

with a spacing of the sampling points of $\Delta\lambda = 0.1 \ldots 1\,\text{nm}$ and to use, for example, the difference schemes

$$\frac{dn}{d\lambda} = \frac{n_1 - 8n_2 + 8n_4 - n_5}{12\Delta\lambda}$$
$$\frac{d^2n}{d\lambda^2} = \frac{-n_1 + 16n_2 - 30n_3 + 16n_4 - n_5}{12\Delta\lambda^2}$$
$$\frac{d^3n}{d\lambda^3} = \frac{-n_1 + 2n_2 - 2n_4 + n_5}{2\Delta\lambda^3} \tag{2.96}$$

For higher accuracy it is recommended not to use the finite difference method but to apply a Sellmeier dispersion fit formula and calculate the derivatives analytically. In particular, the first three derivatives are given by

$$n(\lambda) = \sqrt{n_0^2 + \sum_j \frac{A_j\lambda^2}{\lambda^2 - \lambda_j^2}} \tag{2.97}$$

$$\frac{dn}{d\lambda} = -\frac{1}{n}\sum_j \frac{A_j\lambda\lambda_j^2}{\left(\lambda^2 - \lambda_j^2\right)^2} \tag{2.98}$$

$$\frac{d^2n}{d\lambda^2} = -\frac{1}{n^3}\left[\sum_j \frac{A_j\lambda\lambda_j^2}{\left(\lambda^2 - \lambda_j^2\right)^2}\right]^2 + \frac{1}{n}\sum_j \frac{A_j\lambda_j^2 \cdot \left(\lambda_j^2 + 3\lambda^2\right)}{\left(\lambda^2 - \lambda_j^2\right)^3}$$
$$= -\frac{1}{n^3}\left(\frac{dn}{d\lambda}\right)^2 + \frac{1}{n}\sum_j \frac{A_j\lambda_j^2 \cdot \left(\lambda_j^2 + 3\lambda^2\right)}{\left(\lambda^2 - \lambda_j^2\right)^3} \tag{2.99}$$

$$\frac{d^3n}{d\lambda^3} = -\frac{2}{n}\frac{d^2n}{d\lambda^2}\cdot\frac{dn}{d\lambda} + \frac{1}{n^2}\left(\frac{dn}{d\lambda}\right)^3 + \frac{12\lambda}{n}\sum_j \frac{A_j\lambda_j^2 \cdot \left(\lambda_j^2 + 3\lambda^2\right)}{\left(\lambda^2 - \lambda_j^2\right)^3} \tag{2.100}$$

In the following Table 2.4, for some arbitrary glasses and wavelengths, the results of both methods are given and compared. It is seen that the error strongly depends on the wavelength and the glass. The shorter wavelengths are more critical than the longer ones and the high dispersive flint-glasses are more critical than the low dispersion crown-glasses. Furthermore, it is seen that the slope and the curvature as well as the third derivative is much smaller for longer wavelengths. In the data of the refractive index and the first derivative, usually only very small errors or differences are seen.

2.3.4.3 GVD Properties of Glasses

If the second derivative $d^2n/d\lambda^2$ is plotted as a function of the first derivative $dn/d\lambda$, a behaviour similar to the P–ν dependence of the partial dispersion and the normal-line behaviour is seen. Most of the glasses follow a linear functional dependence. In Figure 2.36 the glasses of the Schott catalogue are shown for three wavelengths: λ = 400, 650 and 1060 nm. As expected, the second derivative shrinks with a larger wavelength, thus the linear curves are then shorter and have a changed slope. Accordingly, the relation of the slope with the wavelength can be written in first

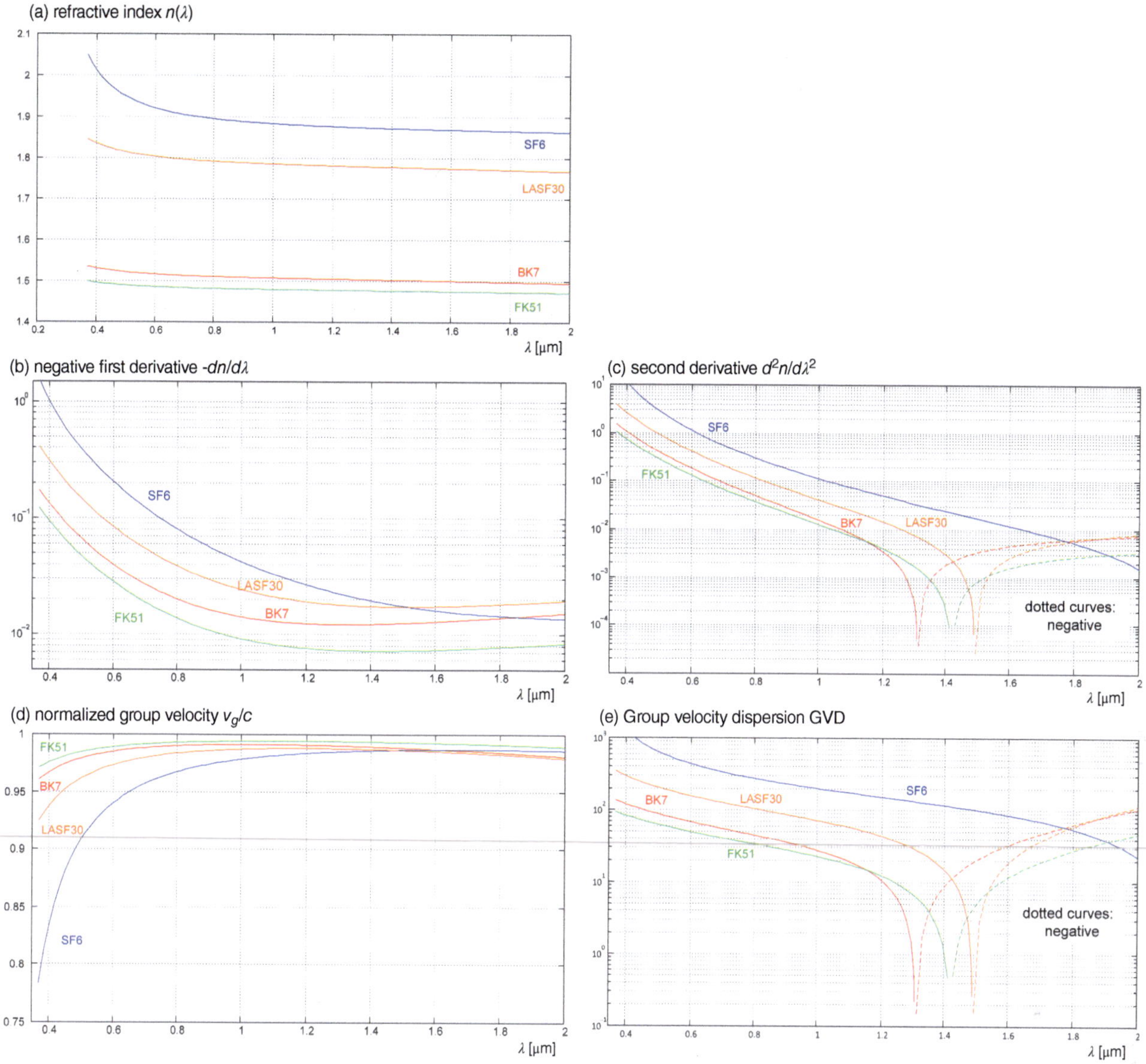

Figure 2.35 Index of refraction (a) and first (b) and second (c) derivative (d) normalized group velocity v_G/c and (e) GVD value D with respect to wavelength as a function of the wavelength for SF6, BK7, FK51 and LASF30. The dotted parts of the curves in (c) and (e) represent negative values, but are plotted positive due to the logarithmic scale.

approximation as

$$\frac{d^2n}{d\lambda^2} \Big/ \frac{dn}{d\lambda} = -\frac{3}{\lambda} \tag{2.101}$$

If a one-term Sellmeier formula in the form

$$n(\lambda) = \sqrt{n_0^2 + \frac{A_1\lambda^2}{\lambda^2 - \lambda_1^2}} \tag{2.102}$$

is assumed, it can be shown that the second derivative can be expressed as

$$\frac{d^2n}{d\lambda^2} = -\frac{3}{\lambda}\frac{dn}{d\lambda} \cdot \left[1 + \lambda\frac{dn}{d\lambda}\frac{3n^2+1}{3n(n^2+1)}\right] \tag{2.103}$$

For small values of the first derivative, the second term in the brackets can be neglected and the approximated linear relationship with $1/\lambda$ is obtained. According to the dispersive behaviour of the typical glasses it becomes apparent that the high dispersive flint-glasses show the largest deviation from the straight line. As is pointed out in [37], the third derivative also follows a similar rule

$$\frac{d^3n}{d\lambda^3} \Big/ \frac{d^2n}{d\lambda^2} = -\frac{4}{\lambda} \tag{2.104}$$

If the GVD constant D is plotted as a function of n and ν for the the Schott catalogue glasses, a pattern like the one in Figure 2.37 is obtained. It is seen that the high dispersive flint-glasses have quite larger values of D. The representation projected onto the ν axis shows a clear correlation of D with ν. Since ν is in principle an inverted slope description, it is obvious that the D–ν function looks like a hyperbolic $1/x$ curve. This representation corresponds to the discrepancy described in Section 2.2.8.2 between the derivative or finite difference-based dispersion formulation.

Table 2.4 Values for the derivatives of the index for some glasses at different wavelengths calculated by numerical or quasi-analytical means, respectively.

Glass	Wavelength in [μm]	Method	n	$dn/d\lambda$ in [1/μm]	$d^2n/d\lambda^2$ in [1/μm^2]	$d^3n/d\lambda^3$ in [1/μm^3]
FK51	0.400	Sellmeier	1.49662	−0.094840	0.7395417	−8.202876
		numerical			0.7515616	−8.306233
	0.550	Sellmeier	1.48779	−0.036049	0.1875307	−1.461961
		numerical			0.1892777	−1.472450
	0.075	Sellmeier	1.48305	−0.015814	0.0493186	−0.291143
		numerical			0.0496559	−0.292089
	1.060	Sellmeier	1.47959	−0.008462	0.0091549	−0.050128
		numerical			0.0092517	−0.050139
BK7	0.400	Sellmeier	1.53085	−0.133163	1.0536217	−12.13215
		numerical			1.0767885	−12.31378
	0.550	Sellmeier	1.51852	−0.050318	0.2581671	−2.060121
		numerical			0.2615018	−2.078242
	0.075	Sellmeier	1.51184	−0.022708	0.0656830	−0.402026
		numerical			0.0663652	−0.403509
	1.060	Sellmeier	1.50669	−0.013210	0.0105287	−0.068916
		numerical			0.0107604	−0.068892
LASF30	0.400	Sellmeier	2.08993	−0.725846	7.1404044	−117.9451
		numerical			7.6445868	−119.5804
	0.550	Sellmeier	2.02948	−0.221008	1.3166185	−11.82728
		numerical			1.3647535	−12.06713
	0.075	Sellmeier	2.00235	−0.082657	0.3177602	−1.914482
		numerical			0.3245844	−1.940421
	1.060	Sellmeier	1.98624	−0.033656	0.0672422	−0.298128
		numerical			0.0683828	−0.299695
SF6	0.400	Sellmeier	2.01238	−1.021616	11.353066	−229.5528
		numerical			12.390346	−230.0084
	0.550	Sellmeier	1.93214	−0.277776	1.7410398	−16.71723
		numerical			1.8209091	−17.07301
	0.075	Sellmeier	1.89892	−0.098029	0.4011168	−2.454861
		numerical			0.4112379	−2.495940
	1.060	Sellmeier	1.88055	−0.036006	0.0865685	−0.366033
		numerical			0.0879473	−0.369105

2.4 Absorption and Transmission

2.4.1 Complex Index of Refraction

Materials with a complex refractive index lead to absorption of optical radiation. The corresponding quantities can be formulated in different ways. Principally, the relation between the complex refractive index and the absorption constant is a formal one. However, one possible formulation variant separates the refractive index into a real and an imaginary part

$$\tilde{n} = n_r + i \cdot n_i \tag{2.105}$$

Another common way of writing this, using the attenuation coefficient κ, is

$$\tilde{n} = n \cdot (1 + i \cdot \kappa) \tag{2.106}$$

Furthermore, a third formulation is possible, which uses the absorption coefficient α in the form of the Lambert–Beer law

$$I(z) = I_0 \cdot e^{-\alpha \cdot z} \tag{2.107}$$

This law describes the decrease in the radiation intensity with the propagation coordinate z. The following relations

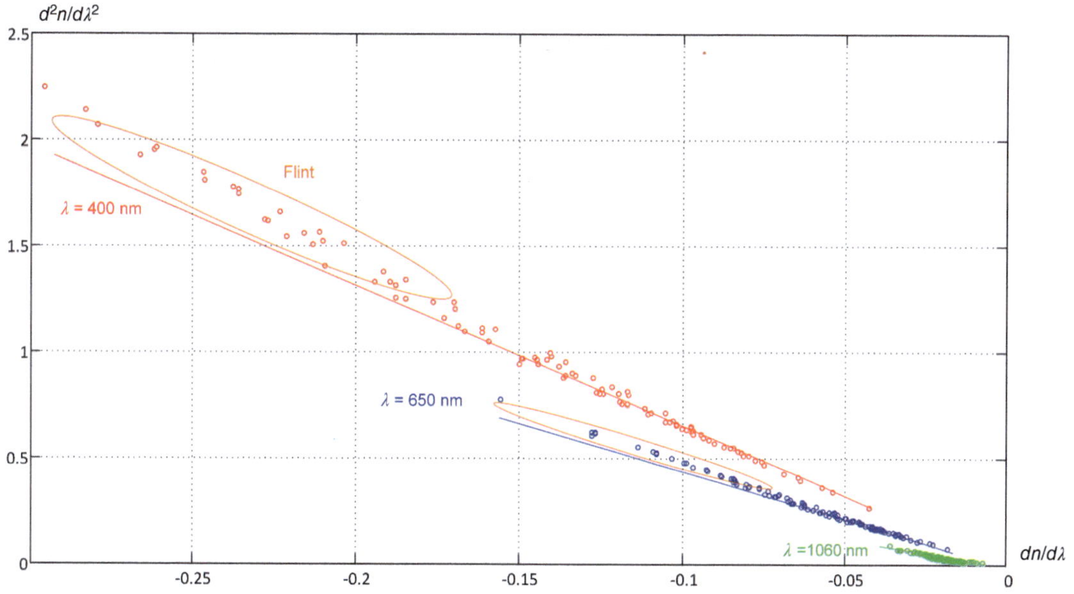

Figure 2.36 Second derivative vs first derivative for the Schott's glasses for three different wavelengths $\lambda = 400$, 650 and 1060 nm showing an approximately linear behaviour.

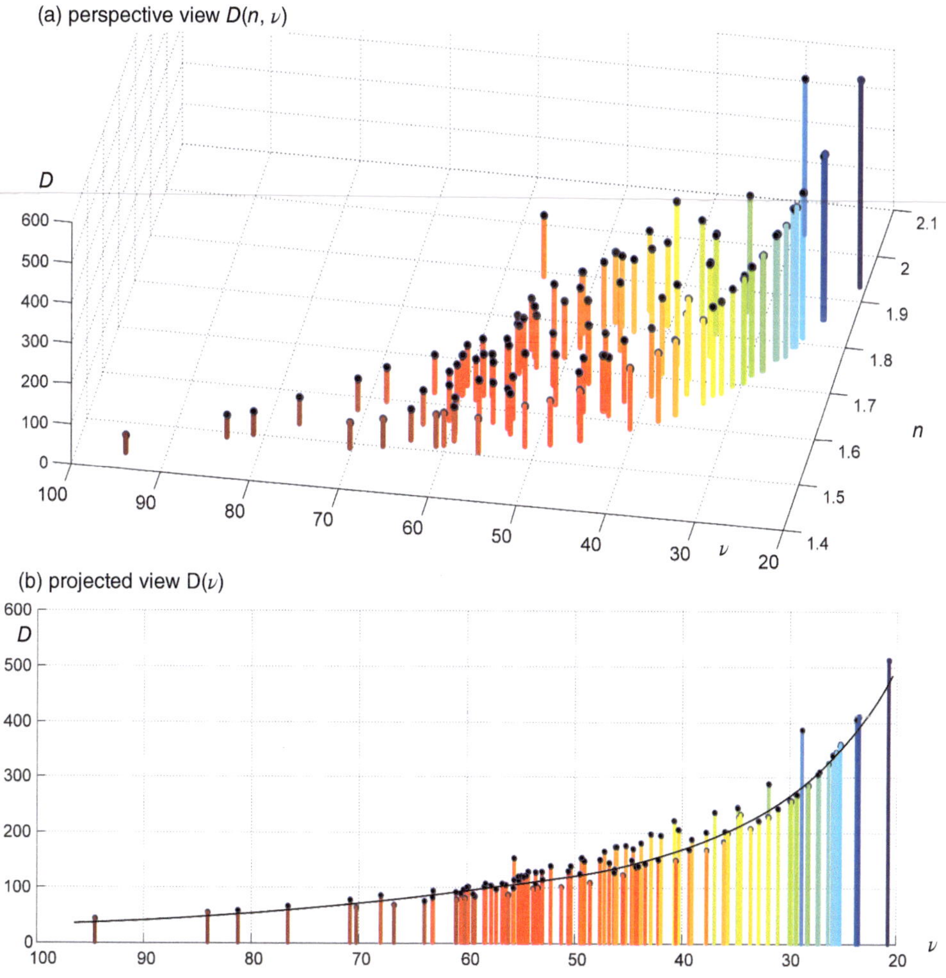

Figure 2.37 GVD constant D as a function of n and ν in (a) a perspective and (b) a projected representation.

are valid to convert between these options

$$\alpha = 2k_0 \cdot n \cdot \kappa = \frac{4\pi \cdot n \cdot \kappa}{\lambda_0}, \quad n_i = n \cdot \kappa = \frac{\lambda_0 \cdot \alpha}{4\pi} \tag{2.108}$$

2.4.2 Lambert–Beers Law

If a material has an absorption constant α the light is damped during propagation through the medium. The Lambert–Beer law Eq. (2.107) gives the decrease in the light intensity as a function of the propagation distance z. The decrease in the intensity with z is shown in Figure 2.38. This law is only valid for the following assumptions.

- The absorbing atoms do not interact.
- There are no non-linear effects, the absorption itself does not change the constant α.
- There is no saturation.
- The material is homogenous, i.e. α does not depend on the location.

It has to be considered further that the formulation refers to the intensity in the medium (reference I_0) and not to the light incident to the sample, so Fresnel effects at the entrance interface are excluded. It should also be mentioned that again effects with values $\alpha < 0$ follow this law.

If the absorption coefficient changes with the position, the Beer law can be written in the more general form

$$I(z) = I_0 \cdot e^{-\int \alpha(z)\, dz} \tag{2.109}$$

Prominent examples for this case are:

- biological samples with different types of tissue;
- gain in laser resonators with saturation;
- second harmonic generation with depletion.

2.4.3 Transmission of a Component

The absolute transmission of a sample is defined as the ratio of the transmitted intensity I_{out} to the incident intensity I_{in}

$$T = \frac{I_{out}}{I_{in}} \tag{2.110}$$

This definition is a technical approach to characterize a component. For the definition of the internal transmission, the Fresnel reflection losses at the entrance and the exit surface are subtracted leaving only the volume effects in the sample. In total, three effects are overlaid according to Figure 2.39.

$$I_{out} = I_{in} \cdot T_{entr} \cdot e^{-\alpha \cdot t} \cdot T_{exit} \tag{2.111}$$

In particular, effects caused by multiple reflection at plane-parallel plates are not considered.

The internal transmission is based on an absorption effect in the volume and on samples with standard thicknesses of $t = 25$ mm. Then, conversion to an arbitrary thickness of the glass can be carried out using the following formulas.

$$T_d = \frac{I_i}{I_{i0}} = e^{-\alpha \cdot t}, \quad T_d = (T_{25\,\text{mm}})^{\frac{t}{25\,\text{mm}}} \tag{2.112}$$

Here α is the absorption factor of the material. The logarithm of the reciprocal transmission is referred to as the optical density D (not to be confused with the use of D in the previous sections).

$$D = \log_{10}\left(\frac{1}{T}\right) \tag{2.113}$$

The measurement of the absorption of a material can be carried out by determining the transmission of a light bundle. Here the light passes a thin plane-parallel plate. On the other side the transmitted power of the ray is measured. Assuming that the plate of thickness t is surrounded by air, the total transmitted power consists of an interference of multiple beams at the two interfaces and the absorption in the medium. For the evaluation of a measurement generally, two measurements at different plate thicknesses $t_{1/2}$

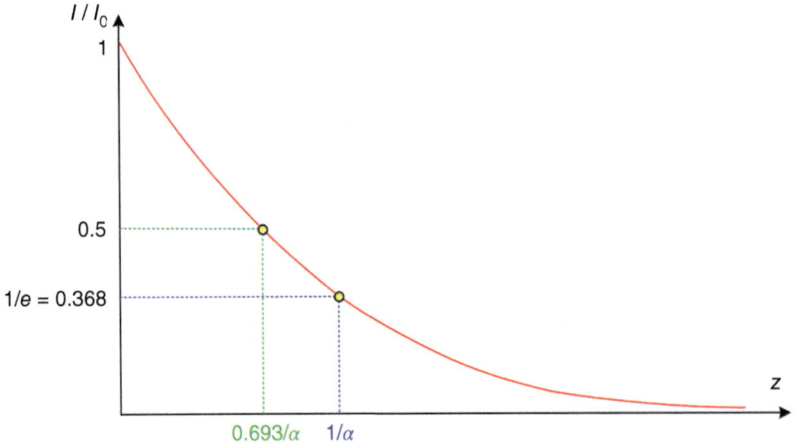

Figure 2.38 Lambert–Beer law.

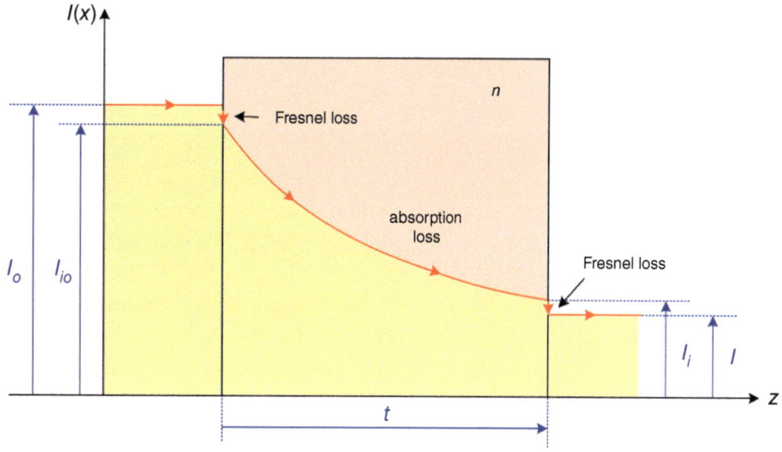

Figure 2.39 Transmission and internal transmission of a plane plate with thickness t and the refractive index n.

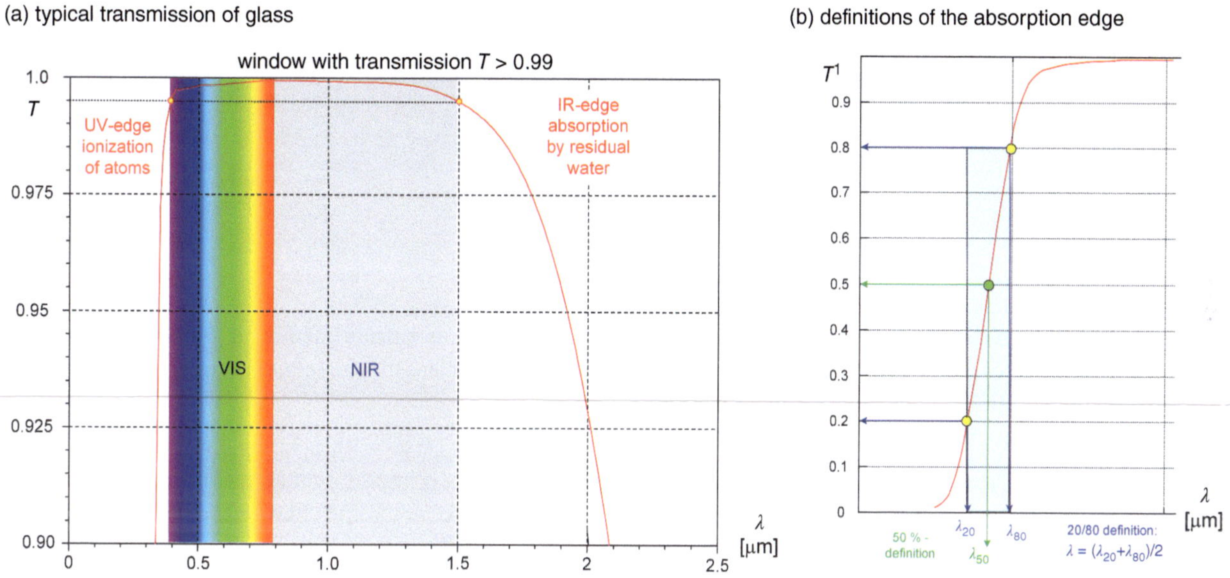

Figure 2.40 Principal path of a glass transmission curve for sample thickness $t = 25$ mm for glass (a) and definition of the absorption edge (b).

are necessary in order to eliminate the surface effects. By using Eq. (2.111) we get

$$\alpha = -\frac{1}{t_2 - t_1} \ln \frac{I_{out2}}{I_{out1}} \tag{2.114}$$

if the equation is formulated for both thickness values and divided through each other. The entrance and exit transmissions are then eliminated.

2.4.4 Transmission of Glasses

The spectral dependence of the transmission of a material in general shows typical windows of high values and corresponding decays to shorter and longer wavelengths. The decrease is due to an absorption inside the material and often shows typical steep edges, mainly found towards the short UV wavelengths. The specification of an absorption edge in the transmission curve is often carried out using the threshold values at $T = 0.5$ or the mean of the wavelengths with transmission of $T = 0.2$ and $T = 0.8$ respectively. Figure 2.40 (a) shows a representative curve for a glass with a steep edge towards the UV and a relaxed decay in the infrared. It should be noticed that the scale is only the upper part of the transmission interval. In part (b) of the figure the definition of the absorption edge is illustrated.

Glasses as dielectric materials are usually transparent in the visible range, the near IR and near UV. A slight residual absorption, however, can be found in all real glasses. The reasons for this residual absorption are as follows.

- Intrinsic absorption: this is based on special atomic vibration states in SiO_2.
- Impurities: displaced impurity atoms in the glass matrix cannot be completely avoided in practice and can lead to absorption bands.

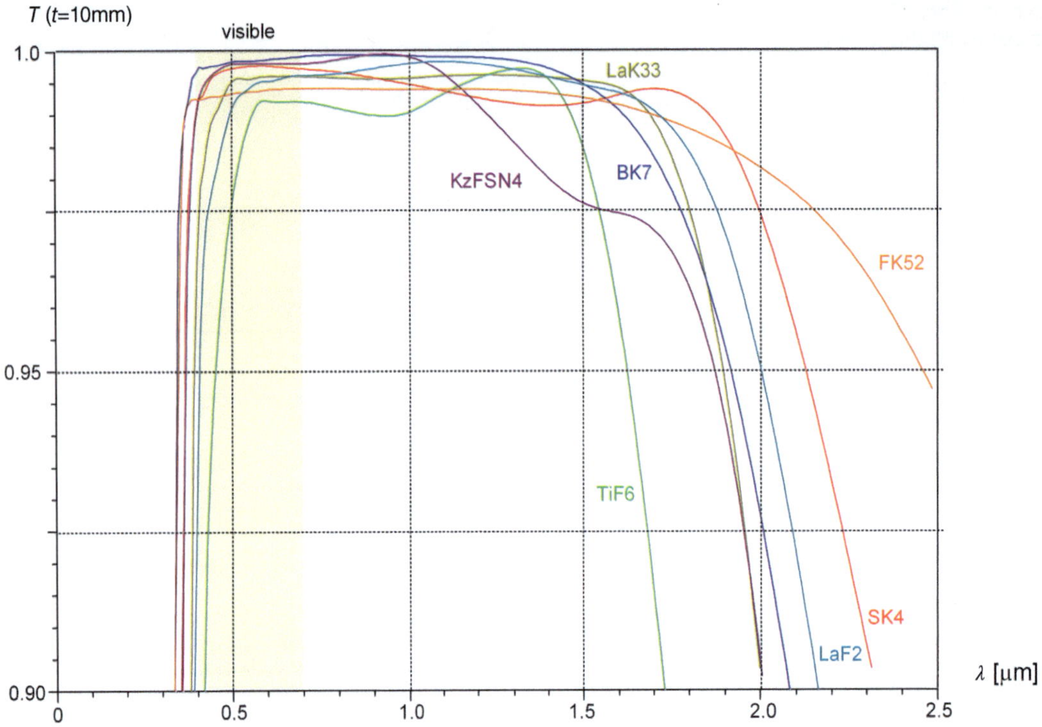

Figure 2.41 Transmission curves of some glasses in the visible and near infrared.

- Scattering: the causes of scattering can be bubbles, contamination or the inclusion of other material in the refractive index. In addition to a non-ideal melting process, the effect of cosmic radiation can also form scattering centres.

With reference to their transmission properties, the optical glasses typically show a UV absorption edge which is between about 300 nm and 400 nm depending on the type of glass. At the long-wavelength side they show good transmission far into the IR range, here larger differences between the various glasses observed. This behaviour is shown in Figure 2.41.

2.5 Thermal Properties

2.5.1 Thermal Expansion

The thermal properties of optical materials are essentially determined by three parameters:

- expansion;
- temperature dependence of the refractive index;
- change of index due to stress and strain.

For complete assembled systems, in addition the expansion properties of the mountings and the housing are relevant; they determine the position of the optical components inside the instrumental reference [48–50].

The thermal expansion coefficient β is defined as the relative change of length with temperature, measured in 1/grd

$$\beta(T) = \frac{1}{l} \cdot \frac{dl}{dT} \tag{2.115}$$

This quantity is important for tensions in mountings and cemented components and changes the lens dimensions in an axial as well as a lateral direction. Plotting the relative expansion $\Delta l/l$ of the glass material vs the temperature, the slope is just equal to the linear expansion coefficient. Such a plot example is shown in Figure 2.42. The expansion values for BK7 are indicated for two different temperature ranges. T_g is the so-called transformation temperature of the glass. This lies between 400 °C and 750 °C depending on the type of glass. About 100 °C below the transformation temperature, the slope of the curve – and thus the expansion coefficient – begins to change noticeably. For the glasses, the expansion coefficient is about 10^{-5} for moderate temperatures. In the range between 0 °C and approximately 300 °C the behaviour can be considered approximately linear. But according to the values of the slope in two temperature ranges in the figure, it is seen that for higher precision smaller T-windows must be used.

For some materials, the expansion coefficients are collected in Table 2.5 together with the mechanically relevant Young modulus.

In Table 2.6 some more essential parameters are listed, which are relevant in the case of thermal simulations.

Figure 2.42 Path of the expansion coefficient with the temperature for a typical glass.

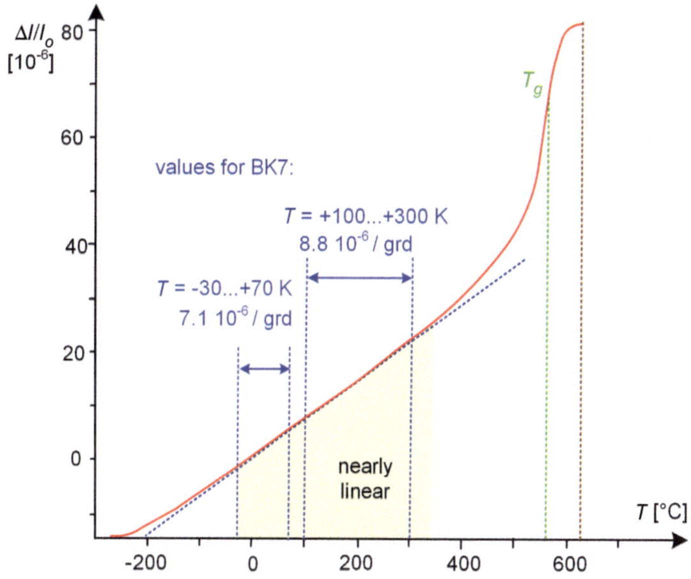

Table 2.5 Thermal and mechanical parameters of certain optical materials and metals. Zerodur is a trademark of Schott.

Material Type	Material	Expansion β in 10^{-6} grd^{-1}	Young's modulus E in 10^3 N/mm^2
Glasses	BK7	7.1	81
	KzFSN4	4.5	60
	LF5	9.1	59
	F2	8.2	58
	LaK9	6.3	110
	PK51A	12.7	73
	SF10	7.5	64
	SK16	6.3	89
Plastics	Acrylic	65	
	CR 39	100	
	Polycarbonate	69	
Crystals	Quartz	0.55	72
	Zerodur™	0.05	91
	Diamond	0.8	1050
	Silicon	2.7	131
	ZnSe	7.1	70
	Sapphire	5.6	400
Metals	Aluminium	24	6800
	Beryllium	12	28 700
	Invar	1.0	141
	Steel	14.7	193

These are thermal conductivity, which is essential for time-dependent processes, and absorption for the wavelength 1060 nm. The absorption is important to know in the case of high-energy beam guiding systems, which are heated not by environmental heat sources but by the signal beam from inside.

Table 2.6 Auxiliary thermal parameter for some materials for $\lambda = 1060$ nm.

Material	Thermal conductivity κ in *W/m/grd*	Absorption α in 1/m
N-BK7	1.114	0.12
N-KzFS4	0.840	0.16
LF5	0.866	0.080
F2	0.780	0.080
N-LaK9	0.908	0.201
N-PK51	0.650	0.241
N-SF10	0.960	0.402
N-SK16	0.818	0.201
Fused silica	0.840	0.00050

2.5.2 Index Change

The second major effect of optical properties with the temperature is the temperature dependence of the refractive index. In reality, this effect is mainly the dominant one in the temperature changes of an optical system in comparison to the impact of the expansion of lenses in optical systems. The refractive index is dependent on the temperature. For temperature changes that are not too large (usual ranges for applications in optical instruments between $T = -30° \ldots +100°$) this dependence can be approximated by a linear correlation

$$n(T) = n_0 + \frac{dn}{dT} \cdot \Delta T = n_0 + \mu \cdot \Delta T \tag{2.116}$$

with the thermo-optical constant μ of the refractive index

$$\mu = \frac{dn}{dT} \tag{2.117}$$

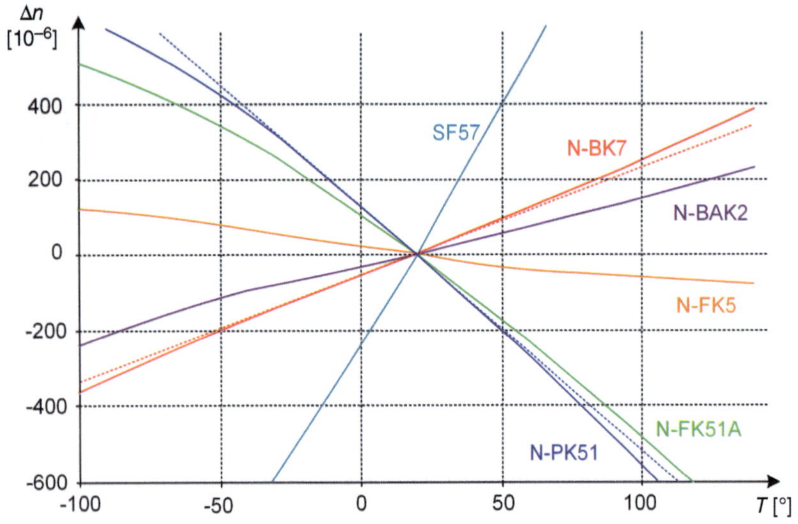

Figure 2.43 Behaviour of *dn/dT* for some glasses, modified from [4]. The dotted lines illustrate the deviation from exact linearity.

This quantity is important for lateral gradients in heated lenses that are cooled by the mechanical mountings of the system. There are additional focal powers and aberrations in optical systems with high energy density and noticeable material absorption. Typically, it also depends on the absolute value of the temperature and the wavelength. Figure 2.43 shows the behaviour of dn/dT in a certain range of temperature for some glasses. It can be seen that there are positive and negative trends and that for most of the glasses inside an interesting range of temperature, the functional dependence is rather well approximated by a linear function.

According to the glass vendors, a more accurate empirical dependence of the refractive index on the wavelength and the temperature is described by the function

$$\frac{dn(\lambda,T)}{dT} = \frac{n^2-1}{2n} \cdot \left(D_0 + 2D_1\Delta T + 3D_2\Delta T^2 + \frac{E_0 + 2E_1\Delta T}{\lambda^2 - \lambda_k^2} \right) \quad (2.118)$$

which included in first approximation the wavelength dispersion also. Here the coefficient D_0 describes the mostly used dominant linear change, rescaled from μ. As an example, for BK-7 and N-PK51 the corresponding parameters (as usual, the reference temperature is 20° C) are given in Table 2.7. It is seen that the dominant parameter is D_0, this means the linear behaviour is the dominating impact.

It should be noted for very accurate calculations and critical compensations that the parameters of the temperature dependence are referenced against vacuum or air. The air itself has a remarkable temperature dependence of the index.

Table 2.7 Temperature parameters of BK7 and N-PK51 for 546 nm for 20° C reference and ΔT in grd.

Parameter	BK7	N-PK51
D_0	$1.86\ 10^{-6}$	$-1.980\ 10^{-5}$
D_1	$1.31\ 10^{-8}$	$-6.060\ 10^{-9}$
D_2	$-1.37\ 10^{-11}$	$1.600\ 10^{-11}$
E_0	$4.34\ 10^{-7}$	$4.160\ 10^{-7}$
E_1	$6.27\ 10^{-10}$	$5.010\ 10^{-10}$
λ_k	0.17	0.134

2.5.3 Thermo-Optical Coefficient

2.5.3.1 Introduction

Corresponding to the effect of a homogenous temperature change on the focal power of a lens, the so-called opto-thermal expansion coefficient is defined as the relative

Table 2.8 Opto-thermal expansion coefficient of certain materials.

Material	Coefficient χ in 10^{-6} grd^{-1}
TiF6	20.94
N-SF57	9.43
BK1	3.28
LaKN9	0.32
N-PK52A	0.08
BAK4	−0.23
KzFS1	−2.89
ZnSe	−28.24
Silicon	−64.1
Germanium	−85.19
Fused silica	−19.5

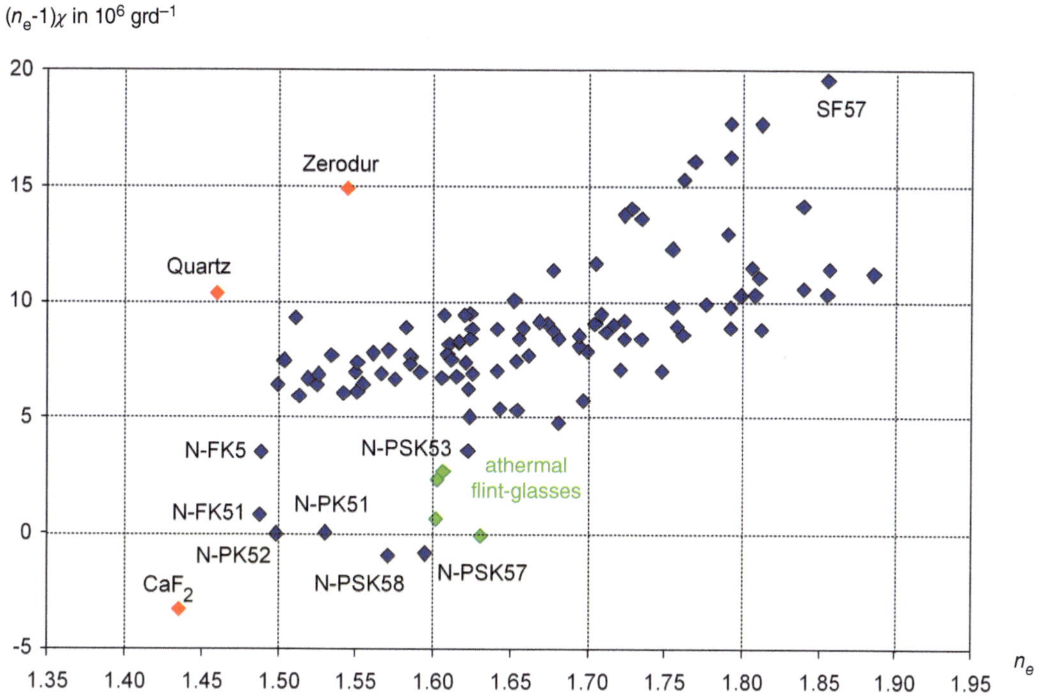

Figure 2.44 Thermo-optical constant as a function of the refractive index [4].

change of the focal length of a thin lens

$$\Delta f = \chi \cdot f \cdot \Delta T \tag{2.119}$$

and is given by the material parameters as

$$\chi = \beta - \frac{1}{n-1} \cdot \frac{dn}{dT} \tag{2.120}$$

It is used as an indicator of optical influence. The first contribution gives the geometrical expansion, the second the physical impact of the T-dependence of the index. In Table 2.8 this quantity is given for some materials at room temperature $T = 20\,°\mathrm{C}$ and wavelength λ_e, Figure 2.44 shows the value scaled by (n_e-1) for some materials. It can also be seen that negative values of χ occur (yellow shaded range in the figure and in Table 2.8), which gives the option for compensating effects in athermalization.

2.5.3.2 Temperature-Dependence of Dispersion*

(*) The content of this section is quite specialized and for advanced reading. Knowledge of this section is not necessary for the following chapters. Beginners are recommended to skip this section.

In the case of more critical applications with strong variations of temperature, it can be necessary to express the refractive index as a function of wavelength and temperature. This can be done, for example, by defining a Sellmeier formula with temperature dependent coefficients [50] in the form

$$n(\lambda, T) = \sqrt{n_0^2(T) + \sum_{j=1}^{j_{\max}} \frac{A_j(T) \cdot \lambda^2}{\lambda^2 - \lambda_j^2(T)}} \tag{2.121}$$

with the temperature dependent parameters

$$n_0(T) = \sum_{m=0} n_{0m} \cdot T^m, \quad A_j(T) = \sum_{m=0} A_{jm} \cdot T^m,$$
$$\lambda_j(T) = \sum_{m=0} \lambda_{jm} \cdot T^m \tag{2.122}$$

which are expressed by a Taylor expansion. This is a more general and more exact approach than Eq. (2.118). For synthetic fused silica as an example, in [48] the coefficients are given, which are fitted onto measurements. As a result, a dependence as shown in Figure 2.45 is obtained. It can be seen that the temperature related variation is smaller than the λ-dispersion and reaches in this case approx. $\Delta n = 0.0015$ over a range of $\Delta T = 200$ K. But as seen in part (b), the difference of the index is not constant, which means that the temperature and the wavelength dependencies are not decoupled and therefore cannot be factorized.

2.6 Other Optical Materials

2.6.1 Crystal Materials for IR and UV

For the spectral ranges beyond the visible in UV and IR, special materials are necessary in order to obtain acceptable transmissions [51–58]. These are mostly crystals.

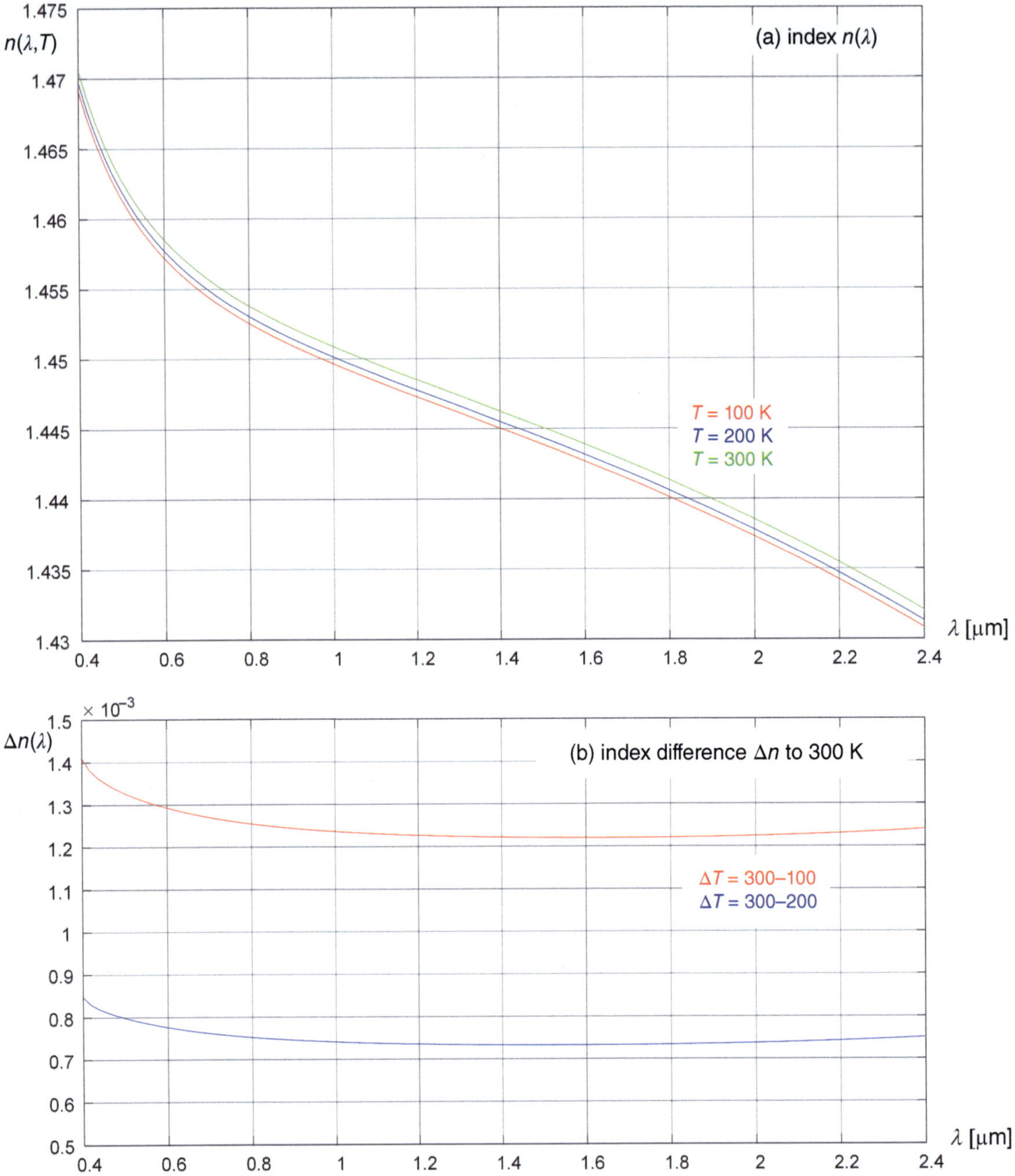

Figure 2.45 Refractive index of fused silica (a) as a function of wavelength for three temperatures and (b) differences of $T = 100$ K and $T = 200$ K values to $T = 300$ K.

Often, however, they are produced in a polycrystalline configuration in order to obtain material with isotropic optical properties. Moreover, semiconductor and special chalcogenide glasses are also used in IR applications. Beneath the optical data for index, dispersion and transmission there are often unwanted properties found, like birefringence or water solubility.

Many materials for the IR range show very large refractive indices, considerably simplifying the optical design in this range regarding the aberration corrections. On the other hand, there are only a few cheap and easily processable materials, which makes correction over large spectral ranges more difficult. This is shown exemplary in Figure 2.46.

Additionally, due to extremely high Fresnel-reflection losses, the coating of the materials plays a particular role.

The wavelength range right above the visible spectrum of $\lambda = 0.8$ μm …1.5 μm is referred to as near infrared. In this

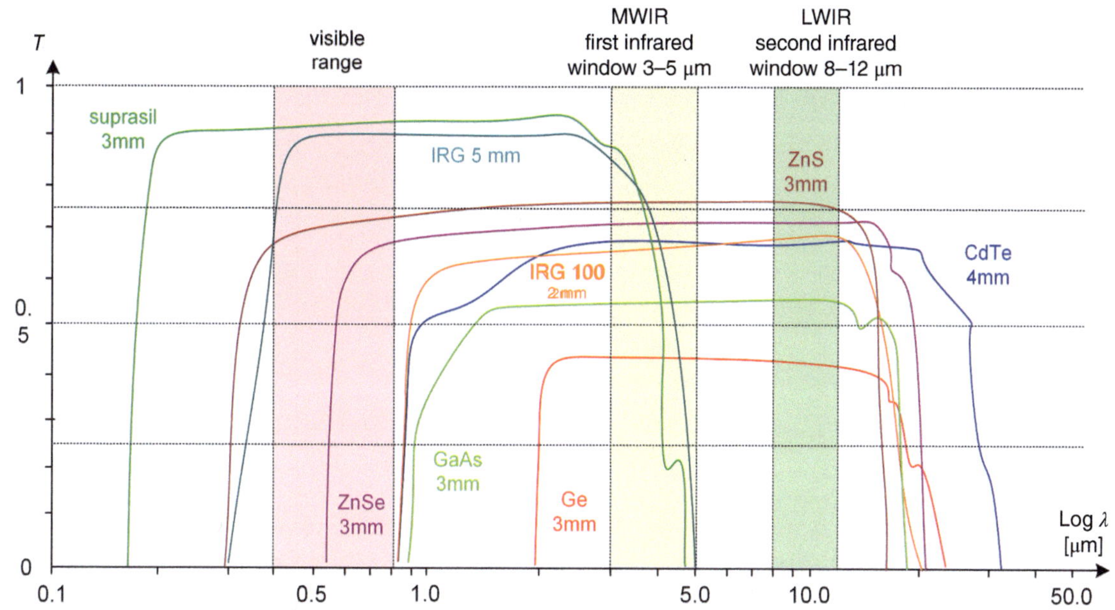

Figure 2.46 Transmission of infrared materials, showing the characteristic windows of the atmosphere and bio-medical applications.

spectral range, most of the glasses are still very transparent such that there is no need to use special materials.

Terrestrial remote sensing applications require transfers over long distances through air. Two spectral windows in the far infrared are of interest for these systems. The first air-transmission window, MWIR, lies between $\lambda = 3$ μm and 5 μm. The second more distant one is the LWIR at $\lambda = 8$–12 μm. In these spectral windows, water absorption is strongly reduced.

Some concrete numbers for refractive indices, dispersion data and useful absorption-free spectral ranges are given in Table 2.9. The definition of the Abbe number is more complicated in the IR window ranges; the numbers are only raw values for orientation here.

The following Figure 2.47 shows the map of refraction index vs Abbe number of several materials for the middle infrared region $\lambda = 3 \ldots 5$ μm. The index is given for 4 μm, the Abbe number is defined as

$$\nu = \frac{n_{\lambda=4\,\mu\mathrm{m}} - 1}{n_{\lambda=3\,\mu\mathrm{m}} - n_{\lambda=5\,\mu\mathrm{m}}} \tag{2.123}$$

The partial dispersion according to the theory explained in Section 2.2.7.5 is shown in Figure 2.48.

Table 2.9 Important UV and IR materials. The quantities are defined for each main spectral application range. The dots give the spectral ranges of the materials, where usage can be recommended. In the case of the empty fields, more detailed information is not available.

Material	Refractive index	λ-range (μm)	UV	IR	ν	P
MgF_2	1.389	0.12–9.0	•	•		
ZnS	2.25	0.4–14.5		•	12.831	0.271
CaF_2, calcium fluoride	1.42	0.12–11.5	•	•	47.53	0.397
ZnSe	2.44	0.5–22.0		•	34.20	0.291
MgO	1.69..0.1.737	0.28–9.5		•	53.4	
CdTe	2.70	0.9–31.0		•		
Diamond	2.3757	0.25..0.3.7, 6.0...	•	(•)	2387	0.469
Germanium	4.003	2.0..0.15		•	701.42	0.556
Silicon	3.433	1.2..0.15		•	420.16	0.523
BaF_2	1.474	0.18..0.12	•	•	81.7	0.651
SiO_2, quartz	1.544	0.15..0.4.0	•	(•)	69.9	0.653
Al_2O_3, sapphire	1.769	0.17..0.5.5	•	(•)	6.626	0.650

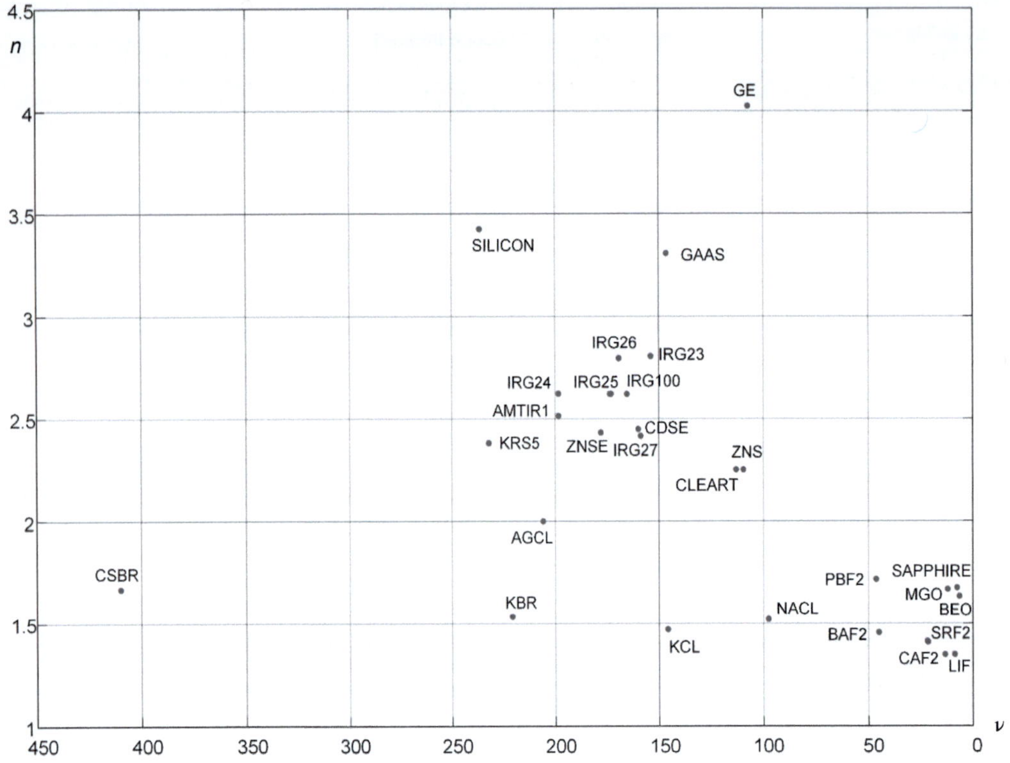

Figure 2.47 Index vs Abbe number map of infrared materials for $\lambda = 3$–5 μm.

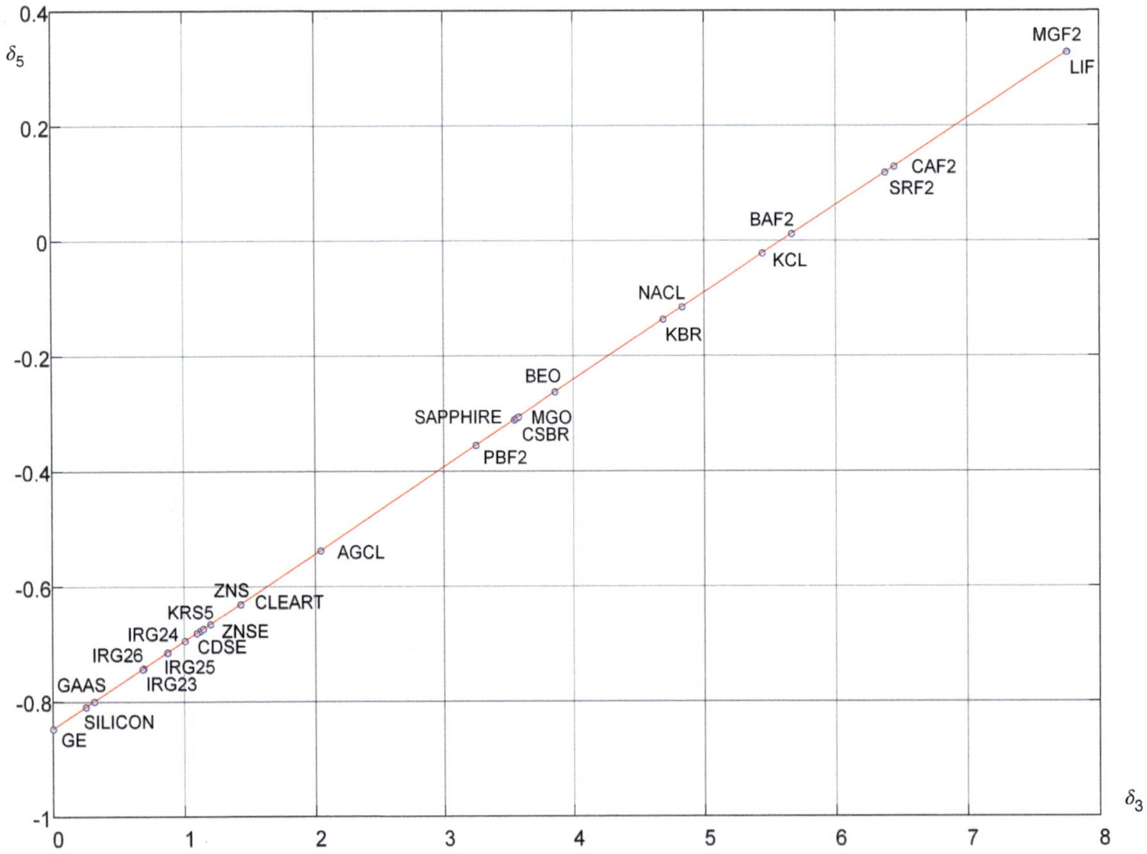

Figure 2.48 δ_3 vs δ_5 correlation of infrared materials for $\lambda = 3$–5 μm according to the partial dispersion description in Section 2.2.7.5.

2.6.2 Plastics

In optical systems plastics are increasingly used for components, mainly in cost-critical consumer applications. Beside their advantages, however, plastics also have some properties which are disadvantageous or which lead to certain limitations [59–63]. In Table 2.10 some of the most important general properties of plastics are compared with those of glass.

Besides these parameters for plastics there are also numerous properties and behaviours that are completely different from those of crystals or glasses. Often they are difficult to describe in a quantitative way, but they are still very important for use in optical systems. These are the following.

- Stress induced birefringence during processing.
- Generation of local inhomogeneities of the refractive index in die casting.
- Water intake (swelling): change of shape (up to 4%) and decrease in the refractive index.
- Electro-static charge.
- Ageing due to cold forming, polymerization, opalescence, yellowing.
- Strong thermal variation of the refractive index,
- Limiting temperature (above the transition temperature the material is destroyed) 100…120 °C.
- For an increased abrasive hardness and for the prevention of charging and swelling, special coatings may have to be applied with low deposition temperatures.
- During the cooling process significant changes occur in the volume caused by shrinking. There are two different types of plastics:
  - thermo sets, shrinking 0.4%…0.7%
  - thermo plasts, shrinking 4%…14%.

For comparison, the plastics can also be represented in a n–ν glass diagram, see Figure 2.49. It can be seen that there is an overlap region with the glass areas, but the plastics extend the available regions towards smaller refractive indices and Abbe numbers. However, some further plastics exist with a large dispersion of about $\nu = 20$, which are not included in the diagram because they have a tendency to strong yellowing and therefore are not used very often.

The values of the refractive index given in the literature refer to the e/d-line or ν for the secondary colours F/C or F'/C'. Therefore, the exact positions of the points vary or are reproduced slightly inaccurately in this diagram.

Table 2.11 gives the refractive index and the Abbe number for some plastic optical materials [9, 10]. Due to the influence of exact chemical compositions and process

Table 2.10 Typical properties of plastics.

Property	Unit	Typical range for plastics	Typical range for glass
Refractive index n		1.49…1.66	1.44...2.05
Dispersion ν		25…60	20...90
Uniformity of the refractive index		10^{-3}…10^{-4}	10^{-4}...10^{-6}
Temperature dependence of the refractive index	10^{-6}*K^{-1}	−100...−160	−10...+10
Vickers hardness	N/mm^2	120...190	3000...7000
Thermal expansion	10−6*grd^{-1}	70...100	5...10
Thermal conductivity	$Wm^{-1grd-1}$	0.15...0.23	0.5...1.4
Internal transmission in the green range		0.97...0.993	> 0.999
Stress – optical coefficient	10^{-12} Pa^{-1}	40	3
Stress – birefringence		5*10^{-5}...10^{-3}	0
Density	g/cm^3	1.05...1.32	2.3...6.2
Water intake	%	0.1...0.8	0

Table 2.11 Index and dispersion of important plastic materials.

Material	Index n	Abbe ν	Trade name
CR-39	1.499	58	
Spectralite	1.537	47	
Polycarbonate	1.586	30	Lexan, Macrolon
RLX Lite	1.555	36	
Ormex	1.558	37	
Finalite	1.600	42	
MR-6	1.597	36	
MR-7	1.660	32	
PMMA	1.492	57	
COC (Cylo olefin copolymer)	1.533	57	Topas
Polystyrol	1.590	31	
O-Pet	1.607	27	
NAS	1.564	33	
PEI	1.658	18	
TPX	1.467	52	
Acrylic	1.492	55	
COP (Cyclo-olefin polymer)	1.525	56	Zeonex
	1.523	52	ARTON
Polystyrene	1.591	31	
Pyrex	1.474	65	TPX
Polysulfone	1.634	23	
Polyester	1.607	28	OKP-4
	1.642	22	OKP-1
	1.661	20	OKP-A2
SAN	1.567	35	

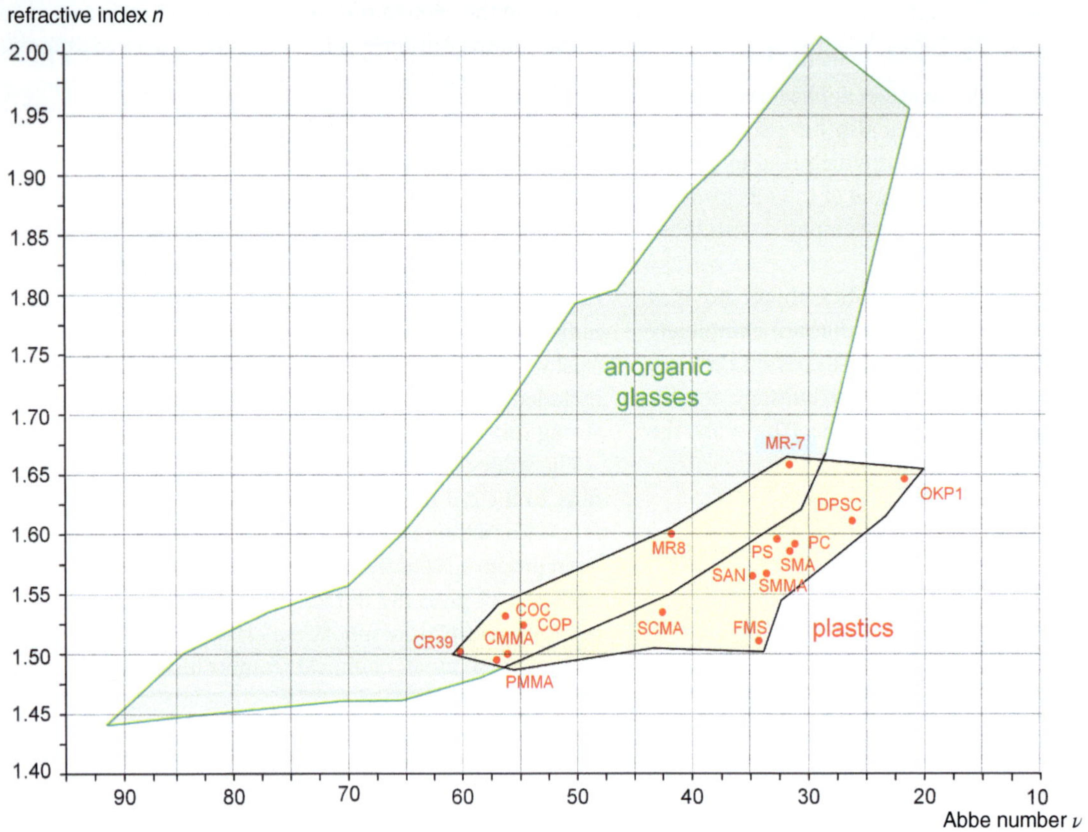

Figure 2.49 Glass diagram with the representation of the real glass ranges and plastics.

parameters during moulding, the accuracy and reproducibility is low. Furthermore, it should be clear that there are ongoing developments for new materials, so the validity and actuality of this overview is limited.

2.6.3 Water

Water has a refractive index of about 1.34 with an Abbe number of $\nu = 55.8$. The dispersion curve is shown in Figure 2.50 [64–66]. It must also be observed that, particularly in the infrared spectral ranges, water has few very pronounced resonances leading to a strong absorption and to large changes of the refractive index. This is seen in part (b) of the figure with two resonances with absorption at approximately 3 μm and 6 μm wavelength. The properties of water are of particular interest in cases of biomedical applications in tissue and propagation in remote sensing applications in humid atmospheres.

At approximately $\lambda = 1$ μm, a turning point is seen in the dispersion curve in the figure (a). This means for $\lambda > 1$ μm the curvature is negative and the curve is bent to the next absorption band.

Biological tissue also has refractive indices in this order, depending on the exact composition. The dispersion is very similar to that of water. In biomedical applications usually the value $n = 1.335$ is used.

Interestingly, the refractive index grows rapidly for shorter wavelengths. In immersion DUV lithography at $\lambda = 193$ nm, the value $n = 1.437$ is used; see also Figure 2.50 (b).

2.6.4 Technical Liquids*

(*) The content of this section is quite specialized and for advanced reading. Knowledge of this section is not necessary to understand the following chapters. Beginners are recommended to skip this section.

A large number of organic liquids are used in, e.g. forensic medicine and analytical chemistry [67, 68]. These have certain optical properties making them interesting for applications in classical optical systems. The most notable characteristic is their unusual position in the glass diagram and their dispersion as shown in Figure 2.51.

Notably the anomalous dispersion properties of these and similar liquids inspired several researchers to design super achromatic objectives and refracting telescopes. Additionally, an amazing number of patents were granted relating to the use of liquids for colour correction.

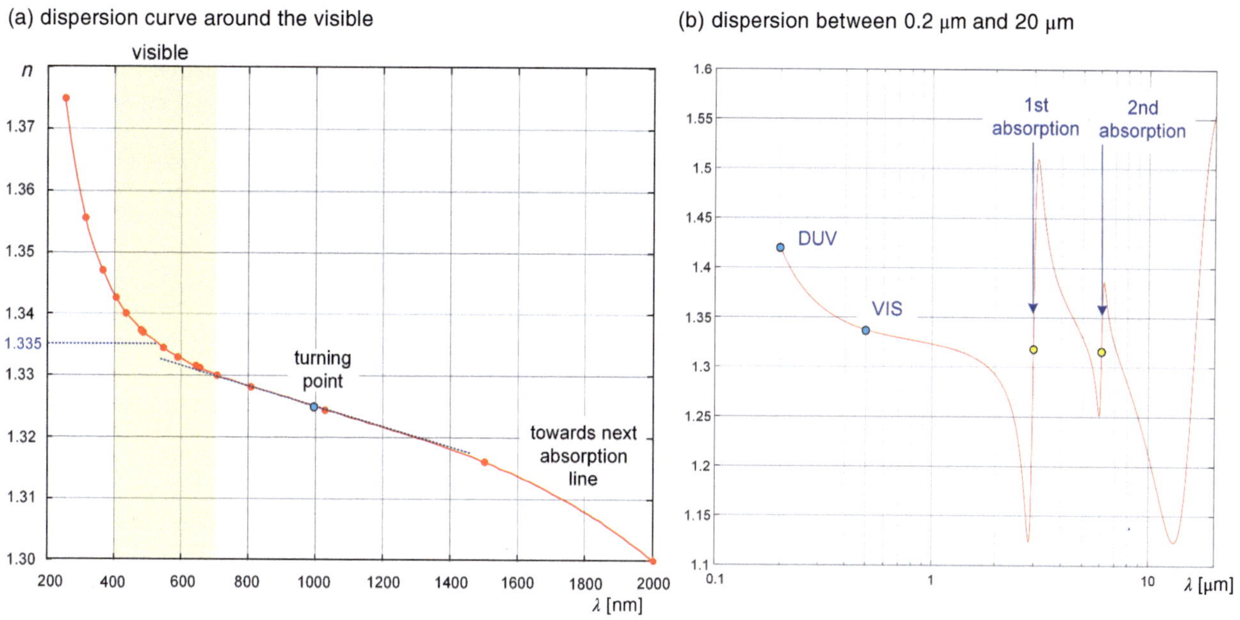

Figure 2.50 Dispersion of the refractive index of water. (a) shows the dispersion around the VIS range, (b) covers the range from 0.1 μm to 20 μm wavelength.

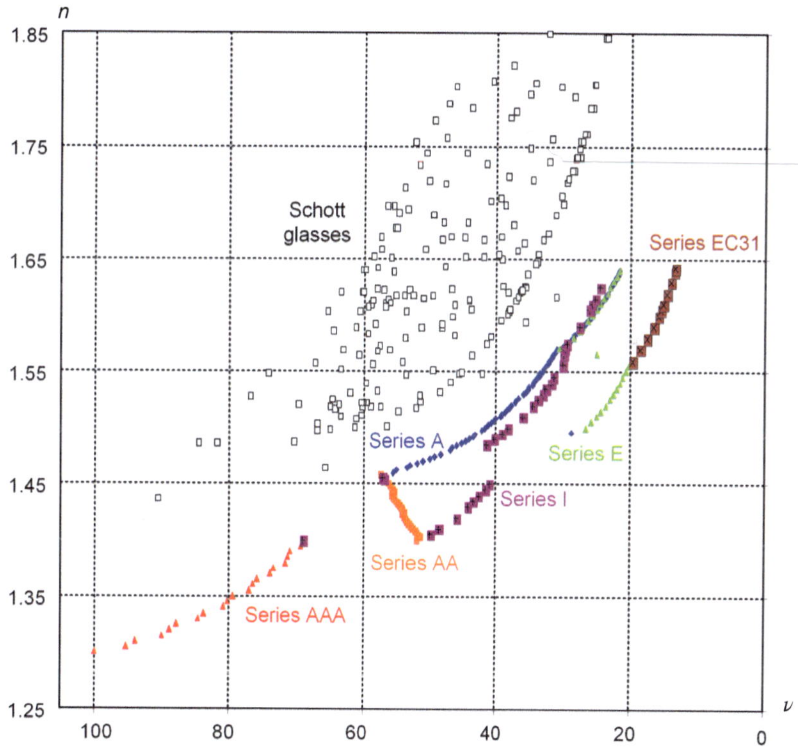

Figure 2.51 Commercial optical liquids company compared with Schott glasses in a n–ν diagram. The data for the liquids are based on data sheets provided by Cargille. Adapted from [69].

Interestingly, use of liquids dates back to the eighteenth century where rather exotic liquids were used as a replacement for a missing flint-glass at that time, to reduce the unavoidable secondary spectrum in achromates. Systems have been successfully built using liquids. This technique became obsolete with glass development and the availability of a broad selection of glass for colour correction.

The advent of liquids with higher stability in the twentieth century, for example from the company Cargille, gave rise to a renaissance of liquids in optical systems. From

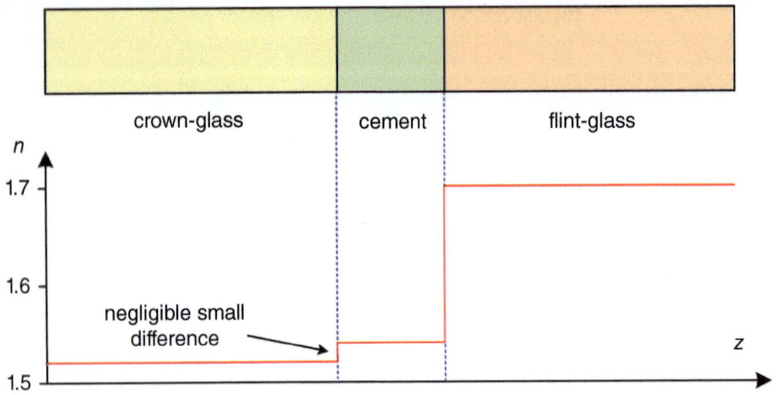

Figure 2.52 Path of the refractive index in a cemented achromate, schematic.

the optical design point of view, liquids can be favourably utilized in the following

- Reducing secondary spectra and/or correcting longitudinal colour for three wavelengths (apochromatism).
- Athermalization due to high linear thermal expansion and high dn/dT.
- Replacement for conventional (hard) optical cements to compensate between glasses with large differences in thermal expansion.
- Building optical systems that are less sensitive to tolerances.

Despite these interesting aspects, the potential pitfalls of using liquids should be carefully taken into consideration.

- Since most of the liquids are based on organic substances, long-term stability is an issue. As a general guideline, it appears that stability increases with decreasing index of refraction.
- Liquids may crystallize after a period of 6–12 months. In particular, this applies to the Cargille series M, H, EH, EC, FH, GH. Conversely, the Cargille series A, AA and AAA seems to be very stable for years.
- Some liquids are toxic and require special handling provisions.
- Almost all liquids are suspectable to ultraviolet (UV) radiation. Significant transmission losses were observed after 2–4 days direct sunlight exposure. However, this effect is suppressed when the liquid is protected from UV-radiation, for example, by a Schott GG375 filter glass, 3 mm thick.
- Liquids exhibit very high thermal expansions that require special precautions for mounting lenses in order to compensate for large varying pressure differences with temperature.

Further technical liquids and their refractive indices are listed in Table 2.12.

Table 2.12 Refractive index of some liquids.

Medium	n_d
Ethylalcohol	1.362
Gasoline	1.440
Benzene	1.501
Glycine	1.455
Methyliodide CH_2J_2	1.742
Carbon sulphate	1.628
Cedar wood oil	1.51
Liquid oxygen	1.221
Liquid hydrogen	1.11

2.6.5 Immersion Liquids

Immersion oils are used in microscopy in order to increase the object-side numerical aperture. They are chemically based on chlorine paraffins, polyphene ether, phthal acid ester and di-benzyl ether. Their refractive indices are in the range of $n = 1.46\ldots1.60$, in extreme cases up to $n = 2.0$ depending on the exact composition. The standard oil has a refractive index of $n_e = 1.518$, and thus lies very near to that of BK7 and K5 as the usual microscopical cover glasses. The Abbe number of this oil is $\nu = 44$, so the optical data are different from the usual cover glass material K5. This is sketched in Figure 2.52. Additionally, glycerine with $n = 1.456$ is used. Extreme oils can have very low refractive indices of $n = 1.29$ or very large values $n = 1.70$, respectively [68, 70].

In microscopy, the use of immersion oils with a refractive index of approximately $n = 1.5$ gives a 50% higher resolution. For more highly refracting materials the effect is even stronger. The properties of the immersion oils lead to difficulties in the UV range. Here and also in the blue spectral range, absorption can be observed. Therefore, particularly in the field of fluorescence microscopy, with excitation wavelengths in the blue range, the use of immersion objectives is more critical.

Table 2.13 Immersion oils. In the case of empty fields, more detailed information is available.

Liquid	Refractive index n_d	Abbe number ν	Remark
Water	1.3311	56.0	
Glycerine	1.474	60.5	
Type A	1.5150	41.3	
Type B	1.5150	43.2	
Type FF	1.4790	57.0	for fluorescence
Type HF	1.5150	43.8	for fluorescence
Glycerol	1.450	58	
Parafin oil	1.480		
Cedarwood oil	1.515		
Bromonaphthalene	1.6585	21	
Methylen iodide	1.742	20	

In the infrared range the oils are usually transparent up to a wavelength of approximately 1.5 μm and do not cause problems. Another difficulty is the relatively strong temperature dependence of the refractive index. With $dn/dT = 0.0004\ \text{grd}^{-1}$ it is significantly higher than the values of glass.

In Table 2.13 some refractive indices and Abbe numbers of immersion liquids are collected [19].

Recently immersion systems have also been used in micro-lithography. In this application only water is used; although in the UV range absorption is present, the effects are small in the very thin layers.

2.6.6 Optical Cements

Optical cements are used for glueing optical components together. Since cements are organic compounds, the following problems have to be considered.

- Stability at higher temperatures, especially the deformation temperature.
- Absorption, especially in the UV range.
- Absorption at high power densities; usually cemented components cannot be used in the range of high energy laser radiation.

The optical cements have refractive indices in the range of 1.45–1.55. For the classical cemented achromate-configurations with an optical crown and a flint-glass, there is usually a very small difference in the refractive index to the low-refracting crown. The cement and its finitely thick layer then almost function as part of the crown lens.

For diameters up to approximately $D = 30$ mm, the thickness of the cement can be assumed to be of the order of 0.01 mm. The cement material can mostly be neglected in the raytracing if in the ideal case the radii of curvature of both lenses are equal and no wedge from misalignment during gluing occurs. Here the cement lens has a nearly perfect meniscus shape. In critical systems such as objectives in microscopes, however, the finite thickness has to be taken into account as a slightly thicker crown lens. In this way surface errors in the crown lens as well as its form tolerance are relatively uncritical. In addition, coating the lenses is not necessary because of the very small difference in the refractive indices.

If the cementing process is not carried out in an expert way, significantly thicker or even wedge-like cement layers can occur and create spherical aberration and coma in the front group of a high numerical aperture microscopic lens.

In practice, the most important cement processed under high temperature is Canada balsam with $n = 1.53\ldots1.55$ depending on the exact composition. Cements that can be used at room temperature are epoxy resins which are hardened by means of UV radiation.

As well as the optical parameters, the thermal-mechanical properties of the cements also play an important role. The cement layer should lie between the lenses and be free of bubbles. It has to stick and harden without any stress.

At the same time, however, the cement must have such elasticity properties that during temperature variations it will compensate for the effects of the different thermal expansion coefficients of the glasses.

References

1 Schott, Technical Information, 2023.

2 Bach, H. and Neuroth, N. (1995). *The Properties of Optical Glass*. Springer.

3 Bass, M. (ed.) (1995). *Handbook of Optics*, Vol II, Chapter 33. McGraw Hill.

4 Hartmann, P. (2014). *Optical Glass*. SPIE Press.

5 Born, M. and Wolf, E. (1999). *Principles of Optics*. Cambridge University Press.

6 Chandrasekhar, S. (1957). The optical rotatory dispersion of quartz. *Proceeding of Indian Academy of Science* 45: 147.

7 Zimmer, H.-G. (1970). *Geometrical Optics*. Springer.

8 Pernick, B.J. (1989). Simultaneous nonlinear regression analysis for n- and k-dispersion data. *Applied Optics* 28: 607.

9 Geffcken, W. (1965). Die Teildispersion von Gläsern Teil II. *Optica Acta* 12: 275.
10 Hartmann, P. (2005). Personal communication, data sheets and diagrams of Schott company, with permission.
11 Sloan, T.R. (1970). Analysis and correction of secondary color in optical systems. *Applied Optics* 9: 853.
12 Turunen, J. and Wyrowski, F. (ed.) (1997). *Diffractive Optics*. Akademie Verlag.
13 Rayces, J. and Rosete-Aguilar, M. (1998). Differential equation of the normal glass dispersion. *Proceedings of SPIE* 3482: 332.
14 Rayces, J. and Rosete-Aguilar, M. (1999). Differential equation for normal glass dispersion and evaluation of the secondary spectrum. *Applied Optics* 38: 2028.
15 Li, C.-L. and Sasian, J. (2014). Adaptive dispersion formula for index interpolation and chromatic aberration correction. *Optics Express* 22: 1193.
16 Robb, N. and Mercado, R.I. (1983). Calculation of refractive indices using Buchdahls chromatic coordinate. *Applied Optics* 22: 1198.
17 Robb, P. (1985). Selection of optical glasses. *Proceedings of SPIE* 554: 60.
18 Sigler, R. (1986). Glass selection for airspaced apochromats using the Buchdahl dispersion equation. *Applied Optics* 25: 4311.
19 Reardon, P. and Chipman, R. (1989). Buchdahl's glass dispersion coefficients calculated from Schott equation constants. *Applied Optics* 28: 3520.
20 Forbes, G. (1984). Chromatical coordinates in aberration theory. *Journal of the Optical Society of America A* 1: 344.
21 Herzberger, M. (1959). Colour correction in optical systems and new dispersion formula. *Optica Acta* 6: 197.
22 Chipman, R. and Reardon, P. (1989). Buchdahl's glass dispersion coefficients calculated in the near infrared. *Applied Optics* 28: 694.
23 Tatian, B. (1984). Fitting refractive index data with the Sellmeier dispersion formula. *Applied Optics* 23: 4477.
24 Hoffmann, H.J., Jochs, W.W., and Westenberger, G. (1992). Use of the Sellmeier dispersion formula for optical glasses. *Proceedings of SPIE* 1780.
25 Shannon, R.R. (1996). Spectral plots for optical glass selection. *Optical Engineering* 35: 2995.
26 Sacks, J., Münz, H., and Bentley, J. (2023). Microscope objective design and analysis using new theory of secondary axial color. *IODC* 3 (896): 100.
27 H. Münz, (2021). *A New Theory of Induced Paraxial Second-Order Colour*, EOSAM Proc.
28 de Albuquerque, B., Sasian, J., de Sousa, F., and Montes, A. (2012). Method of glass selection for color correction in optical system design. *Optics Express* 20 (13): 592.
29 Serna, P. and Revuelta, J. (2022). Optical glass selection for color corrected broadband instrumentation: an overview. *Applied Optics* 61: A50.
30 Münz, H. and Peschka, M. (2024). Principal component analysis of refractive index spaces: A model-free approach to color analysis and color correction. *SPIE Conference Proceedings* 13019.
31 Krishna, K.S. and Sharma, A. (1996). Chromatic aberrations of radial gradient-index lenses. I. Theory. *Applied Optics* 35: 1032.
32 Krishna, K.S. and Sharma, A. (1996). Chromatic aberrations of radial gradient-index lenses. II. Selfoc lenses. *Applied Optics* 35: 1037.
33 Tsuchida, H., Nagaoka, T., and Yamamoto, K. (1992). On the design of the optical system using GRIN materials. *Proceedings of SPIE* 1780: 456.
34 Tsuchida, H., Nagoaka, T., and Yamamoto, K. (2000). Design of microscope objectives that use radial gradient-index rods. *Optik* 111: 497.
35 Tsuchida, H., Ogasawara, S., and Yamamoto, K. (2000). Characteristics of a lens system using low-dispersive radial gradient-index material. *Optical Review* 7: 337.
36 Ryan-Howard, D.P. and Moore, D.T. (1985). Model for the chromatic properties of gradient-index glass. *Applied Optics* 24: 4356.
37 Weiner, A. (2009). *Ultrafast Optics*. Wiley.
38 Diels, J.-C. and Rudolph, W. (1996). *Ultrashort Laser Pulse Phenomena*. Academic Press.
39 Gao, W. and Milster, T. (2018). Strehl ratio for optical systems with ultrafast illumination. *Optics Express* 26: 18028.
40 Li, L. and Yin, A. (2011). Research on the nonlinear pulse propagation by numerical analysis. *Optik* 122: 1195.
41 Bor, Z. and Racz, B. (1985). Dispersion of optical materials used for picosecond spectroscopy. *Applied Optics* 24: 3440.
42 O'Shea, D. (2006). Group velocity dispersion using commercial optical design programs. *Applied Optics* 45: 4740.
43 Miks, A. and Novak, J. (2007). Distortion of ultrashort light pulses in pptical systems. *SPIE Conference Proceedings* 6609.
44 Bor, Z. (1989). Distortion of femtosecond laser pulses in lenses. *Optics Letters* 14: 119.
45 Vaughan, A. (1996). Lens design using group indices of refraction. *Journal of Modern Optics* 43: 1683.
46 Akturk, S., Gu, X., Bowlan, P., and Trebino, R. (2010). Spatio-temporal couplings in ultrashort laser pulses. *Journal of Optics* 12: 093001.
47 Bor, Z. and Racz, B. (1985). Group velocity dispersion in prisms and its application to pulse compression

and travelling wave excitation. *Optics Communication* 54: 165.

48 Leviton, D. and Frey, B. (2006). Temperature dependent absolute refractive index measurements of synthetic fused silica. *SPIE Conference Proceedings* 6273.

49 Steps, H. (1950). Über die Änderung des Brechungsexponenten von optischem Glas mit der Temperatur. *Zeitschrift für Glaskunde* 23: 125.

50 Hoffmann, H.J., Jochs, W.W., and Westenberger, G. (1990). A dispersion formula for the thermo-optic coefficient of optical glasses Proc. *SPIE* 1327: 219.

51 Loewenstein, E.V., Smith, D.R., and Morgan, R. (1973). Optical constants of far infrared materials 2: crystalline solids. *Applied Optics* 12: 398.

52 Savage, J.A. (1985). *Infrared Optical Materials and their Antireflection Coatings.* Adam Hilger.

53 Kokorina, V.F. (1996). *Glasses for Infrared Optics.* CRC Press.

54 Harris, D.C. (1999). *Materials for Infrared Windows and Domes.* SPIE Press.

55 Palik, E.D. (1985). *Handbook of Optical Constants of Solids.* Academic Press.

56 Heaney, J.B., Alley, P.W., and Bradley, S.E. (1991). Catalogue of infrared and Cryo-optical properties for selected materials. *Proceedings of SPIE* 1485: 140.

57 Buerckner-Koydl, D. (1992). Interpolation of indices of refraction extended range UV–IR. *SPIE Conference Proceedings* 1780.

58 Nikogosyan, D.N. (1997). *Properties of Optical and Laser-Related Materials.* Wiley.

59 Schaub, M. (2009). *Plastic Optical Materials.* SPIE Press.

60 Bäumer, S. (ed.) (2010). *Handbook of Plastic Optics.* Wiley-VCH.

61 Lytle, J.D. (1990). Status and future of polymeric materials in imaging systems. *Proceedings of SPIE* 1354: 388.

62 Schreyer, G. (1961). Kunststoffe in der Optik. *Kunststoffe* 51: 569.

63 Waxler, R.M., Horowitz, D., and Feldman, A. (1977). Optical and physical parameters of Plexiglas 55 and Lexan. *Applied Optics* 18: 101.

64 Huibers, P.D. (1997). Models for the wavelength dependence of the index of refraction of water. *Applied Optics* 36: 3785.

65 Rusk, A.N., Williams, D., and Querry, M.R. (1971). Optical constants of water in the infrared. *Journal of the Optical Society of America* 61: 895.

66 Djurisic, A.B. and Stanic, B.V. (1998). Modeling the wavelength dependence of e index of refraction of water in the range 200 nm to 200 my. *Applied Optics* 37: 2696.

67 Neil, I.A. (1995). *Liquid Optics Create High-Performance Zoom lens*, 132. Laser Focus World.

68 Beyer, H. (1988). *Handbuch der Mikroskopie.* VEB Verlag Technik.

69 Cargille Laboratories Inc (2005). *Specialty Optical Liquids Catalogue.* Cargille.

70 Laskar, J., Kumar, P., Herminghaus, S. et al. (2016). High refractive index immersion liquid for superesolution 3D imaging using sapphire. Based aplanatic numerical aperture increasing lens optics. *Applied Optics* 55: 3165.

3

Geometrical Optics

3.1 Introduction

Geometrical optics is a ray based approximation of light propagation and is the most important concept to calculate optical systems. Although the assumption of neglecting all diffraction effects sounds strange, the validity of geometrical optics is quite good in most applications. Due to the huge difference in computational burden in comparison to diffraction related calculation methods, it is quite attractive to use this approach in technical optics. The schemes of geometrical optics and raytrace are therefore nowadays still in widespread use and there are many extensions, generalizations and work arounds to enable use as long as possible. Still, today the computing time in lens design is more than 99% governed by raytrace. But there are limitations and the nature of approximation should be known in order to have a critical view on results and use this approximation meaningfully.

In particular, the paraxial approximation of raytrace is essential to understand as it serves as a reference for all aberration calculations and many terms of definition are based on it. In recent times, more sophisticated optical systems containing diffractive elements, free formed surfaces or gradient index media – as well as micro-optical and segmented components – are used, which makes the raytrace more challenging. This also requires a generalization of paraxiality. It is essential to carefully look at the definition of references and the range of validity of geometrical optics in these cases. Therefore, the understanding of raytrace is an important issue and described here in detail, but from a less theoretical and more practical point of view.

There is a great deal of literature available about geometrical optics, some of the recommended books are [1–13].

3.2 Law of Refraction

3.2.1 Introduction

The most important physical effect in geometrical optics is the refraction at a dielectric interface between two media with different indices. This change of ray direction governs all the light ray behaviour in optical systems in this kind of approximation. Diffraction effects are then excluded by definition; nevertheless, some simple optical elements like gratings or diffractive surfaces can be treated by ray equivalent formulas from the diffraction based results of the grating equation.

Snell's law describes the change of the ray direction at a local plane interface. If the interface has a normal unity vector $\vec{e}$, the incoming ray has the direction unity vector $\vec{s}$ and the outgoing ray is characterized by $\vec{s}'$, the incidence angles i and i' are given by the scalar products in the form

$$\cos i = \vec{s} \cdot \vec{e}, \quad \cos i' = \vec{s}' \cdot \vec{e} \tag{3.1}$$

The new ray direction is then given by the law of refraction in the form

$$n \cdot \sin i = n' \cdot \sin i' \tag{3.2}$$

Here, an additional assumption for this scalar formulation is the property that the three unity vectors $\vec{s}$, $\vec{s}'$ and $\vec{e}$ lie in one plane: the plane of incidence. The derivation of this relationship can be easily obtained by requesting a fulfilment of Fermat's principle of shortest ray travelling time or by forcing the electromagnetic field boundary conditions of the Maxwell equations at the interface. Figure 3.1 visualizes the corresponding geometry.

The derivation of the law of refraction by using the Fermat principle is illustrated in Figure 3.1 part (b). The optical path length sum from the point P until Q is given by

$$s_{OPL} = n \cdot \sqrt{y_1^2 + x_1^2} + n' \cdot \sqrt{y_2^2 + x_2^2} \tag{3.3}$$

The correct position of the point O to fulfil the law of refraction is determined by the relation between the coordinates x_1 and x_2 for a constant difference L The path length should have a minimal value, if the correct refraction location O of the ray is realized. Therefore, we have $L = x_1 - x_2$

Lens Design for Imaging: Fundamentals of Optical Systems, First Edition. Herbert Gross.

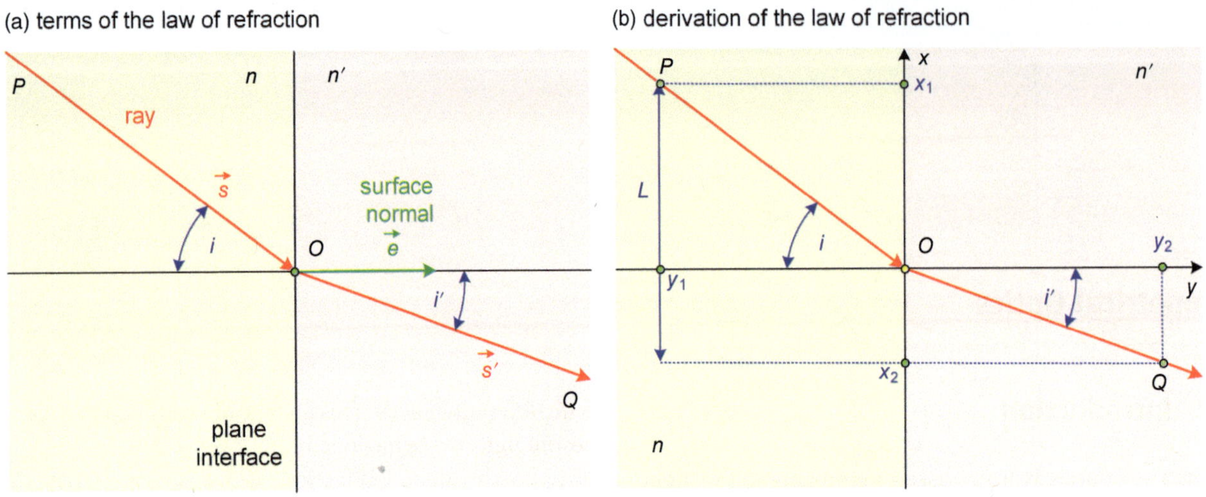

Figure 3.1 Terms of the law of refraction at a plane dielectric interface (a) and (b) derivation by the Fermat principle.

and by eliminating x_1 and optimizing x_2 we get

$$s_{OPL} = n \cdot \sqrt{y_1^2 + (L - x_2)^2} + n' \cdot \sqrt{y_2^2 + x_2^2}$$

$$\frac{\partial s_{OPL}}{\partial x_2} = n \cdot \frac{-2(L - x_2)}{2\sqrt{y_1^2 + (L - x_2)^2}} + n' \cdot \frac{2x_2}{\sqrt{y_2^2 + (x_1 - L)^2}} = 0 \quad (3.4)$$

with the incidence angles

$$\sin i = \frac{L - x_2}{\sqrt{y_1^2 + (L - x_2)^2}}, \quad \sin i' = \frac{x_2}{\sqrt{y_2^2 + x_2^2}} \quad (3.5)$$

we get directly the law of refraction

$$n \cdot \sin i = n' \cdot \sin i' \quad (3.6)$$

If in a general case this relation is written in vectorial form, we get the law of refraction as

$$\vec{s}' = \frac{n}{n'} \cdot \vec{s} + \left(-\frac{n}{n'} \cdot \vec{e} \cdot \vec{s} + \sqrt{1 - \left(\frac{n}{n'}\right)^2 \cdot [1 - (\vec{e} \cdot \vec{s})^2]} \right) \cdot \vec{e} \quad (3.7)$$

which automatically fulfils the coplanarity condition for $\vec{s}$, $\vec{s}'$ and $\vec{e}$. If in particular we set $n' = -n$, this equation changes into the form

$$\vec{s}' = \vec{s} - 2 \cdot \vec{e} \cdot (\vec{e} \cdot \vec{s}) \quad (3.8)$$

and now describes the reflection of light at the interface, where in particular we have $i' = i$.

From a more general physical point of view, an incoming ray is split into a reflected and a refracted part. The energetic splitting ratio is governed by the Fresnel equations (see Section 3.3). This scenario is sketched in Figure 3.2. From the physical point of view, the polarization of the incoming field is additionally essential. This is discussed in Section 3.3 in detail.

Since the sine function in the law of refraction is limited between +1 and −1, some special cases must be considered.

- For grazing incidence $i = 90°$, the ray cannot enter the second medium, independent of the refractive indices.
- If the ray comes from the denser medium with larger refractive index, we have $i' > i$. Since i' is also limited by 90°, there is a maximum angle i for the incoming ray, which allows for a transmission into the second medium. For larger angles, there is no transmission, this case is called TIR (TIR); the corresponding limiting incoming incidence is then given by

$$\sin i = \frac{n'}{n} \quad (3.9)$$

The various cases of i and i' relationships together with the above-mentioned limiting effects are illustrated in Figure 3.3 for several values of the refractive index ratio n/n'. It should be noted that the physical effect at the interface is governed by the index ratio, not by the difference. Figure 3.4 illustrates the transition from the regular refraction to the case of total internal refraction by growing incidence angle i.

3.2.2 Ray Bending

If the law of refraction is considered more from a system point of view in a global coordinate system and is not local in the intersection point of a single surface, the angles of the incoming ray, the outgoing ray and the surface normal can be formulated against the local optical z axis of the surface.

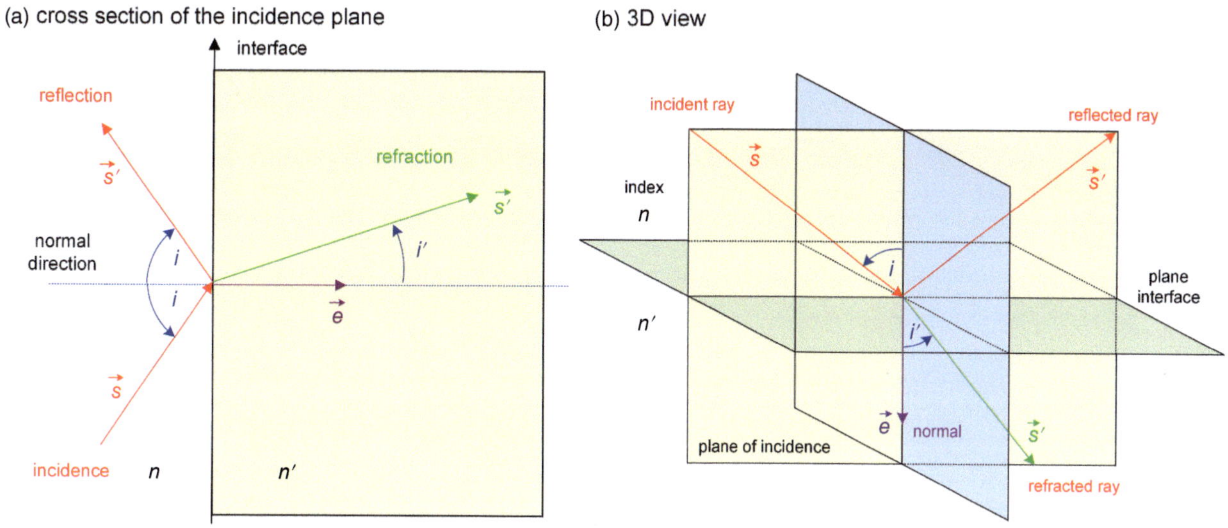

Figure 3.2 Refraction and reflection at a dielectric plane interface. (a) as cross section and (b) in three-dimensional view. The unity vectors $\vec{s}, \vec{s}'$ and $\vec{e}$ lie in the plane of incidence.

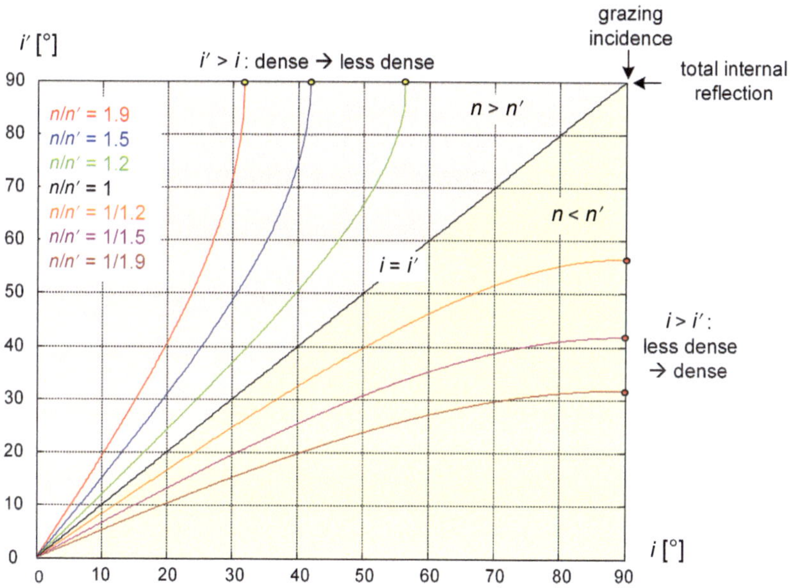

Figure 3.3 Dependence of the outgoing incidence angle i' from the incoming value i as a function of the index ratio n/n'.

The law of refraction then reads [9]

$$n \cdot \sin(\theta_N - \theta_{in}) = n' \cdot \sin(\theta_N - \theta_{out}) \tag{3.10}$$

and the angle with the local unity vector of the surface is related to the local surface slope as a derivative of the surface eq. $F(y)$ in one lateral dimension by

$$\tan \theta_N = -\frac{1}{F'(y)} \tag{3.11}$$

see also Figure 3.5.

If now in a practical task the incoming and outgoing ray angles are pre-given, the required normal surface angle is given by

$$\cos \theta_N = \frac{n_r \cdot \cos \theta_{out} - \cos \theta_{in}}{\sqrt{n_r^2 - 2n_r \cos(\theta_{in} - \theta_{out}) + 1}} \tag{3.12}$$

with the relative index ratio $nr = n'/n$. All the angles have to lie in a range between $-\pi/2 \ldots 0 \ldots + \pi/2$. If a diagram of $i = \theta_N - \theta_{in}$ vs $\delta = \theta_{out} - \theta_{in}$ is plotted, Figure 3.6 is obtained. For a given incoming and outgoing direction, the incidence angle can be changed inside a certain range depending on the index ratio.

The deflection angle δ has a different sign depending on whether the ray is being refracted from the thinner

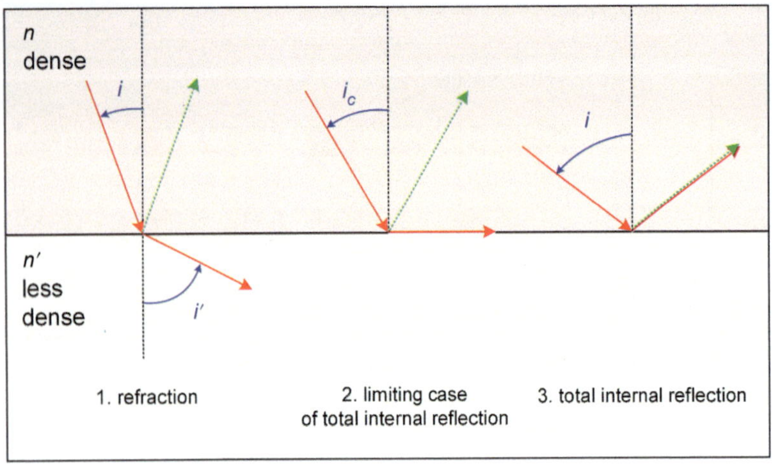

Figure 3.4 Change of incidence angle i for $n > n'$ and transition between refraction, and total internal reflection.

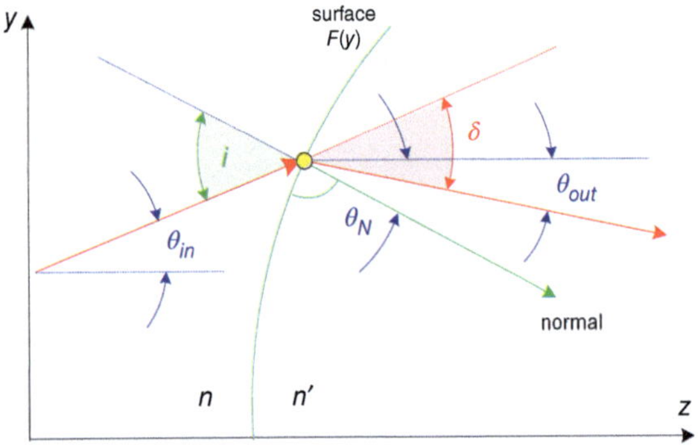

Figure 3.5 Refraction at a surface with angles in a system coordinate system. It should be noticed that θ_{in} is negative here.

into the denser medium or vice versa. In the first case the corresponding curves $\delta(i)$ end at the point of TIR. This is shown in Figure 3.7, which is a clearer picture but equivalent to Figure 3.6.

3.2.3 Description of Refraction in the *k*-Space

The refraction at a dielectric interface and the case of total reflection can also be described in the Fourier space by the wave numbers k. Inserting a plane wave in the form

$$E = E_o \cdot e^{i(\vec{k}\cdot\vec{r}-\omega\cdot t)} \tag{3.13}$$

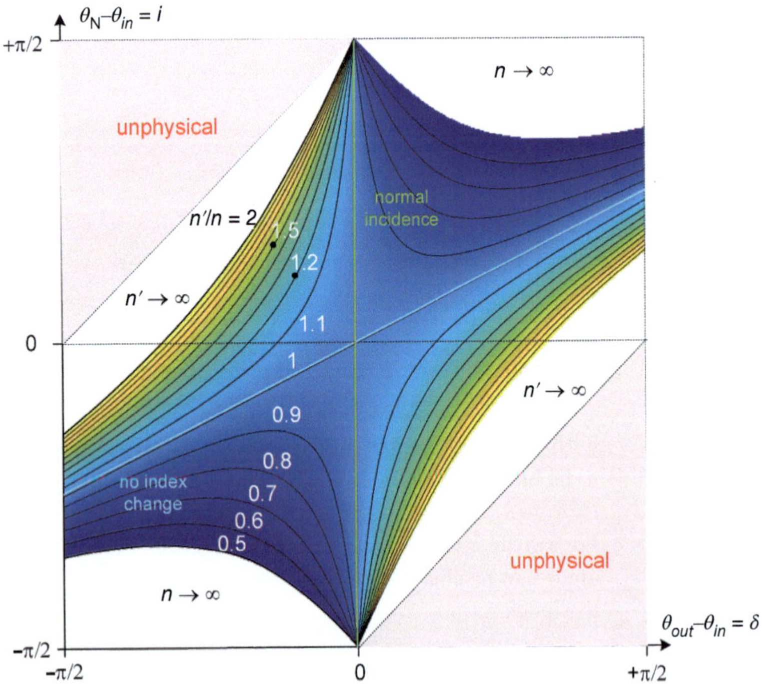

Figure 3.6 Incidence i vs bending angle δ for rays with the index ratio n as parameter. The colour coding and level numbers are the index ratios $n_r = n'/n$.

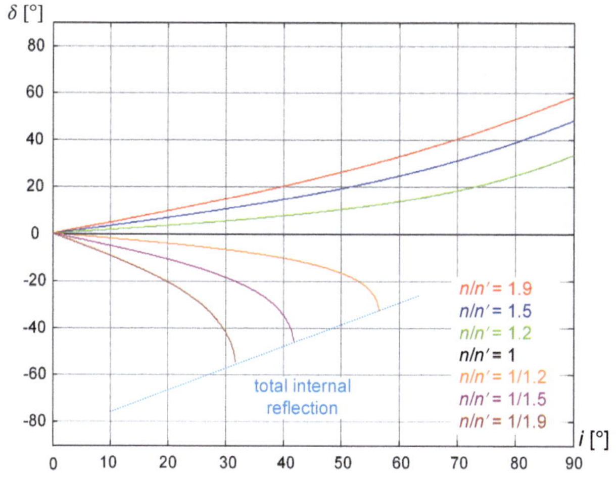

Figure 3.7 Deflection angle δ vs incidence angle i for refraction at an interface for different ratios of the refractive indices.

into the Helmholtz wave equation, the differentiations can be carried out and this yields the equation

$$k_x^2 + k_y^2 + k_z^2 = (n \cdot k)^2 \tag{3.14}$$

This equation describes the propagation in the k-space and is referred to as the Ewald sphere. Using this equation for an interface between the two media n_1 and n_2 leads to

$$k_{1x}^2 + k_{1y}^2 + k_{1z}^2 = (n_1 \cdot k)^2 \tag{3.15}$$

$$k_{2x}^2 + k_{2y}^2 + k_{2z}^2 = (n_2 \cdot k)^2 \tag{3.16}$$

In a two-dimensional representation, these are two circles whose radii are given by the refractive indices respectively. The corresponding geometry is shown in Figure 3.8 (a) in the plane $k_x = 0$. The continuity condition at the interface for the refraction of the ray indicates that the y components are invariant.

$$k_{1y} = k_{2y} = k_y \tag{3.17}$$

If the transition occurs from the denser into the thinner medium, the circle of the incident k-vector is larger. If the incident angle and therefore the y-component of k exceeds a critical value, the continuity condition above can no longer be fulfilled and total reflection occurs. This is shown in Figure 3.8(b); in the purple cases, the y-component of k has then no equal value in the less dense medium.

3.3 Fresnel Formulas

3.3.1 Amplitude Coefficients

If a plane wave falls onto a dielectric surface, beneath the law of refraction to determine the direction of the new wave, the problem is more physically complex [3, 11]. In reality a part of the energy is reflected and the exact division of the field amplitude depends on the refractive indices, the incidence angle and the polarization of the wave. Furthermore the effect of TIR must be considered in more detail and the case of absorbing media changes the behaviour. All these aspects are described by the so-called Fresnel formulae and are explained in this section.

The amplitude coefficients are complex quantities. They describe how the corresponding components of the electric field E and the magnetic field B are transformed at reflection and refraction, respectively. For the electrical field E the amplitude reflection factor is defined as,

$$r_E = \frac{E_r}{E_i} \tag{3.18}$$

with the incident field E_i. An analogous relation holds for the refracted field amplitudes. In particular, there are different factors for p- and s-polarization. Due to the properties of the field, the coefficients are complex quantities too. For the component of the E-field perpendicular to the incident plane (TE for transverse electric, σ- or

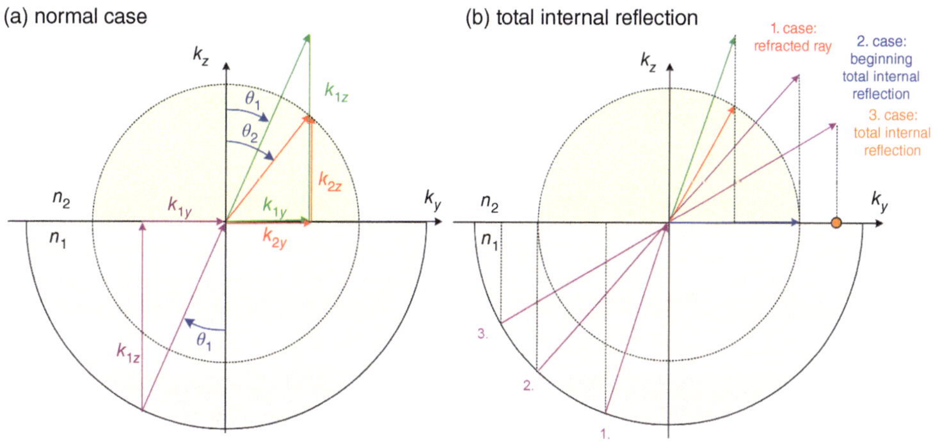

Figure 3.8 Description of the refraction in the k-space: (a) usual case and (b) case of total internal reflection.

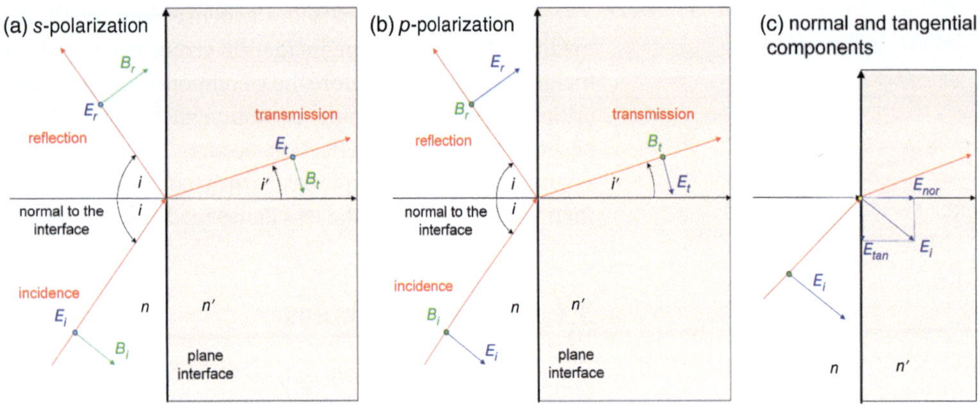

Figure 3.9 Refraction and reflection of an incoming ray at an interface for (a) s-polarization, (b) p-polarization. In (c) the field component decomposition is again sketched.

s-polarization) the coefficient reads with subscript s, for the field components of E lying in the incident plane (*TM* for transvers magnetic, π- or p-polarization) with subscript p. The amplitude ratios are then given by the famous Fresnel equations for reflection as

$$\begin{aligned} r_{Es} &= -\frac{\sin(i-i')}{\sin(i+i')} = \frac{n\cdot\cos i - \sqrt{n'^2 - n^2\cdot\sin^2 i}}{n\cdot\cos i + \sqrt{n'^2 - n^2\cdot\sin^2 i}} \\ &= \frac{n\cdot\cos i - n'\cdot\cos i'}{n\cdot\cos i + n'\cdot\cos i'} \\ r_{Ep} &= \frac{\tan(i-i')}{\tan(i+i')} = \frac{n'^2\cdot\cos i - n\cdot\sqrt{n'^2 - n^2\cdot\sin^2 i}}{n'^2\cdot\cos i + n\cdot\sqrt{n'^2 - n^2\cdot\sin^2 i}} \\ &= \frac{n'\cdot\cos i - n\cdot\cos i'}{n'\cdot\cos i + n\cdot\cos i'} \end{aligned} \tag{3.19}$$

The geometry of these cases is seen in Figure 3.9. The magnetic field is perpendicular to E and is therefore oriented in both cases in the other direction respectively. By analogy, for the refracted transmission component of the field amplitude, we have

$$\begin{aligned} t_{Es} &= \frac{2n\cdot\cos i}{n\cdot\cos i + n'\cdot\cos i'} \\ &= \frac{2n\cdot\cos i}{n\cdot\cos i + \sqrt{n'^2 - n^2\cdot\sin^2 i}} \\ &= \frac{2n\cdot\cos i}{n\cdot\cos i + n'\cdot\cos i'} \\ t_{Ep} &= \frac{2n\cdot\cos i}{n'\cdot\cos i + n\cdot\cos i'} \\ &= \frac{2n'\cdot n\cdot\cos i}{n'^2\cdot\cos i + n\cdot\sqrt{n'^2 - n^2\cdot\sin^2 i}} \\ &= \frac{2n\cdot\cos i}{n'\cdot\cos i + n\cdot\cos i'} \end{aligned} \tag{3.20}$$

This means that in total we have four cases for two polarization directions and transmission vs reflection. The three different formulations are only rearrangements of the variables and are equivalent. In practice the second representation often makes most sense, where the known indices n and n' as well as the given incidence angle i are used. The equations are easily derived by analysing the geometry of the projection and taking the boundary condition of the E field for the Maxwell equations into account. In particular the tangential E component in the interface plane must be continuous, the normal component perpendicular changes according to the refractive index step with index n_o or dielectric constant ε, see Figure 3.9(c).

$$E_{1\tan} = E_{2\tan}, \quad \varepsilon_1\cdot E_{1nor} = \varepsilon_2\cdot E_{2nor}, \\ n_o = \sqrt{\varepsilon_r} = \frac{c_0}{c} \tag{3.21}$$

For the magnetic field, the conditions with magnetic permeability can be formulated as

$$B_{1nor} = B_{2nor}, \quad \frac{B_{1\tan}}{\mu_1} = \frac{B_{2\tan}}{\mu_2} \tag{3.22}$$

Figure 3.9 shows the two geometries and notations for parallel and perpendicular polarization at the plane interface with the corresponding orientations of the fields E and B in parts (a) and (b). The plane of incidence is shown; the dots represent the field components perpendicular to this plane. The index i indicates the incident fields, r the reflected and t the transmitted (refracted) case respectively. The dependence of the amplitude coefficients as a function of the incidence angle i according to the Fresnel formulas are shown in Figure 3.10.

From this graphical presentation of the Fresnel equations, some specific properties can be observed.

- There is a specific incident angle i_B with a vanishing reflection coefficient of the parallel polarization, this is the Brewster angle. The reflected light is then fully s-polarized.
- For grazing incidence $i \to 90°$, the transmission vanishes.
- For an incidence perpendicular to the interface plane with $i = 0$, the coefficients have the same amount for both polarizations. In the case of reflection a right

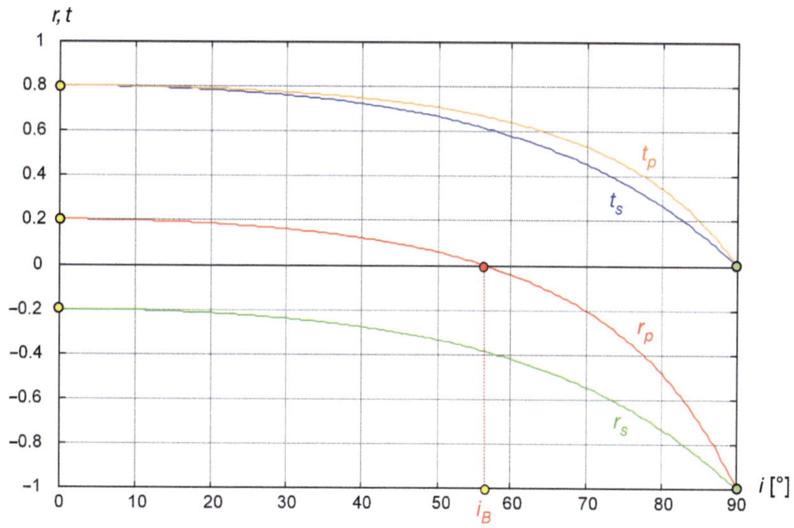

Figure 3.10 Dependence of the amplitude coefficients for refraction and reflection on the incidence angle, for the refractive indices: $n = 1.0/n' = 1.3$.

handed coordinate system is forced after the reflection by an inversion. The change in the coordinate system implies a change of sign.

According to the law of energy conservation, and neglecting any absorption, the following relations are valid for the Fresnel coefficients

$$t_s - r_s = 1, \quad t_p \cdot \frac{\cos i'}{\cos i} + r_p = 1 \tag{3.23}$$

The correction factor with the ratio of the cos functions takes into account the change of the relative cross-sectional area during transmission for a beam of finite width. The corresponding geometry is shown in Figure 3.11. Therefore, it should be noted that, for the calculation of an interface transition, there is energy conservation but the intensity or energy density is not constant. The finite beam diameter changes according to

$$D_r = D_i, \quad D_t = \frac{\cos i}{\cos i'} \cdot D_i \tag{3.24}$$

Additionally, the so-called Stokes relations are valid, connecting the corresponding coefficients for the amplitudes of the Fresnel equations by comparing the cases if the role of the incident and reflected/transmitted rays is interchanged. For the reversal, the coefficients are denoted by primed letters. The following relations are then obtained [3]

$$r'(i') = -r(i), \quad t'(i') \cdot t(i) = 1 - r^2(i) \tag{3.25}$$

In particular the sign switch of the reflected light means that the phase is changed by π in the case of reflection.

3.3.2 Reflectivity and Transmittivity

The definition of reflectivity and transmittivity for the radiant power of the *TM* and *TE* cases and their relation to the corresponding amplitude coefficient are given by

$$R = \frac{P_r}{P_i} = |r^2|, \quad T = \frac{P_t}{P_i} = \frac{n' \cdot \cos i'}{n \cdot \cos i} \cdot |t^2| \tag{3.26}$$

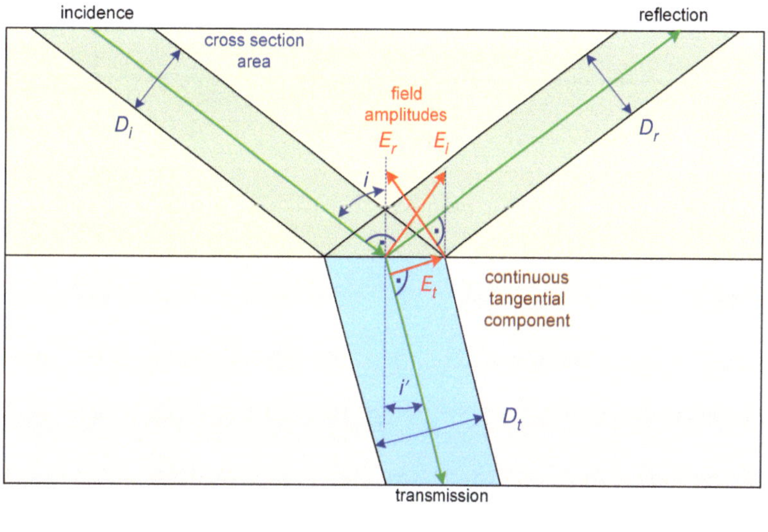

Figure 3.11 Refraction and reflection of an incoming beam with finite width D_i. The change of the incidence angle in case of refraction generates a modified bundle diameter D_t.

From the law of energy conservation, we have the relation, if absorption is neglected

$$R + T = 1 \tag{3.27}$$

Simple squaring the amplitude coefficients gives the intensity-related coefficients $R^{(I)}$ and $T^{(I)}$. They are not additive due to the change in the cross-sectional area. As a function of the incident angle, the formulas are explicitly written in the following way directly from the Fresnel equations

$$R_s^{(I)} = \frac{I_{rs}}{I_i} = \frac{\sin^2(i-i')}{\sin^2(i+i')}, \quad R_p^{(I)} = \frac{I_{rp}}{I_i} = \frac{\tan^2(i-i')}{\tan^2(i+i')} \tag{3.28}$$

$$T_s^{(I)} = \frac{I_{ts}}{I_i} = \frac{\sin 2i' \cdot \sin 2i}{\sin^2(i+i')},$$
$$T_p^{(I)} = \frac{I_{tp}}{I_i} = \frac{\cos 2i \cdot \sin 2i'}{\sin^2(i+i') \cdot \cos^2(i-i')} \tag{3.29}$$

Figure 3.12 shows the coefficients for the power transfer as well as the intensity. It is seen that only the power related coefficients in (a) fulfil the law of preservation with a sum of 1; in both cases of polarization the curves are exactly symmetric around the line with value 0.5. In the case of the intensity in (b), the change in the cross section area of a ray beam gives a sum different from 1. This means the intensities are not additive.

In the special case of a normal ray incidence with $i = 0$, one obtains for the reflectivity the simple and in practice often-used formula

$$R_{s/p} = \left(\frac{n-n'}{n+n'}\right)^2 \tag{3.30}$$

In particular for boundaries between glass and air with $n' = 1$ the reflectivities for some refractive indices are listed in Table 3.1. The case for $n = 1.5$ and $n' = 1$ is typical for many air–glass combinations. The formula (3.30) then delivers the result 0.04. Then there is exactly 4% transmission loss. It can also be seen that, for the large refractive indices $n > 2$ common for the materials used in the infrared range, quite significant residual reflections occur for uncoated components. On the other hand, in case of plastic materials with low index $n = 1.4 \dots 1.46$, the Fresnel losses are quite small.

Table 3.1 Reflectivity at normal incidence for different refractive indices.

n	R
1.4	2.778 %
1.5	4.0 %
1.8	8.16 %
2.4	16.96 %

For the transmittivity we have correspondingly

$$T_{s/p} = \frac{4n \cdot n'}{(n+n')^2} \tag{3.31}$$

this is valid for intensity as well as power due to the value $i = 0$ and a vanishing projection effect.

3.3.3 Transmission of Systems

The total transmission of a complete optical system depends on the number of surfaces and their coatings. In case of uncoated surfaces, the Fresnel equations are directly applicable with the material parameters of the substrate glass. When R is the residual reflection of a single surface and N is the number of surfaces, and assuming for

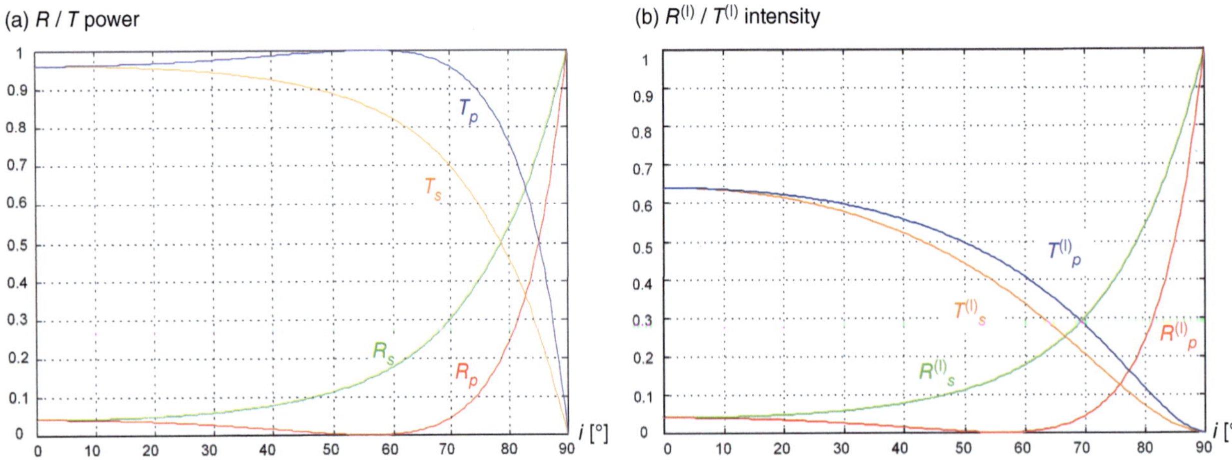

Figure 3.12 Reflection and transmission coefficients for (a) power and (b) intensity as a function of the incidence angle for parallel and perpendicular polarization.

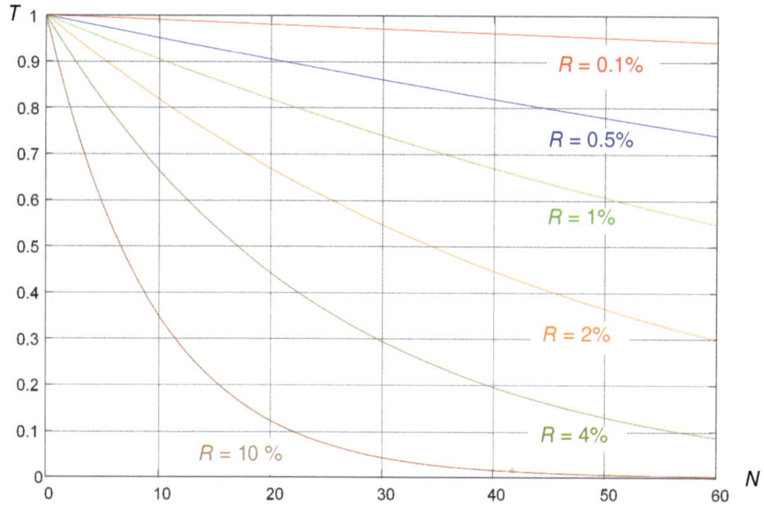

Figure 3.13 Total transmission of an optical system as a function of the number of surfaces N and the reflectivity R.

simplicity that all surfaces have the same R, we have

$$T_{tot} = (1 - R)^N \tag{3.32}$$

Therefore, to get a high transmission in a system with a large number of surfaces, it can be seen that it is extremely desirable to have coatings on the surfaces which have very low residual reflectivities R. The dependence of the total transmission on the number of surfaces and the reflectivity is shown in Figure 3.13. Typical multilayer coatings with high performance reach $R = 0.1\%$, but with a more complicated dependence on incidence and wavelength. If a refractive index of $n = 1.5$ is assumed, from this dependency it is seen that for $N = 10$ corresponding to 5 lenses the overall transmission is $T = 0.67$, for 15 lenses we get $T = 0.3$.

In reality the conditions are more complicated due to several reasons.

- The rays of a bundle intersect a surface under different incidence angles; the case of a fully concentric penetration is more an exception for curved surfaces. For a surface, the averaged value over all incidence angles is relevant for the transmission.
- The surfaces are not always met perpendicular but for an off-axis field point under a mean obliquity angle. This means that the transmission depends on the field point position.
- The angle conditions change from surface to surface.
- The surfaces are coated with a more complicated behaviour than the Fresnel equations describe.

In practice, residual reflectivity issues are often observed in systems with strongly curved surfaces. In such cases the coating only works sufficiently well in the central region near the optical axis. In the outer regions of the components one observes significantly reduced transmission due to large incident angles. Then, a decrease can be observed in the intensity in the outer region of the lens cross section, which is very often the reason for asymmetries in the brightness at field points off the axis.

To get a better understanding about the quantitative influence of the first two effects, a simple paraxial model can be used. Let us assume the transmission formula (3.29)

$$T = \frac{\sin 2i' \cdot \sin 2i}{\sin^2(i + i')} \tag{3.33}$$

to be valid and let's consider only one spherical surface with radius of curvature R (not to be confused with the reflectivity here) and a transition from index n into air. The incidence angle in paraxial approximation is given for the ray height y and incident collimated ray bundle with diameter $2a$ in the x and the y plane respectively for a ray angle u only in the meridional plane by

$$i_x = \frac{x}{R}, \quad i_y = \frac{y}{R} - u, \quad i^2 = i_x^2 + i_y^2 \tag{3.34}$$

This means we consider a collimated ray bundle with inclination in the y–z plane by an angle u. If small incidences are assumed, the law of refraction reads $i' = n\,i$ and a Taylor expansion for the transmission T, if the oblique direction is assumed to be only in y, gives

$$T(x,y) \approx \frac{\left[2ni - \dfrac{8n^3i^3}{6}\right] \cdot \left[2i - \dfrac{8i^3}{6}\right]}{\left[i(n+1) - \dfrac{8(n+1)^3 i^3}{6}\right]^2}$$

$$\approx \frac{4n}{(n+1)^2} \cdot \frac{1 - \dfrac{2}{3}(n^2+1)i^2}{1 - \dfrac{1}{3}(n+1)^2 i^2}$$

$$\approx \frac{4n}{(n+1)^2} \cdot \left[1 - \frac{(n-1)^2}{3} \cdot i^2\right]$$

$$= \frac{4n}{(n+1)^2} \cdot \left[1 - \frac{(n-1)^2}{3} \cdot \left[\left(\frac{x}{R}\right)^2 + \left(\frac{y}{R} - u\right)^2\right]\right] \tag{3.35}$$

If this local transmission is averaged over the circle area with radius a, we get the expression

$$\overline{T} = \frac{1}{\pi a^2} \int_{y=-a}^{a} \int_{x=-\sqrt{a^2-y^2}}^{+\sqrt{a^2-y^2}} T(x,y)\,dx\,dy$$
$$= \frac{4n}{(n+1)^2} \cdot \left[1 - \frac{(n-1)^2}{8} \cdot \left(\frac{a}{R}\right)^2 - \frac{(n-1)^2}{3} \cdot u^2\right] \tag{3.36}$$

This formula shows that due to a finite radius of curvature and a finite incidence u in this simple case the total transmission is reduced for growing angle u of the incoming collimated ray bundle. For $n = 1.5$ and some values of u, Figure 3.14 shows some results as a function of the ratio a/R, which corresponds to the centre angle at the surface. It is seen by inspecting the stretched scale of the transmission axis that the oblique incidence and the curvature have only a small impact of a few percent. But this situation can be different in cases of coatings with a larger i-dependence. It is well known that in the case of microscopic objective lenses with a high numerical aperture in the front group strong effects are observed, which finally give rise to an apodization and a decreased effective numerical aperture.

Using the example of a simple lens imaging, this effect can be understood qualitatively as is shown in Figure 3.15. Shown is (a) a lens with aperture angle $u = 15°$ and a field angle $w = 27°$ in a 1 : 1 imaging setup. The distribution of the incident angles of all the rays inside a ray bundle and the transmission for the axis bundle and the off-axis bundle are shown in Figure 3.15(b) and (c).

The residual reflections in the outer regions also show a special behaviour depending on the wavelength. Therefore, this can lead to coloured reflection images which in addition are polarization dependent. The coatings of lenses are not only applied in real systems to improve the transmission but can also be used to balance the colour equilibrium in the transmission.

3.3.4 Total Internal Reflection

The angle of TIR of glasses typically lies in the range between 33° and 43°. Its dependence on the refractive index can be seen in Figure 3.16. So if TIR is a desired effect, it is better to use a flint-type SF glass with high refractive index; if TIR is an undesirable effect, low index crown-glasses are more comfortable.

In the case of TIR, the Fresnel equations are still valid. The amplitude coefficient of the transmission r_E becomes imaginary. This means that the wave is damped in the new medium with the refractive index n'. In this case the field can be written as

$$E_t(z) = E_e(z) \cdot e^{-\gamma \cdot z} = E_e(z) \cdot e^{-\frac{z}{\delta}} \tag{3.37}$$

with the real damping constant

$$\gamma = \frac{2\pi}{\lambda} \cdot \sqrt{n^2 \cdot \sin^2 i - n'^2} \tag{3.38}$$

or the penetration depth

$$\delta = \frac{1}{\gamma} = \frac{\lambda}{2\pi \cdot \sqrt{n^2 \cdot \sin^2 i - n'^2}} \tag{3.39}$$

Here it is assumed that z is the local propagation coordinate of the wave. So if the light wave is damped

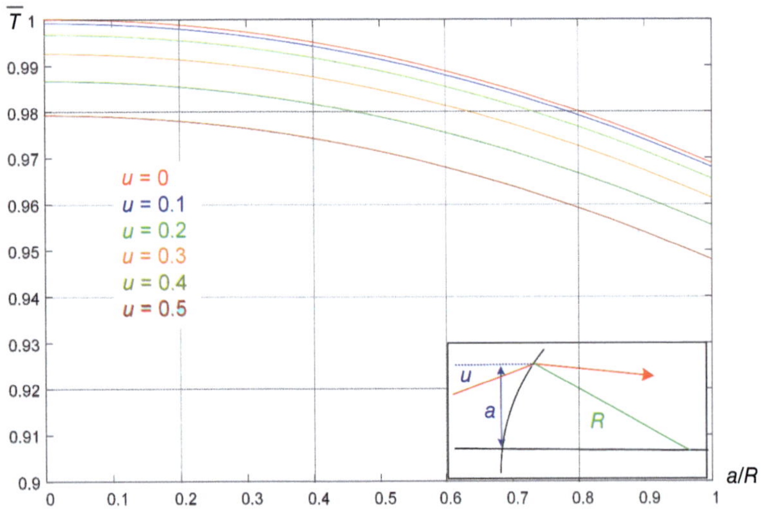

Figure 3.14 Averaged transmission of a collimated ray bundle at a spherical surface with a transition from index n into air as a function of curvature a/R and main incidence angle u.

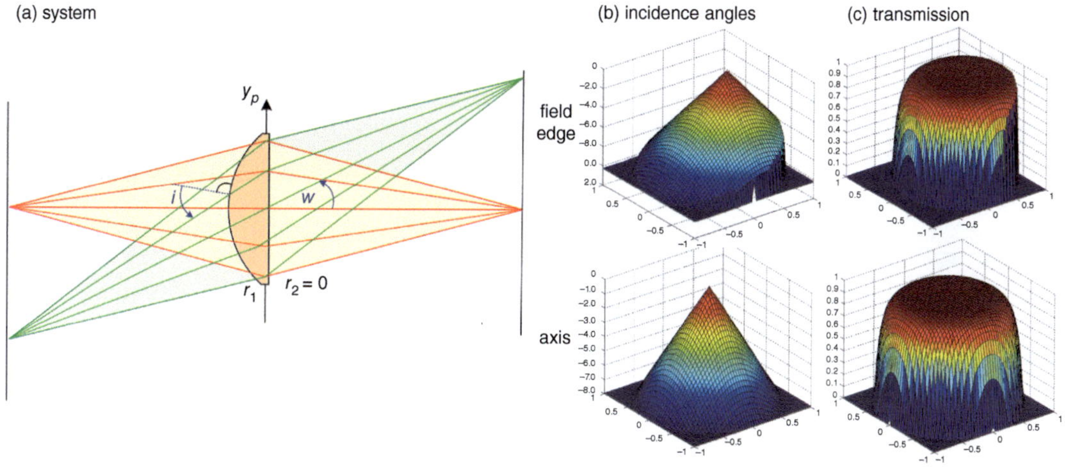

Figure 3.15 Asymmetric distribution of the incident angles for field imaging at strongly curved surfaces. In (b) the incidence angles are given in °, (c) illustrates the transmission as a result of a numerical raytracing based calculation for the axis and in the outermost field point. In (a) the angle i indicates the incidence of an arbitrary ray for illustration, the green ray bundle is tilted by the field angle w.

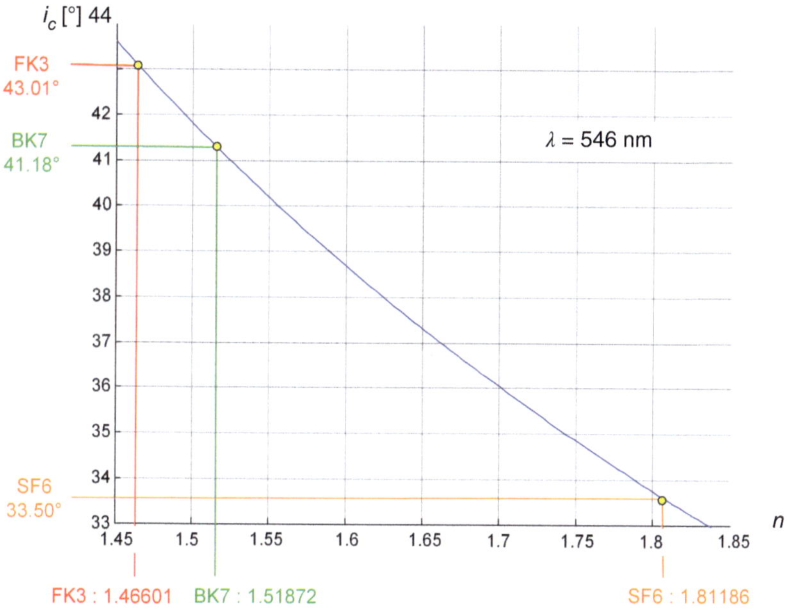

Figure 3.16 Angle of total internal reflection of glasses as a function of the refractive index.

exponentially with the characteristic depth δ, an evanescent wave occurs which penetrates into the new medium for a certain length δ. This is illustrated in Figure 3.17.

By calculating the transmitted wave, it is found that a wave in the new medium runs along the interface with an amplitude damped in the z direction. This effect is illustrated in Figure 3.18. In the interface plane the light intensity has its maximum value for the angle of TIR $i = i_c$. For smaller and also for larger incident angles, the values of the intensity decrease.

If under TIR conditions the second medium has a complex refractive index like a metal, the surface wave penetrating the second medium is partially absorbed. Thus the TIR withdraws energy from the reflected wave. Then the reflectivity is less than 1. At weak absorption can be described by a complex refractive index

$$n' = n'_r + i \cdot n'_i \tag{3.40}$$

For the parameter γ we then have

$$\gamma = \gamma_r + i \cdot \gamma_i \tag{3.41}$$

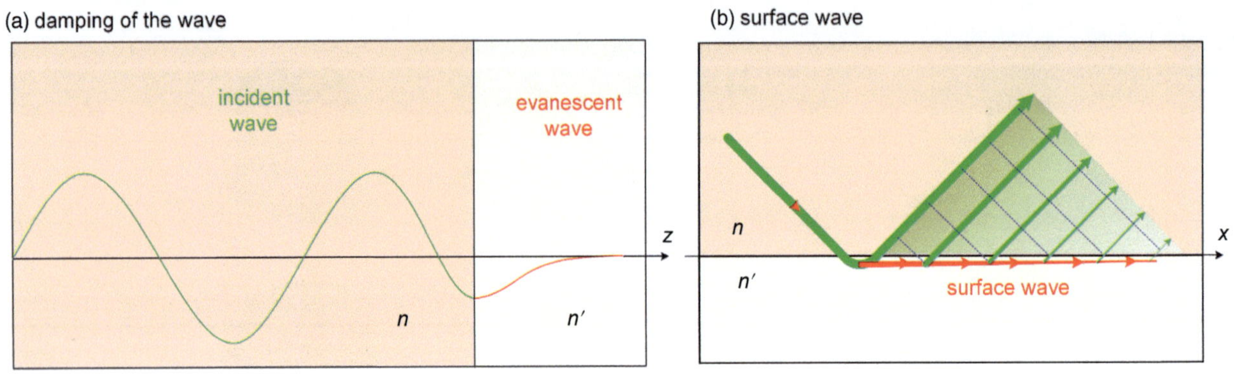

Figure 3.17 Evanescent wave, (a) schematic plot of the damping, (b) surface wave.

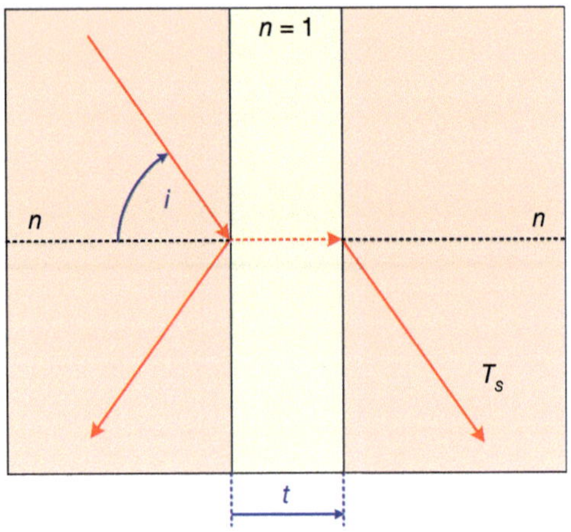

Figure 3.18 Frustrated total reflection and tunnel wave.

with approximately

$$\gamma_r = \frac{\sqrt{n^2 \cdot \sin^2 I - n_r'^2}}{n \cdot \cos i} = -\frac{n_r'^2 \cdot n_i'}{\sqrt{n^2 \cdot \sin^2 i - n_r^2}} \tag{3.42}$$

Then the reflectivities for both polarizations can be calculated by

$$R_s = 1 + \frac{4\gamma_i}{1+\gamma_r^2}, \quad R_p = 1 + \frac{4\gamma_i \cdot \left(\frac{n}{n_r'}\right)^2}{1 + \gamma_r^2 \cdot \left(\frac{n}{n_r'}\right)^4} \tag{3.43}$$

If the second thin medium is air, then the total reflection can be disturbed by inserting an optical component within a distance shorter than the penetration depth $t < \delta$. This causes a tunnelling effect of the light, which can be used for measurement purposes. This effect is called frustrated or disturbed TIR. Consider a sandwich structure consisting of two equal glasses with refractive indices n and a narrow air gap of thickness t in between, as shown in Figure 3.18. For sufficiently small dimensions of t in the range of the wavelength a frustrated total internal reflection (FTIR) occurs, caused by light crossing the gap by means of evanescent waves.

When i is the incident angle and t/λ the normalized gap, we have with the abbreviation

$$\psi = \frac{2\pi \cdot t}{\lambda} \cdot \sqrt{n^2 \cdot \sin^2 i - 1} \tag{3.44}$$

for the transmissions of the tunnel wave

$$\frac{1}{T_s} = 1 + \frac{(n^2-1)^2 \cdot \sinh^2\psi}{4n^2 \cdot \cos^2\psi \cdot (n^2 \cdot \sin^2 i - 1)} \tag{3.45}$$

$$\frac{1}{T_p} = \frac{(n^2-1)^2 \cdot \sinh^2\psi \cdot (n^2 \cdot \sin^2 i - 1)}{4n^2 \cdot \cos^2\psi \cdot (n^2 \cdot \sin^2 i - \cos^2 i)^2} \tag{3.46}$$

depending on the polarization. Here it must be noted that, in the small gap, multiple reflections and interferences occur which must be included in the transmission calculation. It must be mentioned that an additional phase shift takes place in these cases.

In the following Figure 3.19 the dependence of T_s as a function of the normalized gap width t/λ is shown for a

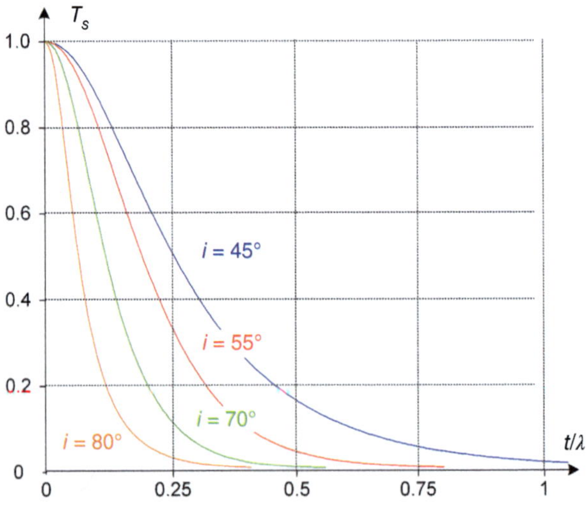

Figure 3.19 Transmission of the frustrated total reflection under s-polarization as a function of the incident angle and the normalized gap width t/λ.

refractive index of $n = 1.5$ and incident angles 45°/55°/70° and 80°.

Using the approach for the evanescent wave, the typical gap distance for a significant light transmission through the sandwich structure is given by

$$t = \frac{\lambda}{2\pi \cdot \sqrt{n^2 \cdot \sin^2 i - n'^2}} \tag{3.47}$$

3.3.5 Reflection at Metals

In the generalized Maxwell description, materials are characterized by the three parameters ε (dielectric constant), μ (magnetic permeability) and σ (electrical conductivity). In particular, the conductivity leads to a damping and absorption of electromagnetic radiation by moving electrons. With frequency ν of the radiation, the following relation with the complex refractive index is valid (here j is the imaginary unit) [3]

$$n = n_r + j \cdot n_i = \sqrt{\mu \cdot (\varepsilon + j \cdot 2\nu \cdot \sigma)} \tag{3.48}$$

respectively, the real and imaginary parts

$$n_r = \frac{1}{\sqrt{2}} \cdot \sqrt{\mu\varepsilon + \sqrt{\mu^2\varepsilon^2 + \frac{4\mu^2\sigma^2}{\nu^2}}},$$

$$n_i = \frac{1}{\sqrt{2}} \cdot \sqrt{-\mu\varepsilon + \sqrt{\mu^2\varepsilon^2 + \frac{4\mu^2\sigma^2}{\nu^2}}} \tag{3.49}$$

In the extreme case of a very strong conductivity, the imaginary part of the complex refractive index dominates and one gets approximately

$$\alpha \approx \frac{\sqrt{4 \cdot \nu \cdot \mu \cdot \sigma}}{c} = 2\sqrt{\frac{\mu \cdot \sigma}{c \cdot \lambda_0}} \tag{3.50}$$

are alternatively

$$n_i = n\kappa = \sqrt{\frac{\sigma}{2\varepsilon_0 \cdot \omega}} \tag{3.51}$$

This simplified relation is also referred to as the Drude equation. The complement of the absorption constant corresponds to a penetration depth of the radiation into the material. At a length of $t = 1/\alpha$ the intensity decreases to $1/e = 36\%$ of the original value. In metals this penetration depth of the field is called the skin depth

$$t_{skin} = \frac{1}{\alpha} = \frac{1}{2} \cdot \sqrt{\frac{c \cdot \lambda_0}{\mu \cdot \sigma}} \tag{3.52}$$

Usually metals show a very large absorption. Therefore, in the approximation

$$\left| \mathrm{Im} \frac{n'}{n} \right| \gg 1 \tag{3.53}$$

the Fresnel equations can be simplified for reflection at a metal after inserting the complex refractive index [3]. In this way the following reflectivities can be found

$$R_s = \frac{\left(\cos i - \frac{n'}{n}\right)^2 + \left(\frac{n' \cdot \kappa'}{n}\right)^2}{\left(\cos i + \frac{n'}{n}\right)^2 + \left(\frac{n' \cdot \kappa'}{n}\right)^2},$$

$$R_p = \frac{\left(1 - \frac{n'}{n} \cdot \cos i\right)^2 + \left(\frac{n' \cdot \kappa'}{n} \cdot \cos i\right)^2}{\left(1 + \frac{n'}{n} \cdot \cos i\right)^2 + \left(\frac{n' \cdot \kappa'}{n} \cdot \cos i\right)^2} \tag{3.54}$$

Further, the phase angles are obtained from the amplitude reflection coefficients depending on the polarization as

$$\tan \delta_s = \frac{\frac{2n' \cdot \kappa'}{n} \cdot \cos i}{\left(\frac{n'}{n}\right)^2 \cdot (1+\kappa)^2 - \cos^2 i},$$

$$\tan \delta_p = \frac{\frac{2n' \cdot \kappa'}{n} \cdot \cos i}{\left(\frac{n'}{n}\right)^2 \cdot (1+\kappa)^2 \cdot \cos^2 i - 1} \tag{3.55}$$

From this a relative phase shift between the two polarization components follows

$$\tan(\delta_p - \delta_s) = \frac{\frac{2n' \cdot \kappa'}{n} \cdot \cos i \cdot \sin^2 i \cdot \left[1 + \left(\frac{n'}{n}\right)^2 \cdot (1+\kappa^2)\right]}{\left[1 + \left(\frac{2n' \cdot \kappa'}{n}\right)^2 + \left(\frac{n'}{n}\right)^4 \cdot (1+\kappa^2)^2\right] \cdot \cos^2 i - \left(\frac{n'}{n}\right)^2 \cdot (1+\kappa^2) \cdot (1+\cos^4 i)} \tag{3.56}$$

So, for reflection at metals, linearly polarized light becomes elliptically polarized. By a particular choice of the incident angle and the azimuth of the linear polarization, it can be arranged that the reflected light is just polarized circularly. For this the angle difference has to be 90°. The corresponding angle is called the principal angle of incidence.

For silver, the material parameters at $\lambda = 587$ nm are: $n' = 0.18$, $\kappa = 20.222$. Using these values in Figure 3.20, the components for the reflectivity and the phase angles of the polarization are shown. For the angle of incidence $i = 76.084°$, the relative phase shift is 90°.

For wavelengths which are not too short there is an approximation for the reflectivity of non-magnetic metals. According to Rubens and Hagen the following simplified formula is valid, which also can be derived from the

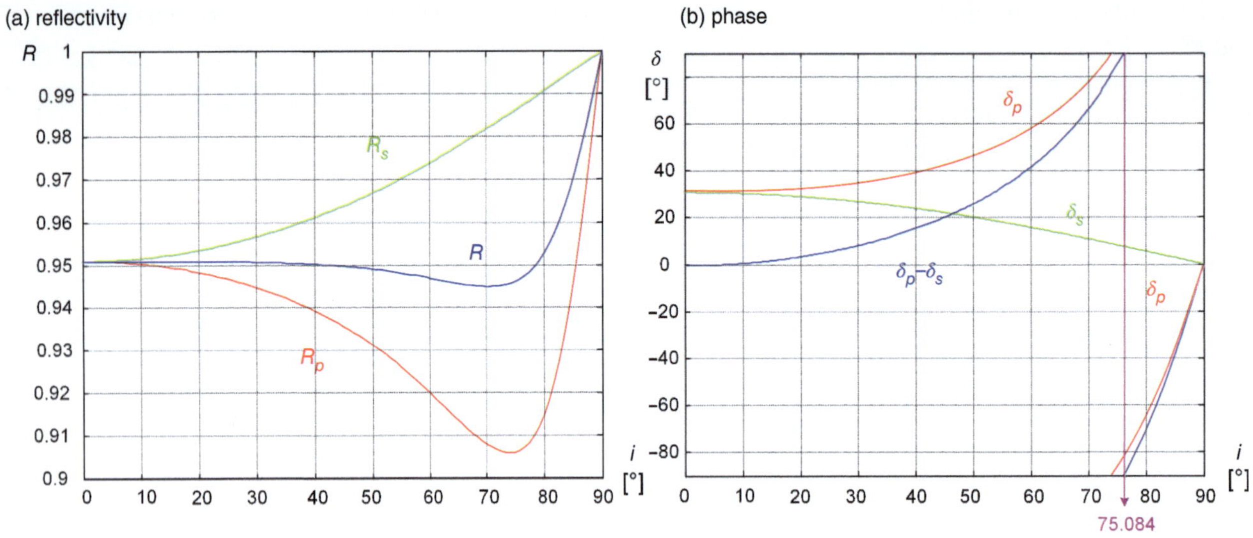

Figure 3.20 Reflection at silver for $\lambda = 587$ nm with (a) the reflectivities and (b) the phase angles.

elementary oscillator model of the atomic absorption

$$R = \frac{\frac{2\sigma}{\nu} + 1 - 2 \cdot \sqrt{\frac{\sigma}{\nu}}}{\frac{2\sigma}{\nu} + 1 + 2 \cdot \sqrt{\frac{\sigma}{\nu}}} \approx 1 - 2 \cdot \sqrt{\frac{\sigma}{\nu}} \tag{3.57}$$

where the last approximation is valid for $\sigma/\nu \ll 1$.

3.4 Raytrace

3.4.1 Introduction

Raytracing is the basis for a pure geometrical modelling of optical systems [14–24]. Apart from the law of refraction, no other physical-optical relations are necessary. In this sense raytracing simplifies the effect of an optical system by using the geometrical-optical approximation of a very small wavelength and neglecting all diffraction effects. Nevertheless, raytracing is among the most fundamental algorithms of all optical calculations and simulations for real systems.

Raytracing uses single rays for the description of the light. There is no transverse extension connected to a ray, so that diffraction effects cannot be considered in geometrical optics.

Even in systems where wave-optical approaches are used to simulate certain diffraction effects, the transport of light through compact systems is reproduced by means of raytracing. By converting the ray data into a wave, a physical-optical propagation can follow, if a weight of the rays is defined that describes the amplitude of the field at the corresponding location and the optical path difference gives the phase of the field. This is referred to as the so-called hybrid method.

Additionally, the modelling of illumination tasks belongs to the application field of raytracing. Here a weight factor is assigned to each ray, representing the transferred energy. This is an advantage also for the calculation of, e.g. point image distributions where the ray weights describe a non-uniform illumination of the system pupil. This so-called apodization is important for the correct determination of the light distributions in the focus area.

Raytrace is the easiest way to describe the propagation behaviour of light. A ray is a representative of a plane wave and in media of constant refractive index it is a straight line. In an optical system, the rays are deflected at the interfaces of the mirrors, lenses or other components. The laws of reflection and refraction are applied to find the new ray directions. The system itself is described in the simplest case by spherical surfaces with radius of curvature R or curvature $c = 1/R$. The distances t between the vertex points determine the axial length of the system's geometry. The materials are described by their refractive indices n and the Abbe number.

The raytrace is performed by the following two major steps.

- Transition from one surface to the next one and determination the intersection point. In the case of second-order surfaces, this point can be calculated analytically swiftly and accurately.
- Perform the refraction/reflection in the intersection point of a surface and find the new ray direction. For this the local normal vector of the surface must be computed first.

This scheme is illustrated in Figure 3.21.

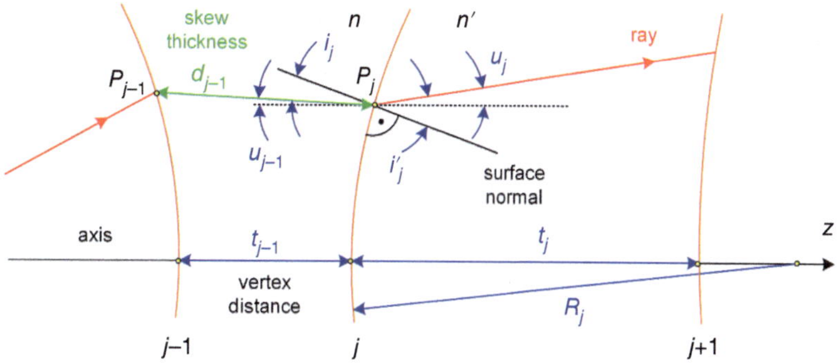

Figure 3.21 Scheme of raytrace in one transverse dimension.

The algorithm of the calculation mainly follows this scheme; the mathematical set of formulas can be found in the next sections. A more detailed workflow consists of the following steps.

1) Transport of a ray over a distance.
   1.1 In a homogeneous medium: the straight line equation in a general vector form.
   1.2 In an inhomogeneous medium: integration of the eikonal equation with a known refractive index function $n(x,y,z)$. As the cut-off condition, the determination of the intercept has to be carried out for each step of the integration. It is reasonable to adapt the increment in the medium in order to adjust the widths of the interval to the size of the local gradient of the refractive index. A possible method for the integration of the corresponding differential equation is the Runge–Kutta method. For a detailed discussion of the eikonal, see Chapter 13.
2) Calculation of the intersection point of the ray with the next interface between two media.
   2.1 If the shapes of the surfaces can be described by linear or quadratic equations, the intercept can be determined analytically. This is fast and exact and therefore desirable.
   2.2 If the shapes of the surfaces are of higher order (aspherical or a surface with a free formed shape, maybe without a closed analytical description), the intercept has to be determined by means of a numerical iteration method. This slows down the speed of raytrace considerably.
3) Calculation of the change in the ray direction in the intersection point with the surface.
   3.1 In the point of intersection, the local normal unity vector of the surface must be calculated.
   3.2 Refracting surface: Snell's law of refraction has to be applied.
   3.3 Reflecting surface: the law of reflection is used.
   3.4 Diffractive elements: in the frame of the local theory of gratings, the grating equation for transmission or reflection is valid.
   3.5 Scattering surface: using the Monte-Carlo method the new direction of the ray is determined by a non-deterministic statistical law according to the scattering characteristics. Then the resulting direction of the ray is broadened into an angular interval.

Figure 3.22 illustrates the logical workflow for the determination of the point of intersection and of the new direction of the ray in more detail. In particular the pitfalls of TIR and failure to find an intersection point or vignetting due to a finite size of a surface are important.

In order to provide a general model description, currently vector raytracing is used for a three-dimensional geometry. For historical reasons a larger number of special sets of formulas exists, which, in practice, can be applied more easily or quickly if calculation speed matters, as in the early days of lens design. The practical impact of these approaches is not significant for today's optical design software. However, the sets of formulas convey some understanding of the different approaches for raytracing and possible approximations and simplifications.

For a description of the ray, the sets of formulas use the following alternatives for a particular intersection point of the ray with a surface.

- The height of the intersection at the surfaces (y or h) as well as the z-value (sag).
- The normal from the surface vertex to the ray (Q).
- The ray height in the vertex plane (y or h)

Finally, it depends on the assumptions of the raytrace set whether the ray height y or h is defined in the vertex plane or exact at the surface. For the direction of the ray they use the following.

- The angle with the optical axis (u).
- The back focal length or intersection length with the optical axis (L, s).

The formula sets using the back focal length have a basic disadvantage. In the case of rays running parallel or nearly

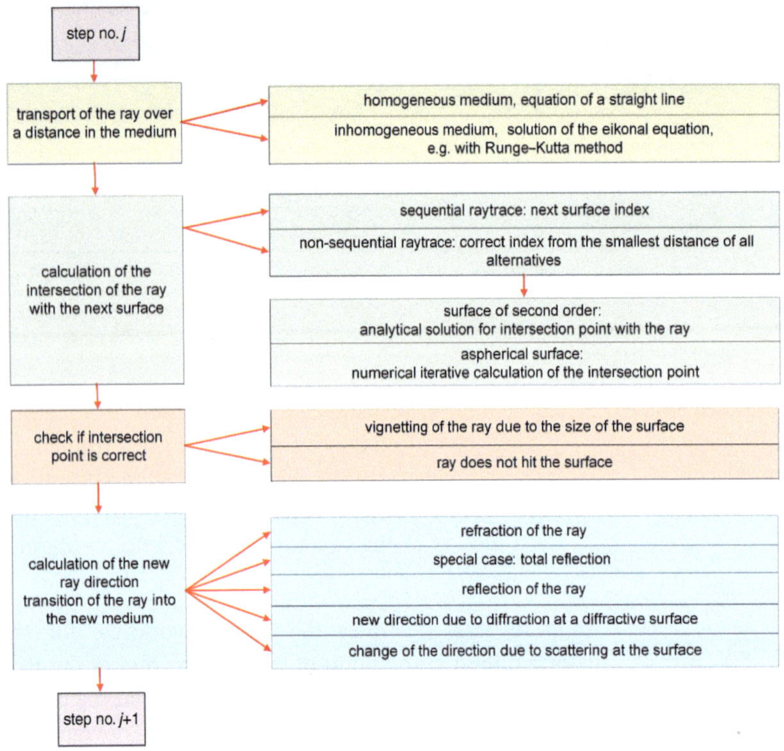

Figure 3.22 Principal workflow of a raytrace. The different red arrows indicate alternative operations depending on the conditions, only one of the variants is realized.

parallel to the axis, they have to take a special case without intersection points into account by using modified formulas. For the recognition of this special case, a threshold criterion for the angle of the ray with the optical axis is necessary, which can depend on the accuracy of the computer if programmed in a disadvantageous way.

The principal limitation of the simplified formula sets used in past pre-computer times is the fact that only rays in the meridional cross section plane are taken into account. In a further simplification the refraction at the surfaces is computed only in paraxial approximation.

In the nomenclature used in the following, the simple terms indicate the region in front of the surface while the primed quantities indicate the region behind the surface.

The different sets of formulas are as follows.

- Paraxial formulas, for reference of ideal imaging and first order estimates.
- Meridional formulas, for circular symmetry in one cross section plane only containing the z axis.
- Vectorial formulas, general three-dimensional systems and arbitrary surface shapes – state of the art today.
- Differential formulas for parabasal rays, transfer of a neighbourhood; this propagates a ray density and therefore is a great advantage in illumination computations. In addition, it directly describes astigmatism.

There are some more specialized aspects, which should be supported in a general raytrace. These are as follows.

- Spherical surfaces or higher orders than second: numerical iterative calculation of intersection points is required (see Section 3.4.7).
- Gradient index media: eikonal differential equation, Runge–Kutta numerical stepwise (see Section 3.4.8).
- Diffractive elements: local grating equation (see Section 3.4.10).
- Non-sequential raytrace: illumination and stray light (see Section 3.4.11).
- Scattering surfaces: stochastic Monte-Carlo decision for new direction.
- Photometric correct raytracing: transfer of relative weighting factor corresponding to the ray density or ray tubes (see Section 3.4.9).
- Polarization raytrace: transfer of a Jones vector on a ray (discussed in a later Chapter).
- Geometrical approximated edge diffraction: ray deviation depends on edge distance.

3.4.2 Paraxial Formulas

The simplest set of raytrace uses the paraxial approximation of small ray angles and small height at the surfaces. For a detailed discussion of paraxiality, see Section 3.5. In this case, all trigonometric functions and surface equations can be simplified by an expansion until first order and the full set of formulas becomes linear. In particular the quadratic sag or Δz-values are neglected and the coordinates of the intersection points can be assumed to lie in the vertex plane. The paraxial set of formulas has the following description parameters.

- Height y of the intersection point of the ray with the surface in the vertex plane.
- Angle of inclination of the ray u.
- Distance t between the surface vertex points.
- Curvature c of the surfaces. It is more comfortable to use the curvature instead of the radius R, as this avoids the singular case of a plane surface with $R \to \infty$.

The procedure consists of two steps according to the general scheme.

- Transition to the next surface

$$y_j = y_{j-1} - t_{j-1} \cdot u'_{j-1} \tag{3.58}$$

the angle of incidence is then

$$i_j = c_j \cdot y_j - u_j \tag{3.59}$$

- The angle of refraction in paraxial approximation of Snell's law

$$i_j{}' = \frac{n_j}{n_j{}'} i_j \tag{3.60}$$

the new ray angle is obtained as

$$u_j{}' = u'_{j-1} + i_j - i_j{}' = y_j c_j - i_j{}' \tag{3.61}$$

The paraxial $y-u$ method can also be represented very clearly in a matrix scheme. This method is described in Section 3.6. Figure 3.23 shows the conventions and terms of the paraxial raytrace at a single surface. It should be emphasized in particular that due to the approximation, the intersection point P does not exactly lie on the surface, but in the vertex plane.

3.4.3 Meridional *Q–U* Method*

(*) The content of this section is quite specialized and for advanced reading. Knowledge of this section is not necessary to understand the following chapters. Beginners are recommended to skip this section.

In case of an exact raytrace in one transverse plane only, there is a way to avoid the special cases of infinite intersection lengths with the optical axis and to create a direct connection to the solution of the paraxial approximation. For this the length Q from the surface vertex V normal to the rays is used instead of the y coordinates perpendicular to the axis [12], see Figure 3.24.

In the paraxial special case, this set of normal formulas has the same structure as the meridional set using y. In the paraxial approximation for small inclination angles, the distances y and Q cannot be distinguished. The transition equation has a more simple form in the finite case since, for the normal, the sine can always be directly applied.

$$Q_j = Q'_{j-1} - t_{j-1} \cdot \sin u_{j-1} \tag{3.62}$$

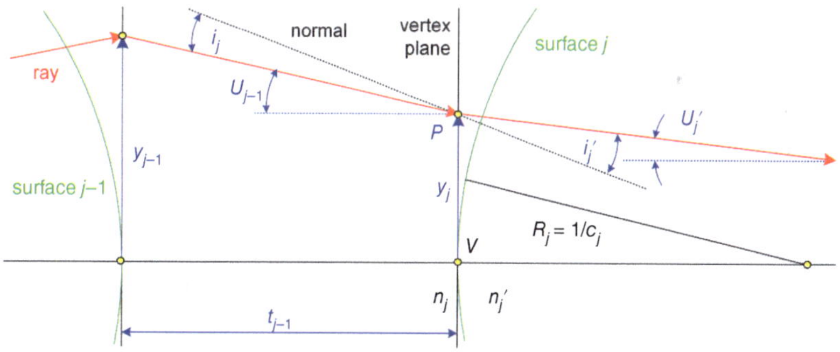

Figure 3.23 Paraxial raytrace at a refracting spherical surface.

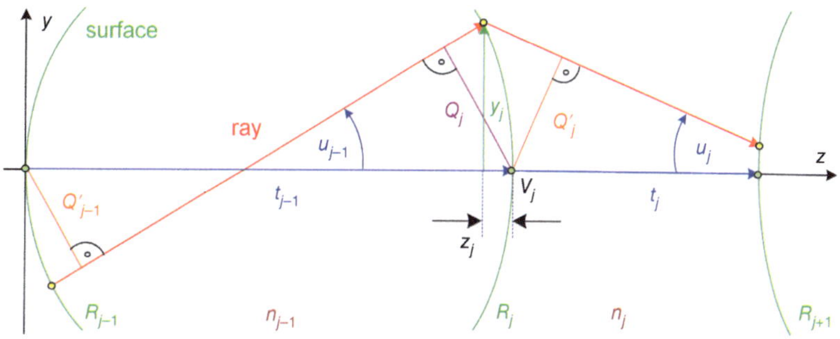

Figure 3.24 Terms and indices for raytracing in the set of formulas using the normal length Q.

The refraction at a spherical surface is described by the following equations

$$\sin i_j = c_j \cdot Q_j - \sin u_{j-1}$$
$$n' \sin i_j' = n \sin i_j$$
$$u_j = u_{j-1} + i_j - i_j' \tag{3.63}$$

The conversion of the normal at the refraction at the surface is given by

$$Q_j' = Q_j \cdot \frac{\cos i_j' + \cos u_j}{\cos i_j + \cos u_{j-1}} \tag{3.64}$$

From the normal Q the ray height y perpendicular to the axis at the surface and the sagittal height z can be determined by using the equations

$$y_j = Q_j \cdot \frac{\cos(u_{j-1} + i_j)}{\cos i_j + \cos u_{j-1}}, \quad z_j = Q_j \cdot \frac{\sin(u_{j-1} + i_j)}{\cos i_j + \cos u_{j-1}} \tag{3.65}$$

For raytracing using the intersection length s with the optical axis we have the relation

$$Q = s \cdot \sin u \tag{3.66}$$

In particular, the final calculation of the back focal length can be determined by this equation in the image domain.

The error pitfall conditions in this set of formulas emerge from the determination of the intersection point with the surface of the second order as well as from the law of refraction or reflection respectively.

- $|\sin i| > 1$ the ray misses the surface.
- $|\sin i'| > 1$ TIR occurs.

The starting calculation of the formula set follows from the height of the object, the distance and the size of the entrance pupil.

3.4.4 Three-dimensional Case

In the general case of a system with three-dimensional geometry and surfaces of second order, the intersection points with the direction of the ray can always be determined analytically too. Modern computers make it possible to apply a general three-dimensional formulation of the set with vector components rather quickly. Therefore, as state of the art, this general case is by far the most important one today [14–17]. A single ray usually is defined in an optical system by fixing its intersection coordinates in the field and in the pupil plane respectively. The coordinates in both planes are normalized, they are denoted by H_x, H_y in the object and P_x, P_y in the entrance pupil plane. This definition can also be used in the special case of infinity distances of object or entrance pupil, in these cases the coordinate interpretation is switched to angle definitions. The scheme to fix one ray in the object domain is illustrated in Figure 3.25. For the exact definition of the entrance pupil, see Chapter 5.

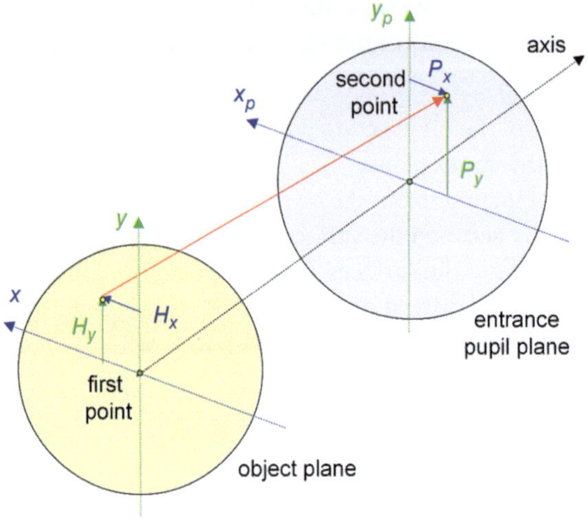

Figure 3.25 Definition of an initial single ray by two points using the normalized coordinates H_x, H_y in the object field plane and P_x, P_y in the entrance pupil plane.

In the general three-dimensional case all relations are formulated vectorially; the basic idea is the same as before. In addition, it is possible to also include tilts and decentring of components. The scheme and the notations are illustrated in Figure 3.26.

The equations to trace a ray are listed in the following. The vector for the direction of the ray is written as

$$\vec{s}_j = (s_{xj} \; s_{yj} \; s_{zj})^T \tag{3.67}$$

and is defined as a unit vector with length 1. In the local coordinate system of the surface no. j the unit normal vector of the direction in the intersection point of the rays perpendicular to the surface is given by

$$\vec{e}_j = \left(e_{xj} \; e_{yj} \; e_{zj}\right)^T \tag{3.68}$$

The position vector of the intersection point of the ray with the surface is

$$\vec{r}_j = \left(x_j \; y_j \; z_j\right)^T \tag{3.69}$$

By inserting the ray vector with skew distance d_{sj} between two intersection points

$$\vec{r}_j = \vec{r}_{j-1} + d_{sj} \cdot \vec{s}_j \tag{3.70}$$

into the quadratic equation of the surface of second order as a conic section with parameters c and κ (for details, see Section 4.6)

$$z(x, y) = \frac{c \cdot (x^2 + y^2)}{1 + \sqrt{1 - (1 + \kappa) \cdot c^2 (x^2 + y^2)}} \tag{3.71}$$

for the skew thickness d_{sj} a quadratic equation follows

$$H_j \cdot d_{sj}^2 + 2F_j \cdot d_{sj} - G_j = 0 \tag{3.72}$$

Figure 3.26 Scheme of raytrace in three dimensions.

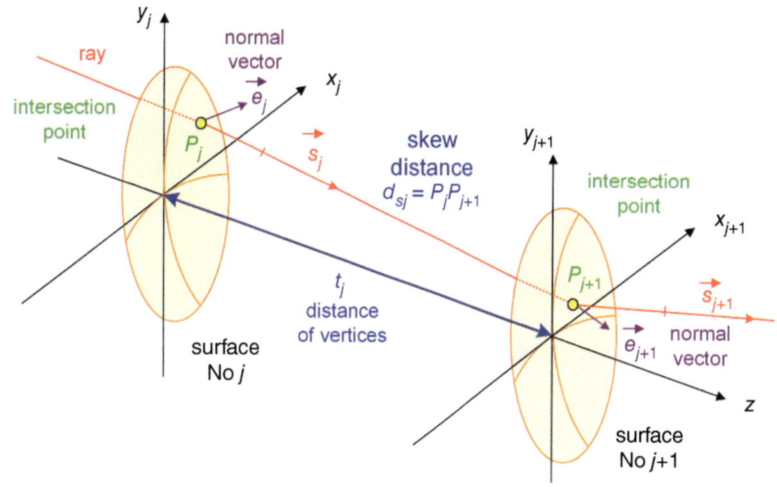

The parameters describing the shape of the surface are as follows (see [21]):

- For a spherical surface the coefficients read

$$H_j = -c_j, \quad G_j = c_j\left(x_j^2 + y_j^2 + z_j^2\right) - 2z_j,$$
$$F_j = s_{zj} - c_j(x_j s_{xj} + y_j s_{yj} + z_j s_{zj}) \tag{3.73}$$

the normal unit vector is given as

$$\vec{e}_j = \left(-c_j x_j \quad -c_j y_j \quad 1 - c_j z_j\right)^T \tag{3.74}$$

- In the case of a conic section with rotational symmetry and conic parameter κ we have

$$H_j = -c_j\left[s_{xj}^2 + s_{yj}^2 + (1+\kappa_j)s_{zj}^2\right]$$
$$G_j = c_j\left[x_j^2 + y_j^2 + (1+\kappa_j)z_j^2\right] - 2z_j$$
$$F_j = s_{zj} - c_j[x_j s_{xj} + y_j s_{yj} + (1+\kappa_j)z_j s_{zj}] \tag{3.75}$$

the unit vector is

$$\vec{e}_j = \left(-c_j x_j \quad -c_j y_j \quad 1 - c_j(1+\kappa_j)z_j\right)^T \tag{3.76}$$

The correct solution of the quadratic equation for the skew thickness is (see Section 3.4.6)

$$d_{sj} = \frac{G_j}{F_j + \sqrt{F_j^2 + H_j G_j}} \tag{3.77}$$

Inserting the skew thickness into Eq. (3.70) delivers the new intersection point with the surface with the index j. In the more general case of a numerically iterative computation of the intersection point (see Section 3.4.7), the formulas for $\vec{e}$ and d_{sj} have to be replaced accordingly. The angle of incidence i_j follows from the scalar product

$$\cos i_j = \vec{s}_j \cdot \vec{e}_j \tag{3.78}$$

and the law of refraction reads in cos-formulation

$$\cos i_j' = \sqrt{1 - \left(\frac{n_j}{n_{j+1}}\right)^2 (1 - \cos^2 i_j)} \tag{3.79}$$

and the law of reflection as a special case is

$$\cos i_j' = -\cos i_j \tag{3.80}$$

Using this with an auxiliary quantity, the so-called refraction invariant,

$$\Gamma_j = n_{j+1}\cos i_j' - n_j \cos i_j \tag{3.81}$$

the new direction of the ray follows as

$$\vec{s}_{j+1} = \frac{n_j}{n_{j+1}}\vec{s}_j + \frac{\Gamma_j}{n_{j+1}}\vec{e}_j \tag{3.82}$$

If the angle of incidence i_j is not explicitly necessary, the vector law of refraction

$$n'\vec{s}' = n\vec{s} + \vec{e}\cdot\left[\sqrt{n'^2 - n^2 + (n\vec{e}\cdot\vec{s})^2} - n\vec{e}\cdot\vec{s}\right] \tag{3.83}$$

or in the case of reflection

$$\vec{s}' = \vec{s} - 2\vec{e}\cdot(\vec{e}\cdot\vec{s}) \tag{3.84}$$

as discussed in Section 3.2.1 can be directly applied instead of the steps by using the Eqs (3.79)–(3.83).

For the description of the spatial geometry of a system, a coordinate system is usually used in which the optical axis is in the z direction and the y axis is in the meridional plane. The sagittal plane is determined by the x–z axis. There is generally an absolute coordinate system valid globally which is fixed in the object or start plane, respectively. In each surface vertex point of the system there is a local coordinate system which can be converted according to the inclination and position of the surface.

Rotational angles are chosen to be positive for clockwise rotation. This property, however, changes with every reflection, reverting a right-handed coordinate system into a left-handed system and vice versa. The position of a surface can be determined by the three following parameters.

- Shift in the z-direction by v_z (in the centred case by the vertex distance t of the surfaces).
- Shift in the y-direction (meridional decentring) by v_y.
- Shift in the x-direction (sagittal decentring) by v_x.

By analogy there are three tilt angles determining the exact geometrical orientation of a surface, these are as follows.

- Rotation around the x-axis (inclination in the meridional plane) by the angle α.
- Rotation around the y-axis (inclination in the sagittal plane) by the angle β.
- Rotation around the z-axis (azimuthal rotation around the apex axis) by the angle γ.

In this definition the angles do not correspond to the Euler angles, which refer to the tilted axes y' and z'. The Euler definition of the angles, however, is possible in theory but not very useful in practice. For the exact interpretation of the shifts and rotations of a surface, the sequence of the application of certain operators is important. Formally, the most general approach of the operators can be written in the following sequence. R and T are translation and rotation matrices before the surface operator S to describe the refraction or reflection applied and transformed from the current coordinate system into the local one of the surface, after the refraction R' and T' rotate and translate again and go into the next transfer or global coordinate system

$$\vec{r}\,' = \underline{T}' \cdot \underline{R}' \cdot \underline{S} \cdot \underline{R} \cdot \underline{T} \cdot \vec{r} \tag{3.85}$$

This is sketched in Figure 3.27. It should be emphasized that first the translation and then the rotation is performed to reach the local coordinates, afterwards the global coordinates are obtained by first rotating and finally shifting. But in principle another sequence of the operators R and T is also possible.

If the ray direction vectors $\vec{s}$ and $\vec{s}\,'$ as well as the refractive indices n and n' are known, the set of equations can be used to reconstruct the surface normal in a fully vectorial scheme as follows

$$\begin{aligned}
a &= \sin(i - i') = |\vec{s} \times \vec{s}\,'| \\
\tan i &= \frac{n'a}{n - n'\cos(\arcsin a)}, \quad \sin i = \frac{\tan i}{\sqrt{1 + \tan^2 i}}, \\
\cos i &= \sqrt{1 - \sin^2 i} \\
\sin i' &= \frac{n}{n'} \sin i, \quad \cos i' = \sqrt{1 - \sin^2 i'} \\
\vec{e} &= \frac{n'\vec{s}\,' - n\vec{s}}{n'\cos i' - n\cos i}
\end{aligned} \tag{3.86}$$

This set of formulas can be used, if, for example, an asphere should be calculated, which exactly performs a refraction with pre-given ray direction. Another example is the determination of the orientation of mirrors for a given new optical axis direction in three dimensions. In principle it is kind of inverting the task of the ray-trace: the ray data are known and the surface should be determined.

3.4.5 Pitfalls and Special Cases

In reality, there are several special error cases for the ray-trace according to the equations. These are as follows.

- The intersection point lies outside of the real diameter (size) of the surface, but is mathematically possible.
- No intersection point exists, the ray fails the sphere or other surface.
- Total internal reflection occurs.
- Negative edge thicknesses of lens centres or edges are unphysical.
- The intersection point is lying on a spherical hemisphere on the rear side. This case is not necessarily an error, but requires special treatment.

These situations are illustrated in Figure 3.28. It should be noted that the negative edge thickness is not really a conflict for the raytrace, which usually works fine in this case. But it is an unphysical geometry and should be avoided in

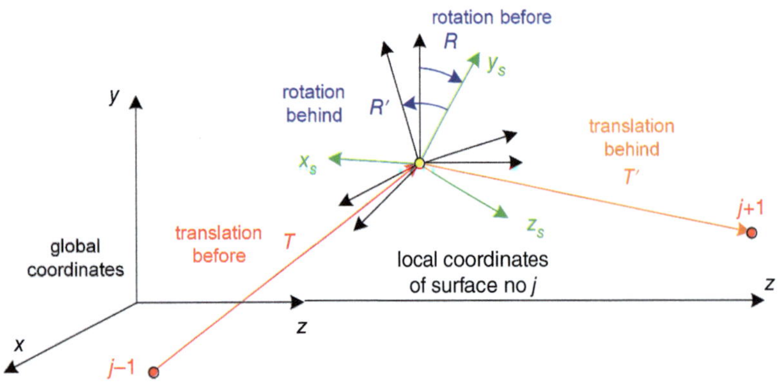

Figure 3.27 Arbitrary geometry with tilt and translation transforms before and after the surface operator.

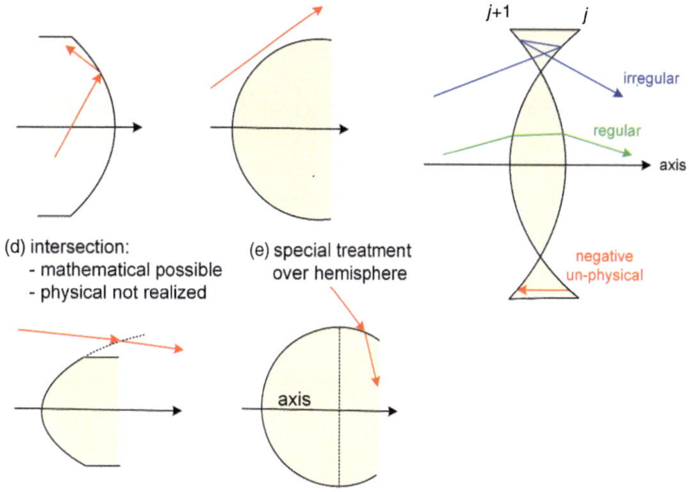

Figure 3.28 Failure, pitfalls and special treatment in raytrace.

reality. But negative thickness values are important and are an often used option to describe reflection at mirrors, and are therefore not simply forbidden in the algorithms.

It has to be considered that there are usually two solutions for the calculation of the intersection point of a path with a surface of second order. Only one of them is correct. In standard cases the sign of the surface radius indicates the correct solution of the two resulting mathematical intersection points. A radius is positive, if the centre C lies to the right of the vertex. For this parameter the two solutions can be distinguished as shown in Figure 3.29. There are, however, exceptions [18–20].

- If the surface extends above the hemispherical meridian. This option requires a special control of the intersection point selection in the implementation.

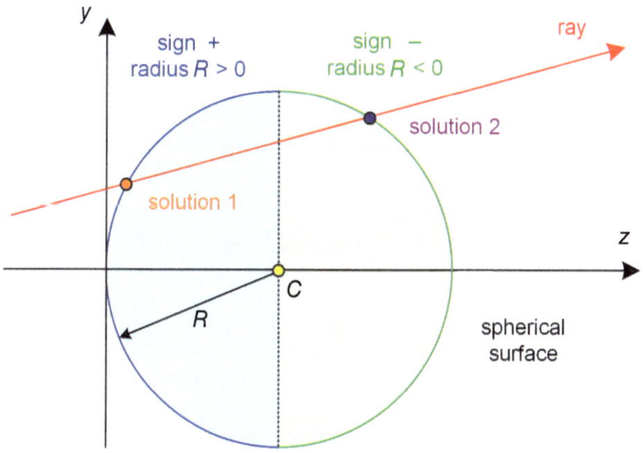

Figure 3.29 Selection of the correct solution and intersection point.

- If a non-sequential raytracing is considered, where the incident ray can come from any side.

Therefore, the design software usually provides an explicit choice between the two possible solutions as indicated in the figure.

3.4.6 Modelling of Mirrors

The modelling of mirror surfaces is a special issue in the raytrace algorithms. The problem is that a mirror reflection converts a right-handed into a left-handed coordinate system. Left-handedness is an unwanted approach; in case of polarization modelling this is an intrinsic source of errors. Therefore, it is recommended to keep right-handedness in any case.

One opportunity to keep the raytrace equations formally correct at a mirror is to give the thickness t_j after the mirror a negative sign and in case of a second mirror again a positive sign. This makes a certain sense as in case of Figure 3.30(a), if the z-component of the ray direction unity vector switches its sign too. But in the case of part (b) of the figure with a kind of grazing incidence, the negative thickness must be selected as well, although the s_z-component stays positive. Therefore, this mathematical trick to maintain the validity of the equations of raytrace is only formal and a physical interpretation is artificial.

From the viewpoint of the equations, a negative refractive index after the mirror also solves the problem; nevertheless, the physical interpretation is not better. In any case it should be taken care of in the calculation of optical path lengths in the case of negative thicknesses or indices. Unfortunately, this selection must be allowed and negative

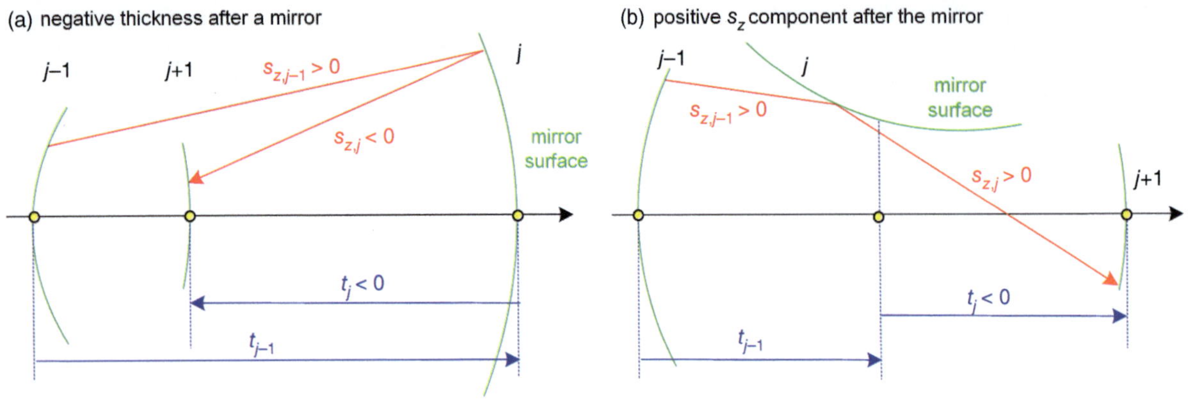

Figure 3.30 Negative thickness t_j after a mirror surface by (a) sign switch of s_{jz} and (b) without sign switch.

edge thicknesses cannot be detected as an unphysical error by the algorithm.

3.4.7 Ray Intersection with Surfaces of Higher Order*

(*) The content of this section is quite specialized and for advanced reading. Knowledge of this section is not necessary to understand the following chapters. Beginners are recommended to skip this section.

The calculation of the intersection point of a ray with an aspherical surface or a surface of higher than second order cannot be solved analytically. This has to be carried out in a general numerical-iterative way. The following scheme has proven to be a very stable algorithm in practice, see Figure 3.31.

- Calculate the intersection point Q_0 with the vertex plane.
- Approximate solution Q_1 on the surface as a z-projection with the same values for x, y.
- Alternatively, it is also possible to use the intercept with the base of the conic section of an asphere or an osculating sphere as the starting point for the calculation of the intersection point, and this can be determined analytically. Often this approach is more advantageous than the other method, in particular for rather steep surfaces with large sag.
- Intercept Q_1' of the ray with the tangential plane in Q_1 with the same x, y values.
- Second approximate solution Q_2 on the surface with invariant x, y.
- Continue iteration, until the desired accuracy is reached.

As a convergence criterion for this iterative improvement of the coordinates of the intersection between the ray vector and the surface, a measure of tolerance for the required accuracy has to be determined. This can be the difference of the absolute distances between the point coordinates

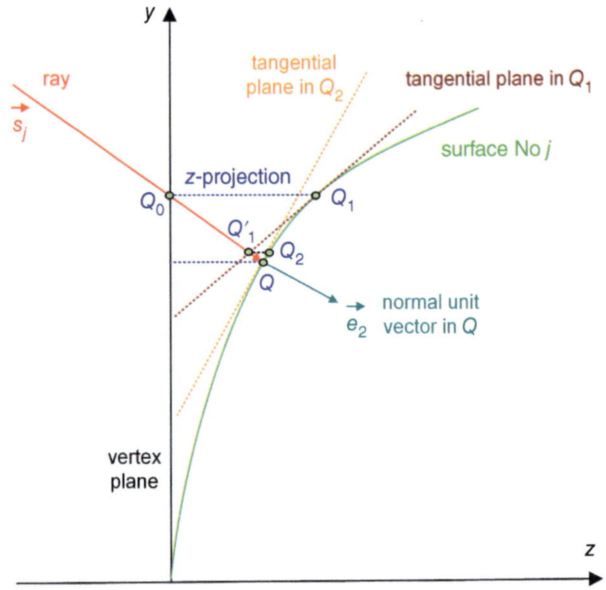

Figure 3.31 Schema of the numerical iterative calculation of the ray intersection with the surface. For an explanation of the various iteration steps, see the text.

of two successive steps of the iteration. With the coordinates of the new intersection point, the normal vector in this point on the surface can be calculated. It should be mentioned that in the case of very strong aspheres or freeform surfaces with steep slopes and especially turning points in the cross section profile, it may be the case that this algorithm fails. What is then observed is a divergent behaviour or oscillations.

The set of equations for this algorithm is listed here.

- Intersection point Q_0 with the vertex plane

$$\vec{r}_0 = \begin{pmatrix} x_0 & y_0 & z_0 \end{pmatrix}^T = \begin{pmatrix} x_{j-1} - \frac{s_{xj}}{s_{zj}} z_{j-1} & y_{j-1} - \frac{s_{yj}}{s_{zj}} z_{j-1} & 0 \end{pmatrix}^T \tag{3.87}$$

- Approximated solution Q_1 on the surface as a z-projection

$$\vec{r}_1 = \begin{pmatrix} x_1 & y_1 & z_0 \end{pmatrix}^T = \begin{pmatrix} x_0 & y_0 & f(x_0, y_0) \end{pmatrix}^T \tag{3.88}$$

- Intercept Q_1' of the ray with the tangential plane in Q_1:

$$z_1' = s_{zj} \frac{\vec{e}_1 \cdot (\vec{r}_1 - \vec{r}_0)}{\vec{e}_1 \cdot \vec{s}_j}, \quad y_1' = \frac{s_{yj}}{s_{zj}} z_1' + y_0,$$

$$x_1' = \frac{s_{xj}}{s_{zj}} z_1' + x_0 \tag{3.89}$$

- Second approximate solution Q_2 on the surface

$$\vec{r}_2 = \begin{pmatrix} x_2 & y_2 & z_2 \end{pmatrix}^T = \begin{pmatrix} x_1' & y_1' & f(x_1', y_1') \end{pmatrix}^T \tag{3.90}$$

etc. With the coordinates of the new intersection point, the normal vector in this point on the surface can be calculated. Furthermore with

$$t_{sj} = \sqrt{(x_j - x_{j-1})^2 + (y_j - y_{j-1})^2 + (z_j - z_{j-1})^2} \tag{3.91}$$

the skew thickness along the ray can be determined, here all coordinates have to be given in the same global coordinate system. In the case of a mirror, the sign of t_{sj} must be switched accordingly.

In the special case of a segmented piecewise definition of surface equations, like spline surfaces or Fresnel ring sub apertures, one step before the intersection point calculation is the determination of the right patch, segment or sub aperture. This usually makes the raytrace more complicated and requires careful handling if the ray meets the surface near a segment border.

3.4.8 Gradient Index Media

Gradient index (GRIN) media are materials with a locally variable refractive index $n(x,y,z)$. In this kind of media the light rays are no longer straight lines but have curved paths. There are components with intended profile distributions of the refractive index which realize special effects, but in reality it is also necessary to model GRIN media when simulating inhomogeneous distributions of the refractive index in glasses or in the atmosphere.

For raytracing in media with non-uniform distributions of the refractive index, the eikonal equation of a light ray has to be solved iteratively by a numerical method [25–28], see Figure 3.32.

The eikonal ray equation can be written in the form [26].

$$\frac{d^2\vec{r}}{dt^2} = n \cdot \nabla n = \vec{D} = \begin{pmatrix} n\frac{\partial n}{\partial x} & n\frac{\partial n}{\partial y} & n\frac{\partial n}{\partial z} \end{pmatrix}^T \tag{3.92}$$

The path parameter along the curved ray is defined as a line integral

$$t = \int \frac{dr}{n} \tag{3.93}$$

and is calculated numerically with the step width Δt. The vector for the optical direction is defined by

$$\vec{T} = n \cdot \vec{s} = \begin{pmatrix} n \cdot s_x & n \cdot s_y & n \cdot s_z \end{pmatrix}^T \tag{3.94}$$

One advantageous opportunity is to solve the eikonal equation numerically by means of the Runge–Kutta method to fourth order for differential equations. Here the raytracing is determined step-wise, leading to correspondingly long computation times. With the auxiliary parameters

$$\vec{A} = \Delta t \cdot \vec{D}(\vec{r}_j),$$

$$\vec{B} = \Delta t \cdot \vec{D}\left(\vec{r}_j + \frac{\Delta t \cdot \vec{T}_j}{2} + \frac{\Delta t \cdot \vec{A}}{8}\right),$$

$$\vec{C} = \Delta t \cdot \vec{D}\left(\vec{r}_j + \Delta t \cdot \vec{T}_j + \frac{\Delta t \cdot \vec{B}}{2}\right) \tag{3.95}$$

with D as defined in Eq. (3.93), the new position vector and the new direction of the ray after a step are given by

$$\vec{r}_{j+1} = \vec{r}_j + \Delta t \cdot \left(\vec{T}_j + \frac{\vec{A} + 2\vec{B}}{6}\right),$$

$$\vec{T}_{j+1} = \vec{T}_j + \frac{\vec{A} + 4\vec{B} + \vec{C}}{6} \tag{3.96}$$

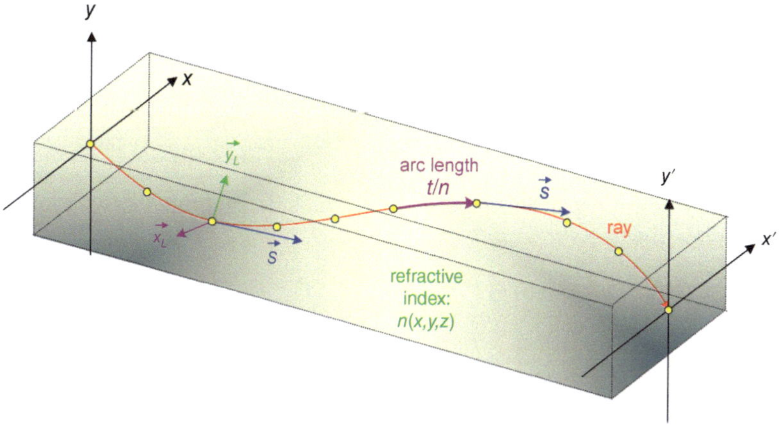

Figure 3.32 Raytracing in gradient-index media. $\vec{s}$ is the direction unity vector of the ray, $\vec{b}$ and $\vec{c}$ are the corresponding orthogonal local transverse vectors.

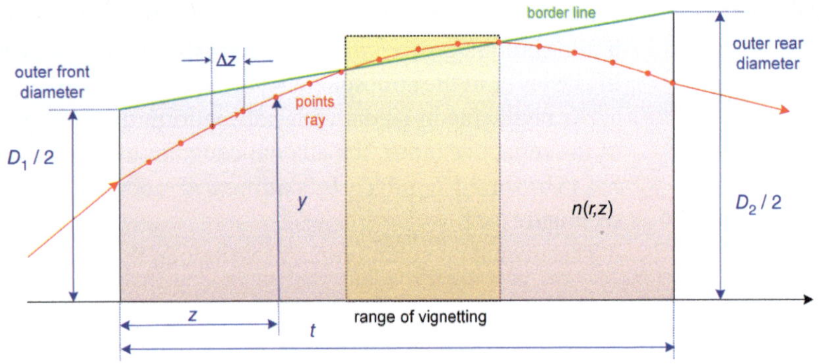

Figure 3.33 Boundary of the diameter inside a GRIN medium; the red dots indicate the discrete propagation steps.

The accuracy of the numerical computation depends on the chosen step width Δt. In order to achieve an optimum between accuracy and computing time a variable control of the step width is usually adaptively implemented in the calculation algorithms. It uses small steps in glass ranges with large gradients of the refractive index and large steps in less changeable regions. The criterion for the adjustment of the step width is the change in the direction per step, which should be small. The order of the error is in this method is $\varepsilon \propto \Delta t^4$. If the ray direction vectors of two successive steps are considered, the change of the ray angle is given by

$$\cos\delta = \vec{s}_j \cdot \vec{s}_{j+1} \tag{3.97}$$

A meaningful criterion to adapt the step width is to limit the angle change

$$\delta(t) < \delta_{thresh} \tag{3.98}$$

with an appropriate value for the maximum allowed change δ_{thresh} in the range of $10^{-2}\ldots10^{-5}$ rad.

When carrying out a raytracing program, the following classes of refractive index function are important where r is the transverse radial coordinate r. Often used index profile functions are (see also Chapter 5.5.3.7) as follows.

- A Taylor expansion into r and z, $n_{0,\lambda}$ is the basic refractive index of the material for the main wavelength

$$\begin{aligned} n(r,z) = {} & n_{0,\lambda} + c_1 r^2 + c_2 r^4 + c_3 r^6 + c_4 r^8 \\ & + z\cdot(c_5 + c_6 r^2 + c_7 r^4 + c_8 r^6 + c_9 r^8) \\ & + z^2\cdot(c_{10} + c_{11} r^2 + c_{12} r^4 + c_{13} r^6 + c_{14} r^8) \\ & + z^3\cdot(c_{15} + c_{16} r^2 + c_{17} r^4 + c_{18} r^6 + c_{19} r^8) \end{aligned} \tag{3.99}$$

- A low power expansion separated in r and z depending on the wavelength:

$$\begin{aligned} n(r,z) = {} & n_{0,\lambda} + c_{1,\lambda} r^2 + c_{2,\lambda} r^4 + c_{3,\lambda} r^6 + c_{4,\lambda} r^8 \\ & + c_{5,\lambda} z + c_{6,\lambda} z^2 + c_{7,\lambda} z^3 \end{aligned} \tag{3.100}$$

In addition some other simple analytical profile functions are common, which provide exact solutions for special requirements. A problem for the practical correction of optical systems is often the unknown spectral variation of the parameters with the wavelength describing the gradient of the refractive index. The derivatives in Eq. (3.93) are in practice calculated by finite differences, in the case of simpler index distribution functions maybe also analytically.

Another special problem is the vignetting of the ray inside the medium (see also Section 5.5.3.7). If it is assumed that the outer boundary has the shape of a cylinder or a truncated cone, it may happen that due to the bent ray path the ray violates this border in between the component and is again possible at the end. This scenario is sketched in Figure 3.33. To avoid this pitfall the ray height y inside the gradient index lens must be checked every step.

Figure 3.34 shows a raytrace in a quadratic index profile, where in first approximation a periodic refocusing can be observed. If the ray bundles of the axis field point (orange) and the off-axis field point (red) are compared, it is seen that the image point is displaced along z. This means that the system suffers from field curvature. Figure 3.35 shows some rays with skew entrance direction in a three-dimensional view. They are following the path of a helix in a quadratic index medium. More details are also presented in Section 4.9.

3.4.9 Differential Raytrace, Parabasal Rays and Ray Tubes

Differential rays are rays very close to a reference ray. They are also called parabasal rays. The reference ray itself can be calculated real with aberrations. This allows us to calculate information about changes and derivatives of ray parameters. This is important for certain aberrations like astigmatism, in illumination issues and in symmetry-free geometries without a defined straight axis. For many components and surface types, the transfer of differential rays can be given analytically resulting in fast raytracing [29, 30]. They can also be numerically calculated as

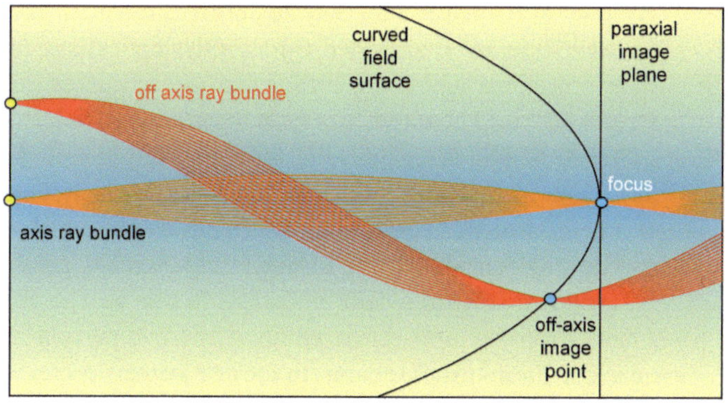

Figure 3.34 Raytracing in a gradient index lens in one cross section.

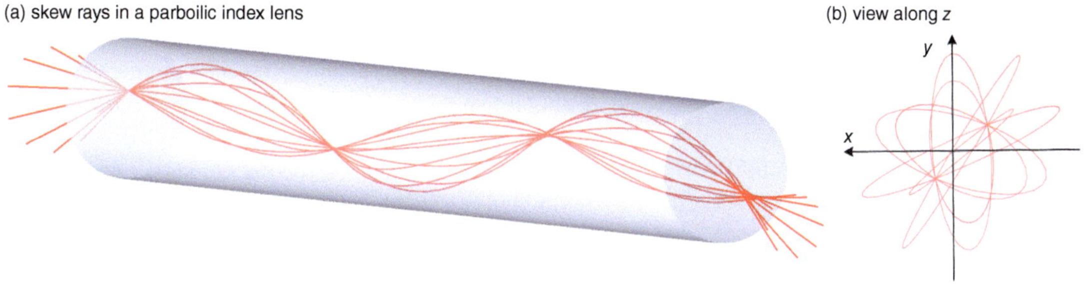

Figure 3.35 Raytracing in a gradient index lens in three dimensions with skew rays and helix-shaped paths.

Figure 3.36 General differential ray, terms of use.

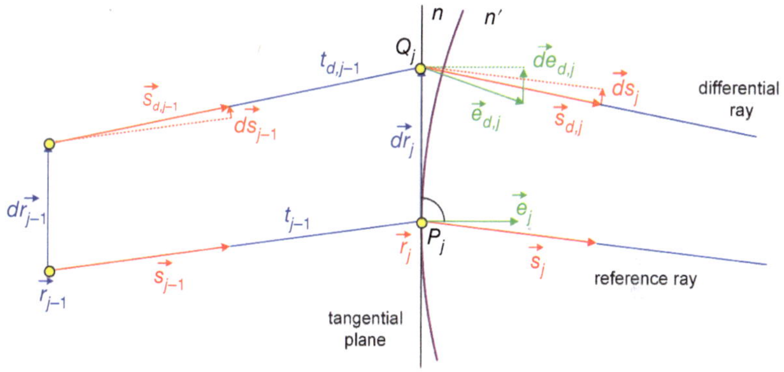

being very close to the reference ray as finite difference. However, especially in GRIN media and at statistical surfaces, raytracing can only be computed numerically.

The idea is to use the representation of a light ray as a straight line as

$$\vec{r}_j = \vec{r}_{j-1} + t \cdot \vec{s}_{j-1} \tag{3.101}$$

and the law of refraction reads with $\mu = n/n'$

$$\begin{aligned}\vec{s}_j &= \mu \cdot \vec{s}_{j-1} \\ &\quad + \left(-\mu \cdot \vec{e}_j \cdot \vec{s}_{j-1} + \sqrt{1 - \mu^2 \cdot \left[1 - (\vec{e}_j \cdot \vec{s}_{j-1})^2\right]}\right) \cdot \vec{e}_j \\ &= \mu \cdot \vec{s}_{j-1} + \Gamma_j \cdot \vec{e}_j\end{aligned} \tag{3.102}$$

For the differential ray, the same equations are valid and are denoted with an additional subscript d and the small changes, see Figure 3.36

$$\begin{aligned}&\vec{r}_{d,j} = \vec{r}_j + d\vec{r}_j, \quad \vec{s}_{d,j-1} = \vec{s}_{j-1} + d\vec{s}_{j-1}, \\ &\vec{e}_{d,j} = \vec{e}_j + d\vec{e}_j, \quad t_{d,j-1} = t_{j-1} + dt_{j-1}\end{aligned} \tag{3.103}$$

Now it can be assumed that the differential ray meets the surface not in the point P_j, but in Q_j, which in first approximation lies in the tangential plane onto the surface in P_j. This has the consequence that

$$\vec{e}_j \cdot d\vec{r}_j = 0, \quad d\vec{e}_j \parallel d\vec{r}_j \tag{3.104}$$

Now the raytrace equations as well as the spherical surface equation with curvature c_j are differentiated and give

$$d\vec{r}_j = d\vec{r}_{j-1} + t_{j-1}d\vec{s}_{j-1} + \vec{s}_{j-1}dt_{j-1} \tag{3.105}$$

$$d\vec{s}_j = \mu \cdot d\vec{s}_{j-1} + d\Gamma_j \cdot \vec{e}_j + \Gamma_j \cdot d\vec{e}_j \tag{3.106}$$

When c_{j1} and c_{j2} are the principal curvatures of the surface along the main azimuthal directions $\vec{u}_1, \vec{u}_2$ a rotation has to be carried out around the azimuthal angle θ. Then the effective radius of curvature can be determined with the azimuthal angle θ_j

$$\cos^2\theta_j = \frac{(\vec{u}_1 d\vec{r}_j)^2}{d\vec{r}_j^2}, \quad \sin^2\theta_j = \frac{(\vec{u}_2 d\vec{r}_j)^2}{d\vec{r}_j^2}$$

$$c_{jn} = c_{j1} \cdot \cos^2\theta_j + c_{j2} \cdot \sin^2\theta_j \tag{3.107}$$

Then for the differential change in the normal direction of the surface it follows that

$$d\vec{e}_j = -c_{jn} \cdot d\vec{r}_j \tag{3.108}$$

Now Eq. (3.106) is multiplied with $\vec{e}_j$ and (3.105) is used. The remaining equation resolved for dt_{j-1} and gives

$$dt_{j-1} = -\frac{\vec{e}_j \cdot (d\vec{r}_{j-1} + t_{j-1}d\vec{s}_{j-1})}{\vec{e}_j \cdot \vec{s}_{j-1}} \tag{3.109}$$

As the last unknown we get from (3.103) the derivative of Γ_j

$$d\Gamma_j = -\frac{\mu\Gamma_j \cdot (\vec{s}_{j-1} \cdot d\vec{e}_j + d\vec{s}_{j-1} \cdot \vec{e}_j)}{\Gamma_j + \mu\vec{s}_{j-1}\vec{e}_j} \tag{3.110}$$

Finally, all equations can be inserted into (3.102, 3.103) to give the changed data of the new differential ray in Eq. (3.104) as a function of the parameters of the reference ray and the previous surface data of the differential ray as

$$\begin{aligned}
dt_{j-1} &= -\frac{\vec{e}_j \cdot (d\vec{r}_{j-1} + t_{j-1}d\vec{s}_{j-1})}{\vec{e}_j \cdot \vec{s}_{j-1}} \\
d\vec{r}_j &= d\vec{r}_{j-1} + t_{j-1}d\vec{s}_{j-1} - \vec{s}_{j-1}\frac{\vec{e}_j \cdot (d\vec{r}_{j-1} + t_{j-1}d\vec{s}_{j-1})}{\vec{e}_j \cdot \vec{s}_{j-1}} \\
d\vec{s}_j &= \mu \cdot d\vec{s}_{j-1} - \vec{e}_j\frac{\mu\Gamma_j \cdot (d\vec{s}_{j-1} \cdot \vec{e}_j - \vec{s}_{j-1} \cdot c_{jn} \cdot d\vec{r}_j)}{\Gamma_j + \mu\vec{s}_{j-1}\vec{e}_j} \\
&\quad - \Gamma_j \cdot c_{jn} \cdot d\vec{r}_j \\
d\vec{e}_j &= -c_{jn} \cdot d\vec{r}_j
\end{aligned} \tag{3.111}$$

So all small changes, which are valid for the differential ray in comparison with the reference ray, for transition and refraction of the differential ray are known. These formulas describe the corresponding raytracing method for a spherical surface as an example.

In the three-dimensional case, the differential rays have a particular significance. Consider the following scenario. Around a reference ray two adjacent differential rays on each side are transported through the optical system in two sections which are orthogonal to each other, see Figure 3.37(a). These five rays form a spatial narrow beam or ray tube which changes its form in the sense of an astigmatic deformation as it passes through the system. If an elliptical (or rectangular) area is always assigned to the rays by considering the distances between the neighbouring rays, then the deformed areas in this way always follow from the intersection points of the five rays on all surfaces. From this the change in the energy density can be determined. Therefore, the transport of such ray tubes is suitable for the modelling of energy transport calculations for illumination or stray light tasks. In principle three rays are sufficient, if the initial values of the parabasal rays are selected appropriately. But practical experience shows that the principle works more robustly and delivers better results if five rays are used. In particular at the borders of the free diameters of a surface or in case of segmented surfaces with discrete patches, the use of four parabasal rays is recommended. If then by vignetting two rays are truncated, at least the remaining two parabasal rays together with the reference ray can be used to calculate the transformed area element.

Then the local power density is known at the intersection point of the centre ray from the size of the included area. So it is not necessary to determine it by integration over many single rays. The concept of a ray tube is sketched in Figure 3.37(a), the transfer of a ray tube through an optical systems is illustrated in Figure 3.37(b).

The calculation of ray tubes becomes difficult for very strongly structured pupils (e.g. illumination condenser with lens array), in GRIN media and for diffractive surfaces. Some more details about parabasal rays are found in Chapters 8 and 9 and Volume 2, Chapter 16.

3.4.10 Diffractive Surfaces

3.4.10.1 Introduction

If an optical system contains components with microstructured surfaces like gratings or diffractive elements, the raytrace must be generalized for this case. Physically, diffraction effects take place and the light behaviour at the surface is more complicated. But for daily standard applications and in particular for optimization, a simplified model can be used which doesn't need the time-consuming rigorous diffraction models. In the point of ray intersection, it can be assumed that the grating equation is locally valid here and changes the new ray direction in addition to the local slope of the substrate. After taking these two effects into account, the raytrace can proceed as usual. But this

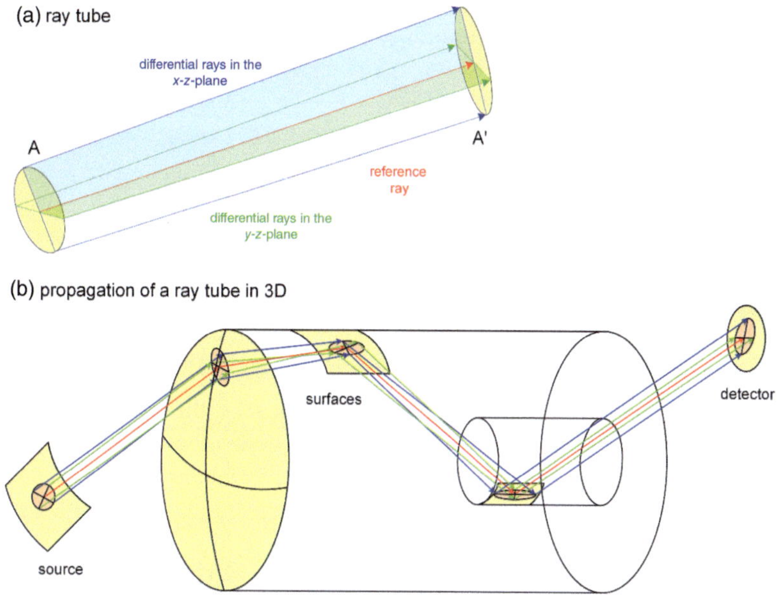

Figure 3.37 General three-dimensional ray tube (a), formed by five rays and (b) transport of a ray tube through an optical system.

simplification assumes two conditions, which must be fulfilled to achieve a valid result.

- Only one diffraction order m of the grating is considered, perturbing spurious light contributions in unwanted diffraction orders are neglected.
- It is assumed that the ideal grating equation valid, which means, for one unique new ray direction, a huge number of grooves must be illuminated to figure out this one direction by the interference function. This hidden assumption may fail in case of very narrow ray bundles.

Therefore, beneath the more complicated procedure to find the new ray direction, the conditions for the validity must be checked [31]. More detail can be found about this issue in Chapter 4.11.

3.4.10.2 Plane Linear Gratings

The easiest case of a diffraction based surface is a linear grating. We assume here without loss of generality that the grooves are located in the local coordinate system of the surface along the x direction, which means that the main deflection is performed in the y direction. The grating substrate has a spherical radius of R. The geometry is sketched in Figure 3.38(a). The variables and vector definitions at the grating are explained in Figure 3.38(b). The diffraction order is denoted here by the integer m. If a ray is considered that starts in Q, strikes the grating in P and is reflected with

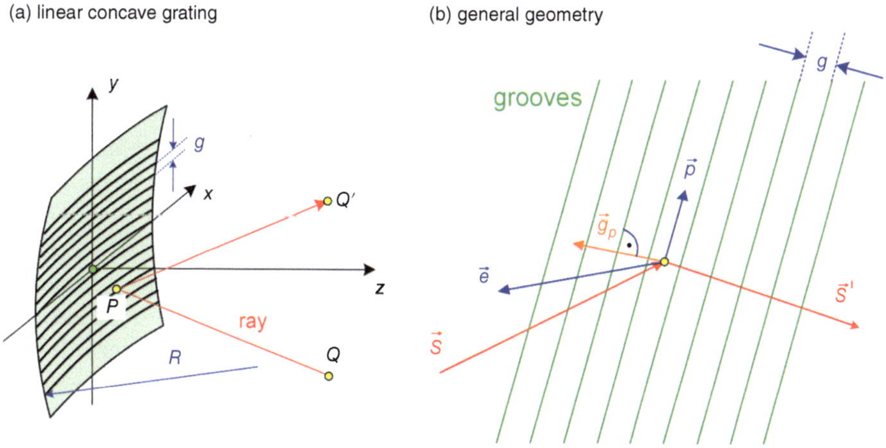

Figure 3.38 Geometry of (a) a linear concave grating and (b) a general grating and the ray trace vectors at a surface with local grating vector g.

the grating diffraction effect into Q', the Fermat principle can be used to find the ray tracing eqs (3.32)

$$s'_{jx} = x \cdot A - s_{jx}, \quad s'_{jy} = y \cdot A - s_{jy} + \frac{m\lambda}{g} \tag{3.112}$$

with the grating constant g and the abbreviation

$$A = -\frac{\cos i}{R} - \frac{m\lambda y}{R^2 g} - \frac{1}{R}\sqrt{\left(\cos i + \frac{m\lambda y}{Rg}\right)^2 + \frac{2s_y \lambda y}{g} - \left(\frac{m\lambda}{g}\right)^2} \tag{3.113}$$

x, y and i are the parameters that are obtained by the raytrace onto the substrate. In the special case of a plane grating $1/R = 0$, we have $A = 0$ and the formulae simplifies to

$$s'_{jx} = -s_{jx}, \quad s'_{jy} = -s_{jy} + \frac{m\lambda}{g} \tag{3.114}$$

3.4.10.3 Generalized Diffractive Surfaces

In a more general case, the normal unit vector in the point of ray intersection is given as $\vec{e}$ and can be calculated from the surface derivatives as [31]

$$\vec{e} = \frac{1}{\sqrt{1 + z_x^2 + z_y^2}} \begin{pmatrix} -z_x \\ -z_y \\ 1 \end{pmatrix} \tag{3.115}$$

The diffractive structure is described by a phase function Φ of the substrate surface, and it gives the local grating groove spatial frequency width as

$$\vec{G} = \frac{1}{2\pi} \nabla\Phi(x, y, z) = \frac{1}{2\pi} \begin{pmatrix} \Phi_x \\ \Phi_y \\ \Phi_z \end{pmatrix} \tag{3.116}$$

with grating period $g = 1/|G|$. The component of the vector $\vec{G}$ perpendicular to the groves is called $\vec{G}_p$ and is given by

$$\vec{G}_p = \vec{G} - (\vec{G} \cdot \vec{e}) \cdot \vec{e} \tag{3.117}$$

This geometry is sketched in Figure 3.41. The vector $\vec{p}$ lies in the incidence point along the grooves. The cartesian components of the vector $\vec{G}_p$ are then given by

$$\begin{aligned} G_{px} &= \frac{1}{2\pi}\left[\Phi_x + \frac{z_x}{1 + z_x^2 + z_y^2} \cdot (-z_x\Phi_x - z_y\Phi_y + \Phi_z)\right] \\ G_{py} &= \frac{1}{2\pi}\left[\Phi_y + \frac{z_y}{1 + z_x^2 + z_y^2} \cdot (-z_x\Phi_x - z_y\Phi_y + \Phi_z)\right] \\ G_{pz} &= \frac{1}{2\pi}\left[\Phi_z - \frac{z_z}{1 + z_x^2 + z_y^2} \cdot (-z_x\Phi_x - z_y\Phi_y + \Phi_z)\right] \end{aligned} \tag{3.118}$$

which is then normalized to $\hat{\vec{G}}_p$. The generalized law of refraction for the substrate and diffraction by the local grating equation of the micro-structure reads then

$$\vec{s}' = \frac{n}{n'} \cdot \vec{s} + \frac{m\lambda}{n'g} \cdot \hat{\vec{G}}_p + \gamma \cdot \vec{e} \tag{3.119}$$

with the abbreviation

$$\gamma = -\frac{n}{n'}\vec{s} \cdot \vec{e} + \sqrt{1 - \left(\frac{n}{n'}\right)^2 \cdot \left[1 - (\vec{e} \cdot \vec{s})^2 + 2\vec{s} \cdot \hat{\vec{G}}_p \frac{m\lambda}{ng} + \left(\frac{m\lambda}{ng}\right)^2\right]} \tag{3.120}$$

This equation combines the refraction or reflection at the substrate with the additional ray deflection due to the local grating with period g in the diffraction order m.

3.4.11 Non-Sequential Raytrace

In conventional raytracing the rays are traced sequentially through an optical system. This means that the refraction at the surfaces have to go through the indices $1, 2, \ldots j, j+1, j+2, \ldots$ in exactly this sequence. This kind of raytracing makes sense for normal tasks in optical design. However, in illumination systems or for the calculation of straylight or general false light, the rays often take a path which is not the same for all the rays within a bundle. So rays in the centre of the bundle can pass through a lens with a small diameter, while the rays in the outer area will miss the lens. Another typical example is the generation of stray light, where scattering at mechanical parts can also be involved [33–36].

For the calculation of this so-called non-sequential raytracing, the algorithm itself has to check after each surface operation which optical component or surface is hit next. Numerically this means that the intersection points with all conceivable surfaces have to be calculated first in order to be able to choose the closest one. Correspondingly such a raytracing algorithm requires a large amount of computational effort. According to the nature of the task this geometrical approach only makes sense if the transverse dimensions of the components are known and can be well described. Also the geometry has to be described in a general way suitable for three dimensions. Some typical examples are collected for illustration in Figure 3.39. Figure 3.39(a) shows a RXI collimator for LEDs with two different ray paths for inner rays (blue) and outer rays (green), (b) shows a prism, where inside a ray bundle the condition of TIR is fulfilled only for some of the rays. Figure 3.39(c) illustrates a Fresnel lens

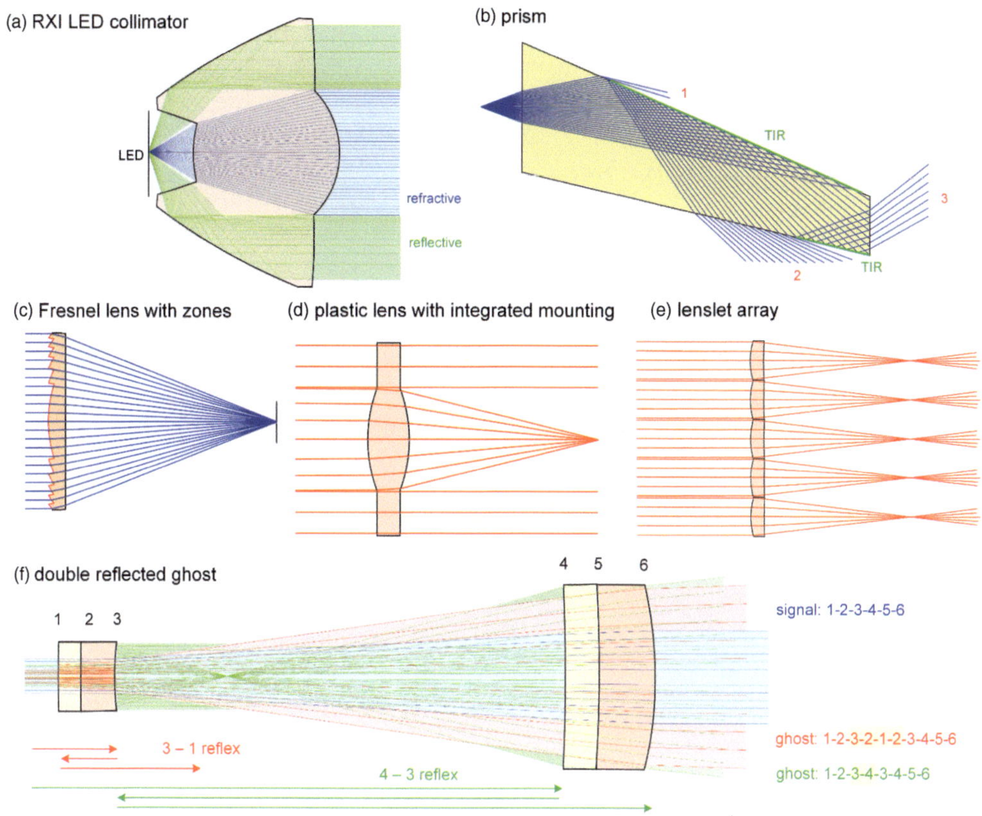

Figure 3.39 Non-sequential raytracing at (a) a RXI-collimator (stands for refractive-reflective incidence), (b) at a prism with total internal reflection for a part of the rays in the bundle, (c) for a Fresnel lens with ring segments, (d) at a plastic lens with a transparent mounting, (e) at a lenslet array and (f) in a lens system due to imperfect coatings as ghost ray bundles.

with segmented rings. Mathematically every ring defines an own surface and channel of propagation, therefore the component is modelled as a non-sequential surface. In (d) a plastic lens is shown, where the mounting for fixation is moulded together with the lens itself. In (e) a lenslet array illustrates the impact of every lens on the rays individually. (f) is an example of the splitting of the ray bundle on regular surfaces of the system due to a non-ideal coating. In this case, reflection of the rays takes place with a certain probability rather than refraction and creates ghost images.

In illumination systems, non-sequential raytrace is often used. Figure 3.40 shows as an example a pumping setup for a cylindrical solid-state laser rod in a cross section. The light of two water cooled flash lamps (green) is collected in a water cooled solid state laser rod in the centre. All the components are cylindrical, only one lateral cross section is shown. This arrangement is used to optimize the transfer efficiency for the pump light, achieving simultaneously a good uniformity inside the laser rod. As is seen in the figure, several different possibilities for the rays occur: splitting, absorption, scattering, multiple reflection and usual refraction and reflection.

3.4.12 Special Ray Types

In imaging applications several rays have a special meaning and name. In particular in the meridional as well as in the sagittal section, there are rays that are important for the characterization of an optical system and its aberrations. Considering an off-axis point object, the plane containing the optical axis and the point object is called a meridional or tangential plane. The definition of the sagittal plane already uses the concept of the chief ray which will be explained below. The chief ray starts in an off-axis field point P of the object plane and meets the entrance pupil in the centre at Q_0. The sagittal section is perpendicular to the meridional plane and contains the chief ray, which encloses a field angle w with the optical axis.

The limiting rays of an aperture ray bundle radiating from a point object on the axis P_0 through the edge of the pupil at Q and lying in the meridional plane are called rim or marginal rays. They form an aperture angle u with the

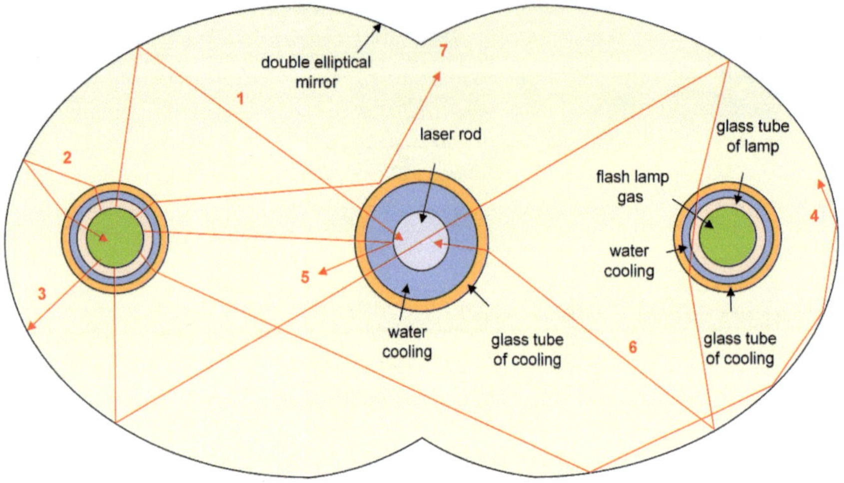

Figure 3.40 Non-sequential raytracing in an illumination system. A double elliptical mirror collects the light of two cylindrical flash lamps into a laser rod as pump light.

optical axis. The limiting rays that radiate from an off-axis point object and lie in the meridional section are called the upper and lower coma rays, respectively. The upper coma ray starts in P and meets the pupil in Q. The sagittal coma rays start from the off-axis point object P, go through the rim of the pupil at Q_s and travel in the sagittal plane of the object space. Within the image space or in the system they do not necessarily intercept the chief ray again. Figure 3.41 shows the various ray types described here.

Usually at every refraction, the sagittal plane of a chief ray changes, see the green and blue sagittal planes in Figure 3.42. It should be noted that these definitions mostly make sense only in circular symmetric optical systems. In general, three-dimensional geometries usually the meridional rays do not remain in this one plane.

Skew rays are all the rays that do not lie in the meridional section. Generally skew rays start from any point object in the x-y starting plane under any angle. These

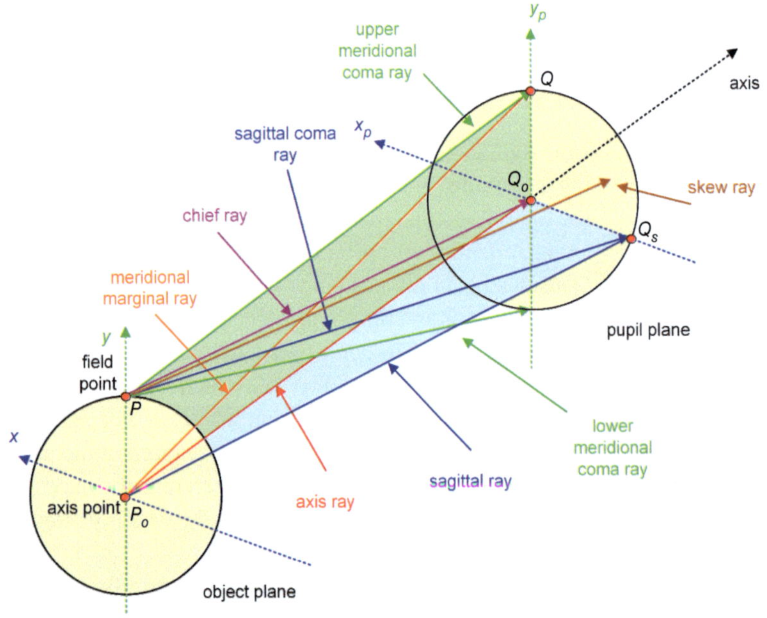

Figure 3.41 Definition of special rays to characterize an optical system. Shown are the rays between a finite object plane and the plane of the entrance pupil.

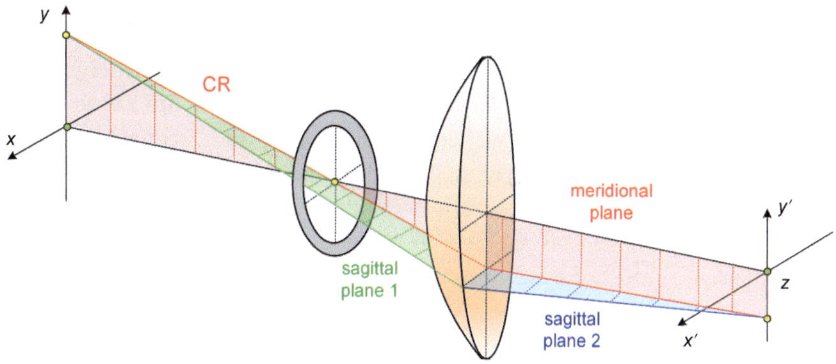

Figure 3.42 Definition of the tangential and the sagittal plane in an optical system for and off-axis field point. It should be noticed that the sagittal plane of the coma ray changes at the lens (from green to blue).

Table 3.2 Different particular rays.

Name	Initial point in object plane	Intersection point in the entrance pupil plane	Remark
Marginal ray	Axis point	Outer edge in y	Meridional plane
Sagittal marginal ray	Axis point	Outer edge in x	Sagittal plane
Chief ray	Outer field in y	Centre	Meridional plane
Tangential coma rays	Outer field in y	Outer edge in $\pm y$	Meridional plane
Sagittal coma rays	Outer field in y	Outer edge in $\pm x$	Skew
Skew ray	Arbitrary x/y	Arbitrary x/y	

rays usually do not intercept the chief ray, either. Their image point is defined as the diapoint and it is situated where they traverse the meridional plane. This is discussed in Section 5.9.1.

A bunch of rays lying in one plane are usually called a ray fan. This selection of rays is found in the calculation of aberration curves. The special properties of the special rays are again collected in Table 3.2.

3.4.13 Ray Aiming

3.4.13.1 Introduction

Pupil aberrations play a role in raytracing in the subject of the so-called ray aiming. In order to trace a selected ray, such as a marginal, a chief or a coma ray, through an optical system, then for fast ray tracing, the entrance pupil is very often used to select the initial data of the rays. As the entrance pupil is defined as the paraxial image of the aperture stop through that part of the system which is in front of the stop, it may occur that rays selected in this way don't meet the real stop inside the system as desired. The reasons are usually large pupil aberrations or large angle conditions, the conjugation between stop and entrance pupil is in this case not well described by the paraxial model. To enforce an exact intersection of marginal and chief ray at the edge and in the centre of the stop respectively, small changes of the initial ray data are adapted. This is in practice a numerical optimization step and is called ray aiming [37, 38]. In total, for an optical system there are different cases for possible aiming.

- Aiming the chief ray to intersect the optical axis intersection in the stop centre point.

  The object position must be invariant. In case of a finite entrance pupil, the chief ray angle is changed; in case of a telecentric entrance pupil, the initial chief ray height is changed.
- Aiming the marginal ray to achieve the right size of the stop.

  The object position is invariant. In case of a finite object, the initial marginal ray angle is changed; in case of an afocal object, the initial marginal ray height is changed.

  If it is not clear which of the surfaces limits the light flux the most, it is not possible to address one surface to aim on. This kind of aiming therefore must take all relevant surfaces into account, which needs a more refined algorithm.
- Aiming the coma rays to meet the stop edges.

  The object position must be invariant. In case of a finite entrance pupil, the coma ray angle is changed; in case of a telecentric entrance pupil, the initial coma ray height is changed.

These six cases are sketched in Figure 3.43.

If, for example, the system part between entrance pupil and stop suffers from spherical aberration, the chief ray initial parameters must be changed to meet the stop centre exactly. This special case is sketched in Figure 3.44 and corresponds to Figure 3.43(a). The stop and the pupil plane are conjugated only in paraxial approximation; for the real chief ray heights and angles a discrepancy occurs. This effect is a result of pupil imaging aberrations and is discussed in more detail in Volume 2, Chapter 17. It should be noted that this aiming changes the entrance pupil position as it can be seen in the figure. In this special case, initial object point and the final point in the stop centre are known and the aiming should determine the one ray, connecting these two points.

Beneath the chief ray aberrations, the coma rays may additionally suffer from pupil aberrations. This means the coma ray initial angles must be changed to intersect the stop plane at the edges. If the pupil aberrations suffer from distortion too, the marginal ray angles must be adapted for

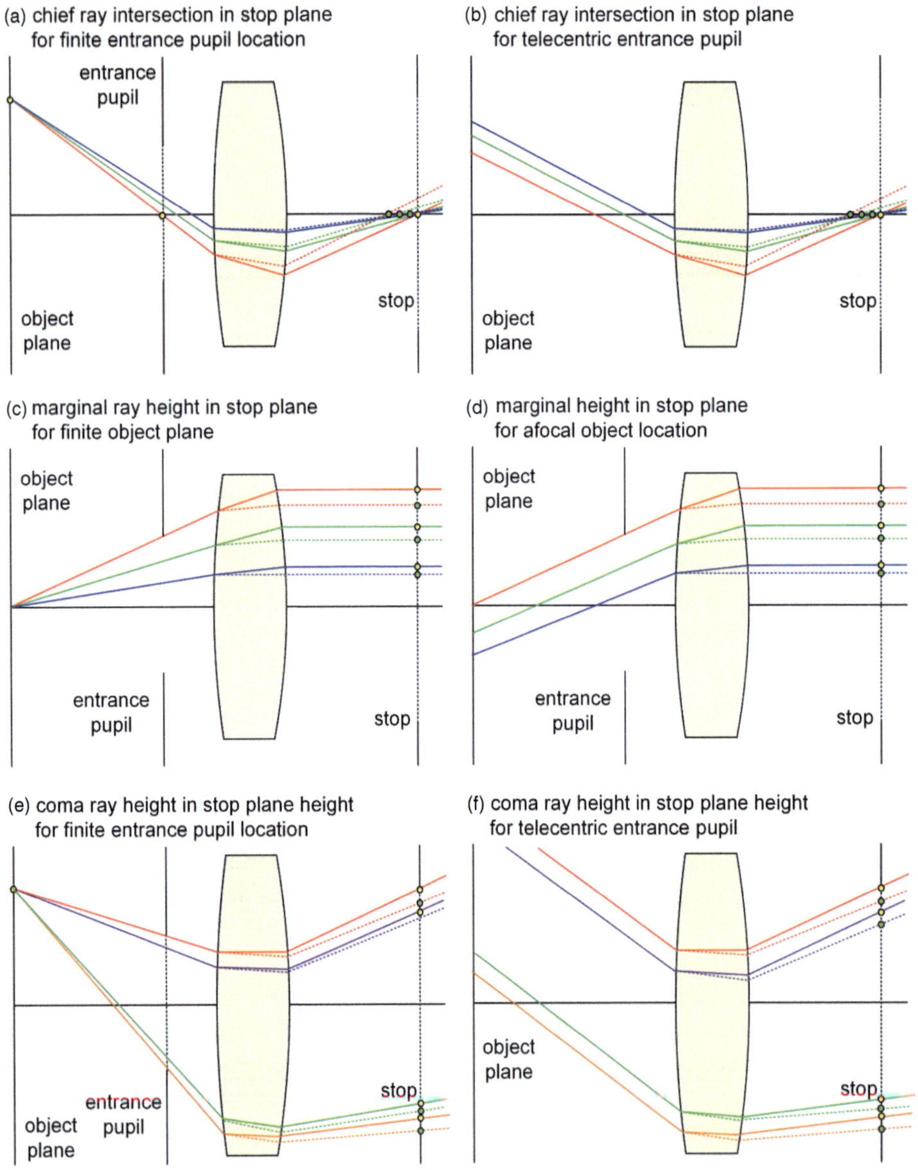

Figure 3.43 Practical cases for aiming of rays in case of pupil aberrations; the dotted rays indicate the wrong paraxial ray paths, the yellow dots the right aimed real rays. The cases (a) and (b) show the chief ray aiming; in (c) and (d) the marginal aiming is shown and the coma rays are indicated in (e) and (f). The infinity cases are always found in the second column.

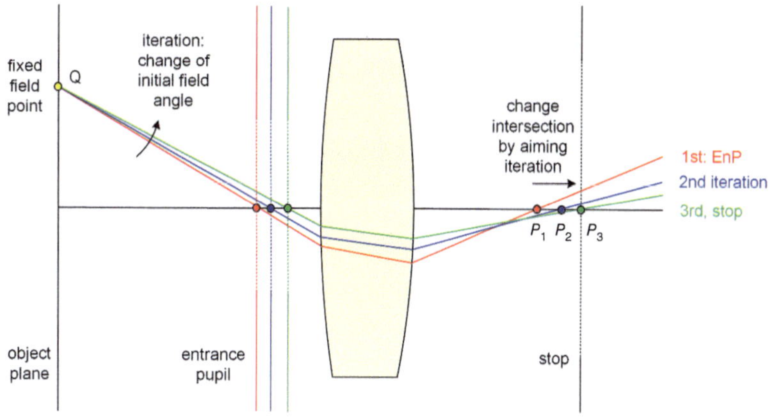

Figure 3.44 Aiming of a chief ray to meet the centre of the stop in P_3. The red case indicates the start situation, the blue is an intermediate result and the green ray represents the final correct chief ray and entrance pupil position.

an equidistant grid in the stop plane corresponding to a uniform illumination at the diaphragm.

If the aperture in a system is not explicitly specified in the object domain, the option 'float by stop size' in the optical design tools means that the effective size of the aperture cone is determined by aiming the outermost coma rays, which are able to pass the system for given lens sizes. The corresponding chief ray can in this case be defined as the ray with the middle angle as the axis of the cone. Figure 3.45 shows this principle of general aperture calculation. This usually has the consequence that for every field height, the effective size of the aperture cone is slightly different and the entrance pupil position depends on the field height. It should be kept in mind that the photometric implications of aiming the marginal or coma rays to achieve a uniform sampling in the stop plane automatically provokes a non-uniform sampling in the angle domain of the object space. Therefore, to get quantitative results, a modified weighting profile of the rays at the start is needed. A change of the initial aperture is then the consequence of a 'float by stop size'. Often, in practice, these details of photometric implications of aiming are neglected.

In the case of the coma aiming problem, it may happen that another factor other than the stop surface limits the aperture dominantly, see the blue upper coma ray in Figure 3.45. Therefore, in this case a simple point-to-point ray search is not enough and all surfaces must be checked for the most restricting one. The flux limiting stop-effect is then distributed along z. As indicated in the figure, it may happen in this case that the calculated chief ray as the middle ray between the coma rays does not intersect the stop at the centre. In this case there occurs a certain discrepancy between the paraxial definition of the chief ray and a physically meaningful interpretation of the chief ray as the ray that represents the centroid of energy in the ray bundle.

It should be mentioned that it is not only aberrations that create a requirement for aiming. In the case of wide-angle systems, the difference between the sine and the tan function of the chief ray angle can also have a corresponding consequence.

Therefore, a special procedure is needed to find the rays that represent the light path more accurately than the linear paraxially approximated ones. In principle, apart from unique cases, two conditions can be required on a ray – maybe points or angles. Then exactly one ray fulfils these two conditions. The aiming is in reality a highly non-linear optimization task and is performed iteratively by numerical methods. Figure 3.46 illustrates this idea for a ray in a more general case. The ray should go through two arbitrary points in a system, which are indicated as green here. The starting height x and the initial angle u are the unknown variables to be found by aiming. In reality, generally the object point is one of the two points the ray should meet and is then known.

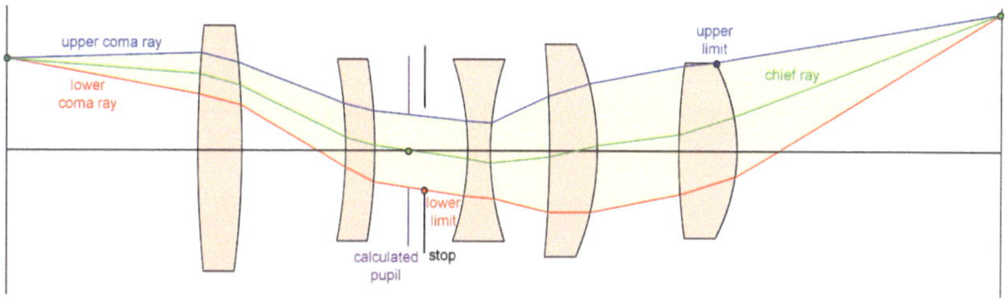

Figure 3.45 Searching the right upper and lower coma rays and definition of the chief ray as the centre of the light cone.

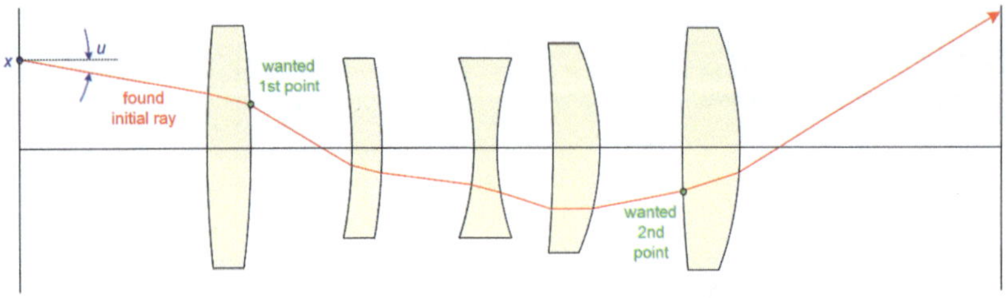

Figure 3.46 General search for a ray through two selected points.

The infinity case works correspondingly by changing the ray heights, keeping the angles the same.

3.4.13.2 Paraxial Aiming Approach

If in more complicated optical systems many rays are needed to perform the quality calculation, the iterative aiming of the rays slows down the computational speed considerably. Therefore, different options to accelerate the aiming process are proposed in the literature and are realized in commercial programs. These options are as follows.

- Real ray aiming:
  general and most accurate case, necessary for chief ray aiming, non-linear and slow.
- Paraxial ray aiming:
  marginal or chief ray sampling in a paraxial plane entrance pupil according to the tan-function. This approximation can solve the large-angle problem, but not the pupil aberrations. In particular. if the flux limiting distributes at different surfaces in $+y$, $-y$ and x-direction, this is a recommended approximation.
- Aplanatic/Hamiltonian ray aiming:
  marginal ray sampling in the aplanatic model for large numerical apertures assumes an entrance pupil sphere on axis. For off-axis field points this corresponds to Hopkins' Hamiltonian approach of a tilted entrance pupil sphere.

For the chief rays, it is always assumed that the object is a plane surface. In the case of an object sided telecentric system, the angle remains constant, but the chief ray height is adapted to aim the rays.

For the fractional sampling of the aperture, the situation is more complicated. The entrance pupil is paraxially assumed to be a plane, but for larger apertures, mostly the sine condition is fulfilled which means the pupil surface is a sphere. Therefore, in these cases the so-called aplanatic sampling on axis or Hamiltonian approach for off-axis object points is more accurate. In the Hamiltonian case, the arcs on the pupil sphere have constant lengths for fractional sampling. The cone is asymmetric, so the upper and lower coma rays do not lie symmetrically around the chief ray. These cases are sketched in Figure 3.47. It should be noted that for small numerical apertures, the paraxial and the aplanatic fractional pupil sampling is practically identical. For an afocal system, the marginal ray height is changed correspondingly.

The ray aiming in the paraxial approximation can be calculated analytically rather simply. If we assume between the object plane and the stop plane a paraxial *ABCD* segment as is indicated in Figure 3.48, the desired final coordinate x_2 is obtained by resolving the matrix equation

$$\begin{pmatrix} x_2 \\ u_2 \end{pmatrix} = \begin{pmatrix} A & B \\ C & D \end{pmatrix} \cdot \begin{pmatrix} x_1 \\ u_1 \end{pmatrix} \tag{3.121}$$

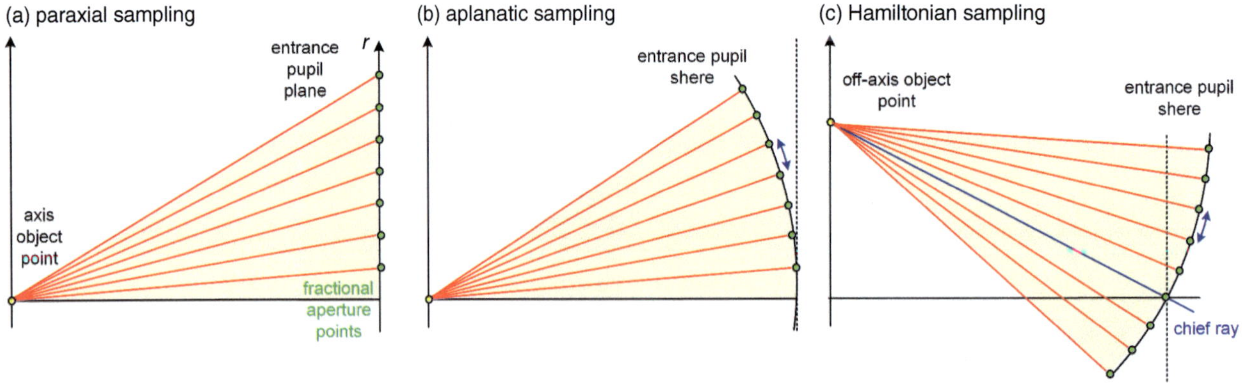

Figure 3.47 Paraxial (a), aplanatic (b) and Hamiltonian (c) aiming principles for the initial aperture aiming. It should be noted here that the full aperture cone is not shown.

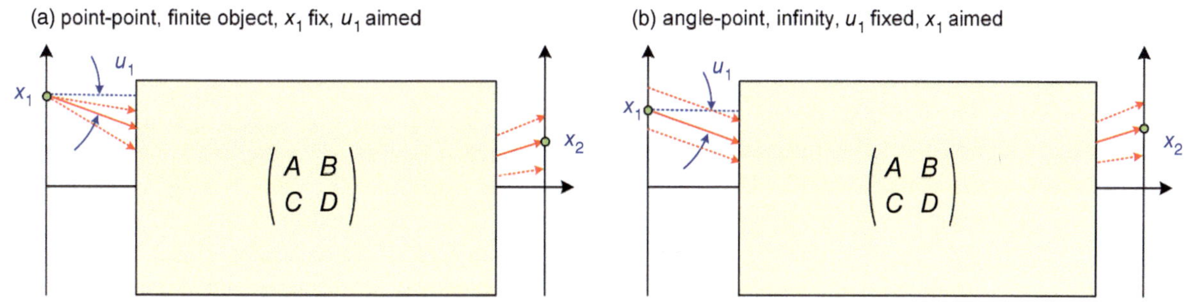

Figure 3.48 Paraxial ray aiming in case of (a) finite initial point or (b) infinity case with constant direction.

in the finite or infinity case respectively for a required x_2 as

$$u_1 = \frac{x_2 - Ax_1}{B}, \quad x_1 = \frac{x_2 - Bu_1}{A} \tag{3.122}$$

3.4.13.3 Real Ray Aiming

One practical problem with real ray aiming in reality for more complicated systems is ray failure. If in the non-linear optimization to find a boundary point by aiming the derivative is calculated, an extrapolation of the ray angle may not pass the system; therefore, a switch between left- and right-handed derivative calculation can be necessary. This means that more highly sophisticated algorithms are needed to achieve robust aiming. Of course this has slower aiming as a consequence. Furthermore, in general systems free of symmetry, there is not only one marginal, chief of coma ray, but also the problem is in its transverse direction two-dimensional. To find the full light flux passing the system, the problem must be solved in x and y simultaneously.

In particular systems without a central chief ray, like asymmetric systems or cases with central obscuration, ray aiming of the coma rays can be problematic because it is not possible to easily generate initial values for the optimization. Usually paraxial rays can serve as an initial guess, but it may be that in the general case they are not meaningful.

If a system has several lenses with mounting boundaries and diaphragms, which limit the light cone, the boarder and the cross section of the aperture cone can be rather complicated. To know it more accurately one possibility is to start with the chief ray, which is assumed to lie inside the aperture, and to aim rays in a radial direction

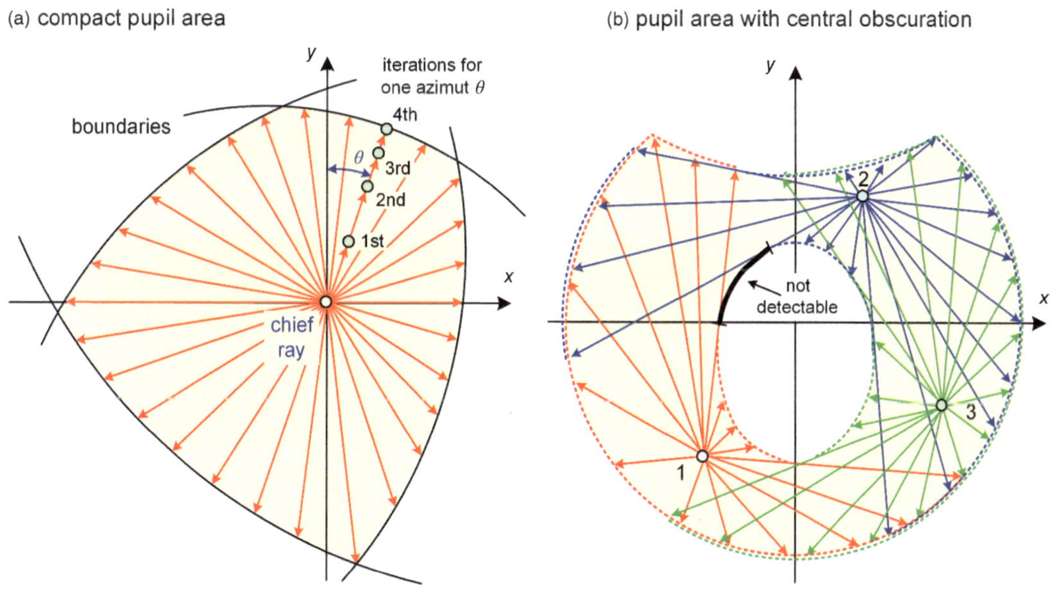

Figure 3.49 Finding the outer boundary of the aperture cone and field point footprint by aiming the radial outside. In (a) the ray pupil height is growing starting with the chief ray in the steps 1-2-3-4, until truncation is detected. In the more complicated case in (b) the same procedure is performed from the starting points 1, 2 and 3, delivering the coloured border sections. But the black arc is not found here.

into a discretized grid of azimuthal angles. This idea is sketched in Figure 3.49(a). An azimuthal sampling with angle θ creates for every angle an iterative scheme to find the maximum size of the initial aperture to pass the system, see the sketched green dots as intermediate results. Finally the boundary is approximated by a polygon line constituted by the end points of the outermost rays, which are not vignetted.

In the case of a central obscuration, the algorithmic problem to find the exact boarders is still more challenging. First it may happen that there is no chief ray in the energetic centroid. Then one of the problems is to find at least one ray that passes the system. But as can be seen in Figure 3.49(b), the scan or azimuthal sampling of the boundaries needs more clever logic and often more than one passing ray. In the constellation shown here, the black part of the inner border cannot be found by any of the three reference points 1–3.

In the practical calculation, the aperture superposed onto the chief ray is increased step by step as is shown in Figure 3.50 and in Figure 3.51 in one transverse cross section. The difference of the ray intersection point h_j at surface no. j and the half free diameter $D_j/2$

$$\Delta h_j = D_j/2 - h_j \tag{3.123}$$

should not be negative. The largest aperture value u_{max} which makes one value $\Delta hj = 0$, gives the outermost ray. To make the algorithm more stable it is a good idea to trace a comb of rays as is shown in Figure 3.50 until the sign of Δhj changes. Usually the curve $\Delta hj(u)$ is non-linear, but for considerably small increments Δu the two points embracing the zero can be used to make a linear or quadratic interpolation to find the zero rather accurately. This searching for the maximum aperture or scanning of the border has to be fulfilled for all the red, blue and green rays in Figure 3.49(b), which makes clear that the computational burden can be large. But it should be mentioned here that maybe the rays beyond the limiting border cannot be traced due to ray failure; then, this interpolation approach is not working.

There are some more comments on ray aiming. In the paraxial approximation, the change of the ray heights behaves linearly and the calculation is quite easy and non-iterative. In many cases this can be used to generate comfortable initial guess values for a more

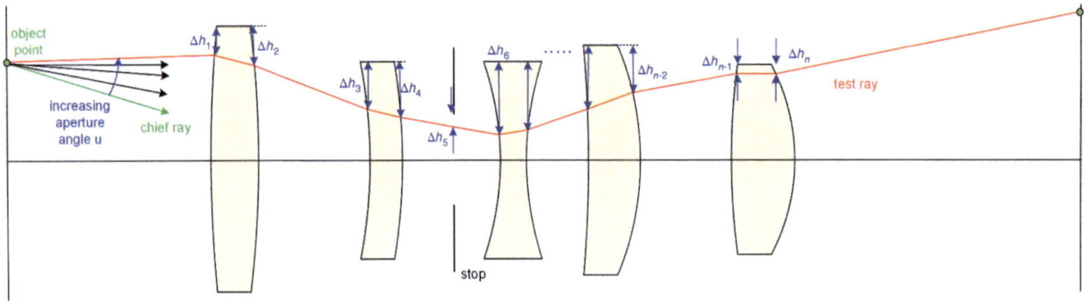

Figure 3.50 Aiming criteria to find the largest ray angle that passes the system.

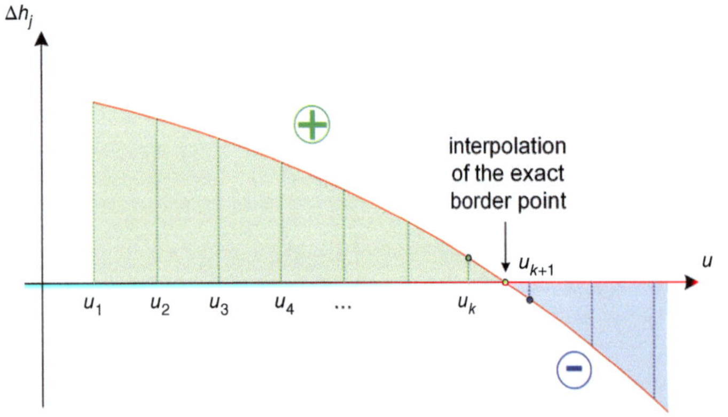

Figure 3.51 Aiming algorithm to search the zero crossing point at the surface number j for the height difference Δh_j.

rigorous non-linear calculation with real raytracing. But in systems with aspheres, free forms of general three-dimensional geometry, the prediction of paraxial optics breaks down rather quickly and the pre-calculations may fail. Furthermore, in special systems, one problem may be that no chief or reference ray is known and the biggest issue is to find at least one ray that passes the system without vignetting or other reasons for failure. This problem can be solved in practice by random number generation of rays and a simple trial and error logic to find at least one successful representative. It should also be noted that the above argumentation is in a certain sense simplified, because in three dimensions the intersection points of a ray with a surface has two coordinate values x_j and y_j and the surface boundary may not have simply a circular shape with one diameter but a complicated shape. It may also happen that a surface has an over-hemispheric shape, which makes the simple diameter check more complicated. Therefore, real algorithms should be programmed very carefully to cover most of the real world scenarios.

Systems with more complex pupil shapes and central obscurations are also a challenge for ray aiming. Most of the commercial design software tools fail in these cases of more complicated aperture shapes.

3.4.13.4 General Formulation of Greynolds*

(*) The content of this section is quite specialized and for advanced reading. Knowledge of this section is not necessary to understand the following chapters. Beginners are recommended to skip this section.

Recently, there have also been more physical concepts proposed in literature [37] to solve ray aiming in optical systems. This approach uses Fermat's principle to find the ray connecting two points in a system as a non-linear system of equations, which can be solved by the well-known algorithms of linear algebra. This approach connects the law of refraction at every surface, therefore all intersection points are found in one shot. In the reference some more examples are discussed and practical insights into the subject of aiming are given.

If in paraxial approximation the refraction at a surface no. j and the two transitions before and after the surface are considered, we have the three equations

$$h_j = h_{j-1} + u_{j-1} \cdot t_{j-1}, \quad n_j u_j = n_{j-1} u_{j-1} - h_j F_j, \quad h_{j+1} = h_j + u_j \cdot t_j \tag{3.124}$$

where F is a function describing the surface power. If here the angles $u_j - 1$, u_j are eliminated, the equation

$$-\frac{n_{j-1}}{t_{j-1}} h_{j-1} + \left(\frac{n_{j-1}}{t_{j-1}} + \frac{n_j}{t_j} - F_j \right) h_j + \frac{n_j}{t_j} h_{j+1} = 0 \tag{3.125}$$

is obtained. If in total $j = 1 \ldots N$ surfaces are considered and it is assumed that the initial height h_0 and the final value h_{N+1} are known and all thickness values t_j are non-zero, a tridiagonal system of equations has to be solved in one transverse dimension

$$\begin{pmatrix} \frac{n_0}{t_0} + \frac{n_1}{t_1} - F_1 & -\frac{n_1}{t_1} & & & \\ \ldots & \ldots & \ldots & & \\ & -\frac{n_{j-1}}{t_{j-1}} & \frac{n_{j-1}}{t_{j-1}} + \frac{n_j}{t_j} - F_j & -\frac{n_j}{t_j} & \\ & & \ldots & \ldots & \ldots \\ & & & -\frac{n_{N-1}}{t_{N-1}} & \frac{n_{N-1}}{t_{N-10}} + \frac{n_N}{t_N} - F_N \end{pmatrix} \begin{pmatrix} h_1 \\ \ldots \\ h_j \\ \ldots \\ h_N \end{pmatrix} = \begin{pmatrix} \frac{n_0}{t_0} h_0 \\ \ldots \\ 0 \\ \ldots \\ \frac{n_N}{t_N} h_{N+1} \end{pmatrix} \tag{3.126}$$

to find the heights h_j. This system can be solved by the well-known non-iterative algorithms and delivers all heights as a solution. In principal this is physically a concatenation of all surface operations fulfilling the law of refraction as constraints. If, here, the elimination of all intermediate h_j is performed, again the simple solution of the *ABCD* approach above is obtained for the price that all intermediate values are not known.

In case of a real raytrace, the system becomes non-linear, has a bandwidth of four and a dimension of 4 N for the coordinates x,y,z and the Lagrange multiplier. This is in principle also possible and can be solved, but the numerical conditions are not very comfortable; speed and convergence are not satisfactory for practical cases of usual complexity, see [37]. This basic approach can also be formulated for angle-point requests and general shapes of surfaces, but the mathematical solution conditions may differ.

3.4.13.5 Examples

Figure 3.52 shows a practical example of the consequence of the ray aiming in a photographic lens, which is shown in part(a) of the figure. The enlarged insets in (b) and (c) show the lack of exact intersection points in a more resolved way.

A second, stranger example is given in Figure 3.53(a), which shows a wide-angle lens to demonstrate the impact

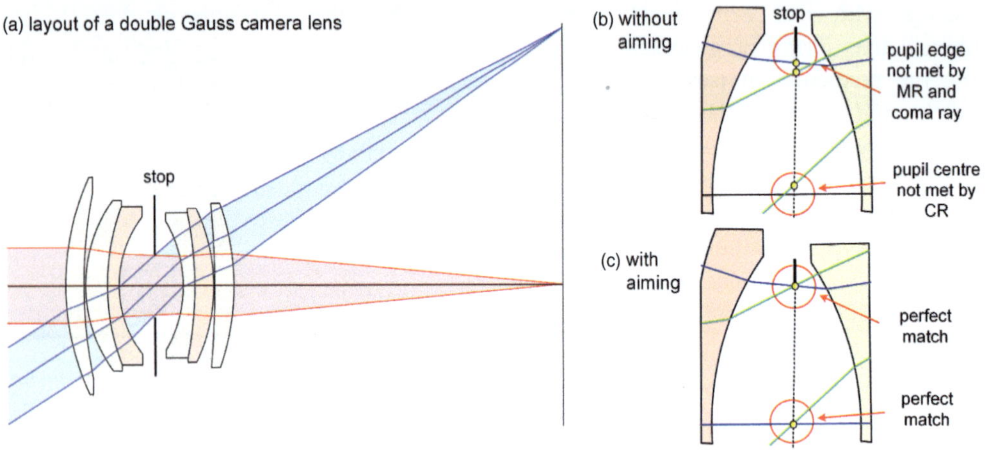

Figure 3.52 Aiming of a chief and coma ray to meet the real stop in the centre and edge.

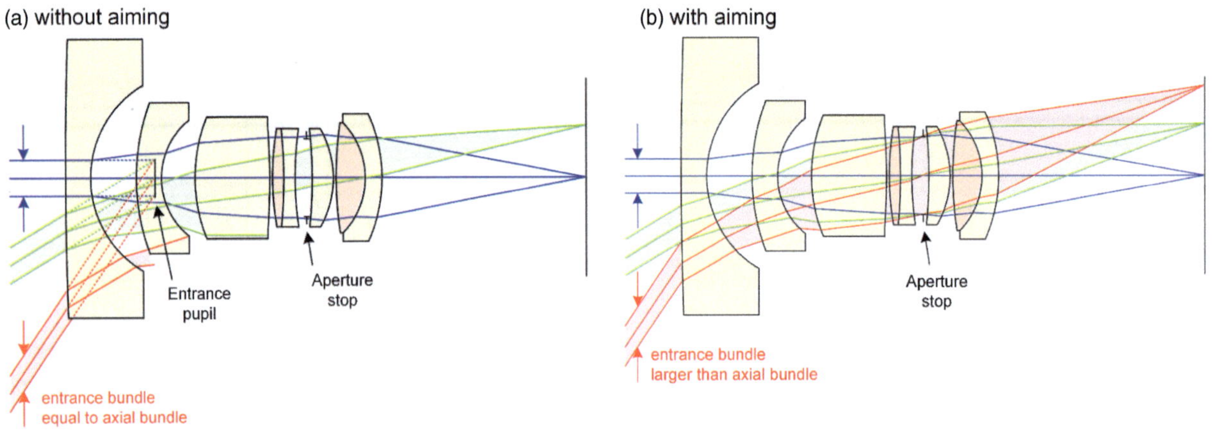

Figure 3.53 Pupil aberration and ray aiming. (a) Ray aiming towards the paraxial entrance pupil, (b) ray aiming towards the real aperture stop.

of aiming. None of the rays from the maximum oblique bundle, when aimed towards the entrance pupil, can pass through the lens. In such systems with large pupil aberrations the rays to be traced must be aimed towards the real aperture stop instead of the paraxial defined entrance pupil as shown in Figure 3.53(b). This is more time consuming but solves the problem.

3.4.14 Complex Rays*

(*) The content of this section is quite specialized and for advanced reading. Knowledge of this section is not necessary to understand the following chapters. Beginners are recommended to skip this section.

3.4.14.1 Introduction

The raytrace approach has been widely used due to its large validity range and simple calculation scheme. But it is a huge simplification of the physical description of light, and therefore many attempts have been made to increase its validity range for more physically related effects. To do this, in particular a better understanding of the approximations behind is needed. The extension of the so-called complex rays is one approach to extend the applicability of rays and will be quickly sketched here. More comprehensive theory can be found in the literature [39–50].

The complex source point theory is strongly connected to the approach of parabasal rays, the geometrical theory of diffraction and the theory of caustics. Its main application areas are scattering at metallic and arbitrary shaped obstacles, radio wave propagation and antenna radiation, evanescent wave, diffraction at curved surfaces and in inhomogeneous media, whispering gallery modes, creeping rays and ray manifolds. It is furthermore also possible to extend the formalism to vectorial waves and polarization as well as time dependent fields.

If a light field is considered as scalar and without a time dependence, the Helmholtz wave eq. [42]

$$\Delta E(\vec{r}) + n^2 k^2 E(\vec{r}) = 0 \tag{3.127}$$

describes the propagation. If the complex field is separated into amplitude A and phase Φ as

$$E(\vec{r}) = A(\vec{r}) \cdot e^{i\Phi(\vec{r})} \tag{3.128}$$

two coupled equations can be obtained by separating the real and imaginary part for the functions $A(r)$ and $\Phi(r)$. If lossy media are included, the eikonal Φ becomes complex and, due to the complex refractive index, not only the phase but also the energy density is included in the eikonal.

In the asymptotic limit of very small wavelengths or large wave vectors k, the so-called Debye series expansion (also called Wentzel-Kramer-Brillouin or WKB ansatz)

$$A(\vec{r}) = \sum_{m=0} A_m(\vec{r}) \cdot \frac{1}{(ik)^m} \tag{3.129}$$

is used for the amplitude function. Then a system of coupled equations for the coefficients $Am(r)$ is obtained, which can be used in different levels of approximation. If this separation is worked out, in zero order one of the obtained equation of motion is the eikonal equation for the phase transport

$$|\nabla\Phi(\vec{r})|^2 = n^2(\vec{r}) \tag{3.130}$$

for constant $A = A0$. This describes the classical raytrace and geometrical optics. In the next order approach of truncation, the series of Eq. (3.130) the equation

$$|\nabla\Phi(\vec{r})|^2 = n^2(\vec{r}) - \frac{\Delta A}{k^2 A} \tag{3.131}$$

can be derived. The interpretation of this equation shows that in generalization of the simple eikonal equation, there is a connection to a change in the lateral amplitude or energy density along the ray path, described by the Laplace operator on the A-function. This allows us to describe the intensity changes in the vicinity of a ray according to the level of approximation. As a simple example of this formulation, the ray giving the axis of a Gaussian beam allows for the lateral apodization description in extension to the pure geometrical ray picture.

This means that beyond the real propagation scheme of the eikonal equation in geometrical optics, moving to a complex formulation gives ‘more physical rays’. Therefore this theory is also called the ‘complex ray’ approach. This allows for a more physical raytrace as taking the energetic Fresnel equations at an interface into account, the change of energy density due to convergent or divergent beam tubes describes effects at caustics and absorption. In principle, due to a complex index of refraction, in lossy media raytrace in any case enters the complex domain and beneath the phase the energetic conditions are also involved. The extension to complex values removes the singularity at caustics and gives a natural physical picture. In a different formulation this means that the wave equations describe the propagation of rays also in situations with formally complex coordinates.

By including the lateral coordinates, the local character of the rays is changed. This is then naturally connected to the geometrical theory of diffraction as a lateral interaction of the rays with a boundary surface as well as general scattering problems. If the imaginary coordinate is introduced and certain spectral or angle constraints are used, a variety of rays, a whole ray bundle or manifold is described, its envelope gives in parabasal approximation the light field of the Gaussian beam around the ray. This means that a complex ray can represent a ray manifold [46].

Interestingly, by taking more than just the first correction into account, this approach is able to describe Gaussian beams beyond the paraxial approximation for higher angles or aberrations [45]. If the strict locality of rays is removed, the wave vector allows for a small lateral component. This means that the wave vector

$$|\vec{k}| = \sqrt{k_x^2 + k_y^2 + k_z^2} = \sqrt{k_t^2 + k_z^2} \tag{3.132}$$

is expanded in a quadratic approximation as

$$k_z = k_o + \frac{k_t^2}{2k_o} \tag{3.133}$$

with the transverse component k_t. This spectral compact representation defines a Fresnel-type volume around the ray or beam tube axis. The ray then changes to a ray tube with a finite cross section and a certain divergence.

The extension to a lateral coordinate and removing the strict locality of real rays opens the opportunity to describe diffraction effects. If the Kirchhoff theory of diffraction and wave propagation is solved approximately by the stationary phase or saddle-point method, the complex ray picture is obtained.

3.4.14.2 Gaussian Beams with Complex Source Point

If a classical spherical wave of the form

$$E = \frac{e^{ikR}}{R} \tag{3.134}$$

is considered and the source point as the centre has complex coordinates

$$R^2 = (x - ix_0)^2 + (y - iy_0)^2 + (z - iz_0)^2 \tag{3.135}$$

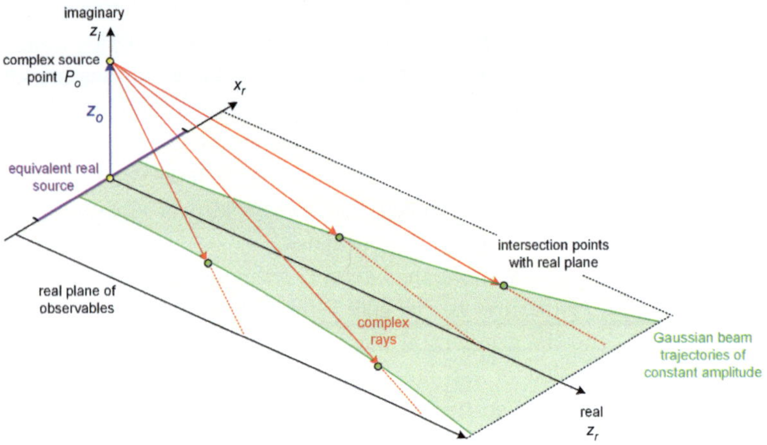

Figure 3.54 Gaussian beam as intersection of a ray manifold of complex rays with the real observable plane $z_i = 0$.

by insertion, the field is in this case given by

$$E = \frac{e^{ikz}}{1 + iz/z_0} \cdot e^{\frac{ik(xx_0+yy_0)}{z_0+iz}} \cdot e^{-\frac{k(x^2+y^2)}{2(z_0+iz)}} \cdot e^{-\frac{ikz\left(x_0^2+y_0^2\right)}{2z_0(z_0+iz)}} \tag{3.136}$$

which describes a Gaussian beam with a lateral offset and an angular inclination. In the special case $x_0 = y_0 = 0$ the complex offset is only present in the z direction, the solution is then aligned and the offset parameter z_0 is exactly equal to the Rayleigh range of the Gaussian beam (see Chapter 8 for more details about Gaussian beams).

To visualize this concept, Figure 3.54 illustrates the geometry for just one lateral coordinate x. All complex rays start from the imaginary point $z_i = z_o$. Their intersection with the plane of real z as possible observable plane marks the curves of constant amplitude of the Gaussian beam with hyperbolic shape [48].

If several complex rays arrive at the same observation point, they have to be summed up and the complex rays describe the interference. This is exactly seen at a caustic. Here the described method has also a strong connection to the method of Gaussian beam decomposition of arbitrary fields.

3.4.14.3 Tracing Complex Rays

As mentioned above, the complex ray includes the energy density propagation in the vicinity of the central geometrical ray. Therefore, in the case of a raytrace at a dielectric interface, the change due to the Fresnel formulas as well as the change in curvature of the local wavefront must be taken into account in addition to the new ray direction, which is geometrically governed by the Fermat principle or the law of refraction. This means that a complex raytrace corresponds to a differential raytrace. In principle not only the geometrical ray direction, but also the change in the physical quantity of the wave vector k has to be transformed with two components as well as the radii of curvature, as in the case of an astigmatic or differential raytrace. Two additional coordinate transforms are necessary to perform the change from the principal curvature system of the incoming wave into the surface incidence plane system with normal vector and the outgoing principal curvature system (see Volume 2, Chapter 16 for more details). Furthermore, along a propagation in a homogeneous medium the radius of curvature in both principal planes changes geometrically according to the skew distance t_j as [47]

$$R'_j = R_j + t_j \tag{3.137}$$

or in a continuous way for a more general astigmatic beamlet the amplitude due to divergence changes as

$$A(z) = A_0 \cdot \sqrt{\frac{R_x R_y}{(z - R_x) \cdot (z - R_y)}} \tag{3.138}$$

Furthermore, the length of the k-vector changes according to the Fresnel equations for the surface transition or corresponding to the properties of a coating at this surface under consideration. This geometry is sketched in Figure 3.55.

To follow the physical properties of the complex ray, the phase shift or optical path difference along the travelling ray and due to the Fresnel formulas at a surface must be considered to correct for the energetic splitting relation accurately. This is necessary to keep the phase correct and to describe interference effects appropriately.

3.5 Paraxial Approximation

3.5.1 Introduction

Monochromatic aberrations in general have their origin in the non-linear behaviour of the sine function in the law of refraction and the non-quadratic behaviour of spherical

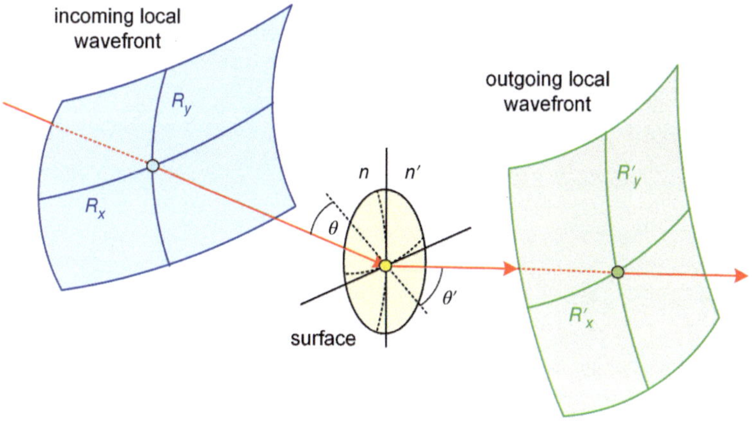

Figure 3.55 Raytrace with local wavefront transform for complex rays.

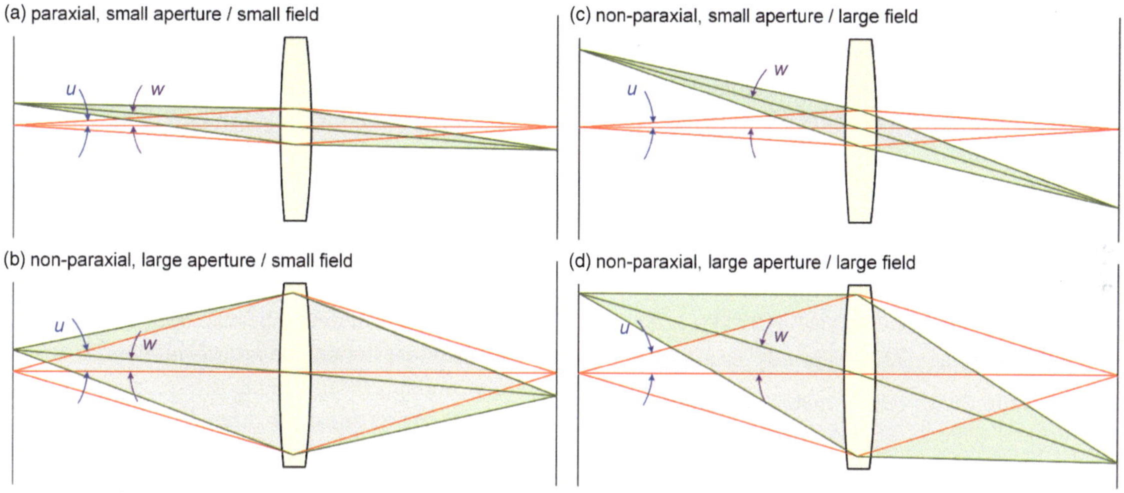

Figure 3.56 Paraxial approximation cases with marginal ray angle u and chief ray angle w. The part (a) is a complete paraxial system, (b) only the marginal ray is paraxial, (c) only the chief ray is paraxial and (d) there is no paraxiality at all.

surfaces [1, 13]. Since the chief ray as well as the marginal ray are essential to characterize an optical system for the size of field of view and aperture, at least these two rays must be taken into account to assess the paraxial properties of a system. A fully paraxial system is obtained if all rays are near to the optical axis with small ray heights and small incidence angles at all surfaces. If only the chief or the marginal ray are considered, the system cannot be fully paraxially approximated. These various cases are illustrated in Figure 3.56.

Sometimes a formulation of the paraxiality definition is found where all the rays are 'near the axis'. This is phenomenological and simplified description for rotational symmetric systems and not true in general, because in three-dimensional cases, the axis itself can have large incidence angles. The physical condition for linearity of ray transfer is the more rigorous criterion and is based on the refraction and the sine function.

The general law of refraction reads

$$n \cdot \sin i = n' \cdot \sin i' \tag{3.139}$$

If the sine function is expanded here into a Taylor series, we get

$$\sin i = i - \frac{i^3}{6} + \frac{i^5}{120} - \frac{i^7}{5040} + \frac{i^9}{362880} - \ldots \tag{3.140}$$

The first term here is the linear one, describing the paraxial approximation of the refraction process. The next higher term is of third order, is non-linear and mostly describes the primary aberrations (third order in transverse aberrations). If the paraxial approximation is used, the law of refraction simply reads

$$n \cdot i = n' \cdot i' \tag{3.141}$$

In the paraxial approximation in addition to the law of refraction also the surface equations are Taylor expanded

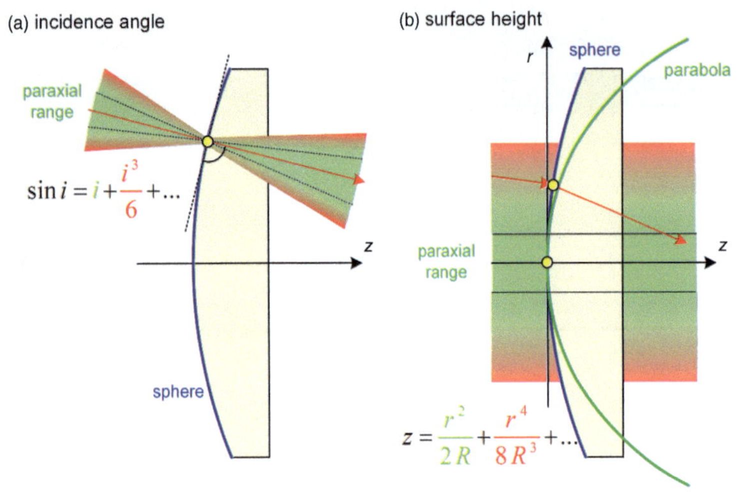

Figure 3.57 Paraxial approximation as a function of incidence and ray height.

until the second order. For the raytrace the slope is essential, the derivative of the second order approximated surface is then linear. Therefore, paraxiality is in general a function of ray height and incidence, as is indicated in Figure 3.57. If a spherical surface can be approximated with sufficient accuracy by a parabolic shape, the paraxiality condition of the surface is fulfilled.

If these approximations are used, this has the following consequences for the properties of optical systems.

All optical imaging conditions become linear (Gaussian optics); calculation with *ABCD* matrix calculus is possible.

No aberrations occur in optical systems.

There are no truncation effects due to transverse finite sized components.

The sag of optical surfaces (difference in z between the vertex plane and real surface intersection point along the optical axis) can be neglected, which means that the ray heights are measured in the vertex plane.

All waves are plane or parabolic; the phase term exponent of waves is quadratic.

$$E(x) = E_0 \cdot e^{-\frac{i\pi x^2}{\lambda R}} \tag{3.142}$$

This is not a contradiction to the linearization, because the ray direction is obtained by the derivative of the phase function and is therefore linear. In particular it should be mentioned that one consequence of the quadratically restricted lateral wave change is separable in x and y due to

$$\begin{aligned} E(x,y) &= E_0 \cdot e^{-\frac{i\pi r^2}{\lambda R}} = E_0 \cdot e^{-\frac{i\pi(x^2+y^2)}{\lambda R}} \\ &= E_0 \cdot e^{-\frac{i\pi x^2}{\lambda R}} \cdot e^{-\frac{i\pi y^2}{\lambda R}} = E(x) \cdot E(y) \end{aligned} \tag{3.143}$$

The paraxial approximation is of importance for every system, because the paraxial range serves as a reference for ideal system conditions. In particular aberrations are measured against this perfect case. In addition the paraxial range is the fundament for the definition of many system properties like the focal length, the principal planes, the magnification and the pupil.

In principle, every approximation produces a certain error. Therefore, the question 'can it be allowed to calculate/consider this as paraxial' is ill posed. The correct procedure should first define what the necessary accuracy is and then decide if in paraxial approximation the results are accurate enough. Figure 3.58 is a representation of the various expansion sums for three different truncation cases of the sine function with a linear ordinate scale. If an acceptable error of 10^{-4} is assumed, this means we get the following limiting incidence angles depending on the selection of approximation. Figure 3.59 shows the contributions of the Taylor series expansion terms in a logarithmic scale, which gives better quantitative information about residual errors. The deviation of the various approximations are obtained for maximum angles i as:

- linear, paraxial: $i < 5°$
- third order: $i < 24°$
- fifth order: $i < 52°$

This means that the paraxial approximation makes sense if all incidence angles are smaller than 5°. It should be noticed that the value of 10^{-4} as a relative error of the sine function is rather arbitrary and must be defined depending on the specific requirements of an application. In real systems the required accuracy is generally much higher.

The remaining error for truncation after the third, fifth, seventh and ninth order term in the sum of the Taylor expansion of the sine function are graphically illustrated in Figure 3.60 in a logarithmic scale. In contrast to Figure 3.59

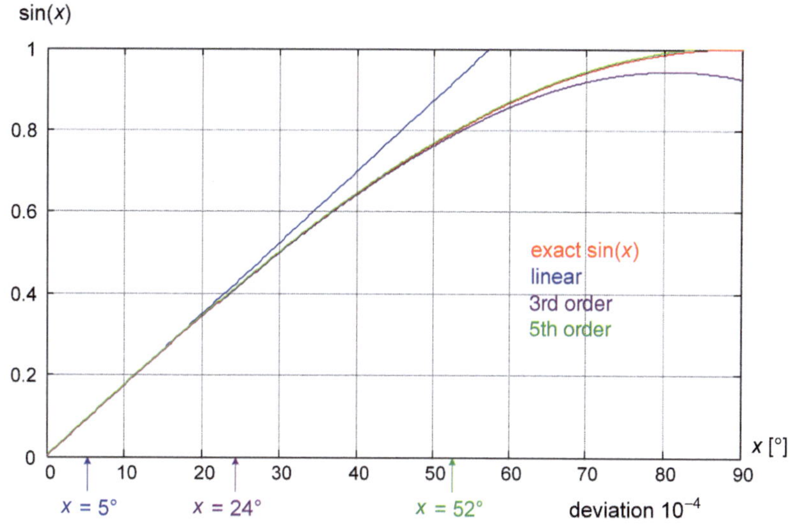

Figure 3.58 Paraxial approximation accuracy for an accepted error of 10^{-4} for one single refraction. The exact curve of the sine function is shown in red, the first, third and fifth approximation according to the Taylor expansion are depicted in blue, purple and green.

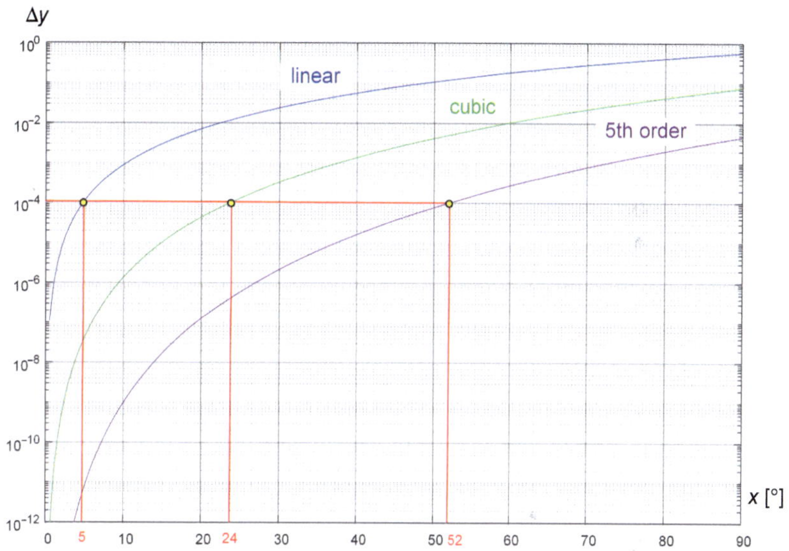

Figure 3.59 Error of the first, second and third approximation of the Taylor expansion of the function $y(x) = \sin(x)$.

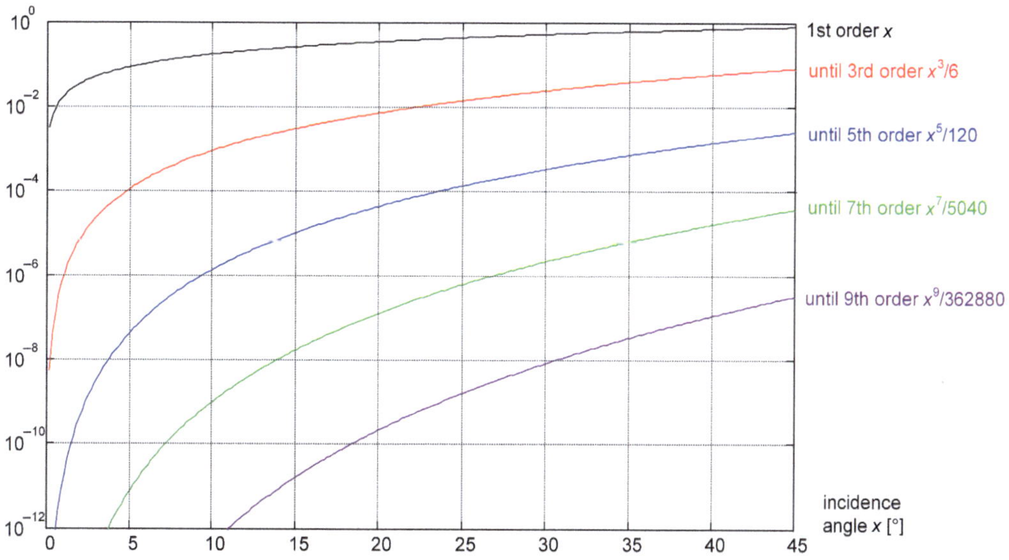

Figure 3.60 Relative error of the various truncation levels in the Taylor expansion of the sine function as a function of the angle x.

the angle i is considered only until 45° and the terms of ninth order included. This kind of separation of orders can also help in reality to estimate what range of validity is obtained for the primary aberrations. They are of third order in transverse deviations and are directly related to the second term of the expansion, but may be magnified by succeeding system surfaces.

The relative error of the paraxial approximation of the law of refraction can be defined as (i, i' are the real, I' is the paraxial angle here)

$$\varepsilon = \frac{i' - I'}{I'} = \frac{\dfrac{n \cdot i}{n'}}{\arcsin\left(\dfrac{n \cdot \sin i}{n'}\right)} - 1 \tag{3.144}$$

with the approximated angle I' in a more quantitative manner.

3.5.2 Abbe Invariant

3.5.2.1 Paraxial Definition

For a paraxial refracting spherical surface, according to Abbe, an invariant can be determined. It is a quantity constant before and after the refraction at a surface for a paraxial ray. The following relation between the intersection lengths on the object side and the image side is valid independent of the inclination u of the ray.

$$Q = n \cdot \left(\frac{1}{R} - \frac{1}{s}\right) = n' \cdot \left(\frac{1}{R} - \frac{1}{s'}\right) \tag{3.145}$$

According to the geometric relations as shown in Figure 3.61, we have the common height h, the angle sum in the triangles PCQ and $P'QC$ and the law of refraction in point T, the following equations are obtained

$$\begin{aligned} h &= -s \cdot u = -s' \cdot u' = R \cdot \varphi \\ 180 &= (180 - i) + u + \varphi \\ 180 &= (180 - \varphi) + i' + u' \\ n \cdot i &= n' \cdot i' \end{aligned} \tag{3.146}$$

With these equations, φ and u can be eliminated on the primed and unprimed side from the first two equations and inserted in the last equation. We immediately get the refraction invariant

$$\begin{aligned} A = n \cdot i &= n \cdot u \cdot \left(1 - \frac{s}{R}\right) = n' \cdot i' \\ &= n' \cdot u' \cdot \left(1 - \frac{s'}{R}\right) \end{aligned} \tag{3.147}$$

and by dividing by h with the first equation of (3.146) the Abbe invariant Q.

The invariant is an alternative formulation of the lens imaging equation. It states that the relation between the intersection length s on the object side and on the image side s' is independent of the incident height and the cone angle of the ray. This is also the form for the imaging equation, which can be easily obtained from (3.145) by re-arranging the terms as

$$\frac{n'}{s'} - \frac{n}{s} = \frac{n' - n}{R} = F \tag{3.148}$$

with the focal power F. This means that the Abbe invariant fixes the paraxial image location.

3.5.2.2 Generalized Abbe Invariant*

(*) The content of this section is quite specialized and for advanced reading. Knowledge of this section is not necessary to understand the following chapters. Beginners are recommended to skip this section.

It is shown in the previous section that the classical paraxial Abbe invariant Q corresponds to the ideal lens maker's formula.

For finite apertures, this definition can be generalized according to [24]. If p and p' are the lengths along the ray in the object and the image space from the axis points P and P' to the surface intersection point T (see Figure 3.62), this formulation reads for the extended real Abbe invariant Q_r

$$Q_r = n \cdot \left(\frac{1}{R} - \frac{1}{s}\right) \cdot \frac{s}{p} = \frac{n \cdot (s - R)}{p \cdot R} = \frac{s}{p} \cdot Q \tag{3.149}$$

With the geometry relations of the triangles C-T-P' (yellow) and Y-T-C (blue) used in the law of refraction in point

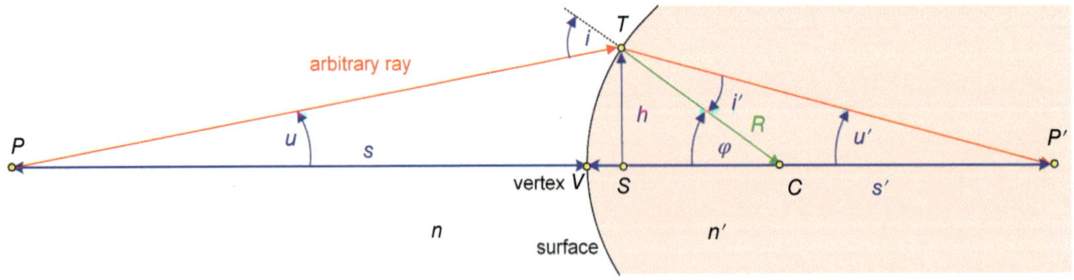

Figure 3.61 Ideal paraxial imaging by a single refracting surface.

T we get

$$h_r = R\sin\varphi = p'\sin u'$$
$$\frac{\sin i'}{\sin u'} = \frac{s'-R}{R}$$
$$n'\sin i' = n'\sin u'\left(\frac{s'}{R}-1\right) = n'\frac{R}{p'}\sin\varphi\left(\frac{s'}{R}-1\right)$$
$$= R\sin\varphi\cdot\frac{s'}{p'}n'\left(\frac{1}{R}-\frac{1}{s'}\right)$$
$$= R\sin\varphi\cdot Q_r \tag{3.150}$$

which gives the same on the object side. Therefore, Q_r is an invariant. In the real case, according to Abbe, an expansion into the powers of the central angle φ can be made as

$$Q_r = Q + q_2\cdot\varphi^2 + q_4\cdot\varphi^4 + \ldots \tag{3.151}$$

with the paraxial invariant Q. Here q_2 describes the third order aberration contribution, q_4 the part of the fifth order and so on. These coefficients are also invariant during refraction for every order, therefore the equations

$$q_2 = q_2', \quad q_4 = q_4' \tag{3.152}$$

are valid. This series representation finally is a power expansion of the longitudinal spherical aberration, see also Volume 2, Chapter 18 for more details. In the second order we get the expression

$$Q_r^{(2)} = Q + \frac{Q^2r^2}{2ns}\cdot\varphi^2 \tag{3.153}$$

and the invariance

$$Q_r^{(2)} = Q + \frac{Q^2R^2}{2ns}\cdot\varphi^2 = Q_r'^{(2)} = Q + \frac{Q^2R^2}{2n's'}\cdot\varphi^2 \tag{3.154}$$

holds. The corresponding terms of this more general definition of the Abbe invariant are illustrated in Figure 3.62. In contrast to the paraxial invariant Q, Q_r depends on the ray height or angle φ at the lens.

3.5.3 Superposition of Rays*

(*) The content of this section is quite specialized and for advanced reading. Knowledge of this section is not necessary to understand the following chapters. Beginners are recommended to skip this section.

In the paraxial approximation the raytrace relation becomes linear. Due to this fact, a general superposition principle is valid. This means if the initial values (y_1, u_1) and (y_2, u_2) are known for two arbitrary rays, every linear combination of these two rays with parameters

$$y_3 = a\cdot y_1 + b\cdot y_2, \quad u_3 = a\cdot u_1 + b\cdot u_2 \tag{3.155}$$

is also a valid ray. Therefore, if two independent rays are calculated, the paraxial properties of the system are fully determined, every paraxial system has only two independent rays. In reality it makes sense to select the chief and marginal ray as special characteristic rays. For example, by adding and subtracting the marginal ray from the chief ray, the upper and lower coma rays (UC and LC respectively) are obtained

$$y_{UC} = y_{CR} + y_{MR}, \quad u_{UC} = u_{CR} + u_{MR}$$
$$y_{LC} = y_{CR} - y_{MR}, \quad u_{LC} = u_{CR} - u_{MR} \tag{3.156}$$

In particular in three dimensions paraxial rays in the x–z and y–z plane respectively can also be overlaid to get a skew ray. If the two rays (y_1, v_1) and (x_2, u_2) are known, an arbitrary skew ray is described by

$$\vec{r} = \begin{pmatrix}x\\y\end{pmatrix} = a\cdot\begin{pmatrix}0\\y_1\end{pmatrix} + b\cdot\begin{pmatrix}x_2\\0\end{pmatrix},$$
$$\vec{w} = \begin{pmatrix}u\\v\end{pmatrix} = a\cdot\begin{pmatrix}0\\v_1\end{pmatrix} + b\cdot\begin{pmatrix}u_2\\0\end{pmatrix} \tag{3.157}$$

with corresponding constants a and b. The advantage of this property is also used in paraxial ray aiming. Since the linear equations can easily be solved for the scaling parameter a and b it is, for example, possible to determine the initial ray

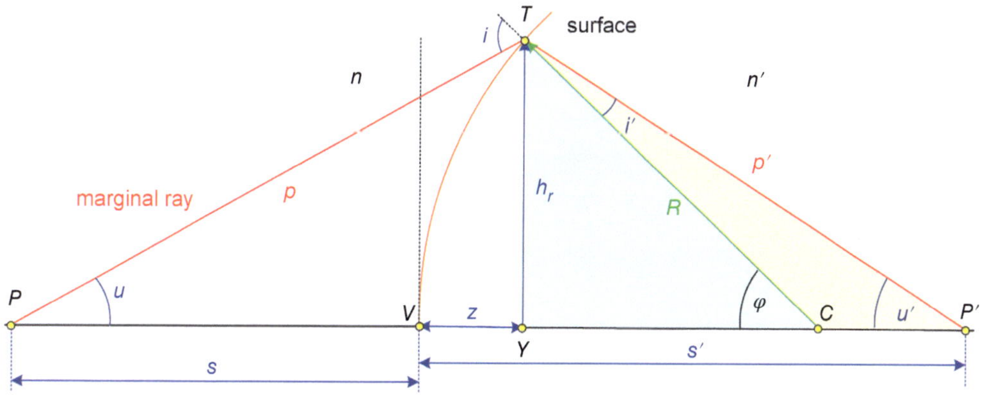

Figure 3.62 Definition of the generalized Abbe invariant for finite aperture angles u.

data if the ray heights at two arbitrary planes are required, see Section 3.4.13.2.

3.5.4 Generalized Paraxiality*

(*) The content of this section is quite specialized and for advanced reading. Knowledge of this section is not necessary to understand the following chapters. Beginners are recommended to skip this section.

In reality it may happen that the classical definition of a paraxial regime fails. In particular two cases are of larger practical relevance.

- Central obscuration.

  Systems using mirrors are often characterized by a central obscuration. As an example, Figure 3.63 shows a Schwarzschild objective with a ring-shaped pupil. The central part with the optical axis is obstructed and rays in the near vicinity of the axis don't exist and are not able to pass the system. Therefore, a paraxial range cannot be defined in the usual way. In principle rays near to the axis can be calculated, although they are truncated by the free diameters. It makes sense to use them as a reference for the evaluation of aberrations.

  Alternatively a ray in the centre of the ring corresponding to the small pupil drawing in Figure 3.63 can be used as reference, but the classical interpretation of paraxial properties is not valid; this ray has finite incidence angles at the surfaces.
- Off-axis systems

  Another type of system without a simple paraxial range is a general three-dimensional system without straight axis, as is shown in Figure 3.64 for a three-mirror anastigmatic telescope system. An optical axis ray can be defined, which represents the energetic centroid of the supported light cone passing the system. But this ray has finite incidence angles at the mirror surfaces and is not fulfilling the condition of a perfect linear aberration-free reference ray. This means that the reference ray has to be calculated real and may suffer from distortion, but the aperture and additional field points are paraxially approximated respective to this central ray as reference, so they are parabasal rays of the axis ray.

  If the ray is used as a reference and a downscaled marginal ray according to part (b) of Figure 3.64 is considered, all the classical properties of a small deviation from the reference concerning resolution are obtained. But the geometrical aberrations of the axis ray are included in the reference and, depending on the correction of the system, astigmatism on axis can occur. A similar consideration can be used to define a parabasal ray, which is obtained as a ray in the near vicinity of the chief ray.

3.5.5 Collinear Transform

If the trigonometric functions in the real refraction model are linearized for small angles, the paraxial framework is obtained. But it is also possible to define a model with finite sized angles which is a priori linear. This approach is called the collinear transform. Consequently a general linear mathematical geometrical transformation [5, 10] of the form

$$x' = \frac{F_1}{F_0}, \quad y' = \frac{F_2}{F_0}, \quad z' = \frac{F_3}{F_0} \tag{3.158}$$

with the linear expression

$$F_j = a_j x + b_j y + c_j z + d_j, \quad j = 0, 1, 2, 3 \tag{3.159}$$

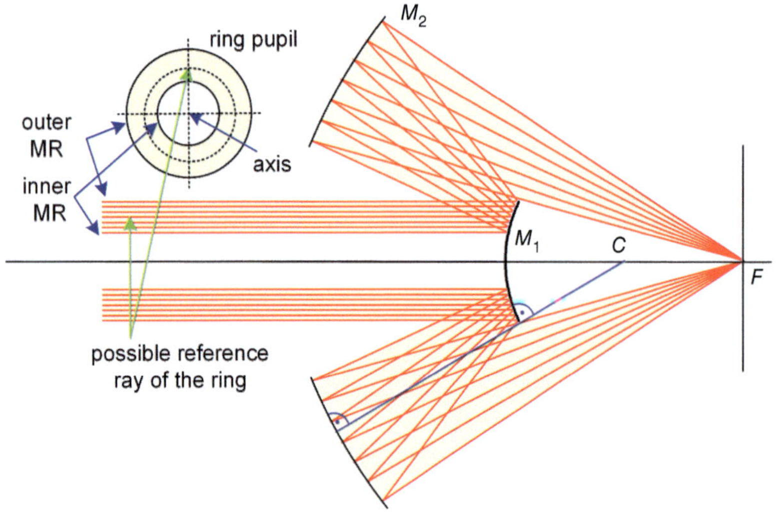

Figure 3.63 Schwarzschild objective lens with central obscuration and ring shaped pupil area.

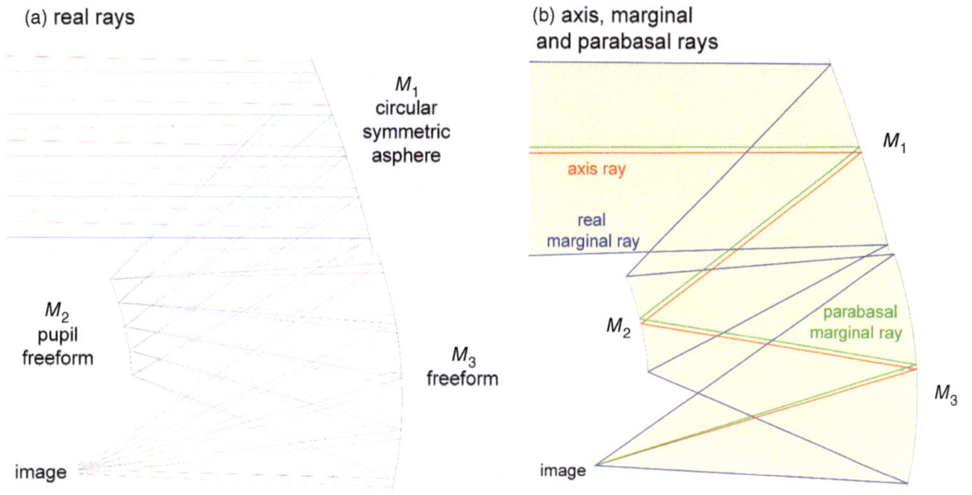

Figure 3.64 Obscuration-free three-mirror anastigmatic telescope with off-axis geometry. (a) shows a real raytrace, (b) the axis ray (in red) and a parabasal ray on it (green).

is considered and it is called a linear collineation. x, y, z are the coordinates of an object point, x', y', z' are those of a point image. After solving the above system of equations for the object coordinates, an analogous expression is obtained

$$x = \frac{F'_1}{F'_0}, \quad y = \frac{F'_2}{F'_0}, \quad z = \frac{F'_3}{F'_0} \tag{3.160}$$

with

$$F'_j = a'_j x' + b'_j y' + c'_j z' + d'_j, \quad j = 0, 1, 2, 3 \tag{3.161}$$

The equations can thus be inverted easily. Therefore, the point object and point image are also called a conjugated pair of points. The collineation can map points into points, lines (rays) into lines or planes (object, image) into planes. It is a general mathematical central projection mapping. This transform is quite general and can be simplified in reality if certain constraints or symmetries are given. This means in particular that this approach is also applicable in the case of general three-dimensional optical systems with rotated or shifted axes under skew conditions. The quantities F and F' describe planes. The planes $F_0 = 0$ and $F'_0 = 0$ are imaged into infinity and thus are the focal planes of the system. Since in this strictly linear imaging model planes are transferred to other planes, so also lines which are the intersection curves of two planes will transfer into other lines.

Until now, the transform was a purely mathematical construction. Now in a second step it is applied onto optical systems and the terms and properties are interpreted for this application.

In optics, rotational-symmetric systems play an important role. With the corresponding constraints and limitations and assuming that the y and the y' axes as well as the x and x' axes are parallel to each other, the equations above can be simplified and with the z axis acting as the axis of symmetry, they become

$$x' = \frac{a_1 x}{c_0 z + d_0}, \quad y' = \frac{b_1 y}{c_0 z + d_0}, \quad z' = \frac{c_3 z + d_3}{c_0 z + d_0} \tag{3.162}$$

If only the ratios of the constants are taken into account this case is described by only four parameters. This means if the unit magnification principal planes are selected as coordinate origin, the equations reduce to

$$x' = \frac{x}{c_0 z + 1}, \quad y' = \frac{y}{c_0 z + 1}, \quad z' = \frac{c_3 z}{c_0 z + 1} \tag{3.163}$$

and only two independent parameter remain. Now the lateral magnification m and the focal lengths read

$$m = \frac{x'}{x} = \frac{y'}{y} = \frac{1}{c_0 z + 1}, \quad f' = \frac{c_3}{c_0}, \quad f = \frac{1}{c_0} \tag{3.164}$$

where the condition of the focal point is to image into infinity. By insertion and elimination, immediately the well-known paraxial equations

$$m = \frac{1}{1 - z/f} = 1 - \frac{z'}{f'}, \quad \frac{z'}{f'} - \frac{z}{f} = 1 \tag{3.165}$$

are obtained. If the origin of the coordinates is alternatively selected in the focal points, the Newton equations of imaging

$$m = \frac{z'}{f'} = \frac{f}{z}, \quad z \cdot z' = f \cdot f' \tag{3.166}$$

are found. In contrast to the interpretation of the principal points as locations of unity magnification, the nodal points

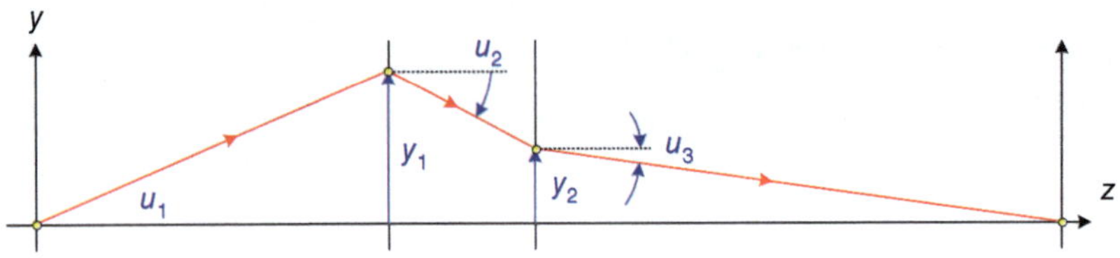

Figure 3.65 Linear collineation with ray bending at planes.

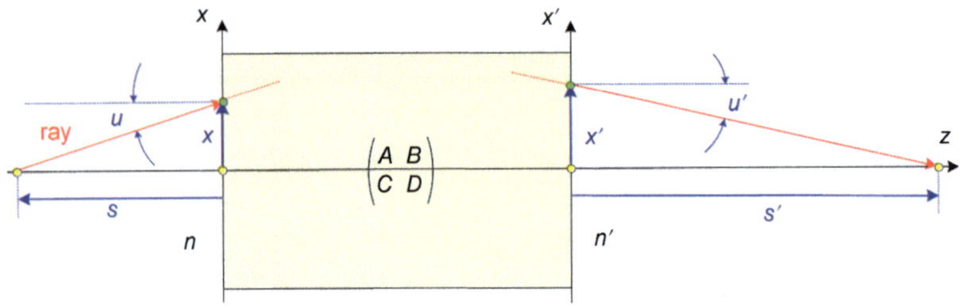

Figure 3.66 Terminology of an *ABCD* system in matrix calculus for transfer through a paraxial partial system.

are characterized as the locations of unit angle magnification or unchanged direction. Thus, paraxial optics can be described completely by a linear collineation.

In the matrix calculus (see next section for details), due to the linearity in cartesian coordinates the tan-function must be used to get mathematically correct results in the case of finite size angles u. This means that the linear collineation reads for finite angles

$$\begin{pmatrix} x' \\ \tan u' \end{pmatrix} = \begin{pmatrix} A & B \\ C & D \end{pmatrix} \cdot \begin{pmatrix} x \\ \tan u \end{pmatrix} \tag{3.167}$$

This means that in this approach the slopes and not the angles are the second parameter in the vectors of the equation, see Figure 3.65.

3.6 Matrix Calculus

3.6.1 Introduction

In a paraxial approximation, optical systems behave linearly with reference to the transport of rays and they can be described by a linear collineation. This is valid for geometrical rays as well as for Gaussian beams. In this model when there are no aberrations of higher order and no truncation effects, propagation problems can be described in a linear matrix form, which has become common particularly for Gaussian beams in laser optics. Therefore, the same *ABCD* system matrix with 2×2 dimensions can be applied to different descriptions of paraxial rays [51–67].

If a geometrical ray with position coordinate x in the starting plane and an inclination angle u with the reference axis is considered, it is transformed by a paraxial optical system into a target plane, the data for the ray behind the system can be calculated linearly in the form

$$x' = Ax + Bu$$
$$u' = Cx + Du \tag{3.168}$$

In matrix form this can be written as

$$\begin{pmatrix} x' \\ u' \end{pmatrix} = \begin{pmatrix} A & B \\ C & D \end{pmatrix} \cdot \begin{pmatrix} x \\ u \end{pmatrix} = \underline{M} \cdot \begin{pmatrix} x \\ u \end{pmatrix} \tag{3.169}$$

with the so-called *ABCD* matrix M of the system, see Figure 3.66. For special arrangements the elements of the matrix M have a simple meaning. In optics of course the same is valid by selecting y instead of x as usual to describe the rays in a meridional plane. But due to common literature, in this chapter x is used as transverse coordinate.

The matrix between conjugated image planes is always characterized by $B = 0$. This means, for example, that the image height x' depends only on the object height x and is independent of the angle u. In other words, a whole cone with many angles forms the same image point. This is the classical understanding of imaging, see Chapter 5. Then the lateral magnification is given by

$$A|_{B=0} = x'/x = m \tag{3.170}$$

For an afocal system, the condition $C = 0$ has to be fulfilled. The angle magnification follows from the ratio of the chief ray angles at the start in the pupil plane with $x = 0$

$$D|_{C=0} = u'/u = \Gamma \tag{3.171}$$

For a ray incident parallel to the axis with $u = 0$, the refractive power of the system is given by

$$F = \frac{1}{f} = \frac{u'}{x} = C \tag{3.172}$$

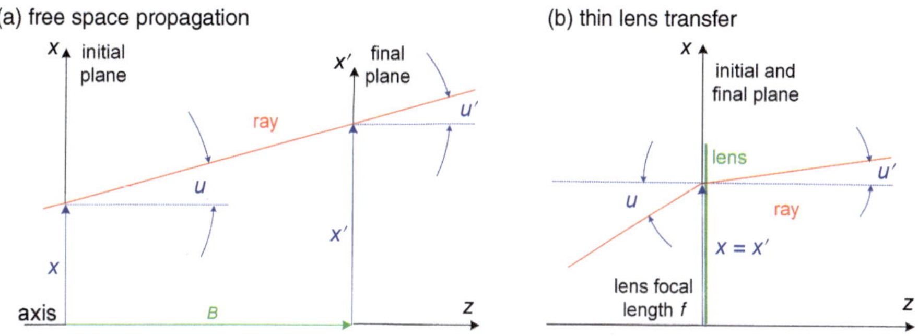

Figure 3.67 Ray operation in matrix calculus for (a) free space propagation as transition into another reference plane and (b) thin lens transfer only impacting the angle.

The simplest examples for the application of the matrix are sketched in Figure 3.67. In (a) a simple free space transition is illustrated with constant angle u and in (b) a thin lens, which changes only the angle and lets the coordinate x be the same.

With s and s', being the intersection lengths of a paraxial marginal ray measured relative to the reference planes of an $ABCD$ system respectively, the change of the distances by means of the $ABCD$ matrix elements is given with (the sign convention for s should be noticed here, see Figure 3.66)

$$s = -x/u, \quad s' = x'/u' \tag{3.173}$$

by dividing the first equation of (3.168) through the second we get

$$s' = \frac{A \cdot s - B}{C \cdot s - D} \tag{3.174}$$

The depth magnification follows from this by differentiation

$$m_z = \frac{ds'}{ds} = -\frac{1}{(C \cdot s - D)^2} \tag{3.175}$$

If a system is located in media n, n' and two additional distance transfers s (negative) and s' are included before and after the $ABCD$ reference planes respectively, the matrix

$$\begin{pmatrix} x' \\ u' \end{pmatrix} = \begin{pmatrix} 1 & s' \\ 0 & 1 \end{pmatrix} \cdot \begin{pmatrix} A & B \\ C & D \end{pmatrix} \cdot \begin{pmatrix} 1 & -s \\ 0 & 1 \end{pmatrix} \cdot \begin{pmatrix} x \\ u \end{pmatrix}$$
$$= \begin{pmatrix} A + s'C & B - sA + s'D - ss'C \\ C & D - sC \end{pmatrix} \cdot \begin{pmatrix} x \\ u \end{pmatrix} \tag{3.176}$$

or

$$x' = (A + s'C) \cdot x + (B - sA + s'D - ss'C) \cdot u$$
$$u' = C \cdot u + (D - sC) \cdot u \tag{3.177}$$

is obtained. Now the cardinal elements of the paraxial system are determined from this equation as follows.

The principal plane positions are found by the condition $x = x'$ independent on u. This gives

$$A + s'C = 1, \quad B + sA + s'D + ss'C = 0 \tag{3.178}$$

which can be resolved with Eq. (3.188) below as

$$s_P = \frac{D - n/n'}{C}, \quad s'_{P'} = \frac{1 - A}{C} \tag{3.179}$$

where the sign of s is switched according to the understanding of the ansatz and a location of the principal plane relative to the entrance reference plane of the $ABCD$ segment. It should be clear that $s'_{P'}$ is measured relative to the exit plane of the segment, s_P relative to the entrance.

- The nodal points are found by the condition $u' = u$ independent on x and x'. Therefore the conditions

  $$D + sC = 1, \quad B + sA + s'D + ss'C = 0 \tag{3.180}$$

  are found and give the nodal plane locations.

  $$s_N = \frac{1 - D}{C}, \quad s'_{N'} = -\frac{A - n/n'}{C} \tag{3.181}$$

- The focal point F' is found by the relation $x' = 0$ for $u = 0$ independent of x and for F we have similar $x = 0$ and $u' = 0$ for all u

  $$A + s'C = 0$$
  $$D + sC = 0 \tag{3.182}$$

therefore, the locations of the focal points are

$$s'_{F'} = -\frac{A}{C}, \quad s_F = -\frac{D}{C} \tag{3.183}$$

The focal lengths and the focal power are given by

$$f' = -\frac{n}{n'C}, \quad f = \frac{1}{C}, \quad F = -C \tag{3.184}$$

If the indices n, n' before and behind the system are identical, the formulas can be simplified with $n/n' = 1$.

3.6.2 Properties of the Matrices

Some important properties of the matrices are listed in the following [51].

- The determinant of the matrix, where n and n' are the refractive indices in front of and behind the system

  $$\det \underline{M} = AD - BC = \frac{n}{n'} \tag{3.185}$$

- When we have equal media, then $AD - BC = 1$. Since this constant relation exists between the matrix elements, there are only three independent parameters for the paraxial system in this case.

  This relation can be interpreted as the validity of the Lagrangian invariant (see Section 3.7). If for a given system the transfer of the marginal and the chief ray are calculated by the same matrix, we get

$$\begin{aligned} y'_M &= Ay_M + Bu_M, \quad u'_M = Cy_M + Du_M \\ y'_C &= Ay_C + Bu_C, \quad u'_C = Cy_C + Du_C \end{aligned} \tag{3.186}$$

- The Lagrange invariant reads in the initial and the final plane

$$L = y_C u_M - y_M u_C, \quad L' = y'_C u'_M - y'_M u'_C \tag{3.187}$$

- By inserting the transfer equation in the primed version of L we get

$$\begin{aligned} L' &= y'_C u'_M - y'_M u'_C \\ &= (Ay_C + Bu_C) \cdot (Cy_M + Du_M) - (Ay_M + Bu_M) \\ &\quad \cdot (Cy_C + Du_C) \\ &= AD \cdot (y_C u_M - y_M u_C) - CD \cdot (y_C u_M - y_M u_C) \\ &= L \cdot (AD - BC) = L \end{aligned} \tag{3.188}$$

  which is only valid if $AD\text{-}BC - 1$.
- The cascading of several systems (note the order) has the resulting matrix

$$\underline{M} = \underline{M}_k \cdot \underline{M}_{k-1} \cdot \ldots \cdot \underline{M}_2 \cdot \underline{M}_1 \tag{3.189}$$

- This means that the individual matrices must be simply multiplied and the total system can again be described in the paraxial approximation by just three parameters.
- The inversion of a system (inversion of the sequence of components) has the matrix

$$\begin{pmatrix} A & B \\ C & D \end{pmatrix}_{\rightarrow} = \frac{n'}{n} \begin{pmatrix} D & B \\ C & A \end{pmatrix}_{\leftarrow} \tag{3.190}$$

- Embedding a system into a medium of refractive index n gives

$$\begin{pmatrix} A & B \\ C & D \end{pmatrix}_{medium} = \begin{pmatrix} A & nB \\ \frac{C}{n} & D \end{pmatrix}_{air} \tag{3.191}$$

- A series of m equal matrices (Silvester's theorem [52, 53]) reads

$$\underline{M}_m = \begin{pmatrix} A & B \\ C & D \end{pmatrix}^m = \frac{1}{\sin\theta} \cdot \begin{pmatrix} A\sin(m\theta) - \sin[(m-1)\theta] & B\sin(m\theta) \\ C\sin(m\theta) & A\sin(m\theta) - \sin[(m-1)\theta] \end{pmatrix} \tag{3.192}$$

  the abbreviated form being

$$\cos\theta = \frac{A+D}{2} \tag{3.193}$$

- This relation is particularly useful for laser resonators and lens ducts.
- The inverse matrix describing the backward propagation of a ray through the same system in air by inversion of the light direction (which should not be confused with 3) is given by

$$\underline{M}^{-1} = \begin{pmatrix} D & -B \\ -C & A \end{pmatrix} \tag{3.194}$$

- This matrix corresponds to the system matrix of the linear Eqs (3.168) if they are resolved to x and u and primed and unprimed coordinates are interchanged.
- Systems which have the same effect in both z directions have certain symmetry properties leading to the fact that the system matrix becomes unimodular with $A = D$. With

$$A = D = \cos\theta \tag{3.195}$$

  and

$$\chi = \sqrt{-\frac{B}{C}} \tag{3.196}$$

  the matrices of such systems can generally be written in the form

$$\underline{M} = \begin{pmatrix} A & B \\ C & D \end{pmatrix} = \begin{pmatrix} \cos\theta & \chi\cdot\sin\theta \\ -\frac{1}{\chi}\cdot\sin\theta & \cos\theta \end{pmatrix} \tag{3.197}$$

- For systems of this kind the Silvester theorem can be written in a simplified form

$$\begin{pmatrix} \cos\theta & \chi\cdot\sin\theta \\ -\frac{1}{\chi}\cdot\sin\theta & \cos\theta \end{pmatrix}^m = \begin{pmatrix} \cos(m\theta) & \chi\cdot\sin(m\theta) \\ -\frac{1}{\chi}\cdot\sin(m\theta) & \cos(m\theta) \end{pmatrix} \tag{3.198}$$

- This means such systems are characterized by only two independent parameters.
- The eikonal or optical path length difference of an $ABCD$ system can be written as

$$L = \frac{Ax_1^2 - 2x_1x_2 + Dx_2^2}{2B} \tag{3.199}$$

- This result can be obtained if the $ABCD$ system is decomposed, as described in the next section, into two transitions and a lens in between with ray height x, the path difference is calculated as

$$L = \sqrt{(x-x_1)^2 + z_1^2} + \sqrt{(x-x_2)^2 + z_2^2} - z_1 - z_2 \tag{3.200}$$

- The roots are paraxial expanded to the quadratic term and the intermediate ray height x is given by

$$\begin{aligned} x &= x_1 + z_1 u_1 = x_2 - z_2 u_2 = x_1 + \frac{D-1}{C} u_1 \\ &= x_2 - \frac{A-1}{C} u_2 \\ u_2 &= C x_1 + D u_1 \end{aligned} \tag{3.201}$$

- where the expressions for z_1 and z_2 are used. If from these relations x and the angles u_1 and u_2 are eliminated, the relation above is obtained.

3.6.3 Matrices of Simple Components

In the following, some elementary ray matrices are given in explicit form.

- Transition over the distance L:

$$\underline{M} = \begin{pmatrix} 1 & L \\ 0 & 1 \end{pmatrix} \tag{3.202}$$

For a plane-parallel plate with refractive index n and thickness t

$$\underline{M} = \begin{pmatrix} 1 & \frac{t}{n} \\ 0 & 1 \end{pmatrix} \tag{3.203}$$

- A thin lens with focal length f':

$$\underline{M} = \begin{pmatrix} 1 & 0 \\ -\frac{1}{f'} & 1 \end{pmatrix} \tag{3.204}$$

- Refraction at a spherical interface with changing refractive indices $n \to n'$ and radius of curvature R:

$$\underline{M} = \begin{pmatrix} 1 & 0 \\ \frac{n-n'}{n'R} & \frac{n}{n'} \end{pmatrix} \tag{3.205}$$

For a plane interface

$$\underline{M} = \begin{pmatrix} 1 & 0 \\ 0 & \frac{n}{n'} \end{pmatrix} \tag{3.206}$$

- A medium with a parabolic radial profile for the refractive index in the form

$$n(r) = n_o \left(1 - \frac{1}{2} n_{r2} r^2\right) \tag{3.207}$$

with constant n_{r2} and length L

$$\underline{M} = \begin{pmatrix} \cos(\sqrt{n_{r2}}L) & \frac{1}{\sqrt{n_{r2}}} \sin(\sqrt{n_{r2}}L) \\ -\sqrt{n_{r2}} \sin(\sqrt{n_{r2}}L) & \cos(\sqrt{n_{r2}}L) \end{pmatrix} \tag{3.208}$$

The approximation for a weak gradient effect $L \cdot \sqrt{n_{r2}} \ll 1$ leads to

$$\underline{M} = \begin{pmatrix} 1 - \frac{1}{2} n_{r2} L^2 & L \\ -n_{r2} L & 1 - \frac{1}{2} n_{r2} L^2 \end{pmatrix} \tag{3.209}$$

- A plane grating with period g between the grooves and of order m

$$\underline{M} = \begin{pmatrix} \frac{\cos\theta_o}{\cos\theta_i} & 0 \\ 0 & \frac{\cos\theta_i}{\cos\theta_o} \end{pmatrix} \tag{3.210}$$

with the grating equation

$$g \cdot (\sin\theta_o - \sin\theta_i) = m\lambda \tag{3.211}$$

- A reflective grating with radius of curvature R and again the grating equation

$$\underline{M} = \begin{pmatrix} \frac{\cos\theta_o}{\cos\theta_i} & 0 \\ -\frac{1}{R} \cdot \left(\frac{1}{\cos\theta_o} + \frac{1}{\cos\theta_i}\right) & \frac{\cos\theta_i}{\cos\theta_o} \end{pmatrix} \tag{3.212}$$

- A spherical mirror surface of radius R:

$$\underline{M} = \begin{pmatrix} 1 & 0 \\ \frac{2}{R} & 1 \end{pmatrix} \tag{3.213}$$

- A retro-reflector or triple mirror

$$\underline{M} = \begin{pmatrix} -1 & 0 \\ 0 & -1 \end{pmatrix} \tag{3.214}$$

- A phase-conjugating mirror

$$\underline{M} = \begin{pmatrix} -1 & 0 \\ 0 & -1 \end{pmatrix} \tag{3.215}$$

- A thick lens in air

$$\begin{aligned} \underline{M} &= \begin{pmatrix} 1 - \frac{n-1}{n} \cdot \frac{t}{R_1} & \frac{t}{n} \\ -(n-1) \cdot \left(\frac{1}{R_1} - \frac{1}{R_2}\right) - \frac{(n-1)^2 \cdot t}{n \cdot R_1 R_2} & 1 + \frac{n-1}{n} \cdot \frac{t}{R_2} \end{pmatrix} \\ &= \begin{pmatrix} 1 - \frac{n-1}{n} \cdot \frac{t}{R_1} & \frac{t}{n} \\ -F & 1 + \frac{n-1}{n} \cdot \frac{t}{R_2} \end{pmatrix} \end{aligned} \tag{3.216}$$

- An afocal telescope with length $L = f_1 + f_2$ and angle magnification Γ

$$\Gamma = -\frac{f_1}{f_2} \tag{3.217}$$

with Kepler $\Gamma > 0$ and Galilei $\Gamma < 0$

$$\underline{M} = \begin{pmatrix} \frac{1}{\Gamma} & L \\ 0 & \Gamma \end{pmatrix} \tag{3.218}$$

- An afocal telescope with focal lengths f_1 and f_2 and telescopic magnification Γ and distances t_1 and t_3 in front of

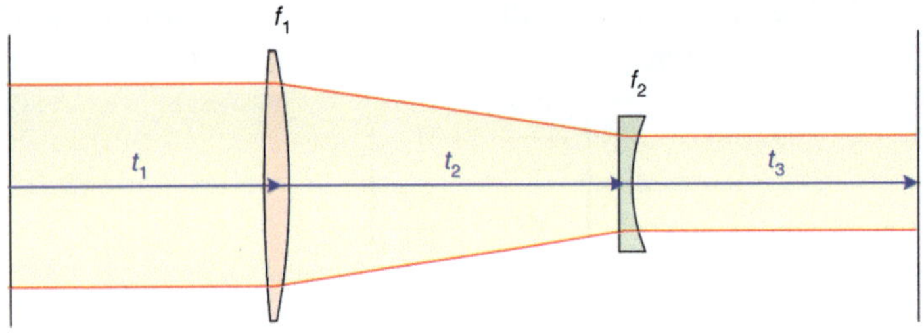

Figure 3.68 Afocal Galilean telescope.

and behind the system (see Figure 3.68)

$$\underline{M} = \begin{pmatrix} -\frac{f_2}{f_1} - t_3 \cdot \left(\frac{1}{f_1} - \frac{1}{f_2}\right) & f_1 + f_2 - t_1 \cdot \frac{f_2}{f_1} - t_3 \cdot \frac{f_1}{f_2} \\ 0 & -\frac{f_1}{f_2} \end{pmatrix} \tag{3.219}$$

In the special case where the starting and target planes are directly in front of and behind the components respectively, $t_1 = t_3 = 0$ the matrix is reduced to Eq. (3.202).

- An afocal Keplerian telescope with focal lengths f_1 and f_2 and setting $B = 0$ in Eq. (3.203) gives in particular for

$$t_1 = f_1, \quad t_3 = f_2 \tag{3.220}$$

a pure magnifier with a diagonal matrix

$$\underline{M} = \begin{pmatrix} -\frac{f_2}{f_1} & 0 \\ 0 & -\frac{f_1}{f_2} \end{pmatrix} = \begin{pmatrix} \Gamma^{-1} & 0 \\ 0 & \Gamma \end{pmatrix} \tag{3.221}$$

We have here a classical Keplerian telescope, starting and ending in the focal planes of the first and second lens respectively, see Figure 3.69. The system is conjugated now as is indicated by the green marginal ray. This is now a classical 4-f-system with a magnification m # −1. By using one negative lens and choosing a Galilean type, the magnifier has $\Gamma > 0$, the system becomes virtual imaging.

- A Fourier transforming setup of a thin lens, starting in the front focal plane and ending in the rear focal plane has the pure anti-diagonal matrix:

$$\underline{M} = \begin{pmatrix} 0 & f \\ -\frac{1}{f} & 0 \end{pmatrix} \tag{3.222}$$

3.6.4 Decompositions of an *ABCD* Matrix

3.6.4.1 Iwasawa Decomposition

In the case of a system with the same medium on both sides $n = n'$ and an arbitrary *ABCD* matrix, the arrangement can be decomposed into three parts [68–71] with different possibilities. Two of these options are given by Iwasawa and can be easily interpreted by optical systems. They are described here.

- For $C \neq 0$ a sequence: transition/lens/transition is described by

$$\underline{M}_{I1} = \begin{pmatrix} A & B \\ C & D \end{pmatrix} = \begin{pmatrix} 1 & z_1 \\ 0 & 1 \end{pmatrix} \cdot \begin{pmatrix} 1 & 0 \\ -\frac{1}{f} & 1 \end{pmatrix} \cdot \begin{pmatrix} 1 & z_2 \\ 0 & 1 \end{pmatrix} \tag{3.223}$$

with the parameters

$$z_1 = \frac{A-1}{C}, \quad z_2 = \frac{D-1}{C}, \quad f = -\frac{1}{C} \tag{3.224}$$

see Figure 3.70.

- For $B \neq 0$ an equivalent sequence of the form lens/transition/lens can be found

$$\underline{M}_{I2} = \begin{pmatrix} A & B \\ C & D \end{pmatrix} = \begin{pmatrix} 1 & 0 \\ -\frac{1}{f_2} & 1 \end{pmatrix} \cdot \begin{pmatrix} 1 & z \\ 0 & 1 \end{pmatrix} \cdot \begin{pmatrix} 1 & 0 \\ -\frac{1}{f_1} & 1 \end{pmatrix} \tag{3.225}$$

with

$$f_1 = -\frac{B}{A-1}, \quad z = B, \quad f_2 = -\frac{B}{D-1} \tag{3.226}$$

see Figure 3.71.

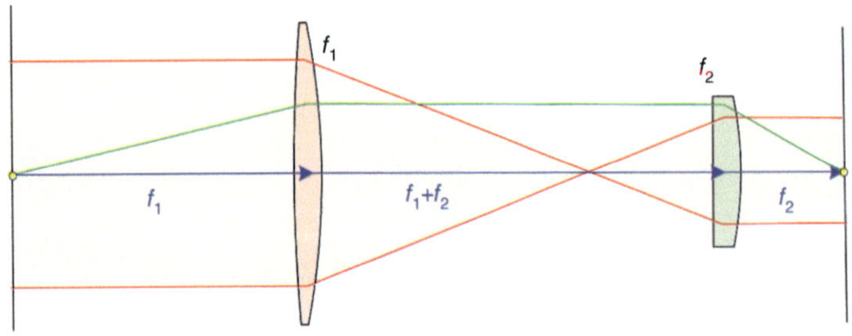

Figure 3.69 Afocal Keplerian telescope versions between conjugated planes as pure magnifier with $B = 0$.

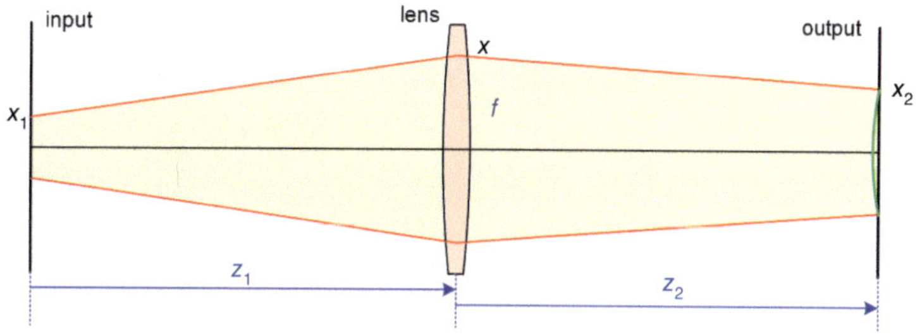

Figure 3.70 Equivalent system consisting of two transitions and a lens.

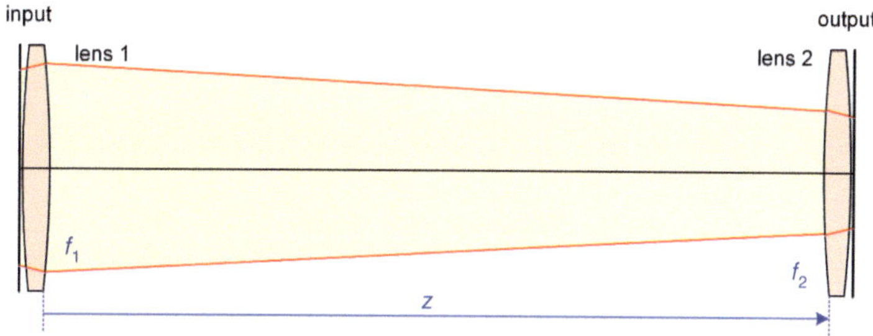

Figure 3.71 Equivalent system consisting of a transition and two lenses.

In a general case, both decompositions are possible. In particular, in the two special cases a restriction exists:

- imaging setup $B = 0$ only I_1 is possible
- afocal setup $C = 0$ only I_2 is possible.

3.6.4.2 Alternative Decompositions

The Iwasawa forms are only special possibilities and easy to interpret. The following options are alternative decompositions [70]. In Eq. (3.211) every line gives one opportunity of decomposition.

$$\begin{aligned}\underline{M} = \begin{pmatrix} A & B \\ C & D \end{pmatrix} &= \begin{pmatrix} 1 & 0 \\ C/A & 1 \end{pmatrix} \cdot \begin{pmatrix} A & 0 \\ 0 & 1/A \end{pmatrix} \cdot \begin{pmatrix} 1 & B/A \\ 0 & 1 \end{pmatrix} \\ &= \begin{pmatrix} 1 & B/D \\ 0 & 1 \end{pmatrix} \cdot \begin{pmatrix} 1/D & 0 \\ 0 & D \end{pmatrix} \cdot \begin{pmatrix} 1 & 0 \\ C/D & 1 \end{pmatrix} \\ &= \begin{pmatrix} B & 0 \\ D & 1/B \end{pmatrix} \cdot \begin{pmatrix} 0 & 1 \\ -1 & 0 \end{pmatrix} \cdot \begin{pmatrix} 1 & 0 \\ A/B & 1 \end{pmatrix} \\ &= \begin{pmatrix} 1 & 0 \\ D/B & 1 \end{pmatrix} \cdot \begin{pmatrix} 0 & B \\ -1/B & 0 \end{pmatrix} \cdot \begin{pmatrix} 1 & 0 \\ A/B & 1 \end{pmatrix} \\ &= \begin{pmatrix} -1/C & -A \\ 0 & -C \end{pmatrix} \cdot \begin{pmatrix} 0 & 1 \\ -1 & 1/A \end{pmatrix} 0 \cdot \begin{pmatrix} 1 & D/C \\ 0 & 1 \end{pmatrix} \\ &= \begin{pmatrix} 1 & A/C \\ 0 & 1 \end{pmatrix} \cdot \begin{pmatrix} 0 & -1/C \\ C & 0 \end{pmatrix} \cdot \begin{pmatrix} 1 & D/C \\ 0 & 1 \end{pmatrix} \end{aligned} \tag{3.227}$$

It should be mentioned that not every one of these possible decompositions can be used in every case. Depending on vanishing matrix elements the versions with a zero in the denominator are not meaningful. If the expressions of the principal planes of an *ABCD* segment in air

$$s_P = \frac{D-1}{C} = \frac{D}{C} + f, \quad s'_P = -\frac{A-1}{C} = -\frac{A}{C} - f \tag{3.228}$$

are used, the last version can be written as

$$\begin{aligned}\underline{M} = \begin{pmatrix} A & B \\ C & D \end{pmatrix} &= \begin{pmatrix} 1 & A/C \\ 0 & 1 \end{pmatrix} \cdot \begin{pmatrix} 0 & -1/C \\ C & 0 \end{pmatrix} \\ &\cdot \begin{pmatrix} 1 & D/C \\ 0 & 1 \end{pmatrix} \\ &= \begin{pmatrix} 1 & -s'_P \\ 0 & 1 \end{pmatrix} \cdot \begin{pmatrix} 1 & -f \\ 0 & 1 \end{pmatrix} \cdot \begin{pmatrix} 0 & f \\ -1/f & 0 \end{pmatrix} \\ &\cdot \begin{pmatrix} 1 & f \\ 0 & 1 \end{pmatrix} \cdot \begin{pmatrix} 1 & s_P \\ 0 & 1 \end{pmatrix} \end{aligned} \tag{3.229}$$

This form corresponds to a so-called UAU decomposition, U stands for upper triangular matrix and A for anti-diagonal. The interpretation of this extended decomposition according to Arsenault [70] is:

- transition into front principal plane;
- transition into front focal plane;
- transfer into rear focal plane by a Fourier sub-system;
- transition into rear principal plane;
- transition into final plane.

The interpretation of the five matrices is illustrated in Figures 3.72 and 3.73(a).

The first decomposition of Eq. (3.211) is in contrast a LDU (L means lower diagonal and D means diagonal) decomposition with a lower triangular Matrix L, see Figure 3.73(b).

$$\begin{aligned}\underline{M} &= \begin{pmatrix} A & B \\ C & D \end{pmatrix} \\ &= \begin{pmatrix} 1 & 0 \\ C/A & 1 \end{pmatrix} \cdot \begin{pmatrix} A & 0 \\ 0 & 1/A \end{pmatrix} \cdot \begin{pmatrix} 1 & B/A \\ 0 & 1 \end{pmatrix} \end{aligned} \tag{3.230}$$

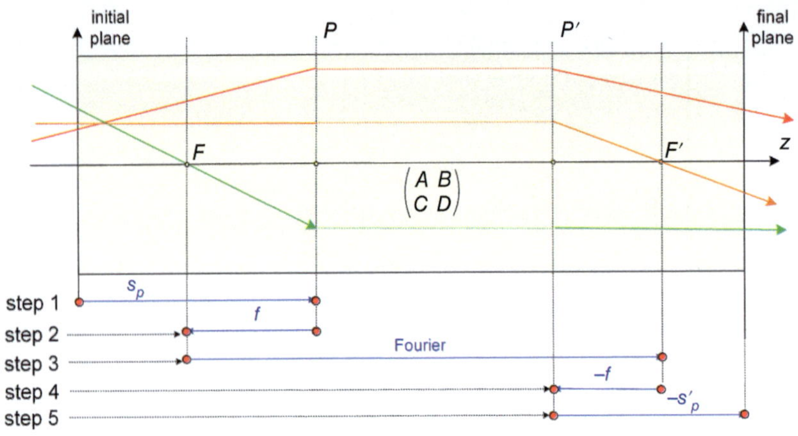

Figure 3.72 Interpretation of the decomposition of Eq. (3.213).

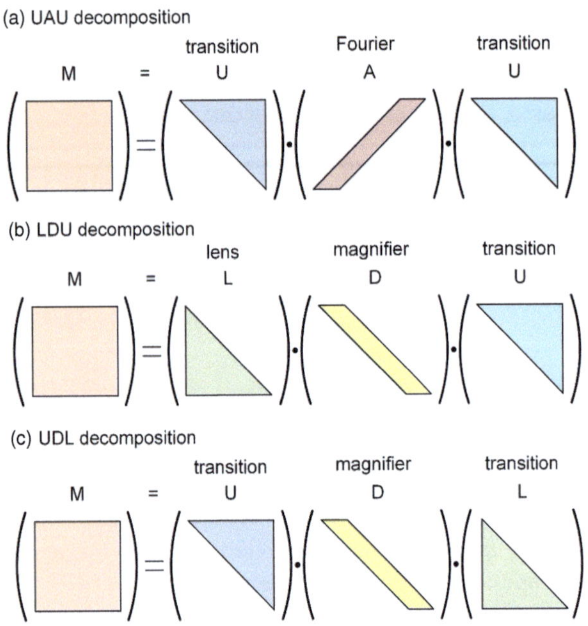

Figure 3.73 (a) UAU and (b) LDU and (c) UDL-decomposition options of an *ABCD* matrix.

Here the first matrix describes a thin lens, the second a magnification or scaling and the last a transition. This corresponds to the group theoretical Wei–Norman decomposition, illustrated in Figure 3.73(b). The second version in the above equation is a UDL decomposition

$$\underline{M} = \begin{pmatrix} A & B \\ C & D \end{pmatrix} = \begin{pmatrix} 1 & B/D \\ 0 & 1 \end{pmatrix} \cdot \begin{pmatrix} 1/D & 0 \\ 0 & D \end{pmatrix} \cdot \begin{pmatrix} 1 & 0 \\ C/D & 1 \end{pmatrix} \tag{3.231}$$

see Figure 3.73(c).

3.6.5 Matrices for Generalized Geometries

3.6.5.1 Two-dimensional 3 × 3 Matrices for Plane-Symmetric Systems*

(*) The content of this section is quite specialized and for advanced reading. Knowledge of this section is not necessary to understand the following chapters. Beginners are recommended to skip this section.

Optical systems without circular symmetry allow a quasi-paraxial description and the use of *ABCD* matrices too, if certain extensions are made. The easiest generalization allows us to include tilt and decentring in the direction of x and can be described by 3×3 matrices. With the help of the 3×1 description vector [61, 65]

$$\vec{r} = \begin{pmatrix} x \\ u \\ 1 \end{pmatrix} = \begin{pmatrix} \vec{e} \\ 1 \end{pmatrix} \tag{3.232}$$

3×3 matrices of the form

$$\underline{M} = \begin{pmatrix} A & B & E \\ C & D & F \\ 0 & 0 & 1 \end{pmatrix} \tag{3.233}$$

with

$$\begin{aligned} x' &= Ax + Bu + E \\ u' &= Cx + Du + F \end{aligned} \tag{3.234}$$

can be defined, which enable the description of decentring and tilting in one coordinate plane. E describes a decentring v and F a tilt through the angle φ in the transversal x–z coordinate plane, see Figure 3.74. Here $\vec{G}$ is the vector containing the centring errors.

$$\vec{G} = \begin{pmatrix} E \\ F \end{pmatrix} = \begin{pmatrix} v \\ \varphi \end{pmatrix} \tag{3.235}$$

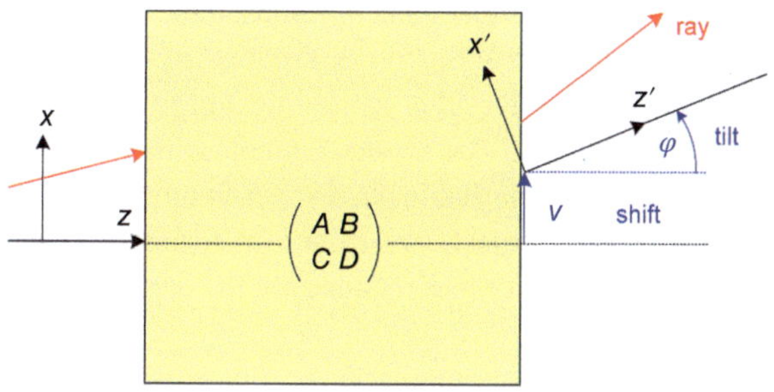

Figure 3.74 Paraxial ray transport for the non-centred case.

Using this abbreviation one gets the following representation for the ray transformation

$$\vec{r}' = \begin{pmatrix} \underline{M} & \vec{G} \\ \vec{0}^T & 1 \end{pmatrix} \cdot \vec{r} \tag{3.236}$$

The determinate of the extended 3×3 matrix is again $AD\text{-}BC = n/n'$ as in the 2×2 case. For inversion of the system, the 3×3-matrix has to be inverted. The condition with retro-reflector matrix and unity matrix $\underline{I}$

$$\underset{\leftarrow}{\underline{M}} \cdot \underline{M}_{Retro} \cdot \underset{\rightarrow}{\underline{M}} = \underline{I} \tag{3.237}$$

which leads to (assuming $AD\text{-}BC = 1$ and therefore equal media in front of and behind the system)

$$\underset{\leftarrow}{\underline{M}} = \begin{pmatrix} D & B & BF - DE \\ C & A & AF - CE \\ 0 & 0 & 1 \end{pmatrix} \tag{3.238}$$

For the sequencing of two 3×3-matrices we have in analogy to the 2×2 case simply

$$\underline{M} = \underline{M}_2 \cdot \underline{M}_1 \tag{3.239}$$

with

$$\begin{aligned} A &= A_2A_1 + B_2C_1, & B &= A_2B_1 + B_2D_1 \\ C &= C_2A_1 + D_2C_1, & D &= C_2B_1 + D_2D_1 \\ E &= A_2E_1 + B_2F_1 + E_2, & F &= C_2E_1 + D_2F_1 + F_2 \end{aligned} \tag{3.240}$$

Special 3×3 matrices for simple setups are

- Decentering by v

$$\underline{M} = \begin{pmatrix} 1 & 0 & v \\ 0 & 1 & 0 \\ 0 & 0 & 1 \end{pmatrix} \tag{3.241}$$

- Tilt by the angle φ

$$\underline{M} = \begin{pmatrix} 1 & 0 & 0 \\ 0 & 1 & \varphi \\ 0 & 0 & 1 \end{pmatrix} \tag{3.242}$$

- *ABCD* system decentred by the value v

$$\underline{M} = \begin{pmatrix} A & B & v \cdot (1 - A) \\ C & D & -v \cdot C \\ 0 & 0 & 1 \end{pmatrix} \tag{3.243}$$

- *ABCD* system tilted by the angle φ

$$\underline{M} = \begin{pmatrix} A & B & -\varphi \cdot B \\ C & D & \varphi \cdot (1 - D) \\ 0 & 0 & 1 \end{pmatrix} \tag{3.244}$$

- Matrix of a medium with a non-rotational-symmetric gradient profile of length L with the function of the refractive index

$$n(x) = n_0 \cdot (1 + n_1x - n_2x^2) \tag{3.245}$$

gives [62]

$$\underline{M} = \begin{pmatrix} \cos(L\sqrt{2n_2}) & \frac{1}{\sqrt{2n_2}} \sin(L\sqrt{2n_2}) & \frac{n_1}{2n_2}[1 - \cos(L\sqrt{2n_2})] \\ -\sqrt{2n_2} \sin(L\sqrt{2n_2}) & \cos(L\sqrt{2n_2}) & \frac{n_1}{\sqrt{2n_2}} \sin(L\sqrt{2n_2}) \\ 0 & 0 & 1 \end{pmatrix} \tag{3.246}$$

- Matrix of a divergent axicon with a tilting angle θ of the cone surface ($\theta = 0$ indicates a plane surface, see Chapter 4 for the discussion of an axicon) and the auxiliary quantity [63]

$$\eta = \arcsin(n \sin \theta) - \theta \tag{3.247}$$

is given by

$$\underline{M} = \begin{pmatrix} 1 - \tan \theta \cdot \tan \eta & 0 & 0 \\ 0 & 1 & \pm\eta \\ 0 & 0 & 1 \end{pmatrix} \tag{3.248}$$

with signs valid for the upper and lower half of the x-coordinate plane, respectively.

For this component the determinant is

$$\det \underline{M} = 1 - \tan \eta \cdot \tan \theta \tag{3.249}$$

i.e. not equal to one. So an axicon describes a non-linear system which changes the quality of an incident beam. This seems to be a contradiction, but here the ray path is formally introduced into the matrix form although the system is not paraxial. So the assumption of the *ABCD* approach is violated.

- Matrix of a convergent axicon with a tilting angle θ of the cone surface and thickness L in the middle

$$\underline{M} = \begin{pmatrix} 1 - \tan\theta \cdot \tan\eta & 0 & \pm L \cdot \tan\eta \\ 0 & 1 & \pm\eta \\ 0 & 0 & 1 \end{pmatrix} \tag{3.250}$$

  with the signs valid for the upper and lower half, respectively.

3.6.5.2 Centered 4 × 4 Matrices

In the more general case of an astigmatic and non-isotropic system but with an axis ray and all components aligned on it, a paraxial matrix representation for the calculation of ray transformations can be formulated by a 4 × 4 approach. The formulation with 4 × 4 matrices here refers to a reference ray in a general geometry. The reference ray itself can pass through an optical system in an arbitrary way with aberrations. In this sense, the matrix transfer describes a vicinity around a reference or base ray. The application of this approach can be used, for example, to formulate parabasal or differential rays in optical systems and the propagation for general astigmatism (GA), see Figure 3.75 for illustration.

The vector of coordinates and directions necessary for the description of a system (version I of the notation) [53, 67] is given by

$$\vec{r} = \begin{pmatrix} x & y & u & v \end{pmatrix}^T \tag{3.251}$$

The linear transformation through an optical system can be written in the form

$$\vec{r}\,' = \underline{M} \cdot \vec{r} = \begin{pmatrix} A_{xx} & A_{xy} & B_{xx} & B_{xy} \\ A_{yx} & A_{yy} & B_{yx} & B_{yy} \\ C_{xx} & C_{xy} & D_{xx} & D_{xy} \\ C_{yx} & C_{yy} & D_{yx} & D_{yy} \end{pmatrix} \cdot \vec{r} \tag{3.252}$$

The single matrix elements with indices xx and yy take effects in the x and y cross sections into account, respectively, and have the same meaning as in the one-dimensional case. The mixed elements with indices xy and yx couple both coordinate planes to each other. Systems where these parts vanish are called orthogonally astigmatic, the two coordinate directions are decoupled in this case.

The 4 × 4 matrices are symplectic, i.e. there are additional six conditions for the 16 matrix elements. Thus, only 10 elements are independent. In the mathematical formulation with the unit or identity matrix $\underline{I}$ and the zero matrix $\underline{O}$ the condition for a symplectic matrix $\underline{M}$ is

$$\underline{M} \cdot \begin{pmatrix} \underline{O} & \underline{I} \\ -\underline{I} & \underline{O} \end{pmatrix} \cdot \underline{M}^T = \begin{pmatrix} \underline{O} & \underline{I} \\ -\underline{I} & \underline{O} \end{pmatrix} \tag{3.253}$$

Consequently. there are three different cases. Beneath the general case two additional setups with remaining symmetry have a smaller number of parameters.

- The general case with ten parameters: general astigmatism (*GA*) and full matrix

$$\underline{M}_{GA} = \begin{pmatrix} A_{xx} & A_{xy} & B_{xx} & B_{xy} \\ A_{yx} & A_{yy} & B_{yx} & B_{yy} \\ C_{xx} & C_{xy} & D_{xx} & D_{xy} \\ C_{yx} & C_{yy} & D_{yx} & D_{yy} \end{pmatrix} \tag{3.254}$$

- The case with six parameters without x-y couplings, respectively: symmetrical orthogonal astigmatism (*SA*)

$$\underline{M}_{SA} = \begin{pmatrix} A_x & 0 & B_x & 0 \\ 0 & A_y & 0 & B_y \\ C_x & 0 & D_x & 0 \\ 0 & C_y & 0 & D_y \end{pmatrix} \tag{3.255}$$

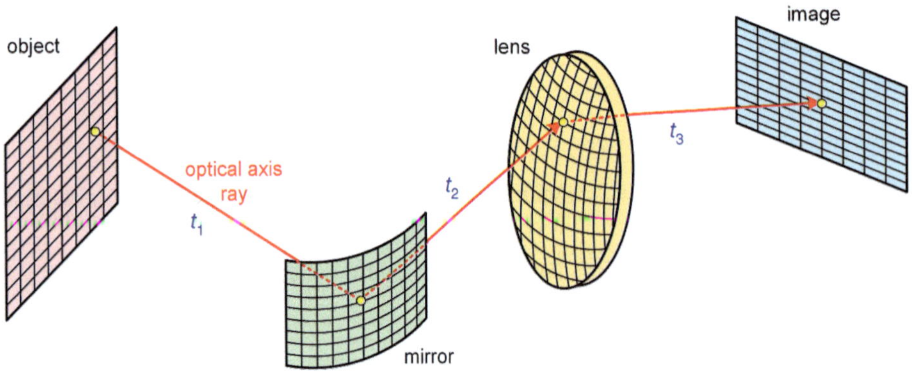

Figure 3.75 General geometry of a system with optical axis ray.

- Rotational-symmetric case with three parameters: stigmatic (ST)

$$\underline{M}_{ST} = \begin{pmatrix} A & 0 & B & 0 \\ 0 & A & 0 & B \\ C & 0 & D & 0 \\ 0 & C & 0 & D \end{pmatrix} \tag{3.256}$$

The 4×4 matrix formalism can be written in a simplified way. This creates a greater similarity to the 2×2-formalism using submatrices and subvectors as elements

$$\begin{pmatrix} \vec{r}' \\ \vec{u}' \end{pmatrix} = \begin{pmatrix} \underline{A} & \underline{B} \\ \underline{C} & \underline{D} \end{pmatrix} \cdot \begin{pmatrix} \vec{r} \\ \vec{u} \end{pmatrix} \tag{3.257}$$

where $\vec{r}$ represents the pure position vector consisting of an x and y component and $\vec{u}$ contains the angles u and v. By converting the matrix equation above, the following equation results, which allows the calculation of the angles from the coordinates

$$\begin{pmatrix} \vec{u} \\ \vec{u}' \end{pmatrix} = \begin{pmatrix} -\underline{B}^{-1}\underline{A} & \underline{B}^{-1} \\ \underline{C} - \underline{D}\underline{B}^{-1}\underline{A} & \underline{D}\underline{B}^{-1} \end{pmatrix} \cdot \begin{pmatrix} \vec{r} \\ \vec{r}' \end{pmatrix} \tag{3.258}$$

Another alternative is to write the ray vector in the following form sorting the variables in the two cross section planes

$$\vec{r} = \begin{pmatrix} x & u & y & v \end{pmatrix}^T \tag{3.259}$$

Then the matrix M has a coordinate related block form

$$\vec{r}' = \underline{M} \cdot \vec{r} = \begin{pmatrix} A_{xx} & B_{xx} & A_{xy} & B_{xy} \\ C_{xx} & D_{xx} & C_{xy} & D_{xy} \\ A_{yx} & B_{yx} & A_{yy} & B_{yy} \\ C_{yx} & D_{yx} & C_{yy} & D_{yy} \end{pmatrix} \cdot \vec{r} = \begin{pmatrix} \underline{M}_{xx} & \underline{M}_{xy} \\ \underline{M}_{yx} & \underline{M}_{yy} \end{pmatrix} \cdot \vec{r} \tag{3.260}$$

3.6.5.3 Special 4 × 4 Matrices

In the following, some special 4×4 matrix forms are given.

- Rotation matrix M_R describing a rotation with reference to the azimuth by an angle φ

$$\underline{M}_R = \begin{pmatrix} \cos\varphi & \sin\varphi & 0 & 0 \\ -\sin\varphi & \cos\varphi & 0 & 0 \\ 0 & 0 & \cos\varphi & \sin\varphi \\ 0 & 0 & -\sin\varphi & \cos\varphi \end{pmatrix} \tag{3.261}$$

- Matrix of a cylindrical lens with focal length f_x rotated by an angle φ to the azimuth

$$\underline{M}_{Z\phi} = \begin{pmatrix} 1 & 0 & 0 & 0 \\ 0 & 1 & 0 & 0 \\ \dfrac{-\cos^2\varphi}{f_x} & \dfrac{-\sin 2\varphi}{2f_x} & 1 & 0 \\ \dfrac{-\sin 2\varphi}{2f_x} & \dfrac{-\sin^2\varphi}{f_x} & 0 & 1 \end{pmatrix} \tag{3.262}$$

- Matrix of a plane plate with the thickness t rotated by an angle φ

$$\underline{M}_{t\varphi} = \begin{pmatrix} 1 & \dfrac{t \cdot n^2 \cdot \cos^2\varphi}{(n^2 - \sin^2\kappa)^{\frac{3}{2}}} & 0 & 0 \\ 0 & 1 & 0 & 0 \\ 0 & 0 & 1 & \dfrac{t}{\sqrt{n^2 - \sin^2\varphi}} \\ 0 & 0 & 0 & 1 \end{pmatrix} \tag{3.263}$$

The plate has an optical path length of size

$$L = \frac{n \cdot t}{\sqrt{n^2 - \sin^2\varphi}} \tag{3.264}$$

and displaces the ray laterally by the value

$$y = t \cdot \sin\varphi \cdot \left(\frac{\cos\varphi}{\sqrt{n^2 - \sin^2\varphi}} - 1 \right) \tag{3.265}$$

Specifically in the case of the Brewster condition

$$n = \tan\varphi \tag{3.266}$$

we have

$$L = \frac{\sqrt{n^2 + 1} \cdot t}{n} \tag{3.267}$$

and

$$\underline{M}^B_{t\phi} = \begin{pmatrix} 1 & \dfrac{t \cdot \sqrt{n^2+1}}{n^4} & 0 & 0 \\ 0 & 1 & 0 & 0 \\ 0 & 0 & 1 & \dfrac{t \cdot \sqrt{n^2+1}}{n^2} \\ 0 & 0 & 0 & 1 \end{pmatrix} \tag{3.268}$$

- Matrix of a dispersion prism with path length L in glass, basic angle α and incident angle θ_1

$$\underline{M}_{\text{Pr is}} = \begin{pmatrix} \dfrac{\cos\theta_2\cos\theta_4}{\cos\theta_1\cos\theta_3} & \dfrac{L\cdot\cos\theta_1\cos\theta_4}{n\cos\theta_2\cos\theta_3} & 0 & 0 \\ 0 & \dfrac{\cos\theta_1\cos\theta_3}{\cos\theta_2\cos\theta_4} & 0 & 0 \\ 0 & 0 & 1 & \dfrac{L}{n} \\ 0 & 0 & 0 & 1 \end{pmatrix} \tag{3.269}$$

with the auxiliary angles (incident and exit angles at the surfaces):

$$\begin{aligned} \sin\theta_2 &= \frac{\sin\theta_1}{n} \\ \theta_3 &= \alpha - \theta_2 \\ \sin\theta_4 &= n\sin\theta_3 \end{aligned} \tag{3.270}$$

- Matrix of a toric lens with refractive powers F_x and F_y rotated by an angle φ in the azimuth

$$\underline{M}_{\text{T}\phi} = \underline{M}_D^{-1}\cdot\underline{M}_T\cdot\underline{M}_D$$
$$= \begin{pmatrix} 1 & 0 & 0 & 0 \\ 0 & 1 & 0 & 0 \\ -F_x\cos^2\varphi - F_y\sin^2\varphi & (F_y - F_x)\cos\varphi\,\sin\varphi & 1 & 0 \\ (F_y - F_x)\cos\varphi\,\sin\varphi & -F_x\sin^2\varphi - F_y\cos^2\varphi & 0 & 1 \end{pmatrix} \tag{3.271}$$

- Matrix of a tilted spherical surface with refractive indices n_1 and n_2, main radii of curvature R_{H1}, R_{H2} and incident angles θ_1 and θ_2, [53]

- Matrix of a combination of three closely arranged cylindrical lenses with focal lengths f, $f/2$ and f rotated by the angles φ, $-\varphi$ and φ in the azimuth

$$\underline{M} = \begin{pmatrix} 1 & 0 & 0 & 0 \\ 0 & 1 & 0 & 0 \\ \dfrac{-4\cos^2\varphi}{f} & 0 & 1 & 0 \\ 0 & \dfrac{-4\sin^2\varphi}{f} & 0 & 1 \end{pmatrix} \tag{3.274}$$

This is a system with variable cylindrical effect; the astigmatism can be adjusted by changing the angle φ.

3.6.5.4 Decomposition of a 4 × 4 Matrix

If a general 4 × 4 matrix is considered, it is meaningless to perform a classical LDU or UAU decomposition. But it makes sense from the view of the optical interpretation to realize a block-decomposition; for example, the special LDU-type

$$\underline{M}_{LDU} = \begin{pmatrix} A_{xx} & A_{xy} & B_{xx} & B_{xy} \\ A_{yx} & A_{yy} & B_{yx} & B_{yy} \\ C_{xx} & C_{xy} & D_{xx} & D_{xy} \\ C_{yx} & C_{yy} & D_{yx} & D_{yy} \end{pmatrix} = \begin{pmatrix} \underline{A} & \underline{B} \\ \underline{C} & \underline{D} \end{pmatrix}$$
$$= \begin{pmatrix} \underline{I} & \underline{0} \\ \underline{CA}^{-1} & \underline{I} \end{pmatrix}\cdot\begin{pmatrix} \underline{A} & \underline{0} \\ \underline{0} & \underline{A}^{-1} \end{pmatrix}\cdot\begin{pmatrix} \underline{I} & \underline{BA}^{-1} \\ \underline{0} & \underline{I} \end{pmatrix} \tag{3.275}$$

$$\underline{M}_{\text{Re}\,fr} = \begin{pmatrix} \dfrac{\cos\theta_2}{\cos\theta_1} & 0 & 0 & 0 \\ 0 & 1 & 0 & 0 \\ \dfrac{n_1\cos\theta_1 - n_2\cos\theta_2}{n_2R_{H1}\cdot\cos\theta_1\cos\theta_2} & 0 & \dfrac{n_1\cos\theta_1}{n_2\cos\theta_2} & 0 \\ 0 & \dfrac{n_1\cos\theta_1 - n_2\cos\theta_2}{n_2R_{H2}} & 0 & \dfrac{n_1}{n_2} \end{pmatrix} \tag{3.272}$$

- Matrix of a tilted spherical mirror with main curvature radii R_{H1}, R_{H2} and tilt angle θ_1

$$\underline{M}_{mirror} = \begin{pmatrix} 1 & 0 & 0 & 0 \\ 0 & 1 & 0 & 0 \\ \dfrac{2}{R_{H1}\cos\theta_{12}} & 0 & 1 & 0 \\ 0 & \dfrac{2\cos\theta_1}{R_{H2}} & 0 & 1 \end{pmatrix} \tag{3.273}$$

or the UAU form

$$\underline{M}_{LDU} = \begin{pmatrix} A_{xx} & A_{xy} & B_{xx} & B_{xy} \\ A_{yx} & A_{yy} & B_{yx} & B_{yy} \\ C_{xx} & C_{xy} & D_{xx} & D_{xy} \\ C_{yx} & C_{yy} & D_{yx} & D_{yy} \end{pmatrix} = \begin{pmatrix} \underline{A} & \underline{B} \\ \underline{C} & \underline{D} \end{pmatrix}$$
$$= \begin{pmatrix} \underline{I} & \underline{AC}^{-1} \\ \underline{0} & \underline{I} \end{pmatrix}\cdot\begin{pmatrix} \underline{0} & -\underline{C}^{-1} \\ \underline{C} & \underline{0} \end{pmatrix}\cdot\begin{pmatrix} \underline{I} & \underline{DC}^{-1} \\ \underline{0} & \underline{I} \end{pmatrix} \tag{3.276}$$

with zero matrix $\underline{O}$ and unity diagonal matrix $\underline{I}$. In the generalized case of two transverse dimensions, in addition an azimuthal rotation matrix comes into the game. This idea can be used to transform the focal power sub-matrix C into diagonal form with its eigenvalues as focal powers under a certain angle φ. If the general matrix is considered in a rotated coordinate system, we get the matrix

$$\underline{M}_{\varphi} = \underline{M}_R^{-1} \cdot \underline{M} \cdot \underline{M}_R$$
$$= \begin{pmatrix} \cos\varphi & -\sin\varphi & 0 & 0 \\ \sin\varphi & \cos\varphi & 0 & 0 \\ 0 & 0 & \cos\varphi & -\sin\varphi \\ 0 & 0 & \sin\varphi & \cos\varphi \end{pmatrix} \begin{pmatrix} A_{xx} & A_{xy} & B_{xx} & B_{xy} \\ A_{xy} & A_{yy} & B_{xy} & B_{yy} \\ C_{xx} & C_{xy} & D_{xx} & D_{xy} \\ C_{xy} & C_{yy} & D_{xy} & D_{yy} \end{pmatrix} \begin{pmatrix} \cos\varphi & \sin\varphi & 0 & 0 \\ -\sin\varphi & \cos\varphi & 0 & 0 \\ 0 & 0 & \cos\varphi & \sin\varphi \\ 0 & 0 & -\sin\varphi & \cos\varphi \end{pmatrix}$$
$$= \begin{pmatrix} A_{xx}\cos^2\varphi + A_{yy}\sin^2\varphi - A_{xy}\sin 2\varphi & \frac{1}{2}(A_{xx} - A_{yy})\sin 2\varphi + A_{xy}\cos 2\varphi & \dots & \dots \\ \frac{1}{2}(A_{xx} - A_{yy})\sin 2\varphi + A_{xy}\cos 2\varphi & A_{xx}\sin^2\varphi + A_{yy}\cos^2\varphi + A_{xy}\sin 2\varphi & \dots & \dots \\ C_{xx}\cos^2\varphi + C_{yy}\sin^2\varphi - C_{xy}\sin 2\varphi & \frac{1}{2}(C_{xx} - C_{yy})\sin 2\varphi + C_{xy}\cos 2\varphi & \dots & \dots \\ \frac{1}{2}(C_{xx} - C_{yy})\sin 2\varphi + C_{xy}\cos 2\varphi & C_{xx}\sin^2\varphi + C_{yy}\cos^2\varphi + C_{xy}\sin 2\varphi & \dots & \dots \end{pmatrix} \quad (3.277)$$

The block parts with B and D have the same structure. It is seen that with a special angle every block can be made diagonal individually. In particular for the most interesting case of the C-block, this angle reads

$$\tan 2\varphi_C = \frac{-2C_{xy}}{C_{xx} - C_{yy}} \quad (3.278)$$

With this rotation angle, the matrix for the C-block reads

$$\underline{M}_{\varphi} = \begin{pmatrix} \dots & \dots & \dots & \dots \\ \dots & \dots & \dots & \dots \\ \frac{C_{xx}+C_{yy}}{2} + \sqrt{\left(\frac{C_{xx}-C_{yy}}{2}\right)^2 + C_{xy}^2} & 0 & \dots & \dots \\ 0 & \frac{C_{xx}+C_{yy}}{2} - \sqrt{\left(\frac{C_{xx}-C_{yy}}{2}\right)^2 + C_{xy}^2} & \dots & \dots \end{pmatrix} \quad (3.279)$$

The two powers in the diagonal form of the C-block are now the eigenvalues of the system

$$C_{p1/2} = \frac{C_{xx}+C_{yy}}{2} \pm \sqrt{\left(\frac{C_{xx}-C_{yy}}{2}\right)^2 + C_{xy}^2} \quad (3.280)$$

corresponding to the principal powers of the system.

3.6.5.5 General 5 × 5 Matrices*

(*) The content of this section is quite specialized and for advanced reading. Knowledge of this section is not necessary to understand the following chapters. Beginners are recommended to skip this section.

If we combine the concept of the 3 × 3 matrix idea with tilt and decentring (also taking into account a non-centred reference system or reference ray) with the general 4 × 4 matrix formulation (where the two coordinate directions are distinguished), one obtains the 5 × 5 matrix representation which describes the most general case. Defining the vector of position and direction necessary for the description of a ray as a five vector

$$\vec{r} = (x \;\; y \;\; u \;\; v \;\; 1)^T \quad (3.281)$$

a linear transformation through an optical system can be written in the form

$$\vec{r}' = \underline{M} \cdot \vec{r} = \begin{pmatrix} A_{xx} & A_{xy} & B_{xx} & B_{xy} & E_x \\ A_{yx} & A_{yy} & B_{yx} & B_{yy} & E_y \\ C_{xx} & C_{xy} & D_{xx} & D_{xy} & F_x \\ C_{yx} & C_{yy} & D_{yx} & D_{yy} & F_y \\ 0 & 0 & 0 & 0 & 1 \end{pmatrix} \cdot \vec{r} \quad (3.282)$$

In analogy with the 3 × 3 case the matrix elements E and F describe the decentring and tilting in the corresponding coordinate directions. In the block representation with the relevant submatrices and subvectors, this formulation can also be expressed as

$$\underline{M} = \begin{pmatrix} \underline{A} & \underline{B} & \vec{E} \\ \underline{C} & \underline{D} & \vec{F} \\ \vec{0} & \vec{0} & 1 \end{pmatrix} \tag{3.283}$$

3.7 Helmholtz–Lagrange Invariant

3.7.1 Introduction

In paraxial imaging theory there are some system invariants that have the same value in front, within and behind the system, i.e. they are constant independent of the position inside the system.

The most important of these constants is the Helmholtz–Lagrange invariant L that corresponds to the conservation of energy. The invariant according to Helmholtz–Lagrange which is also called etendue of space-bandwidth product is given in one transverse dimension with image height y and aperture angle u by

$$L = n \cdot u \cdot y = n' \cdot u'y' = y \cdot NA = y' \cdot NA' \tag{3.284}$$

L is a paraxial quantity and gives a measure of how much light and information, respectively, can be transported through a system. Therefore, it is also called the linear geometrical light throughput. The derivation is carried out very easily at a single lens or for a single refracting surface. Inserting the invariant into the definition of the lateral magnification one obtains the magnification as a function of the aperture angles u and u'

$$m = \frac{y'}{y} = \frac{n \cdot u}{n' \cdot u'} \tag{3.285}$$

By re-arrangement of the primed and unprimed variables on the same side, Eq. (3.268) is obtained. Figure 3.76 illustrates the derivation of the invariant for (a) a single surface and (b) a single thin lens, the pupil is assumed to be at the lens. For the surface case, the required equations are obtained from the geometry and the refraction and read

$$h = -u \cdot s = -u' \cdot s', \quad w = \frac{y}{s} = \frac{y'}{s'} \tag{3.286}$$

so we get by elimination of h and w

$$m = \frac{y'}{y} = \frac{n \cdot u}{n' \cdot u'} = \frac{n}{n'} \cdot \frac{s'}{s} \tag{3.287}$$

In the point Q the law of refraction for the marginal ray gives the relation between u and u', the special chief ray crossing the centre of curvature C is not refracted and delivers the image height y'. This corresponds to the assumption of the pupil location at C. In case of the lens in Figure 3.76(b), the pupil is assumed to be at the lens and the nodal ray is used to determine y'.

In Figure 3.77 the corresponding invariance behaviour is illustrated for a composed two-lens system. A larger field height y results in smaller aperture angle u and vice versa. Furthermore it is shown that in the so-called phase space

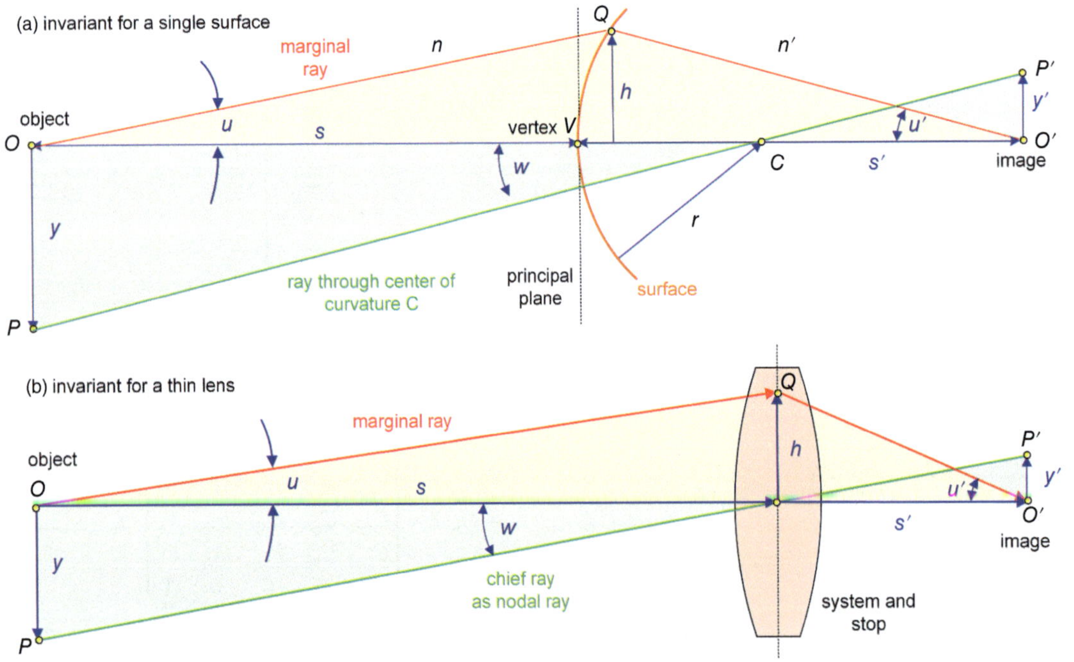

Figure 3.76 Helmholtz–Lagrange invariant for (a) a single surface and (b) a single thin lens.

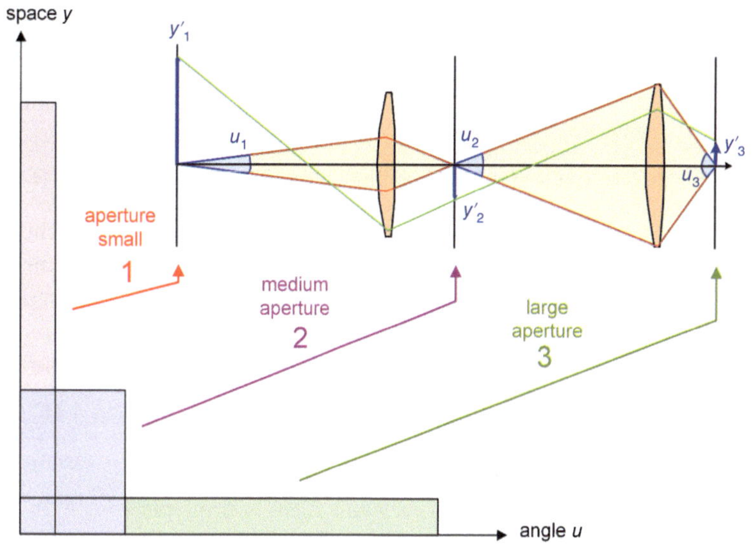

Figure 3.77 Helmholtz–Lagrange invariant for a cascaded system made of two lenses and interpretation in the phase space.

as a representation of the light as a function of the spatial width y (height of the chief ray here) and the angular extension (numerical aperture angle of the marginal ray u) the invariant can be interpreted as a constant rectangular area (see Chapter 10 for more details). Every paraxial lens or system only changes the shape of the rectangle, keeping the area size constant.

The interpretation of the Lagrange invariant can be formulated as:

- conservation of energy;
- constant transfer of information;
- invariant space-bandwidth product.

It should be noted that the simple formulation as a product of numerical aperture and field size assumes conjugated planes for the definition of the ray heights y and y'. This means that the y, y' are ideal image heights and are defined in an image plane only. In the case of vignetting or truncation of rays, the conservation of energy and therefore the invariance of L is not necessarily fulfilled. In a more general context, several generalizations of the simple definition of L make sense. These are the following.

- Definition in an arbitrary z plane.
- Extension to two transverse dimensions.
- Arbitrary cross section areas of the ray bundle.
- Arbitrary light energy density distributions across the light bundle needs to be generalized for the preservation of energy. This approach requires the definition of an energy density in the phase space, which is the Wigner function. Details about this approach are explained in Chapter 10. This extension generalizes the geometrically described energetic definition to a more physical one.
- Validity to non-paraxial larger angles u; this is related to the Abbe sine condition.

It should be mentioned here that the Lagrange invariant in case of a matrix $ABCD$ description reads

$$AD - BC = \frac{n}{n'} \tag{3.288}$$

If p is denoted as the distance between object and entrance pupil according to Figure 3.78, it is possible to re-arrange the Lagrange invariant with

$$NA = n \cdot \frac{D_{EnP}}{2p}, \quad w = \frac{y_{obj}}{p} \tag{3.289}$$

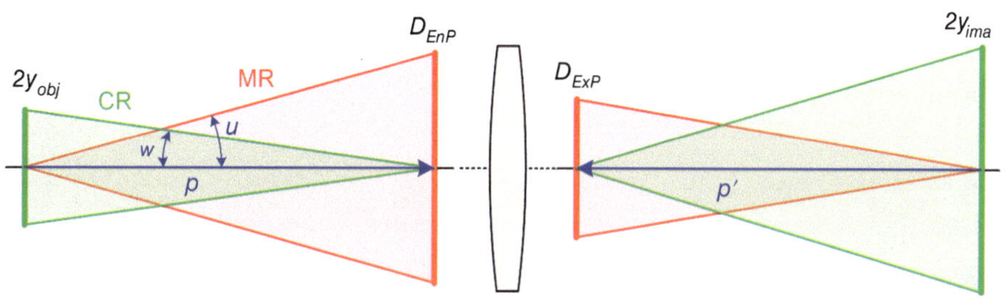

Figure 3.78 Field and pupil planes for a single lens system.

in the form

$$L = y \cdot NA = y_{obj} \cdot n \frac{D_{EnP}}{2p} = y_{obj} \cdot n \frac{w \cdot D_{EnP}}{2y_{obj}}$$
$$= n \cdot w \cdot \frac{D_{EnP}}{2} \quad (3.290)$$

This means first that the value of the Lagrange invariants for the object and the pupil imaging are the same; secondly this expression can be used to quantify the Lagrange invariant in case of an afocal system by using the field angle w instead of the field height y and the size of the entrance pupil D_{EnP} instead of NA, see Figure 3.78.

3.7.2 Lagrange Invariant for Arbitrary z-Positions

In the classical definition of the Helmholtz–Lagrange invariant, the values of field sizes y and y' are in principle measured at different conjugate positions inside the optical system. To get the opportunity to determine the value of L at any arbitrary location, the change of marginal and chief ray height must be taken into account simultaneously. For a general z position the Lagrange invariant in the geometry of Figure 3.79 the triangles SPB, ABO′ and SQA must be investigated for an arbitrary coordinate z. By eliminating the angles and formulating the invariant by the ray heights $y_{CR}(z)$ and $y_{MR}(z)$ only, the expression

$$L = n' \cdot [u' \cdot y_{CR}(z) - w' \cdot y_{MR}(z)] \quad (3.291)$$

is obtained. If $z = 0$ in the image plane, with $y_{MR} = 0$ the usual definition is obtained as a special case. For the second special choice $z = p$ the afocal version of the invariant is obtained with $y_{CR} = 0$.

3.7.3 Generalized Etendue Definitions*

(*) The content of this section is quite specialized and for advanced reading. Knowledge of this section is not necessary to understand the following chapters. Beginners are recommended to skip this section.

The next generalization to characterize an optical system by the Helmholtz–Lagrange invariant is to take two lateral dimensions into account. If double plane symmetry of the system is assumed, L can be written with the cross section area A and the squared numerical aperture or by using the solid angle Ω as [72]

$$L_{parax} = A \cdot NA^2, \quad L_{parax} = A \cdot \Omega \quad (3.292)$$

Figure 3.80 shows the corresponding definition of the terms. In detail it should be distinguished between the cases, where field size and aperture are circular symmetric, squared in shape or maybe rectangular. This depends on the properties of the system. The formulation with A and Ω is the most general one.

This definition still assumes paraxiality and a uniform illumination across the ray bundle cross section. If the cross

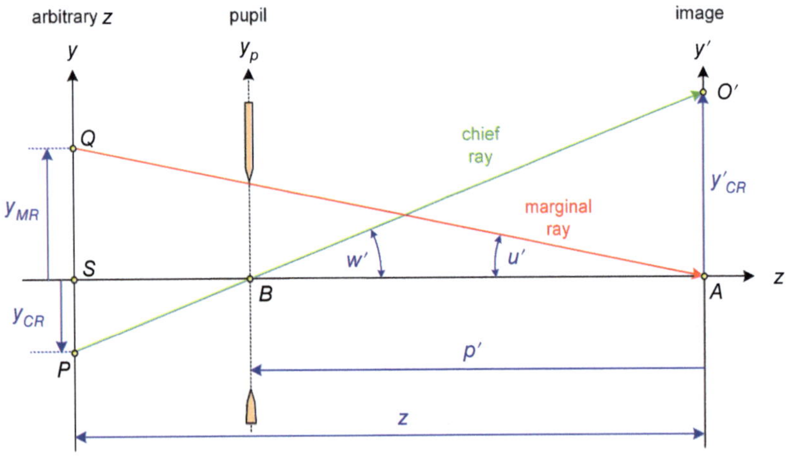

Figure 3.79 Definition of the Helmholtz–Lagrange invariant for a system at an arbitrary position in the system.

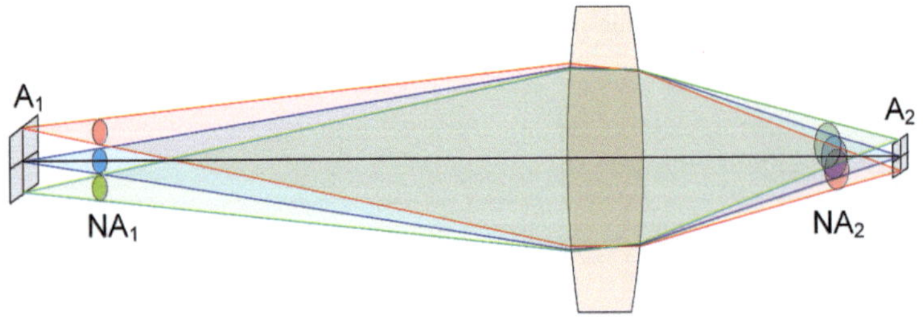

Figure 3.80 Helmholtz–Lagrange invariant in the two-dimensional case.

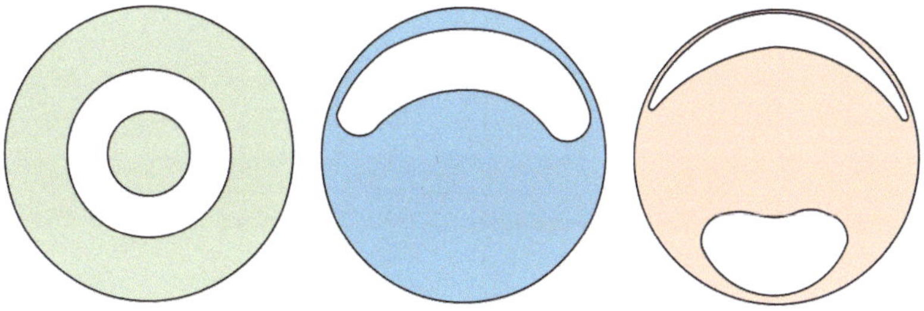

Figure 3.81 Helmholtz–Lagrange invariant for more general cases of a complicated intensity distribution.

section area suffers from a binary distribution and a more complicated boundary, as is shown in Figure 3.81, with the help of the variables [72]

$$p = n \cdot \cos(u), \quad q = n \cdot \cos(v) \tag{3.293}$$

as ray direction cosines multiplied by n, the area and the angle integration can be defined numerically and the Lagrange invariant reads

$$L_{geo} = \iint \iint dx\, dy\, dp\, dq \tag{3.294}$$

This is the most general pure geometrical formulation of the invariant as an integral over the phase-space volume, see Chapter 10.

If now the intensity distributions in the spatial as well as in the angle domain are non-homogeneous, an additional distribution function $W(x,y,p,q)$ must be used, which corresponds to an arbitrary energy density. Then we end up with the most general more physical definition

$$L_{phys} = \iint \iint \int W(x, y, p, q)\, dx\, dy\, dp\, dq \tag{3.295}$$

This equation now corresponds to the phase space integration over the Wigner density function W and is a quite general formulation of the space-bandwidth product and the conservation of energy. See Chapter 10 for more details.

One more aspect to be generalized is the case when the aperture angle u is no longer small and the paraxial approximation is not valid anymore. In this case, the so called sine condition states that the relation

$$n \cdot y \cdot \sin u = n' \cdot y' \cdot \sin u' \tag{3.296}$$

is valid for finite angles, if the system is corrected appropriately and the pupil surfaces are spherical in shape, see a more detailed discussion in Volume 2, Chapter 17. The meaning of the sine condition is a fulfilment of the conservation of energy, also locally for arbitrary sub-apertures. But equation is not necessarily fulfilled as the global conservation of energy, which is a basic physical law. The sine condition needs for s special correction of an optical system; in particular, as a consequence, the magnification is constant for all ranges of the aperture cone. In the wording of correction, the spherical aberration as well as sagittal coma must be corrected. The sine condition reads

$$m' = \frac{n \cdot \sin u}{n' \cdot \sin u'} \tag{3.297}$$

According to Figure 3.82 the interpretation is that the magnification is the same in the green as well as in the orange sub-aperture.

3.7.4 Smith's Cosine Invariant*

(*) The content of this section is quite specialized and for advanced reading. Knowledge of this section is not necessary to understand the following chapters. Beginners are recommended to skip this section.

The so-called cosine invariant of Smith can be obtained from a general setup according to Figure 3.83, if the ray path

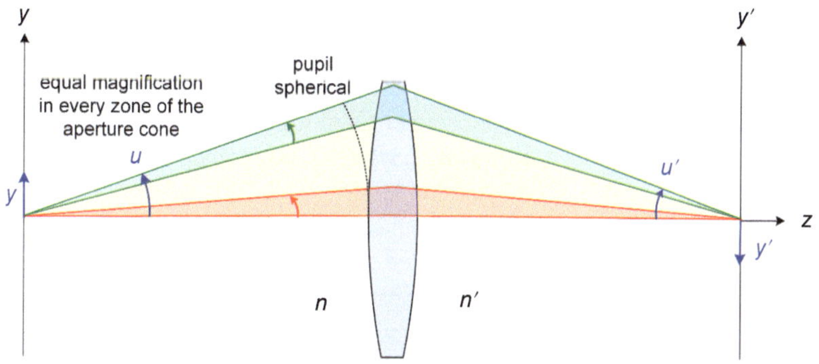

Figure 3.82 Illustration of the sine condition.

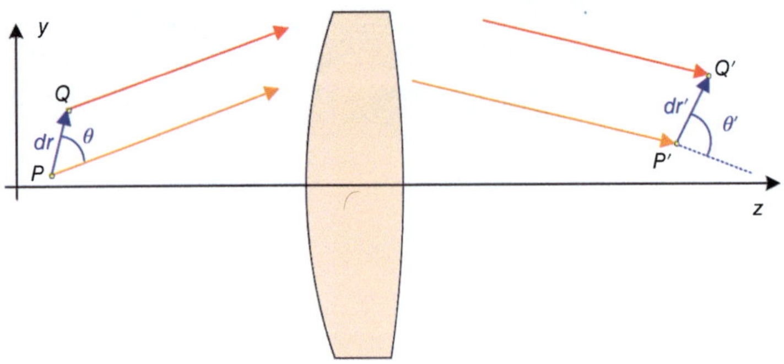

Figure 3.83 Geometry for the definition of the cosine invariant.

from two conjugated points P to P' is changed by a small vector dr with angle θ to the ray with the conjugated points Q, Q'. The optical path difference for both ray alternatives is calculated as follows. From the optical path difference in an optical system

$$\delta L = n'\vec{s}' \cdot d\vec{r}' - n\vec{s} \cdot d\vec{r} \tag{3.298}$$

we immediately get the so-called cosine invariant in the case of an imaging arrangement using the Fermat principle of invariant optical path in both cases.

$$\begin{aligned} n \cdot \vec{s} \cdot d\vec{r} &= n' \cdot \vec{s}' \cdot d\vec{r}' \\ n \cdot dr \cdot \cos\theta &= n' \cdot dr' \cdot \cos\theta' \\ n \cdot \cos\theta &= n' \cdot \delta m \cdot \cos\theta' \end{aligned} \tag{3.299}$$

with the differential magnification $\delta m = dr'/dr$. This is the general invariance condition of Smith. It is a differential invariant; a more rigorous derivation can be found in Chapter 13.

The cos-invariant is a special form of the Fermat principle and therefore is valid also under quite general conditions. For example, in a transverse gradient index medium, the eikonal equation can be formulated in the form

$$\frac{d}{ds}\left(n \cdot \frac{d\vec{r}}{ds}\right) = \nabla n \tag{3.300}$$

The corresponding geometry is depicted in Figure 3.84.

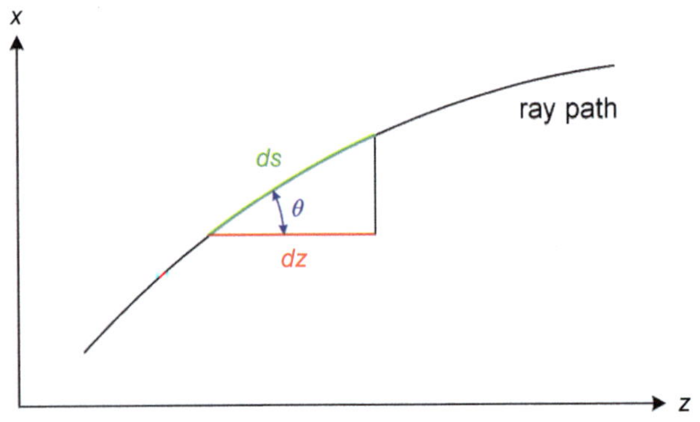

Figure 3.84 Interpretation of the cosine invariant for ray propagation in a lateral gradient index medium.

If only the z component is considered, we get

$$\frac{d}{ds}\left(n \cdot \frac{dz}{ds}\right) = \frac{dn}{dz} \tag{3.301}$$

For a transverse inhomogeneous medium, the derivative dn/dz vanishes and we get the invariant of propagation

$$\begin{aligned} \frac{d}{ds}\left(n \cdot \frac{dz}{ds}\right) &= \frac{dn}{dz} = 0 \\ \frac{d}{ds}(n \cdot \cos\theta) &= 0 \\ n(x,y) \cdot \cos\theta(x,y) &= m = const. \end{aligned} \tag{3.302}$$

or

$$m = n(x,y) \cdot \cos\theta(x,y) = \frac{n(x,y)}{\sqrt{1 + x'^2 + y'^2}} \tag{3.303}$$

3.8 Delano Diagram*

(*) The content of this section is quite specialized and for advanced reading. Knowledge of this section is not necessary to understand the following chapters. Beginners are recommended to skip this section.

3.8.1 Definition

The Delano diagram is a simple representation of the ray paths in optical instruments. With this tool the path of the chief and the marginal ray through the system is plotted in a suitable way. These diagrams are quite helpful for the design of complex compound systems since they provide an overview and allow for a quick estimation of various properties. So usually the Delano diagram is used for more qualitative conceptual planning. Correspondingly, it is mostly used for paraxial considerations, but this is not a necessary limitation [73–79].

In the $y - \bar{y}$ or Delano diagram the height y of the marginal ray is plotted versus the height $\bar{y}$ of the corresponding chief ray. At each surface or lens the diagram line shows a deflection from which the ray deflecting impact of the component can be seen. The simple scheme of a Delano diagram for a system consisting of three lenses is illustrated

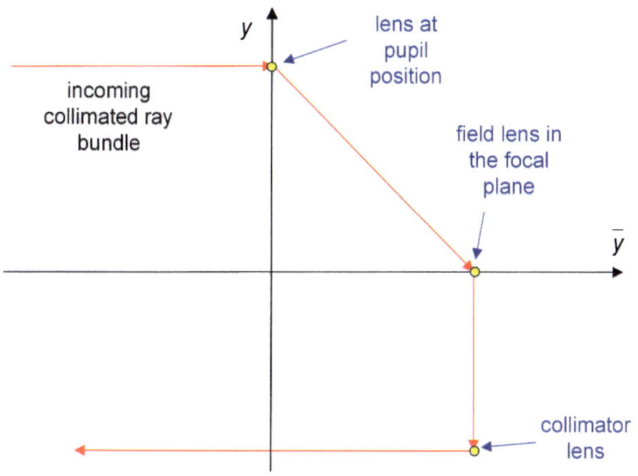

Figure 3.85 Example: Delano diagram of a system with three lenses.

in Figure 3.85. Figure 3.86 reproduces the corresponding y–z spatial domain lens cross section where the chief, the marginal ray and their heights which vary with z are drawn as arrows. In the Delano diagram a positive refractive power causes a deflection of the ray path towards the origin with a clockwise orientation. The greater the refractive power, the larger is the angle of deflection. Figure 3.87 shows this for illustration. However, it has to be observed that the angle of deflection for a fixed focal length also depends on the size of the system and the position of the chief ray, therefore a quantitative evaluation is not trivial.

3.8.2 Properties of the Delano Diagram

Delano diagrams have some very remarkable properties which contribute to their importance. On the other hand, there are also some characteristics which are less straightforward.

- The intersection points with the y-axis are the positions of stops. By extending the initial line the entrance pupil is found as intersection with the y-axis, by back-following the final line the exit pupil is obtained, see Figure 3.88.
- The intercept with the $\bar{y}$-axis is the position of object and image as well as intermediate images (planes of field of view).

Figure 3.86 Optical system for the Delano diagram in Figure 3.86.

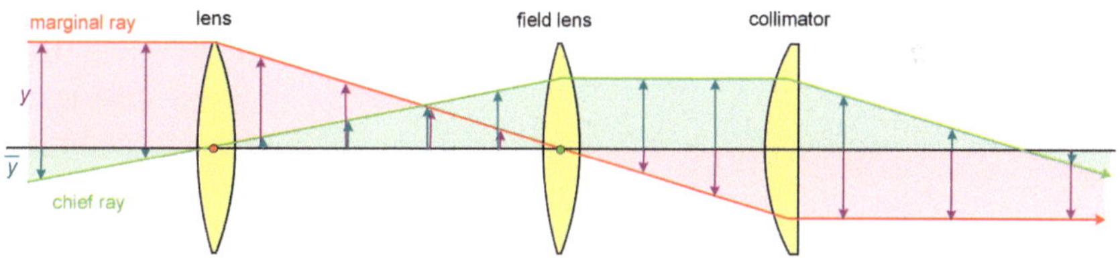

Figure 3.87 Deflection angle of the ray in the Delano diagram for different values of refractive power. The positive powers generate a clockwise, the negative powers a counterclockwise, bending of the Delano curve.

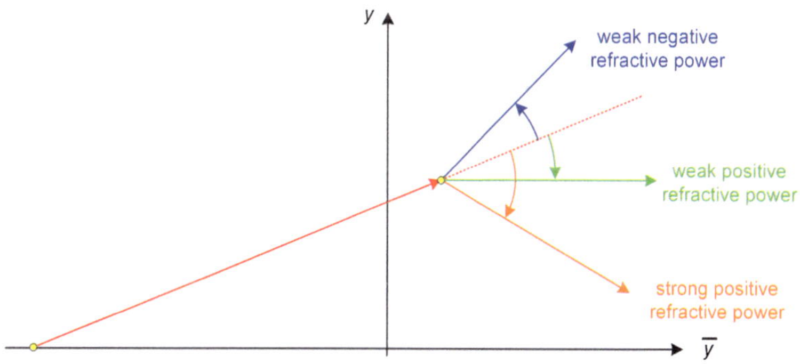

Figure 3.88 Positions of stops and pupils in the Delano diagram.

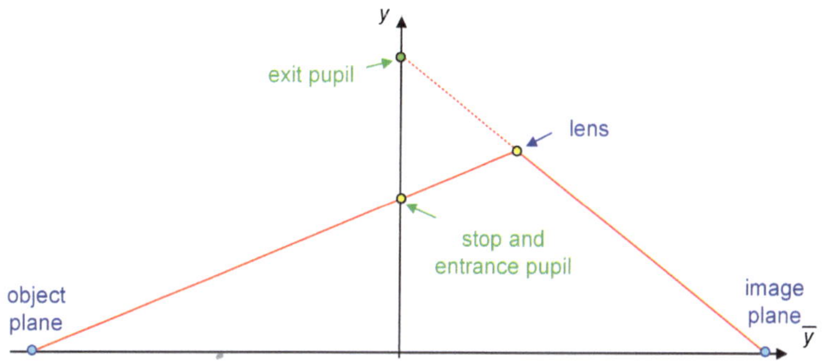

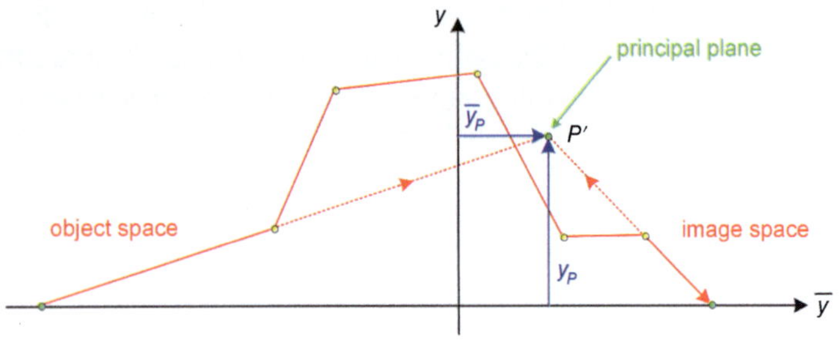

Figure 3.89 Principal plane P' in the Delano diagram.

- Lines parallel to the $\bar{y}$-axis indicate a constant height y of the marginal ray, corresponding to the collimated parallel path of the marginal ray.
- Vertical line with constant height $\bar{y}$ of the chief ray represents a parallel chief ray (telecentricity).
- The sum of the y and $\bar{y}$ coordinate of any point describes the height of the coma ray in the system; see also the next section.
- Positive lenses or components are described by polygon points concave to the origin.
- If the first and the last line segments of a system are extended, the intersection point of these two lines gives the principal plane P' and the heights of the object- and image-side marginal y_p and chief ray $\bar{y}_p$ at this location, see Figure 3.89.

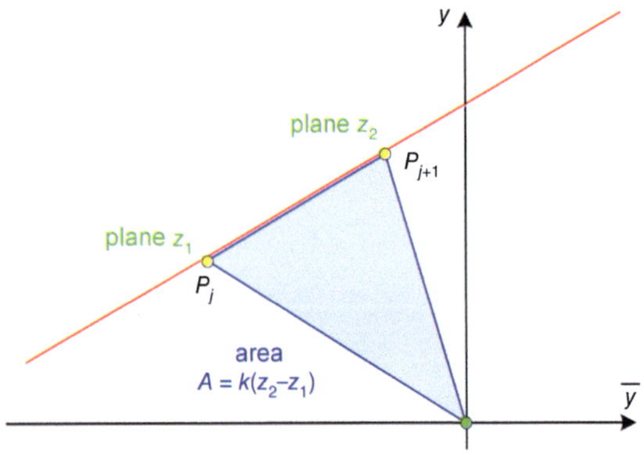

Figure 3.90 z distances in the Delano diagram.

- If the triangle between the two points P_j and P_{j+1} on the diagram and the origin is drawn, the area of the triangle is proportional to the distance $z_2 - z_1$ of the two components along the optical axis inside the optical system, see Figure 3.90 for illustration. The proportionality factor k includes the Lagrange invariant of the system. Therefore, this relation changes if the system is stopped down or the size of the field is modified.
- A shift in the z position of a stop with a fixed size causes a shearing of the points by a fixed angle φ as indicated in Figure 3.91. The same effect can also be obtained by a shift of all the system points in the direction of the $\bar{y}$ axis by a value proportional to y. In this way the tangents of the angle φ remain the same for all points.
- A shift in the object and image position, where the chief ray angle is kept the same, moves all points on lines in the vertical direction of the y axis. The amount of the shifts is proportional to the values of $\bar{y}$.
- Planes that conjugate with each other must always lie on a line through the origin. Such a line through the origin intercepts the Delano curve in a complete set of conjugate points, see Figure 3.92. The conjugate line through the intercept of the object and image ray gives the principal point P' of the system
- In the Delano diagram there is no difference between thin lenses and mirrors.
- The line parallel to the object segment through the origin intercepts the image ray in the rear focal plane F'. The line parallel to the image segment through the origin intercepts the object ray in the front focal plane F, see Figure 3.93.

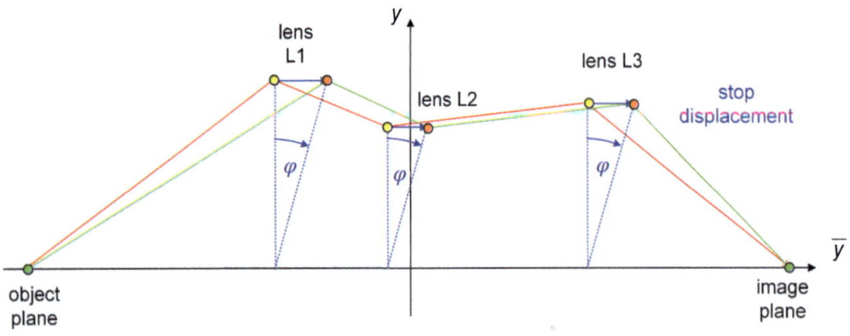

Figure 3.91 Change in the Delano diagram for a shift in the stop position within a system.

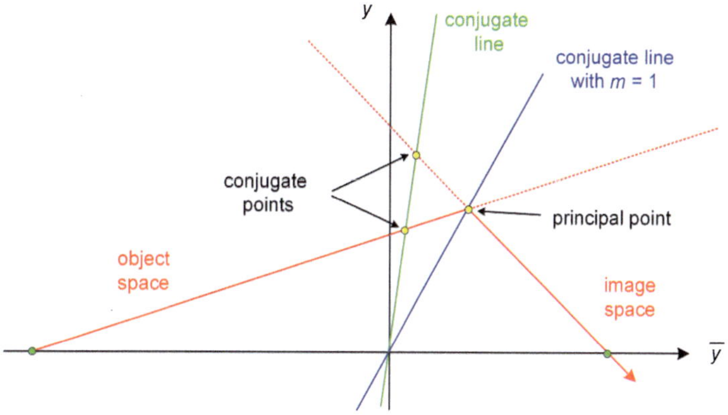

Figure 3.92 Conjugated points in the Delano diagram.

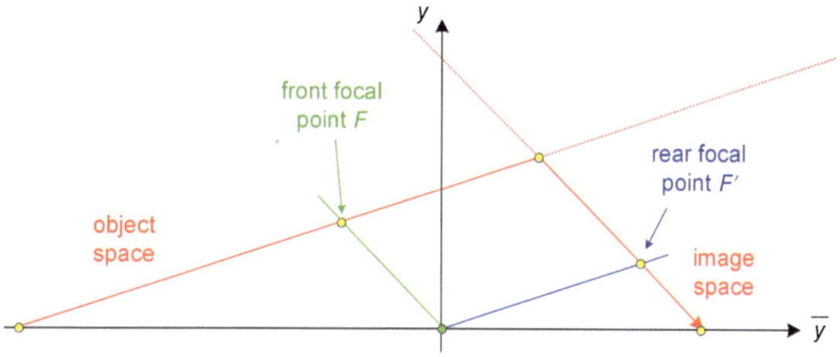

Figure 3.93 Focal points in the Delano diagram.

3.8.3 Vignetting

If a surface of the system on a line segment of the Delano diagram is considered, the height of the marginal and chief ray at this position can be obtained by definition of the diagram. However, in a rotational-symmetric system there is in fact a whole circular cone around the chief ray due to the aperture, see Figure 3.94 in the spatial domain. In paraxial approximation the sum of the heights of the marginal and chief rays gives the coma ray height, which is the largest of all rays.

In the Delano diagram the coma ray corresponds to a line

$$D/2 = |y| + |\bar{y}| \tag{3.304}$$

which is a diagonal line under 45°. It has to be drawn through the relevant point on the diagram line assigned to the system surface (shown in red in the figure) under inspection. Then the circle through the two intersection points with the two diagram axes has to be constructed (see Figure 3.95). This circle yields the free minimum diameter at the system surface where no vignetting occurs for the coma ray [76, 77].

By analogy, the required diameter of a component can immediately be obtained from any Delano diagram by drawing the 45° diagonal line through the point of the corresponding component. This is shown in Figure 3.96.

3.8.4 Examples

For illustration, in the following, some Delano diagrams of selected simple systems are shown. Here the meaning and capabilities of this kind of representation are demonstrated.

- Three-lens triplet system.
- Telephoto and retrofocus lens.
- Galilean and Keplerian telescope. In the case of the Keplerian system, two options are overlaid. If a field lens is used (solid path line), the system diameter at the last lens is changed considerably.

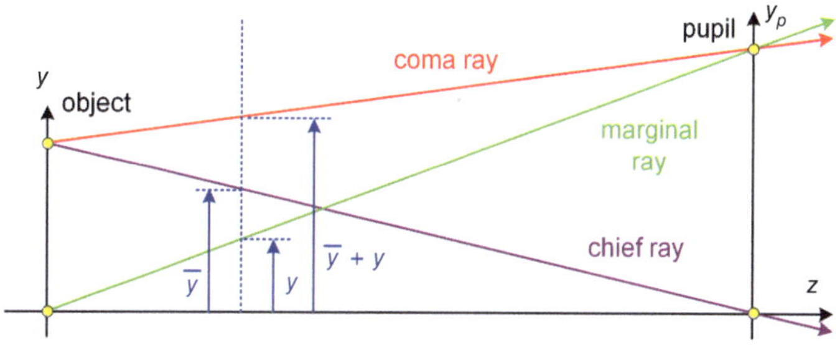

Figure 3.94 Maximum height of the coma ray in red in the spatial domain.

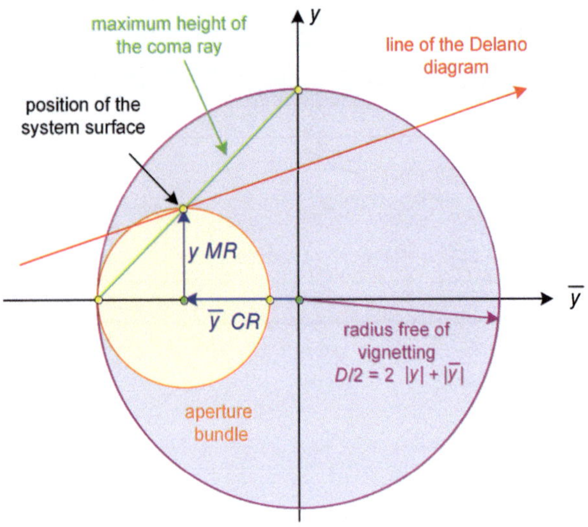

Figure 3.95 Plot of the diameter free of vignetting in the Delano diagram.

- Lithographic refractive projection lens.

  The system considered here is shown in Figure 3.102 in the spatial domain. It is a refractive system with magnification of $m = -0.25$ and three bulges of lenses. In the Delano diagram in Figure 3.103, the bulges, the pupil, the magnification can be clearly seen. The green dotted lines are 45° lines and indicate the maximum diameters to avoid vignetting (bulges) as well as minimum diameters (waists) as tangent to the curve.

In the usual Delano diagram it is easy to detect the largest common light bundle diameters or heights by inspecting the distance of the 45°-diagonals to the origin. In case of the angles, it is not so trivial to get information about the smallest angle $u_{\max} = u_{MR} + u_{CR}$.

Figure 3.104 indicates for another lithographic lens shown in (a) beneath the Delano diagram in (b) the ray heights of marginal and chief rays as well as the sum of both in (c). Here only the positive heights are used, the bulges and waists of the lens can be clearly seen. In the part (d) the angles of marginal ray, chief ray and the sum of the absolute values are depicted. This diagram is far more complicated. It can be seen that the smallest angle sum is obtained in the pupil with the largest ray heights and the switch of the sizes at the waists can be recognized.

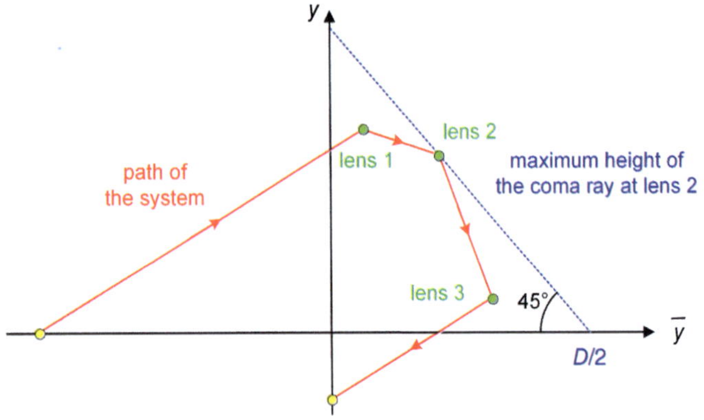

Figure 3.96 Plot of vignetting in the Delano diagram, required dimension of a component.

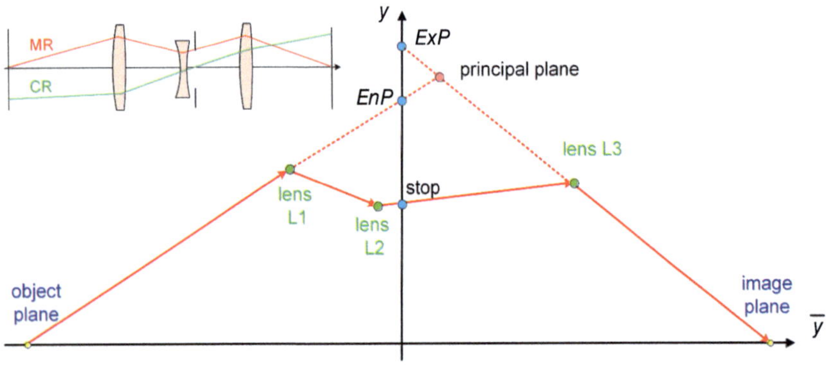

Figure 3.97 Delano diagram of a triplet.

- Microscope.
- Three-mirror reflective telescopes.

In Figure 3.101 several different approaches of three-mirror telescopes are shown as Delano diagrams. Except for the Offner system, all of them have a finite image location.

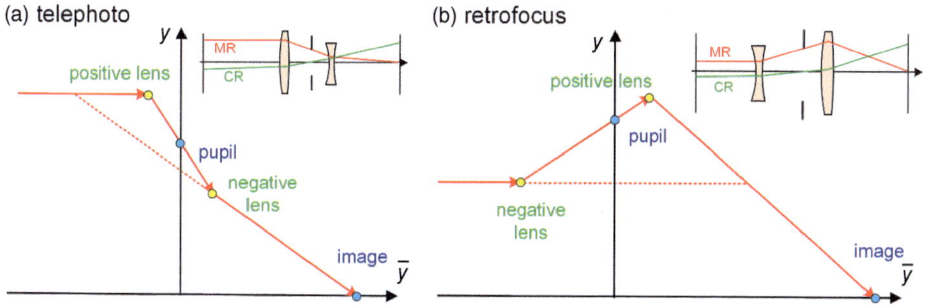

Figure 3.98 Delano diagram of a telephoto lens.

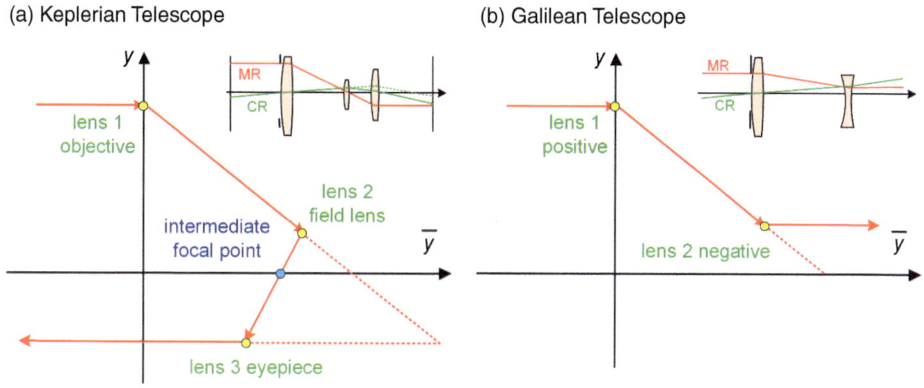

Figure 3.99 Delano diagram of (a) a Keplerian telescope with and without field lens and (b) Galilean telescope.

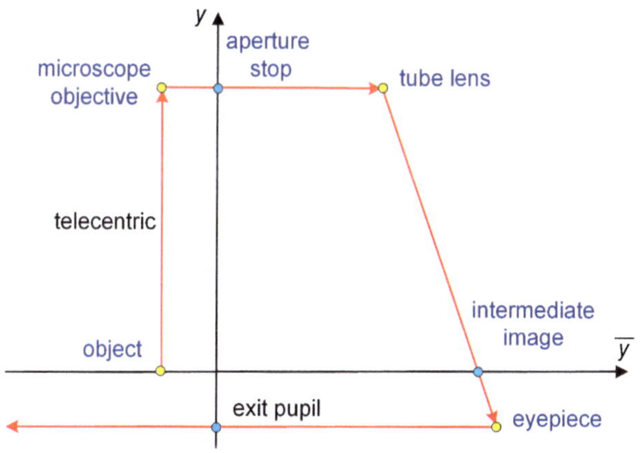

Figure 3.100 Delano diagram of a microscope.

The smallest $u_{\max}$ in the general case is not necessarily at the pupil, because here we have a double telecentric 4-f-equivalent system.

Unfortunately, there is no simple reciprocal behaviour and relation between ray heights and angles.

3.9 Gaussian Brackets*

(*) The content of this section is quite specialized and for advanced reading. Knowledge of this section is not necessary to understand the following chapters. Beginners are recommended to skip this section.

3.9.1 Introduction

The scheme of Gaussian brackets is a special mathematical calculus that allows an easy formal handling of special problems of ray based inversion issues. In particular, the optical effect of a cascaded series of paraxial components can be written in terms of Gaussian brackets quite easily [12, 82–95].

If the a_j are an ordered set of quantities and [] denotes the Gaussian bracket and the starting values are defined as

$$[a_1] = a_1$$
$$[a_1, a_2] = a_1 a_2 + 1 \tag{3.305}$$

the generalized Gaussian bracket is given by the recurrence formula

$$[a_1, a_2, \dots a_n] = [a_1, a_2, \dots a_{n-1}] \cdot a_n + [a_1, a_2, \dots a_{n-2}] \tag{3.306}$$

which gives the next elements as

$$[a_1, a_2, a_3] = a_1 a_2 a_3 + a_1 + a_3$$
$$[a_1, a_2, a_3, a_4] = a_1 a_2 a_3 a_4 + a_1 a_2 + a_1 a_4 + a_3 a_4 + 1$$
$$[a_1, a_2, a_3, a_4, a_5] = a_1 a_2 a_3 a_4 a_5 + a_1 a_2 a_3 + a_1 a_2 a_5 + a_1 a_4 a_5 + a_3 a_4 a_5 + a_1 + a_3 + a_5 \tag{3.307}$$

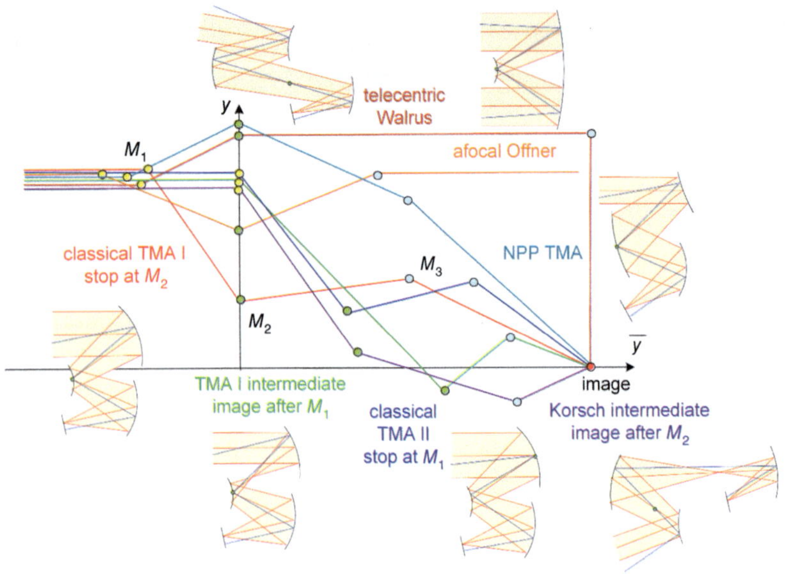

Figure 3.101 Delano diagram of three-mirror reflective telescopes.

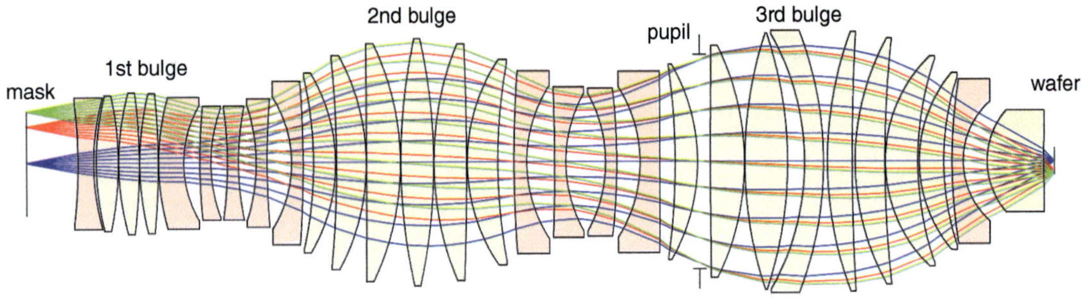

Figure 3.102 Optical layout of the considered lithographic projection lens. Adapted from [80].

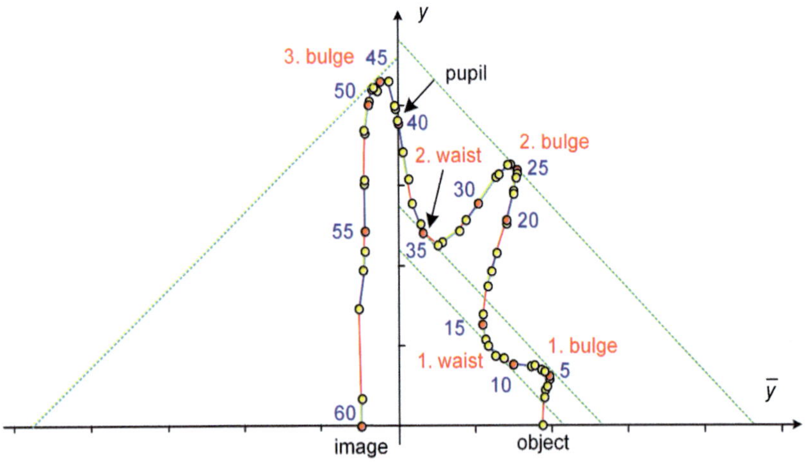

Figure 3.103 Delano diagram of the lithographic lens. Adapted from [81]. The blue numbers count the surfaces.

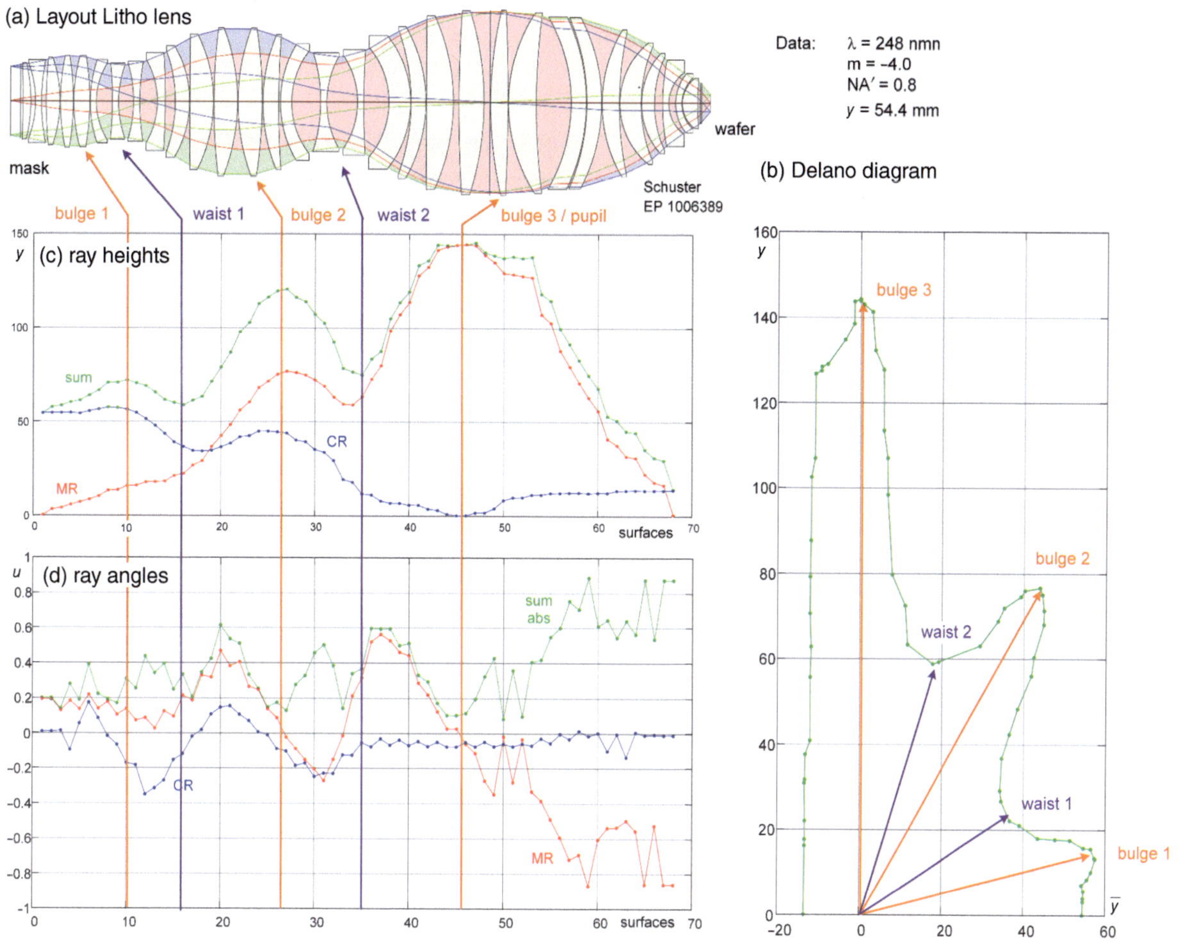

Figure 3.104 (a) Lithographic lens with (b) Delano diagram, (c) ray heights over the surfaces and (d) angles over the surfaces.

Gaussian brackets are linear in all of their elements. As a consequence, a bracket can be developed linearly by factoring out one element in the form

$$[a_1, a_2, \dots a_n] = C_1 \cdot a_k + C_2$$
$$C_1 = [a_1, a_2, \dots a_{k-1}] \cdot [a_{k+1}, a_{k+2}, \dots a_n]$$
$$C_2 = [a_1, a_2, \dots, (a_{k-1} + a_{k+1}), \dots a_n] \tag{3.308}$$

It should be mentioned that in this decomposition, the terms $C_{1/2}$ do not any longer contain the value of a_k. This property can be used to give the derivative of a Gaussian bracket as

$$\frac{\partial}{\partial a_k}[a_1, a_2, \dots a_n] = C_1 = [a_1, a_2, \dots a_{k-1}] \cdot [a_{k+1}, a_{k+2}, \dots a_n] \tag{3.309}$$

One more property is the validity of the so-called determinant formula

$$[a_1, a_2, \dots a_n] \cdot [a_2, a_3, \dots a_{n-1}] - [a_1, a_2, \dots a_{n-1}] \cdot [a_2, a_3, \dots a_n] = 1 \tag{3.310}$$

3.9.2 Alternative Representations

A clearer representation can be obtained when the Gaussian bracket is written with the abbreviation $G_{j,n}$

$$G_{j,n} = [a_j, a_{j+1}, \dots a_n] \tag{3.311}$$

In this notation, the recurrence relation reads

$$G_{j,n} = G_{j,n-1} \cdot a_n + G_{j,n-2} \tag{3.312}$$

and the expansion can be formulated as

$$G_{jn} = G_{j,k-1} \cdot G_{k+1,n} \cdot a_k + G_{j,k-2} \cdot G_{k+1,n} + G_{j,k-1} \cdot G_{k+1,n} \tag{3.313}$$

One more important formula is the recurrence relation

$$\begin{vmatrix} G_{in} & G_{jn} \\ G_{im} & G_{jm} \end{vmatrix} = (-1)^{m-j+1} G_{i,j-2} G_{m+2,n} \tag{3.314}$$

It is also possible to write Gaussian brackets in the form of a determinant in a simple tridiagonal form as

$$G_{k,n} = \begin{vmatrix} a_k & -1 & 0 & 0 & \dots & 0 & 0 \\ +1 & a_{k+1} & -1 & 0 & \dots & 0 & 0 \\ 0 & +1 & a_{k+2} & -1 & \dots & 0 & 0 \\ 0 & 0 & +1 & a_{k+3} & \dots & 0 & 0 \\ \dots & \dots & \dots & \dots & \dots & \dots & \dots \\ 0 & 0 & 0 & 0 & \dots & a_{n-1} & -1 \\ 0 & 0 & 0 & 0 & \dots & +1 & a_n \end{vmatrix} \tag{3.315}$$

From the mathematical point of view, Gaussian brackets have also a direct relationship to continued fractions. This means

$$\frac{[a_1, a_2, a_3, \dots, a_n]}{[a_2, a_3, \dots, a_n]} = a_1 + \cfrac{1}{a_2 + \cfrac{1}{a_3 + \dots \cfrac{1}{a_n}}} \tag{3.316}$$

3.9.3 Relation to *ABCD* Formalism

There is a very tight relation between the Gaussian brackets and the classical *ABCD* matrices of paraxial optics. It is well known that the ray transfer in a paraxial system can be written as (see Section 3.6)

$$\begin{pmatrix} x' \\ u' \end{pmatrix} = \begin{pmatrix} A & B \\ C & D \end{pmatrix} \cdot \begin{pmatrix} x \\ u \end{pmatrix} \tag{3.317}$$

If a paraxial optical system with refractive powers F_j and $(n-1)$ distances t_j between them is considered, the matrix elements of the *ABCD*-matrix can be written in terms of the system parameters in the form of Gaussian brackets as

$$\begin{aligned} A_n &= [F_1, -t_1, F_2, -t_2, \dots, F_{n-1}, -t_{n-1}] \\ B_n &= [-t_1, F_2, -t_2, \dots, -t_{n-2}, F_{n-1}, -t_{n-1}] \\ C_n &= [F_1, -t_1, F_2, -t_2, \dots, -t_{n-1}, F_n] \\ D_n &= [-t_1, F_2, -t_2, \dots, -t_{n-1}, F_n] \end{aligned} \tag{3.318}$$

the focal powers *Fj* can be the expressions of a single refractive surface or the value of a thin lens. With the recurrence relations

$$\begin{aligned} A_n &= C_{n-1} \cdot t_{n-1} + A_{n-1} \\ B_n &= -D_{n-1} \cdot t_{n-1} + B_{n-1} \\ C_n &= A_n \cdot F_n + C_{n-1} \\ D_n &= B_n \cdot F_n + D_{n-1} \end{aligned} \tag{3.319}$$

the building structure of a cascaded system can be observed. This equation allows us to obtain the new matrix of a system when one component is added.

According to the calculation of the focal power of a paraxial system by

$$F = \frac{1}{f} = \left.\frac{u'}{x}\right|_{x=0} = -C \tag{3.320}$$

the power as a composition of the single components can be expressed by Gaussian brackets as

$$F = [F_1, -t_1, F_2, -t_2, \dots, -t_{n-1}, F_n] \tag{3.321}$$

If in a more general case the *ABCD* elements are written for a sub-system from the indices k to n as

$$\begin{aligned} A_{kn} &= [F_k, -t_k, F_{k+1}, -t_{k+1}, \dots, F_{n-1}, -t_{n-1}] \\ B_{kn} &= [-t_k, F_{k+1}, -t_{k+1}, \dots, -t_{n-2}, F_{n-1}, -t_{n-1}] \\ C_{kn} &= [F_k, -t_k, F_{k+1}, -t_{k+1}, \dots, -t_{n-1}, F_n] \\ D_{kn} &= [-t_k, F_{k+1}, -t_{k+1}, \dots, -t_{n-1}, F_n] \end{aligned} \tag{3.322}$$

The expansion theorem allows us to write the system matrix explicitly as a function of the elements with index j and $k \le j < n$ as

$$\begin{aligned} A_{kn} &= A_{kj}B_{jn}F_j + C_{k,j-1}B_{jn} + A_{kj}A_{j+1,n} - C_{kj}A_{j+1,n}t_j \\ &\quad + A_{kj}A_{j+1,n} + C_{kj}B_{j+1,n} \\ B_{kn} &= B_{kj}B_{jn}F_j + D_{k,j-1}B_{jn} + B_{kj}A_{j+1,n} - D_{kj}A_{j+1,n}t_j \\ &\quad + B_{kj}A_{j+1,n} + D_{kj}B_{j+1,n} \\ C_{kn} &= A_{kj}D_{jn}F_j + C_{k,j-1}D_{jn} + A_{kj}C_{j+1,n} - C_{kj}C_{j+1,n}t_j \\ &\quad + A_{kj}C_{j+1,n} + C_{kj}D_{j+1,n} \\ D_{kn} &= B_{kj}D_{jn}F_j + D_{k,j-1}D_{jn} + B_{kj}C_{j+1,n} - D_{kj}C_{j+1,n}t_j \\ &\quad + B_{kj}C_{j+1,n} + D_{kj}D_{j+1,n} \end{aligned} \tag{3.323}$$

From these expansions, the derivatives to the system parameters F_j and t_j are directly calculated by

$$\begin{aligned} &\frac{\partial A_{kn}}{\partial F_j} = A_{kj}B_{jn}, \quad \frac{\partial B_{kn}}{\partial F_j} = B_{kj}B_{jn}, \quad \frac{\partial C_{kn}}{\partial F_j} = A_{kj}D_{jn}, \\ &\frac{\partial D_{kn}}{\partial F_j} = B_{kj}D_{jn}, \quad \frac{\partial A_{kn}}{\partial t_j} = -C_{kj}A_{j+1,n}, \\ &\frac{\partial B_{kn}}{\partial t_j} = -D_{kj}A_{j+1,n}, \quad \frac{\partial C_{kn}}{\partial t_j} = -C_{kj}C_{j+1,n}, \\ &\frac{\partial D_{kn}}{\partial t_j} = -D_{kj}C_{j+1,n} \end{aligned} \tag{3.324}$$

3.9.4 Applications

There are many practical applications of Gaussian brackets in paraxial lens design. Some of the most important ones are briefly sketched here.

3.9.4.1 Tolerancing

The influence of tolerances inside the system is described by small changes of one parameter, thickness or focal power. The expansion formula allows then the simple calculation of the impact on the complete system. For example, with the help of the expression of the total focal power

$$F = [F_1, -t_1, F_2, -t_2, \dots, -t_{n-1}, F_n] \tag{3.325}$$

by expanding the expression to the thickness with index k

$$\begin{aligned} F &= [F_1, -t_1, F_2, -t_2, \ldots, -t_{k-1}, F_k] \\ &\quad \cdot [F_{k+1}, -t_{k+1}, \ldots, -t_{n-1}, F_n] \cdot t_k \\ &\quad + [F_1, -t_1, \ldots, -t_{k-1}, (F_k + F_{k+1}), -t_{k+1}, \ldots, F_n] \end{aligned} \tag{3.326}$$

and the linear dependence on t_k, the impact of a tolerance change of the value of t_k is directly obtained.

3.9.4.2 Achromatization

Due to the very simple calculation of derivatives with the help of Gaussian brackets, the calculation of a system achromatization by the wavelength derivative benefits from this advantage. The invariance of the image plane location is given by $B = 0$. Therefore the derivative

$$\frac{\partial B_{1n}}{\partial \lambda} = \frac{\partial B_{1n}}{\partial F_j}\frac{\partial F_j}{\partial \lambda} = B_{1j}B_{jn}\frac{\partial F_j}{\partial n_j} n_j'(\lambda) = \frac{n_j'(\lambda)}{R_j} B_{1j}B_{jn} \tag{3.327}$$

with spherical radius of curvature R_j can be used to require a wavelength independence. Assuming F_j is the power of a single refractive surface, the B can now be expressed with

$$B_{kn} = [-t_k, F_{k+1}, -t_{k+1}, \ldots, -t_{n-2}, F_{n-1}, -t_{n-1}] \tag{3.328}$$

as

$$\begin{aligned} \frac{\partial B_{1n}}{\partial \lambda} &= \frac{n_j'(\lambda)}{R_j}[-t_1, F_2, -t_2, \ldots, -t_{j-2}, F_{j-1}, -t_{j-1}] \\ &\quad \cdot [-t_j, F_{j+1}, -t_{j+1}, \ldots, -t_{n-2}, F_{n-1}, -t_{n-1}] \end{aligned} \tag{3.329}$$

If this condition is formulated for all surfaces $j = 1\ldots n$ a linear system of equations is obtained for the terms n'/R and allows the calculation of an achromatic compensation.

3.9.4.3 Zoom Systems

In the representation of the paraxial description of zoom systems, the Gaussian brackets allow for a formal and easy formulation of the imaging condition during movement. Therefore, in the literature, the use of Gaussian brackets for the calculation of the paraxial system layout of zoom systems is common practice [92–97]. The calculation of the shifting element inside a zoom system works very much the same as in the tolerancing; usually the goal is to get the same image plane again, but a modified magnification. Therefore by coupling the zoom distances a system of two or more zoom equations is obtained, which can be directly solved. In *ABCD* terms the corresponding requirements are

$$m = A, \quad B = 0 \tag{3.330}$$

If for simplicity a simple mechanically compensated zoom system is considered with two axial moving lenses j and k by the shift values x_j and x_k, the conditions can be formulated as

$$\begin{aligned} m &= [F_1, -t_1, F_2, -t_2, \ldots, -t_{j-1} + x_j, F_{j,}, -t_j - x_j, \ldots, \\ &\quad -t_{k-1} + x_k, F_{k,}, -t_k - x_k, \ldots, F_{n-1}, -t_{n-1}] \\ 0 &= [-t_1, F_2, -t_2, \ldots, -t_{j-1} + x_j, F_{j,}, -t_j - x_j, \ldots, \\ &\quad -t_{k-1} + x_k, F_{k,}, -t_k - x_k, \ldots, F_{n-1}, -t_{n-1}] \end{aligned} \tag{3.331}$$

The second equation is the so-called zoom eq. $Z(x_j, x_k) = 0$. By expanding the Gaussian brackets, a quadratic equation for the movements x_j, x_k is obtained. Furthermore, this equation can be expressed by the magnification term m. By inspecting the derivatives, critical and singular points of the movement can be determined. The corresponding rearrangements of the formulas are not complicated, but tedious and long expressions are obtained if this general formulation is used.

It is also easy to require additional properties with more degrees of freedom for movements like constant position of the principal planes, constant location of the pupil or others.

The Gaussian brackets can also be used for the variants of zoom systems, which use variable power components at fixed positions [92–94]. In this case the Gaussian brackets must be decomposed after the power variables F_j instead of the distances.

In reality, the Gaussian brackets help to simplify the initial conditions in a clear way. For getting concrete results, the bracket terms are then always resolved into the system parameter.

References

1 Mahajan, V.N. (1998). *Optical Imaging and Aberrations, Part I, Ray Geometrical Optics*. Bellingham: SPIE Press.

2 Ditteon, R. (1998). *Modern Geometrical Optics*. New York: Wiley.

3 Born, M. and Wolf, E. (1999). *Principles of Optics*. Cambridge: Cambridge University Press.

4 Lin, P. (2017). *Advanced Geometrical Optics*. Springer.

5 Herzberger, M. (1980). *Modern Geometrical Optics*. Krieger.

6 Mahajan, V.N. (2014). *Geometrical Optics*. SPIE Press.

7 Greivenkamp, J. (2004). *Field Guide to Geometrical Optics*. SPIE Press.

8 Mouroulis, P. and Macdonald, J. (1997). *Geometrical Optics and Optical Design*. New York: Oxford University Press.
9 Cionek, M. and Barbano, E. (2023). Iterative assembly: a quick algorithm for generating aspherical collimating lenses in bidimensional optical systems. *Optics Continuum* 2: 1416.
10 Walther, A. (1995). *The Ray and Wave Theory of Lenses*. Cambridge University Press.
11 Klein, M. (1970). *Optics*. Wiley.
12 Zimmer, H.-G. (1967). *Geometrical Optics*. Springer.
13 Haferkorn, H. (2003). *Optik*. Weinheim: Wiley VCH.
14 Zhang, Y.-W., Chang, M.-J., Yang, C.-W., and Tu, H.-C. (1988). Automatic design of optical systems using a set of general skew ray tracing formulas. *Applied Optics* 27: 1574.
15 Spencer, G.H. and Murty, M.V. (1962). General ray-tracing procedure. *Journal of the Optical Society of America* 52: 672.
16 Feder, D.P. (1951). Optical calculations with automatic computing machinery. *Journal of the Optical Society of America* 41: 630.
17 Stavroudis, O. (1972). *The Optics of Rays, Wavefronts and Caustics*. Academic Press.
18 Freeman, D.E. (1990). A simple method for finding the valid ray-surface intersection. *SPIE Conference Proceedings* 1354: 200.
19 Howell, B.J. and Wilson, M.E. (1982). Automatic ray-surface intersection method. *Applied Optics* 21: 2184.
20 Cheatham, P.S. (1980). Alternate surface intersection points. *SPIE Conference Proceedings* 237: 142.
21 Welford, W.T. (1986). *Aberrations of Optical Systems*. Bristol: Adam Hilger.
22 Chen, J.-L. (1983). Ray tracing of Fresnel systems. *Applied Optics* 22: 560.
23 Glassner, A. (ed.) (1989). *Ray Tracing*. San Francisco: Morgan Kaufmann Publishers.
24 von Rohr, M. (1904). *Die Bilderzeugung in Optischen Instrumenten*. Springer.
25 Sharma, A. (1985). Computing optical path length in gradient-index media: a fast and accurate method. *Applied Optics* 24: 4367.
26 Sharma, A. and Ghatak, A.K. (1986). Ray tracing in gradient-index lenses: computation of ray-surface intersection. *Applied Optics* 25: 3409.
27 Puchalski, J. (1994). Numerical determination of continuous ray tracing: the four-component method. *Applied Optics* 33: 1900.
28 Sharma, A., Kumar, D.V., and Ghatak, A.K. (1982). Tracing rays through graded-index media: a new method. *Applied Optics* 21: 984.
29 Redding, D.C. and Breckenridge, W.G. (1990). Linearized ray-trace analysis. *SPIE Conference Proceedings* 1354: 216.
30 Evans, E.D. and Collins (1991). Expressions for ray-associated electromagnetic fields through optical systems. *Journal of the Optical Society of America A* 8: 841.
31 Lindlein, N. and Schwider, J. (1993). Local wave fronts at diffractive elements. *Journal of the Optical Society of America A* 12: 2563.
32 Welford, W. (1962). Tracing skew rays through concave diffraction gratings. *Optica Acta* 9: 389.
33 Klein, J.E. (2001). Challenges and problems in non-sequential ray tracing. *SPIE Conference Proceedings* 4442: 60.
34 Abernathy, M. (1996). Nonsequential raytracing: enlightened software for illumination engineering. *Optics News* 11: 22.
35 Frasch, H.-J. and Steinmetz, G. (2002). Monte Carlo methods in raytracing software. *SPIE Conference Proceedings* 4769: 55.
36 Freniere, E.R., Gregory, G.G., and Hassler, R.A. (1999). Edge diffraction in Monte Carlo ray tracing. *SPIE Conference Proceedings* 3780: 151.
37 Greynolds, A. (2017). Ray-pathing: a generalized extension and possible alternative to conventional ray aiming. *SPIE Conference Proceedings* 10590: 13.
38 Sun, H. (2017). *Lens Design*. CRC Press.
39 Chapman, S., Tew, R., Lawry, J., and Ockendon, J. (1997). On the theory of complex rays. *Society for Industrial and Applied Mathematics Review* 41 (10): 1137.
40 Duan, Q., Onofri, F., Han, X., and Ren, K. (2023). Numerical implementation of three-dimensional vectorial complex ray model and application to rainbow scattering of spheroidal drops. *Optics Express* 31: 34980.
41 Chang, S. (2011). Evolution of a complex-source-point spherical wave through a symmetric optical system with spherical aberration. *Optik* 122: 2094.
42 Couture, M. and Belanger, P.-A. (1981). From Gaussian beam to complex -source-point spherical wave. *Physical Review A* 24: 355.
43 Deschamps, G. (1971). Gaussian beam as a bundle of complex rays. *Electronics Letters* 7: 684.
44 Einziger, P. and Felsen, L. (1982). Evanescent waves and complex rays. *IEEE Transactions on Antennas and Propagation* 30: 594.
45 Felsen, L. (1985). Novel ways for tracking rays. *Journal of the Optical Society of America A* 2: 954.
46 Kravtsov, Y. and Orlov, Y. (1999). *Caustics, Catastrophes and Wave Fields*. Springer.
47 Kravtsov, Y. (2005). *Geometrical Optics in Engineering Physics*. Alpha Science.

48 Tap, K. (2007). Complex Source Point Beam Expansions for some Electromagnetic Radiation and Scattering Problems. Thesis, Ohio State University.

49 Kravtsov, Y., Forbes, G., and Asatryan, A. Theory and applications of complex rays. *Progress in Optics* 391.

50 Keller, J. and Streifer, W. (1971). Complex rays with application to Gaussian beams. *Journal of the Optical Society of America* 61: 40.

51 Gerrard, A. and Burch, J.M. (1994). *Matrix Methods in Optics*. New York: Dover.

52 Brouwer, W. (1964). *Matrix Methods in Optical Instrument Design*. New York: Benjamin.

53 Wang, S. and Zhao, S. (2000). *Matrix Optics*. Springer.

54 Kloos, G. (2007). *Matrix Methods for Optical Layout*. SPIE Press.

55 Shamir, J. (1999). *Optical Systems and Processes*. Bellingham: SPIE Press.

56 Tovar, A.A. and Casperson, L.W. (1995). Generalized Sylvester theorems for periodic applications in matrix optics. *Journal of the Optical Society of America A* 12: 578.

57 Arsenault, H.H. and Macukow, B. (1983). Factorization of the transfer matrix for symmetrical optical systems. *Journal of the Optical Society of America* 73: 1350.

58 Hodgson, N. and Weber, H. (1997). *Optical Resonators*. Berlin: Springer.

59 Ananev, Y.A. and Bekshaev, A.Y. (1989). Matrix method of analysis of misaligned optical systems with astigmatic elements. *Optics and Spectroscopy* 66: 531.

60 Nazarathy, M., Hardy, A., and Shamir, J. (1986). Misaligned first-order optics: canonical operator theory. *Journal of the Optical Society of America A* 3: 1360.

61 Shaomin, W. (1985). Matrix methods in treating decentred optical systems. *Optical and Quantum Electronics* 17: 1.

62 Shaomin, W. (1982). Propagational problem of optical beams through an asymmetric inhomogeneous medium. *Scientia Sinica* 25: 72.

63 Zhao, D. (1999). The ray transfer matrices for axicons. *Optik* 110: 493.

64 Tache, J.-P. (1987). Ray matrices for tilted interfaces in laser resonators. *Applied Optics* 26: 427.

65 Arsenault, H.H. (1980). A matrix representation for non-symmetrical optical systems. *Journal of Optics* 11: 87.

66 Rodionov, S.A. (1981). Matrix technique of Gaussian optics near an arbitrary ray. *Optics and Spectroscopy* 50: 531.

67 Harrigan, M. (2002). General beam propagation through nonorthogonal optical system. *SPIE Conference Proceedings* 4832: 379.

68 Torre, A. (2005). *Linear Ray and Wave Optics in Phase Space*. Elsevier.

69 Ozaktas, H., Zalevsky, Z., and Kutay, M. (2001). *The Fractional Fourier Transform*. Wiley.

70 Liu, X. and Brenner, K.-H. (2008). Minimal optical decomposition of ray transfer matrices. *Applied Optics* 47: E88.

71 Simon, R. and Mukunda, N. (1998). Iwasawa decomposition in first-order optics: universal treatment of shape-invariant propagation for coherent and partially coherent beams. *Journal of the Optical Society of America A* 15: 2146.

72 Lerner, S. (2006). Etendue and optical system design. *SPIE Conference Proceedings* 6338.

73 Delano, E. (1963). First-order design and the y - ybar diagram. *Applied Optics* 2: 1251.

74 Shack, R.V. (1991). Analytic system design with pencil and ruler – the advantages of the y-ybar-diagram. *SPIE Conference Proceedings* CR 39: 127.

75 Flores-Hernandez, R. and Stavroudis, O.N. (1996). The Delano y-ybar-diagram applied to optical system synthesis and optimization, y-ybar diagram point displacements. *Optik* 102: 171.

76 Besenmatter, W. (1977). Analyse des Vignettierungsverhaltens der Vario-Objektive mit Hilfe des Delano-Diagramms. *Optik* 48: 289.

77 Schwab, M. and Herkommer, A. (2008). The Delano diagram, a powerful design tool. *SPIE Conference Proceedings* 7100. https://doi.org/10.1117/12.797672.

78 Flores-Hernandez, R. (1996). A New Optical Design Method Based on the Delano Diagram. Thesis, University of Guanajuato.

79 Lopez, F. (1973). The Application of the Delano Y-Y Diagram to Optical Design. Thesis, University of Arizona.

80 Kotaro, Y. (1998). Projection Optical System. Jap. Pat. 11006957.

81 Schuster, K.-H. (1999). Mikrolithographisches Reduktionsobjektiv, Projektionsbelichtungsanlage und -Verfahren, EP 1006389.

82 Tanaka, K. (1981). Lagrange-Helmholtz invariant and optical invariants in terms of Gaussian brackets. *Optik* 58: 351.

83 Tanaka, K. (1983). Gaussian brackets and their application to paraxial theory. II. Paraxial theory in terms of generalized Gaussian constants. *Optik* 64: 89.

84 Tanaka, K. (1983). Gaussian brackets and their application to paraxial theory. I. Gaussian brackets and generalized Gaussian constants. *Optik* 64: 13.

85 Tanaka, K. (1982). On Gaussian brackets and Gaussian constants. *Optik* 62: 211.

86 Tanaka, K. (1986). Paraxial theory in optical design in terms of Gaussian brackets. *Progress in Optics XXIII* 63.

87 Yuan, X. and Cheng, X. (2014). Lens design based on lens form parameters using Gaussian brackets. *SPIE Conference Proceedings* 9272. https://doi.org/10.1117/12.2073422.

88 Herzberger, M. (1943). Gaussian optics and Gaussian brackets. *Journal of the Optical Society of America* 33: 651.

89 Liu, Y., Yang, B., Gu, P. et al. (2020). 50x five group inner focus zoom lens design with focus tunable lens using Gaussian brackets and lens modules. *Optics Express* 28: 29098.

90 Pegis, R. and Peck, W. (1962). First-order design theory for linearly compensated zoom systems. *Journal of the Optical Society of America* 52: 905.

91 Herzberger, M. (1952). Precalculation of optical systems. *Journal of the Optical Society of America* 42: 637.

92 Miks, A. and Novak, P. (2014). Paraxial analysis of three component zoom lens with fixed distance between object and image points and fixed position of image-space focal point. *Optics Express* 22: 15571.

93 Miks, A. and Novak, P. (2018). Paraxial design of four-component zoom lens with fixed position of optical center composed of members with variable focal length. *Optics Express* 26: 25611.

94 Miks, A. and Novak, J. (2013). Three component double conjugate zoom lens system from tunable focus lenses. *Applied Optics* 52: 864.

95 Tanaka, K. (1983). Allgemeine Gausssche Theorie eines mechanischen kompensierten Zoom-Objektivs: 3. Kritischer Punkt und singulärer Punkt einer Zoom-Gleichung. *Optics Communication* 45: 231.

96 Tanaka, K. (1979). Allgemeine Gaussche Theorie eines mechanisch kompensierten Zoom-Objektivs. *Optics Communication* 29: 138.

97 Tanaka, K. (1982). Paraxial analysis of mechanically compensated zoom lenses. 1: four-component type. *Applied Optics* 21: 2174.

4

Optical Components

4.1 Overview

Optical components are used to modify the geometrical paths or physical parameters of light. According to their modification impact on the properties of light, different types of components can be distinguished. In principle there are three major groups of components, which change the geometrical, the spectral or the polarization properties respectively. These are as follows.

1) Wavefront management:
   1.1 beam shaping;
   1.2 beam modification, splitting;
   1.3 beam guiding.
2) Wavelength management:
   2.1 angular dispersion;
   2.2 filtering;
   2.3 modification.
3) Polarization management:
   3.1 angular dispersion;
   3.2 filtering;
   3.3 modification.

In Figure 4.1 an illustration of the structured ordering of the components' working principles goes into more detail and gives some examples. For most of the tasks, several component realizations exist. For optical design applications, the category 'Ray/Wavefront Management by Beam Redirection' is by far the most important one.

4.2 Single Refractive Spherical Surface

A single refracting spherical surface with radius R and refractive indices n and n' in front of and behind the surface is the basic refraction module for a lens. In principle, a lens is a combination of two refractive surfaces. In the case of an imaging setup, the following imaging equation relates for a single surface the location s of the object to the location s' of the image in paraxial approximation.

$$\frac{n'}{s'} - \frac{n}{s} = \frac{n' - n}{R} = F \tag{4.1}$$

where F is the refractive power as the inverse value of the focal length. See the next chapter for the derivation of this equation. The central ray through the centre of curvature C of the surface is easily used for the graphical determination of the image position as shown in Figure 4.2; this ray impinges perpendicularly on the surface and was not deflected in point Q. As a second ray, an arbitrary ray can be selected or the special vertex ray.

The individual focal lengths of the surfaces are

$$f' = \frac{n'R}{n' - n}, \quad f = -\frac{nR}{n' - n} \tag{4.2}$$

In the special case of a spherical mirror surface with formally $n' = -n$ it follows that

$$f = f' = \frac{R}{2} \tag{4.3}$$

Spherical surfaces are usually given by the following equation, where the coordinate origin is fixed by the surface vertex point V

$$R^2 = y^2 + (R - z)^2 \tag{4.4}$$

Solving for z we obtain

$$z = r \cdot \left(1 - \sqrt{1 - \left(\frac{y}{R}\right)^2}\right) \tag{4.5}$$

here the right solution is selected according to the sign convention for spherical radii. An equation that in practice is very useful for estimation can be obtained by approximating the sphere to a parabola with the same curvature in the vertex

$$z_p \approx \frac{y^2}{2R} \tag{4.6}$$

This simple equation is obtained by a Taylor expansion of the root in Eq. (4.5) for $|y/R| << 1$. With this, the sagitta or sag z_p can be easily calculated for heights y that are not too great. Figure 4.3 shows the corresponding geometry.

Using the curvature c of the spherical surface, the surface equation can be written in a more stable form, which

Lens Design for Imaging: Fundamentals of Optical Systems, First Edition. Herbert Gross.

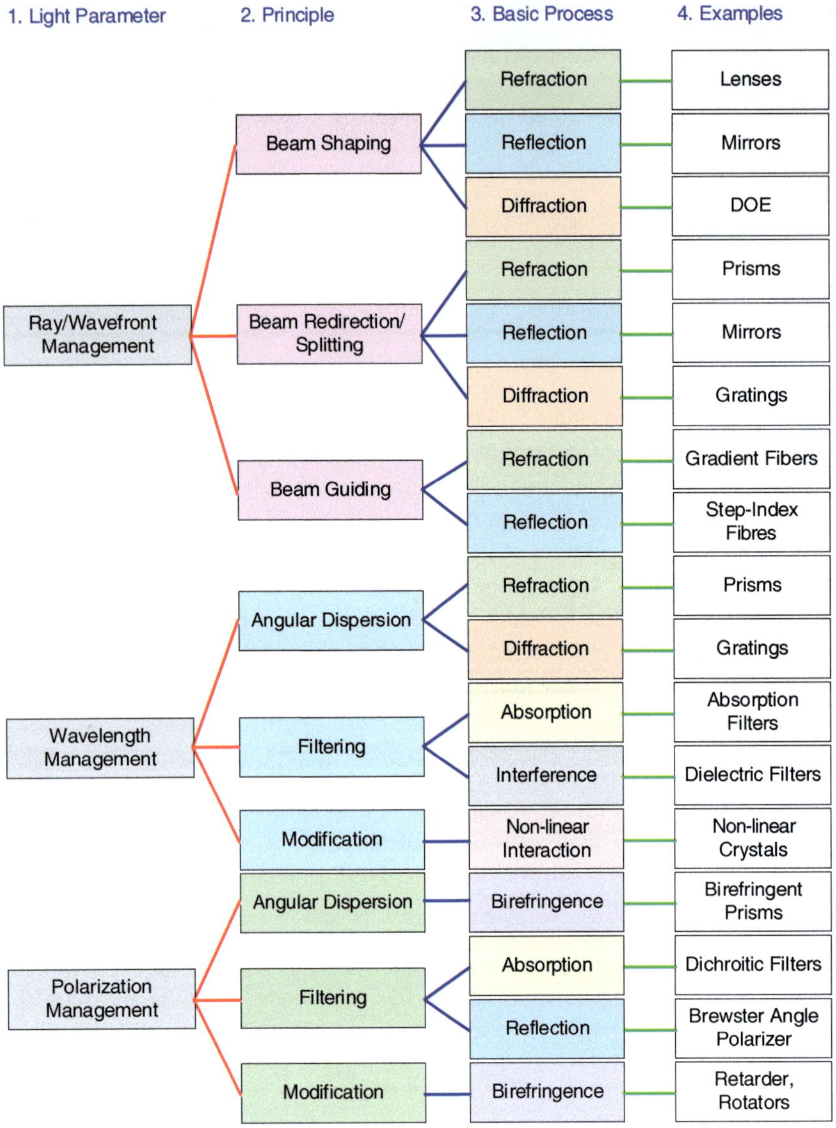

Figure 4.1 Overview on optical components sorted after their physical working principles [1].

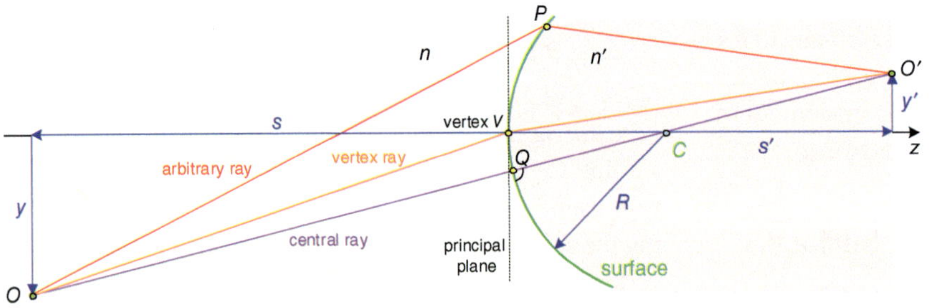

Figure 4.2 Imaging by refraction at a single surface.

in particular includes a plane surface $c = 0$ without the need of a singularity case distinction

$$z = \frac{c \cdot y^2}{1 + \sqrt{1 - c^2 y^2}} \tag{4.7}$$

which is the usual vertex-referenced equation for spherical surface description.

4.3 Plane Plates

4.3.1 Introduction

In reality, there are many components of optical instruments that have special tasks or functionalities, which are from the viewpoint of aberrations only plane plates. Depending on the application it may be that they are

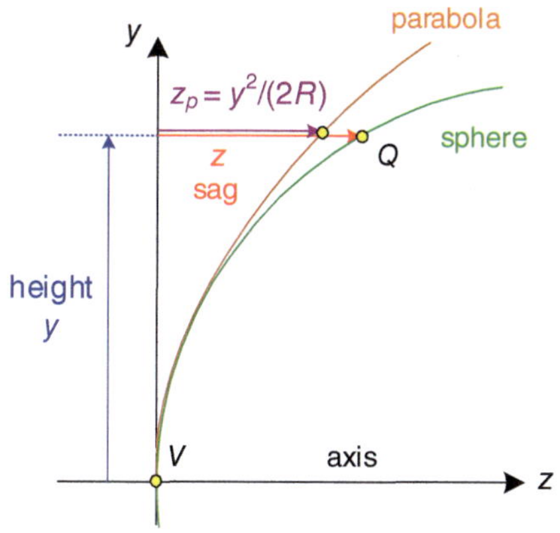

Figure 4.3 Calculation of the sagitta of a spherical surface.

transmitted perpendicularly or under a skew angle. Some possibilities for orthogonal use are as follows.

- Beam splitter cubes.
- Reflection prisms which are entered perpendicularly.
- Cover glasses in microscopy or for protection in front of a sensor.
- Spectral filter glasses.
- Windows for vacuum, cryogenic chambers or liquid samples.

If the optical axis is transmitting the entrance surface obliquely, additional other effects occur. Examples are:

- beam splitter plates;
- Dove or Littrow prisms;
- Brewster plates.

4.3.2 Perpendicular Plates

If a plane plate is employed in practice, generally it is used in transmission in the best case perpendicularly. Furthermore, to reduce aberrations as far as possible, collimated light is used, which is only possible for one field point. In reality, a small field angle must be taken into account, violating the orthogonality for off-axis field points. Figure 4.4 shows two representative examples for cubes or prism for beam splitting. In (a) a classical cube is shown with physical beam splitting by a corresponding coating, in (b) splitting is realized in a Köster prism. The latter generates a lateral shifted part of a beam and is used as in shearing interferometers.

Most of the well-known reflective prisms (see Section 4.10) are used with the optical axis and the axis ray bundle transmitting the entrance and exit surface perpendicularly. One prominent exception is the Dove prism, see Section 4.10.

4.3.2.1 Beam Displacement

A plane-parallel plate of thickness t and refractive index n is equivalent to an air gap with a thickness t/n from an ideal paraxial point of view. A glass plate will shift the image-side paraxial focus forward by (see Figure 4.5(a))

$$\Delta s' = \frac{n-1}{n} t \approx \frac{t}{3} \tag{4.8}$$

(assuming $n = 1.5$ for the last approximate expression) thus increasing the intercept distance.

The exact relationship for the modification of the intersection distance in terms of a finite aperture angle u is

$$\Delta s' = \frac{t}{n} \cdot \left(n - \frac{\cos u}{\cos u'} \right) = t \cdot \left(1 - \frac{\cos u}{\sqrt{n^2 - \sin^2 u}} \right) \tag{4.9}$$

where u' denotes the reduced aperture angle in the medium of the plate. The insertion of a plane-parallel plate into the optical path at an angle θ results in a parallel offset of the optical axis by an amount $\Delta y'$. The same effect is observed if a finite field angle is considered. With the incidence angle of $i = \theta$ on axis one obtains according to Figure 4.5(b)

$$\Delta y' = t \cdot \sin\theta \cdot \left(1 - \frac{\cos\theta}{\sqrt{n^2 - \sin^2\theta}} \right) \tag{4.10}$$

For small inclination angles, the last expression can be approximated by

$$\Delta y' = \frac{n-1}{n} \cdot t \cdot \sin\theta = \Delta s' \cdot \sin\theta \tag{4.11}$$

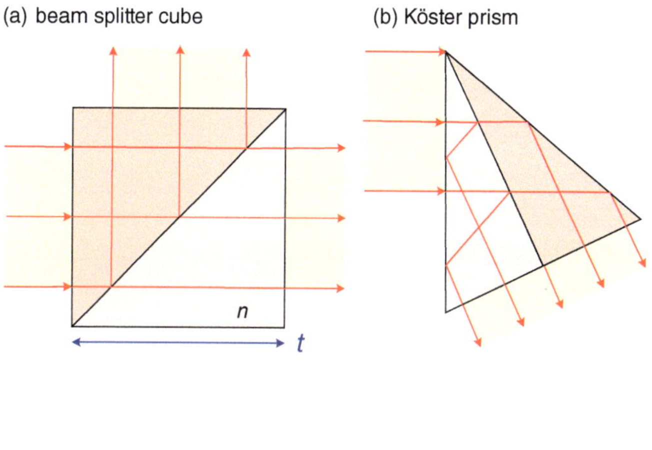

Figure 4.4 Beam splitter (a) cube and (b) Köster prism.

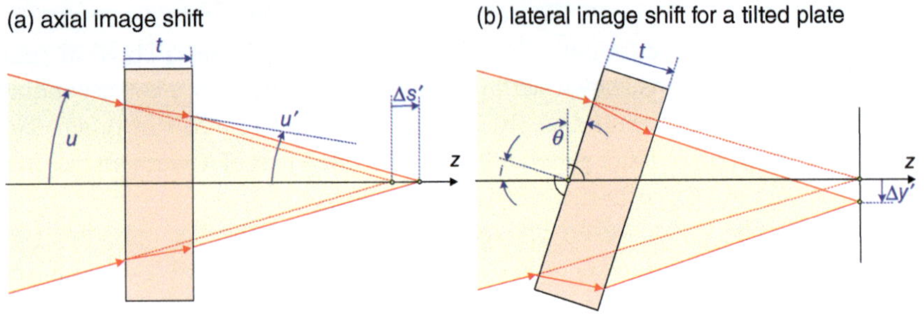

Figure 4.5 Image displacement by a plane-parallel plate in the case of convergent light (a) in axial and (b) in lateral direction for a tilt.

4.3.2.2 Aberrations

If a beam is collimated the insertion of a plane-parallel plate does not introduce any aberrations [2] if the ray bundle impinges the plate perpendicularly (see Volume 2 for more details about aberrations). In contrast, for convergent or divergent beams one has to take into account the aberrations caused by the plane-parallel plate. This effect cannot be neglected, particularly for high numerical apertures, such as for cover glasses in microscope objectives of high numerical aperture. The reflection prisms are used in instrumental optics for beam deflection and to change the desired image orientation. They are inserted mainly into regions of the beam path where the axial pencil of light is collimated or where it possesses a sufficiently small numerical aperture.

For the spherical aberration of a plane-parallel plate, one can obtain the exact spherical longitudinal aberration from

$$\Delta s'_{sph} = \frac{t}{n} \cdot \left(1 - \frac{\cos u}{\cos u'}\right) = t \cdot \left(\frac{1}{n} - \frac{\cos u}{\sqrt{n^2 - \sin^2 u}}\right) \tag{4.12}$$

In the third-order Seidel approximation, the last expression leads to

$$\Delta s'_{sph} = \frac{t \cdot (n^2 - 1) \cdot \sin^2 u}{2n^3} \tag{4.13}$$

Since the axial image displacement directly depends on the index of refraction, a plane-parallel plate introduces a longitudinal chromatic aberration of third order with magnitude

$$\Delta s'_{CHL} = \frac{t \cdot (n - 1)}{n^2 \cdot \nu_e} \tag{4.14}$$

For a finite angle w of the chief ray, the effect of a plane-parallel plate on the field pencil is similar to that of a tilted plate, yielding in third order

$$\Delta y'_{CHV} = \frac{t \cdot (n - 1) \cdot \sin w}{n \cdot \nu_e} \tag{4.15}$$

for the transverse chromatic aberration. Astigmatism also occurs in the field. As a longitudinal aberration this is given by

$$\Delta s'_{ast} = \frac{t}{\sqrt{n^2 - \sin^2 w}} \cdot \left(\frac{n^2 \cos^2 w}{\sqrt{n^2 - \sin^2 w}} - 1\right) \tag{4.16}$$

which can be approximated in third order by

$$\Delta s'_{ast} = \frac{t \cdot (n^2 - 1) \cdot \sin^2 w}{n^3} \tag{4.17}$$

We note that the difference in the back focal lengths is independent of the front focal length, according to the formulae for the astigmatism of plates that are tilted with respect to the field. This is true also for the case of infinity (i.e. in a telecentric configuration) where this difference exhibits the same value.

The following expression holds for the third-order sagittal coma

$$\Delta s'_{coma} = \frac{t \cdot (n^2 - 1) \cdot \sin w \cdot \sin^2 u}{2n^3} \tag{4.18}$$

The aberrations caused by a plane-parallel plate inserted into the beam path are in general independent of its z position within the beam. This is a consequence of the fact that all aberration formulae for a plane-parallel plate contain only the angles and there is no dependence on the beam height. A prominent application of this feature are high-NA microscopic objective lenses with water immersion for biomedical applications. They can sharply image in the depth of the sample despite the cover glass spherical aberration.

4.3.3 Tilted Plates

Tilted plane-parallel plates are used in practice to achieve a lateral image displacement and in particular to obtain a precise alignment, whereas the angle of rotation or inclination should not be chosen to be too large in order to avoid additional residual aberrations (see Figure 4.6).

Very often in real instruments a beam splitter is realized as a tilted thin plate as is sketched in Figure 4.7(a). The reason for this is that it is easier to realize a coating for the amplitude splitting on a glass–air surface than inside the cube parts on a cemented surface, as in Figure 4.4(a).

Figure 4.6 Lateral image displacement by a tilted plane-parallel plate.

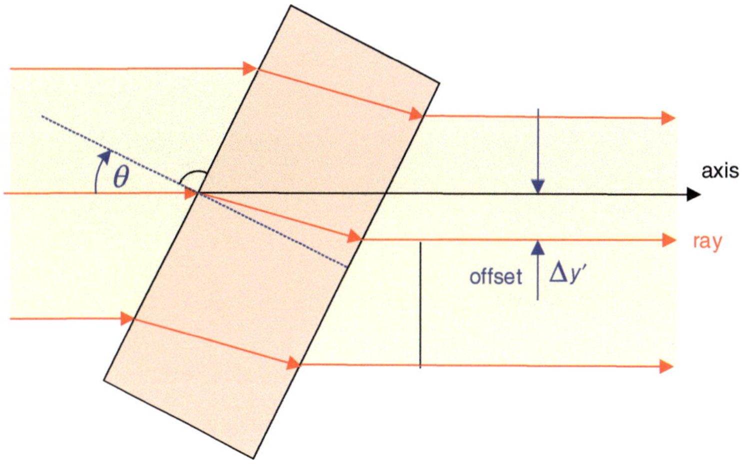

Figure 4.7 Occurrence of skew transmission through a plane plate in (a) and a double image (ghost) by reflection at a plane-parallel plate in (b).

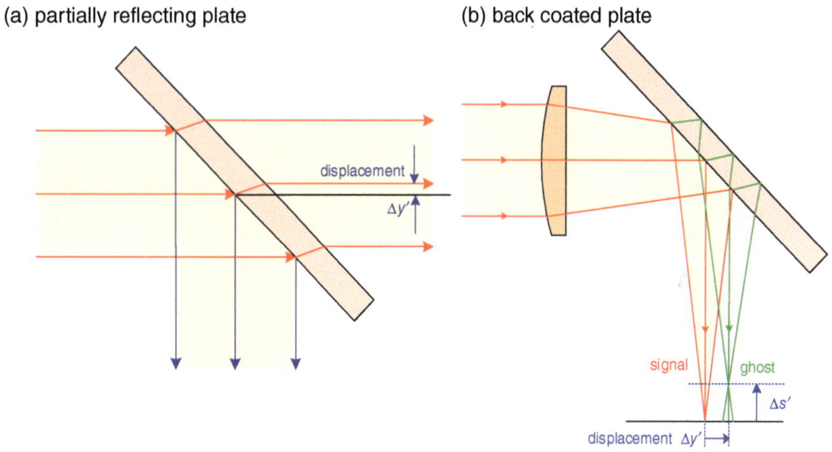

A thin plane-parallel plate with imperfect coatings could lead to spurious double images when used in a non-parallel beam. The reflection at the back surface of the plate has only a slightly shifted back focal distance and thus yields, in the case of small numerical apertures with large depth of focus, an almost sharply defined ghost image with a small lateral displacement $\Delta y'$. The occurrence of such a double image is illustrated in (Figure 4.7(b)).

4.3.4 Non-Orthogonal Prisms

There are a small number of prisms where the axis does not enter normal to the surface. The two most prominent ones are the Dove and the Littrow prism. The unfolding of the Dove prism, in order to generate the tunnel diagram, transforms it into a tilted plate. The two prism types are sketched in Figure 4.8. Due to the ray refraction, the Dove prism is dispersive and the output height changes with wavelength. The ray angle is the same for all wavelengths, but the lateral shift depends on the wavelength.

In the case of the Littrow prism this dispersive effect is used by intention, and the incidence is therefore selected to be very steep. This prism has the property of a retroreflector; it can be considered as a special case of the Dove prism with 90° incidence on the reflective surface.

4.3.5 Plane-parallel Plate in a Convergent Beam*

(*) The content of this section is quite specialized and for advanced reading. Knowledge of this section is not necessary to understand the following chapters. Beginners are recommended to skip this section.

A tilted plane-parallel plate in a convergent beam path introduces spherical aberration, astigmatism, coma, longitudinal chromatic aberration and a chromatic difference in the magnification. Coma varies linearly with the tilt angle, while the dependence of the astigmatism is quadratic [3]. Often this setup is found for beam splitters; therefore, in the figures the reflected ray path is shown in addition in green, see Figure 4.9.

The consideration of the aberrations caused by a plane-parallel plate in a convergent beam and their correction can be simplified and adapted to conventional analysis by taking a different perspective based on the

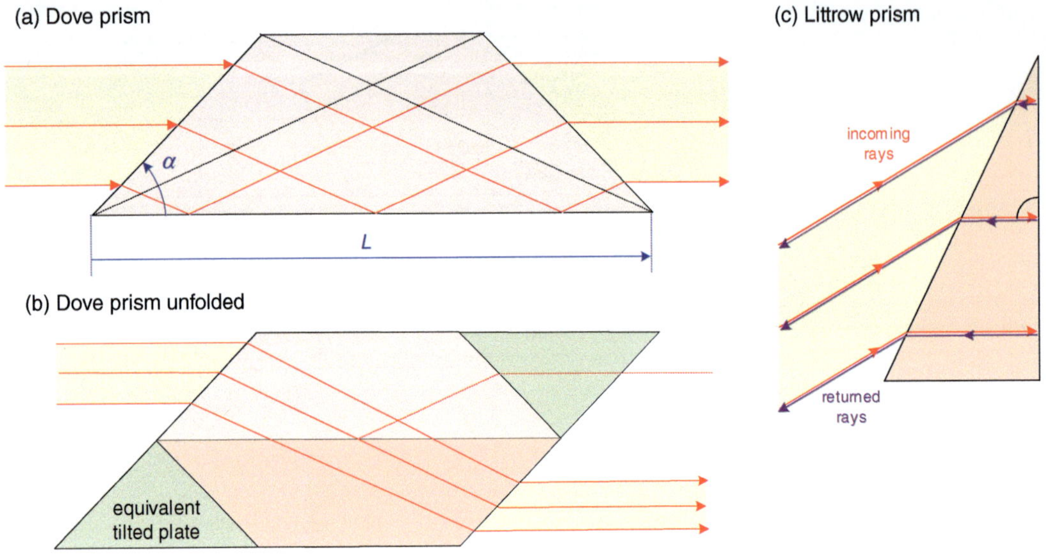

Figure 4.8 Beam paths for (a) a Dove prism and an (b) unfolded Dove as well as (c) a Littrow prism.

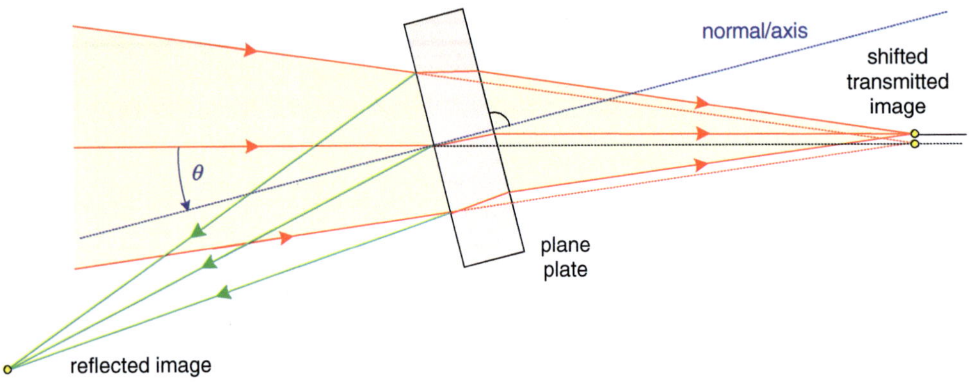

Figure 4.9 Image displacement by a plane-parallel plate centrally located but tilted in a convergent beam.

line connecting the reflected and transmitted images, instead of the direction of the central ray. The new scenario is equivalent to normal imaging and is effective for an off-axis sub-aperture only. This idea is illustrated in Figure 4.10.

In this consideration the plane-parallel plate produces an over-corrected spherical aberration. Correction of this error is possible through an additional lens with a spherical under-correction effect. Figure 4.11 demonstrates this approach.

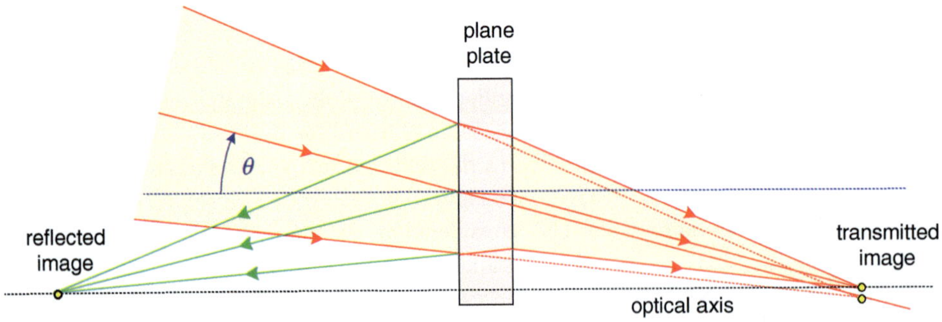

Figure 4.10 Image displacement by a plane-parallel plate from a shifted point of view with a different optical axis.

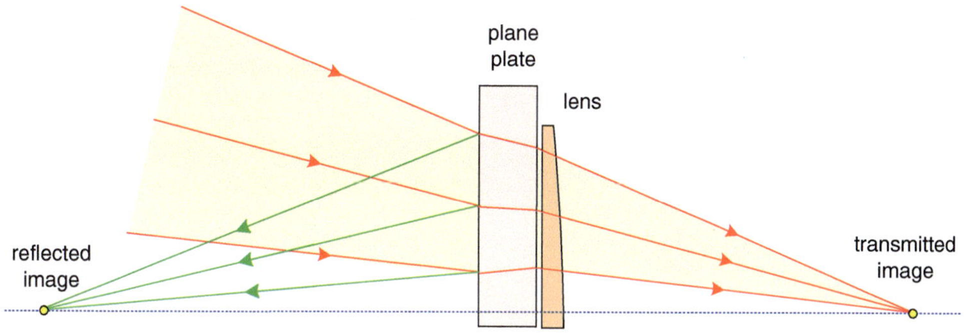

Figure 4.11 Image displacement by a plane-parallel plate in convergent light with an auxiliary lens.

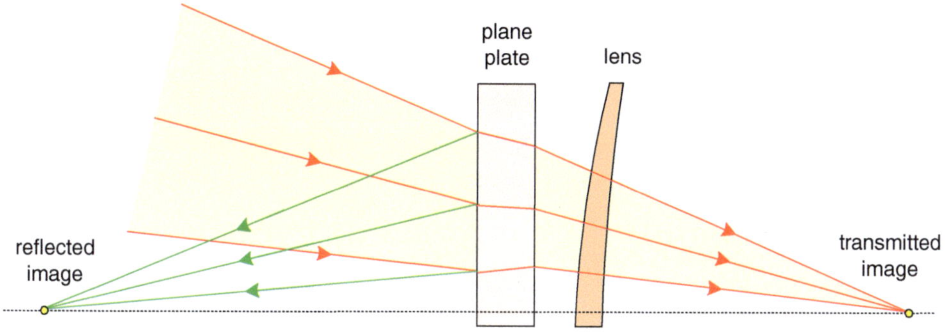

Figure 4.12 Image displacement by a plane-parallel plate in convergent light with an optimized auxiliary lens.

In order to accomplish the spherical correction, the additional lens should be formed through a suitable and specific bending technique. At the same time, the two components should be appropriately separated in order to correct the longitudinal chromatic aberration. This corrected arrangement is shown in Figure 4.12.

Moreover, through the choice of different materials for the two components and optimization of the bending, one can achieve improved coma and astigmatism corrections for a finite image field. Even to the third order, however, this possibility can be realized only approximately.

4.4 Lenses

4.4.1 Notations

The most important optical components to shape the beam geometry of light are lenses. They are used for imaging, beam concentration and general shaping. A lens in the classical sense is a combination of two spherical interfaces with a medium in between which has a different refractive index from either or both of the outer media. The connecting line of the two centres of curvature determines the optical axis of symmetry. The primary parameters of defining a lens are as follows.

- Radii of curvature R_1, R_2 of the surfaces.
- Thickness t measured along the optical axis, which defines the two vertex points.
- Diameter D.
- Refractive index n of the medium.

The definitions of the various quantities are shown in Figure 4.13 and are summarized in Table 4.1.

For practical purposes, the surfaces are mostly defined in a local coordinate system with the origin in the vertex point. The paraxial approximation means that the spherical surface is approximated by the parabola with the same vertex radius of curvature.

For a lens, the meaning of focal point F' is defined as those points where an incoming axis parallel ray intersects the optical axis behind the lens. On the other side, a ray starting at F leaves the lens parallel to the optical axis.

Physically the ray deflection at a lens takes place on both outer surfaces. But this double ray deflection can be considered to take place only once at the principal plane P'. The principal planes define in general those planes where an incoming ray has the same height as the outgoing ray. If this property is used, an incoming ray seems to be deflected only once at the back principal plane, see Section 4.4.3. It should be emphasized that the focal lengths f and f' are measured from the principal plane P and P' respectively. This means that the incoming axis parallel ray seems to be refracted only once; the point Q' is the apparent point of ray bending. This geometry is illustrated in Figure 4.14.

In particular for imaging applications, special properties of lenses and their relative positioning to the object must

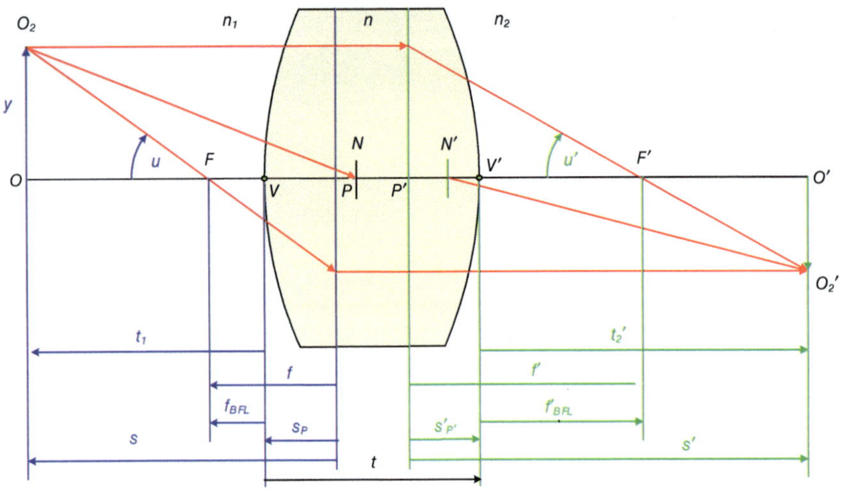

Figure 4.13 Definition of the main quantities at a thick lens with spherical surfaces in different media on both sides. The object side quantities are shown in blue, the image side in green.

Table 4.1 Terms and definitions at a lens.

Quantity	Meaning/name or base point	Remarks
P, P'	Principal planes	
F, F'	Focal points	
V, V'	Vertex points of the surfaces	
N, N'	Nodal planes or base point	If the lens is embedded in the same medium, N and N' are identical to P, P'
f, f'	Focal lengths	They are different if the media before and after are different, measured relative to principal planes
s, s'	Intersection lengths of a ray	Measured relative to vertex points V, V'
R, R'	Radii of curvature	
c, c'	Curvatures of surfaces	$c = 1/R$
t	Thickness along the optical axis	
f_{BFL}, f'_{BFL}	Back focal lengths	Measured relative to vertex points V, V'
O, O'	Object and image points	
s_P, s'_P	Distances OP, $O'P'$	
n	Lens material refractive index	
n_1, n_2	Embedded material indices	

be considered. In some cases, the image is virtual, which means that the image cannot be gathered by a screen. In particular, positive and negative lenses must be distinguished; they have positive or negative focal power and collect or diverge the light. For the construction of the image, special rays can be used that strike the focal or the nodal points of the lens respectively. This is shown in Figure 4.15 for positive and negative lenses. A ray coming parallel to the optical axis becomes a focal ray; an incoming focal ray becomes a parallel ray and a nodal point ray through the centre of a thin lens and has an invariant direction.

4.4.2 Focal Length

The focal length is the most important characteristic quantity of a lens. According to the explanation in the previous section, it is defined as the distance between the focal point and the principal plane base point. Finally, it characterized the focusing properties of a lens. If a ray impinges a lens parallel to the optical axis in a height y, the image side focal length is given by f' in the opposite direction and we get the value of f. u and u' are the angles of the bent rays against the optical axis on both sides respectively.

$$f = \frac{y'}{\tan u}, \; f' = \frac{y}{\tan u'} \tag{4.19}$$

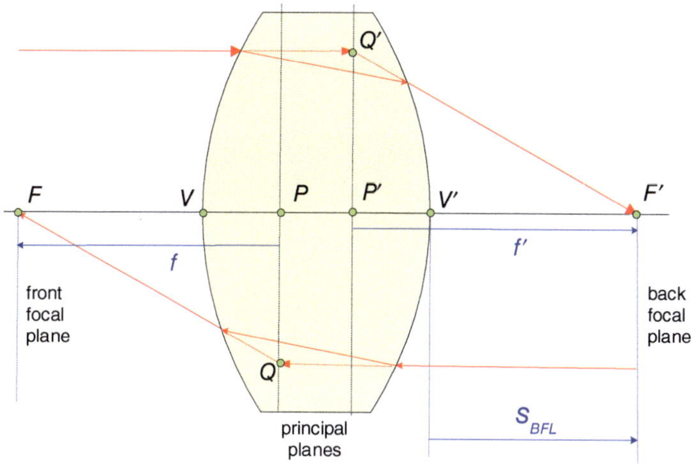

Figure 4.14 Focal points and principal planes of a lens.

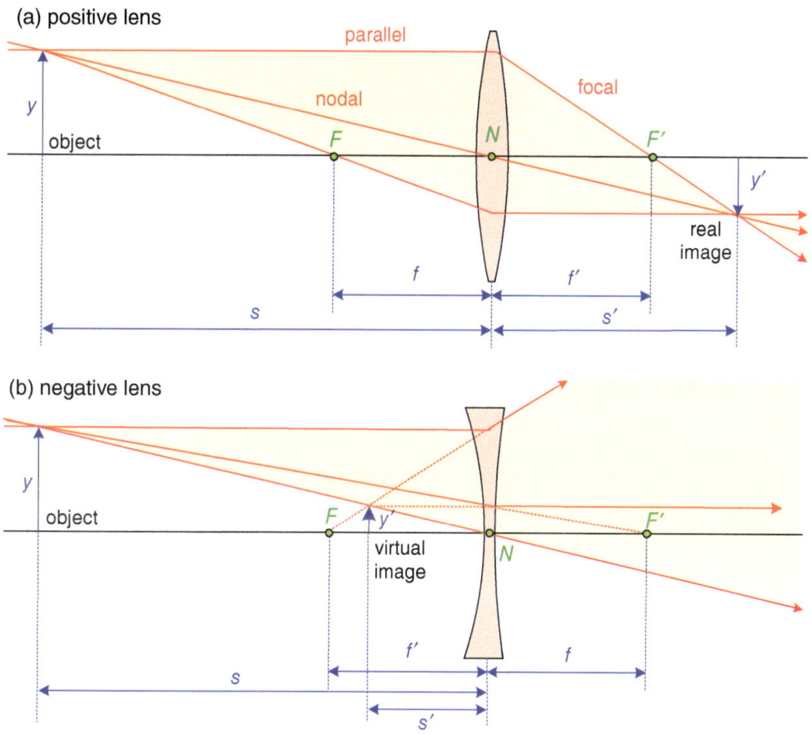

Figure 4.15 Image construction by thin lenses.

The power of a lens is the inverted value of the focal length, corrected by the refractive index. For the power there is only one value and it is given by

$$F = -\frac{n}{f} = +\frac{n'}{f'} \tag{4.20}$$

If the lens is embedded in the same medium, the two focal lengths are the same, only the sign is opposite. A collecting lens has $f' > 0$ and $f < 0$. Usually in daily work the lenses are located in air; in this special case we get the simplified $F = 1/f'$ and the focal length is simply denoted by f.

It should be noted that the focal length is a paraxial definition and that the focal lengths are measured against the principal planes and not relative to the vertex points V; the distance between V and focal point F' is called the back focal length.

In the most general case, a lens consists of one material with refractive index n and different radii. If the refractive index of the medium in front of the lens is n_1 and the medium behind the lens has a refractive index n_2 then with the thickness t of the lens for the focal length we have

$$f' = \frac{nn_1R_1R_2}{-nR_1(n-n_2) + nR_2(n-n_1) + t(n-n_1)(n-n_2)} \tag{4.21}$$

If the media on both sides of the lens are equal, i.e. $n_1 = n_2 = n'$, the equations above can be simplified in the

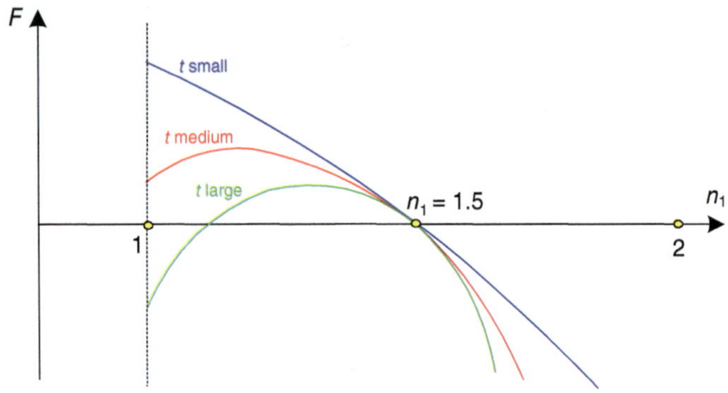

Figure 4.16 Change of power of a lens with $n = 1.5$ in a surrounding medium with index $n_1 = n_2$.

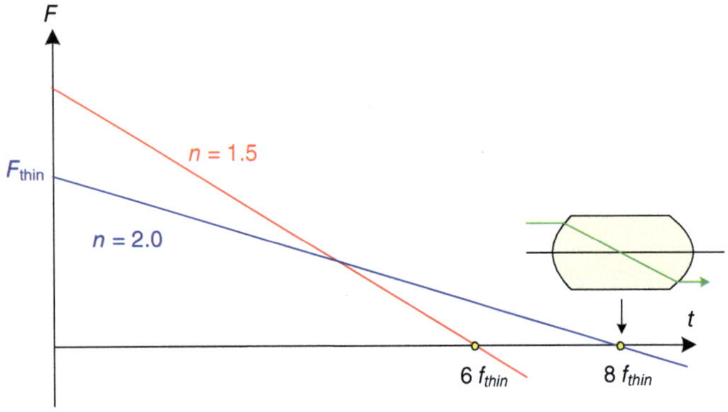

Figure 4.17 Change of power with thickness and afocal rod lenses in air.

following way. The focal length has the value

$$f' = \frac{nn'R_1R_2}{(n-n')\cdot[-nR_1+nR_2+t(n-n')]} \tag{4.22}$$

If the focal power is drawn as a function of the index of the surrounding medium, a behaviour like the one in Figure 4.16 is obtained for different values of the thickness t.

An alternative formulation uses the focal power F, which is the inverted focal length and is given for a lens in air as

$$F = \frac{1}{f'} = (n-1)\cdot\left(\frac{1}{R_1}-\frac{1}{R_2}\right)+\frac{(n-1)^2\cdot t}{nR_1R_2} \tag{4.23}$$

It can be seen that the focal power changes linearly with the thickness. If, in particular, the thickness t can be neglected, the very often used formula for a thin lens focal power and focal length is obtained by simply removing the last term

$$F_{thin} = \frac{1}{f'_{thin}} = (n-1)\cdot\left(\frac{1}{R_1}-\frac{1}{R_2}\right) = (n-1)(c_1-c_2) \tag{4.24}$$

For the special case of a thin lens in air with one plane surface (plano convex lens, PC) or a symmetric form with $R_2 = -R_1$ (biconvex lens, BC), we get the simpler versions for the focal length

$$f'_{PC} = \frac{R}{n-1},\quad f'_{BC} = \frac{R}{2\cdot(n-1)} \tag{4.25}$$

If a symmetric lens is considered with $c = c_1 = -c_2$, the focal power vanishes for the thickness

$$t_o = \frac{2nR}{(n-1)} = 4n\cdot f_{thin} \tag{4.26}$$

This large thickness corresponds to a rod lens as indicated in Figure 4.17, for the refractive index $n = 1.5$ one obtains $t = 6R$.

The locations of the principal planes in the case of a thick lens are given relative to the vertex by distances of

$$s_P = \frac{-t\cdot R_1}{(n-1)\cdot t+n\cdot(R_2-R_1)},$$
$$s'_P = \frac{-t\cdot R_2}{(n-1)\cdot t+n\cdot(R_2-R_1)} \tag{4.27}$$

If a lens is thin and is located in different media, the term with the thickness t can be neglected and

$$f' = \frac{n_1R_1R_2}{-R_1(n-n_2)+R_2(n-n_1)}$$
$$F = (n-n_2)\cdot c_1-(n-n_1)\cdot c_2 \tag{4.28}$$

is obtained. If the media are the same $n_1 = n_2 = n'$, the formula simplifies to

$$f' = \frac{n'R_1R_2}{(n-n')(-R_1+R_2)}$$
$$F = (n-n')\cdot(c_1-c_2) \tag{4.29}$$

It should be noted that the formulation of the indices 1 and 2 is not symmetric.

If a lens is symmetrical and can be considered as being relatively thin, i.e. $t \ll r$, the principal points are symmetrical at

$$s'_{P'} = -s_P = -\frac{t}{2n} \tag{4.30}$$

If the material of the symmetrical lens $R_1 = -R_2 = R$ in air has a refractive index $n = 1.5$, then the equations for the focal length are particularly simple

$$f' = R \tag{4.31}$$

and the principal points are symmetrical at

$$s'_{P'} = -s_P = -\frac{t}{3} \tag{4.32}$$

An overview of all the versions of focal length formula are summarized in Table 4.2.

4.4.3 Principal Planes and Surfaces

As already mentioned, the definition of the principal planes is that an arbitrary incoming ray is transferred from P to P' by the same height h. In particular, an incoming ray parallel to the axis seems to be deflected only once with a bending point on the back principal plane P'. This geometry is sketched in Figure 4.18.

A graphical construction of rays can be done easily with the help of the principal planes. This basic construction of the principal planes is a paraxial one. If the lens is non-paraxial and the outgoing ray angle u cannot be considered to be small, the 1 : 1 transfer of the ray heights is no longer valid. But the basic idea can be generalized; this means there is a non-planar surface for which this requirement is fulfilled, at least in case of rotational symmetry. In particular, a system that fulfils the sine condition is characterized by a principal surface that is a sphere. Figure 4.19 shows this case of a single lens; Figure 4.20 shows a high NA microscopic lens with the point of apparent ray bending in the object space. This means that for a corrected sine condition the principal surfaces are spheres; P' is here indicated by the blue dots for a discrete number of marginal rays. A detailed discussion of the sine condition is found later in Volume 2, Chapter 17. The bending of the principal surfaces is only apparent on the high-NA side, for a collimated ray bundle it remains a plane.

But, unfortunately, the concept of principal surfaces breaks completely down if an off-axis object point is considered. For object points far from the axis, the apparent locations are complicated surfaces that may consist of two branches. This is sketched in Figure 4.21. But this formal correct construction of a principal surface as the location of the apparent ray deflection makes only limited sense in practice.

Table 4.2 Collection of focal lengths formulae.

Thickness	Indices	Curvatures	Focal length/focal power
Thick	General $n_1 - n - n_2$		$f' = \frac{nn_1R_1R_2}{-nR_1(n-n_2) + nR_2(n-n_1) + t(n-n_1)(n-n_2)}$
	$n_1 = n_2$		$f' = \frac{nn'R_1R_2}{(n-n')\cdot[-nR_1 + nR_2 + t(n-n')]}$
	In air $n_1 = n_2 = 1$		$F = \frac{1}{f'} = (n-1)\cdot\left(\frac{1}{R_1} - \frac{1}{R_2}\right) + \frac{(n-1)^2\cdot t}{nR_1R_2}$
Thin	General $n_1 - n - n_2$		$f' = \frac{n_1R_1R_2}{-R_1(n-n_2) + R_2(n-n_1)}$ $F = (n-n_2)\cdot c_1 - (n-n_1)\cdot c_2$
	$n_1 = n_2$		$f' = \frac{n'R_1R_2}{(n-n')(-R_1+R_2)}$ $F = (n-n')\cdot(c_1 - c_2)$
	In air $n_1 = n_2 = 1$		$f' = \frac{R_1R_2}{(n-1)(-R_1+R_2)}$ $F = \frac{1}{f'} = (n-1)\cdot\left(\frac{1}{R_1} - \frac{1}{R_2}\right) = (n-1)(c_1 - c_2)$
		Symmetrical biconvex or biconcave	$f'_{BC} = \frac{R}{2\cdot(n-1)}$
		One side plane, plane-convex or plane-concave	$f'_{PC} = \frac{R}{n-1}$

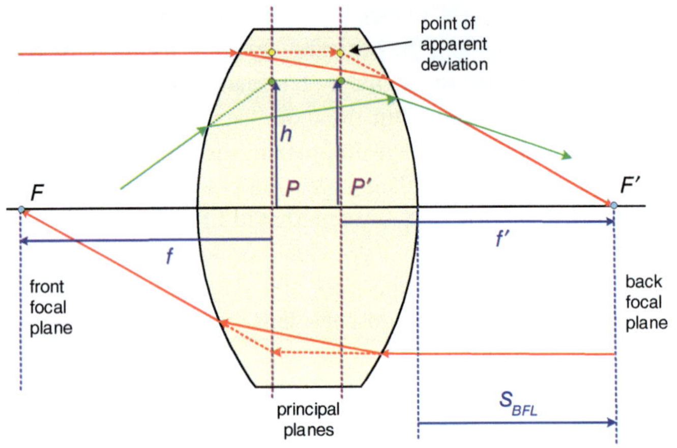

Figure 4.18 Principal plane definition of a lens. The green ray is an arbitrary ray, the red one impinges the lens parallel to the optical axis.

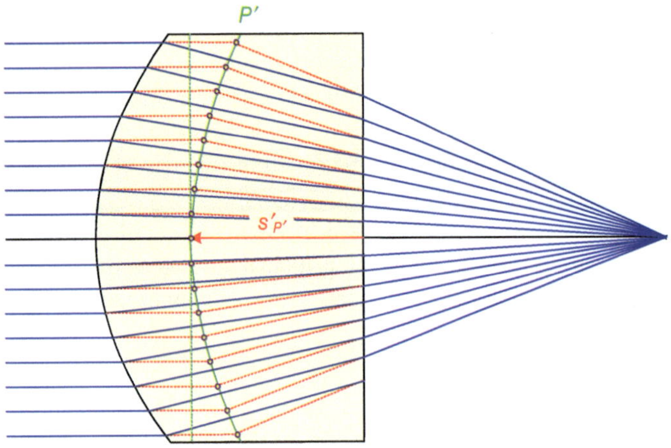

Figure 4.19 Spherical principal surface for a sine-condition corrected lens with large aperture angle.

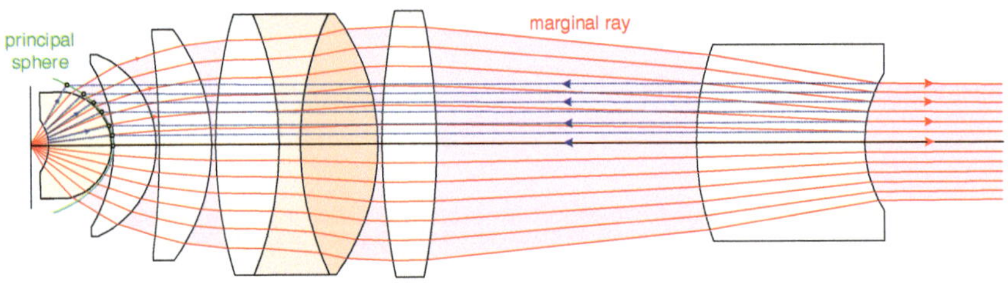

Figure 4.20 Spherical principal surface for a sine-condition corrected lens microscopic lens [4] on the object side, indicated by blue dots.

4.4.4 Lens Shape and Bending

There are different possible shapes for lenses,we distinguish between:

- biconvex/biconcave, symmetric or not;
- plane convex/concave, one surface plane;
- meniscus, both surface radii with the same sign to the left or the right side respectively.

Convex means bending outside and concave corresponds to a hollow surface. In particular, for meniscus shaped lenses it should be noted that the principal planes can be located outside of the lens body, see Figure 4.22.

Changing the shape of a lens is of particular interest in lens design. If the shape is modified by keeping the focal length constant, the paraxial construction of a multi-lens system is preserved, but the distribution of the incidence angles at a lens is changed. This basic idea is of great importance in correcting aberrations. To come to a more quantitative description, the so-called lens bending parameter X is defined as

$$X = \frac{R_1 + R_2}{R_2 - R_1} = \frac{c_1 + c_2}{c_1 - c_2} = \frac{\alpha + 1}{\alpha - 1} \tag{4.33}$$

with the ratio of the radii

$$\alpha = \frac{R_2}{R_1} \tag{4.34}$$

see Figure 4.23 for illustration.

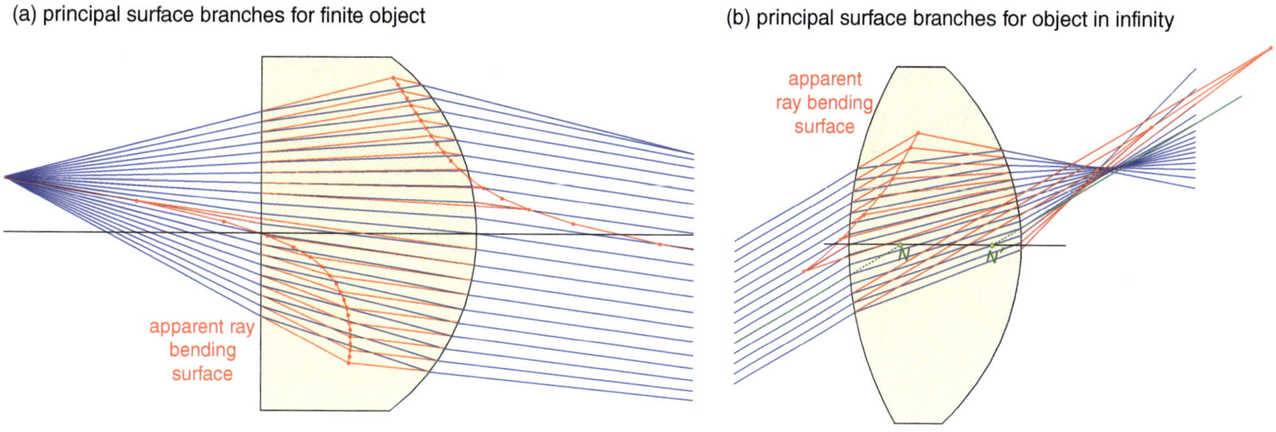

Figure 4.21 General principal surfaces for an off-axis ray bundle for an object (a) in finite distance and (b) in infinity.

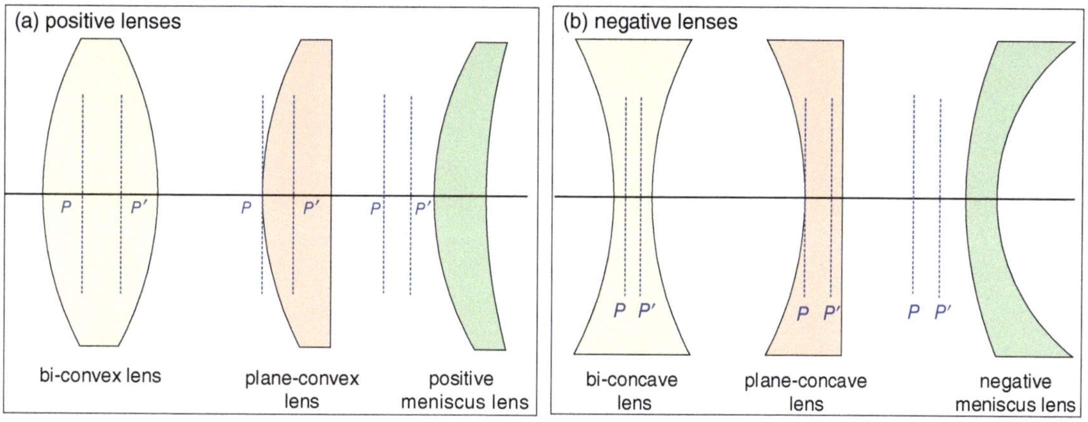

Figure 4.22 Principal planes of different lens shapes.

Figure 4.23 Lens bending X and corresponding shapes.

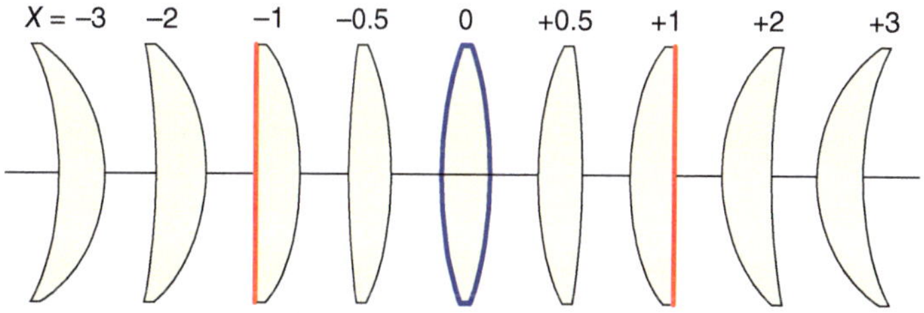

If, for a thin lens, X is changed and the focal length is preserved, the shape and the radii values are determined. Beneath the change of the incidence angles the principal planes of the lens also move relative to the lens position. This is visualized in Figure 4.24. An incoming ray parallel to the optical axis is focused into the focal point in all cases, but the ray paths and the incidence angles at both surfaces are different in every constellation depending on X. In the figure, the principal planes P and P' are kept constant; the lens shapes belonging to the X-values are coloured.

If the power F and the bending X of a thin lens are given, the corresponding radii of curvature can be determined by

$$c_1 = \frac{F \cdot (X+1)}{2(n-1)}, \quad c_2 = \frac{F \cdot (X-1)}{2(n-1)}$$
$$R_1 = \frac{2(n-1)f}{X+1}, \quad R_2 = \frac{2(n-1)f}{X-1} \tag{4.35}$$

These relations show that F and X is an equivalent pair of parameters to describe a thin lens with the radii R_1 and R_2 and in particular that the bending is independent from the power F. For a lens with finite thickness t, the relation

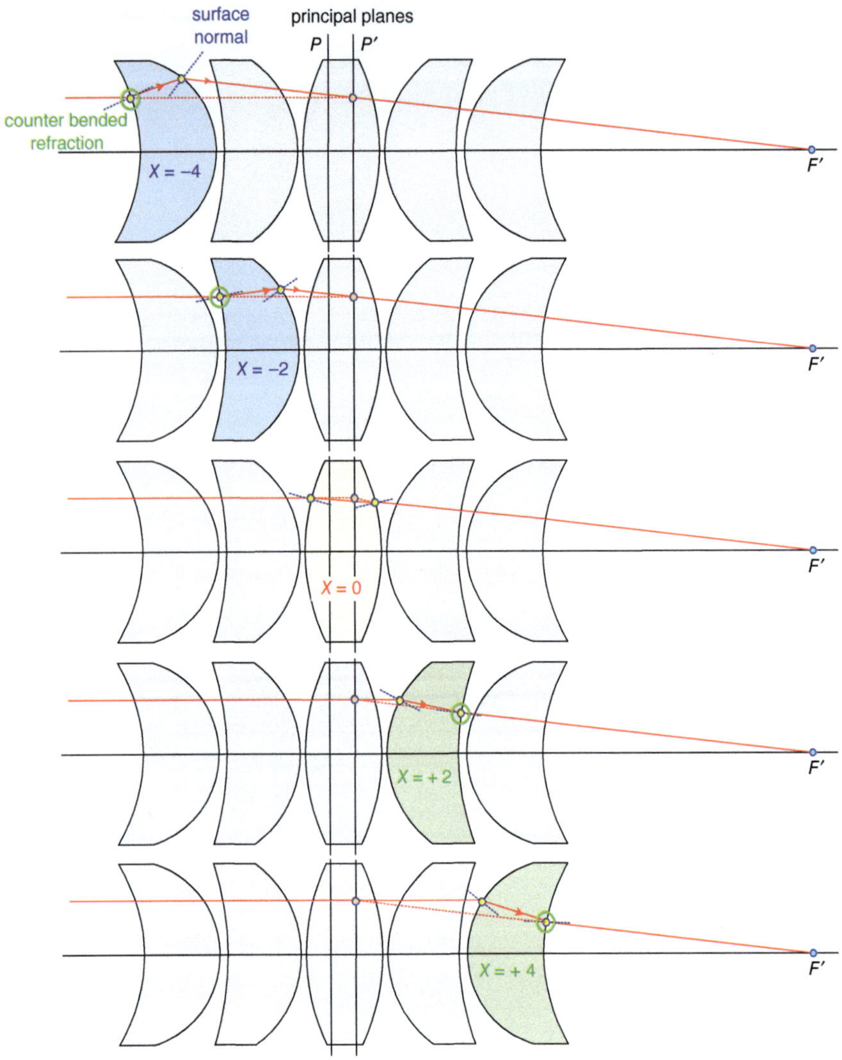

Figure 4.24 Shown here is a lens with constant focal length and different bending values X. All the lenses are drawn simultaneously, keeping the position of the principal plane constant. The coloured lenses belong to the X-values.

between the bending, the focal length and the thickness are given by

$$R_1 = \frac{(n-1)\left[f + \sqrt{f^2 + (X^2-1)\frac{f\cdot t}{n}}\right]}{X+1} \approx \frac{2(n-1)f}{X+1} + \frac{(n-1)(X-1)t}{2n}$$
$$R_2 = \frac{(n-1)\left[f + \sqrt{f^2 + (X^2-1)\frac{f\cdot t}{n}}\right]}{X-1} \approx \frac{2(n-1)f}{X+1} + \frac{(n-1)(X+1)t}{2n} \quad (4.36)$$

Here the last approximated expressions are obtained as a linear truncated Taylor expansion for small thickness values t.

In principle it is easily possible to visualize the various shapes of a single lens in a common diagram according to [5]. This is shown in Figure 4.25 in a normalized manner. According to the re-arranged formula of the focal power in the form

$$\frac{tF}{n-1} = tc_1 - tc_2 + \frac{n-1}{n}\cdot tc_1 \cdot tc_2 \quad (4.37)$$

the diagram has the scaled axes tc_1 and tc_2. The various shapes are indicated by lines, colours and sketches. Thin lenses are near to the origin, special values of the lens power are indicated by red lines, while the blue lines show the location of symmetric, anti-symmetric and concentric shapes.

If a lens is bent while its thickness and focal length are kept constant, the positions of the principal planes change relative to the vertex positions.

A second important parameter for a lens is the relative behaviour of the marginal ray change at the lens. The

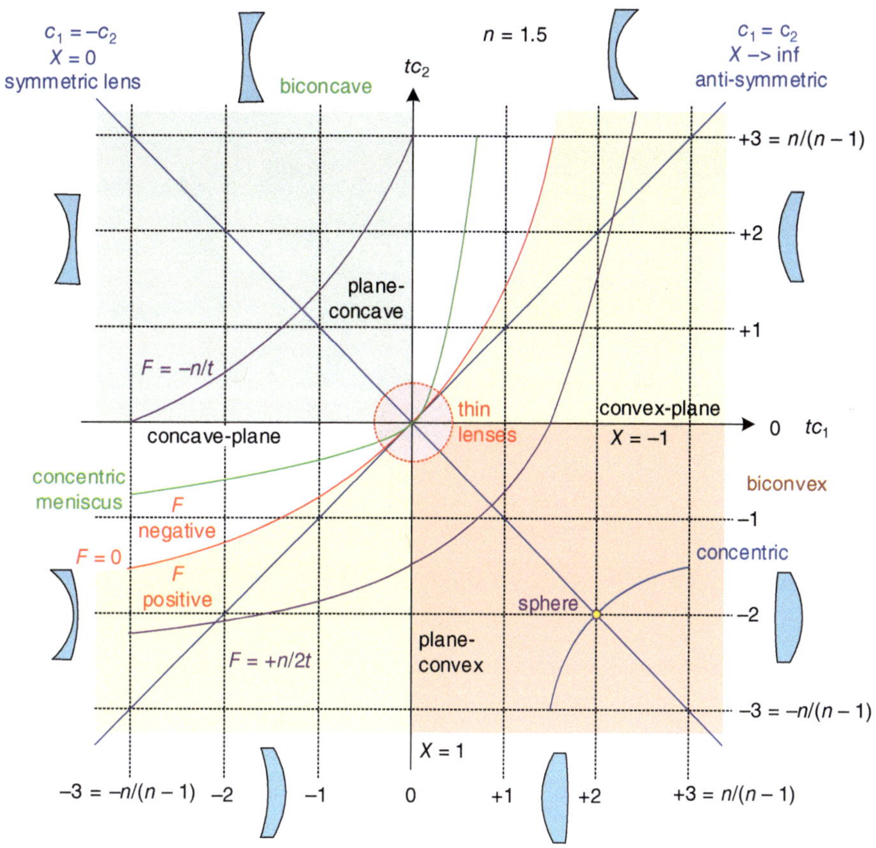

Figure 4.25 Normalized diagram for various lens shapes.

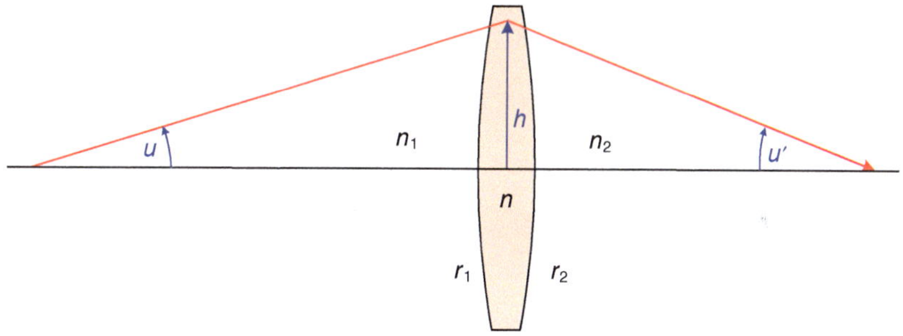

Figure 4.26 General case of a lens embedded into different indices of refraction.

so-called conjugation parameter M indicates how the marginal ray angles u and u' before and behind the lens change, see Figure 4.26. The definition is similar to the bending

$$M = \frac{u + u'}{u' - u} = \frac{1 + m}{1 - m} \tag{4.38}$$

with magnification m. If M, the height h at the lens and F are given, the angles can be determined by

$$u = \frac{hF \cdot (M - 1)}{2}, \quad u' = \frac{hF \cdot (M + 1)}{2} \tag{4.39}$$

4.4.5 Generalized Bending*

(*) The content of this section is quite specialized and for advanced reading. Knowledge of this section is not necessary to understand the following chapters. Beginners are recommended to skip this section.

If a lens in different surrounding media is considered, the well-known formulas for the bending are no longer valid and meaningful and a generalization is needed. Let's assume that the lens material has the refractive index n, the medium before has index n_1 and behind we have n_2.

A feasible definition of the generalized bending parameter is in this case

$$X^* = \frac{(n-n_1)\cdot c_1 + (n-n_2)\cdot c_2}{(n-n_1)\cdot c_1 - (n-n_2)\cdot c_2} = \frac{(n-n_1)\cdot c_1 + (n-n_2)\cdot c_2}{F} \quad (4.40)$$

The conjugation parameter should be formulated in the following way with the ray angles u_1 and u_2

$$M^* = \frac{n_2 u' + n_1 u}{n_2 u' - n_1 u} = \frac{1+m}{1-m} \quad (4.41)$$

From these definitions we get

$$c_1 = \frac{F\cdot(X^*+1)}{2(n-n_1)}, \quad c_2 = \frac{F\cdot(X^*-1)}{2(n-n_2)} \quad (4.42)$$

and

$$u = \frac{hF\cdot(M^*-1)}{2n_1}, \quad u' = \frac{hF\cdot(M^*+1)}{2n_1} \quad (4.43)$$

These equations are relevant for individual lenses of cemented systems.

4.4.6 Modified Bending Definition*

(*) The content of this section is quite specialized and for advanced reading. Knowledge of this section is not necessary to understand the following chapters. Beginners are recommended to skip this section.

The Abbe or refraction invariant for thin lens in air reads at the front and the rear surface respectively

$$Q_1 = \frac{1}{R_1} - \frac{1}{s_1}, \quad Q_2 = \frac{1}{R_2} - \frac{1}{s_2'} \quad (4.44)$$

with intersection lengths $s = s_1$ and $s' = s_2'$ for the lens. A modified version of bending can now defined as the sum of both surface contributions

$$\overline{X} = \frac{1}{R_1} - \frac{1}{s} + \frac{1}{R_2} - \frac{1}{s'} = Q_1 + Q_2 \quad (4.45)$$

If in this equation the focal power F and the imaging equation for the intersection lengths s and s' as well as the classical definitions of X and conjugation parameter M

$$F = (n-1)\left(\frac{1}{R_1} - \frac{1}{R_2}\right), \quad h = -us = -u's'$$
$$X = \frac{R_1 + R_2}{R_2 - R_1}, \quad M = \frac{u+u'}{u'-u} \quad (4.46)$$

are used and the variables R_1, R_2, u, u', s and s' can be eliminated and the relation

$$\overline{X} = F\cdot\left[\frac{X}{2(n-1)} - M\right] \quad (4.47)$$

is obtained. $\overline{X}$ is now a property that is not only describing the shape of the lens, but also includes via M the relative change of the ray path. If the refractive index is assumed to be $n = 1.5$, in the brackets of the formula the difference X-M occurs. The interpretation according to the meaning of the refraction invariant is a relative strength of refraction. If a stronger bending X is given and the ray path is simultaneously changed, the ray deviation work of the lens is still small, so a synchronous change of X and M does not increase the ray bending of the lens. This behaviour is sketched in Figure 4.27 by the diagonal green dotted line. The strong ray refraction along the anti-diagonal line at the end provokes large aberrations, these are the 'counter bent' lenses for corresponding correction purposes.

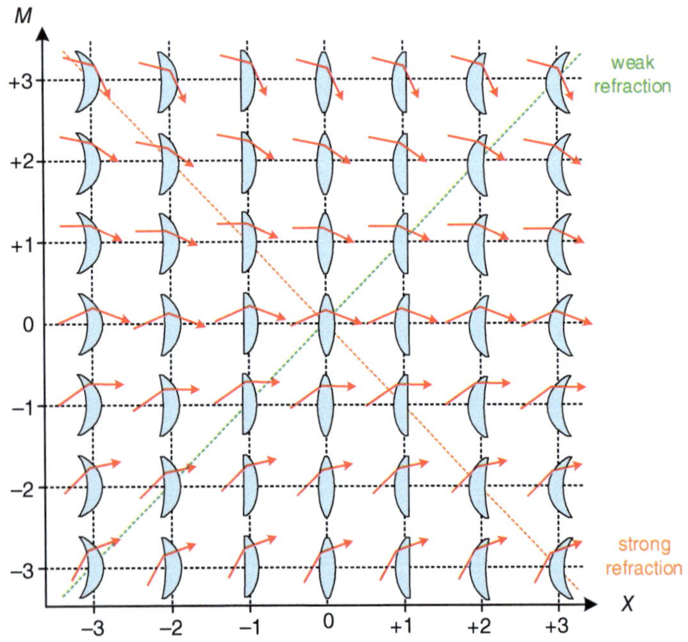

Figure 4.27 Combined effect of lens bending and ray deflection.

4.4.7 Thick Lenses

In practice, the simpler formula of a focal length of a thin lens in air is preferably used. The question is now: what are the conditions that allow for this approximation and what errors occur in this case?

When using a thick lens in practice the question is often raised regarding which thickness t can be considered as 'thin'. If the radii of curvature of the lens are relatively large compared with the thickness

$$|c_1 \cdot t| \ll 1, \quad |c_2 \cdot t| \ll 1 \tag{4.48}$$

the change in height for a ray incident parallel to the axis within the lens can be estimated by assuming that the refracting effect of the first surface is located in the front vertex plane and then the angle of the ray can be used in order to calculate its height in the back vertex plane. In paraxial approximation the relation is

$$y_2 = \left[1 - \frac{n-1}{n} \cdot t \cdot c_1\right] \cdot y_1 \tag{4.49}$$

The corresponding geometry is sketched in Figure 4.28. So the relative height change of the ray is

$$\frac{\Delta y}{y_1} = \frac{n-1}{n} \cdot t \cdot c_1 = \frac{n-1}{n} \cdot \frac{t}{R_1} \approx \frac{1}{3} \cdot \frac{t}{R_1} \tag{4.50}$$

with the approximation of $n = 1.5$ in the last formulation. If this value can be neglected, the lens can be considered as thin. Since the heights of the rays at a lens are of significant importance for the generated aberrations at the surface, this is a rather critical parameter. In the case of a thin lens, the assumption is always that the ray heights are the same at both surfaces.

In the discussion of a paraxial approach for an optical system, usually thin lenses are considered first. For the practical realization of such a system, however, these have to be replaced by real lenses of finite thickness. Consider a single thin lens with the refractive power F. Its two surfaces have the curvatures c_1 and c_2. From a general point of view, the thick lens has three geometrical degrees of freedom, the thickness t and the two curvatures $c_{1/2}$, the thin lens in contrast has only the two parameters $c_{1/2}$. From the viewpoint of a paraxial substitution, it is obvious that the location of the first principal plane as well as the focal length of the lens should be invariant. This means that one more degree of freedom allows for one more requirement.

If a thin lens layout is transferred into a real setup with thick lenses, the main idea is to split the principal plane of a thin lens into the two separated principal planes of a thick lens. The refractive index is kept constant to avoid any complications with chromatical behaviour. Both configurations are compared in Figure 4.29.

The first option to exchange the ideal lens for a real thick one is as follows. The radii $R_{1/2}$ remain the same. The location of the first principal plane P is chosen to be identical to the position of the thin lens. The distances between the rear vertex point and the rear principal plane and the corresponding front points are given by [6, 7]

$$s'_P = -\frac{n-1}{n} \cdot \frac{f \cdot t}{R_1}, \quad s_P = -\frac{n-1}{n} \cdot \frac{f \cdot t}{R_2} \tag{4.51}$$

The separation between the two principal planes and the stretching of the system is given by

$$\begin{aligned}\Delta z &= t - s_P + s_{P'} = t - \frac{n-1}{n} t \cdot f \cdot \left(\frac{1}{R_1} - \frac{1}{R_2}\right) \\ &= \frac{t(n-1)}{n}\end{aligned} \tag{4.52}$$

and is independent from the bending X.

In the second meaningful approach to replace the thin lens, it is requested that the marginal ray as well as the chief ray with stop at the principal plane P are invariant and only shifted by Δz. The chief ray condition fixes the plane P and the marginal ray angle fixes a constant focal power $F = 1/f$. The solution to this problem is not unique, one uncertainty is unresolved, which is the shape or bending of the lens. This open degree of freedom can in reality be used to force one more Seidel aberration to be invariant for the thick lens. This can be the bending X of the lens, which ensures a constant spherical aberration and coma (for stop at the principal plane). If the ratio of the radii of curvature $R_{1/2}$ of the thin lens and $R_{th1/2}$ of the thick lens are constant

$$\alpha = \frac{R_{th2}}{R_{th1}} = \frac{R_2}{R_1} \tag{4.53}$$

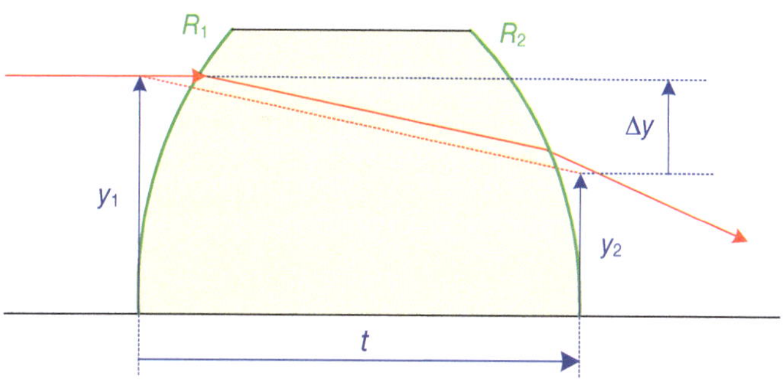

Figure 4.28 Change of the ray height at a thick lens.

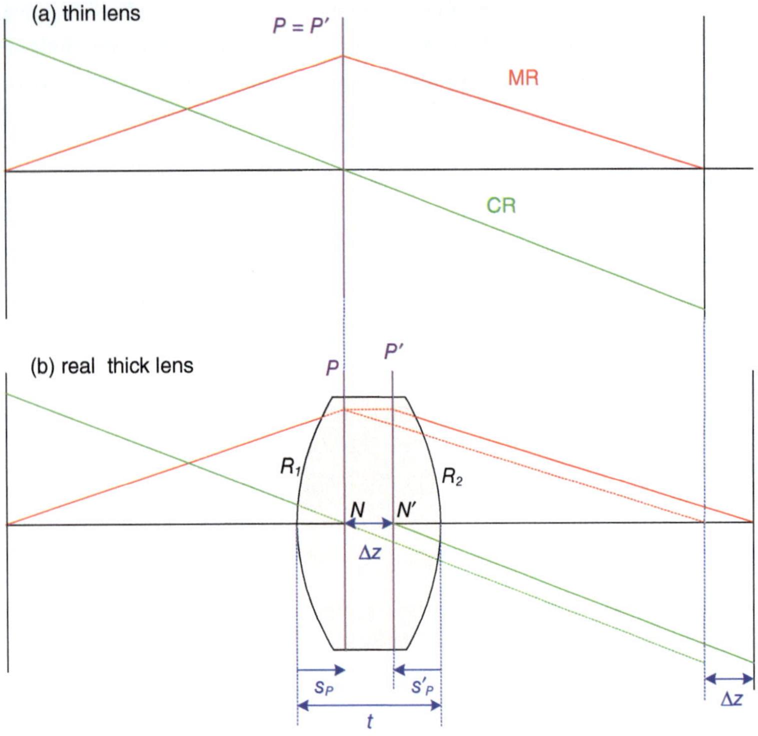

Figure 4.29 Substitution of a thin lens (a) by a thick lens (b).

due to

$$X = \frac{R_1 + R_2}{R_2 - R_1} = \frac{\alpha + 1}{\alpha - 1} \tag{4.54}$$

the bending is therefore constant. If in addition the focal power must be invariant, which means

$$F = (n-1)\left(\frac{1}{R_1} - \frac{1}{R_2}\right) = (n-1)\left(\frac{1}{R_1} - \frac{1}{R_2}\right) + \frac{t(n-1)^2}{nR_1R_2} \tag{4.55}$$

which gives the new radii of the thick lens

$$R_{th1} = r_1 + \frac{(n-1)(1-\alpha)}{2\alpha F} \cdot \left[1 - \sqrt{1 + \frac{4t\alpha F}{n(\alpha-1)^2}}\right]$$
$$R_{th2} = r_2 + \frac{(n-1)(1-\alpha)}{2F} \cdot \left[1 - \sqrt{1 + \frac{4t\alpha F}{n(\alpha-1)^2}}\right] \tag{4.56}$$

The principal planes are then shifted to

$$s_{p1} = \frac{\alpha - 1}{2\alpha F} \cdot \left[1 - \sqrt{1 + \frac{4t\alpha F}{n(\alpha-1)^2}}\right]$$
$$s'_{p1} = \frac{\alpha - 1}{2F} \cdot \left[1 - \sqrt{1 + \frac{4t\alpha F}{n(\alpha-1)^2}}\right] \tag{4.57}$$

and P is then not invariant as in the first change option above. The advantage of this selection is a better coincidence of the third order aberrations by exchanging the thin lens for a thick one.

The option considering the change of the Seidel aberrations is discussed in more detail in Volume 2, Chapter 14.

The equivalent substitution of a thin by a thick lens presented here is only a paraxial consideration.

4.4.8 The Ideal Lens Dilemma*

(*) The content of this section is quite specialized and for advanced reading. Knowledge of this section is not necessary to understand the following chapters. Beginners are recommended to skip this section.

Usually, a lens is called ideal if it generates a perfect image. This means that all the rays starting at one object point intersect in one image point only. Mathematically, the linear collineation can be considered to model ideal systems, this is schematically shown in Figure 4.30(a) for an axis object point. Unfortunately, real optical components are not able to realize an imaging behaviour like this. As a special case, an aplanatic lens with a stigmatic point imaging on axis and a corrected imaging for small fields shows slightly different behaviour. In the image plane, a perfect image point can be achieved, but looking at Figure 4.30(b) it can be seen that in this case the principal surfaces are spheres. In the case of a sphere, the ray heights at the principal surface grow with the sine of the aperture angle. Mathematically, the tan-function of the collineation and the sine-function of the aplanatic case are not compatible. This has the consequence that the ray paths are not identical in both cases. From the modelling point of view, three possible solutions exist for this dilemma [8, 9].

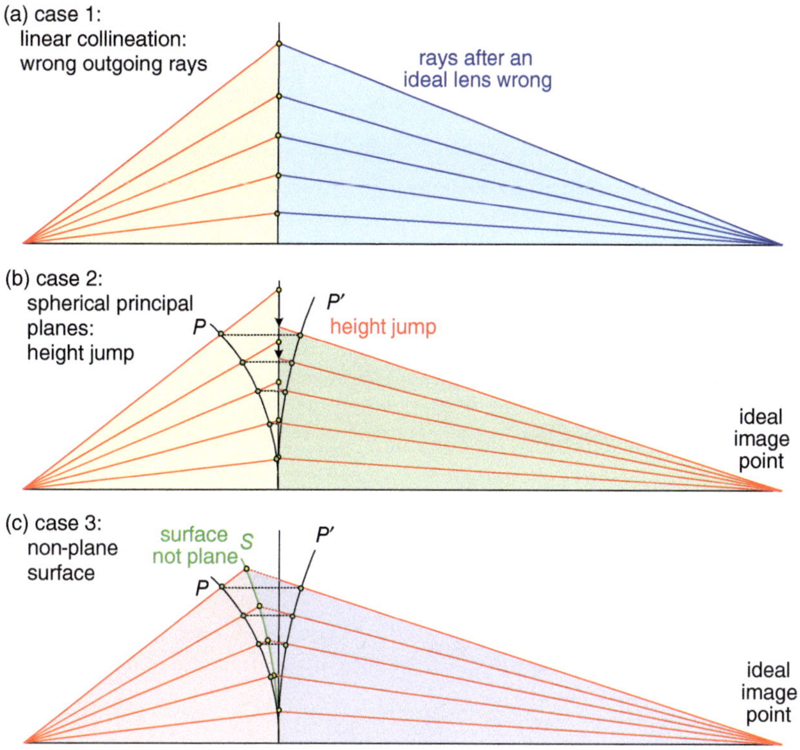

Figure 4.30 Comparison of the possibilities with (a) linear collineation for the blue rays and a wrong outgoing ray direction, (b) a height jump at the surface and (c) a curved surface.

- The ray height at the vertex plane of the component is the same for incoming and outgoing rays; the consequence in this case is an incorrect ray path behind the lens, it is only correct in the image plane. See Figure 4.30(a) and outgoing blue rays. This means that the predicted numerical aperture is wrong too.
- The ideal lens is represented by a plane and it is accepted that the rays make a height jump in this plane to gain the correct outgoing height. The layout looks somewhat unphysical in this case. See the outgoing red rays in Figure 4.30(b). It should be noted that the outgoing aperture angles in the image domain are not consistent with the artificial ideal lens.
- The correct ray paths behind the lens are requested and the synthetic intersection points with the incoming rays are used to construct a curved synthetic ideal lens surface S. This case is illustrated in Figure 4.30c by the red outgoing rays, which intersect the incoming rays at the green surface S.

An ideal lens is a synthetic construct, which is hard to realize in a computational algorithm consistently. An ideal lens is obtained if the principal planes coincide, the refractive index goes to infinity and the focal power remains constant. This means that it is physically impossible to request an ideal behaviour simultaneously for:

- a plane which represents the ideal lens;
- the ray path behind the lens, in particular the ray angles;
- the image plane, to guarantee a perfect stigmatic imaging on and off-axis.

Therefore, the properties that are requested to be 'ideal' must be defined.

A perfect lens should in addition image an off-axis point perfectly. The conditions to gain a perfect lens are:

- The sine condition is fulfilled

$$m = \frac{\sin u}{\sin u'} \tag{4.58}$$

- The imaging is ideal, an object point generated an image point and the optical path difference through the lens is zero

$$\frac{1}{z_i} - \frac{1}{z_o} = \frac{1}{f} \tag{4.59}$$

The magnification m reads (here z is used for the distances instead of s to avoid a confusion with the direction unity vectors)

$$m = \frac{y_i}{y_o} = \frac{z_i}{z_o} \tag{4.60}$$

If all these conditions are simultaneously fulfilled, the coma rays automatically cannot intersect in the plane of the lens in one point, therefore they make a step in height. This does not satisfy the usual behaviour of the ray at the principal plane [10].

If a single perfect lens is considered as shown in Figure 4.31, $\vec{s}$ are the ray unit vectors, the chief ray through

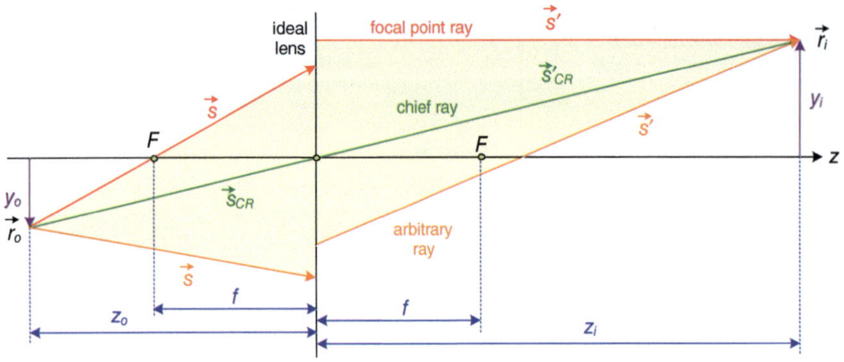

Figure 4.31 Ideal lens magnification and ray split at the principal plane.

the centre of the lens fulfils the condition

$$\vec{s}'_{CR} = \vec{s}_{CR} \tag{4.61}$$

The direction of an arbitrary ray is calculated by the relations, applying the magnification onto the parabasal angle differences to fulfil the sine condition

$$s'_x = s_{x(CR)} + \frac{s_x - s_{x(CR)}}{m}, \quad s'_y = s_{y(CR)} + \frac{s_y - s_{y(CR)}}{m},$$
$$s'_z = \sqrt{1 - s'^2_x - s'^2_y} \tag{4.62}$$

if the object and image have finite distances.

Finally, a problem with the physical consistency exists in the calculation of the optical path lengths [10]. This is sketched in Figure 4.32 for an object point on axis. The change of the optical path length at the lens for an arbitrary ray is given with the optical path length (OPL) of the axis ray for the thin lens in air

$$s_{OPL}^{(axis)} = z_i - z_o = (m-1)\cdot z_o \tag{4.63}$$

with negative z_o. This value is also expected for an ideal lens for all marginal rays. Geometrically and due to the ideal imaging equations above the relations

$$z_i = \frac{f\cdot z_o}{f + z_o}$$
$$\sin u'_2 = \frac{z_o}{z_i}\cdot \sin u = \frac{\sin u}{m}, \quad \tan u'_1 = \frac{z_o}{z_i}\cdot \tan u = \frac{\tan u}{m} \tag{4.64}$$

are found. In the case of the version with a correct unique intersection point Q_1 at the lens position (blue ray in the Figure 4.30(a)) the sum of the geometrical paths of the marginal ray as OQ_1O'

$$s_{OPL}^{(1)} = \frac{-z_o}{\cos u} + \frac{z_i}{\cos u'_1} = \frac{-z_o}{\cos u} + mz_o\sqrt{1+\tan^2 u'_1}$$
$$= \frac{-z_o}{\cos u} + z_o\sqrt{m^2 + \tan^2 u} \tag{4.65}$$

If the correct path after the lens is requested (red ray in the Figure 4.30(b)), the optical path difference is $OQ_1 + Q_2O'$ and gives

$$s_{OPL}^{(2)} = \frac{-z_o}{\cos u} + \frac{z_i}{\cos u'_2} = \frac{-z_o}{\cos u} + \frac{m^2 z_o}{\sqrt{m^2 - \sin^2 u}} \tag{4.66}$$

Both lengths are not compatible and are not identical to the path length along the axis. A plane perfect lens can simply not follow Fermat's principle; it's a non-physical artificial construction. This shows that the calculation of the wave aberration and the application of Fermat's principle at an ideal lens is problematic and requires special treatment in computation software. Mainly in commercial optical design software the mode of calculation can be changed, depending on the requirements.

Figure 4.33 shows a lens for high numerical aperture (NA) with two ideal surfaces and finite thickness as calculated by a commercial optical design program. As can be seen, the solution case 1 is realized here in the layout, which means that the rays after the lens have the wrong angle. In addition, the ray heights inside and after the lens are wrong; they are correct earlier in the image plane.

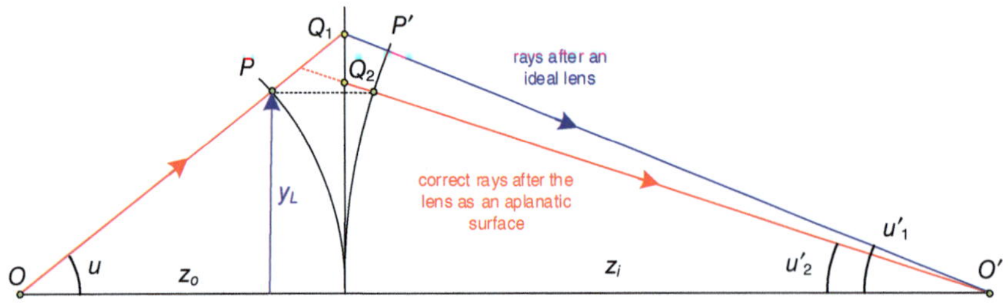

Figure 4.32 Calculation of the optical path length at an ideal lens.

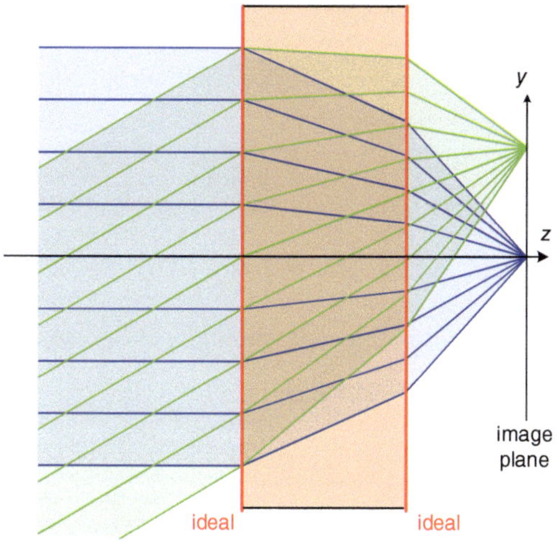

Figure 4.33 Ideal lens for high numerical aperture.

4.5 Mirrors

If mirrors are used in optical systems as components, no wavelength dependence occurs for the ray paths. More precisely, only the physical properties like reflectivity and polarization depend on the wavelength, see Figure 4.34 and the discussion of the Fresnel formulas in Section 3.3.

In general mirror systems have special properties in modelling as well as in real optical systems that must be taken into account. Mathematically, mirrors invert the coordinate system. This means that a left-handed system is transformed into a right-handed one. This is an issue in modelling; therefore, in the simulation tools several opportunities are used to reverse this effect. To get a finally correct raytrace after a mirror, the following work arounds are possible to rescue the desirable right-handed coordinate system.

- Distances behind the mirror are declared to be negative. In the case of a normal incidence, this is obvious; for other cases this seems artificial, although the raytrace then works correctly. This corresponds to a negative z component of the direction vector of the ray $s_z < 0$, see Section 3.4.6.
- The refractive index is set negative behind the mirror; this solution also seems to be unphysical and is only a formal solution.

Due to the usual reflection law, at a centred mirror the optical axis is bent backwards. This has the undesirable effect that the outgoing ray bundles can conflict with the incoming ones. To avoid obscuration effects, the ray bundles must be deflected in a way such that they are separated after a certain propagation distance. For a more clear and transparent drawing of the system layout, the so-called tunnel diagram is often used. In this approach, the system is unfolded which means that the system is turned around the vertex plane of the surface; this in turn means that the optical axis is once more oriented to the right direction. In particular, in the case of reflective prisms with several reflections, this is a meaningful way to make the ray paths clearer. Figure 4.35 shows this scheme for a simple spherical mirror as well as a prism with two reflections.

If curved mirrors are used, the simplest case is a spherical mirror with radius of curvature R. The focal length is then $f = R/2$. Curved mirrors also generate aberrations; on axis we have spherical aberration, and for off-axis field points

Figure 4.34 Reflection at a mirror surface.

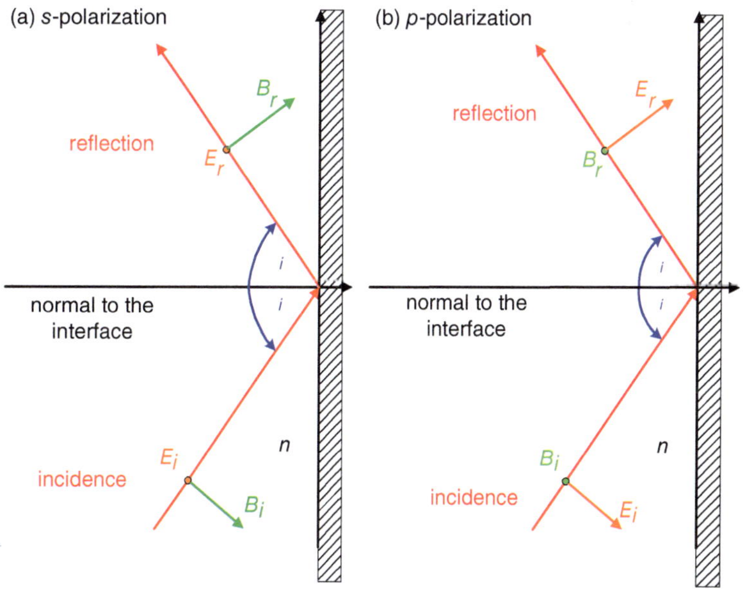

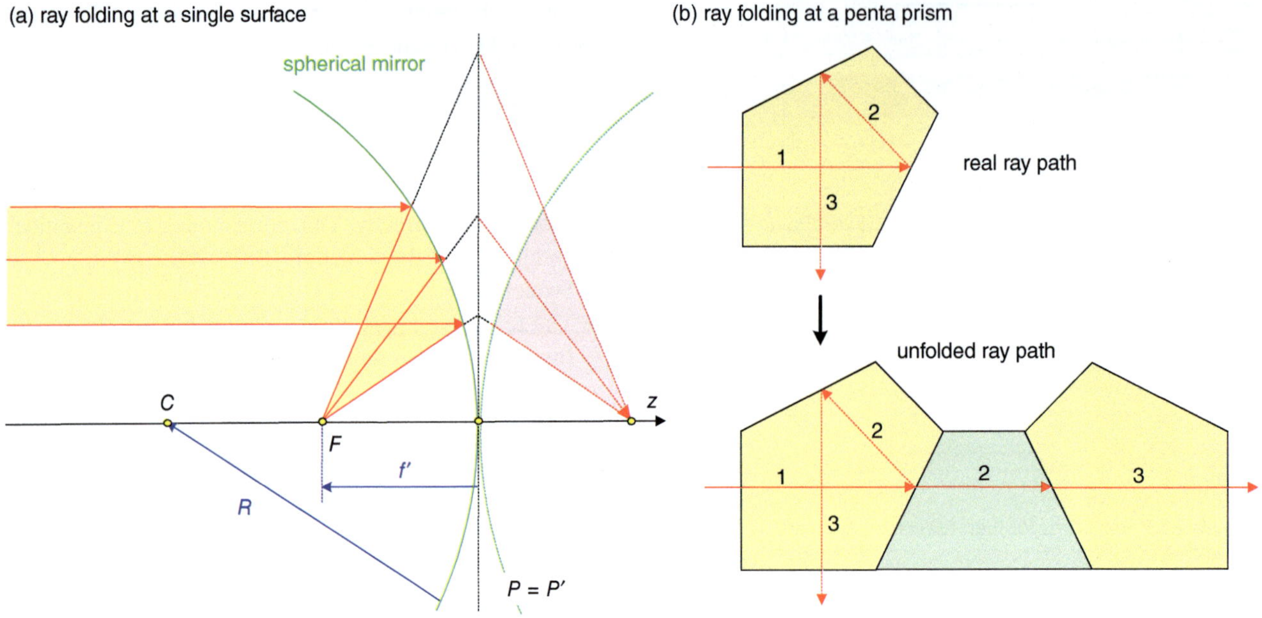

Figure 4.35 Reflection at mirror surfaces by unfolding the ray path for (a) a single surface, the unfolded ray bundle is shaded in red, and (b) a penta prism with two reflexions.

even stronger aberrations occur depending on the stop position.

In the special case of conic sections, particular setups exist that have a perfect focusing capability, at least for one field point. A parabolic mirror with the equation

$$z = \frac{x^2 + y^2}{2R} \tag{4.67}$$

perfectly focuses incoming collimated light if it is parallel to the optical axis. For finite field angles, large astigmatism and coma occur. This setup is therefore used in illumination systems to collimate the light or in astronomical telescopes where the field of view is rather small. Figure 4.36 shows the imaging setup with a curved parabolic mirror with the stop in the centre of curvature C. The axis point P_o is stigmatically perfect focused, the off-axis field point P suffers from astigmatism.

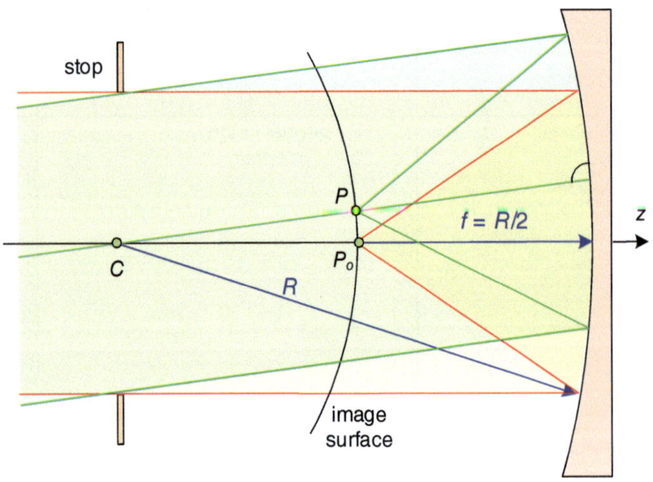

Figure 4.36 Imaging with a curved mirror for the special stop position in the centre of curvature C.

4.6 Aspheres

4.6.1 Introduction

Traditionally the surfaces of lenses have spherical shapes. They are easy to manufacture and there are reasons from aberration theory why this shape is useful. But the ray bending properties are usually not perfect, and therefore lenses with aspherical surfaces make sense in special situations inside optical systems. Figure 4.37 illustrates the cross section of an asphere. The deviation from the spherical shape is described by the so-called sag, which is the surface coordinate $z(r)$ in the coordinate system of the vertex point on the optical axis. Here the term 'asphere' assumes rotational symmetry, therefore the lateral coordinate r is used instead of x or y. The radius of curvature R is

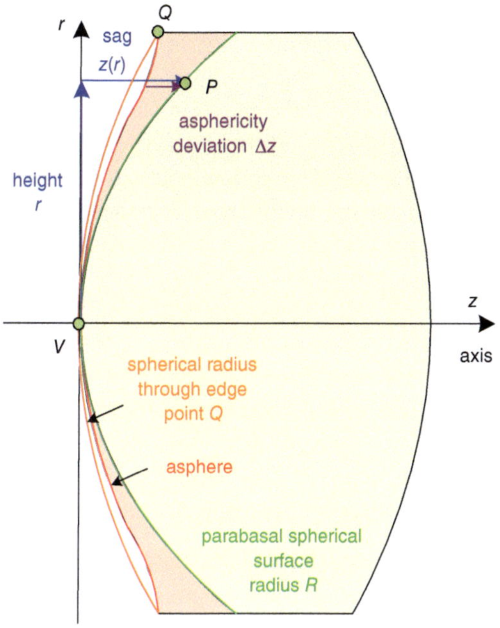

Figure 4.37 Asphere surface profile.

used to avoid confusion. The well-known relation

$$r = \sqrt{x^2 + y^2} \tag{4.68}$$

is valid. If the asphere is drawn by a cross section profile as shown in Figure 4.37, two spherical radii are important. One is the parabasal radius which approximates the asphere near the axis as an osculating surface and the second is the radius meeting the aspherical profile at the outer edge. This second radius is essential for the manufacturing of the surface, because first a spherical raw component is produced with this radius, and in the second step the exact aspherical profile is achieved by removing the additional material. What is important for the manufacturing properties of the asphere is the z difference between the asphere and the parabasal sphere. This value $\Delta z(r)$ determines the asphericity, it influences the measurability and the correction potential of the surface as well as the amount of material removal and therefore the length of grinding time. In particular due to fabrication kinematics, turning points in the aspherical profile are extremely undesirable.

The function $z(r)$ defines the special shape in one cross section. Figure 4.38 shows the spherical aberration (see Chapter 13) due to simple spherical lenses, which illustrate how to correct the ray bending by an additional angle $\delta(r)$ by aspherizing the surface profile appropriately. In the shown cross section view a kind of prismatic wedge re-directs the rays and compensates for the deviation of the outer rays, which is too strong for a spherical surface.

Aspheres are more complicated to fabricate in comparison to a spherical surface, but under certain conditions they give the opportunity to build a system with fewer lenses of a comparable quality. For an exact description of the aspherical profile curve, several opportunities exist. Mainly an expansion is used, and the number of terms gives the additional number of degrees of freedom; but often this is also the reason for a certain waviness, which is undesirable.

The goals for a selection of a specific mathematical representation to describe $z(r)$ are as follows.

- Good numerical conditions for a fast and accurate calculation of the asphere in the raytrace.
- Only a few parameters are necessary for an efficient aberrations correction.
- The expansion terms are orthogonal to guarantee a good convergence in the numerical optimization. Here two types of orthogonality can be selected: orthogonality in space or in slope.

4.6.2 Conic Surfaces

4.6.2.1 Introduction

A special simple case of an aspherical surface is a conic section. The equation of this second order function is given in vertex plane formulation in r or x/y

$$z(r) = \frac{cr^2}{1 + \sqrt{1 - (1+\kappa)c^2r^2}},$$
$$z(x,y) = \frac{c \cdot (x^2 + y^2)}{1 + \sqrt{1 - (1+\kappa) \cdot c^2(x^2+y^2)}} \tag{4.69}$$

Besides the radius of curvature in the vertex point $R = 1/c$, one more degree of freedom with the conical parameter κ is introduced, which allows us to select the special cases of conic section types according to Table 4.3.

The corresponding change in the shape of the conical surface cross section as a function of κ is illustrated in

Figure 4.38 Aspherical correction to improve the image quality on axis.

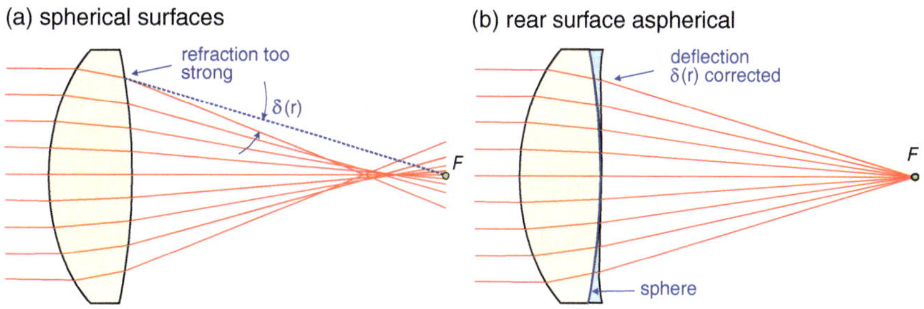

Table 4.3 Influence of the conical parameter κ on the shape of a conic section.

Parameter	Surface shape
$\kappa = 0$	Sphere
$\kappa = -1$	Paraboloid
$\kappa < -1$	Hyperboloid
$\kappa > 0$	Oblate ellipsoid (disc)
$0 > \kappa > -1$	Prolate ellipsoid (cigar)

Figure 4.39. It can be seen that for negative large amounts of κ the largest curvature occurs at the rim, while for large positive κ the centre has the largest curvature.

In general, the rational expression of the conic section in vertex representation can be expanded as $z(r)$ into a polynomial series according to

$$z = \frac{1}{2}cr^2 + \frac{1+\kappa}{8}c^3r^4 + \frac{(1+\kappa)^2}{16}c^5r^6 + \ldots \tag{4.70}$$

The first term gives the paraxial focal point position, the second term corresponds to primary spherical aberration of fourth order in waves, the third term to secondary spherical aberration of sixth order and so on. The fourth order term can be included in the primary aberration Seidel formulas (see Volume 2, Chapter 13). It can be seen that in this formulation, only the parabola with $\kappa = -1$ is of second order, the other types contain higher orders. The second order in z and r is only achieved in the implicit surface equation.

In the simplest case $\kappa = -1$ a parabola is obtained with the quadratic polynomial expression

$$z = \frac{1}{2}c\,r^2 = \frac{r^2}{2R} \tag{4.71}$$

According to the properties of a parabola, collimated light coming parallel to the axis is focused perfectly into the focal point F (see Figure 4.40(a)). For an ellipsoid, a finite object point F_1 is perfectly imaged into the focal point F_2. In the case of a hyperboloid, this finite imaging is virtual, it can be used to shorten the image intersection length. It should be noted that according to the Taylor expansion above, κ does not influence the paraxial properties and in particular the focal length of the surface.

It should be known that all these stigmatic ideal imaging conditions are valid for one pair of conjugated object and image points on axis only. The conic sections do not provide a finite imaging field of view, they are not aplanatic and show a huge coma for finite field sizes.

Depending on the type of conic section, the following imaging properties will apply according to Figure 4.40.

- Parabolas focus the paraxial rays into the focal point.
- Ellipses focus the rays from one focal point into the other one.
- Rays from one focal point of a hyperbola appear to come from the other focal point.

All of these properties are stigmatic in the sense of geometric optics and therefore are free of aberrations. For the stigmatic image between the focal points of a conic section, the back focal lengths vary over the solid angle of the numerical aperture through a very large range. Therefore, the sine condition (see Volume 2, Chapter 17) for these images is not adequately fulfilled. This in turn explains the extremely large aberrations for the field points of imaging by conics.

Thus, imaging systems using a conic section are only useful for very small fields. Consequently, the main areas of

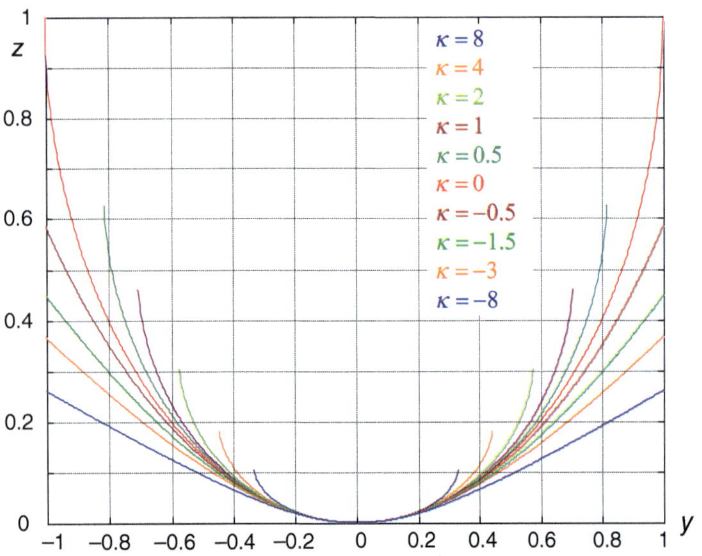

Figure 4.39 Cross section of a conic section as a function of the parameter κ.

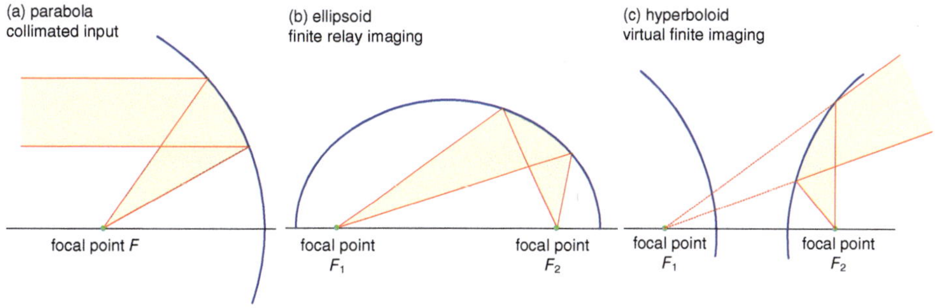

Figure 4.40 Imaging with conic section mirrors (a) from infinity by a parabola, (b) finite by an ellipsoid and (c) virtual by a hyperboloid.

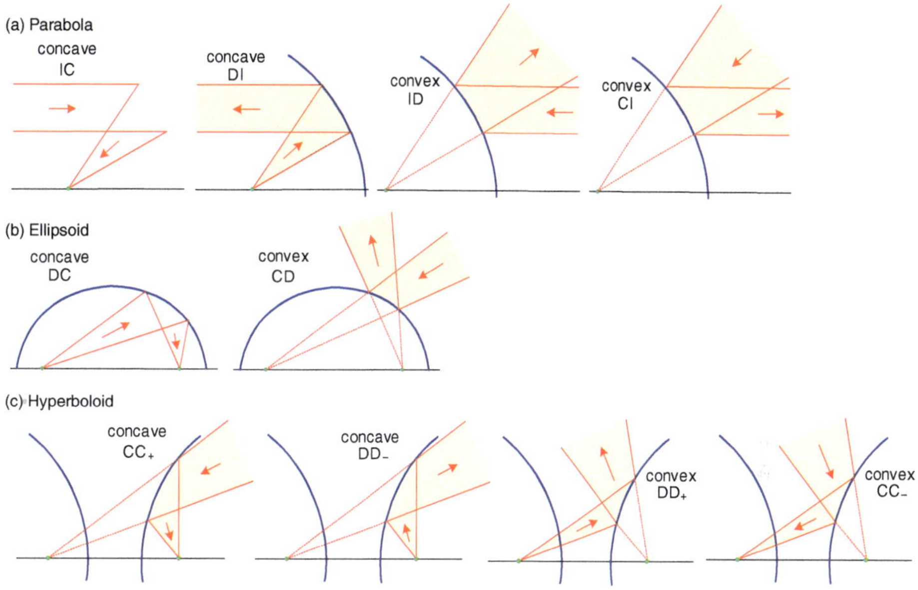

Figure 4.41 Possible ray bundle setups of a single conic mirror with (a) a paraboloid, (b) an ellipsoid and (c) a hyperboloid. Here I stands for infinity, C for convergent and D for divergent. A + means: the type is strengthened, a − means the type is relaxed.

application are astronomical telescopes and illumination collectors.

Conical mirrors can be used on their convex as well as on their concave side, depending on whether the imaging is real or virtual. This means that in total 10 combinations are obtained; these cases and properties are illustrated in Figure 4.41.

Conic sections can also be meaningful in lens design as refractive surfaces. This is seen in the Section 4.6.3.

4.6.2.2 Parabolic Mirrors

In the case of a parabolic mirror, all paraxially incident rays are focused to the focal point F as shown in Figure 4.42. This way an ideally corrected image is obtained for objects at infinite distance. The equation for the paraboloid is

$$z = \frac{c \cdot r^2}{2} = \frac{r^2}{2R} \tag{4.72}$$

where the vertex radius is R. The focal length of the mirror is

$$f' = \frac{R}{2} \tag{4.73}$$

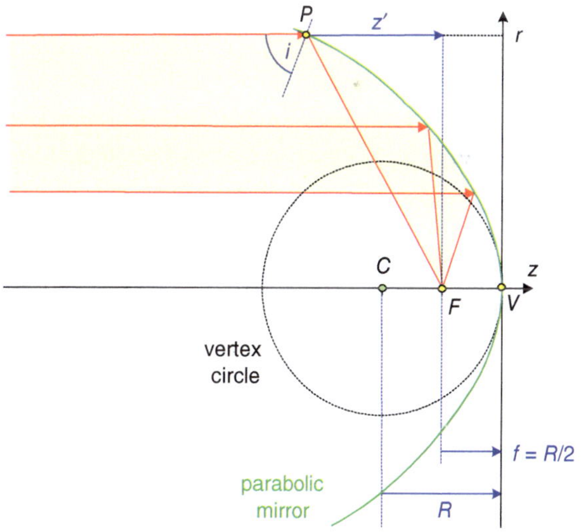

Figure 4.42 Path of the marginal rays for a parabolic mirror.

At a height r the angle of incidence in point P is

$$\tan i = \frac{r}{R} \tag{4.74}$$

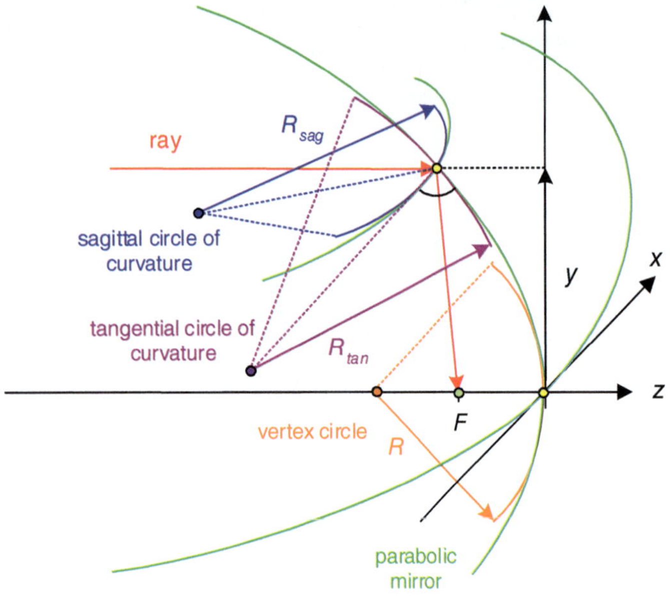

Figure 4.43 Local radii of curvature at the parabolic mirror.

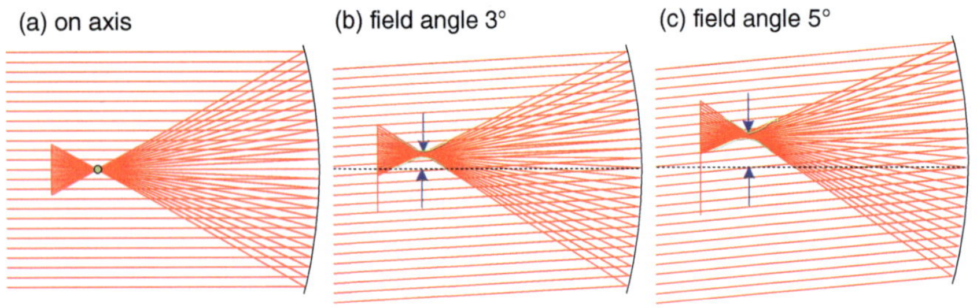

Figure 4.44 Parabolic mirror, showing the caustic for different field angles w with (a) axis, (b) moderate field angle $w = 3°$ and (c) larger field angle $w = 5°$.

and the intersection with the axis for an incoming parallel ray is at a distance z' along z (see Figure 4.42)

$$z' = \frac{R}{2} + \frac{r^2}{2R} = \frac{R}{2\cos i} \cdot \sqrt{1 + \left(\frac{r}{R}\right)^2} \tag{4.75}$$

Chief rays incident under an angle w to the optical axis show astigmatism and coma. At a height r a rotation-symmetric parabolic mirror has two different main radii of curvature

$$R_{sag} = R \cdot \sqrt{1 + \left(\frac{r}{R}\right)^2} \quad R_{\tan} = R \cdot \left[1 + \left(\frac{r}{R}\right)^2\right]^{\frac{3}{2}} \tag{4.76}$$

$$c_{\tan} \cdot c^2 = c_{sag}^3$$

Consequently, the mirror suffers from astigmatism for finite field sizes. The focal power is different in the plane of the incident ray and perpendicular to it. This is shown in Figure 4.43. For the two curvatures the above relation is obtained with $c = 1/R$ in the vertex.

So, for a direction of incidence that is inclined to the optical axis by an angle w, there will be a splitting of the back focal lengths. The astigmatism has the magnitude

$$\Delta s' = s_{\tan} - s_{sag} = 2s' \cdot w \cdot \tan i = \frac{2s' \cdot w \cdot y}{R} \tag{4.77}$$

Figure 4.44 shows the generation of the caustic of the coma for a parabolic mirror, with increasing angle w of the chief ray. Thus the violation of the sine condition can be seen directly.

A off-axis paraboloidal mirror tilted by an angle θ is shown in Figure 4.45. On the optical axis a toroidal surface is obtained.

4.6.2.3 Ellipsoidal Mirror

In the case of the ellipsoidal mirror, all rays coming from the focal point F are focused into the second focal point F'.

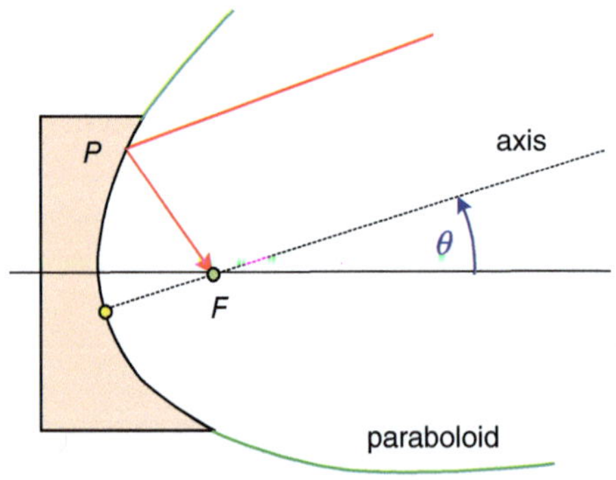

Figure 4.45 Off-axis paraboloidal mirror with tilted axis.

For any coordinate r the curvatures in the tangential and sagittal direction are

$$c_{tan} = \frac{c}{[1-(1+\kappa)\cdot r^2c^2]^{3/2}},$$
$$c_{sag} = \frac{c}{[1-(1+\kappa)\cdot r^2c^2]^{1/2}} \tag{4.78}$$

According to Figure 4.46 it can be seen that the paraxial radius of curvature R of an ellipsoid on axis can be smaller on axis as a prolate ellipsoid ($0 < \kappa < -1$) or larger in the upper point Q as an oblate ellipsoid ($\kappa > 0$).

The lack of fulfilling the sine condition and a stigmatic imaging property only on axis for exactly two finite conjugate points is in the case of an ellipsoid similar as for the paraboloid. This is illustrated in Figure 4.47 for a finite yellow square area. If the aperture rays belonging to the blue ray bundle are used, the magnification is $m = s'/s = 2$, in the green case we get in contrast $m = 1/6$. This means that the magnification changes by an order of magnitude over the full aperture cone and does not allow for a sharp imaging of a finite sized object.

4.6.2.4 Hyperboloidal Mirror*

(*) The content of this section is quite specialized and for advanced reading. Knowledge of this section is not necessary to understand the following chapters. Beginners are recommended to skip this section.

Hyperboloidal mirrors shift the visual source point of a divergent radiation coming from the focal point to the other focal point F' as shown in Figure 4.48. Hyperboloidal mirrors are applied only rarely in optics. One special application is the use in x-ray telescopes of the Wolter-type, here the shortening of the overall telescope length is essential due to the very small grazing incidence angles. This is discussed in more detail in the next section.

4.6.2.5 Grazing Incidence Telescopes

An important application of conic section mirror surfaces in imaging systems are grazing incidence mirror systems in so-called Wolter telescopes. In this case, the focal properties of conic sections are directly used to have a perfect performance on axis. The grazing incidence is in particular used for very short wavelengths in the x-ray regime, because here the transmission of coatings is critical

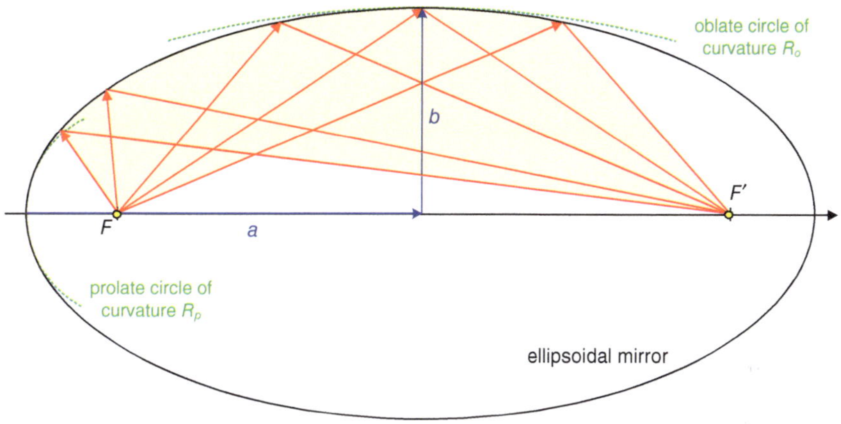

Figure 4.46 Terms and path of the rays for the ellipsoidal mirror.

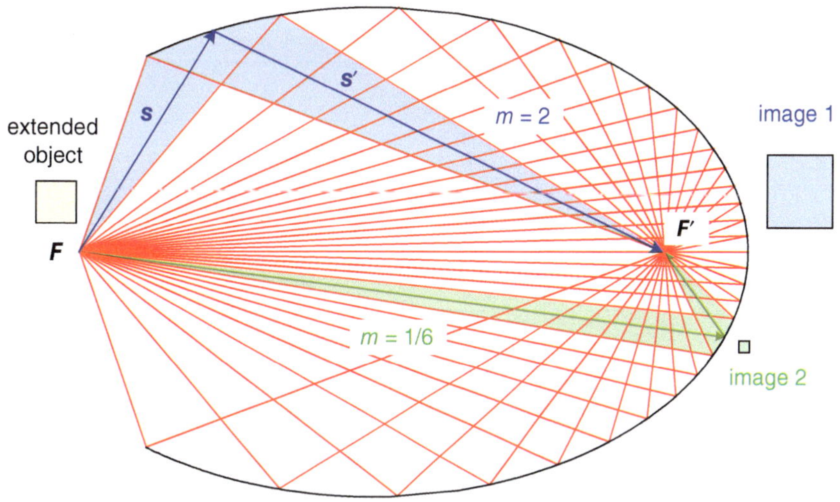

Figure 4.47 Variable magnification of the image for an ellipsoidal mirror and the constant sum of the intersection lengths ($s + s' =$ const.)

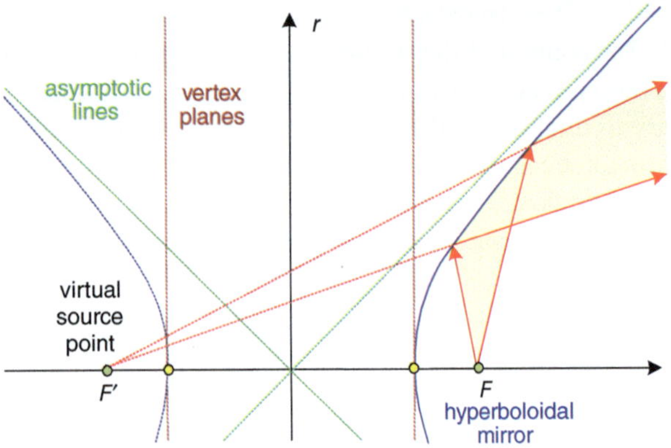

Figure 4.48 Hyperboloidal mirror.

and depends strongly on the incidence angle. If the total internal reflection is used by the grazing incidence, a lossless reflection is obtained. The drawback of these types of systems is the very small aperture cone forced by the incidence condition, which means that the throughput and the Helmholtz–Lagrange invariant are quite small. Furthermore, the ray bending angles are small and a strong focusing requires several surfaces. As one opportunity to overcome this problem, the conic mirror systems are nested with several shells to enlarge the effective usable area of the light bundles.

In total, there are three different Wolter telescope types:

- Wolter type I:
  paraboloid and concave hyperboloid to shorten the final focus distance and system length;
- Wolter type II:
  paraboloid and convex hyperboloid to achieve a tele system with large focal length;
- Wolter type III:
  paraboloid and ellipsoid to achieve a retrofocus system with larger numerical aperture.

These principal layouts are sketched in Figure 4.49.

An example for a Wolter type I telescope is shown in Figure 4.50 that is intended as an x-ray telescope. First a parabolic mirror creates a focus point with usually with a quite large focal length f. To shorten the system for a satellite application, a second mirror with hyperbolic shape is used. In the realized system of the figure, four shells are used to increase the aperture of the system for astronomical applications with critical signal to noise ratio (SNR) conditions. This shows the optical system of the so-called Roentgen satellite (ROSAT) mission [11].

4.6.3 Cartesian Ovaloids*

(*) The content of this section is quite specialized and for advanced reading. Knowledge of this section is not

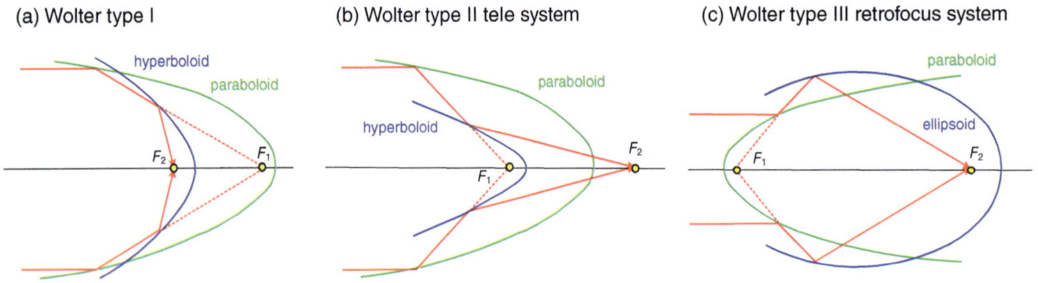

Figure 4.49 Grazing incidence two-mirror telescopes of the three Wolter systems of (a) type I, (b) type II and (c) type III.

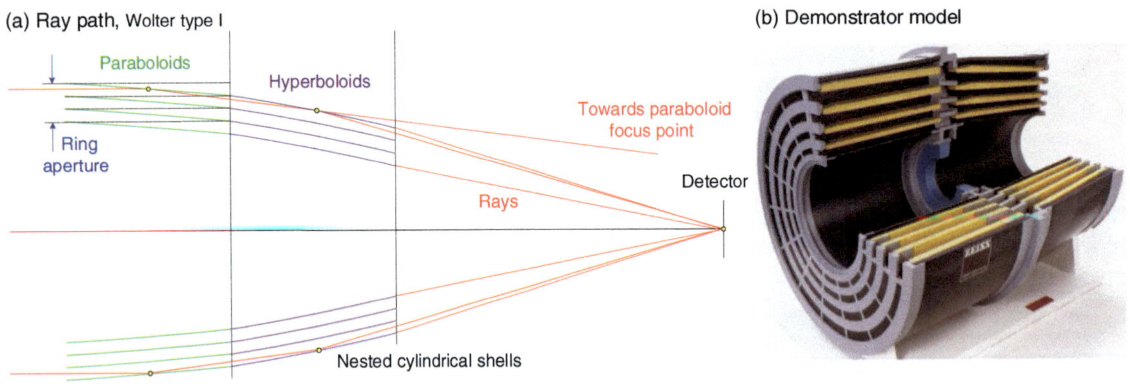

Figure 4.50 Grazing incidence two-mirror telescope of the Wolter I type. (a) ray path, (b) demonstrator model with permission of Zeiss.

necessary to understand the following chapters. Beginners are recommended to skip this section.

4.6.3.1 Introduction

A general Cartesian ovaloid is a surface $z(r)$ for given values of n, n', s, s' with stigmatic imaging properties for a conjugated point pair on axis. The surface is refractive and looks similar to an ellipsoid, but has some aspherical contributions of fourth order, see Figure 4.51. By applying the Fermat principle on the setup shown in the figure, we get

$$n \cdot \sqrt{r^2 + (z - s)^2} + n' \cdot \sqrt{r^2 + (s' - z)^2} = s'n' - ns \tag{4.79}$$

Here it should be noticed that $s < 0$, as it lies left to the surface. This equation describes a general Cartesian oval and is of fourth order, if the square roots are resolved [12–17].

In the case of finite object and image locations, the equation of a Cartesian or Descartes Ovaloid can be written in the implicit form as an equation of fourth order

$$A \cdot z^4 + B \cdot z^3 + C(r) \cdot z^2 + D(r) \cdot z + E = 0 \tag{4.80}$$

with the coefficients

$$\begin{aligned}
A &= (n^2 - n'^2)^2 \\
B &= 4(n^2 - n'^2)(n'^2 s' - n^2 s) \\
C &= 4(n'^2 s' + n^2 s)^2 + 2r^2(n^2 + n'^2)^2 \\
&\quad - 4nn'ss'(n^2 + n'^2) - 4n'^2 n^2(s^2 - 4ss' + s'^2 + 2r^2) \\
D &= (n'^2 s' + n^2 s)[4r^2(n^2 + n'^2)^2 - 8nn'ss' \\
&\quad - 4n'^2 n^2(-2s^2 s + 2ss'^2 + 2r^2 s - 2r^2 s')] \\
E &= (n^2 + n'^2)^2 r^4 - 4nn'ss'(n^2 + n'^2)r^2 \\
&\quad + 4n'^2 n^2 s^2 s'^2 - 4n'^2 n^2(s^2 s'^2 + 2r^2 s^2 + r^2 s'^2) + r^4
\end{aligned} \tag{4.81}$$

In some special cases, the fourth order equations simplify to a quadratic one for the conic section ellipsoid or hyperboloid respectively. These are as follows.

- If the object is located in infinity, a conic section of second order is obtained, depending on $n > n'$ or $n < n'$ an ellipsoid or a hyperboloid occurs.
- If the imaging condition fulfils $ns = n's'$ and therefore is aplanatic, the ovaloid degenerates to a sphere

An ovaloid can also be written as a special type of a so-called super-conic, see Section 4.6.4.4 below.

Figure 4.52 illustrates why this type of surface is called ovaloid; in contrast to a conic, there is no symmetry plane

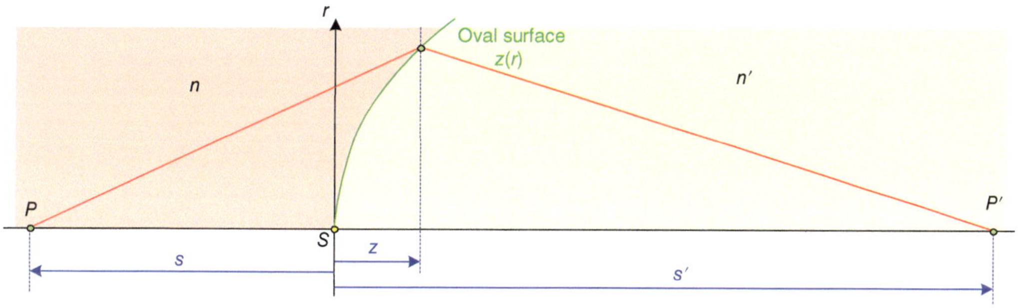

Figure 4.51 Marginal ray path at an ovaloid surface.

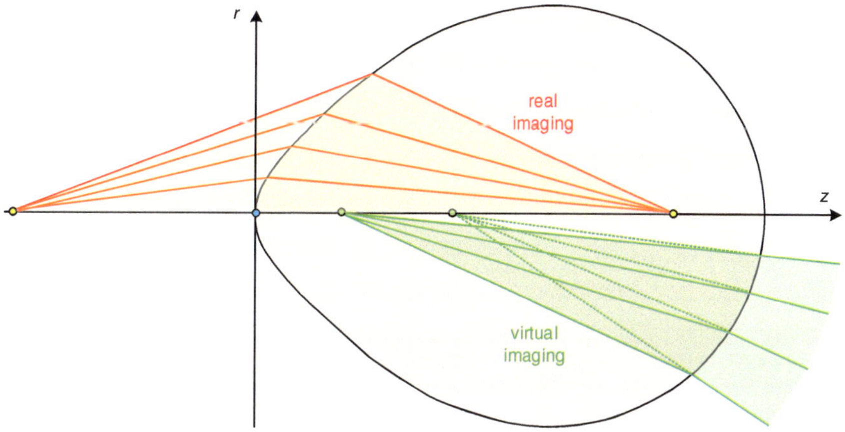

Figure 4.52 Real and virtual imaging with a refractive ovaloid surface.

perpendicular to z. Furthermore, the figure shows that an ovaloid can be used for real as well as virtual imaging.

The fourth order equation of Descartes ovaloid can also be formulated analytically by the corresponding mathematical substitution techniques of Cardano. It is then shown that in general the curve has a solution landscape with several branches that are used depending on the relation $n > n'$ or $n < n'$ and the property of the imaging to be real or virtual in object or image. Due to the squaring, some unphysical solutions come out with negative optical path differences, which are mathematically possible but must be out-selected in reality.

4.6.3.2 Special Case Infinity

If the object point on axis is in infinity, the Fermat condition reads (referenced to the front vertex plane, formally with $s = 0$)

$$n \cdot z + n' \cdot \sqrt{r^2 + (s' - z)^2} = s'n' \tag{4.82}$$

In this case the equation can be re-arranged into the form

$$\frac{\left(z - \frac{n's'}{n+n'}\right)^2}{\left(\frac{n's'}{n+n'}\right)^2} + \frac{r^2}{s^2 \cdot \left(\frac{n'-n}{n'+n}\right)} = 1 \tag{4.83}$$

This is a conic section with the parameters

$$a = \frac{n's'}{n+n'}, \quad b = \sqrt{\left|\frac{n'-n}{n'+n}\right|}s' \tag{4.84}$$

$$R = \frac{n'^2 s'}{(n'+n)\sqrt{|n'^2 - n^2|}}, \quad \kappa = \frac{n'^2}{(n'-n)^2} - 1 \tag{4.85}$$

From a more practical view, a single refractive surface can seldom be applied and it makes more sense to think of a component with two refractive surfaces. To maintain the advantages of the Cartesian surfaces, the other surface is shaped concentric to the marginal rays and therefore has no impact. Depending on the choice of the Cartesian surface being the front or the rear surface, two cases appear.

In the first case, the front surface is the conic, the second surface is concentric and does not affect the rays anymore. In the second case, the front surface is plane and without impact, the second bends the rays. The two cases are shown in Figure 4.53. In part (a) the aspherical front surface is an ellipsoid, in case two shown in part (b) the rear surface is hyperbolic in shape.

But if now a finite field of view is introduced and the incoming ray bundle is tilted by a field angle w, the spot size behaves as the diagram in Figure 4.54 shows. The growing coma is a similar effect as for the case of conical mirrors not fulfilling the sine condition (for details see Volume 2, Chapter 17). The size of the spot diameter grows linear with w, a clear indicator for coma aberrations and, as can be seen, the spot size shows nearly pure coma. This means the surface is not aplanatic.

4.6.3.3 Special Case Aplanatism

In the special case that the optical path length is zero (defining s with sign), we get from this equation for the right side of Eq. (4.80)

$$s'n' = ns \tag{4.86}$$

and the left side becomes

$$n \cdot \sqrt{r^2 + (z-s)^2} + n' \cdot \sqrt{r^2 + (n/n's - z)^2} = 0$$

$$(n/n')^2 \cdot (r^2 + z^2 + s^2 - 2zs) = (r^2 + z^2 + (n/n')^2 s^2 - 2zsn/n')$$

$$z^2(n^2/n'^2 - 1) - 2zs[(n/n')^2 - n/n'] + r^2(n^2/n'^2 - 1) = 0$$

$$\left[z - \frac{sn}{n+n'}\right]^2 + r^2 = \left[\frac{sn}{n+n'}\right]^2 \tag{4.87}$$

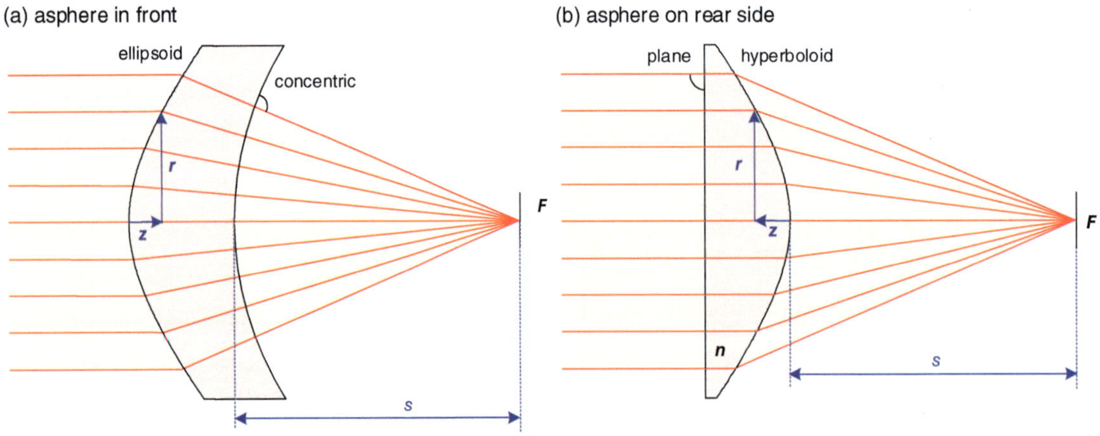

Figure 4.53 Perfect stigmatic focusing asphere as conic section on one side of a lens. The front surface is elliptical as shown in case (a) with a concentric rear surface. The rear surface is a hyperbolic as shown in (b) with a plane front surface. Only the conic surface is deflecting the rays.

Figure 4.54 Spot diagram and spot diameter for the aspheres of Figure 4.53 for a growing field angle w.

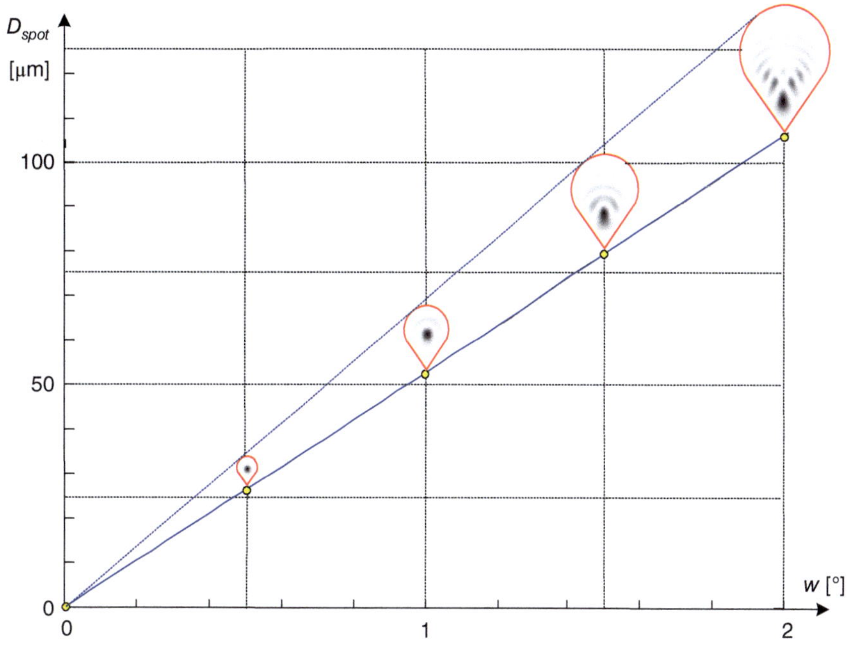

It is seen that the oval surface in particular becomes a sphere with the radius

$$R = \frac{sn}{n + n'} \tag{4.88}$$

This is the special interesting aplanatic spherical surface, well known in aberration theory. If the simple relation of paraxial optics

$$m = \frac{y'}{y} = \frac{ns'}{n's} \tag{4.89}$$

is combined with the sine condition for finite angles in the form

$$m = \frac{n \cdot \sin u}{n' \cdot \sin u'} \tag{4.90}$$

the following equation is obtained, which is equivalent to the sine condition

$$s \cdot \sin u = s' \cdot \sin u'$$

$$\Delta(s \cdot \sin u) = 0 \tag{4.91}$$

This can be easily interpreted geometrically as the location of an aplanatic surface by constructing the semi-circles according to Figure 4.55 and the following receipt. The object point P, the image point P' and the vertex point of the surface V are given.

- Construct the half circle between the object point P and the vertex point V (green as example in the figure).

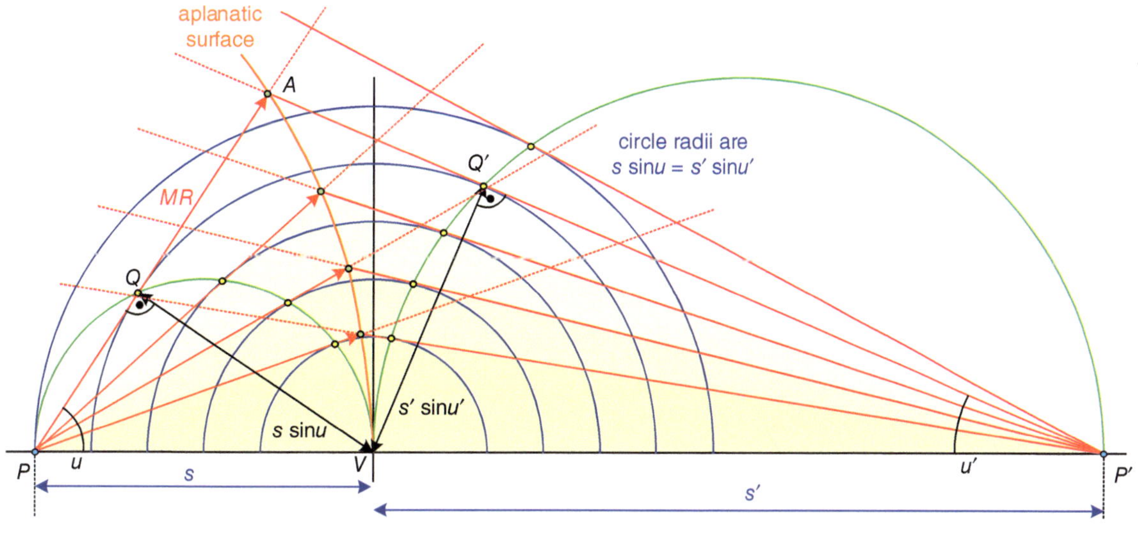

Figure 4.55 Graphical construction of a general aplanatic surface shape.

- Construct an arbitrary marginal ray.
- Construct the half circle around the vertex point V, which is tangential to the marginal ray in point Q.
- Construct the marginal ray in the image space as tangential to this circle too in point Q'.
- The two marginal rays intersect in the point A on the aplanatic surface.

This construction does not contain the refractive indices and usually cannot be fulfilled by real values of n and n'. This corresponds to the statement that a real aplanatic imaging with spherical surfaces is not possible.

4.6.4 Asphere with Polynomial Expansion

4.6.4.1 Traditional Taylor Expansion

A general definition of an aspherical surface has historically been used in lens design from the following equation. The conical basic shape of Eq. (4.69) is supplemented here with a Taylor polynomial expansion term. The coefficients a_m describe the expansion, M is the maximum number of terms.

$$z(r) = \frac{cr^2}{1+\sqrt{1-(1+\kappa)c^2r^2}} + \sum_{m=0}^{M} a_m r^{2m+4} \tag{4.92}$$

In practical usage of this expression the purpose is to fulfil the second order function of the surface in the best case by the basic shape curvature c and conic constant κ. Mostly a sphere with $\kappa = 0$ is used. The polynomial expansion then determines the deviation from the sphere, also called asphericity. It is recommended to have only a small deviation, to help relaxed sensitivity and centring tolerances as well as the manufacturing of the surface. The number of terms M usually is in the range of 2–5. It should be noted that there are only even powers included and the lowest power is r^4. Figure 4.46 illustrates the behaviour of the power terms normalized on the radius. This means that the expansion starts with spherical aberration equivalent correction on axis and has no paraxial focusing contribution and describes only rotational symmetric smooth surfaces without a singularity on axis.

Some drawbacks of this description make the application complicated. These are the following critical properties:

- There is no normalization of the lateral coordinate. This means that every constant has an individual unit. Furthermore, the orders of magnitude of the coefficients are strongly dependent on the absolute size of the diameter of the surface and therefore on the maximum value of the radial coordinate r. This may cause problems in the series convergence for large r-values.
- The Taylor polynomial terms are not orthogonal. This has the consequence that the terms depend on each other and the optimization of the asphere in lens design is not very stable.
- The coefficient values are not directly interpretable and cannot be directly toleranced. As can be seen from Figure 4.56, the maximum slope of the terms occurs at the boundary and grows with the power m.

In practical lens design, according to the additional coefficients, an asphere offers more degrees of freedom for the correction. But an asphere is not the solution for every correction problem. Usually, one aberration type can be corrected to great advantage, while some others become worse. If, for example, an asphere is positioned near the pupil of a system, spherical performance can be achieved nearly perfectly for an object point on axis, but the coma correction for off axis field points is decreased.

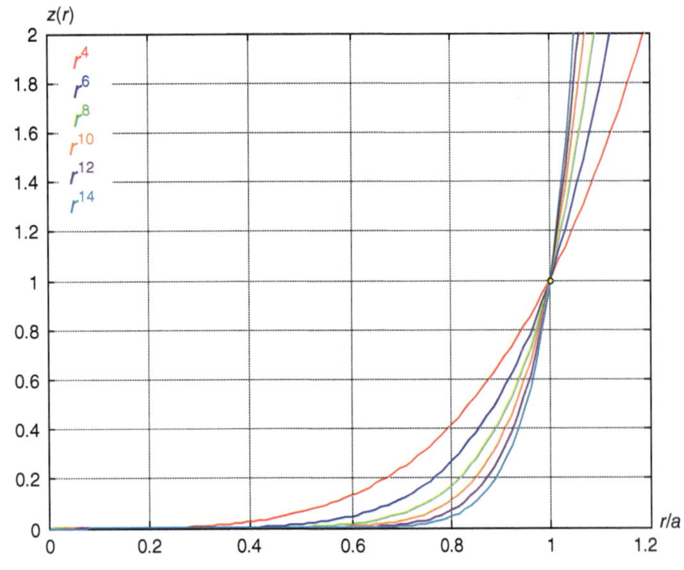

Figure 4.56 Sag contribution of the individual Taylor power terms on a normalized radial coordinate r/a.

4.6.4.2 Strong Forbes Aspheres

Recently two new aspherical representations have been proposed by G. Forbes to overcome the shortcomings mentioned before [18–21]. The first type is the so-called strong Forbes aspheres (or Q-type asphere Q^{con}). They are based on orthogonal Jacobi polynomials and are orthogonal in space. To solve the scaling problem, they are defined on a normalized radius r_{max}. Furthermore, the power $\bar{r}^4$ is extracted out of the polynomials. One more advantage is that the expansion coefficients a_m can be directly toleranced, because they contribute as a sum of squares to the root mean square (rms) deviation from the desired shape. The asphere representation can be written as

$$z_{con}(r) = \frac{cr^2}{1+\sqrt{1-(1+\kappa)c^2r^2}} + \bar{r}^4 \cdot \sum_{m=0}^{M} a_m Q_m^{con}(\bar{r}^2);$$
$$\bar{r} = \frac{r}{r_{max}} \tag{4.93}$$

Figure 4.57 shows the lowest order curves of the scaled polynomials $Q_m{}^{con}$. It is seen that they are oscillatory and have the largest slopes at the boundary, but also contribute some considerable sag-values for smaller radii in contrast to the old-fashion asphere terms in Figure 4.56.

In the following, we write for simplicity r instead of $\bar{r}$. In principle, an extrapolation of the polynomial expression is possible, but values above 1 are strongly increasing. The value at the boundary is always $Q_m{}^{con} = 1$. As a consequence of the normalization in the optimizing of an asphere inside a system, the value of r_{max} must be fixed before the coefficients are optimized or are left as a variable.

There is a relation between the Jacobi polynomials $P_m{}^{(n,j)}$ that is well known in mathematics and the strong Forbes aspheres of the form

$$Q_m^{con}(r) = P_m^{(0,4)}(2r-1) \tag{4.94}$$

The orthogonalization can be written in the spatial domain as

$$\int_0^1 r^4 Q_j^{con}(\mathrm{r}) Q_k^{con}(\mathrm{r}) 2\pi \mathrm{r}\,\mathrm{dr} = \mathrm{N}_{jk}\delta_{jk} \tag{4.95}$$

Some of the lower order polynomials are explicitly given in the following equations

$$\begin{aligned}
Q_0^{con}(r) &= 1\\
Q_1^{con}(r) &= 6r - 5\\
Q_2^{con}(r) &= 28r^2 - 42r + 15\\
Q_3^{con}(r) &= 120r^3 - 252r^2 + 168r - 35\\
Q_4^{con}(r) &= 495r^4 - 1320r^3 + 1260r^2 - 504r + 70\\
Q_5^{con}(r) &= 2002r^5 - 6435r^4 + 7920r^3\\
&\quad - 4620r^2 + 1260r - 126\\
Q_6^{con}(r) &= 8008r^6 - 30030r^5 + 45045r^4 - 34320r^3\\
&\quad + 13860r^2 - 2772r + 210\\
Q_7^{con}(r) &= 31824r^7 - 136136r^6 + 240240r^5 - 225225r^4\\
&\quad + 120120r^3 - 36036r^2 + 5544r - 330\\
Q_8^{con}(r) &= 125970r^8 - 604656r^7 + 1225224r^6\\
&\quad - 1361360r^5 + 900900r^4 - 360360r^3\\
&\quad + 84084r^2 - 10296r + 495
\end{aligned} \tag{4.96}$$

The polynomials can be constructed by the following recurrence formula which follows the properties of the Jacobi polynomials

$$Q_n^{con}(r) = \frac{(2n+3)[(n+2)(n+1)(2r-1)-4]}{n(n+4)(n+1)} \cdot Q_{n-1}^{con}(r) - \frac{(n-1)(n+3)(n+2)}{n(n+4)(n+1)} \cdot Q_{n-2}^{con}(r) \tag{4.97}$$

4.6.4.3 Mild Forbes Aspheres

The second recommended type are the mild aspheres of Forbes, they are formulated differently [18–21]. They are called the best-fit representation. The deviations from the

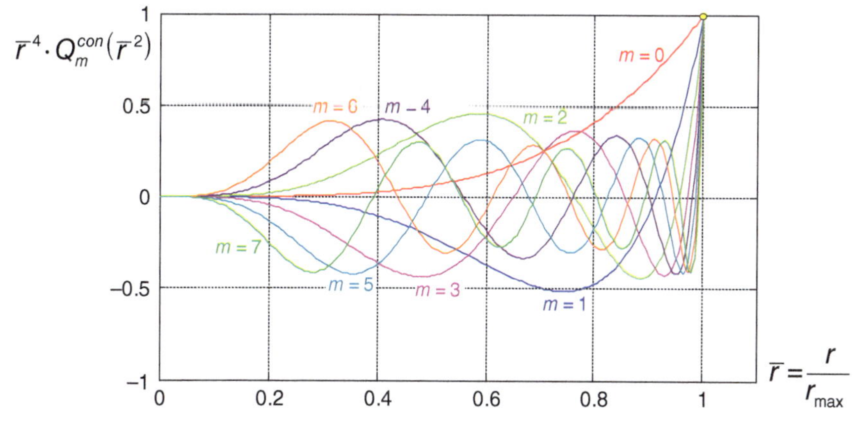

Figure 4.57 Lower order strong Forbes asphere profile forms of the terms.

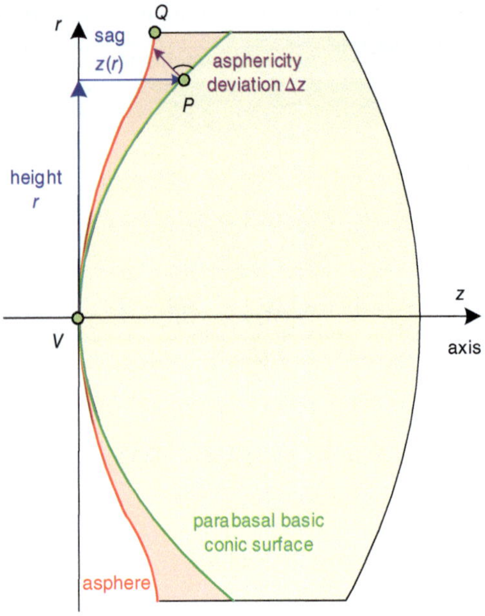

Figure 4.58 Forbes mild asphere surface profile.

basic shape sphere are directly superposed perpendicular to the local position r on the surface, see Figure 4.58.

Therefore a pre-factor is needed in front of the polynomial sum to guarantee the direction as a function of r. This factor has the consequence that the functions are no longer real polynomials but rational functions and a conversion into polynomials can only be achieved approximately. The functions are defined as

$$z_{bfs}(r) = \frac{cr^2}{1+\sqrt{1-c^2r^2}} + \frac{\bar{r}^2(1-\bar{r}^2)}{\sqrt{1-c\bar{r}^2}} \sum_{m=0}^{M} b_m Q_m^{bfs}(\bar{r}^2) \tag{4.98}$$

and are also based on a normalized radius $r_{\max}$. Some lower orders are shown in Figure 4.59. This representation is in particular very comfortable if the overall curvature and slope at the edge is very strong and a large difference occurs between the surface-perpendicular and the z-direction.

The corresponding slopes of the aspherical profile

$$S_m^{bfs}(\bar{r}) = \nabla_r z_{m'}(\bar{r}) = \frac{d}{d\bar{r}}\left[\bar{r}^2(1-\bar{r}^2)\cdot Q_m^{bfs}(\bar{r}^2)\right] \tag{4.99}$$

are shown in Figure 4.60.

There is one more interesting property of the mild aspheres. They are slope orthogonal. In contrast to the usual space orthogonality, the gradients are orthogonal according to the following equation

$$\int_0^1 \frac{\nabla_r z_m(\bar{r})\cdot\nabla_r z_{m'}(\bar{r})}{\sqrt{\bar{r}^2(1-\bar{r}^2)}}\, 2\pi\,\bar{r}\,d\bar{r} = N_{nm}\cdot\delta_{mm'} \tag{4.100}$$

This property gives very good conditions for optimization of ray aberrations. It is further seen by experience that the Forbes aspheres allow a better result in correction with a smaller number of terms and the optimization convergence is much faster and more robust. This means that in practice they are recommended to be used. Similar to the strong polynomials, here the rms of the slope error is directly related to the coefficients with the equation

$$\left\langle\left[\frac{1}{r_{\max}}\sum_m b_m S_m^{bfs}(\bar{r})\right]^2\right\rangle = \frac{1}{r_{\max}^2}\sum_m b_m^2 \tag{4.101}$$

with the area average symbol <>. This means that the coefficients of the best fit sphere Forbes polynomials can be directly used for tolerancing. Some of the lower order polynomials without the pre-factor in front of the sum in Eq. (4.98) are explicitly given by

$$Q_0^{bfs}(r) = 1$$

$$Q_1^{bfs}(r) = \frac{1}{\sqrt{19}}(13-16r)$$

$$Q_2^{bfs}(r) = \sqrt{\frac{2}{95}}[29-4r(25-19x)]$$

$$Q_3^{bfs}(r) = -\sqrt{\frac{2}{2545}}\{207-4r[315-r(577-320r)]\}$$

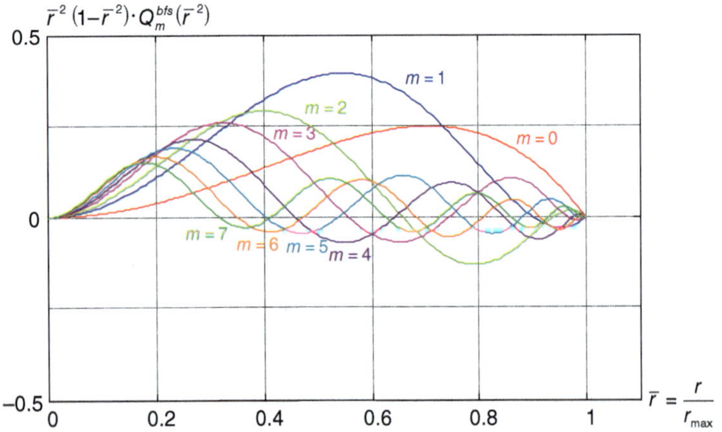

Figure 4.59 Lower order mild Forbes asphere terms.

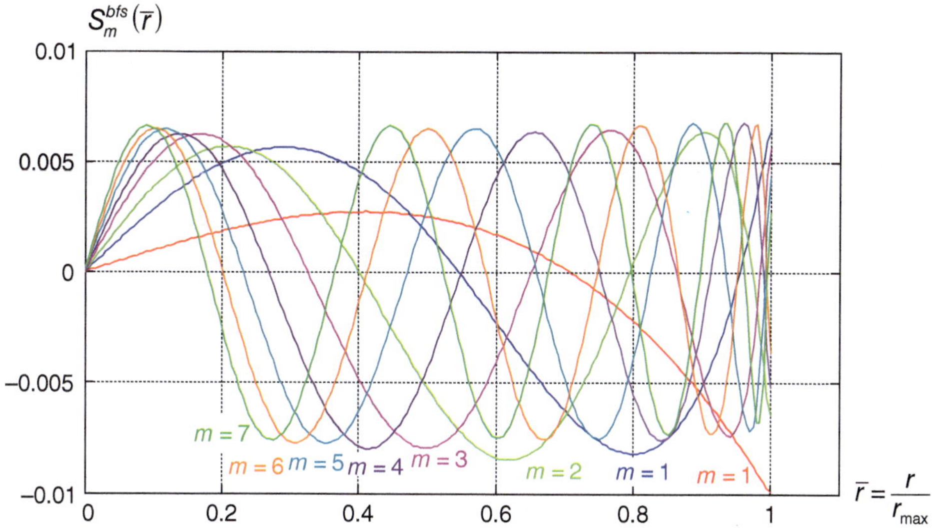

Figure 4.60 Lower order mild Forbes asphere slopes S_m.

$$Q_4^{bfs}(r) = \frac{1}{3\sqrt{131831}}(7737 - 16r\{4653 - 2r[7381 - 8r(1168 - 509r)]\})$$

$$Q_5^{bfs}(r) = \frac{1}{3\sqrt{6632213}}[66657 - 32r(28338 - r\{135325 - 8r[35884 - r(34661 - 12432r)]\})] \tag{4.102}$$

4.6.4.4 Superconical Surfaces

The classical conical section has the explicit representation

$$z(r) = \frac{cr^2}{1 + \sqrt{1 - (\kappa + 1)c^2r^2}} \tag{4.103}$$

as discussed before and is the solution of the implicit equation of second order

$$(\kappa + 1)cz^2 - 2z + cr^2 = 0 \tag{4.104}$$

According to a proposal from [22–24] this equation can be generalized by defining the three-dimensional distance s between the surface point and its vertex as a cord length s

$$s^2 = r^2 + z^2 = x^2 + y^2 + z^2 \tag{4.105}$$

in the form with two even order expansion terms inside the parentheses, the generalization reads

$$Kcz^2 - 2(1 + b_1s^2 + b_2s^4 + \ldots) \cdot z + (c + c_1s^2 + c_2s^4 + \ldots) \cdot s^2 = 0 \tag{4.106}$$

which is called a super-conic asphere. In particular there are some special cases, see Table 4.4

According to the practical experience mentioned in the literature, the super-conic implicit representation has in particular great advantages in the case of surfaces with very high NA. The formulation with s instead of r gives much more stable and improved optimization results and the implicit form allows also for a straightforward modelling of an over hemisphere.

Table 4.4 Special cases of superconics.

Relevant parameter	Vanishing parameter	Shape/surface type
$c, K = \kappa + 1$	All other zero	Conic
K, b_1, c, c_1	All other zero	Cartesian ovaloid
Rational function $z(s)$	$K = 0$	For four expansion terms a quartic Bezier spline is obtained

4.7 Freeform Surfaces

4.7.1 Introduction

In general in this book surfaces are called freeform surfaces if they have reduced symmetry properties; a remaining plane or double plane symmetry may be allowed, but no circular symmetry. This surface type can have therefore a large number of degrees of freedom and is used to correct symmetry-free optical systems with the corresponding higher orders.

If a freeform surface is to be used, there are several criteria for an appropriate mathematical description of the surface. In general, the following criteria are relevant.

- Efficient and robust calculation.
- Covering most application needs.
- Simple and robust import and export of data.

- Direct relation to tolerancing.
- Fit of measured data easily possible.
- Simple extension of the optical used free area to a larger area for fixing the mounting.
- Comfortable robust, accurate and fast raytrace possible.

For an efficient raytrace, the following points are essential.

- Fast calculation of intersection points.
- Fast calculation of local slopes.
- Optimization:
  - significant description of usual surfaces with only a few parameters;
  - fast convergence in optimization.

There are various types of surface descriptions found in the literature to describe optical surfaces. These are as follows.

- Smooth surfaces, analytical for aberration correction.
- Multi-functional surfaces (multi-apertures, segmentation, splitting of bundles).
- Modelling of local deviations of real surfaces (manufactured surfaces).
- Point clouds from measurements, typically noisy.
- Piecewise continuous surfaces (Fresnel surfaces, arrays, segmented mirrors ...).
- Surface equations with full-area support or describing only local deviations.

The usual available mathematical options for the freeform description are as follows.

- Extended polynomials:
  classical non-orthogonal monomial representation, two-dimensional Taylor expansion.
- Zernike surfaces:
  only useful for (nearly) circular pupils and low orders.
- Splines:
  localized description, hard to optimize, good for manufacturing characterization.
- Generalized Forbes polynomials:
  promising new approach, two types, strong relation to tolerancing.
- Radial basis functions:
  non-orthogonal or orthogonal possible, local description approach, good for local effect description.
- Wavelets:
  not preferred for smooth surfaces, only feasible for tolerancing.
- Fourier representation:
  classical description without assumptions, but not well adapted to aberrations.

In reality confirmed by practical experience it is recommended to compose a freeform in lens design into a basic shape that delivers and separates the quadratic parabasal part, and a second part that contains the higher correction orders; this is shown in Figure 4.61.

A general possibility to formulate a freeform surface is therefore as follows [26]:

$$z(x,y) = z_{basic}(x,y) + \frac{B(x,y)}{P(x,y)} \sum_{j}^{j\max} b_j F_j(x,y) \tag{4.107}$$

The meaning of the various terms in this equation is as follows.

- The basic shape z_{basis} defines the paraxial properties of second order.

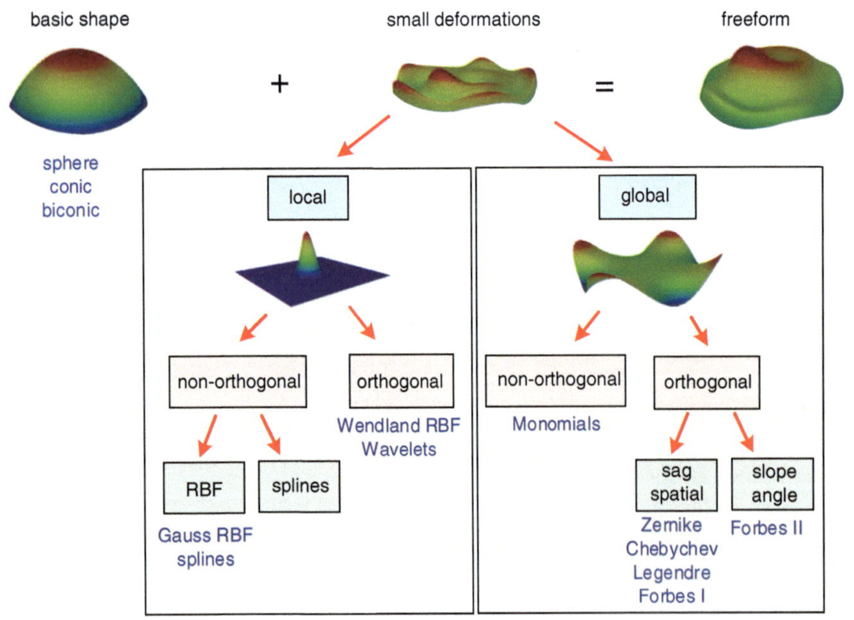

Figure 4.61 General scheme of the possible descriptions of freeform surfaces [25] / with permission of Dennis Osche.

The approach is similar to an asphere, but it is allowed that the basic shape is anamorphic.

- The boundary function $B(x,y)$ forces defined values of the function in the centre or on the boundary.
- The projection factor $P(x,y)$ fixes the property if the higher order contribution are added along the optical z-axis or perpendicular to the surface. The second opportunity makes sense if the shape is rather steep.
- The polynomial function $F(x,y)$ describe the higher order corrections with expansion coefficients b_j.

A more concrete formulation example for a circular symmetric boundary is

$$z(x,y) = \frac{cr^2}{1+\sqrt{1-c^2r^2}} + \frac{\frac{r^2}{a^2}\cdot\left(1-\frac{r^2}{a^2}\right)}{\sqrt{1-c^2r^2}} \cdot \sum_{n=0} b_n F_n\left(\frac{r^2}{a^2}\right) \quad (4.108)$$

Here a is the outer circle radius giving the free diameter as $2a$ and is used for a normalized radial coordinate. Furthermore, the type of functions can be selected in different classifications:

- global polar based;
- global Cartesian based;
- local supporting;
- in particular describing high frequency regular ripple;
- boundary circular or rectangular;
- orthogonality in space or slope domain or no orthogonality at all.

The various options proposed in the literature are again summarized in the following Table 4.5.

In reality, some concrete selected functional descriptions are given in a short overview here.

1) Extended polynomials in x, y: monomial Cartesian product of powers, non-orthogonal. As a basis function, an anamorphic conic is assumed

$$z(x,y) = \frac{c_x x^2 + c_y y^2}{1+\sqrt{1-(1+\kappa_x)c_x^2x^2-(1+\kappa_y)c_y^2y^2}} + \cdot \sum_{n,m=2} a_{n,m} \cdot x^n y^m \quad (4.109)$$

2) Zernike expansion, orthogonal on circle

$$z(x,y) = \frac{c_x x^2 + c_y y^2}{1+\sqrt{1-(1+\kappa_x)c_x^2x^2-(1+\kappa_y)c_y^2y^2}} + \cdot \sum_{j=0} c_j \cdot Z_j(x,y) \quad (4.110)$$

3) Extended Forbes asphere (Q polynomials) Slope orthogonality on circle is guaranteed, is related to tolerancing [27]

$$z(x,y) = \frac{cr^2}{1+\sqrt{1-c^2r^2}} + \frac{\frac{r^2}{a^2}\cdot\left(1-\frac{r^2}{a^2}\right)}{\sqrt{1-c^2r^2}} \cdot \sum_{n=0} a_n^0 Q_n^0\left(\frac{r^2}{a^2}\right) + \sum_{m=1}\left(\frac{r}{a}\right)^m \times \sum_{n=0}\left[a_n^m \cos(m\theta) + b_n^m \sin(m\theta)\right] \cdot Q_n^m\left(\frac{r^2}{a^2}\right) \quad (4.111)$$

4) Legendre polynomials, orthogonal on a rectangle

$$z(x,y) = \frac{c_x x^2 + c_y y^2}{1+\sqrt{1-(1+\kappa_x)c_x^2x^2-(1+\kappa_y)c_y^2y^2}} + \cdot \sum_{j=0} c_j \cdot L_j(x,y) \quad (4.112)$$

Table 4.5 Properties of freeform surfaces.

	Global definition	Local definition	Polar based	Carte-sian based	Sag orthogonal	Gradient orthogonal	Boundary circ	Boundary rect.
Zernike	x		x		x		x	
Zernike differences	x		x			x	x	
Generalized Forbes	x		x			x	x	
Off-axis aspheres	x		x					
Monomials	x			x				
Chebychev	x			x	x			x
Legendre	x			x	x			x
Fourier	x			x	x			x
Splines		x		x				x
Radial basis functions		x	x					
Wavelets		x		x	x			

5) Chebychev polynomials, orthogonal on a rectangle

$$z(x,y) = \frac{c_x x^2 + c_y y^2}{1 + \sqrt{1 - (1+\kappa_x)c_x^2 x^2 - (1+\kappa_y)c_y^2 y^2}} + \cdot \sum_{n,m} a_j \cdot T_n(\overline{x}) T_m(\overline{y}) \quad (4.113)$$

6) Fourier expansion: More meaningful in the case of rectangular regions, orthogonal

$$z(x,y) = \frac{c_x x^2 + c_y y^2}{1 + \sqrt{1 - (1+\kappa_x)c_x^2 x^2 - (1+\kappa_y)c_y^2 y^2}} + \sum_{n,m} B_{nm} \cdot Re\left[e^{-ik_{xn}x - ik_{ym}y}\right] \quad (4.114)$$

7) Radial basis functions: Representation by an expansion into local shifted Gaussian functions, which in general are not orthogonal

$$z(x,y) = \frac{c_x x^2 + c_y y^2}{1 + \sqrt{1 - (1+\kappa_x)c_x^2 x^2 - (1+\kappa_y)c_y^2 y^2}} + \sum_{n,m} a_{nm} \cdot e^{-\left(\frac{x-x_n}{w_x}\right)^2 - \left(\frac{y-y_n}{w_y}\right)^2} \quad (4.115)$$

8) Radial basis functions: Piecewise polynomials of Wendland, Wu, Gneiting, can be made orthogonal

$$z(x,y) = \frac{c_x x^2 + c_y y^2}{1 + \sqrt{1 - (1+\kappa_x)c_x^2 x^2 - (1+\kappa_y)c_y^2 y^2}} + \sum_{n,m} a_{nm} \cdot \left(\frac{r}{w}\right)^m P_m(r) \quad (4.116)$$

9) Cubic spline: Polynomial representation inside finite sized patches with index j, k defined. Very usual are polynomials of third order (other powers are possible)

$$z_{j,k}(x,y) = \sum_{m=0}^{3} \sum_{n=0}^{3} a_{jkmn} \times (x - x_j)^m \times (y - y_k)^n \quad (4.117)$$

Other spline variants are of course also possible.

4.7.2 Basic Shape

The possible options for the selection of the basic shape are as follows [26].

- Best fit sphere (for minimizing the non-spherical contribution).
- Circular conic section (for extracting the circular contribution) and capture of the paraxial behaviour.
- Biconic (to cover low order astigmatism and keep it out of the expansion).

This part should describe the basic parabasal behaviour; it makes sense not to mix this with the higher order corrections. Often this shape also serves as the basis of easy manufacturing, the higher order freeform terms are limited and more restricted from a technological points of view considering the absolute size of the sag as well as change of slope.

The lower orders of the freeform surface are given by the separated basic shape term. It dominates the parabasal properties of the surface near the optical axis. In the simplest case a circular symmetric sphere with curvature $c = 1/R$ is selected.

$$z_{basic}^{(sphere)}(r) = \frac{cr^2}{1 + [1 - c^2 r^2]^{1/2}} \quad (4.118)$$

According to the proposal of Forbes [27] for the Q-polynomials a sphere is used and fitted over the free area, touching the surface in the centre.

$$z_{basic}^{(bfs)}(r) = \frac{c_{bfs} r^2}{1 + \left[1 - c_{bfs}{}^2 r^2\right]^{1/2}} \quad (4.119)$$

The higher order freeform is not contributing in the centre in this case. This approach can be generalized to a circular symmetric conic surface

$$z_{basic}^{(conic)}(r) = \frac{cr^2}{1 + [1 - (1+\kappa)c^2 r^2]^{1/2}} \quad (4.120)$$

with curvature c described in the vertex point and conical constant κ. If in particular for mirror systems an astigmatism on axis is an issue, this representation can be extended to an anamorphic conic section with different parameters in the x- and y-cross section. This toroidal basic shape can be written as a biconic

$$z_{basic}^{(biconic)}(x,y) = \frac{c_x x^2 + c_y y^2}{1 + \sqrt{1 - (1+\kappa_x)c_x{}^2(x^2+y^2) - (1+\kappa_y)c_y{}^2(x^2+y^2)}} \quad (4.121)$$

4.7.3 Boundary and Projection Factor

The boundary function $B(x,y)$ or $B(r)$ forces as an additional constraint the surface values in the centre and on the wanted boundary. Depending on the footprint of light on the surface, this allows us to control the surface values outside the free area. In practice, an additional surface part is required for the positioning of the surface in the mechanical design and locating reference marks. In most cases, due to the higher orders, an extrapolation of the polynomial functions beyond the free diameter is critical. As a boundary curve, simple geometries are recommended. Figure 4.62 gives two simple examples of boundary functions for a square domain $B(\overline{x},\overline{y})$ and a circular case with centre fixation $B(\overline{r})$ [26].

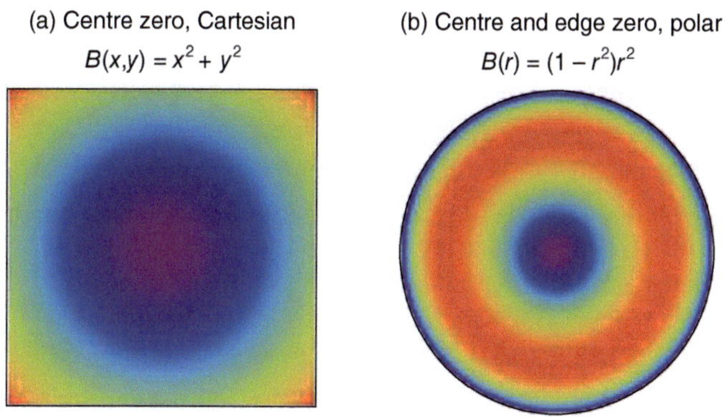

Figure 4.62 Possible boundary constraints of freeform surfaces. (a) shows a Cartesian geometry with a zero in the centre, (b) is a circular symmetric geometry with a zero in the centre and the outer boundary.

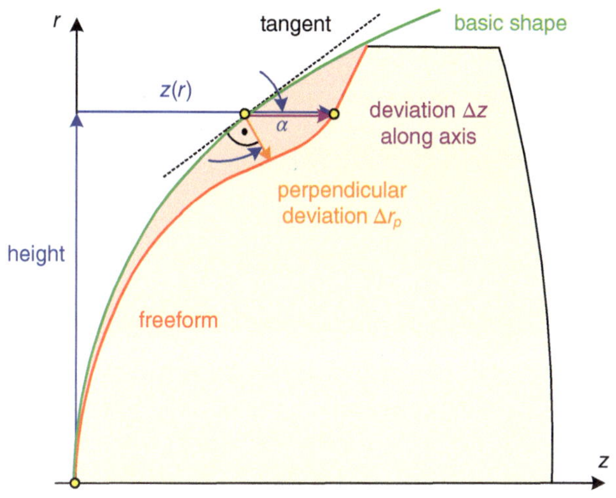

Figure 4.63 Projection factor for the correction contribution along z or perpendicular to the surface.

As is sketched in Figure 4.63 the correction of the basic shape by higher orders can take different directions. To control this property, a projection factor $P(x,y)$ can be defined. In the simplest case the deviation is chosen as Δz parallel to the optical axis. If a surface is strongly bent with large slopes near to the outer boundary, it is more comfortable to select the deviation perpendicular to the surface, as shown in the figure in orange. The corresponding difference Δr_p is then much smaller than Δz and describes material removal by grinding and processing time more accurately. The corresponding projection function has the consequence that the surface description is no longer a polynomial but a rational function. Therefore, surfaces with this representation can only be converted to a polynomial with finite accuracy depending on the number of terms.

As is seen in the figure, the projection factor is given with the slope angle α as $1/\cos(\alpha)$. In general, this expression can be written

$$P(x,y) = \frac{1}{\cos\alpha} = \left[1 + \left(\frac{dz_{basic}}{dx}\right)^2 + \left(\frac{dz_{basic}}{dy}\right)^2\right]^{-1/2} \tag{4.122}$$

and it depends on the selected basic shape function. In practice it is meaningful to avoid linear tilt terms in the basic shape and in the polynomial correction part. Then the local z-axis remains perpendicular to the surface. A tilt of the freeform surface can be included by a rotation of the coordinate transform between the global coordinates and the local coordinates of the surface.

4.7.4 Polynomial Functions

4.7.4.1 Introduction

Finally, the expansion term is expressed as a sum of functions, mostly polynomials with global support on the defined active area. The scaling usually is normalized on the half diameter $a = D/2$; therefore, the coefficients are independent of the diameter. Most of the materials shown here are based on [26]. The expansion polynomials with coefficients b_n can be:

- orthogonal in sag (Zernike, Legendre, Chebyshev);
- orthogonal in slope (Forbes);
- boundary circular/rectangular;
- different weighting functions (uniform, emphasized boundary ...).

The differences between the usual polynomials for this purpose are relatively small in their practical consequence. Most of the versions are defined on normalized coordinates, which means that the user has to decide what size of surface area should be supported. But this normalization improves the numerical conditioning and comparability of different cases. Only the simple monomials are not scaled in transverse coordinates.

One of the major differences is the geometry of the definition area. Some of the polynomials are defined on a circular area and are formulated in polar coordinates. If the light footprint on the surface is 'more rectangular' in shape, which means that the aspect ratio $D_{max}/D_{min} \gg 1$ is large, the corresponding descriptions offer a significantly better modelling. All versions except the monomials are orthogonal in space or in gradient. This delivers comfortable conditions in modelling and fitting given data or optimization of the parameters in optical design. Therefore, choosing orthogonal polynomials is strongly recommended. If the first two mode shapes are fixed, the higher order polynomials can be forced to be orthogonal by using the classical Gram–Schmidt procedure. This can also be done for individual cases of special illumination weightings or pupil boundaries. One more difference between the various possibilities is the weighting function of the orthogonality inner product. The polynomials with an increased weight in the outer region are able to describe the higher order aberrations at the rim of the pupil usually slightly better.

Furthermore, the sorting of the orders of the polynomials is defined differently. Originally, all the two-dimensional functions had two ordering indices, but for more easy handling they are mostly mapped onto a one-dimensional running index. So, the coefficients become a vector.

Except in the case of a used projection factor all the polynomial representations described above can be converted into each other exactly.

The selection of the highest used order of the polynomials is a non-trivial question. On the one hand, higher orders allow for a better overall correction; on the other hand, high orders create a larger sensitivity of the surface in position accuracy and are not easy to fabricate.

The sorting of the polynomial indexing in the best case corresponds to the aberration order. But the opportunity to compare polar and Cartesian descriptions is different. A conversion between different boundary formulations is also not easily possible. Most efficient is a numerical re-fitting. In Figure 4.64 some of the most often used functional sets of polynomials are visualized. Their definition and properties are discussed in the following sections.

4.7.4.2 Monomials

A simple Taylor expansion into separated x- and y-powers is called a monomial representation. It is easy to program, the biggest disadvantage is the lack of orthogonality. In Cartesian coordinates the description reads:

$$z(x,y) = z_{basic}(x,y) + \sum_{m,n}^{M,N} a_{mn} x^m y^n \quad \text{with } m, n = 0, 1 \ldots M, N \tag{4.123}$$

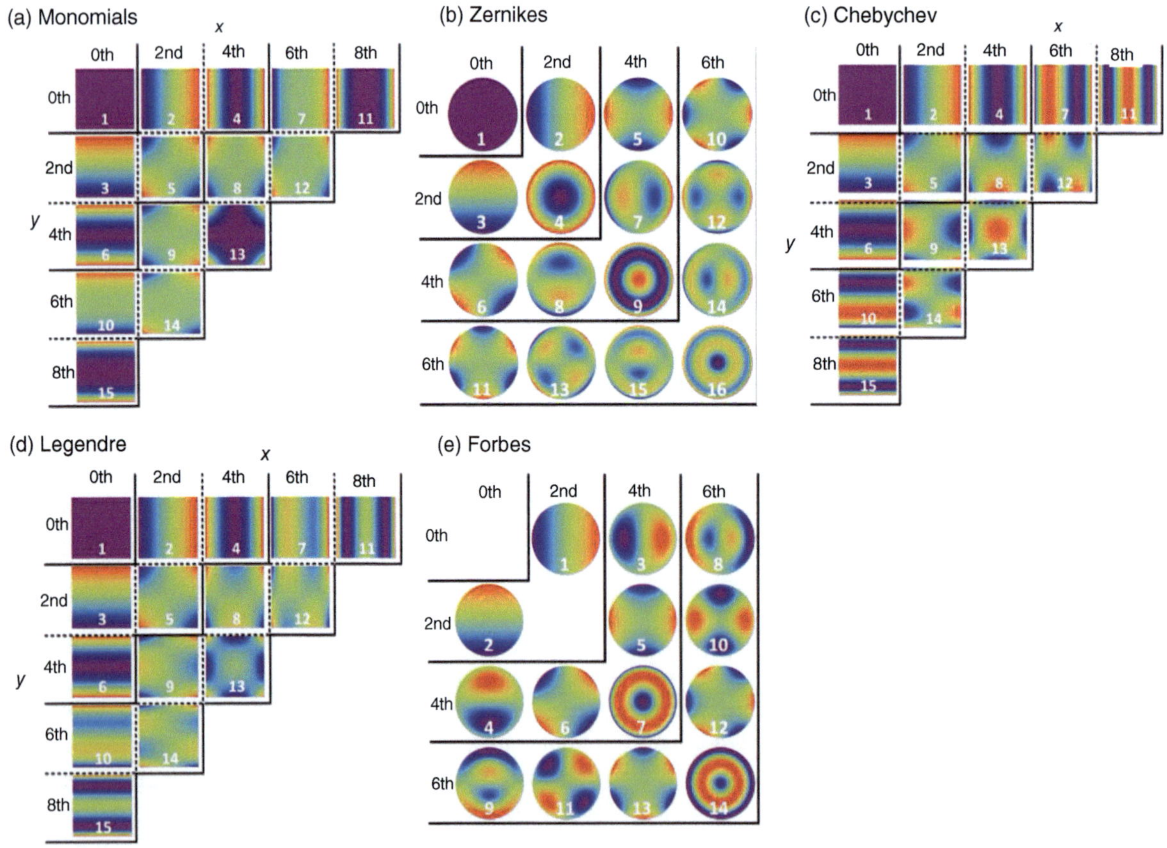

Figure 4.64 Different polynomial function sets (a) monomials and polynomials of (b) Zernike, (c) Chebychev, (d) Legendre, (e) Forbes [26] Anika Brömel / CC BY 4.0.

This form corresponds to the historically used circular symmetric asphere representations discussed in Section 4.6 and it suffers from all the drawbacks discussed there.

4.7.4.3 Zernike Polynomials

The Zernike polynomials are defined in polar coordinates. The functions are spatially orthogonal for a circular domain and a constant weighting function. The advantage of using this representation is the usual description of wave aberrations by Zernikes. Unfortunately, different definitions and conventions for Zernike polynomials are in use. The most common are the standard definition and Fringe convention. The latter is specifically introduced to describe wavefront errors in optical metrology. The functions can be written as.

$$z(\bar{r},\varphi) = \sum_{m,n} a_n^m Z_n^m(\bar{r},\varphi) \quad \text{with } m,n = 0,1\ldots\text{N},\ \bar{r} = \frac{r}{r_{\max}} \tag{4.124}$$

Zernike functions are decomposed by a product of a radial function $R_n^m(\bar{r})$ with an azimuthal part $\Phi^m(\varphi)$ scaled by a normalization factor $N(m, n)$. In the standard convention the Zernike functions are normalized to the RMS, in Fringe convention they are normalized to the peak-to-valley deviation

$$Z_n^m(\bar{r},\varphi) = N(m,n)\,R_n^m(\bar{r})\,\Phi^m(\varphi) \tag{4.125}$$

The radial index n and the azimuthal index m follow special selection rules, they must be both odd or both even and it is always $|m| \le n$. In the Fringe convention it is

$$N(m,n) = 1 \tag{4.126}$$

the radial part for both conventions is

$$\begin{aligned} R_n^m(\bar{r}) &= \sum_{k=0}^{\frac{n-|m|}{2}} (-1)^k \binom{n-m}{k} \binom{n-2k}{\frac{n-m}{2}-k} \bar{r}^{n-2k} \\ &= \sum_{k=0}^{\frac{n-|m|}{2}} (-1)^k \frac{(n-k)!}{k!\left(\frac{n-|m|}{2}-k\right)\left(\frac{n+|m|}{2}-k\right)} \bar{r}^{n-2k} \end{aligned} \tag{4.127}$$

and the azimuthal part

$$\Phi^m(\varphi) = \begin{Bmatrix} \sin(m\varphi) & m > 0 \\ \cos(m\varphi) & m < 0 \\ 1 & m = 0 \end{Bmatrix} \tag{4.128}$$

with the angle φ. This angle is measured against the x-axis. It should be mentioned that due to the angle part, the Zernike functions are not simple polynomials; the usual wording 'Zernike polynomials' is not precise. Zernike polynomials can be developed with the orthogonality over a unit circle by using

$$\int_0^{2\pi}\int_0^1 Z_n^m(r,\varphi)\,Z_{n'}^{m'}(r,\varphi)\,r\,\mathrm{d}r\,\mathrm{d}\varphi = \frac{\pi(1+\delta_{m,0})}{2(n+1)}\delta_{m,m'}\delta_{n,n'}, \tag{4.129}$$

with m'-n' even and $R_n^m(1) = 1$. In practical applications it is easier to have only one running index instead of two. Therefore, a mapping of n and m onto a single index j is common. But the various conventions use different sorting schemes. One major benefit of the Zernike polynomials to describe optical surfaces is the direct relation to the wavefront description if the surface is located in the pupil. Many more details about Zernike polynomials are presented in Volume 2, Chapter 15, which deals with wave aberrations.

4.7.4.4 Chebyshev Polynomials

Chebyshev two-dimensional polynomials are comfortable to use on rectangular domains. They can be described as Cartesian products of the one-dimensional Chebyshev polynomials. The resulting set is spatial orthogonal on a unit square. These are the polynomials of the first kind; other definitions are possible. Chebychev polynomials can be expressed as

$$\begin{aligned} z(x,y) &= z_{basic}(x,y) + \sum_{m,n}^{M,N} a_{mn} T_m(\bar{x})\,T_n(\bar{y}) \\ &\text{with } m,n = 0,1\ldots M,N \\ \bar{x} &= \frac{x}{x_{\max}},\ \bar{y} = \frac{y}{y_{\max}} \end{aligned} \tag{4.130}$$

The one-dimensional terms $T_m(\bar{x})$ and $T_n(\bar{y})$ respectively use the weighting function $(1-x^2)^{-1/2}$:

$$\int_{-1}^{1} \frac{T_m(x)\,T_{m'}(x)}{\sqrt{(1-x^2)}}\,dx = \begin{cases} \frac{1}{2}\pi\,\delta_{mm'} & \text{for } m \neq 0, m' \neq 0 \\ \pi & \text{for } m = m' = 0 \end{cases} \tag{4.131}$$

with the Kronecker delta δ_{mm}. This corresponds to a strong weighting of the outer edges $x = \pm 1$. The polynomial terms can be generated by recurrence with two initial terms and the following recurrence relation in one dimensional

$$\begin{aligned} T_0(x) &= 1 \\ T_1(x) &= x \\ T_{m+1}(x) &= 2xT_m(x) - T_{m-1}(x) \end{aligned} \tag{4.132}$$

The first terms of the Chebyshev first kind can be seen in Figure 4.64(c).

4.7.4.5 Legendre Polynomials

Another option for rectangular areas is to use Legendre polynomials. They are defined similarly to the Chebyshev polynomials as Cartesian products of one-dimensional functions but use a different weighting.

$$z(x,y) = z_{basic}(x,y) + \sum_{m,n}^{M,N} a_{m,n} L_{mn}(\overline{x},\overline{y})$$
$$= z_{basic}(x,y) + \sum_{m,n}^{M,N} a_{m,n} P_m(\overline{x}) P_n(\overline{y})$$
$$\text{with } m,n = 0,1\dots M,N \qquad (4.133)$$

The one-dimensional functions $P_m(\overline{x})$ are orthogonal with a constant weighting function

$$\int_{-1}^{1} P_m(x)\, P_{m'}(x)\, dx = \frac{2}{2m+1}\delta_{mm'} \qquad (4.134)$$

The lower order terms and the recurrence formula reads

$$P_0(x) = 1$$
$$P_1(x) = x$$
$$(m+1)P_{m+1}(x) = (2m+1)xP_m(x) - mP_{m-1}(x) \qquad (4.135)$$

4.7.4.6 Forbes Freeform Surfaces

Forbes or Q-polynomials for the description of freeform surfaces are extended functions of the corresponding circular symmetric aspheres; see Section 4.6.4.3. The basis is the 'mild-asphere' approach and they are gradient-orthogonal and are defined for circular boundaries. The mathematical description uses the concept of the 'best-fit' basic shape, a boundary function and a projection factor. The mild-asphere set is fixed at the centre and the outer boundary function to achieve the 'best-fit' basic shape. The surface sag at the centre of the freeform is only defined by the basic shape. With this definition, a limitation of the slopes is achieved and this generates a manufacturing-friendly surface.

The complete surface description is therefore

$$z(r,\varphi) = z_{basic}^{(bfs)} + z_{Mild-Asphere}(\overline{r}) + z_{Q-Poly}(\overline{r},\varphi)$$
$$= z_{basic}^{(bfs)} + \frac{1}{\sqrt{1-c^2r^2}} \cdot \left[\overline{r}^2 \cdot (1-\overline{r}^2) \cdot \sum_{n=0} a_n^0 Q_n^0(\overline{r}^2)\right]$$
$$+ \frac{1}{\sqrt{1-c^2r^2}}$$
$$\cdot \left[\sum_{m=1} (\overline{r})^m \sum_{n=0} \left[a_n^m \cos(m\varphi) + b_n^m \sin(m\varphi)\right] \cdot Q_n^m(\overline{r}^2)\right]$$
$$\text{with } m,n = 0,1\dots M,N \qquad (4.136)$$

The trigonometric functions in the square bracket are automatically orthogonal for different azimuthal orders. The radial part with the boundary function follows the orthogonality condition:

$$\int_0^1 \nabla[z_{Q-Poly}(\overline{r},\varphi)] \cdot \nabla[z_{Q-Poly}(\overline{r},\varphi)] \frac{d\overline{r}}{\sqrt{1-\overline{r}^2}}$$
$$= \begin{cases} \pi & n = n' \\ 0 & \text{otherwise} \end{cases} \qquad (4.137)$$

Q-polynomials $Q_n^m(\overline{r}^2)$ were specifically designed to be orthogonal in gradient, the mean square gradient of the normal departure from the best-fit basic shape can then be expressed as the sum of the squares of the coefficients in the departure:

$$\langle |\nabla \delta_{nd}(\overline{r},\varphi)|^2 \rangle = \left\langle \left(\frac{\partial \delta_{nd}}{\partial \overline{r}}\right)^2 + \frac{1}{\overline{r}^2}\left(\frac{\partial \delta_{nd}}{\partial \varphi}\right)^2 \right\rangle$$
$$= \sum_{m,n} \left[\left(a_n^m\right)^2 + \left(b_n^m\right)^2\right] \text{ with } b_n^0 = 0$$
$$(4.138)$$

4.7.5 Functional Systems with Local Support

4.7.5.1 Introduction

In lens design, and in particular in the development of freeform systems, there are in principle two different situations that need to describe a freeform surface.

- During the optimization of the system:
  here mostly it is desirable to have not too many parameters, the surfaces should be smooth and polynomials with global area support are preferred. A localized description with a very weak coupling and a huge number of variable parameters is mostly not very comfortable; steep localized deviations are created, which are hard to manufacture.
- After a surface or a component is manufactured it is strongly recommended to re-import the real surface shape into the simulation tool to verify the performance and specification of the complete system. Then it may happen that the measuring data are corrupted by noise, have stochastic deviations and maybe localized errors. In this case the description with many parameters and a stronger localized description makes sense to capture all effects. If the stiffness of the model surface can be adapted to smooth the noisy points and maintain the relevant deviations, the best approach is obtained.

This shows that in particular for real surfaces the need for localized descriptions is obvious. In literature two major approaches are proposed: radial shifted base functions (RBFs) and splines. Both have pros and cons that will be briefly discussed in the next sections. For more rigorous mathematical theory, it is recommended to look into the huge amount of literature.

In addition, a description by wavelets is proposed in literature, but there is not a lot of experience reported and therefore no discussion is presented here.

4.7.5.2 Radial Basis Functions*

(*) The content of this section is quite specialized and for advanced reading. Knowledge of this section is not necessary to understand the following chapters. Beginners are recommended to skip this section.

RBF-like Gaussian modes can be used to describe a deviation locally) [28, 29]. The amplitude and the grid choice of the sampling centres of the RBFs are the variables to optimize a representation. In the simplest case, equidistant transverse shifted Gaussian functions are superposed according to

$$z(x,y) = \sum_n c_n \cdot e^{-\frac{(x-x_n)^2+(y-y_n)^2}{w^2}} \tag{4.139}$$

Figure 4.65 illustrates this idea in one and two dimensions. Although mathematically the Gaussian mode functions never reach a zero level, practically a clip level can be defined and then the modes have a local support area. This allows for a sparse representation and is a great advantage for numerical computations.

In this simple approach, the functions are defined due to their fast convergence only on a locally restricted range. They are not orthogonal, but can be orthogonalized depending on the selected mode functions. In particular pure Gaussian mode functions are never orthogonal. One crucial question is the sampling grid, which defines the centres and the width of the individual functions. If the width w is small relative to the sampling distances Δx, Δy, the overlap vanishes and not every value can be realized between the centres. If the width is large, there is a comfortable overlap, but it is not possible to model sharp edges and steep slopes. In the best case, the individual centres are equidistantly sampled, but in practice this is not a very critical request. The obvious Cartesian grid has the disadvantage that the distance is longer in the diagonal directions. Therefore, hexagonal, Halton-type statistical or Fibonacci grids are preferred and give the best results in reality. Figure 4.66 shows the fit of a real statistical surface with different sampling geometries as a function of the number N of used RBFs. It can be seen that the differences between hexagonal, Fibonacci and statistical grids are only small and a growing number of degrees of freedom of course give better results. The primitive Cartesian grid is roughly worse by a factor of 2.

If the number N is chosen on a given region of interest, the distances between the RBF centres is also changed. Therefore, the best result is obtained if N and w are adapted. Figure 4.67 shows the dependence of the fit rms

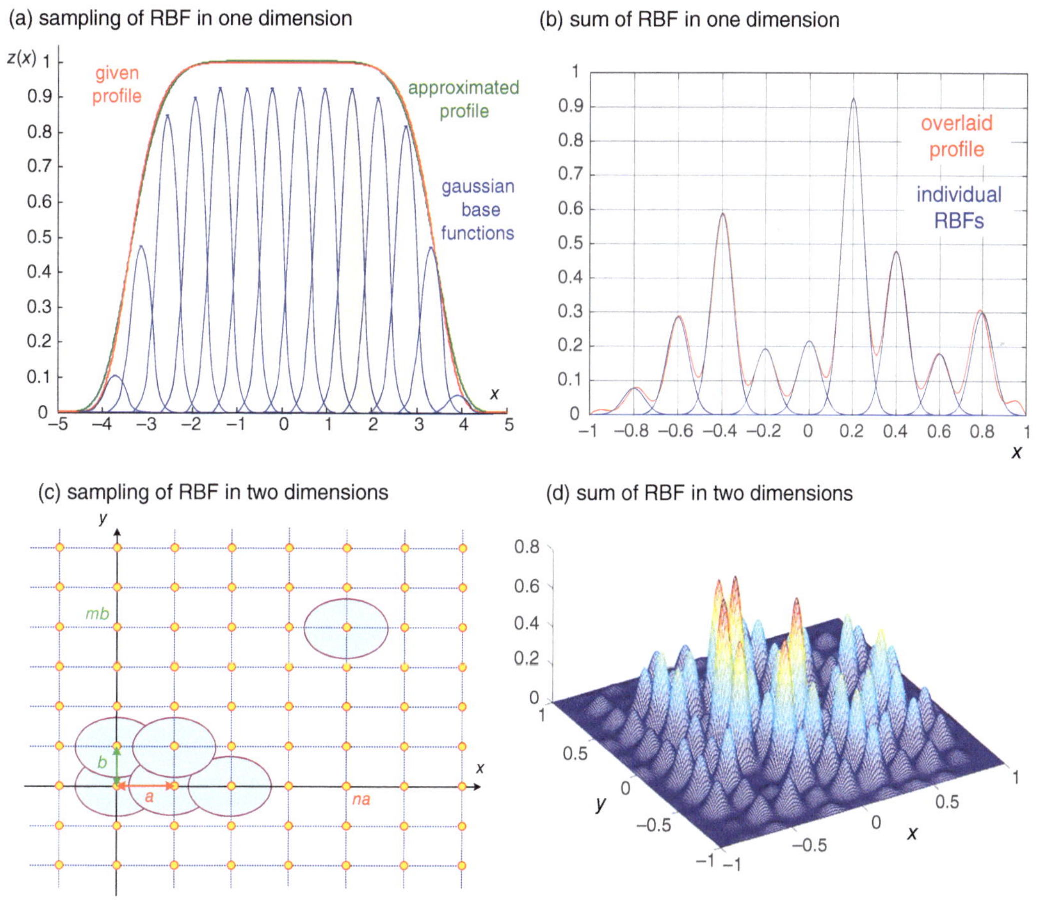

Figure 4.65 Principle of radial shifted base functions for (a) sampling and (b) overlay in one dimension and (c), (d) in two dimensions. The diagrams are schematic and the units arbitrary.

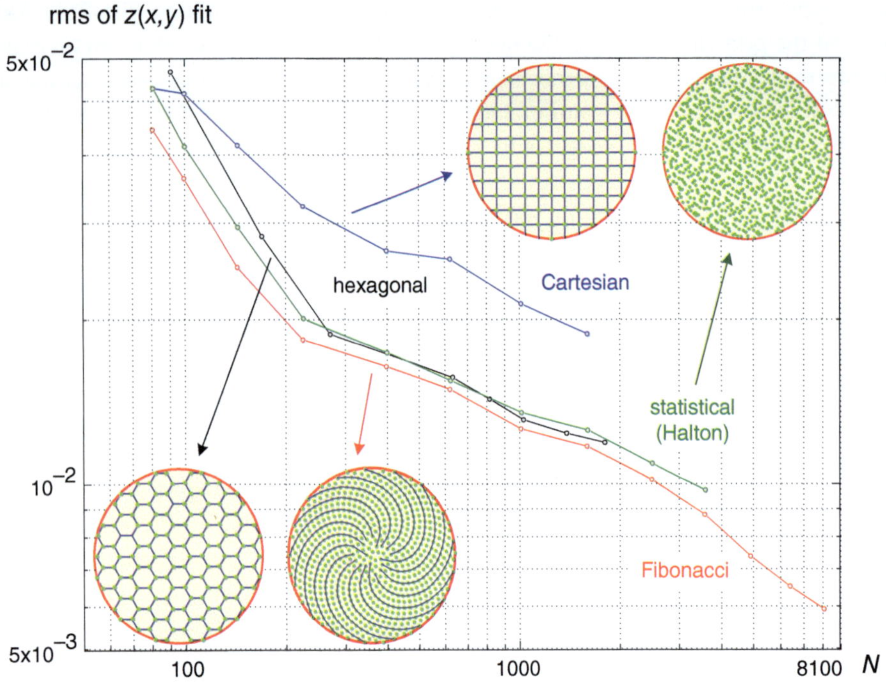

Figure 4.66 Influence of function numbers and sampling geometry for RBF functions for a statistical surface.

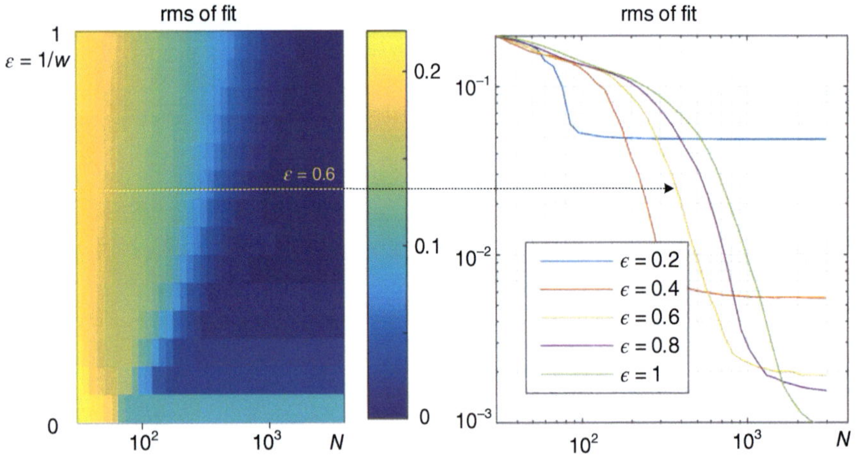

Figure 4.67 Dependence of accuracy on the shape parameter and the number of RBFs [28] by fitting a Zernike Z_{43} surface.

of a Zernike term no. 43 surface on $\varepsilon = 1/w$ and N as error rms(ε,N) and as a cross section for selected values of ε. It can be seen that a large N and smaller w make sense. Of course, the exact quantitative behaviour depends on the spatial frequency content of the surface to be described.

In Figure 4.68 a one-dimensional fit of a super-Gaussian profile with different exponents $m = 2$, 12, 40 and 200 with $N = 128$ radial basis functions is performed and the resulting rms of the model fit is considered. The width w of the Gaussian profiles is scaled by the centre distance Δx of the Cartesian equidistant grid. It can be seen that depending on the maximum slope of the profile, a value of $w = 0.75 \ldots 0.96\ \Delta x$ gives the best match. If the desired profile is smooth with no steep gradients, as the case $m = 2$, larger w are not really a disadvantage. But in the case of steep slopes, w should be in the best case in the range of 0.74 to fill the gap between the sampling points and simultaneously to model the slope appropriately. Therefore, the adaptive optimization of w makes sense in this case.

This simulation shows that higher spatial frequencies, as in the case of the steep slope for large m values of the super-Gaussian functions, reduce the accuracy and need for smaller values of w and therefore automatically for larger mode function numbers N.

In reality, there are several crucial details in the numerical realization of RBF functions on an equidistant grid. If the simple and preferred hexagonal grid is used, the boundary is also hexagonal, which mostly does not correspond to the real border curve. The parameters of the hexagonal grid must therefore be selected in a way to capture all relevant

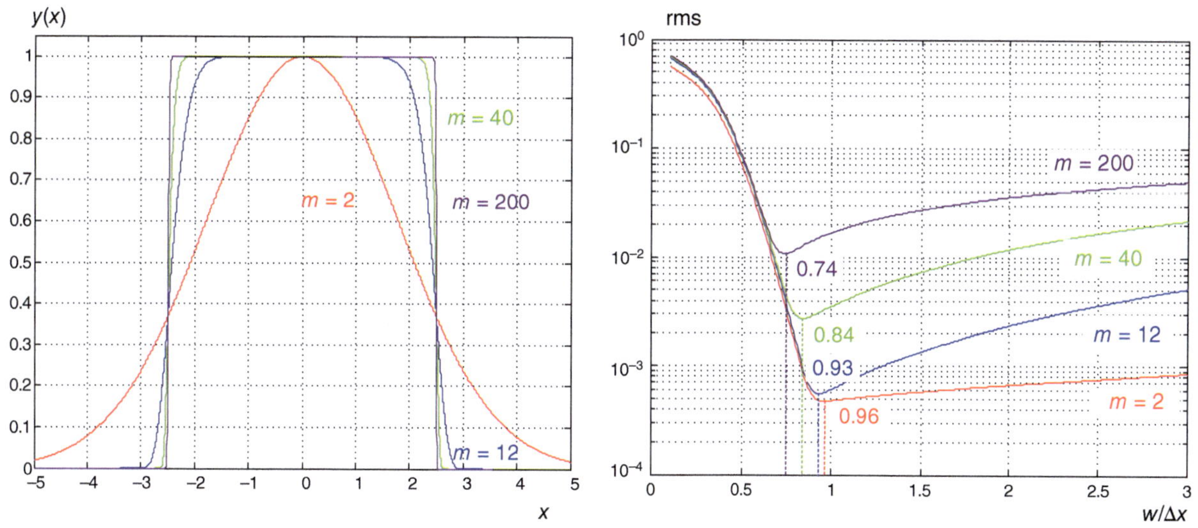

Figure 4.68 Fit of super-Gaussian profiles (a) and accuracy of RBF representation (b) with different exponents m as a function of the width/distance ratio.

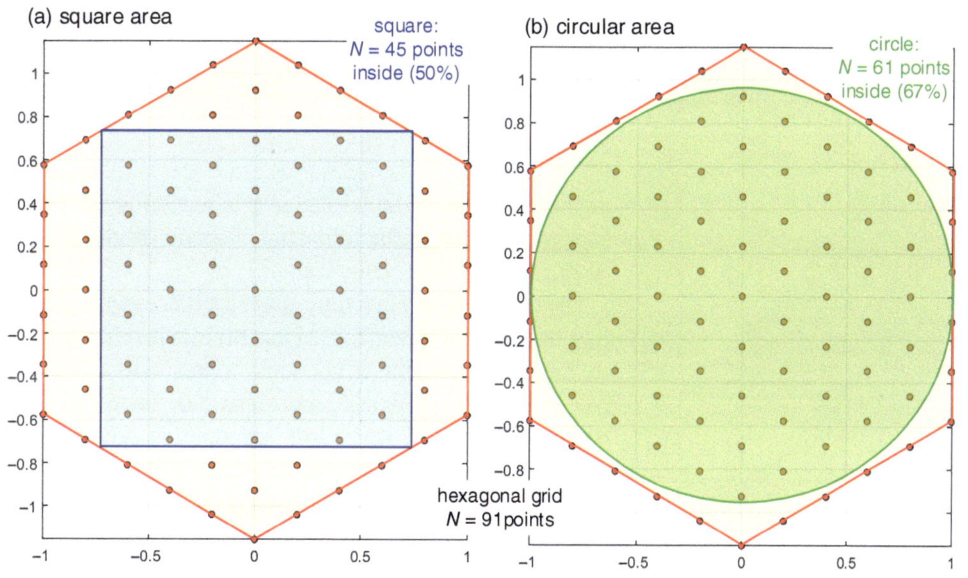

Figure 4.69 Hexagonal grid and the problem of the boundary for filling (a) a square or (b) a circle.

regions. Figure 4.69 shows as an example a hexagonal grid with 91 points and a circular and square area respectively, which has to be sampled. It can be seen that the fill factor is not really convincing if the full hexagon is considered.

Figure 4.70 shows for illustration the result of a RBF fit onto a Gaussian profile on a hexagonal grid with 919 grid points. The rms deviation of this approximation is in the range of 10^{-6}.

There are also more refined techniques proposed in the literature to improve the performance in the sense of complexity vs accuracy of RBF functions. One approach suggests distributing and locally adapting the sampling points according to the changes and gradient values of the function to approximate. Furthermore, it makes sense to allow for different widths of the basis functions. Figure 4.71 shows an example of a non-uniform grid; in the ranges of the peaks, the basis functions are sampled more densely.

It should be clear that Gaussian basis functions are not the only alternative for radial shifted functional systems. Another wide class of functions are piecewise defined polynomials. Figure 4.72 shows some exemplary function types [29]. These basic functions typically have the mathematical representation as a product of one polynomial, which acts as a window function for the finite support area

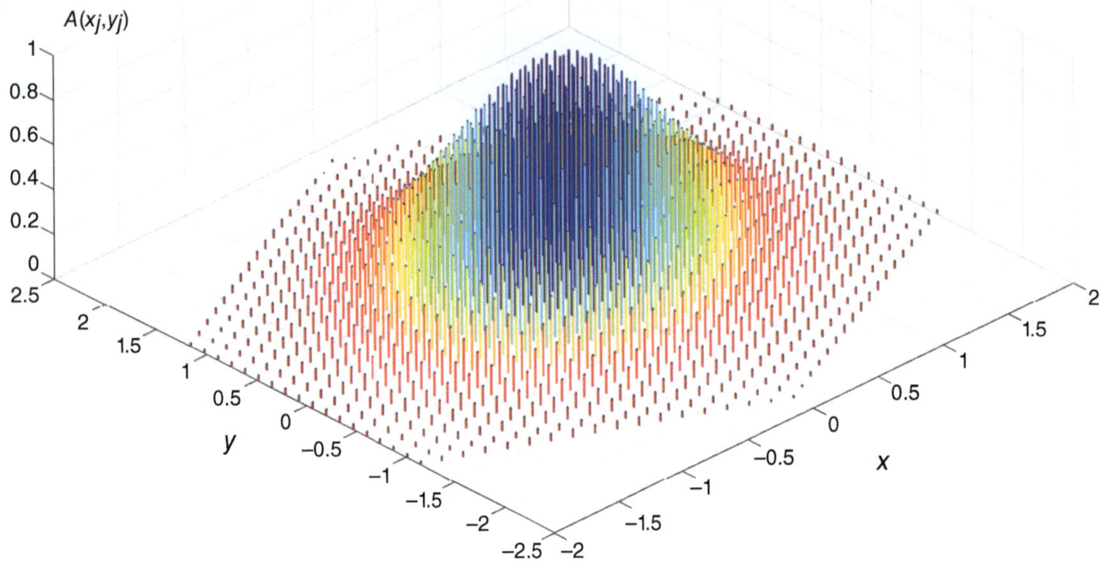

Figure 4.70 RBF decomposition of a Gaussian profile into 919 RBFs of Gaussian shape. Shown are the amplitudes of the RBFs on the hexagonal grid.

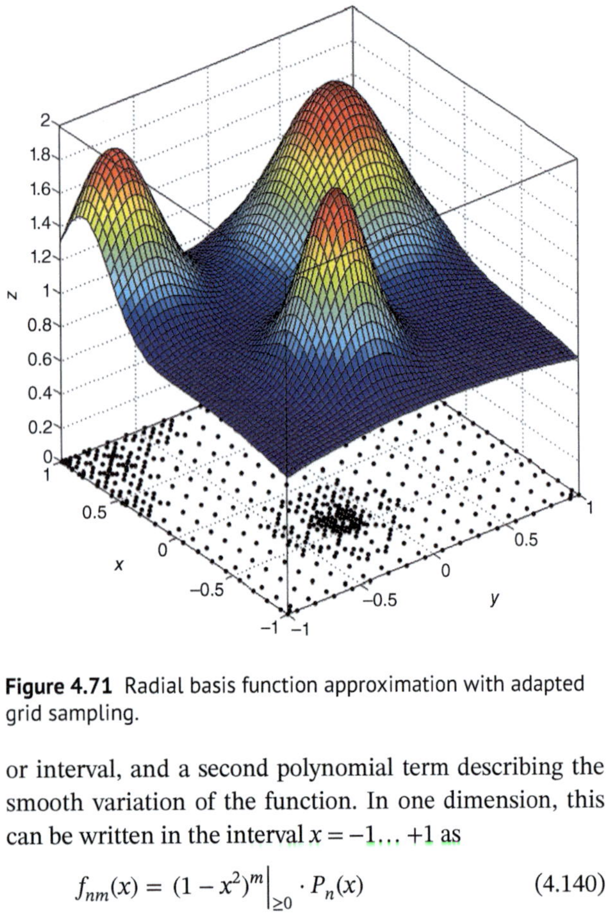

Figure 4.71 Radial basis function approximation with adapted grid sampling.

or interval, and a second polynomial term describing the smooth variation of the function. In one dimension, this can be written in the interval $x = -1 \ldots +1$ as

$$f_{nm}(x) = (1 - x^2)^m\Big|_{\geq 0} \cdot P_n(x) \tag{4.140}$$

where the first factor is the window function. In particular there are various definitions and proposals made by Wendland, Wu and Gneiting. Some of them are illustrated in the figure; it should be noted that negative values of the basis functions are also possible and by selecting the polynomial function orthogonality can also be achieved.

4.7.5.3 Splines*

(*) The content of this section is quite specialized and for advanced reading. Knowledge of this section is not necessary to understand the following chapters. Beginners are recommended to skip this section.

The description of arbitrary shaped surfaces with splines has been well known for a long time and is used a lot in mechanical engineering. This approximation can be performed in one or two dimensions. The basic idea is to divide the area of surface into segments or area patches and to define in every patch a polynomial function locally. This means that a lot of parameters are available. At the boundaries between the patches, the functions must fulfil the constraints of continuous functions in value, slope and maybe second derivative to guarantee a smooth model surface [30, 31].

Figure 4.73 shows an example for illustration. In each element the surface is described analytically by a simple cubic polynomial equation. Therefore, the parameters of description are only valid locally in a relatively small area. A large number of parameters is necessary to describe the whole surface.

In particular, local deformations limited to a small area of the surface can usually be described using splines; a local description is possible without any problem.

There are many known spline representations that have different analytical descriptions, different properties at the boundary lines, a different form for the patches and also different smoothing properties. In practice, very often the

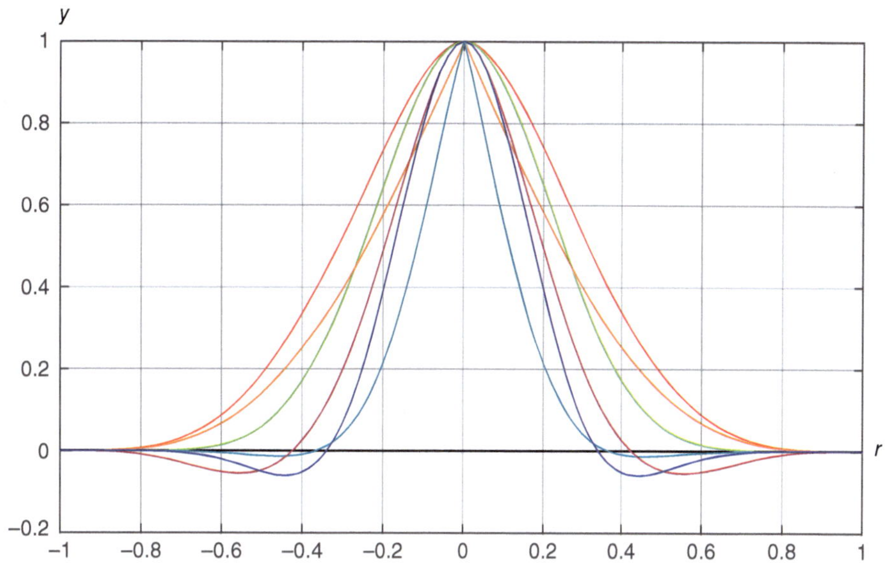

Figure 4.72 Piecewise defined polynomials of different type as basis functions restricted on the interval $r = -1 \ldots +1$.

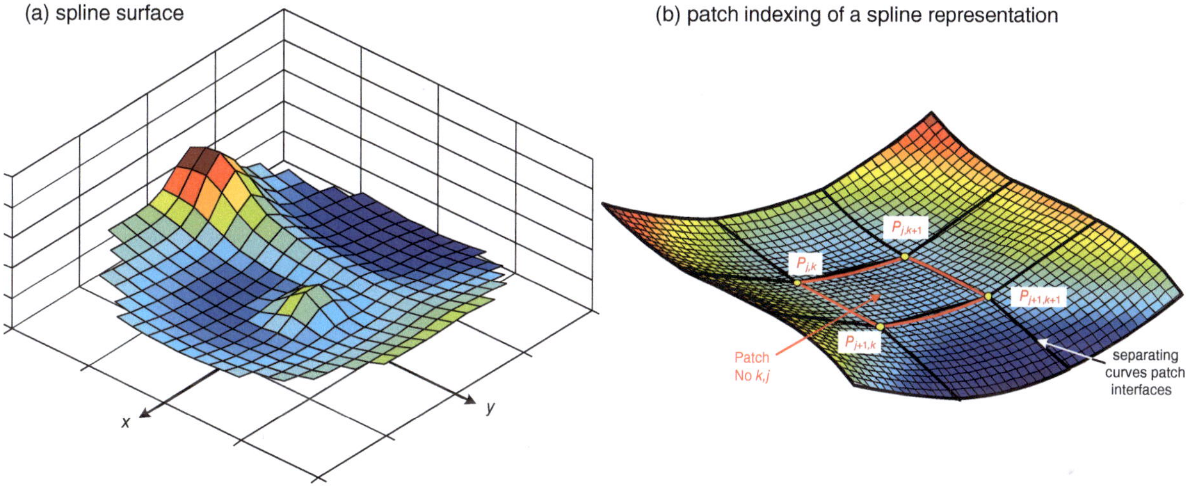

Figure 4.73 Freeform surface (a) modelled as spline, (b) indexing of the patches and smooth transitions between the sub-apertures.

surface is first given at a finite number of discrete points. In a second step, one looks for a surface that interpolates between these sampling points. One possibility is to use interpolating splines. These represent the surface exactly at the supporting points. If there are measurement errors or uncertainties it is reasonable to choose compensating splines with pre-given stiffness. They allow for the possibility of correction of statistical errors by their smoothing effect.

The most important types of splines are:

- Cubic interpolating splines.
  Within a patch the height of the sagitta is described by a two-dimensional cubic polynomial.

$$A_{j,k}(x,y) = \sum_{m=0}^{3}\sum_{n=0}^{3} a_{jkmn} \cdot (x - x_j)^m \cdot (y - y_k)^n \quad (4.141)$$

Altogether 16 parameters are necessary to describe one patch.

- Bezier splines.
  The Bernstein polynomials $B_m(x,\mu)$ of order m resulting from the binomial theorem form a polynomial base system. From this, a so-called Bezier polynomial can be obtained by superposition. For, e.g. a representation to the third order, the Bezier polynomial is written in one dimension as

$$B(x) = b_0 \cdot B_0(x) + b_1 \cdot B_1(x) + b_2 \cdot B_2(x) + b_3 \cdot B_3(x) \quad (4.142)$$

The coefficients of the superposition are the Bezier control points and they determine the trajectory of the curve in one dimension. For a finite interval, according to the boundary between two patches of a surface representation, a curve is obtained that approaches the given tangent at the end points. In between it has a curved form determined by the control points. In two dimensions Bezier spline surfaces behave in an analogous way for given control points as well as given gradients in the vertices of the basic area of the patch.

- Non-uniform rational basis splines (NURBS).
  If a curve in one dimension is not described by a simple polynomial but by a ratio of two polynomials, i.e. a rational function, then rational splines are obtained. If, furthermore, a non-uniform sampling of the control nodes is allowed, then NURBS are generated. For third order we have in one dimension the equation

$$B(x) = \frac{b_0 \cdot B_0(x) + b_1 \cdot B_1(x) + b_2 \cdot B_2(x) + b_3 \cdot B_3(x)}{c_0 \cdot B_0(x) + c_1 \cdot B_1(x) + c_2 \cdot B_2(x) + c_3 \cdot B_3(x)} \tag{4.143}$$

More types of spline representation, such as Hermite or Akima splines, can be found in the literature [30, 31]. A special form is the representation of a surface by rotational symmetric splines. Here only the radial dependence is fixed.

By corresponding control and without the request of smooth slopes at the borders between the patches it is also possible to simulate steps and kinks in a surface by this approach; for example, Fresnel surfaces can be modelled with high accuracy.

The calculation of the raytracing highlights an important problem for a spline surface. First, the right mesh patch for the calculation of the intersection point has to be found. For a ray running towards an edge of the patch, problems can occur if the patch has to be changed again. The spline representation is valid only within the patches and cannot be extrapolated or extended into the neighbouring patch.

4.7.6 Technological Constraints and Manufacturability*

(*) The content of this section is quite specialized and for advanced reading. Knowledge of this section is not necessary to understand the following chapters. Beginners are recommended to skip this section.

If a freeform surface is calculated in the optimization of lens design, usually additional constraints must be taken into account to ensure that the surface can be realized. The exact values and numbers of these constraints depend on the technological approach of manufacturing. If diamond turning is used, the boundary conditions are defined by the following.

- A maximal difference in sag (dynamical range of the machine Δz) as a deviation from the rotational symmetric shape, usually in the range of 5–20 mm.
- If the rotation of the diamond turning tool is considered, a sag difference gives rise to an azimuthal acceleration. These values should be limited to allow for comfortable speeds of processing.
  Figure 4.74 shows as an example the azimuthal acceleration map of a sample, where two surfaces of a TMA mirror are fabricated on a common substrate. This means, the transitions must be smooth and guarantee a small acceleration.

In total it must be clear that for fabricating a freeform, all practical steps of realization must be strongly connected to the design. This is valid for the following.

- The metrology of evaluating the accuracy of the profile after manufacturing. These data must be re-imported into the design software after a corresponding modelling of the point cloud data.
- The mechanical design to consider the positioning of the surface inside the system. Corresponding fiducials are needed to measure and reference the position of the substrate. In particular it is also necessary to generate an

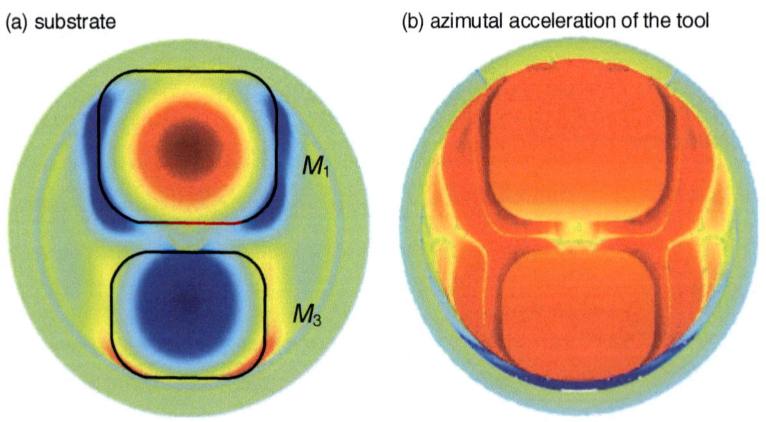

Figure 4.74 Manufactured double freeform surfaces on one substrate with (a) sag and (b) azimuthal acceleration distribution [32] / Optica Publishing Group.

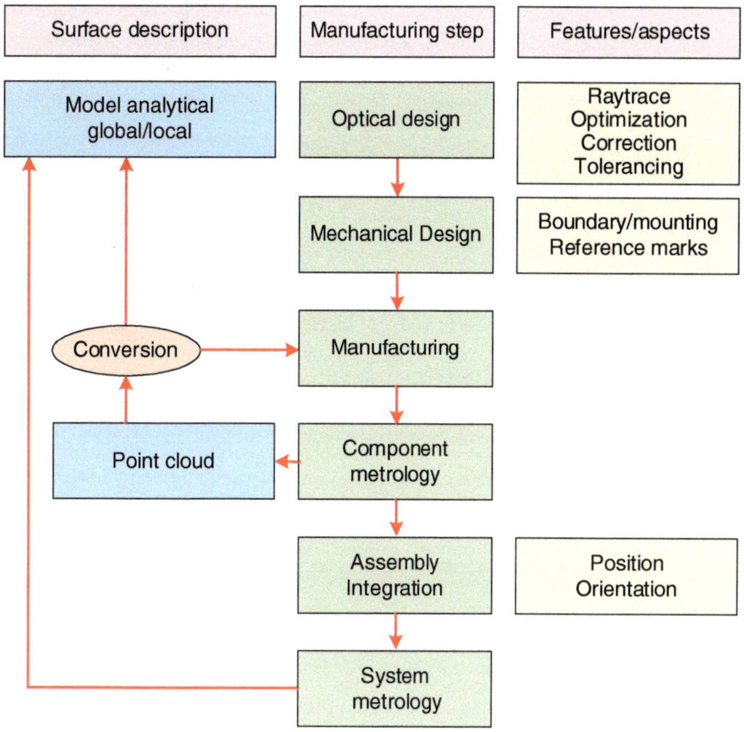

Figure 4.75 Workflow and connection of manufacturing steps for freeform systems.

overflow of the surface outside the required optical area, which guarantees the accuracy.

- Adjustment and assembly of the components in the complete system during the realization of the prototype.

Figure 4.75 shows a corresponding workflow, which demonstrates the important control of the interfaces between the various steps and data exchanges.

The special technologies of single point removal of material in grinding, like diamond turning, creates in reality very characteristic form deviations. To classify and quantify these errors the Fourier spectra are used to separate different ranges of spatial frequencies; in a double logarithmic scaling the absolute values squared of the Fourier coefficients of the surface are used, this is the so-called power spectral density representation. As is illustrated in Figure 4.76, the three ranges of low, middle and high frequency perturbations can be distinguished and show significant signatures of shape deviations. Their impact onto the points spread function and the image quality are quite different. This is discussed in more detail in a later chapter.

4.8 Special Component Types

4.8.1 Cylindrical Lenses

4.8.1.1 Introduction

If a lens has a cylindrical shaped surface, the focusing effect only occurs in one of the transverse cross sections. As a result, the ray paths in the x–z and the y–z cross sections are different. As is sketched in Figures 4.77 and 4.78, lateral focal lines can be observed in the cross section by focusing with a single cylindrical lens. This is one of the practical applications of cylindrical lenses in beam shaping and illumination. In contrast to the simple illustration, real cylindrical lenses are, due to manufacturing reasons, mostly plano-convex [33, 34].

In imaging applications, cylindrical lenses can be used to generate different focal lengths and magnifications in x and y. For illustration, Figure 4.79 shows a simple setup with two crossed cylindrical lenses, which forms a so-called anamorphic optical system. The difference in magnification in x and y allows for a re-shaping of a rectangular object field to modify the aspect ratio. This, for example, can be used to adapt an object field of square geometry to a rectangular sensor area.

4.8.1.2 Aspherical Cylindrical Lenses*

(*) The content of this section is quite specialized and for advanced reading. Knowledge of this section is not necessary to understand the following chapters. Beginners are recommended to skip this section.

As is well known, single lenses always suffer on axis from spherical aberration. For applications in laser beam guiding, often the cylindrical lenses are used to focus a collimated beam into a line focus or to collimate an astigmatic beam to a circular cross section. Correspondingly, the line width as a criterion of quality can be far from being

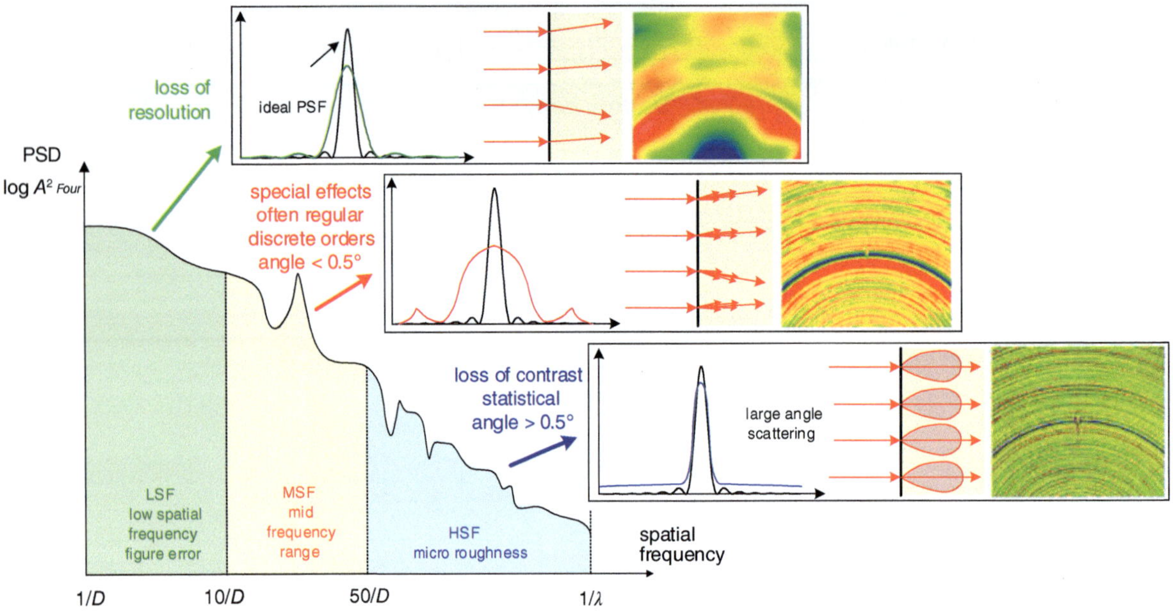

Figure 4.76 Power spectral density and typical consequences onto the point spread function and the angle distributions.

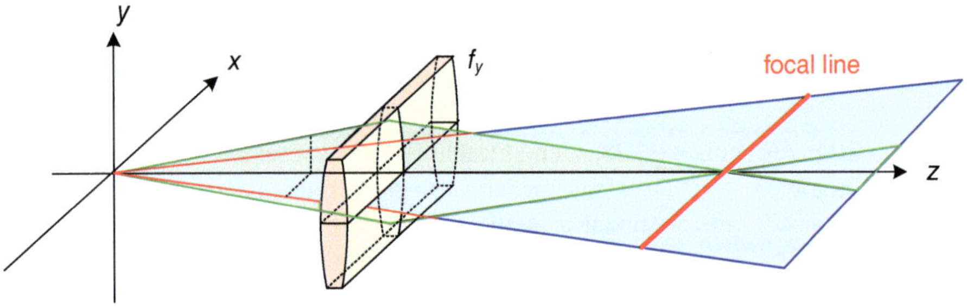

Figure 4.77 Focusing properties of a cylindrical lens as a relay in the y–z-section only.

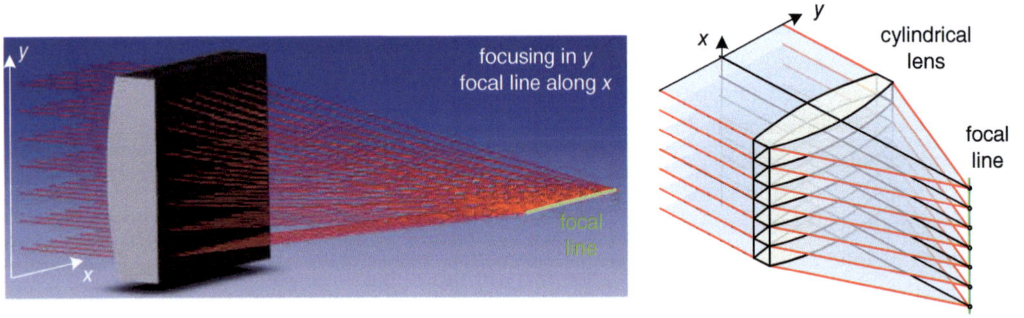

Figure 4.78 Focusing by a cylindrical lens for incoming collimated light. The ray bending is effective in the y–z-cross section only; in total a transverse focal line is generated, which is oriented along x.

diffraction limited if simple spherical cylindrical lenses are used. Since cylindrical lenses are, due to technical reasons, mostly produced with one plane surface, in addition the degree of freedom of bending cannot be used to improve the line quality. Therefore, in particular for larger numerical aperture angles, cylindrical lenses are available with aspherical surfaces on one side, see Figure 4.80.

Unfortunately, in some commercial optical design programs the performance representation of cylindrical systems is not well supported. This means that in a spot diagram it is not possible to select an anamorphic scaling with different resolution in x and y. In the case of wave aberrations, only circular and plane reference waves are possible, but not cylindrical spherical waves. Therefore,

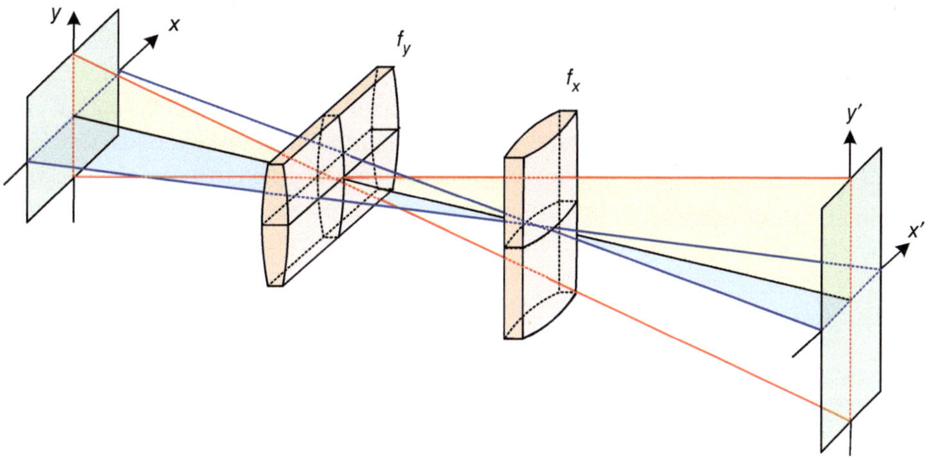

Figure 4.79 Anamorphic imaging by two crossed cylindrical lenses schematically shown by a change of the rectangular cross sections in the object and image plane.

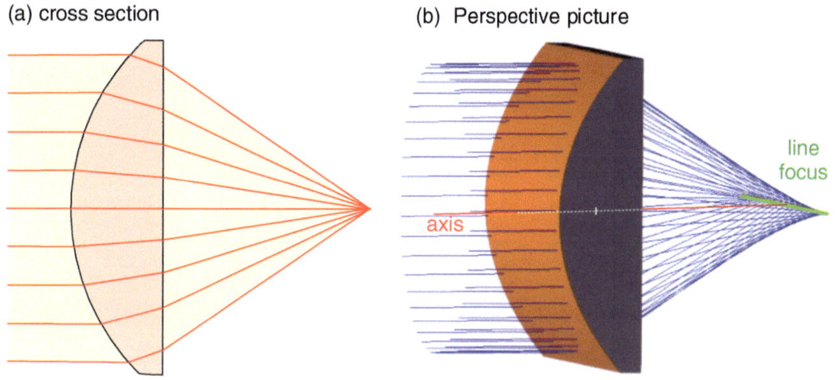

Figure 4.80 Aspherical cylindrical lens with numerical aperture of *NA* = 0.55 as (a) cross section and (b) three-dimensional view.

in practice it is sometimes cumbersome to analyse and correct cylindrical systems.

4.8.1.3 Toroidal Lenses*

(*) The content of this section is quite specialized and for advanced reading. Knowledge of this section is not necessary to understand the following chapters. Beginners are recommended to skip this section.

The easiest paraxial toroidal surface is given by a simple quadratic parabolic shape in each cross section separately. The equation reads

$$z = \frac{1}{2}\left(\frac{x^2}{R_x} + \frac{y^2}{R_y}\right) \tag{4.144}$$

A so-called spherocylindrical surface can be defined by bi-spherical cross sections with the different parameters as

$$z = \frac{c_x x^2 + c_y y^2}{1 + \sqrt{1 - (c_x x^2 + c_y y^2)}} = \frac{(c_x \cos^2\theta + c_y \sin^2\theta)\cdot r^2}{1 + \sqrt{1 - (c_x \cos^2\theta + c_y \sin^2\theta)^2 \cdot r^2}} \tag{4.145}$$

in polar coordinates r,θ or Cartesian ones x, y respectively. It is toroidal in shape and it can be seen that in this case the curvature is a function of the azimuthal angle.

$$c(\theta) = c_x \cos^2\theta + c_y \sin^2\theta \tag{4.146}$$

A more general toric conic section (bi-conic) has a double plane symmetry in x–z and y–z cross section. With the two parameters $c_{x,y}$ and $\kappa_{x,y}$ in every cross section, the surface equation is given by

$$z = \frac{c_x x^2 + c_y y^2}{1 + \sqrt{1 - (1+\kappa_x) c_x^2 x^2 - (1+\kappa_y) c_y^2 y^2}} \tag{4.147}$$

If this expression is expanded until the fourth order, we get

$$z = \frac{1}{2}\left(\frac{x^2}{R_x} + \frac{y^2}{R_y}\right) + \frac{1}{32}\left[(1+\kappa_x)\frac{x^4}{R_x^3} + \left(\frac{1+\kappa_x}{R_x} + \frac{1+\kappa_y}{R_y}\right)\frac{x^2 y^2}{R_x R_y} + (1+\kappa_y)\frac{y^4}{R_y^3}\right] + \ldots \tag{4.148}$$

According to the second mixed term inside the parentheses there is in fourth and higher order no decoupling possible in x and y. An alternative equation for a toric surface is

$$z = R_y - \sqrt{\left(R_y - R_x + \sqrt{R_x^2 - x^2}\right)^2 - y^2} \qquad (4.149)$$

According to the definition, first a circular arc with radius R_x is rotated around the y axis, then a rotation with centre point distance R_y is performed around the x-axis. An expansion of this equation to the fourth order gives the sag

$$z = \frac{1}{2}\left(\frac{x^2}{R_x} + \frac{y^2}{R_y}\right) + \frac{1}{8}\left(\frac{x^4}{R_x^3} + \frac{2x^2y^2}{R_x^2 R_y} + \frac{y^4}{R_y^3}\right) + \ldots \qquad (4.150)$$

It can be seen that this surface is also not fully symmetric between x and y in the mixed term in this representation.

As a conclusion, there are several opportunities to define toric surfaces. All these options are similar in the paraxial quadratic sag part, but have differences in the fourth order terms. Therefore, a more general equation can be written as

$$z = \frac{1}{2}\left(\frac{x^2}{R_x} + \frac{y^2}{R_y}\right) + \frac{1}{8}\left(\frac{x^4}{R_3^3} + \frac{2x^2y^2}{R_4^3} + \frac{y^4}{R_5^3}\right) \qquad (4.151)$$

with arbitrary values for the parameters R_3, R_4 and R_5.

If the x-axis is considered as the axis of revolution of a toroidal surface, there are two types of torus shapes as can be seen in Figure 4.81. If the other radius of curvature around the y-axis is $R_y > R_x$, the torus surface corresponds to a barrel inside, in the case $R_y < R_x$ a doughnut lies inside the surface.

4.8.1.4 Toroidal Ring Lens*

(*) The content of this section is quite specialized and for advanced reading. Knowledge of this section is not necessary to understand the following chapters. Beginners are recommended to skip this section.

A toroidal ring lens is a special component that is able to generate a ring focus out of a homogeneously filled circular aperture [35]. The lens is a body of revolution which is obtained if a circle sector is rotated around the optical axis. Figure 4.82(a) shows the building of the toroidal ring lens. Figure 4.82(b) illustrates the generation of a circular focal line as illustrated in Figure 4.83 in one cross section. If the ring lens diameter is exactly equal to the double ring width, there is no hole in the centre and therefore no loss in this focusing setup. With the help of an additional Fourier lens, a line focus along the z-axis can be created, see Figure 4.84. The ring lens is then identical working as an axicon, see Section 4.8.3, if $D_{in} = 2D$ and there is no inner hole.

4.8.1.5 Combinations of Cylindrical Lenses*

(*) The content of this section is quite specialized and for advanced reading. Knowledge of this section is not necessary to understand the following chapters. Beginners are recommended to skip this section.

Cylindrical lenses break the symmetry of an optical system. In a paraxial consideration of cylindrical lenses it therefore makes sense to take the azimuthal rotation of the lenses into account. A paraxial description requires a 4 × 4 *ABCD* matrix formalism, see Section 3.6.5.2.

If two cylindrical lenses with the same focal lengths f are rotated in opposite direction around the z-axis, a toroidal system occurs. This is sketched in Figure 4.85. The resultant matrix reads

$$\underline{M} = \begin{pmatrix} 1 & 0 & 0 & 0 \\ 0 & 1 & 0 & 0 \\ -\frac{1}{f_x} & 0 & 1 & 0 \\ 0 & -\frac{1}{f_y} & 0 & 1 \end{pmatrix} = \begin{pmatrix} 1 & 0 & 0 & 0 \\ 0 & 1 & 0 & 0 \\ -\frac{2\cos^2\varphi}{f} & 0 & 1 & 0 \\ 0 & -\frac{2\sin^2\varphi}{f} & 0 & 1 \end{pmatrix} \qquad (4.152)$$

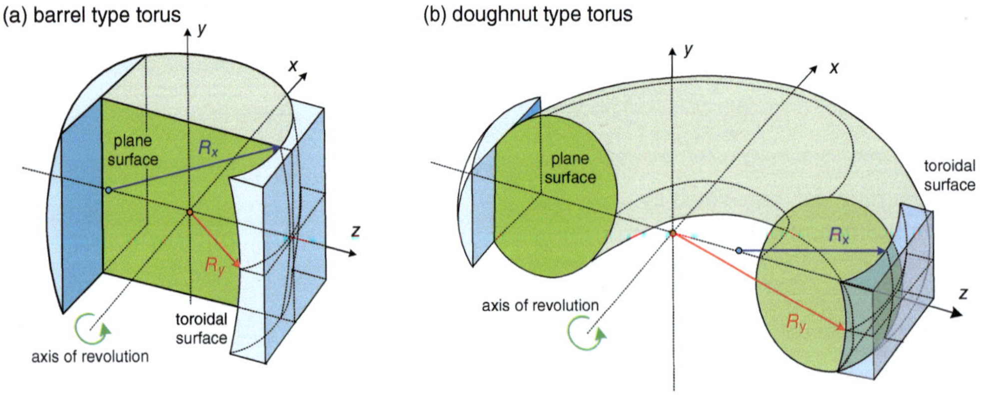

Figure 4.81 Toroidal lens of the (a) barrel type or (b) doughnut type.

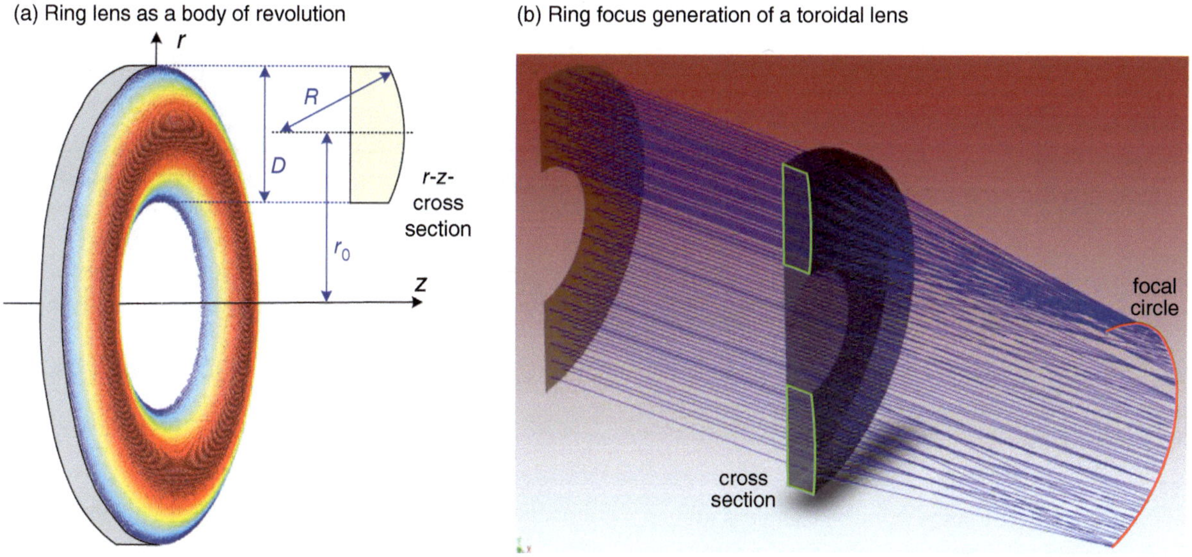

Figure 4.82 Toroidal ring lens lens (a) as a body of revolution with (b) the focusing behaviour to generate a focal circle ring.

Figure 4.83 Cross section of a toroidal lens, which creates a ring focus.

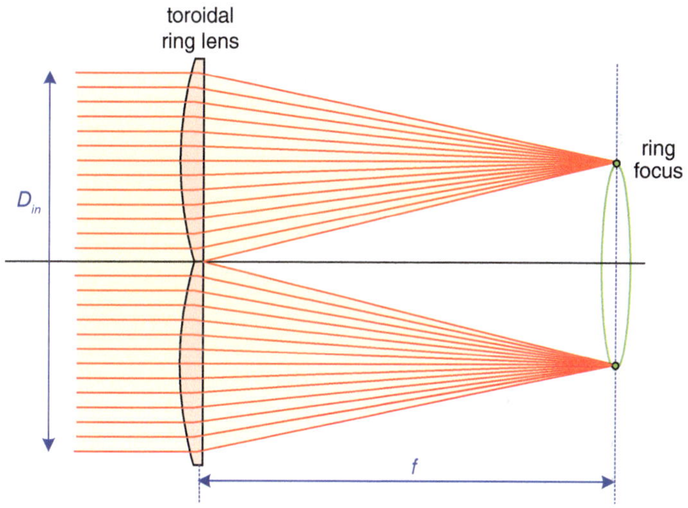

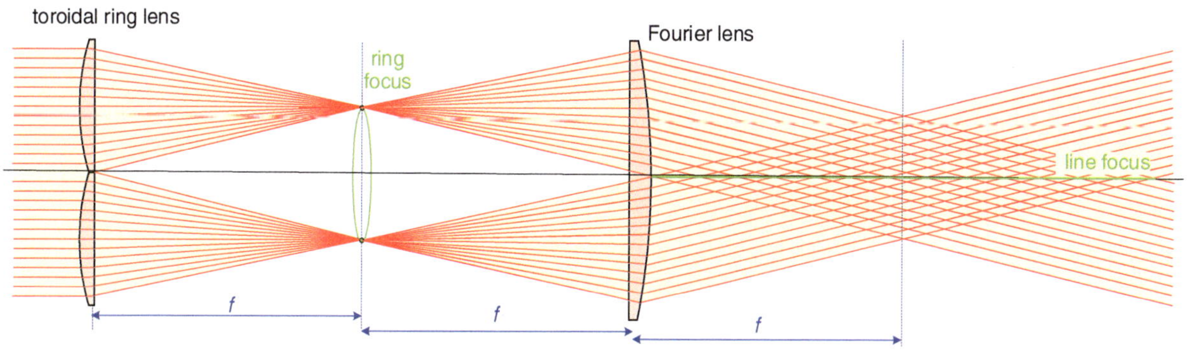

Figure 4.84 Toroidal ring lens with Fourier lens creates an axial line focus.

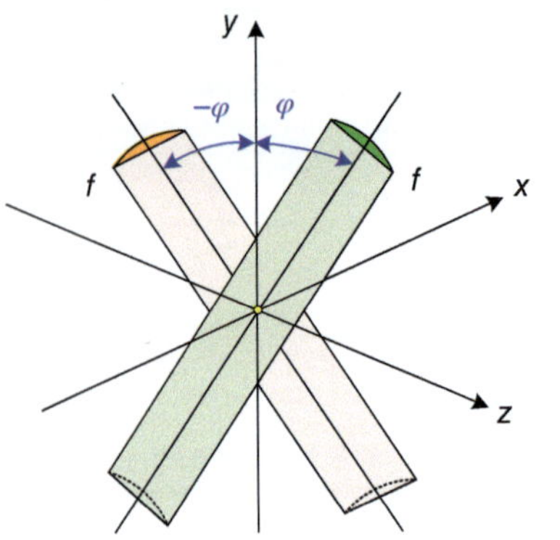

Figure 4.85 Two opposite rotated cylindrical lenses to generate a variable toroidal system.

Table 4.6 Focal length components of a cylindrical lens pair rotating in opposite directions.

Angle φ [°]	Focal length f_x	Focal length f_y	Comment
0	$f/2$	inf	Cylindrical lens in x
22.5	$0.586f$	$3.41f$	Toroidal lens, stronger focusing in x
45	f	f	Circular symmetric lens
67.5	3.41 f	$0.586f$	Toroidal lens, stronger focusing in y
90	inf	$f/2$	Cylindrical lens in y

Depending on the value of the azimuthal angle φ we get different focal power components in the x and y-direction as illustrated in Table 4.6. This means that we have a component with variable astigmatism.

As a second example the very ingenious idea of S. Bernet [36] is presented. The basic approach is to generate a flexible focal length by using only rotating lenses and not moving components as in a traditional zoom system. This kind of module can serve as a component for a zoom system without shifting lenses, which is mechanically more comfortable. The principle of the system is sketched in Figure 4.86. In the first group a doublet of toroidal lenses with opposite focal lengths $\pm f_{C1}$ in the x and y-cross section is established; both lenses are rotated in opposite direction around the z-axis by the angle $\pm\varphi_1$. This is visualized in Figure 4.87(a) for clarity. The next component is a circular symmetric lens with focal length f_L. After a distance $z = f_L$ a similar system is established in reverse order but with focal lengths f_{C2} and azimuthal rotation angles $\pm\varphi_2$.

If this system is calculated in the paraxial approximation, a circular symmetric optical power is obtained and the focal power f is given by $-1/C_{xx}$. The 4 × 4 matrix has no off-diagonal elements, but there are not the same values for A and D which means that the principal planes in x and y-cross section are different. The approach to find this solution is to allow all parameters to be different for this kind of setup and then to look for the conditions of

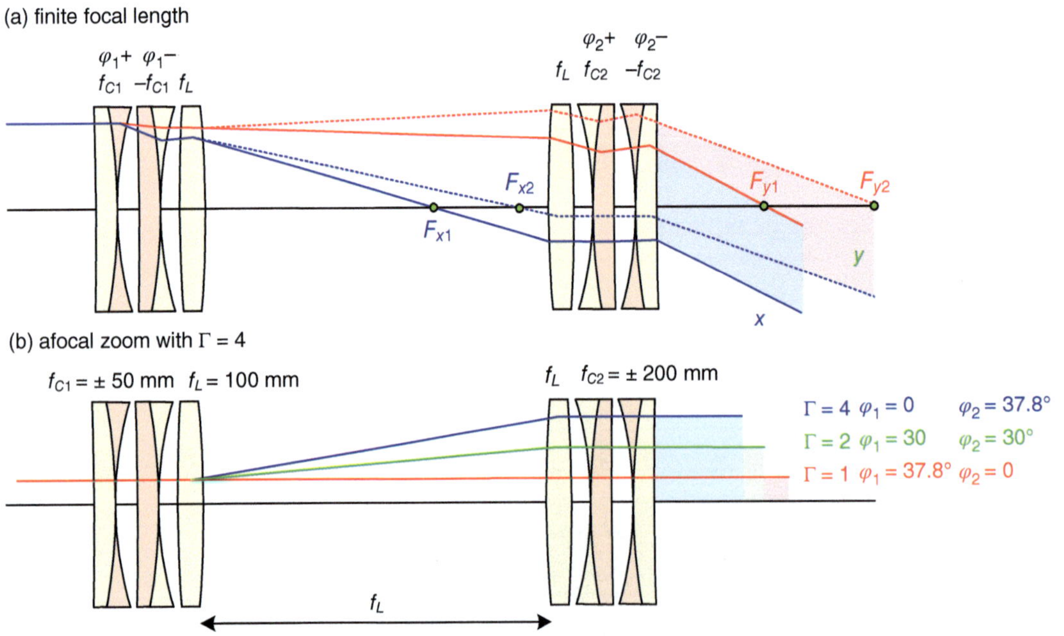

Figure 4.86 System of six thin lenses to create a variable power with (a) finite power relay and (b) afocal setup with factor $\Gamma = 4$.

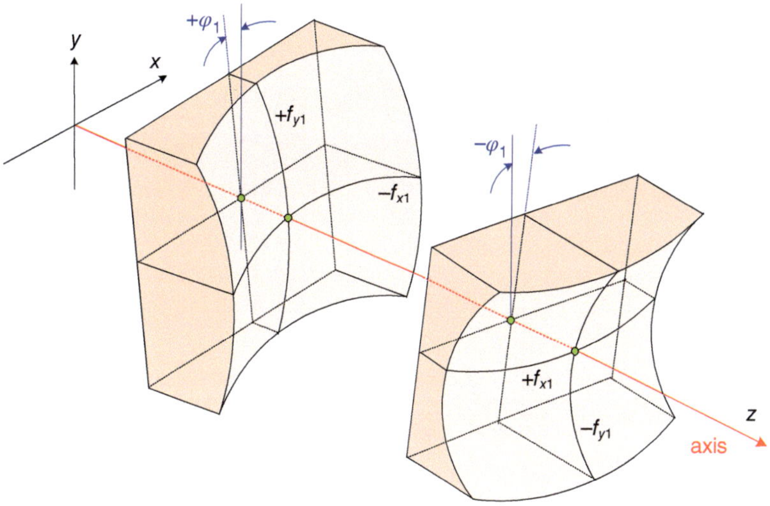

Figure 4.87 System of Figure 4.86 in a perspective view to illustrate the reverse oriented rotation of the toroidal lens pairs.

$C_{xx} = C_{yy}$ and $C_{xy} = 0$.

$$A_{xx} = -A_{yy} = \frac{2f_L}{f_{C1}}\cos(2\varphi_1)$$

$$B_{xx} = B_{yy} = f_L$$

$$C_{xx} = C_{yy} = \frac{1}{f_L} - \frac{4f_L}{f_{C1}f_{C2}}\cos(2\varphi_1)\cos(2\varphi_2)$$

$$D_{xx} = -D_{yy} = \frac{2f_L}{f_{C2}}\cos(2\varphi_2) \qquad (4.153)$$

It is seen that in case of A and D that a switch in the sign occurs, which means that the outgoing beam has the same angle, but different focal points.

If a more meaningful system is desired, the parameters should fulfil certain additional conditions. If, for example, an afocal zoom system is requested with a factor from $\Gamma = 1 \ldots \Gamma_{max}$, the additional constraints

$$f_{c1} = f_L/\sqrt{\Gamma_{max}},\ f_{c2} = f_L \cdot \sqrt{\Gamma_{max}}$$

$$\varphi_1 = 0 \ldots \frac{1}{2}\arccos\frac{1}{2\sqrt{\Gamma_{max}}},$$

$$\varphi_2 = \frac{1}{2}\arccos\frac{1}{4\cos(2\varphi_1)} \qquad (4.154)$$

with the condition $C_{xx} = 0$ gives the desired afocal zoom setup. As an example, Figure 4.86(b) shows a case with $\Gamma_{max} = 4$.

4.8.2 Fresnel Lenses

4.8.2.1 Introduction

Fresnel lenses are lenses with a radial surface structure. If a thick plane-spherical single lens is considered with short focal length, the curved surface $z(y)$ (mostly aspherical) is wrapped in z-direction, keeping the local slopes constant. This is sketched in Figure 4.88(a),(b). Thus, the direction of ray deviation remains the same [37–48]. This reduction of the sagitta is in the case of a circular symmetric lens performed in ring-shaped areas, so that in the mean the component has a homogeneous fairly small thickness. The Fresnel surface itself has a sawtooth-shaped contour in the cross section. The cross section of a Fresnel lens and the principle of thickness reduction is shown in Figure 4.88b.

In practice, the refracting active surface areas are mostly approximated by linear pieces so that the surface consists of a number of conical ring zones. The principle of this approximation is shown in Figure 4.86(c). But it should be clear that the z-shift of the ray deflection points due to the wrapping, as well as the approximation of the zones by linear profiles, generates certain errors in imaging.

Advantages of Fresnel lenses compared with conventional lenses are:

- weight reduction;
- reduction in the construction length compared with equivalently thick aspherical devices for large apertures (for example see Figure 4.88(b));
- reduction of the light paths in the material and improved transmission for absorptive materials;
- multi-functional lenses (e.g. bi-focal, beam-splitter etc.) are possible.

On the other hand, there are the following disadvantages:

- significant aberrations for a finite field of view;
- stray light due to unwanted ray paths with the non-active surface parts at the steps;
- aberrations due to the changed ray bending points by the wrapping process;
- residual aberrations due to the linear approximations of the active parts of the surface mostly found in practice;

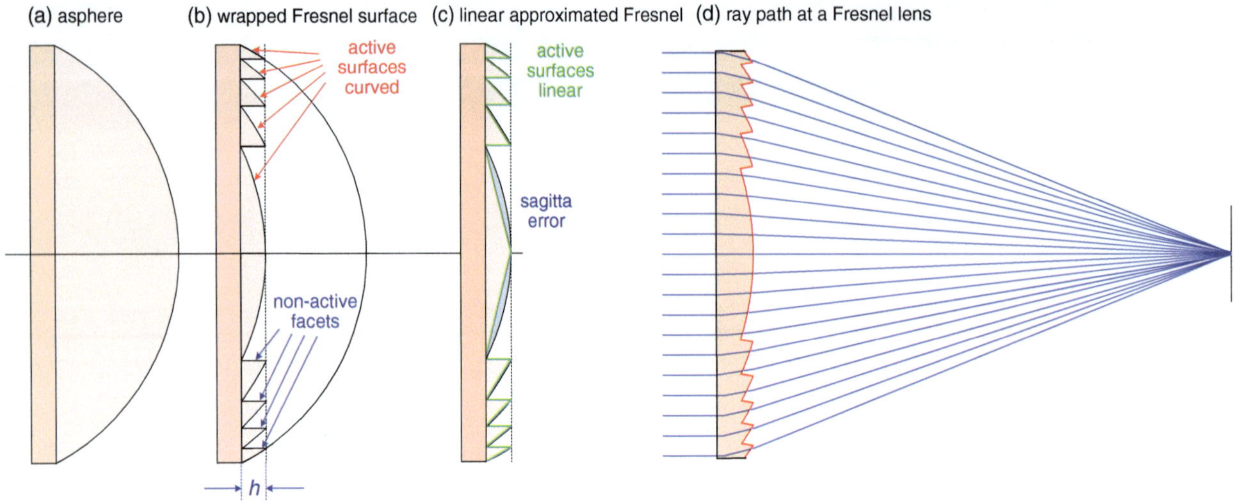

Figure 4.88 Principle of the Fresnel lens with (a) equivalent thick asphere, (b) with wrapped surface, (c) approximated linear ring contours and (d) raytrace.

- interferences in the focus caused by refraction, since Fermat's principle is not satisfied.

Some of the drawbacks of simple Fresnel lenses are addressed in more refined design proposals in the literature, some of them are as follows.

- Aspherical substrates and using reflective Fresnel surfaces for improved resolution and colour performance [41].
- More complicated groove shapes to improve the resolution [42].
- Using double-sided components to get more degrees of freedom for correction [43].
- Using freeform substrate shapes for special illumination applications [47].
- Constructing more complicated side-wall geometries to improve straylight and truncation issues [45].
- Using cylindrical lenses for special applications [47].

4.8.2.2 Basic Equations

With the aperture angles u and u' in front and behind the Fresnel surface, respectively, and n, n' the corresponding refractive indices, the angle of inclination φ of the active ring surface has to fulfil the condition (see Figure 4.89)

$$\tan\varphi = \frac{n' \cdot \sin u' - n \cdot \sin u}{n' \cdot \cos u' - n \cdot \cos u} \tag{4.155}$$

In most cases, with the use of a Fresnel lens, correction of the spherical aberration is required. If the axial point is to be imaged through the single active rings without spherical aberration, then with the ray height y and the intersection lengths s and s', respectively, we find the condition

$$\tan\varphi = \frac{\frac{n'}{s'} \cdot \sqrt{1+\frac{y^2}{s^2}} - \frac{n}{s} \cdot \sqrt{1+\frac{y^2}{s'^2}}}{n' \cdot \sqrt{1+\frac{y^2}{s^2}} - n \cdot \sqrt{1+\frac{y^2}{s'^2}} u} \tag{4.156}$$

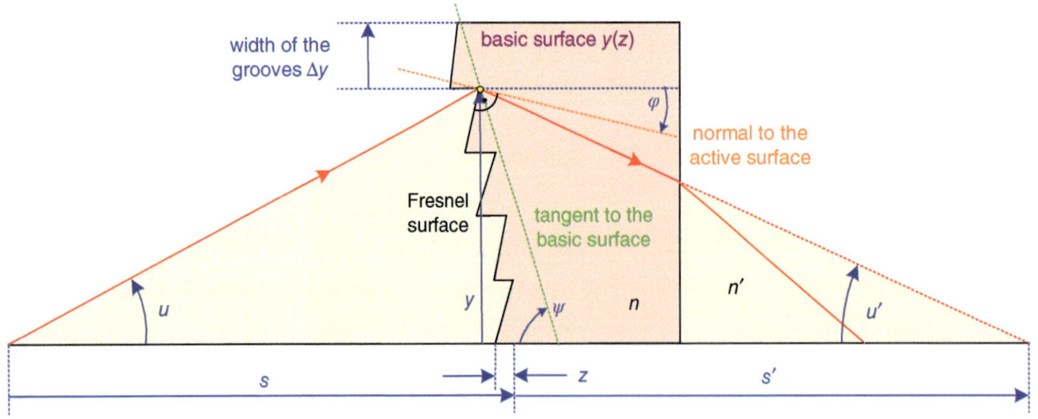

Figure 4.89 Calculation of ray deviation at a Fresnel lens.

In paraxial linear approximation for small aperture angles this equation becomes

$$\tan \varphi = \frac{n'}{n' - n} \cdot \frac{y}{f'} \tag{4.157}$$

where f' is the focal length of the Fresnel lens.

4.8.2.3 Properties of Fresnel Lenses

In practice, Fresnel lenses are usually applied instead of aspherical devices for large apertures. Therefore, there are mainly large deviating angles in the outer zones. The limitation of the maximum possible deviation is then determined by the total internal reflection at the active rings of the surface. With $n > n'$ for a Fresnel lens oriented to the object side, the maximum at the edge is

$$\cos (u' - u)_{max} = \frac{n'}{n} \tag{4.158}$$

In the special case of a collimated incident wave coming from infinity ($u = 0$) this becomes

$$\tan u'_{max} = \sqrt{\left(\frac{n}{n'}\right)^2 - 1} \tag{4.159}$$

If the angle's steepness is too large, total internal reflection can take place, depending on the value of the refractive index. Figure 4.90 shows the limit of total internal reflection in dependence of the height of the ray y and the back focal length s for $n = 1.5$.

From this, the maximum usable diameter of a Fresnel lens for a given s, s' results in

$$2y_{max} = -\frac{s' - s}{\tan (u' - u)_{max}} + \sqrt{\left(\frac{s' - s}{\tan (u' - u)_{max}}\right)^2 - 4s' \cdot s} \tag{4.160}$$

Fresnel lenses do not fulfil Fermat's principle. Therefore, they form only a quasi-geometric image and for finite field angles some issues arise. The phase differences between the single-ring segments are arbitrary and depend on the selected wrapping thickness. Usually, they are not equivalent to multiples of 2π, as for a Fresnel zone plate. Therefore, Fresnel lenses do not image in an aplanatic way. They have a good spherical correction only on axis, but significant coma aberrations for off-axial object points.

Although the reduction of the thickness of a thick lens can be carried out without changing the inclination of the ray, there is, however, an axial shift Δs connected to this change of form. It disturbs the common intersection of the rays in the image point. This effect occurs especially for larger aperture angles. Figure 4.91 illustrates this effect.

If the condition for the correction of the spherical aberration is met for each individual ray at the corresponding mid-point of the surface, then this offset can be corrected. The Fresnel surface then shows a modified angle of inclination compared with the original surface. This condition or correction, however, can only be complied for one ray at a time over the finitely wide groove with height Δy. For this the individual surface rings have to be realized in a linear approximation as conical surfaces, which is common practice. In this way the so-called macro-images are corrected for the corresponding mid-rays at the individual active surface rings. The micro-image over the small sub-aperture of a zone area, however, stays unchanged. Thus, there is still an aberrated image as shown in Figure 4.91(b). From a physical point of view, the crossing of the beam rings creates interference, if the incoming wave is considered to be coherent. The focused field behind the Fresnel lens suffers therefore from fringes and structured distributions, in particular in the range of the focus. This is demonstrated in a more rigorous calculation in Figure 4.91(c).

If the single rings are formed as a conical surface in a linear approximation, then there is no exact image produced. Within this micro-image the lens acts as an

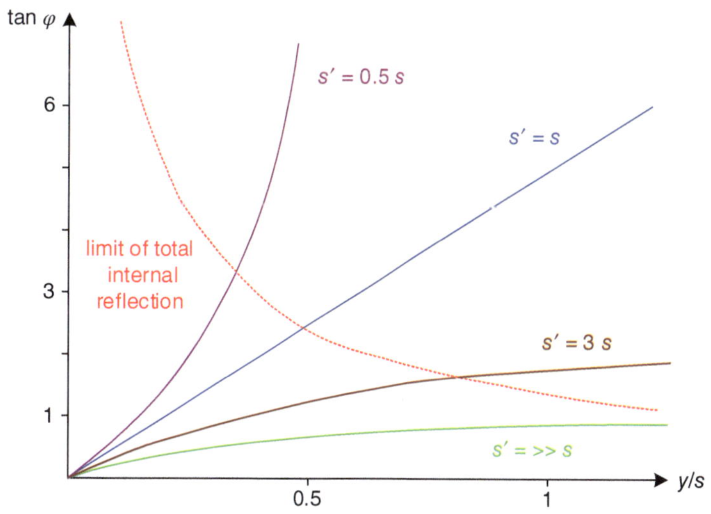

Figure 4.90 Total internal reflection at a Fresnel lens with $n = 1.5$. φ is the inclination angle of the surface part at the height y.

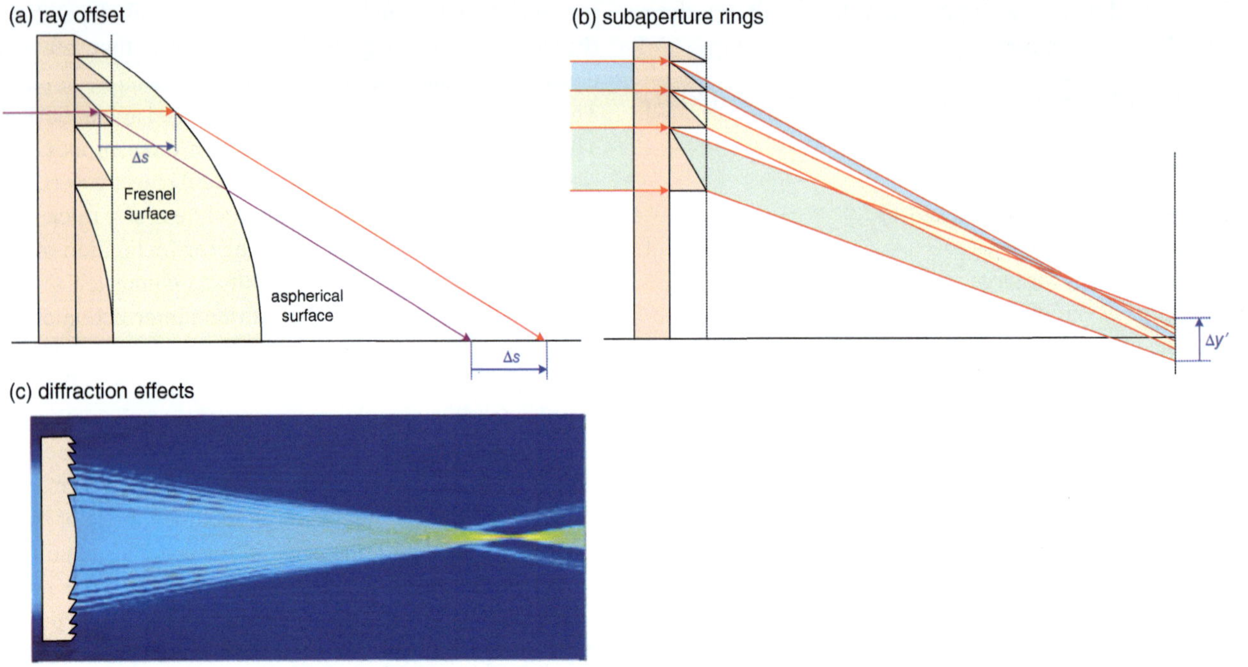

Figure 4.91 Aberration of a Fresnel lens due to an axial offset. (a) illustrates the z-shift of the ray bending point, (b) shows the impact of the linearization of the surface profile sections and (c) demonstrates the diffraction effects due to edges and dead zones.

axicon creating small image circles for each object point. The micro-image is characterized by astigmatism and a very large coma. As the Fermat principle is not met over all sub-apertures, a structured point image, very much widened by interferences, can be observed. These effects result in a correspondingly reduced contrast of the image.

4.8.3 Axicons

4.8.3.1 Geometrical Concept

Axicons are special components, basically used for illumination applications, which have one plane side and one cone surface. Figure 4.92 illustrates the light transformation by this type of component [49, 50]. The tip of the cone is a singular point on axis. The cone surface bends an incoming collimated ray bundle in a rotational symmetric way. This has the consequence that the axicon creates a long axial oriented focal line. The practical use of this line is the major application of axicons. In a simple one-dimensional picture, it can be assumed that the upper part of the axicon acts like a prism, bending the upper half of the ray bundle downwards without focusing effects, see Figure 4.92.

For the length of the line created by an axicon of an incoming beam with full illuminated half diameter $a = D/2$ and final aperture angle θ as seen in Figure 4.92, we get approximately the expression

$$L = \frac{a}{\sin\theta} \cdot \cos\theta \approx \frac{a}{\sin\theta} \tag{4.161}$$

with $\theta = (n\text{-}1)\alpha$ as paraxial relation between axicon angle α and aperture angle θ. If the illumination is performed with a central hole as a ring with width $a = a_{out}\text{-}a_{in}$ to avoid the problems with the singular tip of the axicon, the length of the line focus is the same, only the initial point of the line is moved forward. By combining two axicons or axicons with

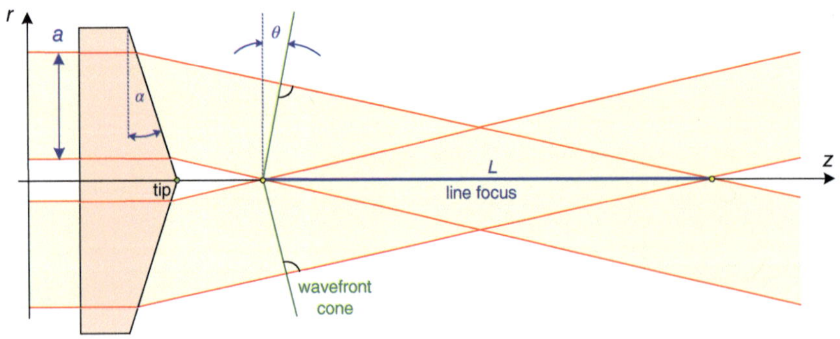

Figure 4.92 Ray path behind an axicon, a line focus is created along the optical axis.

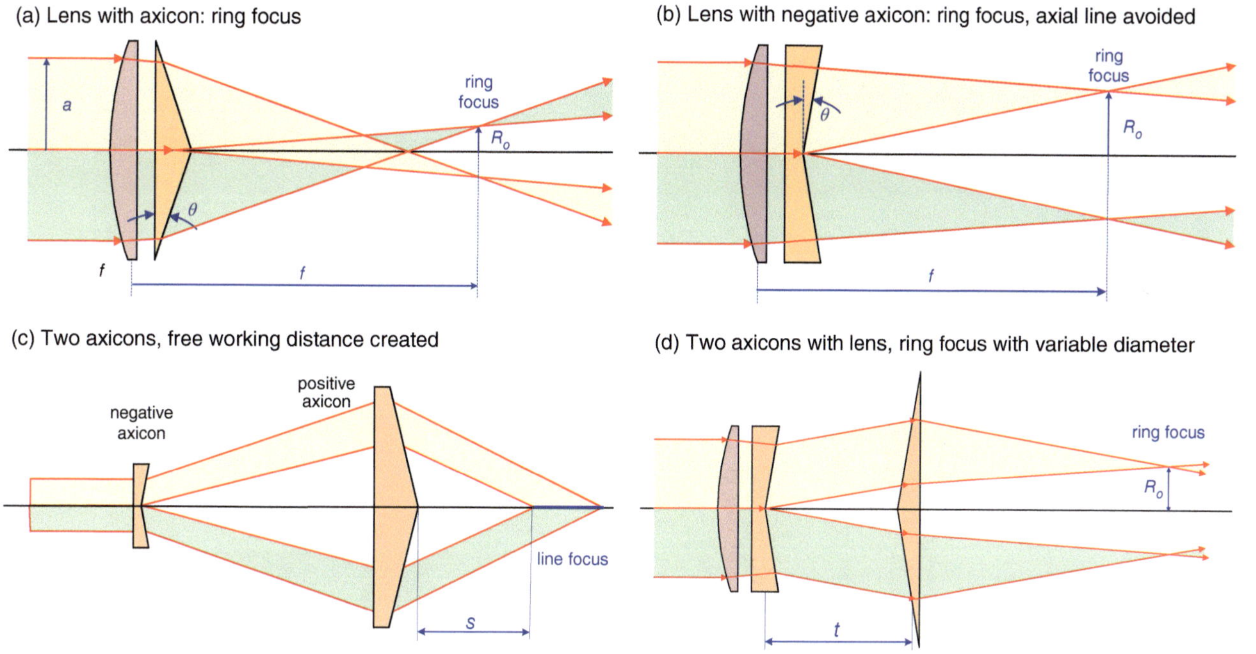

Figure 4.93 Special effects with axicon components (a) ring focus with axicon and lens, (b) ring focus with negative axicon without axial focal line but ring focus, (c) line focus with finite working distance s of the line focus, (d) two axicons with variable distance with lens creates ring focus with variable radius R_o.

lenses, special effects can be obtained. Some examples are shown in Figure 4.93; here the working distance to the line focus as well as variable ring foci radii R_o are addressed.

4.8.3.2 Physical Description*

(*) The content of this section is quite specialized and for advanced reading. Knowledge of this section is not necessary to understand the following chapters. Beginners are recommended to skip this section.

If a plane wave penetrates an axicon, behind the component the phase has a conic shape. This is exactly what is the characteristic property of a Bessel beam. A classical general Bessel beam can be written in the form

$$E_m(\mathrm{r}, \varphi, \mathrm{z}) = \mathrm{E}_0 \cdot J_m(k_r r) \cdot \mathrm{e}^{im\varphi} \cdot e^{-ik_z z} \tag{4.162}$$

In particular the basic mode is obtained for $m = 0$ as a circular symmetric beam as

$$E(r, z) = E_0 \times J_0(k_r r) \cdot e^{-ik_z z} \tag{4.163}$$

This field has a plane phase as a consequence of the overlaid upper and lower part in the interference range of the line focus. There is no spreading of the profile with z, it describes a 'non-diffracting' beam with a pronounced ring structure. This is exactly the radial profile after an axicon; the rings can be understood as the interference of the upper and lower part of the cone phase according to Figure 4.94(c). The result of a diffraction calculation is shown in Figure 4.94. It is seen that the raw geometrical picture above more or less delivers the envelope of the intensity distributions along z, shown in Figure 4.94(b).

There is a pronounced Fresnel ripple along the z-direction and the mean intensity is not really changing linearly with z. In practical applications it is mostly desirable to achieve a more uniform intensity along the line. This can be obtained by an appropriate radial aspherical phase profile along the axicon instead of the linear axicon shape. In the case of these so-called Gaxicons, a redistribution of energy takes place, see Figure 4.95.

4.8.4 Variable Components

4.8.4.1 Introduction

In lens design it is very attractive for many applications to be able to smoothly change the focal length, magnification or field of view of a system. For this purpose, one or two variable components are required [51–65]. There are many very special ideas, working principles, actuator methods and technical realizations proposed in literature. Figure 4.96 sketches a small and non-complete overview of some major approaches. Beneath the classical idea of axially moving lenses in a traditional zoom system, Alvarez lens pairs with coupled lateral movements realize a similar functionality. Deformable mirrors, pixelized arrays and liquid lenses are additional opportunities to smoothly change component parameters.

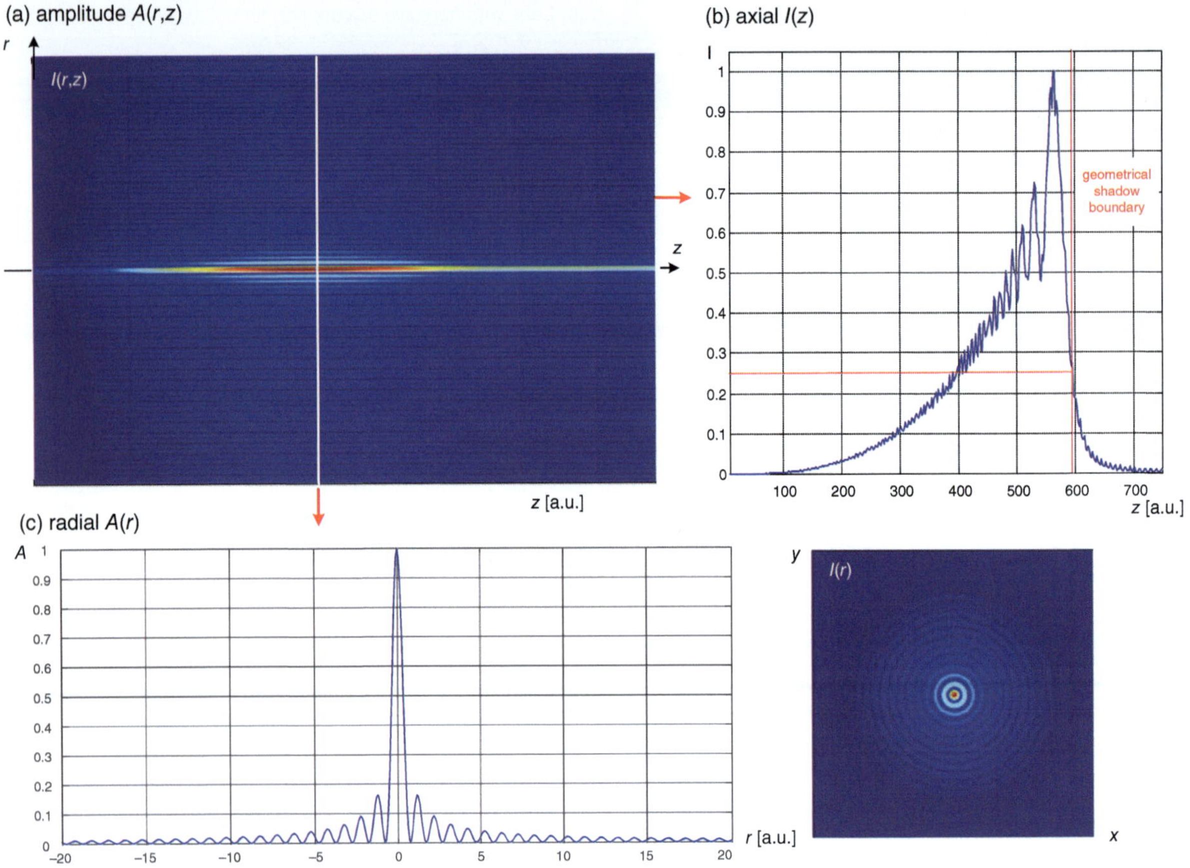

Figure 4.94 Intensity distribution after an axicon, (a) amplitude $A(r,z)$ (b) shows the intensity $I(z)$ on axis and (c) the radial profile as amplitude $A(r)$.

The criteria of the usefulness of a variable component are:

- speed of switching;
- dynamic range of focal power;
- maximum slope, sensitivity and accuracy of adjustment;
- spatial resolution, wavefront quality;
- size/diameter;
- spectral dependence, dispersion;
- polarization dependence;
- further performance criteria like temperature dependence, power loss.

Here in particular the components are briefly introduced that change their focal length without macroscopic movements by modification of the refractive index, the shape of the lens surfaces or small transverse shifts. The field of variable components is still in progress, possible parameters and technological realizations are improved continuously, therefore only qualitative descriptions are presented here. Some major aspects and principles are discussed, but in particular the large variety of technological issues is not presented in detail.

4.8.4.2 Electrowetting Liquid Lenses

If a solid material, a liquid phase and a vapour are building a droplet on the solid plate according to Figure 4.97, the contact angle θ of the liquid is according to Youngs' law given by

$$\cos\theta = \frac{\sigma_{sol-vap} - \sigma_{sol-liq}}{\sigma_{liq-vap}} \tag{4.164}$$

This is a simple balance of molecular forces and surface tension to minimize the surface energy in equilibrium with interface energy $\sigma_{j\text{-}k}$ between the different material pairs. The gravity effects are neglected here, in micro-dimensions they can usually be assumed to be very small. If the diameters of the liquid lenses are growing in diameter, the gravity effects become more and more relevant.

If now the liquid is located on an isolating thin layer above an electrode and in addition a voltage is applied, the energetic conditions change according to Lippmann's equation

$$\cos\theta = \cos\theta_o + \frac{\varepsilon}{2t\sigma_{liq-vap}} \cdot V^2 \tag{4.165}$$

with the thickness t of the isolating dielectric with liquid permittivity ε and interface energy σ and voltage V.

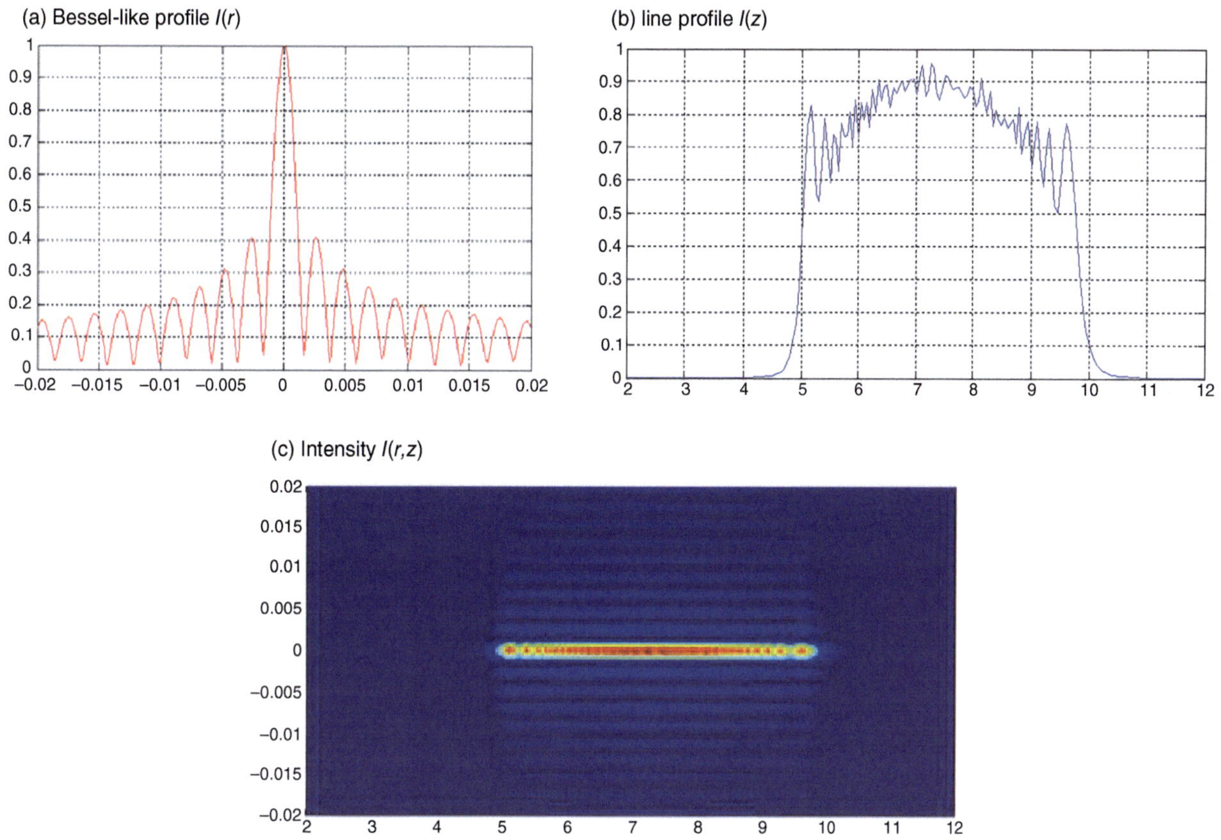

Figure 4.95 Intensity after a Gaxicon with improved uniformity along z. The units at the axes are arbitrary.

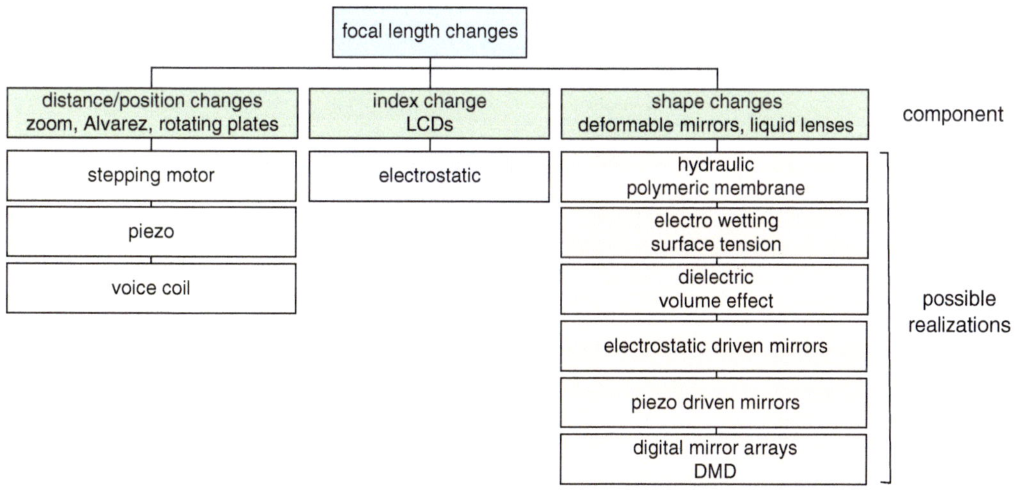

Figure 4.96 Working principles (green) and real actuator options for variable components.

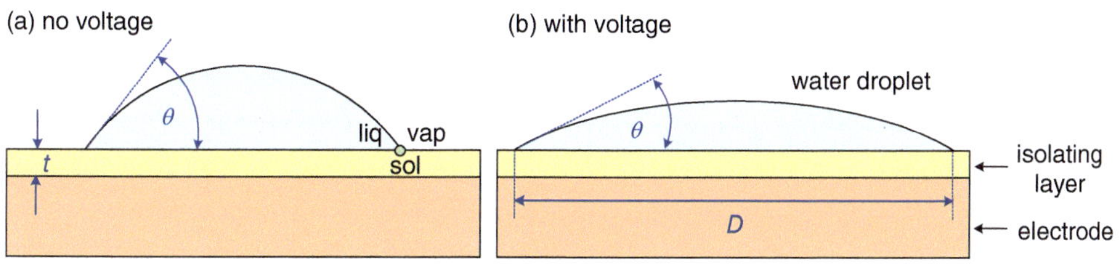

Figure 4.97 Voltage driven contact angle as electrowetting effect.

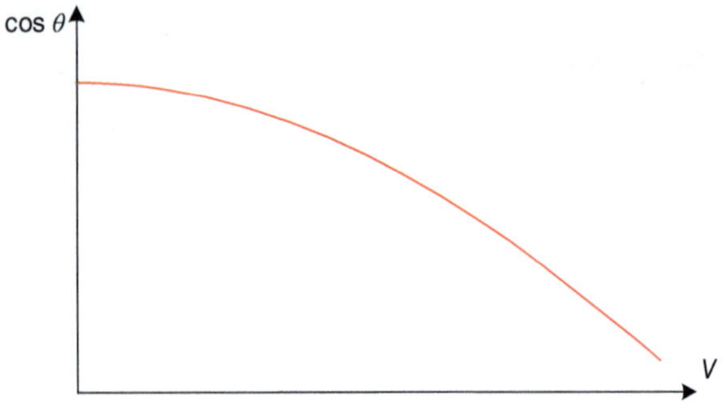

Figure 4.98 Typical contact angle change with voltage.

Figure 4.98 shows the quadratic change of the angle with V schematically.

If in first approximation the surface is considered to be a minimal spherical surface, the focal length of the arrangement is then obtained as

$$f = \frac{D}{2(n_{liq} - n_{vap)}) \cdot \sin\theta} \tag{4.166}$$

with the diameter D of the lens. In reality, an environmental vapour material is not very comfortable. Therefore, in the useful technical realizations, the systems look more like Figure 4.99. As the liquid medium, oil is selected, as a second medium water; the whole liquid is sealed between glass windows. Furthermore, the shapes of the electrodes are optimized to guarantee a good useful free diameter of the lens. Typical values of the main optical parameters are diameters $D = 2 \ldots 10$ mm and focal length ranges $f = 50 \ldots 200$ mm or $2 \ldots 8$ mm depending on the working point and geometry. But these values are continuously improved by more tricky techniques over time.

Figure 4.100 shows the issue with vanishing hysteresis effects in the lens as well as the problem of temperature dependence. In reality a small hysteresis cannot be avoided completely. The intermediate surface between the two liquids should not have an impact on gravity, so the lenses can be used in horizontal as well as vertical or skew orientation. This requires a coinciding mechanical density of the oil and water liquid, which is usually only obtainable at one temperature due to different dependencies.

The switching behaviour with a certain time delay due to a movement of the liquids is illustrated in Figure 4.101(b), the problem of wavefront performance is sketched in Figure 4.101(a). The optical quality depends on the diameter, refractive index difference as well as the temperature.

4.8.4.3 Electrophoretic Liquid Lenses*

(*) The content of this section is quite specialized and for advanced reading. Knowledge of this section is not necessary to understand the following chapters. Beginners are recommended to skip this section.

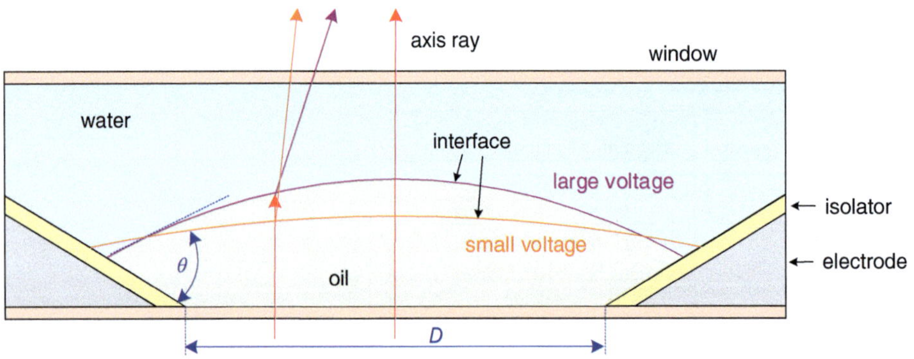

Figure 4.99 Electrowetting lens.

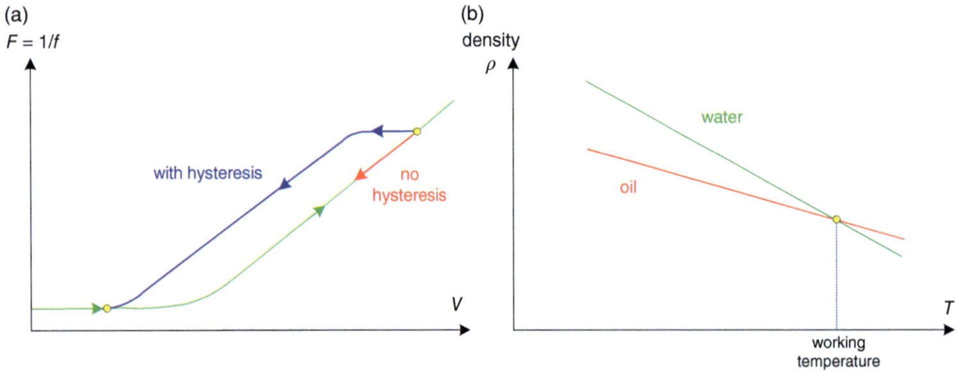

Figure 4.100 Liquid lens (a) power vs voltage with or without hysteresis respectively and (b) density change with temperature.

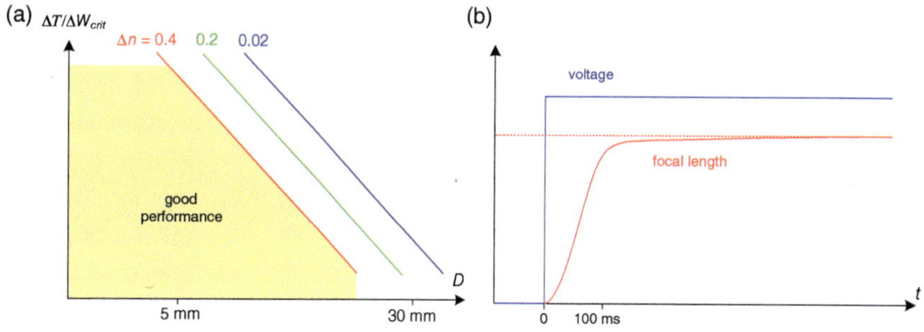

Figure 4.101 Liquid lens (a) performance dependence on diameter and index difference and (b) time behaviour of the switching process.

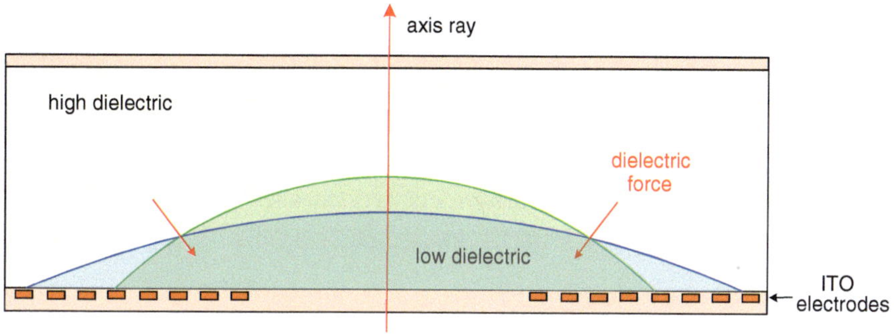

Figure 4.102 Electro-phoretic effect to change the shape of a lens surface between two liquids.

Another actuation principle is the so-called electrophoresis [51]. This is a volume effect; a voltage changes the dielectric forces on two dielectric liquids with different dielectric constants like oil and alcohol. The principle is sketched in Figure 4.102. A system of distributed electrodes acts on both liquids depending on the voltage or field strength in different spatial ranges. The complete system physically minimizes the energy balance and reshapes the interface between both liquids. The dependence of ε on the temperature is an issue also in this kind of device. Typical values are voltages of 200 V, focal length ranges of $f = 12 \ldots 34$ mm are typical.

4.8.4.4 Hydraulic Membrane Liquid Lenses

A completely different approach to vary the focal length of a liquid is a hydraulic change of material balance inside a chamber with a pressure driven change of the liquid volume, see Figure 4.103. The surface of the liquid usually is stabilized by a polymeric membrane and as a surrounding medium another liquid is used to get a sealed off chamber like the approach in the previous sections, as shown in Figure 4.104 [53].

For the liquid lenses of this technology many proposals and realizations are found in the literature considering the pressure actuation, geometry, selection of materials and

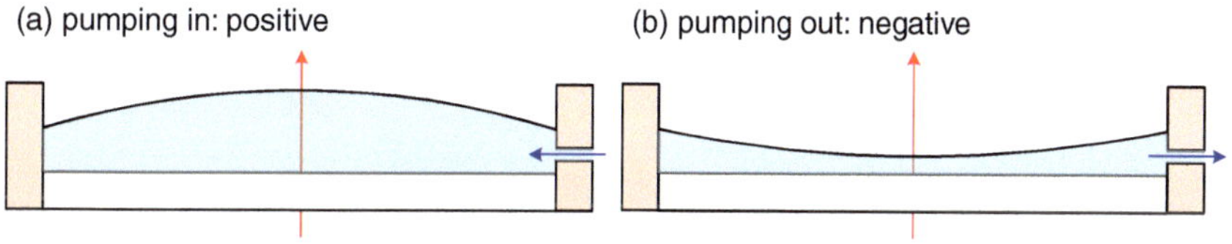

Figure 4.103 Pressure driven liquid lens with positive or negative power respectively.

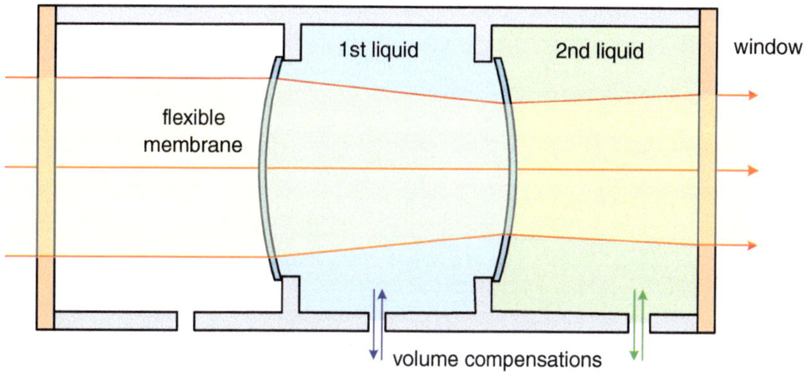

Figure 4.104 Liquid lens doublet driven by pressure with elastomeric interface membranes.

more [53]. This technology also allows for a combination of two chambers to construct a doublet lens, see Figure 4.104.

4.8.4.5 Liquid Crystal Lenses

A very old technology to generate a changing focal power is to use liquid crystals. Inside these liquid phases very long molecules are oriented in different possible ways if an applied voltage generates a selection of orientation due to the overall energy of the polar molecules. This principle is sketched in Figure 4.105. By the re-orientation, the anisotropic refractive index changes in the direction of the optical axis. By a clever arrangement of the geometry of the electrodes, special refractive index variations over the free diameter can be achieved.

Due to the basic principle, liquid crystals are unfortunately strongly polarization dependent. By adding a reflective mirror, reflective variable mirror devices can also be constructed.

4.8.4.6 Deformable Mirrors

Deformable thin membrane metal mirrors have been used in adaptive optics for a long time. The actuators can be electrostatic electrodes (see Figure 4.106) or piezo elements, moving the mirror mechanically. The number of actuators, the flexibility of the membrane and the geometrical sampling arrangement strongly influence the correctability and dynamic range of a mirror. A theory based on the bending of plates due to lateral forces can be used to calculate the so-called influence functions. These functions describe the shape of the mirror if only one actuator is activated. The functions are usually different due to the location of the acting point on the mirror and boundary effects of the plate. Figure 4.107 shows an example of an influence function [54–61].

One of the drawbacks of deformable mirrors is typically the dynamic range. Only rather small radii of curvature can be realized, mostly only in one direction (push or pull, both are not possible) and therefore one sign of the generated power. If the membrane is rather thin and a lot of actuators are applied, larger slope values can be realized and higher orders of aberrations corrected. But the stability of the plate and permanent power consumption are then a problem; therefore, the deformable mirrors are mainly used for smooth low order effects.

Figure 4.108 shows an example of a random perturbed wavefront, which is corrected by an adaptive mirror with hexagonal grid pattern and 127 actuators. The remaining print through of the actuator pattern and the boundary effects are clearly observed.

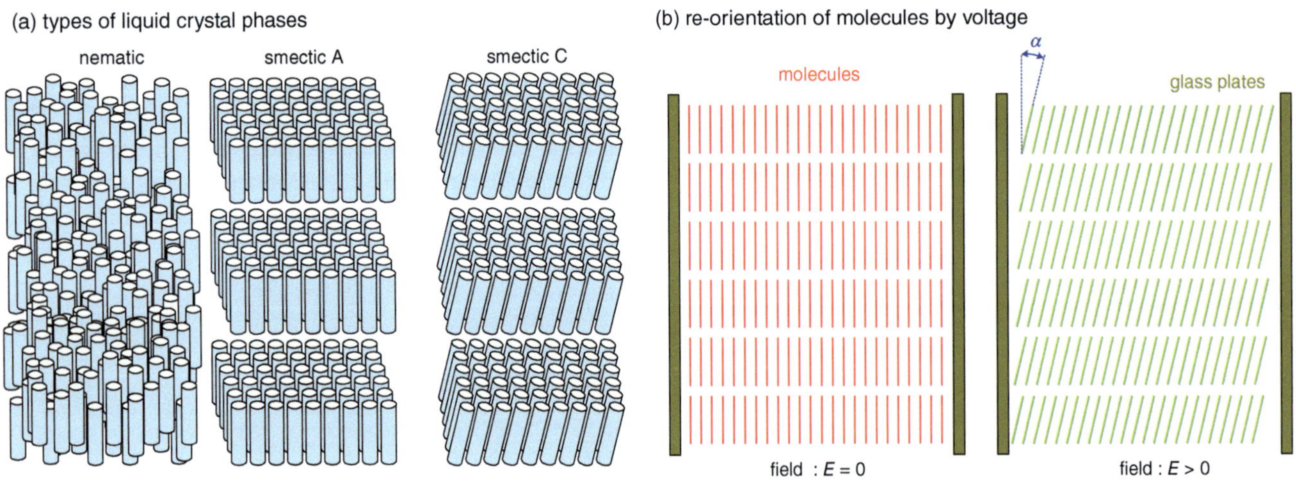

Figure 4.105 Molecular structure in a liquid crystal with (a) different phases of the molecule orientation and (b) switching behaviour.

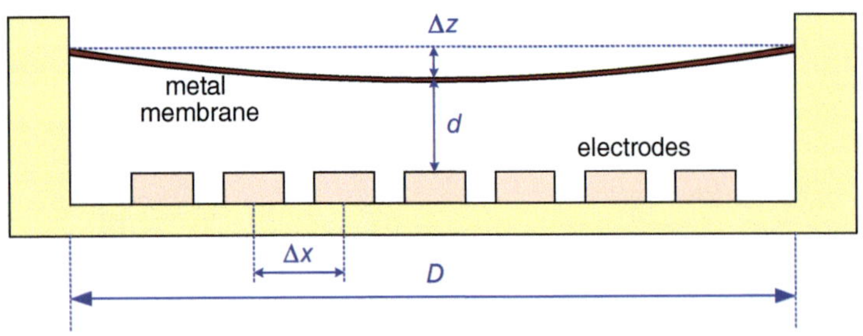

Figure 4.106 Electrostatic membrane mirror.

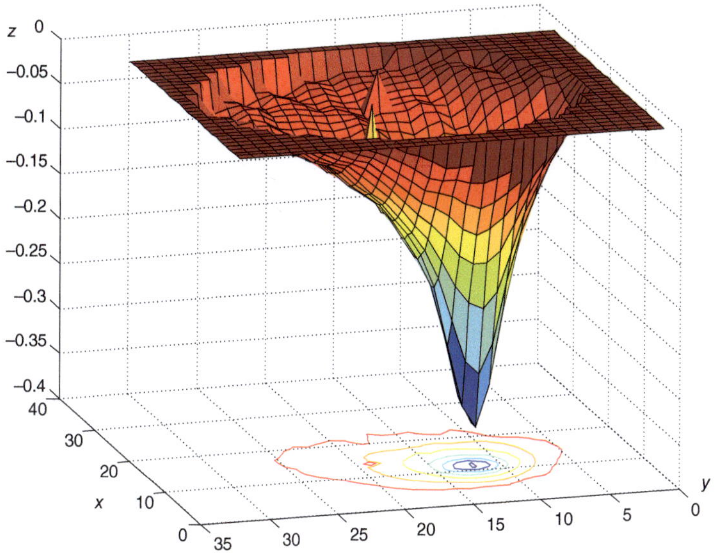

Figure 4.107 Influence function of a single piezo actuator, calculated by finite element calculation of the mechanical plate bending.

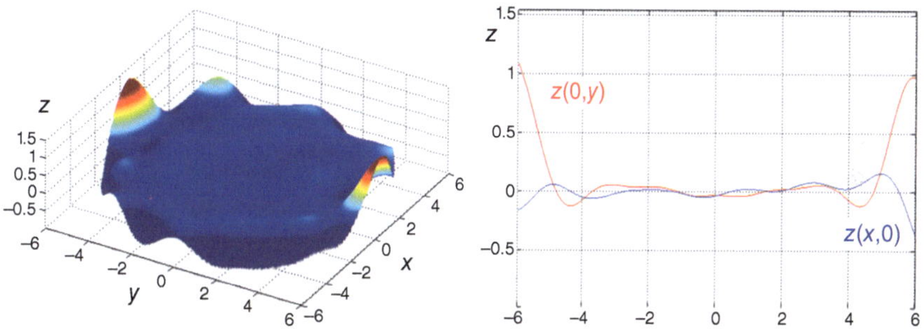

Figure 4.108 Wavefront after correcting adaptive mirror in two dimensions and as cross sections. The axes units are arbitrary.

4.8.4.7 Alvarez Lenses

An Alvarez plate pair is a system made of two thin glass plates which have a special freeform shape as a cubic polynomial with strength A with the sag according to the eq. [62–65]

$$z(x,y) = A \cdot \left(x \cdot y^2 + \frac{1}{3} \cdot x^3\right) \tag{4.167}$$

If the two plates are considered in the simplest model, they can be assumed to be thin and close together. Both plates can be shifted laterally to generate a special desired phase function. In the original system, a shift in x-direction creates a defocus, a shift in y creates an astigmatism. This can be used in practice to compensate for the aberrations of the human eye in ophthalmology. In the corresponding devices, the plates can be shifted in x and y-direction with arbitrary amounts. So a desired combination of focus and astigmatism can be adapted (Figure 4.109).

If, now, two shifts by Δx in opposite directions are performed on the two components respectively with an inverse effect (the step of refractive index has different sign, if both freeform surfaces are oriented to one another), we get the

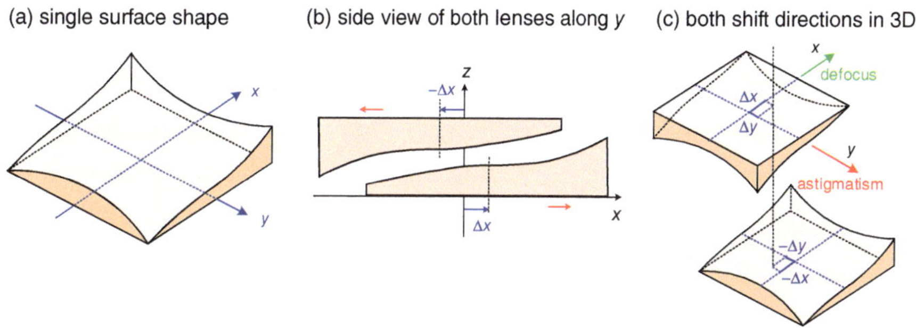

Figure 4.109 Illustration of the Alvarez principle with (a) the shape of a single surface, (b) the counter movement of the lenses in projected view and (c) the two shift directions for generation of defocus and astigmatism in three dimensions.

influence of an x-shift as a defocus

$$\begin{aligned} z^{(x)}(x,y) &= z(x+\Delta x, y) - z(x-\Delta x, y) \\ &= A \cdot \left[(x+\Delta x)\cdot y^2 + \frac{1}{3}\cdot(x+\Delta x)^3\right] \\ &\quad - A \cdot \left[(x-\Delta x)\cdot y^2 + \frac{1}{3}\cdot(x-\Delta x)^3\right] \\ &= 2A\cdot\Delta x\cdot[x^2+y^2] + 2A\cdot\Delta x^3 \\ &\approx +2A\cdot\Delta x\cdot[x^2+y^2] \end{aligned} \tag{4.168}$$

where the second term is neglected as of higher order in the small shift Δx. This corresponds to a paraxial defocus effect that is linearly growing with A and Δx. If a similar shift in y-direction is performed, we get a 45°-astigmatism as

$$\begin{aligned} z^{(y)}(x,y) &= z(x, y+\Delta y) - z(x, y-\Delta y) \\ &\approx -4A\cdot\Delta y\cdot x\cdot y \end{aligned} \tag{4.169}$$

4.8.5 Lenslet Arrays*

(*) The content of this section is quite specialized and for advanced reading. Knowledge of this section is not necessary to understand the following chapters. Beginners are recommended to skip this section.

4.8.5.1 Introduction

Due to growing progress in manufacturing and the needs of miniaturization, lenslet arrays are used in many applications [52, 66, 67]. There are applications found in imaging, illumination and metrology. The arrays divide a beam or bundle cross section area into several sub-apertures or channels. In every individual channel, the well-known geometrical or physical laws for the shaping of light are valid. If the channel widths are quite small, diffraction effects become more and more dominant and the lateral resolution is decreased due to a small numerical aperture in every channel. On the other hand, due to small angles and a parabasal vicinity in every channel, aberrations play only a minor role and the depth of focus in every channel is very comfortable. This kind of multi-aperture segmentation has several beneficial properties, but also some drawbacks.

From the viewpoint of efficiency, it makes sense to cover the area of the beam without dead zones. This means that the channel cross sections can be rectangular, hexagonal or triangular to get a fill factor of nearly 1. Figure 4.110 illustrates the problem of circular cross section and an overall relative loss of signal 1-π/4 = 0.21. On the other hand, in the case of vanishing dead zones, a cross talk between the channels may happen, which can be a problem depending on the application.

Figure 4.111 illustrates the working principle of an array of lenses, in every channel the corresponding lens focuses the light and creates a focus pattern.

From the viewpoint of manufacturing, several opportunities have been proposed in recent times to realize array lens lets with small dimensions. These are, for example:

- moulding;
- thermal re-flow'
- photolithographic techniques, grey tone;
- three-dimensional printing;
- photo-polymerization;
- embossing/printing from a metal master.

4.8.5.2 Special Lens Arrays

Simple spherical lenslets on a plane substrate with Cartesian segmentation is just one option for an array. Modified realizations, which can make sense depending on the application, can be as follows.

- Other tessellation of the area filling structure like hexagonal sub-apertures.

  This gives constant distances from the lenslet centres and less anisotropy of the boundary diffraction.
- Statistical variation of shape of the sub-apertures, the size of the sub-aperture areas and the depth (sag) of the lenslets changes randomly.

  This gives better and more uniform mixing effects in the case of illumination application.
- Aspherical or freeformed surfaces with individual tilt angles (chirped arrays).

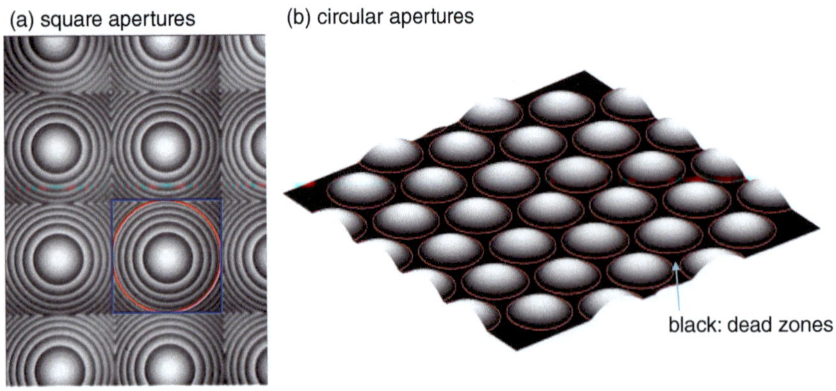

Figure 4.110 Array structures with (a) square sub-aperture shapes or (b) circular cross sections.

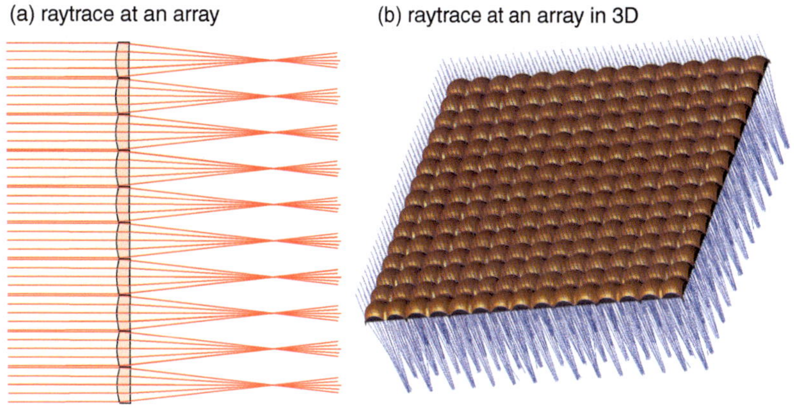

Figure 4.111 Raytrace in array structures illustrated in (a) as a cross section and in (b) in a three-dimensional view.

This approach takes care of bias-field angles in the outer zones.

- Cylindrical array.
- Arrays with curved substrate.
- Diffractive lenslets.

Some of these generalized arrangements are illustrated in Figure 4.112.

4.8.5.3 Matrix Calculation of Arrays

If an array consists of identical elements, its geometrical properties can be described by a matrix of tilted and decentred elements [67]. The tilt and decentre values can depend on the realization of the array envelope according to the parts (b) or (d) in Figure 4.112.

If a simple array on a plane substrate is considered from one lenslet to the next, we get the geometry of Figure 4.113.

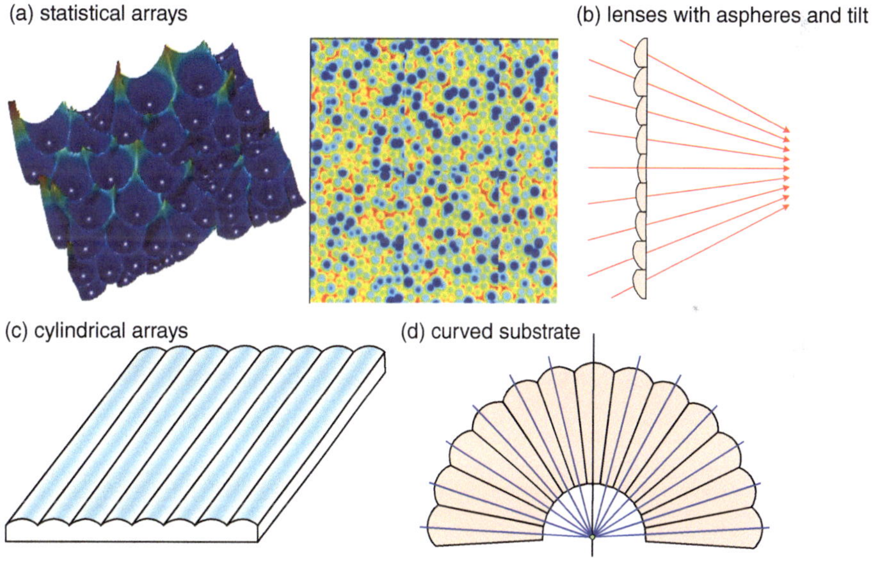

Figure 4.112 Modified array geometries. (a) shows a statistical array in three dimensions and as a colour-coded height profile, (b) shows an array with tilt angles depending on the position, (c) is an array of cylindrical lenses and (d) illustrates an array on a curved substrate.

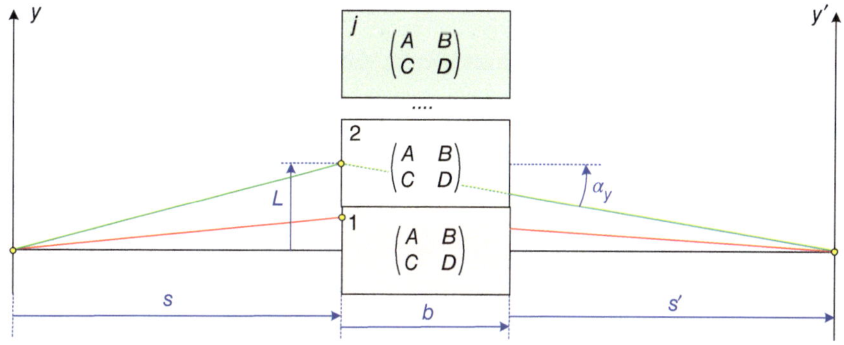

Figure 4.113 Matrix description of a lenslet array in case of imaging.

The shift between the channels is L; let's assume an imaging setup with matrix $ABCD$ for one thick lenslet, an object distance s and an image distance s'. The matrix can now be written as

$$M = \begin{pmatrix} A & 0 & B & 0 & v_x \\ 0 & A & 0 & B & v_y \\ C & 0 & D & 0 & \alpha_x \\ 0 & C & 0 & D & \alpha_y \\ 0 & 0 & 0 & 0 & 1 \end{pmatrix} \tag{4.170}$$

with the offsets v and the rotations angles α. If only a shift is assumed, we have

$$v_x = j \cdot L, \; v_y = m \cdot L, \; \alpha_x = \alpha_y = 0 \tag{4.171}$$

for the lens with indices j, m. If in addition the array substrate is curved as indicated in the figure, the rotation angles are

$$\alpha_x = \frac{v_x}{s' + B} = \frac{j \cdot L}{s' + B}, \; \alpha_y = \frac{v_y}{s' + B} = \frac{m \cdot L}{s' + B} \tag{4.172}$$

If an arbitrary initial ray is started, the indices of the impinged channel are given by

$$j = \text{Int}\left[\frac{x_1 - s \cdot u_1}{L} - \frac{1}{2}\right], \; m = \text{Int}\left[\frac{y_1 - s \cdot v_1}{L} - \frac{1}{2}\right] \tag{4.173}$$

with Int[..] as the smallest integer of the argument. By inserting in the case without rotated array elements the matrix

$$M = \begin{pmatrix} A & 0 & B & 0 & L \cdot \text{Int}\left[\frac{x_1 - s \cdot u_1}{L} - \frac{1}{2}\right] \\ 0 & A & 0 & B & L \cdot \text{Int}\left[\frac{y_1 - s \cdot v_1}{L} - \frac{1}{2}\right] \\ C & 0 & D & 0 & 0 \\ 0 & C & 0 & D & 0 \\ 0 & 0 & 0 & 0 & 1 \end{pmatrix} \tag{4.174}$$

is obtained. It is seen that the matrix elements depend on the initial ray data: the system is no longer linear. In the case with rotations the matrix is still more complicated.

4.8.5.4 Applications

There are many very special applications of array components. In illumination applications the so-called fly-eyes condenser is an important disposal of arrays. The simple version uses only one array, which divides the incoming beam with inhomogeneous intensity distribution into sub-apertures. With an additional condenser lens all channels are transported into a plane where the various channel beams exactly overlap, but come from different locations in the original beam and have different directions after the condenser. This has the impact of a homogenization of the profile in the spatial domain in one plane, see Figure 4.114(a). But the angle space is not homogeneous and the uniform illumination field is changed in case of defocusing. In the more comfortable variant, which is illustrated in Figure 4.114(b), a second aligned array in the focal plane of the first array acts as a field lens in every channel and images the channel boundaries into the final field plane, as can be seen in the red shaded ray bundle. This improves the homogenization and reduces the diffraction ripple at the outer edges considerably. All angle diffraction effects coming from the first array are refocused sharply in the final destination plane, as can be seen in the figure: the illumination fields uniformity is better and the edges are sharp.

The basic idea of homogenization can also be realized by reflective array mirrors, which make sense if high energy densities or very short wavelengths are used. This principle is sketched in Figure 4.115.

A second very prominent application of mostly Cartesian lenslet arrays is a Hartmann–Shack wavefront sensor. The principle of this device is sketched in Figure 4.116. A two-dimensional lenslet array divides the incoming wave into small sub-apertures according to the size of the lens areas. Every lenslet focuses their part of the wave onto a detector and generates a spot. If the wave is tilted in the sub-aperture by the angle u, the spot is shifted in lateral direction by $x = f \tan(u)$. The tilt angles are directly related to the size of the shift components in x and y. Therefore, the local tilt of the wave can be calculated by evaluating the centroid position of the spots. By integrating the gradient fields, the full wave $W(x,y)$ can be reconstructed.

The full setup with collimating beam and showing illumination and detection ray path is illustrated in Figure 4.117. This measurement setup can be used for the metrology of surfaces as well as free propagating wave fronts. A drawback of this kind of sensor is the dispersion in the refractive lenslets and possible cross talk between the channels. Depending on the application, the number of lenslets, the focal lengths, the fill-factor and the sensor pixel size are critical parameters to optimize dynamic range as well as sensitivity and accuracy.

A third application of arrays for imaging is the so called plenoptical or light-field camera [68]. Here the ray bundle is divided into sub-apertures and every sub-aperture creates a special pattern in the sensor plane. The raw intensity distribution is far from being comfortable due to the rupture of the light field. The real image is calculated and reconstructed by digital means. The most interesting property of a plenoptical camera system is the possibility to resharpen the image by calculation for different depths of the object. Therefore, the depth of focus is strongly increased and a focusing of the system is not necessary.

Figure 4.118 sketches one possible layout of a plenoptical camera. The array is placed in the image plane; the sensor

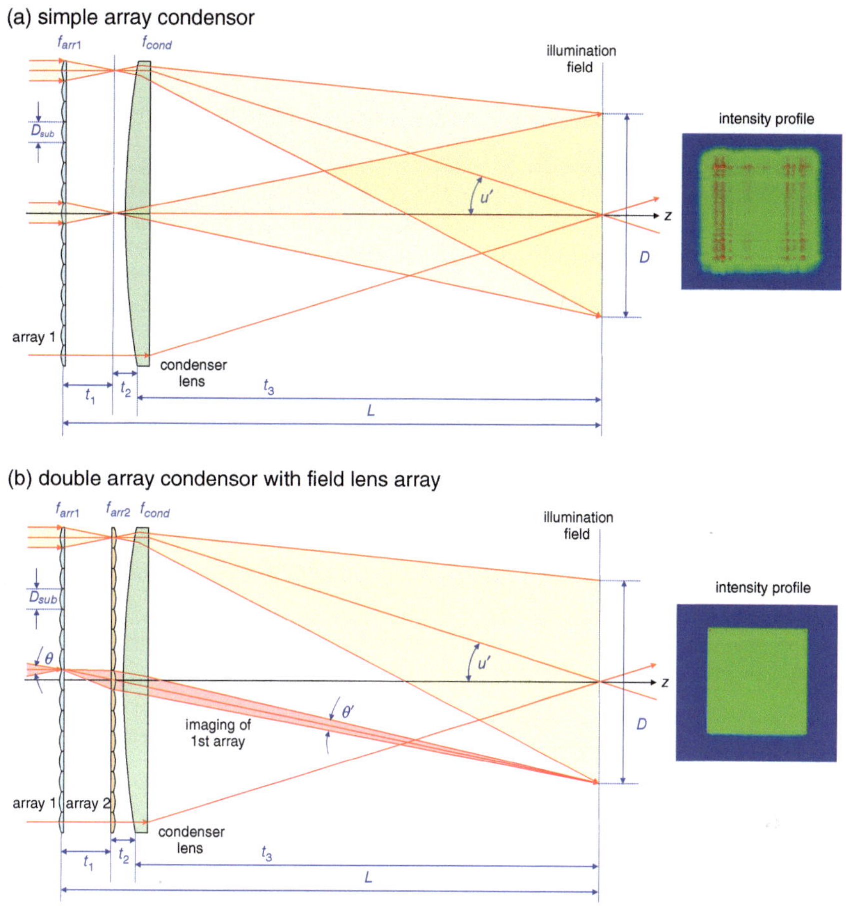

Figure 4.114 Fly-eye condenser with (a) only one array and (b) two arrays with imaging of the first array for illumination homogenization.

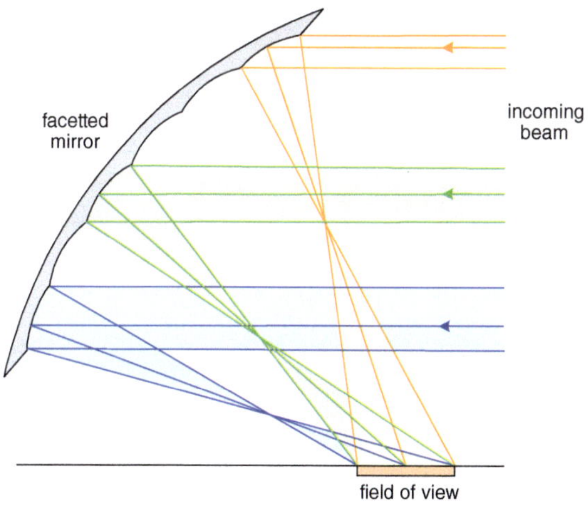

Figure 4.115 Reflective array structure.

is located behind in the focal length distance of the array lenses. The lenslet array mixes the spatial and the angle properties of the rays.

In this special arrangement the array is located in the image plane. The rays in this plane build a rectangular grid. If only the rays through the centres of the lenses are considered, they have the same angles as the original system, but in the focal plane of the array a changed coordinate x occurs. The rectangular pattern of chief ray image points was sheared, as illustrated in Figure 4.119.

If all the light rays are considered that come from one object point, in the classical case without the array coloured lines are obtained as shown in Figure 4.120(a). The angle extent due to the numerical aperture corresponds to the lengths of the lines, the inclined dotted line as an envelope shows the chirp of the angles due to a finite field angle in the image. In part (b) of the figure now the array is included; the lines are tilted. This has the consequence that in every finite width of a pixel, the integration corresponds to a summation over the angle θ in the finite width yellow rectangle; the contributions are coming from several object points. In the sketched case, three object points contribute to the energy in one pixel. This mixing makes the raw image unusual. The intensity distribution can now be deconvolved by digital processing to generate the desired image.

In modified setups, the array is placed in an arbitrary plane between camera lens and image plane, in this case every lens of the array captures in principle the full image information.

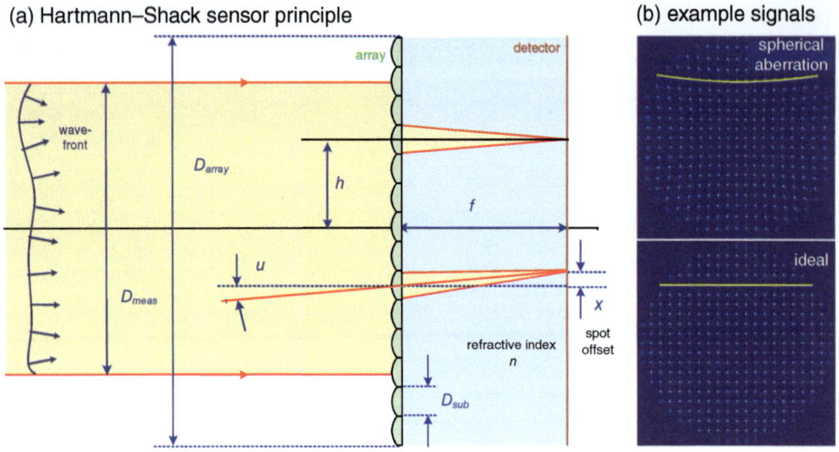

Figure 4.116 Principle of operation of the Hartmann–Shack wavefront sensor (a) and (b) sample signals for illustration.

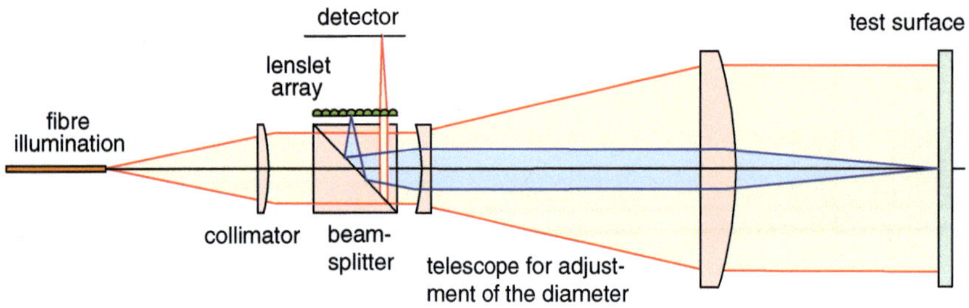

Figure 4.117 Typical setup of optical components testing in reflection by the Hartmann–Shack wavefront sensor.

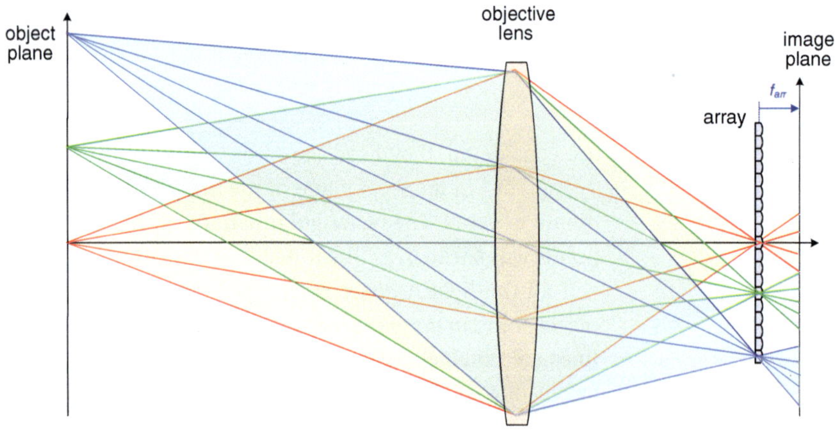

Figure 4.118 Principle of the plenoptical camera.

The angle integration at the sensor location x can be written as

$$I(x) = \int L(x, \theta)\, d\theta \tag{4.175}$$

with the radiance L. If a defocused detail of the object is considered, this equation reads

$$I(x) = \int L(x + z \cdot \theta, \theta)\, d\theta \tag{4.176}$$

with defocus value z. This means that the depth information is coded in the modified lateral spatial distribution via the angle change. If the image entropy is optimized, the unknown depth z can be determined.

In the special setup of the array location in the original image location shown here, the sensor plane is in the focal plane of the array, the relation between the image and the new intensity distribution is simply a Fourier transform and therefore can be reconstructed by an inverse operation. For more details of the imaging mode, the reader is referred to the literature [68] and Chapter 11.

4.8.6 Digital Mirror Device*

(*) The content of this section is quite specialized and for advanced reading. Knowledge of this section is not necessary to understand the following chapters. Beginners are recommended to skip this section.

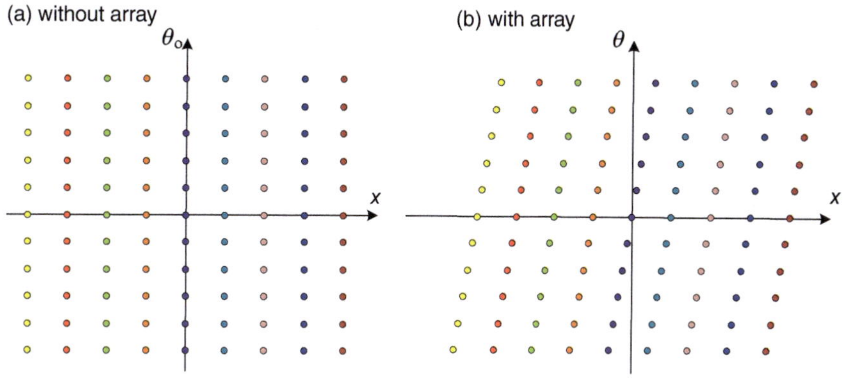

Figure 4.119 Intersection points and angles of the rays going central through the lenslets.

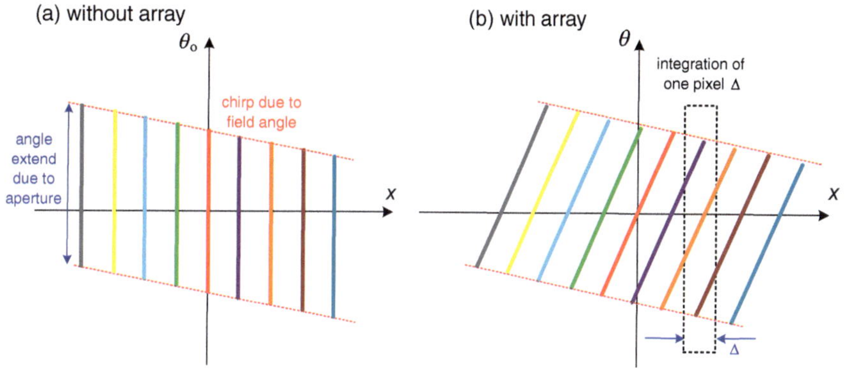

Figure 4.120 Intersection points and angles of the rays from one object point (a) without and (b) with the array.

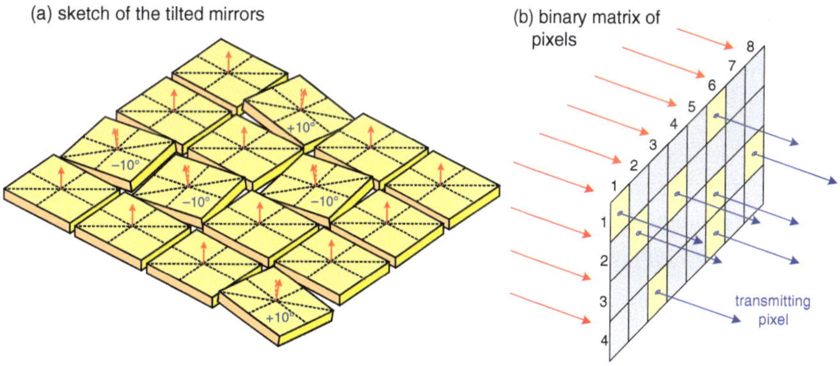

Figure 4.121 (a) Principle of the digital mirror device, and (b) binarization of the light field in one direction, ignoring the reflection.

Spatial light modulators and digital mirror devices (DMDs) are a special type of array structure. A large number of channels, typically 1000 × 1000, discretizes the light field; every channel is processed by one lenslet of an LCD component or a tilt mirror individually. This allows for a well resolved binary multi-aperture system [69–72].

In the case of the so-called DMD, every individual mirror can have three discrete orientations, +10°, 0 or – 10° of the tilt angle. In between these mirrors, for mechanical reasons some dead zone must be accepted; it is in the range of 40% of the area, which is in total approximately 10 mm × 10 mm. One mirror pixel has, for example, a side length of 10 μm. The switching between the angles is done electrically with switching times of 20 ms. The principle geometry is sketched in Figure 4.121.

Over time, different versions of DMDs have been developed; in some cases the turning axis are Cartesian oriented, in some the axes are along the diagonal of the square. Two fixed angles can be realized. For a switched off mirror without voltage, a third indefinite position of the mirrors is observed, see Figure 4.122.

For the characterization of the DMD, the side length L, the number of pixels in one dimension N, the pixel pitch p, the tilt angle α and the mirror size w are useful.

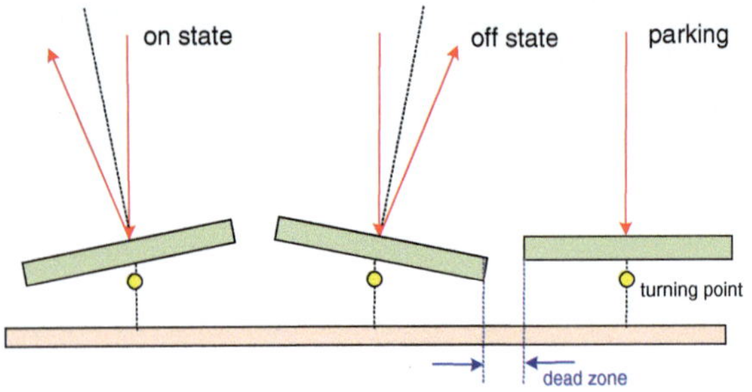

Figure 4.122 The three possible orientations of a mirror.

Typical values of commercial DMDs are (for notation see Figure 4.123) as follows.

N	$1000 \times 768 \ldots 4000 \times 3072$	number of mirrors
α	12°	incidence angle relative to the substrate plane
p	8.6 μm/13.68 μm	pitch size, period of the grid
w	7.6 μm	mirror size

Due to the difference between p and w a dead zone $a = p\text{-}w$ in the range of 1 μm is present, which decreases the efficiency to 61 … 67% in the first diffraction order. As can be seen in Figure 4.123, in the usual case, due to the deflection, a compression of the outgoing beam takes place. The tilt angle of the individual mirrors is 12°, therefore the usual beam deflection is $\alpha = 24°$. This deflection angle results in a diameter compression of

$$\frac{D_{out}}{D_{in}} = \cos 24° = 0.9135 \tag{4.177}$$

in the deflection plane if the incident light meets the DMD plane perpendicular. If a finite incidence angle θ is realized, the lateral compression is

$$\frac{D_{out}}{D_{in}} = \cos(\theta + 24°) \tag{4.178}$$

which causes certain inner obscuration stripes, see Figure 4.123(b).

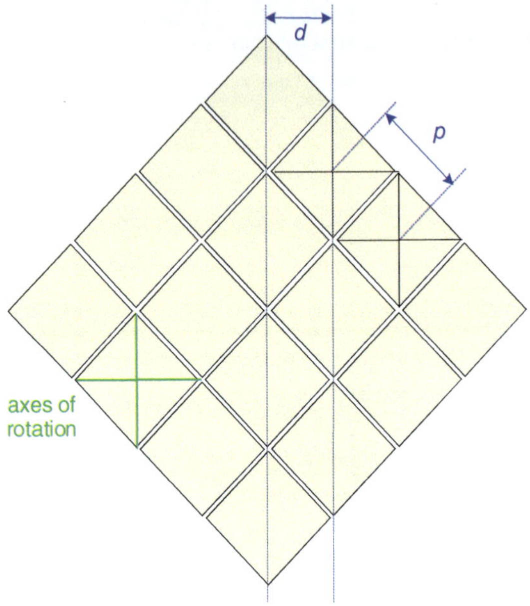

Figure 4.124 Real geometry of the pixel matrix.

A DMD has the structure of an array and therefore shows diffraction effects similar to grating at the pixel boundaries. Due to the geometry, the x and y-direction are nearly perfectly decoupled and therefore the diffraction effects can be considered to be one dimensional. According to the 45° rotated geometry shown in the following Figure 4.124, the real grating pitch is smaller than the pixel pitch as

$$d = p/\sqrt{2} \tag{4.179}$$

So, for a DMD pitch of 13.68 μm, the grating pitch is $d = 9.67$ μm.

According to the geometry in a simple view, it is assumed that the DMD acts as a blazed grating (see Section 4.11). For this, the blaze condition

$$\lambda = \frac{2}{p \cdot m} \sin\theta_B \cos(\theta - \theta_B) \tag{4.180}$$

must be fulfilled with the diffraction order m. But since the wavelength is arbitrary and the angle θ is fixed to be 12°, this

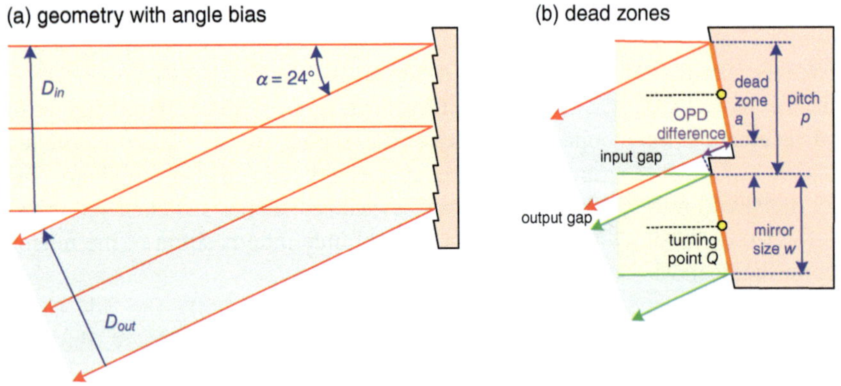

Figure 4.123 The projection effect due to the angle bias is shown in (a) and the obscuration effects are sketched in (b).

Table 4.7 Blaze wavelengths of a DMD.

m	4	5	6	7	8	9	10	11
λ	984	787	656	562	492	437	393	358

condition cannot be fulfilled for arbitrary wavelengths. But on the other way round, this means that there are very special wavelengths for which the DMD can be a blaze grating and works best. Depending on the used diffraction order m, the values are given in the Table 4.7.

If a Fourier optical model of a DMD is established, the transmission function in one dimension can be written as

$$T(x) = \left[\mathrm{rect}\left(\frac{x}{w}\right) \otimes \mathrm{comb}\left(\frac{x}{d}\right)\right] \cdot \mathrm{rect}\left(\frac{x}{L}\right) \quad (4.181)$$

The Fourier transform of this function gives the angle related amplitude in the farfield and the Fraunhofer diffraction pattern with $L = N \cdot d$ as

$$\begin{aligned} \hat{F}[T(x)] &= A(v) \\ &= \left[w \cdot \mathrm{sinc}(w \cdot v) \cdot \sum_m \left(x - \frac{m}{d}\right)\right] \otimes \mathrm{sinc}(N \cdot d \cdot v) \end{aligned} \quad (4.182)$$

with the spatial frequency v. This pattern looks very much the same as for a classical grating. Due to the fixed deflection angle θ_B, there is a phase step between two neighbouring pixels, therefore the intensity is modified by a shifted term as

$$\begin{aligned} I(\varphi) = I_o \cdot \sum_m &\mathrm{sinc}\left[\frac{w}{d} \cdot (m + \sin\theta_{in} + \sin 2\theta_B)\right] \\ &\cdot \mathrm{sinc}\left[N \cdot d\left(\frac{\varphi}{\lambda} - \frac{m}{d}\right)\right] \end{aligned} \quad (4.183)$$

in the deflection plane. In the perpendicular plane, the argument of the first factor is simply $w\, m/d$. This shows that the diffraction pattern is not symmetric in x and y but shows a certain anisotropy. The efficiency in the zeroth order is in the range of 61…67% for commercial devices. This behaviour is sketched in Figure 4.125.

4.9 Gradient Index Lenses

4.9.1 Introduction

In conventional components, it is assumed that the refractive index of the materials is constant inside the lens. But it is also possible to fabricate and use components with a spatially varying refractive index, so-called gradient index lenses (GRIN) [73–79]. The ray paths are then curved in the material and special index distributions are realized to achieve the desired ray deflection. The easiest way to describe the refractive index distribution inside the volume of the component is a Taylor expansion of the function of the volume distribution $n(x,y,z)$. In reality, mostly radial symmetric or axial gradients are found. Possible mathematical descriptions for the spatial distributions are:

- A quite general polynomial expansion with variables x, y, r and z

$$\begin{aligned} n = n_{o,\lambda} &+ c_1\, r + c_2\, r^2 + c_3\, r^4 + c_4\, r^6 + c_5\, r^8 \\ &+ c_6 z + c_7\, z^2 + c_8\, z^3 + c_9\, z^4 \\ &+ c_{10}\, x + c_{11}\, x^2 + c_{12}\, x^3 + c_{13} y + c_{14}\, y^2 + c_{15}\, y^3 \end{aligned} \quad (4.184)$$

- A representation with radial and axial dependence for circular symmetric components, the description has a block form as

$$\begin{aligned} n = n_{o,\lambda} &+ c_1\, r^2 + c_2\, r^4 + c_3\, r^6 + c_4\, r^8 \\ &+ z \cdot (c_5 + c_6\, r^2 + c_7\, r^4 + c_8\, r^6 + c_9\, r^8) \\ &+ z^2 \cdot (c_{10} + c_{11}\, r^2 + c_{12}\, r^4 + c_{13}\, r^6 + c_{14}\, r^8) \\ &+ z^3 \cdot (c_{15} + c_{16}\, r^2 + c_{17}\, r^4 + c_{18}\, r^6 + c_{19}\, r^8) \end{aligned} \quad (4.185)$$

- A square root description with only radial distribution, the parameter c_1 is used for normalization of the radial dependence

$$n = n_{o,\lambda} \cdot \sqrt{\begin{aligned} &1 + c_2\,(c_1\, r)^2 + c_3\,(c_1\, r)^4 + c_4\,(c_1\, r)^6 \\ &+ c_5\,(c_1\, r)^8 + c_6\,(c_1\, r)^{10} \end{aligned}} \quad (4.186)$$

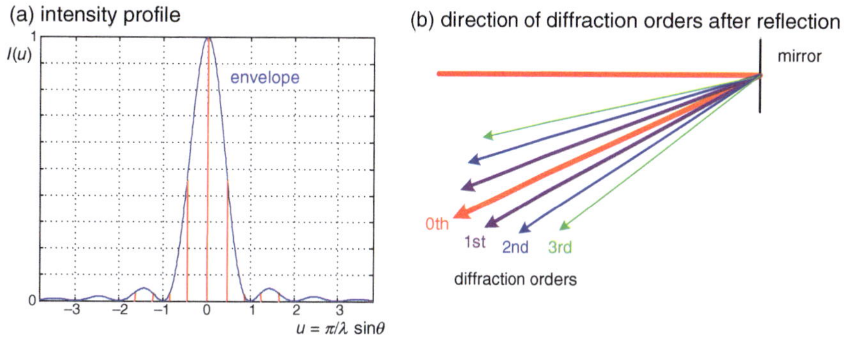

Figure 4.125 Diffraction at a DMD with (a) intensity profile and (b) orientations and strength of the diffraction orders.

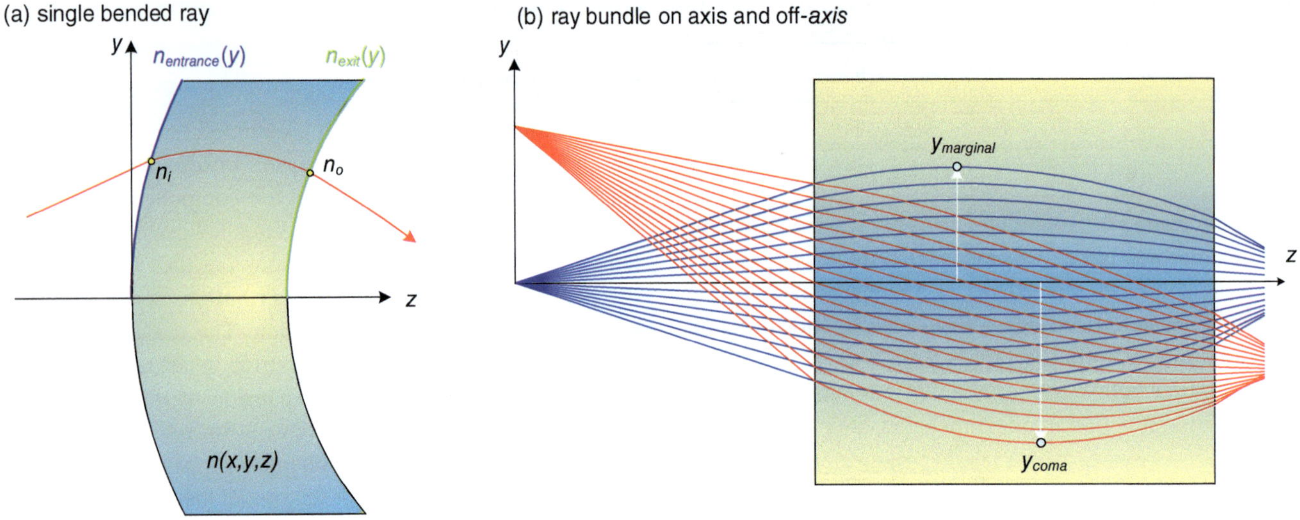

Figure 4.126 Gradient index lens with (a) curved surfaces and (b) a radial profile with imaging functionality.

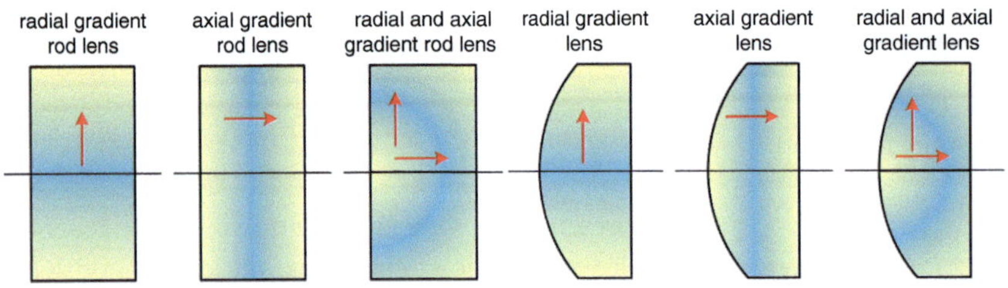

Figure 4.127 Different types of gradient index components with plane or curved surfaces, radial or axial gradients or combinations of these opportunities.

One more degree of freedom is the shape of the outer surfaces. Of course, the manufacturing of the components as well as the packaging of more than one lens is easier for plane outer surfaces, a curvature of entrance and exit surface respectively gives more opportunities. Since the ray bending at the surfaces is governed in this case geometrically by the local slope as well as physically by the local value of the refractive index, this is quite general. One problem in the case of large index variations is the coating, which usually has to be adapted to one value of the refractive index. This condition can for radial gradients only be fulfilled approximately. A component with bent outer surfaces is sketched in Figure 4.126(a), (b) shows the raytrace through a lens with plane entrance and exit surfaces; Figure 4.127 illustrates some combination opportunities between radial index variation, axial index changes and curvatures of the surfaces respectively.

4.9.2 Quadratic Radial Index Lenses

If in particular a quadratic transverse index profile is created with a larger index in the centre, imaging setups can be realized. The profile of a quadratic radial gradient index media is then written in the simple form

$$\begin{aligned} n(r) &= n_0 - n_2 \cdot r^2 = n_0 \cdot (1 - n_r \cdot r^2) \\ &= n_0 \cdot \left(1 - \frac{1}{2} A \cdot r^2\right) \end{aligned} \tag{4.187}$$

where A is a constant and r is the radial coordinate measured from the optical axis. This index profile is of special importance because in the paraxial approximation the wave equation can be solved analytically. Rays follow sinusoidal paths. Refocusing occurs periodically at a characteristic distance which is also known as pitch length p. It is given by

$$p = 2\pi \cdot \sqrt{\frac{n_0}{2n_2}} = \frac{2\pi}{\sqrt{2n_r}} \tag{4.188}$$

A light wave that has passed one complete cycle of the sinusoidal wave has traversed one pitch. Figure 4.128 illustrates this concept for different rod lens lengths.

These lenses can also be used with finite object and image distances as well as off-axis field points. This is sketched in Figure 4.129. But it should be mentioned that these imaging properties are only valid in paraxial approximation. If field of view and numerical aperture have larger values, the usual aberrations arise. If for simplicity the outer surfaces are plane, a gradient component with quadratic index variation is called a Wood lens. If a finite thickness t is considered, depending on the sign of the index difference, positive or negative lenses can be realized. This is shown in Figure 4.129. The principal planes are perfectly separating the lens in parts of $t/3$. For the axis field point also an exact stigmatic imaging is possible for a special radial profile and non-paraxial larger apertures, see Section 4.9.4.

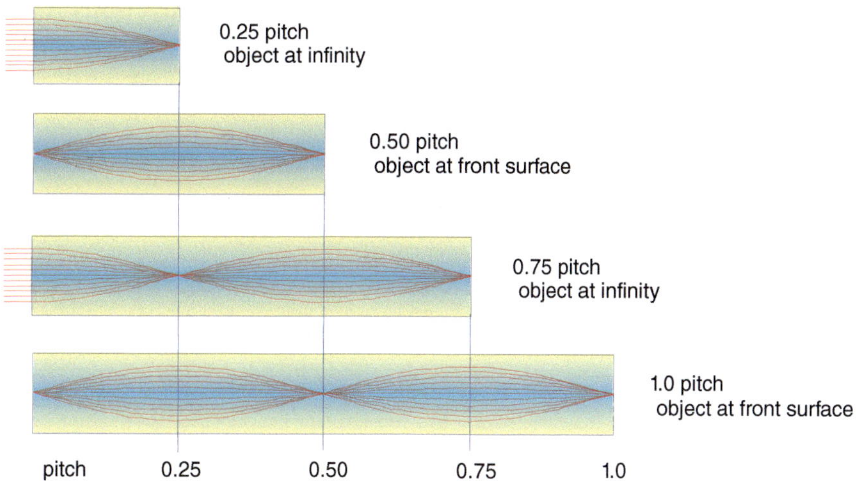

Figure 4.128 Pitch concept of lenses with quadratic radial index profile for different lengths and object locations at the entrance surface or in infinity.

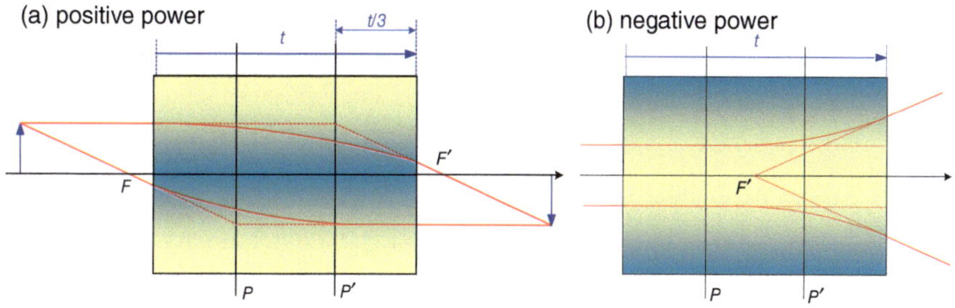

Figure 4.129 Positive and negative Wood lenses with plane surfaces and quadratic radial index profile with (a) positive and (b) negative power.

4.9.3 Axial Gradient Index Lenses*

(*) The content of this section is quite specialized and for advanced reading. Knowledge of this section is not necessary to understand the following chapters. Beginners are recommended to skip this section.

Axial gradients are another possible option for correcting aberrations with index profiles. Figure 4.130 shows a lens with an axial gradient, for illustration. A simple linear axial gradient in a medium is described by the formula

$$n(z) = n_0 + n_{1z} \cdot z \tag{4.189}$$

where the linear coefficient n_{1z} is the gradient of the refractive index in the z-direction.

$$n_{1z} = \frac{\partial n}{\partial z} \tag{4.190}$$

For plane outer surfaces and a collimated paraxial incident ray, a purely axial GRIN lens has no refracting effect. A lens of thickness t with a linear gradient in the refractive index and plane outer surfaces has a focal length

$$f = \frac{1}{n_{1z}} \cdot \ln\left(1 + \frac{n_{1z}t}{n_0}\right) \tag{4.191}$$

So-called 'GRADIUM' glasses (trademark of LightPath Inc.) exhibit an axial refractive index profile only [77–79]. The manufacturing process may be thought of as stacking thin glass plates into a sandwich structure. The refractive index of each plate varies slightly from plate to plate to form a staircase index profile, as shown in Figure 4.131(a) in green. This stack is then thermalized and the discrete index steps are smeared out to a continuous profile by diffusion. Depending on the glass types and the doping, different overall index variations and lengths can be realized, see Figure 4.131(b).

Depending on the base glasses used for composing the index profiles, a maximum axial length z_{max} can be used. Within this profile, the axial position Δz of a lens can be chosen within the limits $0 < z < z_{max}$-t, where t is the thickness of the lens and z is the position of the front surface of the lens. A scheme of this device is shown in Figure 4.132.

Compared with homogeneous glasses, the optical designer now gains an additional degree of freedom by selecting the appropriate index profile and by varying the position of the lens (front vertex of the lens at z) within the profile. z can be made variable in most lens design programs. Typically, the axial gradient is described by a Taylor series of the form

$$n(z, \lambda_{ref}) = \sum_{k=0}^{k_{max}} n_k \cdot \left(\frac{z}{z_{max}}\right)^k \tag{4.192}$$

Some real data of index profiles are shown in Figure 4.131(b). As can be seen, the range of the index reaches a maximum value of 0.3 over a length of 10 mm.

4.9.4 Stigmatic Imaging Gradient Index-Lenses*

(*) The content of this section is quite specialized and for advanced reading. Knowledge of this section is not

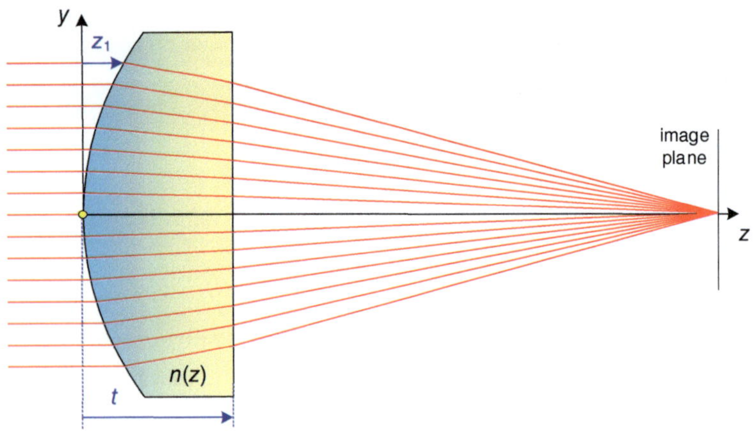

Figure 4.130 Lens with axial gradient of the refractive index.

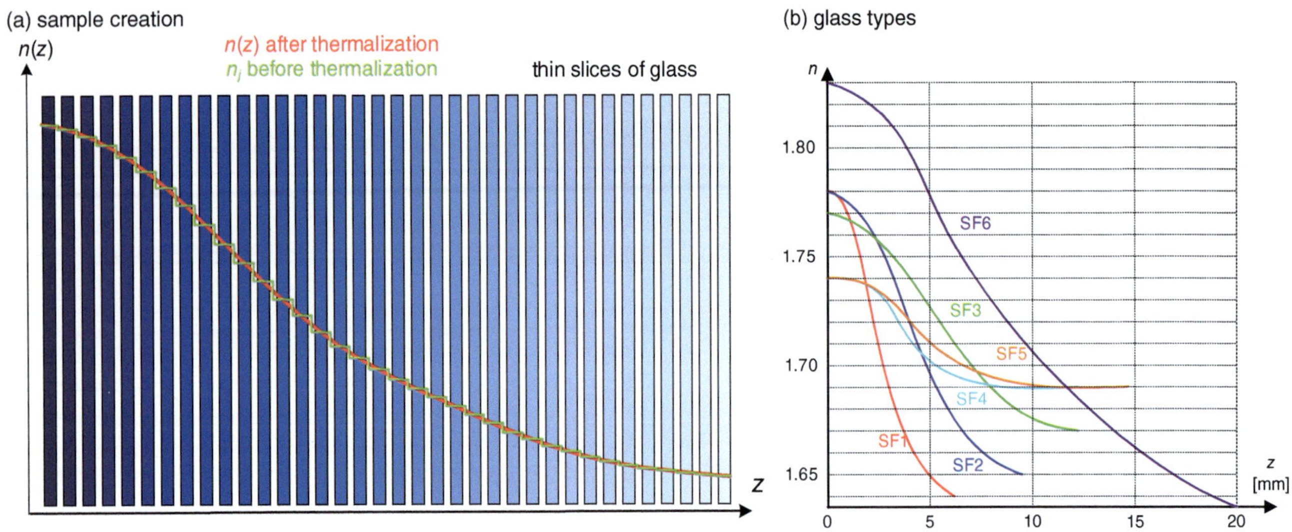

Figure 4.131 GRADIUM material (a) preparation of samples by thin slices with different refractive indices and (b) different glass type profiles and index ranges.

necessary to understand the following chapters. Beginners are recommended to skip this section.

In literature there are some gradient index lenses known that in theory correct the spherical aberration exactly to zero by means of a special distribution of the refractive index. Figure 4.133 illustrates these lens types in comparison.

The lens of Mikaelian (Figure 4.133(a)) is a rod lens with plane outer surfaces of thickness t. Its refractive index has the radial profile

$$n(r) = \frac{n_0}{\cosh\left(\frac{\pi}{2t} \cdot r\right)} \tag{4.193}$$

the lens has the focal length

$$f' = \frac{2t}{\pi \cdot n_0} \tag{4.194}$$

The lens perfectly focuses collimated incoming light into the exit surface plane.

The design of Tarhanov (Figure 4.133(b)) is a plano-convex lens with the axial profile of the refractive index

$$n(z) = \frac{n_0}{\sqrt{1 + 2n_0 \cdot (n_0 - 1) \cdot \frac{z}{R}}} \tag{4.195}$$

where n_0 is the refractive index at the rear vertex. The rear back focal length is

$$s' = -\frac{R}{n_0 - 1} \tag{4.196}$$

The lens of Ilinsky (Figure 4.133c) is corrected also with regard to spherical aberration. It has the shape of a meniscus with thickness t and radii $R_{1/2}$. The gradient of the refractive index is

$$n(x, y, z) = \frac{n_0}{f} \cdot \sqrt{x^2 + y^2 + (z - f)^2} \tag{4.197}$$

The focal length can be calculated from

$$f = \frac{n_0 R_1}{n_0 - 1} \tag{4.198}$$

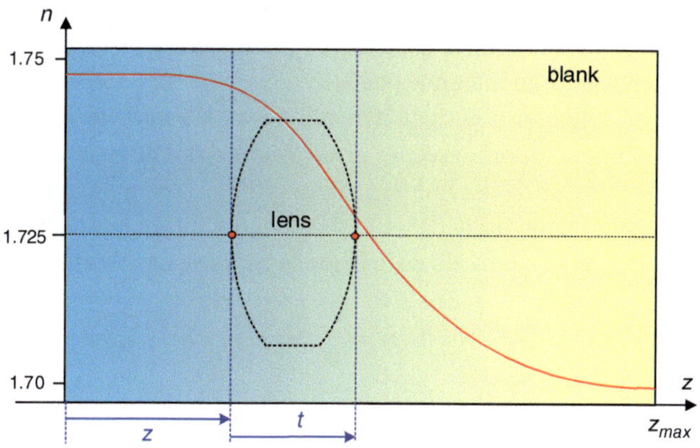

Figure 4.132 Profile of the refractive index in a GRADIUM lens by selecting the position inside the blank.

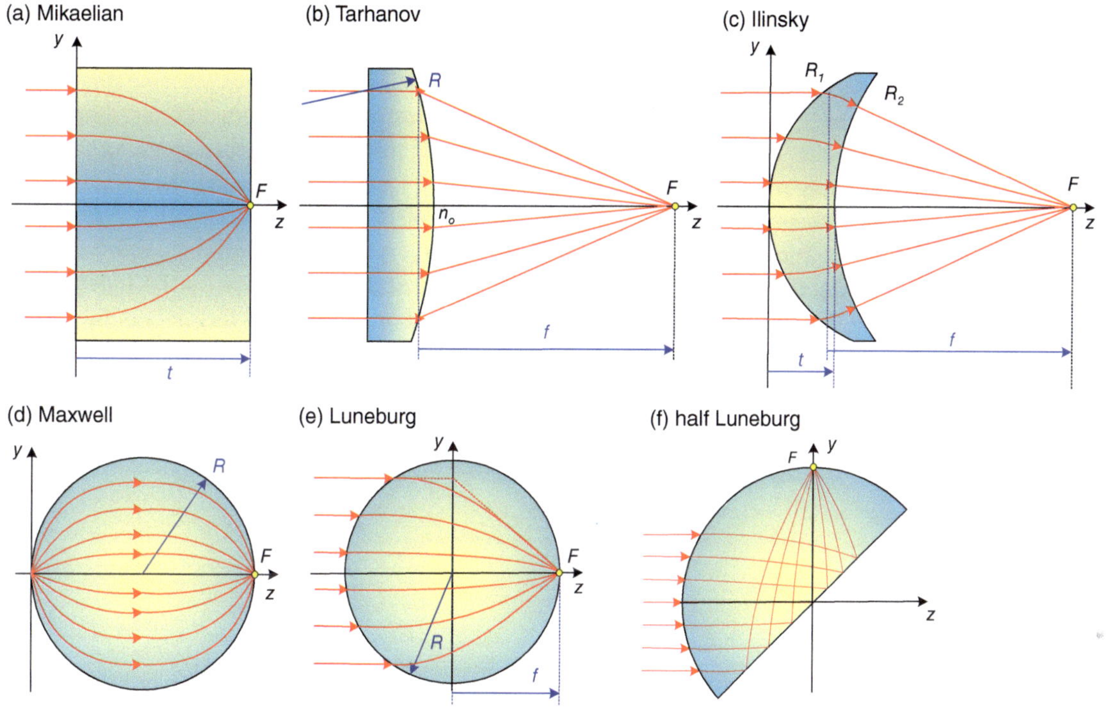

Figure 4.133 Various stigmatic imaging gradient index lenses; for details see the explanation in the text.

and the following correlation has to be fulfilled

$$R_2 = f - t \tag{4.199}$$

In the case of the lens according to Maxwell (Figure 4.133(d)), the object lies in the front vertex of the spherical lens. The image is finite, if the spherical profile of the refractive index follows the formula

$$n(x, y, z) = \frac{2n_o}{1 + \dfrac{x^2 + y^2 + z^2}{R^2}} \tag{4.200}$$

where n_o is the reference refractive index in the vertex point. An object point in the front vertex is imaged into the rear vertex. The paths of light within the lens are circles.

Finally, the spherically corrected gradient lens according to Luneburg (Figure 4.133(e)) is a spherical lens with radius R and spherical gradient

$$n(x, y, z) = \sqrt{2 - \frac{x^2 + y^2 + z^2}{R^2}} \tag{4.201}$$

The focal length of the Luneburg lens is $f = R$. Its focal point lies at the rear vertex of the lens. The object is assumed to be in infinity. The rays within the lens are ellipsoidal curves.

There is an alternative form to the classic Luneburg lens. For this the lens is cut into a half sphere and the cut plane is reflecting (Figure 4.133(f)). In principle, the Luneburg lens can be of advantage if it is used as a collimating optical system for a small almost point-like light source as emitter. In this case the ray paths of Figure 4.133 have to be inverted. For other applications the reflecting rear surface is used as a retroreflector. In the so-called generalized Luneburg lens, the focal point lies outside the lens. In this case the profile of the refractive index can only be calculated numerically. It is given by an integral equation.

The solutions described here are nowadays more or less academic solutions, except the rod lens in (a). The technology of preparing the index gradient profiles is complicated but shows growing success. Perhaps in the future lenses of these types can be realized in practice with reasonable effort and precision.

4.10 Prisms

4.10.1 Reflecting Prisms

There is a large variety of reflecting prisms that are used to redirect the light and to change the image orientation. In most cases, the outer refractive surfaces are used in normal incidence; therefore, for collimated light beams no aberrations occur for the marginal ray [80–84]. Figure 4.134 shows some examples of reflection prisms to deviate or shift ray bundles. It should be noted that the various solutions are only a small selection of all known possible setups. The essentials are mostly the size, the number of reflections, the redirection orientation of the optical axis and the length of the glass path.

Every reflection inverts the image orientation; an even number of reflecting surfaces therefore finally has an erected image. If several internal reflections are used inside a prism, it is possible to modify an image orientation. The main functionalities as well as a simple demonstration are shown in Figure 4.135. In total there are four options for changes of image orientation:

- invariant image orientation;
- reverted image (side reversal);
- reverted image (upside down);
- image inversion.

The function of prisms can be one of the following.

- Bending of the beam path, deflection of the axial ray direction.
- Application in instrumental optics and folded ray paths with parallel off-set or lateral displacement of the axial ray.
- Modification of the image orientation (reversion, inversion, see above).
- Off-set of the image position due to the index of the medium, shift of image position forwards in the propagation direction.

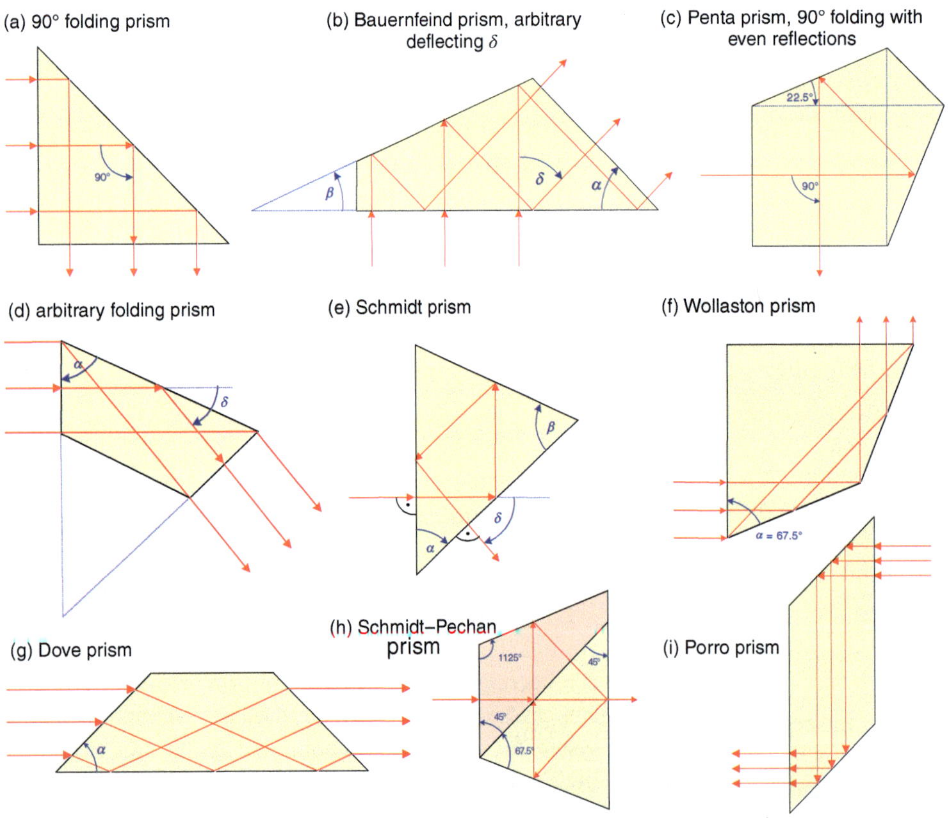

Figure 4.134 Different types of prisms.

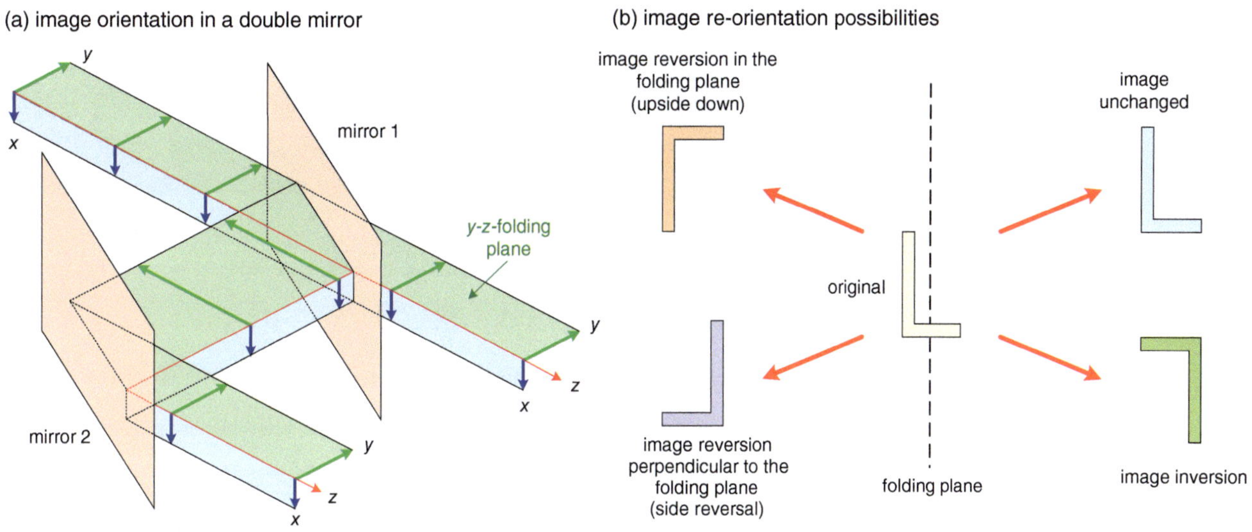

Figure 4.135 Change of image orientation with mirrors (a) shows a two-reflection case, (b) illustrates the possible image orientation options.

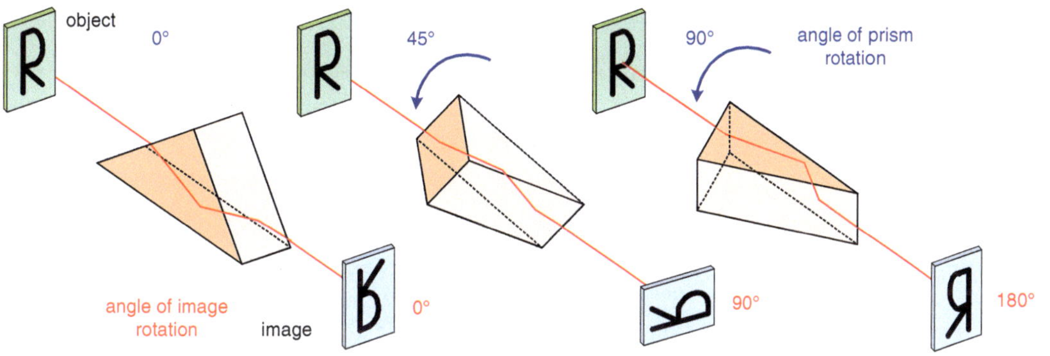

Figure 4.136 Change of image orientation with a Dove prism.

Figure 4.136 illustrates in particular the behaviour of a Dove prism concerning the image orientation. Due to the special setup of this component, a continuous image rotation is possible. In the image plane, the image rotation occurs with twice the angle of the prism rotation. This special feature is used in periscopes. The Dove prism is one of the very few examples where the ray bundle does not enter the prism perpendicularly. This causes special chromatical aberrations.

If the above-mentioned ideal conditions of collimated light impinging perpendicularly are violated and non-collimated beams are used, aberrations can be observed; these are briefly mentioned in Section 4.3. They are:

- astigmatism;
- chromatic aberration;
- spherical aberration in non-collimated beams.

4.10.2 Roof Prisms

4.10.2.1 Description of the Geometry

Roof prisms serve additionally to revert an image in just one direction perpendicular to the folding plane. At the roof surface, the light rays undergo a cross-over. The beam aperture is divided in the centre and the two portions are transposed by reflection at the corresponding roof surfaces. Figure 4.137 shows the beam path in such a 90°-prism where the hypotenuse is substituted by a 90° roof edge. This means, in principle, a roof is a segmented surface with two sub-apertures. The green and the blue part of the collimated ray bundle are interchanged.

Inside the roof prism two reflections take place by two planes containing the roof edge, inclined by 90° to each other. If an incident ray in the roof edge plane is considered, the ray in the middle (red in colour in Figure 4.137(a))

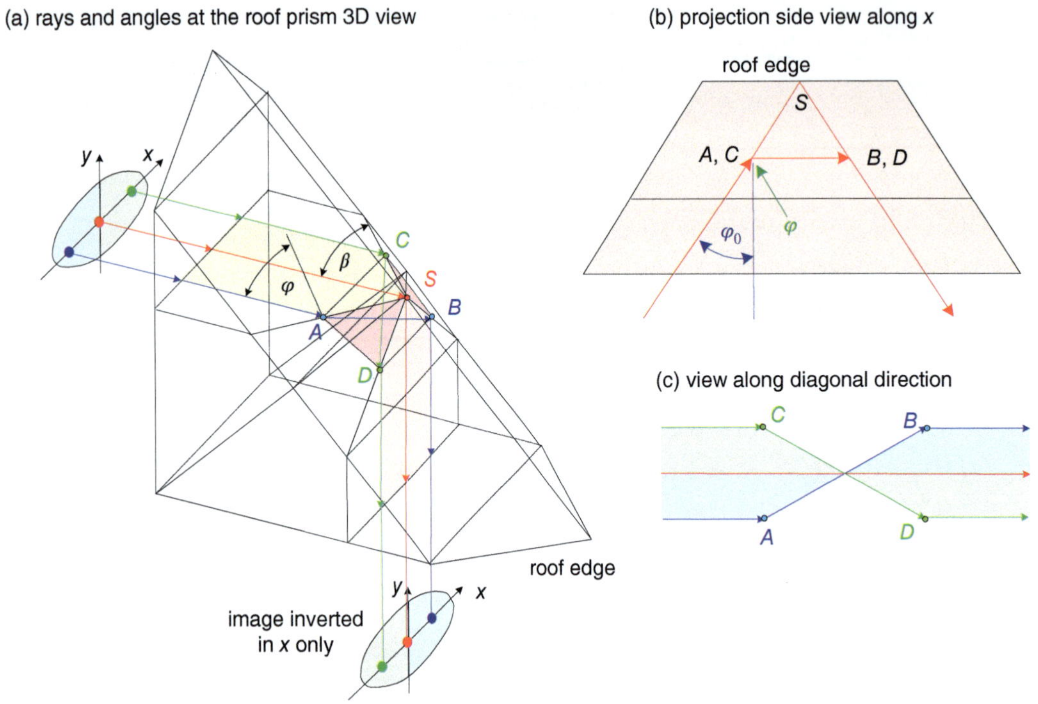

Figure 4.137 Beam path in a right-angled roof prism with (a) three-dimensional view, (b) projected side view and (c) diagonal projection.

will be reflected at point S and directed in the new direction. The two peripheral rays (red and green) strike both roof faces in points A, B and C, D respectively, and will be therefore reflected twice. They will then be deflected in the new direction.

If the incidence angle of the axial beam with the normal to roof edge is φ_0, the actual angle of the beam incident on the roof faces is given by

$$\cos\varphi = \frac{1}{\sqrt{2}} \cdot \cos\varphi_0 \tag{4.202}$$

The projection in Figure 4.137(c) shows the meaning of this angle for the reflection at the roof surfaces of a right-angled reflecting prism. For the most common case of 90° reflection at the roof edge, the resulting incidence angle amounts to 60°. Other examples are included in Table 4.8. In principle, a roof can be applied at nearly every reflecting surface of a prism.

Table 4.8 Commonly used angles φ_0 between the roof edge and the direction of the incidence beam and their respective incidence angles φ of the rays on the roof surfaces.

φ_0	15°	30°	45°	60°	75°
φ	79.4°	69.3°	60°	52.2°	47°

4.10.2.2 Dimensioning of the Size*

(*) The content of this section is quite specialized and for advanced reading. Knowledge of this section is not necessary to understand the following chapters. Beginners are recommended to skip this section.

The substitution of a reflecting surface by a roof edge makes it necessary, in any prism geometry individually, to increase the prism dimensions, which are determined in general by specific geometrical factors. For a circular beam cross-section, the separation e of the beam axis from the roof edge is given by (see Figure 4.138)

$$e = \frac{D}{2\sin\beta} = \frac{D \cdot \sqrt{1+\sin^2\varphi_0}}{2\sin\varphi_0} \tag{4.203}$$

where β is half the angle of the roof edge in a sectioning plane normal to the optical axis and D is the diameter of the beam at the entrance. Therefore

$$\tan\beta = \sin\varphi_0 \tag{4.204}$$

The geometrical factors in this consideration are illustrated in Figure 4.138.

4.10.2.3 Image Formation Problems Related to Roof Edges*

(*) The content of this section is quite specialized and for advanced reading. Knowledge of this section is not necessary to understand the following chapters. Beginners are recommended to skip this section.

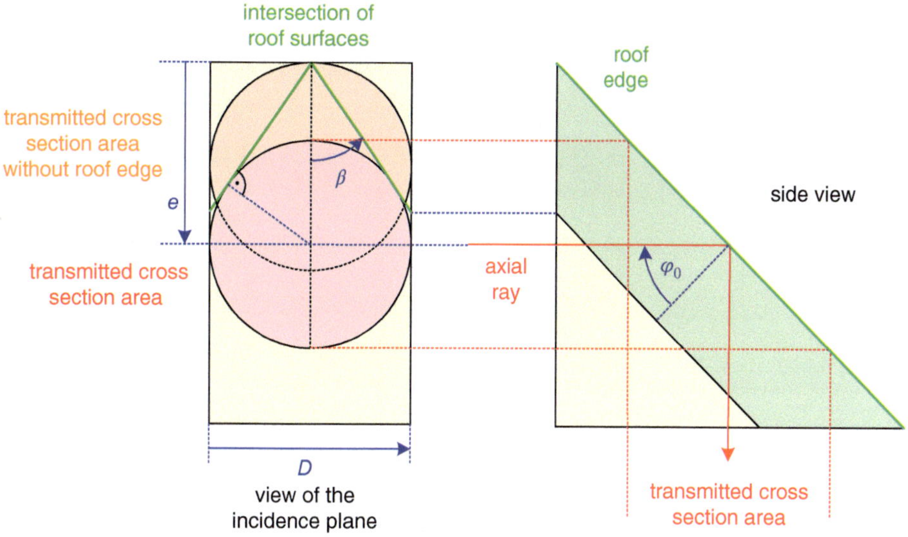

Figure 4.138 Dimensions of the diameter for a right-angled roof prism.

The increased angle of incidence in comparison with the deflection angle makes it possible in most cases to use the roof edges for total internal reflection. In prisms with more than one reflecting surface, the choice of the surface to be substituted by a roof edge is determined by the need to enable total internal reflection and to minimize the efforts required by the increased dimensions necessary in order to guarantee a free transfer through the prism.

On the other hand, total internal reflection produces an elliptic beam polarization. Since the effect is different in the two portions of the beam, the final image quality degrades. As a consequence, a perturbed interference of the pupil halves appears in the point image. Hence, it is a common practice to aluminize the roof faces or to compensate for the depolarization effect by dielectric coatings.

In addition, it should be mentioned here that a roof surface is extremely critical considering the accuracy of the 90° angle. The impact of a tolerance of this value generates a split double image, which is rather critical to observe. The required tolerance depends on the magnification and image distance, but usually in classical tube optics it is in the range of 1″.

4.10.3 Dispersion Prisms

4.10.3.1 Introduction

A dispersion prism has a base angle of α and the light does not necessarily enter the prism perpendicularly. The goal of using a dispersion prism is to split the spectral components of broadband light beam. The component separates the wavelengths by refraction at an interface, see Figure 4.139. Blue is more strongly deflected than red; for a symmetrical ray path of the middle wavelength geometry the angles are given by the formula

$$\varphi = 2 \cdot \arcsin\left(n \cdot \sin\frac{\alpha}{2}\right) - \alpha \approx (n-1) \cdot \alpha \quad (4.205)$$

By calculation of the derivative, the spectral spreading can be determined by

$$\frac{d\varphi}{d\lambda} = \frac{2\sin(\alpha/2)}{\sqrt{1-n^2\sin^2(\alpha/2)}} \frac{dn}{d\lambda} \quad (4.206)$$

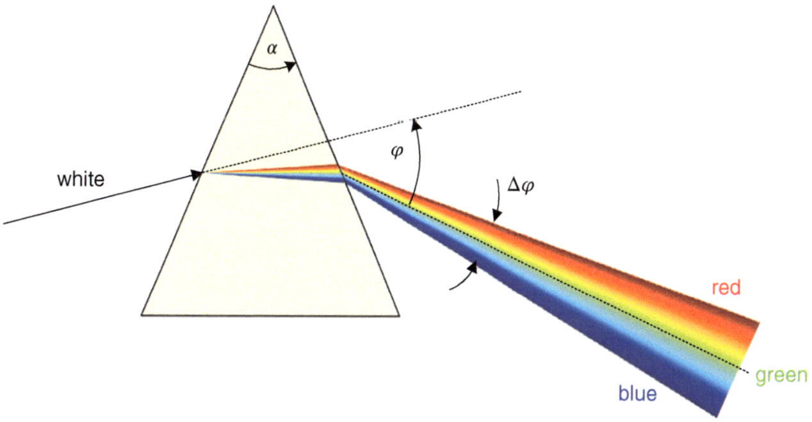

Figure 4.139 Dispersion prism, the dispersion of the colours occurs at both interfaces.

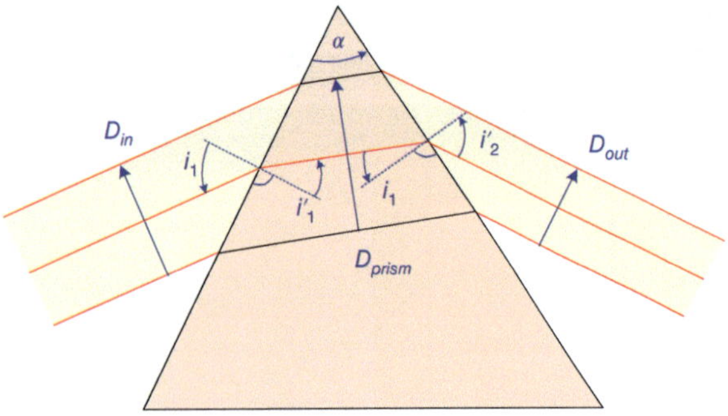

Figure 4.140 The effect of magnification by a prism with a change in the beam cross- section.

An incident beam with a finite width D_{in} is transformed in the general case of a transfer through a dispersion prism to a beam with a different diameter D_{out} if the transfer is not symmetrical. This change of the beam cross-section at the prism exit is described by the so-called prism magnification, which is defined by the ratio of the corresponding diameters, see Figure 4.140.

For a prism angle α and angle of incidence at the front surface i_1 this factor or anamorphism is given by

$$m = \frac{D_{out}}{D_{in}} = \frac{\cos i\prime_1}{\cos i_1} \cdot \frac{\cos i\prime_2}{\cos i_2} = \frac{\sqrt{1 - \frac{\sin^2 i_1}{n^2}}}{\cos i_1} \cdot \frac{\sqrt{1 - \frac{1}{n^2} \cdot \sin^2 \left[\alpha - \arcsin\left(\frac{\sin i_1}{n}\right)\right]}}{\cos\left[\alpha - \arcsin\left(\frac{\sin i_1}{n}\right)\right]} \tag{4.207}$$

The magnification is equal to 1 for a symmetrical beam transit. The scale factor m depends on the angle of incidence at the front surface, the prism angle and the index of refraction. Figure 4.141 shows this dependence of the magnification m on the angle of incidence and the prism angle α for a refractive index $n = 1.5$.

4.10.3.2 Achromatic Prism Pairs*

(*) The content of this section is quite specialized and for advanced reading. Knowledge of this section is not necessary to understand the following chapters. Beginners are recommended to skip this section.

By the use of an achromatic prism pair, the same angle of deviation φ can be achieved for two wavelengths [85]. The two prisms should be made of different materials, possessing different indices of refraction n_1 and n_2 and Abbe numbers ν_1 and ν_2. If a cemented wedge pair, such as the one shown in Figure 4.142(a), is selected for the design, the prism angles should obey the following relations in the small-angle approximation

$$\alpha_1 = \frac{\varphi \cdot \nu_1}{(n_1 - 1) \cdot (\nu_1 - \nu_2)}, \quad \alpha_2 = \frac{\varphi \cdot \nu_2}{(n_2 - 1) \cdot (\nu_2 - \nu_1)} \tag{4.208}$$

The prism basis angles lie opposite to each other and the prism with the higher ν-value made of crown glass has the larger angle. The outgoing beam now has the same angle of deviation φ for two preselected wavelengths. If this applies, e.g. for red and blue, as a secondary chromatic aberration, one obtains a slightly different angle for green. The two beams corresponding to the same angle of deviation have, however, a parallel offset which is a result of the different beam paths in the prisms. The magnitude of this off-set depends on the thickness of the prisms and the corresponding paths in glass, respectively.

If, in addition, the light deflection by the achromatic wedge has to be minimized, the incident and emergent

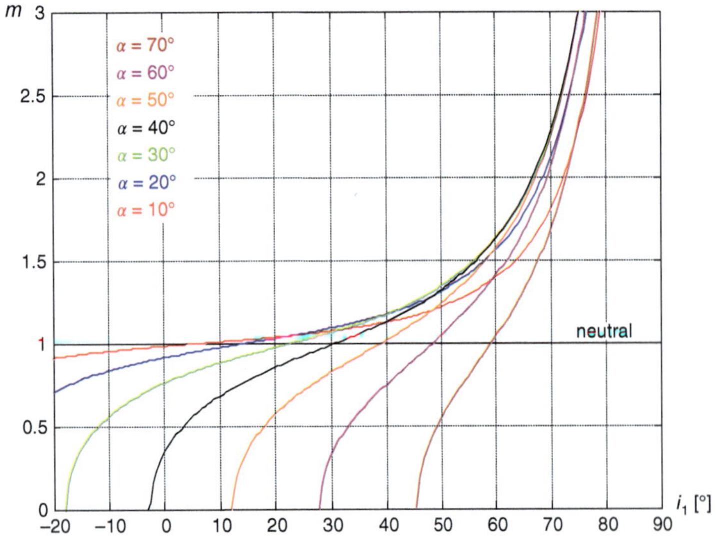

Figure 4.141 Prism magnification as a function of the prism and incidence angles for $n = 1.5$.

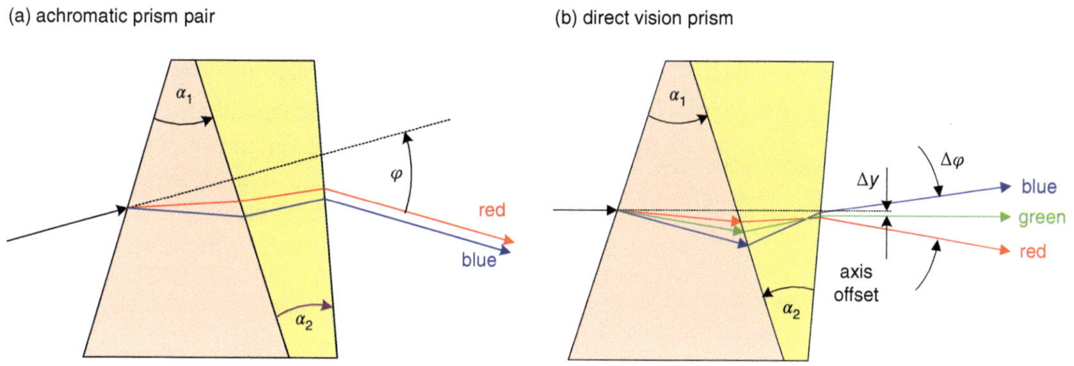

Figure 4.142 (a) Achromatic prism pair and (b) direct-view prism.

beams should propagate in a direction normal to the middle plane of the components for green. This condition leads to an additional requirement on the materials used, which can be expressed as equal values of the parameter

$$x = \frac{n_e}{n_F - n_C} \tag{4.209}$$

for the two glass types. To a good approximation this is satisfied, e.g. for the material combinations LLF3/BaF5, SK19/KF8, SF3/LaSF81, LaSK1/BaF8 (all Schott glasses).

In the case of a so-called direct-vision prism as is shown in Figure 4.142(b), the different wavelengths behind the setup emerge at different angles, but the propagation direction for the central wavelength remains unchanged and blue and red have the same small deviation with opposite sign. This makes it possible to have a linear beam path in a spectrometer without folding the optical axis for the central wavelength. It should be kept in mind, however, that even in the case of normal propagation, the central wavelength will experience a lateral beam offset Δy. The relationships between the corresponding wedge angles α_1 and α_2 of a direct-vision prism and the maximum spectral splitting $\Delta\varphi$ between the red and the blue wavelengths lying at both ends of the spectrum, are given in the thin-prism approximation by

$$\alpha_1 = -\frac{\Delta\varphi_{F'C'}}{n_1 - 1} \cdot \frac{\nu_1 \cdot \nu_2}{\nu_2 - \nu_1}, \quad \alpha_2 = +\frac{\Delta\varphi_{F'C'}}{n_2 - 1} \cdot \frac{\nu_1 \cdot \nu_2}{\nu_2 - \nu_1} \tag{4.210}$$

4.10.3.3 Anamorphic Prism Pair*

(*) The content of this section is quite specialized and for advanced reading. Knowledge of this section is not necessary to understand the following chapters. Beginners are recommended to skip this section.

Anamorphic prism pairs are used to modify the cross-section of collimated beams through the prism magnification factor m. Their most frequent application is the transformation of an elliptical beam cross section into a circular one. Figure 4.143 demonstrates the principle of this arrangement.

The conventional configuration of the anamorphotic prism pair, for the special case of beam compression, includes two identical prisms with the beam at normal entrance incidence to both of them. In this simple case, if θ is the deflection angle of the first prism, α the prism

Figure 4.143 Anamorphic prism pair.

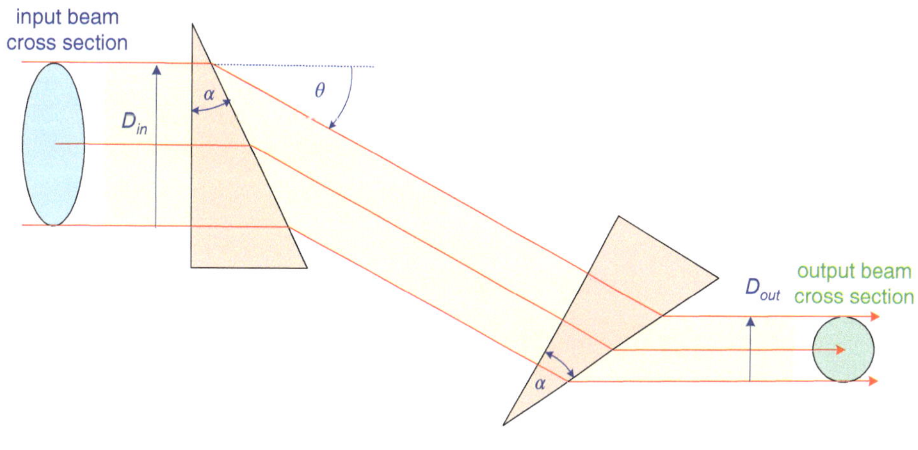

angle and n the refractive index, one obtains for the beam compression in the outgoing beam

$$\frac{D_{out}}{D_{in}} = \frac{1 - n^2\sin^2\alpha}{\cos^2\alpha} \quad (4.211)$$

This arrangement is also called the Brewster telescope. Indeed, it is useful only for monochromatic applications since otherwise rather large chromatic differences in the magnification factor will occur. As can be seen from Figure 4.143, the modification of the cross section is accompanied by increasing angles of incidence and lateral displacement of the beam axis. Both effects complicate the practical application of such systems.

4.10.3.4 Fery Prisms*

(*) The content of this section is quite specialized and for advanced reading. Knowledge of this section is not necessary to understand the following chapters. Beginners are recommended to skip this section.

The Fery prisms, named after their inventor, have curved front and back surfaces (Figure 4.144). They can be employed in a spectrometer in such a way that the refractive input and output surfaces take over the roles of a collimating and a focusing lens, respectively [86–88]. The properties of a Fery prism design for spectrometer purposes are similar to those of the Rowland grating. In particular, the limiting aberration is given by the astigmatism caused by refraction at the oblique surfaces. From a practical point of view, the Fery prisms are problematic because of difficulties related to their manufacturing and centring properties.

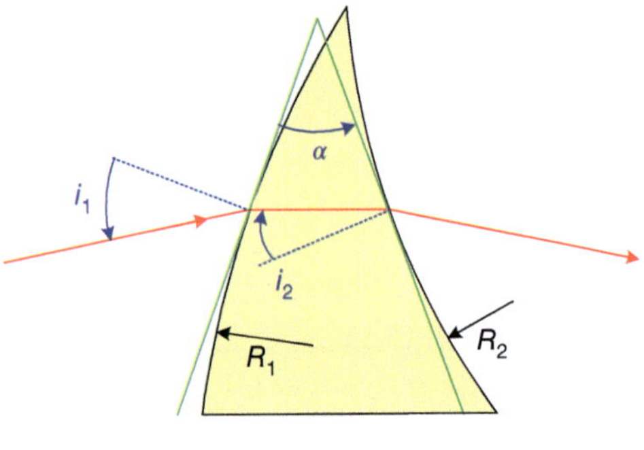

Figure 4.144 Fery prism with curved surfaces.

It is especially interesting that the third-order spherical aberrations vanish in this configuration if both surfaces have equal radii. The astigmatism can be in addition eliminated by an appropriate combination of parameters leaving the residual coma as a limiting factor. A trade-off is also necessary, however, with respect to the dispersive effect, which can require different radii. The most appropriate solutions are given by approaches in which the radii of curvature of the surfaces correspond or are close to the aplanatic case.

Fery prisms are mostly applied in spectrometer setups [89]; in comparison to the classical setups in transmittive or reflective constellation, the lens components can be removed, which simplifies the setup. This idea is sketched in Figure 4.145.

If the lens makers' formula is applied to both surfaces, the thickness of the prism is assumed to be neglectable and in object distance from the first surface is s, the image distance after the prism is given by

$$s' = \frac{R_1R_2s}{R_1R_2 - (n-1)(R_1 - R_2)s} \quad (4.212)$$

In the special case of a symmetric arrangement with $R = R_1 = -R_2$ we get as expected $s' = R/2/(n-1)$. It should further be noted that for the axis ray the dispersion is the same as in the case of a classical prism with angle α. If larger slit lengths or larger apertures are used, it may be a good idea to correct for the aberrations by making the surfaces freeformed.

4.10.4 Risley Prisms*

(*) The content of this section is quite specialized and for advanced reading. Knowledge of this section is not necessary to understand the following chapters. Beginners are recommended to skip this section.

By combining two thin prismatic wedges in series, a beam can be oriented in any direction within a narrow

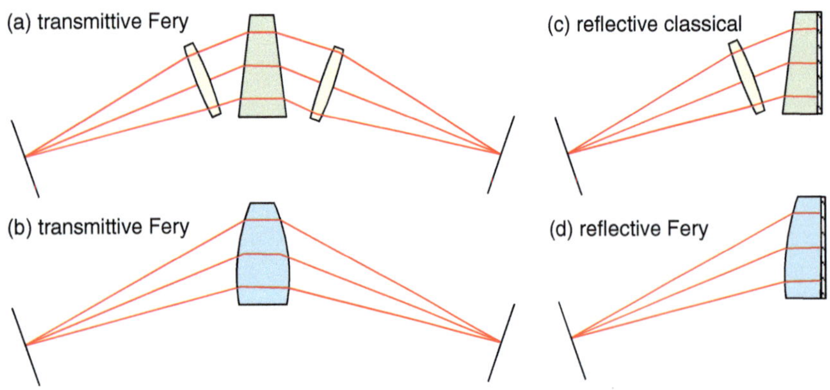

Figure 4.145 Transmittive classical (a) and reflective classical (c) as well as corresponding Fery prism solutions (b) and (d) for spectrometer.

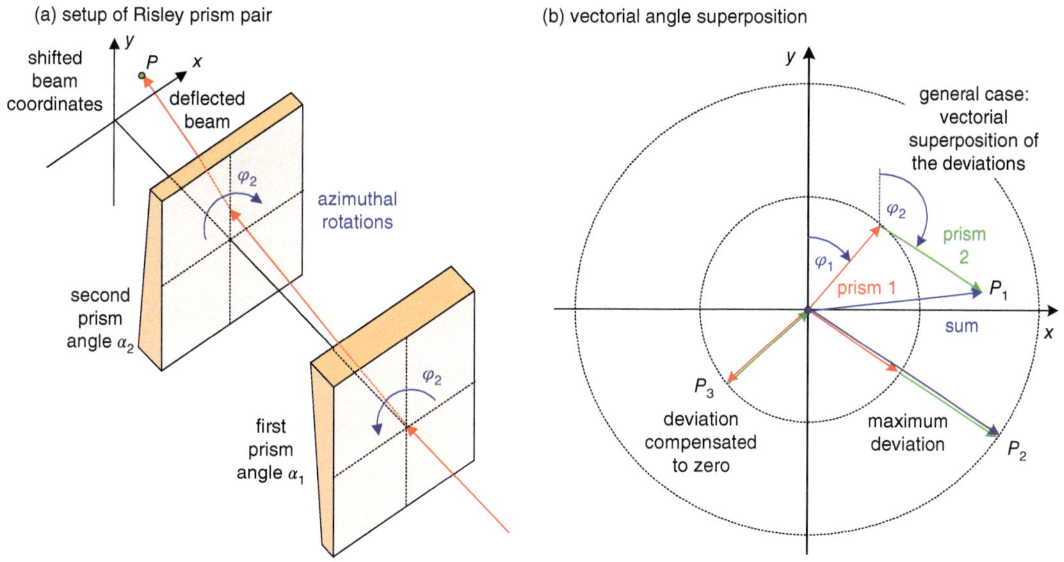

Figure 4.146 Risley Prism (a) arrangement in the rotational scanner and (b) vectorial superposition of the deviations for three combinations.

cone by independently azimuthally rotating the prisms [90–95]. Each prism alone produces a constant angular deviation of the propagation direction, the azimuth of which depends on the rotation position of the prism. The vectorial superposition of the two deviations allows complete scanning of a circular area with a radius twice as great as the angular deviation to be achieved by pure azimuthal rotation of the prisms. Synchronized rotation of the prisms in opposite directions results in linear scanning. Figure 4.146(a) illustrates this so-called Risley prism setup. Figure 4.146(b) shows schematically in a linearized model the vectorial superposition of the two deviations, created by the azimuthal rotation angles φ_1 and φ_2.

If the deviation angles of the prisms are large, depending on the position of the first prism, one obtains different intersection points and as a consequence slightly different deviation angles from the second prism. The linearization and the symmetrical ray transfer is then slightly perturbed. In the vectorial presentation this reflects the fact that the length of the second vector depends on the rotation position of the first vector. In this case one cannot cover the whole area and the exact relationship with the rotation azimuths of the prisms is no longer linear. For more rigorous modelling there are different setups and several approaches discussed in the literature.

From the viewpoint of degrees of freedom of the setup, the following options must be kept in mind.

- Thin vs thick prisms.
- Small vs large wedge angles.
- Same or different materials.
- Small vs larger air gap between the prisms.
- Forcing at least one prism surface perpendicular to the axis or allowing for arbitrary angles.

In principle, four different configurations can be considered. If it is assumed that one prism surface is always perpendicular to the axis, which is denoted by '1', the other deviating surface is '2' respectively. Then according to Figure 4.147, we have the configurations [90]

1) Group A: 21–12 12–12
2) Group B: 21–21 12–21

All final deflections of the four setups in the figure are the same. But as is indicated by the ray paths, in the group B setups, the second prism is used by smaller angles, while in case A the incidence has a certain bias that is not beneficial for the residual aberrations. Therefore, the geometries of case B are preferred.

An exact inversion formula for the most general case doesn't exist, but with some approximations it is possible

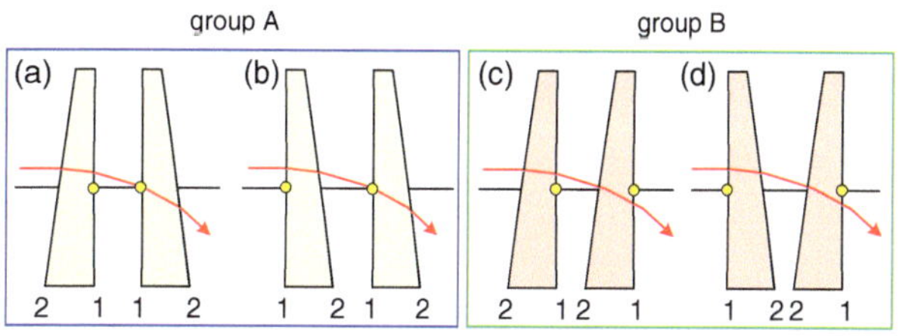

Figure 4.147 Different basic configurations of Risley prisms, the yellow dots indicate the prism surfaces perpendicular to the optical axis, they are denoted by '1'.

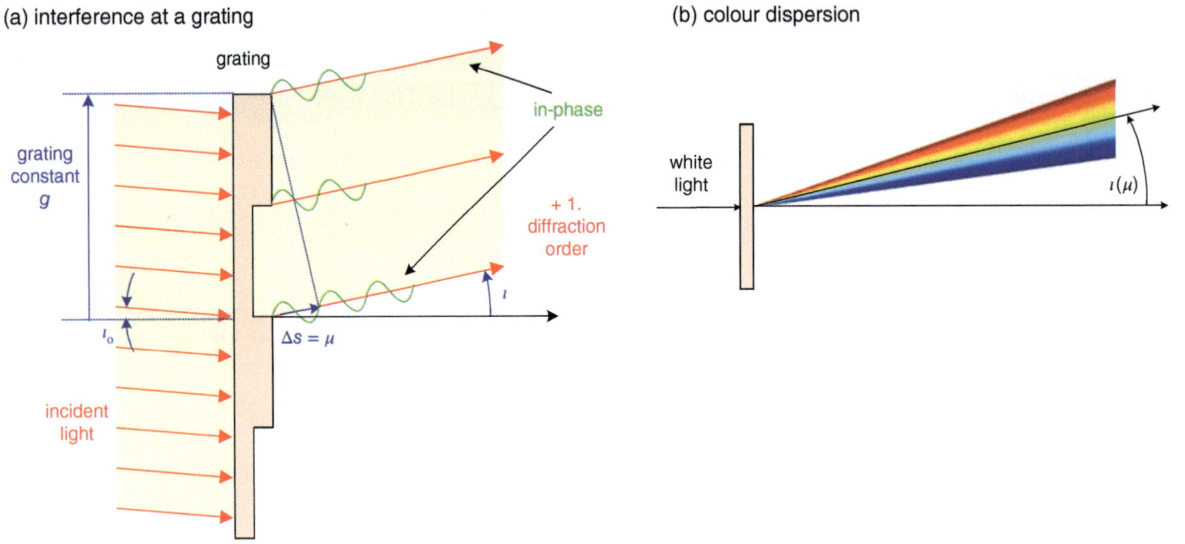

Figure 4.148 Transmittive dispersion grating with (a) path length difference Δs and (b) wavelength dependent deviation $\theta(\lambda)$.

to express the final coordinates or far-field angles as a function of the rotating angles as well as the required system data analytically. There exists a lot of analytical lengthy formulae in the literature for the various configurations. The modelling has various possibilities too, as follows.

- Paraxial vs real raytrace.
- Considering third order approximation for the non-linear effects.
- Neglecting vs considering the finite thicknesses of prisms and the air gap.
- Considering the aimed coordinate point in the far- or in the near-field.
- Direct forward vs inverse calculation of aiming point from the rotation angles.

According to Figure 4.147(b) it is seen that only if the deflection of the two prisms are exactly compensated can the origin point be achieved. This means that some tolerances like wrong wedge angles, mounting errors or other alignment errors give rise to problems in the covering of the full area of the scan.

4.11 Diffractive Elements

4.11.1 Linear Gratings

4.11.1.1 Introduction

Components, which are working by diffraction instead of refraction, are called diffractive optical elements (DOE). A grating disperses a wavelength very similar to a prism, but stronger and with the opposite sign. The grating equation reads for an incoming angle θ_0

$$\sin\theta - \sin\theta_0 = m/g \cdot \lambda \tag{4.213}$$

where g is the period width of the periodic structure and m is the diffraction order [96–98]. This is demonstrated in Figure 4.148. For this device, the red light or the longer wavelength is deviated by larger angles. The energy and the intensity in the various diffraction orders changes with m. The constructive diffraction orders are obtained if the optical path differences Δs is an integer value of the wavelength.

The effect of a dispersion grating on the light can be understood as a combined effect of the slit diffraction with slit width g, describing the impact and phase variation of a single period d, and the interference by the superposition of N periods and a total length $L = N\,d$. If a certain dead zone exists between the free parts, the period is larger than the slit width $d > g$. The corresponding formula for the far-field intensity reads

$$I = N^2 \cdot g^2 \cdot \left[\frac{\sin\left(\frac{\pi\,\theta g}{\lambda}\right)}{\left(\frac{\pi\,\theta g}{\lambda}\right)} \cdot \frac{\sin\left(\frac{N\pi\,\theta d}{\lambda}\right)}{N \cdot \sin\left(\frac{\pi\,\theta d}{\lambda}\right)} \right]^2 \tag{4.214}$$

and is illustrated in Figure 4.149. The first factor in the brackets describes the slit diffraction, the second the interference function as a consequence of the periodicity. If the individual periods have a special linear phase gradient, the relative position of both functions can be shifted. The slit diffraction (shown in blue in the figure) modulates the interference function (in red).

If the second factor in the grating Eq. (4.214) is considered separately, it can be seen that a large number of periods N of the interference function create the constructive maximum of intensity quite sharply. This is seen in Figure 4.150

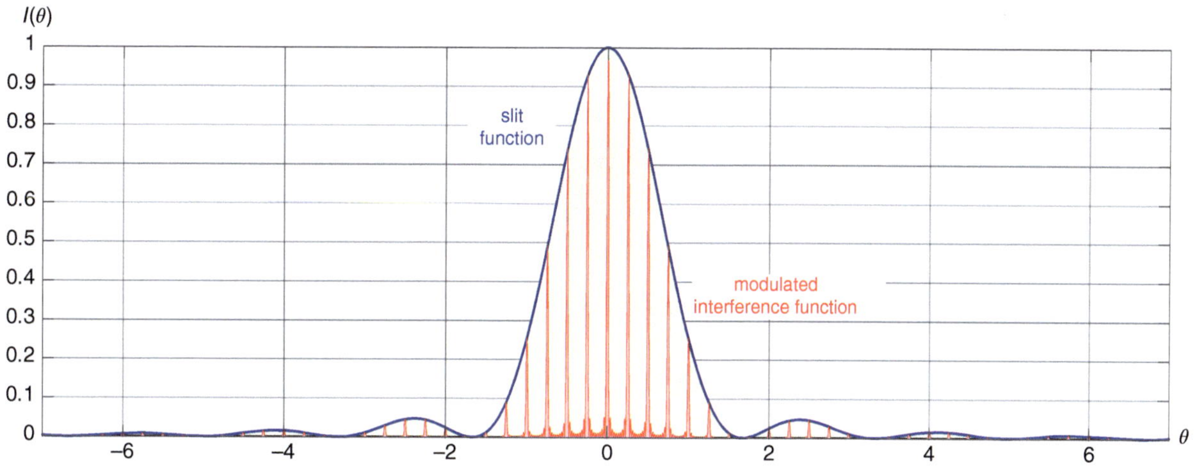

Figure 4.149 Intensity distribution of a grating diffraction as a function of the scaled angle θ. The units of the horizontal axis are arbitrary.

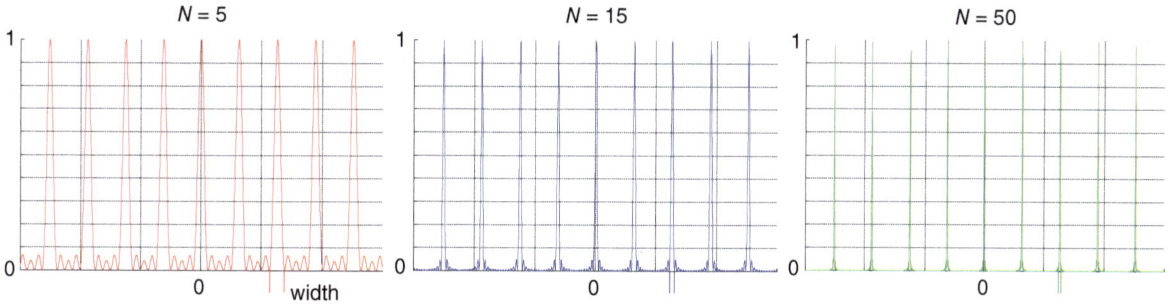

Figure 4.150 Sharpness of the diffraction orders as a function of the number of periods N, the width of every order shrinks with growing number N.

for illustration for several values of N. The interpretation of this physical fact is that the simple ideal grating equation, which predicts only one sharp direction for the location of the diffraction orders, is only valid if an infinity number of periods is involved.

4.11.1.2 Blazed Grating

A special setup is a blazed grating. If the blaze condition is fulfilled, the overlay of the two functional terms in the grating intensity distribution are synchronized in a way that the slit function suppresses all the interference diffraction orders but one. In addition, a linear phase ramp in every period shifts the interference function relative to the slit function. By adjustment of the ramp angle, the zeros of the slit function can be exactly located at the positions of the interference function maxima.

In this basic case, the geometry looks as shown in Figure 4.151. β is the slope angle of one flank, $g = d$ the period length, h_{max} the height difference of the relief at the step and θ the deflection angle of a perpendicular incident ray or wave.

The diffraction order m is defined as the phase difference between the edge rays of one period, scaled in units of the local wavelength, see Figure 4.151(b). It is assumed here that n_{len} is the refractive index of the grating substrate and n_{med} of the surrounding embedding medium. A simple saw tooth profile with a periodic structure generates a deflection in the direction θ. The geometrical and physical relationships for this case are a complete length L of the grating with N grooves in total

$$L = N \cdot g \tag{4.215}$$

with the optical path difference between the rays at the step in the surrounding medium

$$\Delta_{OPD} = (n_{len} - n_{med}) \cdot h_{\max} \cdot \cos\theta \tag{4.216}$$

Phase difference at the step

$$\Phi_{\max} = 2\pi \frac{\Delta_{OPD}}{\lambda} n_{med} \cdot \cos\theta \tag{4.217}$$

Angle of the linear groove face (small)

$$\tan\beta \approx \beta = \frac{h_{\max}}{g} \tag{4.218}$$

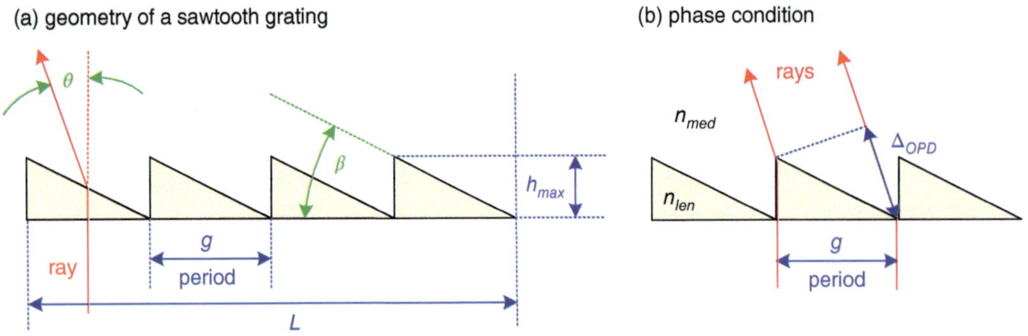

Figure 4.151 Blazed grating (a) terms of geometry and (b) phase condition.

If the blaze condition with order m has to be fulfilled, the formulas are in particular

$$\Delta_{OPD}^{(blaze)} = \frac{m \cdot \lambda}{n_{med}} \tag{4.219}$$

$$h_{\max}^{(blaze)} = \frac{\lambda \cdot m}{n_{med} \cdot (n_{len} - n_{med}) \cdot \cos\theta} \tag{4.220}$$

$$\Phi_{\max}^{(blaze)} = 2\pi \cdot \frac{\lambda \cdot m}{n_{med}} \tag{4.221}$$

$$\beta^{(blaze)} = \frac{h_{\max}^{(blaze)}}{g} = \frac{\lambda \cdot m}{g \cdot n_{med} \cdot (n_{len} - n_{med}) \cdot \cos\theta} \tag{4.222}$$

In the case of a perfectly realized blaze condition for one design wavelength, there is no difference observed in the functionality between a refractive and a diffractive element, one period deviates the light like a corresponding prism. Due to this more physical consideration, a rigid separation between refractive and diffractive components is meaningless in the case of coherent light. If the light has a lateral coherence length of finite size and not all of the pitches can interfere, the behaviour becomes different.

The amplitude transfer function t can be decomposed into the function of one elementary cell (pitch) and the periodic repetition. The periodicity is given by a sequence of delta functions, limited by the overall length L of the grating. Both functions must be convolved.

$$t(x) = \mathrm{rect}\left(\frac{x}{g}\right) \cdot e^{2\pi \cdot i \cdot \beta \cdot x \cdot \frac{n_{len} - n_{med}}{\lambda}} * \mathrm{rect}\left(\frac{x}{L}\right) \cdot \sum_j \delta(x - j \cdot g) \tag{4.223}$$

This means that the cell and the interference can be separated as

$$t(x) = t_{\text{cell}}(x) * t_{\text{inter}}(x) \tag{4.224}$$

The angle distribution or the spatial frequency function is obtained as the Fourier transform of $t(x)$. Due to the well-known rules and concrete transform functions, we get for $\cos(\theta) \approx 1$

$$t(\nu) = \mathrm{sinc}\left[m \cdot \left(\nu \cdot \frac{\lambda}{\beta \cdot (n_{len} - n_{med})} - 1\right)\right] \cdot \sum_j \mathrm{sinc}\left[L \cdot \left(\nu - \frac{j}{m} \cdot \frac{\beta \cdot (n_{len} - n_{med})}{\lambda}\right)\right] \tag{4.225}$$

with the spatial frequency ν and the sinc function defined as

$$\mathrm{sinc}(x) = \frac{\sin(\pi x)}{\pi x} \tag{4.226}$$

According to the Fourier optical basics, the deflection angle is linearly related to the spatial frequency by the equation

$$\theta = \nu \cdot \lambda \tag{4.227}$$

An additional constant phase factor is suppressed in this formula; it has no impact on the result in the case of a simple one-period grating. The interpretation of the formula means as follows.

- The second factor is a comb of sinc functions. This describes the interference effect. The larger the number of grooves $N = L/g$, the more sharply peaked is this comb.
- The total transmission (and also the field after the grating) is then the comb modulated by the cell function (first factor), which is a sinc function too.
- The width of the envelope is governed by the order m, the slope angle β can be used to shift the envelope over the comb.
- By a clever synchronization of m, g and β the blaze case is obtained: the cell transfer function works as an envelope and suppresses all orders but one.

In Figure 4.152 this principle is shown for a fulfilled blaze condition. Part (a) illustrates a perfect blaze case, in (b) it is shown that the cell function has in addition an imaginary part.

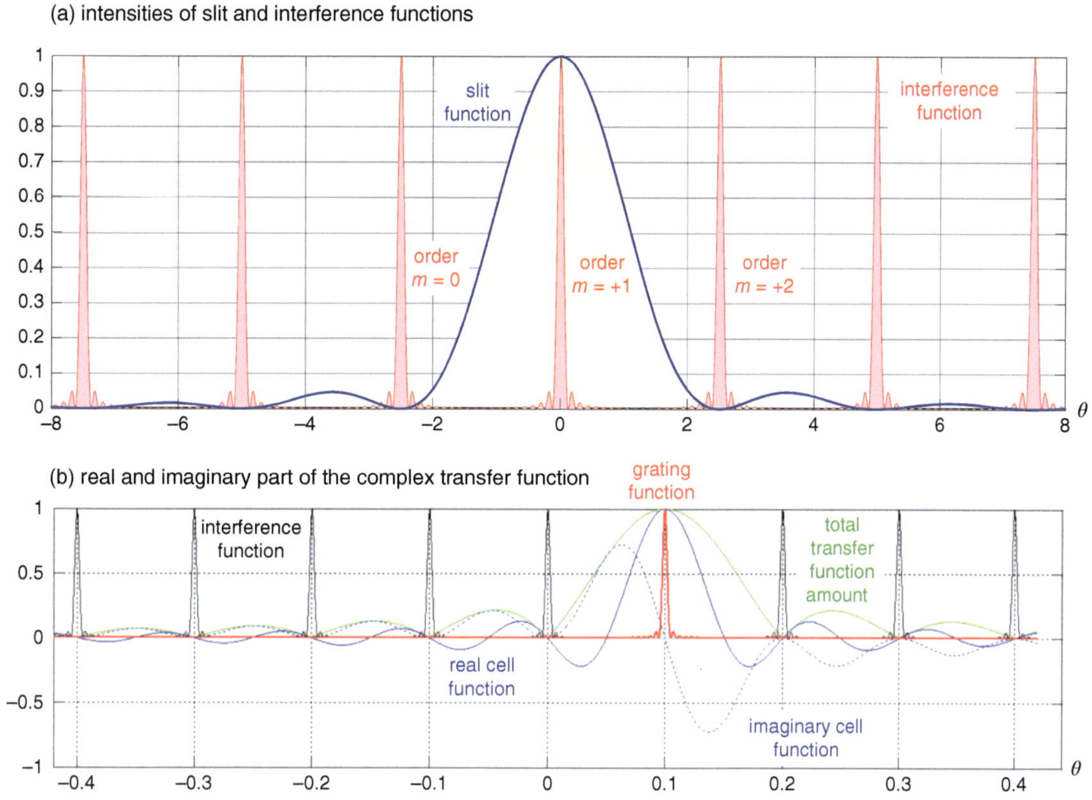

Figure 4.152 (a) Intensities of slit/cell and interference function and (b) cell interference and grating function for a perfect blazed grating. The scales of the angle/frequency axis are different. The horizontal axes have arbitrary scaling here but are proportional to the deflection angle. The units of the horizontal axis are arbitrary.

In general, for a diffractive lens, the angular spreading and therefore the focal point positions are only determined by the period length p and the order m. This first component of the final distribution is the interference function. The relative energy in every focal point is determined by the cell structure only and can be controlled by the phase height or height distribution (for example linear in case of blaze) or a duty cycle as the ratio between feature width inside one cell and the period. Therefore, the second factor of the total distribution is governed by the Fourier transform of the cell profile. This so-called envelope function controls the distribution of energy in the various orders of the interference function.

It should be noted here that the blaze condition is fulfilled only for one selected diffraction order m, one incidence angle and one wavelength.

4.11.2 General Diffractive Elements

4.11.2.1 Introduction

If the basic idea of a grating is used at a surface inside an optical system, the grating equation governs the direction of the outgoing ray. Furthermore, it is possible to orient and modify the grating constant as a function of the spatial position, which means that the grating constant is not fixed but a local parameter $g(x,y)$. With these additional degrees of freedom, monochromatic aberration corrections are possible similar to an asphere.

In recent years, the use of diffractive optical components in optical systems has grown rapidly. The development of physical understanding and the technology to create elements of this type now allows their use in industrial products. In principle, from a simple geometrical point of view, for a single wavelength, diffractive elements have a similar potential in the correction of optical systems as aspheres. It is possible to generate a desired wavefront as the outcome of an element. But according to their working principle, they show a completely different spectral behaviour in comparison with refractive components. Therefore, one benefit in using diffractive components is the possibility of correcting a system for chromatic aberrations, especially in those ranges of the spectrum where only a few transparent materials are available to achromatize. In particular, in the infrared range, the structures of the diffractive elements are coarse and therefore easy to fabricate and the lack of suitable materials to achromatize leads to the widespread use of diffractive elements.

But apart from these benefits in generating degrees of freedom for the correction of the aberrations, there are some drawbacks in the practical use of diffractive elements. The ideal properties, as described below in detail, are valid for one object point or field angle and one wavelength only. If a broad spectrum is used and a finite size of the field of view is taken into account, there are always the following problems to consider.

- Light in unwanted diffraction orders. This stray light is unavoidable, since the ideal blaze condition is only valid for one wavelength and one incidence angle. This effect is discussed under the aspect diffraction efficiency later.
- Stray light coming from the finite range of the incidence angles and the interaction with the non-optically used edges, which are also illuminated by light.
- Stray light that occurs from tolerances of the exact geometry of the complicated micro-structure, non-sharp edges and so on.

The theory and practice of diffractive optics is a complicated field of research and development. There is a large number of publications about the various details and a number of good books dealing with this topic.

Only a basic understanding of the principle and use in imaging optical systems will be presented in this book. The topics that will definitely not be covered here are the following.

- The application of diffractive elements for illumination and beam shaping.
- The technology for manufacturing diffractive elements.
- The rigorous simulation and modelling of diffractive elements by electromagnetic methods.
- The calculation and optimization of real surface topology from the optical requirements.

For these special aspects, the reader is referred to the literature [99–103].

A general diffractive element is a special component very similar to a Fresnel lens, but now with period size in the order of the wavelength. In a simple picture that helps to understand the development of a diffractive element, let's consider a quite thick aspherical lens, as is depicted in Figure 4.153 as a brown line. In the figure only half of the cross section for $r > 0$ is drawn. An asphere of this large thickness is rather bulky and it is desirable to have a component with less weight. The main function of the aspherical contour for a single ray is the slope in the intersection point, it governs the new ray direction by the law of refraction. If now the contour $z(r)$ is simply reduced or wrapped on a component of the thickness $h_{2\pi q}$, see Figure 4.153, the slope at every value of r is identical and the new ray direction for rays coming parallel to the axis is the same. The only difference is a shift of the point of ray bending from P to P' as is seen in the figure. Therefore, the outgoing ray is shifted parallel, and the larger the focal length, the smaller this error. If it can be neglected, the component has nearly the same functionality as before. This idea is identical to a Fresnel lens as described in Section 4.8.2. The thickness $h_{2\pi q}$ is in a certain sense arbitrary. For a later explained argument, it should be in the best case a q times multiple of the wavelength. If $q \gg 1$ and the lens is still of macroscopic thickness, a Fresnel lens or kinoform is obtained, which works purely refractively.

If the thickness now becomes smaller and smaller, according to the slope of the sag $z(r)$ the lateral dimensions also shrink. If they reach the order of magnitude of the wavelength, a diffractive element is obtained. Since

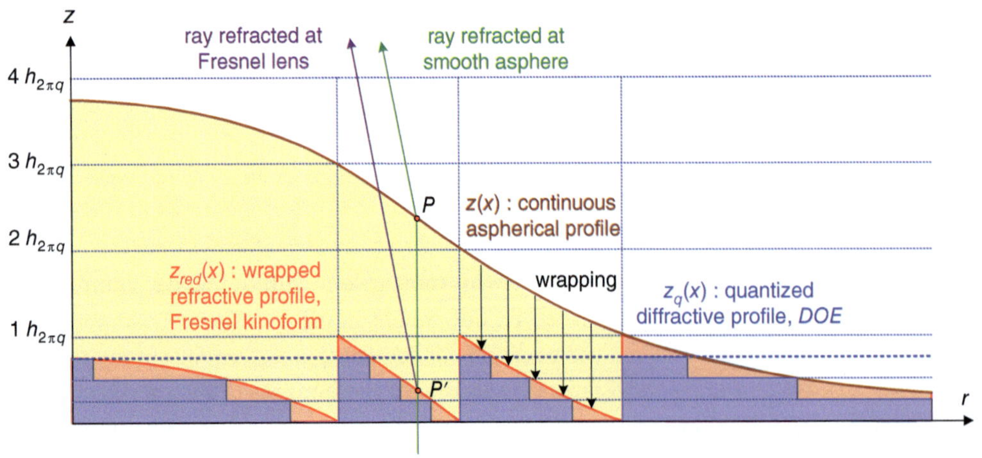

Figure 4.153 Transition from a thick lens over a Fresnel lens to a diffractive element.

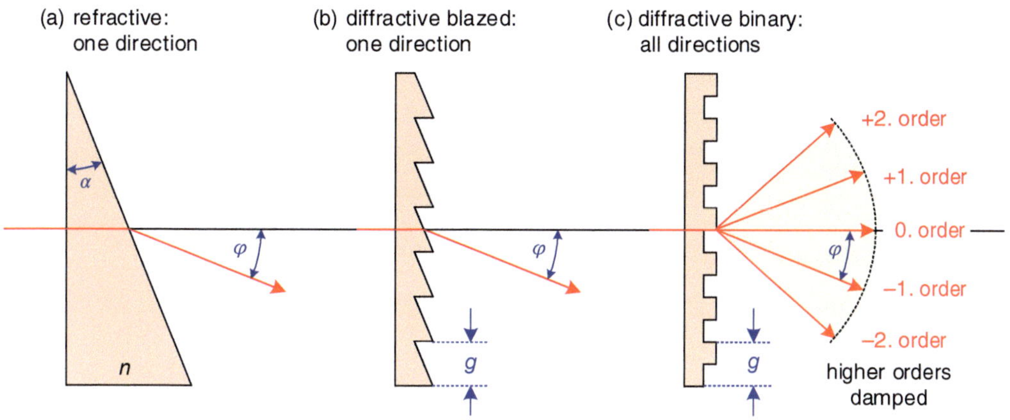

Figure 4.154 Prism, blazed grating or kinoform and binary diffractive element.

the slope is mostly smaller near the optical axis, the lateral structures are coarser here and in the intermediate situation it may happen that the zones near the axis are refractive, while the outer ones work diffractively. If the above-mentioned condition is fulfilled, the zones are always locally blazed and there is a smooth transition between a refractive and diffractive function. In the case of very small structures, it is technologically hard to realize the curved slopes on a very short scale. Therefore, in practice, often the flanks are made linear and, in the extreme case, the smooth contour curves segments are realized by step-like binary parts as indicated in the figure in blue. This structure can be fabricated by lithographic means. Unfortunately, the blazing condition is violated in this case and higher and antisymmetric orders are generated too. This means that the efficiency in the desired order is reduced. These three cases are depicted in Figure 4.154.

The basic physical principle for describing the function of a grating is diffraction. A periodic structure with very small dimensions diffracts the light, depending on the concrete structure of this elementary cell. Furthermore, if light from many cells is superposed, interference is observed.

A similar approach to the blazed grating can be realized with circular symmetry. In this case a focusing diffractive lens is obtained, as shown in Figure 4.155. Now the grating period is not constant and is changed radially according to the following equation in discrete zones.

$$r_m = \sqrt{2\pi f \lambda m}\ , \quad \sin\theta_m = \frac{\lambda}{r_{m+1} - r_m}, \quad \sin\psi_m = \frac{\sin\theta_m}{n} \tag{4.228}$$

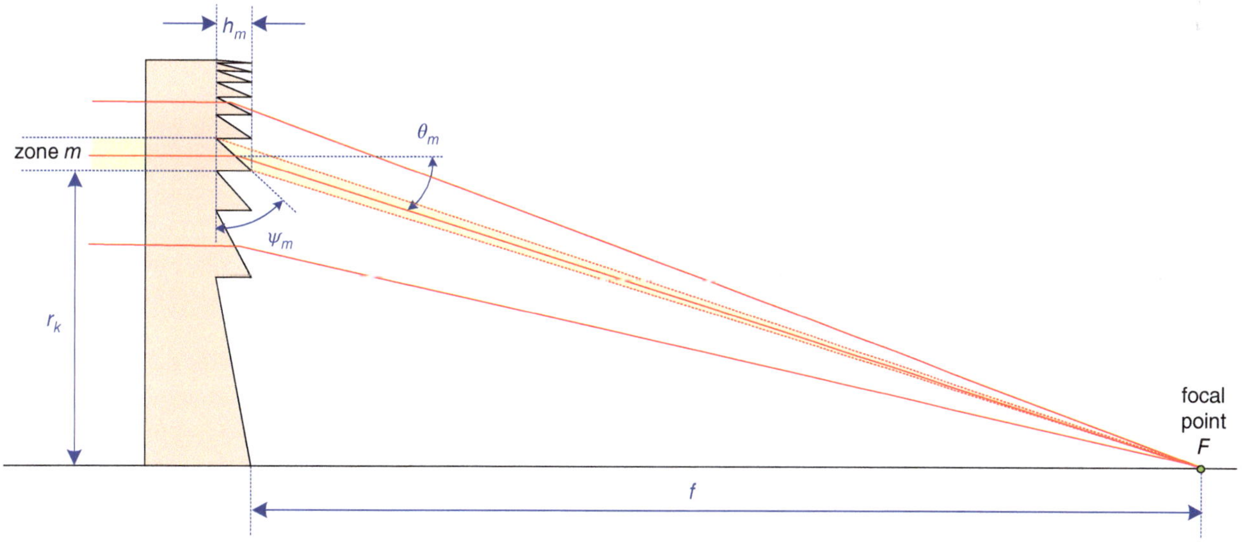

Figure 4.155 Blazed circular symmetric diffractive element.

with the ring radii r_m, the deflection angle θ_m, the refractive index n and the surface ring slope angle ψ_m. f is the focal length of the lens.

According to the realization of the micro-structured surfaces, there are several known types of diffractive elements. The most important ones are:

- Fresnel zone plates (transmission or phase zones);
- binary diffractive elements;
- computer generated diffractive elements;
- blazed diffractive elements.

If a diffractive structure is used under blaze conditions, all the energy is directed into the selected order. Due to the fact that the blaze condition contains the wavelength and the incidence angle, this approach is only valid for certain values of these parameters, in reality mainly for one incidence and one wavelength. If spectral broader light is used, the efficiency of the energy found in the intended direction governed by the grating equation is reduced for deviating wavelengths. This means that the behaviour of the diffractive elements is strongly dispersive. Figure 4.156 shows the change of efficiency in the various diffraction orders with λ. Note the inverted representation of the horizontal wavelength axis.

From the viewpoint of lens design, a diffractive element offers a huge number of degrees of freedom to redirect the light, very similar to an asphere or freeform. A major difference in comparing a DOE with a smooth surface is the dispersion behaviour. The wavelength dependence of the local grating equation is much stronger and has the opposite sign in comparison to the material dispersion for the refraction. Figure 4.157 compares the degrees of freedom for three different types of components.

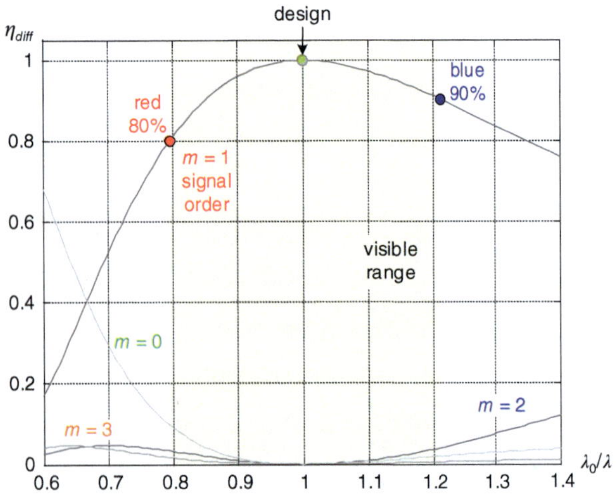

Figure 4.156 Variation of diffraction efficiency with wavelength.

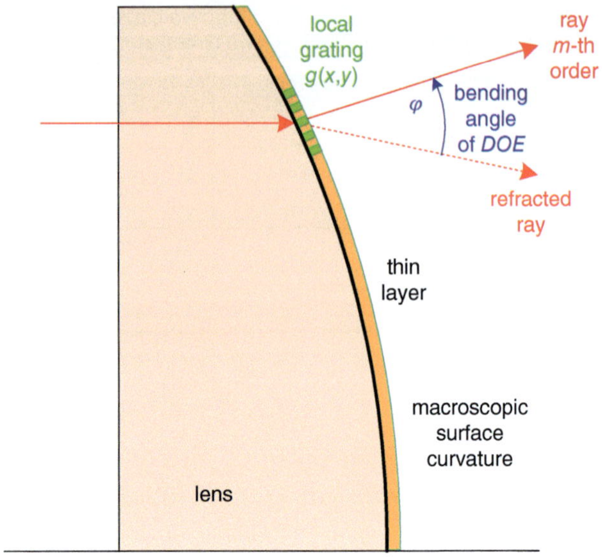

Figure 4.157 Principle of the ray bending locally at a diffractive surface as the overlay of substrate slope and local grating groove density.

- Spherical refractive lens.
  There are two parameters determining the ray bending, the refractive index and the radius of curvature R. If ray bending at a certain height r is requested, for every other radius r the value is fixed.
- Aspherical refractive lens.
  There are several aspherical constants, which can be used to change the surface contour. The local slope $dz.(r)/dr$ governs the ray direction at the height r. Since the sag cross section curve $z(r)$ is uniquely related to the slope as the derivative, the point of ray bending cannot be selected independent of the direction of the outgoing ray. In a certain sense the ray direction for N incoming rays can be selected arbitrarily, if $N+1$ is the number of aspherical coefficients.
- Diffractive structure on an aspherical substrate.
  Very similar to the above. The points of ray bending can be requested at $N+1$ locations arbitrary. But the new ray direction is now given by the slope of the substrate $dz.(r)/dr$ and in addition by the local grating constant $g(r)$. This means that the deflection point and the ray direction are decoupled and can be requested independently from each other. Therefore, a diffractive element with aspherical shape of the substrate offers the maximum degree of freedom to shape the light. What is nevertheless a problem is the assumption of the grating equation: that a very large number of grooves must be illuminated to get sharp diffraction orders and therefore ray directions (Figure 4.158).

The realization of an arbitrary ray bending can be done by smooth surfaces with an asphere or by a micro-structured

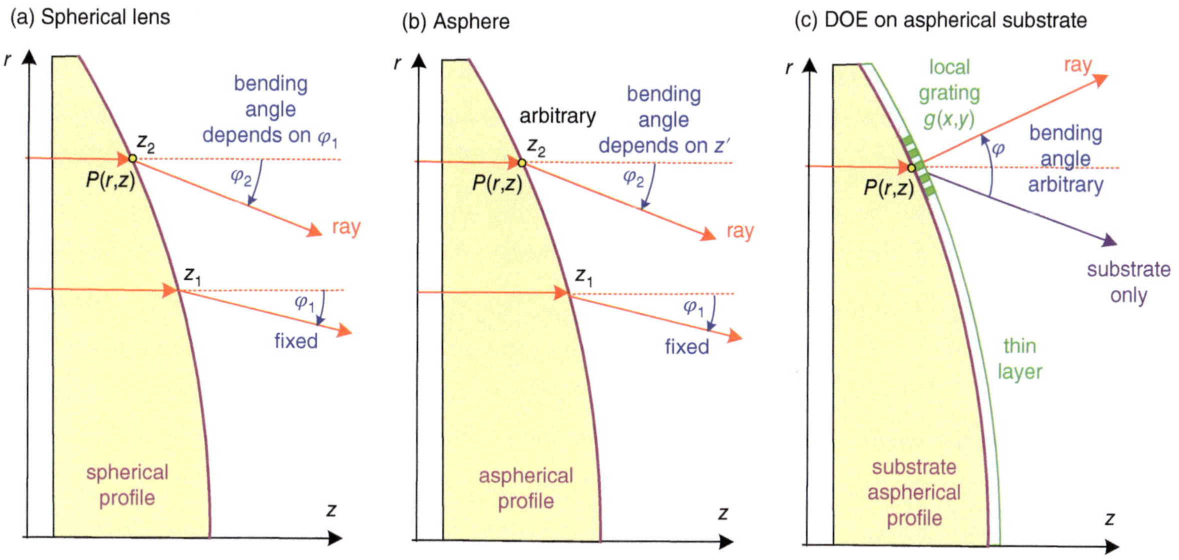

Figure 4.158 Comparison of the degrees of freedom to deflect a ray by (a) a spherical lens, (b) an asphere and (c) a diffractive structure onto an aspherical substrate.

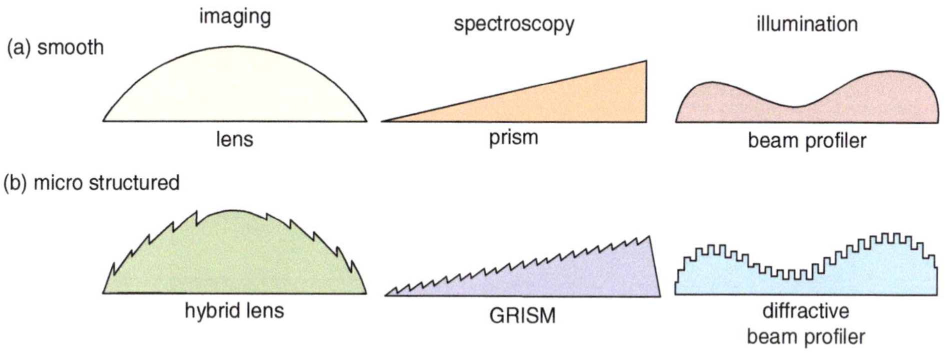

Figure 4.159 Components with (a) smooth and (b) micro-structured surfaces in comparison.

surface such as a DOE. This is illustrated in Figure 4.159. Lenses, prisms or more complicated components can be realized by this approach.

4.11.2.2 Discretization and Quantization*

(*) The content of this section is quite specialized and for advanced reading. Knowledge of this section is not necessary to understand the following chapters. Beginners are recommended to skip this section.

Diffractive elements can be realized as phase or as amplitude structures. From the viewpoint of signal transfer, phase elements are more attractive, because they are in principle lossless. If the manufacturing side is considered, binary structures can be fabricated by lithographic means quite easily. In the case of an amplitude grating, the transmission changes then between 0 and 1; in the case of a phase grating the phase steps are 0 or π. In general, a relation of the form

$$T_{Amp} = \frac{1 + T_{Phas}}{2} \tag{4.229}$$

can be established. This means, both types of DOEs are equivalent, but if the Fourier transform of this equation is calculated

$$\hat{F}[T_{Amp}] = \frac{1}{2}\hat{F}[T_{Phas}] + \frac{1}{2} \cdot \delta(\nu) \tag{4.230}$$

it is seen that there are in the far-field two major differences:

- the overall power is reduced by 50% in case of the amplitude grating, this is well known;
- in the case of an amplitude binary grating, there is an additional zeroth order term, which is described here by the δ-function.

Binary gratings have only real values of the transmission function $T(x)$, therefore the far-field is point symmetric

$$T(\nu_x, \nu_y) = T^*(-\nu_x, -\nu_y) \tag{4.231}$$

and as a consequence, arbitrary far-field distributions cannot be produced.

Binary gratings or DOEs have therefore a 2π quantization in phase, these steps correspond to the classical Fresnel zones. The steps are abrupt, they cannot be fabricated perfectly, finite slopes of the steps are one of the practical problems of binary DOEs. It is also possible to select a higher order of steps, so jumps in the phase of $2\pi m$ with an integer value for p are possible. These so-called super zones have a smaller efficiency due to the higher order. One of the problems in practice is then the very small grating periods in the range of larger slopes near to the boundary of a lens-like system.

An alternative scheme uses constant lateral sizes to avoid very small feature sizes at the boundary of the component. As a consequence, in the centre for a lens-like structure, the phase jumps are very small. The phase jumps then do not follow the 2π discretization scheme and have different properties from classical binary systems. These opportunities are sketched in Figure 4.160. Figure 4.161 shows the wrapping scheme, which couples feature size and phase steps.

4.11.2.3 Types of DOE

There are a large number of different types of diffractive component. They differ in their physical structure and in their technical realization. In the following, only a few of the most important versions are listed.

- Transmission zone plates according to Fresnel. Concentric ring zones, which have different transmittance and select the Fresnel diffraction zones.
- Phase zone plate according to Fresnel.
  Concentric ring zones, which have different phase delays.
- Interferometric zone plates.
  Zone plates, which are fabricated by interferometric methods, the zones have a smooth variation and no sharp edges.
- Binary diffractive elements.
  Realization of a diffractive component by a lithographic fabrication process. They have quantization of the height levels.
- Holographic diffractive elements.
  The diffractive structure is manufactured by a holographic setup, they can be generated by a variation of the refractive index or by a lithographic process.

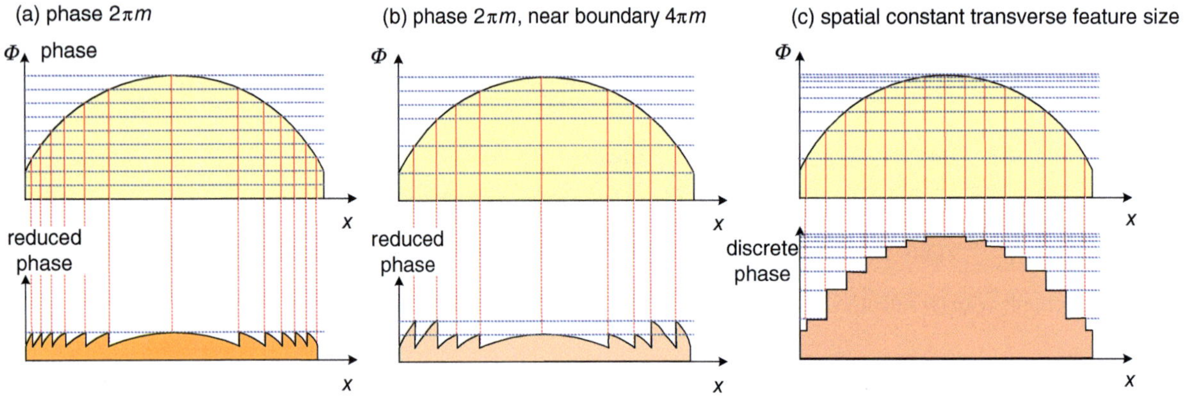

Figure 4.160 Different schemes of discretization, (a) shows the usual $2\pi m$ wrapping, in (b) different orders m are realized to relax the feature sizes at larger radii and in (c) constant feature sizes are used.

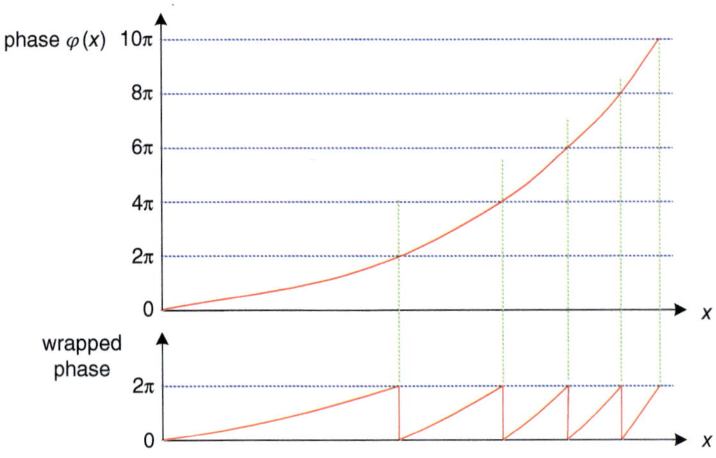

Figure 4.161 Phase wrapping principle by reduction of a smooth profile.

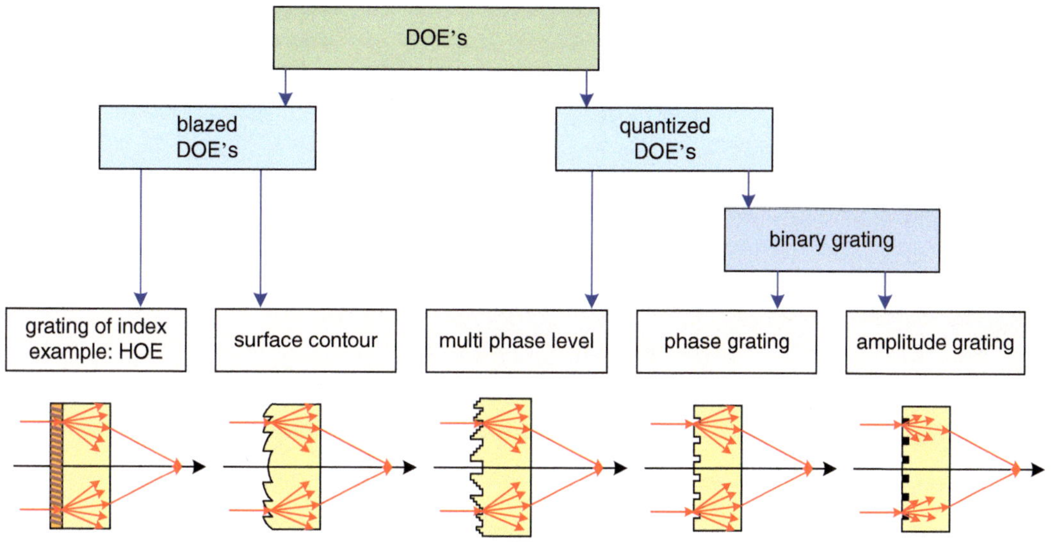

Figure 4.162 Different schemes of discretization and DOE realization.

- Computer-generated diffractive elements.
  The diffractive structure is calculated and fabricated with a grey tone lithography with the help of a computer by laser plotting or electron beam writing.
- Blazed diffractive element.
  The diffractive structure is carried out as a blazed grating for one wavelength and one incident ray direction. This suppresses the unwanted diffraction orders.
- Kinoform lens.
  A general term for a diffractive element, which acts with the fine structure of its surface topology and is produced, in most cases, by single-point diamond turning.
- Dynamically variable diffractive elements.
  The ray bending angles are switchable, realized for example with liquid crystal arrays.

In general, two or more of the above typical features can be found together in a real component. A component that has one surface with a refractive effect and another surface with a diffractive effect is called a hybrid component.

The generation of micro-structured phase can be realized by different physical effects. According to Figure 4.162 this can be done by a surface structuring as well as an index variation. From the viewpoint of realization binary, multi-level, amplitude or phase effects and smooth as well as binarized features are possible.

4.11.2.4 Fresnel Zone Plates

The historically oldest versions of components acting diffractively as lenses are zone plates, according to Fresnel. They can be realized by amplitude or phase jumps. The radial positions following the quadratic behaviour of the path difference according to the Pythagorean law for a flat substrate

$$f^2 + r_p^2 = (f + m\lambda)^2 \tag{4.232}$$

or resolved in paraxial approximation

$$r_m = \sqrt{2m \cdot f \cdot \lambda} \tag{4.233}$$

The focal length is given by

$$f = \frac{r_1^2}{2\lambda} \tag{4.234}$$

The phase differences are 2π and the component works as a lens, if:

- the phase is reversed in sign from ring to ring (or phase-shifted by π in an alternating sequence);
- all even ring numbers are blocked in amplitude.

Both options are possible. This principle is sketched in Figure 4.163.

If the focal intensity is calculated with an amplitude or phase zone plate, small differences are seen. Figure 4.164 shows the transmission profile of a case with $N = 30$ zones as well as the intensity in the focus. For comparison, the Airy pattern is drawn in red; the two profiles for amplitude or phase plates cannot be distinguished in this normalized representation. But it can be seen that the side lobes do not decrease continuously but in an alternating way. The exact behaviour depends on the ring number N. But for practical applications in imaging these small differences can be neglected.

4.11.2.5 Classical Diffractive Lens

For a simple paraxial lens, the width of the grooves changes according to the Fresnel zones as a function of the radius

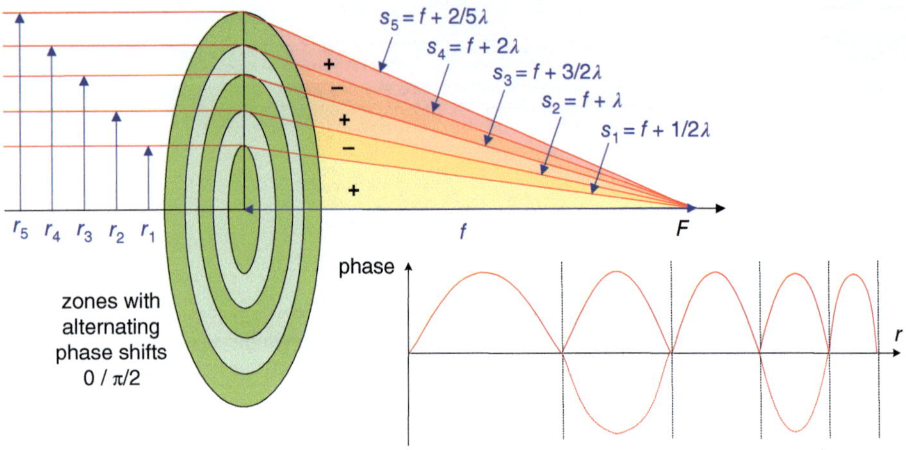

Figure 4.163 Principle of a Fresnel zone plate.

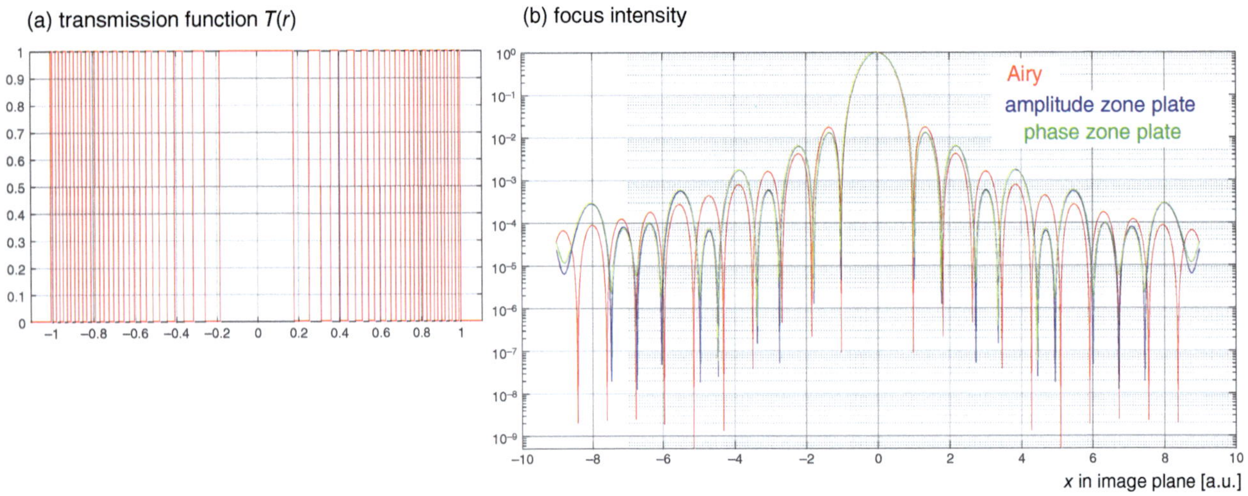

Figure 4.164 Example calculation of a Fresnel zone plate with $N = 30$ zones with (a) transmission function and (b) the focal intensity and comparison of the Airy profile. The units of the horizontal axis are arbitrary.

coordinate as

$$g(r) = \frac{m \cdot \lambda_0}{r} \cdot \sqrt{r^2 + f^2_{\lambda_0,m}} \quad (4.235)$$

At the radius r, the accumulated number of grooves is

$$M(r) = \frac{1}{m \cdot \lambda_0} \cdot \sqrt{r^2 + f^2_{\lambda_0,m}} - \frac{f_{\lambda_0,m}}{m \cdot \lambda_0} \quad (4.236)$$

The wrapped kinoform profile is then given by

$$s(r) = -h_m \cdot [M(r) \cdot mod(1)] \quad (4.237)$$

A classical zonal diffractive lens with focal length f for the design wavelength λ_0 has the exact phase function in order

$$\varphi_{exact}(r) = \frac{2\pi m}{\lambda} \cdot (f - \sqrt{f^2 + r^2}) \quad (4.238)$$

In reality, the phase is wrapped to $2\pi m$, this looks like Figure 4.165. It is seen that the smallest feature size of course occurs at the edge. It can be estimated in paraxial approximation as

$$\varphi(r) = \Delta r_{\min} \cdot \left(\frac{d\varphi}{dr}\right)_{r=D/2}, \quad \Delta r_{\min} = \frac{2\lambda f}{mD} \quad (4.239)$$

If for $r \ll f$ a paraxial approximation is considered, by Taylor expansion we get

$$\varphi_{parax}(r) = -\frac{\pi m \cdot r^2}{f \cdot \lambda} \quad (4.240)$$

A raw estimation with the Rayleigh $\lambda/4$ criterion gives as a condition of validity the size limitation

$$D^2 < 4 \cdot \sqrt{2m\lambda \cdot f^3} \quad (4.241)$$

4.11.2.6 Decomposition into Orders or Zones*

(*) The content of this section is quite specialized and for advanced reading. Knowledge of this section is not necessary to understand the following chapters. Beginners are recommended to skip this section.

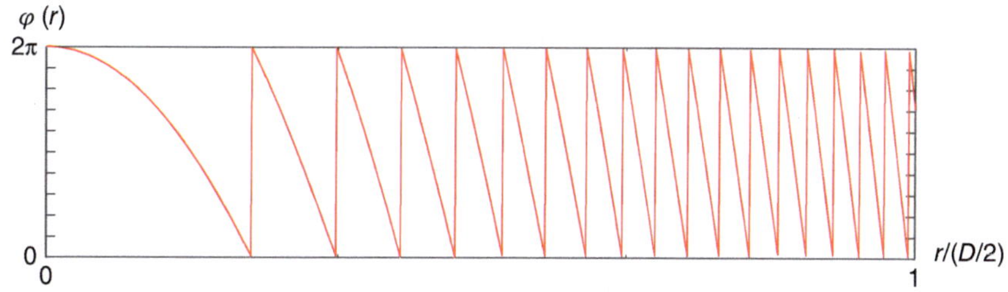

Figure 4.165 Wrapped phase profile of a lens for $m = 1$.

A blazed diffractive lens divides the incoming wavefront into radial zones. In every zone, an individual shape generates a phase jump by an integer multiple of 2π or a physical height jump by an integer multiple of the design wavelength λ_o. Due to these steps in phase or sag, the diffraction orders are different in every zone; in principle, the smooth wavefront of one order or constant optical path length is spatially decomposed in different orders. This means that the DOE effect can be understood by a [99]

- decomposition into orders;
- decomposition into spatial zones.

This is sketched in Figure 4.166. In (a) an ideal blaze condition is valid for all heights r of the component and the wrapping is perfectly matched. If the step height of the DOE is different from $h \neq m\ \lambda$ (in air) as shown in part (b) of the figure, the jumps are slightly incorrect and a dephasing is accumulated by the steps over the height. This means that the phase condition of the blaze is violated the larger the larger r is. The same happens if the wavelengths slightly differ from the design value $\lambda \ \# \ \lambda_0$, the decomposition changes over the radial coordinate r; the zonal decomposition is chirped in this case, see part (c) in the figure.

4.11.2.7 Problems of Real Diffractive Elements

In reality, there are several aspects of diffractive elements that cause deviations from the perfect assumptions of ideal theory. These are as follows.

- The perfect blaze condition can only be fulfilled for one wavelength.
- The ideal blaze condition can only be fulfilled for one incidence wave direction.
- If the blaze condition is violated, a certain amount of the light energy is scattered into unwanted parasitic diffraction orders.

  This causes a reduced signal strength and unwanted light, reducing the contrast of the imaging application.
- The assumption of an ideal thin thickness of the diffractive structure is an idealization. In reality, small transverse effects are observed. For example, light strikes the optically unused facets between the grooves and is therefore scattered into completely unpredictable directions.

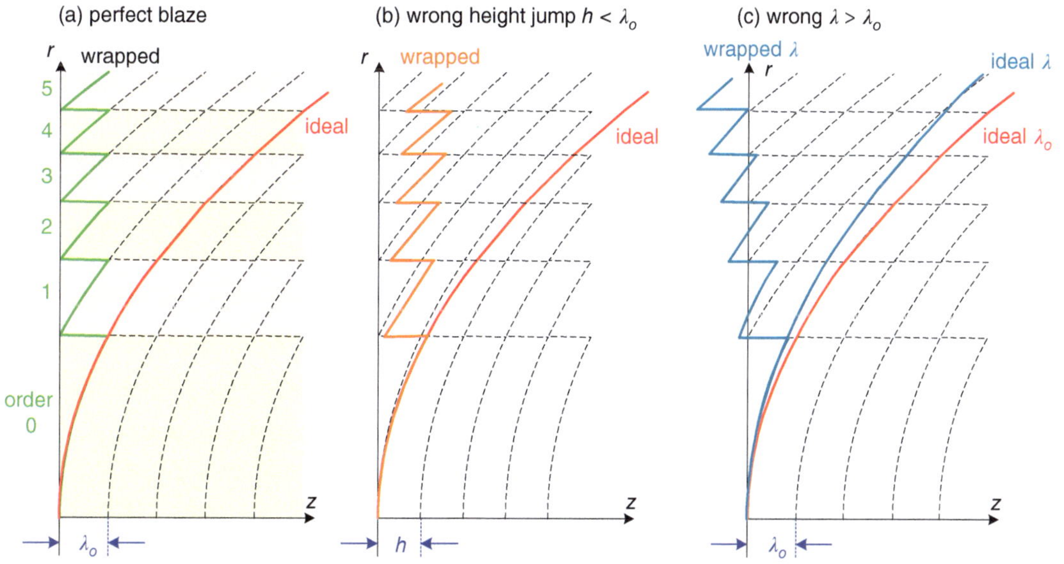

Figure 4.166 Diffractive lens with (a) ideal blaze, (b) incorrect structural height $h < \lambda_o$ and (c) incorrect wavelength $\lambda > \lambda_o$.

Furthermore the finite angles of the outgoing wave can cause small discontinuities and a finite fill factor with gaps.

- If the surfaces are discretized in the relief height, binary structures cause an increase of light power in unwanted orders.
- The tolerances of the diffractive structures have an impact on the performance of the desired light distribution. This can be a wrong height, errors in the slope of the used areas, rounding of edges, finite edge angles.

4.11.3 Modelling DOEs by Raytrace*

(*) The content of this section is quite specialized and for advanced reading. Knowledge of this section is not necessary to understand the following chapters. Beginners are recommended to skip this section.

From a physical point of view, the function of a diffractive element can only be explained by wave optical models. Due to the fact that raytrace is mostly used in lens design, a ray-based equivalence is quite attractive, which allows us to simulate diffractive elements without changing the tool.

According to Sweatt, the effect of a diffractive surface on light with the wavelength λ_0 can be modelled in an approximate manner using a raytracing calculation in a geometrical sense as a lens with an aspherical surface [103]. The fact that the bending of the rays takes place in a thin layer upon the carrier surface is considered by choosing a rather high refractive index in the range of approximately 10^4. Then a small inclination of the surface delivers a rather large bending angle for an incident ray. In principle, the Sweatt model is a special realization that allows for the use of the conventional raytrace algorithms.

The idea of this model is illustrated in Figure 4.167. The optical phase in both setups must be the same, therefore the product

$$\varphi(x,y) = 2\pi \cdot n \cdot z(x,y) = 2\pi \cdot n^* \cdot z^*(x,y) \quad (4.242)$$

must be equal. Since the value of z^* should be very small, the value of n^* must be very large. The Sweatt model is an approximation and a finite thickness remains as long as the refractive index has a finite size. If the refractive index is chosen to be extremely large to overcome these inaccuracies, in practice sometimes problems occur with the accuracy of the raytracing formulas in optical design software.

If λ_0 is the design wavelength of a diffractive structure, the wavelength dependence of the refractive index in this model changes according to

$$n_{eff}(\lambda, m) = m\frac{\lambda}{\lambda_0} \cdot [n(\lambda_0) - 1] + 1 \quad (4.243)$$

to fulfil the grating dispersion behaviour. In this equation m indicates the diffraction order that should be used for the light. The change in the refractive index with the wavelength is quite different in comparison to the material dispersion of optical glasses or crystals. This is the reason for the interesting ability of diffractive components to correct the chromatic aberrations of optical systems. If the freeform sag $z(x,y)$ is represented as a Taylor expansion in two dimensions in the form

$$z(x,y) = k \cdot \sum_{j,m} c_{jm} \cdot x^j \cdot y^m \quad (4.244)$$

the local grating constant in the m-th order of diffraction is then given with the phase Φ by

$$g(x,y) = \frac{2\pi \cdot m}{|\nabla\Phi(x,y)|} = \frac{\lambda \cdot m}{(n-1) \cdot |\nabla z(x,y)|} \quad (4.245)$$

This means that a general diffractive element has spatial varying grating constant.

4.12 Diffusor Plates*

(*) The content of this section is quite specialized and for advanced reading. Knowledge of this section is not necessary to understand the following chapters. Beginners are recommended to skip this section.

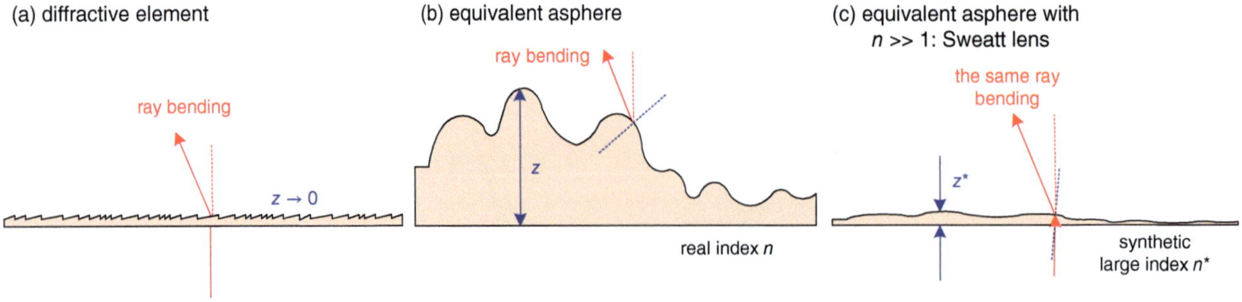

Figure 4.167 Effect of equal ray bending in the surface of the component with a negligible sagitta. The real diffractive element is shown in (a), the equivalent asphere for one order by unwrapping the phase in (b) and the Sweatt-equivalent asphere in (c).

4.12.1 Types of Diffusers

There are different types of diffuser plates found in reality. The main important types are [104–107] as follows.

- Ground glass diffusers.
  One optical surface of a plate is sand blasted or chemically etched and has a random distributed surface height. Mainly Gaussian angle profiles are generated with these devices.
- Bulk diffuser.
  Here suspended bubbles, inclusions or other scatter centres are distributed inside the volume of a plate with finite thickness. As parameters the thickness, the refractive index difference, the centre size and their density are possible. It can be assumed that multiple scattering takes place. In reality a low efficiency with high backscattering characterizes this approach. The best-known system is Opal glass.
- Micro-prism diffuser.
  Small prisms with refractive function redirect the light here. The shape and slope as well as the distribution of the prisms determine the function.
- Diffractive prism diffuser.
  In this case the feature sizes of the elementary prisms are in the range or smaller than the wavelength. The functional principle can be a micro-structured surface or index as a holographic element. Again the elementary cell geometry allows us to generate nearly arbitrary angle distributions.
- Micro-lenslet arrays.
  In this case, small lenses change the angle distribution. As parameters the size, the centres, the focal lengths, the depth, aspherical shapes and the boundaries allow us to modulate the angle distribution nearly arbitrary. If the illumination is coherent and the micro-structures are regular, diffraction patterns and Moire effects can be obtained.

4.12.2 Properties of Diffusers

Consider a diffusing disk in a path of light as shown in Figure 4.168(a). The characteristic of the diffusing disk is a statistical height distribution. As is illustrated in part (a) of the figure, at the diffusing surface scattering creates an enlarged spreading angle θ; a collimated incoming ray bundle becomes divergent after the plate. This characteristic increase in angle usually is observed on axis, where an incoming aperture cone angle u is grown to $u + \theta$, see also part (b) of the figure. If the incoming ray cone is inclined by an angle φ, the situation is more complicated. In general, the main light direction is not exactly φ again, but slightly changed to φ' and the broadening $\theta(\varphi)$ is modified too; this is illustrated in Figure 4.168(c).

In reality, the broadening is not simply a hard geometrical angle value; the intensity drop of a ray bundle or a light cone is smooth. A typical angular characteristic of the diffusing disk is shown in Figure 4.169 in a Cartesian plot for perpendicular light incidence $\varphi = 0$ as well as a polar representation. The plotted intensity $I(\theta)$ in the direction θ to the optical axis is normalized in the far-field to the intensity in the direction of the axis.

The diffusers are characterized by two main properties.

- Angle characteristic, spread angle and distribution inside the cone.
- Transmission efficiency; this describes the relative amount of the incoming light power, which is found in the cone. Mostly backscattering or spurious light at wide angles limit this value.

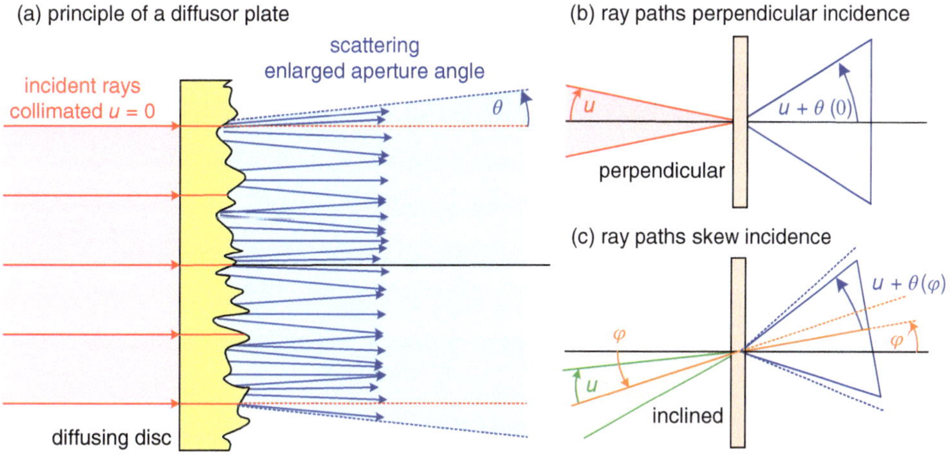

Figure 4.168 Effect of a diffusing disk (a) principle of scattering and (b) geometrical raytrace behaviour for perpendicular incidence and (c) skew incidence φ.

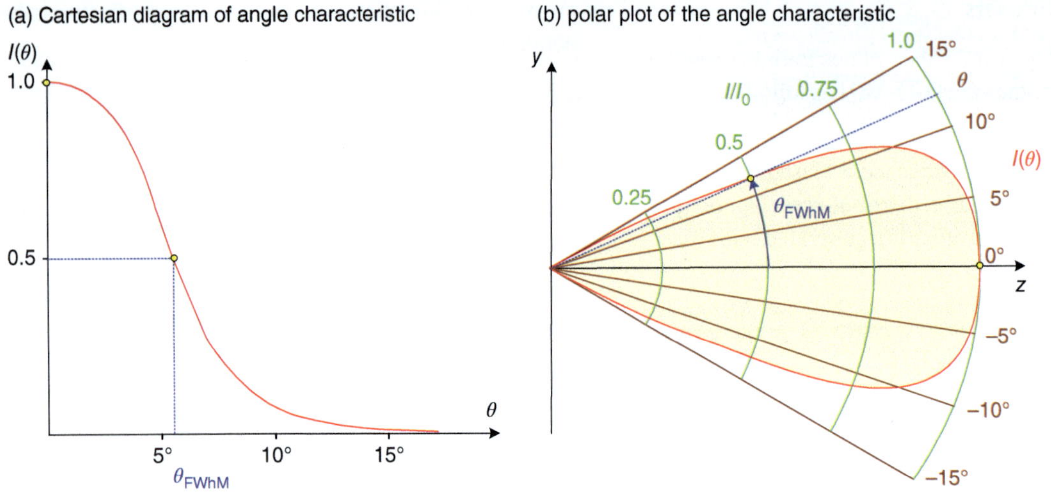

Figure 4.169 Angular characteristic of a diffusing plate in (a) Cartesian and (b) polar representation.

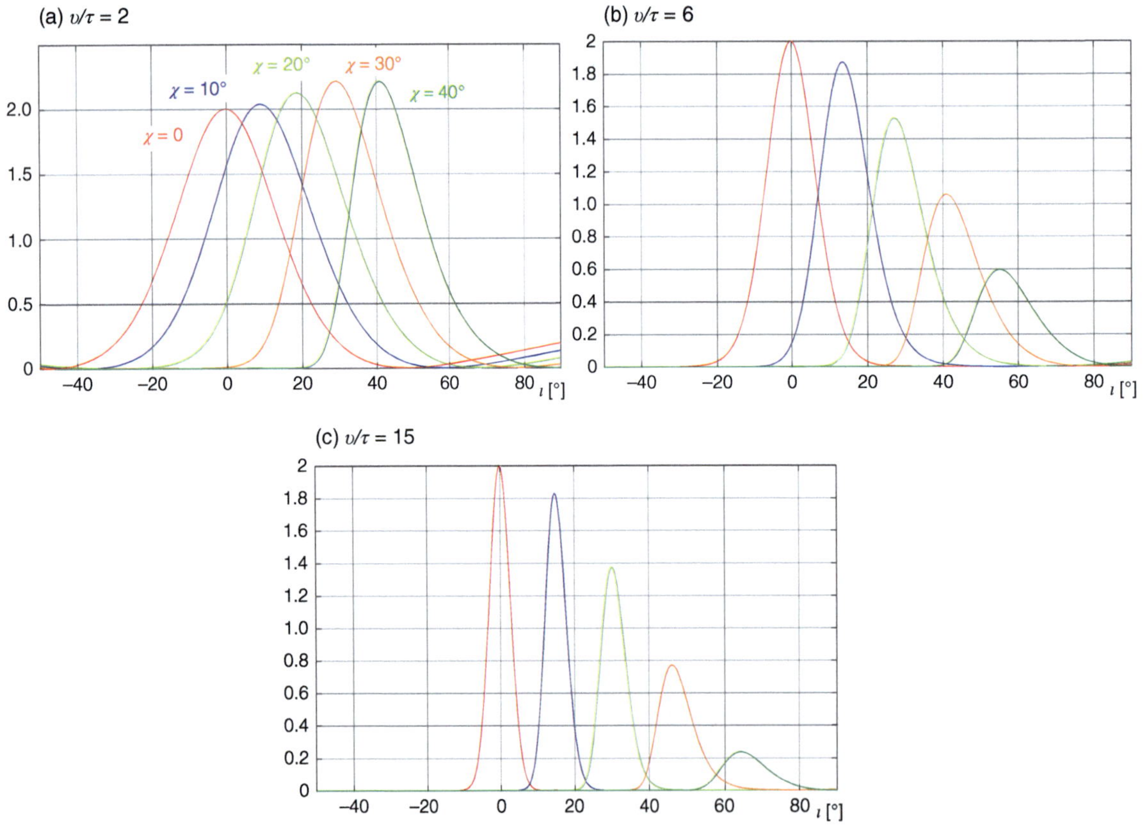

Figure 4.170 Angular characteristic of a diffusing disk as a function of the angle of incidence θ_i and the parameter τ/σ of the statistical parameters for correlation length and roughness. The roughness shrinks from left to right.

4.12.3 Angle Characteristic

In the geometric-optical approximation, which has a high coincidence with a diffraction based model using the Kirchhoff integral, the averaged intensity distribution with regard to the angle behind the diffusing disk can be written as

$$I_\varphi(\theta) = I_0 \cdot \frac{[n\cos(\theta-\varphi)-1]^2}{(n\cos\varphi - \cos\theta)} \cdot e^{-\left(\frac{\sin\theta - n\sin\varphi}{n\cos\varphi-\cos\theta}\cdot\frac{\tau}{2\sigma_{rms}}\right)^2} \quad (4.246)$$

where φ is the angle of incidence, θ the observation angle, n the refractive index, τ the correlation length and σ2rms the variance of the surface height distribution. τ / σrms is a measure of the diffusors surface roughness. This indicatrix of the scattering intensity becomes more asymmetric in form if the angle of incidence φ increases. The indicatrix depends only on the ratio τ/σ of the statistical parameters. The larger the ratio, the more narrowly it becomes concentrated around the refracting angle. Figure 4.170 shows some cases of the dependence of Eq. (4.246) for a refractive index of $n = 1.5$.

The definition of the outgoing direction is not unique and must be defined. In Figure 4.171(a) the change of the intensity peak angle, the centroid angle, is compared with the geometrically expected angle due to the law of refraction. The larger τ/σ, the better this simple geometrical prediction. Furthermore, it is observed in the right part of the figure (b) that the FWhM angle width of the distribution changes with incidence. This gives the modified angle spreading θ in Figure 4.168(c) as a function of φ. This effect means in reality for a diffuser plate in the image plane of an optical system:

- after the diffusor, the aperture angle $u + \theta$ is a function of the chief ray angle $w = \varphi$ and therefore changes with image height y;
- the chief ray w angle after the diffusor changes slightly with the image height, $\Delta w = w' - w$ depends on the image height y too.

This means from the geometrical point of view the simple approach

$$u' = u + \theta, \quad w' = w \quad (4.247)$$

or

$$L = y \cdot \sin(u + \theta) \quad (4.248)$$

is only a zeroth order approximation and must be modified in particular for larger angles of φ or chief ray angle

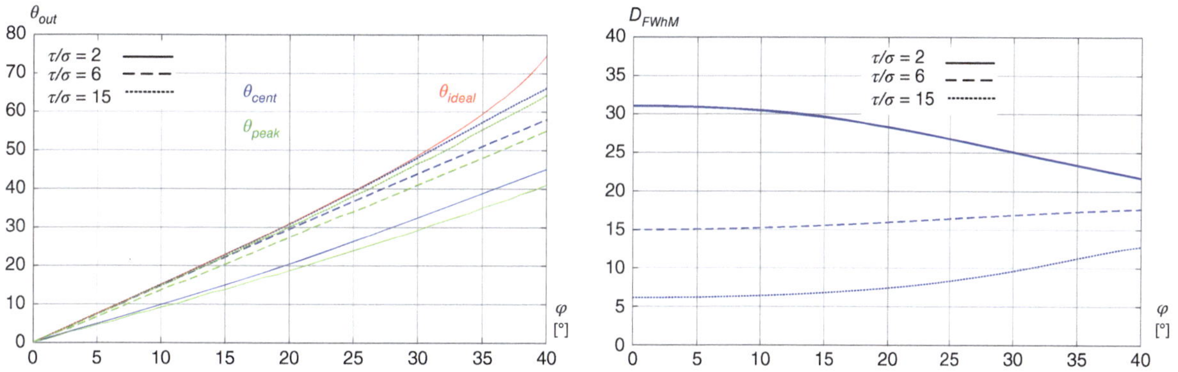

Figure 4.171 Angular characteristic of a diffusing disk as a function of the angle of incidence φ and the parameter τ/σ of the statistical parameters for correlation length and roughness. The blue and the green curves in the left diagram correspond to different definitions of the angle reference as discussed above.

Figure 4.172 Angular characteristic of a lenslet diffuser with varying incidence angle.

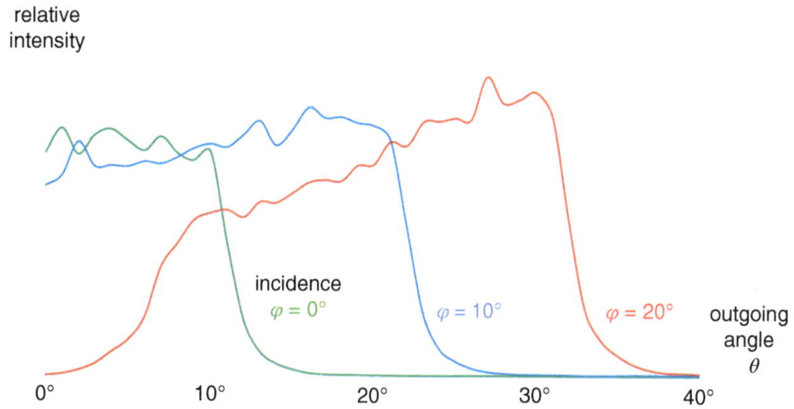

w respectively. In any case, the diffusor enlarges the Lagrange invariant L.

What are not considered in this simple geometrical model of a scatter plate are more complicated changes of coherence.

If a lenslet array is used to generate the desired angle distribution, a skew characteristic is observed for larger incidence angles due to the aberrations coma and astigmatism. As an example, this is sketched in Figure 4.172.

References

1 M. Kempe (2017). Personal communication, with permission.
2 Mahajan, V. (1998). *Optical Imaging and Aberrations*. Part I: SPIE Press.
3 V. J. Doherty and D. Shafer (1980). Simple method of correcting the aberrations of a beamsplitter in convergent light. *SPIE Conference Proceedings* 237: 195.
4 M. Matsubara (1974). Microscope Objective, US Patent 3822931.
5 Giusfredi, G. (2019). *Physical Optics*. Springer.
6 Berek, M. (1970). *Grundlagen der praktischen Optik*. de Gruyter.
7 Herzberger, M. (1944). Replacing a thin lens by a thick lens. *Journal of the Optical Society of America* 34: 114.
8 Walther, A. (1995). *The Ray and Wave Theory of Lenses*. Cambridge University Press.
9 de Meijere, J. and Velzel, C. (1987). Linear ray-propagation models in geometrical optics. *Journal of the Optical Society of America A* 4: 2162.
10 Wilde, J. (2024). Ray tracing model of a perfect lens compliant with Fermat's principle: the cardinal lens. *Applied Optics* 63: 1110.
11 Zeiss Archiv (1989). Schnittmodell des Spiegelsystems des deutschen Röntgenastronomie-Satelliten ROSAT. OMO B 2100/2. With permission.
12 Hsueh, C., Elazhary, T., Nakano, M., and Sasian, J. (2011). Closed-form sag solutions for Cartesian oval surfaces. *Journal of Optics* 40: 168.
13 Stavroudis, O. (1972). *The Optics of Rays, Wavefronts and Caustics*. Academic Press.
14 Luneburg, R. (1964). *Mathematical Theory of Optics*. University of California Press.
15 Braat, J. and Török, P. (2019). *Imaging Optics*. Cambridge University Press.
16 Hazra, L. (2022). *Foundations of Optical System Analysis and Design*. CRC Press.
17 Solva-Lora, A. and Torres, R. (2019). Explicit Cartesian oval as a superconic surface for stigmatic imaging optic systems with real or virtual source or image. *Proceedings of the Royal Society A* 476: 894.
18 Forbes, G. (2007). Shape specification for axial symmetrical optical surfaces. *Optics Express* 15: 5218.
19 G. Forbes and C. Brophy (2008). Asphere, O asphere, how shall we describe thee? *SPIE Conference Proceedings* 7100. https://doi.org/10.1117/12.797770.
20 Forbes, G. (2010). Robust efficient computational methods for axially symmetric optical aspheres. *Optics Express* 18: 19700.
21 Forbes, G. (2010). Robust and fast computation for the polynomials of optics. *Optics Express* 18: 13851.
22 A. Greynolds (2002). Superconic and subconic surfaces in optical design. *Proceedings of the International Society for Optical Engineering*. https://doi.org/10.1117/12.486491
23 A. Greynolds (2021). Hyper-aspheroidal surfaces: Two approaches, *SPIE Conference Proceedings* 12078. https://doi.org/10.1117/12.2603614.
24 Cho, S. (2016). Explicit superconic curves. *Journal of the Optical Society of America A* 33: 1922.
25 D. Ochse (2015). Personal communication, with permission.
26 Brömel, A. (2019). *Development and Evaluation of Freeform Surface Descriptions, Thesis*. FSU University Jena.
27 Menke, C. and Forbes, G. (2013). Optical design with orthogonal representation of rotational symmetric and freeform aspheres. *Advanced Optical Technologies* 2: 97.
28 Stock, J. (2016). *Investigation of Freeform Surfaces under Real Conditions, Master Thesis*. FSU University Jena.
29 Fasshauer, G. (2007). *Meshfree Approximation Methods with MATLAB*. World Scientific.
30 de Boor, C. (1978). *A Practical Guide to Splines*. Springer.
31 Piegl, L. and Tiller, W. (1997). *The NURBS Book*. Springer.
32 Hartung, J., von Lukowicz, H., and Kinast, J. (2018). Theoretical compensation of static deformations of freeform multimirror substrates. *Applied Optics* 57: 4020.
33 Malacara, Z., Malacara-Doblado, D., Malacara-Hernandez, D., and Landgrave, J. (2007). Astigmatic optical surfaces, characteristics, testing and differences between them. *Optical Engineering* 46: 123001.

34 A. Greynolds (2015). Battle of the biconics: Comparison and application of various anamorphic optical surfaces, Paper presented at *Imaging and Applied Optics*, Arlington, Virginia 7–11 June. https://doi.org/10.1364/FREEFORM.2015.FT2B.1.

35 Goodell, J. (1969). Eccentric lenses for producing ring images. *Applied Optics* 8: 2566.

36 Bernet, S. (2021). Combined diffractive optical elements with adjustable optical properties controlled by a relative rotation: tutorial. *Journal of the Optical Society of America A* 38: 1521.

37 Miller, O.E., McLeod, J.H., and Sherwood, W.T. (1951). Thin sheet plastic Fresnel lenses of high aperture. *Journal of the Optical Society of America* 41: 807.

38 Hofmann, C. and Tiedeken, R. (1966). Über die Einsatzmöglichkeiten von Fresneloptik als Bauelement des wissenschaftlichen Gerätebaus. *Feingerätetechnik* 15: 1.

39 Davis, A. and Kühnlenz, F. (2007). Optical design using Fresnel lenses. *Optik and Photonik* 4: 52.

40 J. Egger (1979). Use of Fresnel lenses in optical systems. *SPIE Conference Proceedings* 193: 63.

41 Jimenez-Rodriguez, M., Avendano-Alejo, M., Castillo-Santiago, G., and Huerta-Carranza, O. (2024). Image formation through aspheric concave Fresnel-type mirrors. *Journal of the Optical Society of America A* 41: 2277.

42 Jin, N., Lim, Z., Zhou, G. et al. (2021). Design and fabrication of composite polygonal Fresnel lenses. *Optics Express* 29: 36516.

43 Mo, J., Chang, X., Renqing, D. et al. (2021). Design and fabrication of a double-sided aspherical Fresnel lens on a curved substrate. *Optics Express* 29: 43353.

44 Desnijder, K., Hanselaer, P., and Meuret, Y. (2020). Freeform Fresnel lenses with low number of discontinuities for tailored illumination applications. *Optics Express* 28: 24489.

45 Fujii, T., Goulet, A., Hattori, K. et al. (2015). Fresnel lens sidewall design for imaging optics. *Journal of the European Optical Society - Rapid Publications* 10: 15017.

46 A. Davis (2009). Raytrace assisted analytical formulation of Fresnel lens transmission efficiency. *SPIE Conference Proceedings* 7429. https://doi.org/10.1117/12.823742.

47 Gonzalez, J. (2009). Design and analysis of a curved cylindrical Fresnel lens that produces high irradiance uniformity on the solar cell. *Applied Optics* 48: 2127.

48 Delano, E. (1978). Primary aberration contributions of curved Fresnel surfaces. *Journal of the Optical Society of America* 68: 1306.

49 Dickey, F. (ed.) (2014). *Laser Beam Shaping*. CRC Press.

50 Vetter, C., Steinkopf, R., Bergner, K. et al. (2019). Realization of free-space long-distance self-healing Bessel beams. *Laser and Photonics Reviews* 13: 1900103.

51 Ren, H. and Wu, S.-T. (2012). *Adaptive Lenses*. Wiley.

52 Jiang, H. and Zeng, X. (2013). *Microlenses*. CRC Press.

53 Zappe, H. and Duppe, C. (2016). *Tunable Micro-Optics*. Cambridge University Press.

54 Tyson, R. (1991). *Principles of Adaptive Optics*. Boston: Academic Press.

55 Huang, L., Rao, C., and Jiang, W. (2008). Modified Gaussian influence function of deformable mirror actuators. *Optics Express* 16: 108.

56 Rodrigues, G., Bastaits, R., Roose, S. et al. (2009). Modular bimorph mirrors for adaptive optics. *Optical Engineering* 48: 034001.

57 Claflin, E. and Bareket, N. (1986). Configuring an electrostatic membrane mirror by least-squares fitting with analytically derived influence functions. *Journal of the Optical Society of America A* 3: 1833.

58 Hudgin, R. (1977). Wave-front compensation error due to finite corrector-element size. *Journal of the Optical Society of America* 67: 393.

59 S.K. Ravensbergen, R. Hamelink, P. Rosielle and M. Steinbuch (2009). Deformable mirrors: design fundamentals for force actation of continuous facesheets. *SPIE Conference Proceedings* 7466. http://dx.doi.org/10.1117/12.825832

60 Yaitskova, N. (2009). Adaptive optics correction of segment aberration. *Journal of the Optical Society of America A* 26: 59.

61 Li, E., Dai, Y., Wang, H., and Zhang, Y. (2006). Application of eigenmode in the adaptive optics system based on a micromachined membrane deformable mirror. *Applied Optics* 45: 5651.

62 Lohmann, A. (1970). A new class of varifocal lenses. *Applied Optics* 9: 1669.

63 Barbero, S. (2009). The Alvarez and Lohmann refractive lenses revisited. *Optics Express* 17: 9376.

64 Humphrey, W. (1976). Remote subjective refractor employing continuously variable sphere-cylinder corrections. *Optical Engineering* 15: 287.

65 J. Babington (2015). Alvarez lenses, theory and applications, SPIE Conference Proceedings 9626. http://dx.doi.org/10.1117/12.2193114.

66 Daly, D. (2001). *Microlens Arrays*. Taylor & Francis.

67 Wang, S. and Zhao, D. (2000). *Matrix Optics*. Springer-CHEB.

68 Martinez-Corral, M. and Javidi, B. (2018). Fundamentals of 3D imaging and displays: a tutorial on integral

imaging, light field and plenoptic systems. *Advances in Optics and Photonics* 10: 512.

69 Scholes, S., Kara, R., Pinnell, J. et al. (2020). Structured light with digital micromirror devices: a guide to best practice. *Optical Engineering* 59: 041202.

70 Mitchell, K., Turtaev, S., Padgett, M. et al. (2016). High-speed spatial control of the intensity, phase and polarisation of vector beams using a digital micro-mirror device. *Optics Express* 24: 29269.

71 J. Rice, J. Neira, M. Kehoe and R. Swanson (2009). DMD diffraction measurements to support design of projectors for test and evaluation of multispectral and hyperspectral imaging sensors. SPIE Conference Proceedings 7210. http://dx.doi.org/10.1117/12.808990.

72 Ren, Y.-X., Lu, R., and Gong, L. (2015). Tailoring light with digital micromirror device. *Annalen der Physik* 527: 447.

73 Marchand, E.W. (1978). *Gradient Index Optics*. New York: Academic Press.

74 Gomez-Reino, C., Perez, M.V., and Bao, C. (2002). *Gradient Index Optics*. Berlin: Springer.

75 Greisukh, G.I., Bobrov, S.T., and Stepanov, S.A. (1997). *Optics of Diffractive and Gradient-Index Elements and Systems*. Bellingham: SPIE Optical Engineering Press.

76 Kravtsov, Y.A. and Orlov, Y.I. (1990). *Geometrical Optics of Inhomogeneous Media*. Berlin: Springer.

77 B. V. Hunter, P. Fournier and S. C. Johnston (1999). Improved characterization of GRADIUM Gradient-Index Glasses. *SPIE Conference Proceedings* 343. https://doi.org/10.1117/12.323761.

78 B. V. Hunter and B. Walters (1998). How to design and tolerance with GRADIUM glass. *SPIE Conference Proceedings* 3482: 801.

79 B. V. Hunter, K. E. Moore and P. L. Ruben (1997). Advantages of GRADIUM GK Glasses for polychromatic design. *SPIE Conference Proceedings* 3130:19.

80 Lin, P. (2017). *Advanced Geometrical Optics*. Springer.

81 Vanderwerf, D. (2010). *Applied Prismatic and Reflective Optics*. SPIE Press.

82 Hagen, N. and Tkaczyk, T. (2011). Compound prism design principles I. *Applied Optics* 50: 4998.

83 Hagen, N. and Tkaczyk, T. (2011). Compound prism design principles II, triplet and Jannsen prisms. *Applied Optics* 50: 5012.

84 Hagen, N. and Tkaczyk, T. (2011). Compound prism design principles III, linear in wavenumber and optical coherence tomography prisms. *Applied Optics* 50: 5023.

85 Goncharenko, E.N. and Repinskii, G.N. (1975). The design of achromatic wedges. *Soviet Journal of Optical Technology* 42: 445.

86 Fery, C. (1911). A prism with curved faces for spectrograph or spectroscope. *The Astrophysical Journal* 34: 79.

87 Warren, D., Hackwell, J., and Gutierrez, D. (1997). Compact prism spectrographs based on aplanatic principles. *Optical Engineering* 36: 1174.

88 Morrissey, P.F. (1994). Third-order aberrations of a prism with spherically curved surfaces. *Applied Optics* 33: 2539.

89 Tang, Z. and Gross, H. (2021). Improved correction by freeform surfaces in refractive spectrometer concepts. *Applied Optics* 60: 333.

90 Li, Y. (2011). Closed form analytical inverse solutions for Risley-prism-based beam steering systems in different configurations. *Applied Optics* 50: 4302.

91 Li, Y. (2011). Third-order theory of the Risley-prism-based beam steering system. *Applied Optics* 50: 679.

92 Lu, Y., Zhou, Y., Hei, M., and Fan, D. (2013). Theoretical and experimental determination of steering mechanism for Risley prism systems. *Applied Optics* 52: 1389.

93 Zhao, Y. and Yuan, Y. (2014). First-order approximation error analysis of Risley-prism-based beam steering directing system. *Applied Optics* 53: 8020.

94 Zhao, Y., Chen, Y., Zhu, P. et al. (2018). Limits on field of view for Risley prisms. *Applied Optics* 57: 9114.

95 Lai, S. and Lee, C. (2018). Analytic inverse solutions for Risley prisms in four different configurations for positioning and tracking systems. *Applied Optics* 57: 10172.

96 Hutley, M.C. (1982). *Diffraction Gratings*. London: Academic Press.

97 Loewen, E.G. and Popov, E. (1997). *Diffraction Gratings and Applications*. New York: Marcel Dekker.

98 Kingslake, R. (ed.) (1969). *Applied Optics and Optical Engineering*, vol. V. New York: Academic Press.

99 Wood, A. and Babington, J. (2023). *Diffractive Lens Design*. IOP Publishing.

100 Sinzinger, S. and Jahns, J. (2003). *Microoptics*. Weinheim: Wiley.

101 Turunen, J. and Wyrowski, F. (ed.) (1997). *Diffractive Optics*. Berlin: Akademie Verlag.

102 Herzig, H.P. (ed.) (1997). *Micro-Optics*. London: Taylor & Francis.

103 Sweatt, W.C. (1977). Describing holographic optical elements as lenses. *Journal of the Optical Society of America* 67: 803.

104 G. Odom (2016). Diffusers in Coherently Illuminated, Wide Field of View Applications. Master's thesis, University of Arizona.

105 Garcia-Botella, A., Sanchez-Brea, L., Vazquez-Molini, D., and Bernabeu, E. (1999). Modulation transfer function of translucent rough sheets. *Applied Optics* 38: 5429.

106 Qiantian, P., Rebollo, M., and Gaggioli, N. (1997). Diffusion off light transmitted from rough surfaces. *Journal of Modern Optics* 44: 447.

107 Quintian, O., Rebollo, M., Gaggioli, N., and Raffo, C. (1998). The non-refractive effect in translucent diffusers: a geometrical model. *Journal of Modern Optics* 45: 441.

5

Imaging Systems

5.1 Introduction

5.1.1 Mathematical Description of Optical Systems

If an optical system has to be simulated and optimized, first a mathematical model is needed that describes the system by characteristic parameters. Usually for this purpose the system data as well as the light data must be included; the light data are more or less a description of the illumination source in combination with the interface to the imaging system.

For the system, the following parameters and quantities must be considered.

- Dielectric interface surfaces or mirrors:
  - mathematical modelled surfaces;
  - planes, spheres, aspheres, conics, free shaped surfaces etc.
- Size of components:
  - thickness and distances along the axis;
  - transverse size, circular diameter, complicated boundary contours.
- Geometry of the setup, position and orientation of components:
  - in the special case of rotational symmetry: air distances between the components;
  - general case: 3D, tilt angles, offsets and decenters.
- Materials:
  - refractive indices for all used wavelengths;
  - optional other properties: absorption or transmission, birefringence, nonlinear coefficients, index gradients etc.
- Special aspects:
  - stop position;
  - gratings, diffractive elements;
  - arrays;
  - scattering surface;
  - zoom positions, multi configurations.

For a single spherical surface, the radius of curvature, the diameter, the refractive index and the distance behind the surface are necessary to know; this is illustrated in Figure 5.1.

The main parameters to characterize the light data are:

- wavelength(s), optional with spectral weighting factors;
- aperture, NA, entrance pupil size or F-number;
- field of view as angle or object/image field diameter;
- special data:
  - telecentricity
  - afocal image location.

The complete system is a sequence of surfaces that is indexed in the ascending order from left to right. Figure 5.2 shows the usual indexing scheme, if N surfaces are there. The object gets the index 0, the image plane the index N. The indices of the distances are identical to their preceding surface. In classical imaging systems the rays start in the object plane on the left side and impinge the surfaces one after the other as 0-1-2-…-($N-1$)-N. The optical axis is denoted to be the z-axis of the coordinate system, the main tangential or meridional plane is the y–z plane and the x-coordinate corresponds to the sagittal direction. Every item of data has a sign. Figure 5.3 defines the signs of the length parameter to be positive to the right of the reference and negative to the left. The angles' sign definition is positive in case of a counter-clockwise orientation.

It should be mentioned that the indexing described above can also be carried out differently; in particular the indices of object and image surfaces are not unique.

5.1.2 System Notations

The task of an optical system is mostly to create a good image. Therefore, the relations between object space and image space are essential, independent of the concrete realization of this transfer by the optical system. Therefore, in the mathematical description of optical systems and

Lens Design for Imaging: Fundamentals of Optical Systems, First Edition. Herbert Gross.

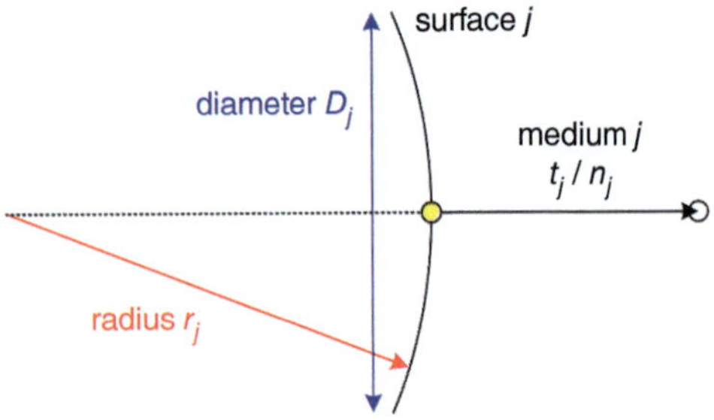

Figure 5.1 Parameter of a single spherical surface with index j.

in particular the formulation of aberration theory, it is enough to consider object and entrance pupil as well as exit pupil and image plane. Figure 5.4 illustrates the usual definitions of the four coordinate systems, the distances of field and pupil planes as well as the lateral aberrations $\Delta x'$, $\Delta y'$ in the image plane. The notation and the meaning of the quantities used here is:

x, y	object coordinates, especially object height;
x', y'	image coordinates, especially image height;
x_p, y_p	*coordinates of entrance pupil;*
x'_p, y'_p	coordinates of exit pupil;
s	distance front principal plane to object;
s'	distance rear principal plane to image;
s_p	distance front principal plane to entrance pupil;
s'_p	distance rear principal plan to exit pupil;
p	distance object to entrance pupil;
p'	distance image to exit pupil;
$\Delta x'$	sagittal transverse aberration;
$\Delta y'$	meridional transverse aberration.

In particular it should be noted here that the intersection lengths s are always referenced to the principal planes in the object as well as image domain. The distances p, p' between field and pupil are in addition important in aberrations theory, but can be expressed easily by s and s_p or s, s'_p respectively.

5.1.3 System Layout

One of the most important control tools for a lens design is the layout of the system with selected rays. It must always be inspected after installing a new system to see whether the geometry and the ray paths are as expected and planned. There are several opportunities to fix the outfit of layouts. For rotational symmetric systems, usually a cross section of the y–z meridional plane is used. This kind of drawing is shown in Figure 5.5(a). The ray bundles for different field points are indicated by different colours, the number of rays per ray bundle is arbitrary. If the control

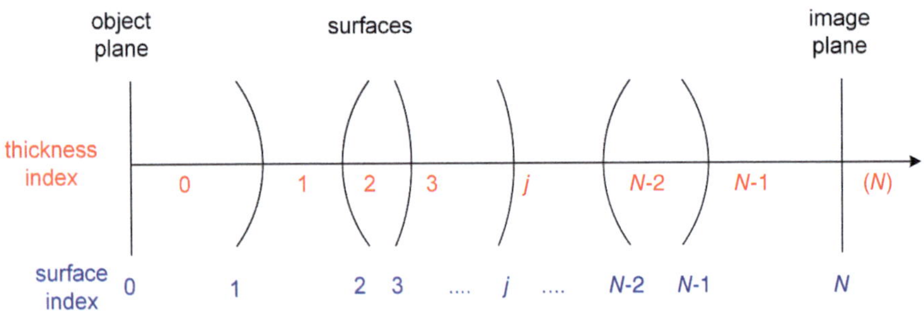

Figure 5.2 Surface indices in an optical system.

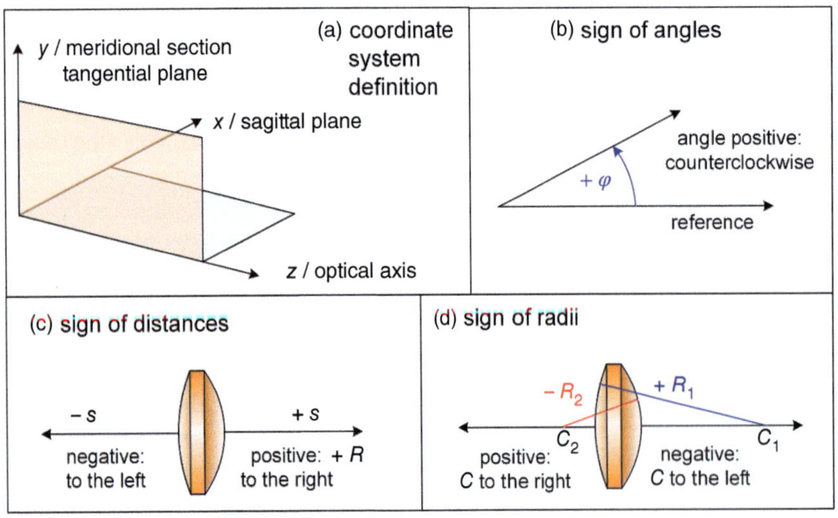

Figure 5.3 Definitions of coordinates and signs.

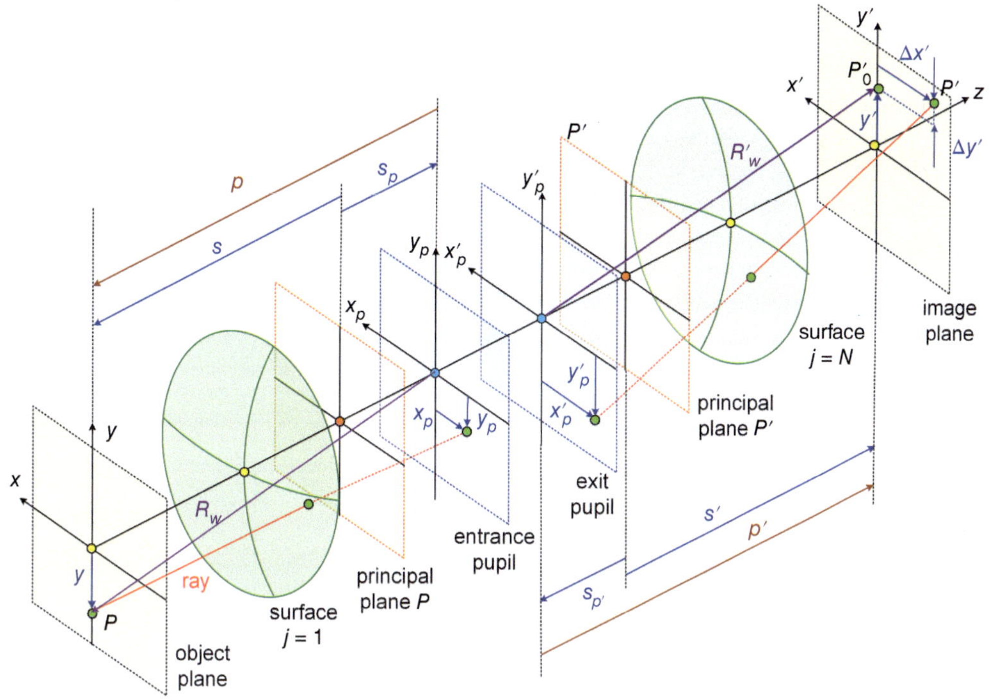

Figure 5.4 Definition of the characteristic system data in field and pupil.

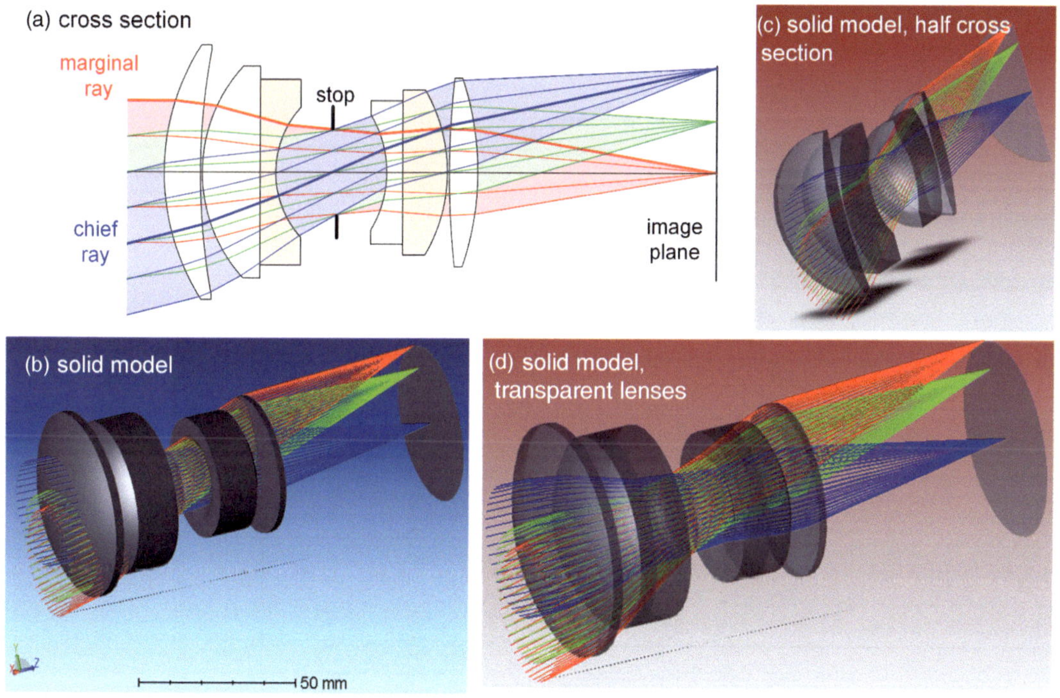

Figure 5.5 Drawing options of a lens layout (a) cross section for circular symmetry, (b) solid model, (c) solid model with half cross section and (d) solid model with transparent components.

of three-dimensional geometries is needed, a solid body model like the one in Figure 5.5(b)–(d) is useful. It depends on the purpose and the setup geometry what kind of options deliver the information best. In Figure 5.5 only some possibilities are illustrated. One of the drawbacks of the solid model version is that due to the perspective and skew viewing angle, the length values are scaled according to the skew projection. In the special case of an anamorphic system with a double plane symmetry around y–z and x–z, it is also meaningful to draw these two separated planes as cross sections.

5.1.4 Terms of Performance Evaluation

In the majority of cases, optical systems for imaging are modelled and designed using raytrace. Many definitions, algorithms and conversions or analysis calculations have been developed over time. Figure 5.6 shows schematically the most important steps, opportunities and calculation algorithms that are used in calculation tools. Most of the quantities as well as the operators are explained in a later section of the text. In total five levels of modelling can be distinguished.

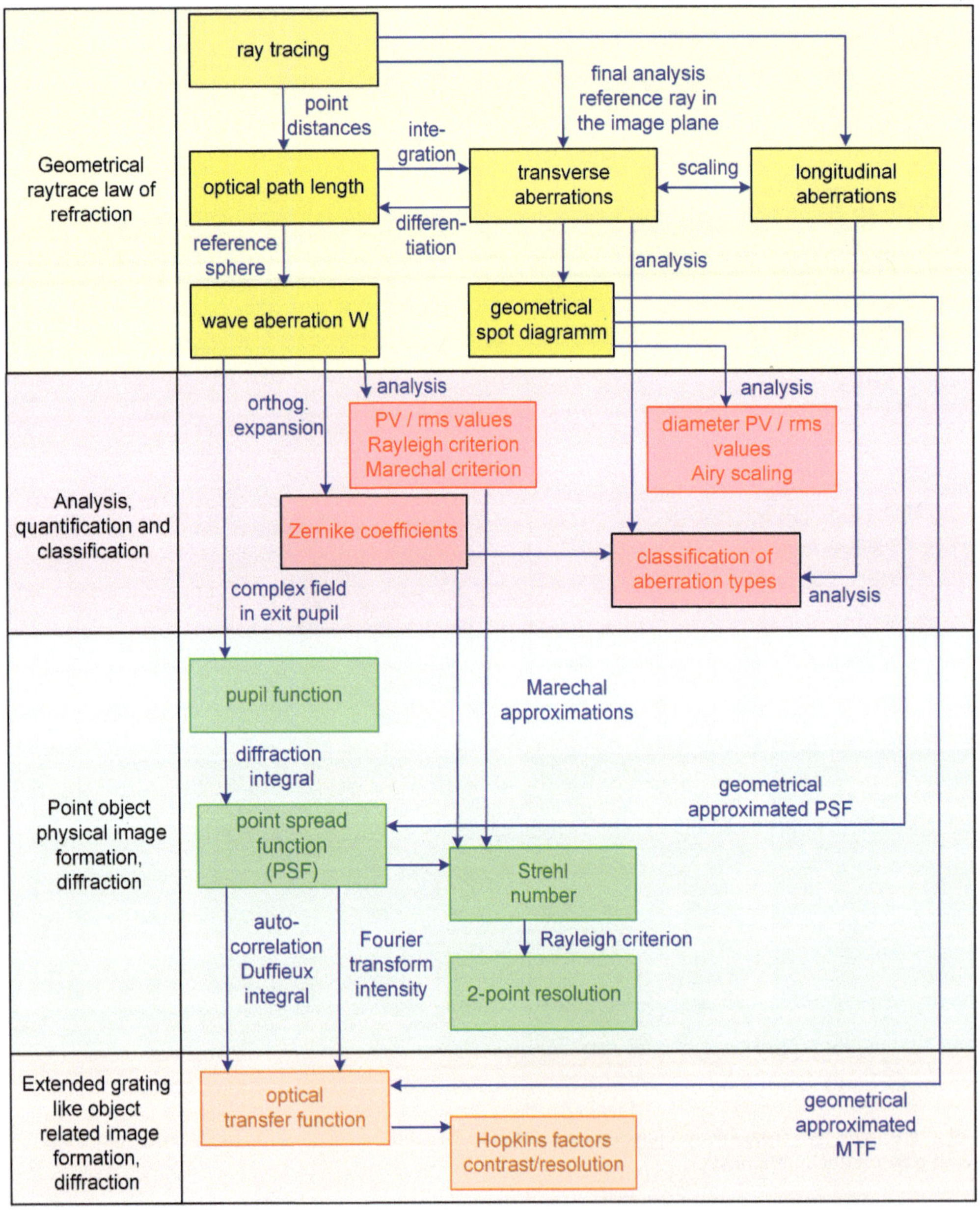

Figure 5.6 Simulation levels of optical systems and relations between the definitions. The blue arrows indicate operators to calculate or analyse the data.

- Yellow layer:
  geometrical raytrace: tracing rays through the system from surface to surface until the image plane is reached, analysis of the resulting numbers of ray intersection point and angle data as well as optical path difference data.
- Red layer:
  equivalent geometrical quantities: classification to make the results easier to interpret and to reduce the huge amount of information from many rays as well as wavefronts to a few significant numbers.
- Green layer:
  physical model: complex pupil function and including diffraction effects in the hybrid model, analysis of resulting numbers, calculation of primary physical quantities like the point spread function, analysis of quality, for example, by the Strehl ratio.
- Orange layer:
  determines secondary physical quantities like optical transfer function and contrast values to assess the performance of the system by Fourier-optical means for extended grating-like objects.

First a raytrace is performed; the information can be coded after getting the data out of this calculation, and the results of many rays are condensed and classified into aberration types or simple numbers like rms values of spot diameters or aberration coefficients. If a geometrical model is sufficient, more complicated numbers can be calculated according to the type of system. If diffraction effects cannot be neglected, the data of the optical path lengths along the rays are used to synthesize a complex field in the exit pupil of a system. Then a calculation of a diffraction integral can be performed to get the point spread function in the image plane. If no point-like objects are considered but an extended field feature like a grating is more application related, an additional Fourier transform delivers the optical transfer function and values of contrast and resolution can be determined. In Figure 5.6, the boxes indicate defined quantities, numbers or functional distributions. They can be used to evaluate the quality of the system or to gather important information. The blue arrows show possible operators or calculation algorithms that allow us to further process or convert the data.

In particular it should be mentioned that for optical systems far above the diffraction limit, a geometrical approximated version for point spread function and transfer function can be used.

5.2 Geometrical Imaging

5.2.1 Introduction

A system is called an imaging system if all the rays emerging from one object point are intersecting, in the best case in one image point. In contrast to an illumination system, in an imaging system a point-to-point relation is required. The size of the field of view is described by the outermost ray: the chief ray. In real non-perfect systems, the rays are collected in a rather small spot with finite lateral extent in the image plane.

The size of the provided light cone or the aperture is described by the marginal ray. It intersects the pupil plane at the largest height (edge). If all rays have small angles against the optical axis, we have paraxial imaging, which perfectly fulfills the above conditions, see Section 3.5. The ray coming from the outermost field point O_2 traversing the optical axis in the pupil centre is the chief ray. This scheme is illustrated in Figure 5.7(b).

In general. a system is called an imaging system if a certain object, mostly a plane structure, is reproduced in an image plane and looks from visual perception to be quite similar. This basic idea is sketched in Figure 5.7(a). In the framework of geometrical optics, the imaging condition means that every ray starting in an object point in a

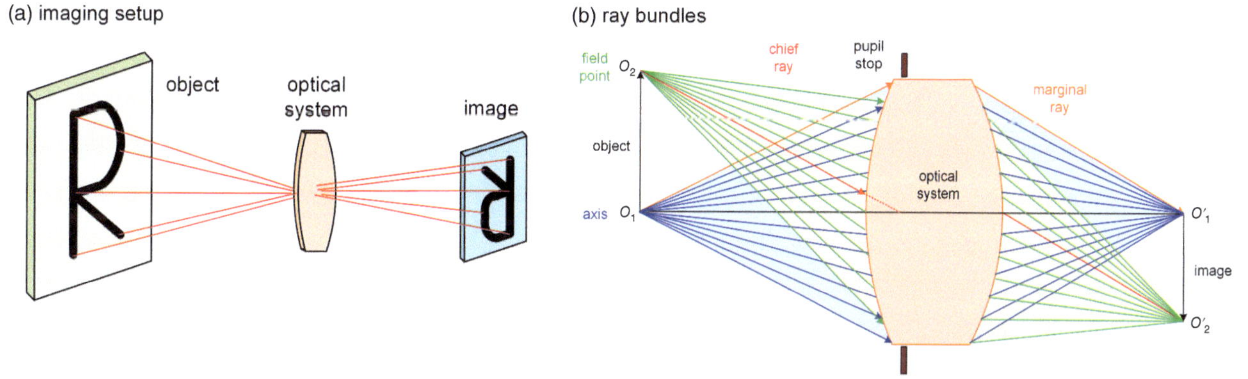

Figure 5.7 Imaging as (a) schematic drawing, and (b) definition of chief ray, marginal ray and pupil.

direction, which is able to pass the optical systems, is again collected in the corresponding image point. In the case of a perfect imaging system, the rays meet in one mathematical point only: the imaging is called stigmatic. This ideal condition is found in reality only in special cases for special points. In general, the ray intersection points in the image plane take different directions and form a blurred spot. The size of this spreading is a measure for the quality and the spatial resolution; it is an effect of the system aberrations. If in addition diffraction effects are included, an additional broadening is observed, see Chapter 6.

First, for simplicity. only a single refracting surface is considered. If s indicates the distance between the object and the position of a refractive surface that separates the indices n and n', the imaging condition formula which gives the location s' of the image reads in paraxial approximation

$$\frac{n'}{s'} - \frac{n}{s} = \frac{n'-n}{R} = \frac{1}{f'} \tag{5.1}$$

If the object lies to the left side of the surface, $s < 0$ is used; $s' > 0$ means that the image lies on the right side of the surface. By using the so-called Abbe invariant Q, this equation can also be written as

$$Q_s = n \cdot \left(\frac{1}{R} - \frac{1}{s}\right) = n' \cdot \left(\frac{1}{R} - \frac{1}{s'}\right) \tag{5.2}$$

which has the form of an invariant. The quantities left to the surface are separated from the primed quantities on the right side. The geometry of this setup, as well as the definition of the quantities. is illustrated in Figure 5.8.

For the derivation of the formulas given above, the following relations are used. In the triangle OCQ the sum of the angles is given by

$$180° = u + \psi + (180° - i) \tag{5.3}$$

Correspondingly we have on the image side in CQO'

$$180° = u' + i' + (180 - \psi) \tag{5.4}$$

The height h in the intersection point Q is given in object as well as image side in paraxial approximation by

$$h = s \cdot u = s' \cdot u' \tag{5.5}$$

Furthermore. the ray meeting the centre of curvature C of the surface is unchanged and therefore

$$w = \frac{y}{R-s} = w' = \frac{y'}{s'-R} \tag{5.6}$$

and finally the law of refraction in point Q gives in paraxial approximation

$$n \cdot i = n' \cdot i' \tag{5.7}$$

By eliminating i, i', u, u' and ψ the imaging Eq. (5.1) is easily obtained.

5.2.2 Lens Makers' Formula

If a lens is considered, the setup of Figure 5.9 is obtained. The imaging is described by the lens makers' formula which reads for a thin lens in air

$$\frac{1}{s'} - \frac{1}{s} = \frac{1}{f} \tag{5.8}$$

and can be derived if the thin lens is constituted by two surfaces and Eq. (5.1) is applied twice. The magnification is defined as the enlargement of the image relative to the object. In paraxial approximation we have

$$m = \frac{y'}{y} = \frac{s'}{s} = \frac{nu}{n'u'} \tag{5.9}$$

which gives a negative value for the simple case illustrated in Figure 5.9.

Figure 5.10 shows a diagram that illustrates the behaviour of the lens makers' formula for different values of the object location. The object as well as the image location distances s, s' are scaled by the focal length f of the lens and are used as the coordinate axes. According to the sign convention, only negative values of s smaller than $-f$

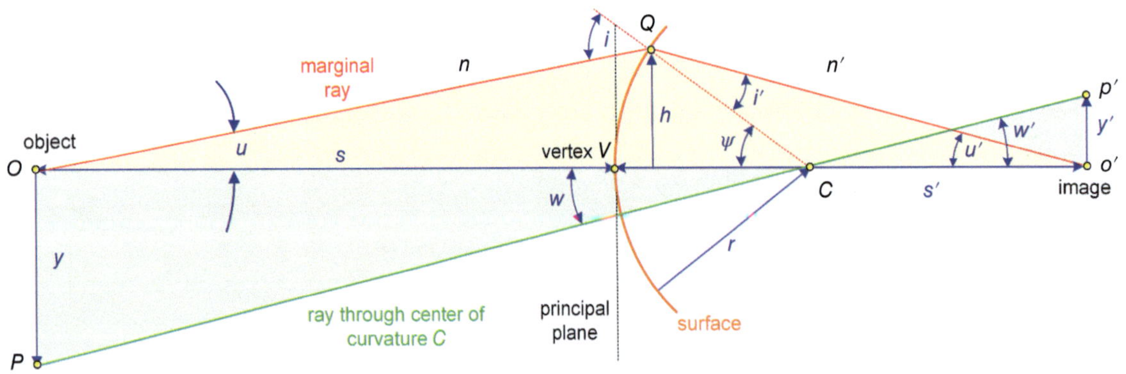

Figure 5.8 Illustration of the imaging condition for a single surface.

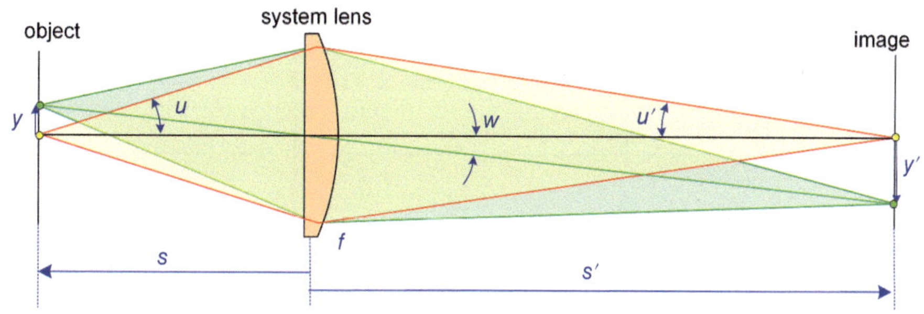

Figure 5.9 Imaging by a single lens.

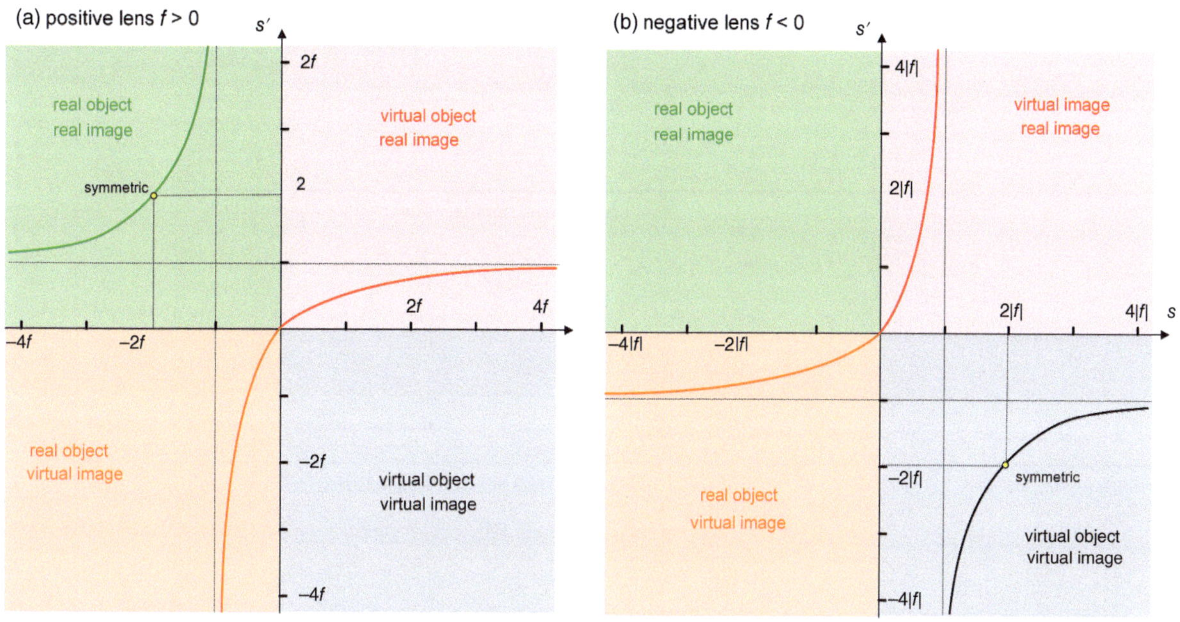

Figure 5.10 Illustration of the lens makers formula for a lens with (a) a positive and (b) a negative focal length.

delivers a real image location $s' > 0$. In total, there are three branches of possible solutions for the object-image location. The real case has already been mentioned; it is shown in green here. The orange branch describes a virtual image and a real object, the red a real image and a virtual object. As is seen in part b) of the figure, a negative lens gives no real image for a real object. For $s = -f$ we get a singular case with the image in infinity. One more interesting case is the symmetrical one: for $s = -2f$ we get $s' = +2f$. Some of these cases are shown in Figure 5.11. Lenses with negative focal lengths are also considered here.

If the position of the object is changed, the location of the image point on axis varies. A negative lens has mostly virtual image locations, only if the incoming ray bundle is strongly convergent might the image position be finite and positive. This last case is realized in tele systems.

If positive focal lengths are considered, the position of the object point is essential for the image location according to the lens makers' formula. If the object is positioned in the front focal point, the image occurs in infinity. If the object is further away than the front focal point, a real image is obtained. If the object is located between the focal point and lens, a virtual image with formally negative intersection length is obtained. Some of the possible cases are sketched in Figure 5.11. In the case of the most often realized setup in part (b) of the figure, the shift of the image follows a simple rule: moving the object shifts the image in the same direction.

If some special rays are selected, it is also possible to find the position and the size of the image by simple graphical means, according to Lister, if the principal planes are known. Figure 5.12 shows this scheme for a positive and a negative lens respectively. The three rays which must intersect in the image point are:

- Ray no. 1:
  ray starting in the object point going parallel to the optical axis. In the image domain, it must strike the optical axis in the back focal point.

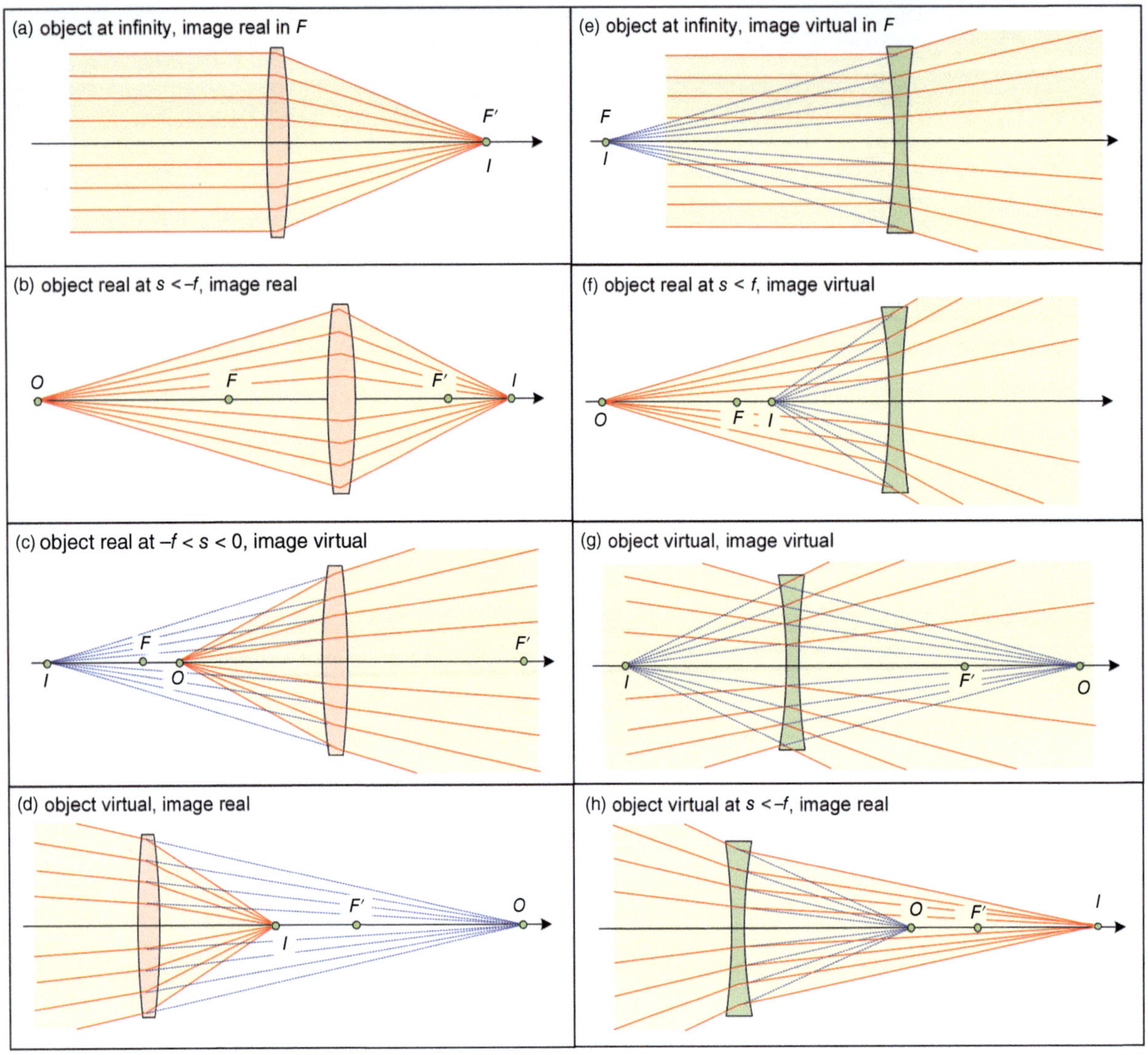

Figure 5.11 Imaging with a paraxial lens, the left column shows the case of a positive f, the right column of a negative focal length. O indicates the object on axis, I the corresponding image point.

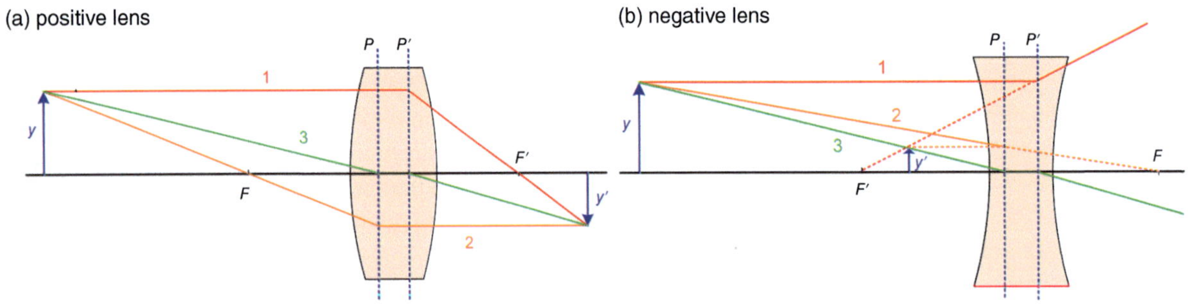

Figure 5.12 Imaging of a paraxial lens by special rays for (a) a positive and (b) a negative focal length.

- Ray no. 2:
  ray starting in the object point going through the front focal point must go parallel to the optical axis behind the system.
- Ray no. 3:
  if the lens is located in the same medium on both sides, the nodal planes coincide with the principal planes. A ray starting in the object point meeting the front nodal point leaves the lens with the same direction.

5.2.3 Paraxial Formulas

A classical one-stage paraxial image formation can be described by five parameters, f', m, s, s' and L. L is the distance between the object and image and is called transfer length or throw. However, two quantities are sufficient for the description of the imaging. So there are in total 30 different equations expressing each of the five parameters using a combination of two others. For $n' = n$, these relations are given in the following Table 5.1. For example, the intersection s can be calculated out of the pairs of known variables s' and f' or s' and m or l and m or L and f' or s' and L or f' and m.

In a completely afocal system, such as a Keplerian or Galilean telescope, both principal planes are at infinity and thus the focal length of such a system is not defined. Consequently, the classical lens equations relating intersection lengths cannot be applied to such systems.

5.2.4 Imaging Transfer Length

For a given focal length of the system and a given magnification m the object–image distance L between the conjugated planes for a thin lens, the relation

$$L = f' \cdot \left(2 - m - \frac{1}{m}\right) \tag{5.10}$$

out of Table 5.1 is interesting. In the case of a thick lens or a lens system, the quantities are referenced to the principal planes and their distance has to then be removed. It is directly seen in this equation that the distance L is unchanged, if m is substituted by $1/m$. Solving this equation for the magnification gives the last equation from the table

$$m = 1 - \frac{L}{2f'} \pm \sqrt{\left(\frac{L}{2f'}\right)^2 - \frac{L}{f'}} \tag{5.11}$$

Table 5.1 Paraxial imaging equations. There is no special ordering scheme in the sequence of the equations.

Quantity to be calculated	Calculation equations for one pair of given quantities		
s	$s = \frac{s'f'}{f'-s'}$	$s = s' - L$	$s = \frac{s'}{m}$
	$s = -\frac{L}{s} \pm \sqrt{\frac{L^2}{4} - f' \cdot L}$	$s = \frac{(1-m)\cdot f'}{m}$	$s = \frac{L}{m-1}$
s'	$s' = \frac{s \cdot f'}{f'+s}$	$s' = s + L$	$s' = \frac{L}{m-1}$
	$s' = \frac{L}{2} \pm \sqrt{\frac{L^2}{4} - f' \cdot L}$	$s' = f' \cdot (1-m)$	$s' = \frac{L \cdot m}{m-1}$
f'	$f' = \frac{s \cdot s'}{s - s'}$	$f' = -\frac{s \cdot (L+s)}{L}$	$f' = \frac{s \cdot m}{1-m}$
	$f' = \frac{s' \cdot (L - s')}{L}$	$f' = \frac{s'}{1-m}$	$f' = -\frac{L \cdot m}{(1-m)^2}$
L	$L = s \cdot (m-1)$	$L = s - s'$	$L = -\frac{s^2}{s+f'}$
	$L = \frac{s'^2}{s'-f'}$	$L = \frac{s' \cdot (m-1)}{m}$	$L = f' \cdot \left(2 - m - \frac{1}{m}\right)$
m	$m = \frac{s'}{s}$	$m = \frac{f'}{s+f'}$	$m = \frac{f'-s'}{f'}$
	$m = \frac{L+s}{s}$	$m = \frac{s'}{s'-L}$	$m = 1 - \frac{L}{2f'} \pm \sqrt{\frac{L}{f'} \cdot \left(\frac{L}{4f'} - 1\right)}$

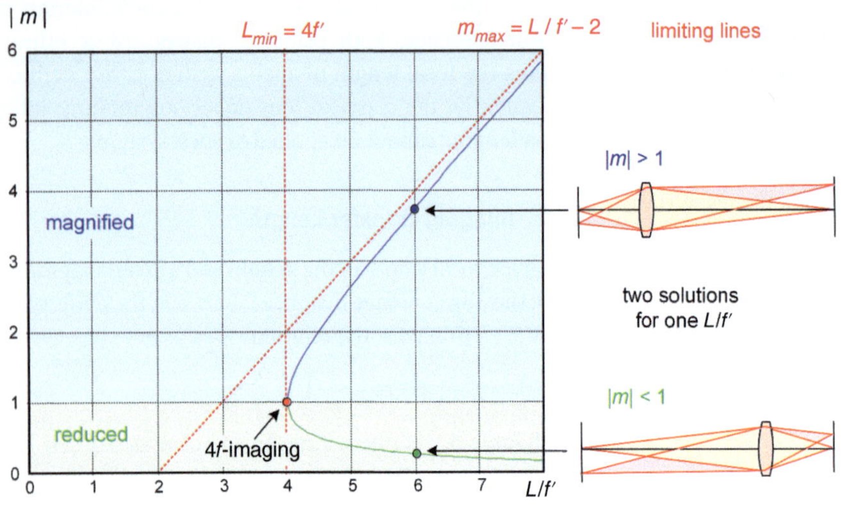

Figure 5.13 Magnification of image formation as a function of the normalized distance between object and image location L/f'.

In the Figure 5.13 this relation is plotted as a function of the normalized ratio L/f'. There are only finite image positions for values $L \geq 4f'$. There are two possible solutions in this case: one magnifies, while the other reduces the image size. Their arrangements are mirror-symmetric.

5.2.5 Principal Planes

5.2.5.1 Principal Planes of Compound Systems

The concept of principal planes is well known for single lenses (see Chapter 4.4) but can be generalized for more complex optical systems. But it should be noted that this idea works only paraxially with the reference on the optical axis; therefore, in three-dimensional geometries and off-axis setups it make no sense. The definition is in any case a surface of a single apparent ray bending or the plane where a ray height is paraxially transported with constant height, see Figure 5.14 or 4.18. The following features and properties should be kept in mind in the general case.

- The principal surface is only a plane in paraxial approximation, in general it is a curved surface.

In particular for large numerical apertures, the principal surface for a corrected sine condition is a sphere.

- The range of validity can be extended to points near the axis, which is called the isoplanatic range.
- In any case, the principal surfaces can be used to find image planes and to define and reference the focal length.

If a more complicated system is analysed, the location of the principal planes strongly depends on the distances. Figure 5.15 depicts a system of two lenses with changing distance t and a fixed image plane. The two principal plane locations are shown in green and purple here. It is seen that they can be located completely outside of the lens positions.

The general construction to find the principal planes is as follows.

- Determine the path of a ray coming parallel to the optical axis.
- Trace the ray in the image space back to the left.
- The intersection point of the incoming and the extended final ray gives the principal plane location.

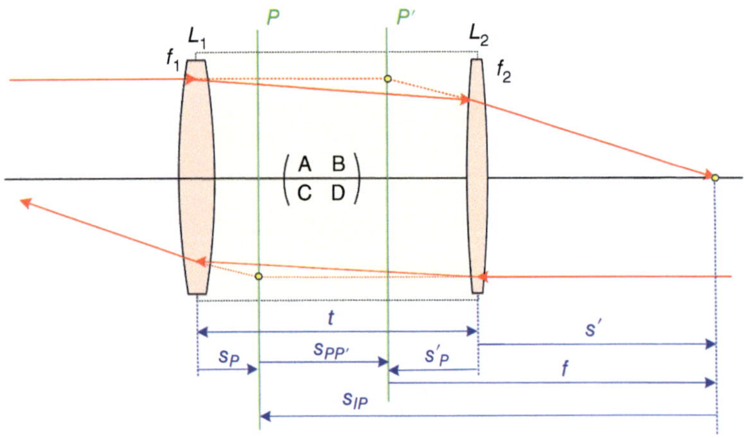

Figure 5.14 Two-lens system schematic with principal planes.

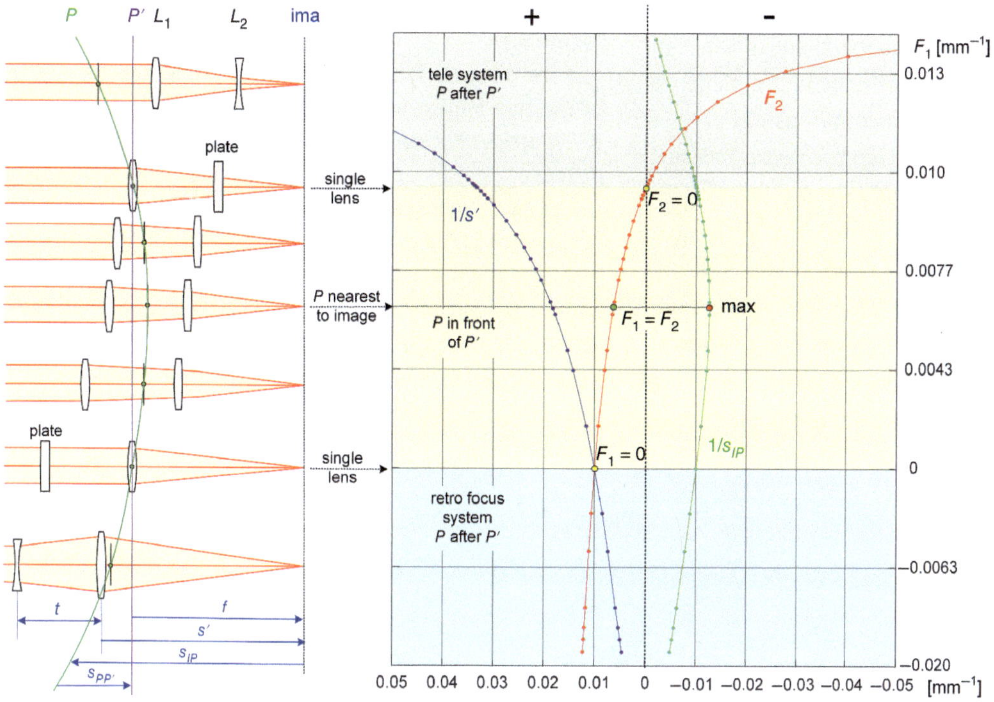

Figure 5.15 Principal planes in a two-lens system for a constant position of the plane P, the image location and the distance t. In the correct scaled diagram on the right side, the vertical axis is not linear.

In the case of multi-lens systems. it should be notid that similar to single lenses, the focal length of a system is referenced against the principal planes. For many lenses, the location of the image sided back principal plane can be a large distance from the last surface.

5.2.5.2 Principal Planes of Two-Lens Systems

If a system of two thin ideal lenses L_1 and L_2 in a distance t in air is considered, the total focal length is given by

$$f = \frac{f_1 \cdot f_2}{f_1 + f_2 - t} \tag{5.12}$$

and the paraxial system matrix reads in a reference plane directly after the lens L_2

$$\underline{M} = \begin{pmatrix} 1 & 0 \\ -1/f_2 & 1 \end{pmatrix} \cdot \begin{pmatrix} 1 & t \\ 0 & 1 \end{pmatrix} \cdot \begin{pmatrix} 1 & 0 \\ -1/f_1 & 1 \end{pmatrix} = \begin{pmatrix} 1 - t/f_1 & t \\ -1/f_1 - 1/f_2 + t/(f_1 f_2) & 1 - t/f_2 \end{pmatrix} \tag{5.13}$$

The final image distance s' as a back focal length, the location s_P of the object sided principal plane relative to lens L_1 and the location s_P' of the image sided principal plane P' relative to L_2 are given by the formulas of the $ABCD$ formalism as

$$s' = \frac{f_2 \cdot (f_1 - t)}{f_1 + f_2 - t} = (f_1 - t)\frac{f}{f_1} \tag{5.14}$$

$$s_P = \frac{D-1}{C} = +\frac{t \cdot f_1}{f_1 + f_2 - t} = \frac{t \cdot f}{f_2},$$

$$s_{P'} = \frac{1-A}{C} = -\frac{t \cdot f_2}{f_1 + f_2 - t} = -\frac{t \cdot f}{f_1}$$

$$s_{PP'} = t + s_P' - s_P = -\frac{t^2}{f_1 + f_2 - t},$$

$$s_{IP} = -s' - s_{PP'} = \frac{t^2 - f_1 f_2}{f_1 + f_2 - t} \tag{5.15}$$

s_{IP} is the distance of P relative to the image plane, s_{PP}' is the distance between the principal planes. The geometry of this two-lens system and the principal planes are illustrated in Figure 5.14.

To illustrate the variation of the principal plane location of P due to the focal power distribution between the two thin lenses, an object is assumed to be at infinity. The focal powers of the lenses are changed, while the position of the image and the principal plane P' are kept at a constant position. This is illustrated in Figure 5.15. The individual lens powers are changed in a way that the total power F is constant with $f = 100$ mm. The distance between the two lenses is invariant with $t = 70$ mm. F_1 is continuously decreasing, F_2 increasing. First the lens L_1 has positive power and F_2 is negative, P lies in front of P' in the z-direction. The system is of the telephoto type in this range. Then a special point is reached with $F_2 = 0$, the first lens alone represents the total power F. Now P lies behind P', we have a classical doublet with both lens powers positive. If F_1 reaches zero, lens L_2 constitutes the total power. For smaller F_1, now the system becomes a retrofocus type; again P' follows P. For approximately $F_1 = F_2$ the principal plane distance of P has the smallest s_{IP} value relative to the image plane. In the figure, for better scaling the values of F instead of f are used and the distances s_{IP} and s' are represented as inverted values.

It is seen that plane P is moving and can be located in front or behind P', dependent on the power ratio. In the case that one of the lenses has no power $F_j = 0$, both planes coincide. If this behaviour is drawn as a function of the powers F_1 and F_2, we get the black curve for F in the Figure 5.16.

If the distance $s_{PP'}$ is shown as a function of the powers of the two lenses, we get the Figure 5.17. Again the path of Figure 5.15 is sketched with the sign switch of the PP' distance.

A positive value of $s_{PP'}$ means that P' lies on the right side of P, which is the usual case.

The following Figure 5.18 shows a case study with a system of constant focal length $f = 100$ mm, but now three different values of the lens distance t and variation

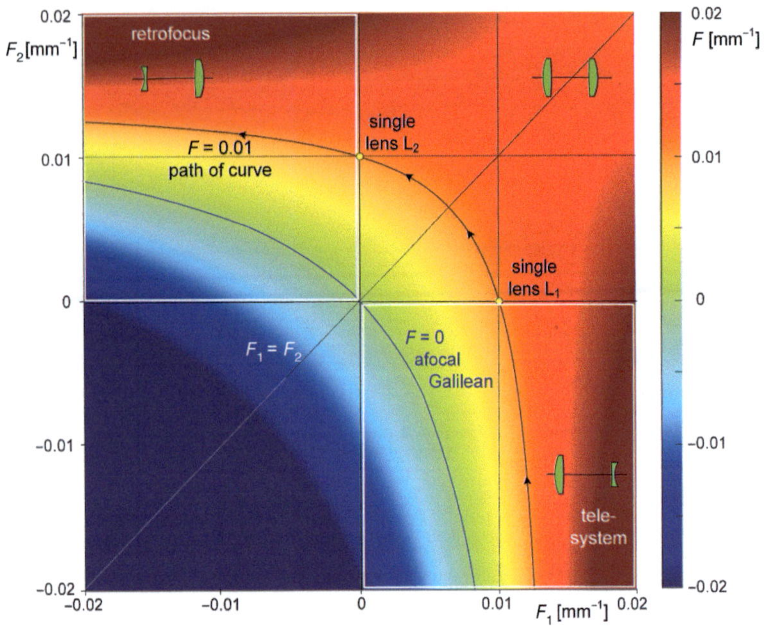

Figure 5.16 Focal power F in a two-lens-system for constant distance $t = 70$ mm.

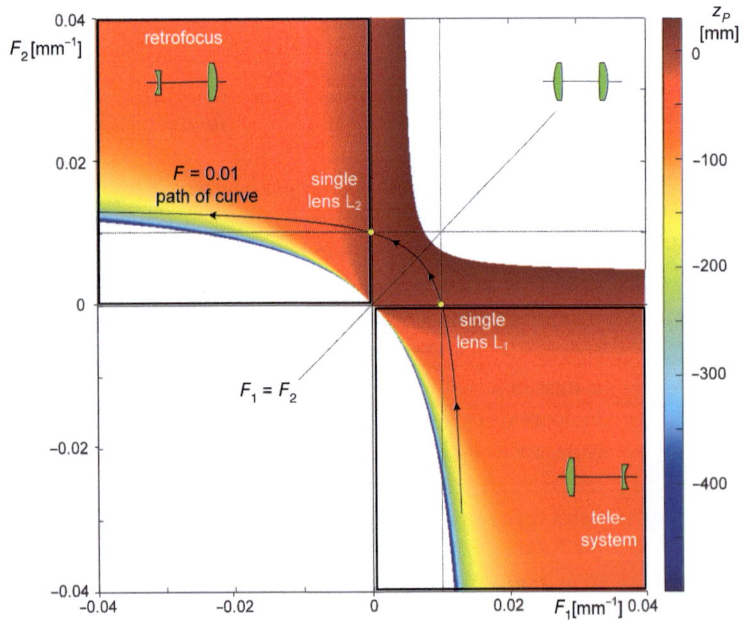

Figure 5.17 Principal plane distance $s_{PP'}$ in a two-lens-system with constant distance $t = 70$ mm. It should be noted that the limits of the shown ranges are doubled in comparison to Figure 5.16.

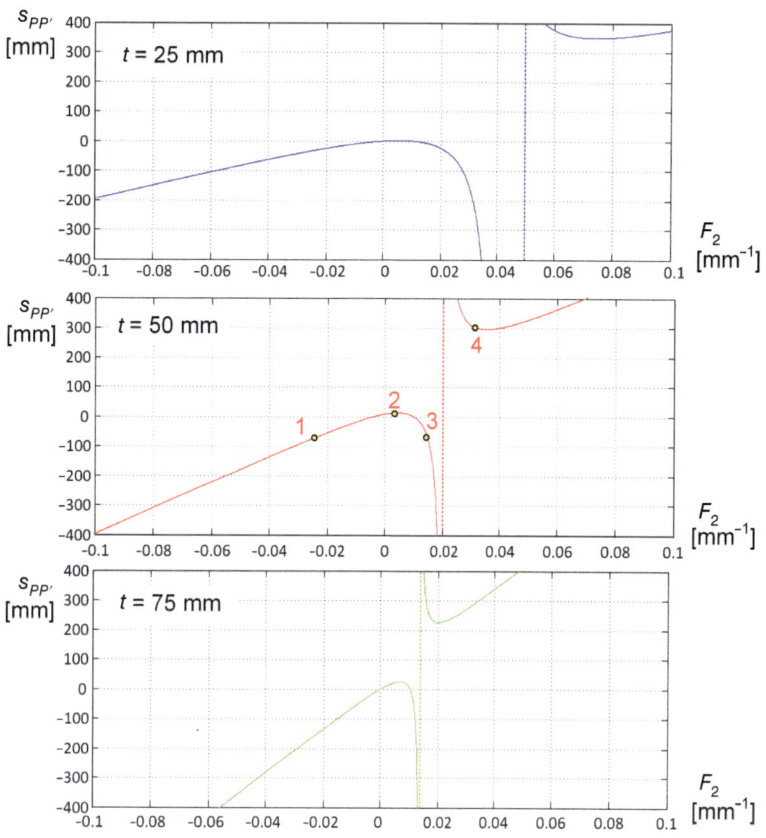

Figure 5.18 Principal plane distance $s_{PP'}$ in a two-lens system for fixed values of $f = 100$ mm and $t = 50$ mm. The red numbers in the middle diagram correspond to the four systems sketched in Figure 5.19.

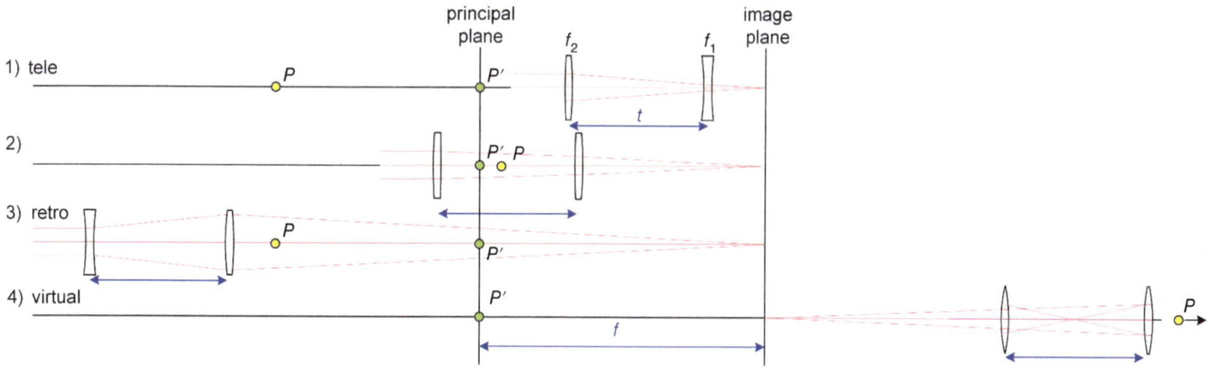

Figure 5.19 Principal planes in a two-lens system for four cases according to the parameter indicated by the four numbers in the central part of Figure 5.18.

of f_1, f_2. The power F_2 is chosen as the dependent variable.

It is seen that there are regions with positive $s_{PP'}$ and regions with negative $s_{PP'}$. Negative values for negative F_2 values correspond to telephoto type systems. The distance is singular for the special condition $f_1 + f_2 = t$.

The case with the red '1' is a telephoto case. For '2' we have a positive $s_{PP'}$. In '3' we have a conventional system. In the point '4' the image location is virtual $z' < 0$.

The following Figure 5.11 shows these four characteristic cases in a lens drawing respectively in a scaled version.

5.2.6 Multi-Lens Systems

If more than one lens constitutes an imaging system, we can consider the image of the first lens to be the object of the second lens and so on. This gives the complete behaviour of a more complex system by simply looking on the full chain of one-lens imaging setups.

If the lenses are thin and close together, the focal power (inverse value of focal length $F = 1/f$) is additive

$$F = \sum_k F_k \tag{5.16}$$

If the distance between the lenses cannot be neglected, the formula reads for the simple case of two lenses in air

$$F = F_1 + F_2 - t \cdot F_1 \cdot F_2 \tag{5.17}$$

The focal length of the combined system is

$$f = \frac{f_1 \cdot f_2}{f_1 + f_2 - t} = -\frac{f_1 \cdot f_2}{e} \tag{5.18}$$

where e is the distance between the two inner focal points F_2 and F_1' (see Figure 5.20). Since both lenses are assumed to be in air, the dash in f_j is omitted here. The back focal length of a ray incident parallel to the axis behind the second (thin) lens is

$$s' = \frac{(f_1 - t) \cdot f_2}{f_1 + f_2 - t} = (f_1 - t)\frac{f}{f_1} \tag{5.19}$$

The principal point P' of a system consisting of two thin lenses can be obtained by graphical construction as indicated in Figure 5.20. Consider a ray incident parallel to the axis that runs through the rear focal point F_1' of the first lens. A ray through the front focal point F_2 of the second lens parallel through the ray between the lenses is parallel after lens 2. The ray direction after the second lens is found to go through the intersection point Q in the back focal plane of lens 2. The position of the rear principal plane P' of the entire system relative to the last surface is then the intersection point S with the prolonged initial ray. The location s_P' relative to lens 2 then follows the equation

$$s_{P'} = \frac{-t \cdot f_1}{f_1 + f_1 - t} = -\frac{t \cdot f}{f_1} \tag{5.20}$$

Alternative formulae for the calculation of a two-component system are as follows.

- For given values of t, f, s and s', the focal length of each lens can be determined from

$$f_1 = \frac{t \cdot f}{f - s'}, \quad f_2 = \frac{-t \cdot s}{f - s' - t} \tag{5.21}$$

- For given values of s, s', t and magnification m the focal lengths are

$$f_1 = \frac{(s-t) \cdot m - s'}{m \cdot t \cdot s}, \quad f_2 = \frac{t - s \cdot m + s'}{t \cdot s'} \tag{5.22}$$

- For given values f_1, f_2, an object to image distance L and magnification m, the distance t follows as a solution of the quadratic equation

$$t^2 - t \cdot L + L \cdot (f_1 + f_2) + \frac{(m-1)^2 \cdot f_1 f_2}{m} = 0 \tag{5.23}$$

and the back focal lengths can be determined from

$$s = \frac{(m-1) \cdot t + L}{(m-1) - t \cdot m/f_1},$$

$$s' = L + s - t = L - t + \frac{(m-1) \cdot t + L}{(m-1) - t \cdot m/f_1} \tag{5.24}$$

Figure 5.21 shows the resulting focal length f of the systems as a function of the distance t between the two lenses for two given focal lengths f_1 and f_2. There are two different regions which have a positive and a negative total refractive power, respectively. They are separated by the special case of an afocal system. The diagonal green line indicates the case where $f = t$ is valid and has two solutions, for which the distance t is equal to one of the focal lengths of the components.

The refractive power of a paraxial system with three lenses having single focal lengths f_1, f_2, f_3 and with refractive indices n_1 and n_2 in the spaces in between and with distances t_1 and t_2 is more general

$$F = F_1 + F_2 + F_3 - \frac{t_1 \cdot F_1 \cdot F_2}{n_1} - \frac{t_2 \cdot F_2 \cdot F_3}{n_2} - \frac{t_1 \cdot F_1 \cdot F_3}{n_1} - \frac{t_2 \cdot F_1 \cdot F_3}{n_2} + \frac{t_1 \cdot t_2 \cdot F_1 F_2 F_3}{n_1 \cdot n_2} \tag{5.25}$$

The refractive power F of a system consisting of k elements with refractive powers F_j or curvatures c_j in case of single surfaces and refractive indices n_j and n_j' where the index $j = 1, 2, \ldots k$ is given by

$$F = \sum_{j=1}^{k} \omega_j^{(\infty)} \cdot F_j = \sum_{j=1}^{k} \frac{h_j}{h_1}(n_j' - n_j) \cdot c_j = \sum_{j=1}^{k} \omega_j^{(\infty)}(n_j' - n_j) \cdot c_j \tag{5.26}$$

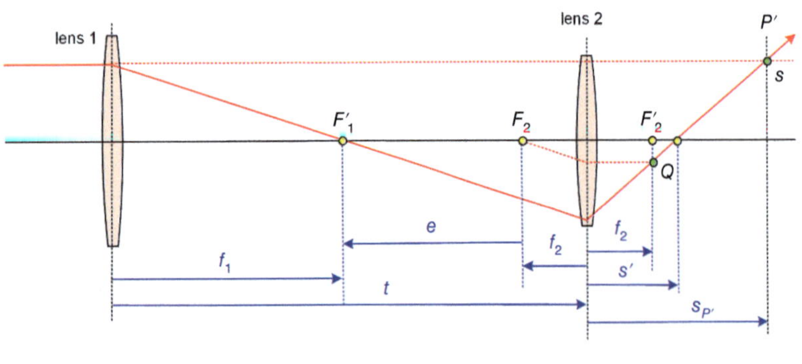

Figure 5.20 Terminology and positions of the principal points of a system consisting of two lenses and scheme for finding the location of P'.

Figure 5.21 Focal length of a two-lens system as a function of the distance, an example with $f_1 = 30$ mm and $f_2 = 50$ mm.

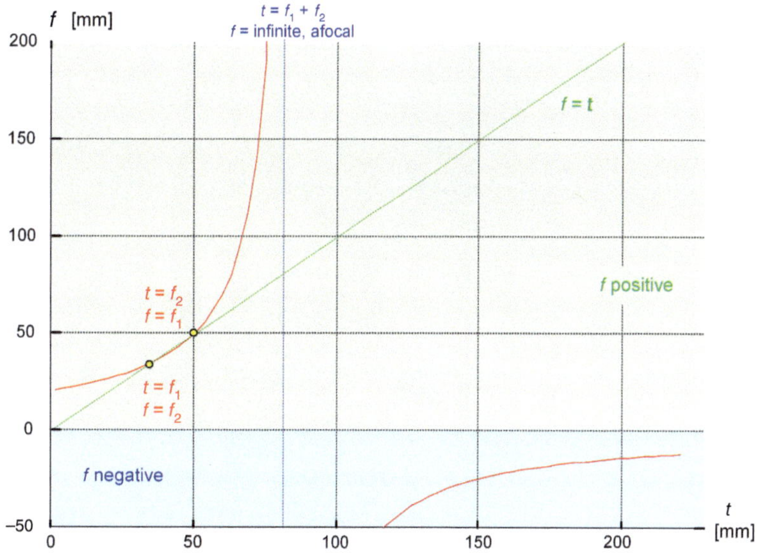

The h_j are the heights of the paraxial ray and

$$\omega_j^{(\infty)} = \frac{h_j}{h_1} \tag{5.27}$$

is the paraxial ratio of the heights. With back focal lengths sj and $s'j$ for the single components, the lateral magnification of the compound system is given by

$$m = \frac{s_1'}{s_1} \cdot \frac{s_2'}{s_2} \cdots \frac{s_k'}{s_k} \cdot \frac{n_1}{n_k'} \tag{5.28}$$

5.3 Magnification, Field of View, Aperture and Vergence

5.3.1 Introduction

The most important properties of an imaging optical system are as follows.

- The field of view: it gives the size of an object, which can be seen in the image plane. In reality, the system mostly doesn't have a field stop and there is no hard boundary to the field. Field stops can be introduced in intermediate images, for instance in a microscope. A special case is if the digital sensor with its finite size limits the field of view. Often, the light cones are shrinking towards the edges of the field of view and the usable field of view has no clear outer edge.
- The aperture, which determines the size of the light cone transporting the information. It is limited by a stop in the system, which can be a real diaphragm, the mounting or the boundary of one of the lenses. This stop is also called the pupil of the system, similar to the iris diaphragm in the human eye. The stop limits the energetic light flux through the system and is responsible for the image brightness and in case of diffraction limited system quality for the achievable resolution.
- The magnification: it is measured for finite object and image location as the ratio of the image to the object size; in the case of an afocal system it is defined as the ratio of the field angles.

Figure 5.22 illustrates the definition of the field and the aperture in an optical system with finite object and image location – a so-called relay system. Both terms are important in the object as well as the image domain. The values can be different in these cases due to a certain magnification of the system. The physical stop inside the system is imaged into the object and image domain respectively; these conjugated planes are the entrance and the exit pupil. The chief ray starts in the off-axis field point in the object plane and meets the centre of the entrance pupil. The marginal ray starts in the centre of the object plane and meets the edge of the entrance pupil. This means that the CR determines the location, the MR the size of the pupil. Entrance and exit pupils are in general not physical diaphragms, they are only images. By definition in the object or image domain respectively, there are no components between object and entrance pupil as well as image and exit pupil respectively. Therefore, these artificial planes are used to describe and fix aperture and field in both domains.

The field of view determines the size of the object area that is provided by the system; it can be defined by the object height y (in case of a finite object location) or by the field angle w.

5.3.2 Numerical Aperture and *F*-Number

The aperture determines the size of the light cone, which is transferred through an optical system. Since every system

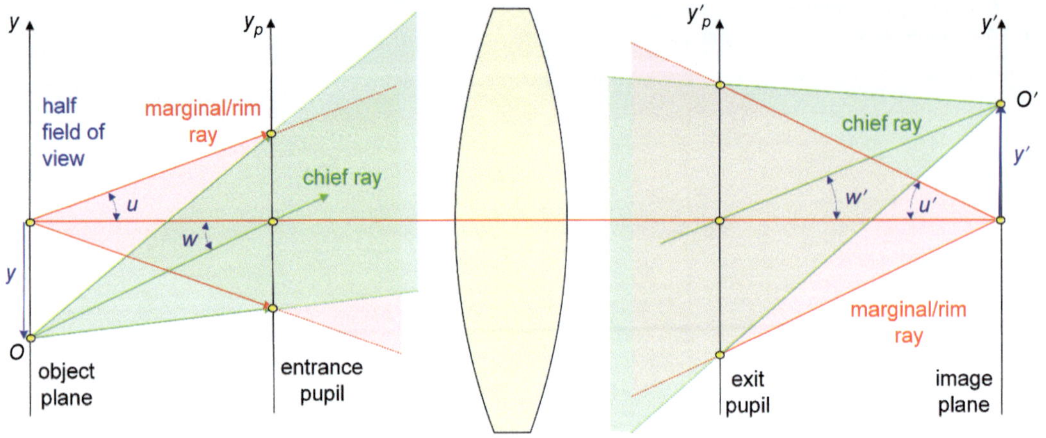

Figure 5.22 Definition of field and aperture, marginal ray (MR) and chief ray (CR).

has a finite lateral size, one component or surface or diaphragm in a system limits the light flux. In reality it is not so trivial to get the value of the open aperture cone. If the stop is positioned inside of the system, it has to be imaged in the object space. This pupil imaging has a magnification and changes the stop diameter to the entrance pupil diameter. Therefore, the real stop and its image in the object as well as the image domain must be considered. This principle problem is illustrated in Figure 5.23. Only in the case of a front stop at or before the first lens is the entrance pupil identical to the stop and the situation is quite simple.

The size of the cone on axis is defined by the marginal ray angle u, which is called the aperture angle. The opening or the size of the light cone which is transmitted by the system can be defined in different ways. One possibility is to fix the size by the numerical aperture, which is the sine of the half angle of the light cone, multiplied by the refractive index.

$$NA = n \cdot \sin u \tag{5.29}$$

This definition is used in microscopy; it fails in cases of an afocal setup like a telescope.

The second opportunity is to define the opening by the F-number; this is usually done for photographic systems. The F-number is the focal length divided by the entrance pupil diameter, assuming an object location in infinity, see Figure 5.24

$$F_{\#} = \frac{f}{D_{EnP}} \tag{5.30}$$

In addition, a more complicated definition exists for the F-number in cases of a finite object location, see next section.

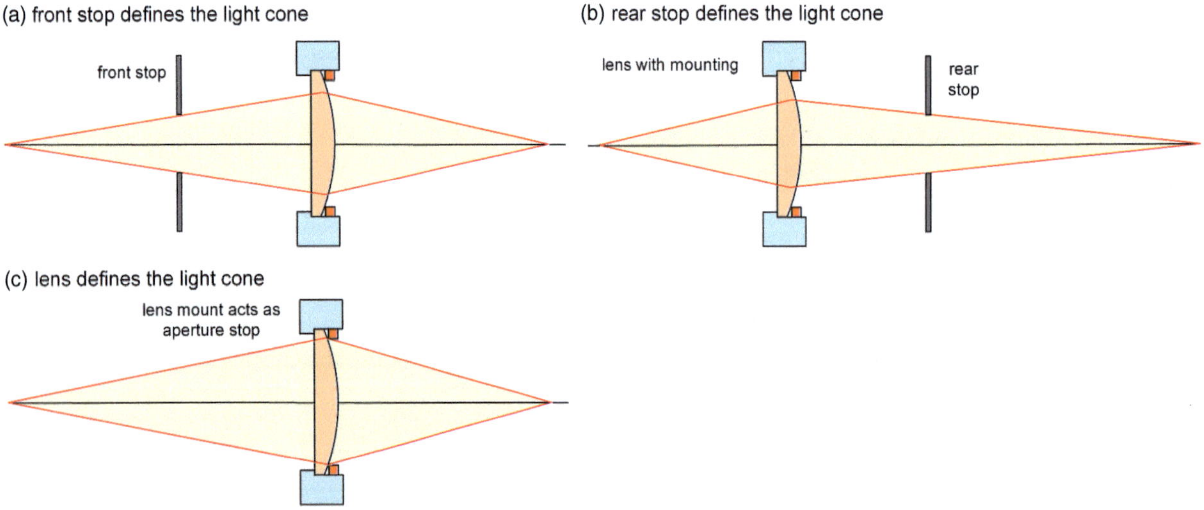

Figure 5.23 Stop and pupil in a simple system of one lens, in (a) the front surface and in (b) the rear stop defines the size of the light cone and (c) the first lens mounting limits the light cone.

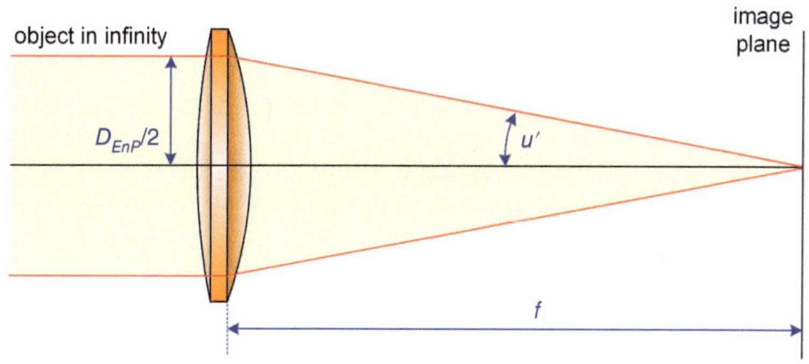

Figure 5.24 Definition of the F-number.

For small aperture angles, an approximated relation between both definitions is given in air by

$$F_{\#} = \frac{1}{2NA} \tag{5.31}$$

It is seen that a large F-number corresponds to a small NA and vice versa.

5.3.3 Generalized *F*-number

The traditional definition of the F-number is useful for camera systems with the object located in infinity. In this case, the diameter of the incoming ray bundles corresponds to the entrance pupil size, which is used in the $F\#$-definition. If the object location is finite, the entrance pupil diameter is not obtained so simply. For corresponding systems, a so-called effective or working F-number is defined. If m is the system magnification, the image location is given by

$$s' = f' \cdot (1 - m) \tag{5.32}$$

In the image domain, the aperture cone angle is related to the exit pupil by

$$\sin u' = \frac{D_{ExP}}{2s'} = \frac{D_{ExP}}{2f'(1-m)} = \frac{m_p D_{EnP}}{2n'f(1-m)} \tag{5.33}$$

with the pupil imaging magnification mp. If these equations are inserted into the relation (5.30), the formulation of the effective F-number is obtained as

$$F_{\#}^{eff} = \frac{1}{2n' \cdot \sin u'} = F_{\#} \cdot \frac{1-m}{m_p} \tag{5.34}$$

The corresponding quantities can be seen in Figure 5.25.

5.3.4 Effective Aperture*

(*) The content of this section is quite specialized and for advanced reading. Knowledge of this section is not necessary to understand the following chapters. Beginners are recommended to skip this section.

If the boundary shape of the aperture in the exit pupil is quite complicated and the aperture angle u of the marginal rays strongly depend on the azimuthal angle θ, according to [1] it is possible to define an averaged effective numerical aperture by the expression

$$NA_{eff} = n \cdot \sqrt{\frac{1}{2\pi} \int \sin^2 u(\theta)\, d\theta} \tag{5.35}$$

Figure 5.26 illustrates the exit pupil shape and the geometry of the light cone with a more complicated cross section area.

5.3.5 Magnification

5.3.5.1 Finite Systems

The magnification is defined as the relative enlargement of the object. If object and image are located in a finite distance, the definition is therefore simply

$$m = \frac{y'}{y} \tag{5.36}$$

see also Figure 5.27. Depending on the sign of y and y', it may be that m is negative.

If a system is paraxially approximated, due to the Helmholtz–Lagrange invariant, the magnification can also

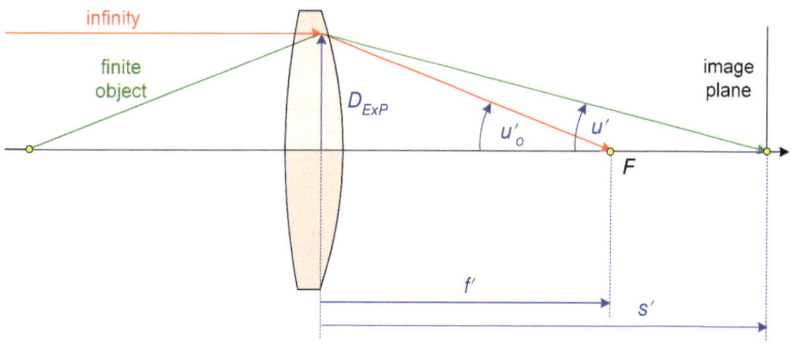

Figure 5.25 Definition of the generalized F-number for finite object locations.

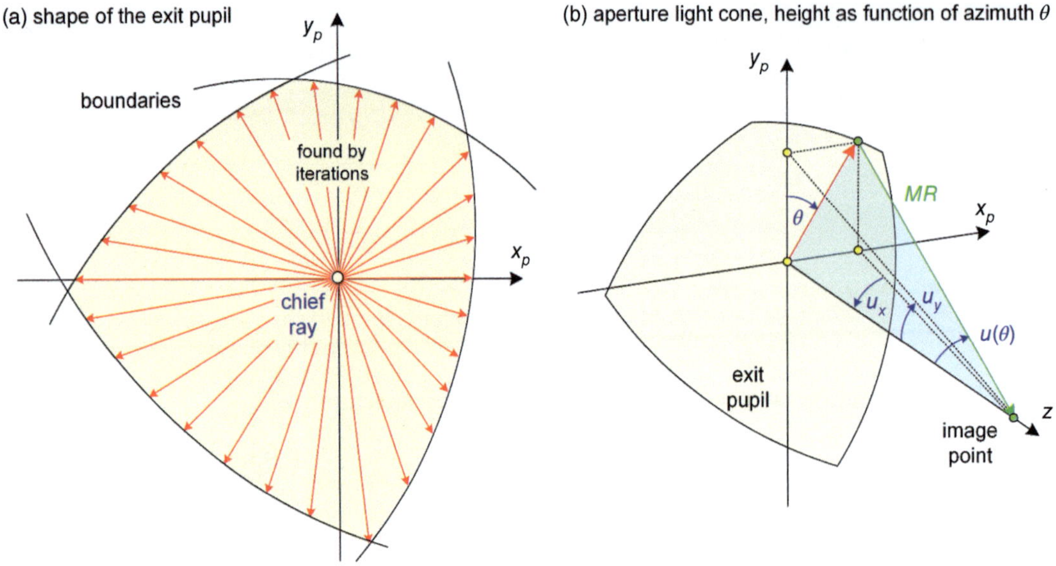

Figure 5.26 Definition of the effective numerical aperture in case of a complicated cross section area of the light cone with the shape of exit pupil and the definition of the azimuthal dependent aperture angle $u(\theta)$.

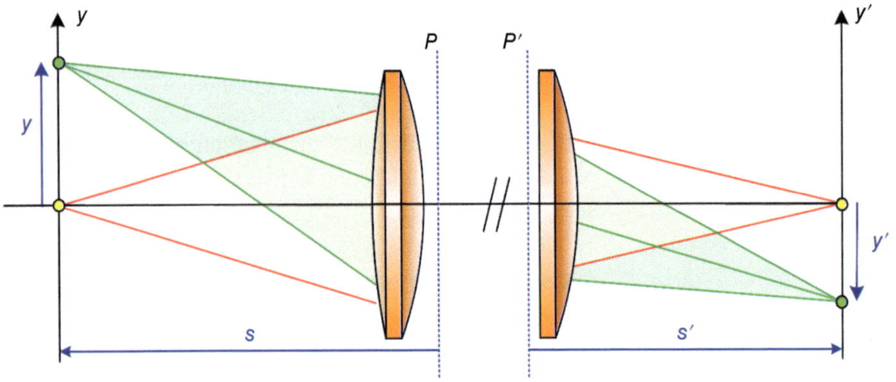

Figure 5.27 Definition of magnification in case of a finite object-image location.

be written by the aperture angles u, u' or with the intersection lengths s, s', which are measured against the principal planes.

$$m = \frac{y'}{y} = \frac{n \cdot u}{n' \cdot u'} = \frac{n \cdot s'}{n' \cdot s} \tag{5.37}$$

Usually, it is assumed that the magnification is determined by the height of an image point, generated by a tangential one-dimensional ray fan. But in the case of asymmetry in the field bundle due to coma, the selection of the ray fan matters. In particular in the definition of the sine condition, therefore, a special so-called sagittal magnification m_s is defined. According to Figure 5.28, the magnification is defined by the y-image height lying in the meridional section, but formed with the aperture rays in sagittal plane only.

$$m_s = \frac{y_s'}{y_s} = \frac{n \cdot u_s}{n' \cdot u_s'} \tag{5.38}$$

This should not be confused with a magnification of lengths lying in the x-direction

$$m_x = \frac{x'}{x} \tag{5.39}$$

If a system suffers from strong distortion, the growth of the image height y' changes non-linearly with the object height, see Figure 5.29(a). This means, for a finite field height y, the relative magnification locally at the field height y is different from the ideal one. Therefore, the local magnification

$$m_{loc}(y) = \frac{\partial y'}{\partial y} \approx \frac{\Delta y_r'}{\Delta y}, \quad m_o = \frac{\Delta y_o'}{\Delta y} \tag{5.40}$$

can be quite different from the paraxial magnification m_o, see Figure 5.29(b).

5.3.5.2 Influence of the Pupil Upon the Magnification

The magnification is defined by comparing the chief ray height or angle in the object and image plane respectively.

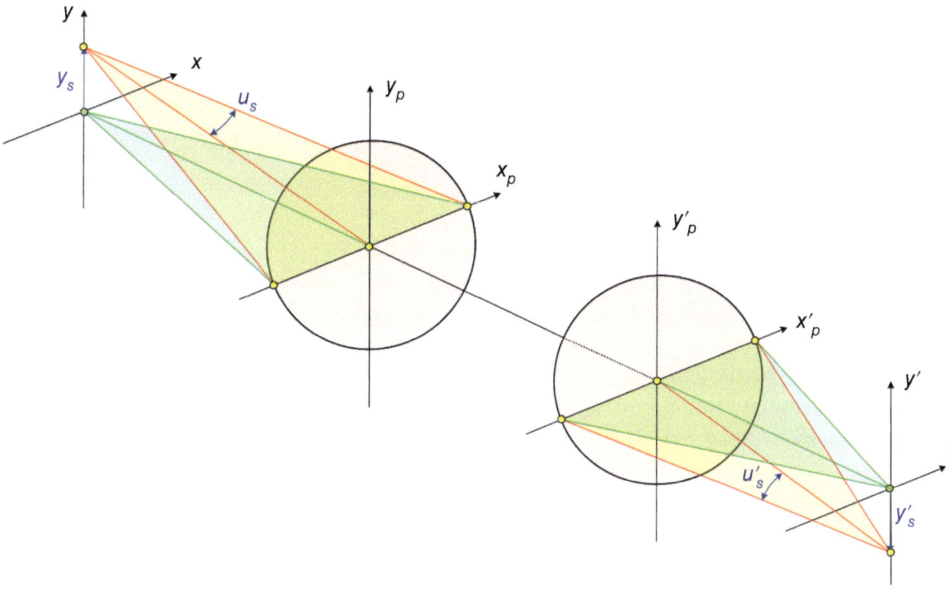

Figure 5.28 Definition of the sagittal magnification.

Figure 5.29 Definition of the magnification. (a) shows the $x'-y'$ image plane in case of a pincushion distortion, (b) illustrates the relation between object y and image height y' in case of distortion and defines the local magnification as a change of the slope.

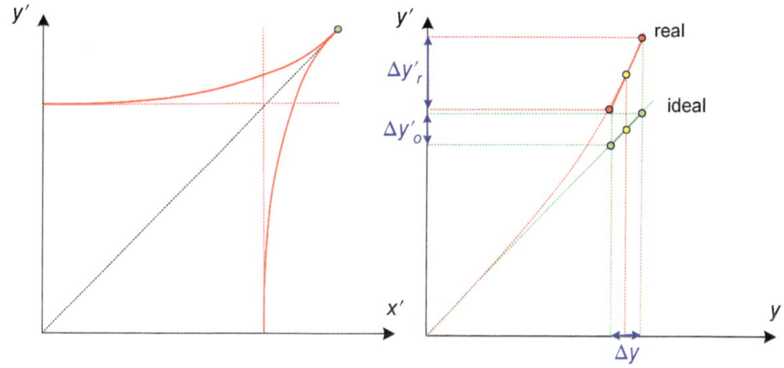

On the other side the chief ray is defined by the stop and pupil inside a system. Therefore, it should be discussed here how the stop position is related to the magnification.

Let s and s_p be the distances of the object and the entrance pupil from the principal plane P, u is the angle of the marginal ray, i.e. the aperture angle and y the size of half the field of view. In the image space the corresponding quantities are primed. With the magnification m for the object we have (see Figure 5.30) in paraxial approximation

$$m = \frac{y'}{y} = \frac{n \cdot u}{n' \cdot u'} = \frac{ns'}{n's} \tag{5.41}$$

The pupil magnification m_p is given by the ratio of the pupil diameters, in case of finite pupil locations we have

$$m_p = \frac{D_{ExP}}{D_{EnP}} = \frac{p' \cdot u'}{p \cdot u} \tag{5.42}$$

Figure 5.30 Image of the field of view and the pupil.

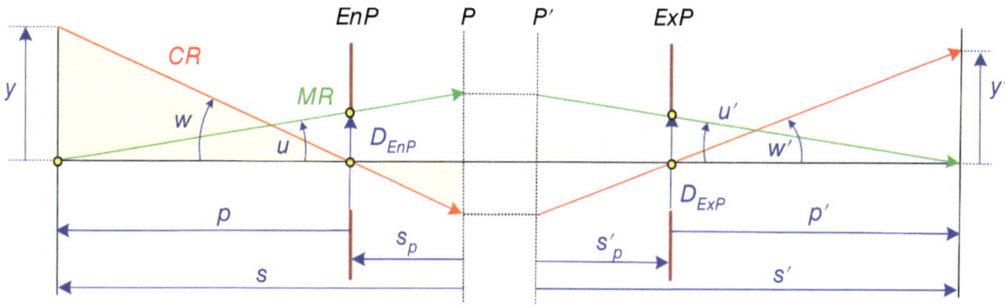

By elimination of the ratio of the angles we get the equation

$$m \cdot m_p = \frac{n \cdot p'}{n' \cdot p} \tag{5.43}$$

This simple relation shows the connection of the field and the pupil magnification. In the case of larger non-paraxial angles, the situation is more complicated. Then the relation depends on whether the sine-condition is fulfilled or not. This is discussed in more detail in Volume 2, Chapter 17.

If the stop is shifted inside the system keeping the aperture constant, the chief ray and the skew ray bundle changes. According to the Lagrange invariant L measured in the object space, this doesn't change the value of L. Therefore, this must be true also in the image domain, so the image height y' stays the same and the magnification m is invariant and independent on the stop position. This is sketched in Figure 5.31(a) schematically and in (b) by a real system raytrace. The corresponding changes in p, D_{EnP} and w can also be directly analysed in the geometry of Figure 5.31(a) and the same result is obtained. If for the pupil imaging the lens makers' formula

$$\frac{1}{s_p'} - \frac{1}{s_p} = \frac{1}{f} \tag{5.44}$$

is used, it is seen that the pupil magnification changes if the stop is moved

$$m_p = \frac{s_p'}{s_p} = \frac{f}{s_p + f} \tag{5.45}$$

and f remains invariant.

If a finite imaging system with general paraxial $ABCD$ matrix and incoming and final intersections s and s' is considered, as is shown in Figure 5.32, it is seen that a simple angle magnification of the chief ray is not meaningful and changes by shifting the pupil. If the matrix from the

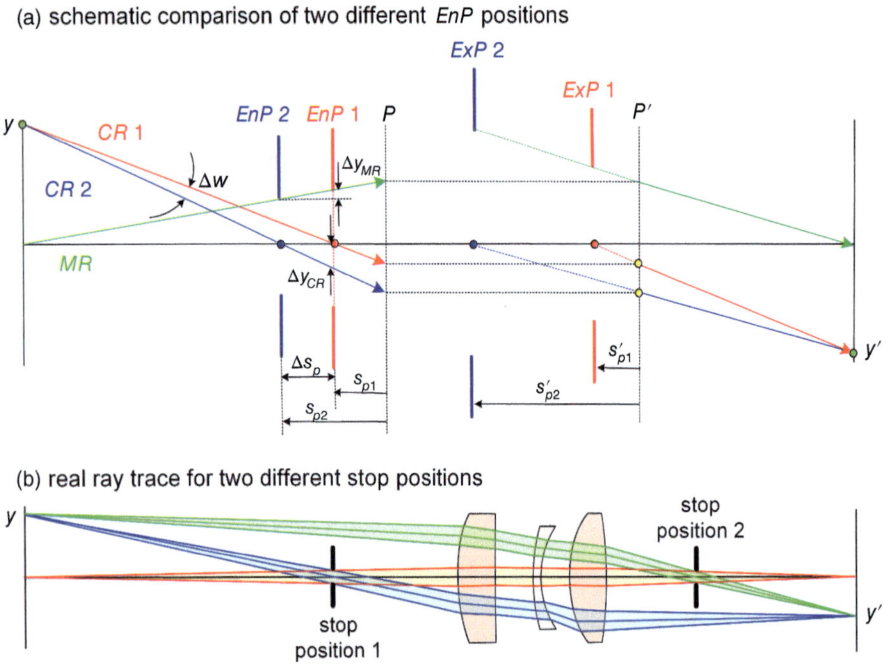

Figure 5.31 Independence of the spatial magnification on the stop position shown in (a) schematically and (b) for a concrete system example.

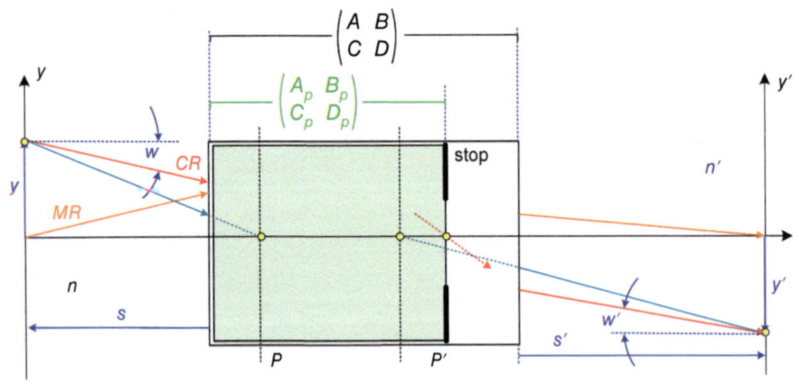

Figure 5.32 Change of the angular magnification with stop position.

ABCD entrance plane to the internal stop is denoted by $A_pB_pC_pD_p$, an angular magnification of the chief ray angles is defined by

$$m_a = \frac{w'}{w} \tag{5.46}$$

The matrices from the object plane to the image and the stop plane respectively reads

$$\underline{M} = \begin{pmatrix} 1 & s' \\ 0 & 1 \end{pmatrix} \cdot \begin{pmatrix} A & B \\ C & D \end{pmatrix} \cdot \begin{pmatrix} 1 & s \\ 0 & 1 \end{pmatrix} = \begin{pmatrix} A+s'C & B+sA+s'D+ss'D \\ C & D+sC \end{pmatrix}$$

$$\underline{M}_p = \begin{pmatrix} A_p & B_p \\ C_p & D_p \end{pmatrix} \cdot \begin{pmatrix} 1 & s \\ 0 & 1 \end{pmatrix} = \begin{pmatrix} A_p & B_p + sA_p \\ C_p & D_p + sC_p \end{pmatrix} \tag{5.47}$$

The condition for the stop location is $y_{CR} = 0$ and gives with the matrix $\underline{M}_p$

$$y'_{CR} = A_p \cdot y + (B_p + s \cdot A_p) \cdot w = 0 \tag{5.48}$$

The calculation of the image sided chief ray angle from the full system matrix $\underline{M}$ reads

$$w' = C \cdot y + (D + s \cdot C) \cdot w \tag{5.49}$$

By eliminating y out of these two equations, the angle magnification is obtained

$$m_a = \frac{w'}{w} = D - C \cdot \frac{B_p}{A_p} \tag{5.50}$$

It is seen that the angle magnification changes with the stop position through the impact of the coefficients A_p and B_p.

5.3.5.3 Magnification of Afocal Systems

For an afocal system the angle magnification is denoted by Γ and follows from the ratio of the chief ray angles before and after the system. In paraxial approximation, we get [2] the so-called telescopic magnification

$$\Gamma = \frac{w'}{w} = \frac{nD}{n'D'} \tag{5.51}$$

If the object and image side are afocal, a telescope is obtained as is demonstrated in Figure 5.33. It is seen in the

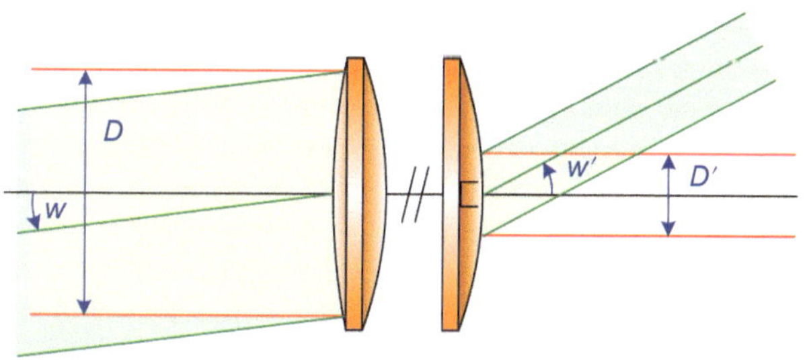

Figure 5.33 Definition of magnification in case of an afocal system.

last expression that the magnification Γ can also be defined by the change of the collimated bundle diameters D and D' respectively, if the refractive index is taken into account. This expression is a simple consequence of the Lagrange invariant. Since the ratio D'/D is the magnification m_p of the exit pupil relative to the entrance pupil

$$m_p = \frac{D'}{D} = \frac{D_{ExP}}{D_{EnP}} \tag{5.52}$$

This finally means that the angle magnification and the pupil magnification are related by

$$\Gamma \cdot m_p = \frac{n}{n'} \tag{5.53}$$

This relation can also be derived directly from the Lagrange invariant.

5.3.5.4 Depth Magnification

If the lens makers' formula is written as

$$s' = \frac{s \cdot f}{s+f} \tag{5.54}$$

the derivative gives the local depth, axial or longitudinal magnification m_z

$$m_z = \frac{ds'}{ds} = \frac{f(s+f) - sf}{(s+f)^2} = \frac{f^2}{(s+f)^2} = m^2 > 0 \tag{5.55}$$

It is seen that m_z is always positive, a movement in the object domain to the right also generates a shift in the image domain to the right, see Figure 5.34. It is well known that the lateral magnification m is constant in an object plane in paraxial approximation. But as is seen from Eq. (5.55), the axial magnification changes with s and is not invariant.

5.3.6 Vergence*

(*) The content of this section is quite specialized and for advanced reading. Knowledge of this section is not necessary to understand the following chapters. Beginners are recommended to skip this section.

If the image location or generally intersection lengths s, s' are considered, the afocal setup is always a singular case and must be especially taken into account. Therefore, in many applications dealing with an afocal situation, it is more convenient to consider the reciprocal value of the intersection length, the so-called vergence. In a more general case this value is multiplied with the refractive index, which is a kind of normalization. If s is the distance to the image position or the centre of the spherical wave in a medium of refractive index n, the vergence is defined as

$$v = \frac{n}{s} \tag{5.56}$$

Positive values of v denote a convergent, negative values indicate a divergent wave with a virtual image location. The afocal case corresponds now simply to the vanishing

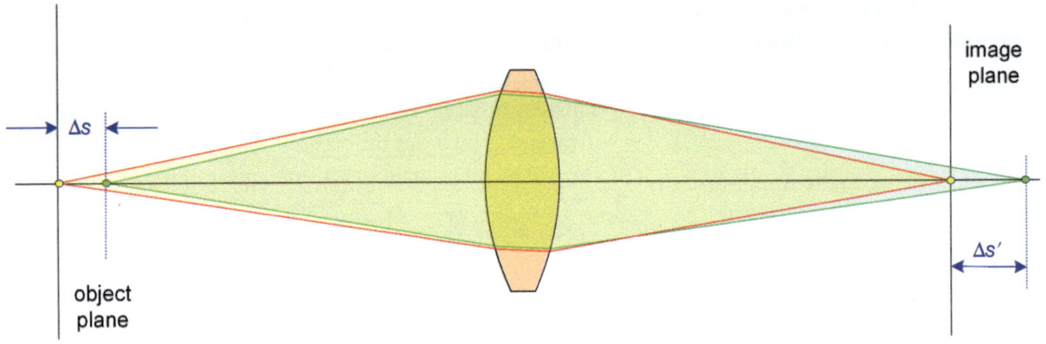

Figure 5.34 Definition of the axial or longitudinal magnification.

vergence $v = 0$. But it should be noted that in the case of a non-circular symmetric system, the vergence must always be referenced on the axis ray, which can be bent too. In practice, the vergence, which has the unit of an inverse length, is mostly scaled in diopters. One diopter is 1 dptr = $1/m$. In particular in ophthalmology this use of the vergence is usual, because the relaxed human eye assumes an afocal infinity object distance.

In the case of an imaging system, the lens makers' formula for the vergence reads simply as

$$v' = v + F \tag{5.57}$$

with the focal power F of a surface or lens. This simple linear relationship corresponds to the Figure 5.10 but is now a simple straight line, as is seen in Figure 5.35.

5.4 Pupil

5.4.1 Introduction

The stop or pupil is a surface or location inside an optical system that limits the light flux and defines the chief ray path and the marginal ray. The chief ray intersects the optical axis in the stop plane and fixes its z-location, this is illustrated in Figure 5.36 for a front stop. The angle of the provided light cone or the aperture is described by the marginal ray. It intersects the stop plane at the largest height (edge) and therefore fixes its size. The half cone angle is the aperture angle u. For the off-axis field point, the throughput of light is described by a similar but asymmetric light cone. It is limited in the meridional plane by the upper

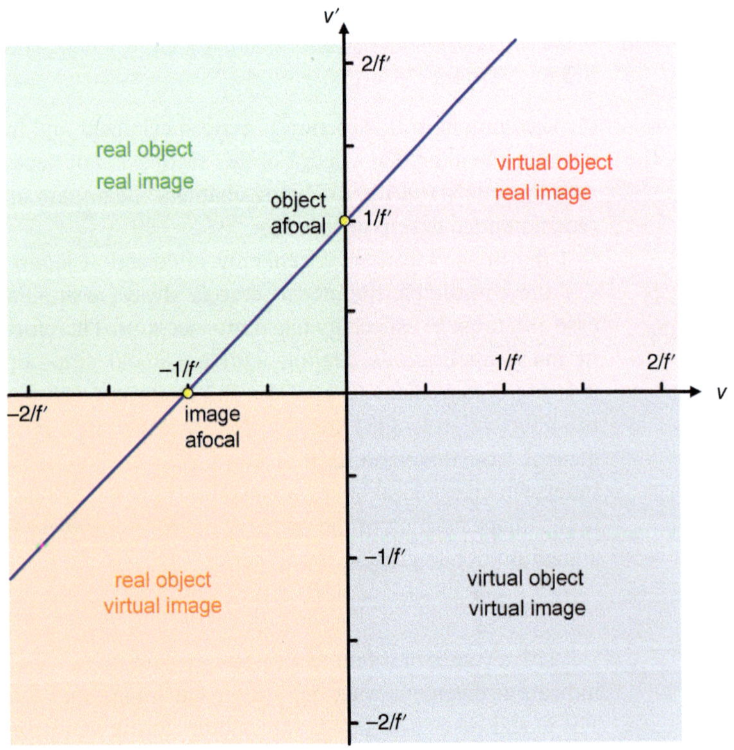

Figure 5.35 Illustration of the lens makers' formula for the vergence.

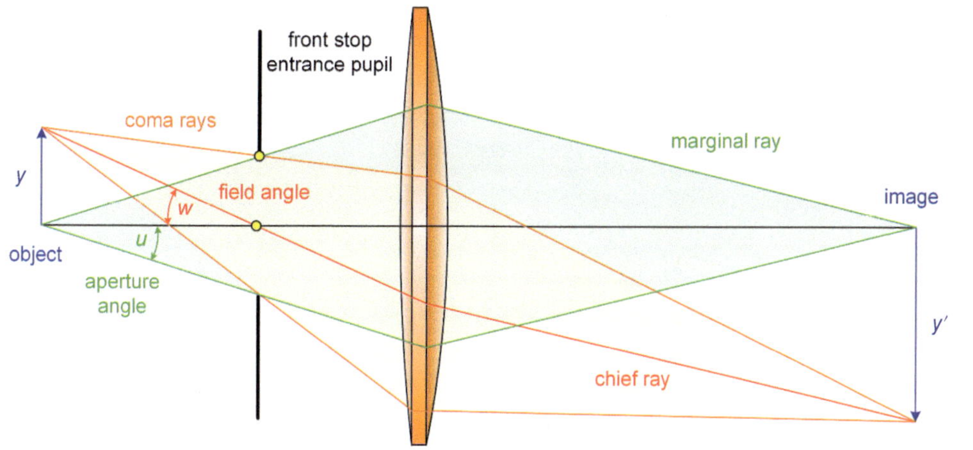

Figure 5.36 Pupil, chief ray and marginal ray in case of a front stop.

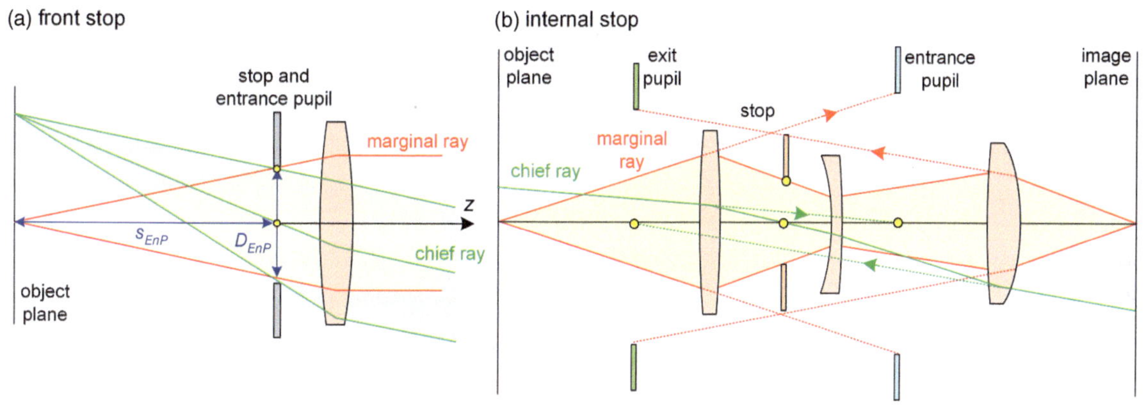

Figure 5.37 Entrance pupil, chief ray and marginal ray in case of (a) a front stop and (b) an internal stop position.

and lower coma ray. For small chief ray angles the chief ray lies in the centre of the oblique light cone. For larger field angles, the skew conditions create a certain asymmetry.

In general, the physical stop limits the light flux and is defined somewhere in the system, mostly by a lens mounting or a diaphragm. In this case, the image of the stop into the object space gives the entrance pupil, the image into the image space gives the exit pupil. This idea is sketched in Figure 5.37(b). The entrance as well as the exit pupil are therefore artificial images of the physical stop; all the three planes are conjugate to each other. It should be noted that the entrance pupil is defined as lying in the object domain, the medium here is assumed to be homogeneous. Therefore, no further components are between the object and the entrance pupil and the marginal and chief ray are not deflected in between. The same happens in the image domain with the image and the exit pupil. Very often the entrance pupil and the exit pupil are virtual images of the stop and their planes cannot be accessed.

The relevance of the system pupil is great. It determines:

- brightness of the image, transfer of energy;
- resolution of details, information transfer for systems with diffraction limited performance;
- image quality, aberrations due to the aperture size;
- image perspective, perception of depth;
- in the case of compound systems, it is necessary that the pupils of the sub-systems match in location and size.

All the three planes of stop, entrance and exit pupil are conjugated to each other. The entrance pupil is found in a real system through the following steps (see Figure 5.38).

- Find the ray starting from the off-axis object point that intersects the optical axis in the plane of the stop.
  This is the chief ray.
- Extend this chief ray in the object space to meet the optical axis, this is the z-location of the entrance pupil.
- Find the marginal ray that meets the stop at the edge.
- Extend the marginal ray in the object space to intersect the entrance pupil plane, this defines the size of the entrance pupil.
- Extend both chief and marginal ray similarly in the image space, this delivers the exit pupil position (chief ray) and size (marginal ray).

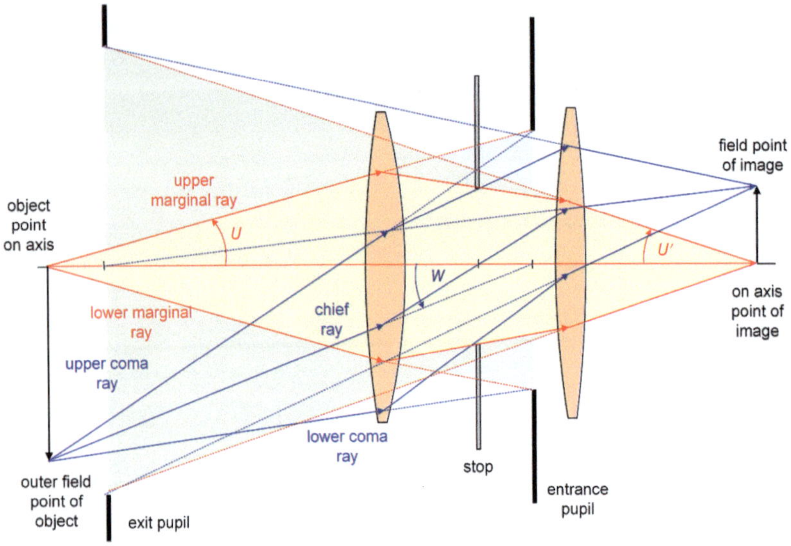

Figure 5.38 Construction of entrance and exit pupil.

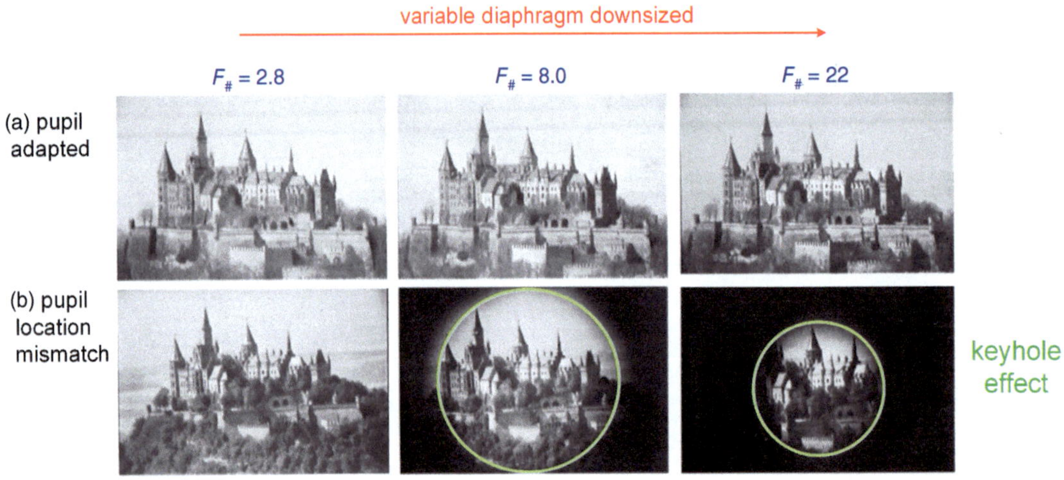

Figure 5.39 Keyhole effect for non-matched pupil locations [3] / with permission of H. Schlemmer. In the upper row (a) the pupil is well matched, in the lower row (b) the pupil is displaced along z and vignetting occurs with a keyhole effect.

5.4.2 Pupil Matching in Composed Systems

If in cascaded systems the pupils of the various parts are not matched, the field of view is reduced and the so-called keyhole effect occurs. Figure 5.39 shows a series of images made by a two-step optical system with increasing F-number. In the first row (a) the pupil locations are matched, the only impact of the changed aperture by stopping down a variable diaphragm is a loss in resolution and brightness. These effects are not really resolved in the images. In the second row (b) the pupil location is not well matched. It is seen that for smaller F-numbers the real field of view is shrinking.

Figure 5.40 shows the keyhole effect of a reduced field of view schematically. The image of the first subsystem works as the object of the second subsystem, the light runs from left to right. In (a) there is a perfect match and the field after the second system has the full size. In case (b) a displacement Δz is introduced. If now the new chief ray is constructed (in orange), which belongs to the full field of the second system, it covers only part of the field of the first system. This means that not the full field is observed. In the lower part of (b) it is visualized that in addition a reduced aperture is obtained in case of the z-mismatch.

But it should be noted that the case here corresponds to finite locations of field and aperture. The discussed consequences in the field are not observed for a telecentric system and the reduction of the aperture is not taking place in an afocal system.

5.4.3 Field Lenses

Field lenses are lenses that are located in or very near to an image plane. Only particular systems with an intermediate

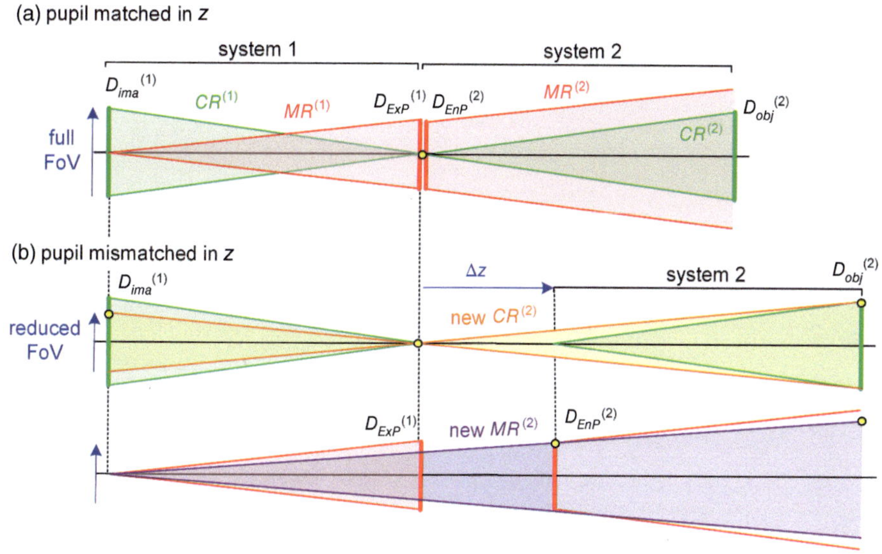

Figure 5.40 In (a) a perfectly matches system concatenation is shown; in (b) a reduced field of view in the case of the concatenation of two systems with z-mismatch of the pupil is obtained as well as a reduced aperture.

image allow this special feature. Since the marginal ray height is zero in a field plane, it is not affected by the field lens. On the other side the chief ray height is maximum here and therefore the chief ray is influenced considerably. If the chief ray path is changed, as a consequence the pupil plane is shifted along the optical axis. Therefore, field lenses allow us to change the pupil location, keeping the field planes invariant and sharp.

Field lenses can be used to match the pupil location to a desired position of a following system, in particular, for instance, to generate telecentric chief ray paths in endoscopes. Furthermore, field lenses can help to compress the chief ray path to smaller diameters to keep the system smaller in size.

Figure 5.41 shows a Keplerian telescope in two different overlaid realizations. The stop is at the first lens. In the first setup in red, no field lens is used, the red rays indicate the exit pupil and the size of the second lens. In the green case a field lens is added in the intermediate image. The size of the second lens can be reduced, the exit pupil is shifted towards the system; all these changes have advantages.

One of the drawbacks of field lenses is their natural property that they are sharply imaged. This means that any defects, scratches, glass bubbles etc., can be directly observed. In particular for a system with small NA and therefore small ray bundle diameters this is critical. It may happen that a dust particle blocks one pixel of the image sensor.

Figures 5.42 and 5.43 show a second example of an endoscopic system with objective, relay system and eyepiece with two intermediate image locations in between. Figure 5.42 illustrates the three subsystems separately, together with the numbers of the field angle sine values at the intermediate images. It is seen that the objective creates a divergent chief ray, although the relay assumes a convergent one. Second, the relay creates a larger divergent field

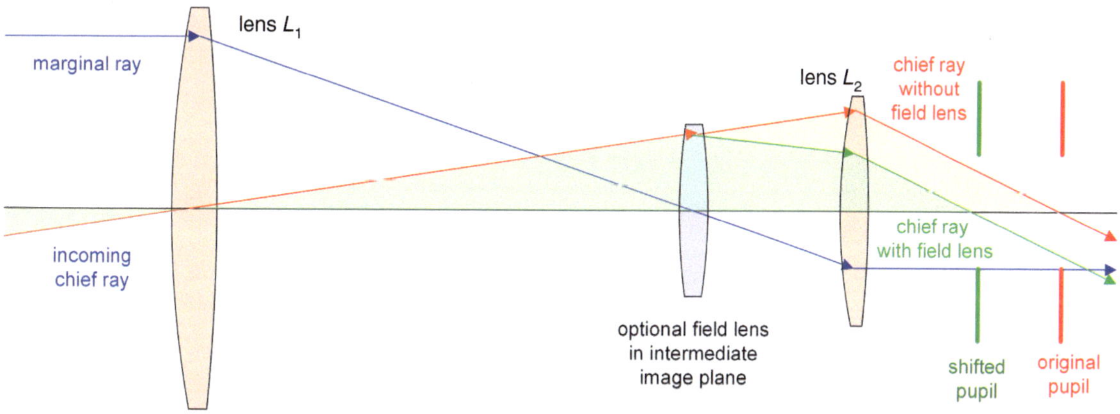

Figure 5.41 Kepler telescope with two cases, drawn overlaid. In the green case a field lens is included, in the red case this lens is removed.

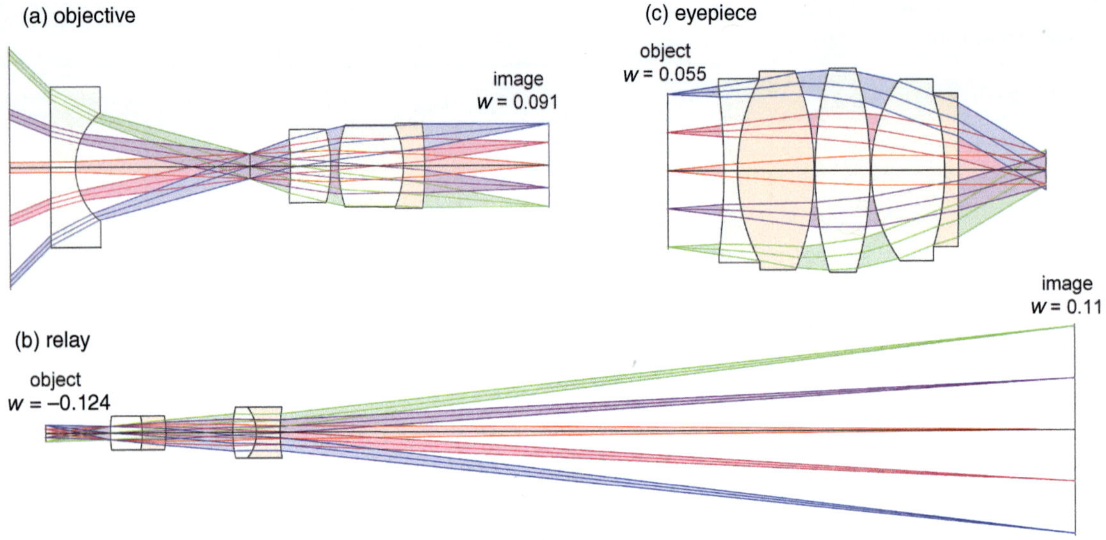

Figure 5.42 Subsystems of a compound endoscope with (a) objective lens, (b) relay and (c) eyepiece. The chief ray angles w are all small, so the sine is approximated by the angle itself.

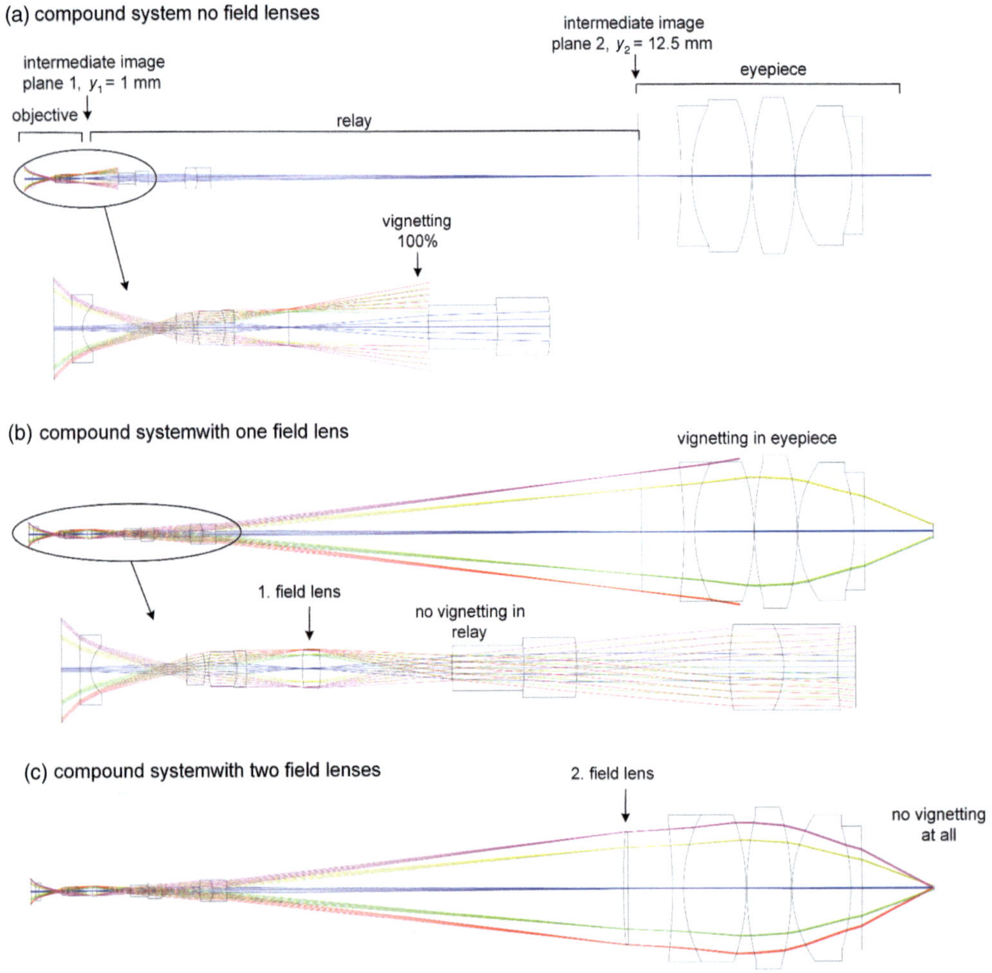

Figure 5.43 Endoscopic system with objective lens, relay and eyepiece and two intermediate images with (a) no field lens, the outer field is totally blocked, (b) one field lens in front of the relay, the outer field is transported but with vignetting and (c) with two field lenses, the full field is observed. Adapted from [4]. In these drawings all the sizes are correctly to scale.

angle as the acceptance of the eyepiece. The three parts are not to scale in the drawing, in reality the intermediate image sizes of $y_1 = 1$ mm and $y_2 = 12.5$ mm are perfectly matching.

In Figure 5.43 the three system parts are concatenated with matching intermediate image plane positions. (a) no field lens is used, the outer field of view is completely blocked, the overall size of the field of view is strongly reduced. In case (b) one field lens is introduced between the objective and relay, this improves the field size, but now a part of the ray bundle is vignetted in the eyepiece. In case (c) two field lenses are used; now the full field is provided.

5.4.4 Pupil and Perspective*

(*) The content of this section is quite specialized and for advanced reading. Knowledge of this section is not necessary to understand the following. Beginners are recommended to skip this section.

According to the equation

$$m = \frac{f'}{s+f'} \tag{5.58}$$

(see Table 5.1), the magnification depends on the object location relative to the imaging lens. This means, at different distance s, the magnification changes. In the usual case of a small compact lens, an object detail with larger distances appears smaller, this is the natural visual impression of the human eye observation. In technical imaging systems consisting of many lenses, it is possible to move the pupil by corresponding design requests. If the pupil is between the object and the principal plane, the so-called natural entocentric perspective is obtained, shown in green in Figure 5.44. If as a special case the exit pupil is shifted to infinity, the chief ray path is parallel to the optical axis, the system and the size of an object appears of the same size independent on the z-location. This telecentric case is discussed in Section 5.6 in more detail.

The last case is a remote exit pupil, lying at a larger distance than the object, see Figure 5.44 in the blue case. Then the object details with larger distances or growing z appear to be larger. This case is called a hypercentric perspective and generates a rather unusual visual perception.

If the yellow object point in the figure is observed instead of the red one, it is seen from the corresponding dotted chief ray angles in the eye that this object with larger distance has a smaller chief ray angle in the entocentric case than usual. In the hypercentric case, the detail with larger distance z generates a larger field angle, which is strange from the experience of visual seeing.

5.5 Vignetting

5.5.1 Introduction

Vignetting is the truncation effect of off-axis ray bundles by lenses, mountings, diaphragms. There is a lot of literature available about this practically important issue [5–15]. If an arbitrary field point is considered, as is shown in Figure 5.45 for a three-lens camera setup, the truncation of the aperture light cone perhaps is not limited by the physical stop, but by lens mountings at different locations. This means, for the off-axis field point Q, the light flux is reduced. This effect is called vignetting. In the figure the real off axis cone is shown in green, in the shown cross section it is limited in the lower part by lens L_1 and in the upper part by lens L_3. If the size of the lenses is larger, the bundle is limited by the stop, the blue cone is obtained if no vignetting takes place. In the case of vignetting, the image intensity is reduced for larger field heights and becomes darker. The light cone is then usually no longer circular in cross section.

Figure 5.46 shows the decrease of the light bundle cross section area over the full field of view. Only in the inner

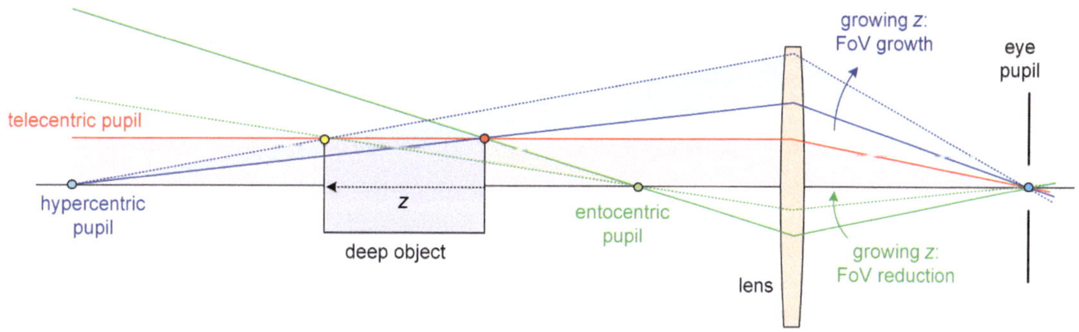

Figure 5.44 Dependence of the visual perspective on the pupil position. The cases of telecentric, entocentric and hypercentric pupil locations are sketched.

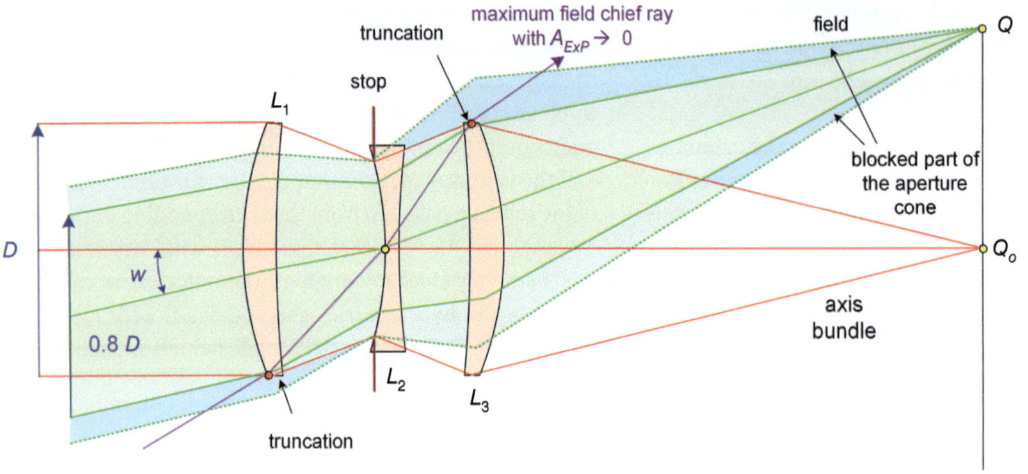

Figure 5.45 Vignetting for a triplet system. The purple chief ray indicates the limiting case of a vanishing exit pupil area. The blue shaded parts of the ray bundle are truncated.

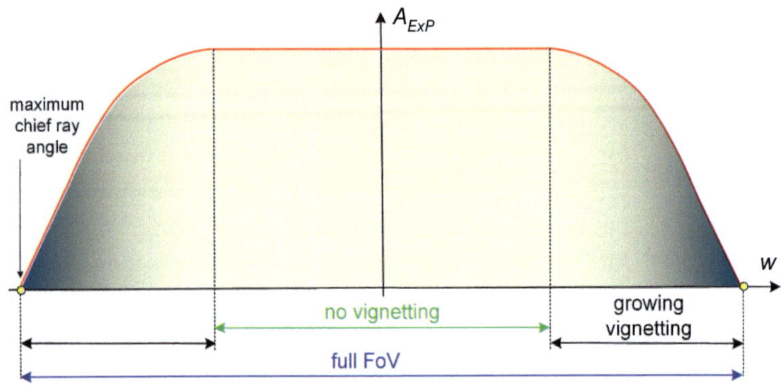

Figure 5.46 Intensity drop at the field boundary due to vignetting.

region is the size constant, for larger field angle the area shrinks. For a theoretical maximum field angle, the pupil area size goes to zero; this is also indicated in Figure 5.45 as a purple chief ray.

The vignetting found in conventional camera systems behaves rotational symmetrically. This means that the smaller cone angles in the outer field points create a darker image. Figure 5.47 shows the corresponding drop of brightness in a real image due to vignetting in the corners of an image. Part (a) shows the full original image without truncation, (b) a partial vignetting case with only 30% remaining brightness in the corners and (c) a case where outside the image is fully blocked and a broad transition range is seen.

But it may also happen that mechanical parts are blocking some light in an asymmetric way, an example of this is seen in Figure 5.48.

If vignetting is considered, several aspects and modelling depths can be considered. The dominant reason usually is a truncation of the light by lens or stop boundaries. Therefore, if a free transmittive area in the entrance pupil plane is considered, the major effects are captured. For the calculation of this area, some cases must be distinguished.

- Finite pupil position with an incoming light cone or the infinity constellation with an incoming cylindrical light tube.
- Calculation of the upper and lower rays in a linear paraxial approach or more exactly by real rays with aberrations.
- Finite chief ray angle vs telecentric arrangement.
- Simple estimation of the pupil area with the tangential and sagittal extensions or more accurate complicated transverse boundary due to the truncation of several surfaces.
- In case of a high-NA system with corrected sine condition, the pupil surface is a sphere and the quantitative evaluation of the pupil area is more complicated (see Volume 2, Chapter 17).

In a more accurate physical model, some further effects have to be taken into account:

- Projection and length effects of the pupil area in case of wide-angle chief ray conditions. This corresponds to

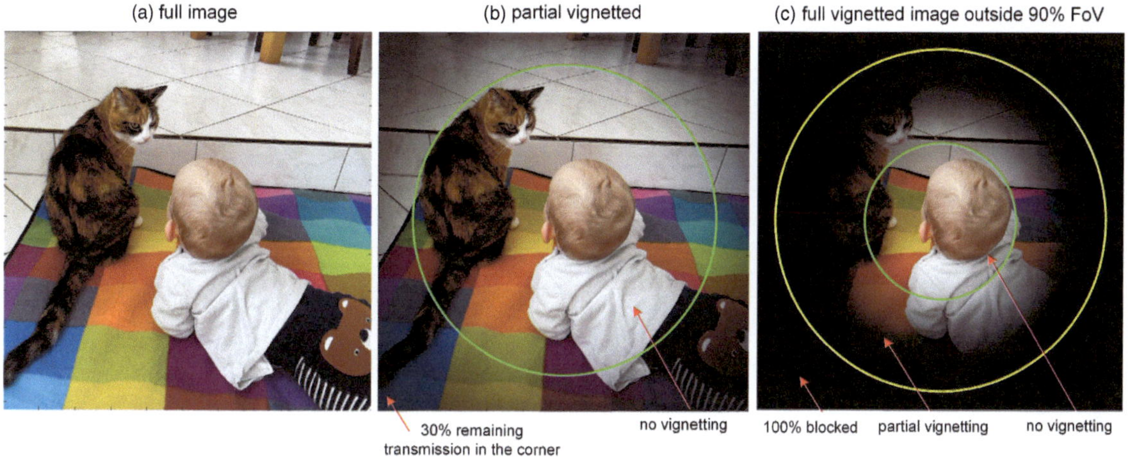

Figure 5.47 Intensity drop due to vignetting in the corners of an image for circular symmetry. In part (b) the 30% reduction of the brightness is hard to detect if the exact intensity distribution in a natural scene is not known.

Figure 5.48 Asymmetrical truncation of the field of view in reality.

the so-called natural vignetting contribution, see next section.

- In the case of pupil aberrations, the distortion of the pupil grid causes a change in the power density.
- The classical distortion of the image influences the brightness in the image due to the corresponding geometrical local grid mapping.
- In the case of a special angle characteristic of the illumination as apodization profiles, the rays must be weighted correspondingly in location and direction.

A vignetting always reduces the effective opening. In circular symmetric systems, the vignetting for off-axis field points is stronger in the tangential than in the sagittal direction. In the case of sensitive applications like microscopy, a vignetting therefore creates an anisotropic effective aperture and correspondingly a change of resolution in S and T orientation.

5.5.2 Photometric Natural Vignetting

The vignetting effects explained in the previous section are a result of the finite size of the components as well as a cone-limiting diaphragm. Due to finite size angles, there is in addition the so-called natural vignetting effect. This phenomenon has its origin in projection effects of tilted planes and changing distances in skew direction, see also Chapter 9.

In Figure 5.49 the image space of a system is illustrated schematically. If w' is the field angle of the chief ray, here it is seen that there are in total three factors influencing the irradiance in the off-axis image point relative to the centre axis point.

- One projection factor $\cos(w')$, because the cross section of the cone is inclined relative to the exit pupil.
- One projection factor $\cos(w')$, because the image plane is inclined similarly.
- The distance from the exit pupil centre to the image point grows by $1/\cos w'$, according to the $1/z^2$ law this creates an additional factor of $\cos^2 w'$.

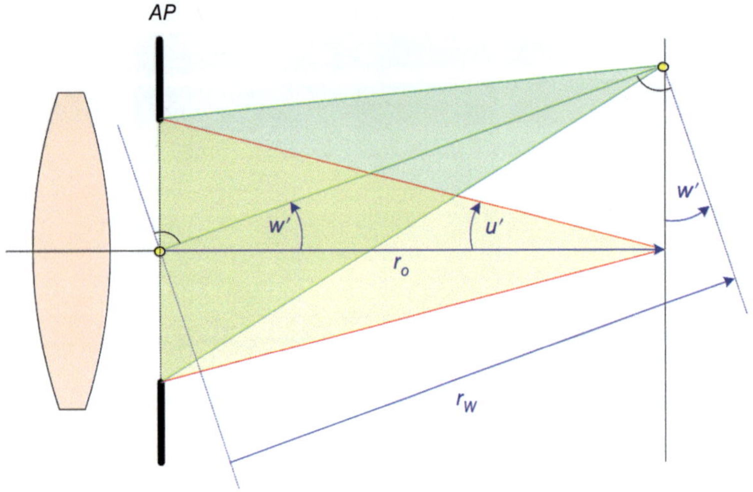

Figure 5.49 Natural $\cos^4 w'$ vignetting for a wide angle system.

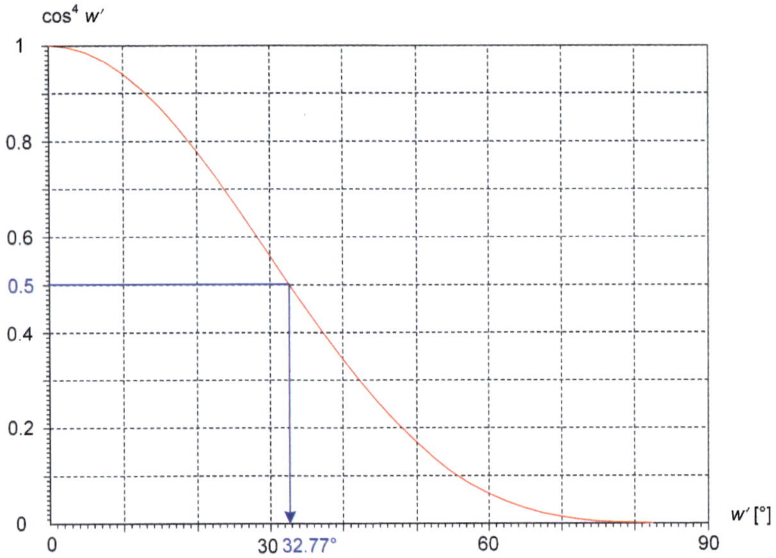

Figure 5.50 $\cos^4 w'$-decay of the irradiance for an off-axis image point with angle w'.

Therefore, in summary the irradiance shrinks by $\cos^4 w'$

$$I(w') = I_0 \cdot \cos^4 w' \tag{5.59}$$

This is the well-known effect of natural vignetting. Since the decrease of the cos-function is rather slow, this effect is only important for wide angle systems. Figure 5.50 shows the principal behaviour of this function. If some of the above used assumptions are violated, the power of the cos-decrease changes. Other examples are a telecentric setup, a monocentric system or a curved sensor. More details of natural vignetting in optical systems are discussed in Chapters 9 and 10. Since the chief ray angles w and w' are connected by the angle magnification, which is constant, this relative drop of irradiance can be similarly formulated by w instead of w'.

5.5.3 Artificial Geometrical Vignetting

5.5.3.1 Introduction

The easiest approach to understanding vignetting is to consider the free area of the entrance pupil for an off-axis field point. Figure 5.51 shows the simplest case of a truncation effect of a circular pupil by one circular boundary. If the position of a field point is moved from the optical axis outwards, usually the vignetting starts to occur on one side of the pupil and grows. The area in this case can be described as the shape of two circle arcs of different radius; only the common area is transitive. This is shown for illustration in Figure 5.51 together with the corresponding curve of the image brightness. The growing asymmetry of the chief ray angle in between the remaining light cone as can be seen in part (a) of the figure should be noted.

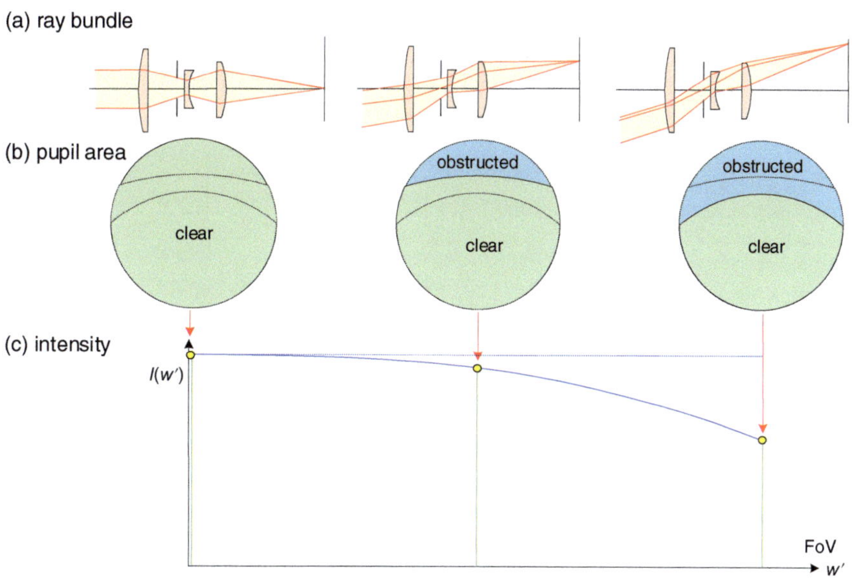

Figure 5.51 Growing vignetting of a system with field angle *w*. The reduction of the exit pupil area corresponds to a decrease of the image brightness, as can be seen in the lower diagram. Shown are (a) the layouts, in (b) the pupil area and in (c) the irradiance in the image for three different field angles.

If in a more general case several surfaces at different z-locations truncate the ray bundles, as a consequence the cross section of the light cone has a more complicated shape. If the boundary of the transmitted area is inspected, there are several arcs, which are the projections of different stops and mountings inside the system into the pupil plane. This means that the transmitted area, shown in yellow in Figure 5.52(a), was constructed by various system surfaces or components. This more general picture already visualizes that the simple elliptical or two-arc approximation is only a coarse approach. Figure 5.52(b) illustrates some of these effects in a real camera lens. In particular, it is seen that a field point with x and y coordinates creates more complicated boundaries. The four arcs in the fourth picture are again illustrated in the schematic drawing, numbering the responsible surfaces.

The geometrical vignetting effects can be considered in the Delano diagram or in the phase space quite illustratively. These forms of representation are discussed in Chapters 3 and 10 respectively.

In lens design it is usually important to know the decay of irradiance for growing field of view quantitatively. The most general approach is an integration over the ray bundle solid angle with weighting coefficients by raytrace.

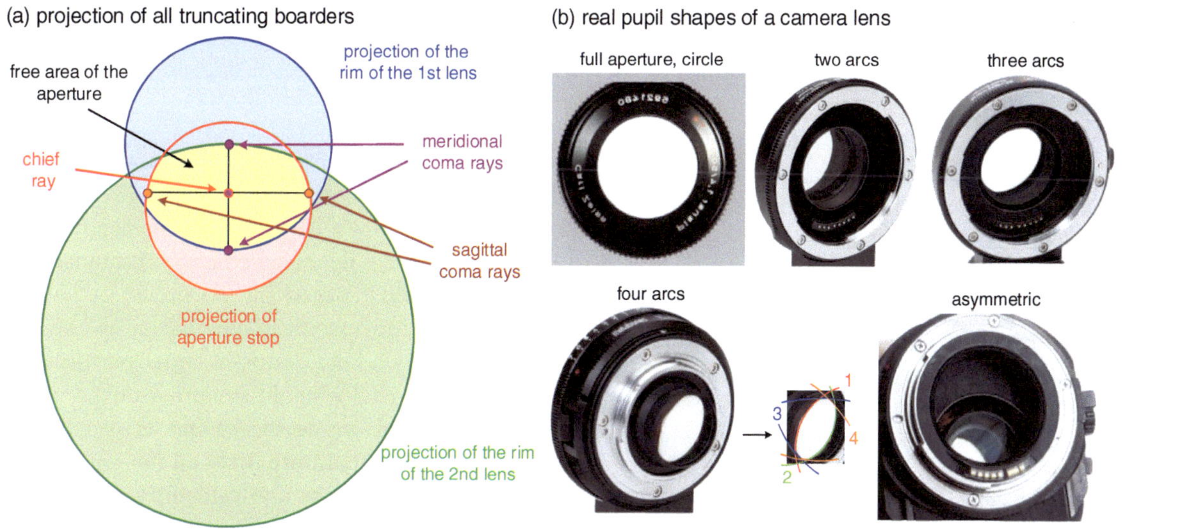

Figure 5.52 General pupil cross section area shown schematically in (a) and illustrated with a real camera lens in (b) [16] / with permission of H. Nasse.

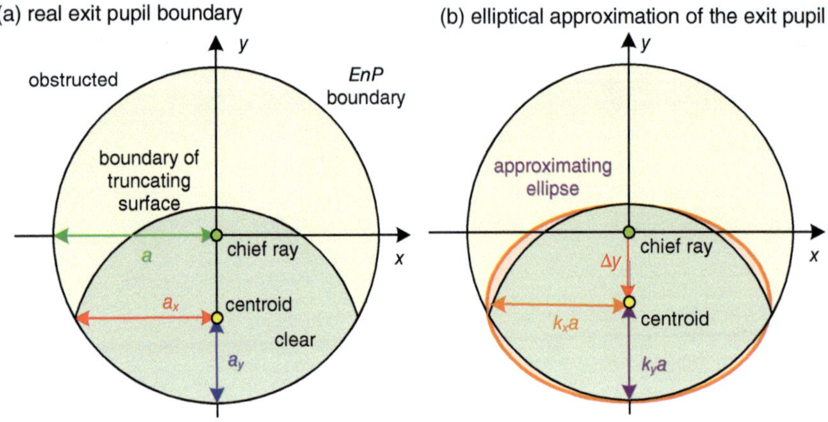

Figure 5.53 Vignetting (a) with anisotropic truncation in x and y and separation between centroid and chief ray and (b) substitution of the clear area by an ellipse.

But these computations are intricate, and therefore several simplified models are proposed in the literature. Some of them are briefly discussed in the following sections.

5.5.3.2 Approximation of the Pupil by an Ellipse*

(*) The content of this section is quite specialized and for advanced reading. Knowledge of this section is not necessary to understand the following chapters. Beginners are recommended to skip this section.

It should be noted that the clear aperture in the right drawing of Figure 5.53(a) is formed by two arcs as discussed before. The reduction factor of the aperture is therefore not the same in the x- and y-direction. If additional components with aberrations are inserted between the stop and the entrance pupil, the shape of the arcs may not be perfectly circular. Therefore, a simple analytical estimation of the pupil area bounded by two circular arcs makes sense, but in reality it is not trivial to judge the accuracy of such a calculation result. If vignetting takes place, usually the truncation occurs on one side first. This has the consequence that the chief ray is no longer representing the centre of gravity. A physically more meaningful choice is then the centroid ray of the clear area of the pupil as is sketched in Figure 5.53(b) [15]. Furthermore, the two-arc area is approximated by an ellipse.

In practical optical design work it is meaningless to start and trace those rays that later in the system disappear through vignetting. Therefore, it makes sense to reduce the initial ray sampling selection in an appropriate way. One simple possibility to achieve this goal is to reduce the initial aperture values by a reduction factor $k < 1$ and centre the clear area onto the centroid and shift or re-orient the chief ray accordingly. This is shown in Figure 5.53(b). According to the discussion above, the reduction factors are typically different in the x- and y-direction. These factors $k_{x/y}$ are defined as

$$a_x = k_x \cdot a; \qquad a_y = k_y \cdot a \tag{5.60}$$

where a is the half diameter of the pupil circle and a_x and a_y are the major axes of the ellipse. According to the area formulas, the overall reduction in energy is then given by

$$\eta = 1 - k_x k_y \tag{5.61}$$

But it should be noted that the substitution of the ellipse only roughly approximates the clear area. But from the viewpoint of a raw estimate, it is a comfortable solution. In reality, the tracing of real rays inside the ellipse may still show truncation due to the imperfect match, but these are very small errors and don't increase the run time considerably. This small discrepancy corresponds to the orange shaded area in Figure 5.53(b).

If the aperture cone is rather large and the pupil is better approximated by a sphere, or if a wide-angle field angle creates rather asymmetric skew conditions, the geometrically more accurate calculation of the vignetting is much more complicated and a simple geometrical model makes only limited sense, because pupil aberrations usually come on top and violate the mathematical relations of the geometry. Some special formulas and analytical descriptions for special assumptions are found in [5, 7, 11, 12].

5.5.3.3 Analytical Approach in the Infinity case*

(*) The content of this section is quite specialized and for advanced reading. Knowledge of this section is not necessary to understand the following chapters. Beginners are recommended to skip this section.

To demonstrate the analytical approaches, one special case should be shown here that is of larger practical relevance. Assume a photographic system with the object in infinity and a stop inside the system as sketched in Figure 5.54(a) [12]. Usually, the first and the last components of this type of quasi-symmetrical setup are the critical surfaces considering vignetting. Here it is assumed that the first and the last surface are those that have to be taken into account. Now these two surfaces as well as the stop

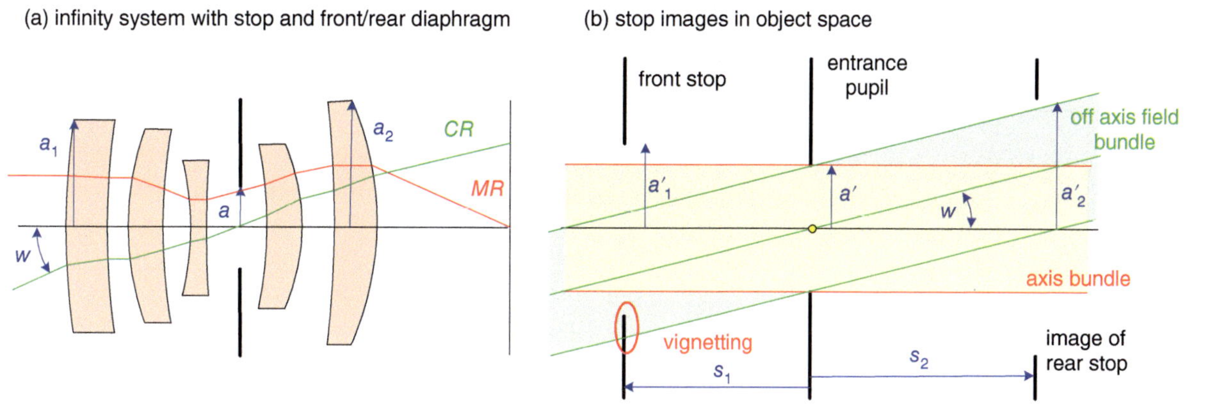

Figure 5.54 (a) Afocal system with internal stop and front/rear diaphragm and (b) images of the three stops in the object space.

are imaged into the object domain. The corresponding half diameter sizes are a_1', a' and a_2' as illustrated in figure part (b). The position of the stop images relative to the entrance pupil are s_1 and s_2.

If the field angle is now growing, two cases have to be distinguished. For small field angle w, the stop radius is still limiting, as is seen in Figure 5.55(a) and (b). For a critical value w_c, the stop just intersects with the other two circular images in one point, this is illustrated in (c). For larger angles w, the stop is no longer actively truncating the beam, see part (d) of the figure.

By inspecting the geometry for the different cases in detail, the vignetting coefficient is found for case (b) to be

$$\begin{aligned} k_b = 1 &- \frac{\arccos(m_1)}{\pi} + \frac{a_1'^2}{\pi a'^2} \arccos \frac{m_1 + \frac{s_1}{2a_1'}\tan w}{a_1'/a'} \\ &- \frac{s_1}{2\pi a_1'} \tan w \cdot \sin(\arccos(\mathrm{m}_1)) \\ &- \frac{\arccos(\mathrm{m}_2)}{\pi} + \frac{a_2'^2}{\pi a'^2} \arccos \frac{m_2 + \frac{s_2}{2a_2'}\tan w}{a_2'/a'} \\ &- \frac{s_2}{2\pi a_2'} \tan w \cdot \sin(\arccos(\mathrm{m}_2)) \end{aligned} \quad (5.62)$$

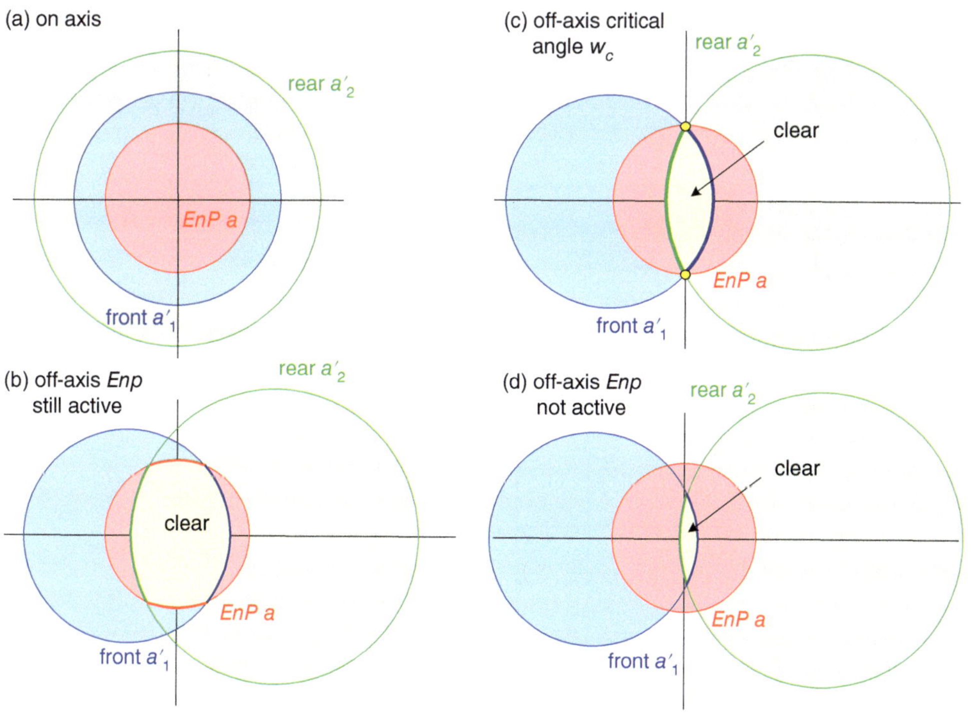

Figure 5.55 Vignetting in an afocal system with middle stop, front and rear truncation. The pupils are shown for a growing field (a) on axis, (b) for a still active stop, (c) the critical field angle and (d) a large field, where the stop is no longer active.

with

$$m_{1,2} = \frac{\left(\frac{a'_{1/2}}{a'}\right)^2 - 1 - \left(\frac{s_{1/2}}{2a'_{1/2}}\right)^2 \tan^2 w}{\frac{s_{1/2}}{a'_{1/2}} \tan w} \tag{5.63}$$

and in case (d)

$$\begin{aligned} k_b = {} & \frac{a_1'^2}{\pi a'^2} \arccos \frac{m}{a'_1/a'} \\ & + \frac{a_2'^2}{\pi a'^2} \arccos \left[\frac{\frac{s_1}{a'_1} + \frac{s_2}{a'_2}}{2a'_2/a'} \cdot \tan w - m \right] \\ & - \frac{a'_1}{2\pi a'} \left(\frac{s_1}{a'_1} + \frac{s_2}{a'_2} \right) \tan w \cdot \sin \left(\arccos \frac{m a'}{a'_1} \right) \end{aligned} \tag{5.64}$$

with

$$m = \frac{\left(\frac{a'_2}{a'}\right)^2 - \left(\frac{a'_1}{a'}\right)^2 + \frac{1}{2}\left(\frac{s_1}{2a'_1} + \frac{s_2}{2a'_2}\right)^2 \tan^2 w}{\left(\frac{s_1}{a'_1} + \frac{s_2}{a'_2}\right) \tan w} \tag{5.65}$$

For illustration, Figure 5.56 shows the vignetting factors for the special case $a'_1 = a'_2 = 1.5$ mm, $a' = 1$ mm for the two values $s_1 = s_2 = 5$ mm and 12 mm respectively. It is seen that there is a starting and an ending point; in between at the critical value of the field angle w case (b) goes over into case (d) with a continuous and smooth transition. The region of the k-change from 0 to 1 is shorter, if the intersection lengths $s_{1/2}$ of the stop images are larger. The dotted lines are the cases where the surface is no longer actively vignetting.

5.5.3.4 Definition of Vignetting Coefficients

A quantitative description of the vignetting is in practice done by a vignetting coefficient. In the literature, there are four different definitions found with the following ideas behind [7, 12]:

- Tangential extension of the pupil area.
  This very often used definition assumes that the area in the pupil plane is roughly determined by the tangential shrink only. If this reduced extension is considered, the pupil area reduction is approximated by a rectangle and the coefficient is given by the major axis D_y in the meridional section as

$$k_T = \frac{D_y}{D} \tag{5.66}$$

- Relative numbers of rays.
  This definition simply integrates the area of the entrance pupil by tracing rays and counting for those that are able to transmit the system. The reference is the number N_i or initial rays similar to the centred axis point.

$$k_{ray} = \frac{N_{tran}}{N_i} \tag{5.67}$$

  This criterion can be evaluated by real or paraxial raytrace and is more time consuming for computation, but is quite general and also applicable for apodized illumination cases.
- Area of footprint in the entrance pupil.
  This criterion is very similar to the second criterion described above. The footprint of rays in the entrance pupil is evaluated and the area determined.

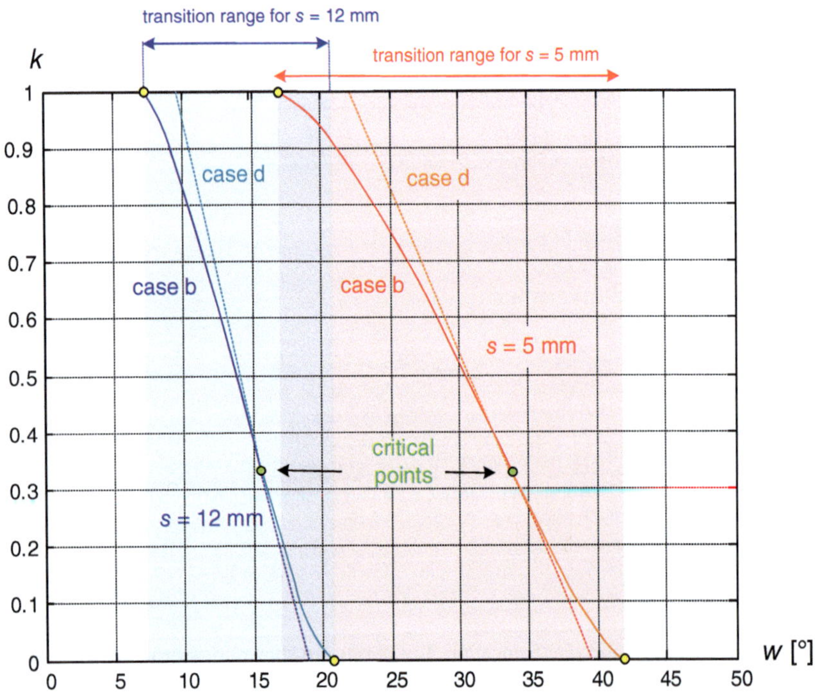

Figure 5.56 Illustration of the vignetting for the analytical model for two cases of $s = 5$ mm (red, orange) and $s = 12$ mm (blue, cyan).

$$k_{AEnp} = \frac{A_{clear}}{A_{Enp}} \tag{5.68}$$

The advantage of this definition is the possibility of giving analytical approximations for simple cases.

- Ratio of the light flux.
  This is the most general, more physical definition and simply relates the transmitted power to the initial one. It is therefore defined as

$$k_{pow} = \frac{P_{trans}}{P_{init}} \tag{5.69}$$

  and can also be used to take the natural vignetting as well as special illumination distributions into account.

Figure 5.57 shows the differences in the dependence of the vignetting factors k_T and k_{AEnP} as a function of the field angle w.

5.5.3.5 Paraxial Vignetting Calculation*

(*) The content of this section is quite specialized and for advanced reading. Knowledge of this section is not necessary to understand the following chapters. Beginners are recommended to skip this section.

In general, the maximum height or diameter limitation of a spherical surface is defined at the surface. This means, due to the finite sag, the height of an incoming ray at the surface is different from its height in the vertex plane, see Figure 5.58. On the other hand, the ray height is defined in paraxial approximation in the vertex plane.

If a system can be considered to be paraxial, the ray heights behave linearly and the calculation of vignetting is simplified. In this case the idea of the phase space corridor can be extended to a case of arbitrary paraxial segments between the diaphragms. Let's assume that the system consists of several $ABCD$ segments with stops of diameter $2a_j$ in between according to Figure 5.59 [6].

If it is assumed that the light signal starts with the height y_o and the angle u_o; after the first segment the coordinate is

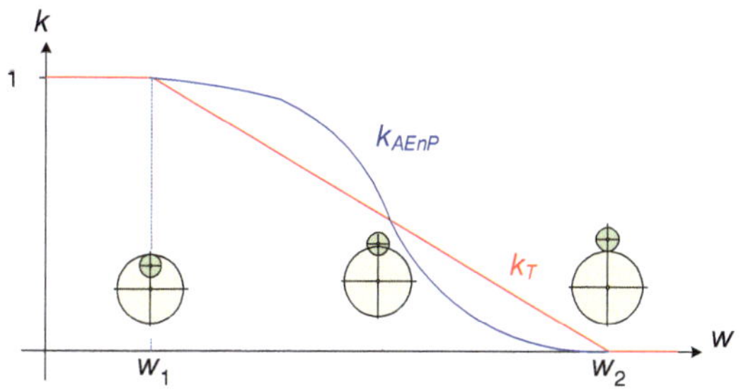

Figure 5.57 Different definitions of the vignetting coefficients as a function of the field angle w.

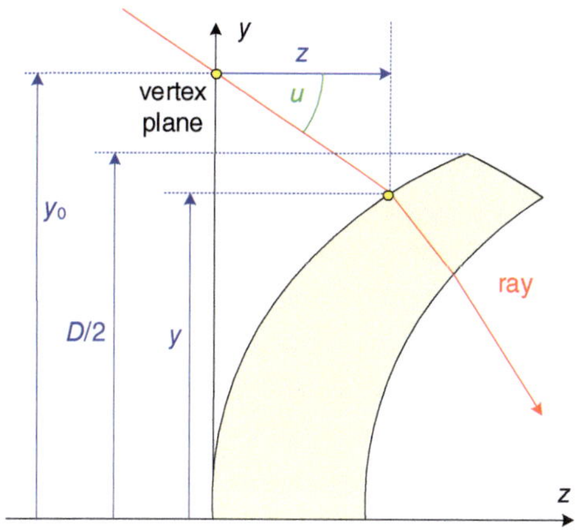

Figure 5.58 Height of an inclined ray at a curved surface y and in the vertex plane y_o.

easily obtained by the $ABCD$ elements of the paraxial segment and compared with the stop radius a_1:

$$y_1 = A_1 y_0 + B_1 u_0 < a_1 \tag{5.70}$$

The same can be done step by step for all compound matrices. In principle the idea is the same as in the traditional geometrical picture: every stop with radius a_j is imaged or transformed into the initial plane. To make the calculation a little bit simpler, only those surfaces must be taken into account that have the chance to truncate the ray bundles.

To demonstrate this approach, Figure 5.60 shows a double Gauss camera lens as an example. The mechanical diameters are fixed in a way that on the axis the stop with index no. 6 is the vignetting surface; for the upper coma ray (TCU), the last surface no. 11 is the vignetting location. The $ABCD$ matrices are calculated to the stop position as M_1 and to surface no. 11 as M_2. The corresponding values are here

$$\underline{M}_1 = \begin{pmatrix} 0.600 & 35.355\ \text{mm} \\ -0.00531\ \text{mm}^{-1} & 1.354 \end{pmatrix},$$
$$\underline{M}_2 = \begin{pmatrix} 0.5779 & 80.706\ \text{mm} \\ -0.010\ \text{mm}^{-1} & 0.3269 \end{pmatrix} \tag{5.71}$$

The stop half diameter is $y_{stop} = 10.23$ mm. With the matrix data, first the marginal ray with the initial height $y^{i}_{MR} = 18$ mm and field angle tan value $w = 0.2867$ corresponding to 16° is calculated.

$$y^{(6)}_{MR} = A_1 \cdot y^{(i)}_{MR} + B_1 \cdot 0 = 10.797\ \text{mm} \tag{5.72}$$

Therefore, the reduction factor is

$$k_{MR} = \frac{y_{stop}}{y^{(i)}_{MR}} = \frac{10.23}{10.797} = 0.947 \tag{5.73}$$

The new marginal ray now passing through the edge of the stop is shown in the figure dotted in red. In the second

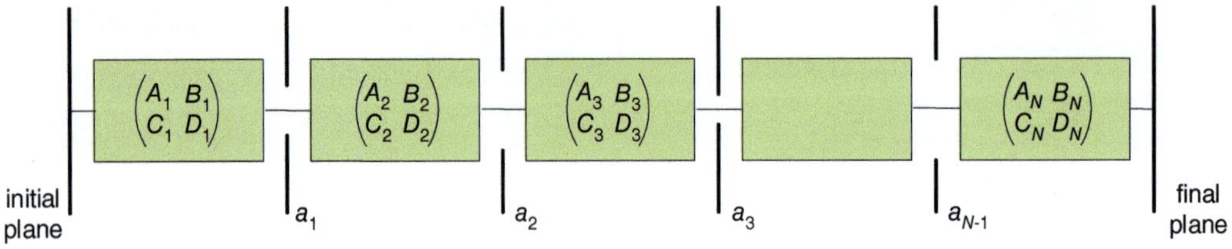

Figure 5.59 Cascaded system of paraxial segments with stops in between.

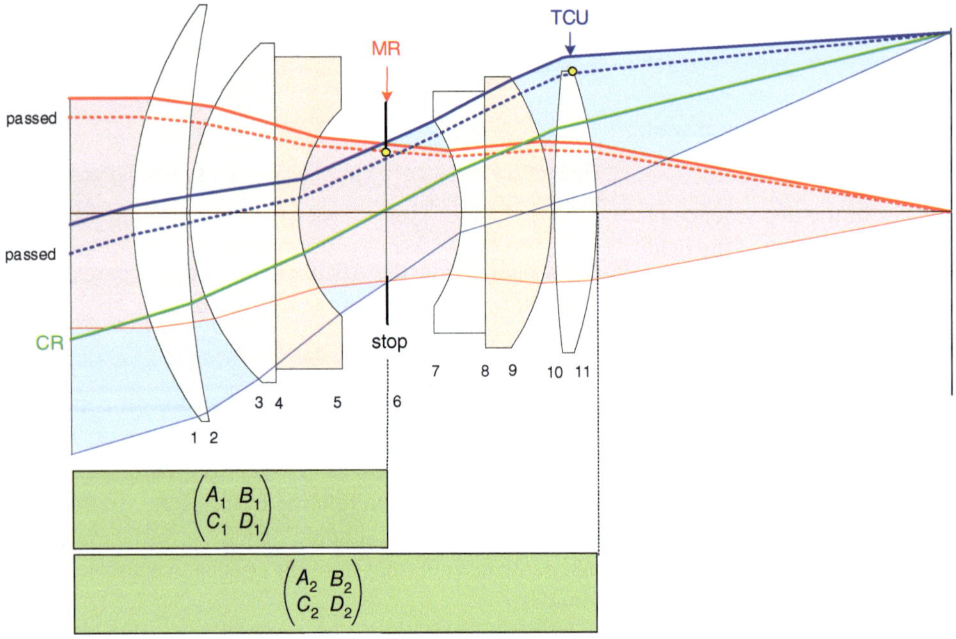

Figure 5.60 Paraxial vignetting calculation in the case of a camera lens with truncation at the surfaces no. 6 (stop) and no. 11 (last surface).

step the initial height of the chief ray to meet the stop centre is calculated

$$y_{CR}^{(i)} = -\frac{B_1 \cdot w_{CR}^{(i)}}{A_1} = -19.765 \text{ mm} \tag{5.74}$$

The half diameter at this surface is $y_{11} = 21.65$ mm. If now the coma ray is calculated, we get

$$y_{TCU}^{(11)} = A_2 \cdot y_{TCU}^{(i)} + B_2 \cdot w_{TCU}^{(i)} = 23.777 \text{ mm} \tag{5.75}$$

To meet the edge of this surface, $y_{TCU}^{(i)}$ must be reduced. By resolving the equation, the value

$$y_{TCU}^{(i)} = \frac{y_{11} - B_2 \cdot w_{TCU}^{(i)}}{A_2} = -5.44 \text{ mm} \tag{5.76}$$

is obtained. This means, the relative aperture between the coma and the chief ray before the system should be

$$\Delta y = y_{TCU}^{(i)} - y_{CR}^{(i)} = -5.44 \text{ mm} + 19.765 = 14.32 \text{ mm} \tag{5.77}$$

and the reduction of the aperture for the coma ray is

$$k_{MR} = \frac{\Delta y}{y_{MR}^{(i)}} = \frac{14.32}{18} = 0.796 \tag{5.78}$$

This ray is shown as blue dotted in the figure.

What is more complicated on the other side is the three-dimensional aspect that for a skew chief ray in y the sagittal plane in x must be investigated too and gives other results. If the paraxial approximation of the ray height is not acceptable in case of larger sag values or angles, a simple approximation can improve the accuracy of the results. If the spherical surface with curvature c is approximated by a parabola in one lateral dimension and we use the relation in the triangle of Figure 5.57

$$y_0 - y = z \cdot \tan u \tag{5.79}$$

we get in third order approximation the corrected height as

$$y = y_0 - \frac{c}{2} \cdot y_0^2 \cdot \tan u \tag{5.80}$$

In this approximation, the same correction is obtained for the x-coordinate with the angle v.

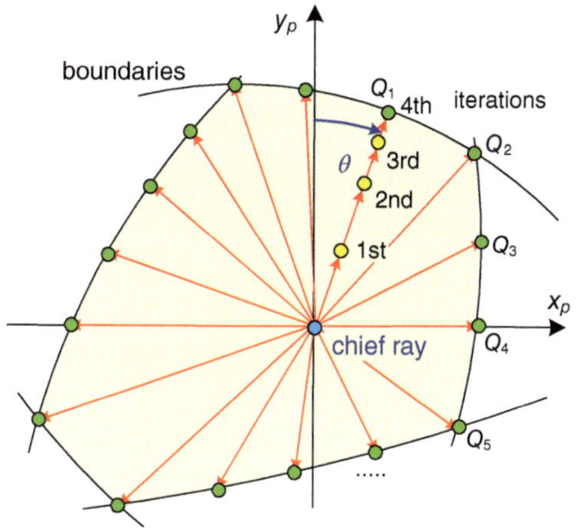

Figure 5.61 Determination of the exit pupil shape in case of a more complicated boundary. The exploration of the boundary shape is sketched by sampling of the azimuthal angle θ. Starting at the chief ray the yellow dots indicate intermediate results of iteration; the green dots are the final points on the boundary.

5.5.3.6 Vignetting for Complicated Pupil Shapes*

(*) The content of this section is quite specialized and for advanced reading. Knowledge of this section is not necessary to understand the following chapters. Beginners are recommended to skip this section.

If there is a more complicated shape of the area of the exit pupil, which allows the rays to pass through, an aiming procedure must be used to determine the boundary line [1, 14]. The process of calculation is sketched in Figure 5.61. The azimuthal angle is sampled according to the star of red arrows in part (a) of the figure. In every azimuthal direction with angle θ a growing angle u is tested for transmission. This is an iterative procedure of aiming, which requires a lot of computer time. Near to the boundary at Q, clever algorithms can be used to accelerate the final convergence. This is equivalent to the aiming problem discussed in Section 3.4.13.3.

5.5.3.7 Vignetting Inside a Gradient Index Lens*

(*) The content of this section is quite specialized and for advanced reading. Knowledge of this section is not necessary to understand the following chapters. Beginners are recommended to skip this section.

A special situation occurs in the case of a gradient index lens with variation of the refractive index. As is well known, the ray path is here represented by a bent curve and is calculated by numerical means in a z-step manner. The corresponding eikonal equation is solved for small step length Δz along the z-direction. As is seen in Figure 5.62, it may happen that the outer boundary of the component is violated in between the lens. Therefore, a vignetting control is needed every step at every position z.

If a ray is touching the border, usually not only a simple truncation of the light takes place, but also a grazing incident setup generates reflected, unwanted ray directions. Usually this is not easy to model exactly in a conventional raytrace-based lens-design program.

5.5.3.8 Vignetting in Phase Space*

(*) The content of this section is quite specialized and for advanced reading. Knowledge of this section is not necessary to understand the following chapters. Beginners are recommended to skip this section.

In general, vignetting is in the geometrical sense an effect that is governed by the ray heights and the angles. Therefore, it is obvious to describe it in the phase space. Chapter 10 explains this approach in more detail. In the space-angle domain the phase space defines a window in the sense of a clear area in one dimension, which allows rays to pass.

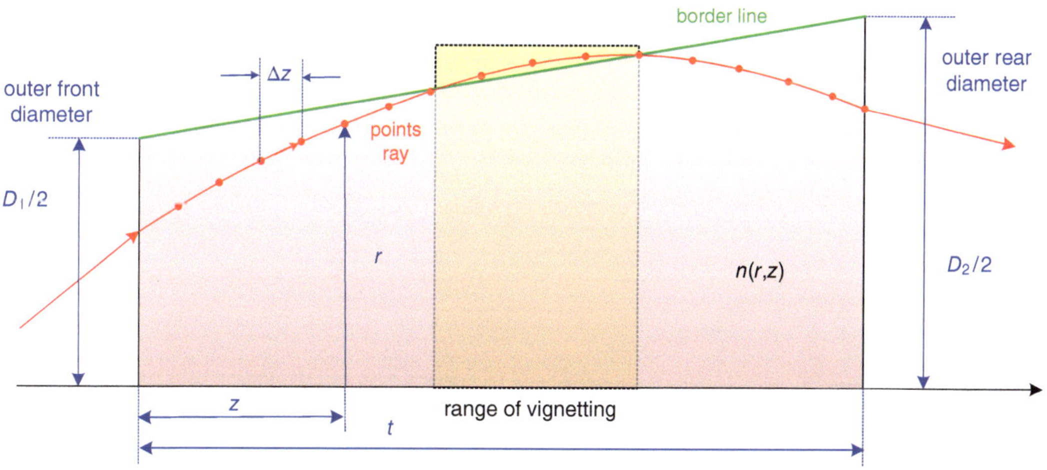

Figure 5.62 Vignetting in a radial gradient index lens in between the lens length.

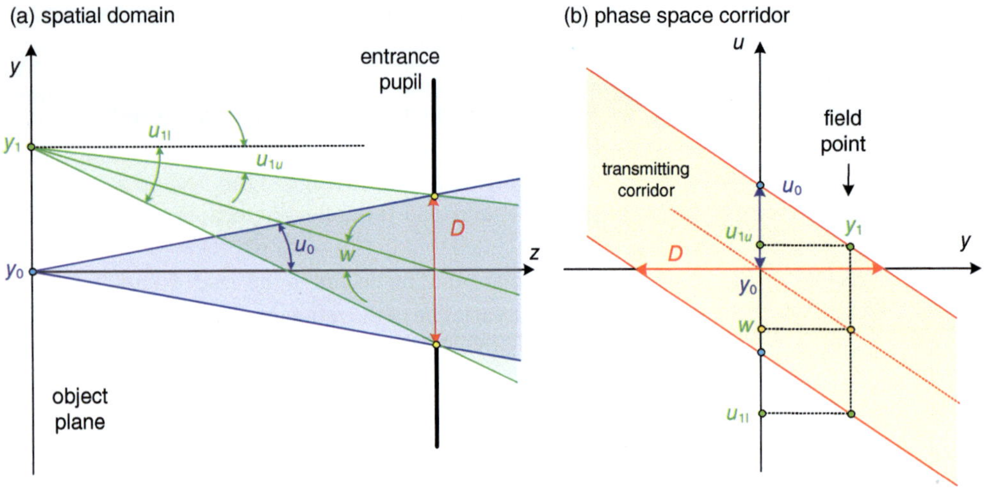

Figure 5.63 Spatial domain (a) and phase space equivalent (b) clear aperture corridor for free space propagation and diameter D of the entrance pupil.

Figure 5.63 illustrates the equivalence between the spatial domain in the image space with a phase-space model. For off-axis object points y the angles u that allow the upper and lower coma ray to pass the entrance pupil change. In the phase space a band shaped corridor is obtained, which contains the allowed y/u variables [13].

The same scheme can be applied if several apertures occur. Figure 5.64 illustrates the phase space corridor for the cases of one, two and three diaphragms and free space propagation between them. For circular symmetric systems, the corridors are all-point symmetric around the origin.

The described paraxial one-dimensional case can be generalized rather straightforwardly into the following aspects:

- Four-dimensional case:
  both dimensions x/y and u/v are taken into account.
- General apertures:
  it is also possible to consider general boundary shapes of the truncating surfaces.
- Aberrations:
  if the transfer of the angles and coordinates is performed by real raytrace, the boundaries of the corridors are no longer straight lines, but are curved due to aberrations.

In the four-dimensional case, similar to the two-dimensional case, a kind of four-dimensional solid body as an allowed corridor for transmitted rays in an optical system can be found [13]. A simple visualization is unfortunately not possible in general, but only for the remaining symmetry. Furthermore, the two spatials and angles are no longer independent and are coupled. The only simple possibility to illustrate this case is a solid body three-dimensional representation in three variables y, u, v for various values of the fourth variable x.

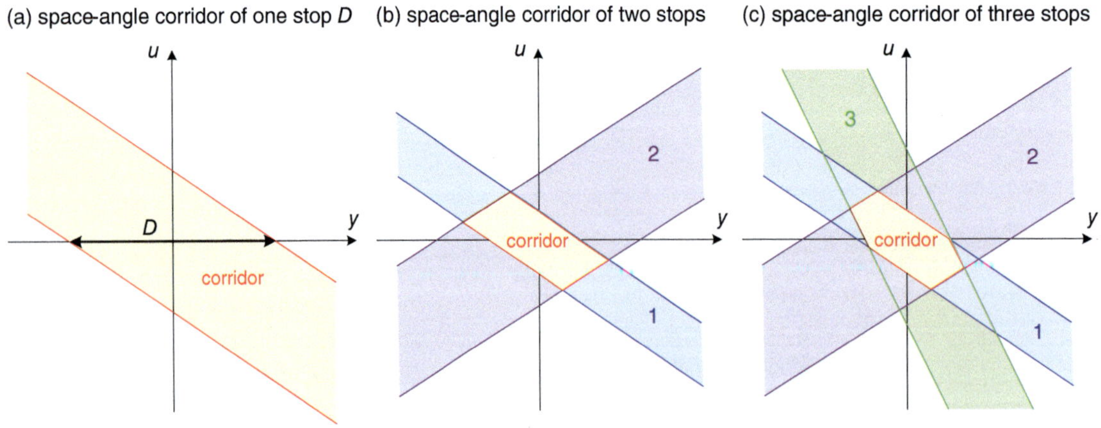

Figure 5.64 Phase space corridor for (a) one, (b) two and (c) three diaphragms.

One more extension of the vignetting model in the phase space is to take photometric distributions and effects into account. Instead of simply calculating the area of clearance, now the integration over the Wigner function as an energy density is performed. This allows us to take into account:

- projection factors like the natural vignetting;
- apodization effects on the surfaces of the system, like imperfect coatings or absorption;
- modelling of the used light source illumination characteristic by an appropriate input Wigner distribution.

Some more details for this approach are discussed in Chapter 10.

5.6 Infinity Cases for Field and Pupil

5.6.1 Overview

The main planes of the system are the object plane, image, entrance and exit pupil. In the most common case, all are assumed to be in a finite distance. In reality, all of these special planes can also move to infinity. If these four planes can be in finite distance or in infinity, in total 16 cases of combination can be found. In reality, seven of these combinations are not physically realizable. It is impossible that the entrance pupil and object or exit pupil and image are in infinity simultaneously. This would mean the marginal and chief ray coinciding, and the Lagrange invariant going to zero. In Figure 5.65 all the remaining nine

case	object	entrance pupil	image	exit pupil	sample systems	
1	finite	finite	finite	finite	relay	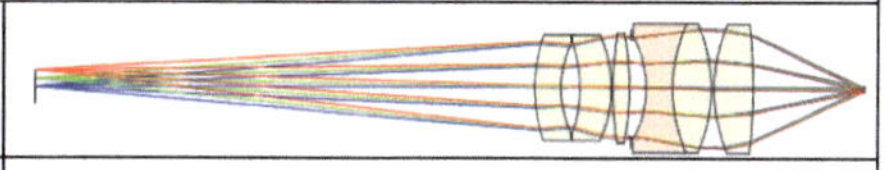
2	infinity afocal	finite	finite	finite	camera lens focusing lens	
3	finite	finite	infinity afocal	finite	eyepiece collimator	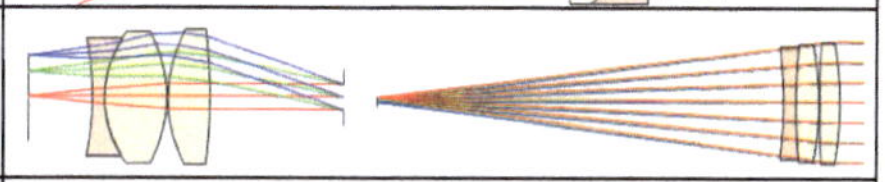
4	infinity afocal	finite	infinity afocal	finite	afocal zoom telescopes beam expander	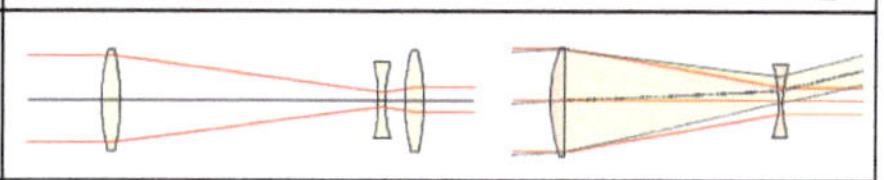
5	finite	infinity telecentric	finite	finite	metrology lens	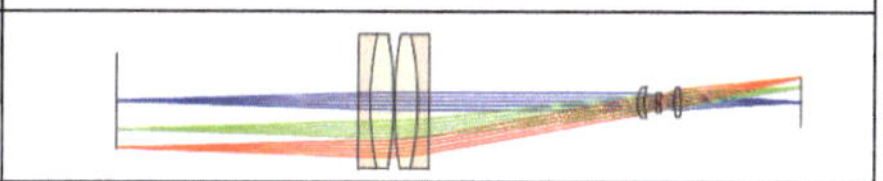
6	finite	finite	finite	infinity telecentric	metrology lens	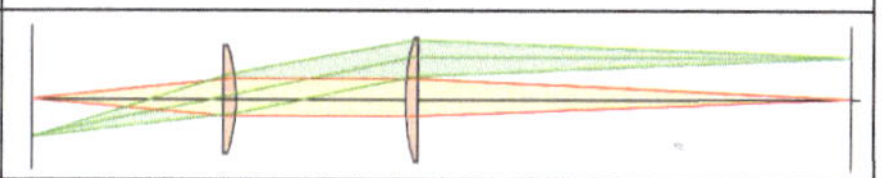
7	finite	infinity telecentric	finite	infinity telecentric	lithographic projection lens 4-f-system	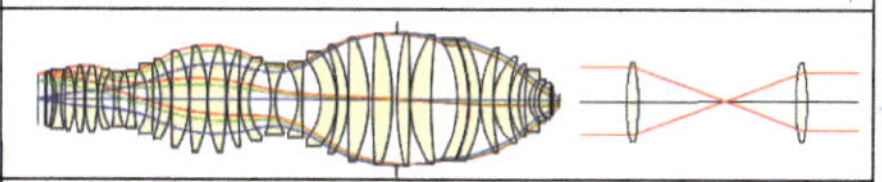
8	infinity afocal	finite	finite	infinity telecentric	metrology lens	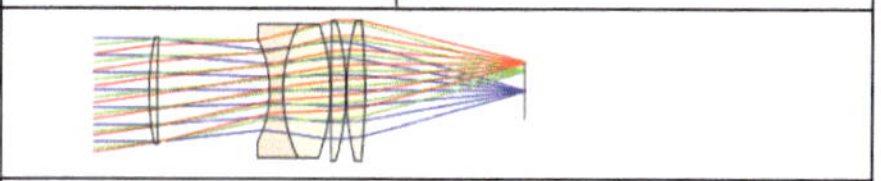
9	finite	infinity telecentric	infinity afocal		microscopic objective	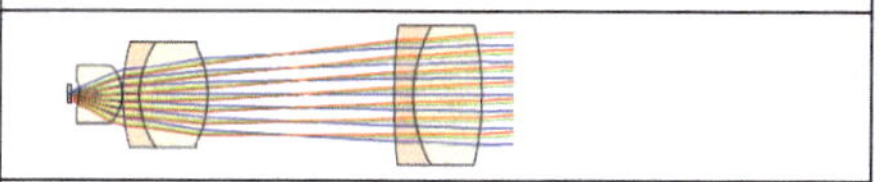
10	infinity	infinity	finite	finite	unphysical combinations	
11	finite	finite	infinity	infinity		
12	infinity	infinity	infinity	finite		
13	infinity	infinity	finite	infinity		
14	infinity	finite	infinity	infinity		
15	finite	infinity	infinity	infinity		
16	infinity	infinity	infinity	infinity		

Figure 5.65 Possible infinity cases of the major planes in an optical system, 9 of the 16 combinations are relevant.

meaningful combinations are listed together with practical examples. It can be seen in the last two columns that all the other combinations make sense and are found in relevant applications.

5.6.2 Afocal Imaging

5.6.2.1 Introduction

If the image is in infinity, the outgoing ray bundles are collimated and the marginal ray runs parallel to the optical axis. Usually, this case must be declared explicitly in a lens design program, because the tool cannot distinguish between small angles for very large distances, aberration induced residual angles and the intended zero angle. If a system image is in infinity, it makes no sense to assess the image quality by lateral aberrations in the spatial domain. In this case, angle or wave deviations are the correct measure to quantify the performance. If wave aberrations are used, the reference sphere simply goes over into a plane reference wave [2].

Figure 5.66(a) shows an eyepiece example system of this type. It is seen that the condition is equivalent to an object located in the focal plane of the system.

If the object is in infinity, incoming collimated ray bundles are obtained as in Figure 5.66(b). Usually in the design model, a distance of infinity is inserted in the data. In this case, the object size cannot be declared in the spatial domain and it is necessary to use field angles.

If the object and image side are simultaneously afocal, a telescope is obtained as is demonstrated in Figure 5.66(c). In this case, the magnification must be defined by the change of the angles or bundle diameters as

$$\Gamma = \frac{\tan w'}{\tan w} = \frac{nD}{n'D'} \approx \frac{w'}{w} \tag{5.81}$$

as discussed in Section 5.3.5.3. This is called the telescopic magnification. As can be seen in the last expression, the magnification can also be defined by the change of the collimated bundle diameters D and D' respectively, if the refractive index is taken into account. The last expression in the equation can be used in the case of small field angles, this is possible typically in astronomical telescopes. According to Figure 5.66(b) the collimated beam diameter D is related to the focal length and the aperture angle u by

$$f \cdot \tan u = y = D/2 \tag{5.82}$$

Since in a both-side afocal system the diameters D and D' are simultaneously the pupil sizes, the relation above can be written as a relation between the angle field and the spatial pupil magnification

$$m_p \cdot \Gamma = \frac{n}{n'} \tag{5.83}$$

which gives 1 if the same media are in both spaces. If an optical system is only one-side afocal, it is not simply possible to characterize it with a spatial magnification m. Only if a second system is used together with the afocal setup to realize a both-side afocal or both-side finite imaging condition is a magnification meaningful. Therefore, for camera objectives no magnification is indicated. In the case of microscope objective lenses. a magnification is used for characterization, but this is only possible if they are used with a fixed focal length tube lens. In this area confusion also often occurs if through the use of

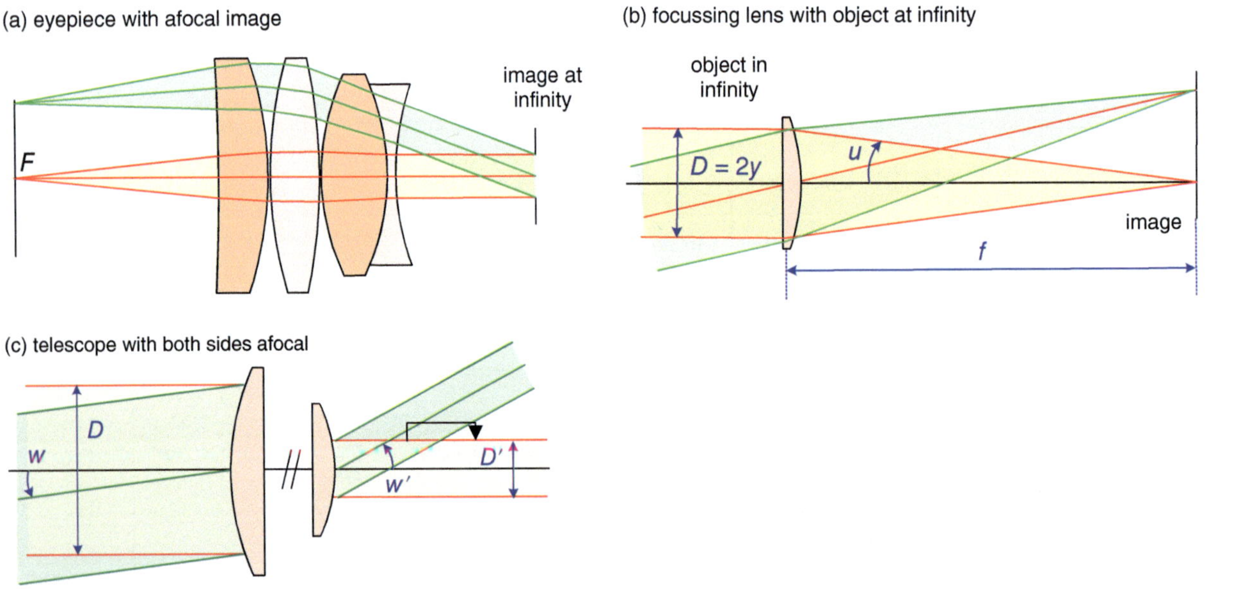

Figure 5.66 Afocal systems on (a) the image side at an eyepiece and (b) the object side and (c) both sides afocal for a telescope.

another tube optic the magnification is different from the lens specification.

5.6.2.2 Relay Imaging with Afocals

A double-sided afocal system must have finite pupil locations due to the non-zero Lagrange invariant. If the pupils and the fields are now interchanged, a finite imaging relay system is obtained. Due to the different angles of chief and marginal ray, maybe the correction and performance of the axis as well as field correction is completely different, but in principle it works at least paraxially. Due to the afocality, now the chief ray is parallel to the axis so the system is both-side telecentric. In addition, in reality the object locations can be changed to modify the magnification. The telecentricity is then destroyed. Figure 5.67 shows for illustration in (a) an afocal Keplerian telescope, in (b) the corresponding relay usage which is then called a 4-f-system and in (c) a relay with a modified magnification be shifting the object.

According to Eq. (5.81) the magnification of a relay keeping the telecentricity is given by the pupil magnification of the afocal. If the object and image are in air, the relation

$$m_p = \frac{1}{\Gamma} \tag{5.84}$$

holds. If a double-sided afocal telescope is used to observe an object with height y in a finite distance z with the human eye, the geometry of Figure 5.68 is obtained. Without the telescope, the viewing angle of the eye is w_o with

$$\tan w_0 = \frac{y}{z + s_0} \tag{5.85}$$

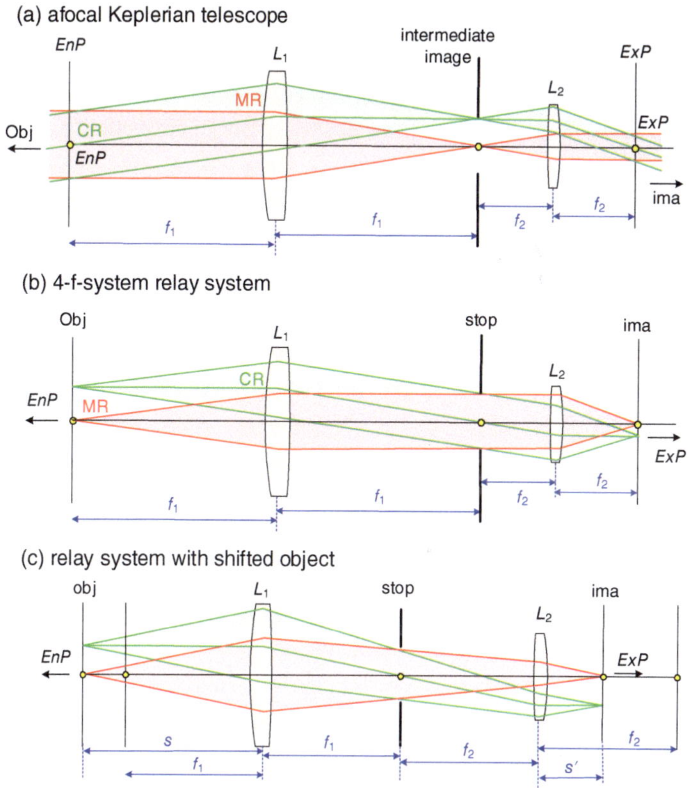

Figure 5.67 Keplerian telescope used in different conjugates: (a) classical afocal use, (b) pupils and fields interchanged to get a relay system and (c) relay variant with changed object location.

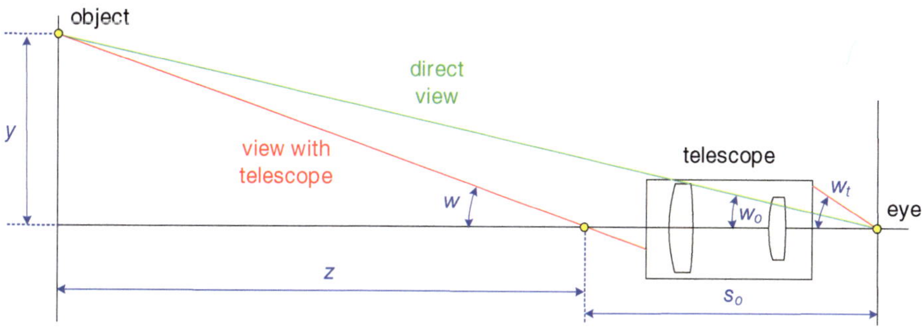

Figure 5.68 Definition of the apparent angle magnification by a visual observation with an afocal telescope.

with the visual natural distance, $s_0 = 250$ mm. If the afocal is used appropriately, the reference point in the observation with telescope is shifted by s_o and the angle magnification Γ gives

$$\tan w = \frac{y}{z}, \quad \Gamma = \frac{\tan w_t}{\tan w} \tag{5.86}$$

This finally results in eliminating y and w in an apparent angle magnification of

$$\Gamma_{app} = \frac{\tan w_t}{\tan w_0} = \Gamma \cdot \left(1 + \frac{s_0}{z}\right) \tag{5.87}$$

which goes to Γ, if $z >> s_0$ as it should be. Often in literature the afocal telescopic magnification is therefore also denoted by $\Gamma\infty$.

5.6.2.3 Applications and Examples

There are many different applications of afocal systems or sub-systems found in technical optics. The following Table 5.2 gives a selected list of well-known applications. Some of these systems are discussed in more detail in volume four of this book from the viewpoint of correction and design aspects.

Table 5.2 Afocal system applications.

Type	Afocality	Examples
Refractive	Double afocal	Keplerian telescope, inverting
		Galilean telescope, upright
		Binocular
		Periscope
		Borescope
		Afocal zoom systems
		Viewfinder
		Laser beam expander
		Power changing attachments
		4-f-system (also used as relay)
		Astronomical field correctors
		Endoscopic relay, 1 : 1
	Single afocal	Collimator
		Microscopic objective
		Eyepiece
		Camera lens
Reflective	Double afocal	Mersenne telescope
		Mirror laser beam expander
		Offner (also used as relay)
	Single afocal	Schwarzschild telescope
Catadioptric	Single afocal	Schmidt telescope
		Bouwer telescope
		Maksutov telescope
		Dyson (also used as relay)

5.6.3 Telecentricity

5.6.3.1 Introduction

A system is called telecentric if the entrance pupil, the exit pupil or both are located in infinity. Telecentricity means that the chief ray is parallel to the optical axis. The stop is positioned in the focal point of the system or a group of the system. In particular, three cases are possible:

- stop in back focal plane: object-side telecentricity;
- stop in front focal plane: image-side telecentricity;
- stop in an intermediate focal plane: both-side telecentricity.

Figure 5.69 shows an example for an object-side telecentric system; this is typical for metrology setups or microscopy objectives. A small defocusing does not change the magnification due to the parallel chief ray path. To get an object-side telecentric chief ray, the stop must be located in the image-side focal point of the system.

The major advantage of telecentricity is the invariance of magnification in the case of defocusing. This is an important property of optical systems in metrology for the measurement of deep samples with finite z-extension. Figure 5.70 illustrates the difference in magnification for an object detail defocused by Δs in comparison of a (a) conventional system and (b) a telecentric system. In the case of telecentricity, only a blur as a result of the defocusing is observed.

A drawback of a telecentric system is the large lateral size of the system. The necessary diameter of the system without vignetting is at least

$$D > 2 \cdot (y_{\max} + s \cdot \sin u) \tag{5.88}$$

with the maximum field height $y_{\max}$. If the chief ray in usual systems goes through the centre of a compact lens system, the collection of all ray bundles is more constricted and the system is smaller in diameter. This is illustrated in Figure 5.71, where both systems are drawn as overlay. In the case of the traditional stop in the system, the objective lens has approximately 1/3 of the size in comparison to the telecentric case for this illustrated example.

In the lens design of a telecentric system, it can be seen that the fulfilment of telecentricity typically is not valid over the complete field of view. This means that due to residual aberrations of the pupil imaging, the telecentricity behaves like an usual aberration and only a tradeoff over all chief ray angles can be achieved. The specification of telecentricity errors is not trivial and the impact of an angle error depends on the focal length and required resolution. Aberration issues of telecentric systems are discussed in a later chapter.

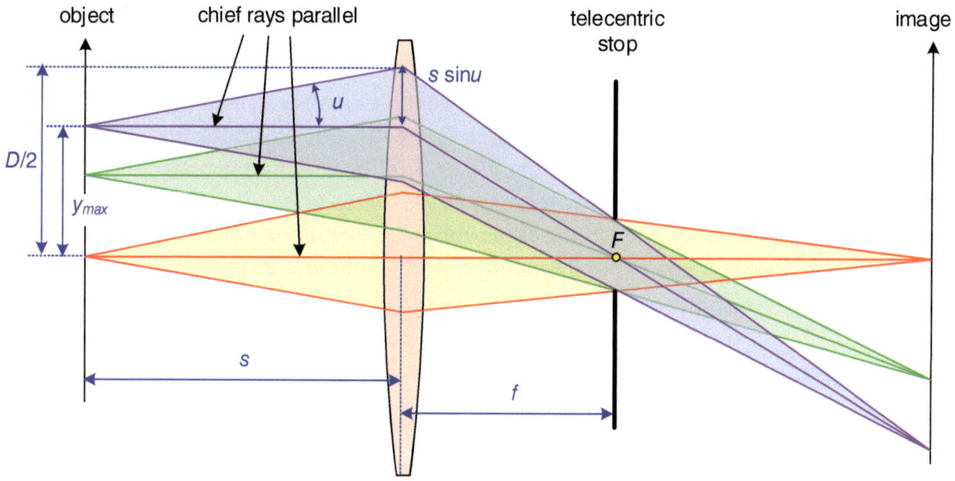

Figure 5.69 Illustration for an object-sided telecentricity.

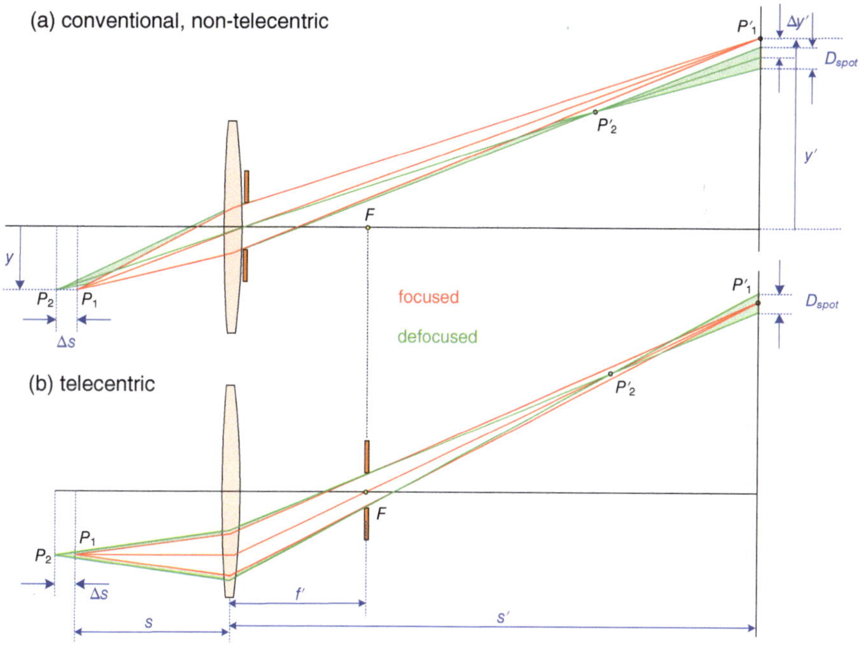

Figure 5.70 Comparison of (a) a conventional measuring system with a change in magnification in the case of defocus and (b) a telecentric system, keeping the image height invariant.

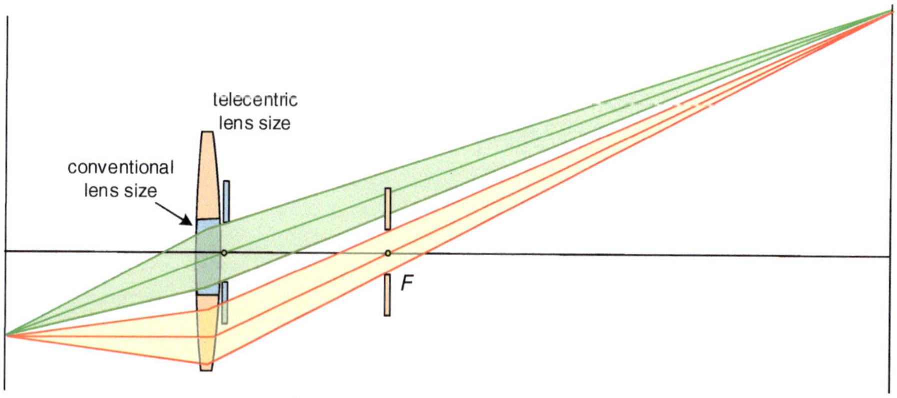

Figure 5.71 Comparison of (a) conventional (green) with (b) telecentric (yellow) measuring system concerning the change in diameter of the lens.

5.6.3.2 Examples

A well-known example for a double-sided telecentric system is a 4-f-system as is shown in Figure 5.72. This arrangement is only possible if the stop is in a common focal point of the front and the rear group respectively. Figure 5.73 shows a refractive lithographic projection system, working in the deep UV wavelength range. It is also double-sided telecentric; the only difference to a simple 4-f-system is the better correction of both lens groups. The purple dotted lenses indicate a paraxial 4-f-substitute of the complex lens system.

The next two examples also show very prominent cases of telecentric systems in reality. In both cases the invariant magnification is not just the reason for this chief ray path selection. Figure 5.74 illustrates a microscopic lens, which is telecentric in the object space. This has two advantages in the application: the magnification is constant in case of volume imaging and secondly the aberrations of the cover

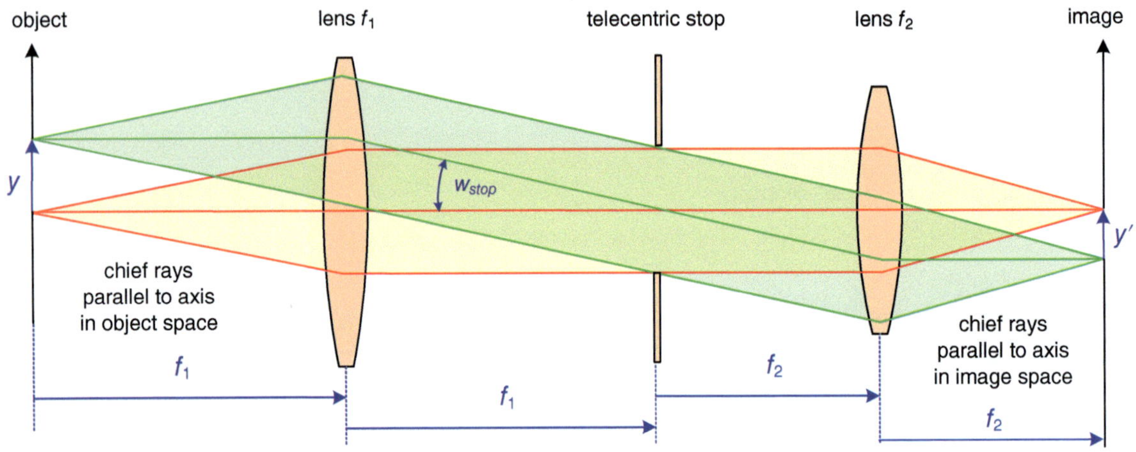

Figure 5.72 Double-sided telecentricity in a 4-f-system.

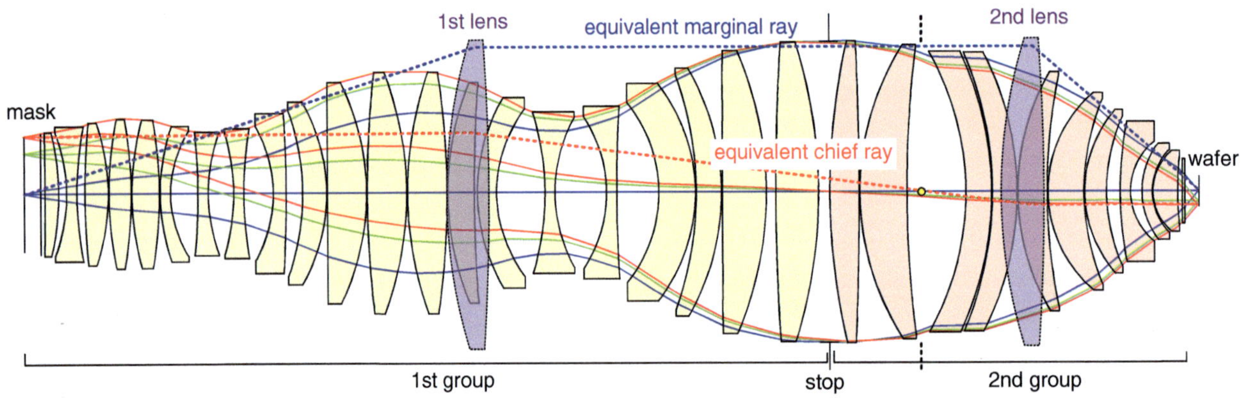

Figure 5.73 Double-sided telecentricity in a lithographic lens [17]. The overlaid purple lenses illustrate the 4-f-substitute.

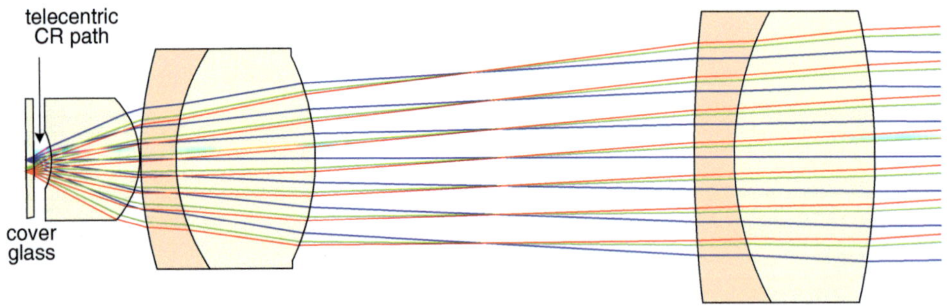

Figure 5.74 Telecentricity in the object space of a microscopic lens.

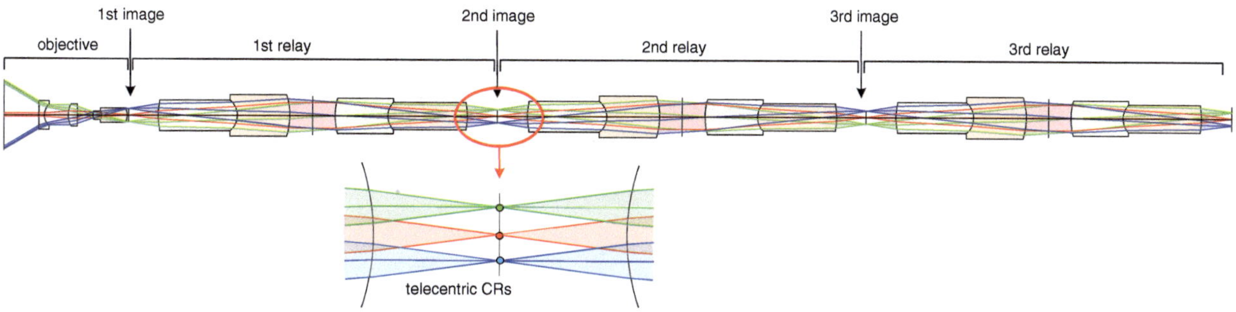

Figure 5.75 Telecentricity in the intermediate images of an endoscope with several relay systems to bridge a large distance.

glass in the case of a large numerical aperture are the same for all field heights, which simplifies the field-independent correction of these effects inside the lens.

Figure 5.75 shows the system of an endoscope with three relay systems to gain a large transfer distance. Every relay has on both sides a telecentric chief ray, otherwise the modularity and flexibility in selecting the number of relays cannot be fulfilled. Therefore, in this case the reason is to avoid vignetting.

5.7 Imaging with Mirrors

5.7.1 Introduction

In the case of using just mirrors, many formulas for calculating the systems and their aberrations have a much simpler form. In particular, in astronomy mirror systems are used extensively due to three reasons [18–30]:

- There is no dispersion and therefore no colour aberrations occur.
- Large diameter components to collect a lot of light is not a technological issue.
- Due to the very small field angles, the problems with astigmatism and coma at mirrors are controllable.

Here, only a short overview about reflective imaging systems in the sense of first order concepts is discussed.

Table 5.3 Types of purely reflective systems.

Object location	Image location	Type	Examples	Application
Infinity	Finite	Telescope	Cassegrain, Schwarzschild, Gregorian	Astronomical observation
Finite	Finite	Relay	Offner	Spectrometer
Infinite	Infinite	Afocal	Mersenne	Laser beam guiding

In principle there are three different relevant constellations considering the conjugations. Mainly the telescopic version is discussed here, see Table 5.3.

It is relatively easy to formulate paraxial equations of multi-mirror systems by a matrix approach to get an overview on magnifications, mirror sizes, pupil location and image position.

Beyond a paraxial investigation, the following aspects mainly influence the concepts.

- Central obscuration:
  to get comfortable properties concerning resolution, contrast and power throughput it is desirable to avoid central obscuration. There are three major approaches to achieve this goal, they are sketched for a two-mirror system in Figure 5.76.

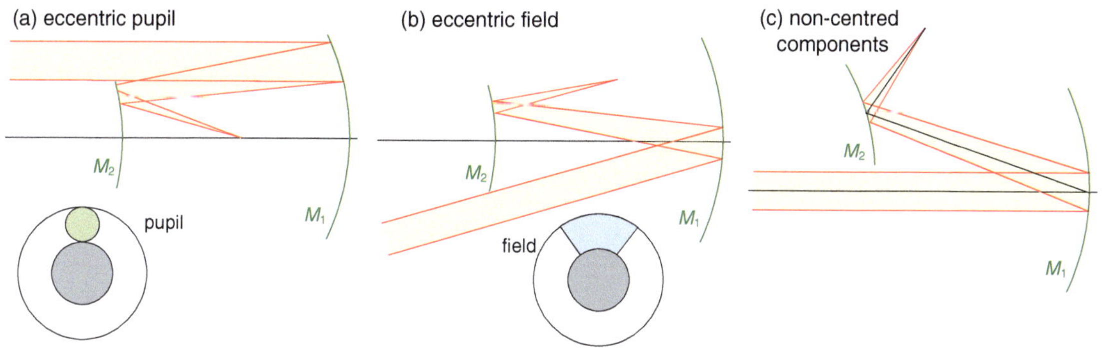

Figure 5.76 Options to avoid central obscuration in case of mirror systems: (a) eccentric pupil, (b) eccentric field of view or (c) non-centred components.

- Linear astigmatism:
  curved mirrors, which are used under skew conditions with finite axis ray incidence, generate a macroscopic linear astigmatism of first order on axis. This affects the resolution a lot and has to be compensated for in real systems.
- Field flattening:
  if only reflective components are used, it is hard to get a flat plane image by correction means and it requires at least one negative power mirror. This is necessary in nearly all practical applications. Details are discussed in a later chapter about the correction of mirror systems.
  In particular for spherical mirrors, the concept of Kutter with the so-called Schiefspiegler approaches addresses this problem.
- Conic mirrors:
  reflective telescopes usually have a large aperture and the correction of spherical aberration of wide aperture cones is a major task. This can be easily done for an axis field point by the use of confocal mirrors: several conic sections are arranged in a way such that their focal points coincide and a stigmatic imaging is possible on axes. This concept is sketched in Figure 5.77 [31, 32]. This means, on axis in the centre of the field the system's image is stigmatic without aberrations at all. This gives a good starting point for the quality correction of finite field sizes.

If in particular two conic sections are combined with the additional requirement that the outgoing focal point of the first mirror is identical with the incoming one of the second mirror, the stigmatic imaging properties are preserved. If the second mirror is rotated around the common focal point, arbitrary orientations of the outgoing ray bundle can be obtained. The various possible arrangements are shown in Figure 5.78, but are restricted here only on an outgoing bundle axis parallel to the incoming one. In total there are three options for the outgoing bundle: it can be afocal, convergent with finite focal point or divergent. Furthermore, the beam shape between the two mirrors can be divergent or convergent with or without an intermediate focal point.

Depending on the mirror power distribution, the two options of a Keplerian approach with two positive mirrors and an intermediate focal point or a Galilean with a negative and a positive mirror can be distinguished. The abbreviations used to characterize the setups in the figure have the following meanings:

P: parabolic mirror, with positive or negative power;
E: elliptical mirror, with positive or negative power;
H: hyperbolic mirror, with negative power;
I: infinite, object and image;
C: convergent ray bundle;
D: divergent ray bundle.

As an example, P + E+/ICC means: the system consists of a positive parabolic and a positive elliptical mirror, the axis ray bundle comes from infinity and is convergent between the mirrors and in the image domain.

- Image plane tilt:
  if mirrors are used under skew and tilted conditions, usually the image plane is tilted and no longer perpendicular to the axis ray. This problem can be removed by a corresponding tilt of the sensor plane. Some more details on this issue can be found in Section 5.8.4.

The discussion of multi-mirror systems on a paraxial basis without considering correction aspects is meaningful for initial system concepts, but aberrations and quality are always an issue. Nevertheless, some basic approaches for centred and decentred systems are discussed here; more detailed correction aspects are found in a later chapter.

5.7.2 Centred Two-Mirror Setups

First circular symmetric two-mirror telescopic systems should be discussed; there are three major constellations, see Figure 5.79:

- Cassegrain type:
  here, a large positive primary mirror M_1 generates convergent ray bundles, a small M_2 creates central

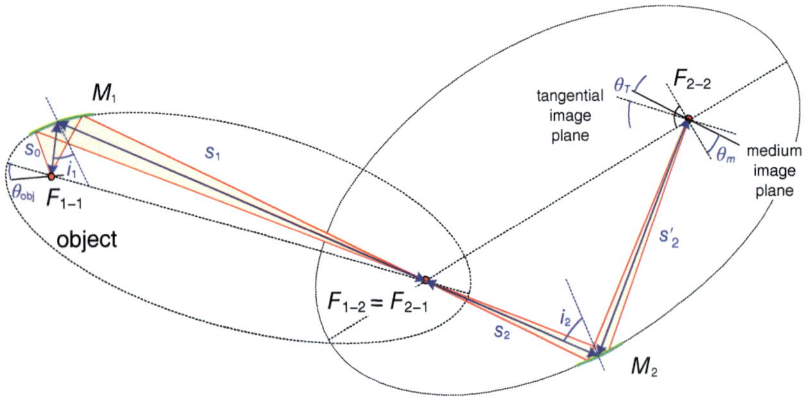

Figure 5.77 Geometry of two confocal conic sections.

Figure 5.78 Two-mirror systems with conical mirrors and (a) afocal, (b) finite real focus and (c) finite virtual focus point.

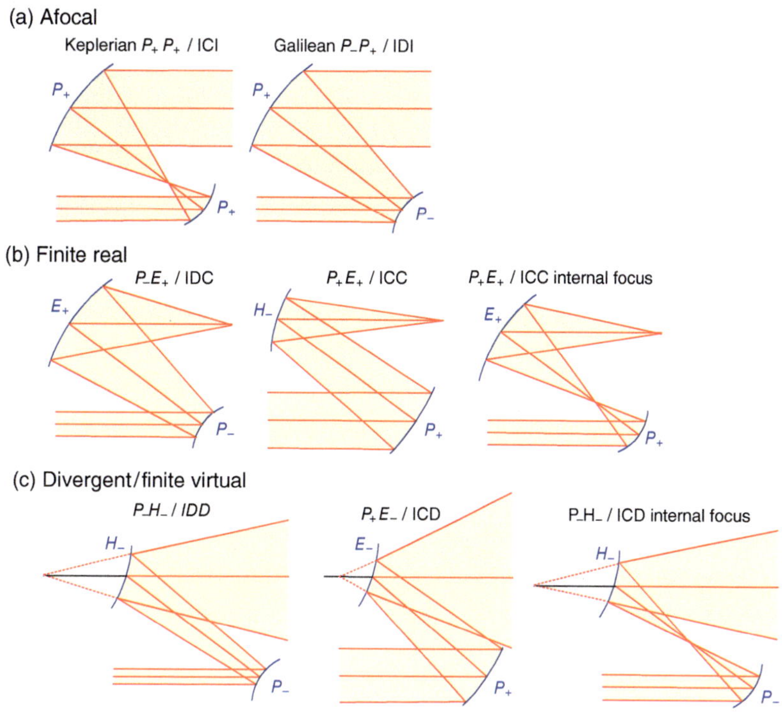

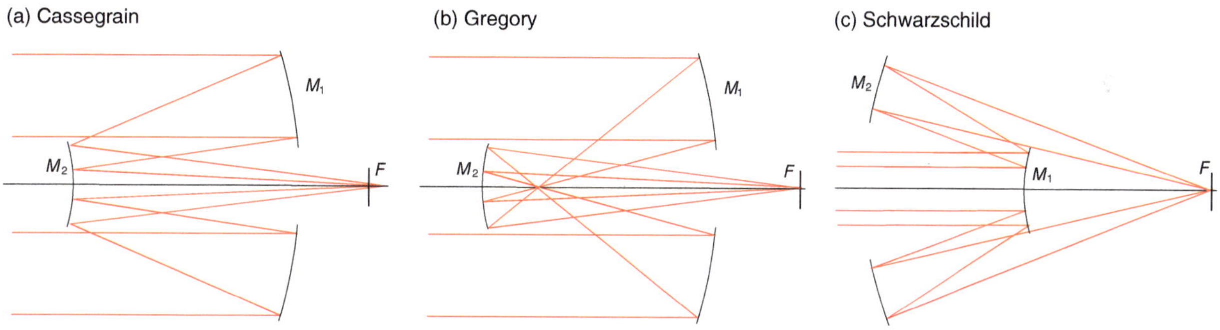

Figure 5.79 Two mirror setups after (a) Cassegrain, (b) Gregory and (c) Schwarzschild.

obscuration and builds a focus, mostly behind the M_1. This arrangement corresponds to a tele-setup. If the Cassegrain layout is built with aplanatic correction, it is also called Ritchey–Chretien telescope.

- Gregory type:
  this setup is similar, but an intermediate focus is generated and both mirrors have positive power. Here a field flattening is impossible.
- Schwarzschild type:
  in this arrangement, the primary mirror is negative and widens the ray bundle, the secondary mirror is then focusing. In this setup a large NA can be achieved, therefore it is found in microscopy or x-ray applications. This geometry corresponds to a retrofocus type system.

In principle there exists a fourth constellation with a plane secondary mirror. This so-called Newton system is not considered here, because it optically corresponds to a single-mirror telescope.

In all three cases a central obscuration is present. For concrete layout equations, mostly parameters for aplanatic and field flatness are selected, which is discussed later. If a mirror system is calculated by paraxial, for example a matrix-based scheme, the sign of the variables should be considered with care. If in a two-mirror system the axis ray bundle or the marginal ray comes from infinity, the first matrix is the M_1, then follows the intermediate distance t_1, then the M_2 and finally the image distance t_2. If the way back after M_1 is described by a negative $t_1 < 0$ and the sign of the spherical mirrors with powers F_j follows the technical convention, the matrix of a two-mirror setup reads [33]

$$\underline{M} = \begin{pmatrix} 1 & t_2 \\ 0 & 1 \end{pmatrix} \cdot \begin{pmatrix} 1 & 0 \\ -F_2 & 1 \end{pmatrix} \cdot \begin{pmatrix} 1 & -t_1 \\ 0 & 1 \end{pmatrix} \cdot \begin{pmatrix} 1 & 0 \\ -F_1 & 1 \end{pmatrix}$$
$$= \begin{pmatrix} 1 + F_1 t_1 - t_2 F_2 - t_1 t_2 F_1 F_2 - t_2 F_1 & -t_1 + t_2 + t_1 t_2 F_2 \\ -F_1 - F_2 - t_1 F_1 F_2 & 1 + t_1 F_2 \end{pmatrix} \tag{5.89}$$

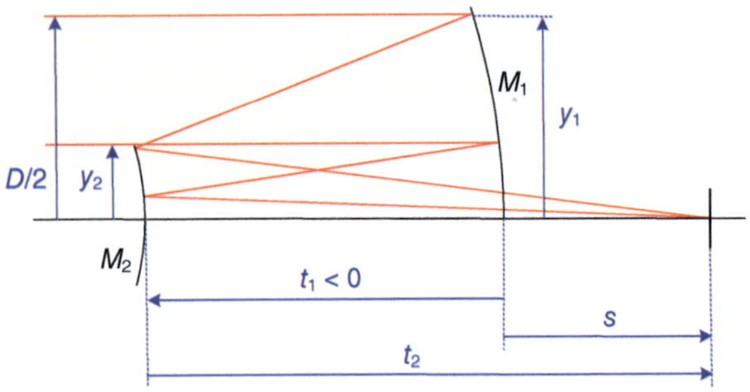

Figure 5.80 Description parameters of a centred two-mirror setup.

If in this convention more mirrors are used, the correct assignment of positive and negative powers is cumbersome. Here we have $F_1 = -2/R_1 = -2_{c1}$ and $F_2 = +2/R_2$. The well-known height ratio between the marginal ray heights at the secondary and primary mirror is now given by (see Figure 5.80)

$$\omega = \frac{y_2}{y_1} = 1 + t_1 \cdot F_1 = 1 - t_1 \cdot 2c_1 \tag{5.90}$$

with the power F_1 of the M_1. The ratio of the inner obscuration due to the M_2 to the incoming bundle diameter is called the central obscuration ε. This value is mostly equal to ω, only in case of the Schwarzschild setup there is a difference, see also Table 5.4. The total power is given by

$$F = F_1 + \omega \cdot F_2 \tag{5.91}$$

and the back focal length as free working distance is given by

$$s = t_1 + t_2 \tag{5.92}$$

where $s_2' = t_2 = L$ can also be interpreted as the length of the system from the secondary mirror to the image.

If now the power F, the incoming diameter D and the condition $s > 0$ (practically desired, but not a necessary optical constraint) and central obscuration $\varepsilon > 0$ are given, the following ranges and setups are possible. There is no unique solution, but certain ranges of possible parameters.

Figure 5.81 illustrates these solution landscapes graphically on a ω scale. For the layout of a two-mirror system, many different sets of formulas exist, depending on the pre-given parameters. These can be the length L, the back focal length s, the incoming diameter $D = 2y_1$, the obscuration ratio ε, the focal length f or the numerical aperture $\sin(u')$. If the sets of equations found in literature are compared, care must be taken to use the same sign conventions for t_1 and R_1. Furthermore, the focal length or the incoming ray height is often normalized.

5.7.3 Obscuration-Free Two-Mirror Systems*

(*) The content of this section is quite specialized and for advanced reading. Knowledge of this section is not necessary to understand the following chapters. Beginners are recommended to skip this section.

If the central obscuration is removed by tilted mirrors, the signs of the tilt angles are one more degree of freedom. Independent of the power distribution of the systems above, three different constellations are possible, see Figure 5.82 [34].

- α-setup:
  here the same sign of the deflection angles (clockwise) generates an overall geometry with overfolding ray bundles.
- u-setup:
  In this geometry the signs are also the same, but by selecting angle value and distances appropriate, the crossing is avoided.
- z-setup
  in this arrangement the first angle is clockwise, the second counter-clockwise, so different signs create a z-type geometry.

From the viewpoint of performance, size, field and aperture support the three options are quite different. In reality

Table 5.4 Essential two-mirror setups.

Name	F_1	F_2	ω-range	Central obscuration parameter ε
Gregory	+	+	$-1 < \omega < 0$	$\varepsilon = \omega = -t_1 \cdot F_1 - 1$
Newton	+	0	0	$\varepsilon = 0$
Cassegrain	+	−	$0 < \omega < 1$	$\varepsilon = \omega = 1 + t_1 \cdot F_1$
Schwarzschild	−	+	$\omega > 1$	$\varepsilon = \dfrac{1}{s \cdot \omega \cdot F} = \dfrac{1}{(t_1 + t_2)(1 + t_1F_1)(F_1 + F_2 + t_1F_1F_2)}$

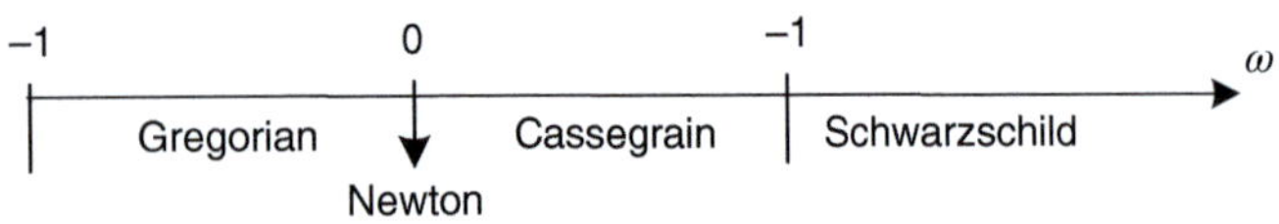

Figure 5.81 Central obscuration of the two-mirror systems.

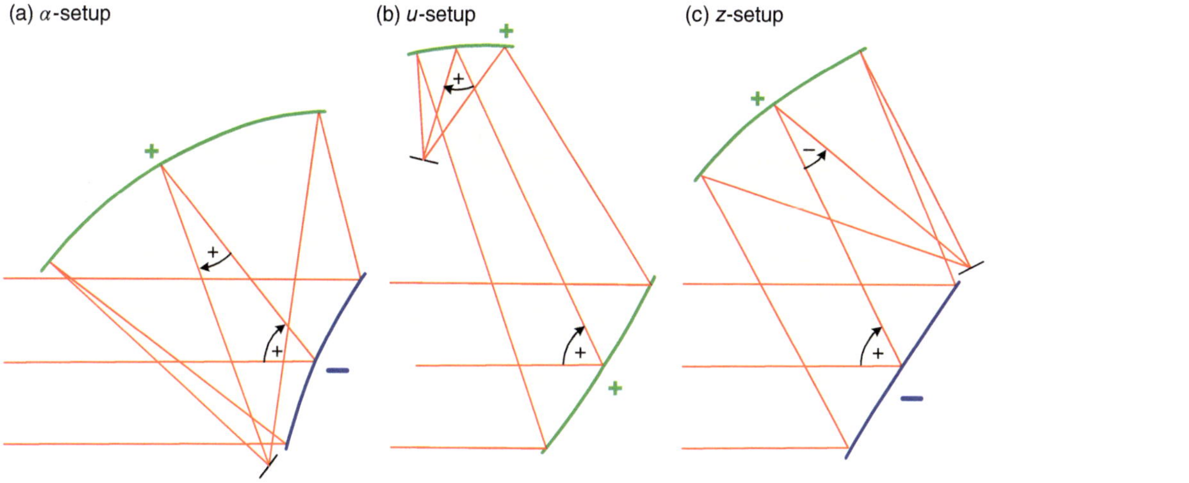

Figure 5.82 Three different geometries for two-mirror obscuration-free constellations: (a) α-setup, (b) u-setup and (c) z-setup.

a good diffraction limited correction can only be achieved if freeform surfaces are used to ensure the performance for a finite field and a large numerical aperture. These aspects are discussed in a later chapter.

5.7.4 Three-Mirror Setups*

(*) The content of this section is quite specialized and for advanced reading. Knowledge of this section is not necessary to understand the following chapters. Beginners are recommended to skip this section.

If a better quality for larger field angles is required in a telescope, three mirrors can be used to gain additional degrees of freedom for correction. An aligned setup with central obscuration is mostly not very practical and tilt angles are introduced to avoid any truncation. But as initial systems and the conceptual values for distances and power distribution between the mirrors, a paraxial layout is in any case of interest. The larger field then needs a better off-axis correction; therefore, flattening as well as aplanatic correction is recommended. Therefore, the three-mirror anastigmates (TMAes) play a major role in the design of modern telescopes, in particular off-axis with fulfilment of the Schiefspiegler condition or using freeform surfaces for correction. These details are discussed in a later chapter.

With the abbreviations [18, 35, 36] of the individual mirror magnifications m_j, the relative marginal ray height ratio at the M_1 and M_1 and the pupil magnifications m_{pj} as

$$m_2 = \frac{s_2'}{s_2}, \quad m_3 = \frac{s_3'}{s_3}, \quad \omega_2 = \frac{y_2}{y_1} \tag{5.93}$$

$$m_{p2} = \frac{-m_2\omega_2}{1+m_2-\omega_2},$$
$$m_{p3} = \frac{m_3}{1-(m_{p2}+1)m_2m_3+m_{p2}m_3} \tag{5.94}$$

a three-mirror telescopic setup can be determined for corrected spherical aberration, coma, astigmatism and field curvature, if the parameters are fulfilling the conditions

$$r_1 = \frac{2f}{m_2m_3}, \quad r_2 = \frac{-2f\,m_{p2}}{m_3(m_2-m_{p2})},$$
$$r_3 = \frac{2f\,m_{p2}m_{p3}}{m_3-m_{p3}} \tag{5.95}$$

$$t_1 = \frac{f(1+m_{p2})}{m_3(m_2-m_{p2})},$$
$$t_2 = \frac{-f\,m_{p2}}{m_3}\left(\frac{m_2+1}{m_2-m_{p2}} + m_{p3}\frac{m_3+1}{m_3-m_{p3}}\right)$$
$$L = s_3' = \frac{f\,m_{p2}m_{p3}(m_3+1)}{m_3-m_{p3}} \tag{5.96}$$

Here it is assumed that the entrance pupil is located at the primary mirror. Figure 5.83 sketches the layout and in Figure 5.84 it is visualized how the three-mirror imaging is decomposed into three parts according to the functionality of the three mirrors.

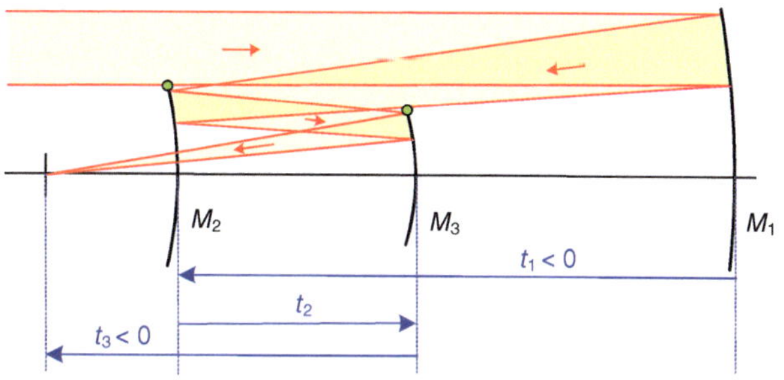

Figure 5.83 Sketch of a centred three-mirror system.

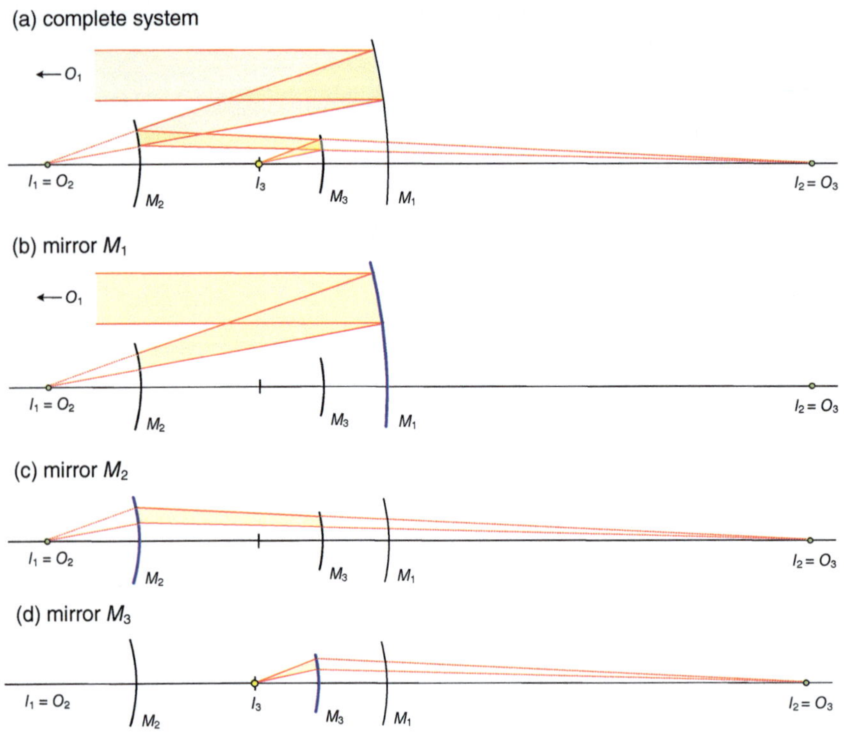

Figure 5.84 Decomposition of a three-mirror system.

5.8 Imaging Without Rotational Symmetry

5.8.1 Anamorphic Systems

5.8.1.1 Introduction

An optical imaging system is called anamorphic if two orthogonal cross sections x–z and y–z have different focal lengths and magnifications, but the same image location. To get these properties, components are needed that break the circular symmetry. Anamorphic systems are therefore double-plane symmetric. In instrumental optics there exist three types of anamorphic system types [37–42]:

- Systems with refractive cylindrical or toroidal surfaces.
- Systems with spherical mirrors under a finite incidence angle or with cylindrical reflective surfaces.
- Systems with prisms.

The first class typically has a straight optical axis. Applications of anamorphic systems are as follows.

- Beam cross section transforms, mostly for laser beams. Here no field of view is present, which simplifies the setups and performance correction considerably.
- Change of the aspect ratio of the field of view due to rectangular scenes or image sensors. This is illustrated in Figure 5.85.

A non-trivial question is whether the object, the pupil or the image is isotropic. This depends on the application and the setup of the stop. In camera applications, often the rectangular CCD sensors are the reason for an anisotropic field of view in the image space. But it is also possible that the imaged object format is anisotropic with a large horizontal width and a smaller height corresponding to a cinema impression. Some of these various options are collected in Figure 5.86 for illustration, but this list is not complete.

The ratio of focal lengths or magnifications in both cross sections is called the anamorphic factor or anamorphism

$$M = \frac{m_x}{m_y} \tag{5.97}$$

If the apertures in the object space are different, an anisotropic resolution appears if the system is diffraction limited. Therefore, if this is unwanted, it can make sense that although the image format is rectangular, the numerical aperture cone is circular. From an optical point of view this means that the anamorphic system can have different values of the etendue in both cross sections. The ratio of the etendue or Lagrange invariant between the two cross sections is denoted by

$$K = \frac{L_x}{L_y} \tag{5.98}$$

The easiest case is obtained if the etendue in both cross sections is the same, and we have $K = 1$. This means that the anamorphic ratio is seen in its inverted value for the imaging of the pupils. If the image, for example, is larger in x-direction, the numerical aperture is correspondingly

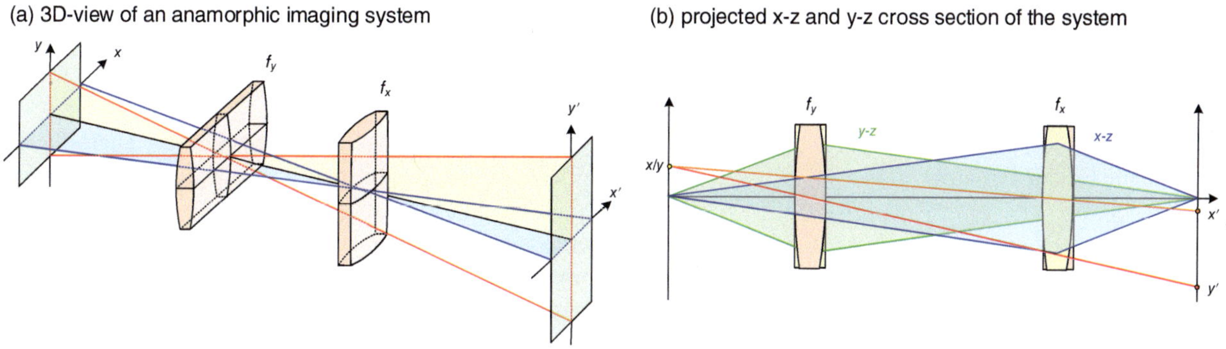

Figure 5.85 Anamorphic imaging by two crossed cylindrical lenses schematically shown by a change of the rectangular cross sections in object and image plane as (a) a three-dimensional view and (b) projected cross sections overlaid.

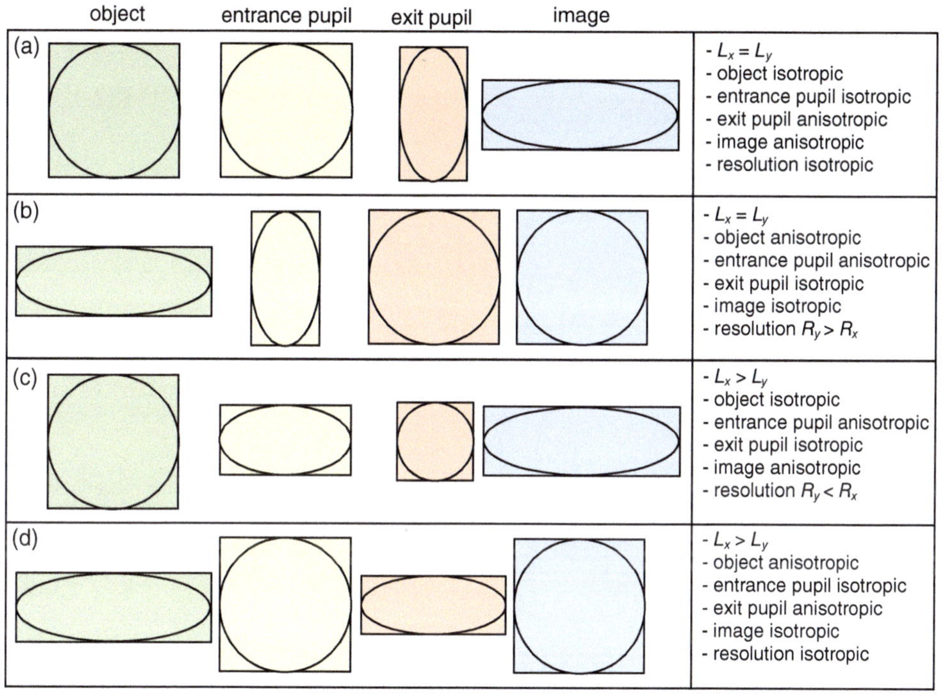

Figure 5.86 Isotropic vs anisotropic cases in object and image space. The resolution is determined by the aperture in the object space. $L_{x/z}$ are the Lagrange invariants in both cross sections.

smaller and the diffraction related resolution changed between x and y. If the object side is isotropic (maybe circular or squared in shape), the entrance pupil is isotropic too, but on the image side we have

$$x = y,\quad NA_x = NA_y$$
$$x' = M \cdot y',\quad NA'_x = \frac{NA'_y}{M} \tag{5.99}$$

which means, the exit pupil is anisotropic; for example, ellipsoidal. Depending on the stop shape and the field limitation, the conditions can also be isotropic on the image side and anisotropic in the object space

$$x = M \cdot y,\quad NA_x = \frac{NA_y}{M}$$
$$x' = y',\quad NA'_x = NA'_y \tag{5.100}$$

These two cases are sketched in Figure 5.86 as (a) and (b). If in addition different etendues are considered, more cases are possible, but the basic idea is the same. In part (c) and (d) of the figure only the case $L_x > L_y$ is sketched. In (c) the image is anisotropic and the aperture in the image space is isotropic, so the entrance pupil is anisotropic. In case (d) the image is isotropic, so the object field and the exit pupil are anisotropic. The mentioned resolution properties are based on the assumption that the system is diffraction limited in both cross sections and therefore the resolution depends on the entrance pupil diameters. In reality it may happen that the performance is dominated by aberrations, and then the resolution depends on these properties.

In plane symmetrical systems, the reason for the anamorphism is due to the projection effect of the incidence angles i_j and i'_j on the optical axis ray. This contribution therefore is paraxial and can be written as

$$M = \frac{m_x}{m_y} = \Pi_{j=1}^{N} \frac{\cos i_j}{\cos i'_j} \tag{5.101}$$

according to the diameter change of a collimated ray bundle for a skew refraction at a plane interface and the Fresnel equation projection factor.

5.8.1.2 Classes of Anamorphic Systems*

(*) The content of this section is quite specialized and for advanced reading. Knowledge of this section is not necessary to understand the following chapters. Beginners are recommended to skip this section.

In general, there are two types of anamorphic systems: the first and most important class is orthogonal and can be separated in the x–z and y–z plane. This means that a ray starting in one of these planes remains there by tracing through the system, and the system is then double plane symmetric. An arbitrary skew paraxial ray can be projected into the x–z and the y–z plane respectively; finally x, u and y, v are independent in the paraxial model.

In the more general case, there is a mixing between x and y due to azimuthal rotated components, which cannot be removed. This means that if a paraxial ray is traced, it doesn't remain in one plane. An example for this case is shown in Figure 5.87 for demonstration. Here a circular symmetric first lens L_1, a toroidal lens L_2 rotated by 45° and a cylindrical lens L_3 working in the y-cross section are assembled. It is seen that the ray fans in the x- and y-cross section respectively leave their plane after the lens L_2 and become twisted.

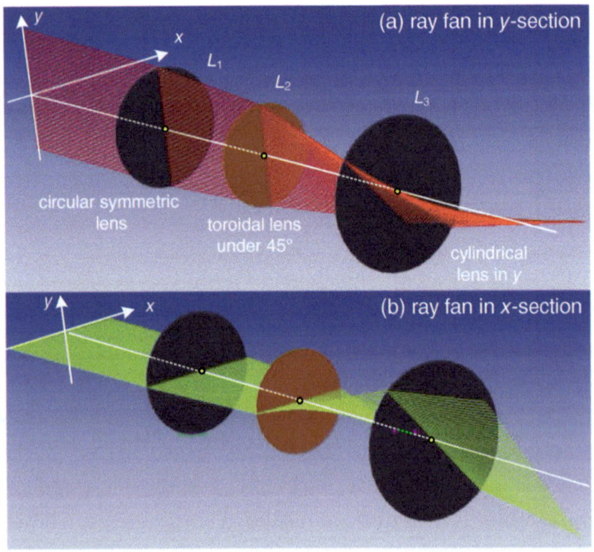

Figure 5.87 System with general non-orthogonal astigmatism with ray fans starting in the (a) y section and (b) x section.

If the components are rotated by arbitrary angles in the azimuthal direction, due to this new degree of freedom, it is always possible to decompose the surfaces into two projection planes and to remove the off-diagonal elements of the corresponding 4×4 $ABCD$ matrix. The conventional way to express the 4×4 matrix reads

$$\begin{pmatrix} x' \\ y' \\ u' \\ v' \end{pmatrix} = \begin{pmatrix} A_{xx} & A_{xy} & B_{xx} & B_{xy} \\ A_{yx} & A_{yy} & B_{yx} & B_{yy} \\ C_{xx} & C_{xy} & D_{xx} & D_{xy} \\ C_{yx} & C_{yy} & D_{yx} & D_{yy} \end{pmatrix} \cdot \begin{pmatrix} x \\ y \\ u \\ v \end{pmatrix} = \begin{pmatrix} \underline{A} & \underline{B} \\ \underline{C} & \underline{D} \end{pmatrix} \cdot \begin{pmatrix} x \\ y \\ u \\ v \end{pmatrix} \tag{5.102}$$

In the special case of orthogonal anamorphism, the x–y coupling off-diagonal elements vanish

$$\underline{M}_{SA} = \begin{pmatrix} A_x & 0 & B_x & 0 \\ 0 & A_y & 0 & B_y \\ C_x & 0 & D_x & 0 \\ 0 & C_y & 0 & D_y \end{pmatrix} \tag{5.103}$$

To show what happens in case of an azimuthal rotation of a system, it is more convenient to write the matrix in a different order with separated x- and y-submatrices as

$$\begin{pmatrix} x' \\ u' \\ y' \\ v' \end{pmatrix} = \begin{pmatrix} A_{xx} & B_{xx} & A_{xy} & B_{xy} \\ C_{xx} & D_{xx} & C_{xy} & D_{xy} \\ A_{yx} & B_{yx} & A_{yy} & B_{yy} \\ C_{yx} & D_{yx} & C_{yy} & D_{yy} \end{pmatrix} \cdot \begin{pmatrix} x \\ u \\ y \\ v \end{pmatrix} = \begin{pmatrix} \underline{M}_{xx} & \underline{M}_{xy} \\ \underline{M}_{yx} & \underline{M}_{yy} \end{pmatrix} \cdot \begin{pmatrix} x \\ y \\ u \\ v \end{pmatrix} \tag{5.104}$$

If the system is orthogonal, this matrix is simplified with the block structure without the mixed elements

$$\begin{pmatrix} x' \\ u' \\ y' \\ v' \end{pmatrix} = \begin{pmatrix} A_{xx} & B_{xx} & 0 & 0 \\ C_{xx} & D_{xx} & 0 & 0 \\ 0 & 0 & A_{yy} & B_{yy} \\ 0 & 0 & C_{yy} & D_{yy} \end{pmatrix} \cdot \begin{pmatrix} x \\ u \\ y \\ v \end{pmatrix} = \begin{pmatrix} \underline{M}_{xx} & 0 \\ 0 & \underline{M}_{yy} \end{pmatrix} \cdot \begin{pmatrix} x \\ y \\ u \\ v \end{pmatrix} \tag{5.105}$$

If this system is rotated by an angle φ we get with the rotation matrix

$$\begin{pmatrix} x' \\ u' \\ y' \\ v' \end{pmatrix} = \begin{pmatrix} \cos\varphi & 0 & \sin\varphi & 0 \\ 0 & \cos\varphi & 0 & \sin\varphi \\ -\sin\varphi & 0 & \cos\varphi & \sin\varphi \\ 0 & -\sin\varphi & -\sin\varphi & \cos\varphi \end{pmatrix} \cdot \begin{pmatrix} x \\ u \\ y \\ v \end{pmatrix} = \underline{M}_{\varphi} \cdot \begin{pmatrix} x \\ u \\ y \\ v \end{pmatrix} \tag{5.106}$$

the new matrix

$$\underline{M}' = \underline{M}_{\varphi}^{T} \cdot \underline{M} \cdot \underline{M}_{\varphi} \tag{5.107}$$

with the abbreviations

$$c = \cos\varphi, \quad s = \sin\varphi \tag{5.108}$$

we have in total the matrix

$$\begin{pmatrix} x' \\ u' \\ y' \\ v' \end{pmatrix} = \begin{pmatrix} c^2\underline{M}_{xx} + s^2\underline{M}_{yy} - cs(\underline{M}_{xy} + \underline{M}_{yx}) & c^2\underline{M}_{xy} - s^2\underline{M}_{yx} - cs(\underline{M}_{xx} - \underline{M}_{yy}) \\ c^2\underline{M}_{yx} - s^2\underline{M}_{xy} - cs(\underline{M}_{xx} - \underline{M}_{yy}) & s^2\underline{M}_{xx} + c^2\underline{M}_{yy} + cs(\underline{M}_{xy} + \underline{M}_{yx}) \end{pmatrix} \cdot \begin{pmatrix} x \\ y \\ u \\ v \end{pmatrix} \tag{5.109}$$

This matrix can be decomposed into an orthogonal part and a second part, which describes the deviation from the orthogonality in the form

$$\begin{pmatrix} x' \\ u' \\ y' \\ v' \end{pmatrix} = [\underline{M}_{ortho} + \underline{M}_{dev}] \cdot \begin{pmatrix} x \\ y \\ u \\ v \end{pmatrix}$$
$$= \left[\begin{pmatrix} c^2\underline{M}_{xx} + s^2\underline{M}_{yy} & 0 \\ 0 & s^2\underline{M}_{xx} + c^2\underline{M}_{yy} \end{pmatrix} + \begin{pmatrix} -cs(\underline{M}_{xy} + \underline{M}_{yx}) & c^2\underline{M}_{xy} - s^2\underline{M}_{yx} - cs(\underline{M}_{xx} - \underline{M}_{yy}) \\ c^2\underline{M}_{yx} - s^2\underline{M}_{xy} - cs(\underline{M}_{xx} - \underline{M}_{yy}) & cs(\underline{M}_{xy} + \underline{M}_{yx}) \end{pmatrix} \right] \cdot \begin{pmatrix} x \\ y \\ u \\ v \end{pmatrix} \tag{5.110}$$

If the non-diagonal matrix is rotated by an angle φ, it gets a diagonal form. By setting the upper right element in $\underline{M}_{dev}$ to zero

$$c^2\underline{M}_{xy} - s^2\underline{M}_{yx} - cs(\underline{M}_{xx} - \underline{M}_{yy}) = 0 \tag{5.111}$$

the relation

$$\tan 2\varphi = \frac{2\underline{M}_{yx}}{\underline{M}_{xx} - \underline{M}_{yy}} \tag{5.112}$$

is obtained and gives the azimuthal direction φ of the principal planes, which guarantees an orthogonal form with decoupled x and y cross sections.

5.8.1.3 Anamorphic Imaging Systems

In the paraxial range of an imaging system, the surfaces can be toric with different focal powers F_x and F_y in both cross sections. A special case is a cylindrical surface, if one of the powers is zero. Usually it is assumed that in the case of imaging systems the object and the image plane are at the same z-location, which means there is no astigmatic separation. Nevertheless, the pupil planes can be at different z-positions, which makes all wave reference conditions and references more complicated.

In general, the location of pupils and field plane locations are independent from each other, which means we have two different marginal rays in x and y as well as chief rays respectively. An arbitrary skew paraxial ray does not remain in one plane during tracing as shown in Figure 5.87, this is different from a circular symmetric system. The corresponding Lagrange invariants in both cross sections can be the same or can have different values. The condition of an anamorphic system, which has the same z-location of object and image, delivers an additional constraint on the system data. In the lens makers' formula formulated in Newton's form, with the distances z and z' of object and image to the focal points and the distances $a_{x/y}$ between first and last surface respectively to the corresponding focal points, this means [37]

$$(a_x - z) \cdot (a'_x - z') = f_x'^2, \quad (a_y - z) \cdot (a'_y - z') = f_y'^2 \tag{5.113}$$

If both equations are fulfilled simultaneously, this gives by subtraction a linear relation between z and z'

$$a_x a'_x - a_y a'_y + z \cdot (-a'_x - a'_y) + z' \cdot (a_y - a_x) = f_x'^2 - f_y'^2 \tag{5.114}$$

If this linear relation is inserted into one of the first equations, a quadratic equation results for z or z' respectively. This means there are always two solutions 1 and 2 for conjugated anamorphic planes. As can be found after a tedious but straightforward calculation, a relationship with inverse magnifications and therefore inverted anamorphic factors [37]

$$M_1 = \frac{m_{1x}}{m_{1y}} = \frac{m_{2y}}{m_{2x}} = \frac{1}{M_2} \tag{5.115}$$

If an anamorphic system has different signs of the magnification in x and y, by rotating the system the image is rotated by double the speed. An example system is shown in Figure 5.88. Two nested Galilean and Keplerian

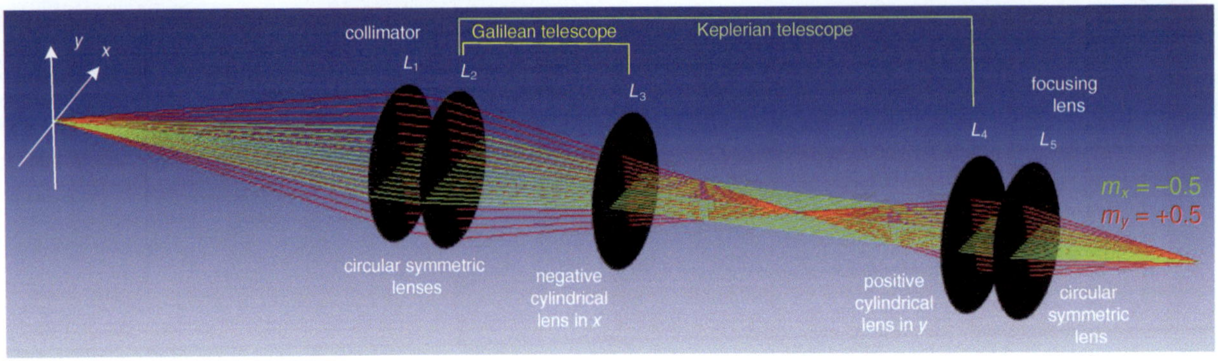

Figure 5.88 System with opposite sign of magnification by nested Galilean and Keplerian telescopes.

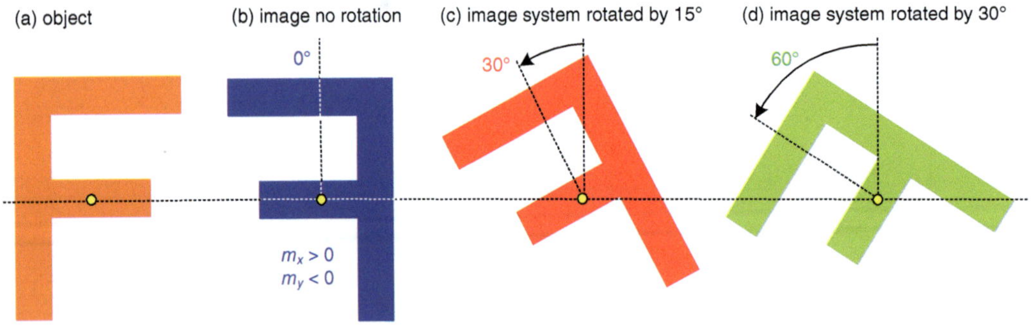

Figure 5.89 System with opposite sign of magnification by nested Galilean and Keplerian telescopes; behaviour of the image in the case of azimuthal system rotation.

telescopes with the same absolute magnification value are realized by cylindrical lenses. The magnifications are in this case $m_x = -0.5$ and $m_y = +0.5$.

Figure 5.89 illustrates the behaviour of the image in this system, if the system is rotated in the azimuthal direction. It is seen that the original capital '*F*' is inverted only in *y*-direction for the unrotated system. If the rotation is applied, the image rotates by the double angle.

5.8.1.4 Refractive Anamorphic Systems

If refractive anamorphic systems are considered, two different types must be distinguished. In the first case, cylindrical or toric components influence the rays in both cross sections differently. According to Figure 5.90(a), in the infinity case a cylindrical Galilean or Keplerian telescope can be used. In the case of finite object locations, a cylindrical so-called Bravais system changes the field vs aperture ratio keeping the image plane invariant. A typical property of this class of anamorphic systems is a straight optical axis.

The second opportunity uses prisms to generate a difference between the *x*–*z* and y–*z* plane. Prismatic anamorphic setups contain usually two prisms according to Section 4.10.3.3 and Figure 4.143. The reason for using two prisms is that the final direction of the optical axis should coincide with the initial one, although a lateral shift cannot be avoided. Only with more than two prisms can this offset also be removed. In imaging applications, prismatic solutions have problems with the aberration correction

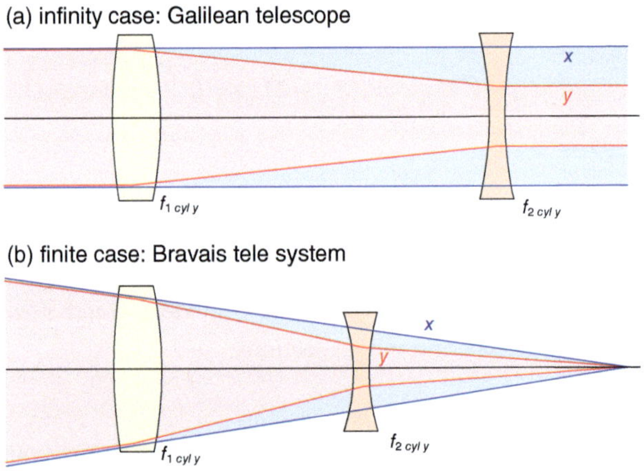

Figure 5.90 Schemes of two cylindrical sub-systems to generate an anamorphic system (a) in the infinity case and (b) the Bravais system for a finite object-image location. All lenses are cylindrical, the *y*–*z* (red) and *x*–*z* (blue) cross sections are overlaid in different colours.

and therefore they are not attractive. The long glass paths are critical and the chromatical aberrations are hard to overcome. In principle achromatic prism pairs can be used, but these systems are complicated and critical in fabrication and handling. The generated astigmatism is in addition a drawback of these attachments; therefore, they are preferably used in collimated beam paths for laser illuminations to change the cross sections.

If, for higher performance requirements, aberrations have to be controlled, it is learned that anamorphic systems are hard to correct because higher orders are not separable in x and y cross sections so easily as the paraxial properties. This means that a system which is performing well in the x- and y-cross section may have huge quality problems under 45°. To get rid of these problems, double-plane symmetric freeform surfaces can be used. These correction issues are discussed in a later chapter.

5.8.1.5 Reflective Anamorphic Systems*

(*) The content of this section is quite specialized and for advanced reading. Knowledge of this section is not necessary to understand the following chapters. Beginners are recommended to skip this section.

In off-axis mirror systems, the astigmatism correction of skew curved mirrors always provokes unwanted anamorphic properties. But it is also possible to use the anamorphism due to its properties [38] as an advantage and a feature. Historically, these systems are called Delrama solutions. Figure 5.91 shows a system made of two cylindrical mirrors, which have their focal power in the x cross section. The system is completely afocal. Figure 5.92 shows corresponding systems with prisms, which suffer from chromatical aberrations in reality, but are as solid-state components more comfortable to manufacture and are more robust in centring sensitivity.

5.8.2 Scheimpflug Imaging

5.8.2.1 Introduction

Usually in optical imaging systems it is assumed that the object and the image plane are perpendicular to the optical axis. If, conversely, the object plane is tilted by an angle θ

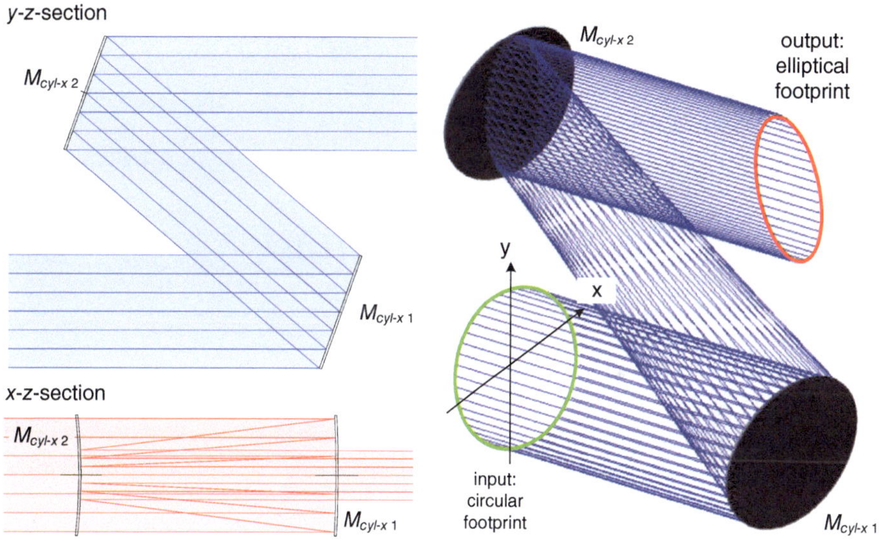

Figure 5.91 Afocal mirror system Delrama type I with an anamorphic factor; shown are the two cross sections and a perspective view.

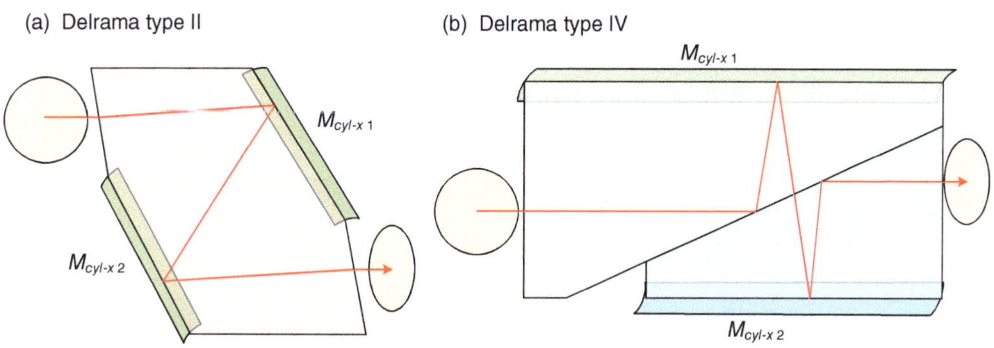

Figure 5.92 Afocal Delrama systems using prisms with cylindrical surfaces and an anamorphic factor; (a) type II with one prism only and (b) type IV with two cemented prisms.

against the axis, there exists a special tilt angle θ' of the image plane that nevertheless allows in paraxial approximation a sharp imaging. This special setup is called the Scheimpflug arrangement. But there are several critical issues with this kind of imaging: a macroscopic keystone distortion cannot be avoided due to the geometry; and the change of magnification over the image plane is complicated and causes in addition special issues of resolution and brightness with a strong anamorphism [43–51].

In Figure 5.93 an optical system with an object plane tilted by the angle θ is shown. Here a sharp image is found only in an image plane with a tilt angle of θ', for which the so-called Scheimpflug condition is fulfilled. This condition states that the extended object and image planes have to intercept with the principal planes of the system in the same height h; see the figure.

To derive the condition for the Scheimpflug setup, let's consider the imaging of one point O on axis and one point A off axis according to the figure. If the principal planes P and P' are considered, the intersection heights of the tilted object and image plane with P and P' are $h = h'$ respectively and the angles are θ and θ'. Relative to the axis point O the point A is shifted along the axis by z. The conditions in the image domain are completely similar. To get the height $y'_\perp$ in the global coordinate system there are two options.

- If the imaging of the off-axis point A with distance s to the principal plane is considered, the magnification in paraxial approximation is given in global coordinates

$$m = \frac{y'_\perp}{y_\perp} = \frac{s'}{s} \tag{5.116}$$

- Alternatively, by using the geometrical relations according to the drawing in the triangle on the object as well as on the image side, we have the relations

$$\tan\theta = \frac{z}{y_\perp} = \frac{s}{h}, \quad \tan\theta' = \frac{z'}{y'_\perp} = \frac{s'}{h'} \tag{5.117}$$

Now $y'_\perp$ is set to be equal from both equations

$$y'_\perp = y_\perp \frac{s'}{s} = \frac{z \cdot h}{s} \cdot \frac{s'}{s} = \frac{z'h'}{s'}$$

$$\frac{z'}{z} = \frac{h}{h'} \cdot \left(\frac{s'}{s}\right)^2 \tag{5.118}$$

Here the lateral and the axis magnification in point A in the global perpendicular coordinates are

$$m_z = m^2 = \frac{z'}{z} = \left(\frac{s'}{s}\right)^2 \tag{5.119}$$

If this is inserted, the necessary condition $h = h'$ is obtained. This is exactly the Scheimpflug condition for sharp paraxial imaging with tilted planes. If the geometrical relations

$$\tan\theta = \frac{s_0}{h}, \quad \tan\theta' = \frac{s'_0}{h'} \tag{5.120}$$

are inserted in the imaging condition on axis, the Scheimpflug equation in the formulation

$$\frac{1}{\tan\theta'} - \frac{1}{\tan\theta} = \frac{h}{f} \tag{5.121}$$

is obtained.

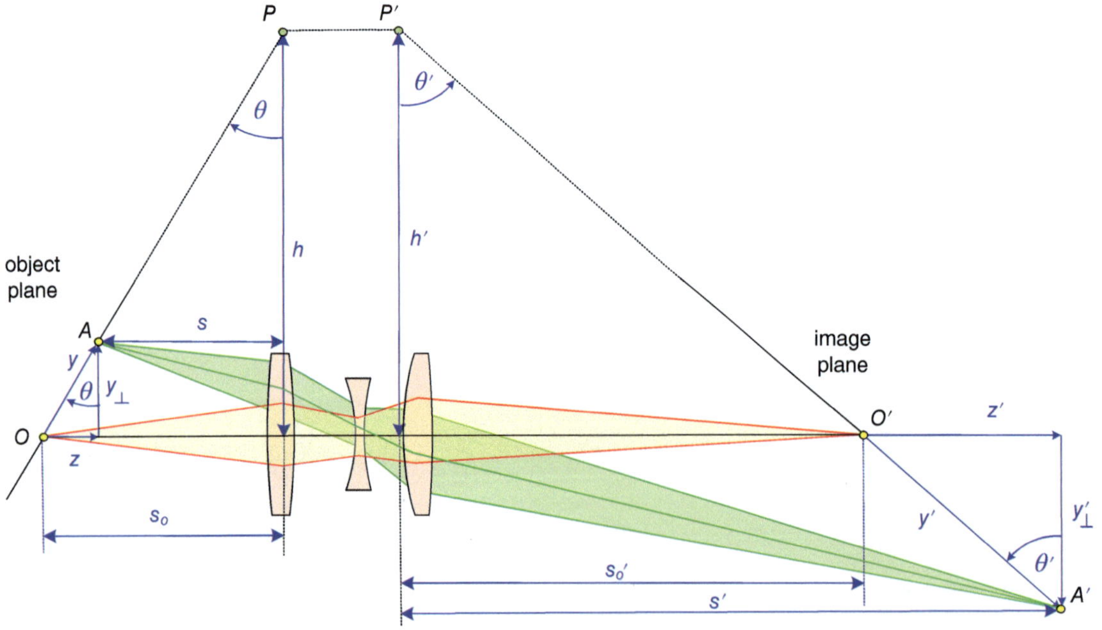

Figure 5.93 Geometry and notations of the Scheimpflug imaging; the necessary condition for sharp imaging is $h = h'$.

5.8.2.2 Keystone Distortion

If the magnification of the Scheimpflug setup is investigated, the x- and the y-direction must be distinguished. For on axis point O we have

$$m_o = \frac{s'_0}{s_0} = \frac{f}{f + s_0} = \frac{\tan\theta'}{\tan\theta} \tag{5.122}$$

In the sagittal x-direction for the off-axis point A, the magnification for the perpendicular coordinate $y'_\perp$ is identical to x and is given by

$$m_x = \frac{x'}{x} = \frac{y'_\perp}{y_\perp} = \frac{s'}{s} \tag{5.123}$$

In contrast to a classical perpendicular imaging setup, the off-axis points like A are simultaneously shifted in z; this always changes the magnification. This has the consequence that m_x depends on y or $y_\perp$ and is not constant. If the imaging equation

$$\frac{1}{s'} - \frac{1}{s} = \frac{1}{f} \tag{5.124}$$

is used to eliminate s' and the relations

$$s = s_0 + z, \quad y_\perp = y \cdot \cos\theta \tag{5.125}$$

are used, the dependence

$$m_x(y) = \frac{f}{f+s} = \frac{f}{f + s_0 + y_\perp \cdot \tan\theta} = \frac{f}{f + s_0 + y \cdot \sin\theta}$$
$$m_x(0) = m_0 = \frac{f}{f + s_0} \tag{5.126}$$

is found. It shows the dependence of the lateral sagittal magnification m_x on y. For the magnification in the tilted y–z section plane of the system, we get with the projection into the tilted plane

$$m_y = \frac{y'}{y} = \frac{y'_\perp \cdot \cos\theta}{y_\perp \cdot \cos\theta'} = m_x(y) \cdot \frac{\cos\theta}{\cos\theta'} \tag{5.127}$$

Correspondingly there is a large keystone distortion and complicated photometric conditions for the image brightness (see Figure 5.94). It should be mentioned that usually distortion is a type of aberration of the optical system of third order or higher. In the case of the Scheimpflug arrangement, the macroscopic keystone distortion results from the tilted plane and is of first order and the strongly changing conditions of object-image distance and magnification. This means it is called distortion, but it is not an inherent error of the system. Figure 5.94 illustrates the magnification conditions for a Scheimpflug system a) schematically and in b) for a real image, where the sharpness conditions are seen for different depths of focus by changing the F#-number.

By substituting the axis magnification m_0 instead of f into the formulas, we get the magnifications in the form

$$m_x = \frac{s \cdot m_0}{s + y \cdot (1 - m_0) \cdot \sin\theta},$$

$$m_y = \left[\frac{s \cdot m_0}{s + y \cdot (1 - m_0) \cdot \sin\theta}\right]^2 \cdot \frac{\cos\theta}{\cos\theta'} \tag{5.128}$$

For the photometric impact of the skew conditions, the local magnification at an arbitrary point x/y is relevant. It is simply given by the area magnification

$$m_{area} = m_y \cdot m_x = m_x^2 \cdot \frac{\cos\theta}{\cos\theta'} \tag{5.129}$$

If the image related incidence angle i' in the meridional plane is drawn as a function of the object side angle i, we get for a thin lens Figure 5.95. With

$$m_o = \frac{\tan\theta'}{\tan\theta}, \quad u' = \frac{u}{m_0}, \quad w' = w \tag{5.130}$$

and the geometry relations

$$i = \theta + w - u, \quad i' = \theta' + w' - u' = \theta' + w - \frac{u}{m_0} \tag{5.131}$$

by insertion we get the image-side incidence angle

$$i' = w - \frac{u}{m_0} + \operatorname{atan}(m_0 \cdot \tan\theta) \tag{5.132}$$

Since w can be positive or negative and θ is always positive, here the dependence is asymmetric.

In Figure 5.96 this dependence is shown for the three magnifications $m_0 = 3,\ 1,\ 0.333$ with field angle $w = 10°$ without an aperture angle. The centre line corresponds to the axis field point, the coloured bands for all the rays inside the finite field. It is seen that in particular for large m_o the incidence angles on the 'long side' of the field can be extraordinarily large. This means, in practice, a selection of large tilt angles θ in combination with a large magnification gives critical large incidences on the image side. This can be seen in Figure 5.95 for the green field bundle in point P'_2.

5.8.2.3 Generalized Scheimpflug Setup*

(*) The content of this section is quite specialized and for advanced reading. Knowledge of this section is not necessary to understand the following chapters. Beginners are recommended to skip this section.

Figure 5.97 shows a somewhat generalized configuration with a lens tilted by φ. In this case we have

$$\frac{s}{s'} = \frac{\tan\theta - \tan\varphi}{\tan\theta' - \tan\varphi} \tag{5.133}$$

This geometry can be considered as a more general Scheimpflug setup with an inclined optical axis running through the lens. In this case the magnifications depend on φ. But by a corresponding coordinate transform in the rotated system of the lens, the original Scheimpflug setup is again obtained but the field of view is then selected to be asymmetric. In principle this arrangement means not only are the conjugated object and image plane tilted, but also the pupil of the system.

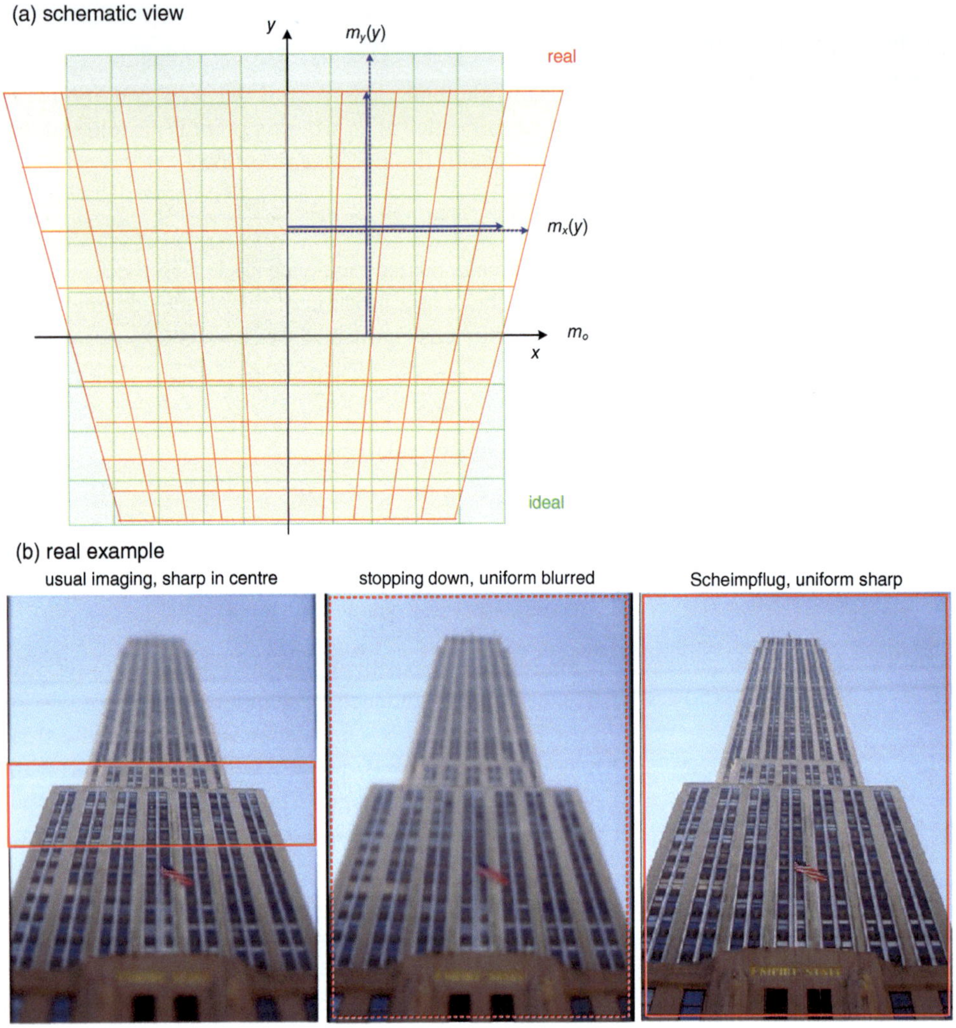

Figure 5.94 Keystone distortion for the Scheimpflug imaging. In (a) the scheme and effect of different magnifications is shown, in (b) a real scene is imaged by conventional imaging in the left column, conventionally stopped down with reduced resolution in the middle and the sharp Scheimpflug case in the right column [52] / with permission of Dennis Ochse.

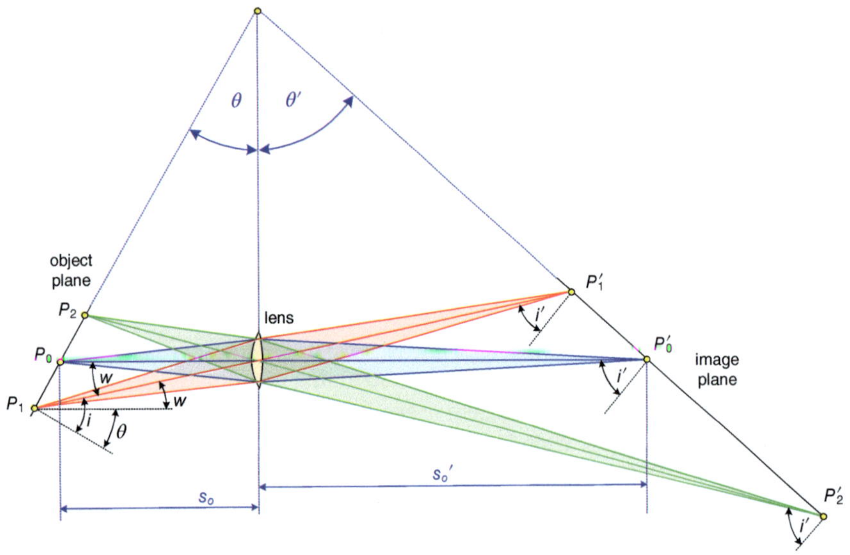

Figure 5.95 Geometry and incidence angles in a Scheimpflug setup for different field points.

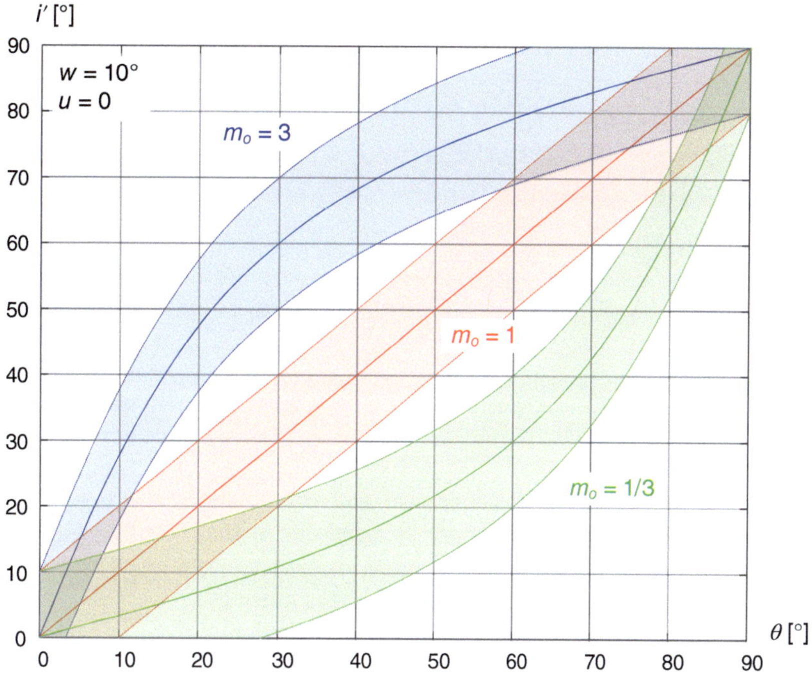

Figure 5.96 Incidence angle dependence on the image side for a Scheimpflug system for field angle $w = 10°$, no numerical aperture and the magnifications on axis of $m_0 = 1/3$, 1 and 3. The coloured bands describe the variation over the field of view.

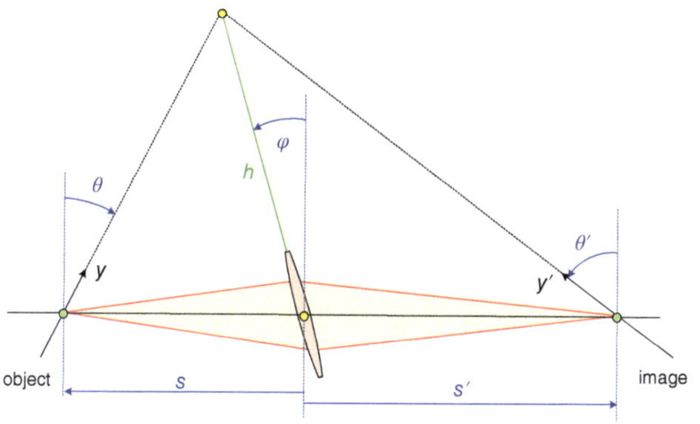

Figure 5.97 Terms for the generalized Scheimpflug imaging.

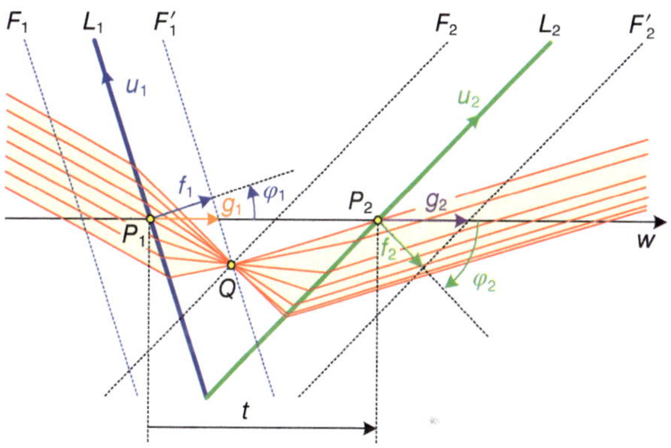

Figure 5.98 Scheme of two tilted lenses in confocal setup of two lenses L_1 and L_2.

5.8.3 General Skew Imaging*

(*) The content of this section is quite specialized and for advanced reading. Knowledge of this section is not necessary to understand the following chapters. Beginners are recommended to skip this section.

In recent times, a more generalized concept to consider paraxial imaging with ideal plane lenses of arbitrary orientation and position [53, 54] has been investigated. If, for example, two lenses are used in a confocal setup and the incoming light is collimated, by knowing the classical imaging conditions, the rays according to Figure 5.98 are obtained. The rays are focused by the lens L_1 into the point Q, which is the intersection point of the focal planes F_1' and F_2. The second lens then collimates the rays again. As is explained in the literature, the classical lens makers' formula can be used

$$\frac{1}{w'} - \frac{1}{w} = \frac{1}{g} \tag{5.134}$$

if skew coordinates u–w according to the figure are used. The z direction is changed into w, which is aligned with the connecting line between the principal points P_1 and P_2. The lateral coordinate u is measured along the lens. Furthermore, the projected focal length g is defined as

$$g = \frac{f}{\cos\varphi} \tag{5.135}$$

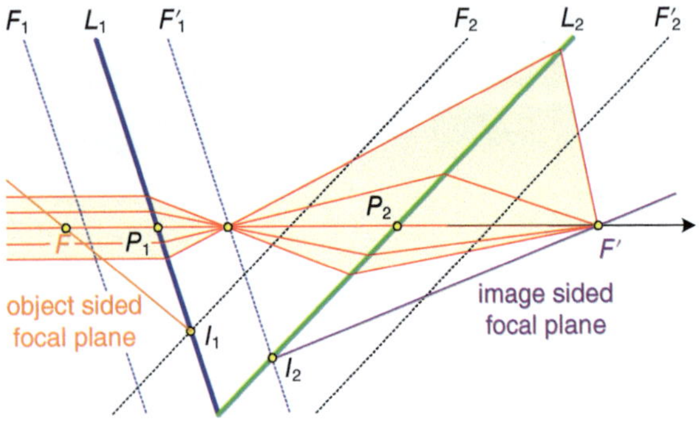

Figure 5.99 Common focal planes of the system constituting two tilted lenses. Here, only the construction of the image sided focal point F is shown by the red rays.

By this clever choice of parameters and coordinate systems, the imaging equations are as simple as in the usual case.

It can be shown that the two lenses can be substituted by a single lens with the resultant focal length very similar to the classical centred case

$$f = \frac{g_1 g_2}{g_1 + g_2 - t} \tag{5.136}$$

with the distance t between $P1$ and $P2$. But in contrast to the usual geometry, the object-side and the image-side focal planes now are inclined against each other according to the drawing in Figure 5.99.

The principal planes of the complete system are found by the conditions that they must be parallel to the focal planes found above and must meet the intersection point of the lens planes. This is shown in Figure 5.100. The principal planes are then also the planes to define the lateral coordinates u and u' of the compound system.

The same scheme can be used if the two lenses are not tilted against each other but are laterally shifted.

All the principles described here can also be used in three dimensions. For more details and further examples and explanations the reader is referred to the literature.

5.8.4 Image Plane Tilt

5.8.4.1 Introduction

There are several practical issues in systems lacking rotational symmetry with the tilt of the image plane. The most prominent examples are [55, 56]

- Scheimpflug systems:
  here the object plane is tilted by intention and the Scheimpflug condition gives the appropriate image plane tilt to gather a sharp image.
- Prism systems:
  in the case of dispersion prisms with non-perpendicular transition of the reference axis ray, an image plane tilt is observed.
- Skew mirror systems:
  if curved mirrors are used to achieve obscuration free imaging, the image plane is typically inclined.

It should be mentioned here that a tilt of the image plane is strongly related to aberrations, in particular astigmatism and coma. Therefore, only some basic and mostly linear relationships are presented here; a more in the depth discussion is found in the second volume about aberration theory.

5.8.4.2 Plane Symmetric Systems and the Scheimpflug Condition*

(*) The content of this section is quite specialized and for advanced reading. Knowledge of this section is not necessary to understand the following chapters. Beginners are recommended to skip this section.

A general rational transformation of the form [57]

$$x' = \frac{a_1 x + b_1 y + c_1 z + d_1}{a_0 x + b_0 y + c_0 z + d_0},$$
$$y' = \frac{a_2 x + b_2 y + c_2 z + d_2}{a_0 x + b_0 y + c_0 z + d_0},$$
$$z' = \frac{a_3 x + b_3 y + c_3 z + d_3}{a_0 x + b_0 y + c_0 z + d_0} \tag{5.137}$$

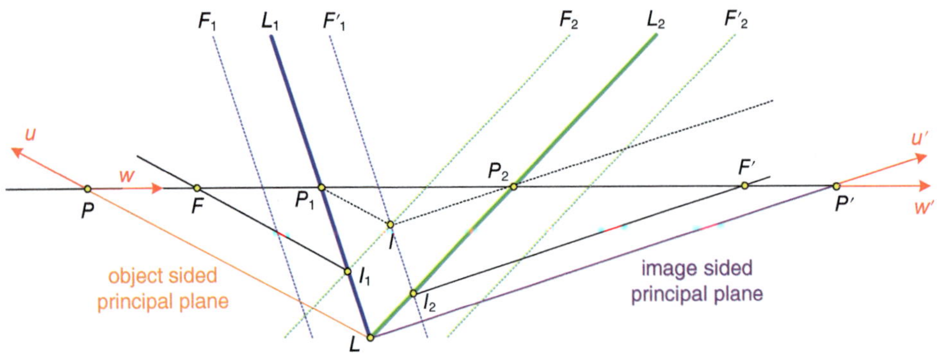

Figure 5.100 Common focal planes of the substitute of two tilted lenses. Here only the construction of the image sided focal point F is shown.

is called a linear collineation, see also Section 3.5.5. Here x, y, z are the coordinates of an object point, x', y', z' are those of a point image. This means in particular that this approach is also applicable in the case of general three-dimensional optical systems with rotated or shifted axes under skew conditions. This means that the effects of image plane tilt can be described in a quasi-parabasal approach with the help of these equations, it is a first order effect. If for simplicity a plane-symmetric system is considered with the y–z plane as mirror symmetric plane, there must be the same result for $-x$ instead of $+x$. This implies the conditions [40, 58]

$$a_0 = a_2 = a_3 = b_1 = c_1 = d_1 = 0 \tag{5.138}$$

If the coordinate origin of the principal planes is selected in x direction we have $a_1 = d_0$. If the origin of the object space is imaged onto the origin of the image domain, the condition $d_2 = d_3 = 0$ follows. If the rotation and tilt in the image space is performed around the x' axis, the condition $c_2 = 0$ must be valid. Normalizing the remaining equation coefficients onto d_o gives for a plane symmetric system with y–z symmetry plane the simplified equations

$$x' = \frac{x}{b_0 y + c_0 z + 1}, \quad y' = \frac{b_2 y}{b_0 y + c_0 z + 1},$$
$$z' = \frac{b_3 y + c_3 z}{b_0 y + c_0 z + 1} \tag{5.139}$$

With the abbreviation

$$z_0' = \frac{c_3 z_0}{1 + c_0 z_0} \tag{5.140}$$

and the ansatz for the tilted object and image plane, located on axis $y = 0$ at z_o and z_0' (see Figure 5.97)

$$z = z_0 - y \cdot \tan\theta, \quad z' = z_0' - y' \cdot \tan\theta' \tag{5.141}$$

by insertion of y', z' and z in the imaging Eq. (5.139) the relation

$$\tan\theta' = \frac{z_0'}{b_2 z_0} \tan\theta + \frac{b_0 z_0' - b_3}{b_2} \tag{5.142}$$

independent of y is obtained. This is a generalized Scheimpflug condition and allows us to calculate θ' out of the system parameter and dependence on θ. With the interpretation of b_2 as image anamorphism M and the abbreviation G as well as the magnification on axis m_o

$$b_2 = M = \frac{m_{y0}}{m_{x0}} = \frac{y'/y}{x'/x}, \quad m_0 = \frac{z_0'}{z_0} = \frac{u'}{u},$$
$$G = u' \cdot \left(b_0 z_0' - b_3\right) \tag{5.143}$$

this equation can be written as

$$M \cdot u' \cdot \tan\theta' = u \cdot \tan\theta + G \tag{5.144}$$

The concrete calculation of M and G is the main key to calculate image plane tilt in real non-circular symmetric systems. It should be known that this formula contains only the geometrical linear contribution as well as the linear astigmatic part of the image plane tilt. As it is pointed out in [56], in case of N surfaces with primary constant coma-contributions W_{31j}, the formula reads with the Lagrange invariant L

$$M \cdot u' \cdot \tan\theta' = u \cdot \tan\theta + G + \frac{2}{L} \sum_{j=1}^{N} \frac{\bar{y}}{y} W_{31j} \tag{5.145}$$

More details and the tight connection of the issue of image plane tilt is discussed in the chapter about aberrations of three-dimensional systems.

5.8.4.3 Tilt of a Skew Spherical Surface*

(*) The content of this section is quite specialized and for advanced reading. Knowledge of this section is not necessary to understand the following chapters. Beginners are recommended to skip this section.

If a spherical surface with radius R, is met by the chief ray under an angle i, it has the anamorphic factor

$$M = \frac{\cos i}{\cos i'} \tag{5.146}$$

The law of refraction reads

$$n \sin i = n' \sin i' \tag{5.147}$$

and the sagittal Coddington equation (see Chapter 16)

$$\frac{n'}{s'} - \frac{n}{s} = \frac{n' \cos i' - n \cos i}{R} \tag{5.148}$$

With the approximated relation (see Figure 5.101)

$$\tan\theta' = \frac{1}{s'} \cdot \frac{\delta s'}{\delta i'} \tag{5.149}$$

and by calculating the total differentials of the previous equations, the expression

$$G = M \cdot u' \tan\theta' = n \sin i \cos i \cdot \left(\frac{\cos i}{n} - \frac{\cos i'}{n'}\right) \cdot \frac{y}{R} \tag{5.150}$$

or

$$\tan\theta' = \frac{n y \sin i \cos i'}{R u'} \cdot \left(\frac{\cos i}{n} - \frac{\cos i'}{n'}\right) \tag{5.151}$$

is found for the image plane tilt. The corresponding geometry is sketched in Figure 5.101.

5.8.4.4 Tilt of a Skew Conical Mirror Surface*

(*) The content of this section is quite specialized and for advanced reading. Knowledge of this section is not necessary to understand the following chapters. Beginners are recommended to skip this section.

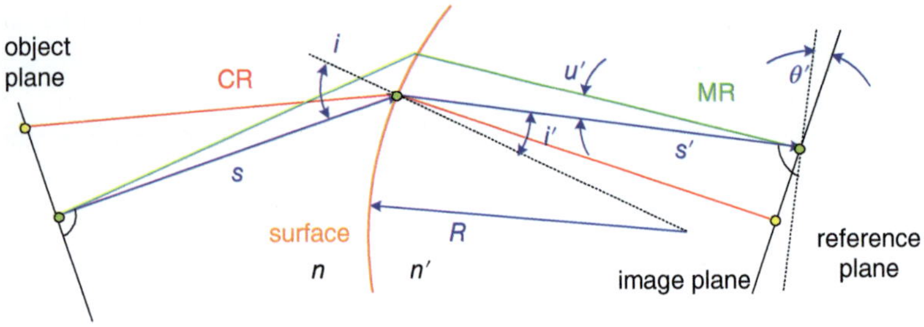

Figure 5.101 Image plane tilt at a spherical skew surface.

Mostly in practice, the image tilt is observed in skew mirror systems. For a tilted mirror the formula (5.150) simplifies to

$$G = \frac{2y \sin i}{R} \tag{5.152}$$

To avoid any consideration about aberrations of higher than first order, it makes sense to think on a conical mirror, which is used as a stigmatic imaging system. The arrangement is seen in Figure 5.102 [31, 32, 55]. It is assumed here that the object plane is tilted by θ_o and the stop is at the mirror. The mirror shape is elliptical and performs a perfect imaging of the axis point from P into P'. The field point Q with field angle w is imaged into Q_s' and Q_t' respectively by tangential and sagittal ray bundles respectively.

If the Coddington equations are used to find the image lengths s_t' and s_s', it is found for small field angles w in linear approximation

$$\frac{1}{s_t'} = \frac{1}{s_0'} + w \cdot \left[\frac{\tan \theta_0}{s_0} + \frac{\sin(2i)}{R} \right]$$
$$\frac{1}{s_s'} = \frac{1}{s_0'} + w \cdot \left[\frac{\tan \theta_0}{s_0} - \frac{\sin(2i)}{R} \right] \tag{5.153}$$

It is seen that the sagittal and the tangential image planes are tilted with different sign, so linear astigmatism cannot be avoided in the case of one mirror. If the axis magnification

$$m_0 = \frac{s_0'}{s_0} \tag{5.154}$$

is used, the individual image plane tilts can be written as

$$\tan \theta_t' = m_0 \cdot \tan \theta_0 - (1 - m_0) \tan i$$
$$\tan \theta_s' = m_0 \cdot \tan \theta_0 + (1 - m_0) \tan i \tag{5.155}$$

If two confocal conical mirrors are used to get a stigmatic sharp image centre (corresponding to the setup of Figure 5.77), by applying the above relation twice the astigmatism can be removed and the image plane tilt

$$\tan \theta' = \frac{s_1}{s_1'} \cdot \frac{s_2'}{s_2} \cdot \tan \theta_0 = \frac{m_{02}}{m_{01}} \cdot \tan \theta_0 \tag{5.156}$$

is obtained. Here the distances in between are measured towards the intermediate common focal point and the relation

$$\frac{s_1 \sin(2i_1)}{R_1} = \frac{s_2 \sin(2i_2)}{R_2} \tag{5.157}$$

must be fulfilled to remove the skew astigmatism. By a consecutive application of this approach for N mirrors the equations and assuming a non-astigmatic object, the condition for corrected linear astigmatism reads

$$(1 + m_{0N}) \cdot \tan i_N + \sum_{j=1}^{N-1} (1 + m_{0j}) \cdot \tan i_j \cdot \prod_{k=j+1}^{N} m_{0k} \tag{5.158}$$

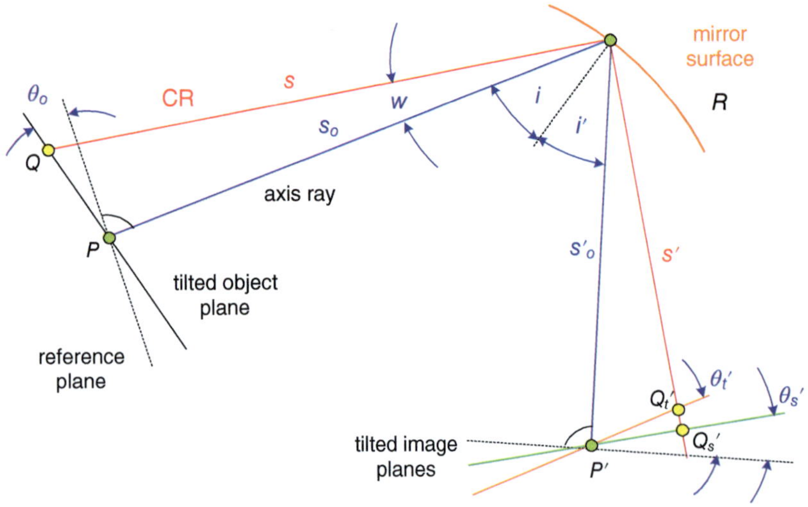

Figure 5.102 Image plane tilt at a spherical skew mirror surface.

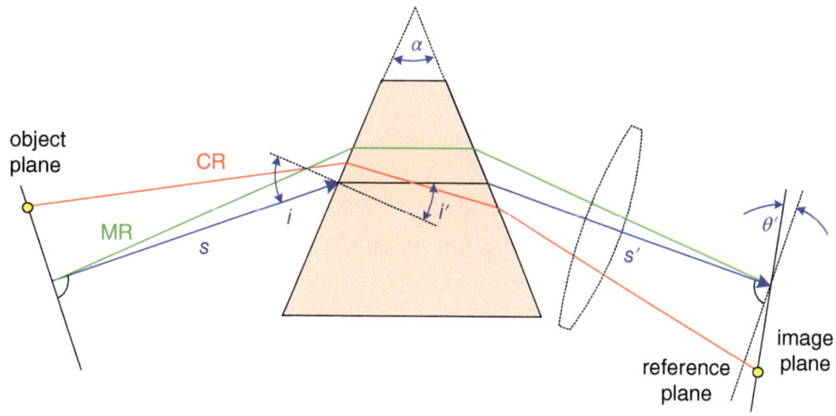

Figure 5.103 Image plane tilt at a symmetrical dispersion prism.

and the resulting image plane tilt is given by

$$\tan\theta' = \tan\theta_0 \cdot \prod_{k=j+1}^{N} m_{0k} \tag{5.159}$$

A more general derivation for plane symmetric systems with arbitrary shaped mirrors in differential approximation is found in [55].

5.8.4.5 Prism Systems*

(*) The content of this section is quite specialized and for advanced reading. Knowledge of this section is not necessary to understand the following chapters. Beginners are recommended to skip this section.

In case of a dispersion prism with angle α, we get the value of $G = 0$ due to the plane surfaces. In the special case of a symmetrical transfer through the prism with the geometrical condition

$$n \sin i = n' \sin\frac{\alpha}{2}, \quad n'u' = nu \tag{5.160}$$

and air around the prism, the image plane tilt is given by

$$\tan\theta' = 2\left(1 - \frac{1}{n}\right) \cdot \sin\frac{\alpha}{2} \tag{5.161}$$

See Figure 5.103 for this arrangement.

5.9 Miscellaneous

5.9.1 Herzberger's Diapoints*

(*) The content of this section is quite specialized and for advanced reading. Knowledge of this section is not necessary to understand the following chapters. Beginners are recommended to skip this section.

In classical geometrical optics, the intersection points of marginal rays with the optical axis are used to define longitudinal aberrations. But in the case of skew rays like sagittal rays in a circular symmetric system, there are mathematically no intersection points with the optical axis. According to the idea of Herzberger [58–60], in circular symmetric systems it makes sense to consider the intersection point of a skew ray with the y–z meridional plane. This is sketched in Figure 5.104 for a circular symmetric system, but the idea can be applied for general symmetry-free systems too. In this case, the sagittal rays with positive and negative x component intersect in the same so-called diapoint D, if the original x-value of the ray is zero. The diapoint D therefore is a kind of sagittal image point. If $\vec{s}' = (s_x' \; s_y' \; s_z')^T$ is the direction unit vector of a ray in the image domain and x', y' are the intersection point coordinates of a skew ray with the image plane, the diapoint coordinates are

$$z_D = -\frac{x' s_z'}{s_x'}, \quad y_D = y' - \frac{x' s_y'}{s_z'} \tag{5.162}$$

If many skew rays are considered in an optical system, the diapoints form a cloud in the meridional plane like a spot diagram. In special cases they lie on a curve or on a straight line. These particular arrangements are connected to the existence of special aberrations. Therefore. diapoints can also be used to characterize aberrations. Unfortunately, they don't exist for meridional rays, which restricts their importance. Since in principle the deviation from a symmetrically located corresponding chief ray in the meridional plane is measured, the diapoints give a special possibility to quantify the asymmetric aberrations coma and distortion.

5.9.2 Canonical Coordinates

5.9.2.1 Axis Case

Canonical coordinates are used for the scaling and normalization of the object- and image-side aperture and field conditions in an optical system. This normalization means that the quality between different systems which are nearly diffraction limited becomes comparable and the influences of wavelength and numerical aperture are no longer an explicit part of the calculation [61–63]. Instead of the common plane entrance and exit pupil, the new

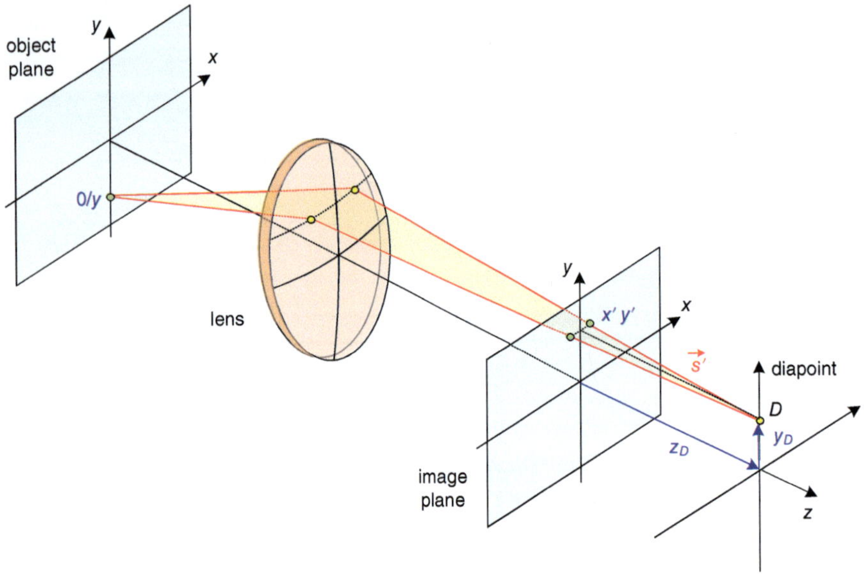

Figure 5.104 Definition of the diapoint D in a circular symmetric system.

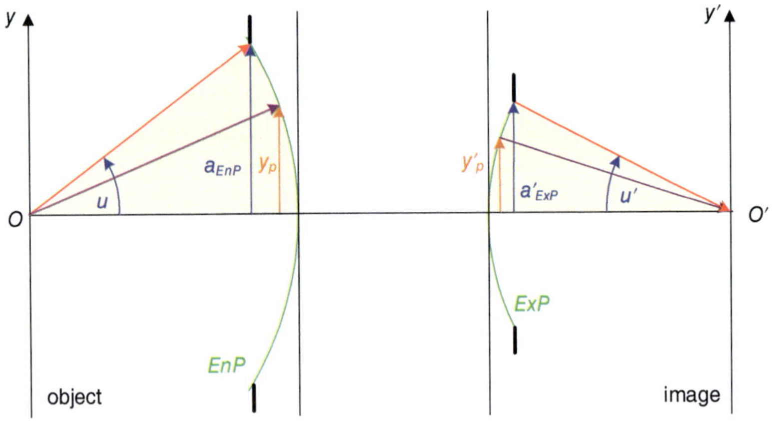

Figure 5.105 Canonical coordinates of an optical system on axis.

discussion is based on pupil spheres around each object and image point, so the fulfilment of the sine condition is assumed (see Chapter 17). The terms are shown in Figure 5.105. For the imaging of a field point on the optical axis, canonical pupil coordinates are defined by a simple normalization. With x_p and y_p as the lateral real pupil coordinates on the sphere of the entrance pupil, the canonical pupil coordinates are given by

$$\bar{x}_p = \frac{x_p}{a_{EnP}},\ \bar{y}_p = \frac{y_p}{a_{EnP}} \tag{5.163}$$

where a_{EnP} is the maximum distance of all points in the clear entrance aperture form the optical axis. This also applies for the exit pupil correspondingly

$$\bar{x}'_p = \frac{x'_p}{a_{ExP}},\ \bar{y}'_p = \frac{y'_p}{a_{ExP}} \tag{5.164}$$

Since the numerical aperture influences the resolution by the Airy point image, the following canonical object coordinates are defined using the real object coordinates x and y

$$\bar{x} = \frac{n \cdot \sin u}{\lambda} \cdot x;\ \bar{y} = \frac{n \cdot \sin u}{\lambda} \cdot y \tag{5.165}$$

Accordingly, in the image plane, it can be determined that the lateral image coordinates have a similar scaling for

$$\bar{x}' = \frac{n' \cdot \sin u'}{\lambda} \cdot x';\ \bar{y}' = \frac{n' \cdot \sin u'}{\lambda} \cdot y' \tag{5.166}$$

Due to this normalization, the entrance and exit pupil as well as object and image plane now become comparable as far as their lateral dimensions are concerned. Figure 5.105 shows the corresponding terms of the canonical coordinates. Thus, in the case of a coma-free imaging for a pupil sphere, we have for the pupil imaging if no aberrations are taken into account

$$\bar{x}_p' = \bar{x}_p, \quad \bar{y}_p' = \bar{y}_p \tag{5.167}$$

In the paraxial case for the field coordinates we have similar

$$\bar{x}' = \bar{x}, \quad \bar{y}' = \bar{y} \tag{5.168}$$

so the equations for the paraxial calculation become very simple.

5.9.2.2 Off-Axis Case

For an off-axis field point the chief ray CR is defined with regard to the upper and lower tangential coma rays (TCU, TCL) so the difference of the direction cosines relative to the chief ray are the same. For the angle of the rays with the optical axis in Figure 5.106, this means that

$$\sin u_{\mathrm{TCU}} - \sin u_{\mathrm{CR}} = -\sin u_{\mathrm{TCL}} + \sin u_{\mathrm{CR}} \tag{5.169}$$

We define the numerical aperture in the meridional section as the sine difference of the outermost coma rays relative to the chief ray

$$NA = n \cdot (\sin u_{\mathrm{CR}} - \sin u_{\mathrm{TCU}}) = n \cdot (\sin u_{\mathrm{TCO}} - \sin u_{\mathrm{CR}}) \tag{5.170}$$

It has the same value in the upper and lower coma bundle due to condition (5.169).

As in the axial case, the entrance pupil is a sphere with the radius R_{EnP} around the object point O, passing through the point Q where the chief ray intersects the optical axis. However, the chief ray OQ generally does not run exactly through the centre of the optical aperture stop, which is defined in a plane perpendicular to the optical axis. But for small values of the field angle this is a good approximation. According to the definition above, the absolute values of the heights y_{TCU} and y_{TCL} have also equal absolute values and are given by

$$y_{\mathrm{TCU}} = R_{EnP} \cdot (\sin u_{\mathrm{TCU}} - \sin u_{\mathrm{CR}}),$$
$$y_{\mathrm{TCO}} = R_{EnP} \cdot (\sin u_{\mathrm{TCO}} - \sin u_{\mathrm{CR}}) \tag{5.171}$$

The parameters of the skew bundle in the tangential section are $u_{\tan}$ and $y_{\tan}$ for the angle and the height, respectively. For reasons of consistency, in the tangential section they are defined for an off-axis point object as

$$u_{\tan} = \frac{\sin u_{\mathrm{TCU}} - \sin u_{\mathrm{CR}}}{\cos u_{\mathrm{CR}}}, \quad y_{\tan} = \frac{y_{\mathrm{TCU}}}{\cos u_{\mathrm{CR}}} \tag{5.172}$$

So the relations

$$y_{\mathrm{TCU}} = R_{EnP} \cdot \cos u_{\mathrm{CR}} \cdot u_{\tan} = y_{\tan} \cdot \cos u_{\mathrm{CR}} \tag{5.173}$$

are valid. The sagittal rays are defined in such a way that they intersect the pupil sphere in the x-z plane. Therefore, for sagittal rays we have $yp = 0$. The sagittal rays passing towards the edge of the pupil form angles with the y-z plane. They are referred to as u_{sag}. The x-coordinate of the intersection point with the pupil is referred to as x_{sag}. Then it is

$$x_{sag} = R_{EnP} \cdot \sin u_{sag} \tag{5.174}$$

The assigned astigmatic parameters of the differential ray u_{sag} and y_{sag} are determined as

$$u_{sag} = \sin u_{sag}, \quad y_{sag} = x_{sag} \tag{5.175}$$

Now the canonical normalized pupil coordinates in the object space are defined as

$$\bar{y}_p = \frac{y_p}{y_{\tan}} \qquad \bar{x}_p = \frac{x_p}{y_{sag}} \tag{5.176}$$

and correspondingly, in the image space

$$\bar{y}_p' = \frac{y_p'}{y_{\tan}'} \qquad \bar{x}_p' = \frac{x_p'}{y_{sag}'} \tag{5.177}$$

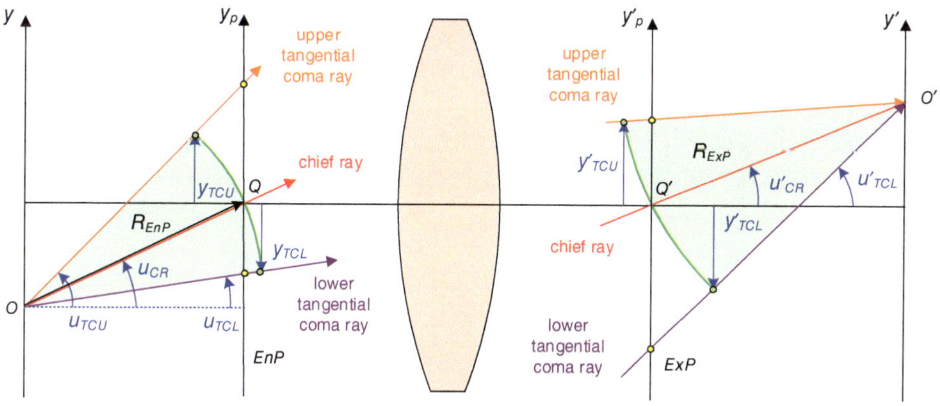

Figure 5.106 Normalized canonical coordinates in oblique bundles for an off-axis field point.

Under the simplifying assumption that the clear pupil form is elliptical in the projection for the skew chief ray, one obtains the equation

$$\left(\frac{x_p}{y_{sag}}\right)^2 + \left(\frac{y_p}{\cos u_{CR} \cdot y_{\tan}}\right)^2 = 1 \tag{5.178}$$

for the pupil in real coordinates under a skew angle. This relation becomes a circle in canonical coordinates

$$\bar{x}_p^2 + \bar{y}_p^2 = 1 \tag{5.179}$$

If the isoplanatic condition is valid with these definitions, the relation

$$\bar{y}_p' = \bar{y}_p \qquad \bar{x}_p' = \bar{x}_p \tag{5.180}$$

is obtained, which corresponds to a unit magnification between the entrance and the exit pupil in canonical coordinates also in the finite field of view.

Often it is reasonable to use the pupil ratios $k_{\tan}$ and k_{sag}. These define a relationship between the tangential and sagittal aperture of the off-axis bundle to the aperture of the axial bundle and are determined in the following way

$$k_{\tan} = \frac{\sin u_{CR} - \sin u_{\tan}}{\sin u}; \quad k_{sag} = \frac{\sin u_{\tan}}{\sin u} \tag{5.181}$$

with the aperture angle u on axis. Corresponding definitions are valid in the image space. The canonical object and image coordinates are defined by analogy with the axial case as

$$\bar{x} = \frac{n \cdot \sin u_{sag}}{\lambda} \cdot x; \quad \bar{y} = \frac{n \cdot (\sin u_{CR} - \sin u_{\tan})}{\lambda} \cdot y \tag{5.182}$$

and in the image space

$$\bar{x}' = \frac{n' \cdot \sin u_{sag}'}{\lambda} \cdot x'; \quad \bar{y}' = \frac{n' \cdot (\sin u_{CR}' - \sin u_{\tan}')}{\lambda} \cdot y' \tag{5.183}$$

In the corrected case similar to the on-axis rays the equation

$$\bar{x}' = \bar{x}, \quad \bar{y}' = \bar{y} \tag{5.184}$$

holds. The canonical coordinates have the advantage that all relations are normalized and have identical forms for the axis range as well as for off-axis field points. Due to this definition, according to Hopkins, a kind of natural scaling is obtained. But the asymmetric geometry in skew ray bundles cannot be avoided by this mathematical trick; as a price the classical chief ray is no longer the centroid ray of the bundle.

5.9.3 Imaging of Curved Objects

If the object surface is curved the image surface will in general be curved too. In the simplest case a spherical curved object surface with a radius of curvature R can be considered in paraxial approximation. If the radius is sufficiently large with an object height y the size of the sagitta can be quadratically estimated by

$$z = \frac{y^2}{2R} \tag{5.185}$$

assuming a parabolic surface shape with vertex radius R. For small z or z' the depth magnification can be approximated according to

$$m_z = \frac{z'}{z} = m^2 \tag{5.186}$$

from this the relation

$$z' = m^2 \cdot z = m^2 \cdot \frac{y^2}{2R} = \frac{y'^2}{2R} \tag{5.187}$$

results. This simple result shows that the image surface is a parabolic surface too with the same curvature radius R' independent of the magnification of the system

$$R' = R \tag{5.188}$$

Figure 5.107 shows the corresponding geometry.

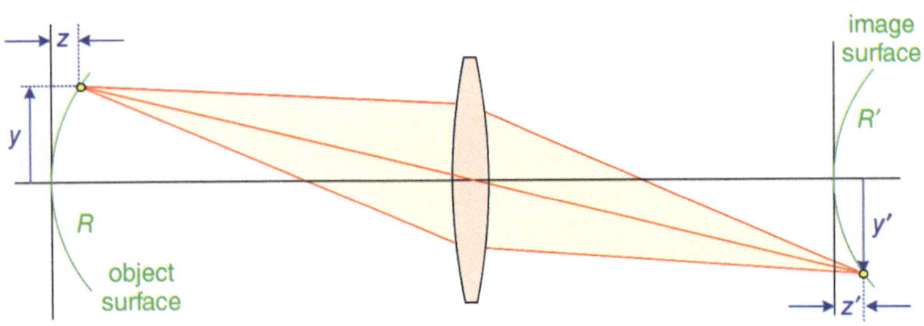

Figure 5.107 Imaging with spherically curved object surface.

References

1 Siew, R. (2005). F/no. and the radiometry of image forming optical systems with non-circular aperture stops. *Proceedings of SPIE* 5867.

2 Wetherell, W.B. (1987). Afocal Lenses. In: *Applied Optics and Optical Engineering*, vol. Vol X, Ch. 3 (ed. R. Shannon and J. Wyant), 110. Academic Press.

3 Schlemmer, H. (2003). Personal communication, with permission.

4 Zügge, H. (2005). Personal communication, with permission.

5 Wynne, C. (1944). Vignetting. *Royal Society of Physics* 56: 366.

6 Ozaktas, H. and Oktem, F. (2013). Phase-space window and degrees of freedom of optical systems with multiple apertures. *Journal of the Optical Society of America A* 30: 682.

7 Duma, V.-F. (2009). Radiometric versus geometric, linear and nonlinear vignetting coefficient Appl. *Optica* 48: 6355.

8 Gitin, A. (2009). Aberrational vignetting and eikonal theory. *Journal of the European Optical Society* 4: 09046.

9 Sands, P. (1973). Prediction of vignetting. *Journal of the Optical Society of America* 63: 803.

10 Besenmatter, W. (1977). Analyse des Vignettierungsverhaltens der Vario-Objektive mit Hilfe des Delano-Diagramms. *Optik* 48: 289.

11 Duma, V.-F. (2008). Vignetting of light beams for objects placed at a finite distance from an optical system. *SPIE Conference Proceedings* 7100.

12 Duplov, R. (2022). Angle-dependent vignetting stop. *Optical Engineering* 61: 065106.

13 Gitin, A. (2012). Phase space description of vignetting in optical systems. *Optics Communication* 285: 2485.

14 Chen, C.-L. (2016). Real Pupils in Optical Systems. Internship report, University of Jena.

15 King, W. (1968). The approximation of vignetted pupil shape by an ellipse. *Applied Optics* 7: 197.

16 Nasse, H. (2003). Private communication, with permission.

17 Schuster, K.-H. (1999). Mikrolithographisches Reduktionsobjektiv, Projektionsbelichtungsanlage und -Verfahren, EP 1006389.

18 Wilson, R. (1996). *Reflecting Telescope Optics I*. Springer.

19 Korsch, D. (1972). Closed form solution for three-mirror telescopes, corrected for spherical aberration, coma, astigmatism and field curvature. *Applied Optics* 11: 2986.

20 Howard, J. and Stone, B. (1998). Imaging a point with two spherical mirrors. *Journal of the Optical Society of America A* 15: 3045.

21 Hannan, P. (1992). General analysis of two-mirror relay systems. *Applied Optics* 31: 513.

22 Schiesser, E., Bahk, S.-W., Bromage, J., and Rolland, J. (2019). Design and alignment of an all-spherical unobscured four-mirror image relay for an ultra-broadband sub-petawatt laser. *Applied Optics* 58: 9514.

23 Bahner, K. (1967). Teleskope. In: *Handbuch der Physik, Band XXIX, Optische Instrumente* (ed. S. Flügge). Springer.

24 Meng, Q., Wang, W., Ma, H., and Dong, J. (2014). Easy-aligned off-axis three-mirror system with wide field of view using freeform surface based on integration of primary and tertiary mirror. *Applied Optics* 53: 3028.

25 Gaivoronskii, S. and Zverev, V. (2012). Versions of mirror-objective composite based on an optical system of Gregory and Cassegrain objectives. *Journal of Optical Technology* 79: 84.

26 Herrera, J., Vazquez, S., Luna, E. et al. (2011). Variant of the anastigmatic telescope with three mirrors for back focal length. *Applied Optics* 50: 1905.

27 Zou, M., Chang, J., Talha, M. et al. (2010). Design and analysis of the four-mirror optical system. *Optik* 121: 1900.

28 Lee, J. and Yu, S. (2009). Analytic design procedure of three-mirror telescope corrected for spherical aberration, coma, astigmatism and Petzval field curvature. *Journal of the Optical Society of Korea* 13: 184.

29 Zverev, V. and Shepelevich, A. (2007). Parametric model of a three-component system of reflective surfaces. *Journal of Optical Technology* 74: 263.

30 Canzek, L. (1980). Systematisierung der Spiegelobjektive nach der Größe der Zentralabdeckung. *Optik* 56: 385.

31 Chang, S. (2015). Linear astigmatism of confocal off-axis reflective imaging systems with N conic mirrors and its elimination. *Journal of the Optical Society of America A* 32: 852.

32 Chang, S., Lee, J., Kim, S. et al. (2006). Linear astigmatism of confocal off-axis reflective imaging systems and its elimination. *Applied Optics* 45: 484.

33 Stavroudis, O. (1967). Two-mirror systems with spherical reflecting surfaces. *Journal of the Optical Society of America* 57: 741.

34 Volatier, J.-B., Duveau, L., and Druat, G. (2020). An exploration of the freeform two-mirror off-axis solution space. *Journal of Physics: Photonics* 2: 014004.

35 Schröder, D. (1987). *Astronomical Optics*. Academic Press.

36 Korsch, D. (1991). *Reflective Optics*. Academic Press.

37 Slyusarev, G. (1984). *Aberration and Optical Design Theory*. Adam Hilger.

38 Bouwers, A. and Blaisse, B. (1955). Anamorphic mirror systems. *Optica Acta* 2: 36.

39 Haferkorn, H. (2003). *Optik*. Weinheim: Wiley-VCH.

40 Sasian, J. (1988). Imagery of the Bilateral Symmetrical Optical System. Thesis, University of Arizona.

41 Yuan, S. (2008). Aberrations of Anamorphic Optical Systems. Thesis, University Arizona

42 Arsenault, H. (1980). A matrix representation for non-symmetrical optical systems. *Journal of Optics* 11: 87.

43 Richter, W. and Mitschunas, B. (1994). Schiefe Abbildung – verallgemeinerte Abbildungsformeln und fotometrische Abschätzung. *Feingerätetechnik & Messtechnik* 102: 165.

44 Forrest, A.K. (1999). Oblique imaging, magnification and image intensity. *Pure and Applied Optics* 1: 687.

45 Richter, W. and Jahn, G. (2000). Some remarks on the skew collinear imaging. *Optik* 111: 215.

46 Bakolias, C. and Forrest, A. (1997). Dark field Scheimpflug imaging for surface inspection. *Proceedings of SPIE* 3029: 57.

47 Hahlweg, C. and Rothe, H. (2008). Utilization of the Scheimpflug principle in scatterometer design. *SPIE Conference Proceedings* 7065.

48 Ott, P. (2005). Imaging with tilted surfaces: an efficient matrix method for the generalized Scheimpflug condition and its application to rotationally symmetric triangulation. *Journal of the Optical Society of America A* 22: 1077.

49 Miks, A., Novak, J., and Novak, P. (2013). Analysis of imaging for laser triangulation sensors under Scheimpflug rule. *Optics Express* 21: 18225.

50 Zhong, Y. (2014). Imaging with Scheimpflug Setup. Masters thesis, University of Jena.

51 Buchroeder, R. (1970). Tilted Component Optical Systems. Thesis, University of Arizona.

52 Ochse, D. (2024). Private communication, with permission.

53 Belin, J. and Courtial, J. (2019). Imaging with two skew ideal lenses. *Journal of the Optical Society of America A* 36: 132.

54 Belin, J., Ferenzi, G., and Courtial, J. (2022). Skew lens image rotator. *Optics Express* 30: 25958.

55 El-Maksoud, R., Hillenbrand, M., and Sinzinger, S. (2014). Parabasal theory for plane symmetric systems including freeform surfaces. *Optical Engineering* 53: 031303.

56 Sasian, J. (1982). Image plane tilt in optical systems. *Optical Engineering* 31: 527.

57 Walther, A. (1995). *The Ray and Wave Theory of Lenses*. Cambridge University Press.

58 Herzberger, M. (1980). *Modern Geometrical Optics*. Krieger.

59 Herberger, M. (1952). The contributions of the single surfaces to the diapoint coordinates. *Journal of the Optical Society of America* 42: 544.

60 Herzberger, M. and Pulvermacher, H. (1968). Finite image error theory based on the diapoint configuration. *Journal of the Optical Society of America* 58: 1100.

61 Hopkins, H. (1985). Image formation by a general optical system. 1: general theory. *Applied Optics* 24: 2491.

62 Hopkins, H. (1985). Image formation by a general optical system. 2: computing methods. *Applied Optics* 24: 2506.

63 Hopkins, H.H. (1983). Canonical and real-space coordinates used in the theory of image formation. In: *Applied Optics and Optical Engineering*, vol. Vol IX, Ch. 8 (ed. R. Shannon and J. Wyant). New York: Academic Press.

6

Diffraction and Point Spread Function

6.1 Diffraction Phenomena

6.1.1 Introduction

Light is represented by an electromagnetic wave. Therefore, it does not simply follow geometrical laws and shows some physical properties; in particular, diffraction effects are observed. These phenomena cannot be understood from the geometrical point of view. Figure 6.1 shows the behaviour of the light after striking a sharp edge in a far distance. It can be seen that some amount of light is found in the geometrical shadow region for $t < 0$, and in the bright region, the intensity shows characteristic oscillations.

A second example is visualized in Figure 6.2. A narrow slit with small width d is illuminated by a monochromatic plane wave. In the geometrical shadow region, typical oscillations are seen. At the angles

$$\sin \alpha_{max} = \frac{\lambda}{d} \cdot \left(m + \frac{1}{2}\right) \tag{6.1}$$

side lobe maxima are found with integer numbers m. The light demonstrates partly constructive and destructive interference due to its wave nature, depending on the path length difference between the outermost edge points

$$\Delta = d \cdot \sin \alpha \tag{6.2}$$

of the slit transition, generating a phase difference

$$\Phi = 2\pi \frac{d}{\lambda} \cdot \sin \alpha \tag{6.3}$$

forming the interference. The scaling factor λ/d is quite characteristic for diffraction phenomena. The relative brightness of the interference maxima decreases with growing order, which means that the diffraction effects are strongest at the geometrical shadow edge.

If a small square or rectangular window is considered as an aperture, Figure 6.3 is obtained. It shows the diffraction pattern behind, in the far-field, which depends on the aspect ratio of the truncating rectangle (shown in yellow). The smaller the rectangular width, the larger the diffraction spreading.

In lens design, there are systems without important diffraction effects, they can be modelled and calculated by geometrical optics with high accuracy. But there are also systems that show phenomena that can only be understood by applying diffraction theory. Therefore, it is important to understand the nature of diffraction and to know the computational approaches and approximations that make the calculation more comfortable. There are many good textbooks about diffraction theory [1–12] and the reader is referred to these resources for further reading, derivations and more rigorous theoretical foundations. There exists in addition a huge number of research papers concerned with improvements of diffraction calculations. It's not the intention here to cite all of them or to discuss all the special tricks and marginal changes of existing algorithms. The goal in this book is more to explain this topic in a pragmatic and descriptive way and to bring the basic concepts and ideas to the reader as an overview. In particular, a special approach following Maggi/Rubinowicz, Miyamoto/Wolf and Keller, which converts the surface integral into a line integral and delivers the so-called geometrical diffraction theory, is not discussed here [5, 8, 13–23].

6.1.2 Fresnel Number

If the diffraction of a wave after a circular stop is considered, several ranges can be distinguished and a quantitative number can be created to measure the relevance and size of diffraction phenomena. Directly behind the stop, the intensity profile suffers from highly oscillating ripples; here the envelope is rather well described by the geometrical approximation with a circular bright area with homogeneous brightness as a classical shadow. In the far-field, only a few side lobes are observed; here, the largest impact of the diffraction is seen and the geometrical prediction completely fails. This is illustrated in Figure 6.4(a) in one transverse dimension for a plane wave impinging on to a stop. If in contrast a convergent spherical wave is considered, the scenario of Figure 6.4(b) is obtained and the

Lens Design for Imaging: Fundamentals of Optical Systems, First Edition. Herbert Gross.

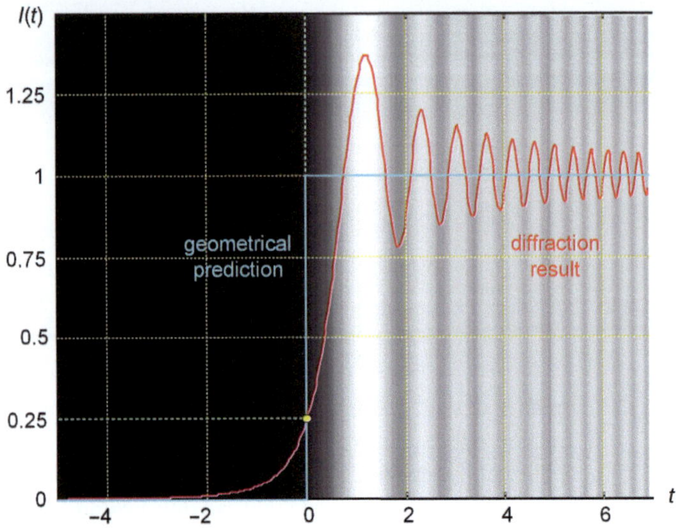

Figure 6.1 Diffraction at an edge, t is an arbitrary scaled lateral coordinate related to the distance of observation z.

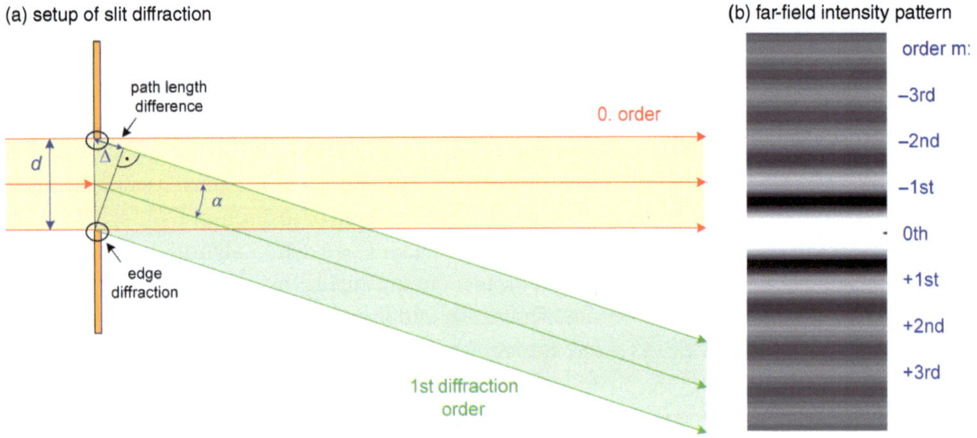

Figure 6.2 Diffraction at a slit, (a) shows the path difference geometry and (b) an exemplary intensity pattern, not to scale with the drawing.

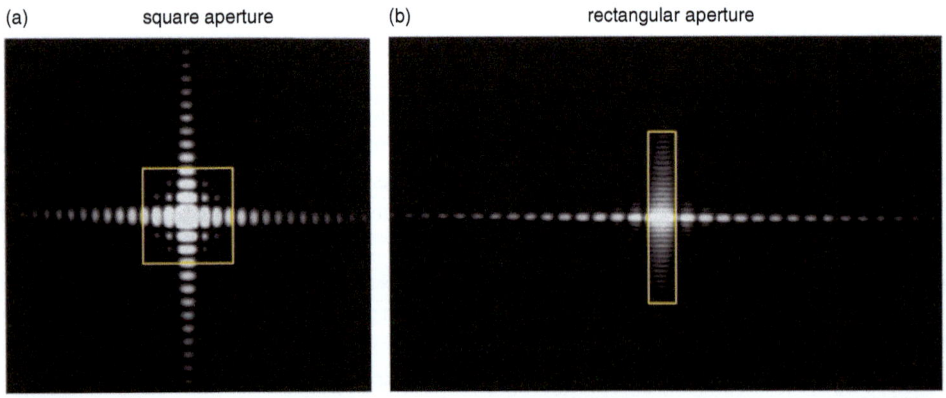

Figure 6.3 Diffraction at a Cartesian aperture with (a) square area and (b) elongated rectangular shape.

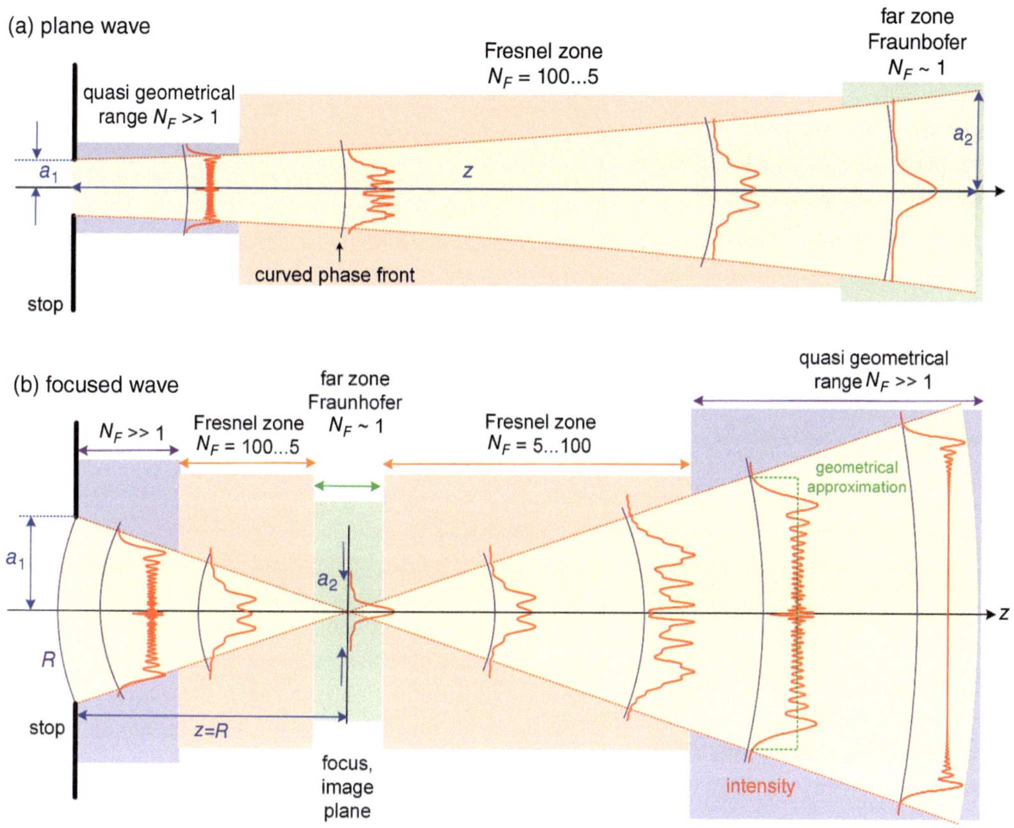

Figure 6.4 Diffraction after a stop for an incident (a) plane wave and (b) convergent spherical wave.

typical ranges change. The far-field is found around the focal point, and in a large distance the geometrical 'shadow throw' is observed again.

Figure 6.4 can be used to discuss the influence of diffraction in more detail and to define a quantitative number for distinguishing the influence of diffraction. In (b), first the light field is focused after the stop as a convergent wave into the geometrical focal plane; behind this position, the wave is again divergent. If the amplitude distribution of the light field is considered, we assume a uniform constant amplitude in the stop plane, although the phase surface is a sphere. In the focal plane the classical point spread function (PSF) is observed. Under perfect conditions, the Airy distribution is created here with a plane phase in a transverse plane of observation. For larger distances, the peaked distribution in the focal plane changes again; the number of oscillations grows, and the envelope increasingly resembles again the tophat profile of the starting plane. The modulation depth of the oscillations is nearly constant, but the spatial frequency of the ripples grows and the oscillation period shrinks with increasing distance. In a moderate distance from the focal point, the so-called Fresnel regime is seen with a considerable ripple structure.

In this context, the Fresnel number is used to distinguish between the geometrical-optical approximated region and cases where the light has to be described by means of wave optics including diffraction. The Fresnel number N_F is in general a measure for the relative strength of diffraction effects during propagation. Roughly speaking, one can distinguish the following regions.

- $N_F \sim 1$ — Fraunhofer far-field, case with strong diffraction influence.
- $N_F = 5 \ldots 100$ — Fresnel case with significant diffraction influence. The numbers are in a certain sense arbitrary, a smooth transition is observed.
- $N_F \gg 1$ — geometrical-optical region with negligible diffraction.

For some simple configurations, the Fresnel number can be clearly expressed in terms of the system data and the wavelength. For example, if a plane wave with wavelength λ is incident on an aperture with radius a and the wave is observed after a distance z, the Fresnel number is defined as

$$N_F = \frac{a^2}{\lambda z} \tag{6.4}$$

One obtains for this configuration shown in Figure 6.4(a) the following explicit physical interpretations of the dimension-free Fresnel number:

- N_F is identical to the number of Fresnel zones in the aperture $2a$ with an optical path difference scaled in $\lambda/2$ between edge and axis, see also Figure 6.17.
- The ratio of the geometrical focusing aperture angle

$$u = a/z \tag{6.5}$$

and the slit diffraction angle

$$\theta_d = \lambda/a. \tag{6.6}$$

therefore,

$$N_F = \frac{u}{\theta_d} = \frac{a/z}{\lambda/a} = \frac{a^2}{\lambda z} \tag{6.7}$$

- The ratio of the focal length f after a focusing lens and the diffraction-limited depth of focus, the so-called Rayleigh unit R_u

$$R_u = \frac{n\lambda}{NA^2} = \frac{\lambda}{n\sin^2 u} \approx \frac{\lambda}{u^2} = \frac{\lambda f^2}{a^2} \tag{6.8}$$

therefore, in the approximation of small angles $\sin u \approx u$ we get

$$N_F = \frac{f}{R_u} = \frac{a^2}{\lambda z} \tag{6.9}$$

- The ratio of the limiting diameter $2a$ of an aperture and the diameter of the diffraction spot

$$N_F = \frac{2a}{\left(\frac{2\lambda}{u}\right)} = \frac{2a}{\left(\frac{2\lambda L}{a}\right)} = \frac{a^2}{\lambda L} \tag{6.10}$$

- A more approximated criteria is that N_F is equal to the number of interference ripples observed in the intensity profile. This is approximately given in the Fresnel regime, see Figure 6.4(a) and (b).

It is clear from these interpretations that the Fresnel number is a measure of the ratio of the geometrical and diffraction effects. Large values of N_F correspond to a quasi geometrical regime, the smaller N_F is, the stronger the diffraction impact.

If a plane and a convergent wave are compared from the viewpoint of near-field, Fresnel range and Fraunhofer far-field, it should be noted that there are the following differences, which are also illustrated in Figure 6.4(b) and are listed in the following table. The far-field is characterized by the property that the phase surface of the wave is plane in the observation point at the distance z. For an initial plane wave, this happens in infinity; for the convergent wave, this is realized in the focal point.

In the more general arrangement with two different apertures of radii a_1 and a_2, respectively, and a separation z between them, the Fresnel number is defined in the case of a plane initial wave as [10], see Figure 6.5(a).

$$N_F = \frac{a_1 a_2}{\lambda z} \tag{6.11}$$

If the incoming wave has a finite radius of curvature R as in Figure 6.5(b), an equivalent Fresnel number can be defined [10]. As can be seen later by discussing the quasi flattening of a diffraction integral in the Fresnel range, a transform with magnification m of the transverse coordinate x and a modified distance z^*

$$m = 1 - \frac{z}{R} = \frac{a_2}{a_1}, \quad x_2^* = \frac{x_2}{m} = x_1, \quad z^* = \frac{z}{m} \tag{6.12}$$

transforms the Fresnel diffraction integral into a completely equivalent form, achieving the identical diffraction patterns. This transform removes the geometrical effects of the finite curvature R, generating a flattened start field showing the pure diffraction effects. As a consequence, the equivalent or effective Fresnel number of the non-collimated setup as in Figure 6.5(b) reads

$$N_F^{(eff)} = \frac{a_1^2}{\lambda z^*} = \frac{a_1^2}{\lambda z} \cdot \left(1 - \frac{z}{R}\right) \tag{6.13}$$

Here the sign of R must be noted, for positive $R = z$ the equivalent Fresnel number becomes zero, which is a singular case coming from geometrical optics where this simple equivalence definition cannot be used.

Table 6.1 Different zones of diffraction effects after a stop.

	Plane wave	Convergent/focused wave radius of curvature R
Fresnel number $N_F = \frac{a_1 a_2}{\lambda z}$ is small, far-field	Large value of distance z	Small value of a_2 near the focal region $z = R$
Location of far-field	In infinity	Near the focal or image plane
Location of quasi-geometrical range	Near the stop for small z	Near the stop for small z and for large ranges $z \gg 2R$
Intermediate Fresnel zone	Only once for moderate distances	Two regions before and after the focal plane
Curvature of the wave in the far field	Asymptotically plane	Exact plane in the focal plane

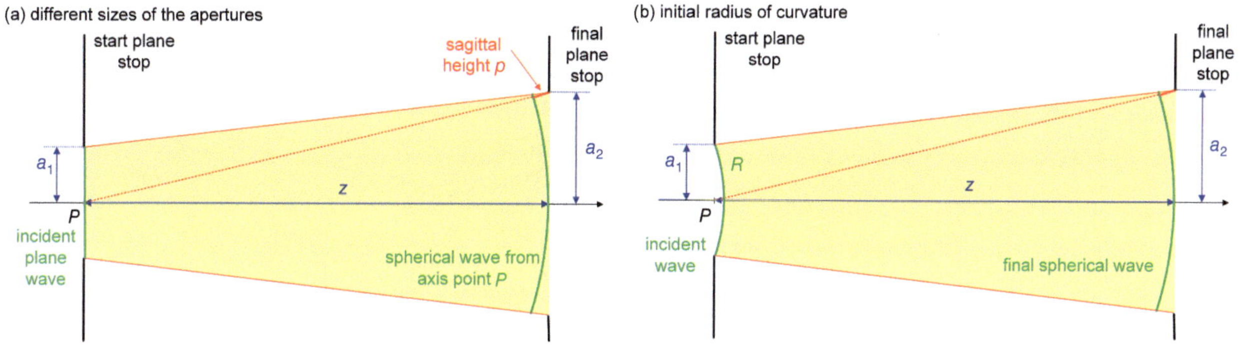

Figure 6.5 Definition of the Fresnel number for different aperture sizes and (a) a plane incoming wave and (b) initial curvature radius *R* of the wave.

6.1.3 Huygens Principle

The generation of diffraction effects and the propagation of waves can be understood heuristically according to the Huygens principle quite simply as follows. Every point on the incoming wave front in the stop plane and everywhere in the propagation regime can be considered as a new secondary source of light that creates a new spherical wave. The superposition of all wavelets generates an envelope, which is the observable new wave after a certain propagation distance. This automatically has the consequence that the light is also able to propagate around the stop edges and is found in the geometrical shadow regions. But in the shadow region, a smaller number of wavelets is superposed and the interference is not always constructive. As a consequence, here we have smaller amplitudes and characteristic oscillation ripple of the amplitude. This principle is sketched in Figure 6.6.

If this principle is applied to the exit pupil of an optical system, the generation of the PSF, the ringing of the intensity and the impact of wave aberrations or apodization can be easily understood, see also Figure 6.7.

Every wavelet can be represented by a small vector or arrow; the length corresponds to the field amplitude, the direction describes the phase of the wavelet and the complex local field E. If first an ideal spherical wave is assumed with uniform intensity, all wavelets are superposed vectorial to achieve the largest overall field intensity in the axis point P, which means that we have a pronounced peak here in the bright focus point.

If we move to an off axis observation point Q in the focal plane, the different distances to the points of the wave in the stop have the consequence of an ongoing phase difference between neighbouring points, represented here as a tilt between adjacent arrows. This sums up to a reduced overall length and intensity. In the special case of the point

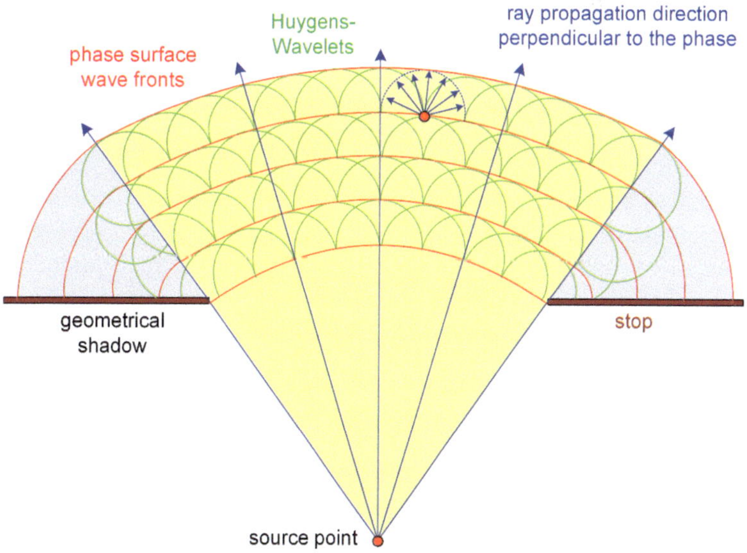

Figure 6.6 Huygens wavelet principle at a stop, every point as the red example dot creates a local spherical wave, which is drawn in green here. The envelopes are the observable red wave fronts.

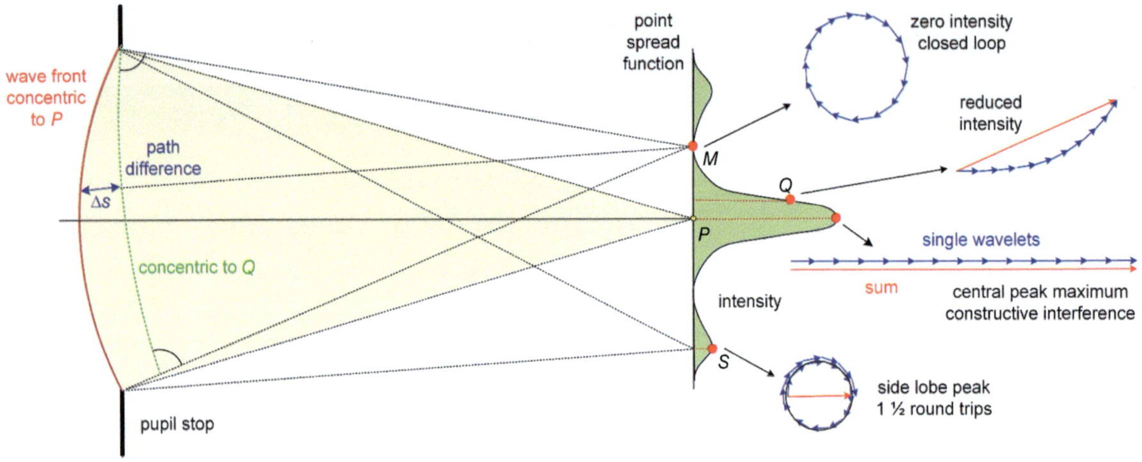

Figure 6.7 Huygens wavelet principle explaining the creation of the PSF. The lengths and numbers of the arrows are not to scale.

M a full round trip of all wavelets is obtained, we just get darkness with a sum of zero; this represents the first dark diffraction ring. The first side lobe is sketched by point S; here the arrows are strongly tilted and form one and a half round trip, building a relative but reduced maximum of intensity.

In addition, the impact of aberrations can be understood in this simple picture. If spherical aberration occurs, the wavelets near the optical axis are constructively superposed, but we get a growing phase difference and therefore arrow tilt near the edge; this corresponds to a reduced peak height on axis and a broadening of the PSF. Furthermore, the impact of an apodization as a changing energy density inside the exit pupil can be modelled in the Huygens picture, the arrows in this case have different lengths with the consequence on the overall summed intensity. This is explained in more detail in Section 6.3.1 and Chapter 18.

The mathematical formulation of the Huygens principle is given by the so-called Kirchhoff diffraction integral

$$E(\vec{r}\,') = \iint E(\vec{r}) \cdot \frac{e^{ik|\vec{r}-\vec{r}\,'|}}{|\vec{r}-\vec{r}\,'|} \cos\theta d^2\vec{r} \tag{6.14}$$

Here the first factor in the integral gives the complex field of the incoming wave in the stop plane. The second factor describes a spherical wave, starting at a point r. The cos factor takes a projection into account and the integral formulations corresponds to a summation over all wavelets. In an optical system, the Huygens principle is more concretely formulated in the form

$$E(x',y') = \iint_{ExP} T(x_p,y_p) \cdot e^{2\pi i W(x_p,y_p)} \cdot e^{\frac{2\pi i}{\lambda R_{ExP}}(x_p x' + y_p y')} dx_p dy_p \tag{6.15}$$

where T describes the amplitude by a transmission function of the system and W describes the wave aberration as a deviation from the ideal spherical wave with radius R_{ExP} in the exit pupil. The exponential factor is now linearly approximated in the far-field equivalent range near the image plane and it is seen that the diffraction integral is now a simple Fourier transform of the field in the exit pupil. This is discussed in more detail in Section 6.2.5.

6.2 Calculation of Diffraction Effects

6.2.1 Introduction

The calculation of diffraction effects in optical systems is not easy and ongoing efforts are being published to reach more accurate results with faster algorithms. In comparison to raytracing, the numerical problems are orders of magnitude more complicated. Therefore, many approaches are made to apply meaningful approximations to improve the efficiency of these modelling computations. In this chapter, only some elementary fundamental methods are discussed to understand the results for the PSF as well as the optical transfer function. The origin of all optical field related calculations are the Maxwell equations. The main point is to simplify the model as far as possible to maintain the required information. Aspects for simplifications are as follows.

- Disregarding magnetic and absorption effects.
- Removing the time dependence; this, for example, is not allowed for in case of modern ultra short pulse simulations.
- Reducing the vectorial field description to a scalar field value.
- Considering only long-range variations and removing the fast oscillations in the short scale of the wavelength.

- Restricting on a monochromatic field and fully coherent conditions.
- Restricting on paraxial conditions.

All these points are physical approximations of the model. As a second class of simplifications, the mathematical algorithms to perform the calculations can be considered. These are, for example, as follows.

- Calculation of the field in only one plane by using the integral formulation instead of the wave equation.
- Restricting on the Fraunhofer far-field or the Fresnel region to avoid the complexity of near-field effects.
- Calculation of equivalent integrals by using clever transforms, which remove the linear and quadratic phase part to improve the sampling issues due to the discretization of the oscillatory integral kernels.
- Improving the accuracy of results by clever weightings of the integration points, in particular at the boundaries of the integration domain to reduce errors of discretization.

There exists a huge amount of literature about the modelling, efficient computation and clever approximation of propagation and diffraction effects, but here only a small number of relevant proposals and options is discussed with the goal of creating a basic understanding for practical use.

6.2.2 Levels of Modelling

There are different levels of modelling the propagation of light waves according to the listings above. The rigorous solution of the Maxwell equations is the most accurate approach, but is mostly impossible to use due to numerical complexity, memory consumptions of the discretization on the λ-scale and resulting long runtime. There are various possible approximations, which reduce the numerical effort. But they can only be accepted if the effects to be searched for allow these simplifications. The first approximation assumes time independence, neglects magnetic effects and changes on the length scale of the wavelength. Next, vectorial effects are neglected. If the angles are considerably small, the small-angle paraxial approximation (Fresnel diffraction) makes sense. The simplest case allows in addition small angles only; then we are in the paraxial domain that e.g. serves as a reference for higher order aberration effects. The next step is a significant and important simplification. If the wavelength goes towards zero, diffraction effects can be neglected and we are in the geometrical-optical regime. This shows that there are several levels of modelling depth to describe an optical system. On the other hand, a calculation should be fast, efficient and accurate. Usually a more accurate result requires deeper physical approaches with fewer approximations. In any case a decision about the selected model is needed. The physical model and mathematical algorithm should be as simple as possible but as complicated as necessary to get the desired results. Figure 6.8 shows certain levels of approximations from top to bottom and right to left, illustrating the impact of accuracy onto the computational burden. If the calculation is made easier and faster, simultaneously certain aspects are lost.

Some further aspects of real complexity are not considered here. One of the essential phenomena in

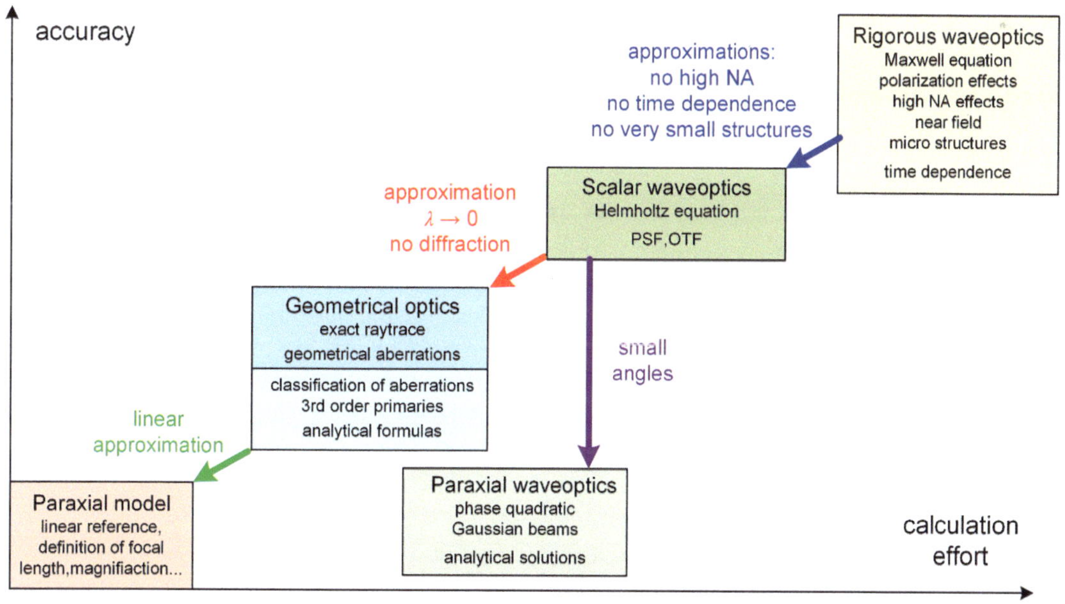

Figure 6.8 Illustration of the trade-off between calculation effort and accuracy; the axes are not to scale.

corresponding optical systems is the occurrence of partial coherent light.

6.2.3 Wave Equation

From the Maxwell equations for the electromagnetic field it is possible derive the wave equation for the electric field $\vec{E}$ in the case of non-conducting media and in the absence of charges and magnetic effects as well as very small feature sizes. This wave equations can be written as

$$\nabla^2 \vec{E} = \mu\varepsilon \frac{\partial^2 \vec{E}}{\partial t^2} \tag{6.16}$$

where ε is the dielectric constant and μ is the magnetic permeability. Using the ansatz

$$\vec{E}(\vec{r}, t) = \vec{E}(\vec{r}) \cdot e^{-i\omega t} \tag{6.17}$$

the time-independent Helmholtz wave equation describing the stationary wave phenomena is obtained

$$\Delta \vec{E} + k^2 \cdot \vec{E} = 0 \tag{6.18}$$

where Δ is the Laplace operator and k is the wave number in a medium with refractive index n

$$k = \frac{\omega}{c} = \frac{2\pi n}{\lambda_o} \tag{6.19}$$

In Cartesian coordinates the time-independent wave equation with the vacuum speed of light c reads

$$\frac{\partial^2 \vec{E}}{\partial x^2} + \frac{\partial^2 \vec{E}}{\partial y^2} + \frac{\partial^2 \vec{E}}{\partial z^2} - \frac{n^2(x, y, z) \cdot \omega^2}{c^2} \vec{E} = 0 \tag{6.20}$$

If the fast oscillating z-dependence on the scale smaller than the wavelength is eliminated by the ansatz (for better readability, the different approximation levels are not distinguished here with different letters or indices)

$$\vec{E}(x, y, z) = \vec{E}(x, y, z)\, e^{-ikz} \tag{6.21}$$

the so-called slowly varying envelope (SVE) approximation for the remaining weak z-dependence

$$\left|\frac{\partial^2 \vec{E}}{\partial z^2}\right| \ll k \cdot \left|\frac{\partial \vec{E}}{\partial z}\right| \tag{6.22}$$

leads to the paraxial wave equation for the transversely polarized scalar field E in Cartesian coordinates

$$\frac{\partial^2 \vec{E}}{\partial x^2} + \frac{\partial^2 \vec{E}}{\partial y^2} - 2\,ik \cdot \frac{\partial \vec{E}}{\partial z} - k^2 \vec{E} = 0 \tag{6.23}$$

This equation describes the wave-optical diffraction phenomena for small angles with the z-axis as the main direction of propagation. Using this approximation one can describe only such diffraction effects for which the changes along the light propagation direction occur on a scale $\Delta z \gg \lambda$.

This form of the wave equation holds if the following conditions are satisfied.

- The diffracting aperture is large in comparison with the wavelength, as only in that case can the influence of currents and charges near the boundaries of the diffracting edge be neglected.
- The field is evaluated at a distance z from the aperture, which is large in comparison with the wavelength. Only when this approximation holds can one neglect the axial field component and near-field effects are excluded.
- There are no large angles of the propagation of phase surface normal direction. This means that high-NA effects with axial field components are not included.

If the field distribution in a volume is necessary to know or if the setup of the wave propagation is not homogeneous with constant refractive index, the wave equation must be solved with corresponding approximations and methods including spatial variations of $n(x,y,z)$. If the field is only of interest in one plane – for example, the image plane of an optical system – the Stokes integral theorem can be used to switch under certain assumptions from the partial differential equation to an integral formulation. This simplifies and accelerates the diffraction calculation a great deal. Cases of relevance where this is not possible are in reality:

- inhomogeneities of the refractive index due to turbulence, tissue variations, striae or index tolerances;
- non-linear effects in special crystals, here the local field influences the propagation; for example, in case of the Kerr effect, second harmonic generation or absorption;
- if the energy density in the whole volume is necessary to know; for example, in thermal simulations with absorption.

In this chapter, only the integral approaches are considered. For the majority of optical system tasks, this resolves the problems.

6.2.4 Green's Theorem*

(*) The content of this section is quite specialized and for advanced reading. Knowledge of this section is not necessary to understand the following chapters. Beginners are recommended to skip this section.

If the complex function $E(r)$ is of interest and is described by the wave equation, then with the help of an auxiliary function G, Green's function, it is possible to express the function values on a closed surface S around a volume V in case of continuous functions as [1, 2]

$$\iiint_V E \cdot \nabla^2 G - G \cdot \nabla^2 E d^3V = \iint_S E \cdot \frac{\partial G}{\partial n} - G \cdot \frac{\partial E}{\partial n} d^2S \tag{6.24}$$

with the partial derivative normal n, outwards oriented to the surface S. This theorem allows us to convert the volume integral into a surface integral. For a clever choice of surface S and the function G it helps to calculate diffraction integrals very successfully.

In the approach of Kirchhoff, Green's function is selected as a spherical wave starting in the point r_o in free space of the form

$$G(\vec{r},\vec{r}_o) = \frac{e^{ik|\vec{r}-\vec{r}_o|}}{|\vec{r}-\vec{r}_o|} = \frac{e^{ikr}}{r} \tag{6.25}$$

A special shape of the surface S is necessary; the source point of the spherical wave must be excluded to remove the singularity here and to guarantee a continuous function. The direction derivative is then given by

$$\frac{\partial G(r)}{\partial n} = \cos\theta_{\vec{n}\vec{r}} \cdot \left(ik - \frac{1}{r}\right) \cdot \frac{e^{ik\vec{r}}}{r} \tag{6.26}$$

and by selecting the surface as an inner sphere around the starting point and an outer sphere, the diffraction integral can be written as

$$E(\vec{r}') = \frac{1}{4\pi}\iint\limits_S \frac{\partial E}{\partial n}\cdot\frac{e^{ikr}}{r} - E(\vec{r})\cdot\frac{\partial}{\partial n}\left(\frac{e^{ikr}}{r}\right) d^2\vec{r} \tag{6.27}$$

of the well-known Helmholtz–Kirchhoff form. Figure 6.10 shows an illustration of the surface S for the special case of diffraction at a screen.

6.2.5 Alternative Formulations of Diffraction Integrals

6.2.5.1 Introduction

In principle, there are some major physical approaches to calculate a diffraction integral. The idea is always to expand the field into basic modes that can be easily propagated. These can be as follows.

- Point source expansion.

  The corresponding spherical waves propagate easily, the superposition corresponds to a conventional point-by-point integration. The Kirchhoff and Sommerfeld formulations are representatives for this approach. The calculations are in principle direct integrations and the numerical effort grows with $N_i^2 N_o^2$.
- Expansion into plane waves.

  Plane waves are also easily propagated. From the numerical point of view, this approach is usually performed by Fourier transforms, because a plane wave corresponds to one Fourier component. This allows typically for much faster calculations with effort $(N \log N)^2$. The drawbacks here are the spatial-frequency coupling, which forces special sampling widths, and the necessity of using equidistant grids for the sampling. Details can be found in the appendix Chapter 12.
- Expansion into special eigenmodes,

  As an eigenmode, the field components like, for example, Gaussian beams for free space propagation or modes in guiding structures, are also invariant in propagation shape and show only a scaling and simple phase change. These techniques are mostly used in systems with boundary conditions like fibres, waveguides or resonators.

All these decomposition techniques can be realized in the spatial as well as in the angle or spatial frequency domain. The most comfortable selection depends on the task.

The Kirchhoff integral calculates the field in an arbitrary plane by integration in the spatial domain, which means that the fields created by infinitesimally small area elements are superposed according to the Huygens principle. From a physical point of view, this corresponds to an expansion of the initial field into spherical waves

$$E(\vec{r}') = \iint \frac{e^{ik|\vec{r}-\vec{r}'|}}{|\vec{r}-\vec{r}'|}\cdot E(\vec{r})\, d^2\vec{r} \tag{6.28}$$

where the inclination factor is neglected here for small angles. In the far-field approximation, the diffraction integral was approximated by the Fraunhofer integral, which mathematically has the form of a Fourier transform. A Fourier transform corresponds to an expansion into plane waves; every Fourier component represents one plane wave with an angle $\theta = \lambda\ \nu$ against the optical axis and can be interpreted by the grating equation for a grating spatial frequency $\nu = 1/g$. In addition. the diffraction integral in Fresnel approximation can be written as a convolution integral as

$$E(\vec{r}') = \hat{F}^{-1}\left[e^{ik_z z}\cdot \hat{F}[E(\vec{r})]\right] \tag{6.29}$$

with the Fourier transform operator notation $\hat{F}$ and can be evaluated by Fourier transform techniques. Physically, spherical as well as plane waves are eigenfunctions of the free space propagation. This means that these elementary wavelets propagate without changing their shape. Both expansion types correspond to a decomposition of the field into elementary modes and are equivalent, but have different practical properties. Figure 6.9 illustrates both approaches again in comparison. The question of which kind of expansion is more comfortable depends only on the symmetry and parameters of the setup to be computed.

In the following, a more intuitive physical description is chosen; for a more rigorous mathematical argument, the reader is referred to the literature.

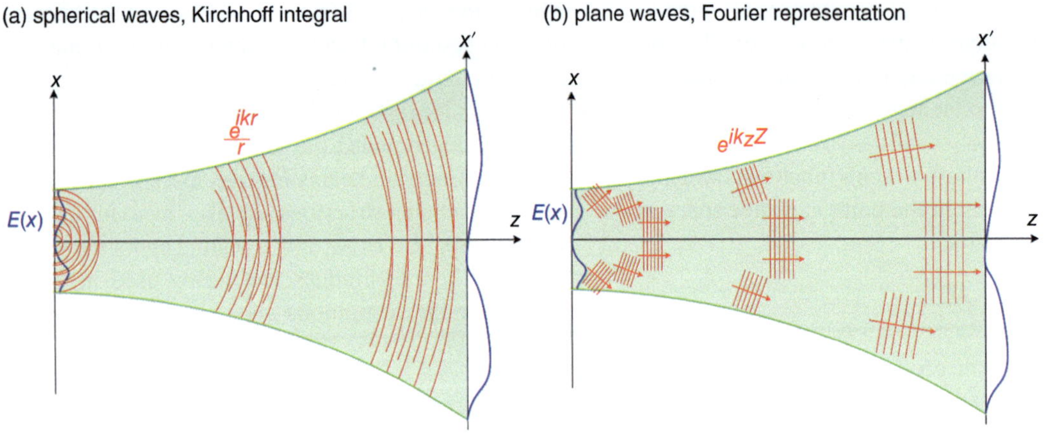

Figure 6.9 Expansion of a field $E(x)$ into (a) spherical waves or (b) plane waves.

6.2.5.2 Kirchhoff Integral

Using Green's theorem one can obtain from the wave equation a diffraction integral that describes the propagation of the field amplitude. One of the most important formulations for this approach is the so-called Kirchhoff diffraction integral. However, it is not rigorous from a mathematical point of view if it is applied in the usual form for the diffraction at a screen, it is actually based on inconsistent assumptions. But the results based on this approach are in most practical cases of high accuracy, and therefore it is in widespread use.

If $L(\vec{r},\vec{r}')$ is the eikonal function of the optical path length from the initial to the observation plane and θ_i and θ_o are the angles of the direction vectors $\vec{r}$ and $\vec{r}'$ for the incident and diffracted rays (see Figure 6.10), it is possible to represent the integral for the computation of the field, following Kirchhoff, in the form

$$E(\vec{r}') = \frac{i}{2\lambda} \iint E(\vec{r}) \cdot \frac{e^{-ik[L(\vec{r},\vec{r}')+|\vec{r}-\vec{r}'|]}}{|\vec{r}-\vec{r}'|} \cdot (\cos\theta_i + \cos\theta_o)\, d^2\vec{r} \tag{6.30}$$

The integral explicitly describes the superposition of weighted spherical waves according to the Huygens principle as outlined above. Accordingly, the wave on a phase surface can be decomposed into separate small wavelets which in turn propagate independently of each other as spherical waves and which have to be superposed in order to obtain the total field at an arbitrary position. The eikonal L describes here the phase of the wave.

When formulating the boundary conditions, the Kirchhoff approach has to be implemented as follows. The space consists of an opaque screen with a diaphragm.

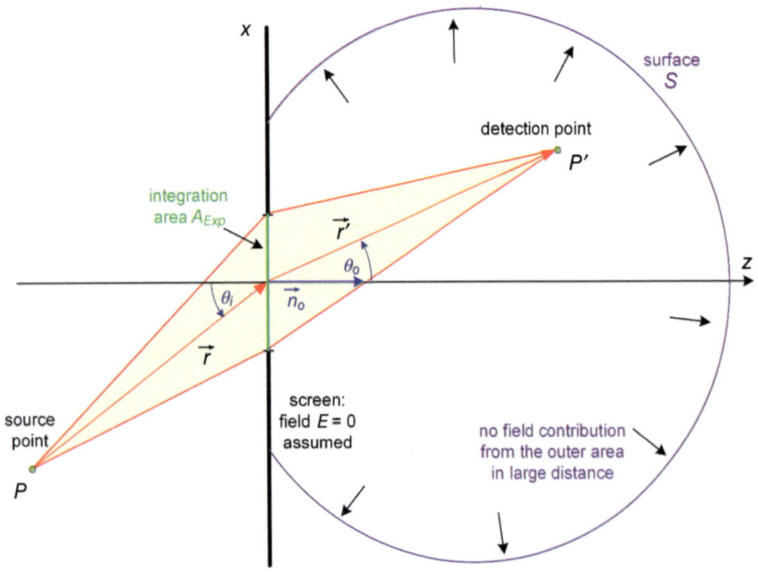

Figure 6.10 Geometry of the Kirchhoff integral.

The field amplitude vanishes at the screen itself, $E = 0$, whereas the amplitude in the open area of the screen is not affected and is considered to have a given value corresponding to the incoming field. One uses for the surface integral a closed surface S in the screen and a hemisphere at infinity. The field contribution of the hemisphere vanishes if the Sommerfeld radiation condition in the form

$$\lim_{r\to\infty} r \cdot \left(\frac{\partial E}{\partial r} - ikE \right) = 0 \tag{6.31}$$

is satisfied. In this approximation, the function E makes a jump at the edge of the diaphragm from zero to E. The assumption that this discontinuous transition does not cause any errors for the field computation in the observation plane is well justified only at sufficiently large distance from the screen. This means that large distances in comparison to the wavelength are assumed; it breaks down in the near-field. As evidenced in practice, this approximation is almost always acceptable as long as computation in very short distances is avoided. Assuming, in particular, that the field distribution has to be computed behind a screen opening and that the distances from the source and detection points to the screen are large compared with the transverse dimensions of the screen, so $\cos\theta_i \approx \cos\theta_0 \approx 1$ holds approximately, one then obtains the simplified form of the Kirchhoff diffraction integral for a constant eikonal function

$$E(\vec{r}\,') = \frac{i}{\lambda} \int_{A_{ExP}} E(\vec{r}) \frac{e^{-ik\,|\vec{r}-\vec{r}\,'|}}{|\vec{r}-\vec{r}\,'|} dF \tag{6.32}$$

where the integral is taken over the area of the opening A_{ExP} in the diffracting screen. In this representation the terms appearing in the integral can be directly associated with the Huygens principle: the first term is the incident field amplitude; the second term is the excited spherical wave.

The physical inconsistency of Kirchhoff's formulation is the fact that the screen outside of the opening also influences and changes the light field in a certain sense, which is neglected here. But all practical calculations show quite good accuracy, so this assumption is mostly allowed without considerably degrading the result.

In principle, the assumption of a step in the field strength at the stop boundary is not necessary. Then the Kirchhoff integral gives an exact solution of the diffraction problem as the Rayleigh–Sommerfeld representations do. But in this case, the inhomogeneities at the material surfaces force a coupling of the field components and the diffraction integral can no longer be formulated for the scalar field amplitude. The Kirchhoff approximation neglects the interaction of the light field with the material at the boundary of the diffracting diaphragm. It is assumed that the unperturbed field is valid for $z = 0+$ in close neighbourhood of the stop.

If high numerical aperture setups are considered, the scalar forms of the diffraction integrals can be applied to every scalar component of a general vectorial field separately.

6.2.5.3 Rayleigh–Sommerfeld Diffraction Integrals

The Kirchhoff integral can be written with the spherical wave as Green's function

$$G(\vec{r}\,', \vec{r}) = \frac{e^{ik|\vec{r}\,'-\vec{r}|}}{|\vec{r}\,'-\vec{r}|} = \frac{e^{ikr}}{r} \tag{6.33}$$

in the form

$$E(\vec{r}\,') = \frac{1}{4\pi} \iint \frac{\partial E}{\partial n} \cdot \frac{e^{ikr}}{r} - E(\vec{r}) \cdot \frac{\partial}{\partial n} \left(\frac{e^{ikr}}{r} \right) d^2\vec{r} \tag{6.34}$$

In Cartesian components this gives

$$\begin{aligned}
E_x(\vec{r}\,') &= -\frac{1}{2\pi} \iint E_x(x,y,0) \cdot \frac{\partial G(\vec{r}\,', \vec{r})}{\partial z} dxdy \\
E_y(\vec{r}\,') &= -\frac{1}{2\pi} \iint E_y(x,y,0) \cdot \frac{\partial G(\vec{r}\,', \vec{r})}{\partial z} dxdy \\
E_z(\vec{r}\,') &= +\frac{1}{2\pi} \iint \left[E_x(x,y,0) \cdot \frac{\partial G(\vec{r}\,', \vec{r})}{\partial x} \right. \\
&\quad \left. + E_y\,(x,y,0) \cdot \frac{\partial G(\vec{r}\,', \vec{r})}{\partial y} \right] dxdy
\end{aligned} \tag{6.35}$$

If the field is known in the initial plane $z = 0$, we have a boundary condition of the Dirichlet type. Then the field can be calculated by the so-called first Sommerfeld diffraction integral in the representation [24, 25]

$$\begin{aligned}
E_I(\vec{r}\,') &= -\iint E(\vec{r}) \frac{\partial G(\vec{r}, \vec{r}\,')}{\partial n} dS \\
&= -\iint E(\vec{r}) \cos\theta_{\vec{n}\vec{r}} \cdot \left(ik - \frac{1}{r} \right) \cdot \frac{e^{ikr}}{r} dS \\
&= -\frac{1}{2\pi} \iint E(x,y) \cdot \frac{\partial}{\partial z} \left(\frac{e^{ik|\vec{r}-\vec{r}\,'|}}{|\vec{r}-\vec{r}\,'|} \right)_{z=0} dxdy
\end{aligned} \tag{6.36}$$

with the simple distance

$$r = |\vec{r} - \vec{r}\,'| \tag{6.37}$$

If now the very near-field in distances in the range of the wavelength are excluded, in the second expression above the second term inside the parentheses can be neglected and the simplified version

$$E_I(\vec{r}\,') = -\frac{i}{\lambda} \iint E(\vec{r}) \cdot \frac{e^{ik|\vec{r}-\vec{r}\,'|}}{|\vec{r}-\vec{r}\,'|} \cdot \cos\theta_d dx\,dy \tag{6.38}$$

as Rayleigh–Sommerfeld's diffraction integral of the first kind is obtained.

If the derivative of the field in Eq. (6.34) is used instead of the value, the second type of Sommerfeld diffraction integral is obtained as

$$E_{II}(\vec{r}') = \iint_S G(\vec{r}, \vec{r}') \frac{\partial E(\vec{r}')}{\partial n} dS = -\frac{1}{2\pi} \iint \left.\frac{\partial E(\vec{r})}{\partial z}\right|_{z=0} \cdot \frac{e^{ik|\vec{r}-\vec{r}'|}}{|\vec{r}-\vec{r}'|} dxdy \quad (6.39)$$

It should be noted here that the average of the two Rayleigh–Sommerfeld formulations of the diffraction integral is exactly identical to the Kirchhoff formulation.

6.2.5.4 Angular Spectrum of Plane Waves

If the spatial formulation of the wave equation and the diffraction integral is converted by Fourier transform into the spatial frequency domain, another class of computational approaches can be found. Every Fourier component can then be physically interpreted as a plane wave, the field E is decomposed into plane waves. In practical problems in optics it is meaningful to decompose the wave vector into a transverse and an axial component by

$$\vec{k} = \vec{k}_t + \vec{k}_z = (\vec{k}_x + \vec{k}_y) + \vec{k}_z, \quad k_z = \sqrt{k^2 - k_t^2} \quad (6.40)$$

In the case that the field is dominant forward propagating, the field can be written as [26]

$$E(x,y,z) = \frac{1}{(2\pi)^2} \iint A(k_x, k_y, z)\, e^{+i(xk_x+yk_y)}\, dk_x\, dk_y \quad (6.41)$$

with the spectral Fourier amplitudes $A(kx,ky)$. The inversion is given by

$$A(k_x, k_y, z) = \iint E(x,y,z)\, e^{-i(xk_x+yk_y)}\, dx\, dy \quad (6.42)$$

and shows that the plane wave spectrum is simply the Fourier transform of the spatial field representation. A plane wave propagates in a homogeneous medium quite simply; the phase grows corresponding to the direction of the wave vector relative to the axis direction z. Therefore, the propagation of a plane wave is given in the spectral domain by

$$A(k_x, k_y, z) = A(k_x, k_y, 0) \cdot e^{-ik_z \cdot z} = A(k_x, k_y, 0) \cdot e^{-ik \cdot z \cdot \sqrt{1-\left(\frac{k_x}{k}\right)^2\left(\frac{k_x}{k}\right)^2}} \quad (6.43)$$

In the Fresnel approximation, the square root can be expanded and truncated after the second order terms and we have

$$A(k_x, k_y, z) = A(k_x, k_y, 0) \cdot e^{-ik\cdot z} \cdot e^{-\frac{iz}{2k}\left(k_x^2+k_y^2\right)} = A(\nu_x, \nu_y, 0) \cdot e^{-ik\cdot z} \cdot e^{-i\pi\lambda z\cdot\left(\nu_x^2+\nu_y^2\right)} \quad (6.44)$$

with the spatial frequencies ν_x and ν_y

$$k_x = 2\pi\nu_x; \quad k_y = 2\pi\nu_y \quad (6.45)$$

we get for the propagation

$$\begin{aligned} E(x,y,z) &= \iint A(\nu_x, \nu_y, 0) \cdot \mathrm{e}^{2\pi i z\sqrt{1/\lambda^2-\nu_x^2-\nu_y^2}} \cdot e^{2\pi i(x\nu_x+y\nu_y)} d\nu_x d\nu_y \\ &= \iint \mathrm{e}^{2\pi i z\sqrt{1/\lambda^2-\nu_x^2-\nu_y^2}} \cdot \left[A(\nu_x, \nu_y, 0) \cdot e^{2\pi i(x\nu_x+y\nu_y)}\right] d\nu_x d\nu_y \\ &= \hat{F}^{-1}\left[\mathrm{e}^{2\pi i z\sqrt{1/\lambda^2-\nu_x^2-\nu_y^2}}\, \hat{F}[E(x,y,0)]\right] \end{aligned} \quad (6.46)$$

This method expresses the wave by a superposition of the spectrum of plane waves (SPW). Every plane wave corresponds to a concrete angle; therefore, the method is also called angular spectrum method (AS).

With this method quite short distances z can be calculated without problems, in contrast to the Sommerfeld integral method. The numerical evaluation of the propagation method typically uses the fast Fourier transform and a convolution approach can be used. The numerical limitations of the discrete method are the sampling constraints in the initial field as well as for the propagator term. In particular, one limitation occurs in the case of long distances z. If the largest angle component reaches the boundary of the spatial computational window, aliasing occurs. Therefore, this method is limited in the propagation distance that comes from numerical reason. There is no physical approximation in this formulation and therefore it is valid for large angles as well. The formulation only assumes negligible evanescent waves. Therefore the AS propagation method is valid for z-values larger than a wavelength.

6.2.5.5 Convolution Formulation

All the diffraction integrals in the formulation of

- Kirchhoff;
- Rayleigh–Sommerfeld;
- AS, spectrum of plane wave expansion;
- Fresnel approximation, see next section;

can all be formulated as filtering in the spectral domain [27]

$$E(x',y') = \hat{F}_{xy}^{-1}[H_{CTF}(\nu_x, \nu_y) \cdot \hat{F}[E(x,y)]] \quad (6.47)$$

or in the spatial domain as convolution

$$E(x',y') = H_{CTF}(x,y) * E(x,y) \quad (6.48)$$

with the coherent transfer function (CTF) $H_{CTF}(\nu)$ as kernel. This is then a description of the wave propagation in the framework of the linear system theory. The transfer function of the exact plane wave spectrum reads

$$H_{CTF}(\nu_x, \nu_y) = e^{2\pi i \cdot z \cdot \sqrt{\frac{1}{\lambda^2} - \nu_x^2 - \nu_y^2}} \tag{6.49}$$

and describes the spherical shape of the phase exactly. It must be noted that in this and the next cases, no separation between the x and the y section is possible in two dimensions, both directions are coupled. After performing the quadratic Fresnel approximation, the x and y direction are decoupled.

The Rayleigh–Sommerfeld formulation uses the exact spherical wave in the exponent but approximates the pre-factor, which describes an apodization effect if the calculation is performed in a plane. This gives the kernel function

$$H_{CTF}^{(RS)}(\nu_x, \nu_y) = \hat{F}\left[\frac{z}{i\lambda(x^2+y^2+z^2)} \cdot e^{\frac{2\pi i}{\lambda}\sqrt{x^2+y^2+z^2}}\right] \tag{6.50}$$

Finally the transfer function according to Kirchhoff is the exact formulation in the scalar case and can be written as

$$H_{CTF}^{(Ki)}(\nu_x, \nu_y) = \hat{F}\left[\frac{z+\sqrt{x^2+y^2+z^2}}{2i\lambda(x^2+y^2+z^2)} \cdot e^{\frac{2\pi i}{\lambda}\sqrt{x^2+y^2+z^2}}\right] \tag{6.51}$$

This convolution structure allows for a fast computation using the Fourier transform. In the special case of the Fresnel approximation, the transfer function reads by a second order expansion of the root in Eq. (6.49)

$$H_{CTF}^{(Fres)}(\nu_x, \nu_y) = e^{i\pi\lambda \cdot z \cdot \left(\nu_x^2+\nu_y^2\right)} \tag{6.52}$$

and corresponds to a quadratic approximation. The exact spherical wave front is approximated by a parabola. This is always true in the paraxial approximation range.

6.2.5.6 Fresnel Approximation

In the majority of practical applications, the Kirchhoff integral must not be evaluated in its general form but instead can be simplified using special approximations [35]. One of the most important approaches is that of Fresnel, which assumes that the lateral dimensions of the diaphragm and the diffraction structures are small in comparison to the axial distances. One even more special approximation is attributed to Fraunhofer and describes the field only in the far region; see next section. A mathematically consistent presentation of the different approximation steps is possible using the concept of the so-called fractional Fourier transform with truncated order [49]. The order is then related to the propagation distance z and is a continuous variable.

For sufficiently large distances from the diffracting aperture, one can expand the distance between the detection and source points in a Taylor series using a quadratic approximation with $z' = 0$

$$r = |\vec{r} - \vec{r}'| = \sqrt{(x-x')^2 + (y-y')^2 + z^2}$$
$$\approx z + \frac{(x-x')^2}{2z} + \frac{(y-y')^2}{2z} + \ldots. \tag{6.53}$$

and the following representation of the Kirchhoff integral in the Fresnel approximation is obtained in Cartesian coordinates:

$$E'(x', y', z) = \frac{ie^{-ikz}}{\lambda z} \iint E(x, y, 0) \cdot e^{-\frac{i\pi}{\lambda z}\left[(x-x')^2+(y-y')^2\right]}\, dx\, dy \tag{6.54}$$

Since the Fresnel integral depends only on the difference of the coordinates, it can be represented as a convolution of the input field E and the transfer function H_{CTF} (impulse response function and CTF in the Fresnel approximation, respectively)

$$H_{CTF}(x, y, z) = \frac{ie^{-ikz}}{\lambda z} e^{-\frac{i\pi}{\lambda z}\cdot(x^2+y^2)} \tag{6.55}$$

$$E'(x, y, z) = E(x, y, z) * H_{CTF}(x, y, z) \tag{6.56}$$

as already mentioned before. Hence, in accordance with the convolution theorem, one can build a physical model and evaluate the wave propagation in the Fresnel approximation as a product in the Fourier spectral domain. This permits the use of the fast Fourier transform, which leads to a substantial reduction in the required computation time.

In terms of geometrical optics, 'paraxial' always stands for the near-axis aberration-free case. The Fresnel integral is also valid, however, for the description of the propagation of wave aberrations of higher order as long as the approximation is fulfilled. In this sense 'paraxial approximation' in wave optics does not mean that only paraboloidal waves can be considered as the initial field, it actually means that only propagation directions with small angles relative to the optical axis are relevant. Therefore, the paraxial approximation of the Fresnel integral allows only for small angles of the connecting lines towards the optical axis, but for non-paraxial aberrated wavefront of the initial field. Only the propagator term in Eq. (6.46) is approximated. Only if the initial field contains large slopes or high spatial frequencies does this computational scheme become critical, because large angles are needed to express the wave appropriately and this can only be done without truncation if the transverse computational grid is extended, which then of course affects the propagator.

The Fraunhofer and the Fresnel approximation of the diffraction integral are the result of truncated Taylor expansions of the eikonal function in the exponent. Therefore, the validity of both approaches is limited. If a change of $\pi/4$ over the aperture is taken as criterion, the

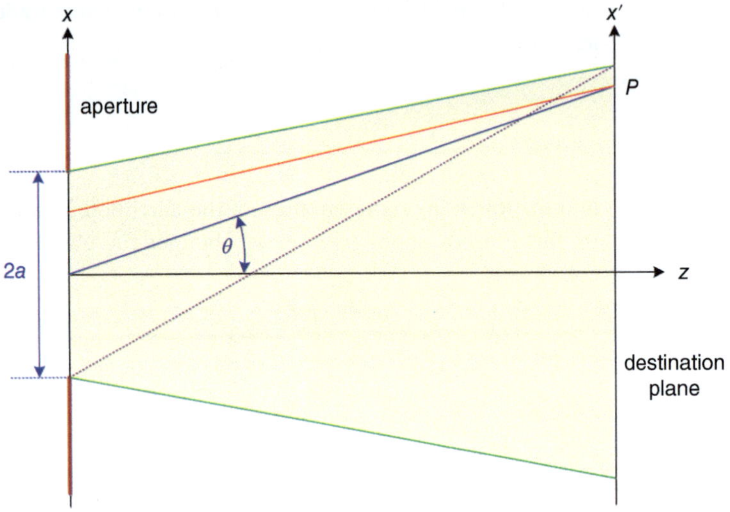

Figure 6.11 Geometry for estimation of the Fresnel approximation in the diffraction calculation.

following estimations for the ranges of application can be determined. The corresponding terms are shown in Figure 6.11.

In the case of the Fraunhofer approximation, the phase varies linear over the aperture with diameter $2a$. The Fraunhofer far-field is valid if the condition

$$r > r_{Fraun} = \frac{2a^2}{\lambda} \tag{6.57}$$

is fulfilled, corresponding to a Fresnel number $N_F \approx 1$. In the case of the Fresnel approximation, the phase over the aperture varies quadratically and the approach can be taken for distances z greater than [28–30]

$$z_{Fres} = \sqrt[3]{\frac{a^4}{2\lambda}} \tag{6.58}$$

But due to the growing distance for off axis points, at a distance $z > z_{Fres}$ only transverse coordinates in the range

$$x < |(2\lambda z^3)^{1/4} - a| \tag{6.59}$$

are allowed in the approximation. Therefore, both approximations fail near to the aperture and have the characteristic validity areas as indicated by the corresponding colours in the illustration of Figure 6.12.

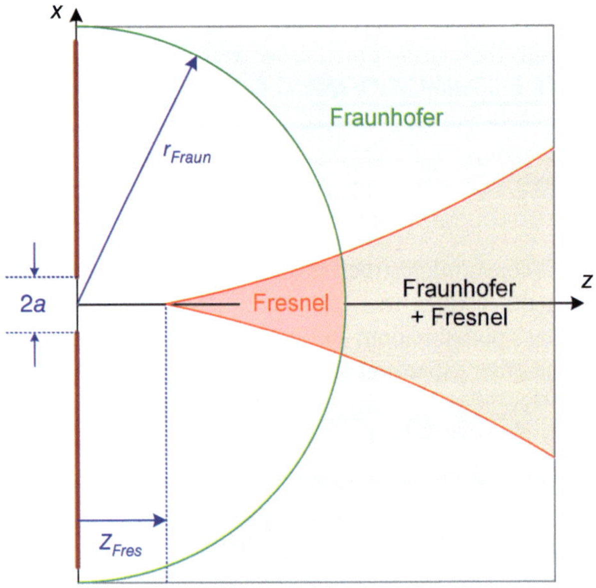

Figure 6.12 Validity range estimations for the Fraunhofer and Fresnel approximation.

6.2.5.7 Fraunhofer Far-field

In the limit of a very large propagation distance z, and small Fresnel number $N_F \sim 1$, it is possible in the Fresnel expansion to neglect the unprimed quadratic terms in the exponent too. If the factor containing the primed quadratic terms is separated and moved outside of the integral, one obtains the Fraunhofer far-field diffraction approximation [31]

$$E'(x',y',z) = \frac{ie^{-ikz}}{\lambda \cdot z} \cdot e^{-\frac{i\pi}{\lambda z}\cdot(x'^2+y'^2)} \iint E(x,y,0) \cdot e^{-\frac{2\pi i}{\lambda z}\cdot(xx'+yy')}\, dx\, dy \tag{6.60}$$

Thus, apart from a quadratic phase factor, the far-field is obtained from the near-field by a Fourier transform. The far-field approximation is justified when the Fresnel number N_F of the transition is of the order of 1 or smaller.

The Fraunhofer far-field approximation is used quite often when optical systems are analysed by means of wave-optic models because, in the Fourier sense, the image plane lies in the far-field of the pupil for a convergent spherical wave with the axial focal point as centre. This means in the case of a convergent wave that the quasi-far-field is located at a finite distance R. Through their focusing effect, lenses have the capability of producing the far-field of the amplitude from the front focal plane in their rear focal plane.

6.2.5.8 Debye Diffraction Integral

The representation of the diffraction integral according to Debye assumes that the field in the focal point of a setup is decomposed only by those plane wave components that fall in the solid angle of the focusing aperture cone, see Figure 6.13 for illustration. If θ_m is the angle of the cone, the general integral of the spectrum of plane waves reads in the frequency space

$$E(x,y,z) = \iint A(k_x,k_y)\cdot e^{i\cdot(x\cdot k_x+y\cdot k_y+z\cdot k_z)}\,dk_x dk_y \tag{6.61}$$

with

$$k_z = \sqrt{k_0^2 - k_x^2 - k_y^2},\quad k_x^2 + k_y^2 < k_0^2\cdot\sin^2\theta_m \tag{6.62}$$

$$\vec{k} = k_0\cdot\begin{pmatrix}\cos\varphi\sin\theta\\ \sin\varphi\sin\theta\\ \cos\theta\end{pmatrix} \tag{6.63}$$

reduces to the Debye integral as a far-field angle integral in spherical coordinates

$$\begin{aligned}E(x,y,z) &= \iint\limits_{k_x^2+k_y^2<k_0^2\cdot\sin^2\theta_m} A(k_x,k_y)\cdot e^{i\cdot(x\cdot k_x+y\cdot k_y+z\cdot k_z)}\,dk_x dk_y\\ &= \frac{i}{\lambda}\int_0^{2\pi}\int_0^{\theta_m} A(\theta,\varphi)\\ &\quad\cdot e^{\frac{2\pi i}{\lambda}(x\cos\varphi\sin\theta+y\sin\varphi\sin\theta+z\cos\theta)}\\ &\quad\cdot\sin\theta d\theta d\varphi\end{aligned} \tag{6.64}$$

As is seen here, the Debye integral also has the general structure of a Fourier integral and therefore allows for a correspondingly fast calculation [32].

In contrast to the Kirchhoff integral, where the field is truncated outside of the screen or stop in the spatial domain, in the Debye case the angle spectrum is truncated by the aperture cone. This limits the spectral bandwidth of the field behind the diffracting screen. In comparison, the Debye approximation forces an abrupt jump of the field in the spectral Fourier domain at the aperture cone angle and therefore contains a similar approximation as the Kirchhoff integral.

The Debye approximation creates problems if the total angle extension of the field is small, as in the case of low Fresnel numbers [31–35].

If $\vec{s}$ is the direction vector of the connecting line between the initial point r on the spherical wave in the screen and the final destination point $\vec{r}'$ in the focal plane, the Debye diffraction integral can also be written in a spatial representation with the solid angle Ω as

$$E(\vec{r}') = \frac{i}{\lambda}\cdot\iint\limits_{\Omega} E(\vec{r})\cdot e^{-ik\cdot\vec{s}\cdot\vec{r}'}\,d\Omega \tag{6.65}$$

This is a purely plane wave representation. It must be noted that in this form the given field $E(\vec{r})$ has to be defined on a sphere and not in a plane. Only those plane waves are taken into account whose wave vector $\vec{k}$ coincides with the calculated geometrical rays in the pupil plane. This procedure allows us to incorporate wave aberrations of an optical system into consideration without problems. In the stationary phase method, the contributing point is located exactly at the corresponding ray intersection point in the pupil.

6.2.5.9 Collins Integral

The diffraction integral of Kirchhoff in the quadratic approximation of Fresnel reads for only one transverse dimension in Cartesian coordinates

$$E(x',z) = \frac{\sqrt{i}\cdot e^{-ikz}}{\sqrt{\lambda z}}\int_{-a}^{a} E(x,0)\cdot e^{-\frac{i\pi}{\lambda z}\cdot(x'-x)^2}\,dx \tag{6.66}$$

This formulation corresponds to the Huygens principle of integrating elementary parabolic waves in the starting

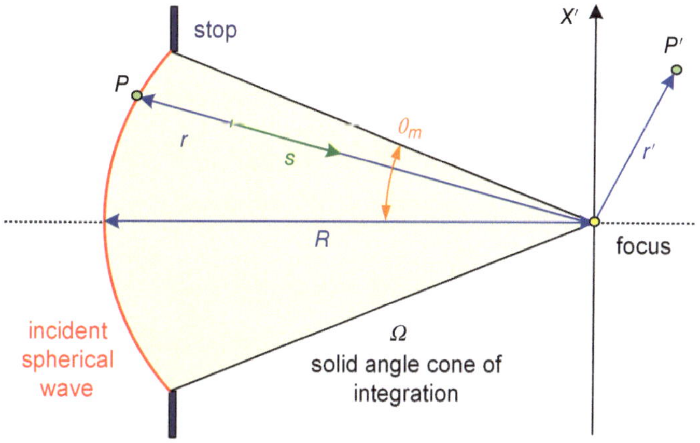

Figure 6.13 Geometry and principle of the Debye integral.

plane $z = 0$ to get the field in the final plane of computation (x', z). The term in the exponent in the second factor of the integrand contains the optical path difference between the two points in the starting and the receiving plane. The above integral describes the propagation of the light in the free space over a distance z with a homogeneous medium.

In practical cases, there are very often optical systems located between the starting and the receiving plane. If these components or subsystems can be considered as corrected and paraxial approximated, the optical path length L or eikonal between two corresponding planes can be easily written in terms of the paraxial system matrix

$$\underline{M} = \begin{pmatrix} A & B \\ C & D \end{pmatrix} \tag{6.67}$$

in the form

$$L = \frac{Ax^2 - 2\,xx' + Dx'^2}{2B} \tag{6.68}$$

see also Section 3.6. Correspondingly, the following generalized Fresnel integral is obtained, which is named after Collins [36–41]

$$E(x') = \frac{\sqrt{i}}{\sqrt{\lambda |B|}} \cdot \int_{-a}^{a} E(x) \cdot e^{-\frac{i\pi}{\lambda B}\cdot(Ax^2 - 2\,xx' + Dx'^2)}\, dx \tag{6.69}$$

Here the elements of the matrix occur in the exponent and constitute the eikonal function. Figure 6.14 shows the principle of the idea; the paraxial system is completely described by the three elements A, B and D of the matrix, the element C can be calculated with the help of the determinant

$$\det(\underline{M}) = \frac{n}{n'} \tag{6.70}$$

The Collins equation for the Fresnel diffraction integral is a paraxial approximation, but it is able to describe the propagation of an initial field with wave aberrations and special intensity profiles through nearly perfect optical systems. Therefore, its range of application is mainly the simulation of laser optical systems, where the radiation has a strong directionality and the Fresnel approximation gives very accurate results. The calculation of optical resonators and laser beam delivery systems are mainly based on the formulation of the Fresnel integral above.

The Collins integral can be written in the form of a convolution integral too by using a straightforward re-arrangement of the terms as

$$E(x') = \frac{1}{\sqrt{\lambda \cdot B_x}} \cdot e^{-\frac{i\cdot\pi}{\lambda\cdot B_x}\cdot(D_x - 1)\cdot x'^2} \cdot \int \left[E(x) \cdot e^{-\frac{i\cdot\pi}{\lambda\cdot B_x}\cdot(A_x - 1)\cdot x^2}\right] \cdot e^{-\frac{i\cdot\pi}{\lambda\cdot B_x}\cdot(x - x')^2}\, dx \tag{6.71}$$

and therefore can be calculated very efficiently with the help of the fast Fourier transform. The part of the integrand that is inside the square brackets corresponds to a flattened field. The numerical computation of the Collins integral is well conditioned if the relation

$$|A| \le \frac{x_{\max}^2 \cdot n}{\lambda \cdot N \cdot |B|} \tag{6.72}$$

is valid according to the sampling theorem with N grid points and a maximum spatial extend of $x_{\max}$. In the special case of exact or nearly conjugate planes, the value of B tends to zero $|B| \to 0$. Here the above equation can be written with the help of the Dirac delta function. Using its definition

$$\delta(x) = \left.\frac{1}{\sqrt{i\pi\varepsilon}} e^{\frac{ix^2}{\varepsilon}}\right|_{\varepsilon \to 0} \tag{6.73}$$

gives for the limiting case $B \to 0$ the equation for conjugate planes

$$E(x') = \frac{1}{A} \cdot e^{\frac{i\pi C}{\lambda A}\cdot x'^2}\, E(x)\,\Big|_{x=\frac{x'}{A}} \tag{6.74}$$

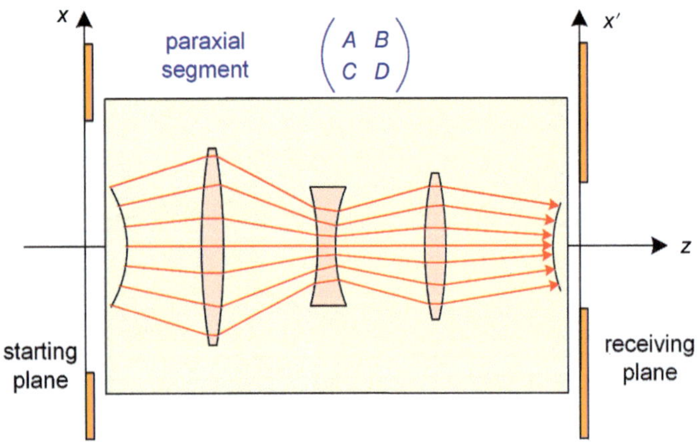

Figure 6.14 Propagation of the field through a paraxial *ABCD* system.

It can be seen from this expression that the transverse field is scaled but conserves its profile. But an additional quadratic phase is superposed on the complex field.

6.2.6 Numerical Computation of Diffraction Integrals

6.2.6.1 Introduction

For several decades researchers have tried to improve the diffraction light propagation models and algorithms for a more comfortable practical application. Correspondingly a large variety of proposals is available in a huge amount of literature about this subject. It is far beyond the scope of this book to give a complete review of all these approaches. But for a good understanding the main ideas and developments are briefly sketched here. Some of the interesting aspects are presented later in the appendix Chapter 12 (chirp-z-transform, non-uniform and semi-analytical Fourier transform, linear canonical transform, fractional Fourier transform).

The binary distinction between Fresnel and Fraunhofer far-field diffraction is artificial. In the illustration of the transition in the phase-space, the far-field transition is a rotation by 90°, a switch between spatial and angular coordinates. But it is also possible to rotate in the phase-space by arbitrary angles, which correspond to a medium distance propagation. Therefore, it was found that a so-called fractional Fourier transform and the linear canonical transform exactly represent this kind of operator [42–50]. Some details about this approach, which 'interpolates' between Fresnel and Fraunhofer diffraction is discussed in Chapter 12.

Most of the practical, used algorithms are based on the fast Fourier transform. The properties of this algorithm are a rigid coupling of the resolution between the spatial and frequency variable and an assumed periodicity beyond the calculation window. This has several practical consequences, and the runtime advantage must be paid for with problems of discretization, sampling and aliasing. This is discussed in Chapter 12. In particular the use of the chirp-z-transform can solve some of these problems [51, 52]. In principle this approach works like a loupe in the frequency domain and therefore allows us to achieve a higher resolution in the angle domain. The price is a higher computational effort; this transform is realized as a convolution. The initial field is multiplied by a chirp function and then Fourier transformed. Mathematically it can also be considered as an implicit sinc-interpolation in the complex domain. This sinc-approach can also be applied directly to calculate the propagation, this is described in Section 6.2.6.5 below [53, 54].

A very useful step to relax the sampling requirements in the case of convergent or divergent waves is a flattening transform [55–60]. In the Fresnel approximation it is possible to transform the integral into a fully equivalent form with a better flatness of the phase and therefore improved sampling conditions. This is described in Section 6.2.6.4 below. A more generalized approach removes in addition any global phase tilt and arbitrary quadratic terms, which gives the so call semi-analytical Fourier transform [61, 62].

Many more proposals can be found in the literature, which discusses the sampling optimization in more detail, together with the two methods just mentioned [63–78]. A simplified discussion is presented in the next sections. A special issue is the discretization errors resulting from the smooth curved boundaries. This is briefly sketched in Section 6.2.6.6 [79, 80].

The fast Fourier transform requires equidistant sampling points. If, due to the propagation, a distorted grid is required, a cumbersome interpolation of the complex field must be performed. Another idea here is to use under special condition the non-uniform Fourier transform [81–85], see again Chapter 12.

Finally, not every field propagation has to be computed by a finite distance transfer via diffraction integral. In many practical situations a component can be considered with high accuracy as a thin phase mask, which mainly influences the phase on an invariant coordinate grid. This can be approximated by the so-called thin-element approximation (TEA). If a finite thickness distorts the coordinate grids only moderately, this approximation can be improved by the local plane interface or local spherical interface approximation [86–88]. This is briefly sketched in Section 6.2.6.7 below.

6.2.6.2 The Phase Oscillation Problem

In practice, the fast oscillating phase term in the diffraction integral causes the major problem of an accurate and fast numerical evaluation. The e^{ikr}-term is mostly responsible for the selection of the necessary number of grid points and consequently determines the computational time. Hence, approximate forms of the diffraction integral have gained practical importance since, with appropriate assumptions, they can be evaluated considerably faster. The practical calculation of diffraction integrals is a cumbersome problem because of the numerical effort required to get accurate results in fulfilling the Shannon sampling theorem. In one dimension, the main structure of the diffraction integral is of the form

$$E'(x') = C \cdot \int_{-a}^{a} E(x) \cdot \frac{e^{-ikr}}{r + r'} dx \tag{6.75}$$

where $r + r'$ is the distance between the starting point $P(x)$ and the observation point $P(x')$ via the point Q in the

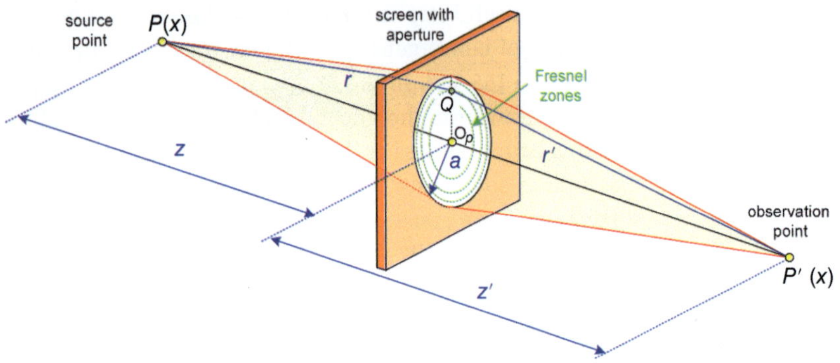

Figure 6.15 Geometry of the diffraction calculation. The field of the source is diffracted by a diaphragm in a screen. The number of Fresnel zones seen by the observation point determines the calculation effort required.

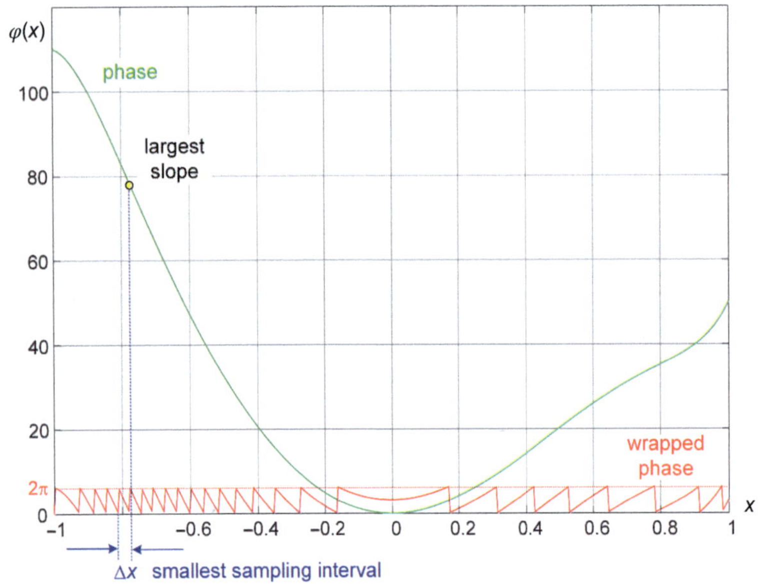

Figure 6.16 Arbitrary phase function $\varphi(x)$ and wrapped version to the interval 2π.

screen plane, see Figure 6.15. The major contribution from the integral comes from the points of stationary phase, as the summation over the oscillating parts results in a vanishing amplitude. Both factors in the above integral contribute to the phase; in general, the initial field $E(x)$ can contain a quadratic defocus distribution and additional higher order variations caused by aberrations. The second factor characterizes the propagation part of the Huygens spherical wavelets and is critical if the Fresnel number of the propagation geometry is high. Figure 6.15 shows the geometry of the corresponding setup, here especially for circular symmetry.

The diffraction integral in the Fresnel approximation in one lateral dimension reads

$$E'(x') = C \cdot \int_{-a}^{a} |E(x)| \cdot e^{2\pi \cdot i \cdot \varphi(x)} \cdot e^{-\frac{i\pi}{\lambda z} \cdot (x-x')^2} \, dx \tag{6.76}$$

is used with the phase of the incoming field φ. For the mathematical calculation, only the distribution of the phase function, which is wrapped to the interval 2π, is relevant. Figure 6.16 illustrates this behaviour for an arbitrary phase curve $\varphi(x)$. The smallest sampling interval Δx of the red wrapped phase is located at the coordinate with the largest slope.

The wrapped phase shows the oscillating behaviour in the form of a zig-zag curve. To describe this distribution sufficiently accurately, the density of the sampling points is determined at the location of the smallest feature size. It can be seen from this figure that the largest derivative or slope of the phase function defines this limit, because the grid points must be necessarily equidistant.

The oscillating behaviour can also be demonstrated by the real part of the integrand. Since this is a smooth function without jumps, a mathematical discussion using this term is much easier.

6.2.6.3 Sampling in Diffraction Calculations

If the geometry of the image space of an optical system is considered to calculate the PSF near to the focal plane of the system, Figure 6.17 is obtained. If the Fresnel approximation is assumed to be accurate enough, Eq. (6.54) governs

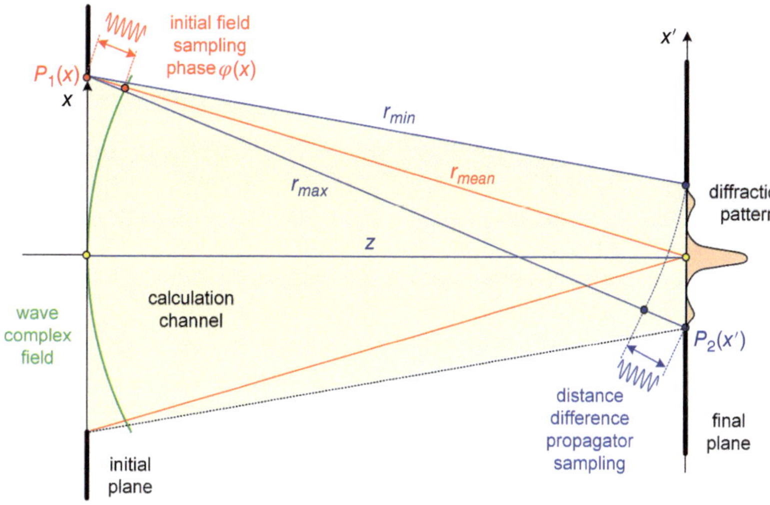

Figure 6.17 Sampling of the diffraction integral, shown in one transverse plane x–z.

the calculation. In the focal point of a perfect system, all the points are in phase and the integral shows good conditioning. It should be emphasized that in the integral two complex factors occur: the initial field and the propagator. The phase term of the propagation (second factor in the integrand) in the best case cancels the phase part of the incoming field (first factor). In the end, for the efficiency of the calculation, only the product matters, corresponding to a sum of both phase contributions. This means a part phase compensation, and therefore improvement can occur. Problems with the sampling arise if:

- the final point is not on axis, for off axis points in the image plane the conditioning becomes worse with larger phase variation due to the skew geometry;
- the final point is not located in the image plane, which means the field should be calculated in a defocused plane;
- the aberration term W is huge and the phase is strongly deviating from the parabolic shape; this is the case for badly corrected systems with a considerable wave aberration. In particular, higher order aberrations usually have a large slope and are therefore hard to sample appropriately.

For the calculation of a diffraction problem, there are several possible numerical algorithms. The optimal choice of the algorithm depends on the computer memory resources, the computing time, the required accuracy and the use of an intelligent pre-calculation, which is able to select the best available tool. There are many calculation schemes proposed in the literature.

In optical systems, the most interesting range is a final destination point near to the image plane. This means that the Fraunhofer approximation is feasible and Fourier techniques can be used. If the FFT is used for reasons of efficiency, the following constraints have to be taken into account.

- The number of sampling points in the initial as well as in the final plane are the same.
- The sampling points are equidistant in both plans individually.
- The discretizations in the start and in the final frequency plane are coupled by

$$\Delta x_p \cdot \Delta \nu = 1/N \tag{6.77}$$

with the total number of sampling points N with spatial frequency $\nu = \lambda f x'$.

If now more concrete, the PSF calculation is considered; this means:

- the sampling in the pupil is $\Delta x_p = D_{Exp}/N$
- the sampling in the image plane is $\Delta x'$. The scaling of a region of interest here is fixed by the Airy diameter of the typical lateral size of the PSF. It is given by

$$D' = f \cdot \frac{\lambda}{D_{Exp}/2} = \frac{\lambda}{NA} = N \cdot \Delta x' \tag{6.78}$$

By combining these equations, it can be found that the spatial resolution in the image plane is given by

$$\Delta x' = \Delta \nu \cdot f \cdot \lambda = \frac{\lambda \cdot f}{N \cdot \Delta x_p} \tag{6.79}$$

This means: to have many significant points in the image plane with small values of $\Delta x'$, N must be large or Δx_p must be large. A coarse sampling of the pupil is critical if stronger variations of the wave aberrations must also be sampled. Furthermore, a coarse sampling provokes larger discretization errors due to the boundary of the pupil; see below. Usually only a small number of significant points is obtained inside the Airy diameter. To overcome this drawback there are two possibilities.

- Sampling a larger area in the pupil plane by zero-padding the range outside the illuminated diameter. Simultaneously the point number N must grow to keep Δx_p constant. The drawback here is that most of the calculation work in this case is shuffling zeros. So the efficiency of calculation is low and the memory requirements are large.
- Use of the chirp-z-transform, which means that an indirect sinc-interpolation improves the density of the obtained points. In this case the integral is formulated as a convolution and the overall computational loading is increased by a factor of approximately 2.5. Some more details are discussed in the mathematical appendix Chapter 12.

According to [2], a good understanding of the sampling problems can be obtained if the bandwidth of the Fresnel propagator is inspected. If the propagator

$$H_{CTF}(x) = \frac{1}{\sqrt{\lambda z}} e^{-\frac{i\pi}{\lambda z} \cdot x^2} \tag{6.80}$$

is Fourier transformed and the sum of the squared spectral amplitudes is calculated, a measure of the normalized bandwidth as a function of the Fresnel number $B(N_F)$ is obtained.

$$B(N_f) = N \cdot \int |\hat{F}[H_{CTF}(x)]|^2 \, dv \tag{6.81}$$

with a normalization factor N for the frequency $v = 0$. If this behaviour is determined, the curve of Figure 6.18 is obtained.

For small and large Fresnel numbers, the asymptotic values are straight lines in the double logarithmic representation; they are found by

$$B(N_F \ll 1) = \frac{1}{\sqrt{4N_F}}, \quad B(N_F \gg 1) = \sqrt{4N_F} \tag{6.82}$$

and are indicated in red and green in the figure. The interpretation is: a large bandwidth and therefore a bad performance is obtained for

- very large Fresnel numbers, where in the near-field the sampling becomes critical due to the spatial discretization;
- very small Fresnel numbers, for large distances the frequency discretization becomes inaccurate.

6.2.6.4 Removal of Parabolic Curvature

In the range of the Fresnel approximation, one important option is to use the equivalence transform to relax the sampling requirements. Since all the integrals have the form of a convolution, the numerical computation can be done by means of the fast Fourier algorithms, as in the previous case. The diffraction integral in the free space Fresnel approximation for one lateral coordinate reads

$$E'(x', z) = C \cdot \int_{-a}^{a} E(x, 0) \cdot e^{-\frac{i\pi}{\lambda z} \cdot (x' - x)^2} \, dx \tag{6.83}$$

It is possible to transform this integral in an equivalent form, but with a rather better conditioning of the sampling problem. The following equation with an arbitrary number M delivers identical results for the diffraction pattern

$$E'(x', z) = C \cdot e^{-\frac{i\pi(|M|-1) \cdot x'^2}{\lambda \cdot M \cdot z}} \cdot \int \left[E(x, 0) \cdot e^{-\frac{i\pi(1-|M|) \cdot x^2}{\lambda \cdot z}} \right] \cdot e^{-\frac{i\pi \cdot M}{\lambda z} \cdot \left(x - \frac{x'}{|M|}\right)^2} \, dx \tag{6.84}$$

where x is the coordinate in the source or pupil plane over which the integration is calculated. The additional second factor in the square brackets describes a defocusing in paraxial approximation. The physical interpretation of this equation is that for every diffraction setup, there are other geometries for which the scaling of the observation plane, a final paraboloidal phase correction and the propagation

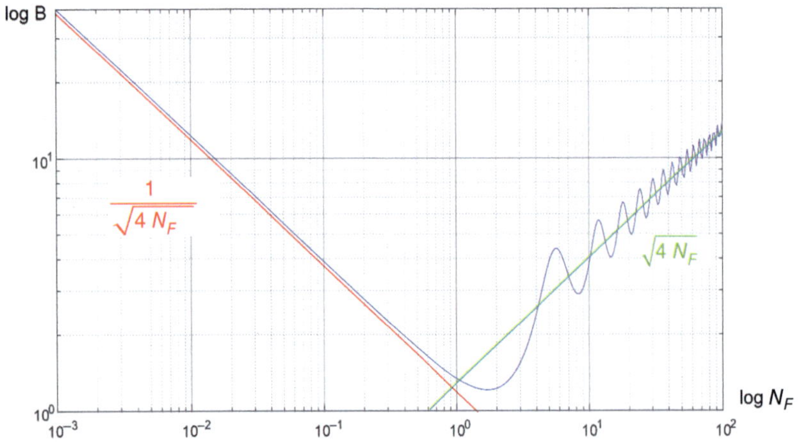

Figure 6.18 Bandwidth of the Fresnel propagator as a function of the Fresnel number.

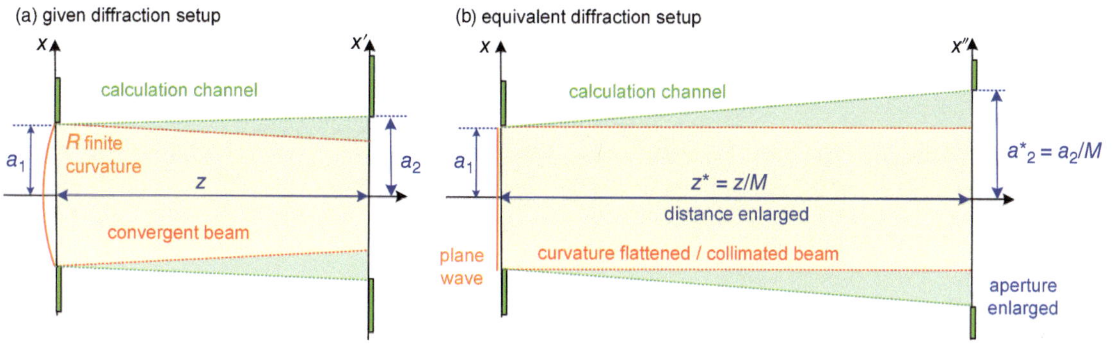

Figure 6.19 Equivalent diffraction setups with scaling of the final aperture and the distance and using a different phase correction. (a) shows the original geometry with a curved initial phase, (b) gives the modified setup with plane starting phase.

distance have to be chosen appropriately, but the field distribution is identical and only scaled. This is illustrated in Figure 6.19. It should be noted that this equivalence of the equations is only valid in paraxial approximation.

The choice of the magnification factor M is arbitrary and can be used to optimize the sampling conditions for the calculation of the integral. If the incoming field has no phase distortions of higher order and is therefore of paraboloidal form, the optimal defocusing corresponds to a flattening of the phase surface. As a result, the phase front becomes plane after the equivalence transformation. If the radius of curvature is R and the propagation distance z, the best choice of the magnification factor M is simply by geometrical means given by [10]

$$M = 1 - \frac{z}{R} = \frac{x'}{x} \tag{6.85}$$

with the sign convention that $R > 0$ corresponds to a convergent wave. The correct propagation distance is then given by z^* and the scaling of the coordinates in the observation plane as x^* with

$$z^* = \frac{z}{M}, \quad x^* = \frac{x}{M} \tag{6.86}$$

Finally, the final field has to be corrected by a geometrical parabolic phase factor with the radius of curvature

$$R' = R - z \tag{6.87}$$

It should be noted that the above formulas are also valid in the case of negative values of M, this happens for distances z greater than R. In this case, an internal focal point is located at the propagation distance and x switches in sign. If propagation in a focal plane is required, the value of M is nearly or exactly zero. In this special case, the equivalence transformation is not suitable in the simple form of Eq. (6.84), this formula is only advantageous for real Fresnel transitions. But Eq. (6.84) is still useful; the arbitrary parameter M is then to be selected appropriately to avoid a shrink of the final x^* coordinate to zero. If the size of the final plane diameter is fixed, M can again be determined by the second form of (6.85).

It should be mentioned that also in the case of a more complicated incoming phase profile the above transform makes sense. The parameter M is then optimized with the goal of a minimum of the residual phase gradients. Then at least a quasi-flattening with more comfortable sampling conditions is obtained. This is discussed for example in [97].

There is another geometrical interpretation of the equivalence transform. The flattening of the field corresponds to a transition into spheroidal coordinates with a centre in the focal point. The Figure 6.20 shows this interpretation. In particular, this picture illustrates the occurrence of an internal focal point with a change in the sign of the coordinate in the observation plane. But this interpretation has to be used with care; in reality, the phase distributions in Fresnel approximation are parabolic, not spherical.

It should be mentioned that this representation of the equivalence transform can also be interpreted as a fractional Fourier transform, which does not correspond to a full 90° rotation in the phase space as in the Fraunhofer case.

The Fresnel number of a setup as a criterion for the relative impact of diffraction is defined for plane waves at the input and in the output plane only. Since the diffraction influence is identical in the case of the equivalence transform in the Fresnel approximation, the Fresnel number remains constant as it should be and has the value

$$N_F = \frac{a_1 a_2^*}{\lambda z^*} = \frac{a_1 a_2}{\lambda z} \tag{6.88}$$

The above explanation uses free-space propagation and is limited to only one transverse dimension. But the idea of the equivalence transform can be extended without problems to two dimensions and setups with paraxial *ABCD* system parts. This means, this method can be applied to

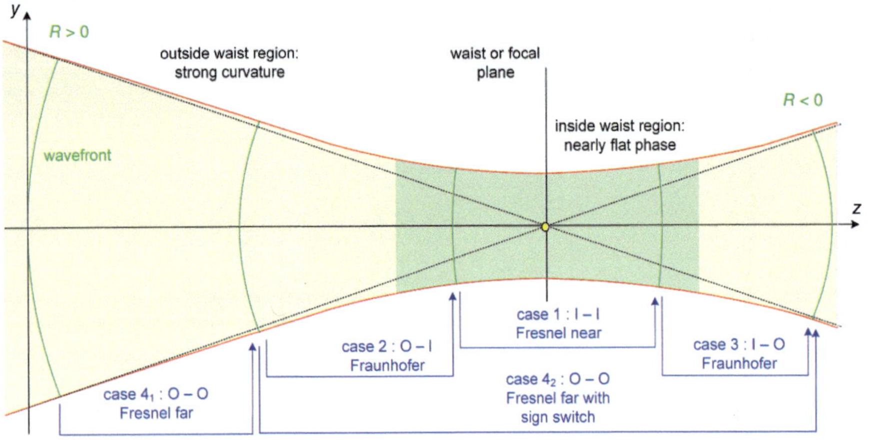

Figure 6.20 Equivalent diffraction setup with scaling of the aperture and the distance, and using a different phase correction, after [55] / with permission of Academic Press.

the Collins integral as well. The equivalence transform then reads in one lateral dimension

$$E(x') = \frac{n_1}{\sqrt{i\lambda_0|B|}} \cdot \sqrt{\frac{n}{n'}} \cdot e^{-\frac{i\pi(|M|\cdot D\cdot n'-n)}{\lambda_0 BM}\cdot x'^2} \cdot \int_{-\infty}^{\infty} E(x) \cdot e^{-\frac{i\pi n\cdot(A-|M|)}{\lambda_0 B}\cdot x^2 - \frac{i\pi M\cdot n}{\lambda_0 B}\cdot\left(x-\frac{x'}{|M|}\right)^2} dx \tag{6.89}$$

where in addition different refractive indices are included.

If a diffraction calculation has to be performed, first of all the optimal conditioning of the given setup has to be analysed. In most cases, the distinction between a Fresnel setup with a near-field transition and a Fraunhofer approximation with far-field conditions has to be made. If a diffraction calculation in the Fraunhofer approximation can to be made, only one Fourier transform is necessary or a simple integration if the computation of the integral is realized directly.

On the other hand, in the Fresnel case, the spectral distribution first has to be calculated, and after propagating this, a transform back to the spatial domain has to be performed.

The scenarios can be selected into special transition cases according to Figure 6.20, depending on whether the initial and final field plane are near to the focal plane or far away, respectively. There are four cases for a diffraction calculation depending on the location of the initial and the final plane. In the ranges outside the focal region, the geometrical curvature is of a considerable size, so an equivalence transformation, according to the previous discussion, is advantageous. Inside the focal region in the range of the focal depth, the curvature is rather small, and there is no real problem in the sampling of the complex field. The separation between the four cases is not rigorous and gives equal results in the overlapping ranges. But a violation of the criteria of optimal calculating ranges results in an increasing computational effort. The four cases are illustrated in Figure 6.20, where the fourth option has two versions. The cases are in more detail as follows [55]:

1) Case I → I

Transition between two planes in the inner region of the focal plane. This is the easiest case and the calculation procedure follows the formula

$$E_{I-I}(x') = \hat{F}^{-1}\{e^{i\pi\lambda\, z v^2} \cdot \hat{F}[\, E(x)]\} \tag{6.90}$$

The phase front is nearly plane and an additional flattening transform is not necessary.

2) Case O → I

Transition from the outer to the inner region of the focal range. This also corresponds to a Fraunhofer transition, but first the flattening phase factor has to be applied. The quadratic factor to correct for the final curvature has only a small effect.

$$E_{O-I}(x') = \frac{1}{\sqrt{\lambda z}} \cdot e^{-\frac{i\pi}{\lambda z}x'^2} \cdot \hat{F}^{-1}\left\{E(x) \cdot e^{-\frac{i\pi}{\lambda z}x^2}\right\} \tag{6.91}$$

According to the relations of the Fourier transform, the scaling of the coordinates in the observation plane can be calculated with the equation

$$x' = \lambda \cdot z \cdot v \tag{6.92}$$

with the spatial frequency v. It should be noticed that in this case only one Fourier transform is needed.

3) Case I → O

Transition from the inner to the outer region of the focal range. This corresponds to a simple Fraunhofer transition into the far-field too; the calculation can be written as

$$E_{I-O}(x') = \frac{1}{\sqrt{\lambda z}} \cdot e^{-\frac{i\pi}{\lambda z}x'^2} \cdot \hat{F}^{-1}\left\{E(x) \cdot e^{-\frac{i\pi}{\lambda z}x^2}\right\} \tag{6.93}$$

Since the relations of the Fourier transform are valid, the scaling of the coordinates in the observation plane can be calculated according to the Eq. (6.92). Also in this case only one transform is needed.

4) Case O → O

For the transition between two outer regions of the focal plane, the flattening idea described above is required here. The Fresnel calculation uses the equivalence transform, with the Fourier transform $\hat{F}$ it follows that

$$E_{O-O}(x') = \frac{e^{-ik\frac{z}{M}}}{\sqrt{|M|}} e^{-\frac{i\pi(|M|-1)}{\lambda Mz}x'^2} \times \hat{F}^{-1}\left\{e^{i\pi\lambda\frac{z}{M}\nu_x^2}\hat{F}\left[E(x)\cdot e^{-\frac{i\pi(1-|M|)}{\lambda z}x^2}\right]\right\} \quad (6.94)$$

The spatial frequency is given here by the sampling theorem

$$\nu_x = \frac{x}{\lambda \cdot z} \quad (6.95)$$

The flattening parameter M has to be optimized according to the criteria described above; it lies in the range

$$M = 1 - \frac{z}{R} \quad (6.96)$$

with the mean radius of curvature R of the field in the starting plane. If M becomes very small for $z \to R$, it makes sense to fix the size of the final computational grid as a multiple of the Airy diameter and determine the magnification M as

$$M = \frac{x'_{max}}{x_{max}} = \frac{K\cdot\lambda/NA}{x_{max}} \quad (6.97)$$

with $K = 4\ldots10$. There are two possibilities in this fourth case which have to be distinguished and are denoted in Figure 6.21 separately:

- in the case 4_1, $|z| < |R|$, $M > 0$
- in the case 4_2, $|z| > |R|$, $M < 0$, here the direction of the coordinate x' is inverted, the propagation runs over an internal focal point, but Eq. (6.91) is directly usable.

Since the magnification M occurs in the denominator of the propagation formula, this formulation is not valid in the range, where $M \to 0$. If this condition is fulfilled, cases two or three have to be applied or the modified M evaluation according to Eq. (6.97) used.

It should be mentioned that in the cascaded two-step Fourier transform procedure for a O–O transition the internal frequency resolution is hidden. Therefore, if the first Fourier transform is performed by a chirp-z transform, the scaling factor s for the internal frequency resolution must be controlled to guarantee that no aliasing occurs in the angle domain.

6.2.6.5 Sinc-based Algorithm*

(*) The content of this section is quite specialized and for advanced reading. Knowledge of this section is not necessary to understand the following chapters. Beginners are recommended to skip this section.

The idea of this alternative propagation method in particular for Fresnel approximated field propagators is an expansion of the initial field into lateral shifted sinc-functions

$$\text{sinc}(x) = \frac{\sin(\pi x)}{\pi x} \quad (6.98)$$

which reads

$$\varphi_m(x) = \text{sinc}\left(\frac{x}{\Delta x} - m\right) \quad (6.99)$$

It is proposed in [53, 54] and more details are found there. In one lateral dimension this means that the field is described by the expansion

$$E(x_i) = \sum_m E_m \cdot \varphi_m(x_i) \quad (6.100)$$

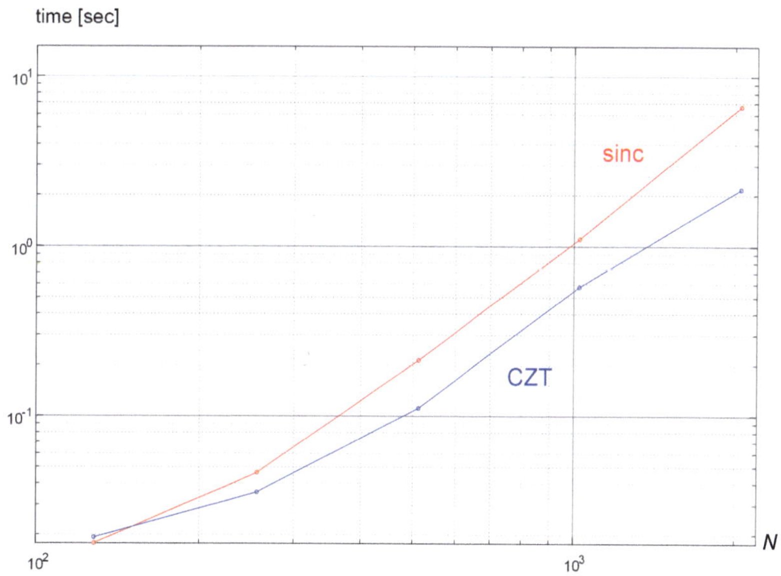

Figure 6.21 Comparison of the runtime between the sinc- and the classical Fourier based algorithm, realized by the chirp-z-transform.

The propagation of the field can be formulated as a convolution with the propagator function, which reads in Fresnel approximation

$$H_{CTF}(x) = \frac{-ike^{ikz}}{2\pi z} \cdot e^{\frac{i\pi}{\lambda z}x^2} \tag{6.101}$$

The field in the final plane after a distance z is then given by the convolution

$$E(x_o) = H_{CTF}(x_o - x_i) * E(x_i) \tag{6.102}$$

This can be expressed by writing the expansion as

$$E(x_o) = \sum_m E_m \cdot \Phi_m(x_i - x_o) \tag{6.103}$$

The integral to get the final expansion function Φ can be written analytically with the help of the Fresnel integrals C and S (see Chapter 12) as

- propagation distance, as long as $z \gg \lambda$ stays in the Fresnel approximation range.

In Figure 6.22 the residual error of the normalized final intensity profile for a Gaussian profile with radius w and a radius of curvature $R = 100$ mm for a wavelength of $\lambda = 1.0$ µm is depicted. There is no flattening performed. The range $w = 0.5\ldots4.0$ mm corresponds to a normalized sag of $\Delta z/\lambda = 1.25\ldots80.0$ which must be sampled. As a consequence, it is seen that:

- of course larger point numbers N help in accuracy;
- the sinc and FFT algorithm are nearly of the same accuracy, sinc is slightly better;
- for small w there is a larger error for both methods due to a near-field model inaccuracy;

$$\begin{aligned}\Phi_m(x) &= \int H_{CTF}(x_i) * \varphi(x_i - x_o)\, dx_i \\ &= \frac{\Delta x}{\pi}\sqrt{\frac{\pi}{\lambda z}}\, e^{\frac{i\pi}{\lambda z}x^2} \cdot \left\{ \begin{array}{l} \left[C\left(\frac{\pi}{2\Delta x}\sqrt{\frac{\lambda z}{\pi}} - x\sqrt{\frac{\pi}{\lambda z}}\right) - C\left(-\frac{\pi}{2\Delta x}\sqrt{\frac{\lambda z}{\pi}} - x\sqrt{\frac{\pi}{\lambda z}}\right)\right] \\ -i \cdot \left[S\left(\frac{\pi}{2\Delta x}\sqrt{\frac{\lambda z}{\pi}} - x\sqrt{\frac{\pi}{\lambda z}}\right) - S\left(-\frac{\pi}{2\Delta x}\sqrt{\frac{\lambda z}{\pi}} - x\sqrt{\frac{\pi}{\lambda z}}\right)\right] \end{array} \right\}\end{aligned} \tag{6.104}$$

Figure 6.21 illustrates a raw estimation of the computational time in seconds as a comparison of the sinc- and the chirp-z-transform based Fraunhofer transform as a function of the pixel number N. It is seen that the sinc approach needs slightly more time, but there is no categoric difference.

In contrast to the usual Fourier based plane wave expansion, the accuracy of this approach is independent of:

- image plane discretization;
- wavelength;
- for large w a stronger phase variation of the complex field must be sampled and creates the same error problems in both methods.

If the propagation distance z is changed from the near-field into the far-field, the sinc algorithm shows much more constant behaviour and small errors for larger distances. The FFT based algorithm definitely has problems with the poor sampling in the spectral domain. Here one of the major advantages of the sinc-approach can be seen, see Figure 6.23.

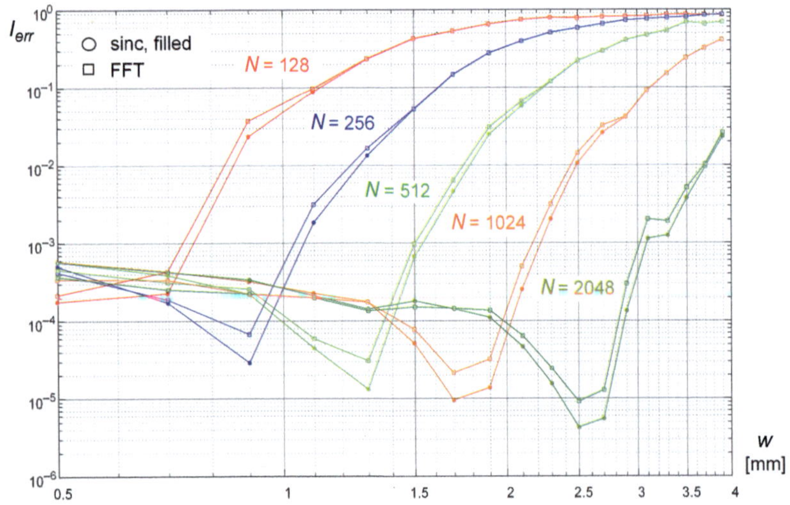

Figure 6.22 Comparison of the residual error of a Gaussian beam propagation using the sinc- and the FFT based calculation respectively as a function of the Gaussian width w for different pixel numbers N with a fixed curvature and wavelength. The data are $R = 100$ mm and $\lambda = 1$ µm.

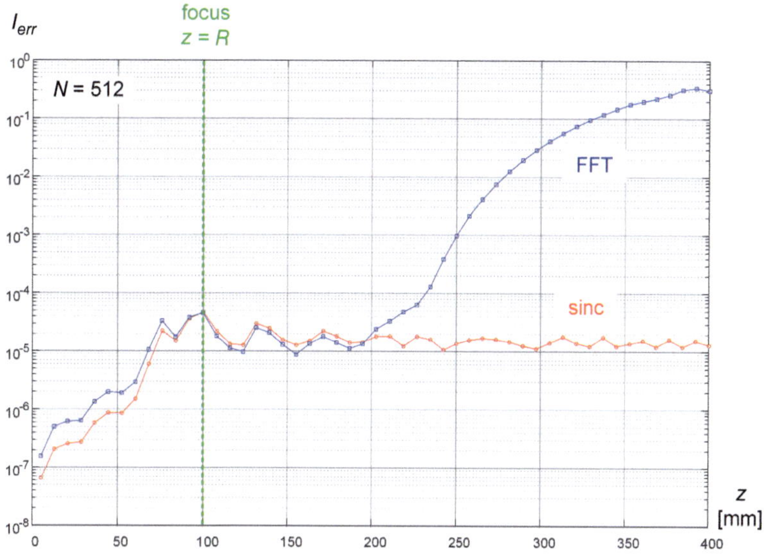

Figure 6.23 Intensity error for the sinc- and the FFT based calculation as a function of the propagation distance z. A smooth Gaussian profile is considered here. The data are $R = 100$ mm and $\lambda = 1$ µm.

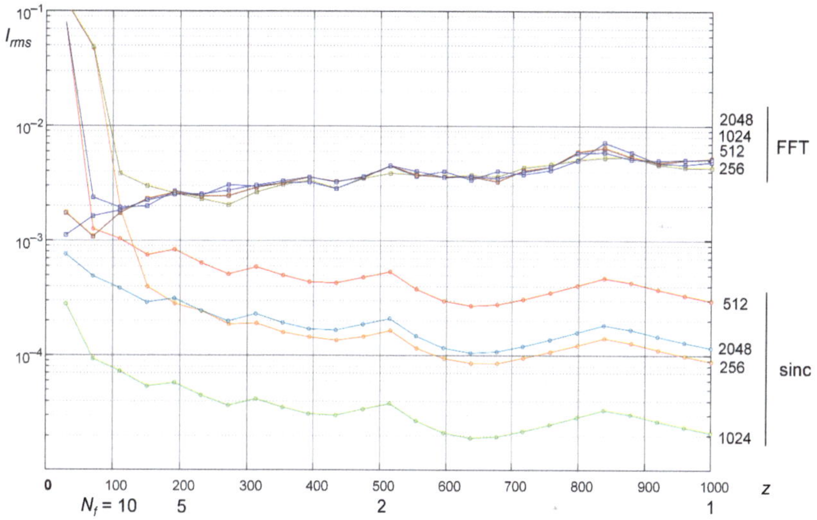

Figure 6.24 Comparison of the sinc- and the FFT-based algorithm for a truncated plane wave with uniform amplitude. The unit of z is arbitrary here.

If the exact calculation of a plane wave diffraction at a square aperture is taken as an example, the sinc algorithm shows much better performance than the FFT-based method. Since the sinc-function is the natural Fourier transform of a slit aperture, a good adaptation can be assumed and explains this behaviour. The accuracy is 1–2 orders of magnitude better and the z-dependence becomes better for larger z. This is illustrated in Figure 6.24. For a $2a = 2$ mm wide screen a wavelength of $\lambda = 1$ µm and the point numbers $N = 256$, 512, 1014 and 2048, the rms of the error as a function of the distance z looks as follows. What is strange here is that:

- the accuracy of the FFT does not depend on N;
- in the case of the sinc method, the accuracy is dependent on N, but shows an interesting waviness, which can be understood as a changed matching of the slit width with an integer pixel size.

Some advantages are obtained (to be proved) by this sinc based approach because of the following.

- The rectangular sampling and the discretization has less influence.
- The restriction onto 2^N sampling numbers is gone.
- The sinc function as a basis is more directly related to the edge diffraction; therefore, in the case of hard clipped intensity profiles, the sinc approach is more accurate.
- For very small Fresnel numbers N_F corresponding to large propagation distances, the sinc-method is much more accurate.

6.2.6.6 Boundary and Sampling Problems*

(*) The content of this section is quite specialized and for advanced reading. Knowledge of this section is not

necessary to understand the following chapters. Beginners are recommended to skip this section.

Until now, the sampling has been considered under the aspect of a dense covering of the wave in the open aperture area, but one more problem occurs through using the fast Fourier transform.

In two dimensions the grid must be Cartesian with equidistant pixel size. Therefore, for a circular boundary the discretization is only very rough and has a corrugated form. This causes some errors in the computation, depending on the geometry of the integration area. Figure 6.25 illustrates this problem. The accuracy can be increased if a suitable weighting of the mesh points near the boundary is introduced [79, 80].

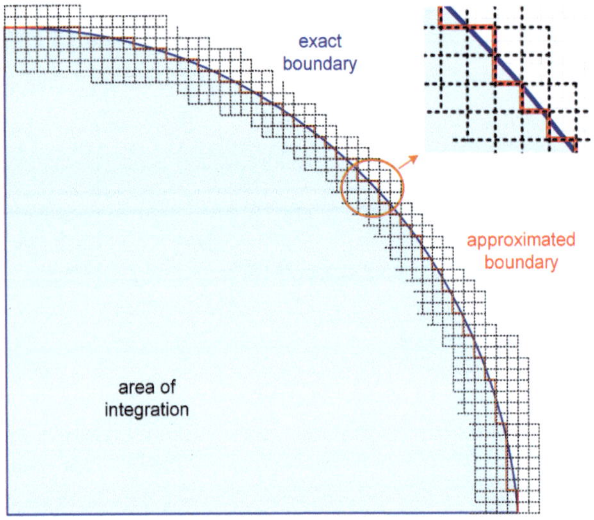

Figure 6.25 Demonstration of the inaccuracy of a smooth curved boundary description in the case of a Cartesian sampling.

6.2.6.7 Phase Masks: TEA-, LPI- and LSI Approximations

Although not directly related to diffraction, when propagating a wave through an optical system not only issues with distance-related transitions occur. Many components can be modelled with good accuracy by a phase mask, changing the phase of field on the same coordinate grid. For example, a thin lens with aberrations under small-angle conditions can be modelled in this way. According to part (a) of Figure 6.26, in the so-called thin-element approximation (TEA) the phase change is then calculated by raytrace in measuring the optical phase difference $\psi(x,y)$ and the component leaves the amplitude distribution and the lateral coordinates x, y unchanged [86–88]. For example, this method can be applied for diffractive elements with good success. The change of the incoming complex field

$$E_{in}(x,y) = A_{in}(x,y) \cdot e^{i\varphi_{in}(x,y)} \tag{6.105}$$

follows then the simple equation

$$E_{out}(x,y) = A_{in}(x,y) \cdot e^{i\cdot[\varphi_{in}(x,y)+\psi_{mask}(x,y)]} \tag{6.106}$$

If the depth of a lens, for example a deep asphere, cannot be neglected, an extension of the above illustrated approach can be used: the local plane interface approach, illustrated in Figure 6.26(b). Here, first the incoming field is decomposed in a plane in front of the component with coordinates x, y into a set of rays, the amplitude describes the ray weights, the phase the direction. Then a raytrace is performed, ending in a second plane behind the component with coordinates x', y'. Due to depth effects and finite ray

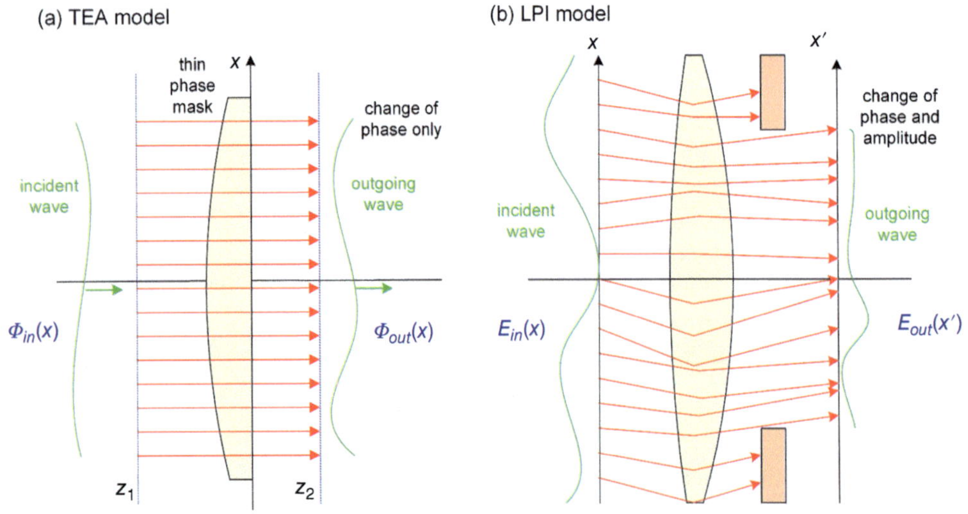

Figure 6.26 Illustration of the thin phase mask approaches with (a) a thin element and (b) a change of the coordinates by a non-thin mask component as local plane interface model.

angles, the coordinates of every ray are changed in general, resulting in a distorted grid. The changes of the resulting pixel size of every ray as a modified distance to their neighbours and therefore changed representing area, corrects the energy density and modifies the new amplitude at x', y'. The ray angles give the new phase gradient components. Finally, to get once again an equidistant grid for the next propagation step, the real two-dimensional distributions of amplitude $A(x', y')$ and phase $\psi(x', y')$ are interpolated on a new grid. This approach is called the local plane interface method (LPI).

In the case of a larger curvature of the incoming phase surface, this approach can also be generalized to the so-called local spherical interface (LSI) method, for details see [87].

6.3 Point Spread Function in Optical Systems

6.3.1 Hybrid Model of the Diffraction Calculation in Optical Systems

If diffraction plays an essential role in an optical system, the basic principles and computational algorithms as described above are used. If this is the case, the diffraction related spreading of the image point is somewhat larger than the spreading due to geometrical aberrations. The system is then called diffraction limited; this case is realized in well corrected setups. Practical examples for this constellation are microscopy, lithography or astronomical instruments. Mostly the field in the focal plane is of interest; therefore, Fraunhofer far-field conditions can be assumed to be acceptable. The complex field is defined in the exit pupil of the system and is determined by the phase corresponding to the wave aberrations W and the amplitude corresponding to the transmission T of the system together with the illumination and maybe a grid distortion. Therefore, the diffraction integral is formulated as

$$E(x', y') = N \cdot e^{-\frac{i\pi}{\lambda R_{ExP}} \cdot (x'^2 + y'^2)} \iint T(x_p, y_p) \cdot e^{2\pi i W(x_p, y_p)} \cdot e^{-\frac{2\pi i}{\lambda R_{ExP}} \cdot (x_p x' + y_p y')} \, dx_p \, dy_p \tag{6.107}$$

with a normalization factor N and the radius of the reference sphere in the exit pupil R_{ExP}. The calculation of this scheme follows in nearly all practical cases the so-called hybrid model. It is assumed that all lateral truncations and all diffraction effects occur in the exit pupil of the system. Every stop or limiting diameter is projected along the ray directions into this plane. The phase is calculated as the wave aberration in the exit pupil too by geometrical raytracing. This scheme is illustrated in Figure 6.27(a).

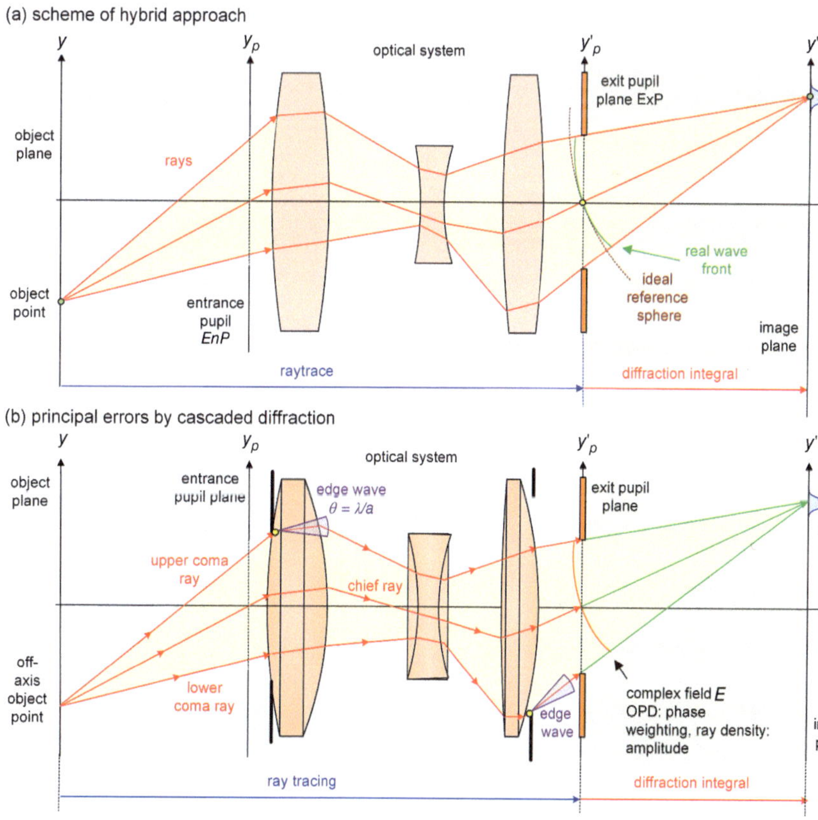

Figure 6.27 (a) Hybrid model of diffraction in an optical system, all diffraction influences are assumed to take place in the exit pupil and (b) inaccuracy of the hybrid model in case of strong diffraction effects inside the system.

In the exit pupil the complex field is constituted by the optical path, length along the rays point by point and the amplitude optional by taking a ray weighting into account. The diffraction effects are then calculated by a Fraunhofer far-field diffraction integration in the image by a free space transition from the exit pupil into the image plane. At the end, in the image plane the PSF is calculated.

This approximation is quite good for most of the optical systems. Only if diffraction in the front group of the system influences the phase and amplitude in the exit pupil due to propagation over a large distance do errors occur. This happens, for example, in beam clean-up telescopes or very large propagation distances in laser systems. The edge diffraction at some early surfaces or diaphragms with diameter $2a$ creates a diffraction wave with angle $\theta = \lambda/a$. This light field is overlaid in the exit pupil to the signal field. The larger the propagation distance from the diffracting surface to the exit pupil, the larger the lateral spreading is. This principal error case is sketched in Figure 6.27(b). Details of cascaded diffraction effects are discussed in Section 6.8.

This model together with the Huygens principle discussed above shows that the PSF can simply be understood as an interference phenomena of the Huygens wavelets in the exit pupil. In addition to the explanation in Section 6.1.3, here two more important factors of influence are discussed.

- If the pupil is apodized with reduced amplitude towards the edge, the arrows lying here are shorter and of reduced energetic weighting. This lowers the central peak height, see Figure 6.28(a). More details are presented about this phenomenon in Section 6.4 below.
- If the system suffers from aberrations, the phase and therefore the orientation of the arrows changes continuously. For spherical aberration, as an example, the slope grows towards the outer pupil points. This reduces the superposed arrow length and therefore the intensity, see Figure 6.28(b). This corresponds to the well-known reduction of the Strehl ratio in the case of aberrations. This is discussed in Volume 2, Chapter 18 in more detail.

6.3.2 Ideal Point Spread Function

6.3.2.1 Introduction

In every optical system there is always a transversal truncation of the ray bundles observed due to the finite size of the components. These boundaries influence the PSF and

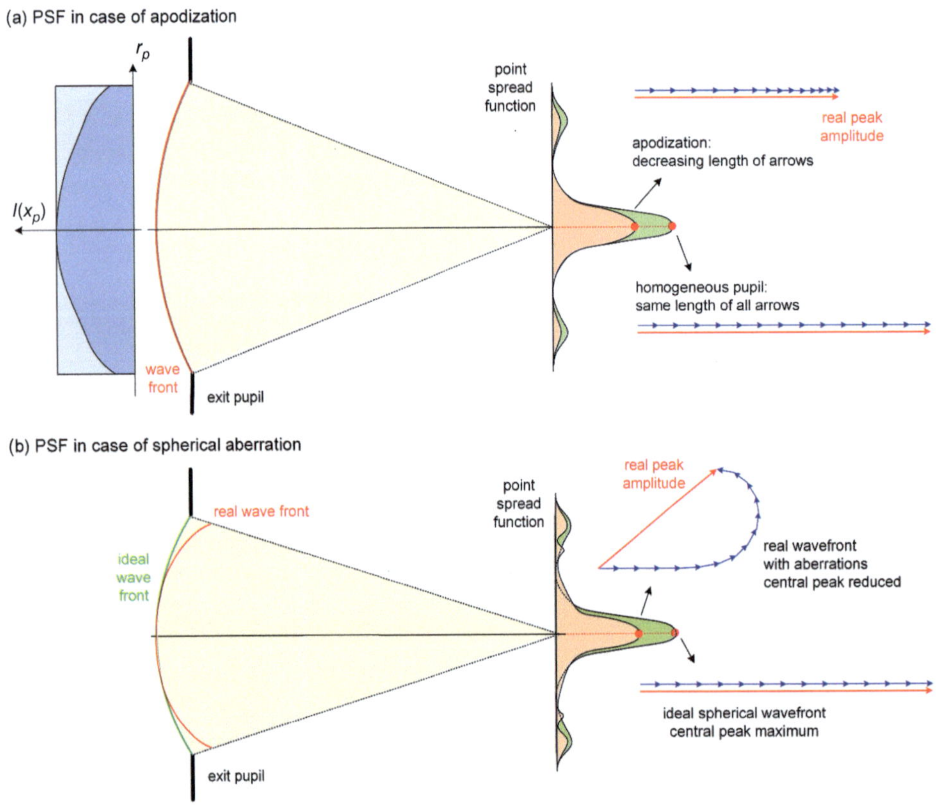

Figure 6.28 Generation of the point spread function in case (a) of apodization and (b) spherical aberration.

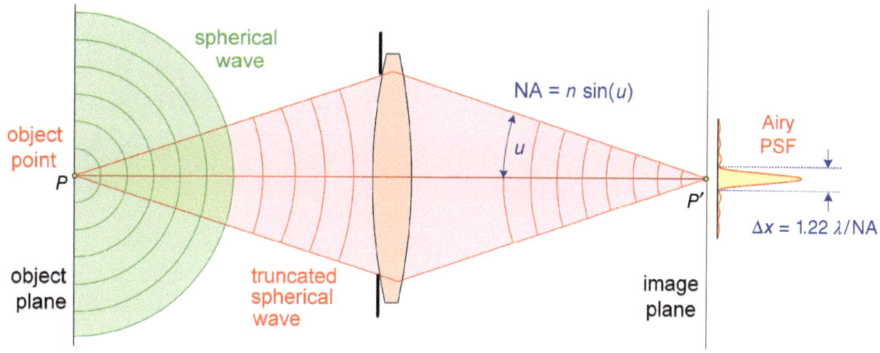

Figure 6.29 Point spread function of an optical system due to diffraction.

cause diffraction effects. As can be seen in Figure 6.29, a self-luminous point in the object plane emits a spherical wave in the half space of growing z-values, shown in green. The system cuts the inner part and transmits only the red wave parts inside the acceptance cone of the aperture. In the image domain, only these supported waves are superposed by interference. This gives rise to a spread intensity distribution in the image plane, although there are in the best case no phase aberrations. The intensity pattern of an ideal but truncated spherical wave for a monochromatic field of constant amplitude is the perfect PSF. Concentric rings are arranged around a dominant central peak according to the grey scale picture in Figure 6.32. This pattern is called the Airy distribution. The first dark ring is used to laterally scale the width of this distribution. The diameter of this dark ring is given by

$$D_{Airy} = \frac{1.22 \cdot \lambda}{NA} = \frac{1.22 \cdot \lambda}{n \cdot \sin u} \tag{6.108}$$

Here $NA = n \sin(u)$ is the numerical aperture in the image domain.

From the viewpoint of the geometrical image point P', the convergent spherical waves of finite solid angle have a constant phase, therefore we have far-field conditions here. This means that the PSF can be calculated by the Fraunhofer diffraction integral. Figure 6.30 shows the complete image domain for one lateral coordinate r in the case of circular symmetry. In addition to the picture in Figure 6.30(b) the axial distribution is included, which means that the intensity distribution is also considered in defocused planes. Near to the image plane a pronounced peak is obtained, in the axial as well as in the transverse direction. The length of this focal caustic is scaled in the lateral direction by D_{airy}, in the axial direction by the Rayleigh length R_u. R_u is responsible for the depth of focus and is given by the equation

$$R_u = \frac{n \cdot \lambda}{NA^2} = \frac{\lambda}{n \cdot \sin^2 u} \tag{6.109}$$

For the special cases of the image plane $z = 0$ and on the optical axis ($x = y = r = 0$) the intensity distributions of the perfect PSF can be expressed analytically. With the help of

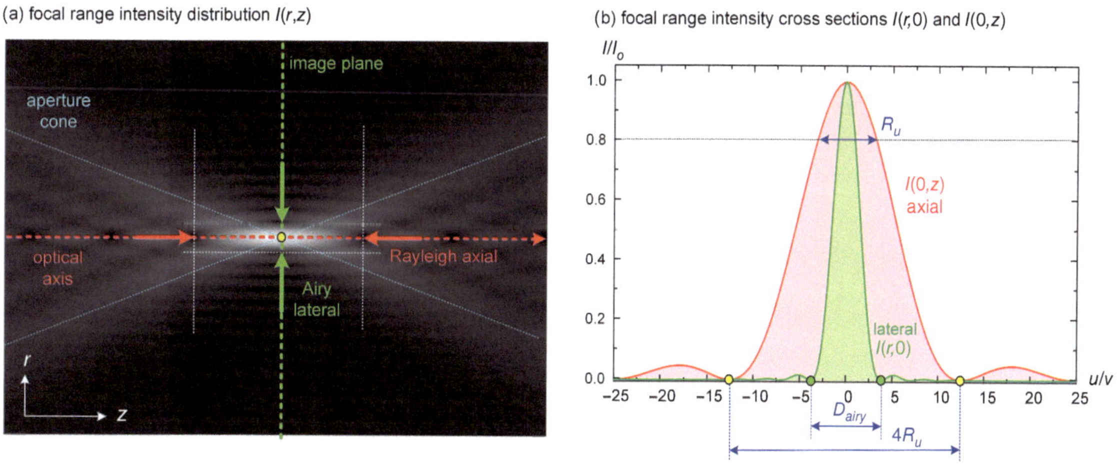

Figure 6.30 Point spread function of an optical system in the focal region (a) as $I(r,z)$ and (b) as cross sections $I(u)$ in red and $I(v)$ in green.

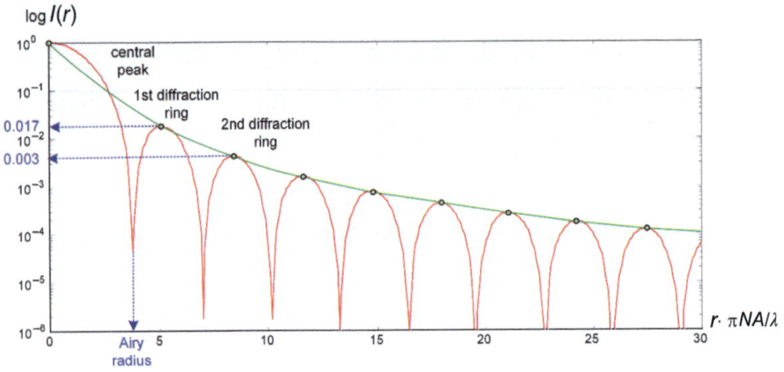

Figure 6.31 Ideal Airy PSF as a logarithmic plot for positive radii only.

normalized canonical coordinates according to

$$v = \frac{2\pi x}{\lambda NA}, \quad u = \frac{2\pi n z}{\lambda NA^2} \tag{6.110}$$

we have in the image plane the intensity as a function of the lateral normalized coordinate as the Airy or Besinc function squared

$$I(0, v) = \left[\frac{2J_1(v)}{v}\right]^2 I_0 \tag{6.111}$$

and along the optical axis the scaled sinc function squared

$$I(u, 0) = I_0 \cdot \left[\frac{\sin(u/4)}{u/4}\right]^2 = I_0' \cdot \text{sinc}^2\left(\frac{u}{4}\right) \tag{6.112}$$

which is also oscillatory. J_1 is the Bessel function of the first kind. In contrast to the Airy width, the distance between the first two zero points along the axis is given by 4 R_u, not simple 1 R_u. Figure 6.30(b) shows these two intensity curves together with the characteristic scaling widths.

The effect of the Airy diameter D_{airy} is the limitation of the lateral resolution in the case of a point like object. This is realized in microscopy or astronomy. It can be seen that there are three primary options to increase the resolution.

- Use of smaller wavelength, this is done in micro-lithography.
- Putting a medium in the object space, this is used in immersion microscopy and lithography.
- Increase the aperture cone angle sin(u). This option is limited because sin(u) <1 and large NA systems are hard to correct for aberrations and the 'ideal' PSF cannot be achieved.

It should be mentioned here that in the scope of a more general physical model, other factors for resolution modifications can be also used; for example, the geometry – which defines the numerical pre-factor in the Airy formula, the coherence or the polarization.

If the ratio of the two scaling numbers D_{airy} and R_u is calculated, we get

$$\eta = \frac{D_{airy}}{4R_u} = 0.3 \cdot \sin u \tag{6.113}$$

This means that for optical systems the focal caustic has always an elongated shape along z, the focal volume is cigar shaped, not disc shaped. The lateral resolution is larger than the axial one. This ratio is independent of the refractive index and the wavelength. The focal volume can be estimated as

$$V \approx D_{airy}^2 \cdot 4R_u = \frac{4n \cdot \lambda^3}{NA^4} \tag{6.114}$$

If in practice the diameter definition is changed to FWhM or something else, the value of this ratio can be different, but from fundamental principles the elongated shape is maintained.

6.3.2.2 Airy Pattern

The Airy distribution as the transverse intensity distribution of the ideal PSF is analysed here in more detail. This quite important pattern is responsible for the image resolution in diffraction limited optical systems and serves as a reference for the correction quality in case of aberrations. To better resolve the side lobes, a logarithmic representation shows the radial structure in Figure 6.31. It can be seen that the relative intensity height of the first

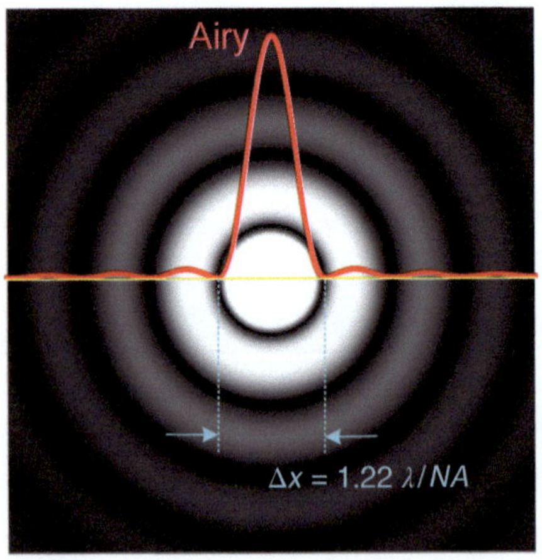

Figure 6.32 Ideal Airy PSF as a grey-scale picture.

diffraction ring is 1.7%. It can be seen that the convergence to intensity zero is poor and the energy is considerably smeared out. Figure 6.32 shows the lateral pattern as grey scale image above 3%.

Another useful representation is shown in Figure 6.33. It is the so-called encircled energy plot that gives the relative amount of enclosed energy for a stop of variable radial width. This shows that the central peak contains 84% of the total energy. The regions with horizontal slope indicate the dark rings, where locally no energy growth is obtained with increasing radius. It is seen that the energy is strongly spread and the limit of the complete energy according to the energy conservation is reached for very large radii.

6.3.2.3 Axial Distribution

The sinc function in Eq. (6.112) governs the axial intensity profile. Figure 6.34 shows this function again. If the 80% drop of the peak intensity is considered, the axial width by numerical calculation is obtained numerically as

$$\Delta z_{diff} = 0.986 \cdot \frac{\lambda}{n \cdot \sin^2 u} \approx R_u \tag{6.115}$$

This gives the diffraction limited depth of focus distance, the so-called Rayleigh unit. The 80% decrease corresponds to the classical approach of the Rayleigh resolution limit in transverse direction to get a diffraction limited system. It should be noticed that in contrast to the Airy diameter,

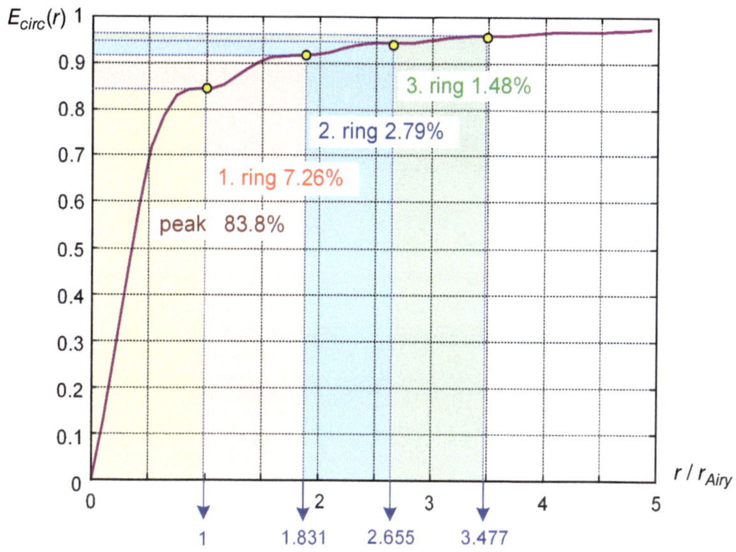

Figure 6.33 Encircled energy of the ideal Airy PSF as a function of a normalized radius.

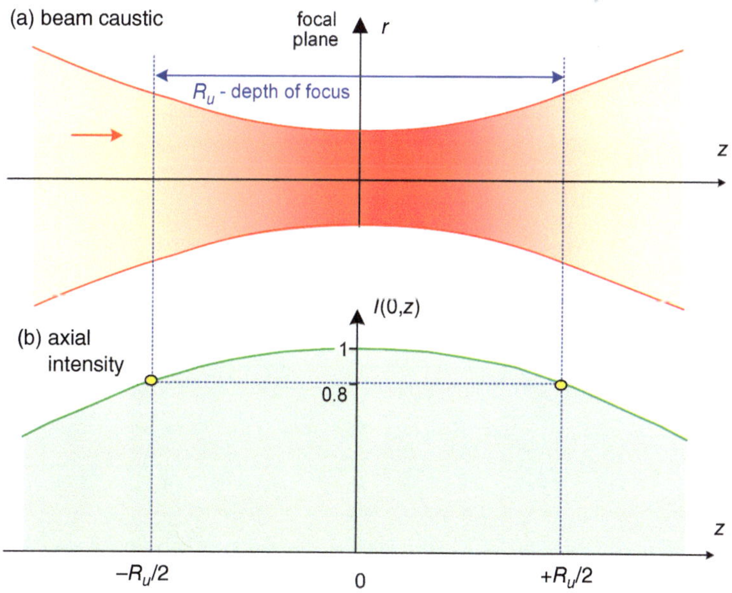

Figure 6.34 Definition of depth of focus in a beam with a caustic shown in (a) and as intensity $I(z)$ on axis in (b).

the depth of focus scales quadratically with the numerical aperture.

The R_u-definition is similar to the Rayleigh length of the Gaussian beam formula

$$z_0 = \frac{\lambda}{\pi n \theta_0^2} \tag{6.116}$$

Due to the Gaussian transverse intensity profile, a different pre-factor $1/\pi$ is obtained here.

If the intensity distribution is considered outside of the perfect image plane, we get the pictures in Figure 6.35. In the upper row, the intensity is renormalized to 1. It can be seen that a spreading with a ring structure takes place, which is perfectly symmetric around the image plane. In the lower row, the energy is conserved due to the spreading and increased area, and the peak height is reduced for the defocusing planes correspondingly.

Similarly to the transverse Airy pattern, the Rayleigh length scales linearly with the wavelength. Therefore, by increasing λ the axial depth grows linearly. This is shown in Figure 6.36 in a series of four cases with a changed wavelength and constant numerical aperture.

The depth of focus defines the extension of a sharp focus along the z-axis. The Rayleigh unit is a measure for this property in case of a diffraction limited system. The depth of focus depends on the numerical aperture (NA), the index n and the wavelength. In contrast to the

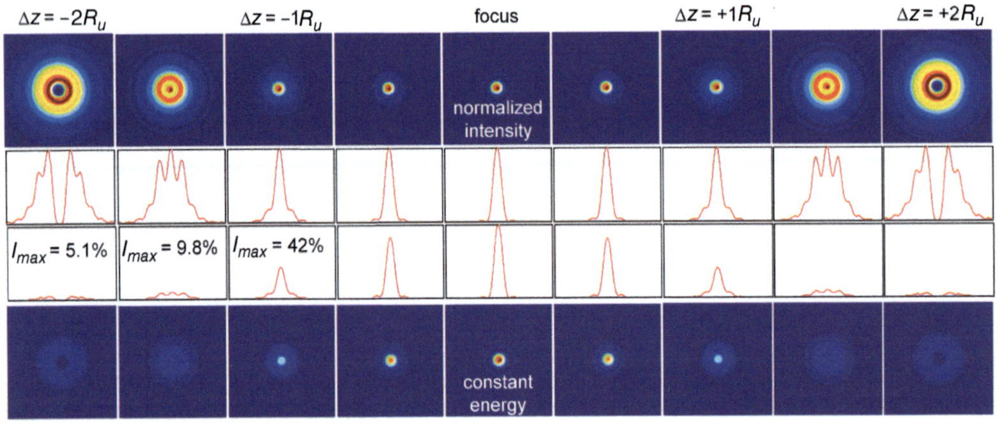

Figure 6.35 Ideal PSF for defocusing in intensity normalized (upper row) and power normalized (lower row) representation.

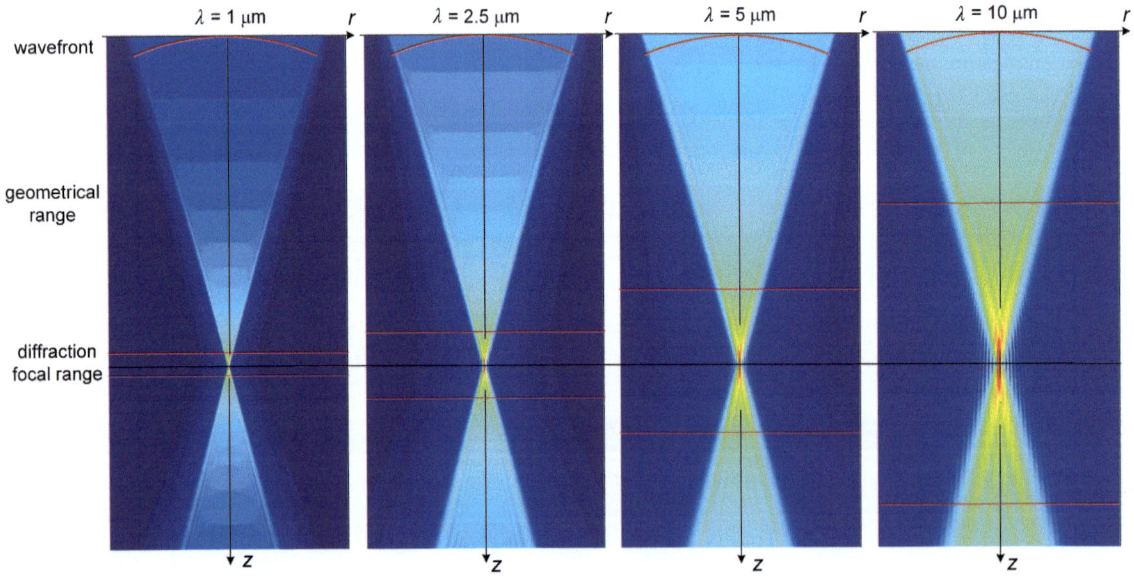

Figure 6.36 Ideal PSF with $I(r,z)$-caustic for different wavelengths. To emphasize the levels of low energy density, the square root of the amplitude is shown.

Figure 6.37 Pictorial comparison of depth of focus in between large NA (a) and small NA (b) with small and large depth of focus respectively. In (c) an image is shown with large depth variation, (d) shows the sharpness map.

Airy diameter, an additional factor of the refractive index occurs. A large aperture results in a small depth of focus, and the small aperture results in a large depth of focus, see Figure 6.37(a) and (b). In (a) with a short depth of focus, only the middle part is sharply imaged. By using a sharpness metric like gradients of intensity, it is possible to create a depth map of an image. This is sketched in part (c) and (d) of the figure. Darker lines in (d) correspond to large gradients and are obtained for the sharp areas of the image, here again the middle part is the sharpest. But it should be noted that the content of the image also influences this map.

The depth of focus is a relevant quantity for:

- accuracy of detector position;
- sharpness range of details in a three-dimensional scene.

6.4 PSF in Case of Apodization

6.4.1 Apodization Effects

Amplitude as well as phase distributions influence the value of the diffraction integral, but the impact of the phase is quite large. If the assumption of a constant pupil illumination is not fulfilled, the PSF and focal caustic changes. If a non-uniform amplitude distribution in the pupil exists, the system is called apodized. From the physical point of view it can be assumed that the amplitude works as an energetic weighting of the phase. Therefore, in the case of considerable aberrations mostly located towards the outer pupil region, a reduced amplitude here helps increase the resolution.

One special practical relevant case of a non-circular homogeneous illumination is a circular ring pupil. This

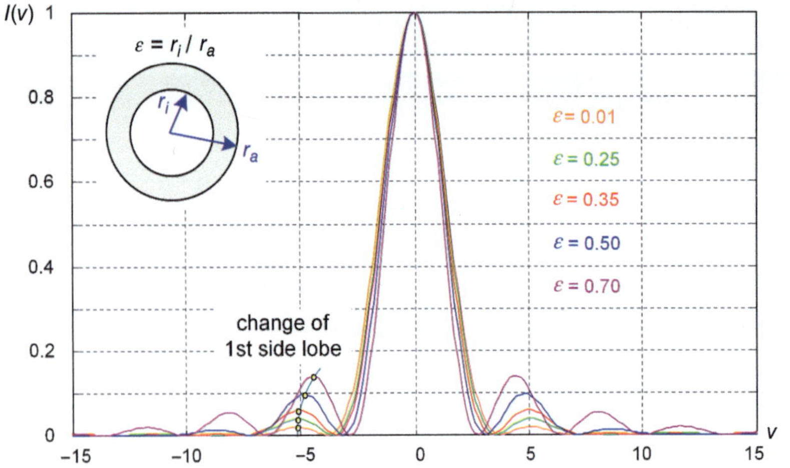

Figure 6.38 Point spread function in the case of a ring pupil for different values of the relative central obscuration values ε.

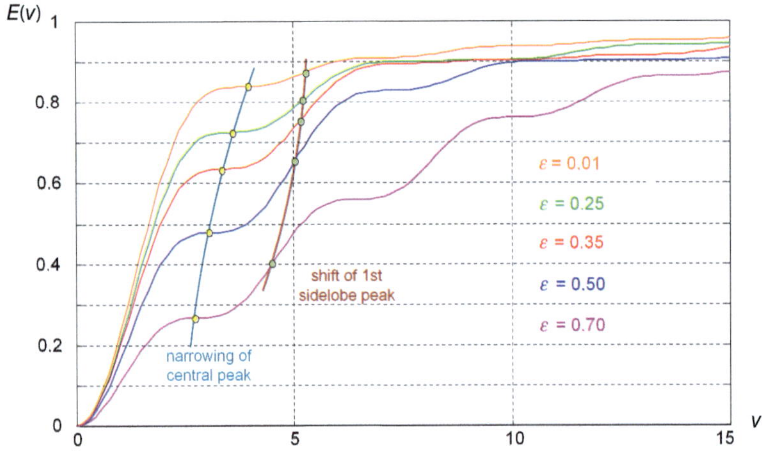

Figure 6.39 Encircled energy functions in the case of a ring pupil for different values of the relative central obscuration ε.

can be found in astronomical telescopes with central obscuration. If the relative size of the inner to the outer radius of the ring pupil is defined as

$$\varepsilon = \frac{a_i}{a_a} < 1 \tag{6.117}$$

it can be seen that the intensity in the image plane without aberrations is a modified Airy pattern with the equation

$$I(v) = \frac{1}{(1-\varepsilon^2)^2}\left[\frac{2J_1(v)}{v} - \varepsilon^2 \cdot \frac{2J_1(\varepsilon v)}{\varepsilon v}\right]^2 \tag{6.118}$$

and the normalized lateral coordinate v. Figure 6.38 shows this distribution for several values of ε. It can be seen that in contrast to the Airy pattern:

- the width of the central pattern is slightly decreased;
- the heights of the side lobes are considerably increased;
- the ring peaks are shifted radially towards the axis.

In Figure 6.39 the encircled energy is plotted for the same cases of central obscuration. In this representation it can be seen much more clearly that a large portion of the energy is distributed into the side lobes. For $\varepsilon = 0.5$ less than 50% of the power is collected in the central peak.

Furthermore, it can be calculated analytically that the depth of focus is increased in the case of a ring pupil. The depth of focus scales with

$$\Delta z = \frac{2\lambda}{n\sin^2 u \cdot (1-\varepsilon^2)} \tag{6.119}$$

The larger the value of ε, the longer the depth of focus; the focal caustic is stretched along z for growing ε values.

The PSFs of annular pupils suffer from large secondary maxima, which may have a diminishing effect on the contrast of an optical image. On the other hand, it is possible to suppress the secondary maxima by an approximately Gaussian shape of the pupil function. In Figure 6.40 for comparison, the ideal case of a completely Gaussian-shaped pupil function is shown as the PSF together with the Airy case and a ring shaped pupil (with $\varepsilon = 0.75$).

Figure 6.40 Radial intensity profiles for different point-spread functions according to a circular, a ring shaped and a Gaussian pupil illumination respectively.

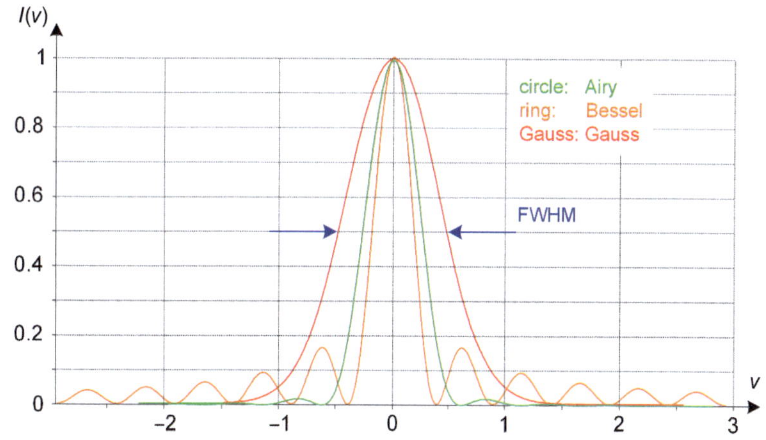

Figure 6.41 Encircled energy for different point-spread functions according to a circular, a ring shaped and a Gaussian pupil illumination. The unit of w is arbitrary here.

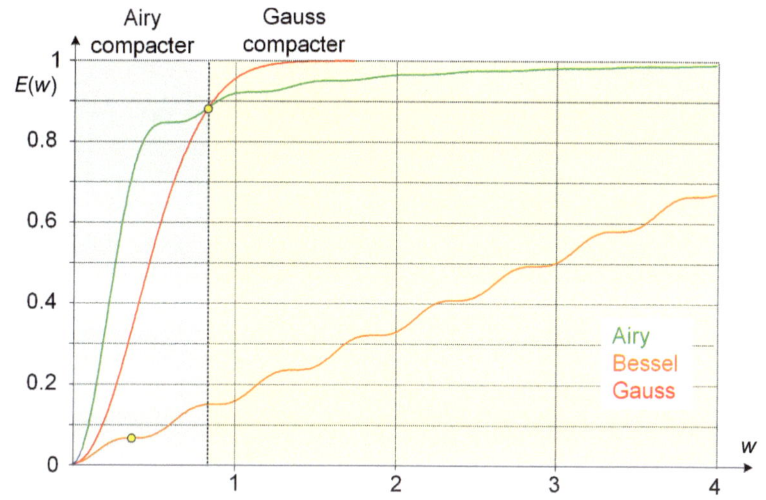

On one side, the Gaussian pupil has a considerably larger diameter in the focal plane if the 50% FWhM threshold is considered as the measure of the width. On the other side, the Gaussian image intensity drops much faster to zero, which means that there is only one concentrated distribution without large spreading and ringing structures. In Figure 6.41 the corresponding encircled energy curves are seen. Here the impact of the large vs small spreading is much more pronounced. For smaller values of $v < 0.8$, the Airy pattern shows a smaller energy-related width, for larger values of v the Gaussian profile it is more compact as indicated in the drawing.

With the Gaussian apodization, the maximum encircled energy of quasi 100% is reached at a smaller radius of the point-spread function compared with the Airy diffraction pattern; while with the latter, more energy is concentrated in the centre of the point image. The Bessel function of an annular pupil has, in contrast, a smaller width of the central peak (e.g. determined by the FWHM), but distributes the energy over a very large area.

In practice, the illumination of coherent optical systems is often realized with lasers operating in a fundamental mode, given by a Gaussian intensity distribution. Particularly in scanning confocal microscopy, when the object is scanned by focused beams, Gaussian beam shapes may be applied instead of plane waves with corresponding Airy point-spread functions. The comparison of the PSF widths is a little bit critical, because in the case of a Gaussian distribution the exit pupil with the profile has in principle a smaller effective aperture if a throughput of at least 99% is requested. The definition of the aperture edge is not this clear as for a circular homogeneous illumination.

It should be mentioned that a 'weak apodization' occurs in optical systems in the case of pupil aberrations. If the imaging from the entrance pupil to the exit pupil suffers from distortion, the energetic angle distribution of the object is modified and a changed amplitude profile is observed in the exit pupil. This is discussed in more detail in Chapter 17.

6.4.2 Super-Gaussian Profile*

(*) The content of this section is quite specialized and for advanced reading. Knowledge of this section is not

necessary to understand the following chapters. Beginners are recommended to skip this section.

To discuss the impact of a pupil apodization more generally, here a non-truncated super-Gaussian profile in the exit pupil of the form

$$I(r) = I_0 \cdot e^{-2\left(\frac{r}{a}\right)^m} \tag{6.120}$$

is considered. If $m = 2$, a classical Gaussian profile is obtained. For larger values of m, the profile becomes more and more rectangular in shape with larger slopes in the decay region. For $m < 1$ the profile is more peaked in the centre with a cusp at $r = 0$. By changing the exponent m, and keeping the formal width a constant, a smooth transition between the Airy and the Gauss PSF is obtained. It should be mentioned here that in this case, the effective numerical aperture changes too; it must be defined explicitly, because there is a certain part of the energy outside $r = a$ in the pupil and no abrupt decay from one to zero is present.

Figure 6.42 shows the change of the profile with m as discussed. Figure 6.43 illustrated the growing maximum slope of the illumination profile at $r = a$ in the pupil.

Figure 6.44 shows the intensity profile in the focus as the Fourier transform or far-field PSF if the super Gaussian profile is assumed in the exit pupil. To emphasize the side

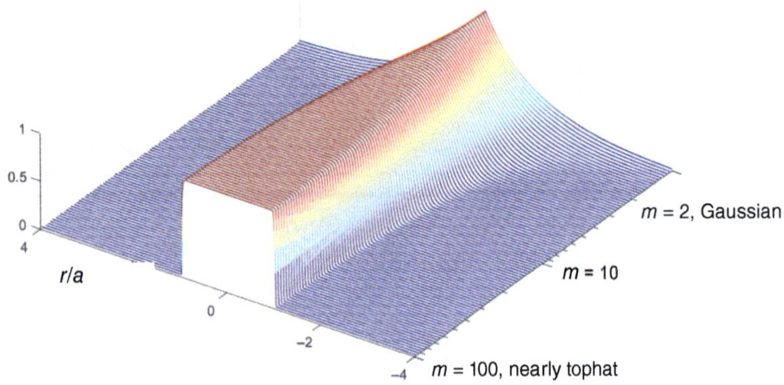

Figure 6.42 Super-Gaussian intensity profile with changing exponent m.

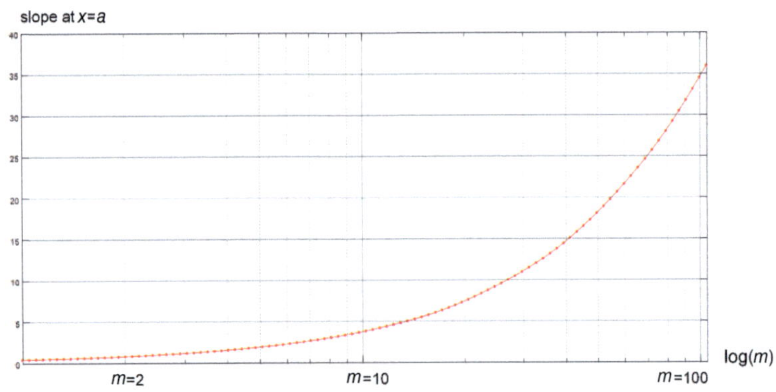

Figure 6.43 Slope of the apodization profile of a pupil with super-Gaussian intensity profile with changing exponent m.

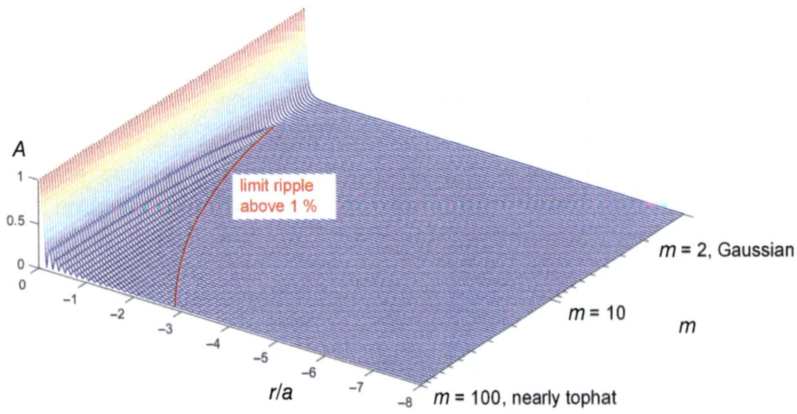

Figure 6.44 Point spread function amplitude of a pupil with super-Gaussian intensity profile with changing the exponent m.

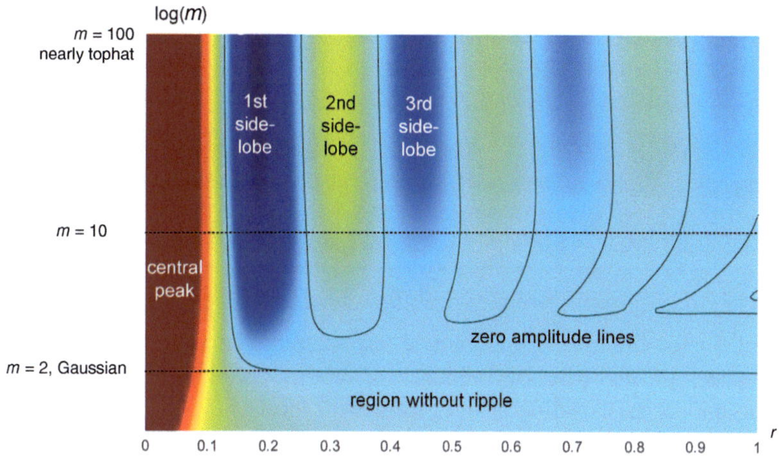

Figure 6.45 Point spread function amplitude of a pupil with super-Gaussian intensity profile with changing exponent m. The amplitude has a sign and switches from side lobe to side lobe, the black lines indicate the zeros. The unit of r is arbitrary here.

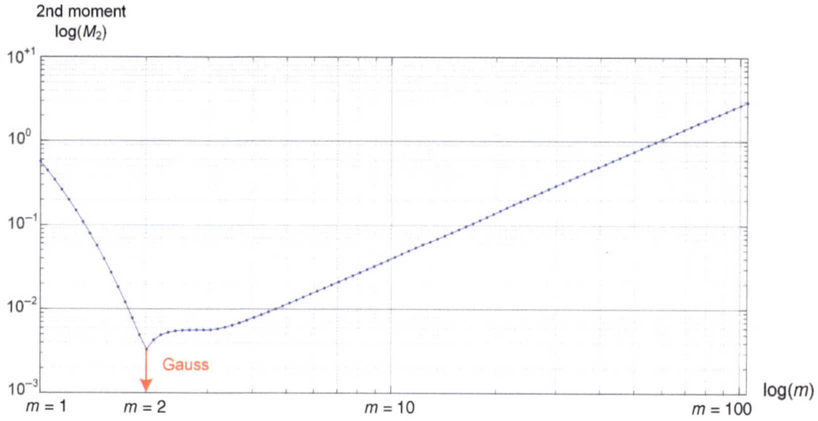

Figure 6.46 Second moment M_2 of the point spread function for a super-Gaussian pupil illumination.

lobes, the amplitude is illustrated here not the intensity. It is seen that with growing m, the side lobes increase in height and the well-known ring system of the Airy pattern occurs. Figure 6.45 shows the same behaviour in an exaggerated representation. As can be is seen, for $m > 2$ ripples occur with first very small side lobe heights, which grow with m. The position of the side lobes is nearly unchanged.

If the quality of the PSF is evaluated by calculating the second moment of the intensity, the result in Figure 6.46 is obtained. It can be seen that the Gaussian profile with $m = 2$ delivers the best result; interestingly for smaller m the moment M_2 also again grows. In principle the effective numerical aperture is growing too and the comparison is somewhat critical. But this graph shows rather clearly that under the assumption of this width criterion, the Gaussian profile delivers the most compact focal spot.

6.5 Focusing at Low Fresnel Numbers*

(*) The content of this section is quite specialized and for advanced reading. Knowledge of this section is not necessary to understand the following chapters. Beginners are recommended to skip this section.

6.5.1 Introduction

If a beam is focused by a very small numerical aperture, or more precisely a small Fresnel number in the range of 1, some of the traditional approximations of the calculation of the diffraction integral concerning the change of the $1/z$ term are not valid inside the focal caustic [89, 90]. This means that the divergence spreading due to diffraction is of the same order of magnitude as the geometrical convergence, forcing a focal point. If the z-dependence is considered, the $1/z$-factor in the diffraction integral matters and its change inside the depth of focus has an impact. In this case it is no longer feasible to move the $1/z$ factor of the diffraction integral as quasi constant outside the integral. It is observed in this case that the geometrical centre of curvature of the spherical focusing wave is no longer coincident with the point of largest intensity on the optical axis, as is shown in Figure 6.47. The reason is that the centre of curvature considers only the phase,

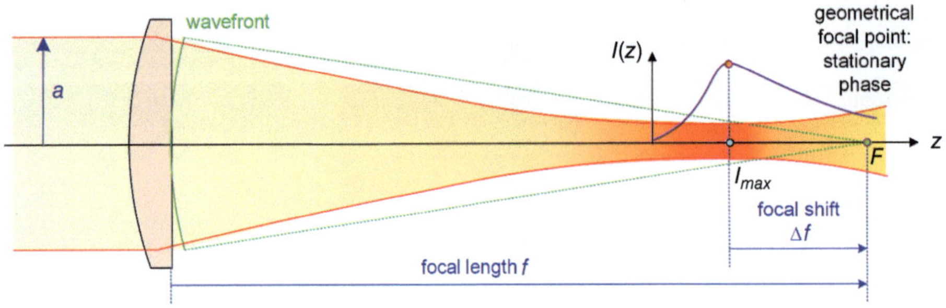

Figure 6.47 Principle of the focal shift effect as a balance between energetic weighting and phase mismatch.

but for the intensity the amplitude also matters and this changes through the $1/z$ influence. If one moves along the optical axis starting at the geometrical focal point F towards the lens, the stationary phase condition is violated and the constructive interference is deteriorated, but the $1/z$ advantage grows. This generates a maximum intensity closer to the system and the so-called focal shift Δf occurs. In the complete focal caustic, furthermore the $1/z$ factor creates an asymmetry of the intensity around the peak location.

Due to the asymmetry, for larger distances from the system, no condition of complete destructive interference in the Huygens principle exists and therefore there are no oscillatory zero points on axis for approximately $N_F < 4$; outside of the focal distance, the intensity $I(z)$ decreases smoothly to zero. On the other side, towards the system, the sequence of zero points grows and they are no longer equidistant. This behaviour is shown in Figure 6.48 for different Fresnel numbers. In Figure 6.49 the full $I(r,z)$ plot illustrates for three different Fresnel numbers the asymmetry along z more clearly also for finite values of the transverse coordinate v. In the case of the left side, for $N_F = 50$ no asymmetry can be resolved; for the right case with $N_F = 7$ it is seen rather clearly.

According to the usual calculation scheme of the axis intensity, the normalized axial coordinate u relative to the focal point is formulated now as

$$u = 2\pi \cdot N_F \cdot \frac{z}{z+f} \tag{6.121}$$

to scale the diffraction integral in an appropriate way. With this abbreviation, the intensity is given by

$$I(u) = I_0 \cdot \left(1 - \frac{u}{2\pi \cdot N_F}\right)^2 \cdot \left(\frac{\sin(u/4)}{u/4}\right)^2 \tag{6.122}$$

Due to the first factor, the intensity is no longer symmetric around $u = 0$. For the asymptotic case of very large Fresnel numbers, the first pre-factor is approximated 1, so the following definition results from Eq. (6.121) for $z \ll f$

$$u_\infty = \frac{2\pi \cdot N_F \cdot z}{f} = \frac{2\pi \cdot z \cdot \sin^2 u}{\lambda} \tag{6.123}$$

which is used in most practical cases. A parameter η can be defined to quantify the asymmetry with the help of the notations of Figure 6.50(a) in the normalized form

$$\eta = \frac{\Delta u_{right} - \Delta u_{left}}{\Delta u_{right} + \Delta u_{left}} \tag{6.124}$$

The definition of the Δu values here follows a FWhM definition of the intensity as indicated in the figure.

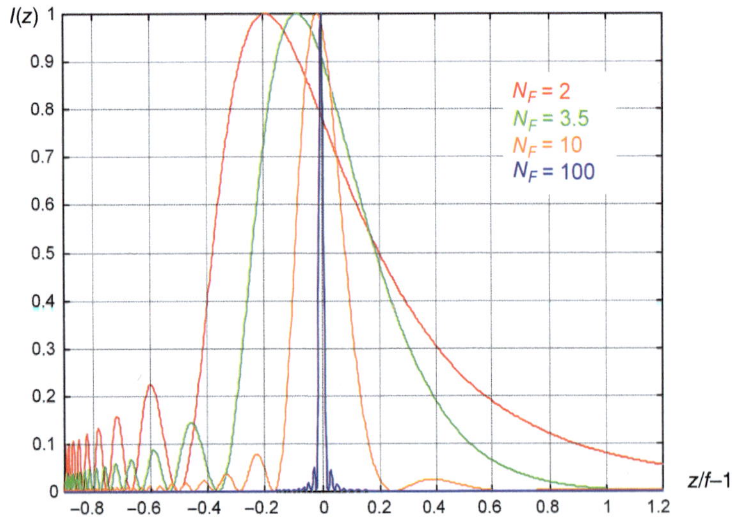

Figure 6.48 Axial intensity $I(z)$ for different Fresnel numbers.

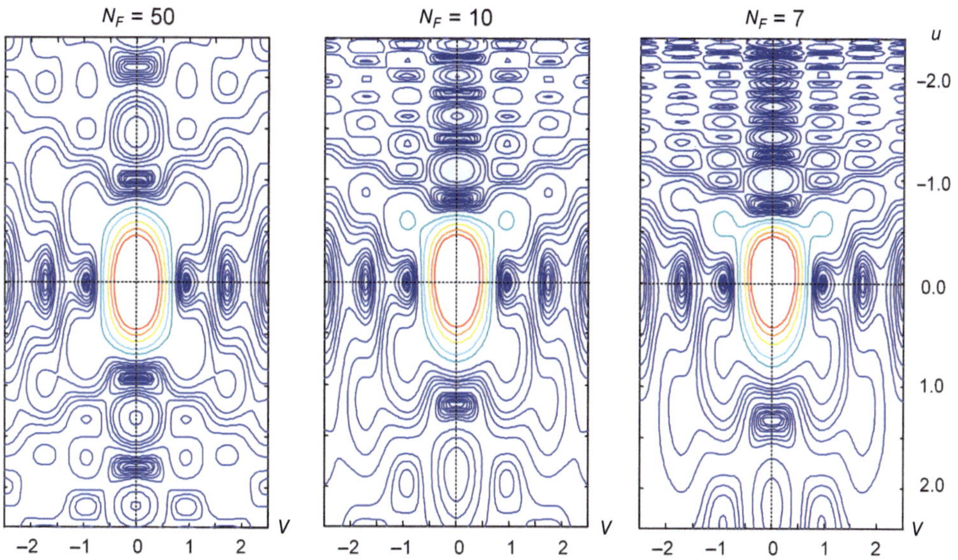

Figure 6.49 Axial intensity distribution for a circular pupil with uniform illumination for the Fresnel numbers $N_F = 7$, 10 and 50. For $N_F = 50$ no asymmetry is seen; for $N_F = 7$ the asymmetry is already significant.

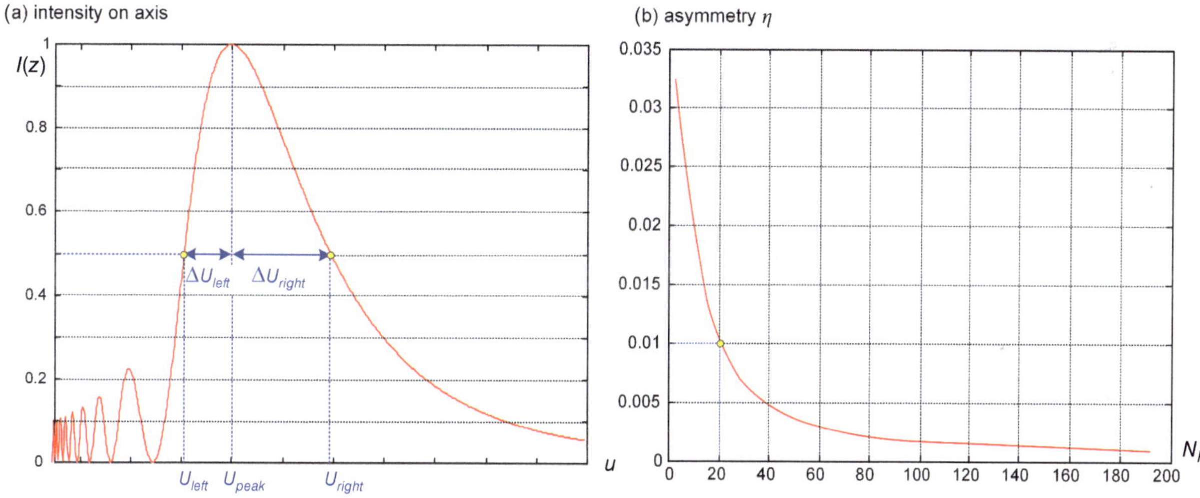

Figure 6.50 Low N_F axial intensity distribution with (a) definition of the asymmetry of the axial intensity distribution and (b) asymmetry parameter as a function of the Fresnel number.

We thus get the dependence of the asymmetry on the Fresnel number as is shown in Figure 6.50(b) as a result of numerical calculations. The asymmetry is larger than 1% for Fresnel numbers below $N_F < 20$.

6.5.2 Focal Shift

As is explained above, the asymmetry of the focal intensity distribution has its explanation in the asymmetric behaviour of the photometric law of distance

$$I(z) \propto \frac{1}{z^2} \tag{6.125}$$

The corresponding focal point must now be defined by more physical means. One meaningful possibility is to look for the maximum intensity on axis, given by the condition

$$\frac{\partial I(0,z)_{foc}}{\partial z} = 0 \tag{6.126}$$

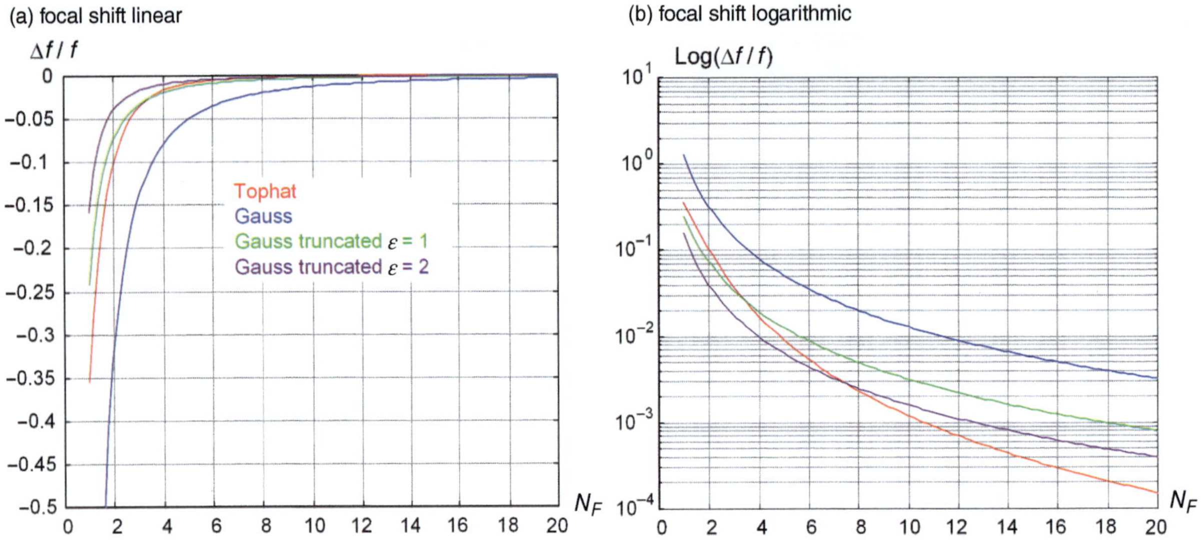

Figure 6.51 Focal shift effect as a function of the Fresnel number in (a) linear and (b) logarithmic representation for the given four different apodization profiles.

This focal shift means that the best location to gather an image is not found in the geometrically predicted focal point. This is an effect that is well known for Gaussian beams. The exact value of the focal shift depends on the intensity profile. If the criterion in (6.126) is used and the analytical formulas are expanded for a uniform pupil illumination, Eq. (6.122) is obtained. Figure 6.51 illustrates the focal shift behaviour by numerical results as a function of the Fresnel number for four different amplitude apodization profiles in linear as well as logarithmic representation. But again, it should be clear that for these different pupil illuminations, the numerical aperture definition is not trivial and the effective values are different in these cases.

There are some empirical analytical approximations, in particular for a uniform illumination proposed in the literature. The formula

$$\frac{\Delta f}{f} = -\frac{1}{1 + 1/12 \cdot (\pi N_F)^2} \tag{6.127}$$

has an error of less than 1% for $NF > 12$, while the formula

$$\frac{\Delta f}{f} = -\frac{1}{1 + N_F} \tag{6.128}$$

has an error of less than 1% for $NF < 0.17$. For Gaussian beam apodization, the focal shift can be calculated analytically exact from the transformation formula as (see Chapter 8)

$$\frac{\Delta f}{f} = -\frac{f \cdot z_0^2}{(f+z) \cdot \left[(f+z)^2 + z_0^2\right]} \tag{6.129}$$

With the help of the beam equivalent Fresnel number N_w, given by

$$N_w = \frac{w^2}{\lambda \cdot f} \tag{6.130}$$

for an incoming beam with the waist position at the lens we get the relation

$$\frac{\Delta f}{f} = -\frac{1}{1 + \pi N_w^2} \tag{6.131}$$

which resembles Eq. (6.127) despite the pre-factor in the denominator.

6.6 Focusing at High Numerical Aperture

6.6.1 Introduction

If a system has a large numerical aperture angle, three effects change the model, which are in the classical approximation neglected for simplicity.

- The transfer of the amplitude from the entrance to the exit pupil is more complicated than in the paraxial regime [91–98]. According to Figure 6.52(a), a geometrical scalar mapping from the pupil vertex plane to the spherical pupil surface changes the size of the area element of a narrow bundle. This results in a first apodization factor, see next section.
- On the nearly spherical surface of the wave in the exit pupil, linear polarization field components behave in their projection differently in the x–z and the y–z plane respectively. Therefore, the field projection onto

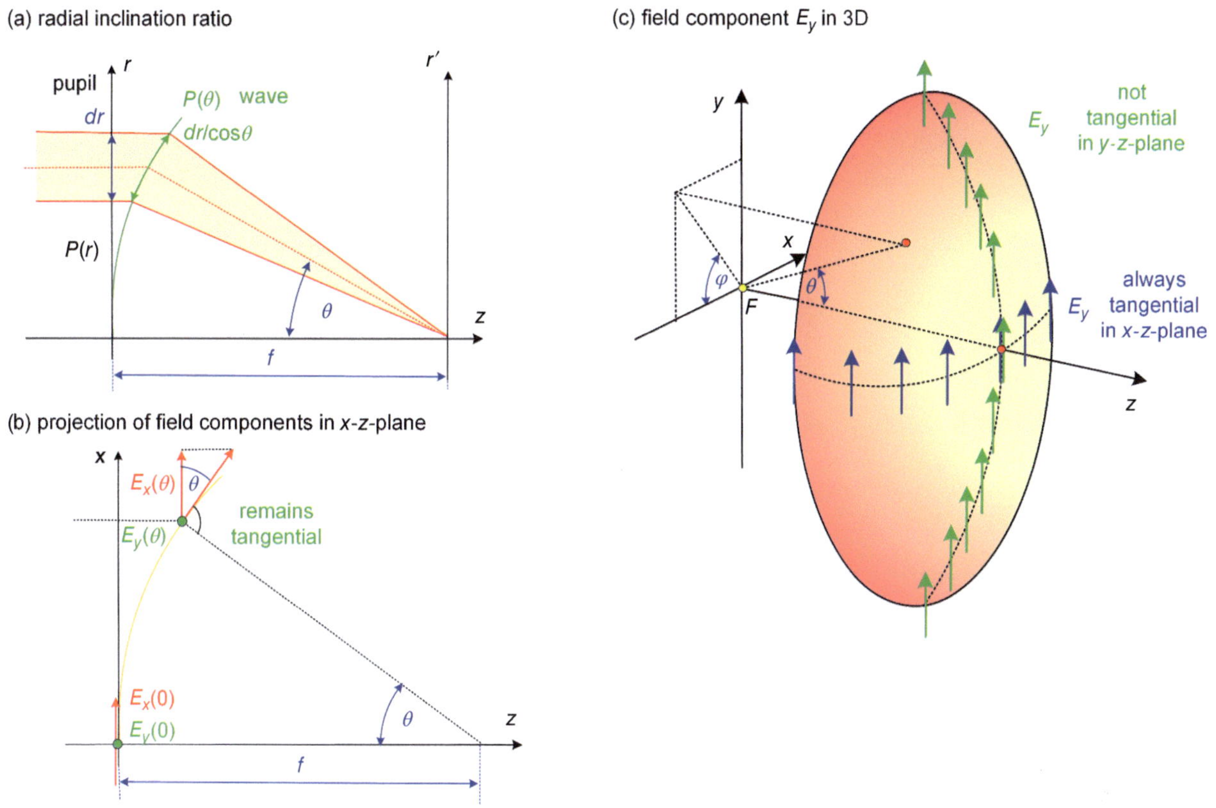

Figure 6.52 Photometric impact of (a) a tilted integration area for the diffraction integral and (b) the difference in projection for linear x-/y-polarization.

the spherical wave has an azimuthal dependence, which looks like astigmatism with a toroidal amplitude apodization. This is sketched in Figure 6.52(b) in one cross section and in Figure 6.52(c) in three-dimensional perspective. This is also described quantitatively in the next section [95, 96].

- The orthogonality of the field to the propagation direction according to the Maxwell equations is locally valid on every ray with an individual angle θ to the optical axis inside the bundle. For the complete cone with a large θ_{max} this has the two consequences that an axial field component E_z occurs and second the decoupling of x and y components is no longer fulfilled. As is sketched in Figure 6.53, this needs a switch between local (L) and global (G) coordinate systems for the field transfer from the object to the entrance pupil, through the system into the exit pupil and from the exit pupil in the image plane [97–103].

It should be mentioned that the mapping from the low-NA entrance pupil to the high-NA exit pupil (or the other way round) depends on the correction of the system. Mostly in lens design the goal is to fulfil the sine condition (see Chapter 17). In this case the exit pupil is spherical in shape. But this is not the only possibility, therefore different classical mapping functions can be distinguished. This is discussed in Section 6.6.3. The vectorial effects of the field propagation finally are treated in detail in Section 6.6.4.

6.6.2 Scalar Apodization Effects

In the high-NA case, a significant difference between the transverse coordinates in the exit pupil plane and the angle domain viewed from the focus point exists. Therefore, two apodization functions must be distinguished: the function $P(r)$ describes the amplitude distribution in the plane perpendicular to the axis as is usual in the case of small apertures using global coordinates. The mapping function can be written with x, y or r. If the wave is considered, the aperture angle θ matters according to the Debye integral and therefore an apodization function $P(\theta)$ is defined in spherical coordinates, which is relevant for the field in the focal point. Now the projection function $g(\theta)$ is defined as

$$g(\theta) = \frac{r}{f} \tag{6.132}$$

with the distance f between the pupil and the image. This function depends on the correction of the system. In the exact conventions of Chapter 3 this should be denoted by

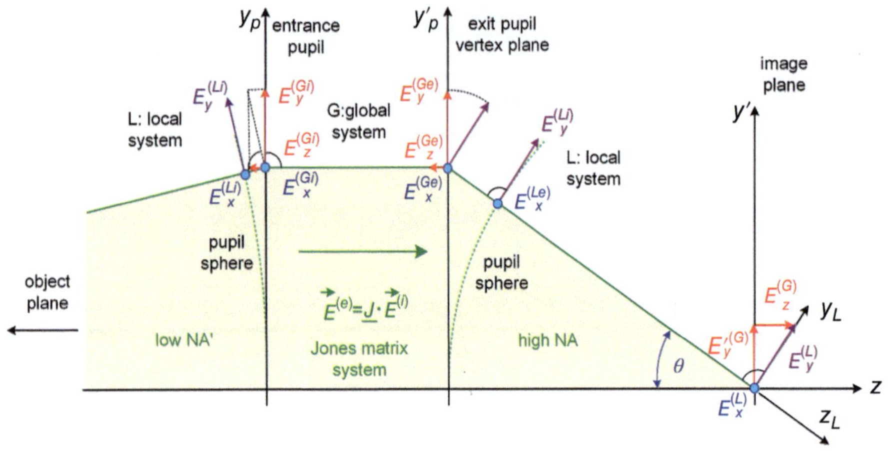

Figure 6.53 Tilted field component in a high-NA focusing geometry.

p, but using f is the usual abbreviation in literature and avoids here a confusion with the geometrical dimensionless amplitude apodization function P. The relation between $P(r)$ and $P(\theta)$ is obtained by forcing the preservation of energy in a narrow ring area element dS in the incoming as well as in the outgoing focused bundle, see Figure 6.52(a).

$$P^2(r) \cdot dS_{in} = P^2(\theta) \cdot dS_{out} \tag{6.133}$$

The amplitude change by the transform is given by

$$A(r) = A_{in}(r) \cdot P(r) \tag{6.134}$$

Now with the area elements

$$\begin{aligned} dS_{in} &= 2\pi r dr = 2\pi f^2 g(\theta) g'(\theta)\, d\theta \\ dS_{out} &= 2\pi f^2 \sin\theta d\theta \end{aligned} \tag{6.135}$$

with derivative $g'(\theta)$ and using Eq. (6.132) by inserting in (6.133) gives the desired relation between $P(r)$ and $P(\theta)$

$$P(\theta) = P(r) \cdot \frac{g(\theta)g'(\theta)}{\sin(\theta)} \tag{6.136}$$

If a spherical pupil shape is assumed for a fulfilled sine condition, we get

$$g(\theta) = \sin\theta, \quad P(\theta) = P(r)\sqrt{\cos\theta} \tag{6.137}$$

This relation can also be directly seen in Figure 6.52(a) by evaluating the tilted area element dr^*

$$dr^* = \frac{dr}{\cos\theta} = \frac{dr}{\sqrt{1-\sin^2\theta}} \tag{6.138}$$

The intensity changes are reciprocal to the area size; therefore, the angular amplitude apodization decreases with growing θ corresponding to the square root of the cosine according to

$$A(\theta) = A(r) \cdot P(r)/P(\theta) = A(r) \cdot \sqrt{\cos\theta} \tag{6.139}$$

In the special case that an initial linear x or y-polarization is assumed, there is an additional scalar effect, which stems from the tilt of the complex field components, as is shown in Figure 6.52(b) and (c). The reason for this vectorial effect is that the property of an electromagnetic field to be orthogonal polarized is valid locally on every ray in a light field. Therefore, every ray in a light cone has a different orientation θ, φ and projection of the field components onto the pupil sphere. It is seen in the figures that the x-field component remains tangential in the shown x–z plane, in the y–z plane a cosθ-projection is needed. This effect changes then during an azimuthal rotation. This azimuthal projection effect of the field causes an additional apodization on the exit pupil sphere corresponding to [95, 96]

$$P_{x-lin}(\theta,\varphi) = \frac{1}{2} \cdot [(1+\cos\theta) - (1-\cos\theta)\cdot\cos 2\varphi] \tag{6.140}$$

with the azimuthal angle φ and a resulting corresponding amplitude $A(\theta,\varphi)$. It should be noted that this effect quantitatively depends on the initial polarization and vanishes for an initial radial or tangential polarization with circular symmetry. The equation shows that the high-NA setup causes a second apodization and a non-constant mapping from the entrance to the exit pupil in the case of a linear initial polarization. Therefore, for imaging conditions with low-NA object side to a high-NA image side, the exit pupil has a non-uniform amplitude according to Figure 6.54 if it can be assumed that the initial energy density is uniform. The illumination has an astigmatism-like shape and causes an elliptical PSF in the image plane for a linear initial polarization which grows with NA.

These scalar apodization effects have some consequences for the usual imaging and performance properties of the system.

- The depth of focus is enlarged and has the equivalent Rayleigh range

$$R_u = \frac{\lambda}{4n\cdot(1-\cos\theta_{max})} \tag{6.141}$$

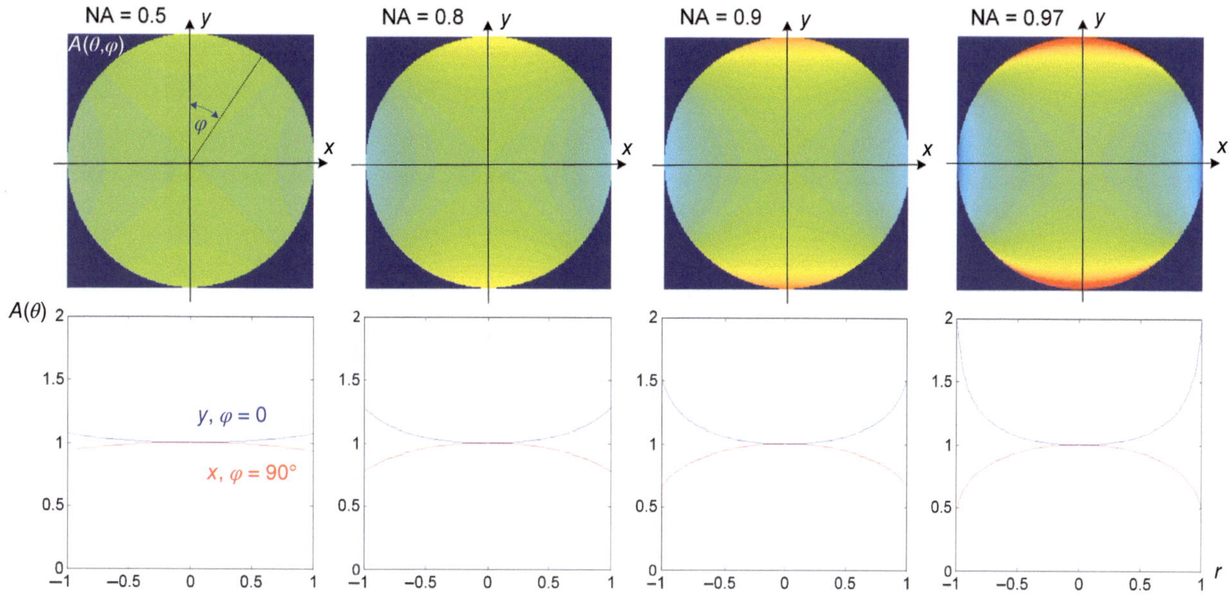

Figure 6.54 Amplitude $A(\theta,\varphi)$ as scalar photometric impact of a high NA focusing in the exit pupil for a uniform initial amplitude and linear x-polarization.

- The Strehl ratio always has to be calculated taking the apodization into account, according to the modified Marechal formula of Lowenthal (see Chapter 18) in global cylindrical coordinates

$$D_S = \frac{\left| \frac{1}{\pi} \int_0^{2\pi} \int_0^1 A(r,\varphi) \cdot e^{2\pi i W(r,\phi)} r\, dr\, d\varphi \right|^2}{\left| \frac{1}{\pi} \int_0^{2\pi} \int_0^1 A(r,\varphi) r\, dr\, d\varphi \right|^2} \tag{6.142}$$

- The best image location has to be changed in the case of residual phase aberrations. The classical predictions are no longer true. If the relative power of the linear polarized components are given by

$$P_x = \int_0^{2\pi} \int_0^1 |A_x(r,\varphi)|^2 r\, dr\, d\varphi,$$
$$P_y = \int_0^{2\pi} \int_0^1 |A_y(r,\varphi)|^2 r\, dr\, d\varphi \tag{6.143}$$

with the total power

$$P = P_x + P_y \tag{6.144}$$

the Strehl ratio must be calculated for every polarization separately, because orthogonal field components cannot interfere

$$D_{Sx} = \frac{\left| \frac{1}{\pi} \int_0^{2\pi} \int_0^1 A_x(r,\varphi) \cdot e^{2\pi i W_x(r,\varphi)} r\, dr\, d\varphi \right|^2}{\left| \frac{1}{\pi} \int_0^{2\pi} \int_0^1 A_x(r,\varphi) r\, dr\, d\varphi \right|^2},$$

$$D_{Sy} = \frac{\left| \frac{1}{\pi} \int_0^{2\pi} \int_0^1 A_y(r,\varphi) \cdot e^{2\pi i W_y(r,\varphi)} r\, dr\, d\varphi \right|^2}{\left| \frac{1}{\pi} \int_0^{2\pi} \int_0^1 A_y(r,\varphi) r\, dr\, d\varphi \right|^2} \tag{6.145}$$

The total Strehl ratio therefore reads as a weighted sum of x and y-field components

$$D_S = \frac{P_x}{P} D_{Sx} + \frac{P_y}{P} D_{Sy} \tag{6.146}$$

The following Figure 6.55 shows some examples and the corresponding differences between the paraxial and the high-NA formulation in axial and transverse direction. First the changes of the lateral intensity are shown as well as the axial intensity profiles in the case of larger aperture angles θ in comparison to the paraxial small angle case. It is seen that the impact of the large aperture is much larger in the axial direction. Figure 6.56 shows in particular the axial case in a better resolved logarithmic representation. Here it is seen that the changes are:

- the side lobe heights are different;
- the location of the side lobes is shifted along z;
- there are no longer exact zero intensity points.

It should be noted that in the case of large angles, the usually used normalized coordinates, which are used in the figure, must now be redefined in contrast to Eq. (6.110) as

$$v = \frac{2\pi}{\lambda} r \sin\theta, \quad u = \frac{8\pi}{\lambda} z \sin^2\frac{\theta}{2} \tag{6.147}$$

Especially the u-definition is now different.

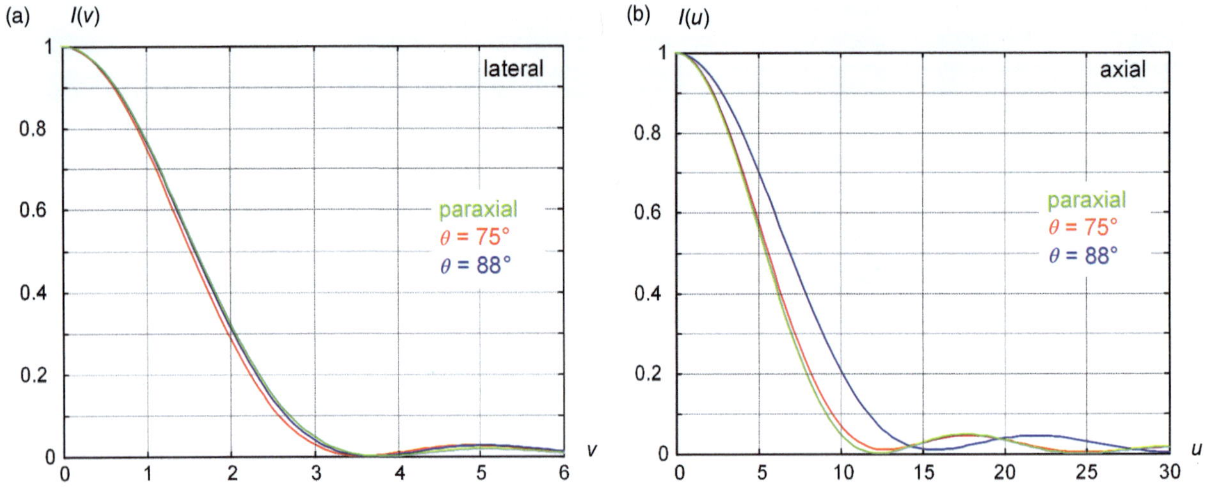

Figure 6.55 Change of the (a) lateral and (b) axial intensity distributions in case of high NA focusing together with the small angle approximation. The used coordinates are the scaled values *u* and *v*.

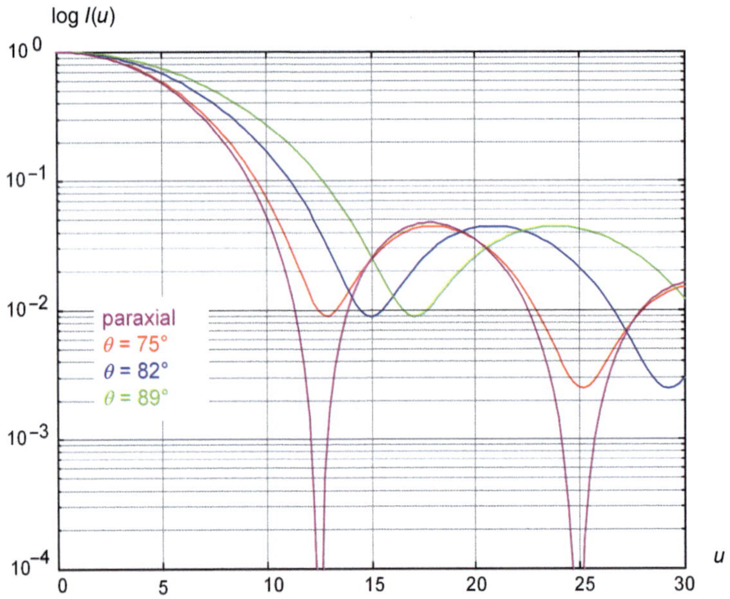

Figure 6.56 Change of the axial intensity distributions in case of high NA focusing together with the small angle approximation in a logarithmic representation.

6.6.3 Natural Mapping Function*

(*) The content of this section is quite specialized and for advanced reading. Knowledge of this section is not necessary to understand the following chapters. Beginners are recommended to skip this section.

In a more general case of high NA the calculation of the PSF is more meaningfully formulated as the Debye integral, which integrates over the angle directions onto the wave in the exit pupil. The projection effects of the area change are then of no concern and no problems arise with the definition of the complex field in a flat reference plane.

In the case of circular symmetry, the scalar diffraction integral according to Debye can be written as [31–34]

$$\begin{aligned} E(u,v) &= \frac{2\pi \cdot i \cdot e^{-ikz'}}{\lambda} \cdot \int_0^{\theta_{max}} A(r)/P(r) \cdot P(\theta,\varphi) \\ &\quad \cdot J_0\left(\frac{v \cdot \sin\theta}{\sin\theta_{max}}\right) \cdot e^{\frac{iu\cdot\sin^2\theta/2}{2\sin^2\theta_{max}/2}} \sin\theta d\theta \\ &= \frac{2\pi \cdot i \cdot e^{-ikz'}}{\lambda} \cdot \int_0^{\theta_{max}} A(\theta,\varphi) \\ &\quad \cdot J_0\left(\frac{v \cdot \sin\theta}{\sin\theta_{max}}\right) \cdot e^{\frac{iu\cdot\sin^2\theta/2}{2\sin^2\theta_{max}/2}} \sin\theta d\theta \end{aligned} \tag{6.148}$$

Table 6.2 Pupil mapping functions and apodization for different geometries.

	Sine condition	Herschel condition	Lagrange condition	Parabolic condition	Helmholtz condition or linear collineation
Spatial mapping function $g(\theta)$	$\sin\theta$	$2\sin\frac{\theta}{2}$	θ	$2\tan\frac{\theta}{2}$	$\tan\theta$
Pupil surface	Sphere			Parabola	Plane
Apodization factor $P(\theta)$	$\sqrt{\cos\theta}$	1	$\sqrt{\frac{\theta}{\sin\theta}}$	$\frac{1}{\cos^2\frac{\theta}{2}}$	$\frac{1}{\cos^{\frac{3}{2}}\frac{\theta}{2}}$

with the pupil mapping functions P and the initial amplitude $A(r)$ for the aperture angle θ and the generalized scaled coordinates according to Born and Wolf in lateral and axial direction as Eq. (6.147). In this formulation, high NA effects are incorporated.

The mapping functions of the pupil coordinates can be changed by the correction of the optical system and the fulfilment of the sine condition or alternative rules and correction schemes. In summary, there are several different possibilities for doing this and influencing the imaging performance properties of the system. The following Table 6.2 shows some of the most prominent cases. $g(\theta)$ describes the geometrical mapping from the entrance to the exit pupil, $P(\theta)$ the corresponding apodization function to correct the amplitude.

All these possible solutions are shown in Figure 6.57 for comparison. For small angles, all these conditions are equivalent, the differences are seen only for larger numerical apertures. The difference in the $r_p - \cos\theta$ relation is clearly indicated, which means all conditions in the Figure 6.57 have the same θ but different ray corresponding heights r_p.

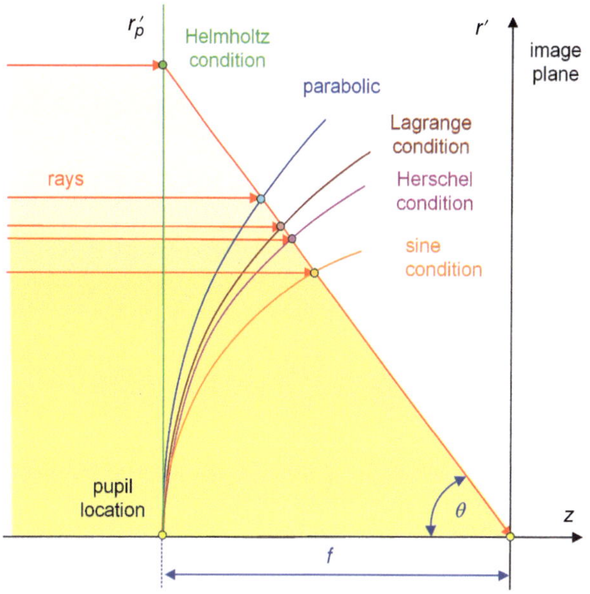

Figure 6.57 Change of the r_p - θ relation for high NA focusing for different mapping conditions.

6.6.4 Vectorial Diffraction Integral

6.6.4.1 Introduction

To compute the focus of electromagnetic fields, we consider an aplanatic optical system with a plane entrance pupil and a spherical exit pupil with fulfilled sine condition. The field distribution in the entrance pupil is described by a distribution of field vectors in pupil coordinates. Two-dimensional field vectors are sufficient because the electromagnetic field is transverse [41]. The geometry and the change of field components is sketched in Figure 6.53; a simplified picture is given in Figure 6.58.

It should be noted that there is a certain contradiction in the classical argumentation: if we allow a phase distribution of the electric field in the entrance pupil, the direction of propagation cannot be purely axial. The only exception is an object in infinity with a collimated input ray bundle on axis. For vector fields this results in a non-vanishing z-component, but for macroscopic pupils (with diameter of at least some mm) and phase errors around a few wavelengths, this component in reality is vanishingly small.

Two-dimensional, normalized dimensionless pupil coordinates are denoted by (p_x, p_y), which are related to the propagation angles of the corresponding plane wave in image space by

$$p_x = n \cdot \sin\theta \cdot \cos\varphi, \quad p_y = n \cdot \sin\theta \cdot \sin\varphi,$$
$$p_z = n \cdot \cos\theta, \quad p_r = \sqrt{p_x^2 + p_y^2} \leq n \cdot \sin\theta_m = NA \tag{6.149}$$

with the refractive index n in image space and the radial component pr. These are the direction unit vector components with the refractive index as an additional factor. The inverse relations are

$$\sin\theta = \frac{1}{n}\sqrt{p_x^2 + p_y^2}, \quad \sin\varphi = \frac{p_y}{\sqrt{p_x^2 + p_y^2}},$$
$$\cos\varphi = \frac{p_x}{\sqrt{p_x^2 + p_y^2}} \tag{6.150}$$

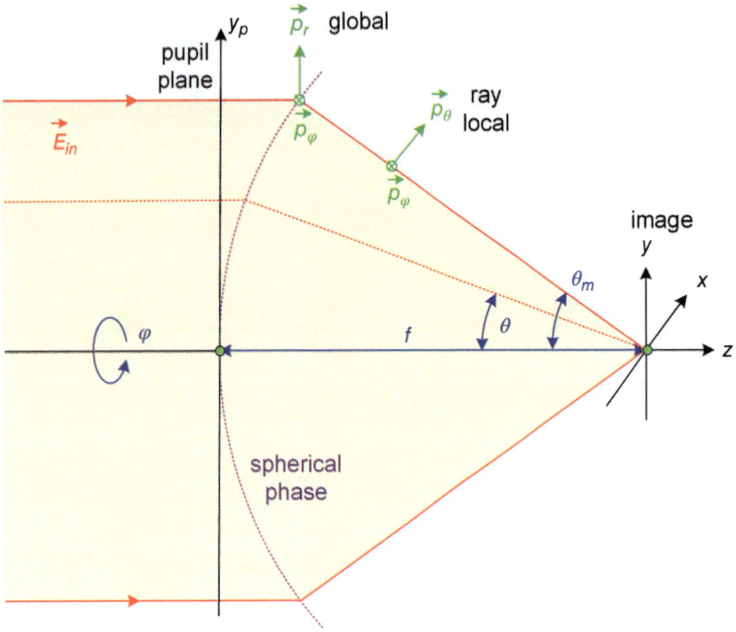

Figure 6.58 High NA focusing geometry.

6.6.4.2 Richards–Wolf Integral

The computation of the focus fields can be done along the lines of the classical paper by Richards and Wolf [99] on polar coordinates for circular pupils. The field components of the PSF follow from the Debye integral for an incoming field, which is decomposed into linearly polarized components in the form

$$\vec{E}(r,z) = \vec{E}_{0x} \cdot e^{-\frac{iu}{4\sin^2\theta_m/2}} \cdot \begin{pmatrix} -i\cdot(I_0 + I_2\cdot\cos 2\varphi) \\ -i\cdot I_2\cdot\sin 2\varphi \\ -i\cdot I_1\cdot\cos\varphi \end{pmatrix}$$
$$\vec{E}(r,z) = \vec{E}_{0y} \cdot e^{-\frac{iu}{4\sin^2\theta_m/2}} \cdot \begin{pmatrix} -i\cdot I_2\cdot\sin 2\varphi \\ -i\cdot(I_0 - I_2\cdot\cos 2\varphi) \\ -i\cdot I_1\cdot\sin\varphi \end{pmatrix} \tag{6.151}$$

where φ is the angle of the azimuth against the x-axis and θ_m is the maximum angle of the full numerical aperture. It is seen that I_1 is responsible for the axial effects and I_2 for the transverse changes. The following three auxiliary functions are used in the above expressions

$$I_0(r,z) = \int_0^{\theta_m} \sqrt{\cos\theta}\cdot\sin\theta\cdot(1+\cos\theta)\cdot J_0(kr\sin\theta)\cdot e^{ikz\cos\theta}\,d\theta$$
$$I_1(r,z) = \int_0^{\theta_m} \sqrt{\cos\theta}\cdot\sin^2\theta\cdot J_1(kr\sin\theta)\cdot e^{ikz\cos\theta}\,d\theta$$
$$I_2(r,z) = \int_0^{\theta_m} \sqrt{\cos\theta}\cdot\sin\theta\cdot(1-\cos\theta)\cdot J_2(kr\sin\theta)\cdot e^{ikz\cos\theta}\,d\theta \tag{6.152}$$

If the normalized coordinates u and v are used, the integrals I_0, I_1 and I_2 can be written in the scaled form

$$I_0(u,v) = \int_0^{\theta_m} \sqrt{\cos\theta}\cdot\sin\theta\cdot(1+\cos\theta)\cdot J_0\left(\frac{v\sin\theta}{\sin\theta_m}\right)\cdot e^{\frac{iu\sin^2\theta/2}{2\sin^2\theta_m/2}}\,d\theta$$
$$I_1(u,v) = \int_0^{\theta_m} \sqrt{\cos\theta}\cdot\sin^2\theta\cdot J_1\left(\frac{v\sin\theta}{\sin\theta_m}\right)\cdot e^{\frac{iu\sin^2\theta/2}{2\sin^2\theta_m/2}}\,d\theta$$
$$I_2(u,v) = \int_0^{\theta_m} \sqrt{\cos\theta}\cdot\sin\theta\cdot(1-\cos\theta)\cdot J_2\left(\frac{v\sin\theta}{\sin\theta_m}\right)\cdot e^{\frac{iu\sin^2\theta/2}{2\sin^2\theta_m/2}}\,d\theta \tag{6.153}$$

In the diffraction computation according to the Debye approximation, each point of the exit pupil is taken as the source of a plane wave with amplitude, phase and polarization prescribed by this point. A straightforward approach to compute the image field is to sum up the vector plane waves corresponding to different pupil points. It can be easily generalized to non-circular pupils of arbitrary polarization states and implemented into optical simulation routines.

In the formulation of Richards–Wolf, the azimuth-integration is done analytically and delivers the Bessel functions. Therefore, this representation cannot be used to calculate arbitrary aberrations in the pupil.

According to Török [95, 100], with the amplitude transmission factors τ_p and τ_s for the corresponding polarization directions and for the incoming field components E_x and

E_y, the field in the exit pupil reads in Cartesian component form

$$\vec{E}(\varphi_p,\theta_p) = \frac{1}{2}\cdot\begin{pmatrix} E_x\cdot[(\tau_s+\tau_p\cos\theta_p) - \cos 2\varphi_p\cdot(\tau_s-\tau_p\cos\theta_p)] - E_y\cdot\sin 2\varphi_p\cdot(\tau_s-\tau_p\cos\theta_p) \\ E_y\cdot[(\tau_s+\tau_p\cos\theta_p) + \cos 2\varphi_p\cdot(\tau_s-\tau_p\cos\theta_p)] - E_x\cdot\sin 2\varphi_p\cdot(\tau_s-\tau_p\cos\theta_p) \\ 2\tau_p\cdot\sin\theta_p\cdot(E_x\cos\varphi_p + E_y\sin\varphi_p) \end{pmatrix}$$
$$= \frac{E_x}{2}\cdot\begin{pmatrix} (\tau_s+\tau_p\cos\theta_p) - \cos 2\varphi_p\cdot(\tau_s-\tau_p\cos\theta_p) \\ -\sin 2\varphi_p\cdot(\tau_s-\tau_p\cos\theta_p) \\ 2\tau_p\cdot\sin\theta_p\cdot\cos\varphi_p \end{pmatrix} + \frac{E_y}{2}\cdot\begin{pmatrix} -\sin 2\varphi_p\cdot(\tau_s-\tau_p\cos\theta_p) \\ (\tau_s+\tau_p\cos\theta_p) + \cos 2\varphi_p\cdot(\tau_s-\tau_p\cos\theta_p) \\ 2\tau_p\cdot\sin\theta_p\cdot\sin\varphi_p \end{pmatrix} \tag{6.154}$$

with the pupil coordinates

$$x_p = r_p\cdot\cos\varphi_p,\quad y_p = r_p\cdot\sin\varphi_p \tag{6.155}$$

The angles $\varphi = \varphi_p$ and $\theta = \theta_p$ are the same as in the Richards–Wolf integral. The diffraction integral now has the form of a double angle integration according to the Debye representation

$$E(r,\varphi,z) = -\frac{ik}{2\pi}\cdot\int_0^{\theta_m}\int_0^{2\pi} E(\varphi_p,\theta_p)\cdot e^{ikr\sin\theta_p\cdot\cos(\varphi-\varphi_p)}\cdot e^{ikz\cos\theta_p}\,d\varphi_p d\theta_p \tag{6.156}$$

In the special case of a homogeneous illumination without further apodization, the integration over the azimuthal angle can be performed by the integrals

$$\int_0^{2\pi}\sin(m\alpha)\cdot e^{ia\cos(\alpha-\beta)}\,d\alpha = 2\pi i^m\cdot\sin(m\beta)\cdot J_m(a)$$
$$\int_0^{2\pi}\cos(m\alpha)\cdot e^{ia\cos(\alpha-\beta)}\,d\alpha = 2\pi i^m\cdot\cos(m\beta)\cdot J_m(a) \tag{6.157}$$

and the θ-integration uses the expressions with the Bessel functions, the auxiliary integrals now become

$$I_0(r,z) = \int_0^{\theta_m}\sqrt{\cos\theta}\cdot\sin\theta\cdot(1+\cos\theta)\cdot J_0(kr\sin\theta)\cdot e^{ikz\cos\theta}\,d\theta$$
$$I_1(r,z) = \int_0^{\theta_m}\sqrt{\cos\theta}\cdot\sin^2\theta\cdot J_1(kr\sin\theta)\cdot e^{ikz\cos\theta}\,d\theta$$
$$I_2(r,z) = \int_0^{\theta_m}\sqrt{\cos\theta}\cdot\sin\theta\cdot(1-\cos\theta)\cdot J_2(kr\sin\theta)\cdot e^{ikz\cos\theta}\,d\theta \tag{6.158}$$

and can be inserted into Eq. (6.151).

6.6.4.3 Cartesian Debye Integral Formulation

In the case of more complicated geometrical conditions, it makes sense to change the selected coordinate system from a cylindrical into a Cartesian one [97, 98, 103]. The propagation vector corresponding to the pupil coordinate (p_x, p_y) with $k_0 = 2\pi/\lambda$ in a medium with refractive index n is

$$\vec{k} = k_0\cdot(p_x\ p_y\ p_z)^T,\quad p_z = \sqrt{n^2 - p_x^2 - p_y^2} \tag{6.159}$$

While the electric field in the pupil coordinates is a two-component phasor, the electric field E_{ExP} in global coordinates in image space usually has three components and can be derived by application of a 3 × 2 transformation matrix T. The transformation matrix follows the change in the direction of the propagation vector. The succession of three rotational matrices with $\varphi - \theta - (-\varphi)$ that provides this change of direction is given by

$$\vec{E}_{ExP} = \begin{pmatrix}\cos\varphi & -\sin\varphi & 0\\ \sin\varphi & \cos\varphi & 0\\ 0 & 0 & 1\end{pmatrix}\begin{pmatrix}\cos\theta & 0 & \sin\theta\\ 0 & 1 & 0\\ -\sin\theta & 0 & \cos\theta\end{pmatrix}\times\begin{pmatrix}\cos\varphi & \sin\varphi & 0\\ -\sin\varphi & \cos\varphi & 0\\ 0 & 0 & 1\end{pmatrix}\begin{pmatrix}E_x\\ E_y\\ 0\end{pmatrix} = T(\theta,\varphi)\cdot\begin{pmatrix}E_x\\ E_y\\ 0\end{pmatrix} \tag{6.160}$$

and yields the transformation matrix

$$T = \begin{pmatrix}\cos^2\varphi\cos\theta + \sin^2\varphi & -\sin\varphi\cos\varphi\cdot(1-\cos\theta)\\ -\sin\varphi\cos\varphi\cdot(1-\cos\theta) & \sin^2\varphi\cos\theta + \cos^2\varphi\\ -\cos\varphi\sin\theta & -\sin\varphi\sin\theta\end{pmatrix} \tag{6.161}$$

It should be noted that the field E has according to Eq. (6.160) three field components. Finally, the diffraction integral in Debye formulation

$$\vec{E}(\vec{r}) = \iint_{EP}\frac{1}{\sqrt{\cos\theta}}T(p_x,p_y)\vec{E}_{ExP}(p_x,p_y)e^{i\cdot\vec{k}\cdot\vec{r}}\,dp_x dp_y \tag{6.162}$$

is obtained for the image plane, which can be implemented as an inverse Fourier transform [15]

$$\vec{E}(x,y,z) = \iint_{EP}\vec{E}(p_x,p_y)e^{i\cdot k_0(xp_x+yp_y+zp_z)}\,dp_x dp_y = \hat{F}^{-1}\left[\vec{E}(p_x,p_y)e^{i\cdot kzp_z}\right] \tag{6.163}$$

with

$$\vec{E}(p_x, p_y) = \sqrt{\frac{n}{p_z}} \cdot T(p_x, p_y)\vec{E}_{ExP}(p_x, p_y) \tag{6.164}$$

The intensity distribution in the focus is the sum of the intensities of the three field components

$$I(\vec{r}) = I_x + I_y + I_z = |E_x(\vec{r})|^2 + |E_y(\vec{r})|^2 + |E_z(\vec{r})|^2 \tag{6.165}$$

It should be emphasized here that the three field components in the focus are all perpendicular to each other and therefore cannot interfere. Therefore, the intensity calculation follows Eq. (6.165).

6.6.4.4 Formulation of Mansuripur

With the unit vector p in the pupil plane and the incident field in the entrance pupil

$$\vec{E}_p = \begin{pmatrix} E_{px} & E_{py} & 0 \end{pmatrix}^T \tag{6.166}$$

the unit direction vector after focusing in the exit pupil reads [97, 98]

$$p'_x = \begin{pmatrix} 1 - \frac{p_x^2}{1+p_z} \\ \frac{-p_x p_y}{1+p_z} \\ -p_x \end{pmatrix}, \quad p'_y = \begin{pmatrix} \frac{-p_x p_y}{1+p_z} \\ 1 - \frac{p_y^2}{1+p_z} \\ -p_y \end{pmatrix} \tag{6.167}$$

The field in the image plane is then given by

$$\begin{pmatrix} E'_x \\ E'_y \\ E'_z \end{pmatrix} = \iint \frac{1}{\sqrt{p_z}} \cdot \begin{pmatrix} 1 - \frac{p_x^2}{1+p_z} & \frac{-p_x p_y}{1+p_z} \\ \frac{-p_x p_y}{1+p_z} & 1 - \frac{p_y^2}{1+p_z} \\ -p_x & -p_y \end{pmatrix} \cdot \begin{pmatrix} \hat{F}[E_{ExP,x}(p_x, p_y)] \\ \hat{F}[E_{ExP,y}(p_x, p_y)] \end{pmatrix} dp_x dp_y \tag{6.168}$$

This expression corresponds directly to the T-matrix formulation of the Debye integral in the previous section as a result of the orientation changes. Explicitly we get the three components according to Mansuripur as

$$E'_x = \iint \frac{e^{2\pi i\cdot\left(xp_x + yp_y + z\sqrt{1-p_x^2-p_y^2}\right)}}{\sqrt[4]{1-p_x^2-p_y^2}} \cdot \left[E_{ExP,x} \cdot \frac{p_x^2\sqrt{1-p_x^2-p_y^2} + p_y^2}{p_x^2+p_y^2} + E_{ExP,y} \cdot \frac{-p_x p_y\left(1-\sqrt{1-p_x^2-p_y^2}\right)}{p_x^2+p_y^2} \right] dp_x dp_y$$

$$E'_y = \iint \frac{e^{2\pi i\cdot\left(xp_x + yp_y + z\sqrt{1-p_x^2-p_y^2}\right)}}{\sqrt[4]{1-p_x^2-p_y^2}} \cdot \left[E_{ExP,x} \cdot \frac{-p_x p_y\left(1-\sqrt{1-p_x^2-p_y^2}\right)}{p_x^2+p_y^2} + E_{ExP,y} \cdot \frac{p_y^2\sqrt{1-p_x^2-p_y^2} + p_x^2}{p_x^2+p_y^2} \right] dp_x dp_y$$

$$E'_z = \iint \frac{-e^{2\pi i\cdot\left(xp_x + yp_y + z\sqrt{1-p_x^2-p_y^2}\right)}}{\sqrt[4]{1-p_x^2-p_y^2}} \cdot \left[E_{ExP,x} \cdot p_x + E_{ExP,y} \cdot p_y\right] dp_x dp_y \tag{6.169}$$

It should be emphasized that the relative size of the vectorial effects according to Eq. (6.164) depends on the angle θ_m, not the numerical aperture, although the equations above use the scaled variables p for convenience.

6.6.4.5 Relative Size of Vectorial Effects

As was seen in the previous sections, a full vectorial calculation is much more laborious than a scalar approach. Therefore, in practice, very often the question arises whether the vectorial model is needed and what the error size is if the scalar model is used. The following figures illustrate the relative size of vectorial effects as a normalized deviation of the intensity in the focal region. In Figure 6.59 the sizes of the integral contributions I_1 for the axial effect and I_2 for the transverse changes according to the Richards–Wolf formulation normalized on I_0 are drawn. The values are plotted versus the aperture angle θ_m and are valid in the ideal image plane. On axis, the changes vanish.

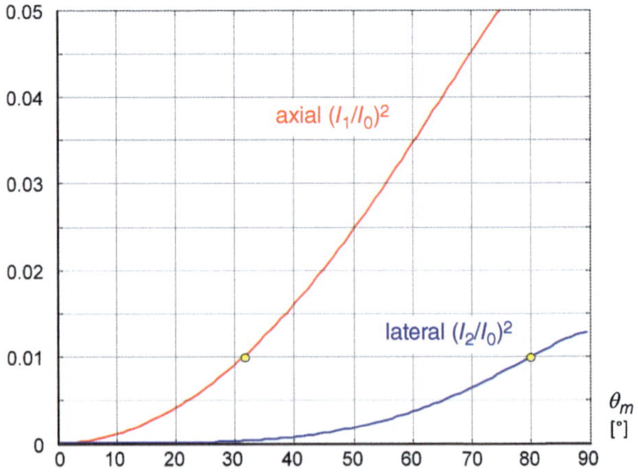

Figure 6.59 Relative size of the correcting Wolf integrals I_1 and I_2 as a function of the aperture angle θ_m in the image plane.

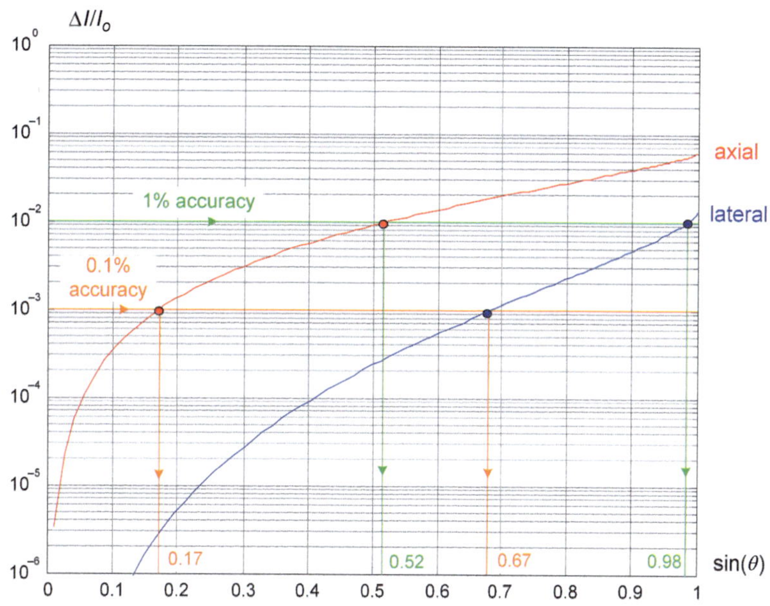

Figure 6.60 Relative *PV* error of the intensity values as a function of the aperture for the scalar paraxial formulation.

The corrections in the transverse direction are smaller than the axial effects. An effect of approximately 1% arises at 30° in the axial direction, and in the transverse direction at 80°.

Figure 6.60 shows the maximum intensity error obtained by the paraxial approach for the axial field component as well as the x–y cross coupling for a linear polarized input field as a function of the sine of the aperture angle. If a certain accuracy is requested, the corresponding maximum value for the critical NA can be determined. For example, an accuracy of 1% relative to the peak value is obtained as long as $\sin(\theta_m) < 0.52$. In Table 6.3 some of these limiting values are summarized numerically.

In principle, three regions can be roughly distinguished concerning the changes due to high numerical aperture; as a rule of thumb:

- $\sin(\theta_m) < 0.4$:the paraxial approximation is accurate better than approximately 0.6%;

Table 6.3 Relative intensity errors of the paraxial approach for various aperture angles for the axial and the cross coupled field strength.

$\sin(\theta_m)$	Error axial	Error lateral
0.2	0.0013	0
0.3	0.0031	0
0.4	0.0057	0.0001
0.5	0.0090	0.0002
0.6	0.0137	0.0005
0.7	0.0193	0.0011
0.8	0.0274	0.0023
0.9	0.0392	0.0048

  – $\sin(\theta_m) = 0.4\ldots0.6$:in this quasi high-NA regime the scalar extension with the apodization correction can be used, and the error made is not larger than approximately 1.5%;
- $\sin(\theta_m) > 0.6$:a vectorial consideration is necessary.

6.6.4.6 Example Calculation

In Figure 6.61 three different polarizations are calculated for a focusing aperture of $\sin(u) = 0.95$ corresponding to an angle of 65°. An initial linear polarization in x, a radial polarization and an azimuthal polarization are visualized for the three field component intensities and the superposed total intensity. It should be noted that the colour scales are different; every plot is renormalized. The following observations are interesting.

- In the linear polarized case, the total intensity is no longer rotational symmetric, the illumination polarization breaks the symmetry of the system.
- For the radial polarization, the z-component is the largest.
- For the azimuthal polarization, there is no z-component occurring (only numerical noise in the figure).
- For the azimuthal polarization, the total intensity is ring-shaped.

6.7 PSF for Compound Systems*

(*) The content of this section is quite specialized and for advanced reading. Knowledge of this section is not necessary to understand the following chapters. Beginners are recommended to skip this section.

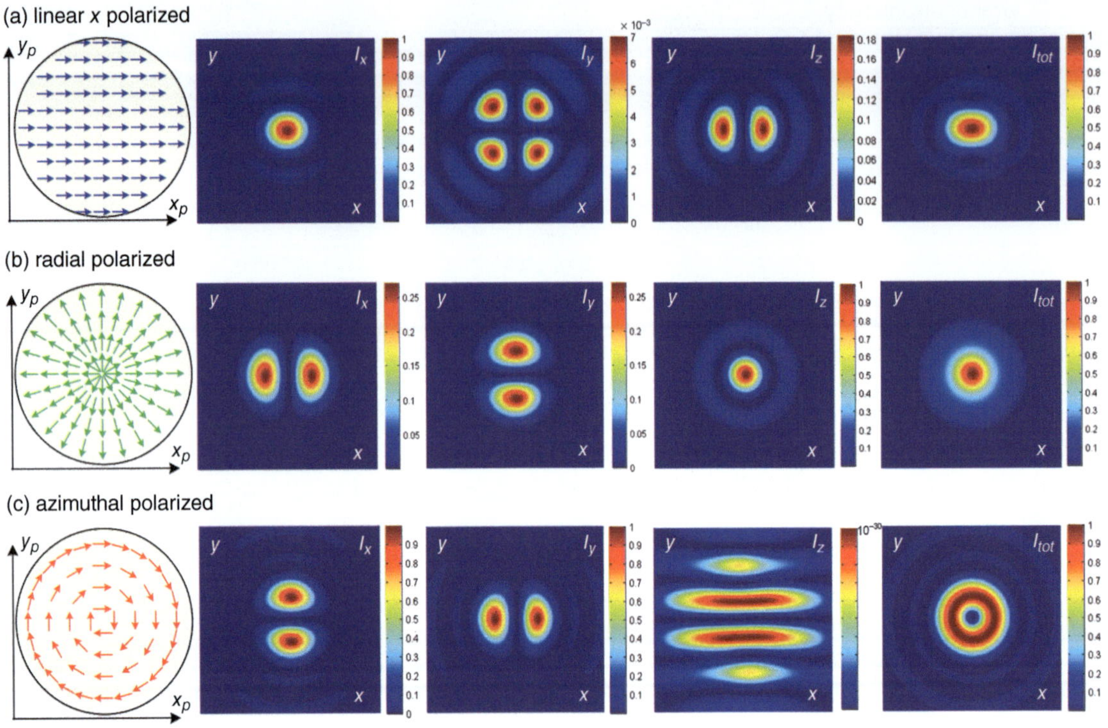

Figure 6.61 Vectorial example calculations with the three field component as well as the total intensity for three different polarizations (a) linear in *x*, (b) radial and (c) azimuthal.

6.7.1 Introduction

If two or more systems are connected and used together, the question is how the PSF of the individual system parts can be overlaid. This is a non-trivial question and depends on the following.

- The interface.
  There is a difference if between both sub-systems only an artificial air surface is assumed or if a diffuser is located in the interface plane, destroying the coherence and the direction memory of the rays.
- The coherence.
  In the fully coherent case the phase aberrations are directly superposed at an air interface and are additive; in the case of a partially coherent or incoherent illumination, the interface plane serves as a new secondary light source.
- The correction.
  If the first system is corrected in the sense that the numerical ray grid is undistorted, the wave aberrations can be directly added. In the case of a considerable distortion of the grid, care must be taken in simply adding the aberration results of the sub-systems, because induced effects occur.

This means that the conditions of coupling are critical and must be inspected carefully. In principle a simple connection of the systems and a calculation of the raytrace in the complete system is always possible, but for reasons of modularity and efficiency, it is often desirable to use the results of the system parts to evaluate the compound system.

6.7.2 Coherent Field Propagation Through a Single Thin Lens

Firstly, the coherent field propagation through a single thin lens is considered. For simplicity, the formulae are formulated with one transverse coordinate only. The field is propagated from the object plane (index *o*) to the image plane (*i*) or in general into an arbitrary plane after the lens. In the plane of the lens, the coordinates are x_p, y_p. The setup is sketched in Figure 6.62.

The propagation of the field from the initial object plane into the plane in front of the lens in Fraunhofer approximation can be written as

$$E(x_p) = -\frac{1}{ikz_o} e^{ikz_o} e^{+\frac{\pi i x_p^2}{\lambda z_o}} \int E(x_o) \cdot e^{+\frac{2\pi i}{\lambda z_o} x_o x_p} dx_o$$
$$= -\frac{1}{ikz_o} e^{ikz_o} e^{+\frac{\pi i x_p^2}{\lambda z_o}} \cdot \hat{F}[E(x_o)]_{\frac{x_o}{\lambda z_o}} \qquad (6.170)$$

where it is assumed that the pupil is located at the thin lens and both planes are Fourier conjugated and it should be taken into account that $z_o < 0$. The transfer of the field

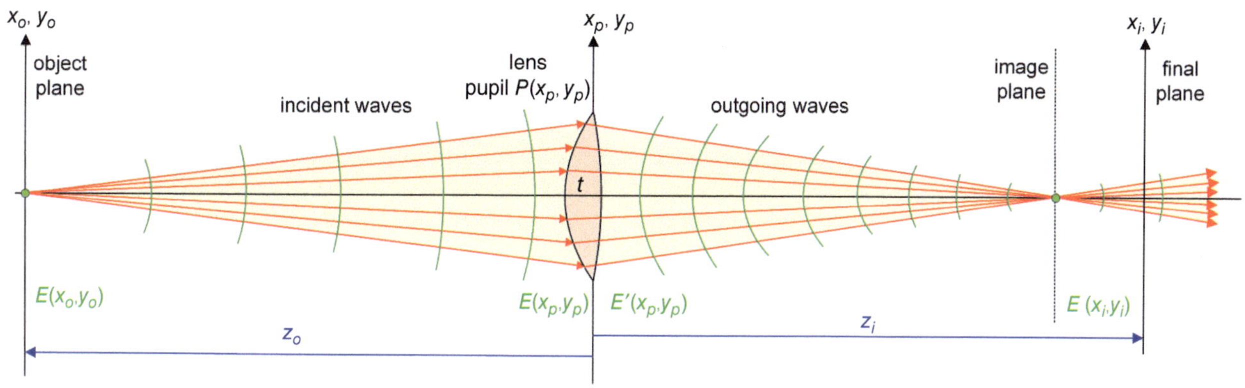

Figure 6.62 Coherent field propagation through a lens.

through the lens with the pupil function P containing transmission part T and phase filtering Φ given by

$$P(x_p) = T(x_p) \cdot e^{2\pi i \Phi(x_p)} \tag{6.171}$$

is simply a multiplication and gives

$$E'(x_p) = E(x_p) \cdot T(x_p) \cdot e^{2\pi i\, \Phi(x_p)} \tag{6.172}$$

In case of a non-perfect lens, we can in particular separate the wave aberration W from the lens power

$$P(x_p) = T(x_p) e^{+\frac{\pi i\, x_p^2}{\lambda f} + 2\pi i W(x_p)} \tag{6.173}$$

The propagation into the final plane again in Fraunhofer approximation can be written as

$$\begin{aligned} E(x_i) &= \frac{1}{ikz_i} e^{ikz_i} e^{-\frac{\pi i x_i^2}{\lambda z_i}} \int E'(x_p) e^{-\frac{\pi i x_p^2}{\lambda z_i}} \cdot e^{-\frac{2\pi i}{\lambda z_i} x_i x_p} dx_p \\ &= \frac{1}{ikz_i} e^{ikz_i} e^{-\frac{\pi i x_i^2}{\lambda z_i}} \cdot \hat{F}\left[E'(x_p) e^{-\frac{\pi i x_p^2}{\lambda z_i}} \right]_{\frac{x_p}{\lambda z_i}} \end{aligned} \tag{6.174}$$

By inserting the fields and eliminating the field in the intermediate planes we have

$$\begin{aligned} E(x_i) = {} & \frac{e^{-\frac{\pi i x_i^2}{\lambda z_i}}}{ikz_i} e^{ikz_o} \frac{1}{ikz_o} e^{ikz_i} \int T(x_p) e^{-\frac{\pi i x_p^2}{\lambda z_i}} \\ & \times e^{+\frac{\pi i\, x_p^2}{\lambda f} + 2\pi i W(x_p)} e^{-\frac{2\pi i}{\lambda z_i} x_i x_p} e^{+\frac{\pi i x_p^2}{\lambda z_o}} \int E(x_o) \\ & \cdot e^{+\frac{2\pi i}{\lambda z_o} x_o x_p} dx_o dx_p \end{aligned} \tag{6.175}$$

In the special case of conjugated planes, the lens makers' formula can be used and with the magnification m

$$\frac{1}{z_i} - \frac{1}{z_o} = \frac{1}{f}, \quad m = \frac{z_i}{z_o} \tag{6.176}$$

we finally get in the image plane

$$\begin{aligned} E(x_i) = {} & \frac{e^{-\frac{\pi i x_i^2}{\lambda z_i}}}{\lambda^2 z_i z_o} \iint E(x_o) T(x_p) \\ & \cdot e^{2\pi i W(x_p)} e^{-\frac{i\pi}{\lambda z_i}(x_i - m x_o) x_p} dx_p dx_o \end{aligned} \tag{6.177}$$

If only one illuminated point is considered, the initial field is given by a delta function

$$E(x_o) = E_o \cdot \delta(x_o) \tag{6.178}$$

and one of the integrals can be resolved. We get then the coherent PSF as the Fourier transform of the CTF, which is a scaled version of the pupil function

$$\begin{aligned} E_{PSF}(x_i) &= \frac{E_o e^{-\frac{\pi i x_i^2}{\lambda z_i}}}{\lambda^2 z_i z_o} \int T(x_p) \cdot e^{2\pi i W(x_p)} \cdot e^{-\frac{i\pi}{\lambda z_i} x_i x_p} dx_p \\ &= \frac{E_o e^{-\frac{\pi i x_i^2}{\lambda z_i}}}{\lambda^2 z_i z_o} \cdot \hat{F}[P(x_p)] = \frac{E_o}{\lambda^2 z_i z_o} \cdot \hat{F}[H_{CTF}(x_p)] \end{aligned} \tag{6.179}$$

In the last version, the defocusing factor is included in the CTF and we get therefore

$$H_{CTF}(\nu) = e^{-\frac{i\pi\, x_i^2}{\lambda z_i}} P(\nu) \tag{6.180}$$

6.7.3 Coherent Propagation Through a Cascaded System

It is seen from Eq. (6.179) that the quadratic phase factor destroys the isoplanatic invariance condition for a coherent transfer by principle, see Figure 6.63.

If two systems are concatenated as illustrated in Figure 6.64, therefore a simple double convolution model cannot be correct in general because the isoplantism for the second stage is not fulfilled. If the intermediate image lies in air and therefore the coherence is preserved in this plane, the second sub-system gets a coherent wave and the phase aberrations can be balanced between the two systems. It is possible that two bad systems in total create a well-corrected doublet system. But a double convolution would in any case broaden the PSF in the final image plane.

Only if the field of view in the intermediate image plane is very small and the quadratic phase factor can be neglected, can the classical convolution approach be

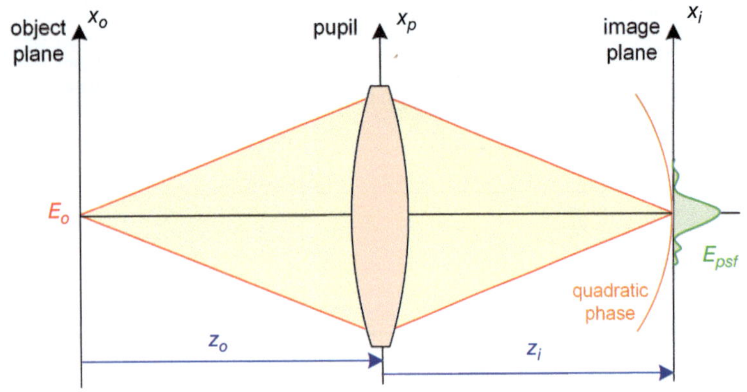

Figure 6.63 Coherent field propagation between two planes in an optical system.

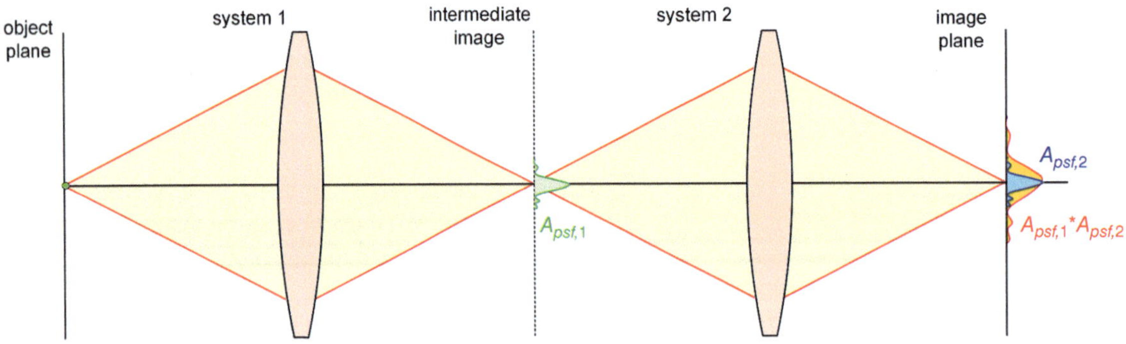

Figure 6.64 Common PSF of a coherent cascaded optical system.

applied twice. With the real apodization function T and the wave aberration W, the CTF can be written for every sub-system as

$$H_{CTF}(\nu) = T(\nu) \cdot e^{2\pi i W} \tag{6.181}$$

In the perfect case with a highly corrected first system and a small FoV, the overlay for the combined systems can be calculated simply as

$$\begin{aligned} T(\nu) &= T_1(\nu) \cdot T_2(\nu) \\ W(\nu) &= W_1(\nu) + W_2(\nu) \end{aligned} \tag{6.182}$$

as the apodization product and the sum of the wave aberrations of the complete system. From this last expression it can be seen that the aberrations are added with their sign and therefore compensation effects for correction the complete system are possible. The solution of the PSF depends only on the sum of the individual aberrations, if the mentioned assumptions are fulfilled.

Therefore, the convolution approach for extended fields is not valid by principle for two concatenated systems. But if only one field point is considered and no extended field of view, the question is how well Eq. (6.182) is fulfilled if the first system is not well corrected. If the residual aberrations of the first system have a considerable size, these deviations cause a modified use of the second system and a non-linear induced effect takes place, the so-called retrace error. Also the Zernike coefficients of the wave aberrations are no longer additive as predicted in Eq. (6.182). This is demonstrated in the following two examples. For a better understanding of the argumentation of the Zernike representation of the wave aberrations, see Chapter 15.

In the following two examples, first two well-corrected subsystems are concatenated. Figure 6.65 shows the layout of both systems and gives the major data of aperture and field of view. The Strehl number for the first system is $D_s^{(1)} = 0.9869$ and for the second system $D_s^{(2)} = 0.8989$.

It should be noted that by cascading the two sub-systems each with negative magnification, the Zernikes with an odd field dependence have to be subtracted. In this case of using the Fringe convention, this has to be done for the running indices j = 2,3,7,8,10,11,14,15. The maximum difference of the Zernike coefficients between the overlay of the individual systems and the total system is $\Delta c = 0.0047$ λ. In this case the additivity is fulfilled to a rather good accuracy and retrace errors can be neglected. This is shown in Figure 6.66, where the sum of the Zernikes as well as the Zernikes for the complete system are plotted for comparison.

If the PSFs are compared, the total system must be worse than the poorer one of the single systems. Figure 6.67 shows this for the example together with the convolution result for the axis point. The green and the orange curve are in the best case identical. The calculation is performed with N = 256 pixel and 16 Zernikes, therefore a small

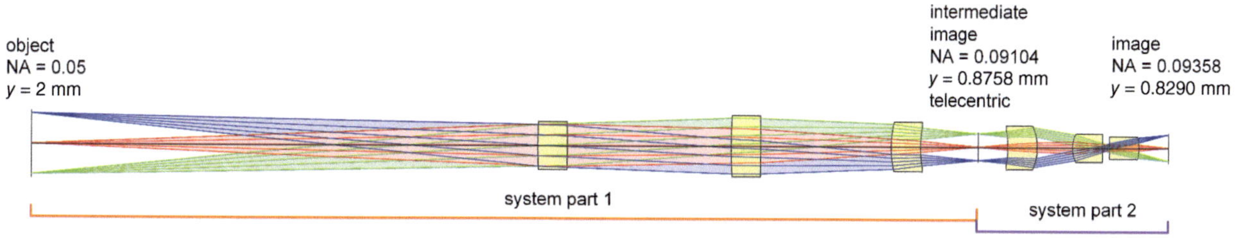

Figure 6.65 Example system with corrected intermediate image.

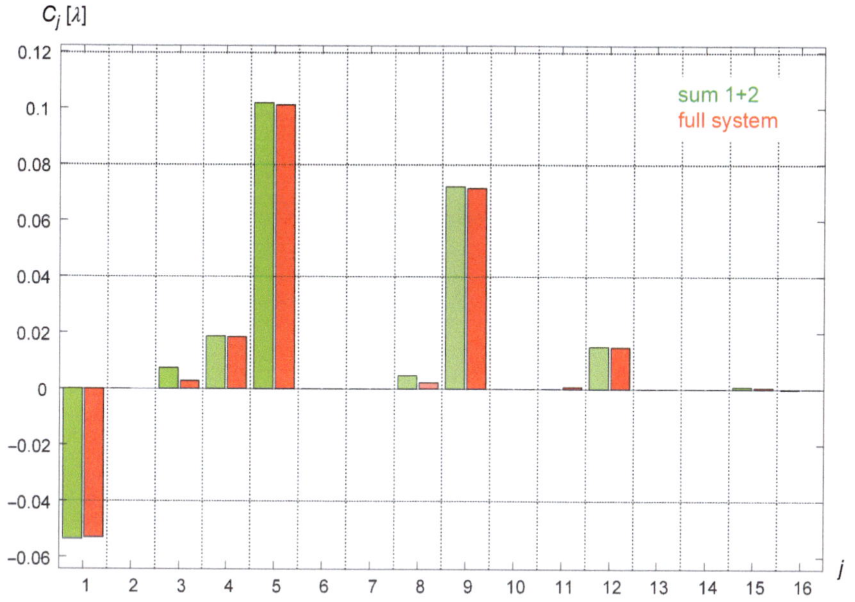

Figure 6.66 Additivity of the Zernikes 1–16 for the example system for the outer field point.

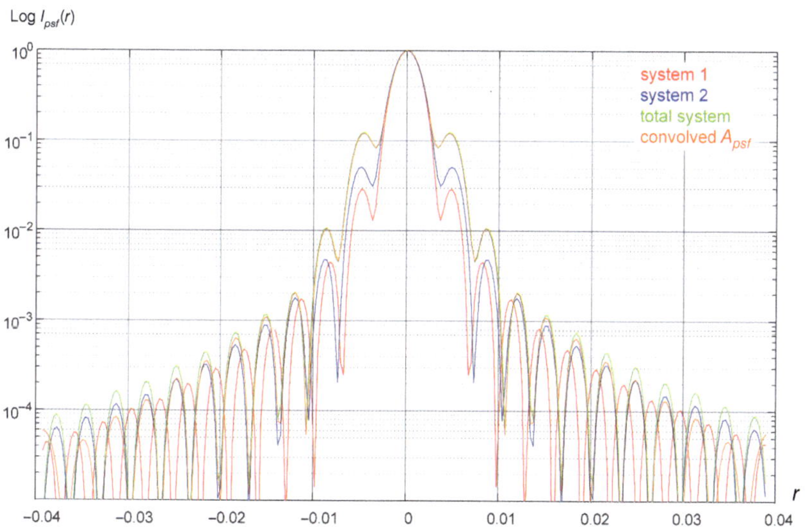

Figure 6.67 Point spread function on axis for the first system, the second system, the total system and the convolution of the PSFs of the first and second systems. The unit of r is arbitrary here.

discretization error may also be there. But the coincidence is really good; some deviations are seen in the order of magnitude of 10^{-3}. It should be noted here that the PSFs are here re-normalized to 1 on axis.

In the second example, the first system has a correction with only Strehl $D_s^{(1)} = 0.64$; the second system is identical to the previous example with $D_s^{(2)} = 0.8989$. The paraxial data are the same as in the first example; Figure 6.68 shows the new system.

In this case the additivity of the Zernikes is fulfilled with less precision, see Figure 6.69, because the residual aberrations of the first system create a larger retrace error in the second part. This means that the rays take a slightly modified path in the second system due to the aberrations of the first system. The uncorrected wave in the intermediate image generates different entry conditions for the second system, provoking modified aberrations.

If the PSFs are considered in Figure 6.70, the red curve now has larger side lobes and it can be seen that, due to the retrace error, in addition the coherent convolution in orange and the total system PSF in green show already significant differences in the range of 8%.

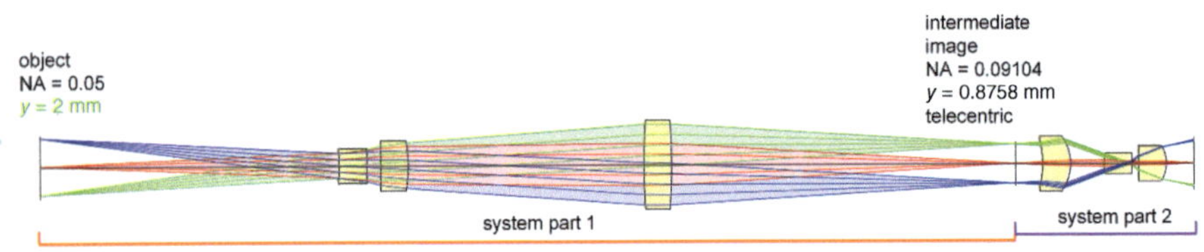

Figure 6.68 Second example system with less corrected intermediate image.

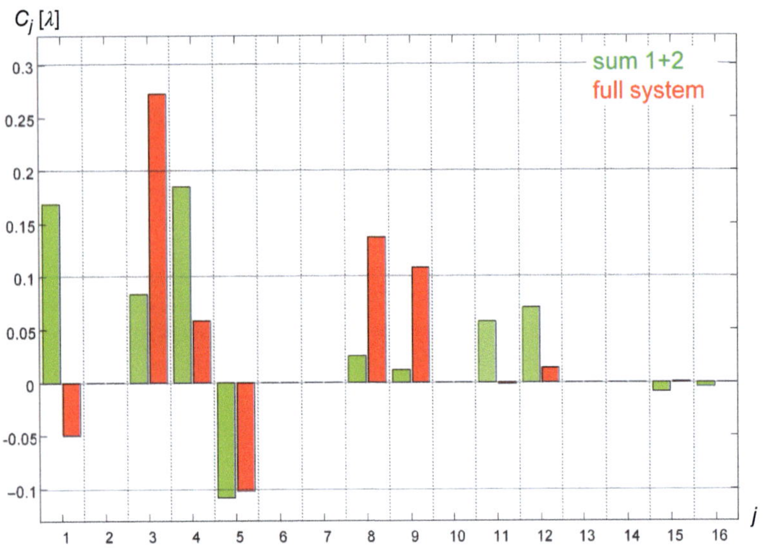

Figure 6.69 Zernikes in the second example system for the outer field point.

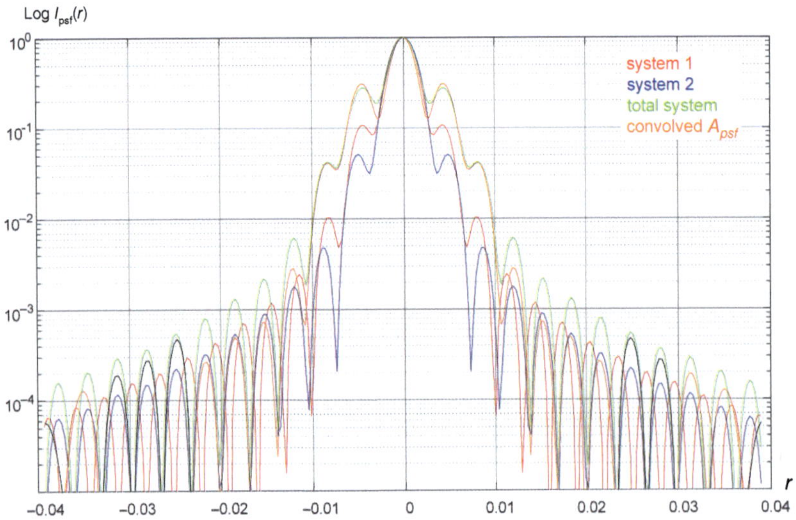

Figure 6.70 Point spread function on axis for the first system, the second system, the total system and the convolution of the PSFs of the first and second system. The unit of r is arbitrary here.

6.8 Cascaded Diffraction in Optical Systems*

(*) The content of this section is quite specialized and for advanced reading. Knowledge of this section is not necessary to understand the following chapters. Beginners are recommended to skip this section.

6.8.1 Introduction

As has been discussed before, every real optical system consists of several lateral truncating diaphragms like stops, mountings or mechanical parts. To simplify the computational scheme, the hybrid approach approximates the diffraction model and assumes all edge diffractions effects take place in the exit pupil only. But it is important to get a quantitative measure of how good this approximation is and which parameters are influencing the accuracy. Therefore, in this section a more sophisticated model of cascading diffraction effects is presented and discussed. Although the hybrid approximation has been used in nearly all classical design tools over the decades, only a small amount of research is found addressing this question [104–112]. We are following here the approach of [104, 105].

If two or more stops are located in a sequence, cascading effects of diffraction occur. For a two edge-stop setup the impact of cascading is illustrated in Figure 6.71. The second edge can be illuminated by primary signal light or by structured light due to the first edge diffraction. If the second stop is illuminated by straylight out of the geometrical shadow range, the impact is quite small. Otherwise, if the second stop gets primary signal light as sketched in Figure 6.71, the perturbed uniformity modifies the energy density at the edge. Due to the strong damping on the geometrical shadow side after an edge, the cases with two or more edges involved create only very low light levels. The size of the effects is decreasing in the corresponding order of interactions. Therefore, in the following we are only discussing the cascaded effect by two stops.

If, for example, a system with three stops is considered, according to Figure 6.72 the following diffraction contributions can be distinguished as edge diffraction effects, perturbing the direct light in green.

- Single diffraction at every direct illuminated stop edge (1 or 3, red and blue rays).
- Diffraction at one stop, the second interaction at a following stop lies inside the bright but modified direct light cascaded double diffraction 1–3 (purple ray).

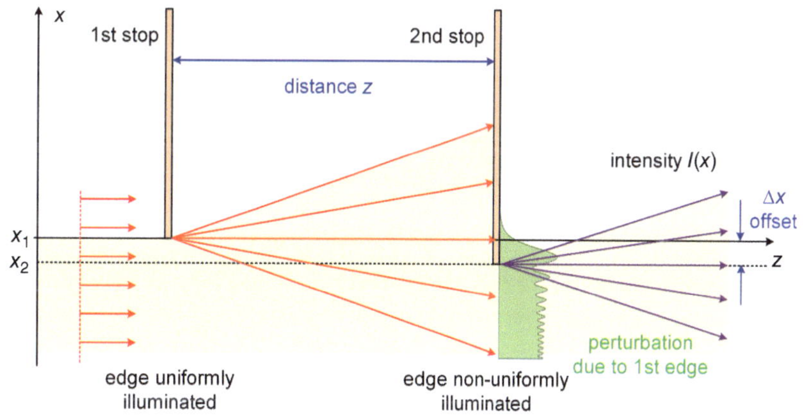

Figure 6.71 Cascaded diffraction, the second stop is illuminated by the light, which is affected by the diffraction at the first stop.

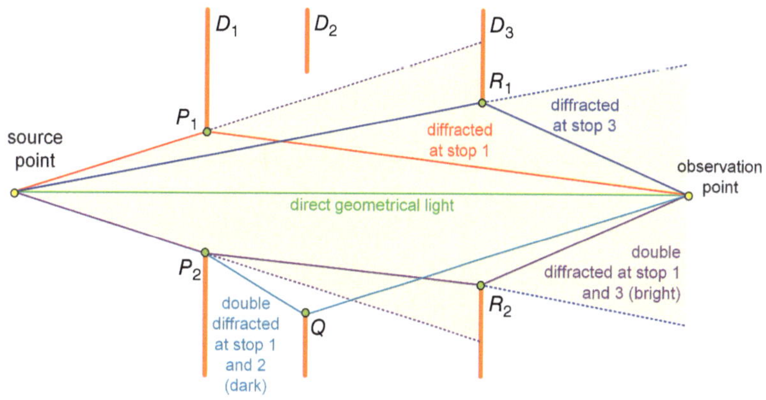

Figure 6.72 Cascaded diffraction in case of three stops; the various cases are discussed in the text.

- Diffraction at one stop, the second interaction at a following stop lies in the geometrical dark but finite illumination cone cascaded double diffraction; here, for example, 1–2 (dark cyan ray).

The main reason for cascaded diffraction effects is the slow convergence of the oscillations on the bright side of the pattern. The amplitude of the maxima is nearly independent from the distance z between the stops, but the widths of the maxima are decreasing. On the other hand, the light damping on the geometrical dark side is quite fast, see Figure 6.1. Therefore, the impact of the interaction between two stops strongly depends on the relative change of the stop sizes and locations. Mainly two scenarios are possible.

- If the second stop is larger in diameter, its edge is located in the shadow region of the first stop and the impact is small.
- If the second stop is smaller, its edge is fully illuminated and the oscillations of the amplitude due to the first stop are of larger influence.

6.8.2 System Model

Every lateral truncating diaphragm (real stops or lens mountings) creates a diffraction effect and a corresponding boundary wave, which spreads the propagating diffraction in the order of magnitude by an angle $\theta = \lambda/a$, see Figure 6.73. If the system is segmented into the parts between the effectively illuminated diaphragms, the wave can be propagated step by step with interfaces at the stops and probably lenses in between. This scheme is sketched in Figure 6.74. For simplicity the system segments between the stops are considered to be paraxial. This makes the wave-optical field propagation between the stops easier and can be calculated in Fresnel approximation with the *ABCD* matrices of the sub-systems. A more accurate

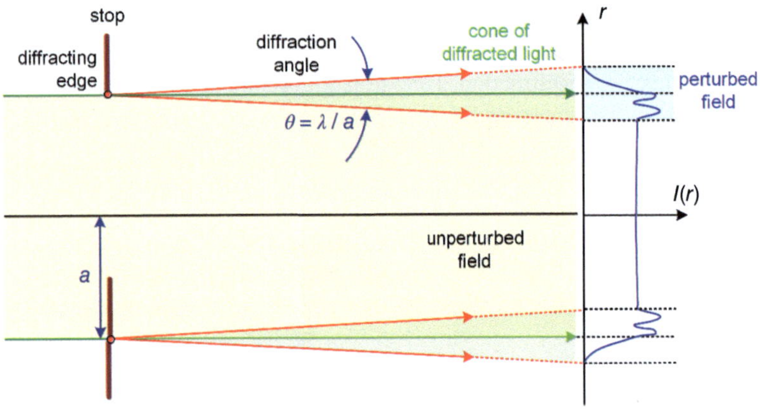

Figure 6.73 Generation of the boundary diffraction wave in an optical system diaphragm. For simplicity, the diffraction cone is drawn here as symmetric; in reality, the perturbation is quite asymmetric. See the comment in the text.

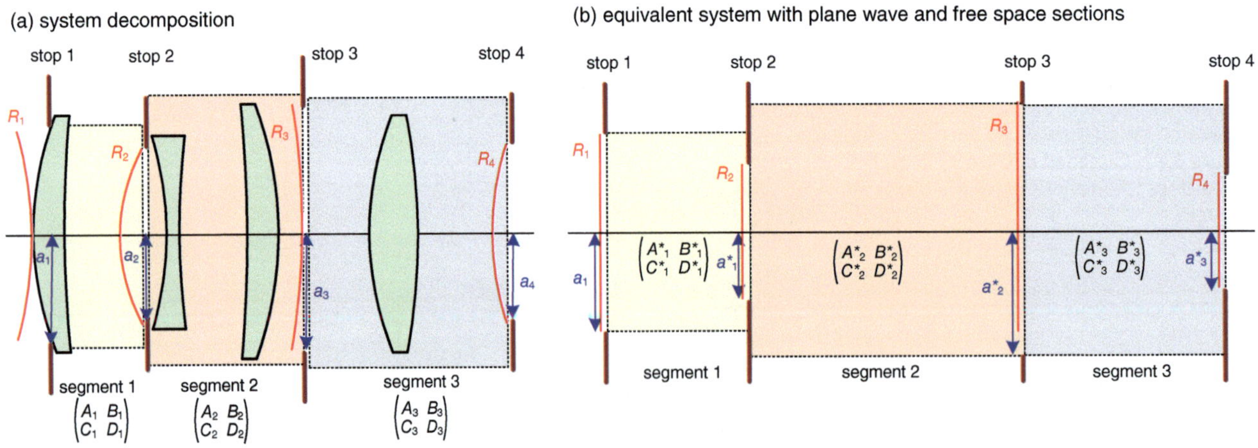

Figure 6.74 Cascaded diffraction concept with (a) decomposition into segments between the stops (b) equivalent free-space system with plane waves and scaled distances and diameters.

calculation taking the aberrations into account is not a principal problem, but lets the computational burden grow and mostly brings only a small improvement of accuracy.

From the viewpoint of diffraction, a flattening transform as explained in Section 6.2.6.4 in every segment between the stops generates a setup with plane waves, laterally scaled stops sizes and modified distances/*ABCD*-matrices without components. The propagation is reduced onto a simple plane wave diffraction under free-space conditions, which is sketched in Figure 6.74(b) with modified matrix elements. If desired and necessary, every internal lens system between the apertures can be taken into account by using Collins' approach in paraxial formulation. If an off-axis field point of the illumination is considered, the geometry must be established under skew conditions around the chief ray as reference.

In the special case that a segment is imaging with $B = 0$, conjugated planes are obtained as seen in Figure 6.75. All angles or directions of the diffracting edge waves are refocused again into one point. The diffraction related impact can in this case easily be adapted from the initial plane to the final plane.

There are now five steps in establishing a numerical calculation model, they are sketched in Figure 6.76.

- The real system is decomposed into segments between active stops.
- Substituting every segment by a paraxial approximation and an *ABCD* matrix.
- Transforming every segment into an equivalent scaled empty one with flattened waves.
- Generating an initial skew wave for off-axis field points, which show a z-distributed truncation and diffraction.
- Re-aligning the light beam around the chief ray by decentring the stops in an equivalent manner. This equivalent system allows for a simpler application of the diffraction integral. In principle the wave is transformed from the chief ray coordinate system into an aligned one on axis.

A systematic error in this model is the simple change from the system into the chief ray coordinate system. In reality, the stops are then tilted by the chief ray angle and become elliptical in shape. The neglected effect is an ellipticity in the range of $\cos(w_{CR})$, which is usually quite small.

If in the various segments the light propagation suffers from aberrations which cannot be neglected, the exact conjugation between pupil planes is perturbed. In this case the simulation can be extended by incorporating additional phase masks, which describe the wave aberration s. But they are collected to appear in one plane only in this case, which is an approximation too. In real systems, the transition between arbitrary planes can exhibit quite large aberrations, because the sub-systems are not well corrected. Therefore, the paraxial propagation approach with the Collins integral is a strong simplification.

In the first order the aberrations can be taken into account if every segment is analysed according to its wave aberrations firstly. For the propagation, the segment is split into two parts with equal length as indicated in Figure 6.77. After the propagation over the first half part, the wave aberrations are superposed. They have then the impact of changing amplitude and phase in the next stop plane, which changes the edge diffraction results quantitatively. The split in the middle is simply an approximation which averages the effects approximately.

6.8.3 Examples

6.8.3.1 Diffraction of a Gaussian Beam

In this example calculation, two separated circular symmetric lenses with the stops at the lenses and a collimated Gaussian beam in between are considered. The wavelength is selected as $\lambda = 1\ \mu m$, the initial beam radius is $w_i = 0.006367\,mm$. The Gaussian beam has a width $w = 10\,mm$ at the first lens L_1. The truncation by the first diaphragm is at $r = w$. After a collimated segment with different lengths z the beam is again focused by the second lens L_2 with focal length $f_2 = 200\,mm$. The system is sketched in Figure 6.78. In the following, three different

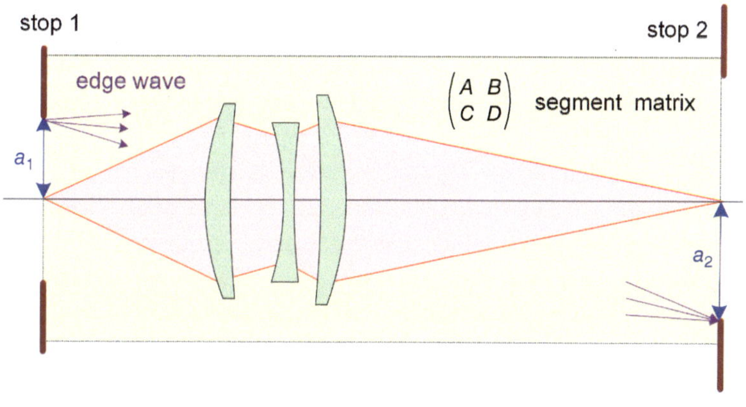

Figure 6.75 Propagation between conjugated planes.

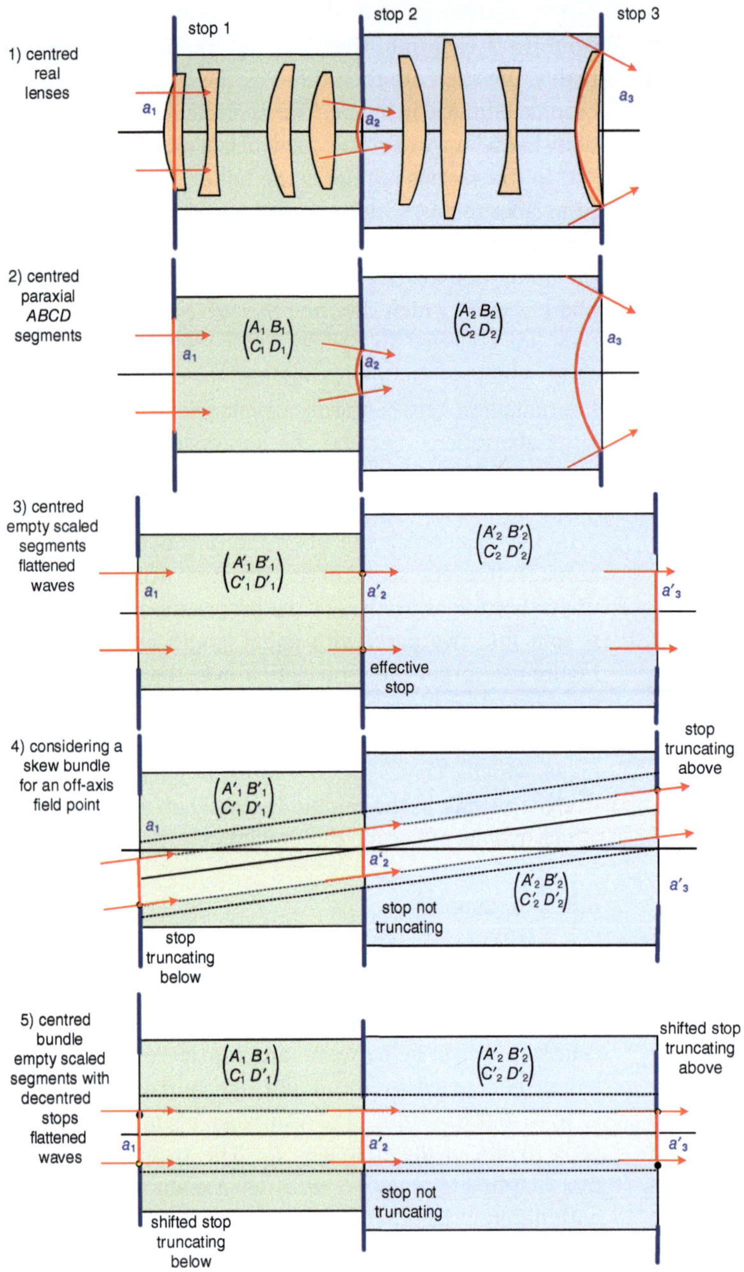

Figure 6.76 Cascaded diffraction workflow of computation in five steps; for the description, see text.

stop constellations are evaluated to examine the effect of cascaded diffraction.

If firstly both stops have the same size $D = 10\,\text{mm}$, at the second stop there is a certain loss of energy and in the focus a degradation is observed due to the impact of the cascaded diffraction. Table 6.4 gives these values as a function of the intermediate distance z. As expected, the larger the distance, the larger the influence at the second stop due to a growing spread.

Figure 6.79 shows the loss at the second stop as a function of the Fresnel number. A small Fresnel number means large diffraction effects and strong impact at the second stop.

The diffraction at the stop L_1 propagates depending on the distance and a change of the final profile in the focal point is observed. The intensity in the final plane as a function of the stop distance z for a full circular symmetric setup with both stops actively truncating is illustrated in Figure 6.80. The changes indicate the impact of the diffraction at the first stop. The intensity Log $I(r)$ as well as the changes are drawn for different Fresnel numbers. It is seen that the changes are below 1% if the Fresnel number is $N_F > 1000$.

If in a first modification of the setup the second stop is used as a central obscuration of the same size, the

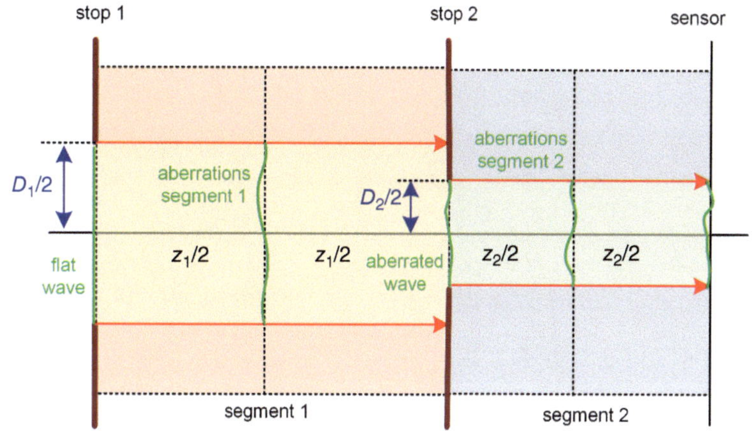

Figure 6.77 Cascaded diffraction in case of residual aberrations, which cannot be neglected.

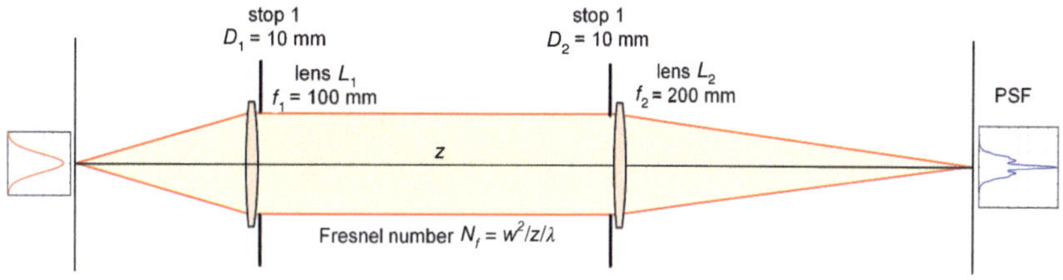

Figure 6.78 Cascaded diffraction of a Gaussian beam.

Table 6.4 Cascaded diffraction of a Gaussian beam. The spot diameter is defined as a second moment or variance of the ray intersection points; see Volume 2, Chapter 13.6 for details.

z [mm]	N_F	Loss second stop in %	Spot D in μm
30	655	0.151	42.5
300	67	0.645	43.2
1000	19	1.24	43.9
3000	5.5	2.24	44.4
6000	2.6	3.30	43.8
12000	1.14	4.85	43.5

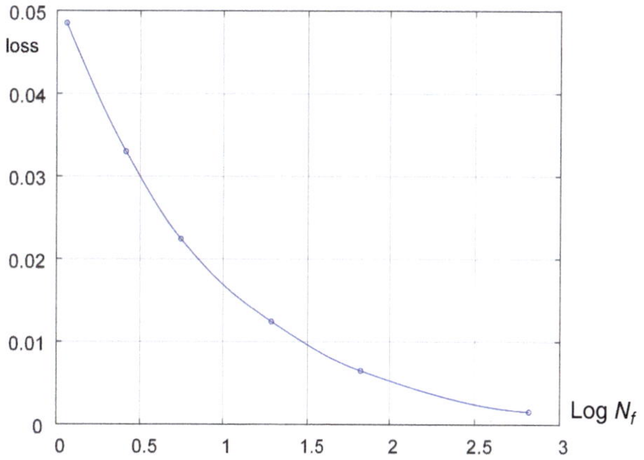

Figure 6.79 Cascaded diffraction of a Gaussian beam, shown is the loss of energy as a function of the Fresnel number.

diffracted light at the first stop is directly seen in the sense of a dark-field setup. The following profiles together with the relative beam power show the dependence on the Fresnel number for different distances z. Geometrically no light is expected after the second stop, see Figure 6.81.

The intensity direct after the second plane with central obscuration is visualized in Figure 6.82. A narrow ring as the contribution of the diffraction spreading is seen in parts (a), (b) and (c). If this light is focused by the lens as before, and the distance z is changed, three cases for z = 30 mm, 1 m and 12 m are shown in parts (d)–(f). It can be seen that if the Fresnel decreases, the amount of light grows by orders of magnitude and the ripple changes slightly.

In a third version, the system setup is modified by modifying the stop shapes. The circular stop is decomposed into two U-shaped half stops with circular arcs, see Figure 6.83; the upper one is fixed in the position of lens L_2 and the lower one at lens L_1. Now the distance z between both groups is changed. This has the effect that the truncating stop is split in z. The lower part of the beam is diffracted at lens L_1, the upper part at L_2. As a consequence, an asymmetry of the diffraction pattern is observed, which is shown in Figure 6.94 for different values of z. The smaller the Fresnel number, the larger is the observed asymmetry of $I(y)$. As expected, the cross section $I(x)$ remains symmetric, see Figure 6.84.

The asymmetry of the z-distributed diffracting stop is seen in the angle spectrum as well as in the focal intensity

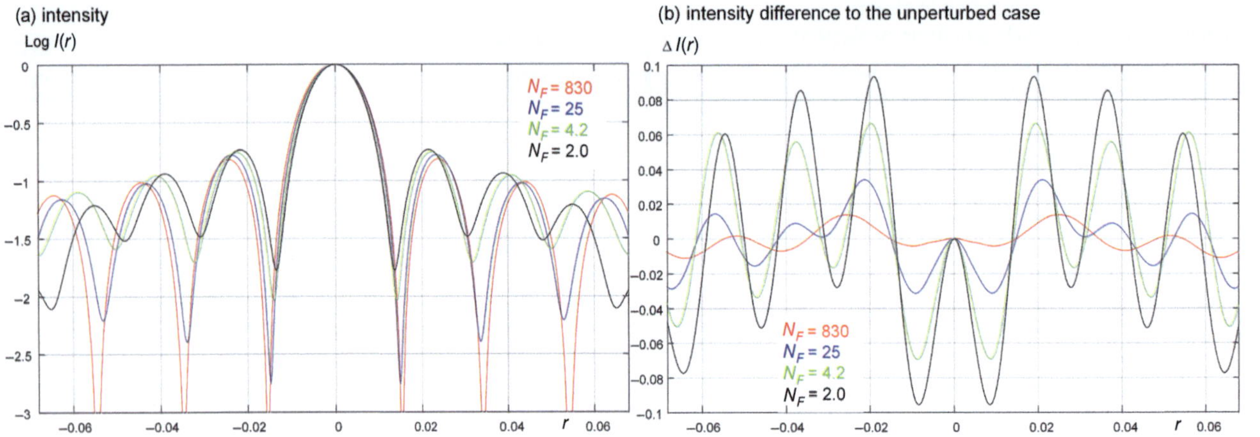

Figure 6.80 Cascaded diffraction, (a) intensity in the focal plane as a function of the Fresnel number in the collimated segment and (b) difference relative to the unperturbed case without any stop. The units of r are arbitrary here.

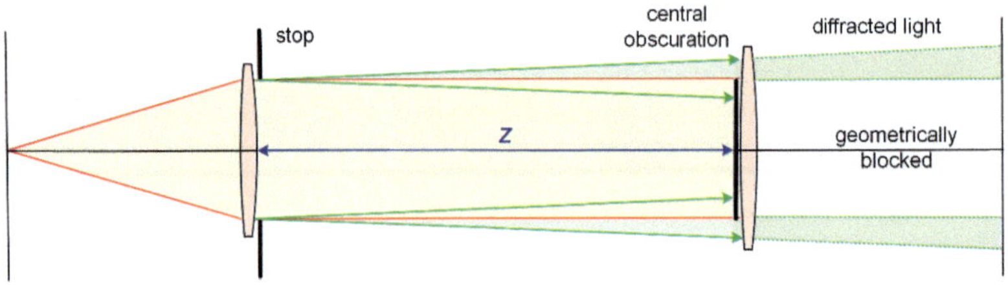

Figure 6.81 Cascaded diffraction with central obscuration in a dark-field setup.

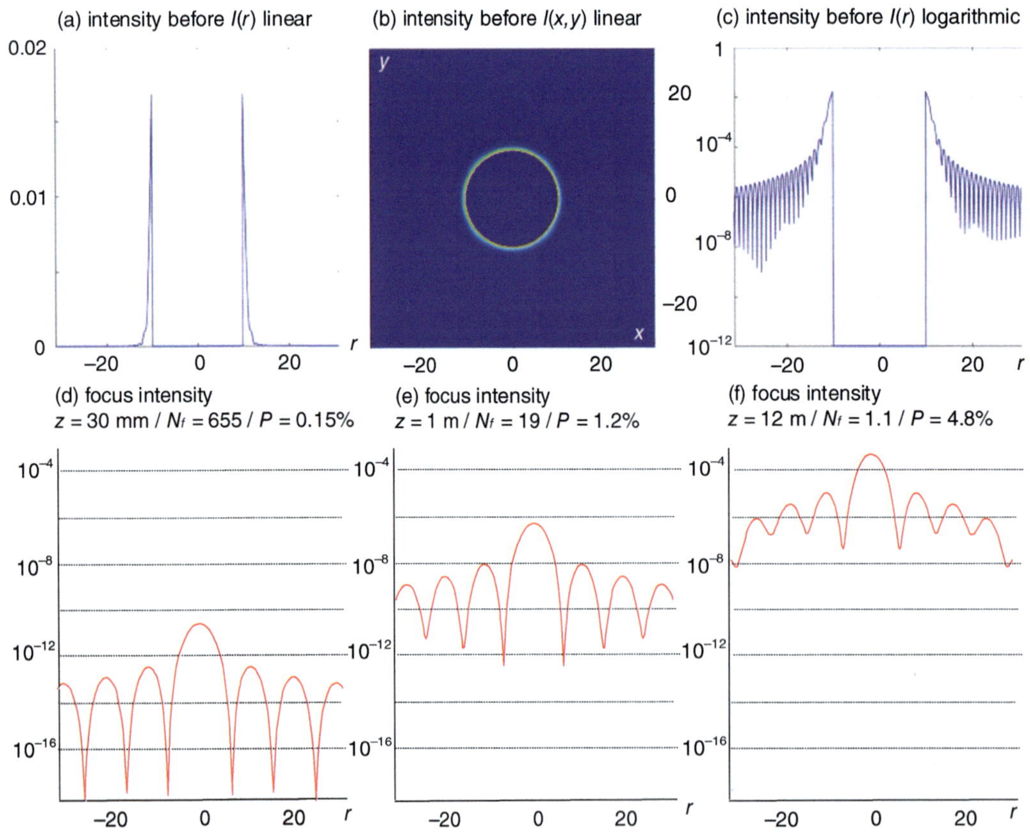

Figure 6.82 Illustration of cascaded diffraction according to the setup of Figure 6.81 with a central obscuration at lens L_2. (a) shows the light diffracted at the stop at the location of the second lens, as a cross section. (b) shows the same in two dimensions and (c) as a logarithmic representation. The parts (d)–(f) show the intensity in the focus of the second lens for three different distances z corresponding to different Fresnel numbers. The units of r are arbitrary here.

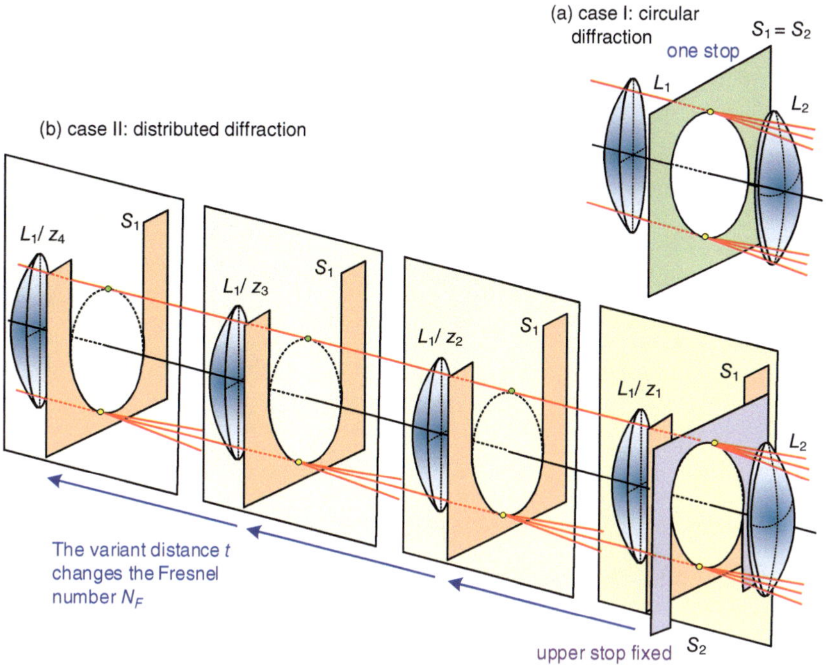

Figure 6.83 Cascaded diffraction with two asymmetric stops and changing the distance z between the stops S_1 and S_2. The lens L_1 and the stop S_1 are moving.

$I(y)$ after lens L_2. This asymmetry grows with distance z. A quantitative representation is seen in Figure 6.85 as the difference of the intensities in comparison to the case of a pure exit pupil diffraction in (a) and as x- and y-cross section in (b). The fields are re-aligned to a common centroid before the difference is calculated.

The maximum and the rms of the intensity anisotropy change exponentially with the Fresnel number. The rms value is roughly one order of magnitude smaller than the peak valley, this is seen in Figure 6.86.

6.8.3.2 Double Gauss Camera Lens

As a last example, a real photographic lens is considered. The diffraction of an off-axis ray bundle is distributed along z. The system layout is sketched in Figure 6.87(a). Considering the truncations by the stop and lens diameter it is seen that:

- the lower edge of the bundle is clipped and therefore diffracted at the first system surface; it has a Fresnel number of $N_F = 3036$ for the transition to the stop if the wave curvature is assumed to be plane at the entrance;
- the upper edge of the bundle is clipped and therefore diffracted at the last system surface; the second segment has a Fresnel number of $N_F = 6148$;
- in the sagittal direction, the stop itself truncates the ray bundle, see the lower part of (a) of Figure 6.87.

The footprints of the ray bundles and the ranges of active diffraction edges are sketched in the lower part of Figure 6.87(a). The Fresnel numbers are defined here in a generalized way by the stop widths and the radius of curvature R_w. The heights are those of the marginal ray.

$$N_F = \frac{h_{MR} h'_{MR}}{\lambda B} \cdot \left(A - \frac{B}{R_w} \right)^2 \tag{6.183}$$

The parameters of the segments for the wavelength $\lambda = 550\,\text{nm}$ are listed in Table 6.5.

If a diffraction related calculation is performed with and without the z-distributed diffraction, a difference of the peak valley of the relative intensity with the amount 3.5 10^{-5} is observed in the y-section, while in the x-section the difference is smaller than 10^{-6}. The calculation is realized with 2048^2 sampling points of the ray bundle. The results are sketched in Figure 6.88. In (a) the footprint of the open pupil area is seen with the four segments of truncation. Part (b) shows the PSF in the image as the logarithm of the intensity. Cross sections of this plot are found in part (c). Although the plots are drawn in a logarithmic scale the asymmetry cannot be resolved. In (d) the difference $I(x)$-$I(-x)$ and similar for $I(y)$ exaggerates the asymmetry in y, which is only of the order of 5 10^{-5} relative. For the calculation of the y-difference, first the peak is aligned to $y = 0$ to remove the effects of distortion. Finally, it is seen that the very large Fresnel numbers before and after the stop reduce the cascaded diffraction effects to this really negligible order of magnitude.

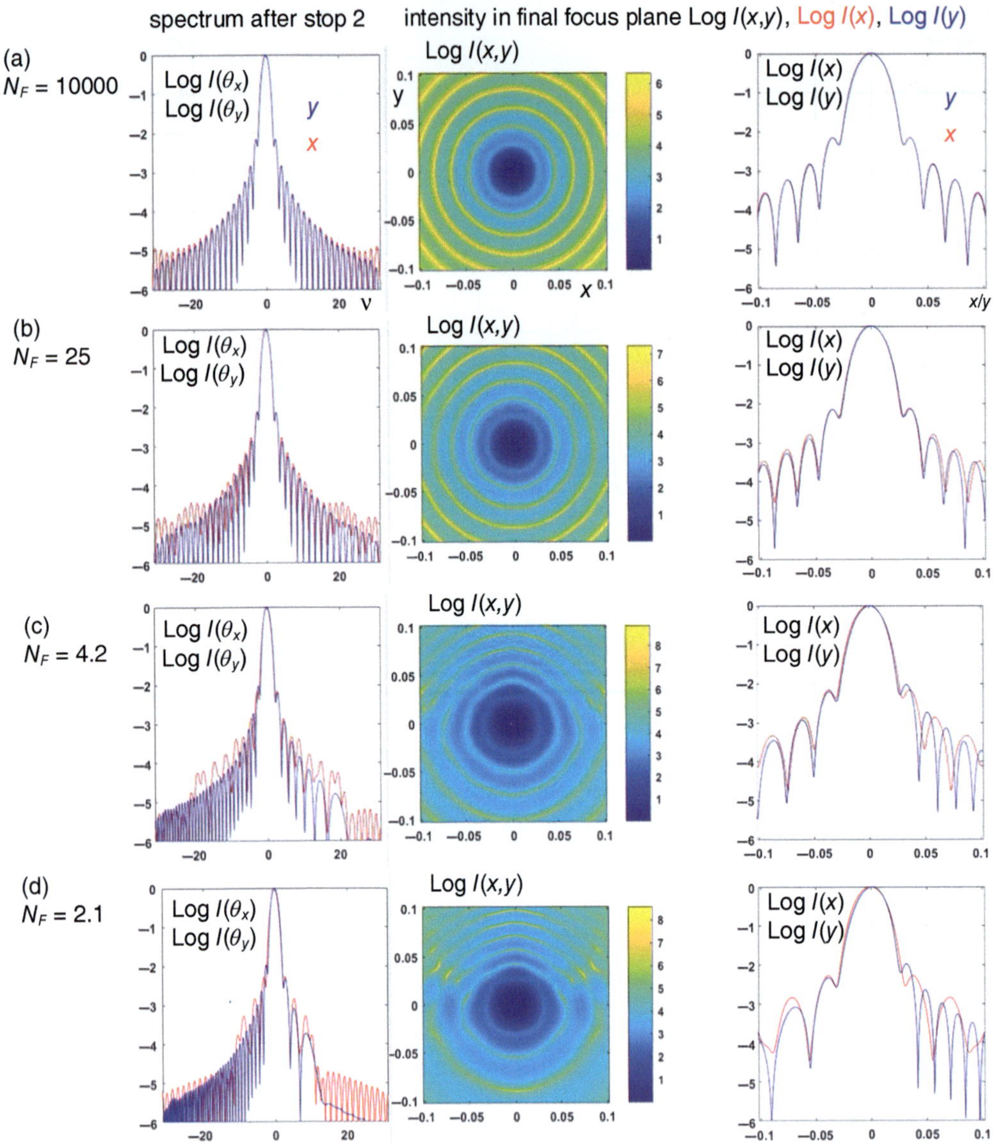

Figure 6.84 Cascaded diffraction, the distance z is increased from (a)–(d), correspondingly shrinking the Fresnel number, and the asymmetry grows. Shown in the first column are the angle spectra after stop two in logarithmic representation. The middle and the right column show the intensities in the focal point in logarithmic representation in two dimensions and as a cross section. The units of r are arbitrary here.

6.9 Miscellaneous

6.9.1 Polychromatic PSF

If a light source is not monochromatic, it can be usually assumed that the spectral components cannot interfere and for every wavelength an individual PSF of the well-known form is generated. In the case of a circular symmetric aberration-free case the Airy function is given by

$$I(r) = \left[\frac{2 \cdot J_1(2\pi/\lambda r \sin u')}{2\pi/\lambda r \sin u'}\right]^2 \tag{6.184}$$

Due to the incoherence, in a broadband point response the intensities are superposed without impact on the phase. This can be written simply as a sum or an integral

$$I_{poly}(r) = \frac{1}{\Delta\lambda}\int_{\lambda_1}^{\lambda_2} S(\lambda) \cdot \left[\frac{2 \cdot J_1(2\pi/\lambda \cdot r \sin u')}{2\pi/\lambda \cdot r \sin u'}\right]^2 d\lambda \tag{6.185}$$

The individual PSF contributions are weighted here with the spectral function $S(\lambda)$ of the illumination source. The diffraction ripples are scaled with the wavelength; therefore, in the overlay the zero locations of the intensity are at different r-locations and the total PSF usually has no zero. This is sketched in Figure 6.89 where for a constant spectral weight $S(\lambda) = \text{const}$, the polychromatic PSF is calculated for different bandwidths $\Delta\lambda/\lambda$ in logarithmic representation.

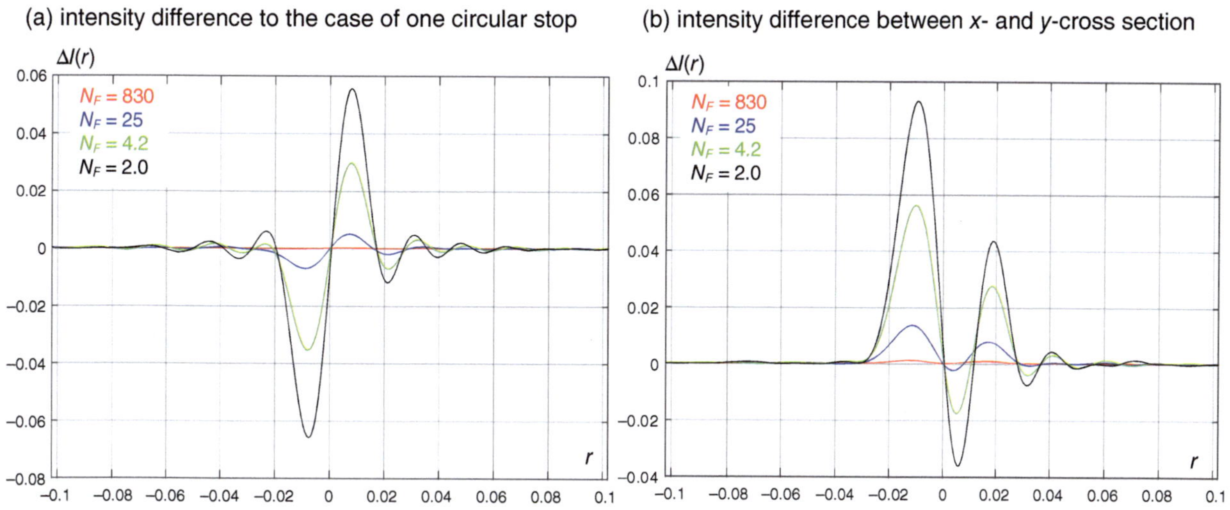

Figure 6.85 Due to cascaded diffraction with z-distributed truncation the intensity changes in comparison to the unperturbed case in the focal plane. Here (a) shows the comparison of intensity differences with the circular symmetric case, (b) gives the difference between x and y-cross section. The units of r are arbitrary here.

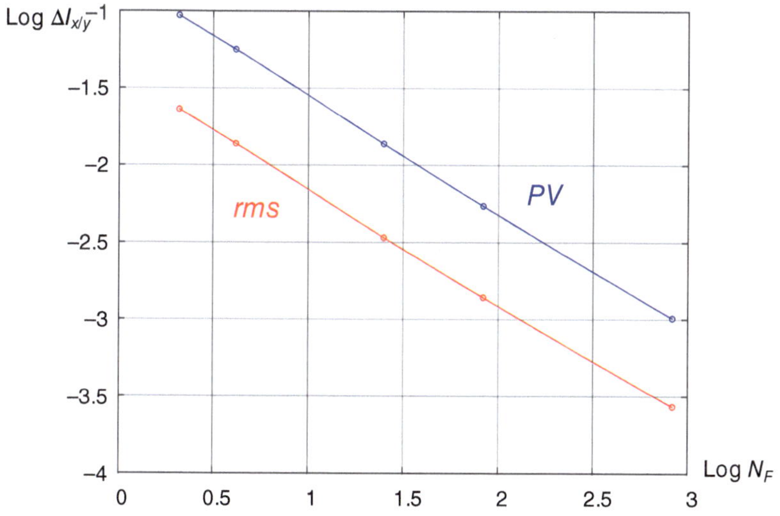

Figure 6.86 Asymmetry of the intensity difference as a function of the Fresnel number due to cascaded diffraction.

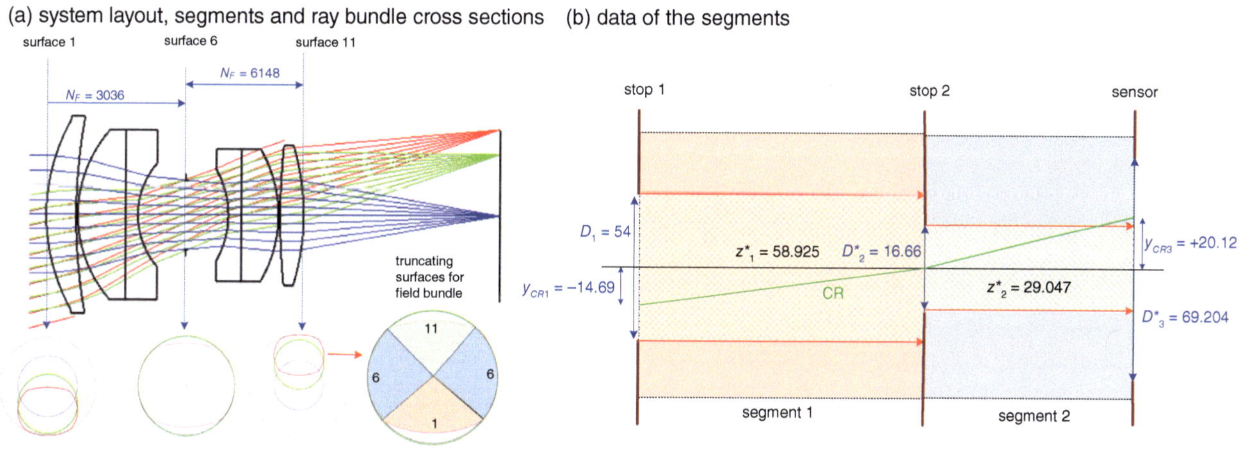

Figure 6.87 Cascaded diffraction in a camera lens; (a) shows the layout with the two segments before and after the stop and the beam bundle cross sections. (b) illustrates the segmentation of the skew ray bundle in the flattened version.

Table 6.5 Data of the camera lens segments of the current example. w_{CR} is the chief ray angle, M is the flattening magnification.

	Surf 1 – 7 (stop)	Surf 7(stop) – 11	Surf 11 – image
$ABCD$	$\begin{pmatrix} 0.5999 & 35.355 \\ -0.00531 & 1.3540 \end{pmatrix}$	$\begin{pmatrix} 1.211 & 27.981 \\ -0.01187 & 0.5514 \end{pmatrix}$	$\begin{pmatrix} 1 & 57.31 \\ 0 & 1 \end{pmatrix}$
$ABCD$	$\begin{pmatrix} 0.5779 & 80.706 \\ -0.0101 & 0.3269 \end{pmatrix}$		
R_w	0	112.95	57.31
h_{MR}	16.65	10.0	9.96
h_{CR}	−14.69	0	+11.63
w_{CR}	14°	23.51°	11.965°
vignetting	1.75%	1.75%	
M	0.600	0.9633	0
N_F	3036	6148	0
D_{stop}	54	20	40
D^*_{stop}	54	33.33	69.204
$B^* = z^*$	58.925	29.047	
h^*_{CR}	−14.69	0	+20.12

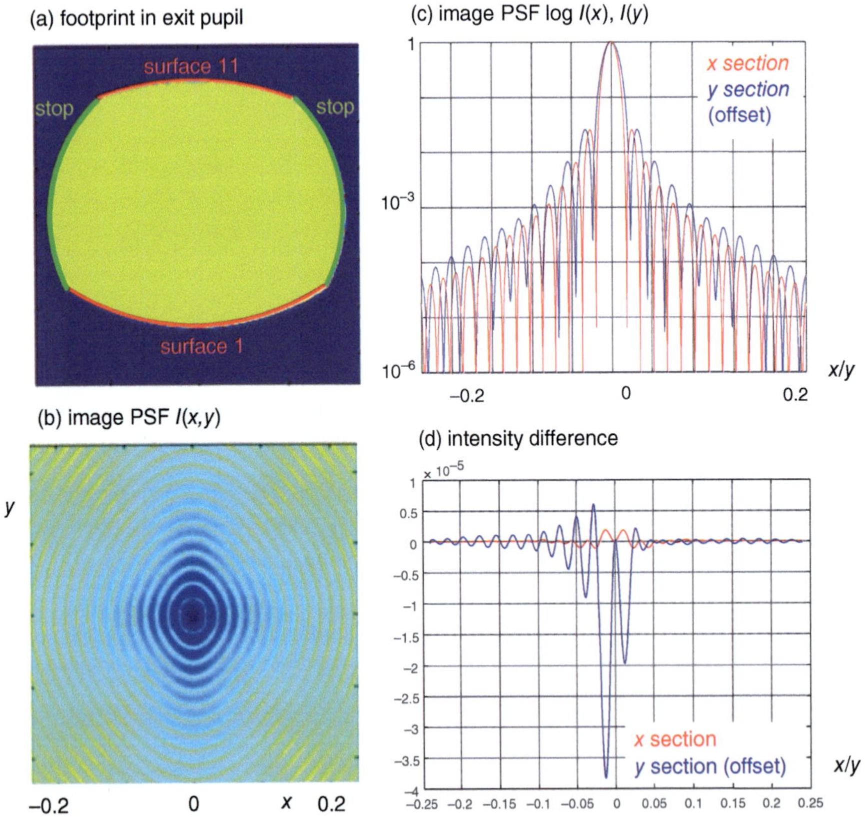

Figure 6.88 Cascaded diffraction in the camera lens for an off-axis field point. (a) shows the pupil area, (b) the PSF logarithmic as $I(x,y)$, (c) the cross sections $I(x)$ and $I(y)$. In (d) the difference between the cross section for plus and minus illustrates the y-asymmetry. The units of x and y are arbitrary here.

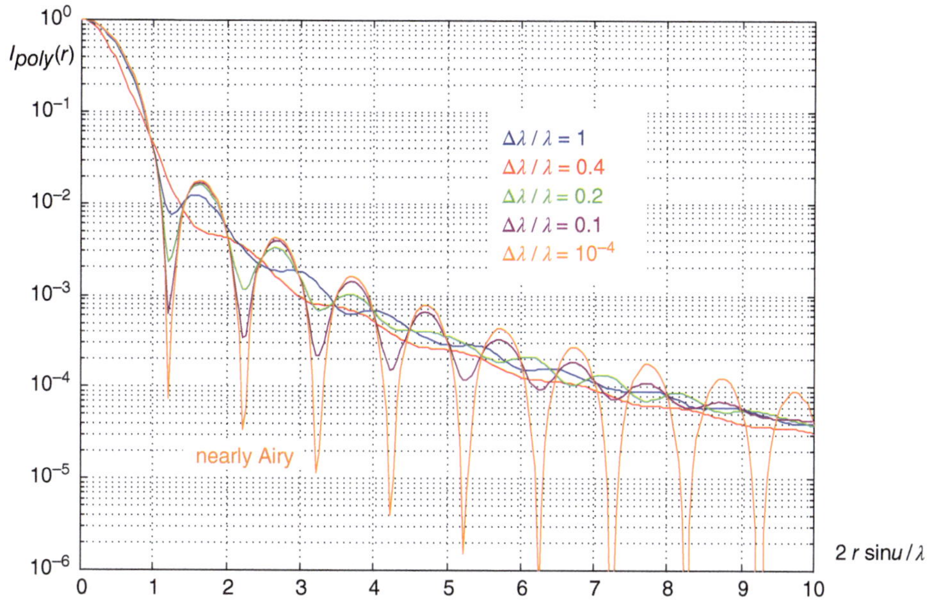

Figure 6.89 Ideal polychromatic point spread function with different bandwidth values $\Delta\lambda/\lambda$.

With growing bandwidth, the modulation depth of the oscillations decreases. The blue case approximately corresponds to the visible light with $\lambda = 0.4\ldots0.7$ μm.

6.9.2 Line of Sight*

(*) The content of this section is quite specialized and for advanced reading. Knowledge of this section is not necessary to understand the following chapters. Beginners are recommended to skip this section.

It can be shown by very fundamental considerations of the wave equation that the centroid of energy inside a beam is always propagating along a straight line [113]. This property is called the line-of-sight principle. In the application of this basic property in optical systems it should be clear that in an arbitrary constellation of phase and amplitude distribution, the location and propagation direction of:

- the centroid of energy;
- the intensity peak – can change during propagation and is mainly of interest in the image plane;
- the chief ray – defined geometrically in the pupil;

of the PSF are different, see the schematic picture of Figure 6.90. While the centroid and chief ray propagate straight, the peak usually moves along a bent curve. Only in special cases do two or all of them coincide. It should also be clarified that the peak and centroid can be defined

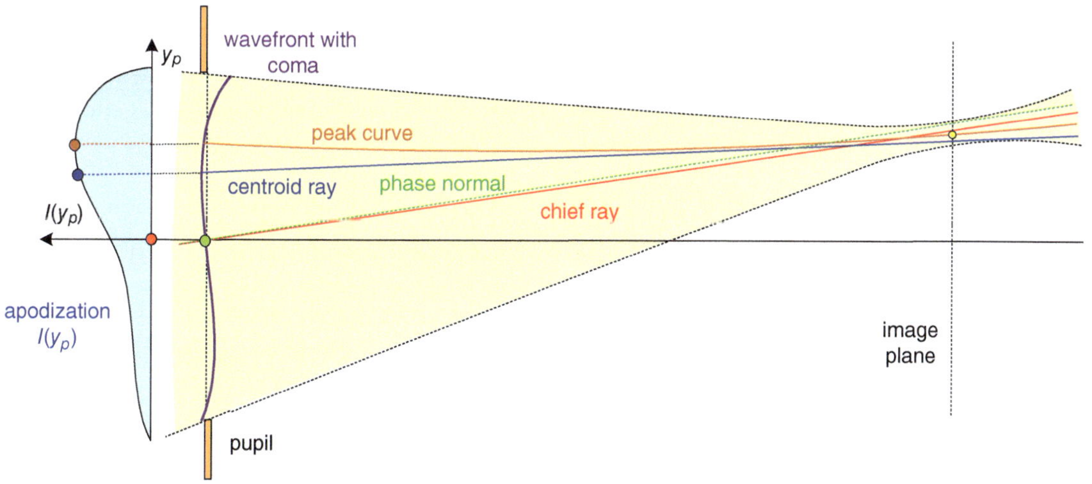

Figure 6.90 Schematic picture of a focusing caustic with the various rays between the exit pupil and the image plane.

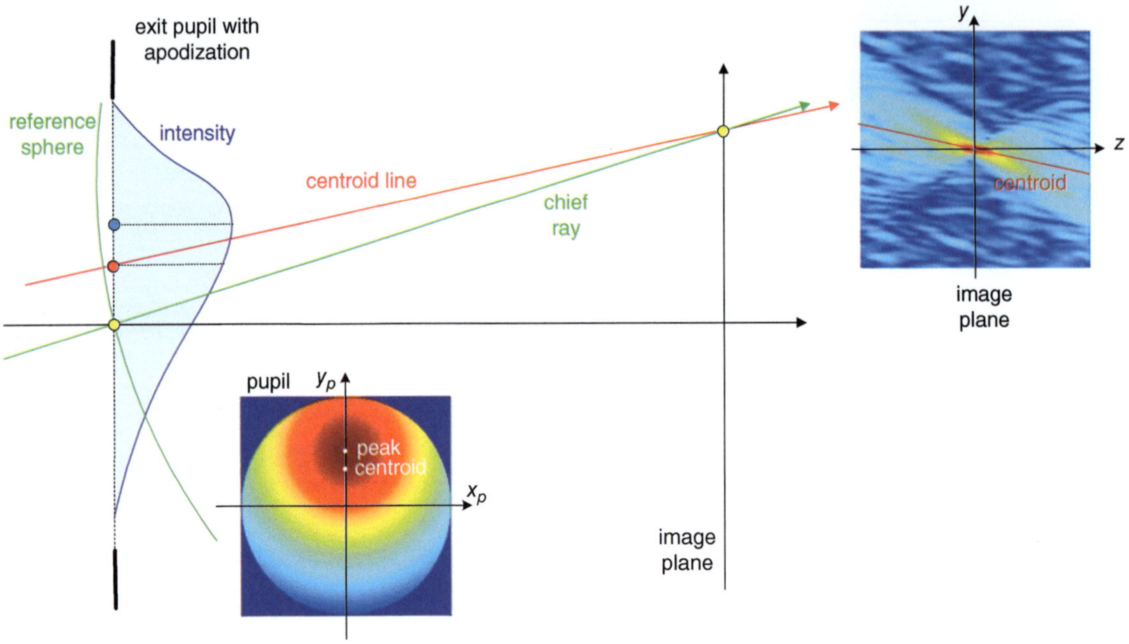

Figure 6.91 Schematic picture of a focusing caustic in the case of an asymmetric apodization.

in the exit pupil or the image plane. The corresponding properties can change during propagation. At an arbitrary location z, the centroid of the intensity profile $I(x,y,z)$ is defined by

$$x_c(z) = \frac{\iint x \cdot I(x,y,z)\,dx\,dy}{\iint I(x,y,z)\,dx\,dy} = \frac{1}{P} \cdot \iint x \cdot I(x,y,z)\,dx\,dy$$

$$y_c(z) = \frac{1}{P} \cdot \iint y \cdot I(x,y,z)\,dx\,dy \tag{6.186}$$

with the beam power P. The line of sight means that the centroid coordinates move along a straight line

$$A \cdot x_c(z) + B \cdot y_c(z) + C \cdot z + D = 0 \tag{6.187}$$

Let's first consider the phase effect. If it is assumed that the exit pupil is uniformly illuminated and therefore no apodization effects are present, only the wave aberrations are important and dominate the propagation behaviour. In an optical system a discrepancy between the centroid ray and the chief ray occurs if an odd function of the phase exists, which means that the wave aberration contains coma-type terms. If only lower order Zernike coma coefficients are considered, the terms c_{nm} with $m = +1$ and -1 are relevant and contribute to the deviation between chief ray and centroid ray as

$$y_c(z) = \frac{2 \cdot z}{D_{ExP}} \cdot \sum_{n=1,3,5,\ldots} \sqrt{2(n+1)} \cdot c_{n,+1},$$

$$x_c(z) = \frac{2 \cdot z}{D_{ExP}} \cdot \sum_{n=1,3,5,\ldots} \sqrt{2(n+1)} \cdot c_{n,-1} \tag{6.188}$$

Higher order n-indices of the Zernikes contribute differently. Of course, higher order odd powers of the trigonometric function also have the same effect.

If an asymmetric exit pupil apodization is assumed and the intensity distribution around the focal caustic is calculated, Figure 6.91 is obtained as a numerical exemplary result. Since the intensity is calculated by the absolute value squared and the image plane is Fourier-conjugated to the pupil, in the case of an even power symmetric phase function the chief ray and the centroid must intersect in the image plane, although they are separated in the pupil.

6.9.3 Extended Zernike Approach*

(*) The content of this section is quite specialized and for advanced reading. Knowledge of this section is not necessary to understand the following chapters. Beginners are recommended to skip this section.

6.9.3.1 Introduction

The classical Zernike expansion is an approximation of the phase distribution in the exit pupil only in the transverse direction. Details of this approach are discussed in

Chapter 15 in depth. According to [95, 96, 114, 115] it is also possible to extend this expansion in the third dimension z. This can be expressed by adding an additional coefficient a_{nm} with z-dependence in the form.

$$E(\bar{r},\varphi,\bar{z}) = 2\cdot a_{00}(\bar{r},\bar{z}) + 4\pi\cdot i\cdot \sum_{n,m\geq 0} c_{nm}\cdot i^m \cdot a_{nm}(\bar{r},\bar{z})\cdot\cos(m\varphi) + 4\pi\cdot i\cdot\sum_{n,m<0} c_{nm}\cdot i^m\cdot a_{nm}(\bar{r},\bar{z})\cdot\sin(m\varphi) \quad (6.189)$$

with normalized coordinates $\bar{r},\bar{z}$. To include the z-coordinate, a quadratic phase term describing defocusing is necessary. A meaningful expansion of the parabolic Fresnel phase term can be found by Bessel functions as

$$e^{i\bar{z}\cdot\bar{r}^2} = e^{\frac{i\cdot\bar{z}}{2}}\cdot\sqrt{\frac{\pi}{\bar{z}}}\cdot\sum_{s=0}^{\infty}(2s+1)\cdot i^s\cdot J_{s+\frac{1}{2}}\left(\frac{\bar{z}}{2}\right)\cdot R_{2s}^0(\bar{r}) \quad (6.190)$$

with the radial Zernike terms Rj. This gives the unknown coefficients a_{nm} as

$$a_{nm}(\bar{r},\bar{z}) = e^{iz}\cdot\sum_{l=1}^{\infty}(-2i\bar{z})^{l-1}\sum_{j=0}^{p} b_{lj}^{nm}\cdot\frac{J_{m+l+2j}(\bar{r})}{l\cdot\bar{r}^l} \quad (6.191)$$

with the abbreviations

$$b_{lj}^{nm} = (-1)^p\cdot(m+l+2j)\cdot\binom{m+j+l-1}{l-1}\cdot\binom{j+l-1}{l-1}\cdot\frac{\binom{l-1}{p-j}}{\binom{q+l+j}{l}} \quad (6.192)$$

and the indices

$$p=\frac{n-m}{2},\quad q=\frac{n+m}{2} \quad (6.193)$$

In this expansion the coordinates are the normalized canonical coordinates. In the transverse direction they are obtained from the real spatial coordinates x', y' by

$$\bar{x} = \frac{2\pi\cdot n\cdot\sin u}{\lambda}\cdot x',\quad \bar{y} = \frac{2\pi\cdot n\cdot\sin u}{\lambda}\cdot y',\quad \bar{r}^2 = \bar{x}^2+\bar{y}^2 \quad (6.194)$$

In the axial direction z the normalization is performed by the Rayleigh unit R_u as

$$\bar{z} = \frac{2\pi}{\lambda}\cdot z'\cdot\left(1-\sqrt{1-n^2\cdot\sin^2 u}\right) \approx \frac{\pi\cdot z'\cdot n^2\cdot\sin^2 u}{\lambda} = \pi\cdot\frac{z'}{R_u} \quad (6.195)$$

where the last two expressions are approximations for small and moderate apertures $\sin(u)$ in Fresnel approximation.

This expansion is now valid in the complete image domain from the exit pupil until the image plane and beyond. If only cos terms are considered, the sum in Eq. (6.197) gives the approximate representation

$$E(\bar{r},\varphi,\bar{z}) = 2\cdot\frac{2J_1(\bar{r})}{\bar{r}} + 4\pi\cdot i\cdot\sum_{n,m\geq 0} c_{nm}\cdot i^m\cdot\cos(m\varphi)\cdot e^{i\bar{z}}\cdot\sum_{l=1}^{\infty}(-2i\cdot\bar{z})^{l-1}\sum_{j=0}^{p} b_{lj}^{nm}\cdot\frac{J_{m+l+2j}(\bar{r})}{l\cdot\bar{r}^l} \quad (6.196)$$

The main advantage of this kind of PSF calculation is that the caustic takes the aberrations into account, which is given by the Zernike coefficients c_{nm} in the exit pupil. No issues with Fourier-related sampling and discretization problems occur and no validity ranges limit the application as long as the series converge to the desired accuracy. The z-expansion is one of the major issues in this approach. The main parameters influencing the convergence properties are:

- the size of the Zernike coefficients c_{nm};
- the defocus size $\bar{z}$;
- The maximum radial order n of the aberrations.

The larger the defocus $\bar{z}_{max}$ is, the more terms of the l-sum are needed in Eq. (6.199). A more comprehensive investigation gives empirically a necessary number

$$l_{max} = 8.5\cdot|\Delta\bar{z}_{max}| + 15 \quad (6.197)$$

for truncation of the series to achieve an accuracy of 10^{-5} in the field amplitude. This nearly linear behaviour is extracted out of numerically calculated values as seen in Figure 6.92.

If the Zernike coefficients are non-zero, the accuracy of the expansion decreases depending on the size of c_j. If the correlation of the calculated PSF with the exact computed profile is used as a quality criterion, the behaviour of Figure 6.93 is obtained for some lower order aberration modes. The error grows with the size of the aberration coefficients c_{nm} and a stronger degradation is seen for coefficients $c_j > 0.2\,\lambda$, but the exact behaviour depends on the aberration type too.

6.9.3.2 Properties of the Extended Zernike Approach

In contrast to all the calculations of diffraction related intensity computations based on Fourier transform or Kirchhoff-integral algorithms, the extended Zernike approach does not primarily depend on discretization. This means that there are no restrictions considering sampling and the approach can be used also for high-frequency perturbations. One of the main advantages is the easy calculation of the field in a broad z-range without being forced

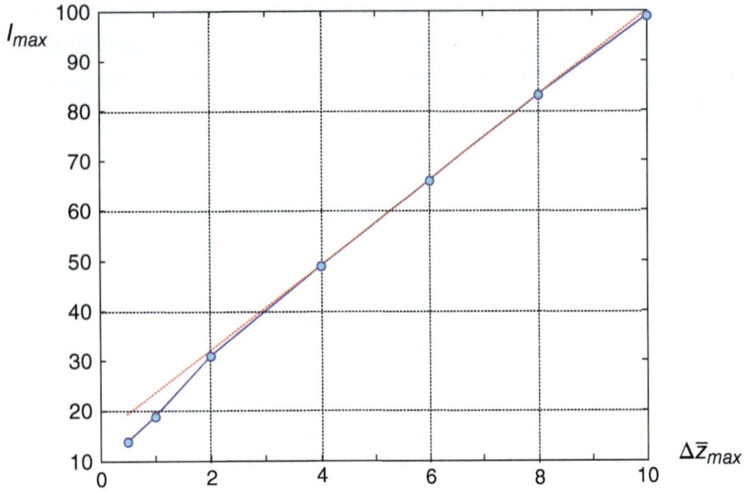

Figure 6.92 Necessary number of terms in the l-sum as a function of the defocus $\bar{z}_{max}$.

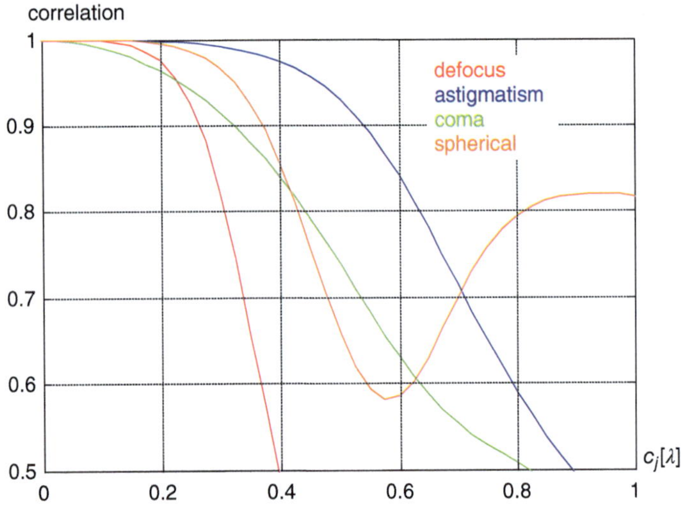

Figure 6.93 Accuracy measured by the correlation of the extended Zernike PSF in comparison to exact numerical calculations as a function of the Zernike coefficient size c_j scaled in λ for some low order aberrations.

to calculate the full algorithm again. By re-ordering the sums in the basic equations and pre-calculating several of the involved functions and indexed coefficients, a complete focal volume can be computed quite efficiently.

There are several possible generalizations and extensions of the extended Zernike approach. These are, for example:

- non-circular pupils;
- non-uniform pupil apodization;
- high-NA conditions, scalar or vectorial.

For these details the reader is referred to the literature. Figure 6.94 illustrates the calculation of a three-dimensional amplitude distribution in case of an optical focusing setup with several Zernike aberrations of the sizes $c_5 = +0.2\,\lambda$, $c_6 = -0.1\,\lambda$, $c_7 = +0.2\,\lambda$, $c_8 = -0.25\,\lambda$, $c_9 = +0.35\,\lambda$. The three-dimensional caustic volume in this example with 256 × 256 × 100 discretization numbers with a not special optimized code, on a normal computer, takes less than 10 seconds runtime.

6.9.4 Encircled Energy

The encircled energy, also referred as 'power in the bucket', is another possibility for measuring the PSF image quality. Figure 6.95 shows the measurement principle of the encircled energy. A beam is truncated by a variable stop and the power of the remaining beam is measured integrally by a detector. For the rotational symmetry case, the energy function is formulated as the integral over the intensity profile distribution as

$$E(r) = \int_0^r I(r)\, 2\pi r\, dr \tag{6.198}$$

There are special applications where the encircled energy is quite well matched to the specification requirements. For example, a satellite transfer system with a huge distance between source and receiver: the gathered power is the major term of quality because it is very small. The distribution of the signal over the entrance aperture of the receiver is of no concern. Therefore, the drop of the

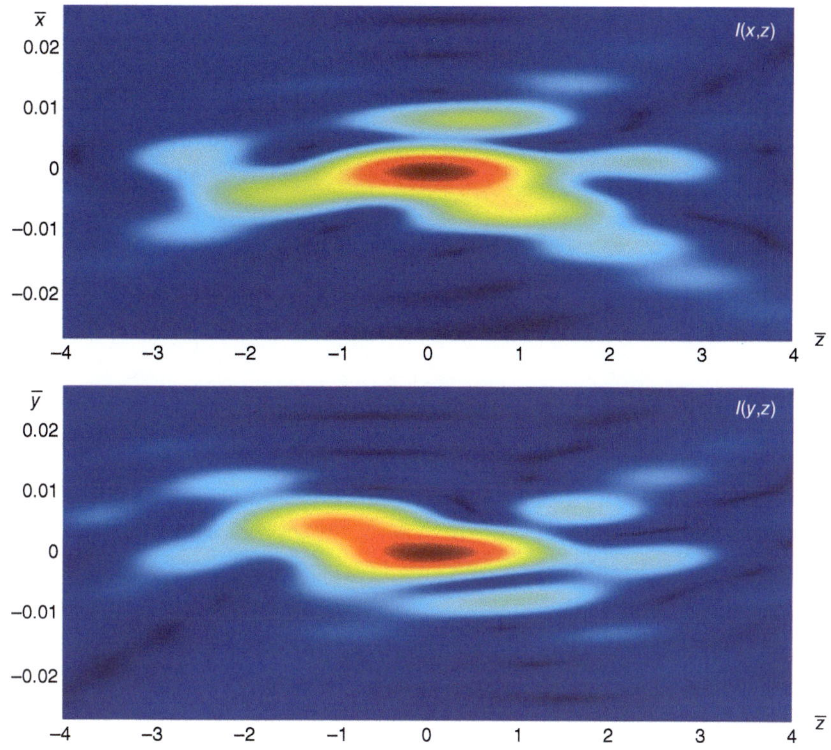

Figure 6.94 Amplitude of the focal region in a system with multiple Zernike aberrations.

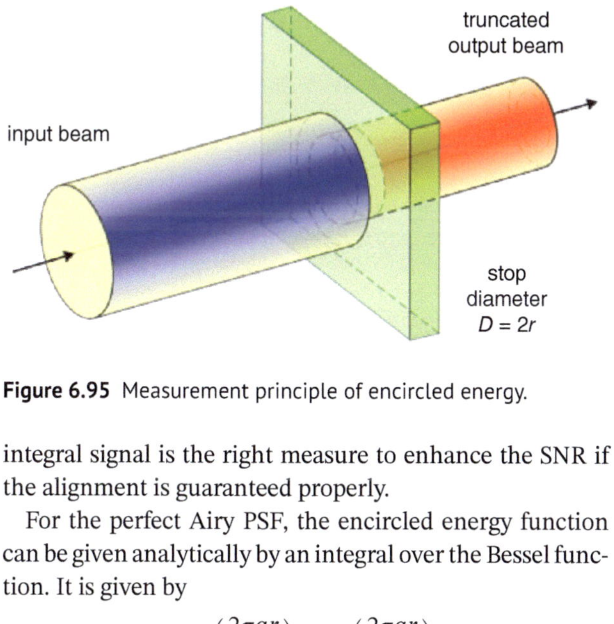

Figure 6.95 Measurement principle of encircled energy.

integral signal is the right measure to enhance the SNR if the alignment is guaranteed properly.

For the perfect Airy PSF, the encircled energy function can be given analytically by an integral over the Bessel function. It is given by

$$E(r) = 1 - J_0^2\left(\frac{2\pi ar}{\lambda R}\right) - J_1^2\left(\frac{2\pi ar}{\lambda R}\right) \tag{6.199}$$

For a threshold of $1/e^2 = 0.135$, the radius is

$$r_{1/e^2} = 0.6739 \cdot r_{Airy} \tag{6.200}$$

with the energy

$$E_{1/e^2} = 0.7659 \tag{6.201}$$

and can be compared with the classical values for Gaussian beams. This function is illustrated in Figure 6.33. According to the Airy function with the zeros with index m at r_m and the condition

$$I(r_m) \sim J_1^2\left(\frac{2\pi a r_m}{\lambda R}\right) = 0 \tag{6.202}$$

the diffraction ring number m carries the relative energy

$$E_m = J_0^2\left(\frac{2\pi a r_m}{\lambda R}\right) - J_0^2\left(\frac{2\pi a r_{m-1}}{\lambda R}\right) \tag{6.203}$$

with this formula the numbers given in Figure 6.33 are found.

If in reality not a circular, but a square diaphragm is used, the integration function looks of course different. But if this is computed quantitatively for the Airy pattern and compared with the encircled function, Figure 6.96 is obtained. The difference is only 7.3%, which seems rather small. This ensquared energy in reality is more related to a square shaped pixel structure of a CCD sensor.

Usually, the calculation of the encircled energy is referred onto the chief ray intersection point in the image plane. But if an asymmetric PSF is obtained in case of coma, the reference point as the midpoint of the integrating circle is quite essential. A PSF with coma splits the chief ray location from the energetic centroid, see Chapter 18. Therefore, the encircled energy curve is changed considerably if the centroid or the chief ray is selected. Figure 6.97 shows the difference of $E(r)$ for a case with $c_8 = 1\,\lambda$ of coma. It is seen that, for example, for $E = 0.8$ the width is increased by nearly a factor

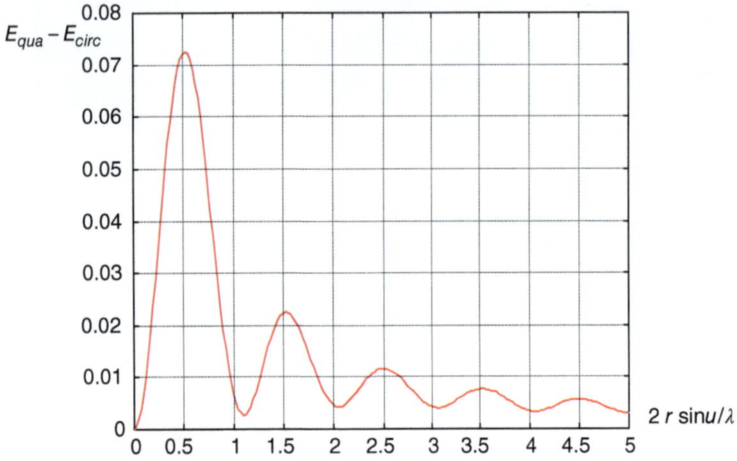

Figure 6.96 Difference between the encircled and the ensquared function for the Airy PSF.

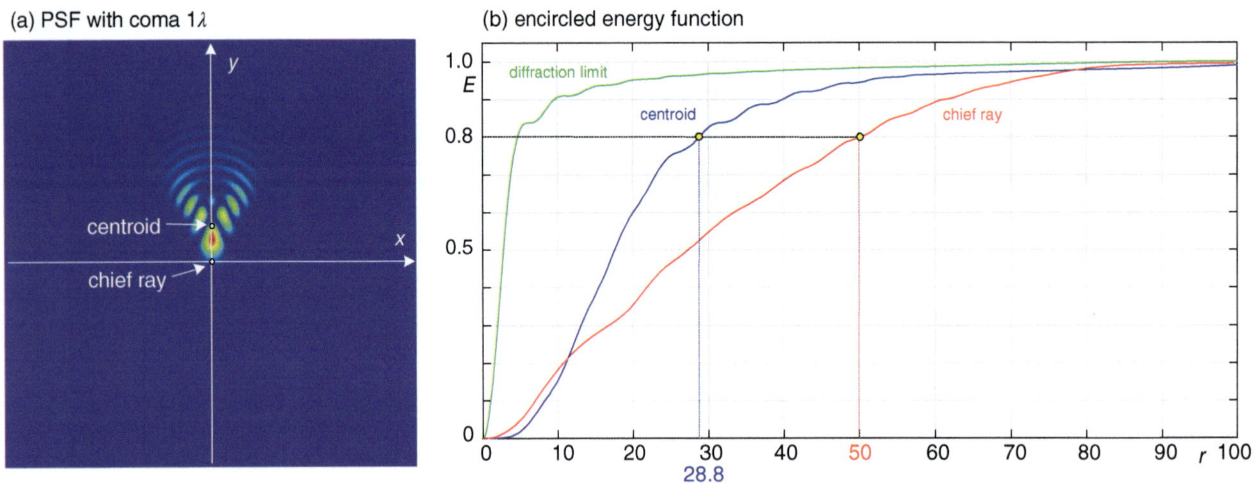

Figure 6.97 In (a) a coma intensity pattern shows the separation of the chief ray and the centroid position of the PSF. (b) gives the encircled energy function for the Airy function and the coma point spread function in the case that the centroid or the chief ray is selected as reference. The unit of r is arbitrary here.

of 2 if the chief ray reference is used. Physically, the centroid is much more meaningful.

6.9.5 Fresnel Edge Diffraction*

(*) The content of this section is quite specialized and for advanced reading. Knowledge of this section is not necessary to understand the following chapters. Beginners are recommended to skip this section.

The usual PSF is a very meaningful and reliable indicator of aberration degradation. But an existing problem with the approach in practice is preparing a point like object and controlling the small amount of energy during detection. Thus the SNR of such a measurement is an issue. Therefore, one attempt to improve this SNR issue is to select a straight edge as an object. In principle, the diffraction pattern of an edge contains the same information if in the case of an anisotropic PSF at least two different orientations of the edge are considered.

From the basic diffraction theory, the pattern after an edge, illuminated by a plane monochromatic wave, is given by an analytical formula containing the so-called Fresnel integrals [2], see Chapter 12 for more details.

$$C(x) = \int_0^x \cos\left(\frac{1}{2}\pi t^2\right) dt,$$

$$S(x) = \int_0^x \sin\left(\frac{1}{2}\pi t^2\right) dt \tag{6.204}$$

The intensity in a certain distance z after the edge in Fresnel approximation then follows by the equation

$$I(x) = \frac{1}{2} \cdot \left[\frac{1}{2} - C\left(\sqrt{\frac{k}{z \cdot \pi}} \cdot x\right)\right] + \left[\frac{1}{2} - S\left(\sqrt{\frac{k}{z \cdot \pi}} \cdot x\right)\right]^2 \tag{6.205}$$

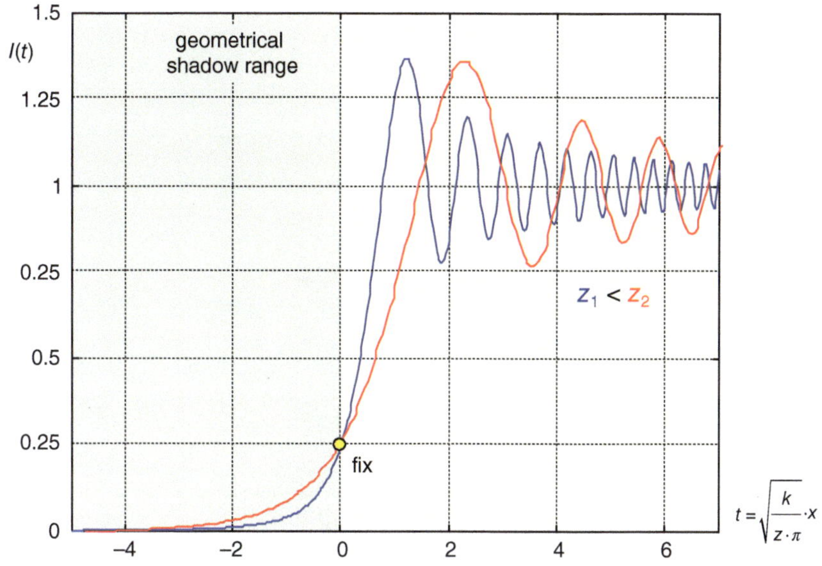

Figure 6.98 Fresnel edge diffraction intensity for two different distances z_1 and z_2.

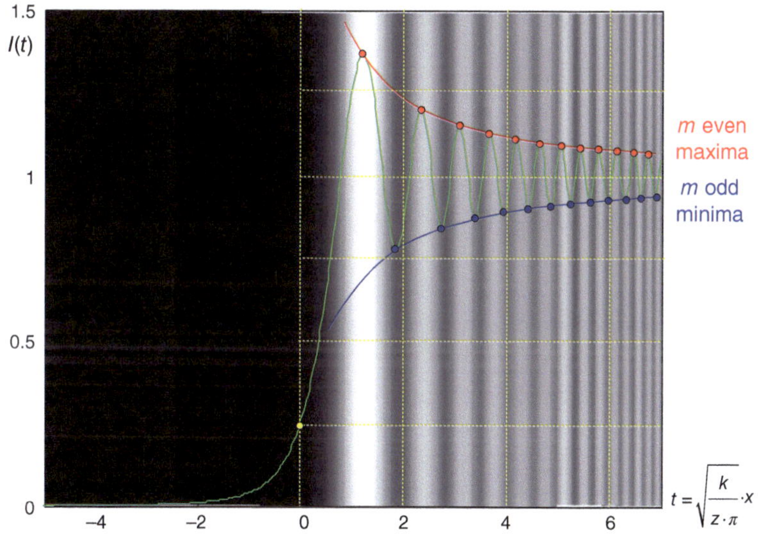

Figure 6.99 Fresnel edge diffraction intensity maxima and minima.

where x is the transverse coordinate perpendicular to the edge. Figure 6.98 shows this function dependent on the normalized lateral coordinate t instead of x. It can be seen that in the geometrical shadow range the intensity drops in a kind of exponential damping. On the bright side, an oscillatory behaviour is seen with a rather bad convergence. The corresponding grey-scale image is seen in Figure 6.99. The peak value is approximately 30% larger than in the unperturbed case. If the detection takes place at a larger distance z, the waviness is rescaled to coarser feature sizes. Interestingly, directly at the geometrical shadow boundary, the intensity is 1/4 of the maximum value. This point is the same for every distance z.

If the envelope on the bright side is considered, the maxima and minima follow the curve

$$\frac{I_m}{I_0} = \left[1 \pm \frac{1}{\pi \cdot \sqrt{4m-1}}\right]^2 \tag{6.206}$$

with the running index m for the maxima (red) and minima (blue) points, see Figure 6.99.

6.9.6 Line Spread Function*

(*) The content of this section is quite specialized and for advanced reading. Knowledge of this section is not necessary to understand the following chapters. Beginners are recommended to skip this section.

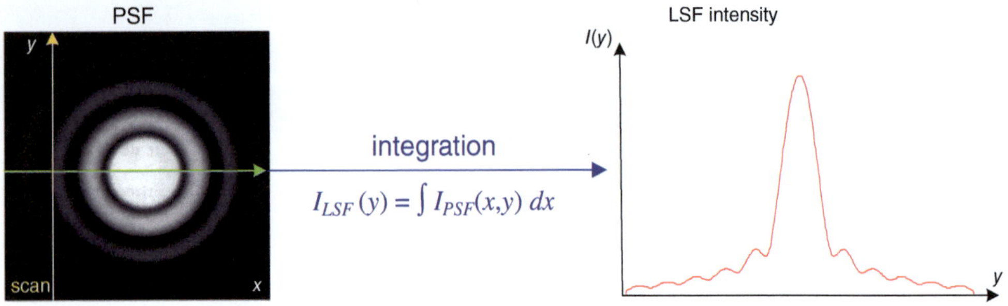

Figure 6.100 Line spread function as the integral over the point spread function.

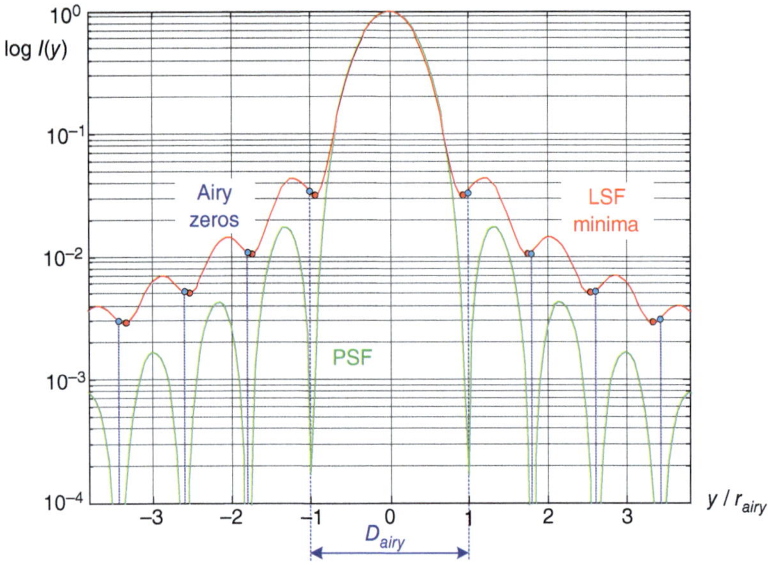

Figure 6.101 Line spread function as the integral over the point spread function and the location shift between zeros and minima.

To overcome the SNR issue of a point source, an alternative possibility is to integrate the PSF in one lateral direction. In this case the so-called ideal line spread function (LSF) is obtained [2, 9].

$$I_{LSF}(x) = \int I_{PSF}(x, y)\, dy \tag{6.207}$$

This idea is illustrated in Figure 6.100. Therefore, as an inversion of this equation, the PSF is the derivative of the LSF. One of the main properties of the LSF is that due to the one-dimensional integration over a circular symmetric pattern, there are no longer zeros of intensity found. Furthermore, it is illustrated in Figure 6.101 that the zero locations of the PSF (blue) no longer coincide with the minima of the LSF (red).

6.9.7 Slit Diffraction*

(*) The content of this section is quite specialized and for advanced reading. Knowledge of this section is not necessary to understand the following chapters. Beginners are recommended to skip this section.

The LSF of the previous section is an academic construct; however, a line of infinitesimal width cannot be realized. If a finite width $2a$ of a narrow slit is considered, a simple model of two overlaid edges can be used and the slit width serves as parameter. This is much more meaningful from the viewpoint of manufacturing and SNR optimization.

In accordance with the Fresnel edge diffraction, in the case of a slit with width $2a$ the field of a slit is therefore given by

$$\begin{aligned} E(x,z) &= \frac{1}{2}\left[C\left(\frac{(x-a)\sqrt{2}}{\sqrt{\lambda\cdot z}}\right) + C\left(\frac{-(x+a)\sqrt{2}}{\sqrt{\lambda\cdot z}}\right)\right] \\ &\quad + i\cdot\left[S\left(\frac{(x-a)\sqrt{2}}{\sqrt{\lambda\cdot z}}\right) + S\left(\frac{-(x+a)\sqrt{2}}{\sqrt{\lambda\cdot z}}\right)\right] \end{aligned} \tag{6.208}$$

With the Fresnel number N_F

$$N_F = \frac{a^2}{\lambda\cdot z} \tag{6.209}$$

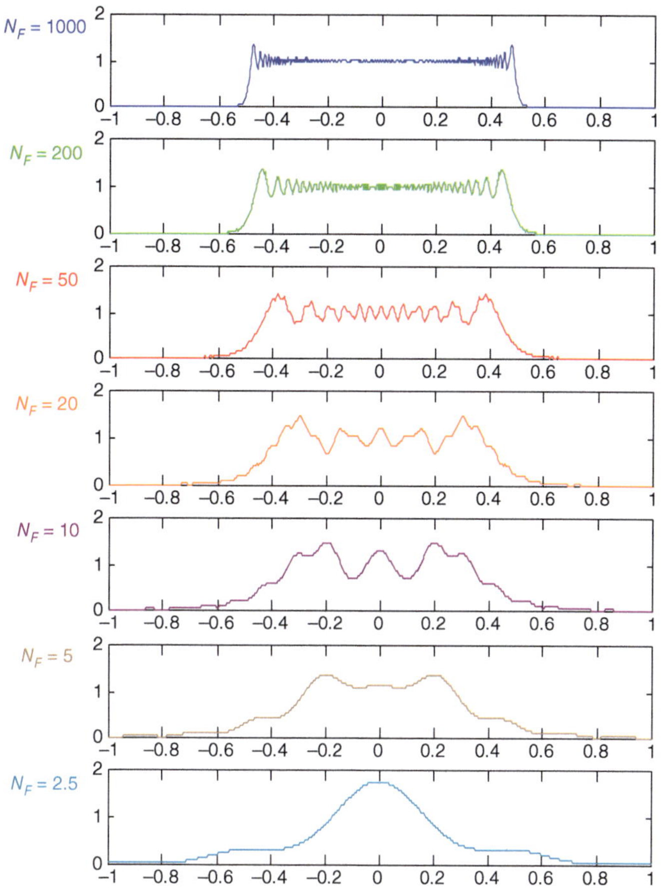

Figure 6.102 Fresnel diffraction of a slit for different Fresnel numbers. The horizontal axis is rescaled here to get a meaningful comparison for every distance.

the intensity in the distance z can be written as

$$I(x,z) = \frac{1}{4}\left[C\left(\left(\frac{x}{a}-1\right)\sqrt{2N_F}\right)+C\left(-\left(\frac{x}{a}+1\right)\sqrt{2N_F}\right)\right]^2 + \left[S\left(\left(\frac{x}{a}-1\right)\sqrt{2N_F}\right)+S\left(-\left(\frac{x}{a}+1\right)\sqrt{2N_F}\right)\right]^2 \tag{6.210}$$

Figure 6.102 illustrates the intensity pattern for some cases of N_F, calculated in re-scaled coordinates. It is seen that with a growing Fresnel number that means a decreasing distance z of observation for constant wavelength and slit width, the pattern resembles a more and more geometrically expected rectangular profile and gets high spatial frequency ripples.

In the far-field as a function of the angle θ, the intensity is given by

$$I(\theta) = I_0\left(\frac{2La}{\lambda R}\right)\left[\frac{\sin(2\pi/\lambda\cdot a\theta)}{2\pi/\lambda\cdot a\theta}\right]^2 = I_0'\cdot \operatorname{sinc}^2\left(\frac{2\pi\cdot a\theta}{\lambda}\right) \tag{6.211}$$

In this function, the relative side lobe maxima with index m are found at the angles

$$\sin\theta_{\max} = \frac{\lambda}{2a}\cdot\left(m+\frac{1}{2}\right) \tag{6.212}$$

With growing index m, the height of the maxima is decreasing as

$$\frac{I_{m,\max}}{I_0} = \frac{1}{\pi^2\cdot\left(m+\frac{1}{2}\right)^2} \tag{6.213}$$

6.10 Field in a Tilted Plane*

(*) The content of this section is quite specialized and for advanced reading. Knowledge of this section is not necessary to understand the following chapters. Beginners are recommended to skip this section.

6.10.1 Introduction

Usually, the diffraction based field propagation is performed between two parallel planes, which are perpendicular on the optical axis. Most of the available algorithms deal with this geometry. In practice, it can be of interest to know the intensity distribution on a tilted plane. Meanwhile, there are several approaches and suggestions regarding this problem found in the literature [116–134].

The main problem of this task is the interpolation of a complex field on a non-linear distorted grid, which can produce considerable errors if the phase variation is strong. The non-linearity is a function of the tilt angle size, in the end a robust algorithm for large angles is required. Potential applications are the calculation of the footprint on a mirror or grating with tilt angles, which cannot be considered to be small. A different range of applications addressed in practice is the field calculation in off-axis digital holography.

In the tilted plane we get a strong linear contribution of the phase variation. Therefore, if the field should be further propagated to a next location, the question is if this strongly varying field can be sampled properly and if it makes sense to use this field for the next step. If a simple estimation is made for a mirror under 45° incidence angle, a wavelength of $\lambda = 632$ nm and a collimated beam with diameter 2 mm, we get a phase variation of approximately 4500 periods across the mirror. This shows that a sampling of the field on the mirror needs 10 000 sampling points in the corresponding cross section, which is not realistic for practical calculations. Nevertheless, the proposed algorithm helps in determining the intensity distribution in the tilted plane.

Most of the suggested algorithmic approaches are general rotations in the frequency space with a clever interpolation scheme of the complex field represented as an angle spectrum. The distortion of the frequency grid for not too large angles can be taken into account if the non-uniform Fourier transform is used [117, 118], see Chapter 12 for details. But this attempt is only efficient if the tilt angles are small.

We are following here the approach suggested in [116], which is a quasi-fast algorithm to deal with this question. The basic idea here is to compute an azimuthal field rotation by three shearing operations, which can be performed with FFT or chirp-z-transform in a fast manner. A remaining tilt around x can be computed also quasi-fast with a split-z-step approach by a one-dimensional decomposition as described in [134].

6.10.2 Analysis of the Geometry

First in this approach, an arbitrary rotation operation is decomposed into three tilts with angles similar to a classical procedure of Euler. In the first tilt, the coordinate system is rotated around the optical z_1-axis by an angle θ. In the second step, the plane is tilted around the new x_2-axis by φ and finally a tilt with the angle γ is performed around the z_3-axis. Figure 6.103 shows the geometry of the system and indicates the corresponding rotations.

The three rotational matrices of the operations are given by

$$\underline{M}_\theta = \begin{pmatrix} \cos\theta & -\sin\theta & 0 \\ \sin\theta & \cos\theta & 0 \\ 0 & 0 & 1 \end{pmatrix},$$

$$\underline{M}_\varphi = \begin{pmatrix} 1 & 0 & 0 \\ 0 & \cos\varphi & -\sin\varphi \\ 0 & \sin\varphi & \cos\varphi \end{pmatrix},$$

$$\underline{M}_\gamma = \begin{pmatrix} \cos\gamma & -\sin\gamma & 0 \\ \sin\gamma & \cos\gamma & 0 \\ 0 & 0 & 1 \end{pmatrix} \tag{6.214}$$

and transform the coordinates as

$$\begin{pmatrix} x_4 \\ y_4 \\ z_4 \end{pmatrix} = \underline{M}_\gamma \underline{M}_\varphi \underline{M}_\theta \cdot \begin{pmatrix} x_1 \\ y_1 \\ z_1 \end{pmatrix} \tag{6.215}$$

The transform of a complex field into the new coordinates by this decomposition now requires two different tasks: a tilt around z, which corresponds to a classical azimuthal rotation and is used twice; and a tilt around the

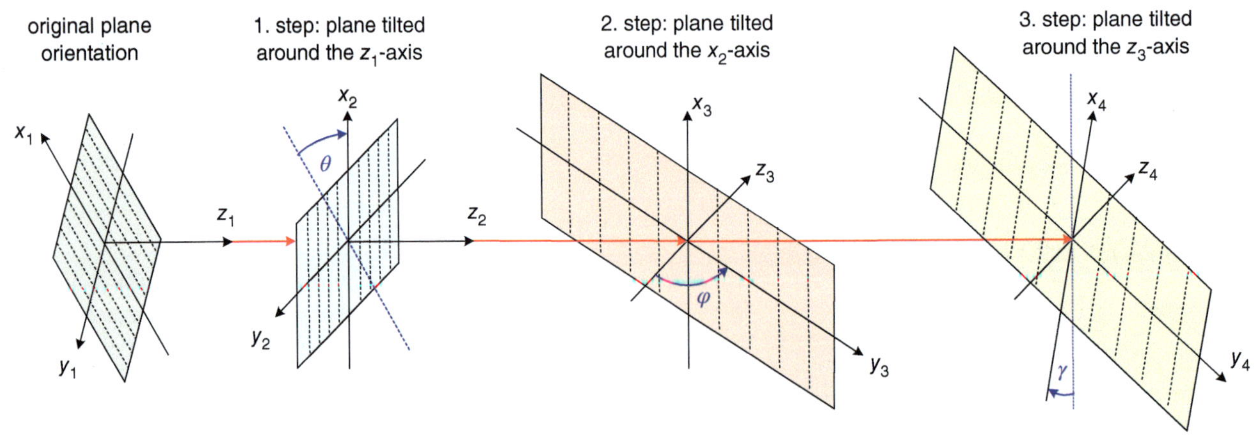

Figure 6.103 Analysis of the geometry and decomposition of the rotation into three tilt steps.

x-axis, which must be computed differently and stretches the coordinate in one direction.

6.10.3 *z*-Rotation by Shear Transform

It is well known that a rotation of an arbitrary two-dimensional distribution can be decomposed into three shearing operations with appropriate values of the shear parameters [135]. The underlying idea is illustrated in Figure 6.104. If the matrix of a z-rotation in the x–y plane is decomposed into three successive shearing matrices with the parameters a, b and c, we get the corresponding values of these parameters to obtain the rotation angle θ as follows. The rotation matrix:

$$\underline{M}_R = \begin{pmatrix} \cos\theta & -\sin\theta \\ \sin\theta & \cos\theta \end{pmatrix} \quad (6.216)$$

is factorized into three shearing transforms. A shear transform with parameters a or b can be performed in x or y-direction respectively as

$$\underline{M}_{ax} = \begin{pmatrix} 1 & a \\ 0 & 1 \end{pmatrix}, \quad \underline{M}_{by} = \begin{pmatrix} 1 & 0 \\ b & 1 \end{pmatrix} \quad (6.217)$$

If three shears are performed in y by the value a, in x by the value b and again in y by c respectively, we have the product matrix

$$\underline{M}_R = \begin{pmatrix} \cos\theta & -\sin\theta \\ \sin\theta & \cos\theta \end{pmatrix} = \begin{pmatrix} 1 & 0 \\ c & 1 \end{pmatrix} \begin{pmatrix} 1 & b \\ 0 & 1 \end{pmatrix} \begin{pmatrix} 1 & 0 \\ a & 1 \end{pmatrix} = \begin{pmatrix} 1+cb & a+b+abc \\ b & 1+ab \end{pmatrix} \quad (6.218)$$

from which we get the shear parameters as

$$b = \sin\theta, \quad a = c = -\tan\frac{\theta}{2} \quad (6.219)$$

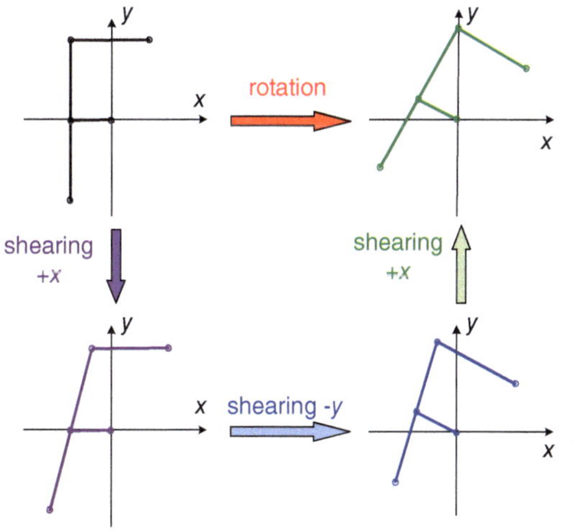

Figure 6.104 Decomposition of a rotation into three successive shear operations.

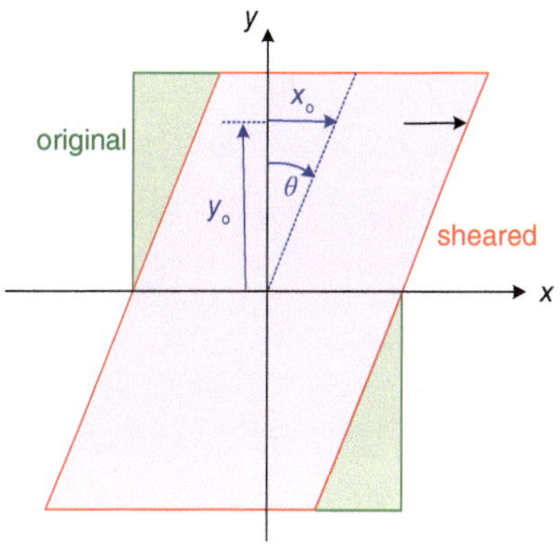

Figure 6.105 Shear in x-direction, keeping the y-coordinate unchanged.

The scheme of this decomposition is sketched in Figure 6.104.

A shear transform for example in x-direction can be considered as a shift operation in x with a y-dependent shear value, see Figure 6.105. For the y-direction it works similarly.

The shear can be calculated quickly with the help of the shift theorem of the Fourier transform. If $f(x)$ is the spatial function and $F(\nu)$ the Fourier spectrum

$$F(\nu) = \int f(x) \cdot e^{-2\pi i x\nu} dx \quad (6.220)$$

$$f(x) = \int F(\nu) \cdot e^{+2\pi i x\nu} d\nu \quad (6.221)$$

we get

$$f(x-x_0) = \int F(\nu) \cdot e^{+2\pi i(x-x_0)\,\nu} d\nu = \int \left[F(\nu) \cdot e^{-2\pi i x_0\,\nu}\right] \cdot e^{+2\pi i x\nu} d\nu \quad (6.222)$$

For every coordinate y_0, the corresponding shift x_0 is determined by

$$x_0 = y_0 \cdot \tan\theta \quad (6.223)$$

and the spectrum is multiplied by a linear phase and back transformed into the spatial domain

$$f(x-x_0) = \hat{F}^{-1}\left\{e^{-2\pi i x_0\,\nu} \cdot \hat{F}[f(x)]\right\} \quad (6.224)$$

The shear in the y-direction works similarly. Figure 6.106 shows for demonstration the performance of this algorithm with the three shear operations. In the example shown here, a given real valued profile with 128 × 128 points is rotated by an angle of 30° in (a), (b) illustrates the recovering by of the original profile by the inverse transform and in (c) 12 successive rotations by 30° deliver the original profile back with high accuracy. But it is clear that the accuracy of this algorithm is lower if a strong curvature of

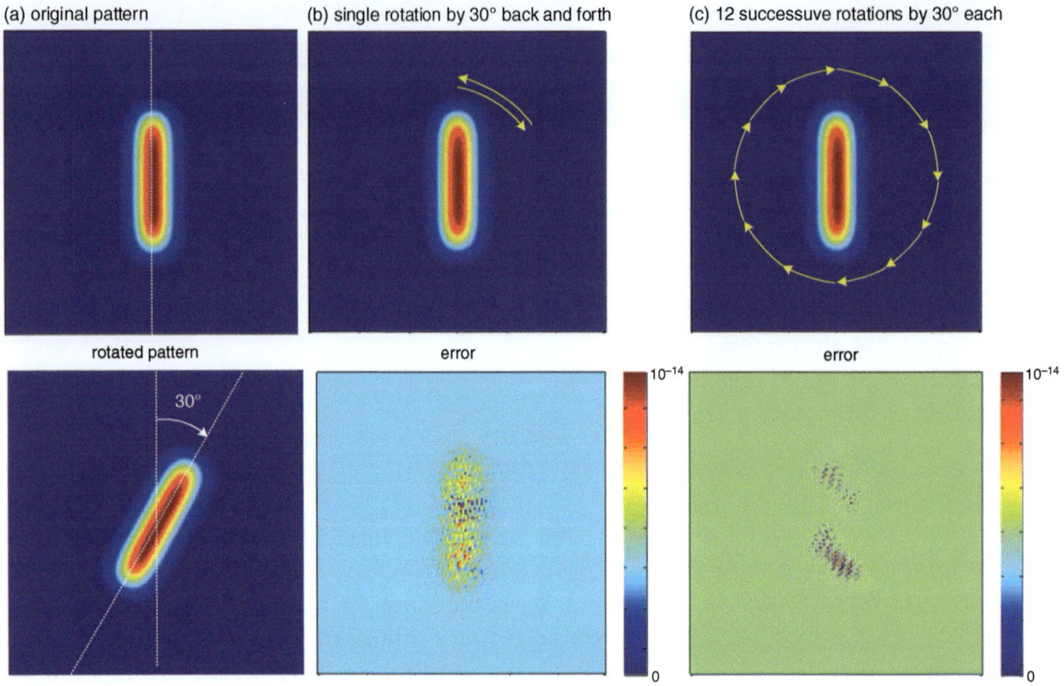

Figure 6.106 Test of the rotating algorithm by rotating a profile. The initial profile on the left side is rotated by 30° for 128 × 128 pixels. Test of the rotating algorithm by rotating and back-rotating a profile. The difference to the initial profile is in the range of numerical precision of the machine. In (a) a single rotation, in (b) single rotation forward and backwards and in (c) 12 successive rotations are performed.

the incoming phase exists and the corresponding sampling is more critical.

6.10.4 Calculation of the *x*-Rotation

In the case of a tilt around x the change of the field can be understood as a propagation over a y-dependent distance z, see the scheme in Figure 6.107. Here the case is illustrated, which in addition includes a z-propagation offset by z_o, but an in-place transform with $z_o = 0$ is possible as well.

Now according to Figure 6.107, the distance z is correlated to the transverse y-coordinate by

$$z(y) = z_0 + y \cdot \tan\varphi \tag{6.225}$$

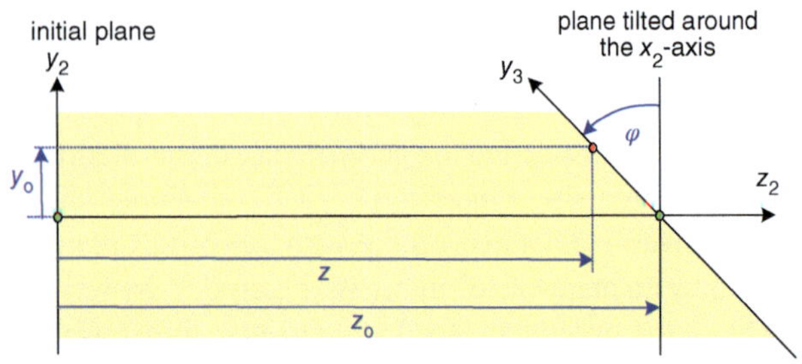

Figure 6.107 Transform of a propagation in z with a tilt around x_2.

and the new y-coordinate is projected onto the tilted plane by

$$y_3 = \frac{y_2}{\cos\varphi} \tag{6.226}$$

For the propagation, the well-known expansion into plane wave is used. First the plane wave spectrum at the location $z = 0$ is calculated by a Fourier transform

$$E(v_x, v_y, 0) = \iint E(x, y, 0) \cdot e^{-2\pi i \cdot (x v_x + y v_y)} \, dxdy \tag{6.227}$$

Generally, the propagation of the plane wave in a homogeneous medium can be calculated in scalar approximation by multiplying a simple phase factor. For small angles, the Fresnel approximation can be used and we get

$$E(v_x, v_y, z) = E(v_x, v_y, 0) \cdot e^{-ikz} \cdot e^{-i\pi\lambda z \cdot \left(v_x^2 + v_y^2\right)} \tag{6.228}$$

By back transforming the field, we get the distribution in the spatial domain at the corresponding position z

$$\begin{aligned} E(x, y, z) &= \iint E(v_x, v_y, z) \cdot e^{+2\pi i \cdot (x v_x + y v_y)} \, dv_x \, dv_y \\ &= \iint E(v_x, v_y, 0) \cdot e^{-ikz} \cdot e^{-i\pi\lambda z \cdot \left(v_x^2 + v_y^2\right)} \\ &\quad \cdot e^{+2\pi i \cdot (x v_x + y v_y)} \, dv_x \, dv_y \end{aligned} \tag{6.229}$$

In the case of the tilted plane, the values of y and z are coupled; especially, the propagation distance z depends on the height y. Therefore, the above calculation must be performed for every y-value separately with adapted propagation distance z. The complete two-dimensional field is than computed line-by-line. If in Eq. (6.229) the value y is considered, we get the corresponding z-value from Eq. (6.225). In Eq. (6.229) we get by separating the integrals in x and y-direction the propagation formula

$$E(x,y,z) = \int e^{+2\pi i \cdot x v_x} \cdot e^{-i\pi\lambda\, z(y)\cdot v_x^2} \times \left[\int E(v_x, v_y, 0)\cdot e^{-ikz(y)} \cdot e^{+2\pi i\cdot y\cdot v_y} \cdot e^{-i\pi\lambda\, z(y)\cdot v_y^2} dv_y\right] dv_x \tag{6.230}$$

Now the inner integral over v_y is no longer a Fourier integral. It is just a summation of the spectrum with a constant factor, a linear contribution and a quadratic chirp. This can be calculated quickly for every value of y. Then the conventional inverse Fourier transform along x is performed with the concrete value of $z(y)$. The constant factor and the chirp depend on the distance z, the slope factor describes the lateral offset for y.

In conclusion, the propagation algorithm scales as the conventional two-dimensional Fourier transform with the number of points. Only the individual multiplications in the formula above cause some larger computational effort. Now the propagation calculation can be performed by calculating N_y times the N_x sampling points with different distances. This slows down the calculation but in principle is quasi-fast. If the Fresnel approximation is not feasible in cases of larger numerical aperture, the formula (6.229) can also be written in a compatible form with the exact propagation kernel as

$$E(x,y,z) = \int e^{+2\pi i \cdot x v_x} \times \left[\int E(v_x, v_y, 0)\cdot e^{iz(y)\cdot \sqrt{1/\lambda^2 - v_x^2 - v_y^2}} \cdot e^{+2\pi i\cdot y\cdot v_y} dv_y\right] dv_x \tag{6.231}$$

which allows us to use the same algorithm.

If in the new tilted y-direction the same grid spacing Δy is desired as in the original mesh to maintain the same resolution of the computational grid in x and y, a resizing of the pixel size must be performed. This can be done with the chirp-z-transform, which allows for a rescaling of the frequency grid in the classical fast Fourier transform, see Chapter 12 for more details. There is an additional scaling parameter $s > 1$ in this transform, which changes the frequency spacing by a corresponding value. If now a function is Fourier transformed with a certain scaling factor s_1 and it is re-transformed with a different factor s_2, the spatial coordinates are stretched by a factor s_1/s_2. In the two-dimensional case, the transform and the corresponding scaling can be selected in only one of the coordinate directions. Therefore, it is possible to adapt the pixel size in y-direction only by properly selecting the values for s_1 and s_2.

6.10.5 Run Time Performance

The various matrix calculations in the propagation routine are scaling with N^4, this finally dominates the run time for the propagation routine. Table 6.6 shows some numbers for a two-dimensional calculation with the traditional FFT and the chirp-z transform in two dimensions together with the time for the rotation operations as a function of the number of points $N_x = N_y$.

If the tilt is performed around the x-axis, the computational effort in y-direction is larger and here we do not see the classical $N \log N$-increase of run time. To see the effect, Figure 6.108 shows the computational time if in one dimension $N = 256$ points are fixed and in that direction

Table 6.6 Run times in seconds of the two routines together with corresponding times for comparison of the pure FFT and the chirp-z-transform routines in two dimensions and the rotation processing.

$N_x = N_y$	Time FFT2	Time czt2	Time rotation	Time propagation
64	0.0001	0.0040	0.050	0.063
128	0.0003	0.0080	0.036	0.026
256	0.0013	0.030	0.083	0.14
512	0.0075	0.094	0.265	1.07
1024	0.031	0.361	0.870	7.0
2048	0.167	1.69	3.42	56.2

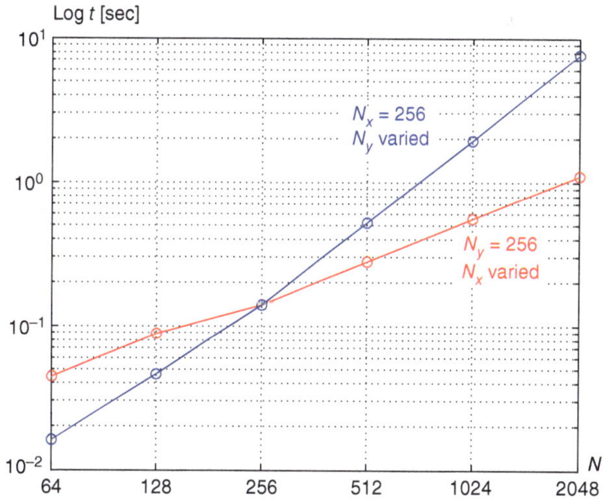

Figure 6.108 Computational time in seconds for different sampling point numbers N in x and y-direction correspondingly in logarithmic representation.

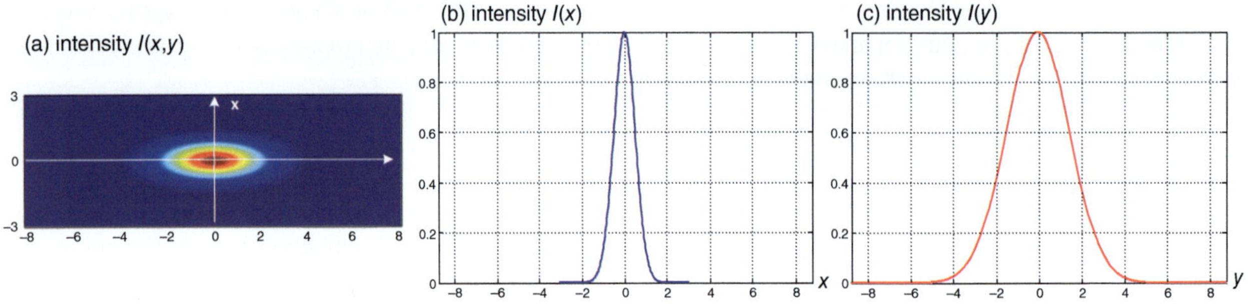

Figure 6.109 Gaussian beam with circular cross section on a plane that is tilted by 70°. The units of x and y are arbitrary.

the number of points is increased in powers of 2 from 64 to 2048 respectively. It is seen that the increase in time is considerably larger in the y-direction. By a rough estimation, the effort grows proportionally with $N_x N_y^2$. This means that the algorithm is only fast in the uncritical section and is slow in the y-section.

6.10.6 Examples

Here some examples of propagation of a field calculation into a tilted plane are presented to demonstrate the proposed algorithm.

The first example shows a circular symmetric Gaussian beam, which falls onto a tilted mirror. The tilt angle is 70°, the enlargement of the y-cross section by $1/\cos(70°) = 2.9$ into an elliptical cross section is clearly seen in Figure 6.109.

The second example shows a nearly tophat profile, which impinges onto a strongly tilted plane. A corresponding application is the beam profile on a Littrow grating, which is used in a high diffraction order. Here the angle is 89.4°. It is seen that the change of the distance z over the grating footprint matters and the Fresnel number from the near edge to the far edge of the plane deforms the pattern and the classical properties of the Fresnel diffraction pattern to an asymmetric distribution, as shown in Figure 6.110.

In a third example, the phase in the tilted plane for small tilt angles is inspected to control the algorithm. The

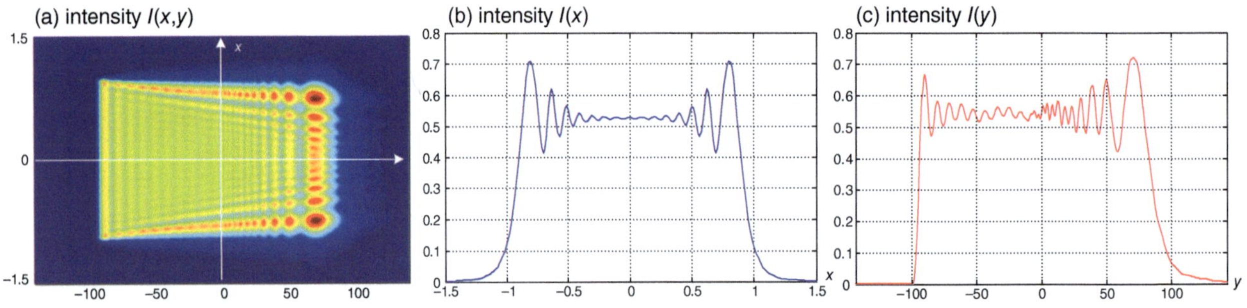

Figure 6.110 Propagation of a collimated beam with nearly rectangular profile onto a plane, which is rotated by an angle of 89.4°. It should be noted that the x- and y-scalings are completely different in these representations. The units of x and y are arbitrary.

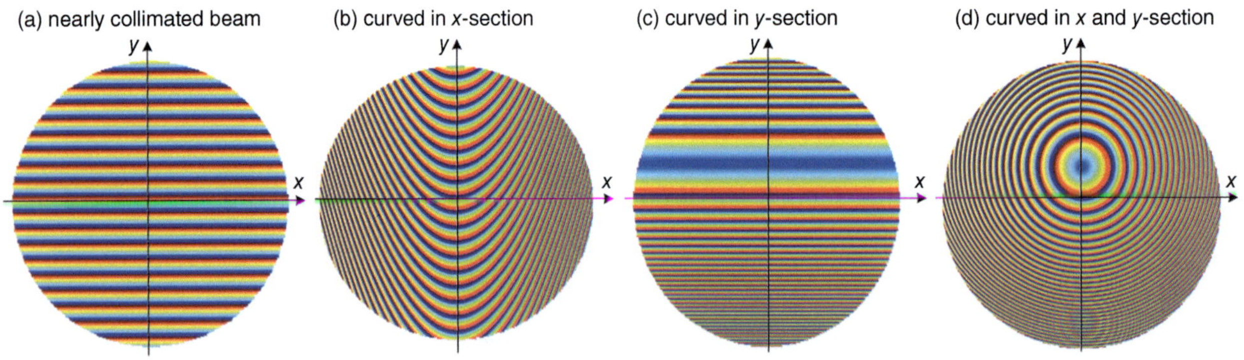

Figure 6.111 Phase in the tilted plane for a small tilt angle in three different cases of the initial beam phase.

following Figure 6.111 shows three cases depending on the initial phase of a Gaussian beam, which is propagated in the near-field onto a slightly tilted plane (0.3°) for different curvatures of the initial wave. In the left part, a collimated case shows the linear wrapped phase distribution according to the tilt only. The parts (b) and (c) illustrate the phase for cylindrical curved phase in x and y respectively and in (d) a combined curvature is applied. As can be seen, although the tilt angle is rather small, the correct sampling of the resultant phase distribution is quite challenging.

The algorithm described here can also be used with the conventional O-O-Fresnel propagator for geometrical convergent or divergent beams as described in Section 6.2.6.4. In this case, it must be taken into account that the magnifications change for the different propagation distances. Since a constant final grid is desired, the curvature for flattening the initial beam must be adapted correspondingly. This causes bad conditioning for convergence or divergence angles that are too large.

References

1 Born, M. and Wolf, E. (1999). *Principles of Optics.* Cambridge: Cambridge University Press.

2 Goodman, J. (2017). *Fourier optics.* Freeman.

3 Barton, G. (1989). *Elements of Green's Functions and Propagation.* Oxford Science Publications.

4 Walther, A. (1995). *The Ray and Wave Theory of Lenses.* Cambridge University Press.

5 Solimeno, S., Crosignani, B., and di Porto, P. (1986). *Guiding, Diffraction and Confinement of Optical Radiation.* Academic Press.

6 Stamnes, J. (1986). *Waves in Focal Regions.* Adam Hilger.

7 Nieto-Vesperinas, M. (2006). *Scattering and Diffraction in Physical Optics.* World Scientific.

8 Ishimaru, A. (1991). *Electromagnetic Wave Propagation, Radiation and Scattering.* Prentice Hall.

9 Reynolds, G., DeVlies, J., Parrent, G., and Thompson, B. (1989). *Physical Optics Notebook: Tutorial in Fourier Optics.* SPIE Press.

10 Siegman, A. (1986). *Lasers.* University Science Books.

11 Gu, M. (2000). *Advanced Optical Imaging Theory.* Springer.

12 Braat, J. and Török, P. (2019). *Imaging Optics.* Cambridge University Press.

13 Ufimtsev, P. (2009). *Theory of Edge Diffraction in Electromagnetics.* SciTech Publishing.

14 McNamara, D., Pistorius, C., and Malherbe, J. (1990). *The Uniform Geometrical Theory of Diffraction.* Artech IIousc.

15 Babic, V. and Buldyrev, V. (1991). *Short-Wavelength Diffraction Theory.* Springer.

16 Keller, J. (1958). *A Geometrical Theory of Diffraction.* McGraw Hill.

17 Maggi, G. (1888). Sulla propagazione libera e perturbata delle onde luminose in un mezzo isotropo. *Annali di Matematica Pura ed Applicata II* 16: 21.

18 Rubinowicz, A. (1966). *Die Beugungswelle in der Kirchhoffsche Theorie der Beugung*, 2nde. Berlin: Springer.

19 Marchand, E.W. and Wolf, E. (1962). Boundary diffraction wave in h domain of the Rayleigh-Kirchhoff diffraction theory. *Journal of the Optical Society of America* 52: 761.

20 Keller, J. (1962). Geometrical theory of diffraction. *Journal of the Optical Society of America A* 52: 116.

21 Felsen, L. and Marcuvitz, N. (1973). *Radiation and Scattering of Waves.* Prentice-Hall.

22 Miyamoto, K. and Wolf, E. (1962). Generalization of the Maggi-Rubinovicz theory of boundary diffraction wave, part I. *Journal of the Optical Society of America* 52: 615.

23 Miyamoto, K. and Wolf, E. (1962). Generalization of the Maggi-Rubinovicz theory of boundary diffraction wave, part II. *Journal of the Optical Society of America* 52: 626.

24 Marathay, A. and McCalmont, J. (2004). On the usual approximation used in the Rayleigh-Sommerfeld diffraction theory. *Journal of the Optical Society of America A* 21: 510.

25 Cheng, N., Wang, C., and Heidrich, W. (2022). HTRSD: hybrid Taylor Rayleigh-Sommerfeld diffraction. *Optics Express* 30: 37727.

26 Kozacki, T. and Falaggis, K. (2015). Angular spectrum-based wave-propagation method with compact space bandwidth for large propagation distances. *Optics Letters* 40: 3420.

27 Li, J., Fan, Z., and Yunchang, F. (2002). The FFT calculation for Fresnel diffraction and energy conservation criterion of sampling quality. *SPIE Conference Proceedings* 4915: 180.

28 Rees, W. (1987). The validy of the Fresnel approximation. *European Journal of Physics* 8: 44.

29 Forbes, G. (1996). Validity of the Fresnel approximation in the diffraction of collimated beams. *Journal of the Optical Society of America A* 13: 1816.

30 Southwell, W. (1981). Validity of the Fresnel approximation in the near field. *Journal of the Optical Society of America* 71: 7.

31 Wang, Z., Baladron-Zorita, O., Hellmann, C., and Wyrowski, F. (2021). Generalized far field integral. *Optics Express* 29: 1774.

32 Leutenegger, M., Rao, R., Leitgeb, R., and Lasser, T. (2006). Fast focus field calculations. *Optics Express* 14: 11277.

33 Sheppard, C. (2000). Validity of the Debye approximation. *Optics Letters* 25: 1660.

34 Wolf, E. and Li, Y. (1981). Conditions for the validity of the Debye integral representation of focused fields. *Optics Communication* 39: 295.

35 Li, Y. (2022). Disguised assumptions and the conditions for validity of Debye integral representation of focused field. *Journal of the Optical Society of America A* 39: C156.

36 Collins, S. (1970). Lens-system diffraction integral written in terms of matrix optics. *Journal of the Optical Society of America* 60: 1168.

37 Wang, S. and Zhao, D. (2000). *Matrix Optics*. Springer.

38 Palma, C. and Bagini, V. (1997). Extension of the Fresnel transform to ABCD systems. *Journal of the Optical Society of America A* 14: 1774.

39 Zhao, C. and Tan, W. (2008). An optical system with aberrations on diffraction integrals written in terms of a generalized ABCD matrix. *Central European Journal of Physics* 6: 895.

40 Li, J. and Li, C. (2008). Algorithm study of Collins formula and inverse Collins formula. *Applied Optics* 47: A97.

41 Baues, P. (1969). The connection of geometrical optics with the propagation of Gaussian beams and the theory of optical resonators. *Opto-Electronics* 1: 103.

42 D. Kelly, J. Sheridan and W. Rhodes (2003). The scaled optical Fourier transform: practical limitations. *SPIE Conference Proceedings* 5182: 149.

43 Shakir, S., Fried, D., Pease, E. et al. (2015). Efficient matrix approach to optical wave propagation and linear canonical transform. *Optics Express* 23: 26853.

44 Healy, J., Kutay, M., Ozaktas, H., and Sheridan, J. (2016). *Linear Canonical Transform*. Springer.

45 Healy, J. and Sheridan, J. (2010). Fast linear canonical transforms. *Journal of the Optical Society of America A* 27: 21.

46 J. Healy and J. Sheridan (2008). Bandwidth, compact support, apertures and the linear canonical transform in ABCD systems. *SPIE Conference Proceedings* 6994. https://doi.org/10.1117/12.780313.

47 Hennelly, B. and Sheridan, J. (2005). Fast numerical algorithm for the linear canonical transform. *Journal of the Optical Society of America A* 22: 928.

48 Hennely, B. and Sheridan, J. (2005). Generalizing, optimizing and inventing numerical algorithms for the fractional Fourier, Fresnel and linear canonical transform. *Journal of the Optical Society of America A* 22: 917.

49 Ozaktas, H., Zalevsky, Z., and Kitay, M. (2001). *The Fractional Fourier Transform*. Wiley.

50 Koc, A., Ozaktas, H., and Hesselink, L. (2010). Fast and accurate algorithm for the computation of complex linear canonical transforms. *Journal of the Optical Society of America A* 27: 1896.

51 Roose, S., Brichau, B., and Stijns, E.W. (1993). An efficient interpolation algorithm for Fourier and diffractive optics. *Optics Communication* 97: 312.

52 Bakx, J.L. (2002). Efficient computation of optical disk readout by use of the chirp z transform. *Applied Optics* 41: 4897.

53 Cubillos, M. and Jimenez, E. (2022). Numerical simulation of optical propagation using sinc approximation. *Journal of the Optical Society of America A* 39: 1403.

54 Cubillos, M. and Jimenez, E. (2022). Diffraction integral computation using sinc approximation. *Applied Numerical Mathematics* 178: 69.

55 Lawrence, G. (1992). Optical Modeling. In: *Applied Optics and Optical Engineering*, Vol XI (ed. R. Shannon and J. Wyant), 125ff. Academic Press.

56 Mas, D., Espinosa, J., Perez, J., and Illueca, C. (2006). Scale corrections for faster evaluation of convergent Fresnel patterns. *Journal of Modern Optics* 53: 259.

57 Kelly, D., Sheridan, J., and Rhodes, W. (2006). Finite-aperture effects for Fourier transform systems with convergent illumination. Part I: 2-D system analysis. *Optics Communication* 263: 171.

58 Mas, D., Perez, J., Hernandez, C. et al. (2003). Fast numerical calculation of Fresnel patterns in convergent systems. *Optics Communication* 227: 245.

59 Mendlovic, D., Zalevsky, Z., and Konforti, N. (1997). Computation considerations and fast algorithms for calculating the diffraction integral. *Journal of Modern Optics* 44: 407.

60 Mas, D., Garcia, J., Ferreira, C. et al. (1999). Fast algorithms for free-space diffraction patterns calculation. *Optics Communication* 164: 233.

61 Asoubar, D., Zhang, S., Wyrowski, F., and Kuhn, M. (2014). Efficient semi-analytical propagation techniques for electromagnetic fields. *Journal of the Optical Society of America A* 31591.

62 Z. Wang (2020). *Fourier Transform Techniques for Fast Physical Modeling*. Doctoral thesis, University of Jena.

63 Shimobaba, T., Matsushima, K., Kakue, T. et al. (2012). Scaled angular spectrum method. *Optics Letters* 37: 4128.

64 Shimobaba, T., Masuda, N., and Ito, T. (2012). Arbitrary shape surface Fresnel diffraction. *Optics Express* 20: 9335.

65 Miora, R., Rohwer, E., Kielhorn, M. et al. (2024). Calculating point spread functions: methods, pitfalls and solutions. *Optics Express* 32: 27278.

66 Abedi, M., Saghafifar, H., and Rahmi, L. (2024). Improvement of optical wave propagation simulations: the scaled angular spectrum method for far-field and focal analysis. *Optics Continuum* 3: 935.

67 Liu, Y., Guo, F., Zhang, Z., and Wu, R. (2024). Wave optical propagation in realistic lens systems through multi-slice decomposition with phase compensation. *Applied Optics* 63: F18.

68 Fujimori, S., Ito, T., and Shimobaba, T. (2024). Rationalized diffraction calculations for high accuracy and high speed with few bits. *Journal of the Optical Society of America A* 41: 303.

69 Wei, H., Liu, X., Hao, X. et al. (2023). Modeling off-axis diffraction with the least-sampling angular spectrum method. *Optica* 10: 959.

70 Zhao, J. (2022). Fast numerical propagation in high-NA imaging using the resampling angular spectrum method. *Optics Express* 30: 41492.

71 Schmidt, J. (2010). *Numerical Simulation of Optical Wave Propagation*. SPIE Press.

72 Onural, L. (2000). Sampling of the diffraction field. *Applied Optics* 39: 5929.

73 J. Mansell, R. Praus and S. Coy (2007). Determining wave-optics mesh parameters for complex optical systems. *SPIE Conference Proceedings* 6675. https://doi.org/10.1117/12.734887.

74 Onural, L. (2007). Exact analysis of the effects of sampling of the scalar diffraction field. *Journal of the Optical Society of America A* 24: 359.

75 Zhang, W., Zhang, H., Sheppard, C., and Jin, G. (2020). Analysis of numerical diffraction calculation methods: from the perspective of phase-space optics and the sampling theorem. *Journal of the Optical Society of America A* 37: 1748.

76 S. Coy (2005). Choosing mesh spacings and mesh dimensions for wave optics simulation. *SPIE Conference Proceedings* 5894: 45.

77 Xiao, J., Zhang, W., and Zhang, H. (2022). Sampling analysis for Fresnel diffraction fields based on a phase-space representation. *Journal of the Optical Society of America A* 39: A15.

78 Koc, A., Ozaktas, H., and Hesselink, L. (2010). Fast and accurate computation of two-dimensional non separable quadratic-phase integrals. *Journal of the Optical Society of America A* 27: 1288.

79 Luchini, P. (1984). Two-dimensional numerical integration using a square mesh. *Computer Physics Communications* 31: 303.

80 Will, S. and Fienup, J. (2020). Algorithm for exact area-weighted antialiasing of discrete circular aperture. *Journal of the Optical Society of America A* 37: 688.

81 Greengard, L. and Lee, J.-Y. (2004). Accelerating the nonuniform fast Fourier transform. *SIAM Review* 46: 443.

82 Fessler, J. and Sutton, B. (2003). Nonuniform fast Fourier transforms using min-max interpolation. *IEEE Transactions on Signal Processing* 51: 560.

83 Kuo, J.-T. and Lee, H.-Y. (2009). Improved accuracy factors for the nonuniform fast Fourier transform (NUFFT) algorithm. *IEEE Microwave and Wireless Components Letters* 19: 3.

84 Shimobaba, T., Kakue, T., Oikawa, M. et al. (2013). Nonuniform sampled scalar diffraction calculation using nonuniform fast Fourier transform. *Optics Letters* 38: 5130.

85 Xiao, Y., Tang, X., Qin, Y. et al. (2016). Nonuniform fast Fourier transform method for numerical diffraction simulation on tilted planes. *Journal of the Optical Society of America A* 33: 2027.

86 Shi, R. and Wyrowski, F. (2019). Comparison of aplanatic and real lens focused spots in the framework of the local plane interface approximation. *Journal of the Optical Society of America A* 36: 1801.

87 H. Lajunen, J. Tervo, T. Vallius, et al. (2003). Local spherical interface approximation in the analysis of the field propagation through modulated interfaces. *SPIE Conference Proceedings* 5182: 15.

88 H. Zhong, S. Zhang, F. Wyrowski and H. Schweitzer (2014). Parabasal thin element approximation for the analysis of the diffractive elements. *SPIE Conference Proceedings* 9131. https://doi.org/10.1117/12.2055116.

89 Li, Y. and Wolf, E. (1984). Three-dimensional intensity distribution near the focus in systems of different Fresnel numbers. *Journal of the Optical Society of America A* 1: 801.

90 Li, Y. and Wolf, E. (1981). Focal shifts in diffracted converging spherical waves. *Optics Communication* 39: 211.

91 Takada, A. and Shibuya, M. (2010). Incoming inclination factor for scalar imaging theory. *Optical Engineering* 49: 023202.

92 M. Shibuya, A. Nishikata, A. Takada and S. Nakadate (2006). Proposing the inclination factor which satisfies the reciprocity theorem in scalar imaging theory and confirming the validity by numerical calculations. *SPIE Conference Proceedings* 6288. https://doi.org/10.1117/12.680207.

93 Guo, H., Chen, J., and Zhuang, S. (2006). Resolution of aplanatic systems with various semi apertures, viewed from the two sides of the diffraction aperture. *Journal of the Optical Society of America A* 23: 2756.

94 Gu, M. (2000). *Advanced Optical Imaging Theory*. Springer.

95 Braat, J. and Török, P. (2019). *Imaging Optics*. Cambridge University Press.

96 A. van de Nes (2005). *Rigorous Electromagnetic Filed Calculations for Advanced Optical Systems*. Doctoral thesis, University of Delft.

97 Mansuripur, M. (1989). Certain computational aspects of vector diffraction problems. *Journal of the Optical Society of America A* 6: 786.

98 Mansuripur, M. (1986). Distribution of light at and near the focus of high-numerical aperture objectives. *Journal of the Optical Society of America A* 3: 2086.

99 Richards, B. and Wolf, E. (1959). Electromagnetic diffraction in optical systems II. Structure of the image field in an aplanatic system. *Proceedings of the Royal Society A* 253: 358.

100 P. Török (2003). *An Imaging Theory for Advanced, High Numerical Aperture Optical Microscopes*. Doctoral thesis, Imperial College London.

101 Ciocci, L., Echarri, R., and Simon, J. (2010). Diffraction in high numerical aperture systems: polarization effects. *Journal of Optics* 12: 12408.

102 Andreas, B., Mana, G., and Palmisano, C. (2015). Vectorial ray-based diffraction integral. *Journal of the Optical Society of America A* 32: 1403.

103 Flagello, D., Milster, T., and Rosenbluth, A. (1996). Theory of high-NA imaging in homogeneous thin films. *Journal of the Optical Society of America A* 13: 53.

104 Gross, H. (2020). Cascaded diffraction in optical systems. Part I: simulation model. *Journal of the Optical Society of America A* 37: 240.

105 Gross, H. (2020). Cascaded diffraction in optical systems. Part II: example calculation. *Journal of the Optical Society of America A* 37: 250.

106 Mout, M., Wick, M., Bociort, F., and Petschulat, J. (2018). Ray-based method for simulating cascaded diffraction in high-numerical aperture systems. *Journal of the Optical Society of America A* 35: 1356.

107 Ferdman, B., Saguy, A., Xiao, D., and Shechtman, Y. (2022). Diffractive optical system design by cascaded propagation. *Optics Express* 30: 27509.

108 Rataj, M. (1984). Convolution formulation of Fresnel diffraction for cascaded optical systems. *Optica Acta* 31: 1237.

109 deVelis, J. and Parrent, G. (1967). Transfer function for cascaded optical systems. *Journal of the Optical Society of America* 57: 1486.

110 Lakshminarayanan, V. and Calvo, M. (1990). Point spread function and modulation transfer function of a photo lens treated as a cascade linear system under the Fresnel regime: application to a Tessar lens. *Optical Engineering* 29: 263.

111 Yurlov, V., Han, K., and Yu, N. (2018). Cascaded Fresnel diffraction theorem and its application. *Optical Engineering* 57: 055108.

112 Calvo, M. and Lakshminarayanan, V. (1991). General transfer function of a compound optical system under linearity conditions in the Fresnel regime. *Optical Engineering* 30: 1503.

113 Mahajan, V. (1985). Line of sight of an aberrated optical system. *Journal of the Optical Society of America A* 2: 833.

114 Braat, J., Dirksen, P., and Jannsen, A. (2002). Assessment of an extended Nijboer-Zernike approach for the computation of optical point spread functions. *Journal of the Optical Society of America A* 19: 858.

115 Braat, J., Dirksen, P., Jannsen, A., and van de Nes, A. (2003). Extended Nijboer-Zernike representation of the vector field in the focal region of an aberrated high-aperture optical system. *Journal of the Optical Society of America A* 20: 2281.

116 Stock, J., Worku, N., and Gross, H. (2017). Coherent field propagation between tilted planes. *Journal of the Optical Society of America A* 34: 1849.

117 Xiao, Y., Tang, X., Qin, Y. et al. (2016). Nonuniform fast Fourier transform method for numerical diffraction simulation on tilted planes. *Journal of the Optical Society of America A* 33: 2027.

118 Chang, C., Xia, J., Wu, J. et al. (2014). Scaled diffraction calculation between tilted planes using nonuniform fast Fourier transform. *Optics Express* 22: 17331.

119 Tommasi, T. and Bianco, B. (1993). Computer generated holograms of tilted planes by a spatial frequency approach. *Journal of the Optical Society of America A* 10: 299.

120 Doumic, M., Duboc, F., Golse, F., and Sentis, R. (2009). Simulation of laser beam propagation with a paraxial model in a tilted frame. *Journal of Computational Physics* 228: 861.

121 Abedin, K. and Rahman, S. (2012). The iterative Fresnel integrals method for Fresnel diffraction from tilted rectangular apertures: theory and simulations. *Optics and Laser Technology* 44: 939.

122 S. Zhang, H. Zhong, D. Asubar, et al. (2012). Tilt operator for electromagnetic fields and its application to propagation through plane interfaces. *SPIE Conference Proceedings* 8550. https://doi.org/10.1117/12.999424.

123 Zhang, S., Asoubar, D., Hellmann, C., and Wyrowski, F. (2016). Propagation of electromagnetic fields between non-parallel planes: a fully vectorial formulation and an efficient implementation. *Applied Optics* 55: 529.

124 Kim, Y.-H., Kim, G.H., Ryu, H. et al. (2015). Exact light propagation between rotated planes using non-uniform sampling and angular spectrum method. *Optics Communication* 344: 1.

125 Onural, L. (2011). Exact solution for scalar diffraction between tilted and translated planes using impulse functions over a surface. *Journal of the Optical Society of America A* 28: 290.

126 Matsushima, K., Schimmel, H., and Wyrowski, F. (2003). Fast calculation method for optical diffraction on tilted planes by use of the angular spectrum of plane waves. *Journal of the Optical Society of America A* 20: 1755.

127 Delen, N. and Hooker, B. (1998). Free-space beam propagation between arbitrarily oriented planes based on full diffraction theory: a fast Fourier approach. *Journal of the Optical Society of America A* 15: 857.

128 Matsushima, K. (2008). Formulation of the rotational transformation of wave fields and their application to holography. *Applied Optics* 47: D110.

129 Yamamoto, K., Ichihashi, Y., Senoh, T. et al. (2012). Calculating the Fresnel diffraction of light from a shifted and tilted plane. *Optics Express* 20: 12949.

130 S. Zhang, H. Zhong, D. Asoubar, et al. (2012). Tilt operator for electromagnetic fields and its application to propagation through plane interfaces. *SPIE Conference Proceedings* 8550. https://doi.org/10.1117/12.999424.

131 Modregger, P., Lübbert, D., Schäfer, P. et al. (2008). Fresnel diffraction in the case of an inclined image plane. *Optics Express* 16: 5141.

132 Delen, N. and Hooker, B. (2001). Verification and comparison of a fast Fourier transform-based full diffraction method for tilted and offset planes. *Applied Optics* 40: 3525.

133 Yu, L., An, Y., and Cai, L. (2002). Numerical reconstruction of digital holograms with variable viewing angles. *Optics Express* 10: 1250.

134 H. Gross (2010). Wave optical simulation of a spectral narrowed resonator. *SPIE Conference Proceedings* 7717. https://doi.org/10.1117/12.853695.

135 Cox, R. and Tong, R. (1999). Two- and three-dimensional image rotation using the FFT. *IEEE Transactions on Image Processing* 8: 1297.

7

Optical Transfer Function

7.1 Spatial Frequency Concept

In Fourier and Abbe's famous idea of understanding imaging in a physical optical picture, the concept of considering spatial frequencies is used. This means the object is assumed to be a grating or a superposition of many gratings with different periods, and the well-known behaviour of the grating diffraction is used to model the image formation [1–5].

A grating separates the incoming light into various diffraction orders according to the grating equation, which reads

$$\sin\theta \approx \theta = \frac{m\,\lambda}{g} = m\,\lambda \cdot \nu \tag{7.1}$$

Here θ is the angle of deflection, g is the grating period and m the diffraction order, which takes integer values $m = \ldots -2, -1, 0, +1, +2, \ldots$ ν is the so-called spatial frequency of the grating. This property is defined as the inverted value of the grating period $g = 1/\nu$. In the following description, the angle θ is mostly assumed to be small. Figure 7.1 illustrates the basic idea of Fourier. If the object is a grating, the ray deflection angle θ is equivalent to a spatial frequency ν or a transverse wave vector component k_t. There are different possibilities to describe these effects, by the angle, the spatial frequency or the wave vector. All of these variables are equivalent and are connected by

$$\theta = \frac{\lambda}{g} = \lambda \cdot \nu = \frac{k_t}{k} \tag{7.2}$$

considering the dominating first diffraction order $m = 1$, see Figure 7.1.

Now, in the understanding of the imaging by an optical system it is assumed that the object is a grating, see Figure 7.2. This simple picture assumes a monochromatic plane wave, impinging perpendicular onto the grating. But the idea can be extended to much more generalized setups. If the deflection into the first grating diffraction order lies inside the numerical aperture cone of an optical system, the three plane waves for the zeroth, +first and −first order are transmitted by the system. In the image space, these three waves are interfering and therefore a corresponding modulation is observed, see Figure 7.2(a). If the grating period is decreased, the diffraction angle increases. If it falls outside of the acceptance light cone, only the zeroth order is transmitted; in the image where there is no modulation or interference seen, this grating period is not resolved. Figure 7.2(b) illustrates this case.

The idea of Fourier converts the spatial quantity g of the grating period into the angle θ or frequency quantity ν. In optical systems, this corresponds to a transition from the object into the pupil of the system. For a classical system, as indicated in Figure 7.3, a plane illumination wave in the object is focused in the back focal plane. Therefore, the various diffraction orders coming from the object diffraction are spatially resolved in the focal plane, which is therefore also called a Fourier conjugated plane. If the diffraction angles are larger, the collimated light bundles are strongly deflected and start to be truncated in the system due to the finite aperture. This shows much more clearly than Figure 7.2 that the pupil size is essential for the resolution of the system and, in this plane, the object spectrum can be observed. The simple ray picture of the diffraction orders neglects the fact that in reality the bundles have a finite diameter; therefore the case of a partial blocking by the system aperture is possible, giving an information and interference in the image plane, but with reduced contrast due to the reduction in energy of the diffracted wave as is shown in Figure 7.2(c).

In a more general case of an object, it is Fourier transformed and every Fourier component corresponds to one plane wave with corresponding strength and direction. Due to the fact that the Fourier theory is linear, every arbitrary object can be considered as a weighted composition of gratings and every grating with one fixed period can be qualified regarding the resolution.

If exactly the two first diffraction orders are supported by the aperture cone, the modulation of the image intensity corresponds to an interference of three plane waves. Due to the energy distribution of the grating diffraction orders, the three waves are not of equal weight. This gives

Lens Design for Imaging: Fundamentals of Optical Systems, First Edition. Herbert Gross.

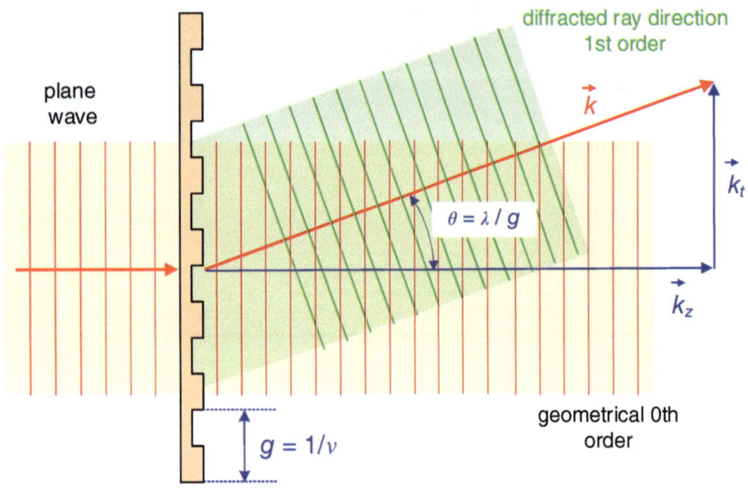

Figure 7.1 Definition of the spatial frequency and first order diffraction of a plane monochromatic wave at a grating.

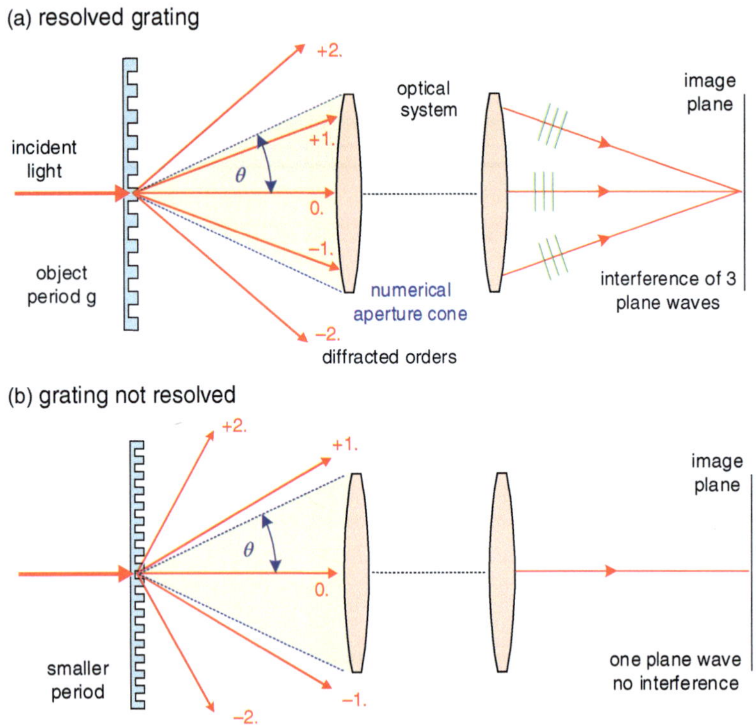

Figure 7.2 Resolution of a grating period by comparing the grating diffraction angles with the numerical aperture cone with (a) a resolved grating and (b) an unresolved case.

the oscillatory profile as it is shown in the first row of Figure 7.4. If more than the minimum number of orders is provided by the system, the interference of the weighted plane wave shows a more accurate image reconstruction. This is seen in the figure for the cases $m = 2, 3, 4$. This means, for the resolution of the period, at least the two first orders are necessary, but a larger number of orders gives a better image quality. This is illustrated in Figure 7.5, here dependent on the maximum number m_{max} of included orders; the rms error of the real image in comparison to the ideal one in the case of a rectangle profile is shown.

If a more complicated scene is now considered, as shown in Figure 7.6, the object must be decomposed into Fourier components. The decomposition into Fourier waves is shown in (c), the filtered spectrum in the image domain depends on the truncation of angles by the numerical aperture. The higher spatial frequencies correspond to smaller feature sizes and steeper edges correspond to larger diffraction angles. The lower spatial frequencies are always supported, the higher frequencies are truncated by the finite aperture; this means that every optical system works as a low pass filter. The narrower the spectrum used for reconstruction corresponding to the coloured ranges in part (c) of the figure, the smoother the imaged details shown in (b). The details in (b) have an offset for better visibility.

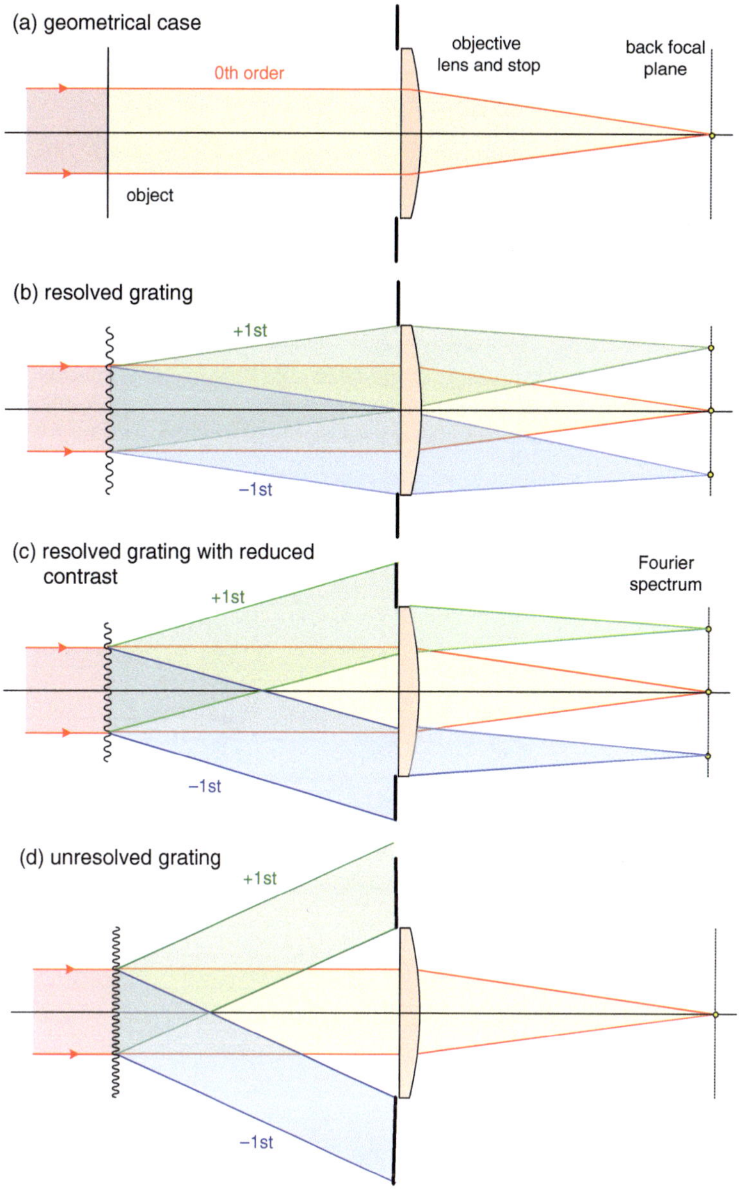

Figure 7.3 Resolution of a grating period by the system pupil size, (a) geometrical 0th order, (b) resolved grating, (c) resolved grating with reduced contrast and (d) unresolved grating.

7.2 Optical Transfer Function

7.2.1 Introduction

The Fourier theory of optical images allows the image formation process to be considered in the formalism of linear system theory. The system has a transfer function with given filter properties which acts upon the light angle or spatial frequency distribution in the object and generates the image distribution. The filtering is performed in the angle or spatial frequency space by the optical system, which corresponds to the systems numerical aperture [1–8].

In practice, incoherent illumination and coherent conditions, respectively, are most often encountered in conventional optical systems. Therefore, the consideration of the incoherent transfer function (OTF, optical transfer function) will be described in the following in more detail. This means the different illuminating object points show no interaction or interference. For simplicity, here only one transverse direction is considered; the extension to both coordinates x and y is straightforward.

The filtering by the optical system means that the object spectrum is changed by the system transfer function. This function is obtained by the Fourier transform of the point spread function intensity in accordance with the computation of the image intensity in incoherent image formation and reads with normalization factor N

$$H_{OTF}(v) = N \cdot \iint I_{PSF}(x) \cdot e^{-2\pi \, iv \, x} \, dx = \hat{F}[I_{PSF}(x)] \quad (7.3)$$

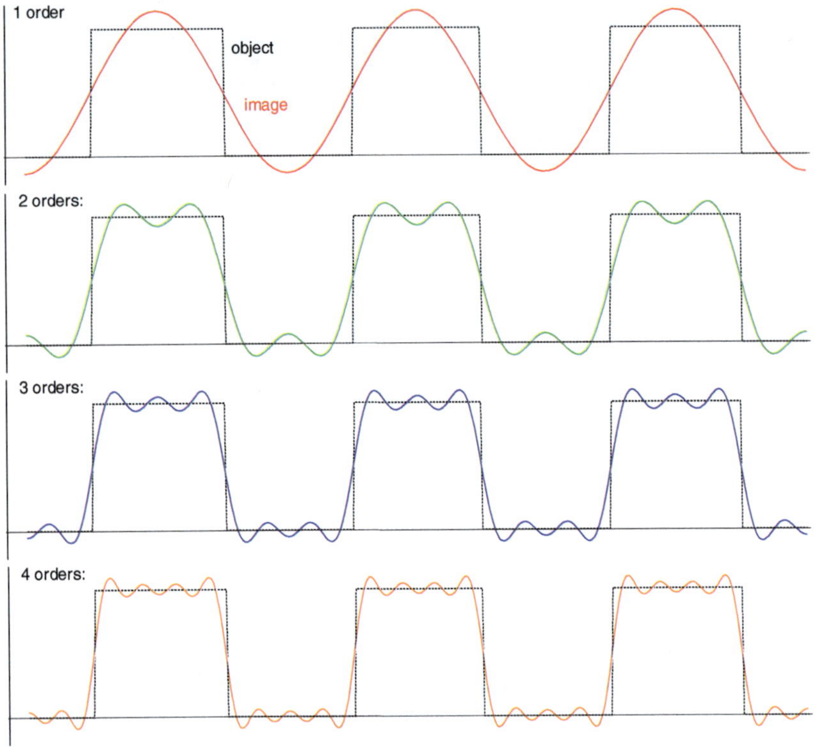

Figure 7.4 Image of a rectangular grating object.

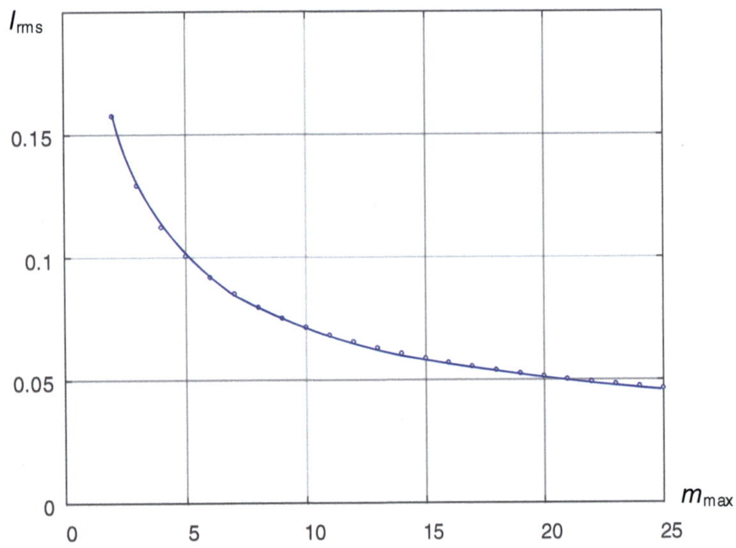

Figure 7.5 Error of the real image intensity of a grating as a function of the number m_{max} of contributing diffraction orders.

The OTF is normalized with respect to the value of the 0th frequency component. Due to the properties of the Fourier transform, the OTF is obtained as a complex function of the spatial frequency ν. The OTF describes the filtering effect of the optical systems. This is changed if aberrations are present or the shape of the pupil boundary or the transmission is altered. The absolute value of the OTF is called the modulation transfer function (MTF) and is real

$$H_{MTF}(\nu) = |H_{OTF}(\nu)| \tag{7.4}$$

It is shown below that this value is exactly the contrast of an imaged grating. According to the finite numerical aperture, the values of the OTF above the maximum cut

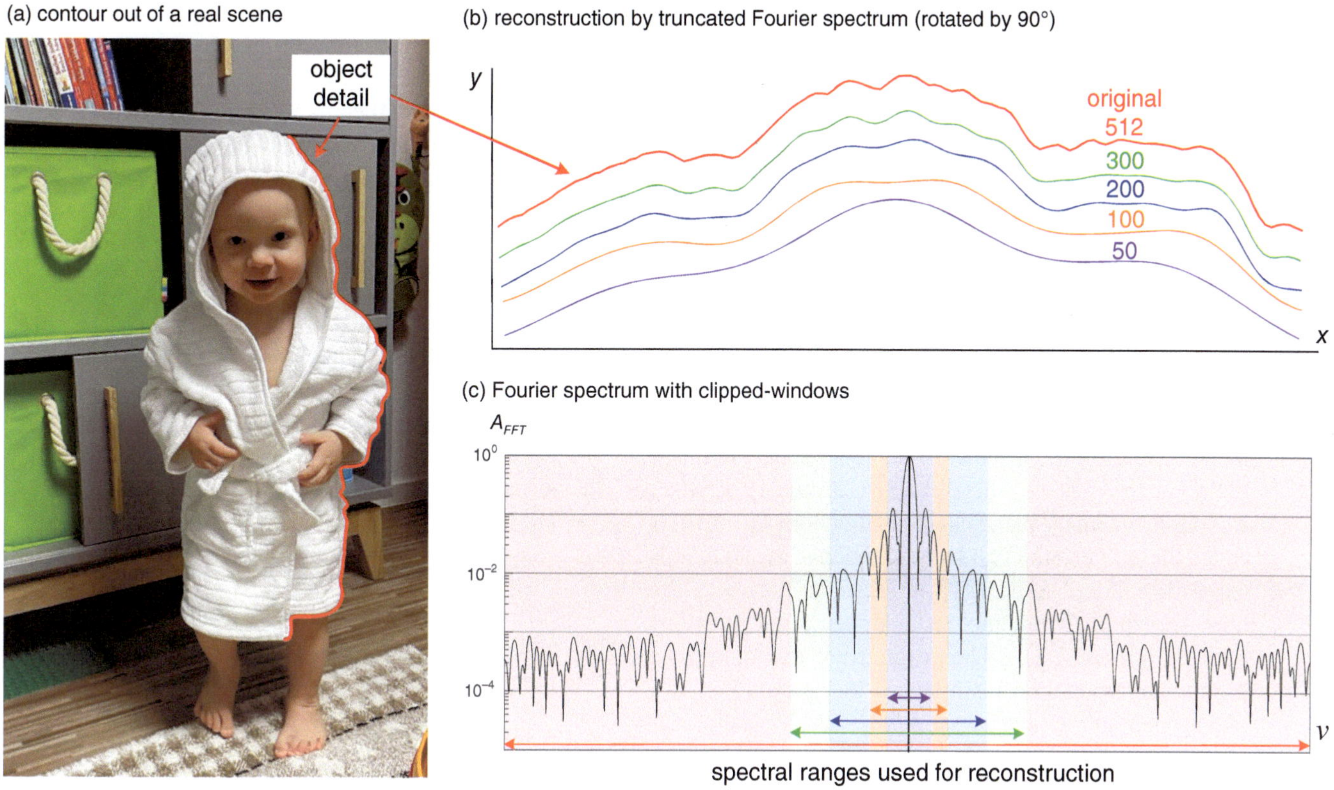

Figure 7.6 Fourier decomposition of an extended scene. The red contour line is considered as a detail in (a); in (b) the real image depending on the width of the used spectrum corresponding to the number of Fourier components included; in (c) the full spectrum and the truncated ranges are seen. The axes are arbitrarily scaled here.

off frequency

$$\nu_{\max} = 2\nu_o = \frac{2na}{\lambda f} = \frac{2n \sin u'}{\lambda} \quad , \quad \nu_o = \frac{NA}{\lambda} \tag{7.5}$$

are always zero. The frequency ν_o is often used for normalization instead of $\nu_{\max}$. The radius of a circular pupil size is denoted by a here. By inspecting this expression, it can be seen that it is approximately (neglecting the pre-factor 1.22) corresponding to a grating period of the Airy pattern radius

$$\nu_{\max} = \frac{2n \sin u'}{\lambda} \approx \frac{2}{D_{airy}} = \frac{1}{r_{airy}} \tag{7.6}$$

If the point spread function is symmetric, due to the properties of the Fourier transform, the OTF is a real function. This means, for coma aberrations, the OTF becomes complex. In general, the complex function can be decomposed into the MTF and the so-called phase transfer function (PTF) in the form

$$H_{OTF}(\nu) = H_{MTF}(\nu) \cdot e^{iH_{PTF}(\nu)} \tag{7.7}$$

Since a linear phase in the pupil plane corresponds to a tilt of the diffraction order according to Figure 7.3, this means the PTF changes the image point location. In turn, this means that the PTF is responsible for the description of the image point position and therefore describes distortion or spatial displacements, while the MTF gives the lateral resolution. But in practical use, the decomposition of Eq. (7.7) is not uniquely used. A switch in the sign corresponding to a contrast reversal can be described by a phase of 180° in the PTF or by a negative contrast in the MTF. Mostly the latter is more intuitive. See also the explanations and the example in Section 7.2.5.

7.2.2 Duffieux Representation of the OTF

In the mathematical theory of Fourier transforms it is well known that the Fourier transform of a squared function E can be calculated as an autocorrelation (ACF) of the Fourier transform of the function itself. If the Fourier transform of E is denoted by P

$$P(\nu) = \hat{F}[E(x)] \tag{7.8}$$

we get therefore

$$\hat{F}[|E(x)|^2] = C_{ACF}[\hat{F}(E(x))] = C_{ACF}[P(\nu)] = \int P(\nu) \cdot P^*(\nu - \nu_o) d\nu_o \tag{7.9}$$

In the case of the OTF this means that the Fourier transform of the point spread function amplitude E is the pupil function P. The OTF is written as the autocorrelation of the systems complex pupil function $P(x_p)$. By a symmetrical choice of the coordinate origin, this reads as

$$H_{OTF}(\nu) = \int P\left(x_p + \frac{\lambda f \nu}{2}\right) P^*\left(x_p - \frac{\lambda f \nu}{2}\right) dx_p \tag{7.10}$$

the pupil coordinate x_p is here rescaled as a spatial frequency. The frequency ν corresponds to an angle $\theta = \lambda\nu$; the spatial coordinate in the pupil is given by

$$x_p = f \cdot \theta = f\,\lambda\,\nu \tag{7.11}$$

Here it is assumed for simplicity that the focal length f is identical to the pupil distance or reference sphere radius R of the system, in more general cases the distance p between exit pupil and image plane has to be used here. Eq. (7.10) is called the Duffieux integral representation of the OTF. This expression describes the OTF in frequency space as the area overlap of two diametrically shifted pupil functions with the offsets $\lambda f\nu/2$. The geometry of this quite nice physical interpretation is illustrated in Figure 7.7 in two lateral coordinates x and y respectively. In a physically clear sense the Duffieux integral, as an extension of the above outlined Abbe model of resolution, infers that the resolution in the form of transferred contrast tends to improve with increasing surface overlap of the system pupils in the zeroth and first diffraction orders of the object spectrum. When the two pupil areas exhibit no common overlap, the resolution for the corresponding spatial frequency is completely absent. The larger the area overlap, the larger the modulation depth, because the interference of the first diffraction orders gets a larger contribution in comparison to the geometrical zeroth order. If the two circles of the pupil in 0th and first order are exactly osculating, the cut-off spatial frequency is obtained, and no interference takes place.

Figure 7.8 illustrates again the relations of the grating imaging for five different grating periods from left to right. In (a) the grating with the feature sizes are seen, (b) shows the pupil overlap of the zeroth with the +first and −first diffraction order in the pupil, (c) gives the changing contrast in the images with constant mean value and (d) finally is the corresponding MTF contrast curve for the five frequency points.

7.2.3 Contrast and Resolution

If an ideal transmission grating is considered that changes the intensity after the component between zero and one, it is imaged in reality by a reduced contrast, which means the maximum and minimum values as well as the steep slopes are degraded. The contrast or visibility is then defined as the normalized intensity difference

$$V = \frac{I_{max} - I_{min}}{I_{max} + I_{min}} \tag{7.12}$$

Figure 7.9 illustrates the changes of a rectangular grating as a result of the low-pass filtered imaging process. It is seen that the perfect structure is changed by three aspects.

- The maxima I_{max} are reduced.
- The minima I_{min} are increased.
- The ideally vertical slopes of the intensity black-white steps are smeared out to a finite gradient.

The interpretation of the optical system filtering properties is again illustrated in Figure 7.10 in a representation of the contrast vs the spatial frequency. The lower range with contrast zero means uniform grey, and the upper boundary with black-white a full modulated image. The larger the spatial frequency (horizontal coordinate), the smaller the contrast (vertical axis) in the image. Only for very small spatial frequencies or coarse structures do we get a full black-white modulation in the image. Furthermore, it is shown in this figure that there exists a contrast sensitivity curve in most real systems, which is marked in green here. This means a sensor needs a certain contrast to accept a

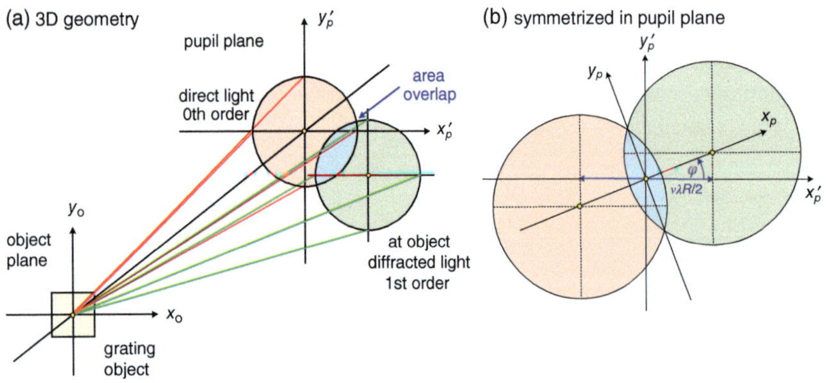

Figure 7.7 Scheme of incoherent imaging with (a) lateral displaced pupil of the zeroth and the first diffraction order of a grating object; the blue overlap area measures the contrast and is drawn in (b) in a usual symmetric arrangement.

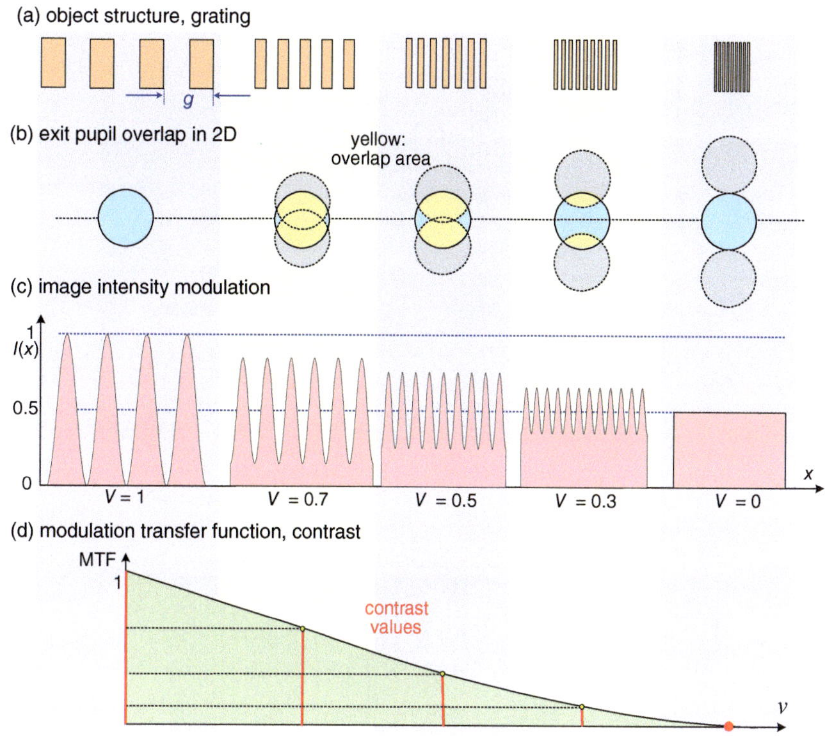

Figure 7.8 Scheme of contrast change by pupil overlay (a) the grating feature, (b) the pupil overlap, (c) the image contrast and (d) the modulation contrast curve. The values are not perfectly to scale here and are only illustrative.

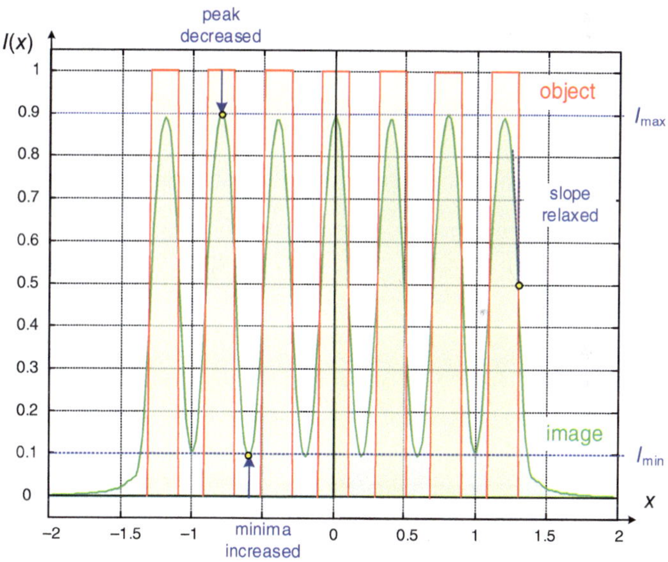

Figure 7.9 Definition of contrast by considering the image of a rectangular transmission grating. The unit of x is arbitrary.

signal. For very small contrast differences, the gap is interpreted as noise or not significant. This behaviour is more pronounced if the spatial frequency is higher; therefore, as can be seen, the green curve grows with ν. This kind of sensor behaviour is typical and is found, for example, in the human eye or for a CCD (cross talk between neighbouring pixels). The interpretation of this aspect means:

- it is always necessary to take the filtering properties of the detector into account, since the OTF is the influence of the optical system only;
- there is a real cut-off frequency due to this detector influence, which may be smaller than the theoretical value by Eq. (7.5). The physical interpretation is: the real contrast must be referenced against the contrast sensitivity curve, not against the zero-axis.

Contrast and resolution are coupled by the MTF behaviour of an optical imaging system, but they must be clearly distinguished as different performance criteria. This is again visualized in Figure 7.11 for a synthetic test pattern with a chirped grating. The visual perception of

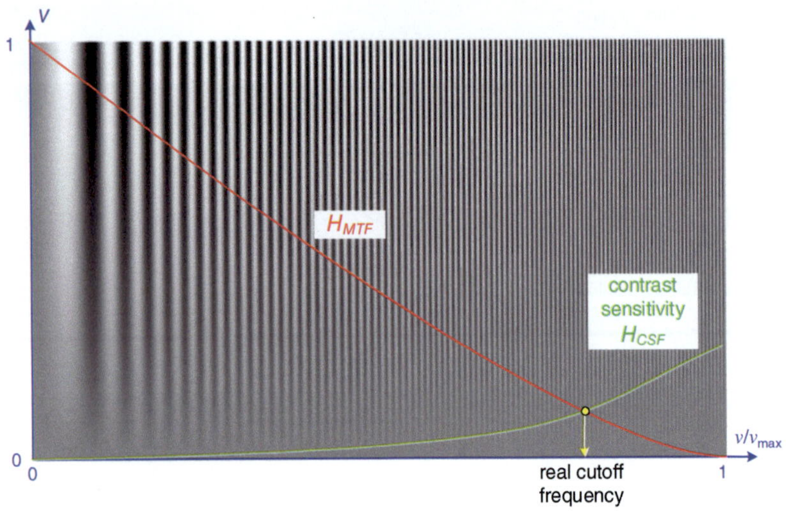

Figure 7.10 Interpretation of MTF and measuring the MTF against a threshold of contrast H_{CSF}.

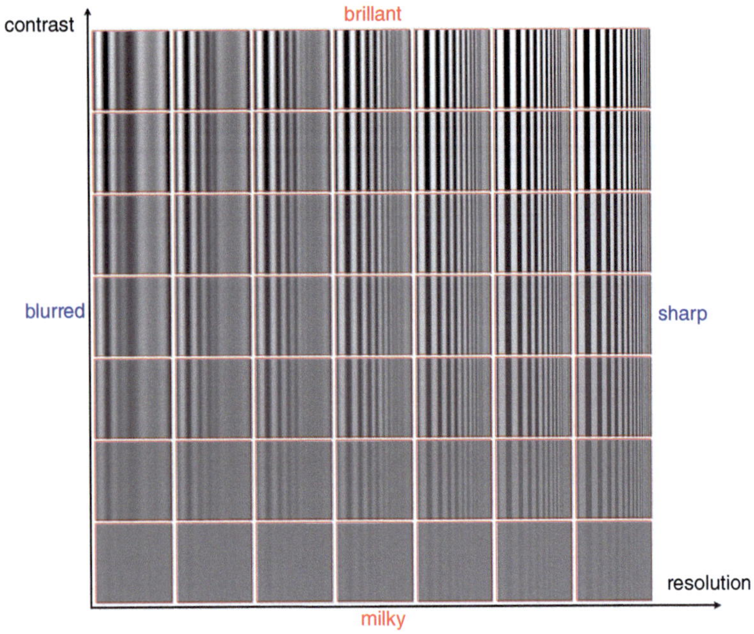

Figure 7.11 Resolution (horizontal) and contrast (vertical) for a test pattern.

a loss in resolution (blurred, left to right) and a loss in contrast (milky, upper to lower cases) is quite different.

The same impression is obtained for a real scene as it is seen in Figure 7.12. Mainly a trade-off between contrast and resolution looks most comfortable. This is exactly what an optical system with its mixed filtering properties does.

A more quantitative description of resolution and contrast can be obtained with the Siemens star. This very ingenious pattern is shown in Figure 7.13. It contains radial lines of decreasing width, representing a continuously changing azimuthal spatial frequency. Outside we see a good black-white contrast, if more inner regions are inspected we see a grey ring, demonstrating the limited resolvable azimuthal spatial frequency of the system under test. Interestingly it is observed that within the first grey ring again a certain contrast is obtained. But more rigorous inspection shows that the colours black and white are inverted here in comparison to the outer range. This effect is called pseudo resolution with contrast reversal. It occurs if the MTF curve has considerable value after a zero crossing point. The nominal negative MTF inverts the contrast. But it should be clear that a zero in the transfer function has the consequence that one spatial frequency is completely missing in the image. The finite contrast for higher frequencies has a certain value for understanding the imaging process, but it is not possible to reconstruct the object for these frequencies due to the gap.

In reality, the achieved resolution of a system is often estimated by inspecting the imaging of specific test charts. These special structures allow for a quantitative contrast

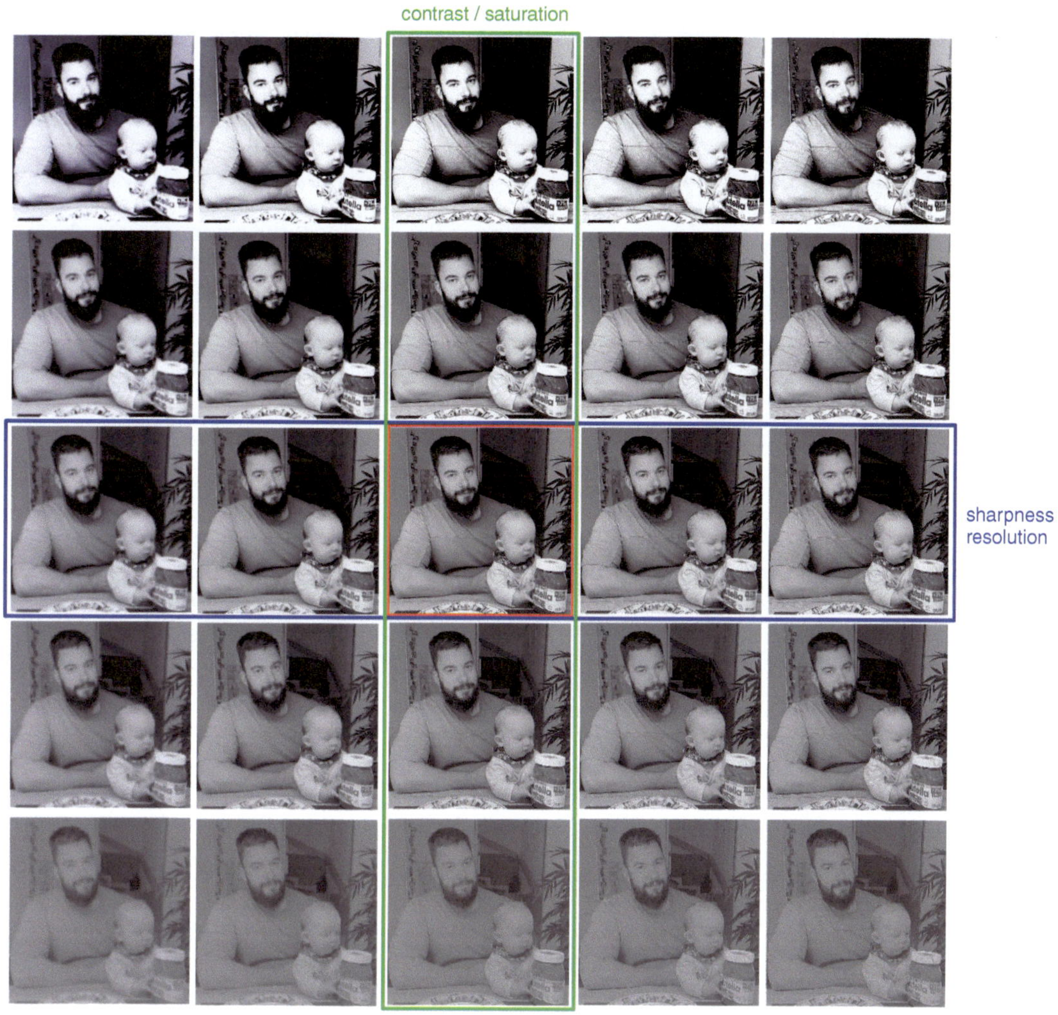

Figure 7.12 Resolution (horizontal) and contrast (vertical) for a real scene, the red marked case gives the best trade-off.

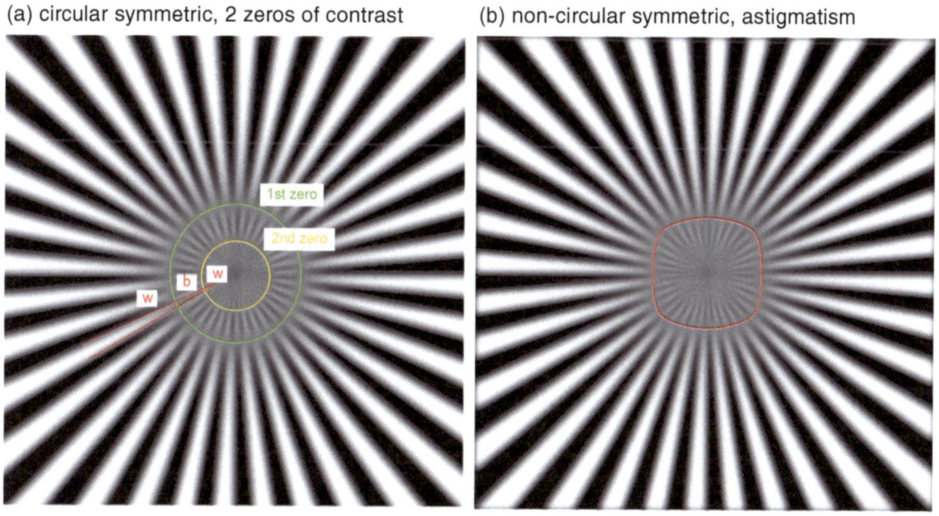

Figure 7.13 Siemens star resolution test plate. The blue arcs indicate a grey with contrast zero; in between a finite reverted contrast is observed.

evaluation of bar pattern or radial stripes with varying width and grey scale in the Siemens star pattern.

7.2.4 Ideal MTF

For a system with a circular pupil and no aberrations, the normalized MTF is given according to the Duffieux representation by the area overlap A of two circles, see picture in Figure 7.14. This function can be calculated analytically and gives the so-called Chinese hat function with normalized variable $\overline{\nu} = \nu/\nu_{max}$. For the ideal point spread function (PSF) of an Airy point spread function, the MTF looks like the curve in Figure 7.14. A decreasing function is obtained, which has the analytical representation

$$H_{OTF}(\nu) = \frac{2}{\pi} \cdot \left[\arccos\left(\frac{\nu}{\nu_{max}}\right) - \left(\frac{\nu}{\nu_{max}}\right)\sqrt{1-\left(\frac{\nu}{\nu_{max}}\right)^2}\right] \quad (7.13)$$

with the maximum cut-off frequency according to Eq. (7.5).

For a better understanding and a clearer interpretation of the meaning of the transfer function, let us make a short calculation. Assume there is a cos-shaped modulated grating with spatial frequency ν_0 given as an object (see Figure 7.15)

$$I_{obj}(x) = c + a \cdot \cos(2\pi\nu_0 x) \quad (7.14)$$

According to the definition, the contrast in the object is therefore

$$V_{obj} = \frac{I_{max} - I_{min}}{I_{max} + I_{min}} = \frac{(c+a)-(c-a)}{(c+a)+(c-a)} = \frac{a}{c} \quad (7.15)$$

The spectrum of the object is given by the Fourier transform

$$\overline{I}_{obj}(\nu) = \widehat{F}[I_{obj}(x)] = c \cdot \delta(\nu) + \frac{a}{2} \cdot \delta(\nu - \nu_0) + \frac{a}{2} \cdot \delta(\nu + \nu_0) \quad (7.16)$$

According to the definition of the transfer function, which is real here due to the symmetry of the object, the spectrum in the image is given by

$$\overline{I}_{ima}(\nu) = \overline{I}_{obj}(\nu) \cdot H_{MTF}(\nu) \quad (7.17)$$

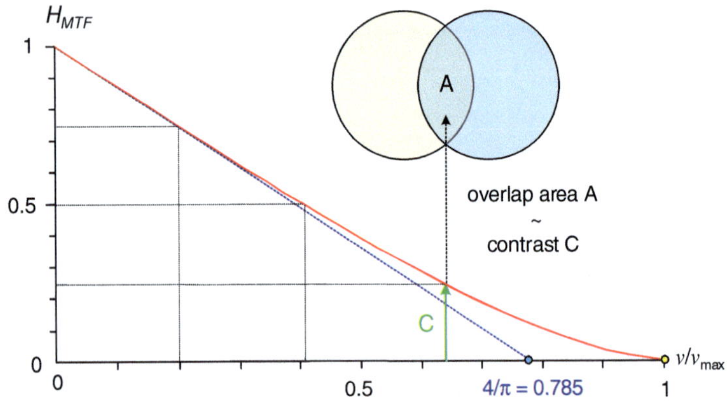

Figure 7.14 Ideal MTF as a function of the normalized spatial frequency.

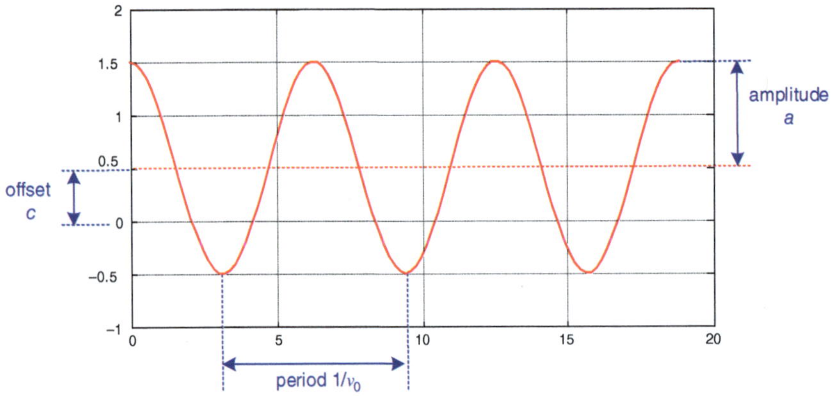

Figure 7.15 Terms of use for the MTF interpretation as contrast. The x units are arbitrary.

Inverting the Fourier transform and inserting the results of the previous steps gives the image intensity as

$$\begin{aligned} I_{ima}(x') &= \hat{F}^{-1}[\bar{I}_{ima}(\nu)] = \hat{F}^{-1}[\bar{I}_{obj}(\nu) \cdot \mathrm{H}_{MTF}(\nu)] \\ &= \hat{F}^{-1}\Big[c \cdot \mathrm{H}_{MTF}(\nu) \cdot \delta(\nu) + \frac{a}{2} \cdot \mathrm{H}_{MTF}(\nu) \\ &\quad \cdot \delta(\nu - \nu_0) + \frac{a}{2} \cdot \mathrm{H}_{MTF}(\nu) \cdot \delta(\nu + \nu_0)\Big] \\ &= c \cdot \mathrm{H}_{MTF}(0) + \frac{a}{2}\mathrm{H}_{MTF}(\nu_0) \cdot e^{2\pi\, i\nu_0 x} \\ &\quad + \frac{a}{2}\mathrm{H}_{MTF}(-\nu_0) \cdot e^{-2\pi\, i\nu_0 x} \\ &= c + a \cdot \mathrm{H}_{MTF}(\nu_0) \cdot \cos(2\pi\, i\nu_0 x) \end{aligned} \tag{7.18}$$

which has the contrast

$$V_{ima} = \frac{a}{c} \cdot H_{MTF}(\nu_0) = V_{obj} \cdot H_{MTF}(\nu_0) \tag{7.19}$$

This shows that the MTF directly modifies the contrast of an object as a factor for the corresponding spatial frequency.

If the ideal MTF is considered, the drop of modulation for small spatial frequencies is linear and can be approximated by the first terms of the Taylor expansion as

$$H_{OTF}(\nu) = 1 - \frac{4 \cdot \nu}{\pi \cdot \nu_{\max}} + \frac{\nu^3}{3\pi \cdot \nu_{\max}^3} - \dots \tag{7.20}$$

The ideal transfer function is often used as a reference to indicate the drop of the real curves due to the residual system aberrations.

7.2.5 Phase Transfer Function*

(*) The content of this section is quite specialized and for advanced reading. Knowledge of this section is not necessary to understand the following chapters. Beginners are recommended to skip this section.

The PTF describes a shift in the image and therefore is equivalent to a distortion effect. If a decomposition of several sinusoidal modulation details in the object is considered, in the image these coherent superposed partial waves suffer from individual phase factors, which finally distort the image [9]. This interpretation of the PTF can be demonstrated as follows. If the Fourier definition of the OTF is inverted by

$$\begin{aligned} I_{PSF}(x) &= \int H_{OTF}(\nu)\, e^{-2\pi\, i\nu x} d\nu \\ &= \int H_{MTF}(\nu)\, e^{iH_{PTF}(\nu)} e^{-2\pi\, i\nu x} d\nu \end{aligned} \tag{7.21}$$

and the PTF is expanded into a Taylor series until the linear term as

$$H_{PTF}(\nu) = i\nu \cdot H'_{PTF}(\nu) + \dots \tag{7.22}$$

we can rearrange the exponent and get

$$\begin{aligned} I_{PSF}(x) &= \int H_{MTF}(\nu)\, e^{i\nu H'_{PTF}(\nu)} e^{-2\pi\, i\nu x} d\nu \\ &= \int H_{MTF}(\nu) \cdot e^{-2\pi\, i\nu\left(x - \frac{H'_{PTF}(\nu)}{2\pi}\right)} d\nu \\ &= I_{PSF}\left(x - \frac{H'_{PTF}(\nu)}{2\pi}\right) = I_{PSF}(x - \Delta x) \end{aligned} \tag{7.23}$$

which illustrates the lateral translation Δx by the derivative of the PTF. The special effect of a contrast reversal is illustrated in Figure 7.16. The derivative of the PTF creates an angle $\Delta\phi$ of the chief ray and an offset Δx in the image plane, which is exactly a half period of a sine modulation. This means the maxima of the sine images between ideal and real chief ray are exactly shifted by $\Delta x = 1/(2\nu)$ due to the distortion.

In classical optical design tasks, the PTF usually has only minor importance. It can be observed that the PTF values can be large although the MTF is small. For illustration, Figure 7.17 shows an example of a lens that suffers from different values of coma. The PTF is scaled in radiant as an angle in the pupil; it is seen that the case of large coma $c_8 = 0.5\,\lambda$ generates larger values of the PTF.

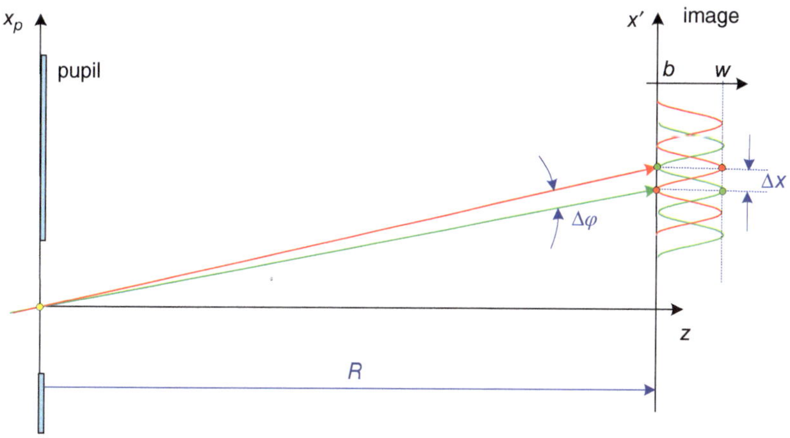

Figure 7.16 Interpretation of the PTF and a contrast reversal.

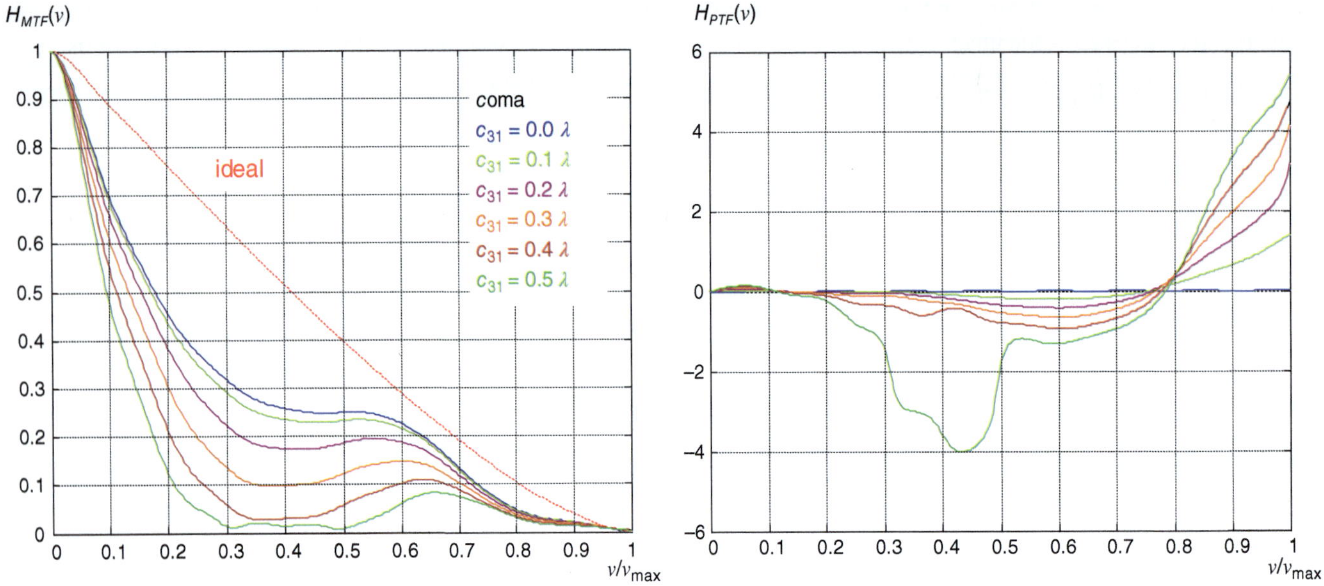

Figure 7.17 MTF and PTF for an optical system suffering from different values of coma (Zernike c_8).

7.2.6 Coherent Transfer Function (CTF)

The classical OTF is defined as the Fourier transform of the PSF intensity. This is meaningful if the illumination of a system is incoherent; by squaring the field in the PSF, the phase information and connection between adjacent object points is lost. If a system is illuminated by a coherent laser source, in a similar manner a coherent transfer function (CTF) is defined as the Fourier transform of the complex field of the PSF; see also Chapter 6.7 [10].

$$H_{CTF}(\nu) = \widehat{F}\,[E_{PSF}(x)] = P\left(\frac{x_p}{\lambda f}\right) \tag{7.24}$$

Since the PSF is calculated in the far-field Fraunhofer approximation in the image plane as the Fourier transform of the complex pupil function, the CTF is identical to the pupil function, disregarding a constant scaling factor.

Corresponding to the definition of the incoherent transfer function, the connection between OTF and CTF is given according to the Duffieux representation by the autocorrelation

$$\begin{aligned} H_{OTF}(\nu) &= C_{ACF}[H_{CTF}(\nu)] \\ &= \int H_{CTF}(\nu) \cdot H^*{}_{CTF}(\nu - \nu_o) d\nu_o \end{aligned} \tag{7.25}$$

As long as the diffraction order falls inside the pupil area, the full coherent contrast $V = 1$ is obtained. If in an ideal model the diffraction order is assumed to be infinitesimally sharp, the contrast coherent breaks immediately down to zero if the spatial frequency exceeds the coherent cut-off frequency

$$\nu_{cut}^{(coh)} = \frac{NA}{\lambda} \tag{7.26}$$

which is not identical to the incoherent cut-off. This behaviour is illustrated together with the incoherent case in Figure 7.18. If both cases are compared, the following differences are observed qualitatively.

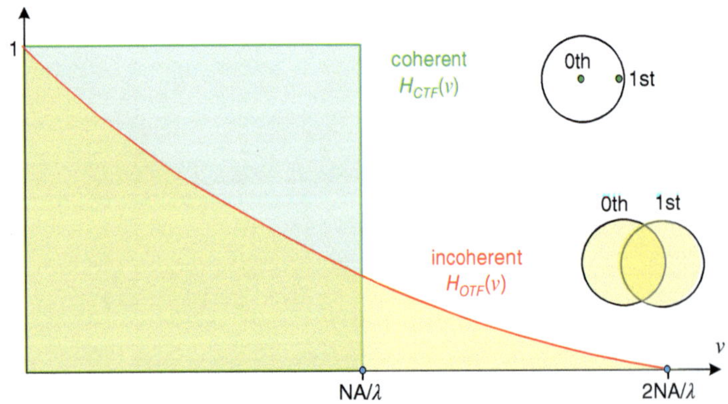

Figure 7.18 Comparison between the incoherent OTF and the coherent CTF.

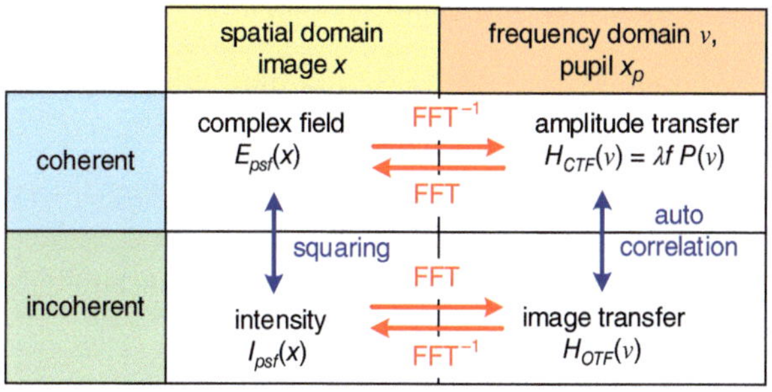

Figure 7.19 Connections between spatial and frequency space in the coherent and incoherent case respectively.

- For $\nu < NA/\lambda$ the contrast is better in the coherent case.
- For $\nu > NA/\lambda$ the contrast is only seen in the incoherent case, but on a rather low level.

An incoherent illumination sends the light in a broad angle range; therefore, the pupil can be assumed to be completely filled with light. This is different in the case of coherent light with a very narrow angle spectrum; although the pupil is large, only a small area is really used. This explains why the contrast decreases abruptly if the first diffraction order reaches the boundary of the pupil. This difference is sketched in Figure 7.18 as an inlet.

But, as is well known, in the coherent case due to interference some ringing and interference effects perturb the image and generate effects at edges, which does not correspond to the natural expectation and experience and does not resemble the object. So, contrast is not the only criterion for a good image.

The relationships of the coherent and the incoherent case together in the spatial as well as the frequency or pupil domain are again summarized in Figure 7.19. The importance of the coherent transfer function is mainly found in microscopy, where often lasers are used for illumination.

7.2.7 Geometrical Approximated Transfer Function (GTF)

The usual definition of the OTF is based on the Fourier model of image formation and therefore always includes diffraction effects. But if a system suffers from huge residual aberrations, the main contributions to the transfer function are obtained at rather low spatial frequencies and geometrical aberrations dominate over diffraction related broadening [11–14]. In these cases, a geometrical approximation of the transfer function is possible and can be calculated by raytrace only. If the OTF (here for simplicity in one lateral dimension) is formulated according to Duffieux as

$$H_{OTF}(\nu_x) = \left(\frac{2R}{k}\right)\cdot\int e^{ik[W(x_p)-W(x_p-\nu\lambda R)]}dx_p \qquad (7.27)$$

with the radius R of the reference sphere, a first order Taylor expansion of the shifted wave aberration W gives with the relation to the transverse aberration $\Delta x'$

$$\frac{\partial W}{\partial x_p} = -\frac{\Delta x'}{R} \qquad (7.28)$$

the expression

$$\begin{aligned} H_{GTF}(\nu_x) &= \left(\frac{2R}{k}\right)\cdot\int e^{ik\left[W(x_p)-\left(W(x_p)-\nu\lambda R\frac{\partial W}{\partial x_p}\right)\right]}dx_p \\ &= \left(\frac{2R}{k}\right)\cdot\int e^{2\pi i\nu\,\Delta x'}dx_p \\ &\approx \left(\frac{2R}{k}\right)\cdot\int\left[1+2\pi i\nu\,\Delta x'-\frac{1}{2}(2\pi\nu\,\Delta x')^2\right]dx_p \end{aligned} \qquad (7.29)$$

with again a Taylor expansion including the second order in the last expression. This expression is called the geometrical optical transfer function (GTF). Due to the strong oscillations for larger shear differences, the integral can be taken over the full pupil area. The approximation with the expansion of the exponential function is not necessary; the integration can also be calculated by a direct integration, but then most of the computational benefit is gone. In this last expression, we can introduce the centroid and the spot rms spreading in the geometrical approximation

$$\begin{aligned} \Delta x_c'^{(CR)} &= \frac{1}{a}\int\Delta x'\,dx_p \\ \Delta r_{rms}^2 &= \frac{1}{a}\int\left(\Delta x'-\Delta x_c'^{(CR)}\right)^2 dx_p \end{aligned} \qquad (7.30)$$

with the pupil radius a and get

$$H_{GTF}(\nu_x) = 1 + i\cdot\pi\nu\cdot\Delta x_c'^{(CR)} - \pi^2\nu^2\cdot\Delta r_{rms}^2 \qquad (7.31)$$

This equation shows that for small spatial frequencies, the drop of the OTF can be calculated by simple raytrace results and transverse aberrations. It should be mentioned here that in the two-dimensional case, the pre-factors of the terms depend on the geometry and are different for a circular than for a square pupil area. Furthermore, in the case of circular symmetry, formally an azimuthal averaging has to be performed.

In this geometrical approximation, the interpretation of both terms is as follows.

- The deviation of the centroid position relative to the chief ray position creates an imaginary contribution and is responsible for a distortion related effect and corresponds to the PTF. It grows linearly with the spatial frequency.

- The broadening of the geometrical spot reduces the contrast and contributes to the MTF. This drop behaves quadratically with the spatial frequency. According to the approximation, this is only valid for small spatial frequencies.

If the geometrical GTF is compared with the diffraction related model, there are some major differences.

- In the GTF the wavelength dependence is weak, only due to dispersion.
- The real part of the GTF starts to decrease quadratically with the horizontal slope; the MTF linearly.
- The quantitative comparison in the sense of a GTF error depends on the system quality and changes with the spatial frequency.

Figure 7.20 shows two examples of a better (W_{rms} = 0.12 λ) and a worse (W_{rms} = 0.40 λ) corrected system for comparison. The system is a double Gauss camera lens with different field points and different $F_{\#}$-numbers. It is seen that the agreement for the worse system is better; and the coincidence is better for lower spatial frequencies. In particular, for the good corrected system the difference in the starting slope is clearly observed.

7.2.8 Connection Between OTF and PSF

By definition the OTF is the Fourier transform of the PSF intensity

$$H_{OTF}(v_x, v_y) = \iint I_{PSF}(x', y') \cdot e^{-2\pi\, i\, (x'v_x + y'v_y)}\, dx'\, dy' \quad (7.32)$$

By inverting this transform, it follows (constant prefactor neglected)

$$I_{PSF}(x', y') = \iint H_{OTF}(v_x, v_y) \cdot e^{2\pi\, i\, (x'v_x + y'v_y)}\, dv_x\, dv_y \quad (7.33)$$

If in this equation the centre coordinate is selected $x' = y' = 0$, it is seen that the integral over the two-dimensional OTF is identical to the peak value of the PSF.

$$I_{PSF}(0, 0) = \iint H_{OTF}(v_x, v_y)\, dv_x\, dv_y \quad (7.34)$$

By normalization it follows the Strehl ratio as the normalized volume below the OTF-surface in the two-dimensional frequency domain

$$D_S = \frac{\iint H_{OTF}(v_x, v_y)\, dv_x\, dv_y}{\iint H_{OTF}^{(ideal)}(v_x, v_y)\, dv_x\, dv_y} \quad (7.35)$$

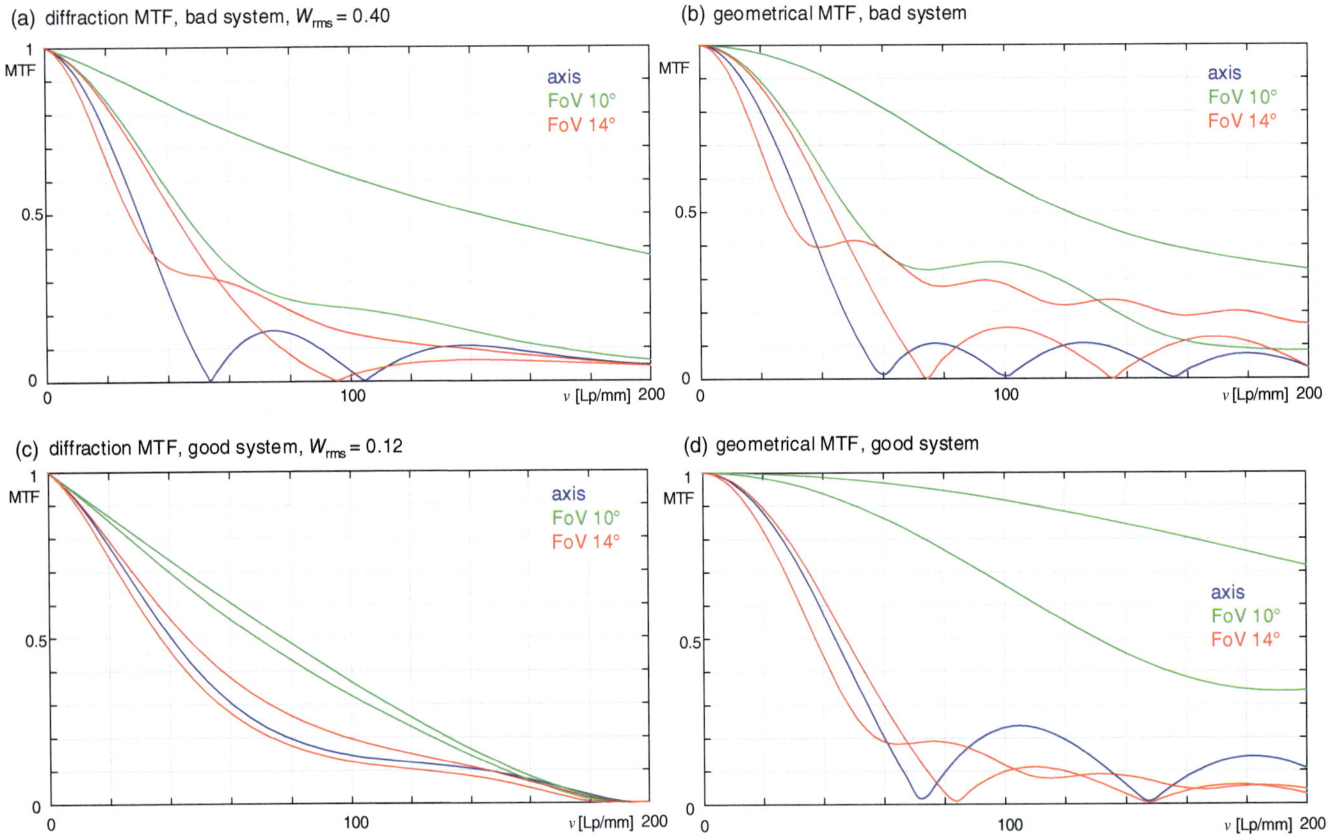

Figure 7.20 Comparison of MTF and GTF for two systems with different correction quality.

If the derivatives of the representation of the OTF according to the frequency coordinates in Eq. (7.32) are calculated and the expression is evaluated again at the origin for the frequencies $\nu_x = \nu_y = 0$ we get

$$\frac{\partial H_{OTF}(\nu_x,\nu_y)}{\partial \nu_x} = \iint I_{PSF}(x',y')\, 2\pi\, ix' \cdot e^{2\pi\, i\, (x'\nu_x + y'\nu_y)}\, dx'\, dy'$$

$$\left.\frac{\partial H_{OTF}(\nu_x,\nu_y)}{\partial \nu_x}\right|_{\nu_x,\nu_y=0} = 2\pi\, i \iint x' I_{PSF}(x',y')\, dx'\, dy' = 2\pi\, i \cdot \langle x' \rangle_I \tag{7.36}$$

This means that the centroid coordinates of the PSF distribution are obtained by the slope component values of the two-dimensional distribution function of the imaginary part of the OTF.

$$\langle x' \rangle_I = \frac{1}{2\pi} \cdot Im \left[\left.\frac{\partial H_{OTF}(\nu_x,\nu_y)}{\partial \nu_x}\right|_{\nu_x,\nu_y=0} \right],$$

$$\langle y' \rangle_I = \frac{1}{2\pi} \cdot Im \left[\left.\frac{\partial H_{OTF}(\nu_x,\nu_y)}{\partial \nu_y}\right|_{\nu_x,\nu_y=0} \right] \tag{7.37}$$

For symmetrical shapes of the PSF, this obviously gives zero. But in the case of coma contributions, the centroid is not identical with the chief ray position and a finite value is obtained.

The imaginary part of the OTF is directly related to the PTF; therefore, again it is seen that the PTF is related to the distortion. The slope of the MTF at the origin is nevertheless linearly decreasing as described in Eq. (7.20).

7.2.9 Three-Dimensional Coherent Transfer Function*

(*) The content of this section is quite specialized and for advanced reading. Knowledge of this section is not necessary to understand the following chapters. Beginners are recommended to skip this section.

In Section 6.7.2 the formula

$$E(x_i) = \frac{e^{-\frac{\pi i x_i^2}{\lambda z_i}}}{ikz_i} e^{ikz_o} \frac{1}{ikz_o} e^{ikz_i} \int T(x_p) e^{-\frac{\pi i x_p^2}{\lambda z_i}} \times e^{+\frac{\pi i\, x_p^2}{\lambda f} + 2\pi i W(x_p)} e^{-\frac{2\pi i}{\lambda z_i} x_i x_p} e^{+\frac{\pi i x_p^2}{\lambda z_o}} \int E(x_o) \cdot e^{+\frac{2\pi i}{\lambda z_o} x_o x_p} dx_o dx_p \tag{7.38}$$

was derived for the field after a lens in an arbitrary distance. Here, only one transverse coordinate x is considered for simplicity, the y direction behaves completely similarly. In the conjugated plane the lens makers' formula is valid and the magnification m can be used according to [10]

$$\frac{1}{z_i} - \frac{1}{z_o} = \frac{1}{f}, \quad m = \frac{z_i}{z_o} \tag{7.39}$$

and delivers in the image plane centre for $x_o = 0$ the PSF

$$E_{psf}(x_i) = \frac{E_o e^{-\frac{\pi i x_i^2}{\lambda z_i}} e^{ik(z_i - z_o)}}{k^2 z_i z_o} \int T(x_p) \cdot e^{2\pi i W(x_p)} e^{-\frac{i\pi}{\lambda z_i} x_i x_p} dx_p = \frac{E_o e^{ik(z_i - z_o)}}{k^2 z_i z_o} \cdot \hat{F}[P(x_p)] = \frac{E_o e^{ik(z_i - z_o)}}{\lambda^2 z_i z_o} \cdot \hat{F}[H_{CTF}(x_p)] \tag{7.40}$$

with the coherent transfer function H_{CTF}, which is a scaled version of the pupil function, here in the conjugated plane to the initial field. In contrast to the formulation of the previous chapter we write s instead of z here to avoid confusion with the later defined defocusing z. The goal is now to achieve a description in three dimensions. To this end, we consider small defocusing values z_o and z_i from the exact conjugation that is denoted by s_o, s_i. In a linear approximation we get

$$\frac{1}{z_o} = \frac{1}{s_o - z} \approx \frac{1}{s_o} + \frac{z}{s_o^2}, \quad \frac{1}{z_i} = \frac{1}{s_i + z'} \approx \frac{1}{s_i} - \frac{z'}{s_i^2} \times \frac{1}{z_i} - \frac{1}{z_o} - \frac{1}{f} = \frac{1}{m^2 s_o^2}(z_i + m^2 z_o) \tag{7.41}$$

and the field reads

$$E(x_i) = \frac{e^{-\frac{\pi i x_i^2}{\lambda s_i}}}{k^2 m s_o^2} \iint E(x_o) T(x_p) \cdot e^{2\pi i W(x_p)} \times e^{ik(z_i - z_o)} e^{-\frac{\pi i}{\lambda m^2 s_o^2}(z_i + m^2 z_o) \cdot x_p^2} e^{-\frac{i\pi}{\lambda s_o m}(x_i - m\, x_o) x_p} dx_p dx_o$$
$$= \frac{e^{-\frac{\pi i x_i^2}{\lambda s_i}} e^{ikz_i}}{k^2 m s_o^2} \int E(x_o) \left[\int T(x_p) e^{2\pi i W(x_p)} \times e^{ikz_o} e^{-\frac{\pi i}{\lambda m^2 s_o^2}(z_i + m^2 z_o) \cdot x_p^2} e^{-\frac{i\pi}{\lambda m s_o}(x_i - m x_o) x_p} dx_p \right] dx_o$$
$$= \frac{e^{-\frac{\pi i x_i^2}{\lambda s_i}} e^{ikz_i}}{k^2 m s_o^2} \int E(x_o) H_{CTF}^{(3D)}(x_i - m x_o, z_i + m^2 z_o)\, dx_o \tag{7.42}$$

The linearization corresponds to an approximation, which makes the PSF invariant to a small z-shift around the conjugated planes. Now, due to a finite depth the integration must also be extended to z_o. If for completeness the y coordinate is included, the three-dimensional convolution version reads

$$E(x_i, y_i) = \iiint E(x_o, y_o, z_o) \times H_{CTF}^{(3D)}(x_i - m x_o, y_i - m y_o, z_i + m^2 z_o)\, dx_o dy_o dz_o \tag{7.43}$$

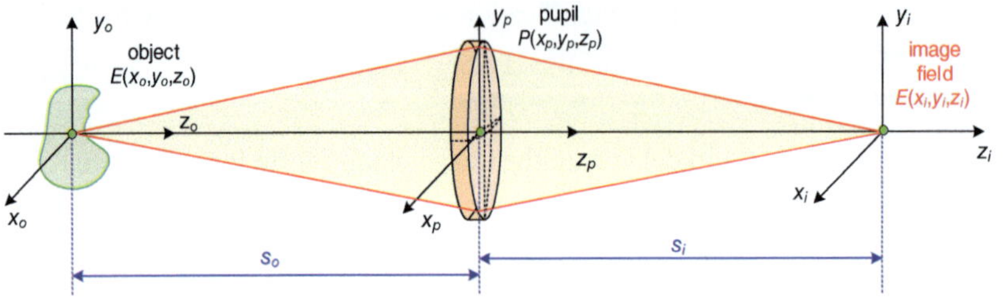

Figure 7.21 Scheme of a three-dimensional imaging system.

where now z_o and z_i are measured against the conjugated planes respectively and z_o is negative. The coherent three-dimensional transfer function is given by

$$H_{CTF}^{(3D)}(x_o, y_o, z_o) = \iint T(x_p) e^{2\pi i W(x_p)} e^{ikz_o} e^{-\frac{\pi i}{\lambda m^2 s_o^2} z_o \cdot \left(x_p^2 + y_p^2\right)} \times e^{-\frac{i\pi}{\lambda m s_o}(x_o x_p + y_o y_p)} dx_p dy_p \tag{7.44}$$

The phase factor e^{ikz_o} takes the phase difference inside the depth of the object into account. The Figure 7.21 illustrates this arrangement of a three-dimensional volume imaging modality. It should be mentioned that in literature mostly the demagnification $M = 1/m$ is used instead of m. This should not be confused.

The intensity in the image is now given as

$$I(x_i, y_i, z_i) = |E(x_i, y_i, z_i)|^2 \tag{7.45}$$

The different magnifications in lateral and axial direction should be noticed, in particular the sign. With the help of the matrix of the demagnifications

$$\underline{M} = \begin{pmatrix} 1/m & 0 & 0 \\ 0 & 1/m & 0 \\ 0 & 0 & -1/m^2 \end{pmatrix} \tag{7.46}$$

the incoherent three-dimensional imaging can be written as a three-dimensional convolution with the object transmission t in vectorial compact form as

$$I(\vec{r}_i) = \int |t(\vec{r}_o)|^2 \cdot \left|H_{CTF}^{3d}(\vec{r}_o + \underline{M} \cdot \vec{r}_i)\right|^2 d^3\vec{r}_o \tag{7.47}$$

This derivation shows that the concept of the OTF can be extended to the three-dimensional case, but the z-coordinate plays a special role.

7.3 Miscellaneous

7.3.1 Non-Isotropic Point Spread Function

If an off-axis field point in an optical system is considered, usually coma or astigmatism create anisotropic PSFs. For primary coma, the ratio of the geometrical diameters of the spot is 2 : 3, therefore in the sagittal direction the resolution and the contrast are improved. In case of astigmatism, it depends on the combination with field curvature and defocus whether the tangential or the sagittal focal line dominates the spreading. This has the consequence that the resolution of a grating-like structure depends on its orientation. To take these azimuthal angle considerations into account, usually T- and S-MTF curves are given for the performance evaluation of optical systems in the field of view for two grating orientations in tangential or sagittal (corresponding to radial) structures. In circular symmetric systems, this effect depends on the location of the off-axis point relative to the optical axis. This is shown in Figure 7.22 and the resolution consequence in Figure 7.23. In very special and critical applications like microlithography, it makes sense to inspect some more feature orientation angles ψ as is sketched in Figure 7.22.

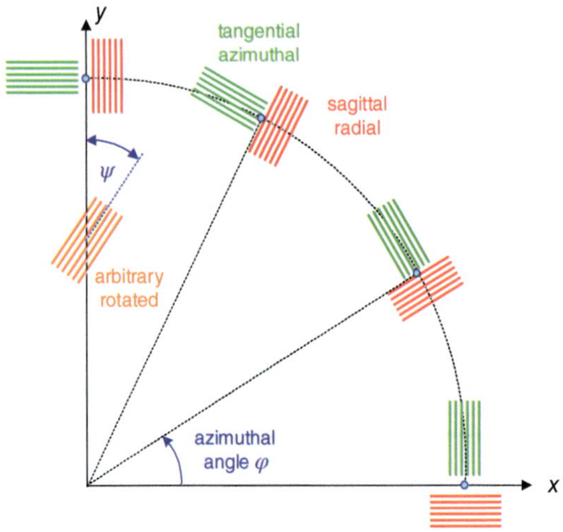

Figure 7.22 MTF dependence on the local orientation of the grating structures in the field of view.

7.3.2 Properties of MTF Curves

Since the transfer function usually is shown in a normalized scaling of the spatial frequency, according to the parameters of the Airy diameter the dependencies of the MTF are somehow hidden. As is shown in Figure 7.24, the cut-off frequency grows with [15]:

- increasing NA;
- decreasing wavelength.

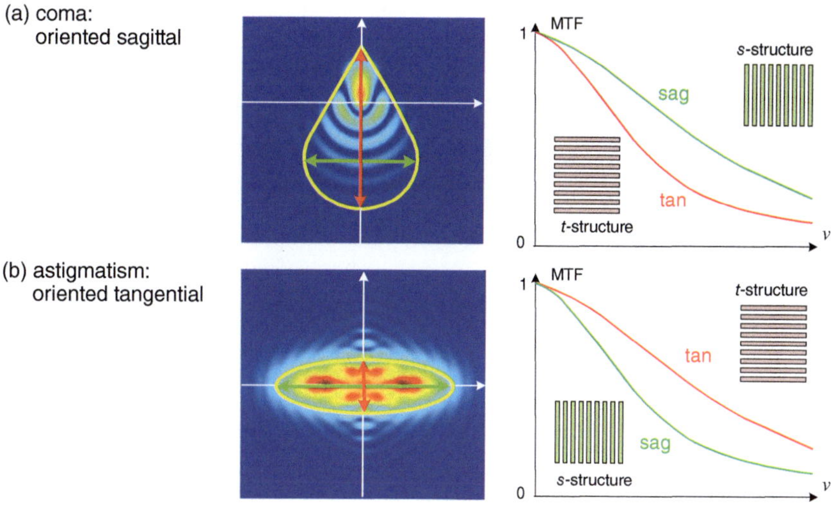

Figure 7.23 PSF for an off-axis field point in case of (a) coma and (b) astigmatism as well as the corresponding MTF curves for S- and T-oriented structures.

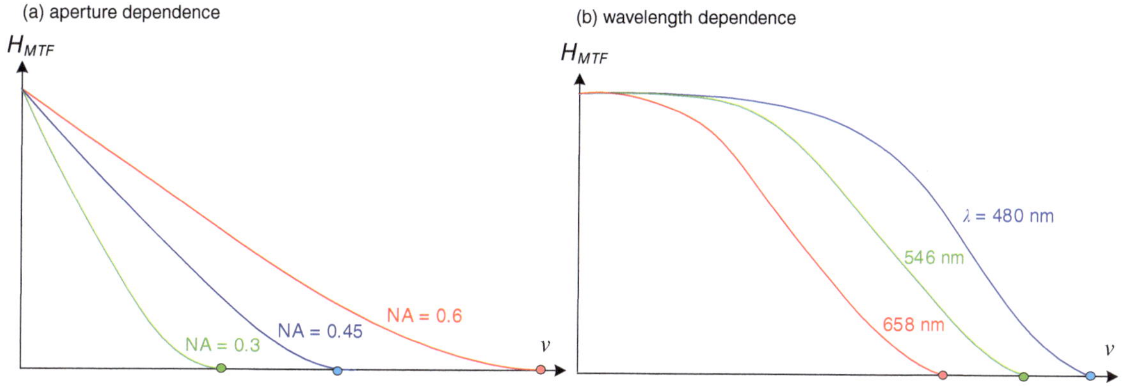

Figure 7.24 Dependence of the MTF cut-off frequency on (a) the numerical aperture and (b) the wavelength.

If the illumination of an optical system is incoherently superposed by a broadband spectrum, the resulting OTF is obtained by a spectral weighted superposition. The normalized spectral emission function $S(\lambda)$ of the source acts as the proper weighting function.

$$H_{OTF}^{(poly)}(\nu) = \int_0^\infty S(\lambda) \cdot H_{OTF}(\nu, \lambda)\, d\lambda \tag{7.48}$$

If, on the other hand, the intensity distribution of the system point spread function is formulated as a polychromatic weighted sum of the spectral components, the OTF can be obtained by the Fourier transform of this intensity according to the definition of the OTF:

$$H_{OTF}^{(poly)}(\nu) = \hat{F}\left[I_{psf}^{(poly)}(x)\right] \tag{7.49}$$

where the normalization factor is suppressed. The additivity of the non-interfering spectral PSF functions together with the linear behaviour of the Fourier transform allow, therefore, both formulations.

Figure 7.25 shows the transfer function of a microscopic lens on the axis for six wavelengths and the polychromatic superposition with equal weighting. The ideal reference curve for the polychromatic case is defined by the shortest wavelength 350 nm. Since the wavelength is a parameter of the scaled frequency axis, the curves do not lie very close to each other, although the correction is nearly diffraction limited. This misleading fact should always be remembered. The coloured dots in the drawing give the cut-off frequencies for the corresponding wavelengths. The strong decrease in the performance for the two shortest wavelengths can be clearly seen; the lens is not really well corrected below 450 nm.

In principle two different types of characteristic MTF curves can be distinguished and inside certain limits provoked by the aberration correction. Finally, it depends on the application which type of correction is more meaningful. Figure 7.26 sketches both alternatives qualitatively.

- The contrast first decreases slowly with a good contrast at low and medium frequencies, then it drops fast to very small levels. In principle the effective usable overall resolution is smaller, although the cut-off frequency is the same. This is the red drawn case 1 in the figure.

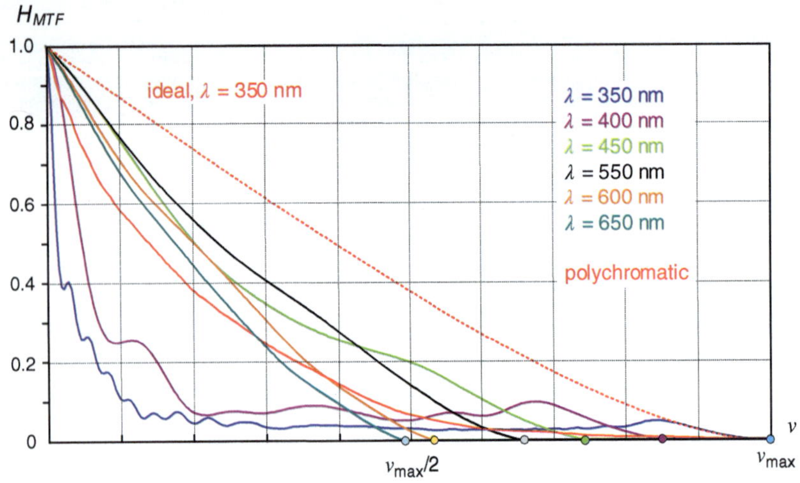

Figure 7.25 Modulation transfer function of a microscope objective for six different wavelengths and superposed polychromatic transfer function with equal weighting. The ideal curve is indicated by a dashed line for the shortest wavelength 350 nm.

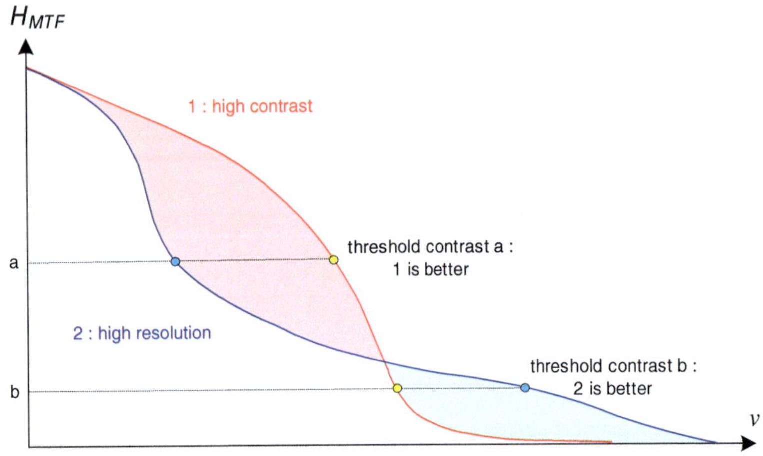

Figure 7.26 Two limiting cases of the MTF behaviour with high resolution or high contrast at medium spatial frequencies and the opposite for larger frequencies.

- The contrast drops fast to a medium value for small spatial frequencies, then it decreases slowly towards the cut-off frequency. A considerable contrast is then seen for larger frequencies. This is the blue case 2 in the figure.

7.3.3 MTF for Defocused Systems

7.3.3.1 Introduction

In reality, it is often of interest to know the behaviour of the OTF if the image plane is defocused. This indicates the decrease in image quality and the loss of resolution for small defocus tolerances. Figure 7.27 shows the variation of the MTF curves for a defocus without other aberrations. If the real part of the OTF is drawn, negative values are possible. It can be seen that there are zero crossing points in the transfer function if the defocus exceeds a certain limit (indicated by the coloured dots). The first zero is shifted to smaller frequencies if the defocus is increased. The yellow band indicates the contrast reversal range. The zero points are the limits for maximum resolution. The finite contrast for frequencies larger than the first zero cannot be used for a correct imaging process, because the frequency components of the zeros are completely absent in the image. Therefore, it is usually called spurious resolution. In this case, the similarity between the object and image cannot be guaranteed beyond the first zero.

Figure 7.28 shows the defocusing in an often used representation. If the system is defocused, special selected spatial frequencies are considered and their drop by deviating from the best image plane are shown. The defocusing is scaled with the Rayleigh unit of depth of focus. It can be seen that the higher frequencies are more sensitive to defocus and that their contrast value in the image plane is still smaller.

7.3.3.2 Approximations of the MTF*

(*) The content of this section is quite specialized and for advanced reading. Knowledge of this section is not necessary to understand the following chapters. Beginners are recommended to skip this section.

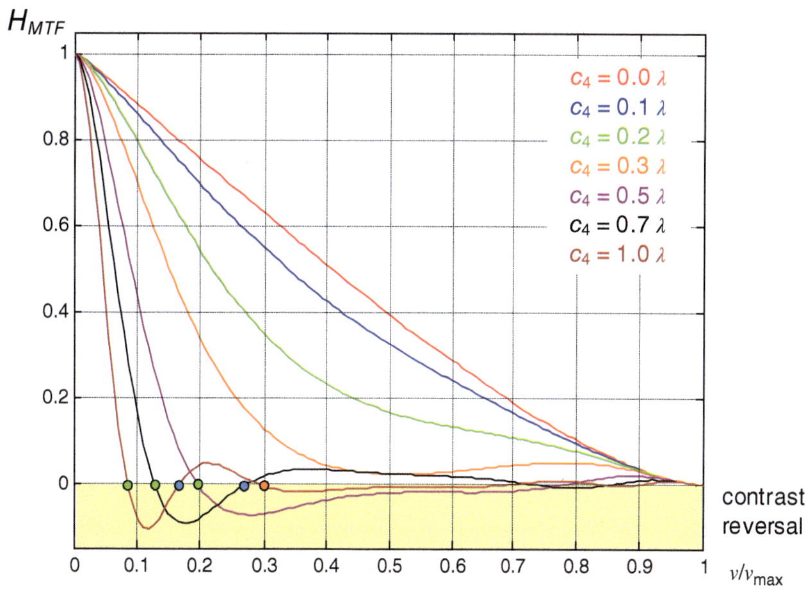

Figure 7.27 Drop in the ideal MTF function for various values of the defocus Δz, given in Zernike coefficients c_4. The real part of the OTF is drawn. The green dots indicate a first zero, the blue a second and the orange a third zero point of the various curves.

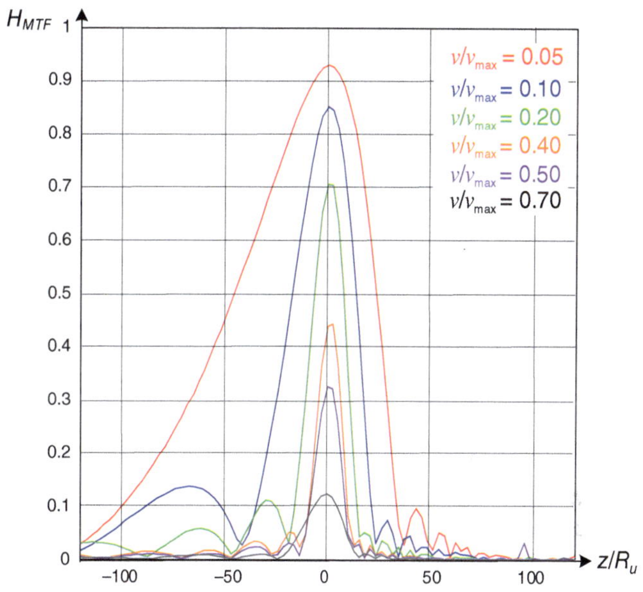

Figure 7.28 MTF function as a function of defocus for different frequencies and some residual aberrations (seen by the asymmetry around the image plane).

For small values of the defocus described by the Zernike coefficient c_4, a simple approximation for small spatial frequencies ν can be written as

$$H_{MTF}^{(Defoc)} = \frac{2 \cdot J_1(2\pi \cdot \nu \cdot c_4)}{2\pi \cdot \nu \cdot c_4} \tag{7.50}$$

with normalized frequency and defocus Zernike coefficient c_4. It is obtained by a Taylor expansion for small ν. According to [14], in case of pure defocus the MTF can be written empirically for all spatial frequencies with good accuracy with errors smaller than 15% in the form [16]

$$H_{MTF}^{(Defoc)} = 2 \cdot (1 - 0.69 \cdot \overline{\nu} + 0.0076 \cdot \overline{\nu}^2 + 0.043 \cdot \overline{\nu}^3) \cdot \frac{J_1[\pi \cdot \nu \cdot c_4 \cdot (2 - \overline{\nu})]}{\pi \cdot \nu \cdot c_4 \cdot (2 - \overline{\nu})} \tag{7.51}$$

The frequency is normalized here to $\overline{\nu} = \nu/\nu_o$ Figure 7.29 shows this approximation for several c_4 values. Eq. (7.50) is a linear approximation of this heuristic formulae.

7.3.3.3 The Ambiguity Function*

(*) The content of this section is quite specialized and for advanced reading. Knowledge of this section is not necessary to understand the following chapters. Beginners are recommended to skip this section.

The Duffieux formulation of the OTF of an optical system reads in one transverse dimension

$$H_{OTF}(\nu_x) = \int P\left(x_p + \frac{\lambda \cdot f \cdot \nu_x}{2}\right) \cdot P^*\left(x_p - \frac{\lambda \cdot f \cdot \nu_x}{2}\right) dx_p \tag{7.52}$$

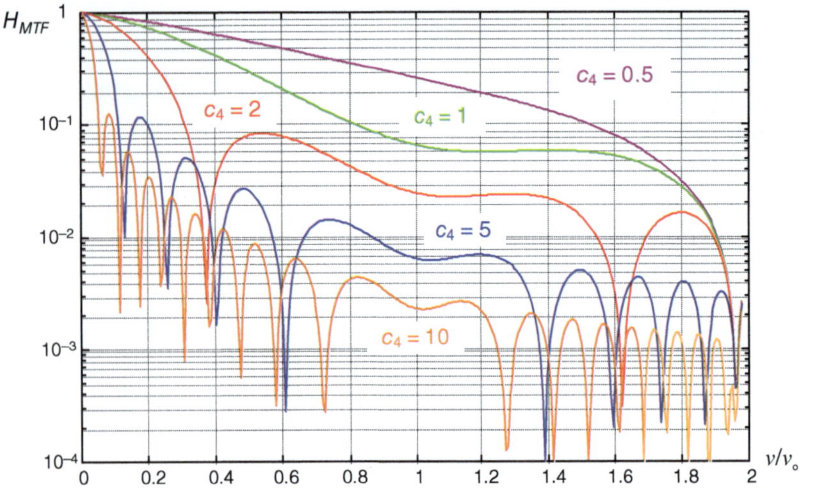

Figure 7.29 Approximation of the MTF for defocusing with different Zernike coefficients c_4 according to Stockseth.

with the pupil function P, which is decomposed in an optical system by a transmission and a phase factor as

$$P(x_p) = T(x_p) \cdot e^{2\pi \cdot i \cdot W(x_p)} \tag{7.53}$$

The spatial frequency v_x relation to the pupil coordinate x_p and the normalization frequency v_0.

$$v_x = \frac{x_p}{\lambda \cdot f}, \quad v_o = \frac{x_{p\max}}{\lambda \cdot f} = \frac{\sin u_{\max}}{\lambda} \tag{7.54}$$

The last approximation assumes small apertures or the fulfilment of the sine condition. In the case of defocusing, the wave aberration can be written as (see Chapter 15)

$$W_{40} = \frac{1}{2\lambda} \cdot \Delta z \cdot \sin^2 u = \frac{\Delta z \cdot x_p^2}{2\lambda \cdot f^2} = \frac{1}{2} \cdot \lambda \cdot \Delta z \cdot v_x^2 \tag{7.55}$$

and the transfer function reads for absent higher order aberrations

$$H_{OTF}(v_x, \Delta z) = \int T\left(x_p + \frac{\lambda \cdot f \cdot v_x}{2}\right) \cdot T^*\left(x_p - \frac{\lambda \cdot f \cdot v_x}{2}\right) \cdot e^{2\pi \cdot i \cdot \frac{\Delta z}{f} \cdot x_p \cdot v_x} dx_p \tag{7.56}$$

If the defocus is defined in a normalized way as a distance d or an angle φ

$$d = \frac{\Delta z}{R_u} = \frac{\Delta z \cdot \sin^2 u_m}{\lambda} = \frac{\Delta z \cdot x_{pm}^2}{\lambda \cdot f^2} = 2W_{40,\max} = \tan \varphi \tag{7.57}$$

it gives

$$\begin{aligned} H_{OTF}(v_x, d) &= \int T\left(x_p + \frac{\lambda \cdot f \cdot v_x}{2}\right) \cdot T^*\left(x_p - \frac{\lambda \cdot f \cdot v_x}{2}\right) \cdot e^{2\pi \cdot i \cdot \frac{\lambda \cdot f \cdot d}{x_{pm}^2} \cdot x_p \cdot v_x} dx_p \\ &= \int T\left(\overline{x}_p + \frac{\overline{v}_x}{2}\right) \cdot T^*\left(\overline{x}_p - \frac{\overline{v}_x}{2}\right) \cdot e^{2\pi \cdot i \cdot d \cdot \overline{x}_p \cdot \overline{v}_x} d\overline{x}_p \end{aligned} \tag{7.58}$$

With the normalized variables

$$\overline{x}_p = \frac{x_p}{x_{p\max}} = \frac{x_p}{a}, \quad \overline{v}_x = \frac{v_x}{v_{x\max}} = \frac{\lambda \cdot v_x}{\sin u'} \tag{7.59}$$

This Fourier transform as a special OTF for defocusing is called the ambiguity function A; it is a polar representation of the OTF, the radial coordinate is the spatial frequency and φ is the polar angle and describes the defocusing d.

$$H_{OTF}(v, d) = A(v, v \cdot \tan \varphi) = A(v, v \cdot d) \tag{7.60}$$

A straightforward extension with two transverse dimensions is easily possible.

For a one-dimensional case, the ambiguity function is the well-known sinc function [17]

$$A(v, d) = \left(1 - \frac{v \cdot \lambda}{2 \sin u}\right) \cdot sinc\left[\pi d \cdot \left(1 - \frac{v \cdot \lambda}{2 \sin u}\right)\right] \tag{7.61}$$

A cross section through this distribution gives the contrast change during defocusing as it is illustrated in Figure 7.30. The angle φ is the angle against the frequency axis and measures the size of defocusing; its value corresponds to the radial spatial frequency. A line usually has several zeros. The zeros of the contrast can be counted along one straight line; they are occurring for the defocus

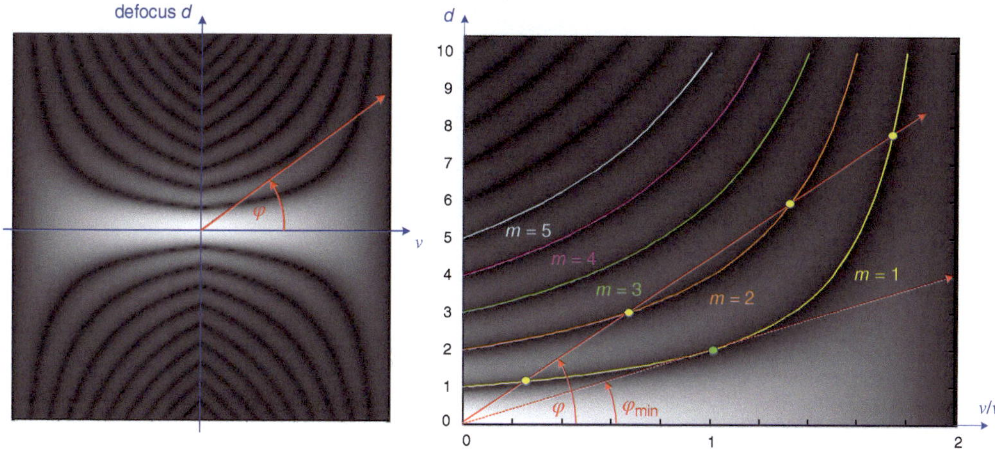

Figure 7.30 Interpretation of the ambiguity function (a) and zeros of the contrast (b).

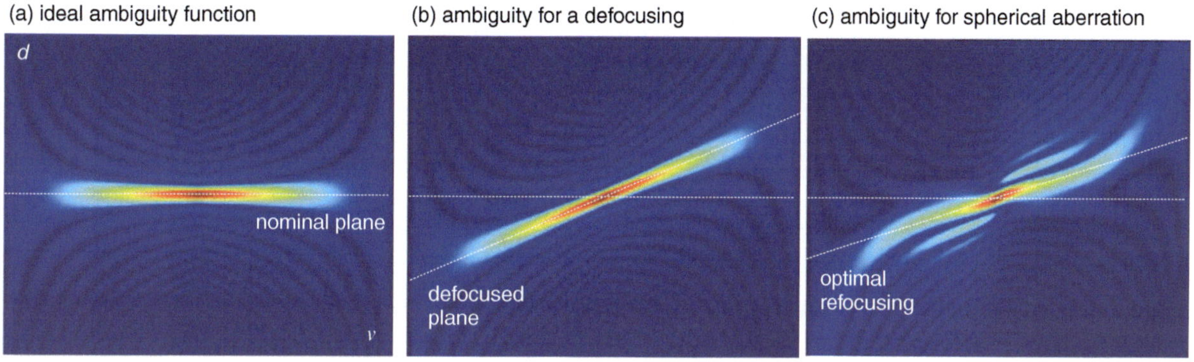

Figure 7.31 Ambiguity function $A(v,d)$ for an (a) ideal system, (b) a system with defocus and (c) a system with residual spherical aberration.

values of the hyperbolic curves

$$d_{zero} = \frac{m}{1 - \frac{\lambda \cdot v}{2 \sin u}} = \frac{m}{1 - \frac{v}{2v_o}} \tag{7.62}$$

where $m = 1, 2, 3, \ldots$ counts the zeros. If the geometric condition for the $m = 1$ curve is constructed for the defocus angle φ, the maximum allowed defocus $\varphi_{\min}$ without contrast reversal is obtained, see Figure 7.30(b) for the yellow curve with the green dot.

If a system has modified properties like residual aberrations, the shape of the ambiguity function changes. This is visualized in Figure 7.31, where an ideal case – a defocused plane and a system with spherical aberration – are sketched. For a defocused plane, obviously the function is sheared in φ-direction; in case of spherical aberration, a cubic deformation takes place and shows that by defocusing out of the ideal plane the contrast can be improved.

7.3.4 MTF for Special Pupil Shapes*

(*) The content of this section is quite specialized and for advanced reading. Knowledge of this section is not necessary to understand the following chapters. Beginners are recommended to skip this section.

7.3.4.1 MTF for a Square Pupil Shape

If the pupil of an optical system has the shape of a square with side length $2a$, the limiting spatial frequency for a finite overlap of the areas in case of a lateral shift is given by

$$v_{\max} = \frac{2a}{\lambda \cdot f} \tag{7.63}$$

The OTF function due to this simple geometry according to Figure 7.32 is in this case a linear triangle function with the analytical representation

$$\begin{aligned} H_{OTF}(v_x, v_y) &= \left(1 - \left|\frac{v_x \lambda f}{2a}\right|\right) \cdot \left(1 - \left|\frac{v_y \lambda f}{2a}\right|\right) \\ &= \mathrm{Tri}\left(\frac{v_x \lambda f}{2a}\right) \cdot \mathrm{Tri}\left(\frac{v_y \lambda f}{2a}\right) \end{aligned} \tag{7.64}$$

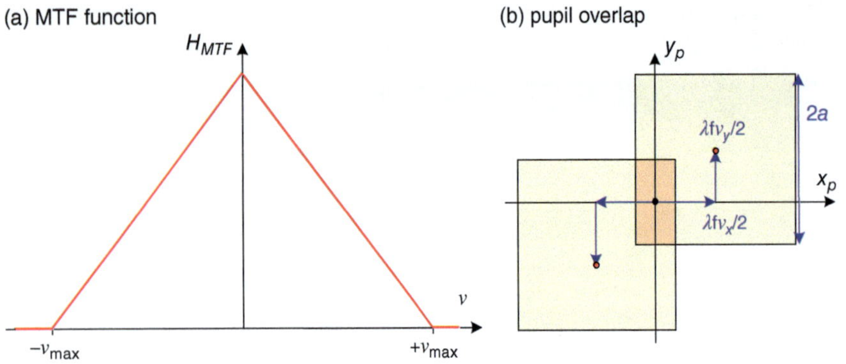

Figure 7.32 Modulation transfer function and pupil overlap for a square aperture.

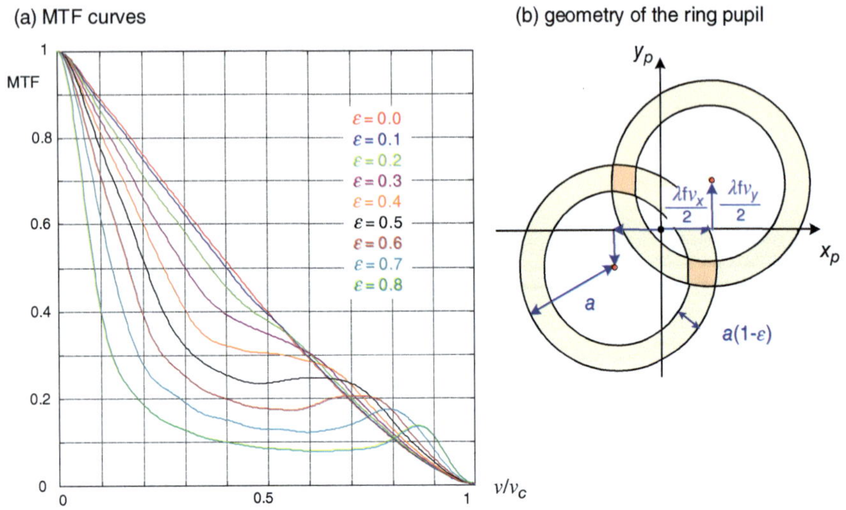

Figure 7.33 Exact calculated modulation transfer function (a) and pupil overlap geometry (b) for a ring aperture with different values of ε.

with the triangle function Tri(). The residual area overlap is now a rectangle; the drop of the MTF curve is therefore linear as shown in Figure 7.32(a).

7.3.4.2 MTF of a Ring Pupil*

(*) The content of this section is quite specialized and for advanced reading. Knowledge of this section is not necessary to understand the following chapters. Beginners are recommended to skip this section.

If a ring pupil is considered with outer radius a and inner radius $\varepsilon\, a$, the maximum overlap of the pupil can in the best case reach the value of the ring area

$$H_{OTF}^{\max}(v) = \pi\, a^2 \cdot (1 - \varepsilon^2) \tag{7.65}$$

(not normalized). Since the maximum spatial frequency depends on the size of the pupil, the resolution and the cut-off spatial frequency of a ring pupil are the same as for a circular area. But due to the geometry of the overlap area in Figure 7.33(b), the contrast first drops fast and then saturates; the value is roughly double the ring width squared. This constancy is better if the ring width $a(1\text{-}\varepsilon)$ is smaller. Later the contrast drops at the cut-off frequency v_c. It should be mentioned that the normalization here gives the wrong picture, since in reality the contrast of the full circle with $\varepsilon = 0$ cannot be exceeded by a ring. The same artefact can be seen in Figure 7.34(b) for the blue curve.

As can be seen from exact diffraction calculations, a narrower ring aperture has a better Strehl ratio in comparison to the reference case of a full illuminated circle aperture. Simultaneously the relative height of the central peak relative to the first side lobe height decreases, and therefore the contrast is lowered. This behaviour is illustrated in Figure 7.34 as a function of the parameter ε.

7.3.4.3 MTF for Systems with Apodization

If a pupil is not uniformly illuminated inside a circular shape, the MTF changes. There are many opportunities for these apodization effects [18]. Here just one often occurring example case should be discussed: a Gaussian illuminated pupil. It is well known that in paraxial approximation a Gaussian profile is maintained during propagation, therefore this functional dependence is also seen in the image as well as for the MTF.

In the case of Gaussian apodization, the boundary of the light cone and the value of the numerical aperture are not clearly defined. Therefore, no unique cut-off frequency

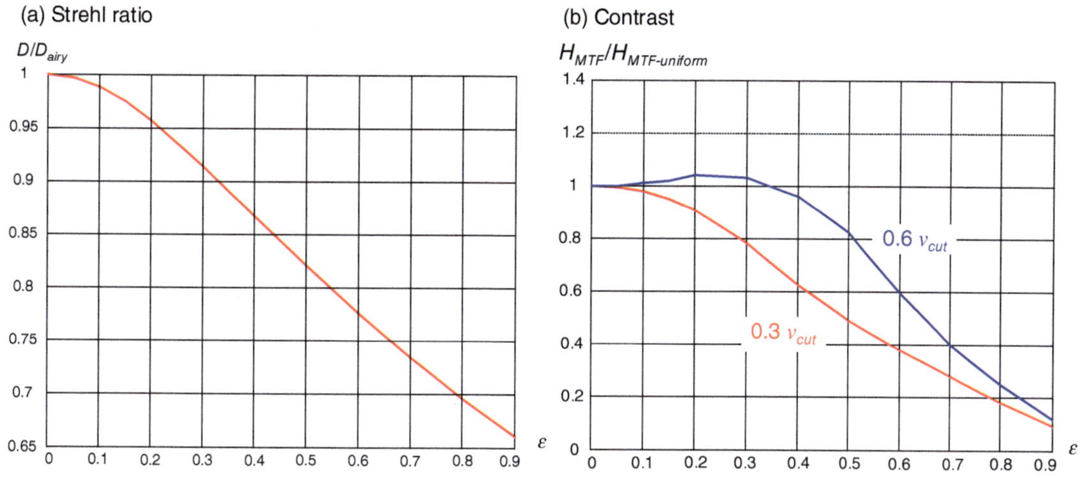

Figure 7.34 Influence of the ring parameter ε as (a) the Strehl ratio behaviour and (b) as contrast for two spatial frequencies. The scale of the axis for the Strehl ratio should be noted.

Figure 7.35 Transfer function in case of a classical uniform pupil as well as a Gaussian apodization.

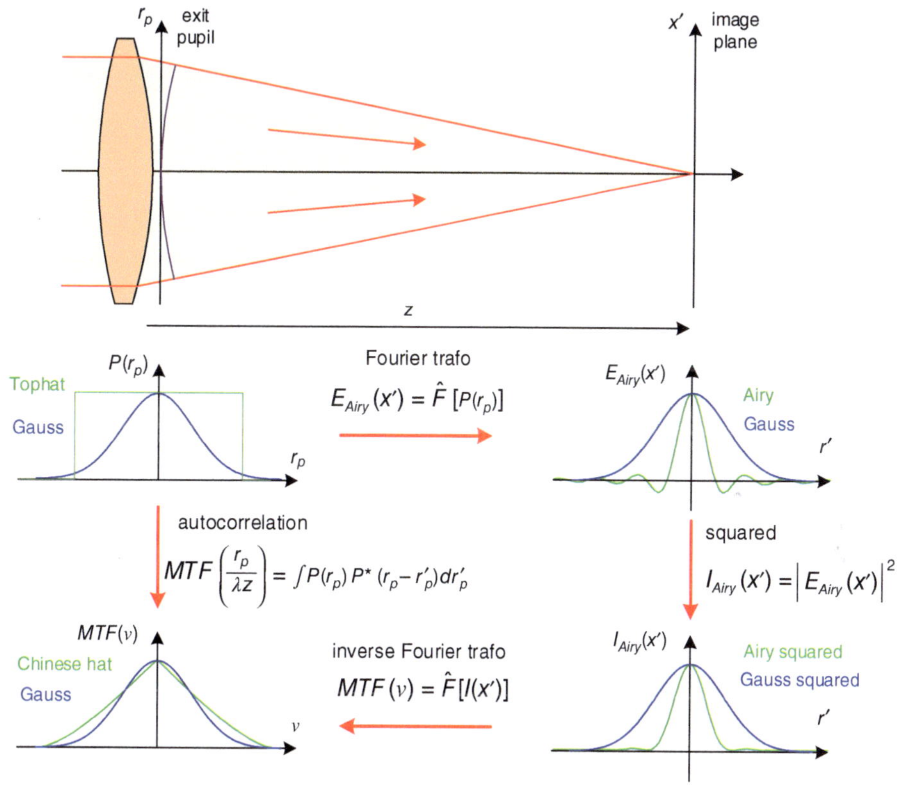

can be defined. In the following Figure 7.35, the Fourier relationships between the pupil (ν or r_p) and the image plane (x') are illustrated for the two cases:

- homogeneous illuminated pupil, Airy pattern in image plane;
- Gaussian illumination, also Gaussian in image plane.

The corresponding mathematical relations of the Fourier transforms can be found in mathematical textbooks. It should be noted that the Fourier relationship of the pattern in the image plane is significantly different, if the amplitude of the Airy function (with negative parts) or the squared function (only positive and sidelobes smaller in peak height) are used. In one case we get in the pupil a tophat; in the other case the Chinese hat function (typical ideal incoherent MTF curve). The corresponding relationships and comparisons are illustrated in Figure 7.35.

7.3.5 Relationships of System Response Functions

To understand the behaviour of an optical system, the response of the system for special illuminated object structures can be used. Depending on the choice of the structure, different response functions are obtained. If the structure is rather general, all these test functions are equivalent; although they have quite different scalings, shapes, sensitivities and parameters. In principle all of these options can be used, but in the practical application it is most useful to select structures and geometries that are similar to the expected imaging features in the application. This then allows the best prediction of image quality of the system.

The useful structures should contain all spatial frequencies to give a more or less complete answer function of the system. The most important ones are compared in the following Table 7.1 together with some remarks about their properties, advantages and shortcomings.

All these simple structures allow for a conversion of the transfer properties. PSF and OTF are the basic approaches and they are discussed in more detail. The other options with edges, lines and slits are more or less integrated versions of a PSF. These give better options for real measurements and the problems of small sources and bad SNR are overcome, but the separation of the residual aberrations is more cumbersome due to the integration. Furthermore, due to the geometry, at least two azimuthal orientations must be considered to get the full response behaviour. This is also true for the OTF with the *S*- and *T*-structures as described above. According to the various definitions, the diffraction related criteria PSF, OTF, ESF and LSF are mathematically connected by different operators and transforms. Figure 7.36 illustrates these links and converting equations; in Table 7.2 the equations are listed.

Table 7.1 Important response functions.

Object structure	Name	Abbr.	Remark
Point	Point spread function	PSF	Hard to measure, low power meaningful for microscopy, astronomy
Edge	Edge spread function	ESF	Easy to realize; two azimuths are needed
Line	Line spread function	LSF	Hard to realize; two azimuths are needed
Slit	Slit spread function		Meaningful for spectroscopy; two azimuths are needed
Grating	Optical transfer function	OTF	Good separability of resolution and contrast meaningful for photography, lithography; two azimuths are needed

Figure 7.37 visualizes two special conversions between ESF, LSF and PSF graphically.

There are also more relationships with the quantitative numbers based on one of these functions; for example, the integrated volume under the OTF function gives the Strehl ratio

$$D_S = \frac{\iint H_{OTF}(v_x, v_y)\, dv_x\, dv_y}{\iint H_{OTF}^{(ideal)}(v_x, v_y)\, dv_x\, dv_y} \tag{7.66}$$

7.3.6 Imaging of Phase Structures*

(*) The content of this section is quite specialized and for advanced reading. Knowledge of this section is not necessary to understand the following chapters. Beginners are recommended to skip this section.

In the case of a classical incoherent image formation framework, the image intensity is obtained as a convolution of the object intensity and the PSF intensity

$$I_{ima}(x) = I_{psf}(x) \otimes I_{obj}(x) \tag{7.67}$$

This means that phase distributions and structures without intensity modulations in the object cannot generate any contrast and therefore they are not seen in the image plane [19–21]. In biological and medical applications, very often the samples are fully transparent and the structures – for example, cells in water solution – are fully transparent and change only the optical path length by different

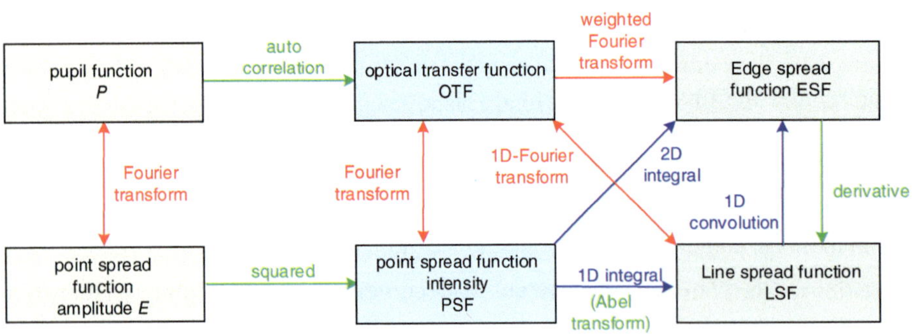

Figure 7.36 Connections between the different diffraction criteria.

Table 7.2 Relations between the diffraction related quality criteria.

Goal	Given	Equation	Remark
ESF	PSF	$I_{ESF}(x) = \int_{-\infty}^{x}\int_{-\infty}^{\infty} I_{PSF}(x_o, y)\, dy\, dx_o$	
	LSF	$I_{ESF}(x) = \int_{0}^{\infty} I_{LSF}(x - m \cdot x_o)\, dx_o$	
	OTF	$I_{ESF}(x) = \frac{1}{2} - \frac{1}{2\pi i} \cdot \int \frac{1}{v_x} H_{OTF}(v_x, 0) \cdot e^{-2\pi i \cdot x \cdot v_x} dv_x$	
LSF	PSF	$I_{LSF}(x) = \int I_{PSF}(x, y) dy$	General case
		$I_{LSF}(x) = 2\int \frac{I_{PSF}(r)}{\sqrt{r^2 - x^2}} \cdot r dr$	Especially for circular symmetry: Abel transform
	ESF	$I_{LSF}(x) = \frac{d\, I_{ESF}(x)}{dx}$	
	OTF	$I_{LSF}(x) = \int H_{OTF}(v_x, 0) \cdot e^{2\pi i \cdot x v_x} dv_x$	
PSF	OTF	$I_{PSF}(x, y) = \iint H_{OTF}(v_x, v_y) \cdot e^{-2\pi i \cdot (x v_x + y v_y)}\, dv_x\, dv_y$	
OTF	ESF	$H_{OTF}(v_x) = \int \frac{d\, I_{ESF}(x)}{dx} \cdot e^{2\pi\, i v_x \cdot x}\, dx = 2\pi\, i v_x \cdot \hat{F}[I_{ESF}(x)]$	
	LSF	$H_{OTF}(v_x) = \int I_{LSF}(x) \cdot e^{2\pi\, i v_x \cdot x}\, dx$	Along the line
	PSF	$H_{OTF}(v_x, v_y) = \iint I_{PSF}(x, y) \cdot e^{2\pi\, i(x v_x + y v_y)}\, dx\, dy$	

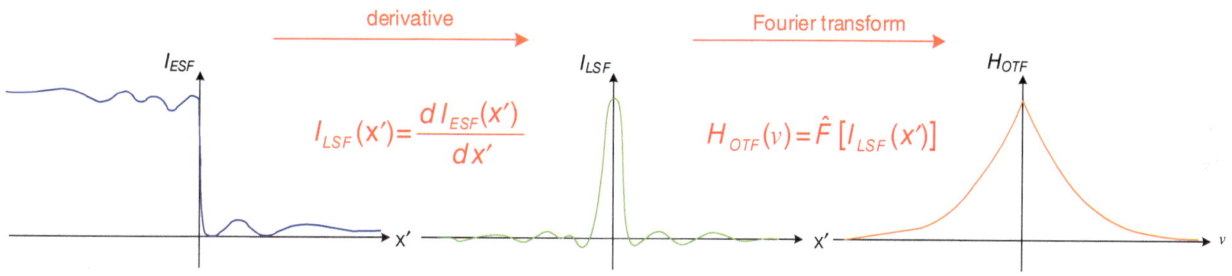

Figure 7.37 Visualization of the relations between ESF, LSF and OTF.

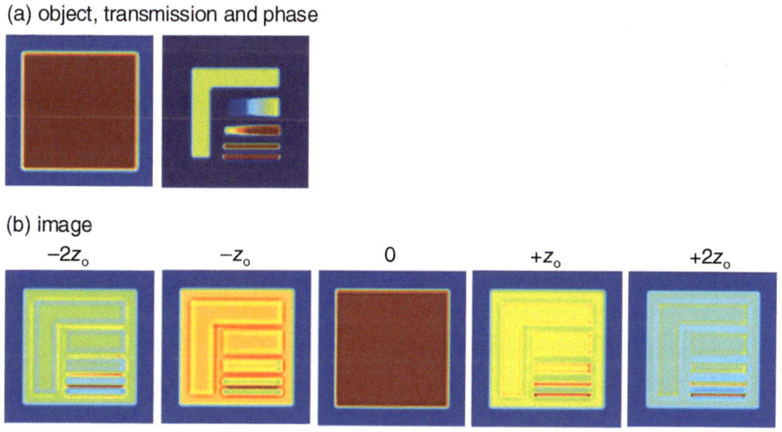

Figure 7.38 Imaging of a pure phase object, (a) transmission and phase and (b) intensity in five different defocused planes.

refractive indices. These phase structures are classically not seen and therefore a phase imaging and contrasting approach is greatly desired. Figure 7.38 shows a pure phase object, in the highly resolved image plane no contrast is observed. But in the defocused planes, the structures are seen though blurred and not symmetric around the image plane.

There are three major groups of methods that allow us to visualize phase structures of an object. These are as follows.

- Using coherent light for illumination, which is phase sensitive.
- Using the fact that the phase changes the propagation direction of the light; therefore, in a certain distance the intensity distribution depends on the initial phase. The change of the intensity with z is coded by the phase and from several planes it can be digitally reconstructed. This approach can also be generalized by a diversification of other form. This is called a digital phase retrieval.
- The classical Fourier image formation is systematically changed so the interference of the partial plane waves finally gives a contrast. This means this model phase can be contrasted if the balance between the diffraction orders is changed or perturbed. In microscopy many different attempts have been developed by this approach over time.

The most important approaches are listed below [22–24]. The main issues in all these methods are the question of lateral resolution and whether the obtained phase 'image' is qualitative or gives the phase values exactly quantitatively. Furthermore, if digital algorithms are involved, the generation of the phase image requires time and therefore is not obtained instantaneously. A quasi-real-time imaging of course is desired. The most important methods to visualize phase structures are as follows.

- Zernike phase contrast: here the first diffraction order is changed in amplitude or phase or both against the other diffractions orders.
- Differential interference contrast: the contrast is generated by creating differences in phase by using polarization effects.
- Modified phase masks: generalized phase variations introduced by masks located in the pupil to create similar effects as in Zernike contrast (Hoffmann, Varel, spiral…).
- Holographic imaging: using a coherent superposition of the object wave with a reference beam.
- Darkfield imaging: suppressing of the geometrical 0th diffraction order
- Interferometric imaging: using coherent illumination and coherent superposition with a reference wave.
- Ptychography: illumination coherent under different directions or positions and tomographic reconstruction.
- Digital phase retrieval: gathering the image in several distances or configurations, digitally recovering the phase by Gerchberg-Saxton, transport of intensity propagation or other algorithms digitally.

7.4 OTF of Cascaded Systems*

(*) The content of this section is quite specialized and for advanced reading. Knowledge of this section is not necessary to understand the following chapters. Beginners are recommended to skip this section.

7.4.1 Introduction

If two optical systems are concatenated, there are different cases from the viewpoint of physics that must be distinguished [22–26].

- If in a real intermediate image plane a diffusing screen is located, the coherence and the phase of the wavefront are destroyed. In this special configuration, the broadening of the PSF by residual phase aberration of both subsystems is superposed and the resulting PSF is worse than the individual PSF-widths by convolution.
- In most real situations, there is no physical screen located in the intermediate image plane. Therefore, for the second system, there is a coherent incoming wave with residual phase perturbations and not an incoherent illuminated object. Phase aberrations of the first systems are superposed by the phase transfer of the second system. If the corrections of both systems are chosen properly, it is possible to get a well performing system; the wave aberrations of the first system can be compensated by the second system because the phase is additive. See also the discussion in Chapter 6.

The following Figure 7.39 shows two limiting cases for illustration with opposite values of the wave aberrations in both sub-systems. In the upper row (a) both individual system MTFs are shown for incoherent illumination. Both systems suffer from a defocus of exactly the opposite value but equal amount. In the lower row (b) on the left side a coherent case illustrates the compensation of the opposite wave aberration to a combined corrected system. On the right side an incoherent case without phase-memory in the intermediate image plane by introducing a diffuser is sketched; here the degradation is stronger for the combined system.

The situation is much more complex if the illumination of the object is partially coherent. This general case is not further considered here.

If the residual aberrations of the first system are quite large, the ray path in the second system shows significant changes as a function of the aberrations of the first system, called raytrace error. In this case of large macroscopic aberrations, both system aberrations can no longer be considered as independent. The linear superposition of the phase is no longer valid and the overall transfer function

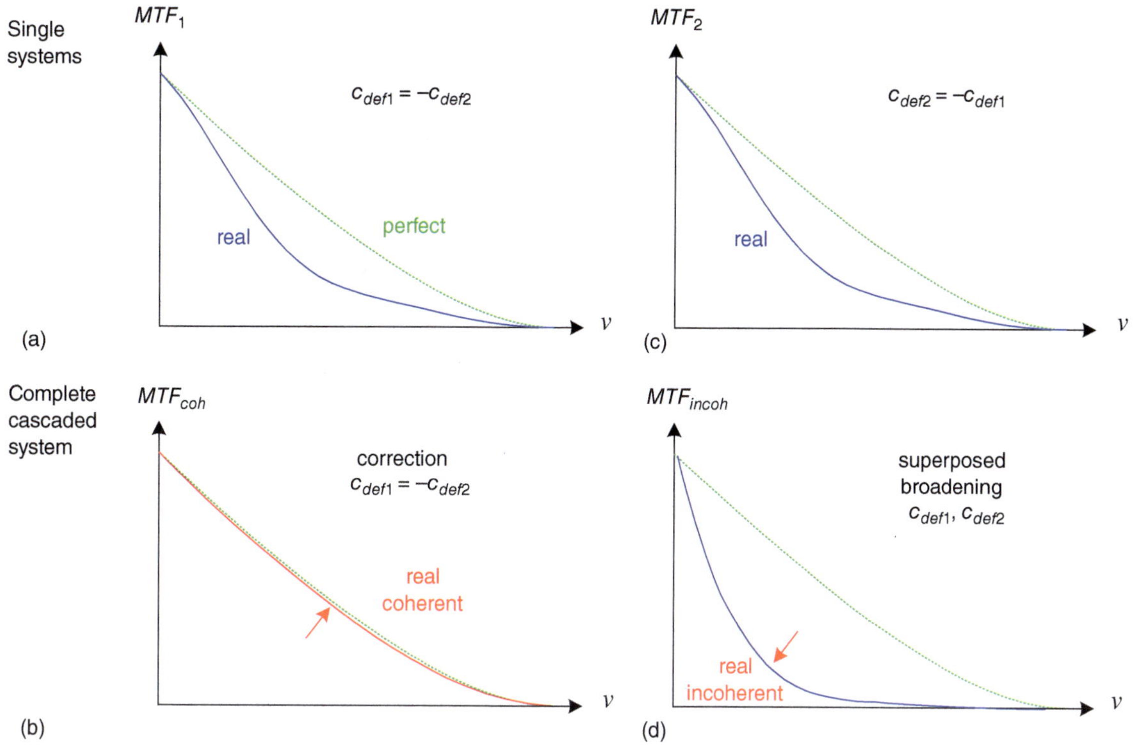

Figure 7.39 Transfer functions for combined systems with the single system MTF curves in (a) and (b). In (c) and (d) the MTF of the complete concatenated system is illustrated in the coherent and the incoherent case respectively.

depends non-linearly on the individual function of the first subsystem. In principle, this effect corresponds to induced aberrations.

7.4.2 Coherent Cascaded Systems

In the case of a simple intermediate air image, the cascading of two systems for coherent illumination by simply multiplying the OTFs is wrong. The question therefore is: if the known incoherent OTFs of both sub-systems can be used to calculate the OTF of the combined system. If the incoherent OTF is written as autocorrelation in the Duffieux form as

$$\begin{aligned} H_{OTF}(v) &= \int P(v) \cdot P^*(v-\mu)\, d\mu \\ &= \int H_{CTF}(v) \cdot H^*_{CTF}(v-\mu)\, d\mu \end{aligned} \tag{7.68}$$

the equation

$$\begin{aligned} H_{OTF}(v) = \int & H^{(1)}_{CTF}(v) \cdot H^{(2)}_{CTF}(v) \cdot H^{(2)*}_{CTF}(v-\mu) \\ & \cdot H^{(2)*}_{CTF}(v-\mu)\, d\mu \end{aligned} \tag{7.69}$$

is obtained for the cascaded system as the correct formula for the overall OTF. If, on the other hand, the OTF functions of both sub-systems are written similar to autocorrelations

$$\begin{aligned} H^{(1)}_{OTF}(v) &= \int H^{(1)}_{CTF}(v) \cdot H^{(1)*}_{CTF}(v-\mu)\, d\mu \\ H^{(2)}_{OTF}(v) &= \int H^{(2)}_{CTF}(v) \cdot H^{(2)*}_{CTF}(v-\mu)\, d\mu \end{aligned} \tag{7.70}$$

and are simply multiplied, the function

$$\begin{aligned} F(v) &= H^{(1)}_{OTF}(v) \cdot H^{(2)}_{OTF}(v) \\ &= \left[\int H^{(1)}_{CTF}(v) \cdot H^{(1)*}_{CTF}(v-\mu)\, d\mu\right] \\ &\quad \cdot \left[\int H^{(2)}_{CTF}(v) \cdot H^{(2)*}_{CTF}(v-\mu)\, d\mu\right] \end{aligned} \tag{7.71}$$

is found, which is not identical to the above correct OTF. It is seen that both system parts are connected due to the common integral, so the simple multiplication of the integrals is suppressing this relationship. The individual incoherent OTF functions do not allow for a direct determination of the performance of the cascaded system.

7.4.3 Incoherent Cascaded Systems

7.4.3.1 Introduction

If the linear system theory of incoherent systems is applied, the image forming process under incoherent illumination conditions in an optical system with a diffuser in between

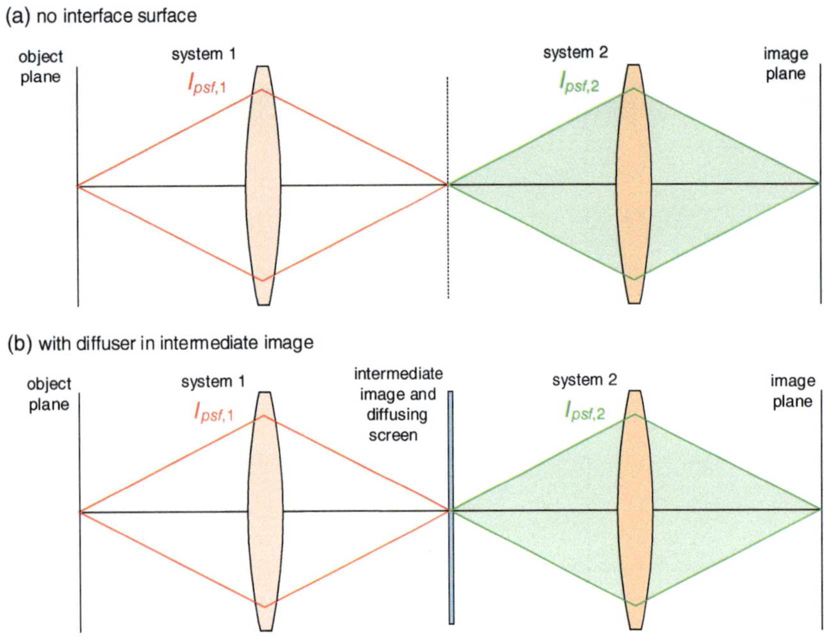

Figure 7.40 Transfer of coupled systems (a) coherent without screen and (b) incoherent with diffusing screen in the intermediate image.

can be described under the assumption of a space-invariant PSF in the spatial domain in the following way (for simplicity, the formalism is formulated here only with one spatial coordinate). In the intermediate image plane the intensity is given by

$$I_{interm-image}(x) = I_{psf1}(x) \otimes I_{object}(x) \tag{7.72}$$

where the operator corresponds to a convolution. Going into the frequency domain (spatial frequency ν) by Fourier transforming both sides of the equation gives the image as a simple product of the corresponding spectra

$$I_{interm-image}(\nu) = I_{psf1}(\nu) \cdot I_{object}(\nu) \tag{7.73}$$

If two system parts are combined in a two-stage image forming system without phase effects, the description of the imaging process corresponds to a successive application of the above equations

$$\begin{aligned} I_{incoher}(x) &= I_{psf2}(x) \otimes [I_{psf1}(x) \otimes I_{object}(x)] \\ I_{incoher}(\nu) &= I_{psf2}(\nu) \cdot [I_{psf1}(\nu) \cdot I_{object}(\nu)] \end{aligned} \tag{7.74}$$

Both equations can be rearranged in the order of the terms according to

$$\begin{aligned} I_{incoher}(x) &= I_{object}(x) \otimes [I_{psf1}(x) \otimes I_{psf2}(x)] \\ I_{incoher}(\nu) &= I_{object}(\nu) \cdot [I_{psf1}(\nu) \cdot I_{psf2}(\nu)] \end{aligned} \tag{7.75}$$

as it is shown in (7.56). This means that the convolution of both PSF intensities forms a new broader PSD, which is responsible for the overall imaging process.

For the validity of this approach the question is: how the phase memory in the intermediate image plane can be interrupted. If a diffusing screen is placed in the intermediate image plane, both sub-systems work based on incoherent imaging theory, see Figure 7.40.

7.4.3.2 Simple Case of Two Gaussian PSFs

In the simple case of two point spread functions with Gaussian shape, the convolution of the previous section can be performed analytically. Let us assume we have two amplitude profiles of the form

$$I_1(x,y) = I_{01} \cdot e^{-2\frac{x^2+y^2}{w_1^2}} \quad , \quad I_2(x,y) = I_{02} \cdot e^{-2\frac{x^2+y^2}{w_2^2}} \tag{7.76}$$

then the convolution is again a Gaussian function with an enlarged width

$$\begin{aligned} I(x,y) &= I_1(x,y) \otimes I_2(x,y) \\ &= \iint I_1(x',y') \cdot I_2(x-x',y-y')\, dx'dy' \\ &= I_{01} \cdot I_{02} \cdot \iint e^{-2\frac{x'^2}{w_1^2} - 2\frac{y'^2}{w_1^2} - 2\frac{x^2}{w_2^2} + \frac{4x\,x'}{w_2^2} - 2\frac{x'^2}{w_2^2} - 2\frac{y^2}{w_2^2} + \frac{4y\,y'}{w_2^2} - 2\frac{y'^2}{w_2^2}}\, dx'dy' \\ &= I_{01} \cdot I_{02} \cdot e^{-\frac{2x^2}{w_2^2} - \frac{2y^2}{w_2^2}} \int e^{\frac{4x\,x'}{w_2^2} - \frac{2x'^2}{w_2^2} - \frac{2x'^2}{w_1^2}} dx' \int e^{\frac{4y\,y'}{w_2^2} - \frac{2y'^2}{w_2^2} - \frac{2y'^2}{w_1^2}} dy' \\ &= I_{01} \cdot I_{02} \cdot e^{-2\frac{x^2+y^2}{w_2^2}} \int e^{\frac{4x\,x'}{w_2^2} - \frac{2x'^2}{w_2^2}} dx' \int e^{\frac{4y\,y'}{w_2^2} - \frac{2y'^2}{w_2^2}} dy' \\ &= \pi I_{01} \cdot I_{02} \cdot \left(\frac{1}{w_1^2} + \frac{1}{w_2^2} \right) e^{-2\frac{x^2+y^2}{w_1^2+w_2^2}} = A_0 e^{-2\frac{x^2+y^2}{w^2}} \end{aligned} \tag{7.77}$$

with

$$w^2 = w_1^2 + w_2^2 \tag{7.78}$$

In the special case of identical initial Gaussian beams, the width is enlarged by $\sqrt{2}$. In any case, the convolution creates a broadening, of which the resultant width is always larger than the largest contribution.

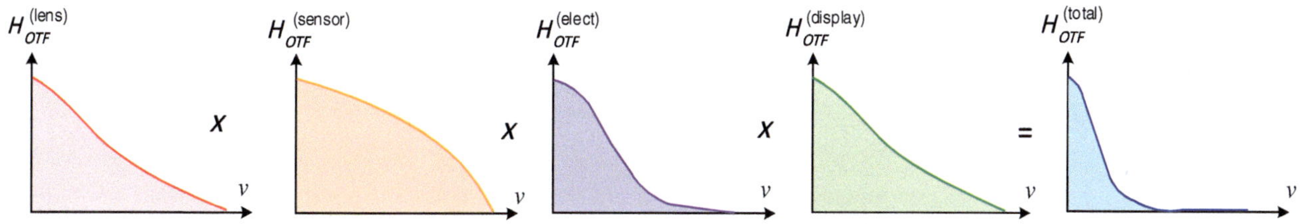

Figure 7.41 Sequence of transfer influences in a signal chain in the sense of a linear system theory.

7.4.4 MTF of a Complete Signal Chain

If the contrast behaviour of a complete signal chain in an optical system together with sensor, electronics and display of the result is considered, in the sense of the linear system theory, every step has an impact on the transfer properties. The individual modules of the signal processing have their individual transfer functions and frequency characteristics. The total behaviour is then finally given by the product of all involved filter functions. This scheme is illustrated in Figure 7.41. Every decrease in contrast reduction is accumulated and the total result usually is worse in comparison to the optical system influence only.

In the classical optical consideration discussed in this chapter, only the first contribution is discussed and, as can be seen, this is not fully describing reality.

References

1 Goodman, J. (2017). *Fourier Optics*. Freeman.
2 Cathey, T. (1974). *Optical Information Processing and Holography*. Wiley.
3 Duffieux, P.M. (1983). *The Fourier Transform and its Applications to Optics*. New York: Wiley.
4 Reynold, G.O., DeVelis, J.B., Parrent, G.B., and Thompson, B.J. (1989). *Physical Optics Notebook*. Bellingham: SPIE Press.
5 Gaskill, J.D. (1978). *Linear Systems, Fourier Transforms and Optics*. New York: Wiley.
6 Williams, T. (1999). *The Optical Transfer Function of Imaging Systems*. IoP Publishing.
7 Boreman, G. (2001). *Modulation Transfer Function in Optical and Electro-Optical Systems*. SPIE Press.
8 Williams, C.S. and Becklund, O.A. (1989). *Introduction to the Optical Transfer Function*. New York: Wiley.
9 Hopkins, H.H. (1984). Image shift, phase distortion and the optical transfer function. *Optica Acta* 31: 345.
10 Gu, M. (2000). *Advanced Optical Imaging Theory*. Springer.
11 Tsai, C.-M. (2015). Evaluation of geometrical modulation transfer function in optical lens system. *Mathematical Problems in Engineering* 863: 201.
12 Barrett, H., Chiu, M., Gordon, S. et al. (1979). Optical transfer function synthesis: a geometrical optics approach. *Applied Optics* 18: 2760.
13 Diaz, J. and Navarro, R. (2023). Geometric optical transfer function and pupil sampling patterns. *Optik* 279 (170): 746.
14 Mahajan, V. (2001). *Optical Imaging and Aberrations II, Wave Diffraction Optics*. SPIE Press.
15 Subbarao, M. (1990). Optical transfer function of a diffraction-limited system for polychromatic illumination. *Applied Optics* 29: 554.
16 Stockseth, P.A. (1969). Properties of a defocused optical system. *JOSA* 59: 1314.
17 Testorf, M., Hennelly, B., and Ojeda-Castaneda, J. (2010). *Phase-Space Optics*. McGraw Hill.
18 Chung, C.S. and Hopkins, H.H. (1989). Influence of nonuniform amplitude on the optical transfer function. *Applied Optics* 28: 1244.
19 Popescu, G. (2011). *Quantitative Phase Imaging of Cells and Tissues*. McGraw Hill.
20 Liu, C., Wang, S., and Veetil, S. (2022). *Computational Optical Phase Imaging*. Springer.
21 Glückstad, J. and Palima, D. (2009). *Generalized Phase Contrast*. Springer.
22 Yurlov, V., Han, K., and Yu, N. (2018). Cascaded Fresnel diffraction theorem and its application. *Optical Engineering* 57: 055108.
23 Calvo, M. and Lakshminarayanan, V. (1991). General transfer function of a compound optical system under linearity conditions in the Fresnel regime. *Optical Engineering* 30: 1503.
24 Rataj, M. (1984). Convolution formulation of Fresnel diffraction for cascaded optical systems. *Optica Acta* 31: 137.
25 deVelis, J. and Parrent, G. (1967). Transfer function for cascaded optical systems. *JOSA* 57: 1486.
26 Lakshminarayanan, V. and Calvo, M. (1990). Point spread function and modulation transfer function of a photolens treated as a cascade linear system under the Fresnel regime: application to a Tessar lens. *Optical Engineering* 29: 263.

8

Gaussian Beams

8.1 Introduction

Gaussian beams are a special eigensolution of the paraxial wave equation. They are often used as model descriptions for laser beams; in particular, the lowest order TEMoo as a fundamental mode. Gaussian beams are paraxial solutions; therefore, all effects perturbing this property – like higher numerical aperture angles, truncation or aberrations – have the consequence that the beams are no longer Gaussian. Therefore, it is a kind of idealized laser beam model. Due to the fact that the influence of diffraction is included, the Gaussian beam shows different behaviour as rays and purely geometrical models. The paraxiality allows for a very simple analytical calculation with the same *ABCD* matrices that are used in paraxial raytrace.

If Gaussian beams are used as illumination sources inside real optical systems, the validity of the hidden assumptions must always be kept in mind. The linear properties mentioned above have the positive effect that the propagation of Gaussian beams can always be performed by a paraxial raytrace equivalent in conventional software to reconstruct the beam parameter [1–6]. There are several generalizations of Gaussian beams proposed in the literature for less idealized conditions. These are for example:

- astigmatic beams with different parameters in two cross sections;
- beams beyond the paraxial approximation of small angles;
- truncated beams;
- beams suffering from spherical aberration;
- partially coherent beams with special profiles of the coherence functions.

These extensions are also briefly discussed here to cover further real conditions.

8.1.1 Definition of Basic Properties

A Gaussian beam is a solution of the Helmholtz wave equation for free-space propagation in a paraxial slowly varying envelope approximation of the Helmholtz wave equation approximation, see Section 6.2.3. The basic form has, in particular, a transverse amplitude variation with a Gaussian profile. This gives solutions with a high directionality as a result of the wave equation beam. There is one small lateral extension, the waist of the beam with radius w_0 and a far field divergence angle θ_0 to describe the mode. Only two parameters of the beam are independent. If the wavelength λ is fixed, only the waist radius w_0, the Rayleigh range z_0 as a depth of the waist region or the far field divergence θ_0 can be chosen arbitrarily. The other two of these three numbers are then fixed and they are related by the equations

$$w_0 = z_0\theta_0 = \frac{\lambda_0}{\pi\theta_0} = \sqrt{\frac{\lambda_0 z_0}{\pi}},$$
$$\theta_0 = \frac{w_0}{z_0} = \frac{\lambda_0}{\pi w_0} = \sqrt{\frac{\lambda_0}{\pi z_0}},$$
$$z_0 = \frac{\pi w_0^2}{\lambda_0} = \frac{w_0}{\theta_0} = \frac{\lambda_0}{\pi\theta_0^2} \tag{8.1}$$

If w_0, λ_0, z_0, θ_0 are the parameters in vacuum, the corresponding values in a medium with refractive index n reads

$$\lambda_n = \frac{\lambda_0}{n}, \quad w_n = w_0, \quad \theta_n = \frac{\theta_0}{n}, \quad z_{on} = n \cdot z_0 \tag{8.2}$$

The relations between the parameters are identical for these terms.

The paraxial propagation of simple coherent Gaussian beams follows analytical formulas. The beam width w is defined as the $1/e^2 = 13.5\%$ intensity level of the peak value, which changes during free space propagation according to a hyperbola

$$w(z) = w_0 \cdot \sqrt{1 + \left(\frac{z - z_T}{z_0}\right)^2} \tag{8.3}$$

as is shown in Figures 8.1 and 8.2. z_T is here the waist position. In the waist location, there is a horizontal tangent corresponding to a plane wave front, the waist w_0 is the smallest possible lateral extension. The Rayleigh range

Lens Design for Imaging: Fundamentals of Optical Systems, First Edition. Herbert Gross.

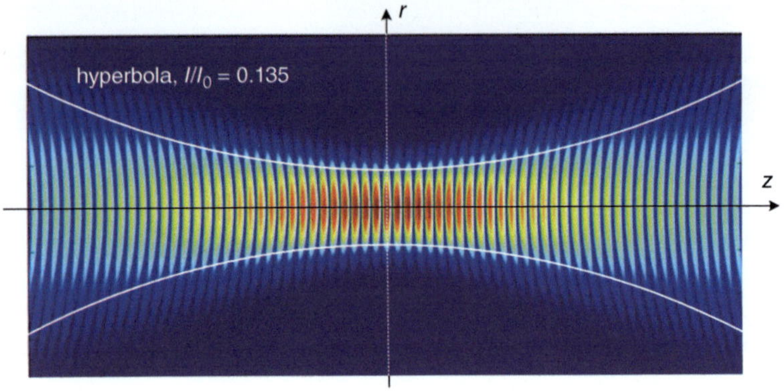

Figure 8.1 Hyperbolic caustic of a Gaussian beam during free space propagation.

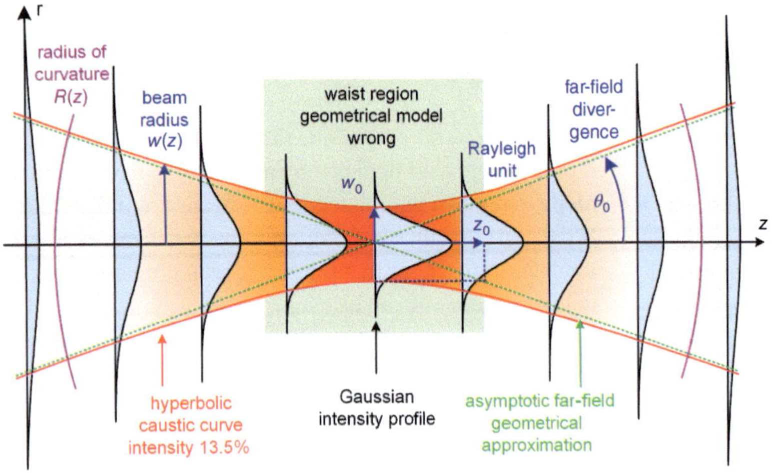

Figure 8.2 Definition of properties at a Gaussian beam and hyperbolic beam radius during propagation.

z_0 defines the axial extension of the waist region, at the distances $+z_0$ or $-z_0$ of the waist position the area is doubled, the beam radius is grown by $\sqrt{2}$. As a consequence of the hyperbolic shape of the envelope, the lateral growth of the beam far away from the waist is linear and can be approximated by the asymptotes. Due to the special property of being an eigensolution of free space propagation, the Gaussian amplitude profile does not change during propagation, only the size is scaled following Eq. (8.3). This is illustrated in Figure 8.2. As a physical solution of the wave equation, the Gaussian beam shrinks in the case of focusing only to the finite waist size, not to a geometrical point. In the far field with the asymptotic limit, the beam behaves like a geometrical ray. The far field divergence angle corresponds then to the slope of the asymptote.

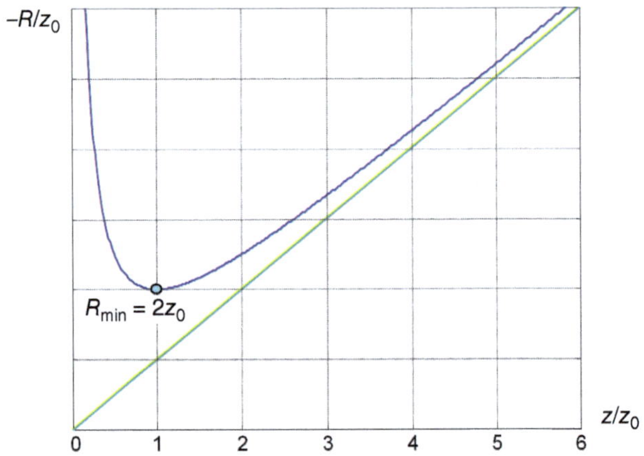

Figure 8.3 Radius of curvature during propagation.

The radius of curvature is parabolic and behaves as

$$R(z) = -(z - z_T) - \frac{z_0^2}{z - z_T} \tag{8.4}$$

This function is shown in Figure 8.3. The sign must be taken with care. In the definition above the geometrical technical definition is used with positive radii to the left of the waist. This is mostly different in the literature concerning laser optics. The radius has its minimal value at $z = z_T$, the wave is plane in the waist and approximates $R = -z$ in the geometrical limit for large values of z. It should be noted that R is never concentric around the waist centre, only asymptotical for large $z \gg z_0$. Furthermore, the ideal curvature in the paraxial approximation is parabolic, not spherical.

At a thin lens the radius of curvature changes according to

$$R' = \frac{-1}{1/f + 1/R} \tag{8.5}$$

corresponding to the lens makers' formula. If in an arbitrary location, and if the radius of curvature R and the beam width w are known, the position of the waist and its radius can be determined from the two equations

$$z_T = \frac{R}{1+\left(\frac{\lambda R}{\pi \cdot w^2}\right)^2}, \quad w_0 = \frac{w}{\sqrt{1+\left(\frac{\pi w^2}{\lambda \cdot R}\right)^2}} \tag{8.6}$$

For large values of z in comparison to z_0, the geometrical optical limiting case is obtained. The nonlinear equations of the beam parameter development are then simplified to a linear dependence as

$$w(z) = w_0 \cdot \frac{z - z_T}{z_0} = \theta_0 \cdot (z - z_T),$$
$$R(z) = -(z - z_T), \quad \theta(z) = \theta_0 \tag{8.7}$$

With the help of the radius of curvature and the waist size, the relation

$$\theta_0^2 = \left(\frac{w}{R}\right)^2 + \left(\frac{\lambda}{\pi w_0}\right)^2 \tag{8.8}$$

can be derived. The interpretation of this equation is that the far-field divergence angle is constituted by a quadratic superposition of the diffraction angle (second term) and the geometrical spreading (first term). If the far field divergence θ_0 and the actual beam radius $w(z)$ are known, the corresponding waist radius size and its position are given by

$$w_0 = \frac{\lambda}{\pi \theta_0}, \quad z_T = \pm \frac{\pi w_0^2}{\lambda} \cdot \sqrt{\left(\frac{w(z)}{w_0}\right)^2 - 1} \tag{8.9}$$

If the actual beam radius $w(z)$ and the local divergence $\theta(z)$ are known, the bundle parameter of the waist can be obtained by the formulas

$$w_0 = \frac{w(z)}{\sqrt{1+\left(\frac{\pi w(z) \cdot \theta(z)}{\lambda}\right)^2}},$$
$$z_T = -\frac{\pi^2 w_0^2 w(z) \cdot \theta(z)}{\lambda^2} \tag{8.10}$$

8.1.2 Amplitude and Intensity Distributions

The complex field of the beam in one lateral dimension r (or x, y in one-dimensional Cartesian representation) of a Gaussian beam is given by

$$E(r,z) = \frac{A_0}{\sqrt{1+\left(\frac{z-z_T}{z_0}\right)^2}} \cdot e^{-\left(\frac{r}{w(z)}\right)^2} \cdot e^{-\frac{i\pi r^2}{\lambda R(z)}} \cdot e^{-\frac{2\pi i z}{\lambda}} \cdot e^{-i \cdot \arctan \frac{z-z_T}{z_0}} \tag{8.11}$$

with waist amplitude A_0. The pre-factor regulates the conservation of energy here: the first factor gives the Gaussian profile, the second term describes the parabolic phase, the third term is the fast exp.(−ikz) phase variation and the last term describes the so-called Guoy phase. The sign switch of the Guoy phase by stepping through the waist should be noticed. The intensity profile can be calculated with the help of the equation

$$I(r,z) = \frac{P}{\pi w_0^2} \cdot \frac{1}{1+\left(\frac{z-z_T}{z_0}\right)^2} \cdot e^{-2\cdot\left(\frac{r}{w(z)}\right)^2} \tag{8.12}$$

where P is the beam power. It is Gaussian in the transverse direction and corresponds to a Lorentzian shape along the axis z. These functions are illustrated in Figure 8.4(a) and (c), (d).

If the phase parts are omitted, the amplitude and the intensity read simply

$$A(r) = e^{-\left(\frac{r}{w}\right)^2}, \quad I(r) = e^{-2\left(\frac{r}{w}\right)^2} \tag{8.13}$$

The encircled energy function can be calculated as

$$E(r) = \frac{\int_0^r I(r')\, 2\pi r'\, dr'}{\int_0^\infty I(r')\, 2\pi r'\, dr'} = 1 - e^{-2\left(\frac{r}{w}\right)^2} \tag{8.14}$$

as is shown in Figure 8.4(b). The total power of a Gaussian beam is given by

$$P = \int_0^\infty I(r)\, 2\pi r\, dr = \frac{\pi}{2} I_o w^2 \tag{8.15}$$

Table 8.1 gives the most important quantitative numbers of a Gaussian beam profile.

The last term in the complex amplitude representation contains the Guoy phase with the value

$$\psi(z) = \arctan\left(\frac{z-z_T}{z_0}\right) \tag{8.16}$$

This phase variation is depicted in Figure 8.5; it changes from −90° to +90° with a sign switch at the waist.

8.1.3 Complex Representation and Collins Chart*

(*) The content of this section is quite specialized and for advanced reading. Knowledge of this section is not

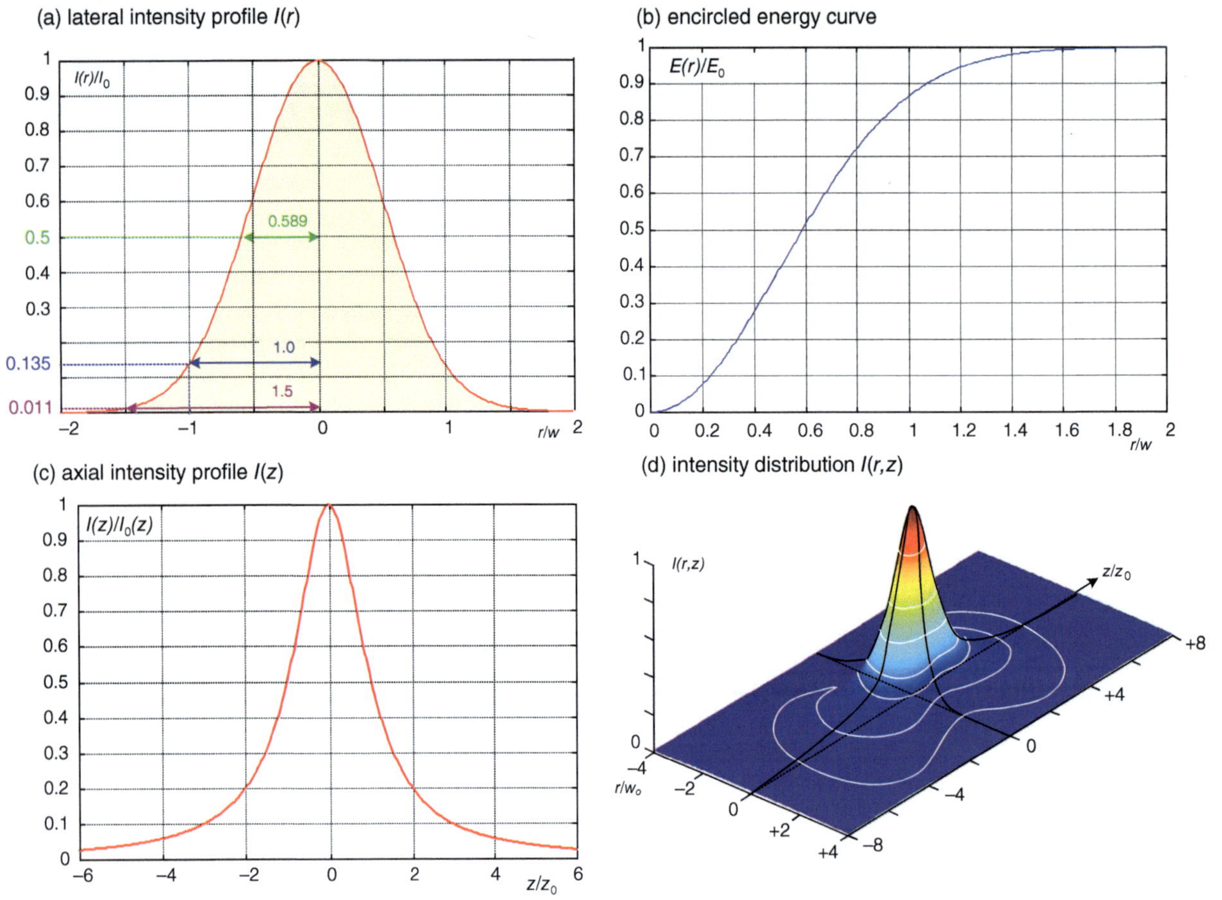

Figure 8.4 Intensity profiles of a Gaussian beam in (a) lateral, (c) axial and (d) both directions as well as the encircled energy in (b).

Table 8.1 Some values of Gaussian beam parameters.

Radius r	Diameter	Amplitude A	Intensity I	Power P
0.58871 w	1.17741 w	0.7071	0.5000	0.5000
w	2 w	0.3679	0.1353	0.8647
1.0730 w	2.146 w	0.3162	0.1000	0.9000
1.5 w	3 w	0.1054	0.0111	0.9889
2 w	4 w	0.0183	0.0003	0.9997

necessary to understand the following chapters. Beginners are recommended to skip this section.

The Gaussian beams can be represented by a complex beam parameter q according to the following equations

$$q = z + i\,z_0, \quad \frac{1}{q} = \frac{1}{R} - \frac{i\,\lambda}{\pi w^2} \tag{8.17}$$

It should be mentioned that this representation is a simple mapping of the parameters w and R into the complex plane, which is comfortable from the viewpoint of mathematics. But there is no deeper physical reason to do this. The parameter q can be propagated through a paraxial $ABCD$ system by the equation

$$q_2 = \frac{Aq_1 + B}{Cq_1 + D} \tag{8.18}$$

If a value of q is given, the physical parameters can be recovered by the formulas

$$\frac{1}{R} = \mathrm{Re}\left[\frac{1}{q}\right], \quad \frac{1}{w^2} = -\frac{\pi}{\lambda}\,\mathrm{Im}\left[\frac{1}{q}\right] \tag{8.19}$$

The Guoy phase of the beam obeys the equation

$$\tan\psi\,(z) = \frac{z - z_T}{z_0} = \arg q(z) = \frac{\pi w^2(z)}{\pi R(z)} \tag{8.20}$$

For a given value of q, the complex field can be expressed more compactly in the form

$$E(x, z) = \frac{E_0}{q(z)} \cdot e^{-\frac{i\pi \cdot x^2}{\lambda \cdot q(z)}} \tag{8.21}$$

With the definition of the complex q-parameter, a Gaussian beam can be represented in a so-called Collins chart.

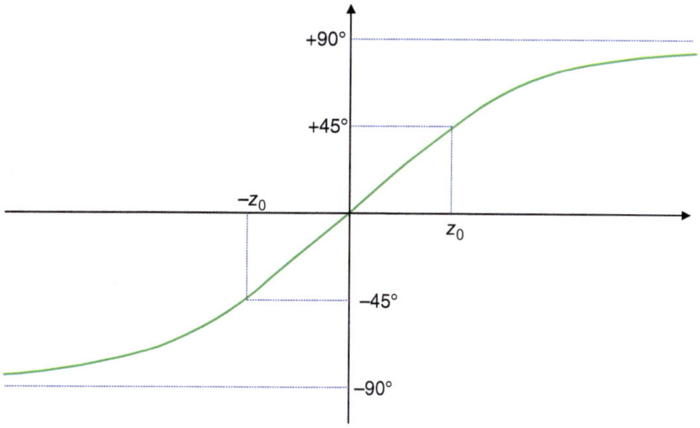

Figure 8.5 Guoy phase as a function of z.

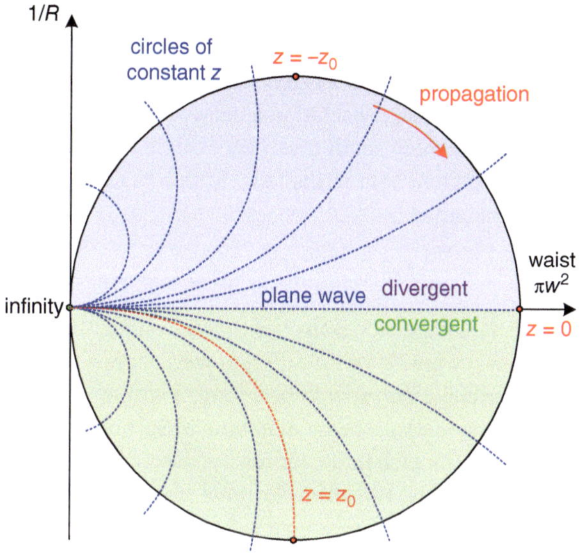

Figure 8.6 Collins chart of a Gaussian beam.

With the real and imaginary part defined by

$$\omega = \mathrm{Im}[q] = \frac{\lambda}{\pi w^2}, \quad c = \frac{1}{R} = \mathrm{Re}[q] \tag{8.22}$$

a beam is represented as a circle. According to Figure 8.6 the waist position $z = 0$ is the right vertex and the infinity point is the left vertex. If the light is assumed to propagate from left to right, from minus to plus infinity, the point on the circle depends on z and makes one round trip in clockwise direction.

If the two equations

$$w(z) = w_0 \cdot \sqrt{1 + \left(\frac{z - z_T}{z_0}\right)^2},$$

$$R(z) = -(z - z_T) - \frac{z_0^2}{z - z_T} \tag{8.23}$$

are used to eliminate z_0 together with the relation between w_0 and z_0 the equation

$$\left(\frac{\lambda}{\pi w^2}\right)^2 + \left(\frac{1}{R} - \frac{1}{2(z - z_T)}\right)^2 = \left(\frac{1}{2(z - z_T)}\right)^2 \tag{8.24}$$

is obtained. This is the equation of a circle in vertex location and describes the Collins chart. The Collins chart is more a tool for visualization than for calculation; but particularly for beam transform and matching problems, it is useful to understand the solutions, see Section 8.2.

8.2 Gaussian Beam Transformation

8.2.1 Transform By a Lens

Gaussian beams can be transformed through a paraxial lens of the focal length f with the help of the modified lens makers' formula

$$\frac{1}{z_T'} - \frac{1}{z_T} \cdot \frac{1}{\left[1 + \dfrac{z_0^2}{z_T\,(f + z_T)}\right]} = \frac{1}{f} \tag{8.25}$$

z_T is the distance of the lens from the initial waist position and is negative, if a real waist is located to the left of the lens. This means that the z-coordinate origin $z = 0$ is selected at the lens; see Figure 8.7. If this equation is resolved for z'_T, we get

$$z_T' = f - \frac{f^2 \cdot (f + z_T)}{(f + z_T)^2 + z_0^2} \tag{8.26}$$

This dependence is visualized in Figure 8.8 (a). It should be noticed that the transformed waist is not exactly located in the classical image plane. Therefore, the waists are not really 'imaged' in the geometrical classical sense; although

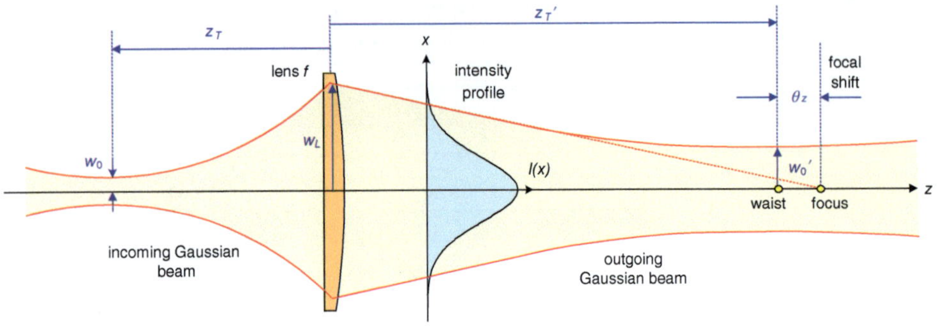

Figure 8.7 Schematic drawing of the Gaussian beam transform at a lens.

in most literature the transform is described like this. Due to the diffraction effects, there occurs a small focal shift. From the above formula, the geometrical limit is obtained by setting $z_0 = 0$. The behaviour of the beam transfer equation is visualized in Figure 8.7. The beam size at the lens is given by

$$w_L = w_0 \cdot \sqrt{1 + \left(\frac{\lambda z_T}{\pi w_0^2}\right)^2} = w_0' \cdot \sqrt{1 + \left(\frac{\lambda z_T'}{\pi w_0'^2}\right)^2} \tag{8.27}$$

If in this equation z_T' is inserted and it is resolved for w_0', the image sided beam parameter are obtained

$$z_0' = \frac{z_0 f^2}{(f + z_T)^2 + z_0^2}, \quad w_0' = \frac{w_0 f}{\sqrt{(f + z_T)^2 + z_0^2}} \tag{8.28}$$

The resulting waist w_0 size follows a Lorentzian-type curve. The ratio of the image-side to the object-side waist size can be interpreted as a kind of magnification, as

$$m_w = \frac{w_0'}{w_0} = \frac{f}{\sqrt{(f + z_T)^2 + z_0^2}} \tag{8.29}$$

This is illustrated in Figure 8.8 (b).

The equation of the Gaussian beam transform resembles the lens makers' formula. It is visualized in Figure 8.8. But there are some significant differences.

- There exists a correction term; due to the diffraction, the image-side waist position is shifted. The waist is in any case nearer to the lens than the geometrical image position.
- The image-side waist position is always finite; it is impossible to move the waist to infinity, this is a singular geometrical case which can never be realized in physical optics.
- The image-side waist size w_0' is always finite and shrinks with a growing divergence angle or aperture.
- In the special case that the initial waist is located in the front focal plane, the image-side waist is located in the rear focal plane. But this is not really a tiny focus, but as in the geometrical expectation a nearly collimated beam with an extremely small divergence angle. The beam is quasi-collimated behind the lens. In this case, the beam waist is obtained by

$$w_0' = \frac{w_0}{z_0} f = \theta_0 f \tag{8.30}$$

- If a special image-side waist position is desired, there are always two solutions with different waist sizes. Only the above special case provides a unique solution. See also the discussion about beam matching in Section 8.2.4.4.
- If the lens is located in the far field of the Gaussian beam with $z_T \gg z_0$, the transform formula simplifies approximately to

$$w_0' = \frac{w_0 f}{z_T} = \frac{\lambda f}{\pi w_{lens}} \tag{8.31}$$

If here w_{lens}/f is interpreted as the numerical aperture, this formula corresponds to the Airy formula. It is seen that the numerical pre-factor 1.22 is changed into $2/\pi = 0.64$ in case of the Gaussian profile.

8.2.2 Beam Transform Through Afocal Telescopes

An afocal telescope of the Kepler or Galilean type with magnification Γ is a common setup to transform a Gaussian beam to another waist size or divergence angle. It changes the beam parameter as

$$w_0' = \frac{w_0}{|\Gamma|}, \quad \theta_0' = \theta_0 \cdot |\Gamma| \tag{8.32}$$

independent of the waist position of the incoming beam relative to the telescope (see Figure 8.10). The length L and the

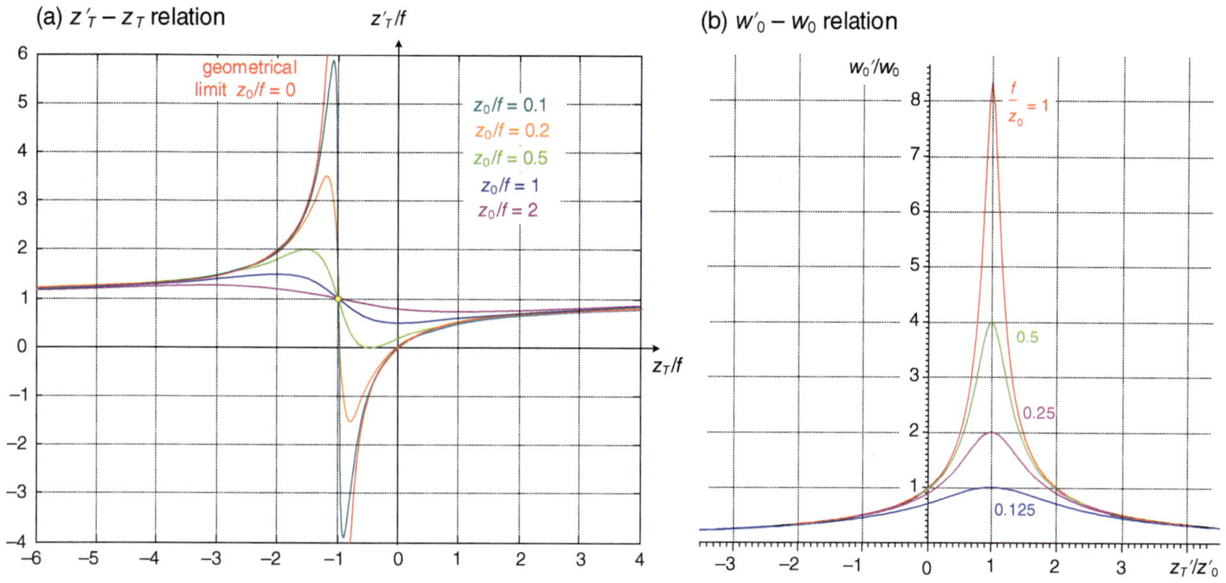

Figure 8.8 Propagation function for the (a) beam waist location and (b) size.

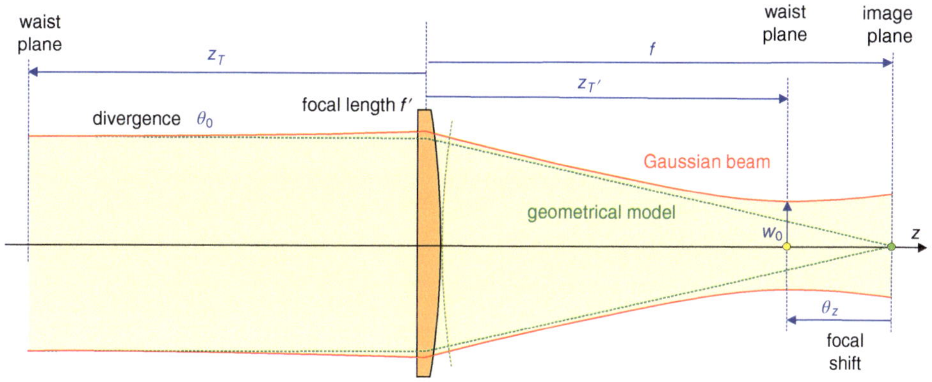

Figure 8.9 Focal shift and image-side waist location.

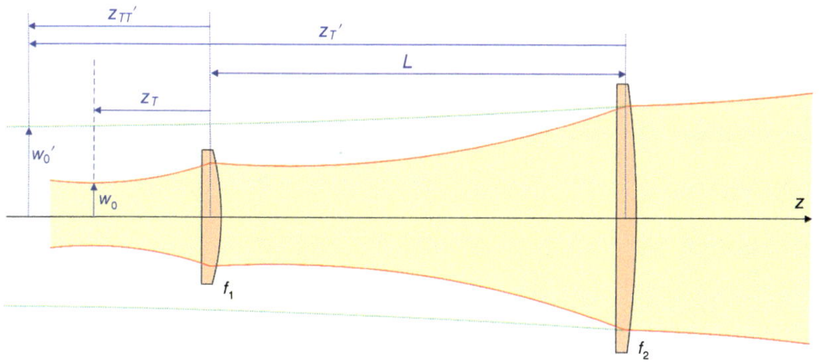

Figure 8.10 Transformation of a Gaussian beam through an afocal telescope.

angle magnification of the telescope are

$$L = f_1 + f_2, \quad \Gamma = -\frac{f_1}{f_2} \tag{8.33}$$

and z_T gives the waist position relative to the first thin lens; the image-side waist position is located at

$$z'_T = L \cdot \Gamma - z_T \cdot \Gamma^2 \tag{8.34}$$

relative to the second lens. In particular for a 1 : 1 transfer with $\Gamma = +1$ for example a 4-f-system, the waists have the distance

$$z_{TT'} = -2L = -4f \tag{8.35}$$

in the forward direction. In the case of a Keplerian telescope, the waist is shifted to the left by $+2\,L$.

8.2.3 Transform by an *ABCD*-Segment

The transformation of a Gaussian beam through a general paraxial *ABCD*-system of any structure can be calculated in real quantities with the equations

$$w' = w\sqrt{\left(\frac{\lambda B}{\pi w^2}\right)^2 + \left(A + \frac{B}{R}\right)^2},$$

$$R' = \frac{\left(A + \frac{B}{R}\right)^2 + \left(\frac{\lambda B}{\pi w^2}\right)^2}{\left(A + \frac{B}{R}\right)\left(C + \frac{D}{R}\right) + D\left(\frac{\lambda B}{\pi w^2}\right)^2} \tag{8.36}$$

see Figure 8.11.

The Guoy phase behind the segment is transformed according to

$$\psi' = \psi + \arctan\left[\frac{\pi w^2}{\lambda}\cdot\left(A + \frac{B}{R}\right)\cdot\left(C + \frac{D}{R}\right) + D\left(\frac{\lambda B}{\pi w^2}\right)\right]$$

$$= \psi + \arctan\left[\frac{\lambda B}{\pi w^2\left(A + \frac{B}{R}\right)}\right] \tag{8.37}$$

The far field divergence angle can be calculated by

$$\theta'_0 = w \cdot \sqrt{\left(C + \frac{D}{R}\right)^2 + \left(\frac{\lambda D}{\pi w^2}\right)^2},$$

$$\theta_0'^2 = C^2 w^2 + 2CD\frac{w^2}{R^2} + D^2\theta_0^2 \tag{8.38}$$

In the special case of a start in the waist plane, the formulas look simpler, as follows

$$w' = \sqrt{A^2 w_0^2 + B^2\theta_0^2}, \quad R' = \frac{A^2 w_0^2 + B^2\theta_0^2}{ACw_0^2 + BD\theta_0^2},$$

$$\theta'_0 = \sqrt{C^2 w_0^2 + D^2\theta_0^2} \tag{8.39}$$

An example of numerical calculation of a beam transport in a simple system of thin lenses is shown in Figure 8.12.

The Collins chart is particularly useful to understand the various solutions of beam transforms and the occurrence of typically two possibilities. Figure 8.13 illustrates the case of

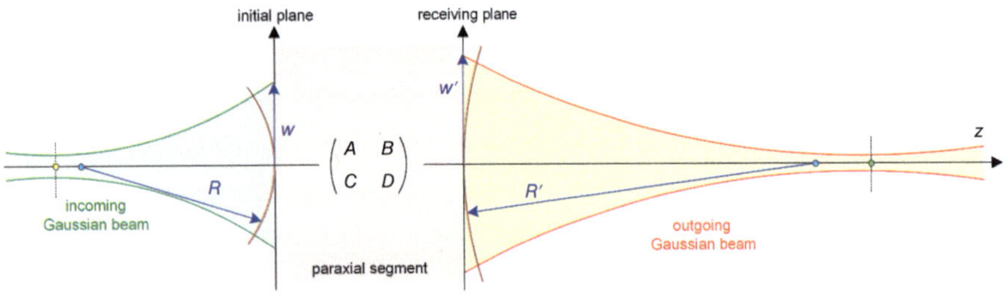

Figure 8.11 Transform of a Gaussian beam by a paraxial *ABCD*-system.

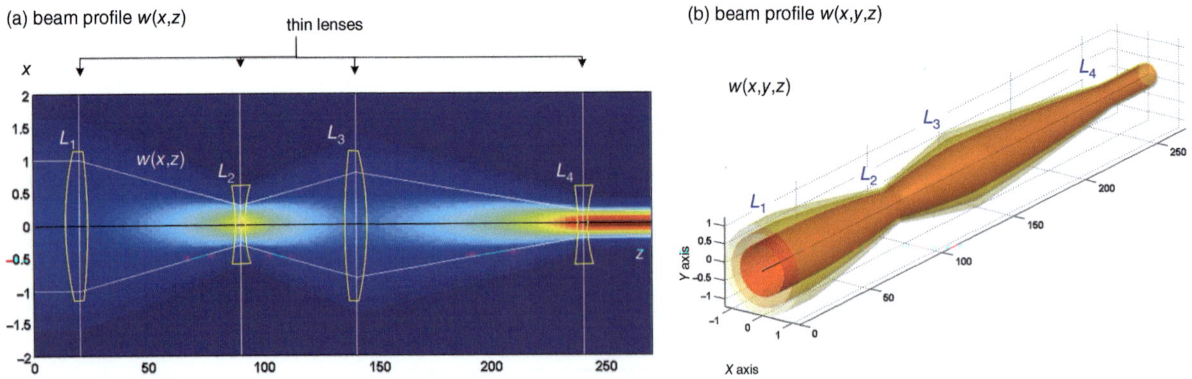

Figure 8.12 Example of a Gaussian beam propagation through a system of four lenses as (a) the intensity variation in the x–z-cross section and (b) as a three-dimensional view of concentric tubes at three levels of constant intensity. The units of r and z are arbitrary.

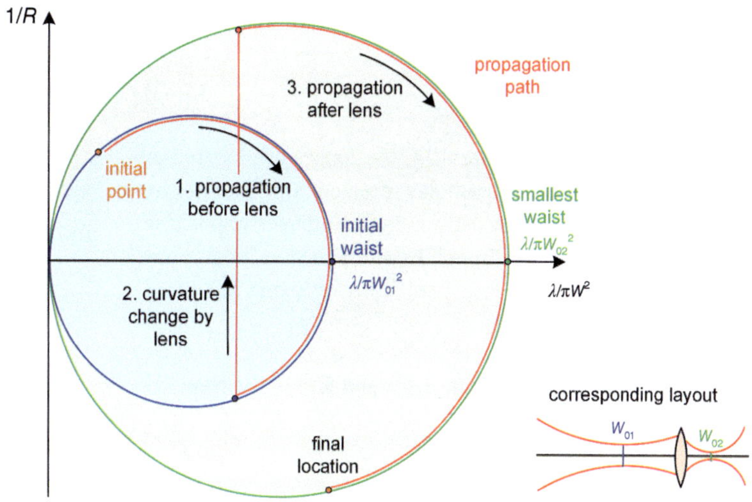

Figure 8.13 Collins chart of a Gaussian beam focused by a lens. First the beam propagates along the blue circle, and after the lens it moves to the green circle.

a beam transform by a single lens. At the thin lens, only the curvature $1/R$ is changed, keeping the beam radius w constant. This means that the z-path moves from one circle to another one with another radius in the vertical direction. The lens can be located in the divergent or convergent part of the initial beam; therefore, the transition is located in the upper or lower part of the circle. Furthermore, a larger waist after the lens corresponds to a smaller circle, of which the diameter is $2\lambda/(\pi w_{00})^2$.

8.2.4 Matching of Beams

8.2.4.1 Introduction

Notation mode matching means that a certain task for given incoming and outgoing Gaussian beams is defined and the system to perform this task is searched for. Gaussian beam propagation hyperbolas are of second order; therefore, in many cases, two solutions exist [1].

Since the Gaussian beams have diffraction effects automatically included, some well known more singular geometrical optical setups do not work due to the boundary conditions of physics. In the following, some classical tasks of beam matching are discussed.

8.2.4.2 Given Incoming and Outgoing Gaussian Beams

In this task, two Gaussian beams in size and relative waist distance $z_{TT'}$ are given. In the special case $w_0' = w_0$ there exists only one solution with a lens of focal length

$$f = \frac{1}{4} z_{TT'} + \frac{z_0}{z_{TT'}} \tag{8.40}$$

The lens has to be positioned symmetrically at

$$z_L = \frac{z_{TT'}}{2} \tag{8.41}$$

from the initial waist. In the general case, there are two solutions for the transforming lens; one solution has a virtual waist position as sketched in Figure 8.14. Since two different hyperbolas are intersecting in any case, this problem has always a solution. The two possible focal lengths are

$$f_{1/2} = \frac{1}{(z_0 - z_0')^2} \cdot \left[-2z_{TT} z_0 z_0' \pm \sqrt{\begin{matrix} (2z_{TT} z_0 z_0')^2 \\ + z_0 z_0' (z_0 - z_0')^2 \left[z_{TT}^2 + (z_0 - z_0')^2 \right] \end{matrix}} \right] \tag{8.42}$$

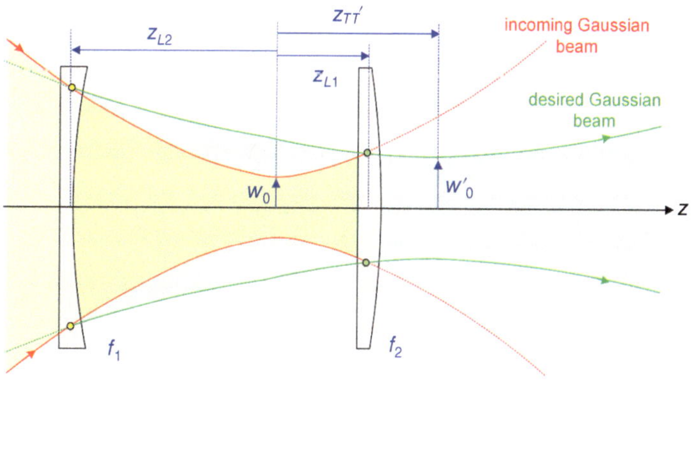

Figure 8.14 Matching of two Gaussian beams. Both solutions with focal lengths f_1 or f_2 are shown simultaneously, although only one lens is needed.

The lens distance from the given Gaussian beam waist is

$$z_{L1/2} = f_{1/2} \pm z_0' \cdot \sqrt{\frac{f_{1/2}^2}{z_0 z_0'} - 1} \tag{8.43}$$

As can be seen in Figure 8.14, the lenses are located at the planes where the hyperbolas intersect and change the phase of the beam only. Mostly one solution needs a positive and one a negative lens. In reality, often one lens position is uncomfortable; in this case, practically only one meaningful technical solution exists.

8.2.4.3 Given Gaussian Beam, Lens and Size of the New Beam

If a Gaussian beam as well as the lens focal length and the desired waist size w_0' of the new beam are fixed, there is only a solution, if the condition

$$f > \frac{w_0' z_0}{w_0} \tag{8.44}$$

is fulfilled. In this case the problem has two solutions with the lens positions

$$z_{L1/2} = -f \pm z_0 \cdot \sqrt{\frac{f^2}{z_0 z_0'} - 1} \tag{8.45}$$

and the new waists at the distances

$$z_{TT',1/2} = 2f \mp (z_0' + z_0) \cdot \sqrt{\frac{f^2}{z_0 z_0'} - 1} \tag{8.46}$$

One of the two solutions always has a virtual waist position. Figure 8.15 shows the possible setups simultaneously.

8.2.4.4 Given Gaussian Beam, Lens and New Waist Position

If the position z of the lens and the new waist position are fixed, a solution is only possible if

$$z_T' < \frac{z_T^2 + z_0^2}{2z_0} \tag{8.47}$$

is valid. The possible two solutions have the focal lengths

$$f_{1/2} = \frac{2zz' - z^2 - z_0^2 \pm \sqrt{\left(2zz' - z^2 - z_0^2\right)^2 + 4z'(z - z')\left(z^2 + z_0^2\right)}}{2(z - z')} \tag{8.48}$$

and the obtained waists are

$$w_{0,1/2}' = \frac{w_0\, |f_{1/2}|}{\sqrt{(f_{1/2} + z)^2 + z_0^2}} \tag{8.49}$$

As can be seen in Figure 8.16, one solution is a real focusing setup, the other delivers the right waist position, but has a quasi-collimated beam as result.

8.2.4.5 Given Gaussian Beam, Lens Location and Size of the New Beam

If beneath the incoming Gaussian beam the lens position z_T and the desired new waist size w_0' are given, there are always two solutions with the focal lengths

$$f_{1/2} = z_T \cdot \frac{1 \pm \sqrt{1 + \left[\left(\frac{z_0}{z_T}\right)^2 + 1\right] \cdot \left[\left(\frac{w_0}{w_0'}\right)^2 - 1\right]}}{\left(\frac{w_0}{w_0'}\right)^2 - 1} \tag{8.50}$$

8.2.4.6 Largest Waist Distance

The waist of a given Gaussian beam cannot be shifted to arbitrary long distances from a lens. There is a certain focal length that delivers the maximum distance. If the focal length is further increased, the waist position 'comes back'. Behind a lens with beam radius w_L the equation

$$w_L^2 = w_0'^2 \cdot \left[1 + \left(\frac{\lambda z_T'}{\pi w_0'^2}\right)^2\right] \tag{8.51}$$

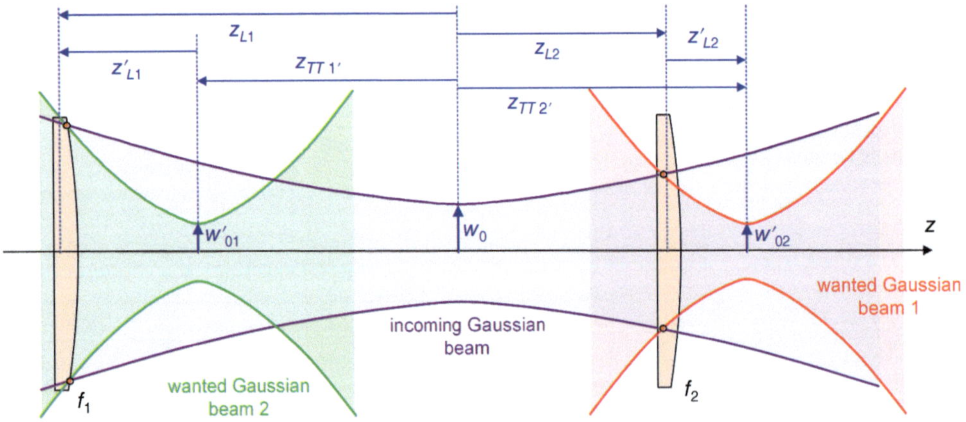

Figure 8.15 Matching of Gaussian beams; again, both possible solutions are drawn overlaid.

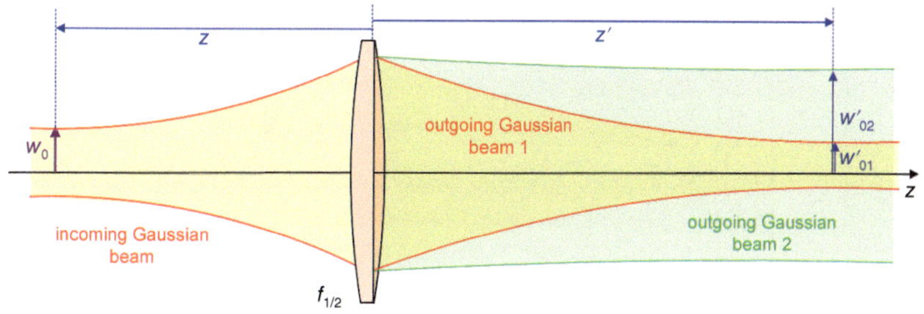

Figure 8.16 Focusing of a Gaussian beam with fixed image-side waist position.

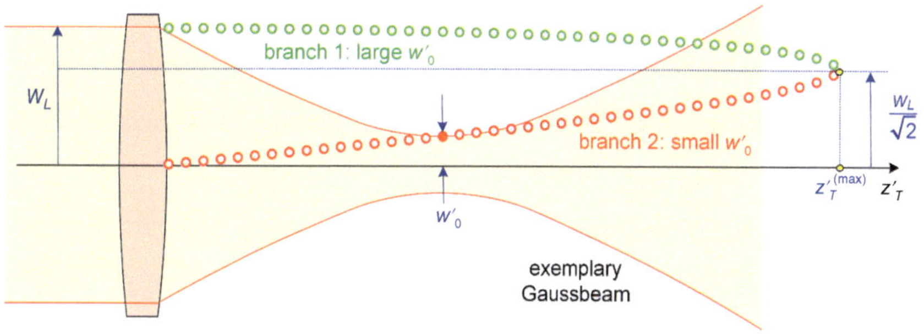

Figure 8.17 Focusing of a Gaussian beam with maximum distance of waist position and two solution branches.

can be used to calculate the derivative

$$\frac{\partial z'_T}{\partial w'_0} = 0 \tag{8.52}$$

which gives

$$z'^{(\max)}_{\mathrm{T}} = \frac{\pi w_L^2}{2\lambda}, \quad w'^{(\max)}_0 = \frac{w_L}{\sqrt{2}} \tag{8.53}$$

If Eq. (8.52) is resolved for w_0, there are always two solutions: one branch with large and one with small w_0.

$$w'_0 = \sqrt{\frac{w_L^2}{2} \pm \sqrt{\frac{w_L^4}{4} - \left(\frac{2\lambda z'_T}{\pi}\right)^2}} \tag{8.54}$$

In Figure 8.17 this situation is illustrated with two solution branches. The maximum possible distance zT' is seen, as well as the value of $w'_0{}^{\max}$.

8.2.4.7 Minimal Beam Size in a Certain Distance

If the aim is to get the minimal waist size in a certain fixed distance z' after the lens with the beam radius w_L, the two conditions

$$w_L^2 = w_0'^2 + \left(\frac{\lambda z'_T}{\pi w'_0}\right)^2,$$

$$w'^2 = w_0'^2 + \left(\frac{\lambda}{\pi w'_0}\right)^2 \cdot (z' - z'_T)^2 \tag{8.55}$$

must be fulfilled and the best z'_T and w'_0 have to be calculated, to minimize $w'_{\min}$, see Figure 8.18.

If z'_T is eliminated, the equation

$$w'^2 = w_0'^2 + \left(\frac{\lambda z'}{\pi w'_0} - \sqrt{w_L^2 - w_0'^2}\right) \tag{8.56}$$

is obtained. In the following, the effective Fresnel number of the beam is used as an abbreviation and is defined as

$$N_{eff} = \frac{\pi w_L^2}{\lambda z'} \tag{8.57}$$

Now the extremum of

$$\frac{\partial w'^2}{\partial w'_0} = 0 \tag{8.58}$$

gives the waist and the waist location by re-inserting as

$$w'_0 = \frac{w_L}{\sqrt{1 + N_{eff}^2}} = \frac{w_L}{\sqrt{1 + \left(\frac{\pi w_L^2}{\lambda z'}\right)^2}},$$

$$w'_{\min} = \frac{w_L}{N_{eff}} = \frac{\lambda z'}{\pi w_L}$$

$$z'_T = \frac{z'}{1 + N_{eff}^{-2}} = \frac{z'}{1 + \left(\frac{\lambda z'}{\pi w_L^2}\right)^2},$$

$$\Delta z = z' - z'_T = \frac{z'}{1 + N_{eff}^2} = \frac{z'}{1 + \left(\frac{\pi w_L^2}{\lambda z'}\right)^2} \tag{8.59}$$

This means, according to Figure 8.18, that the waist should be located slightly before the desired distance z'. This shift enlarges the beam radius, but simultaneously the larger divergence angle decreases the waist. The formulas above give then the optimum encircled energy in the distance z'. In the limiting case of very large z' values, $N_{eff} \to 0$ and in principle the waist should be in the initial plane $z'_T = 0$ and is $w'_o = w_L$.

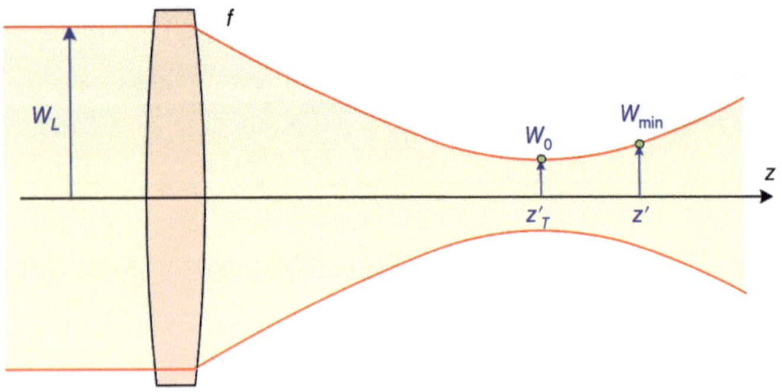

Figure 8.18 Focusing of a Gaussian beam with minimum beam size at a given distance z.

8.2.4.8 Coupling of Two TEMoo-Beams*

(*) The content of this section is quite specialized and for advanced reading. Knowledge of this section is not necessary to understand the following chapters. Beginners are recommended to skip this section.

If two different fundamental modes with Gaussian profile are considered, an important question in practical applications is the coupling efficiency of one mode into the other. In particular this question arises in the launching of a laser mode into a monomode fibre, if the fundamental fibre mode is approximated by a Gaussian beam – see Section 8.8. If the complex fields are E_1 and E_2, the power coupling can be written in general as [7–9]

$$\eta = \left| \frac{\iint E_1(x,y) \cdot E_2^*(x,y)dxdy}{\sqrt{\iint |E_1(x,y)|^2 dxdy} \cdot \sqrt{\iint |E_2(x,y)|^2 dxdy}} \right|^2 \tag{8.60}$$

A certain mismatch of the parameters reduces this value; reasons can be:

- different beam radii w_J;
- Different radii of curvature R_j, which corresponds to a defocusing;
- Relative tilt of the axes by θ;
- Relative offset of the axes v.

See Figure 8.19 for illustration.

If the complex field equation is inserted into the coupling formula above, the integration can be figured out analytically. If the values of tilt angle θ and offset v are assumed to be small in the sense of a tolerance, the approximated result is

$$\eta = \frac{4}{\left(\frac{w_2}{w_1} + \frac{w_1}{w_2}\right)^2 + \left(\frac{\pi w_1 w_2}{\lambda}\right)^2 \cdot \left(\frac{1}{R_1} - \frac{1}{R_2}\right)^2} - \frac{v^2}{2} \cdot \left(\frac{1}{w_{01}^2} + \frac{1}{w_{02}^2}\right) - \frac{2\pi^2\theta^2}{\lambda^2} \cdot \left(w_1^2 + w_2^2\right) \tag{8.61}$$

If no alignment errors are present, $v = 0$ and $\theta = 0$, the two mismatch parameters for size and curvature

$$\sigma = \frac{\pi w_2^2}{\lambda} \cdot \left(\frac{1}{R_1} - \frac{1}{R_2}\right), \quad \varepsilon = \frac{w_{01}}{w_{02}} \tag{8.62}$$

can be defined and the simple formula

$$\eta_{00} = \frac{4}{2 + \varepsilon^2 + \frac{1+\sigma^2}{\varepsilon^2}} \tag{8.63}$$

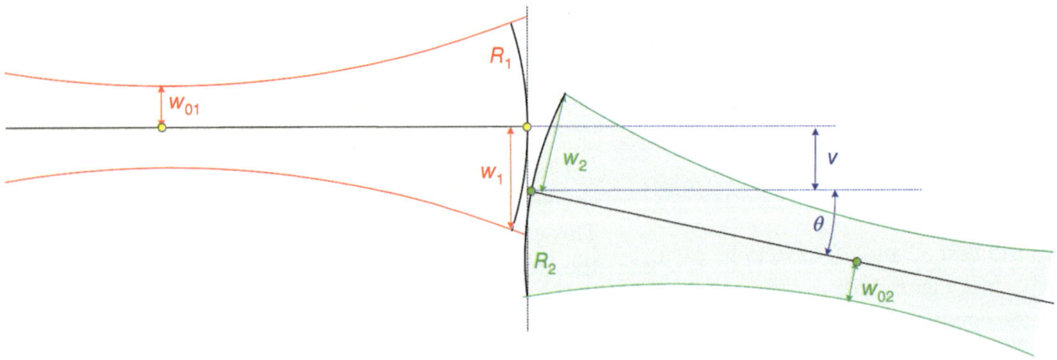

Figure 8.19 Geometry of a two-beam mismatch due to size w, radius of curvature R, tilt θ or offset v.

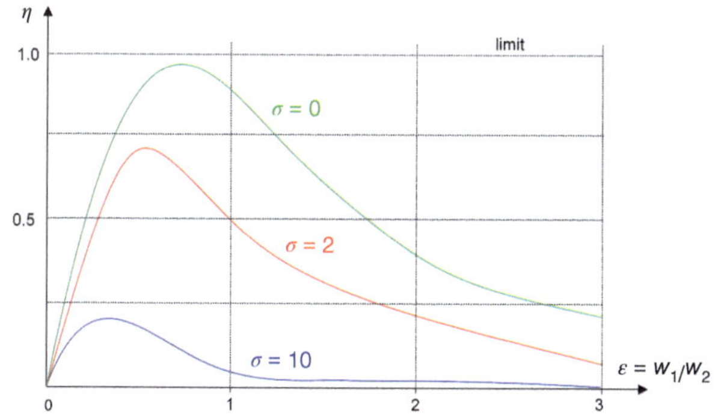

Figure 8.20 Coupling efficiency as a function of ε and σ.

is obtained. The graphical representation for different values of ε and σ is illustrated in Figure 8.20.

In the special case of a coincidence of the waists without any phase discrepancy but just a size mismatch, the formula further simplifies to

$$\eta_{00} = \frac{4}{\left(\dfrac{w_2}{w_1} + \dfrac{w_1}{w_2}\right)^2} \tag{8.64}$$

8.3 Astigmatic Beams

8.3.1 Orthogonal Astigmatic Gaussian Beams

The simplest kind of astigmatic beam has different parameters in the x–z and the y–z cross section [10–12]. There are two independent sets of data for both sections without any mixing effects between them. The complex beam parameters are therefore written as [1]

$$q_{x/y} = z_{x/y} + i \cdot z_{0x/y}, \quad \frac{1}{q_{x/y}} = \frac{1}{R_{x/y}} - \frac{i\lambda}{\pi w_{x/y}^2} \tag{8.65}$$

They can be combined into a matrix or beam tensor of diagonal structure

$$\underline{Q}_{ortho}^{-1} = \begin{pmatrix} \dfrac{1}{q_x} & 0 \\ 0 & \dfrac{1}{q_y} \end{pmatrix} \tag{8.66}$$

Since both waist positions can be different

$$\Delta_T = z_{Tx} - z_{Ty} \tag{8.67}$$

there occurs in general an elliptical beam intensity pattern, which changes along the z-propagation of the beam. For every location z, this ellipse can be transformed into a transverse coordinate system oriented along the major axes of the intensity ellipse by an azimuthal rotation with angle φ. This is shown in Figure 8.21. The rotated matrix is then

$$\underline{Q}^{-1} = \begin{pmatrix} \dfrac{\cos^2\varphi}{q_x} + \dfrac{\sin^2\varphi}{q_y} & \dfrac{1}{2}\sin 2\varphi \cdot \left(\dfrac{1}{q_x} - \dfrac{1}{q_y}\right) \\ \dfrac{1}{2}\sin 2\varphi \cdot \left(\dfrac{1}{q_x} - \dfrac{1}{q_y}\right) & \dfrac{\sin^2\varphi}{q_x} + \dfrac{\cos^2\varphi}{q_y} \end{pmatrix} \tag{8.68}$$

The differences in the imaginary part of the off-diagonal terms

$$\mathrm{Im}\left[\frac{1}{2}\sin 2\varphi \cdot \left(\frac{1}{q_x} - \frac{1}{q_y}\right)\right] = \frac{1}{2}\sin 2\varphi \cdot \left(-\frac{\lambda}{\pi w_x^2(z)} + \frac{\lambda}{\pi w_y^2(z)}\right) \tag{8.69}$$

show that it is possible to let the off-diagonal elements vanish for an appropriate value of the rotation angle φ. In this system, the beam is orthogonal without mixed

Figure 8.21 Focusing of an astigmatic Gaussian beam.

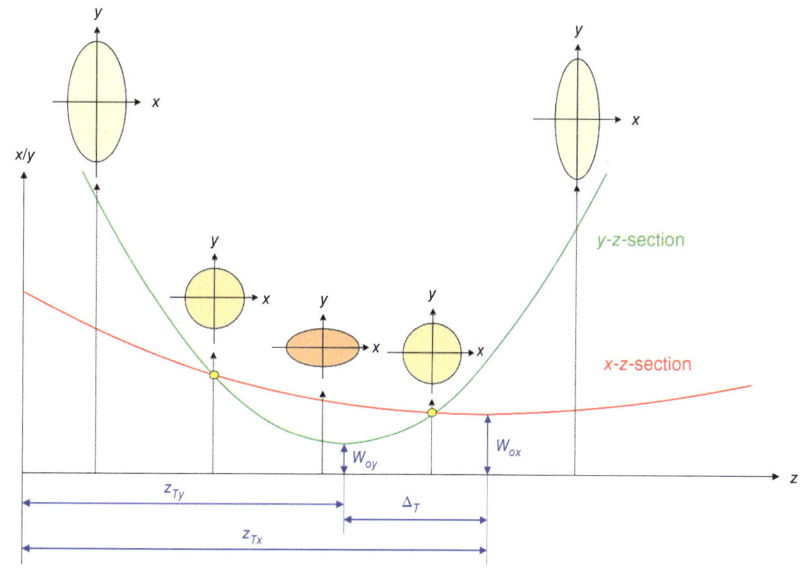

terms. The field can be written as

$$E(x,y,z) = E_0 \cdot e^{-\left[\frac{x^2}{w_x^2(z)}+\frac{y^2}{w_y^2(z)}\right]} \cdot e^{-\frac{i\pi}{\lambda}\left[\frac{x^2}{R_x(z)}+\frac{y^2}{R_y(z)}\right]} \cdot e^{\frac{i}{2}\Psi(z)} \tag{8.70}$$

and can be factorized as

$$E(x,y,z) = E_x(x, z_x, w_{0x}) \cdot E_y(y, z_y, w_{0y}) \tag{8.71}$$

with the Guoy phase

$$\begin{aligned}\Psi(z) &= \frac{1}{2}\arctan\frac{\text{Re}[q_x]}{\text{Im}[q_x]} + \frac{1}{2}\arctan\frac{\text{Re}[q_y]}{\text{Im}[q_y]} \\ &= \frac{1}{2}\left[\arctan\frac{z-z_{Tx}}{z_{0x}} + \arctan\frac{z-z_{Ty}}{z_{0y}}\right]\end{aligned} \tag{8.72}$$

For given values of the complex parameters q_x and q_y, the complex field can be expressed in a plane with removed mixed terms by using the complex q-parameter as

$$E(x,y,z) = \frac{E_0}{\sqrt{q_x(z)q_y(z)}} \cdot e^{-\frac{i\pi x^2}{\lambda q_x(z)}} \cdot e^{-\frac{i\pi y^2}{\lambda q_y(z)}} \cdot e^{\frac{i}{2}\Psi(z)} \tag{8.73}$$

With

$$\underline{Q}^{-1}_{ortho} = \begin{pmatrix} q_x^{-1} & 0 \\ 0 & q_y^{-1} \end{pmatrix} \tag{8.74}$$

(see above) and with the transverse coordinate vector

$$\vec{r}_t = \begin{pmatrix} x \\ y \end{pmatrix} \tag{8.75}$$

the field amplitude can be therefore written as

$$E(x,y,z) = \frac{E_0}{\sqrt{q_x(z) \cdot q_y(z)}} \cdot e^{-\frac{i\pi}{\lambda}\cdot \vec{r}_t^{\,T} \cdot \underline{Q}^{-1}_{ortho} \cdot \vec{r}_t} \tag{8.76}$$

The area of the intensity ellipse is given by

$$A(z) = \pi \cdot w_x(z) \cdot w_y(z) \tag{8.77}$$

In general, there are two z-locations with a circular cross section; see Figure 8.21 for illustration. These are the positions where the two beam hyperbolas intersect each other. These locations can be calculated by

$$z = \frac{\Delta_T}{2} \cdot \frac{w_{0y}^2 + w_{0x}^2}{w_{0y}^2 - w_{0x}^2} \pm \frac{w_{0y}w_{0x}}{w_{0y}^2 + w_{0x}^2} \cdot \sqrt{\Delta_T^2 + \frac{\pi^2}{\lambda^2}\left(w_{0y}^2 - w_{0x}^2\right)^2} \tag{8.78}$$

If an astigmatic Gaussian beam with an initial toric wavefront is focused, the ellipticity w_x/w_y changes in the caustic region as well as the intensity on axis. The larger the divergence angle or numerical aperture is, the better the two focal lines are separated, and the behaviour is more symmetric around the plane in between the focal locations. For small angles, the astigmatic difference comes into the range of the Rayleigh ranges and the intensity on axis becomes more asymmetric. The ellipticity is defined as

$$\varepsilon = \frac{w_x}{w_y} \tag{8.79}$$

and the astigmatic difference is given by Δz. The ellipticity becomes more symmetric if the angles are small. This is illustrated in Figure 8.22, where a constant astigmatic difference of $\Delta z = 1$ mm is assumed and the divergence angles are varied.

Astigmatism and ellipticity are for small differences approximately related by

$$\Delta z = z_0 \cdot \sqrt{1-\varepsilon^2} \tag{8.80}$$

For example, a maximum accepted ellipticity of 0.9 requires a maximum astigmatism of $\Delta z = 0.436\, z_0$.

8.3.2 General Astigmatic Gaussian Beams*

(*) The content of this section is quite specialized and for advanced reading. Knowledge of this section is not necessary to understand the following chapters. Beginners are recommended to skip this section.

If an orthogonal astigmatic beam parameter

$$\underline{Q}^{-1}_{ortho} = \begin{pmatrix} \frac{1}{q_x} & 0 \\ 0 & \frac{1}{q_y} \end{pmatrix} \tag{8.81}$$

is rotated by an angle φ, off-diagonal matrix elements occur

$$\underline{Q}^{-1}_{rot} = \begin{pmatrix} \frac{\cos^2\varphi}{q_x} + \frac{\sin^2\varphi}{q_y} & \frac{1}{2}\sin(2\varphi)\left(\frac{1}{q_x} - \frac{1}{q_y}\right) \\ \frac{1}{2}\sin(2\varphi)\left(\frac{1}{q_x} - \frac{1}{q_y}\right) & \frac{\sin^2\varphi}{q_x} + \frac{\cos^2\varphi}{q_y} \end{pmatrix} \tag{8.82}$$

In a more general case, the intensity as well as the phase consists of a rotated elliptical or toric surface with different azimuthal orientations, respectively. The complex field is then given by

$$E(x,y,z) = E_0 \cdot e^{-\left[\frac{x^2}{w_x^2(z)}+\frac{y^2}{w_y^2(z)}+\frac{2xy}{w_{xy}^2(z)}\right]} \cdot e^{-\frac{i\pi}{\lambda}\left[\frac{x^2}{R_x(z)}+\frac{y^2}{R_y(z)}+\frac{2xy}{R_{xy}(z)}\right]} \cdot e^{\frac{i}{2}\Psi(z)} \tag{8.83}$$

and the principal axis azimuthal angles are different for the intensity and the phase and are given by

$$\begin{aligned}\tan(2\varphi_w) &= \frac{\frac{2}{w_{xy}^2}}{\frac{1}{w_x^2} - \frac{1}{w_y^2}} = \frac{W_{12}+W_{21}}{W_{11}-W_{22}}, \\ \tan(2\varphi_R) &= \frac{\frac{2}{R_{xy}}}{\frac{1}{R_x} - \frac{1}{R_y}} = \frac{C_{12}+C_{21}}{C_{11}-C_{22}}\end{aligned} \tag{8.84}$$

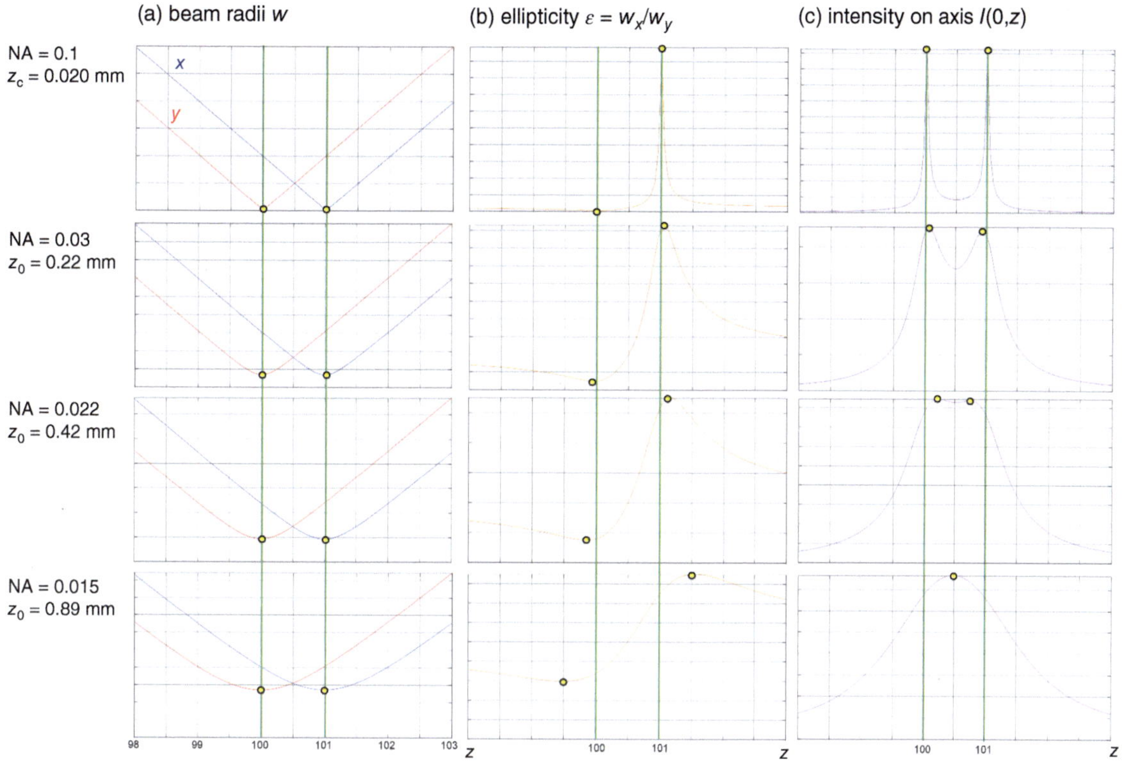

Figure 8.22 Focusing of astigmatic Gaussian beams. To be comparable, all the vertical axes are rescaled and are only relative here. (a) gives the beam radii, (b) the ellipticity and (c) the intensity on axis. The unit of z is arbitrary.

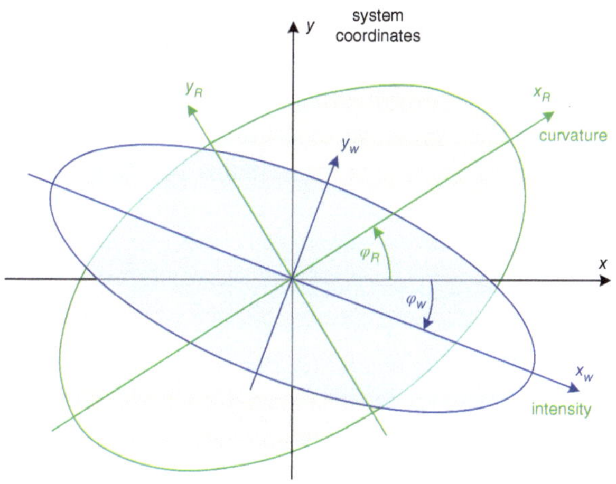

Figure 8.23 Independent rotated ellipses of intensity and curvature.

see also Figure 8.23. The meaning of the matrix elements Wjk and Cjk is explained below. If the coordinate system is rotated by φ_w, the principal axis orientation of the intensity ellipse is obtained

$$|E(\overline{x},\overline{y},z)| = E_0 \cdot e^{-\left[\frac{\overline{x}^2}{\overline{w}_x^2(z)}+\frac{\overline{y}^2}{\overline{w}_y^2(z)}\right]} \tag{8.85}$$

with the principal axes

$$\frac{1}{\overline{w}_x^2} = \frac{\cos^2\varphi_w}{w_x^2} + \frac{\sin^2\varphi_w}{w_y^2},$$
$$\frac{1}{\overline{w}_y^2} = \frac{\sin^2\varphi_w}{w_x^2} + \frac{\cos^2\varphi_w}{w_y^2} \tag{8.86}$$

but the phase ellipse with a toroidal surface is not aligned in this case; this requires a rotation by φ_R.

The general form of an astigmatic beam can now be written as

$$E(x,y,z) = E_0 \cdot e^{-\vec{r}^T \underline{W}(z)\vec{r}} \cdot e^{-\frac{i\pi}{\lambda}\vec{r}^T \underline{C}(z)\vec{r}} \cdot e^{i\,\Phi} \cdot e^{i\,\Psi(z)} \tag{8.87}$$

with the 2×2-matrices of the intensity as well as the curvature ellipse

$$\underline{W}(z) = \begin{pmatrix} \frac{\cos^2\varphi_w}{w_x^2} + \frac{\sin^2\varphi_w}{w_y^2} & \frac{1}{2}\sin(2\varphi_w)\left(\frac{1}{w_x^2} - \frac{1}{w_y^2}\right) \\ \frac{1}{2}\sin(2\varphi_w)\left(\frac{1}{w_x^2} - \frac{1}{w_y^2}\right) & \frac{\sin^2\varphi_w}{w_x^2} + \frac{\cos^2\varphi_w}{w_y^2} \end{pmatrix}$$
$$\underline{C}(z) = \begin{pmatrix} \frac{\cos^2\theta_R}{R_x} + \frac{\sin^2\theta_R}{R_y} & \frac{1}{2}\sin(2\theta_R)\left(\frac{1}{R_x} - \frac{1}{R_y}\right) \\ \frac{1}{2}\sin(2\theta_R)\left(\frac{1}{R_x} - \frac{1}{R_y}\right) & \frac{\sin^2\theta_R}{R_x} + \frac{\cos^2\theta_R}{R_y} \end{pmatrix} \tag{8.88}$$

as well as the Guoy phase

$$\Psi(z) = \frac{1}{2}\arctan\frac{\mathrm{Re}[q_x]}{\mathrm{Im}[q_x]} + \frac{1}{2}\arctan\frac{\mathrm{Re}[q_y]}{\mathrm{Im}[q_y]}$$
$$= \frac{1}{2}\left[\arctan\frac{z - z_{Tx}}{z_{0x}} + \arctan\frac{z - z_{Ty}}{z_{0y}}\right] \quad (8.89)$$

which is unaffected by the rotations. The 2 × 2 beam tensor

$$\underline{Q}^{-1} = \begin{pmatrix} q_{xx}^{-1} & q_{xy}^{-1} \\ q_{yx}^{-1} & q_{yy}^{-1} \end{pmatrix} \quad (8.90)$$

is transformed and propagated by the equation

$$\underline{Q}' = \frac{\underline{A}\underline{Q} + \underline{B}}{\underline{C}\underline{Q} + \underline{D}} \quad (8.91)$$

or

$$\underline{Q}'^{-1} = \frac{\underline{C} + \underline{D}\,\underline{Q}^{-1}}{\underline{A} + \underline{B}\,\underline{Q}^{-1}} \quad (8.92)$$

The components in particular for a free space propagation are given by

$$\frac{1}{q'_{xx}} = \left[\left(1 + \frac{z}{q_{xx}}\right)\left(1 + \frac{z}{q_{yy}}\right) - \frac{z^2}{q_{xy}^2}\right]^{-1} \cdot \left[\frac{1}{q_{xx}}\left(1 + \frac{z}{q_{yy}}\right) - \frac{z}{q_{xy}^2}\right]$$
$$\frac{1}{q'_{yy}} = \left[\left(1 + \frac{z}{q_{xx}}\right)\left(1 + \frac{z}{q_{yy}}\right) - \frac{z^2}{q_{xy}^2}\right]^{-1} \cdot \left[\frac{1}{q_{yy}}\left(1 + \frac{z}{q_{xx}}\right) - \frac{z}{q_{xy}^2}\right]$$
$$\frac{1}{q'_{xy}} = \left[\left(1 + \frac{z}{q_{xx}}\right)\left(1 + \frac{z}{q_{yy}}\right) - \frac{z^2}{q_{xy}^2}\right]^{-1} \cdot \frac{1}{q_{xy}} \quad (8.93)$$

It is seen here that the x and y cross sections are coupled and therefore the propagating is more complex. This means that the physical interpretation of the complex parameters q is not trivial and not obvious. The intensity is given by a rotated ellipse, the curvature by a different rotated ellipse of a phase surface with toroidal shape. Both principal coordinate systems are different and rotate during propagation. In general, it is not possible to align both coordinate systems without special non-circular symmetric components.

If a free space propagation of a general astigmatic beam is considered, a typical picture like Figure 8.24 is obtained. It can be seen in this example that the behaviour is much more complicated. The rotation of the beam causes more complex changes with three extreme intensity radii. It is further noticed that due to the x–y mixing, a unique indication of what is x and what is y is not unique.

8.4 Ray Equivalent of Gaussian Beams

8.4.1 Principle

Gaussian beams are paraxial and therefore can be described by two parameters similar to paraxial rays in a linear approach. The transfer of Gaussian beams through paraxial systems is equivalent to the raytrace calculus and therefore can be determined by the result of a raytrace calculation. This model is only paraxial and three major effects of real propagation are not considered in this approximation [13–18].

- Transverse truncation of the beams.
- Phase aberration of higher order. In the paraxial model only quadratic phase distributions are possible. Thus, defocus and astigmatic effects are incorporated.
- Higher NA, where the paraxiality is violated.

In the ray equivalent propagation, two rays are traced in a system. Any two independent rays are possible, but it is easier to choose two special ray types.

- The divergence ray passes through the centre of the beam waist and has a direction corresponding to the divergence angle of the beam. The beam isophote at $1/e^2$ intensity approximates this ray asymptotically far away from the waist location. This ray corresponds to a marginal ray from the waist centre as the object plane.
- The waist ray starts at the waist radius w_0 and has a vanishing angle with the axis. This corresponds to the chief ray with telecentric geometry of a classical system.

These two special ray selections are illustrated in Figure 8.25.

If the data of these two rays in the image space are used, the parameters of the Gaussian beam can be recovered by the following formulas at every position z:

$$w = \sqrt{x_{wr}^2 + x_{dr}^2}$$
$$w'_0 = \frac{x'_{wr} \cdot \tan u' - x'_{dr} \cdot \tan w'}{\sqrt{\tan^2 u' + \tan^2 w'}},$$
$$z'_T = \frac{x'_{dr} \cdot \tan u' + x'_{wr} \cdot \tan w'}{\tan^2 u' + \tan^2 w'} \quad (8.94)$$

As can be shown, every other two independent rays can be used to determine the parameters of the Gaussian beam [5]; see the schematic shown in Figure 8.26.

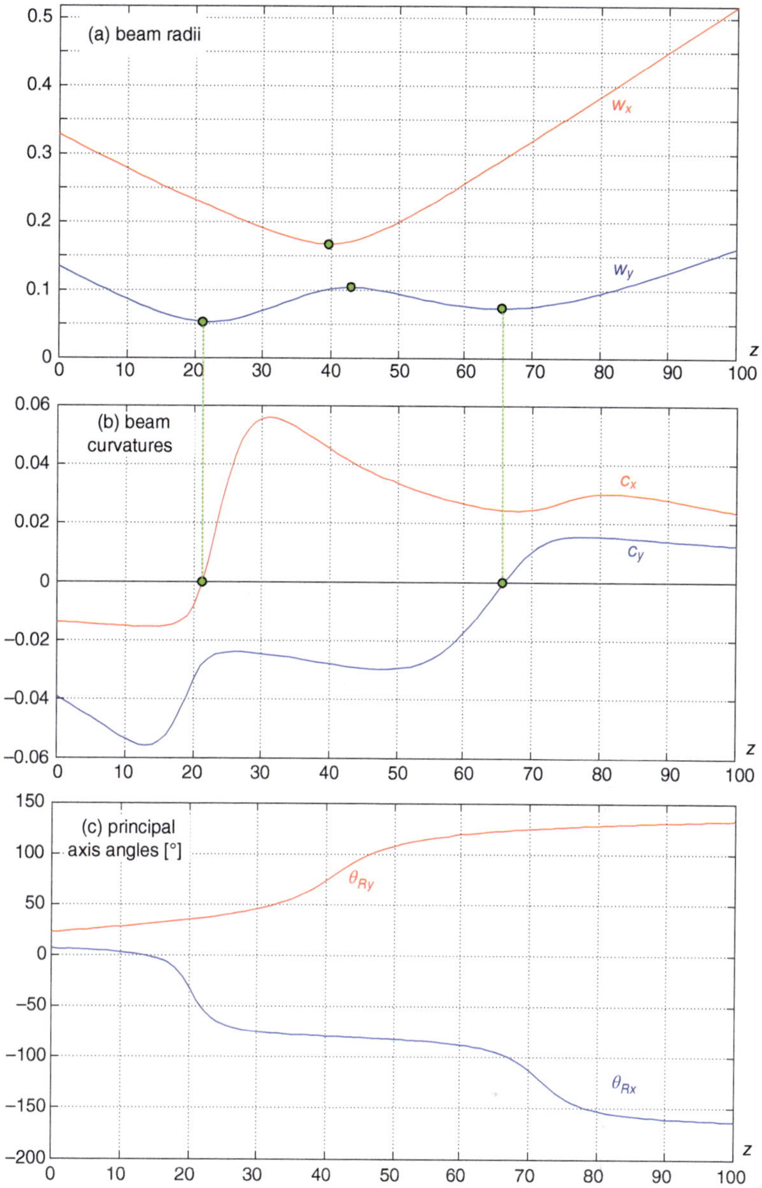

Figure 8.24 Free space propagation of a general astigmatic beam. In (a) the beam radii and in (b) the curvatures in x and y cross section are shown respectively. (c) shows the azimuthal angles of the intensity and curvature ellipse principal directions. The unit of z is arbitrary.

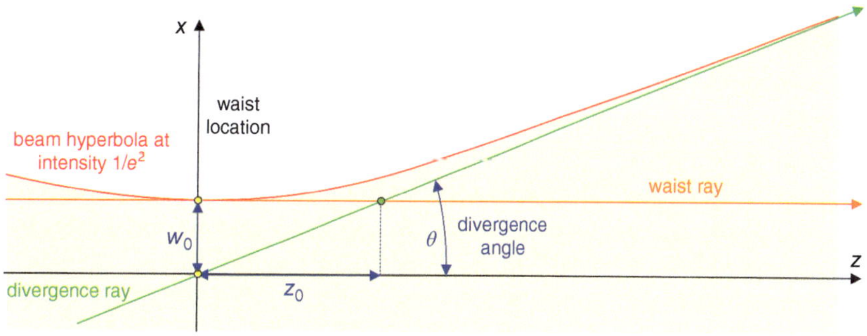

Figure 8.25 Ray equivalent of a Gaussian beam with special waist and divergence ray.

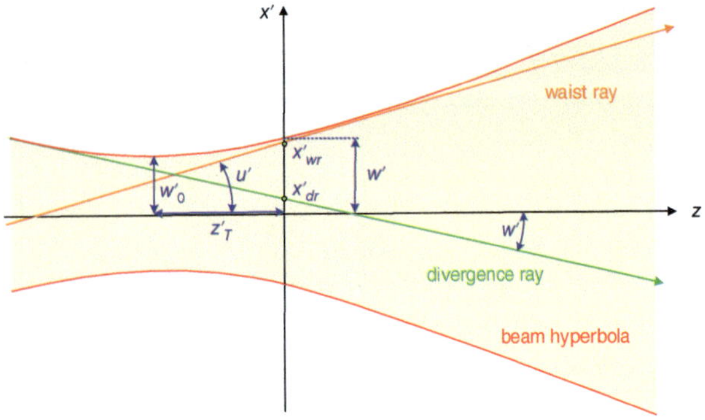

Figure 8.26 Ray equivalent of a Gaussian beam with generalized ray selection instead of a waist and divergence ray.

8.4.2 Ray Equivalent in Three Dimensions*

(*) The content of this section is quite specialized and for advanced reading. Knowledge of this section is not necessary to understand the following chapters. Beginners are recommended to skip this section.

In the more general three-dimensional geometry, the ray equivalent model can be used as well to propagate the Gaussian beam through a parabasal optical system. The centre axis of the beam is used as a centre or base ray of the system and is calculated first by raytrace, maybe real with aberrations. The raytrace must be performed with three coordinate components in a vectorial form:

Axis ray: Intersection point in the final plane: P_0, unity vector of the ray $\vec{s}_0$

Arbitrary ray: Intersection point in the final plane: P_1, unity vector of the ray $\vec{s}_1$

With the difference of the vectors:

$$\vec{r}_{01} = \vec{r}_1 - \vec{r}_0 \tag{8.95}$$

and the orthogonal vector

$$\vec{n} = \vec{r}_{01} \times \vec{s}_0 \tag{8.96}$$

the ray in the vicinity of the base ray with direction $\vec{s}_1$ must be rotated into the plane spanned by $\vec{r}_{01}$ and $\vec{s}_0$:

$$\vec{s}_{1p} = \vec{s}_1 - (\vec{s}_1\vec{n}) \cdot \vec{n} \tag{8.97}$$

see Figure 8.27.

The optical system can be of general geometry without circular symmetrical restrictions. To generate the general 4 × 4 parabasal matrix around the base ray, five rays must be traced through the system. The base ray corresponds to the axis of the Gaussian beam. Two marginal paraxial rays are in infinitesimal neighbourhood of the base ray in the x and the y section. They start in the centre of the waist and have an aperture angle corresponding to the far field divergence of the Gaussian beam. This is sketched in Figure 8.28. The real aperture angle is scaled down to ensure differential behaviour. All the intersection points at the various surfaces of the system must be in close vicinity of the base ray.

In the paraxial vicinity of the base ray, changes of the ray parameters X, Y, U, V are linear and correspond to the general mapping equations

$$\begin{aligned} X' &= X'(X,Y,U,V), \quad Y' = Y'(X,Y,U,V) \\ U' &= U'(X,Y,U,V), \quad V' = V'(X,Y,U,V) \end{aligned} \tag{8.98}$$

If these relations are differentiated, one obtains

$$\begin{aligned} dX' &= \frac{dX'}{dX}dX + \frac{dX'}{dY}dY + \frac{dX'}{dU}dU + \frac{dX'}{dV}dV \\ dY' &= \frac{dY'}{dX}dX + \frac{dY'}{dY}dY + \frac{dY'}{dU}dU + \frac{dY'}{dV}dV \\ dU' &= \frac{dU'}{dX}dX + \frac{dU'}{dY}dY + \frac{dU'}{dU}dU + \frac{dU'}{dV}dV \\ dV' &= \frac{dV'}{dX}dX + \frac{dV'}{dY}dY + \frac{dV'}{dU}dU + \frac{dV'}{dV}dV \end{aligned} \tag{8.99}$$

On the other hand, in the matrix calculus, the equation

$$\begin{pmatrix} x' \\ y' \\ u' \\ v' \end{pmatrix} = \begin{pmatrix} A_{11} & A_{12} & B_{11} & B_{12} \\ A_{21} & A_{22} & B_{21} & B_{22} \\ C_{11} & C_{12} & D_{11} & D_{12} \\ C_{21} & C_{222} & D_{21} & D_{22} \end{pmatrix} \cdot \begin{pmatrix} x \\ y \\ u \\ v \end{pmatrix} \tag{8.100}$$

is valid. By comparison, we get the relationship between the matrix elements and the finite differences in the real

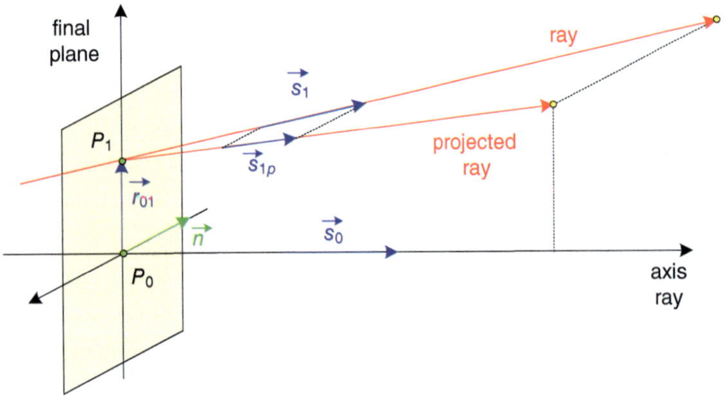

Figure 8.27 Geometry of equivalent rays in the vectorial model.

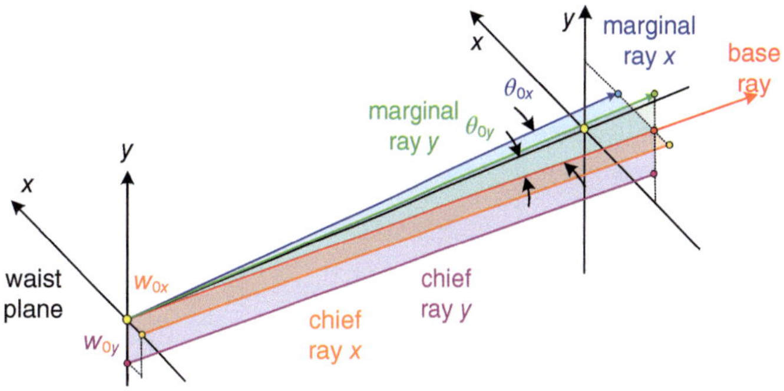

Figure 8.28 Geometry and equivalent rays in the general neighbourhood of the base ray.

raytrace in the formulas as

$$A_{11}=\frac{dX'}{dX},\quad A_{12}=\frac{dX'}{dY},\quad B_{11}=\frac{dX'}{dU},\quad B_{12}=\frac{dX'}{dV}$$
$$A_{21}=\frac{dY'}{dX},\quad A_{22}=\frac{dY'}{dY},\quad B_{21}=\frac{dY'}{dU},\quad B_{22}=\frac{dY'}{dV}$$
$$C_{11}=\frac{dU'}{dX},\quad C_{12}=\frac{dU'}{dY},\quad D_{11}=\frac{dU'}{dU},\quad D_{12}=\frac{dU'}{dV}$$
$$C_{21}=\frac{dV'}{dX},\quad C_{22}=\frac{dV'}{dY},\quad D_{21}=\frac{dV'}{dU},\quad D_{22}=\frac{dV'}{dV} \tag{8.101}$$

In a real implementation, the derivatives can be approximated by finite differences of parabasal rays.

8.4.3 Selecting General Reconstruction Rays*

(*) The content of this section is quite specialized and for advanced reading. Knowledge of this section is not necessary to understand the following chapters. Beginners are recommended to skip this section.

If a Gaussian beam with general astigmatism is considered, four rays are needed to reconstruct the beam after any *ABCD* system. Assume that $w_{0x/y}$ is the waist radii of the intensity ellipse and $\theta_{0x/y}$ is the divergence angles in an arbitrary initial coordinate system. The angles α, β, δ and γ are arbitrary angles of the parabasal rays and p is a scaling factor, which describes the paraxiality around the chief or axis ray. The meanings of the angles are [12]:

α azimuthal angle in the x–y plane for the points/rays one and two;
β azimuthal angle in the x–y plane for the points/rays three and four;
γ angle difference between the point pairs;
δ difference in lengths of the ray distances to the axis ray.

Then the following four rays must be calculated and propagated:

$$\begin{aligned}
x_1 &= pw_{0x}\cdot[\cos\alpha\cos\delta+\sin(\alpha+\gamma)\sin\delta]\\
y_1 &= pw_{0y}\cdot[\sin(\alpha+\gamma)\cos\delta-\cos\alpha\sin\delta]\\
s_{x1} &= -p\theta_{0x}\cdot[\sin\alpha\cos\delta-\cos(\alpha+\gamma)\sin\delta]\\
s_{y1} &= p\theta_{0y}\cdot[\cos(\alpha+\gamma)\cos\delta+\sin\alpha\sin\delta]
\end{aligned} \tag{8.102}$$

$$\begin{aligned}
x_2 &= pw_{0x}\cdot[\cos(\alpha+\pi/2)\cos\delta+\sin(\alpha+\pi/2+\gamma)\sin\delta]\\
y_2 &= pw_{0y}\cdot[\sin(\alpha+\pi/2+\gamma)\cos\delta-\cos(\alpha+\pi/2)\sin\delta]\\
s_{x2} &= -p\theta_{0x}\cdot[\sin(\alpha+\pi/2)\cos\delta-\cos(\alpha+\pi/2+\gamma)\sin\delta]\\
s_{y2} &= p\theta_{0y}\cdot[\cos(\alpha+\pi/2+\gamma)\cos\delta+\sin(\alpha+\pi/2)\sin\delta]
\end{aligned} \tag{8.103}$$

$$\begin{aligned}
x_3 &= -pw_{0x}\cdot[\cos\beta\cos\delta+\sin(\beta-\gamma)\sin\delta]\\
y_3 &= -pw_{0y}\cdot[\sin(\beta-\gamma)\cos\delta-\cos\beta\sin\delta]\\
s_{x3} &= -p\theta_{0x}\cdot[\sin\beta\cos\delta-\cos(\beta-\gamma)\sin\delta]\\
s_{y3} &= p\theta_{0y}\cdot[\cos(\beta-\gamma)\cos\delta+\sin\beta\sin\delta]
\end{aligned} \tag{8.104}$$

$$\begin{aligned}
x_4 &= -pw_{0x}\cdot[\cos(\beta+\pi/2)\cos\delta+\sin(\beta+\pi/2-\gamma)\sin\delta]\\
y_4 &= -pw_{0y}\cdot[\sin(\beta+\pi/2-\gamma)\cos\delta-\cos(\beta+\pi/2)\sin\delta]\\
s_{x4} &= -p\theta_{0x}\cdot[\sin(\beta+\pi/2)\cos\delta-\cos(\beta+\pi/2-\gamma)\sin\delta]\\
s_{y4} &= p\theta_{0y}\cdot[\cos(\beta+\pi/2-\gamma)\cos\delta+\sin(\beta+\pi/2)\sin\delta]
\end{aligned} \tag{8.105}$$

An easier selection defines two left- and two right-handed rays according to the scheme in Figure 8.29 and is obtained for $\alpha=\beta=\gamma=\delta=0$. The special selected rays guarantee that all the parabasal rays are independent.

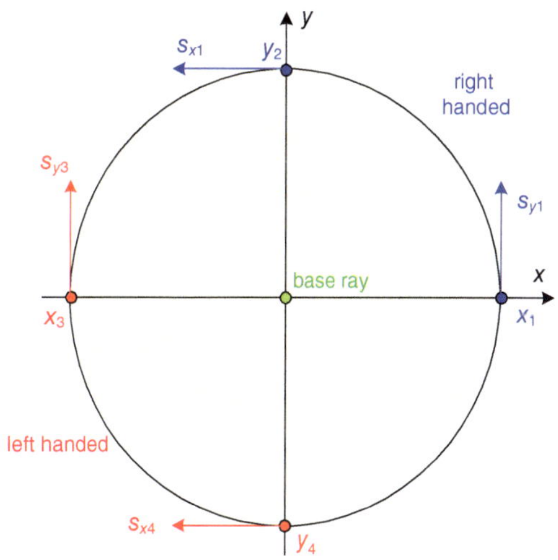

Figure 8.29 Simplest selection of four rays to simulate the general astigmatic Gaussian beam propagation by rays.

The classical waist and divergence angle selection of Greynolds is only one special choice of the angles with $\gamma = \pi/2$ and $\delta = \pi/4$. With the four propagated rays, the corresponding ray values in the final image plane (dashed) allow us to calculate the projection sizes of the intensity ellipse by

$$D_0 = \frac{1}{p\sqrt{2}}\sqrt{x_1'^2 + x_2'^2 + x_3'^2 + x_4'^2},$$
$$D_{\pi/2} = \frac{1}{p\sqrt{2}}\sqrt{y_1'^2 + y_2'^2 + y_3'^2 + y_4'^2}$$
$$D_{\pi/4} = \frac{1}{2p}\sqrt{(x_1' + y_1')^2 + (x_2' + y_2')^2 + (x_3' + y_3')^2 + (x_4' + y_4')^2} \tag{8.106}$$

and the orientation of the intensity ellipse is given by the angle φ_w' as

$$\varphi_w' = \frac{1}{2}\arctan\left(\frac{2D_{\pi/4}^2 - D_0^2 - D_{\pi/2}^2}{D_0^2 - D_{\pi/2}^2}\right) \tag{8.107}$$

In the system of the rotated coordinates

$$x_1'' = x_1' \cos\varphi_w' + y_1' \sin\varphi_w',$$
$$y_1'' = -x_1' \sin\varphi_w' + y_1' \cos\varphi_w'$$
$$s_{x1}'' = s_{x1}' \cos\varphi_w' + s_{y1}' \sin\varphi_w',$$
$$s_{y1}'' = -s_{x1}' \sin\varphi_w' + s_{y1}' \cos\varphi_w' \tag{8.108}$$

(and similarly for rays two, three and four), the intensity ellipse is aligned and the major axes of the intensity ellipse are

$$w_x' = \frac{1}{p\sqrt{2}}\sqrt{x_1''^2 + x_2''^2 + x_3''^2 + x_4''^2},$$
$$w_y' = \frac{1}{p\sqrt{2}}\sqrt{y_1''^2 + y_2''^2 + y_3''^2 + y_4''^2} \tag{8.109}$$

for the data of the phase ellipse, the curvatures along x'' and y'' are calculated by

$$C_x' = \frac{1}{w_x'}\frac{dw_x'}{dz} = \frac{x_1''s_{x1}'' + x_2''s_{x2}'' + x_3''s_{x3}'' + x_4''s_{x4}''}{x_1''^2 + x_2''^2 + x_3''^2 + x_4''^2}$$
$$C_y' = \frac{1}{w_y'}\frac{dw_y'}{dz} = \frac{y_1''s_{y1}'' + y_2''s_{y2}'' + y_3''s_{y3}'' + y_4''s_{y4}''}{y_1''^2 + y_2''^2 + y_3''^2 + y_4''^2} \tag{8.110}$$

With the auxiliary variables

$$\frac{dD_{\pi/4}}{dz} = \frac{1}{2p}\cdot\frac{\begin{array}{c}(x_1'' + y_1'')(s_{x1}'' + s_{y1}'') + (x_2'' + y_2'')(s_{x2}'' + s_{y2}'')\\ + (x_3'' + y_3'')(s_{x3}'' + s_{y3}'') + (x_4'' + y_4'')(s_{x4}'' + s_{y4}'')\end{array}}{\sqrt{\begin{array}{c}(x_1'' + y_1'')^2 + (x_2'' + y_2'')^2\\ + (x_3'' + y_3'')^2 + (x_4'' + y_4'')^2\end{array}}} \tag{8.111}$$

$$u = \arctan\left(\frac{w_y'}{w_x'}\tan\frac{\pi}{4}\right)$$

$$x_t' = w_x' \cos u, \quad y_t' = w_y' \sin u \tag{8.112}$$

$$S_0 = \frac{C_x'}{2}, \quad S_{\pi/2} = \frac{C_y'}{2},$$
$$S_{\pi/4} = \frac{1}{4}(C_x' + C_y') + \frac{1}{2}\,\frac{\sqrt{2}\frac{dD_{\pi/4}}{dz} - x_t'C_x' - y_t'C_y'}{x_t' + y_t'}$$
$$M = \frac{1}{4}(C_x' + C_y'),$$
$$T = \sqrt{(S_0 - M)^2 + (S_{\pi/4} - M)^2} \tag{8.113}$$

the angle of the curvature system relative to the intensity systems is given by

$$\varphi'' = \frac{1}{2}\arctan\left(\frac{(S_0 - M)/T}{(S_{\pi/4} - M)/T}\right) \tag{8.114}$$

so we have

$$\varphi_R' = \varphi'' + \varphi_w' \tag{8.115}$$

and the two principal values of the curvatures are

$$C_x'' = 2(M + T), \quad C_y'' = 2(M - T) \tag{8.116}$$

8.4.4 Numerical Reconstruction of *ABCD* by Parabasal Rays*

(*) The content of this section is quite specialized and for advanced reading. Knowledge of this section is not necessary to understand the following chapters. Beginners are recommended to skip this section.

Let's assume that two rays are traced in a real algorithm: a reference ray and a parabasal ray. The goal is to find with the help of several parabasal rays the 4×4 paraxial matrix on the skew reference ray through an optical system. [18]

The intersection points Q_j and Q_{j+1} of the reference ray have the global coordinates q_j and q_{j+1} respectively, see Figure 8.30. This variable should not be confused with the complex beam parameter q. The skew thickness between these intersections is then

$$d_j = |\vec{q}_{j+1} - \vec{q}_j| \tag{8.117}$$

The matrices for the propagation along the skew distance d_j between surface no. j and $j+1$ are simply given by

$$\underline{M}_p = \begin{pmatrix} A_{xx} & A_{xy} & B_{xx} & B_{xy} \\ A_{xy} & A_{yy} & B_{xy} & B_{yy} \\ C_{xx} & C_{xy} & D_{xx} & D_{xy} \\ C_{xy} & C_{yy} & D_{xy} & D_y \end{pmatrix} = \begin{pmatrix} 1 & 0 & d_j & 0 \\ 0 & 1 & 0 & d_j \\ 0 & 0 & 1 & 0 \\ 0 & 0 & 0 & 1 \end{pmatrix} \tag{8.118}$$

The unit direction vector between Q_j and Q_{j+1} is denoted by t_j. To be consistent with the parabasal approximation, plane surfaces are constructed in Q_j, which are

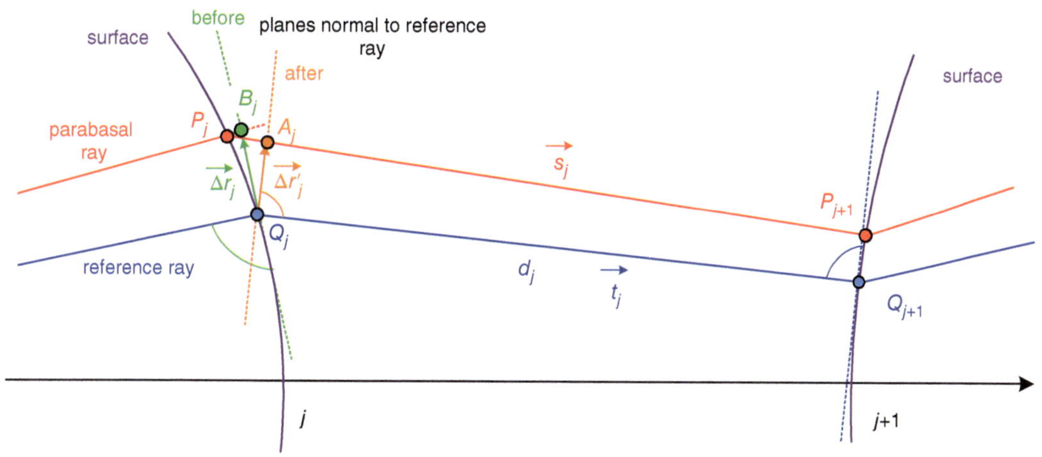

Figure 8.30 Parabasal raytrace around a reference ray.

perpendicular to the incoming and the outgoing ray. The parabasal ray with exact intersection point P_j with vector position r_j meets the plane at B_j before and A_j after the refraction. The differences between Q_jB_j and Q_jA_j are the parabasal ray heights relative to the reference ray. By inserting the corresponding rays into the equations of the tangential planes, the differences are determined to be

$$\Delta\vec{r}_j = \vec{r}_{j-1} + \frac{(\vec{q}_j - \vec{r}_{j-1})\cdot\vec{t}_{j-1}}{\vec{s}_{j-1}\cdot\vec{t}_{j-1}}\cdot\vec{s}_{j-1} - \vec{q}_j,$$
$$\Delta\vec{r}'_j = \vec{r}_j + \frac{(\vec{q}_j - \vec{r}_j)\cdot\vec{t}_j}{\vec{s}_j\cdot\vec{t}_j}\cdot\vec{s}_j - \vec{q}_j \tag{8.119}$$

The matrices for the transition at the surface contain the tilt of the planes as well as the refraction at the surface itself. With the changes in the perpendicular planes above, this can be written as

$$\Delta\vec{R}'_j = \underline{M}_t \cdot \Delta\vec{R}_j \tag{8.120}$$

with the four-vector

$$\Delta\vec{R}_j = (\Delta x_j, \Delta y_j, \Delta u_j, \Delta v_j)^T = (\Delta x_j, \Delta y_j, \Delta s_{xj}, \Delta s_{yj})^T \tag{8.121}$$

If now several different parabasal rays are calculated by finite raytrace, the matrix M_t can be reconstructed out of the ray data changes as

$$\begin{pmatrix} \Delta x'_1 & \Delta x'_2 & .. & \Delta x'_k \\ \Delta y'_1 & \Delta y'_2 & .. & \Delta y'_k \\ \Delta u'_1 & \Delta u'_2 & .. & \Delta u'_k \\ \Delta v'_1 & \Delta v'_2 & .. & \Delta v'_k \end{pmatrix} = \begin{pmatrix} A_{xx} & A_{xy} & B_{xx} & B_{xy} \\ A_{xy} & A_{yy} & B_{xy} & B_{yy} \\ C_{xx} & C_{xy} & D_{xx} & D_{xy} \\ C_{xy} & C_{yy} & D_{xy} & D_{y} \end{pmatrix} \cdot \begin{pmatrix} \Delta x_1 & \Delta x_2 & .. & \Delta x_k \\ \Delta y_1 & \Delta y_2 & .. & \Delta y_k \\ \Delta u_1 & \Delta u_2 & .. & \Delta u_k \\ \Delta v_1 & \Delta v_2 & .. & \Delta v_k \end{pmatrix} \tag{8.122}$$

or in abbreviated matrix notation

$$\underline{D}'_{kx4} = \underline{M}_{4x4} \cdot \underline{D}_{kx4} \tag{8.123}$$

if $k = 4$, the system is exactly solvable. For $k > 4$ the matrix equation is over-determined. This needs more computational time but allows us to get rid of some discretization errors. The solution of this equation can then be written as a non-least-square optimum

$$\underline{M}_t = (\underline{D}' \cdot \underline{D}^T)\cdot(\underline{D}\cdot\underline{D}^T)^{-1} \tag{8.124}$$

This means that by inserting in D and D' the values of the k parabasal probe rays before and after the surface, the transition matrix M_t can be reconstructed.

8.5 Truncated Gaussian Beams

8.5.1 One-dimensional Beam

If a one-dimensional Gaussian beam is truncated in the waist by a slit of diameter $2a$, with the auxiliary parameter

$$b = \frac{w_0^2}{4}\cdot\frac{1 - \frac{i\pi\cdot w_0^2}{\lambda\cdot z}}{1 + \left(\frac{\pi\cdot w_0^2}{\lambda\cdot z}\right)^2} \tag{8.125}$$

the field after this stop in a distance z can be determined by performing the finite integration with the error function $erf()$ and delivers the result [19]

$$E(x,z) = \frac{e^{-ikz}}{\sqrt{-ikz}}\cdot\sqrt{\pi\cdot b}\cdot e^{-\frac{i\pi x^2}{\lambda\cdot z} - \left(\frac{2\pi\cdot x}{\lambda\cdot z}\right)^2\cdot b} \cdot\left[erf\left(\frac{2i\pi\cdot x\cdot\sqrt{b}}{\lambda\cdot z} + \frac{a}{2\sqrt{b}}\right) + erf\left(-\frac{2i\pi\cdot x\cdot\sqrt{b}}{\lambda\cdot z} + \frac{a}{2\sqrt{b}}\right)\right] \tag{8.126}$$

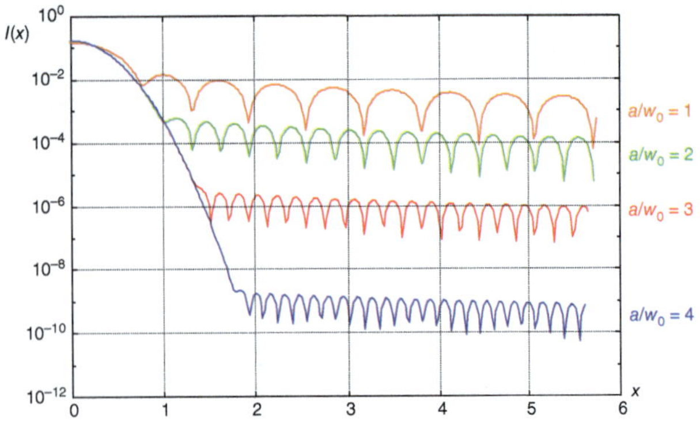

Figure 8.31 Diffraction ripple of a one-dimensional truncated Gaussian beam in the far field. The transverse coordinate x is here given in arbitrary units.

This distribution can be interpreted as a purely unperturbed Gaussian beam with a correction factor. This correction in the brackets of the formula is complex and generates the diffraction ripple by the error function. Figure 8.31 shows the behaviour of this function as intensity in the far field for different values of the normalized truncation a/w_0.

8.5.2 Circular Symmetric Beam

8.5.2.1 Introduction

The problem of a rotational symmetric beam with transverse Gaussian profile but a truncation at a stop with radius a often occurs in practice. It is sketched in Figure 8.32.

The relative truncation ratio is defined as the parameter

$$\varepsilon = \frac{a}{w} \tag{8.127}$$

for large ε nearly no truncation takes place, small $\varepsilon \to 0$ corresponds to a rectangular profile which gives the Airy distribution in the focal plane. The stop truncation gives rise to a power loss and can be described by the truncation of intensity

$$T = \frac{I_{out}}{I_{in}} = 1 - e^{-2\frac{a^2}{w^2}} = 1 - e^{-2\varepsilon^2} \tag{8.128}$$

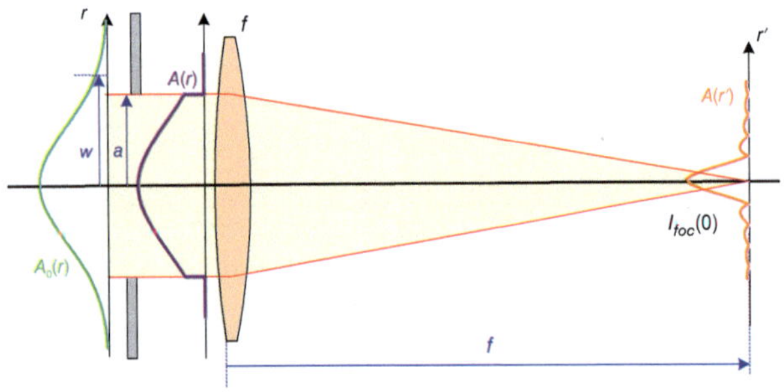

Figure 8.32 Focusing of a truncated circular symmetric Gaussian beam.

The intensity profile after the stop looks in this case as is shown in Figure 8.33 in an intensity normalized representation in (b) and by normalization of the transmitted power in (a).

8.5.2.2 Exact Field Calculation*

(*) The content of this section is quite specialized and for advanced reading. Knowledge of this section is not necessary to understand the following chapters. Beginners are recommended to skip this section.

The diffraction integral of a Gaussian beam truncated in the waist plane can be expressed in terms of the so-called Lommel functions analytically. Here ρ is the normalized pupil coordinate; u and v are the scaled coordinates in axial and lateral direction in the image space defined as

$$\rho = \frac{r}{a}, \qquad u = \frac{2\pi}{\lambda}\left(\frac{a}{f}\right)^2 \frac{z}{1+z/f},$$
$$v = \frac{2\pi}{\lambda}\frac{a}{f}\frac{r}{1+z/f}, \quad v_0 = \frac{2\pi}{\lambda}\frac{a}{f} r \tag{8.129}$$

with the lateral coordinate v_0 in the stop plane. The relevant Fresnel diffraction integral is split into the real and the imaginary part as

$$2 \cdot \int_0^1 J_0(\rho v) \cdot e^{-\frac{i}{2}\cdot u\rho^2}\, \rho\, d\rho = C(u,v) - i \cdot S(u,v) \tag{8.130}$$

with the two-dimensional functions C and S, which are not the classical Fresnel integrals. They read

$$\begin{aligned} C(u,v) &= 2 \cdot \int_0^1 J_0(\rho\, v) \cdot \cos\left(\frac{1}{2}\cdot u\rho^2\right) \rho\, d\rho \\ &= \frac{\cos(u/2)}{(u/2)} \cdot U_1(u,v) + \frac{\sin(u/2)}{(u/2)} \cdot U_2(u,v) \\ &= \frac{2}{v}\sin\frac{v^2}{2u} + \frac{\sin(u/2)}{(u/2)} \cdot V_0(u,v) \\ &\quad - \frac{\cos(u/2)}{(u/2)} \cdot V_1(u,v) \\ S(u,v) &= 2 \cdot \int_0^1 J_0(\rho\, v) \cdot \sin\left(\frac{1}{2}\cdot u\rho^2\right) \rho\, d\rho \\ &= \frac{\sin(u/2)}{(u/2)} \cdot U_1(u,v) - \frac{\cos(u/2)}{(u/2)} \cdot U_2(u,v) \\ &= \frac{2}{u}\cos\frac{v^2}{2u} - \frac{\cos(u/2)}{(u/2)} \cdot V_0(u,v) \\ &\quad - \frac{\sin(u/2)}{(u/2)} \cdot V_1(u,v) \end{aligned} \tag{8.131}$$

Here, $U_{1/2}$ and $V_{0/1}$ are the Lommel functions, which are defined by the Bessel series expansions

$$\begin{aligned} U_n(u,v) &= \sum_{s=0}^{\infty} (-1)^s \cdot \left(\frac{u}{v}\right)^{n+2s} \cdot J_{n+2s}(v) \\ V_n(u,v) &= \sum_{s=0}^{\infty} (-1)^s \cdot \left(\frac{v}{u}\right)^{n+2s} \cdot J_{n+2s}(v) \end{aligned} \tag{8.132}$$

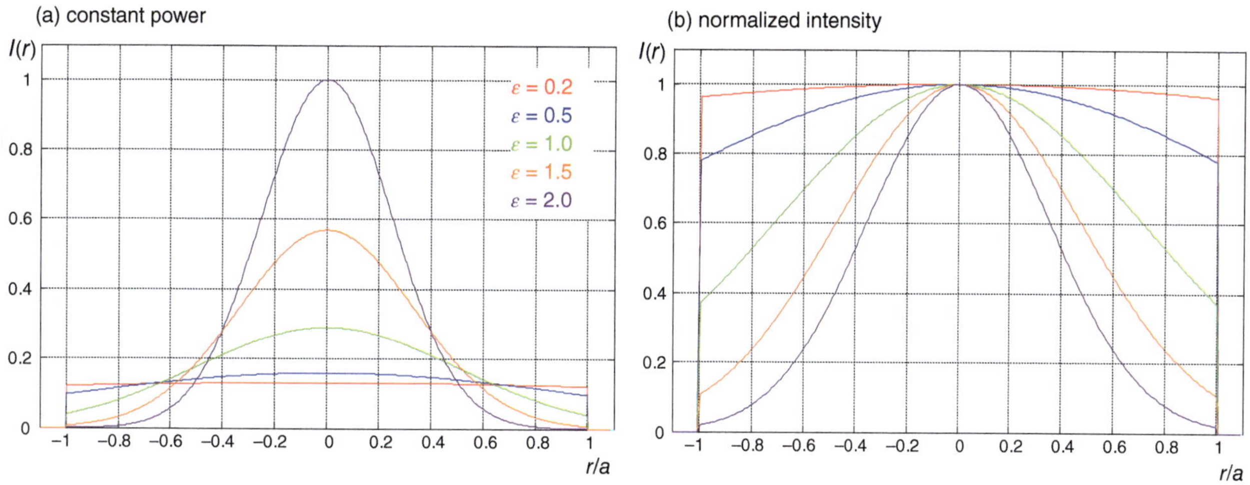

Figure 8.33 Intensity normalized profiles of a truncated circular symmetric Gaussian profile for different truncation ratios and constant stop radius a for (a) constant power P and (b) normalized peak intensity.

In the special case $v = 0$, where only points on the axis are considered, the Lommel functions degenerate to the classical Fresnel integrals. U is the more convenient function for $|u/v| < 1$, which corresponds to the geometrical shadow region, while V is useful for $|v/u| < 1$, which corresponds to the bright region of the light bundle. There are three special cases, which can be expressed by simpler formulae.

- The lateral field in the focal plane of a lens and therefore the far-field is given by the series

$$E(0,v_0) = -\frac{i2\pi a^2 A_0}{\lambda f^2}\cdot e^{\frac{ir^2}{\lambda f}}\cdot e^{-\varepsilon^2}\cdot\sum_{n=1}^{\infty}\frac{J_n(v_0)}{v_0^n}(2\varepsilon^2)^{n-1} = -\frac{i2\pi a^2 A_0}{\lambda f^2}\cdot e^{\frac{ir^2}{\lambda f}}\cdot[C(u-2i\varepsilon^2,v_0)-iS(u-2i\varepsilon^2,v_0)] \tag{8.133}$$

- Along the optical axis for $v = 0$, the intensity is given by

$$I(u,0) = A_0^2\left(1-\frac{u\lambda f}{2\pi a^2}\right)^2\frac{\varepsilon^4}{\varepsilon^4+(u/2)^2}\cdot\frac{\cosh\varepsilon^2-\cos u/2}{\cosh\varepsilon^2-1} \tag{8.134}$$

- The peak intensity, especially in the focal point, is given by

$$I_{foc} = \left(\frac{A_0\,\pi a^2}{\lambda f^2}\cdot\frac{1-e^{-\varepsilon^2}}{\varepsilon^2}\right)^2 \tag{8.135}$$

8.5.2.3 General Properties

Figure 8.34 shows the intensity profiles $I(r)$ in the focal plane of a lens in logarithmic representation for the truncation parameters ε = 1, 2, 3 and 4 as a result of numerical calculations. For values of $\varepsilon < 2$ the well-known diffraction ripples are observed. They resemble the Airy distribution, which is obtained for very small ε in asymptotic behaviour. The ripple height and positions changes and in addition the central lobe increases in diameter with growing ε. For truncation ratios larger than $\varepsilon > 2$, no ripples are seen at least in the range of relative intensities above 10^{-7}. For very large ε, the classical Gaussian profile is observed, which is parabolic in dependence in the logarithmic representation of the figure. In principle this scheme demonstrates the transition from the unperturbed Gaussian case to the Airy pattern if ε is changed from nearly 0 (exact zero makes no sense) to infinity.

If the heights of the first diffraction side lobes are considered, the normalized values behave like the graphic in Figure 8.35. It is seen that in the case $a = w$, which is often realized in practice, the sidelobe has a relative height of 0.5%.

If the peak height of the intensity in the focal plane is calculated and the influence of the truncation considered, there are two different references possible.

- If the incoming full Gaussian beam power P_0 is the reference, the beam intensity can only shrink, see the red curve in Figure 8.36. In asymptotic behaviour we get from the focal intensity formula for small ε the first term of the series-type correction

$$I_{foc} = \left(\frac{A_0\,\pi a^2}{\lambda f^2}\cdot\frac{1-e^{-\varepsilon^2}}{\varepsilon^2}\right)^2 \approx I_0\cdot(1-2\varepsilon^4+\ldots) \tag{8.136}$$

- If the power P after the truncation is considered, the focal point intensity has a maximum value in the case

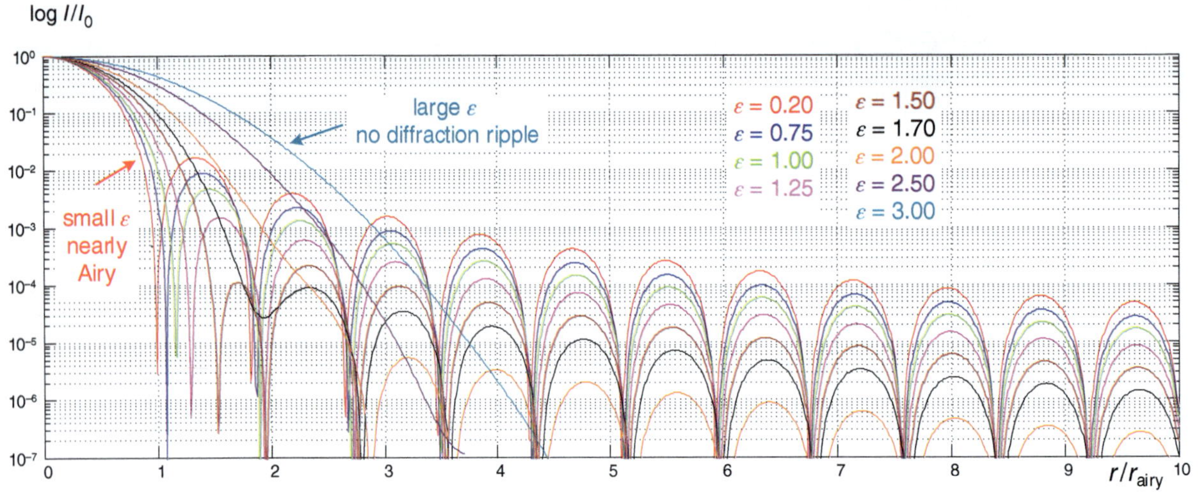

Figure 8.34 Logarithm of the intensity profile $I(r)$ of a focused truncated Gaussian beam for different values of the truncation ratio ε.

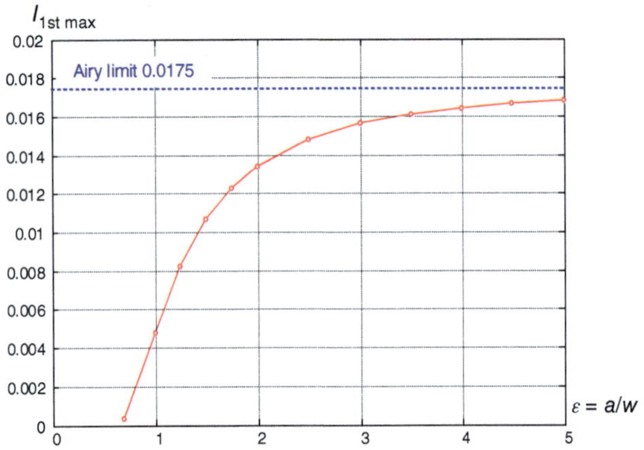

Figure 8.35 Relative height of the first diffraction side lobe of the intensity profile of a focused truncated Gaussian beam for different values of the truncation ratio ε.

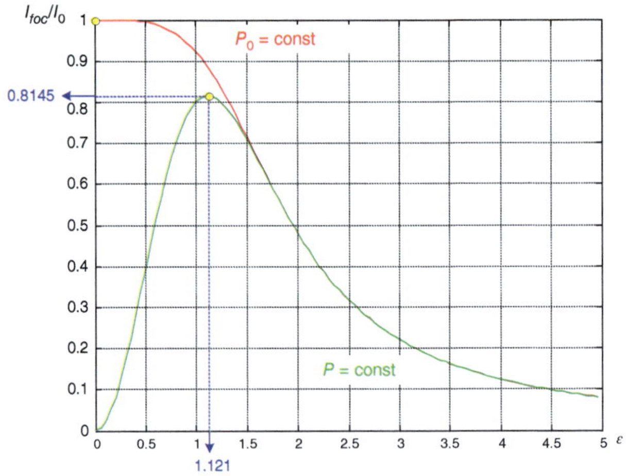

Figure 8.36 Focal intensity peak as a function of ε for the two cases of constant incoming power before the stop P_0 and after truncation P.

of the truncation ratio $\varepsilon = 1.121$. The best value is then $I_{foc}/I_o = 0.8145$ times the peak of the untruncated Gaussian beam. This is represented by the green curve. The optimum comes from the trade-off between effective aperture and transmission at the stop.

The width of the focal distribution is shown in Figure 8.37. It is seen that the relative broadening

$$\eta = \frac{w}{w_{\varepsilon=0}} \tag{8.137}$$

due to diffraction ripple as a 13.5% threshold value of the central peak grows for smaller values of truncation ε, this is also seen in the left central peak of Figure 8.34. The exact numerical behaviour depends on the width definition, in the figure the $1/e2 = 13.5\%$ value as well as the 50% values are compared. For $a = w$ the broadening is roughly 50%.

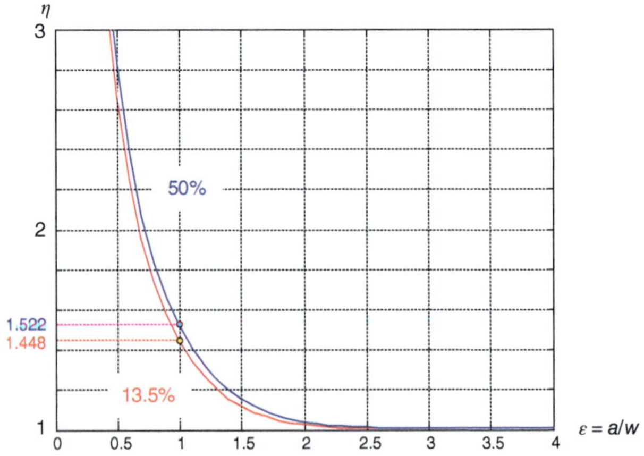

Figure 8.37 Focal peak broadening as a function of the truncation.

The reason for this is the missing energy in the outer range of the stop; the effective NA of the focusing is decreasing.

8.6 Gaussian Beams Beyond the Paraxial Approximation*

(*) The content of this section is quite specialized and for advanced reading. Knowledge of this section is not necessary to understand the following chapters. Beginners are recommended to skip this section.

8.6.1 Introduction

Gaussian beams are paraxial solutions of the wave equation. If one of the assumptions described in 8.1.1 is violated, a starting Gaussian beam is no longer Gaussian during propagation. In this section the first item is discussed. Here in particular the constraint of small angles is considered, which occur in practice; for example, in the case of semiconductor laser sources or high aperture focusing. There is a huge amount of literature and possible calculation schemes and approximations in this field [20–36].

First it should be mentioned that leaving the paraxial range and allowing for larger angles involves vectorial diffraction effects with high/NA aspects. This means that a scalar model is no longer valid; exact transverse polarization like TE, TM, TEM no longer exists. This corresponds of course to the high-NA focusing considerations in Section 6.6. But, similarly as in usual diffraction modelling, there is a range of medium size angles where a better approximation of the paraxial results can be obtained; nevertheless, the vectorial effects are still neglected.

In the scalar models we have to distinguish in elaborating on more general approaches; if the geometry is:

- one dimensional;
- two-dimensional with Cartesian symmetry (Hermite modes), in particular astigmatic beams.
  This is an important case for the application in semiconductor laser sources;
- two-dimensional with circular symmetry (Laguerre modes).

In the literature, there are different approaches proposed to overcome the paraxial approximation.

- Transform of the diffraction integral into prolate spherical coordinates and exact solution of the wave equation.
- Introduce Gaussian beams with complex source points, which are exact.
- Use a Taylor expansion in the sense of a perturbation theory and expansion with the paraxial solution as the lowest order term. Different expansion variables and functions are used and proposed, giving different representations. The expansion coefficients are in some cases simple mathematical terms; in other realizations, lengthy recurrence formulas are obtained.
- Calculate the plane wave expansion using the Fourier transform for the far field representation and simplify and approximate the results.

If large angles are involved, the question becomes how the difference to the classical Gaussian beams is to be referenced. If a Gaussian distribution is assumed in the waist, in the far field the profile is no longer Gaussian. Another viewpoint is to look for the eigensolution of the wave equation for large angles. In this case, the properties are not Gaussian in the near nor the far field, but the profile remains invariant during propagation.

8.6.2 Exact Solution of the Wave Equation

If the wave equation is solved in prolate spherical coordinates, exact scalar solutions can be obtained that are non-paraxial. In the case of a Gaussian amplitude profile, the generalized mode functions [20, 21]

$$E(r,z) = \frac{A_0}{\frac{2\pi}{\lambda}\sqrt{w_0^2+z_0^2}\sqrt{\cos^2\theta + \left(\frac{z-z_T}{z_0}\right)^2}} \cdot e^{-\frac{2\pi}{\lambda}\sqrt{w_0^2+z_0^2}\cdot(1-\cos\theta)} \cdot e^{-\frac{2\pi i z}{\lambda z_0}\sqrt{w_0^2+z_0^2}} \cdot e^{-i\cdot\arctan\frac{z-z_T}{z_0\cos\theta}} \tag{8.138}$$

can be derived with the abbreviation

$$\cos\theta = \sqrt{1 - \frac{r^2}{\left(w_0^2+z_0^2\right)\cdot\left[1+\left(\frac{z}{z_0}\right)^2\right]}} \tag{8.139}$$

with $\theta = \theta_0$ for points on axis $r = 0$. In this generalized solution, the intensity surfaces are again hyperboloids and the phase surfaces are ellipsoids and not paraboloids. The lateral intensity distribution is not a perfect Gaussian profile anymore. The beam hyperbola again reads

$$w^2(z) = w_0^2 \cdot \left[1+\left(\frac{z}{z_0}\right)^2\right] \tag{8.140}$$

The radius of curvature on axis $r = 0$ is given by

$$R(0,z) = z + \frac{z_0^2}{z\cdot\cos^2\theta_0} \tag{8.141}$$

and on axis we have now for not necessarily small angles

$$\tan\theta_o = \frac{w_0}{z_0} \tag{8.142}$$

The Guoy angle is given by

$$\tan\Phi = \frac{z}{z_0 \cdot \cos\theta_0} \tag{8.143}$$

The shape of the beam profile deviates from a perfect Gaussian form, as can be seen in the following section.

8.6.3 Taylor Expansion of the Complex Field

It must be noted that all the approximations of a high aperture scalar field of a Gaussian beam are a little bit critical, because, in reality, the vectorial nature of the field cannot be neglected in the non-paraxial case and the different field components have to be taken into account. In addition, here only the circular symmetric case is considered. With the help of the parameter

$$\varepsilon = \frac{\sqrt{2}}{kw_0} = \frac{\lambda \cdot \sqrt{2}}{2\pi w_0} = \frac{\theta_0}{\sqrt{2}} \tag{8.144}$$

and the abbreviation

$$\xi = \frac{z}{z_0} \tag{8.145}$$

The exact field of a Gaussian beam under non-paraxial conditions can be written as an expansion with the smallness parameter $\varepsilon \ll 1$ in the form [23, 24, 26]

$$E = E^{(0)} + \varepsilon^2 \cdot E^{(2)} + \varepsilon^4 \cdot E^{(4)} + \ldots \tag{8.146}$$

The first term in this equation is the well-known paraxial solution, which reads here by insertion of Eq. (8.145)

$$E^{(0)} = \frac{e^{ikz}}{1 + i \cdot \xi} \cdot e^{-\frac{r^2}{w_0^2 \cdot (1+i\cdot\xi)}} \tag{8.147}$$

The next two correction terms of low orders can be written as

$$\begin{aligned} E^{(2)} &= \frac{i \cdot \xi}{(1 + i \cdot \xi)^2} \cdot L_2\left[\frac{r^2}{w_0^2 \cdot (1 + i \cdot \xi)}\right] \cdot E^{(0)} \\ E^{(4)} &= \frac{3i \cdot \xi}{(1 + i \cdot \xi)^4} \cdot \left\{ L_4\left[\frac{r^2}{w_0^2 \cdot (1 + i \cdot \xi)}\right] + \frac{r^2}{4w_0^2} \right. \\ &\quad \left. \cdot L_3^1\left[\frac{r^2}{w_0^2 \cdot (1 + i \cdot \xi)}\right] \right\} \cdot E^{(0)} \end{aligned} \tag{8.148}$$

The functions $L_n{}^m$ are the Laguerre polynomials with

$$\begin{aligned} L_2(x) &= 1 - 2x + \frac{1}{2}x^2 \\ L_3^{(1)}(x) &= 4 - 6x + x^2 - \frac{1}{6}x^3 \\ L_4(x) &= 1 - 4x + 3x^2 - \frac{2}{3}x^3 + \frac{1}{24}x^4 \end{aligned} \tag{8.149}$$

If the beam radius is evaluated, we get the following expansion [29]

$$w(z) = w_0 \cdot \sqrt{1 + \frac{\theta_0^2}{4} + \left(\frac{z}{z_0}\right)^2 \cdot \left(1 + \frac{3\theta_0^2}{4}\right)} \tag{8.150}$$

This formula shows that the beam width at $1/e^2$ is increasing with the size of the divergence angle.

A polynomial expansion of the field in r and z is given in the literature with the abbreviations

$$f = \frac{1}{1 - i \cdot z/z_0}, \quad \rho = \frac{r}{w_0} \tag{8.151}$$

by the equation

$$E = -iE_0 f \cdot e^{-f\rho^2} \cdot \left\{ \begin{array}{l} 1 + \theta_0^2 \cdot \left[f^2\rho^2 - \frac{1}{4}f^3\rho^4\right] \\ + \theta_0^4 \cdot \left[\frac{1}{8}f^2 - \frac{1}{4}f^3\rho^2 + \frac{15}{16}f^4\rho^4 - \frac{3}{8}f^5\rho^6 \right. \\ \qquad \left. + \frac{1}{32}f^6\rho^8\right] \\ \cdot \left[\frac{3}{16}f^3 - \frac{3}{16}f^4\rho^2 - \frac{9}{32}f^5\rho^4 + \frac{15}{16}f^6\rho^6 \right. \\ \quad \left. - \frac{7}{16}f^7\rho^8 - \frac{1}{16}f^8\rho^{10} - \frac{1}{384}f^9\rho^{12}\right] \end{array} \right\} \tag{8.152}$$

This equation gives the field changes by a θ_0-expansion as a function of r and z. Interestingly the change with z along the optical axis changes only with $\theta_0{}^4$ if in this formula $\rho = 0$ is inserted. Therefore, the z-dependence is much weaker than the change in radial direction.

To get an idea of the absolute impact of this non-paraxiality correction, Figure 8.38 shows in a logarithmic scale the profiles in the waist for different values of the divergence angle. For a better resolution, in part (b) the correction difference is shown. Interestingly for $r < 1.92$ w_0, the intensity is higher; for larger radius values it is smaller than the classical Gaussian profile. It is seen that the correction can reach 25% for $\theta_0 = 0.5$. If the beam radius w is defined as $1/e^2$ of the intensity threshold, the beam width is also changing with growing non-paraxiality. Figure 8.39 shows this behaviour as a function of θ_0.

A more sophisticated analysis of the relative changes makes only limited sense from the viewpoint of practical applications, because the scalar simplification creates an error of considerable size for $\theta_0 = 0.4$ by neglecting the vectorial effects.

8.6.4 Angle Expansion of Higher Order Modes

In paraxial approximation the Gauss–Hermite modes with complex argument can be written in one lateral

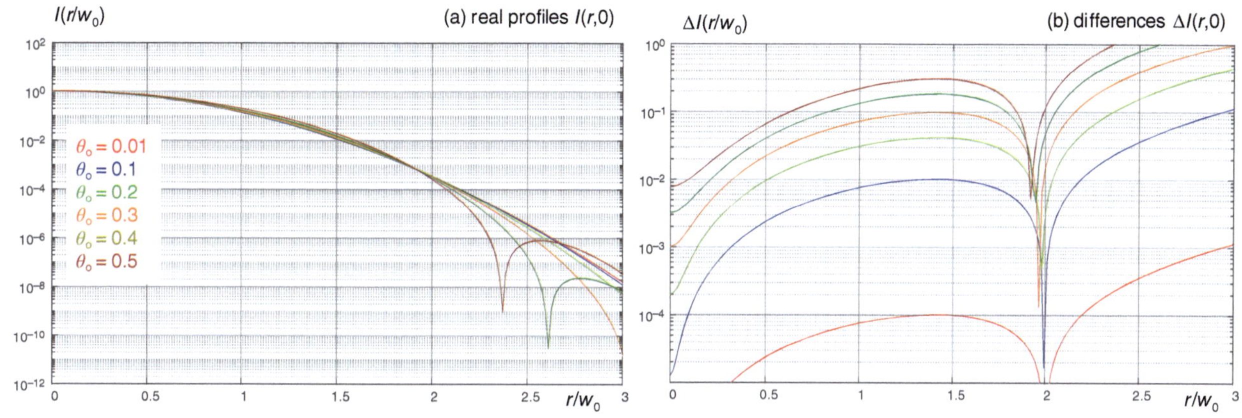

Figure 8.38 Real (a) and correction (b) of the perfect Gaussian intensity profile for different sizes of the divergence angle θ_0.

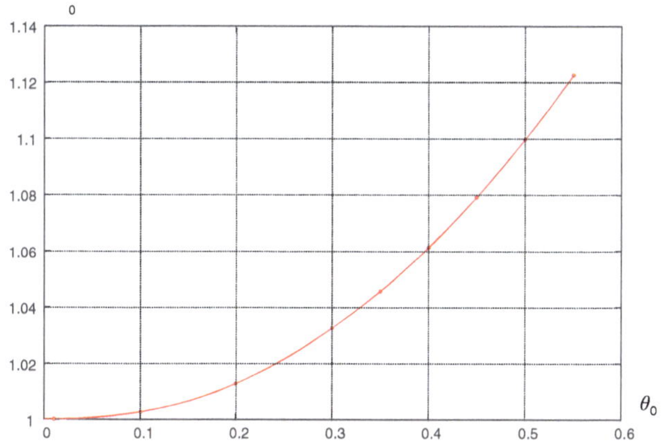

Figure 8.39 Increase of the beam radius as $1/e^2$ width of the corrected profile.

coordinate x as

$$E_m^{(0)}(x,z) = 2^{-m} \cdot \sigma^{\frac{m+1}{2}} \cdot e^{-\frac{\sigma x^2}{w_0^2}} \cdot H_m\left(-\frac{\sqrt{\sigma}\cdot x}{w_0}\right) \tag{8.153}$$

with the complex parameter

$$\sigma = \frac{1}{1 - i\cdot z/z_0} = \frac{iz_0}{q} \tag{8.154}$$

The non-paraxial beams can be written as a Taylor expansion with the divergence angle θ as variable

$$E_m^{(exact)}(x,z) = \sum_{j=0}^{\infty}\left(\frac{\theta}{2}\right)^{2j} \cdot E_m^{(2j)}(x,z) \tag{8.155}$$

The term for $j = 0$ is the paraxial expression

$$E_m^{(0)}(x,z) = \sqrt[4]{\frac{2}{\pi}} \cdot \sqrt{\frac{\sigma}{2^n n! w_0}} \cdot \left(\frac{\sigma}{\sigma^*}\right)^{\frac{m}{2}} \cdot e^{-\frac{\sigma x^2}{w_0^2}} \cdot H_m\left(-\frac{\sqrt{2\sigma\sigma^*}\cdot x}{w_0}\right) \tag{8.156}$$

The first correction term has the form

$$E_m^{(2j)}(x,z) = \sum_{k=0}^{j}\frac{(-1)^{j+k}\cdot(2j)!}{j(k-1)!(j-k)!(j+k)!}\cdot\left(\frac{iz}{z_0}\right)^k \cdot E_{m+2j+2k}^{(0)}(x,z) \tag{8.157}$$

This can be written as an expansion with the transverse derivatives of the basic mode field as correcting terms in the form

$$E_n^{(exact)}(x,z) = \sum_{j=0}^{\infty}\left(\frac{\theta_0}{2}\right)^{2j}\sum_{k=0}^{j}\frac{(-1)^{j+k}\cdot(2l)!}{j(k-1)!(j-k)!(j+k)!} \cdot\left(\frac{iz}{z_0}\right)^k\cdot\left(-\frac{w_0}{2}\right)^{2j+2k}\cdot\frac{\partial^{2j+2k}E_n^{(0)}(x,z)}{\partial x^{2j+2k}} \tag{8.158}$$

8.6.5 Off-axis-Focusing

If a Gaussian beam is focused with a large off-axis shift, large angles are also involved, but only in one Cartesian direction. If u is the numerical aperture angle, a is the offset and w is the beam radius, the diffraction integral can be written as a Debye integral

$$E(r,\theta) = -ikE_0 e^{-\frac{a^2}{w^2}} \cdot \int_0^u e^{iu\frac{\cos\varphi}{\sin^2 u} - \frac{\sin^2\varphi}{w^2}} \cdot J_0\left(kr\sin\varphi\sin\theta - 2ia\frac{\sin\varphi}{w^2}\right) \cdot \sqrt{\cos\varphi}\cdot\sin\varphi\, d\varphi \tag{8.159}$$

The geometry is sketched in Figure 8.40. By numerical evaluation of the field it can be seen that formally the Strehl ratio increases above one as a result of the offset and the large angles in the y cross section. This behaviour is illustrated in Figure 8.41. The unusual values of $D_s > 1$ are an artificial result of the reference here and must be interpreted with care.

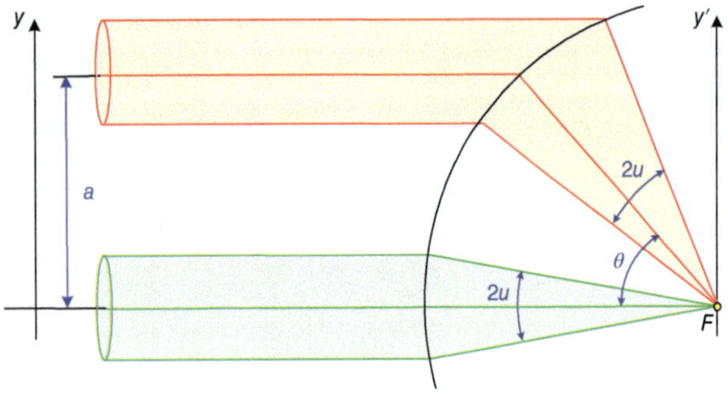

Figure 8.40 Off-axis geometry for focusing of a Gaussian beam.

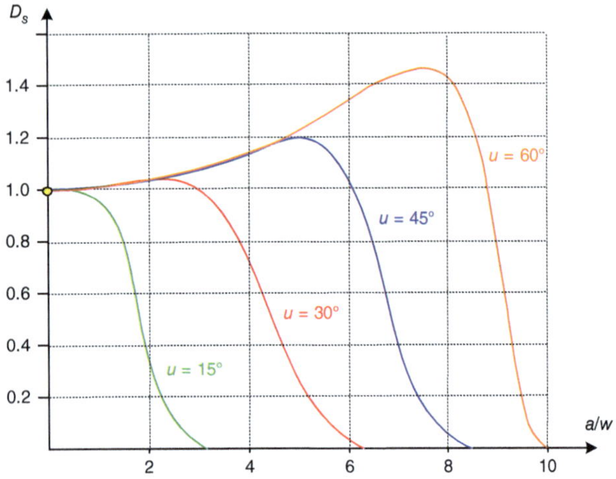

Figure 8.41 Strehl ratio of an off-axis focused Gaussian beam.

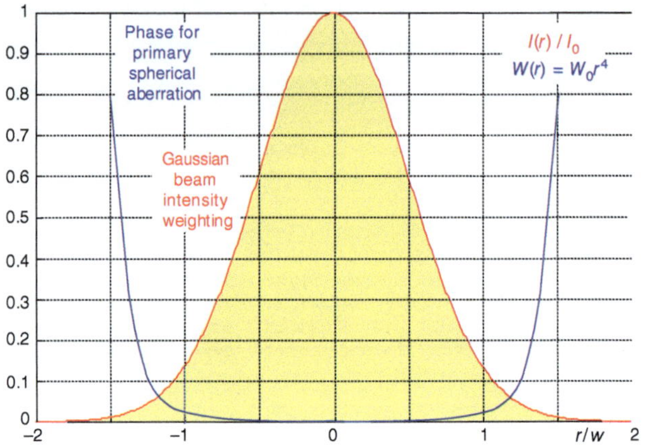

Figure 8.42 Gaussian weighting of primary spherical aberration.

8.7 Gaussian Beam with Spherical Aberration

8.7.1 Introduction

If a Gaussian beam suffers from spherical aberration due to focusing by a non-ideal lens, the paraxiality assumption is violated by the r^4-term of the phase and the beam no longer remains Gaussian [37–46]. From the theoretical point of view, the diffraction integral reads with the spherical aberration coefficient W_{40} as the optical path difference at $r = w$

$$A(r', z) = -\frac{2\pi i A_0}{\lambda z} \int_0^{\infty} e^{-\left(\frac{r}{w}\right)^2} \cdot e^{2\pi i \frac{W_{40}}{\lambda}\left(\frac{r}{w}\right)^4} J_0\left(\frac{2\pi r r'}{\lambda z}\right) r dr \tag{8.160}$$

This integral cannot be solved in analytical form; therefore, only numerical calculations, special cases or approximations in the form of a Bessel function for circular symmetry can be given.

One of the problems in combining Gaussian beams and spherical aberrations as Zernike definition c_9 is the selection of the normalization radius of the aberrations function. Usually $r = w$ is used, but this means that the field contribution outside sees large phase differences and the Zernike polynomials must be extrapolated, but it is only a little weighted by intensity. Therefore, this choice is in a certain sense arbitrary. From a practical point of view, a normalization radius $r = 1.5\,w$ is more appropriate.

It is seen from numerical calculations that a Gaussian apodization changes the point spread function in the image domain, the peak height, the optimal balancing with defocus by inspecting that the Strehl ratio is slightly different in comparison to the top hat illumination. One additional reason for small changes is the focal shift of Gaussian beam relative to the paraxial geometrical image plane.

If only primary spherical aberration is considered, the main phase aberrations occur at the border, where the weighting amplitude of the Gaussian profile is already low. Therefore, the impact of primary aberration on the focus intensity is much smaller in the case of a Gaussian beam in comparison to the Airy profile. This is sketched in Figure 8.42.

8.7.2 Intensity on Axis

With the usual normalized defocusing parameter u and the abbreviations

$$t_1 = \sqrt{\frac{2\pi W_{40}}{\lambda}} \cdot \left(1 - \frac{u\lambda}{4\pi W_{40}}\right),$$
$$t_2 = \sqrt{\frac{2\pi W_{40}}{\lambda}} \cdot \frac{u\lambda}{8\pi W_{40}} \tag{8.161}$$

the amplitude on axis can be expressed by Fresnel integrals as

$$E(u, 0) = \frac{\lambda}{4W_{40}} \left[(C(t_1) - C(t_2))^2 + (S(t_1) - S(t_2))^2\right] \tag{8.162}$$

Another approach for small spherical aberrations uses the auxiliary argument

$$t = \frac{2\pi W_{40}}{1 + \frac{\pi^2 w^4}{\lambda^2 R^2}\left(\frac{R}{z} - 1\right)} \tag{8.163}$$

and expresses the axial intensity at $r = 0$ as the series

$$I(z) = \frac{\frac{\pi^2 w^4}{\lambda^2 R^2}}{1 + \frac{\pi^2 w^4}{\lambda^2 R^2}} \left(\frac{R}{z}\right)^2 \frac{1}{1 + \frac{\pi^2 w^4}{\lambda^2 R^2}\left(\frac{R}{z} - 1\right)^2} \cdot \left[\begin{array}{l} 1 - 4t\sin\theta - 4t^2(6\cos 2\theta - 1) - 48t^3(\sin\theta - 5\sin 3\theta) \\ + 48t^4(3 - 10\cos 2\theta + 70\cos 4\theta + \ldots) \end{array} \right] \tag{8.164}$$

This expansion is valid as long as $W_{40} < \lambda/2\pi$.

8.7.3 Aberration Balancing

To get an idea about the change of the aberration impact in the case of spherical aberrations due to the Gaussian weighting function, let's assume a primary spherical aberration which is compensated by defocusing in the form

$$W(r) = a_4 \cdot r^4 + a_2 \cdot r^2 \tag{8.165}$$

The goal is to calculate the optimum value of a_2 as a function of a_4 for an optimal image plane selection.

If, first, a uniform illuminated pupil as in the classical case is considered, we have

$$\langle W \rangle = \frac{1}{\pi}\int_0^1 (a_4 r^4 + a_2 r^2)\, 2\pi r\, dr = \frac{1}{3}a_4 + \frac{1}{2}a_2$$

$$\langle W^2 \rangle = \frac{1}{\pi}\int_0^1 (a_4 r^4 + a_2 r^4)^2\, 2\pi r\, dr = \frac{1}{5}a_4^2 + \frac{1}{2}a_4 a_2 + \frac{1}{3}a_2^2$$

$$W_{rms} = \sqrt{\langle W^2 \rangle - \langle W \rangle^2} = \sqrt{\frac{4}{45}a_4^2 + \frac{1}{6}a_4 a_2 + \frac{1}{12}a_2^2} \tag{8.166}$$

The optimum defocusing can be found using the criterion

$$\frac{dW_{rms}}{d\,a_2} = \frac{a_4 + a_2}{12\sqrt{\frac{4}{45}a_4^2 + \frac{1}{6}a_4 a_2 + \frac{1}{12}a_2^2}} = 0 \tag{8.167}$$

and therefore we have the well-known result $a_2 = -a_4$ for a uniform illuminated pupil. The Strehl ratio in the homogeneous case can be written analytically in the form

$$D_s = \frac{e^{-\frac{\pi a_2^2}{a_4}}}{8a_4} \cdot \left| erf\left(\frac{\sqrt{\pi}}{i+1} \cdot \frac{a_2 + 2a_4}{\sqrt{a_4}}\right) - erf\left(\frac{\sqrt{\pi}}{i+1} \cdot \frac{a_2}{\sqrt{a_4}}\right) \right|^2 \tag{8.168}$$

If now a Gaussian beam illumination with amplitude

$$A(r) = e^{-(r/w)^2} \tag{8.169}$$

is considered and we use the abbreviation of the relative pupil filling $\varepsilon = a/w$ with pupil radius a, the corresponding calculation gives [42]

$$\langle W \rangle = \frac{1}{\pi}\int_0^1 (a_4 r^4 + a_2 r^2) \cdot e^{-\varepsilon^2 r^2}\, 2\pi r\, dr = \left(\frac{2}{\varepsilon^6}a_4 + \frac{1}{\varepsilon^4}a_2\right) \cdot (1 - e^{-\varepsilon^2})$$

$$\langle W^2 \rangle = \frac{1}{\pi}\int_0^1 \left(a_4^2 r^8 + a_2 r^8\right)^2 \cdot e^{-\varepsilon^2 r^2} 2\pi r\, dr = \left(\frac{24}{\varepsilon^{10}}a_4^2 + \frac{12}{\varepsilon^8}a_4 a_2 + \frac{2}{\varepsilon^6}a_2^2\right) \cdot (1 - e^{-\varepsilon^2})$$

$$W_{rms} = \sqrt{\langle W^2 \rangle - \langle W \rangle^2} = \sqrt{\begin{array}{l} \frac{\varepsilon^2}{30(1 - e^{-\varepsilon^2})} \cdot \left(10a_2^2 + 15a_2 a_4 + 6a_4^2\right) \\ -\frac{\varepsilon^4}{36(1 - e^{-\varepsilon^2})^2} \cdot (3a_2 + 2a_4)^2 \end{array}} \tag{8.170}$$

The minimum criterion gives here the defocusing as

$$a_2 = -\frac{\varepsilon^2 - 3 \cdot (1 - e^{-\varepsilon^2})}{3\varepsilon^2 - 4 \cdot (1 - e^{-\varepsilon^2})} \cdot a_4 \tag{8.171}$$

and the Strehl ratio reads

$$D_s = \frac{1}{8a_4(1 - e^{-\varepsilon^2})^2} \cdot e^{\frac{\varepsilon^4 - 4\pi^2 a_2^2 + 4\pi\varepsilon^2 a_2}{4\pi a_4}} \times \left| erf\left(\frac{i\varepsilon^2 + 2\pi\varepsilon^2 a_2 + 4\pi a_4}{2\sqrt{2\pi i a_4}}\right) - erf\left(\frac{i\varepsilon^2 + 2\pi\varepsilon^2 a_2}{2\sqrt{2\pi i a_4}}\right) \right|^2 \tag{8.172}$$

It is seen that the weighting impact of the Gaussian apodization affects the optimal image plane considerably in the case of spherical aberration.

8.7.4 Numerical Example Calculations

The spherical aberration changes the intensity distribution in the lateral and axial directions. In particular, the axial profile is no longer symmetric around the waist position. Figure 8.43 shows the axial intensity function $I(z)$ for different values W_{40} for the spherical aberration of a Gaussian beam. The transverse reference is here the radius $1.5w$, W_{40} is the optical path difference in waves at the height $r = 1.5w$. The peak is shifted along z and reduced in height. In the case shown here, the Fresnel number is $N_F = 5$; therefore, a difference between positives and values of W_{40} is also observed. This pure diffraction caused asymmetry vanishes for larger Fresnel numbers.

Figure 8.44 shows several cases of the amplitude distribution $A(r,z)$ for Gaussian beams suffering from spherical aberration with Zernike coefficient c_9. Here the Fresnel

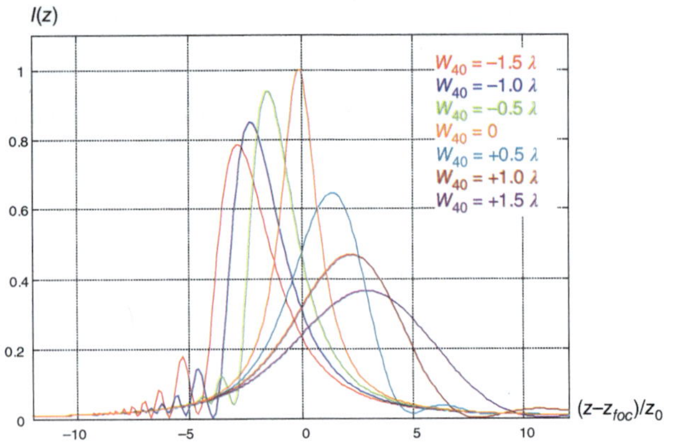

Figure 8.43 Axial intensity of Gaussian beams with spherical aberration W_{40} at 1.5w.

number is $N_F = 320$, which means that negative Zernike coefficients deliver the same profiles inverted around the waist or focal plane. The general behaviour looks very similar to the case of classical ray caustics, but the relative weighting of the aberrations in the outer region is reduced. On the other hand, here the Gaussian beams are not truncated and the Zernike normalization radius is defined at the value $r = 1.5w$. For a better visibility, the amplitude is selected here instead of the intensity. If the intensity is selected, the focal volume looks much more compact and the perturbations around the waist region are hard to observe.

8.8 Single Mode Fibre Coupling with Gaussian Beam*

(*) The content of this section is quite specialized and for advanced reading. Knowledge of this section is not necessary to understand the following chapters. Beginners are recommended to skip this section.

8.8.1 Basic Fibre Parameters

The launching of a fundamental mode Gaussian laser beam into a monomode step index fibre is a quite important task in several applications of lens design. Therefore, it should be discussed here in more detail.

The limiting angle of total internal reflection of a step index fibre is given by

$$\sin\theta = \sqrt{n_1^2 - n_2^2} = n_1 \cdot \sqrt{2\Delta} \tag{8.173}$$

with the indices n_1, n_2 of the fibre cladding and core. If a is the radius of the core, the structural constant is defined as

$$V = \frac{2\pi a}{\lambda} \cdot \sin\theta \tag{8.174}$$

The larger the value of V is, the more modes can propagate inside the fibre without significant loss. As a formal limit of the monomode case, a value of $V = 2.405$ is defined. In any case, the mode field is also extended into the cladding. For smaller values of V, this relative amount grows. The radial field amplitude profile for various values of V is seen in Figure 8.45.

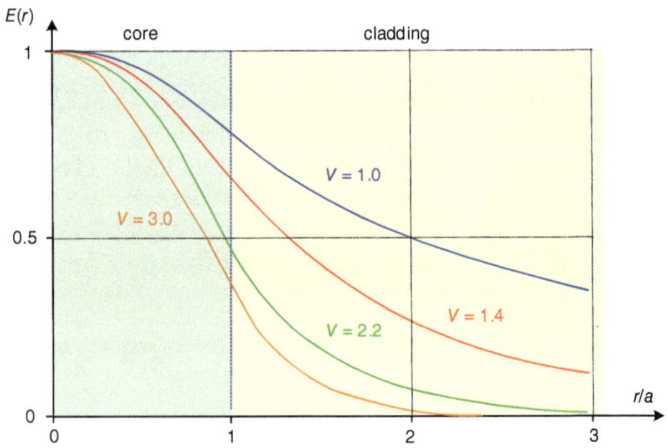

Figure 8.45 Radial mode field amplitude profile for different values of V for a constant peak value.

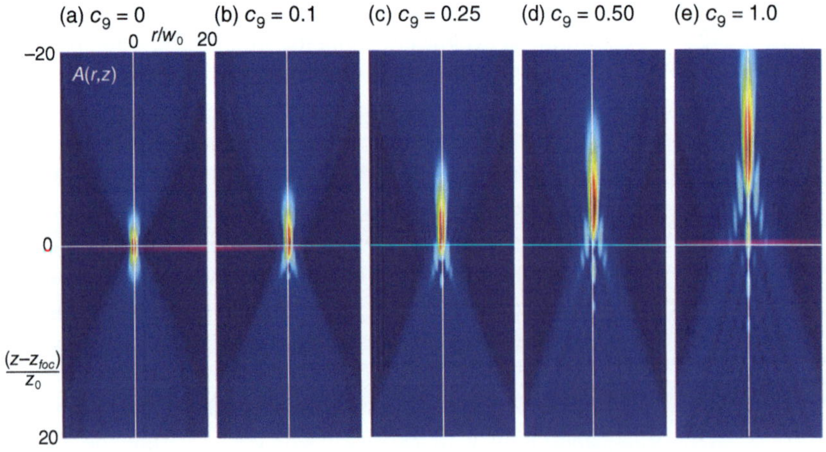

Figure 8.44 Amplitude distribution $A(r,z)$ of Gaussian beams with spherical aberration c_{40} at 1.5w.

The correct field is given by the following equation (in the approximation of: weakly guiding LP-modes [44–46])

$$E(r) = \begin{cases} E_0 \cdot \dfrac{J_0\left(\dfrac{ur}{a}\right)}{J_0(u)} & \text{für } r < a \\ E_0 \cdot \dfrac{K_0\left(\dfrac{wr}{a}\right)}{K_0(w)} & \text{für } r > a \end{cases} \tag{8.175}$$

8.8.2 Fundamental Mode Shape

The fundamental mode can be approximated by a Gaussian profile with the amplitude

$$E_L(r) = E_0 \cdot e^{-\frac{r^2}{w_0^2}} \tag{8.176}$$

To evaluate the residual error, an overlap integral in the form [45]

$$\eta = \left| \frac{\int E_F(r) \cdot E_L^*(r)\, 2\pi \cdot r\, dr}{\sqrt{\int |E_L(r)|^2\, 2\pi \cdot r\, dr} \cdot \sqrt{\int |E_F(r)|^2\, 2\pi \cdot r\, dr}} \right|^2 \tag{8.177}$$

can be used, it is in principle a kind of correlation between the basic fibre mode (EF) and a Gaussian illumination (EG). This is in addition the best coupling efficiency of a TEM00 laser beam into a monomode fibre. The best fit is found with the values $V = 2.405$ for $w0/a = 1.090$ with $\eta = 0.9967$ and 0.33% energy mismatch; see Figure 8.46.

The following Figure 8.47 shows the optimum coupling as a function of the structural parameter V. If the parameter V is selected to be larger at $V = 2.82$, the efficiency can be slightly increased, but then the monomode purity of the fibre is no longer guaranteed.

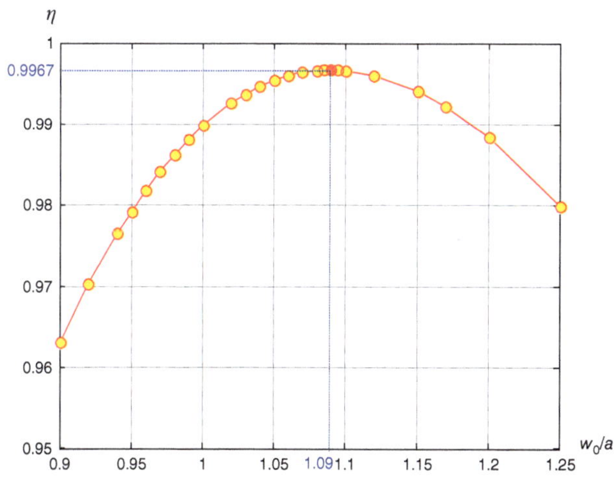

Figure 8.46 Coupling of the fundamental mode with a Gaussian profile as a function of the beam radius for $V = 2.405$.

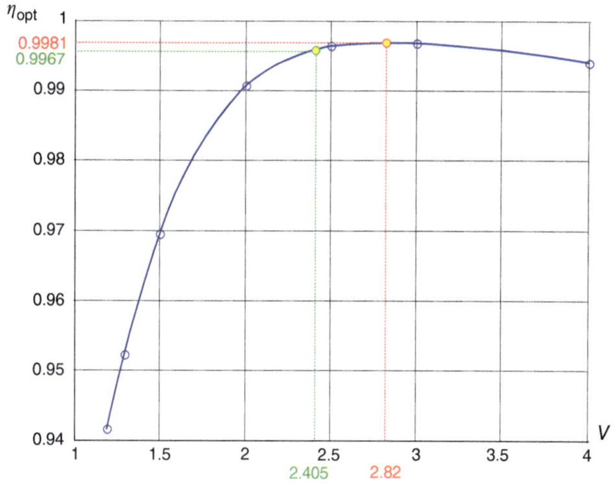

Figure 8.47 Optimal coupling of the fundamental mode with a Gaussian profile as a function of V.

A good approximation (better than 1% in the range $V = 2\ldots2.4$) for the V-dependence is given by the following two alternative empirical approximation formulas

$$\frac{w_0}{a} = 0.65 + 1.619 \cdot V^{-1.5} + 2.879 \cdot V^{-6}$$
$$\frac{w_0}{a} = 0.77 \cdot e^{\frac{2}{V^2}} \tag{8.178}$$

8.8.3 Calculation of Fibre Radiation Mode

If a given fibre radiates in a fundamental mode, the equivalent data of the corresponding Gaussian can therefore be obtained by the following steps. Given here are the fibre data a and $\sin\theta$ as well as the wavelength λ:

- the structural parameter is determined as

$$V = \frac{2\pi a}{\lambda} \cdot \sin\theta \tag{8.179}$$

- the Gaussian beam waist w_o at the fibre exit surface is approximately by Eq. (8.178)

$$w_0 = a \cdot (0.65 + 1.619 \cdot V^{-1.5} + 2.879 \cdot V^{-6}) \tag{8.180}$$

- the far field divergence angle according to the Gaussian beam formulas is given by

$$\theta = \frac{\lambda}{\pi w_0} \tag{8.181}$$

If there is a discrepancy with the measured data, a multimode case or incorrect fibre specification must be assumed.

The most meaningful general definition of a mode diameter uses the second moment definition

$$D_{2nd} = 2w_o = 2 \cdot \sqrt{\frac{2\int r^2 I(r)\, 2\pi r dr}{\int I(r)\, 2\pi r dr}} \tag{8.182}$$

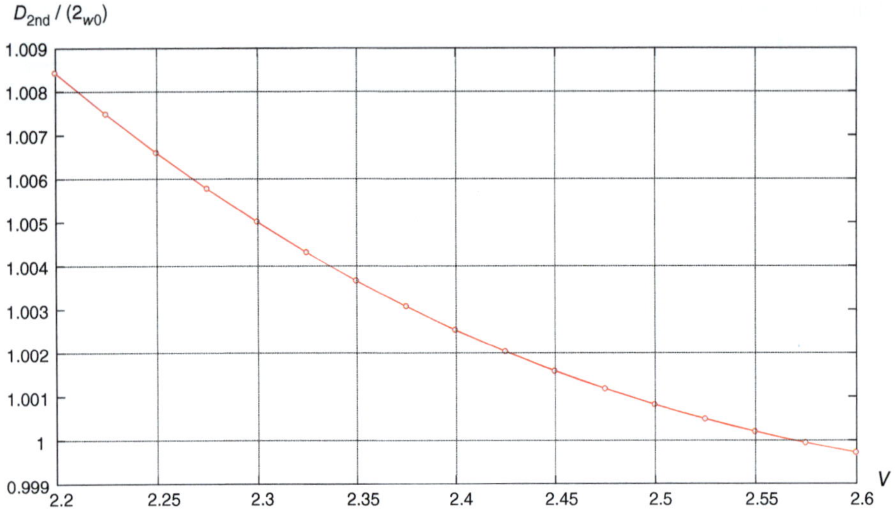

Figure 8.48 Ratio of the diameter with second moment criterion to the Gaussian value profile as a function of V.

In particular, if the mode function is approximated by a Gaussian beam, the result of this general approach is analytically exact $D_{2nd} = 2w_0$.

Due to the good agreement of the fundamental mode with a Gaussian profile (see above) and the various approximations of the exact mode function (V = 2.405, weakly guiding, scala …) in reality the mode field diameter is mostly defined by the equivalent Gauss. If the second moment diameter of the exact LP mode is calculated and normalized on the $2w_0$-value of the equivalent Gaussian, it is seen that the exact diameter is slightly larger for small V values; see the next diagram in Figure 8.48. This behaviour makes sense; for small V a larger contribution of the mode field is in the cladding and extends here with a smaller drop down – see the figure. But it is observed that for the V range shown here, the error is less than 1%.

8.9 Partial Coherent Gauss–Schell Beams*

(*) The content of this section is quite specialized and for advanced reading. Knowledge of this section is not necessary to understand the following chapters. Beginners are recommended to skip this section.

8.9.1 Introduction

There exists a simple type of partially coherent beam that can be analytically described in close analogy to the coherent Gaussian beams. These so-called Gaussian–Schell beams (GSBs) rest upon simple statistical assumptions concerning the coherent basis functions of the coherence function Γ and a Gaussian shape of the cross spectral density function. Their transport properties are fully equivalent with those of the Gaussian beams and their propagation can be described analytically by *ABCD* too. As a consequence of the finite coherence, one more parameter for the description is necessary: the lateral width of the coherence region L_c. In practice, these ideal solutions are useful as a first approximation for the description of partially coherent radiation and will be considered in more detail in the following [7–9, 47–56].

8.9.2 Simple Gauss–Schell Beams

GSBs are partially coherent Gaussian beams with a Gaussian shaped coherence function. The coherence function as the correlation of the complex field E here at two different locations r_1 and r_2 is given by the expression

$$\Gamma(r_1, r_2) = e^{-\frac{r_1^2 + r_2^2}{w_0^2}} \cdot e^{-\frac{(r_1 - r_2)^2}{2L_c^2}} \cdot e^{-\frac{i\pi}{\lambda R}\cdot(r_1 - r_2)^2} \tag{8.183}$$

or as the modified description with the difference $\Delta r = r^1 - r^2$

$$\begin{aligned}\Gamma(r, \Delta r) &= e^{-2\frac{r^2}{w_0^2}} \cdot e^{-\frac{1}{2}\left(\frac{1}{w_0^2} + \frac{1}{L_c^2}\right)\cdot\Delta r^2} \cdot e^{-\frac{i\pi}{\lambda R}\cdot\Delta r^2} \\ &= e^{-2\frac{r^2}{w_0^2}} \cdot e^{-\frac{\Delta r^2}{2w_c^2}} \cdot e^{-\frac{i\pi}{\lambda R}\cdot\Delta r^2}\end{aligned} \tag{8.184}$$

In the literature, sometimes the slightly different scaled versions with the parameters σ_c and σ_I are found

$$\begin{gathered}\Gamma(r_1, r_2) = e^{-\frac{r_1^2 + r_2^2}{4\sigma_I^2}} \cdot e^{-\frac{(r_1 - r_2)^2}{2\sigma_c^2}} e^{-\frac{i\pi}{\lambda R}\cdot(r_1 - r_2)^2} \\ \sigma_c^2 = L_c^2, \quad 4\sigma_I^2 = w_0^2, \quad 2\sigma_c\sigma_I = w_0 L_c\end{gathered} \tag{8.185}$$

The transverse coherence length L_C in the above formula is an additional parameter responsible for the description of the coherence. In the limiting case $L_C \to \infty$, one obtains the conventional coherent Gaussian beams. With decreasing L_C:

- the emitting area radiates more incoherent radiation;
- the beam divergence increases for a constant emitter area;
- the beam depth of focus decreases;
- the value of the beam parameter product becomes larger and the beam quality degrades.

These dependences are shown in the figures below. Besides L_C the following equivalent quantities are commonly used:

- normalized transverse coherence length

$$\alpha = L_c/w_0 \tag{8.186}$$

- auxiliary parameter ε according to

$$\varepsilon = 1 + \frac{1}{\alpha^2} = 1 + \left(\frac{w_0}{L_c}\right)^2 > 1 \tag{8.187}$$

- degree of coherence β

$$\beta = \frac{1}{\sqrt{1+(w_0/L_c)^2}} = \frac{1}{\sqrt{1+\alpha^{-2}}} = \frac{1}{\sqrt{\varepsilon}} < 1 \tag{8.188}$$

In contrast to the classical Gaussian beams, three parameters are necessary for the description of the GSBs. The waist radius w_o and the far field divergence θ_o are now independent parameters for a given wavelength λ. In particular, the depth of focus for the GSBs is shorter in comparison to the coherent Gaussian beams by a factor β

$$z_0' = \frac{z_0}{\sqrt{\varepsilon}} = \beta \cdot z_0 = \frac{z_0}{\sqrt{1+\alpha^{-2}}} < z_0 \tag{8.189}$$

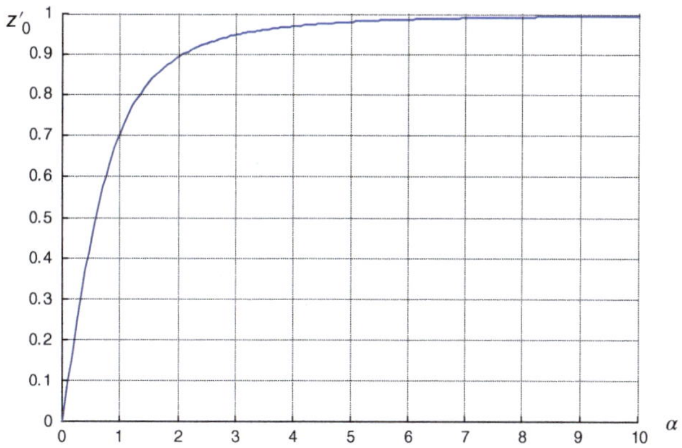

Figure 8.49 Depth of focus as a function of the normalized transverse coherence length α for Gaussian Schell beams.

The parameter z_0' decreases for reduced coherence, which is shown in Figure 8.49.

At a constant far-field divergence θ_0 the waist radius of GSBs increases when L_c decreases; i.e., when the degree of coherence β becomes smaller. Figure 8.50 demonstrates this behaviour for three different values of the parameter β.

If, on the other hand, the waist radius w_o is kept constant, the far field divergence angle grows with decreasing coherence. Figure 8.50 (b) shows some examples of this case. Practically, the meaning of this relationship is that in order to achieve the same size of the focal diameter, the light from a more incoherent source should be focused with a larger numerical aperture.

This dependence of the far-field divergence angle from the coherence can be quantitatively described by the expression

$$\theta_0' = \theta_0 \cdot \frac{1}{\beta} = \theta_0 \cdot \sqrt{1+\left(\frac{w_0}{L_c}\right)^2} \tag{8.190}$$

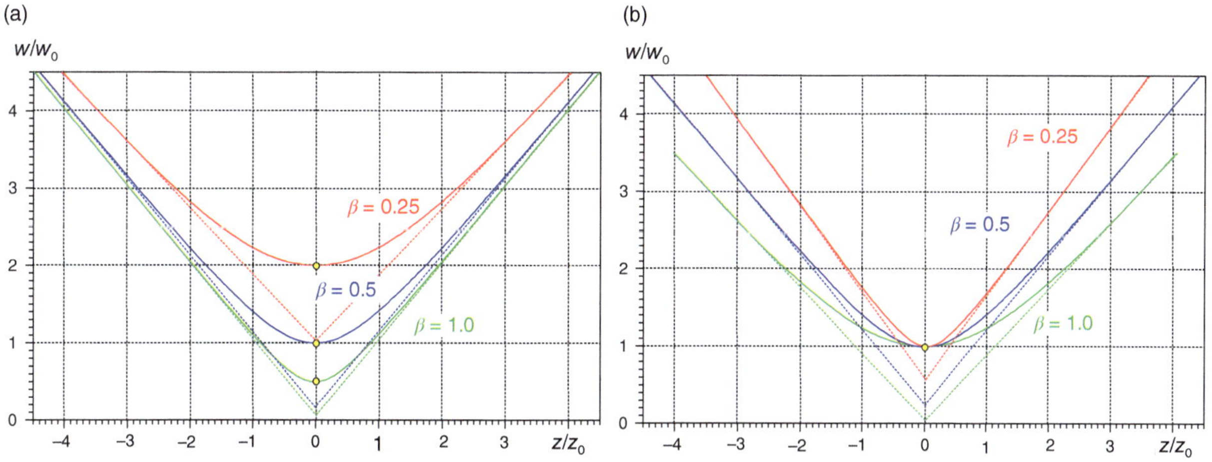

Figure 8.50 Gaussian–Schell beams with (a) beam radius for constant divergence angles and different degrees of coherence in the waist region and (b) for constant waist radius and different degrees of coherence in the waist region.

The coherence length L_c in the waist can be calculated for given λ, w_0 and θ'_0 from the following formula

$$L_c = \frac{w_0}{\sqrt{\left(\frac{\pi w_0 \theta'_0}{\lambda}\right)^2 - 1}} \tag{8.191}$$

8.9.3 Propagation of Gauss-Schell Beams

8.9.3.1 Initial Values

Similar to classical coherent Gaussian beams, Gauss–Schell beams are considered as quasi-paraxial and can be propagated in free space as well as in paraxial *ABCD*-systems in a simple analytical way. The *ABCD* matrices are identical to the Gaussian beam calculus, only one more coherence parameter L_c is needed (in the circular symmetric case) to characterize the beam light.

There are several different possibilities for given beam parameters of Gauss–Schell beams. Therefore, some formulas are needed to extract the primary parameters out of the given data. The following combinations are possible.

- Given λ, w_o and θ'_0:

$$L_c = \frac{w_0}{\sqrt{\left(\frac{\pi w_0 \theta'_0}{\lambda}\right)^2 - 1}} \tag{8.192}$$

- Given $w(z)$, $R(z)$ and $L_c(z)$:

$$\varepsilon = 1 + \left(\frac{w(z)}{L_c(z)}\right)^2, \quad w_0^{GSB} = \frac{w(z)}{\sqrt{1 + \frac{1}{\varepsilon}\left(\frac{\pi w^2(z)}{\lambda R(z)}\right)^2}} \tag{8.193}$$

- Given λ, z'_0 and L_c:

$$w_0 = \frac{\lambda z'_0}{\sqrt{1 - \left(\frac{\lambda z'_0}{L_c}\right)^2}}, \quad \varepsilon = \left(\frac{\pi w_0^2}{\lambda z'_0}\right)^2,$$

$$L_c = \frac{w_0}{\sqrt{\varepsilon - 1}}, \quad \theta'_0 = \frac{w_0}{z'_0} \tag{8.194}$$

- Given λ, w_0 and θ'_0:

$$\varepsilon = \left(\frac{\pi w_0 \theta'_0}{\lambda}\right)^2, \quad L_c = \frac{w_0}{\sqrt{\varepsilon - 1}}, \quad z'_0 = \frac{\pi w_0^2}{\lambda\sqrt{\varepsilon}} \tag{8.195}$$

- Given λ, w_0 and L_c:

$$\varepsilon = 1 + \left(\frac{w_0}{L_c}\right)^2, \quad \theta'_0 = \frac{w_0}{z'_0}, \quad z'_0 = \frac{\pi w_0^2}{\lambda\sqrt{\varepsilon}} \tag{8.196}$$

- Given λ, θ'_0 and L_c:

$$w_0 = \frac{\lambda}{\sqrt{\pi^2 \theta'^2_0 - (\lambda/L_c)^2}},$$

$$\varepsilon = 1 + \left(\frac{w_0}{L_c}\right)^2, \quad z'_0 = \frac{w_0}{\theta'_0} \tag{8.197}$$

8.9.3.2 Free-Space Propagation

GSBs exhibit a value of the degree of coherence β which is between the geometrical-optical incoherent case with $\beta = 0$ and the ideal coherent Gaussian beams with $\beta = 1$. This can also be seen in the image formation behaviour, which is shown in Figure 8.51 for a single lens with focal length f. The image formation or transformation equation for GSBs reads identically to the Gaussian beams as

$$\frac{1}{z'_T} - \frac{1}{z_T} \cdot \frac{1}{\left[1 + \frac{z'^2_0}{z_T \cdot (f + z_T)}\right]} = \frac{1}{f} \tag{8.198}$$

and deviates, similarly to the Gaussian beams, from the geometrical-optical models. In an analogous way, the position of the beam waist at the image-side can be calculated analytically. The beam radius changes similarly to a coherent Gaussian beam with a hyperbolic law and the radius of curvature also behaves the same as for classical Gaussian beams

$$w(z) = w_0 \sqrt{1 + \left(\frac{z}{z'_0}\right)^2} = w_0 \sqrt{1 + \varepsilon\left(\frac{z}{z_0}\right)^2}$$

$$R(z) = -\left(z + \frac{z'^2_0}{z}\right) = -\left(z + \frac{z_0^2}{\varepsilon \cdot z}\right) \tag{8.199}$$

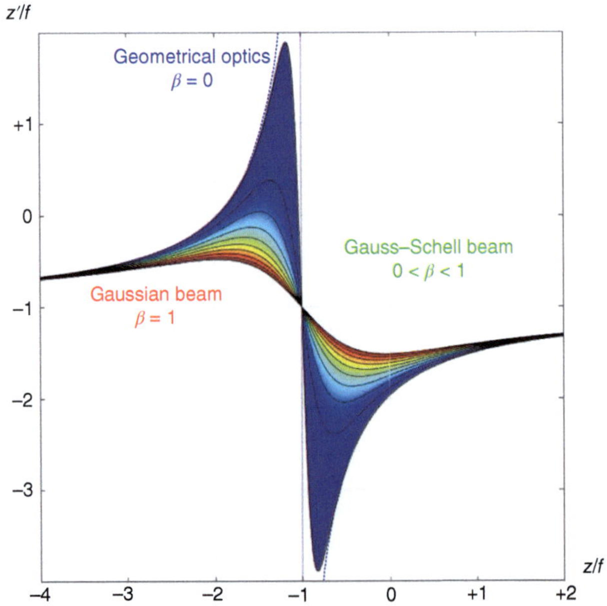

Figure 8.51 Illustrating the image formation properties of Gaussian–Schell beams with object-side and image-side waist positions z_T and z'_T depending on the coherence parameter β.

It should be mentioned that the normalized degree of coherence in the transport of GSBs does not change with the coordinate z

$$\beta(z) = \text{const} \tag{8.200}$$

but $L_c(z)$ scales with w. It should also be noted that Gaussian–Schell model beams are by definition paraxial and the relation (8.198) is valid only in this approximation.

Over time the simple partially coherent GSBs described here were generalized using various beam models. These include twisted beams, flatter or more complex intensity profiles and vector properties.

8.9.3.3 ABCD-Propagation

The easiest way to propagate a GSB is to use the above defined complex q-parameter. It propagates in a paraxial $ABCD$ segment in one dimension as

$$q_2 = \frac{Aq_1 + B}{Cq_1 + D} \tag{8.201}$$

or in the general three-dimensional case with different values in x, y as a matrix by

$$\underline{Q}'^{-1} = \frac{\underline{C} + \underline{D}\,\underline{Q}^{-1}}{\underline{A} + \underline{B}\,\underline{Q}^{-1}} \tag{8.202}$$

The terms $\underline{A}, \underline{B}, \underline{C}, \underline{D}$ are 2×2 submatrices of the general 4×4 matrix of paraxial systems. The mixed terms describe x–y coupling effects in the system.

The change of the explicit beam parameters can be written as

$$w_2^2 = w_1^2 \left[\left(A + \frac{B}{R_1} \right)^2 + \left(\frac{\lambda}{\pi w_1 w_{c1}} \right)^2 B^2 \right]$$

$$L_{c2}^2 = L_{c1}^2 \left[\left(A + \frac{B}{R_1} \right)^2 + \left(\frac{\lambda}{\pi w_1 w_{c1}} \right)^2 B^2 \right]$$

$$R_2 = \frac{\left[\left(A + \frac{B}{R_1} \right)^2 + \left(\frac{\lambda}{\pi w_1 w_{c1}} \right)^2 B^2 \right]}{\left(A + \frac{B}{R_1} \right)\left(C + \frac{D}{R_1} \right) + \left(\frac{\lambda}{\pi w_1 w_{c1}} \right)^2 BD} \tag{8.203}$$

with w_c defined in Eq. (8.192).

$$\frac{1}{w_c^2} = \frac{1}{w^2} + \frac{1}{L_c^2} \tag{8.204}$$

References

1 Siegman, A. (1986). *Lasers*. University Science Books.
2 Goldsmith, P. (1998). *Quasioptical Systems*. IEEE Press.
3 Wang, S. and Zhao, D. (2000). *Matrix Optics*. Springer.
4 Gerrard, A. and Burch, J.M. (1994). *Matrix Methods in Optics*. New York: Dover.
5 Brouwer, W. (1964). *Matrix Methods in Optical Instrument Design*. New York: Benjamin.
6 Kloos, G. (2007). *Matrix Methods for Optical Layout*. SPIE Press.
7 Simon, R. (1985). A new class of anisotropic Gaussian beams. *Optics Communication* 55: 381.
8 Turunen, J. and Friberg, A. (1986). Matrix representation of Gaussian Schell-model beams in optical systems. *Optics Technology* 18: 259.
9 Simon, R. (1985). Generalized rays in statistical wave optics: application to anisotropic Gaussian Schell-model beams. *Journal of Optics* 14: 92.
10 Kochkina, E., Wanner, G., Schmelzer, D. et al. (2013). Modeling of the general astigmatic Gaussian beam and its propagation through 3D optical systems. *Applied Optics* 52: 6030.
11 Rohani, A., Shishegar, A., and Safavi-Naeini, S. (2004). A fast Gaussian beam tracing method for reflection and refraction of general astigmatic Gaussian beams form general curved surfaces. *Optics Communication* 232: 1.
12 Colbourne, P. (2014). Generally astigmatic Gaussian beam representation and optimization using skew rays. *SPIE Conference Proceedings*, 9243. https://doi.org/10.1117/12.2071105.
13 Herloski, R., Marshall, S., and Antos, R. (1983). Gaussian beam ray-equivalent modeling and optical design. *Applied Optics* 22: 1168.
14 Harrigan, M. (2002). General beam propagation through non-orthogonal optical systems. *Proceedings of SPIE* 4832: 379.
15 Greynolds, A. (1985). Propagation of generally astigmatic Gaussian beams along skew ray paths. *Proceedings of SPIE* 560: 33.
16 Greynolds, A. (1986). Vector formulation of the ray-equivalent method for general Gaussian beam propagation. *Proceedings of SPIE* 679: 129.
17 Mansell, J., Suizu, R., Praus, R. et al. (2007). Integrating wave optics and 5x5 ray matrices for more accurate optical system modeling. *SPIE Conference Proceedings*, 6675. https://doi.org/10.1117/12.734849.
18 Tang, Z. (2022). *Improved Correction Methods for Symmetry-Free Systems*. Master's thesis, University of Jena.
19 Horvath, Z. and Bor, Z. (2002). Focussing of truncated Gaussian beams. *Optics Communication* 223: 51.

20 Landesman, B. and Barrett, H. (1988). Gaussian amplitude functions that are exact solutions to the scalar Helmholtz equation. *Journal of the Optical Society of America A* 5: 1610.

21 Landesman, B. (1989). Geometrical representation of the fundamental mode of a Gaussian beam in oblate spheroidal coordinates. *Journal of the Optical Society of America A* 6: 5.

22 Kim, H.-C. and Lee, Y. (1999). Hermite-Gaussian and Laguerre-Gaussian beams beyond the paraxial approximation. *Optics Communication* 169: 9.

23 Lax, M. (1975). From Maxwell to paraxial wave optics. *Physical Review A* 11: 1365.

24 Agrawal, G. and Lax, M. (1983). Free-space wave propagation beyond the paraxial approximation. *Physical Review A* 27: 1693.

25 Kiselev, A. (2017). Time-harmonic Gaussian beams: exact solutions of the Helmholtz equation in free space. *Optics and Spectroscopy* 123: 935.

26 Salamin, Y. (2007). Fields of a Gaussian beam beyond the paraxial approximation. *Applied Physics B: Lasers and Optics* 86: 319.

27 Kiselev, A., Plachenov, A., and Chamorro-Posada, P. (2012). Nonparaxial wave beams and packets with general astigmatism. *Physical Review A* 85: 043835.

28 Agrawal, G. and Pattanayak, D. (1979). Gaussian beam propagation beyond the paraxial approximation. *Journal of the Optical Society of America* 69: 575.

29 Chen, C., Konkola, P., Ferrera, J. et al. (2002). Analysis of vector Gaussian beam propagation and the validity of paraxial and spherical approximations. *Journal of the Optical Society of America A* 19: 404.

30 Couture, M. and Belanger, P. (1981). From Gaussian beam to complex source point spherical wave. *Physical Review A* 24: 355.

31 Duan, K. and Lue, B. (2004). Propagation properties of vectorial elliptical Gaussian beams beyond the paraxial approximation. *Optics and Laser Technology* 36: 489.

32 Lekner, J. (2001). TM,TE and 'TEM' beam modes: exact solutions and their problems. *Journal of Optics A* 3: 407.

33 Seshadri, S. (2008). Fundamental electromagnetic Gaussian beam beyond the paraxial approximation. *Journal of the Optical Society of America A* 25: 2156.

34 Sheppard, C. and Saghafi, S. (1999). Electromagnetic Gaussian beams beyond the paraxial approximation. *Journal of the Optical Society of America A* 16: 1381.

35 Sheppard, C. and Saghafi, S. (1998). Beam modes beyond the paraxial approximation: a scalar treatment. *Physical Review A* 57: 2971.

36 Takenaka, T., Yokota, M., and Fukumitsu, O. (1985). Propagation of light beams beyond the paraxial approximation. *Journal of the Optical Society of America A* 2: 826.

37 Alkelly, A. (2007). Spot size and radial intensity distribution of focused Gaussian beams in spherical and non-spherical aberration lenses. *Optics Communication* 277: 397.

38 Pu, J. and Zhang, H. (1998). Intensity distribution of Gaussian beams focused by a lens with spherical aberration. *Optics Communication* 151: 331.

39 Yoshida, A. and Asakura, T. (1976). Diffraction patterns of off-axis Gaussian beams in the presence of third-order spherical aberration in the optical system. *Optics Communication* 19: 387.

40 Stamnes, J. (1986). *Waves in Focal Regions.* Adam Hilger.

41 Pu, J. and Nemoto, S. (1999). Encircled energy of Gaussian beams focused by a lens with spherical aberration. *Optik* 110: 217.

42 Mahajan, V. (1986). Uniform versus Gaussian beams: a comparison of the effects of diffraction, obscuration and aberrations. *Journal of the Optical Society of America A* 3: 470.

43 Biswas, S. and Villeneuve, J.-E. (1986). Diffraction of a laser beam by a circular aperture under the combined effect of three primary aberrations. *Applied Optics* 25: 2221.

44 Neumann, E. (1988). *Single-Mode Fibers.* Springer.

45 Ghatak, A. and Thyagarajan, K. (1998). *Fiber Optics.* Cambridge.

46 Snyder, A. and Love, J. (1983). *Optical Waveguides.* Chapman and Hall.

47 Bastiaans, M. (1992). ABCD law for partially coherent Gaussian light, propagating through first-order optical systems. *Optical and Quantum Electronics* 24: S1011.

48 Serna, J., Mejias, P., and Martinez-Herrero, R. (1992). Beam quality dependence on the coherence length of Gaussian Schell-model fields propagating through ABCD optical systems. *Journal of Modern Optics* 39: 625.

49 Friberg, A. and Turunen, J. (1986). Algebraic and graphical propagation methods for Gaussian Schell-model beams. *Optical Engineering* 25: 857.

50 Qiang, L. and Shaomin, W. (1990). Transformation of nonsymmetric Gaussian beam into symmetric one by means of tensor ABCD law. *Optik* 85: 67.

51 Lü, B., Feng, G., and Cai, B. (1993). Complex ray representation of the astigmatic Gaussian beam propagation. *Optical and Quantum Electronics* 25: 275.

52 Friberg, A., Tervonen, E., and Turunen, J. (1994). Interpretation and experimental demonstration of twisted Gaussian Schell-model beams. *Journal of the Optical Society of America A* 11: 1818.

53 Simon, R. and Mukunda, N. (1993). Twisted Gaussian-Schell-model beams. *Journal of the Optical Society of America A* 10: 95.

54 Simon, R. and Mukunda, N. (1998). Twist phase in Gaussian-beam optics. *Journal of the Optical Society of America A* 15: 2373.

55 Wang, L. and Zhao, D. (2019). Controllable rotating Gaussian Schell-model beams. *Optics Letters* 44: 735.

56 Lin, Q. and Cai, Y. (2002). Tensor ABCD law for partially coherent twisted anisotropic gauss-Schell model beams. *Optics Letters* 27: 216.

9

Photometry and Radiometry

9.1 Introduction

9.1.1 General Remarks

Photometry, or illumination technology, describes different aspects of the energy flux of optical radiation transport. Basically, there are two systems of units and terms. In objective physical-optical photometry the purely energetic aspects are regarded through MKSA units. They are also called the radiometric quantities. This description does not depend on the kind of detector and is independent on the spectral ranges. In addition to these quantities there is also the photometric unit system that is based on physiological detection by the human eye. The conversion of these quantities can be achieved by means of the sensitivity function $V(\lambda)$ of the photopic eye. Therefore, this approach is in a certain sense subjective.

In photometry the geometrical conditions of area size, direction of the radiation transfer and the solid angle are the most important quantities for considering and quantitatively calculating energy transport. In practice, these conditions are superimposed by the spectral properties of the transitive or reflective materials of the radiation transport system as well as the source radiation and the detector sensitivity. The interested reader is referred to the literature for more details [1–11].

In classical photometry, incoherent radiation is assumed. Therefore, as there are no coherence or interference effects to consider, the radiation density can be calculated purely geometrically. If, however, illuminations by laser have to be taken into account, this approximation may no longer be sufficient and the radiation transfer by means of simple raytracing needs to be replaced by an approach using the Kirchhoff integral [12]. A short description of radiometry in the case of partial coherent light is given in the last section of this chapter.

9.1.2 Definition of the Radiometric Quantities

The most important basic radiometric quantities are defined in the following manner.

- Radiant power:
  the radiant power P is measured in Watts and is the integral power of a source or a light field. It is also called the radiant flux or the light flux. With the help of the Poynting vector S for energy density, derived from the Maxwell equations, the radiant power can be obtained by integrating the density over an area taking into account the vector character of the directed radiation transfer

$$P = \int \vec{S} \cdot d\vec{A} \tag{9.1}$$

- Radiance:
  if the radiant power is not uniformly distributed on a surface area A or a filled solid angle Ω, a density must be defined whose integral again yields the power. The radiance is measured in $W/m^2/sr$ and is defined as

$$L = \frac{d^2P}{\cos\theta \, d\Omega \, dA} \tag{9.2}$$

  where θ is the angle between the surface normal and the direction of radiation.
- Radiant intensity:
  emitting surfaces radiate into a particular solid angle. The larger this angle, the lower the light intensity in the defined direction for a constant power. Therefore, the radiant intensity is defined as the power per solid angle or the area integrated radiance

$$I = \frac{dP}{d\Omega} = \int L \, dA_\perp = \int L \cdot \cos\theta \, dA = \vec{s} \int L \, d\vec{A} \tag{9.3}$$

  and is measured in W/sr. The unit vector $\vec{s}$ indicates the direction of the surface and the final point of calculation.

- Radiant excitance:
  if an emissive surface emits into a finite solid angle, a power per area is generated that is also called the emittance or radiant excitance. It is defined as
  $$E = \frac{dP}{dA} = \int L \cdot \cos\theta \, d\Omega \tag{9.4}$$
  and is measured in W/m^2.
- Illuminance and irradiance:
  if a surface is illuminated passively from different directions in space, a power per area is also generated and is called irradiance. It is defined in the same way as the radiant excitance as
  $$E = \frac{dP}{dA} = \int L \cdot \cos\theta \, d\Omega \tag{9.5}$$
  and is also measured in W/m^2. In everyday language the power per area (i.e., the quantity E) is often incorrectly called the intensity I.
- Radiant exposure:
  To describe the effect of an exposure, the illuminance must be integrated over time. Therefore, the quantity
  $$H = \int E \, dt \tag{9.6}$$
  is defined as the radiant exposure and is measured in Ws/m^2
- Energy:
  the integration of the radiation flux over time yields the transferred energy in Joules
  $$Q = \int P \, dt \tag{9.7}$$

Besides the purely geometrical dependencies, the radiometric quantities in reality also depend on the wavelength. Therefore, corresponding generalizations of the simple definitions are necessary; e.g., the spectral densities of radiometric quantities

$$\begin{aligned} P_\lambda &= \frac{dP}{d\lambda}, \quad P = \int P_\lambda \, d\lambda \\ L_\lambda &= \frac{dL}{d\lambda}, \quad L = \int L_\lambda \, d\lambda \\ I_\lambda &= \frac{dI}{d\lambda}, \quad I = \int I_\lambda \, d\lambda \end{aligned} \tag{9.8}$$

The quantities related to the wavelength also have corresponding units such as W/nm etc. For the exact determination of the radiation transport the spectral dependencies of the material parameters such as transmission, reflectivity, absorption etc. also have to be taken into account. The ratio of the spectral distribution of the luminance received by a detector to the spectral luminance emitted by the source determines the spectral efficiency of the radiation transfer.

$$\sigma(\lambda) = \frac{L_e(\lambda)}{L_s(\lambda)} \tag{9.9}$$

9.1.3 Photometric Quantities

By introducing the special spectral detector properties of the human eye as a sensitivity function, the radiometric quantities can be converted into photometric ones. In addition to this, the units are redefined leading to a conversion factor, the so-called radiometric radiation equivalent

$$K_m = 683 \, \frac{\text{lm}}{\text{W}} \tag{9.10}$$

This is the constant for daylight conditions. A slightly different value is valid for scotopic conditions at night. The conversion of the spectral radiometric power into the visual power is then given by the sensitivity $V(\lambda)$ and the conversion factor as

$$P_V(\lambda) = K_m \cdot V(\lambda) \cdot P(\lambda) \tag{9.11}$$

where the subscript V stands for 'visual'. So the photometric quantities follow from the physical radiation quantities by weighting and evaluating using the eye-sensitivity function $V(\lambda)$. The latter is defined in the spectral window between 380 nm and 780 nm. This distribution is sketched in Figure 9.1 in a peak-normalized representation. The maximum value is found in the green at $\lambda = 555$ nm.

9.1.4 Comparison of Quantities

In the following Table 9.1 the corresponding units, terms and definitions of the radiometric and photometric quantities are compared.

Some typical illuminance values for everyday conditions are listed in Table 9.2. It should be noted that a large dynamic range of $1:10^5$ is covered.

9.1.5 Energy, Power and Photons

According to Einstein, the quantum energy of a photon is given by

$$E = \hbar\omega = h\nu \tag{9.12}$$

where the vacuum speed of light $c = 2.99792\ 108$ m/s and the Planck constant $h = 6.626 \cdot 10^{-34}$ Js. The frequencies of the light are in the range of 1015 Hz. Coherent light oscillates extremely quickly such that it cannot be seen because every detector and even the eye are too slow to detect it and integrate the power over time.

$$\nu = \frac{1}{T} = \frac{c}{\lambda} \approx \frac{3 \cdot 10^8 \frac{\text{m}}{\text{s}}}{\lambda} \approx \frac{300 \cdot 10^{15}\ \text{Hz} \times \text{nm}}{\lambda} \tag{9.13}$$

The circle frequency $\omega = 2\pi/T$ determines the energy of the radiation

$$E = h\nu = \frac{19.865 \cdot 10^{-17}\ \text{J} \times \text{nm}}{\lambda} \tag{9.14}$$

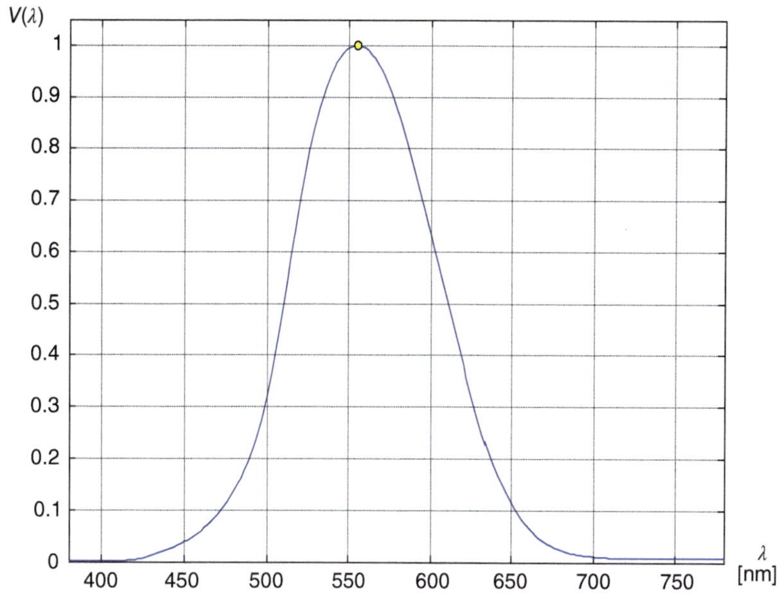

Figure 9.1 Spectral sensitivity $V(\lambda)$ of the human eye during daylight.

Table 9.1 Radiometric and photometric quantities and units.

Quantity	Formula	Radiometric		Photometric	
		Term	Unit	Term	Unit
Energy		Energy	J = Ws	Luminous energy	lm s
Power, radiation flux	P		W	Luminous flux	lumen, lm
Power per area and solid angle	$L = \frac{d^2P}{\cos\theta\, d\Omega\, dA}$	Radiance	W/sr/m^2	Luminance	cd/m^2=stilb
Power per solid angle	$I = \frac{dP}{d\Omega} = \int L\, dA_\perp$	Radiant intensity	W/sr	Luminous intensity	lm/sr=candela
Emitted power per area	$E = \frac{dP}{dA} = \int L \cdot \cos\theta\, d\Omega$	Radiant excitance	W/m^2	Luminous excitance	lm/m^2
Incident power per area	$E = \frac{dP}{dA} = \int L \cdot \cos\theta\, d\Omega$	Irradiance	W/m^2	Illuminance	lux=lm/m^2
Time integral of the power per area	$H = \int E\, dt$	Radiant exposure	Ws/m^2	Light exposure	lux s

or, with the energy given in electron volts,

$$E = \frac{1239.9\text{ eV}}{\lambda}\text{nm} \tag{9.15}$$

The number of photons per second can be calculated from the power of a source as

$$N_{photon} = \frac{\lambda \cdot P}{c \cdot h} = 5.034 \cdot 10^{15} \cdot \frac{\lambda}{\text{nm}} \cdot \frac{P}{\text{W}} \tag{9.16}$$

A source emitting 1000 photons per second in the visible range thus has a power of about 2*f* Watts (1 *f*W = 10^{-15} W). A 30 mW-HeNe laser emits about 10^{17} photons/second. Since the energy of the photons increases with decreasing wavelength, then the number of photons for the same power decreases with the wavelength.

Table 9.2 Typical illuminances.

Illuminance	Description
1 Lux	Just visible
50–100 Lux	Coarse work
100 Lux	Projection onto screen
100–300 Lux	Fine work
1000 Lux	Finest work
100 000 Lux	Sunlight on paper

For example, assuming that in an EUV system, an energy of 15 mW arrives at the wafer and a field of 2 × 26 mm^2 is

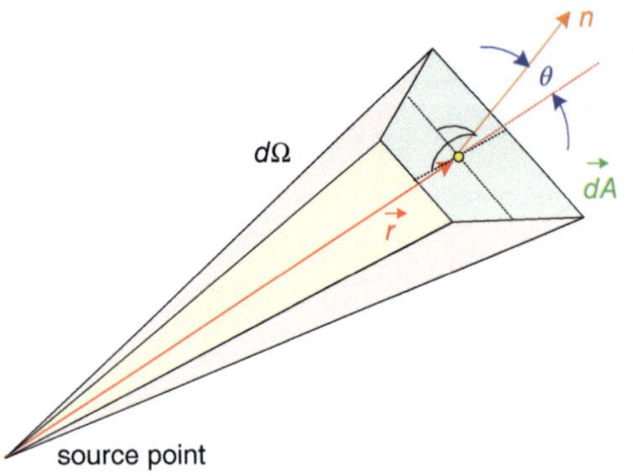

Figure 9.2 Definition of the solid angle element with normal *n* distance *r* and area element *dA*.

illuminated, there are still 2×10^{13} photons/mm^2 in the image field; so per image point with a size of $30 \times 30\,\text{nm}^2$ each there are exactly 17 640 photons. Therefore, for an exposure time of 0.2 s about 3500 photons are contributing to an image point, and the expected noise is $\sqrt{N}/N \sim 1.5\%$.

9.1.6 Solid Angle

A solid angle is defined as the quantity for which a surface element dA whose surface $\vec{n}$ normal encloses an angle θ with the connecting line, appears at a distance r

$$d\Omega = \frac{\cos\theta \cdot dA}{r^2} = \frac{dA_\perp}{r^2} \tag{9.17}$$

So it is the two-dimensional generalization of the classical angle definition, see Figure 9.2.

The unit of measurement of the solid angle is the steradian (sr). The complete space has a solid angle of 4π. For some simple geometrical shapes which are important in optics, the solid angle can be given analytically.

- Cone with half-aperture angle φ corresponds to (see Figure 9.3(a))

$$\Omega = 4\,\pi\,\sin^2\frac{\varphi}{2} = 2\,\pi\,(1 - \cos\varphi) \tag{9.18}$$

- Narrow ring on a sphere with half-aperture angle φ and width $d\varphi$ (Figure 9.3(b))

$$d\Omega = \frac{2\pi\,r \cdot \sin\varphi \cdot r \cdot d\varphi}{r^2} = 2\,\pi \cdot \sin\varphi \cdot d\varphi \tag{9.19}$$

9.2 Lambertian Source

9.2.1 Classical Lambertian Radiator

A radiation field or a light source with a constant radiance L, which does not depend on the observation direction $\vec{s}$, is called a Lambertian radiator.

$$L(\vec{r}, \vec{s}) = L = const \tag{9.20}$$

Thus, the radiant intensity in a direction θ to the surface normal n is given by

$$I(\theta) = L \cdot A \cdot \cos\theta = I_o \cdot \cos\theta \tag{9.21}$$

and has a maximum perpendicular to the surface and vanishes for a gracing angle 90°. Figure 9.4 illustrates a polar diagram for L and I.

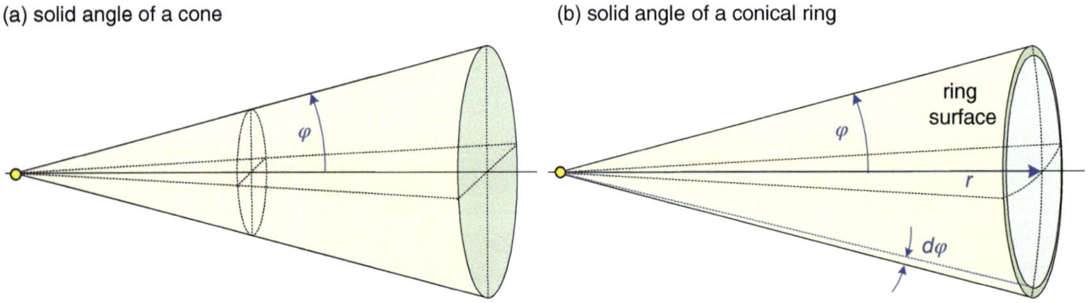

Figure 9.3 Conical ring as a solid angle.

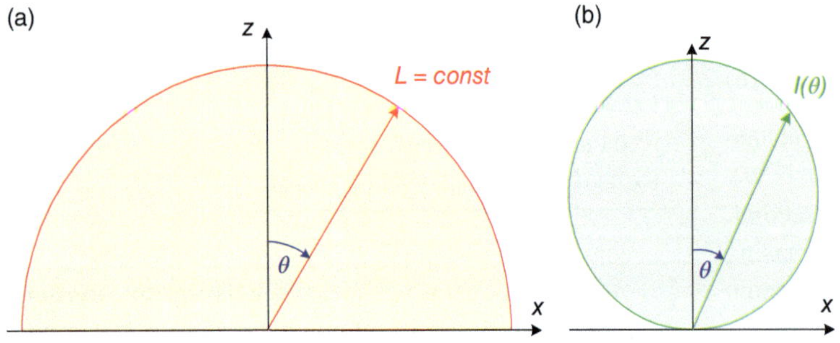

Figure 9.4 Lambertian radiator, shown in (a) radiance L and (b) radiant intensity I representation.

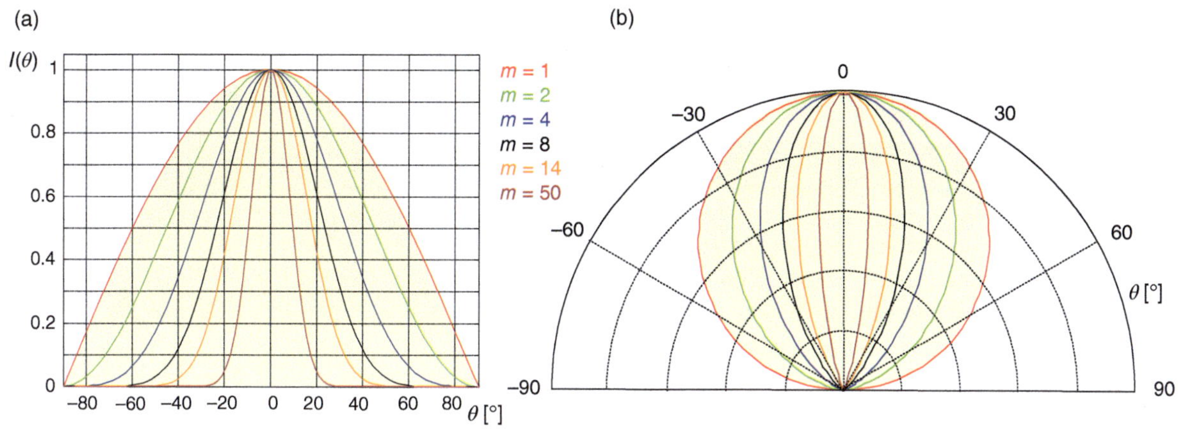

Figure 9.5 Generalized Lambert characteristic, (a) Cartesian representation and (b) polar representation.

A Lambertian radiator often serves as an idealized model for a light source. Real sources, however, often come very close to this mathematical approach. The sun and black body radiators can be approximated by the Lambert model quite well. Illuminated surfaces usually exhibit an intensified radiant emittance in the range of the specular angle of the law of reflection and thus do not fulfil the Lambert condition particularly well.

Integrating over the hemisphere the total radiant power of a Lambertian radiator with $I_0 = LA$ yields

$$P_{Lam} = \int I(\theta)\, d\Omega = L \cdot \int_0^{\pi/2} 2\pi \sin\theta \cdot \cos\theta \, d\theta \, dA = \pi \cdot A \cdot L \tag{9.22}$$

For a cone with half-aperture angle φ this gives

$$P_{Lam}(\varphi) = \pi \cdot A \cdot L \cdot \sin^2\varphi = \pi \cdot I_0 \cdot \sin^2\varphi \tag{9.23}$$

This formula is used particularly in optical systems where $\sin\varphi$ is equal to the numerical aperture on the axis of centred configurations in air.

9.2.2 Generalized Lambertian Radiator*

(*) The content of this section is quite specialized and for advanced reading. Knowledge of this section is not necessary to understand the following chapters. Beginners are recommended to skip this section.

One method of describing the angular distribution of the radiant emittance of light bulbs or light sources in a very simple way is by means of a generalized Lambert indicatrix which can be written in the form

$$I(\theta) = I_0 \cdot \cos^m\theta \tag{9.24}$$

with additional shape parameter m. For $m = 1$ the result is the classical Lambert distribution. $m > 1$ describes a more directed radiation with a dominant concentration in the direction perpendicular to the surface around $\theta = 0$. For integration over the half-space we obtain the power

$$P_{Lam} = \int I(\theta)\, d\Omega = L \cdot A \int_0^{\pi/2} 2\pi \sin\theta \cdot \cos^m\theta \, d\theta = \frac{\pi \cdot A \cdot L}{m+1} \tag{9.25}$$

The shape of this distribution can be seen below in Figure 9.5 for several m-values. In (a) the radiant intensity as a function of the inclination angle θ is shown, (b) gives a polar diagram of the radiant intensity.

Table 9.3 Generalized Lambert characteristic, exponents m for scattering surfaces.

Material	Exponent m
Ceramics unglazed	1.08....1.15
Ceramics with white glaze	1.23
Ceramics with yellow glaze	1.18
Spectralon	1.23

The generalized Lambert law of the angular distribution as given in the form above can also be used for the scattering properties of diffusely reflecting or transmitting surfaces. For important practical cases one finds some typical values of the exponent's m in Table 9.3.

9.3 Radiometric Transfer of Energy

9.3.1 Radiance and Irradiance

The radiance L is the most important quantity for the characterization of a light source. According to its definition it describes the amount of power emitted from a surface element into a solid angle for a particular direction at a given wavelength. In this sense this quantity characterizes the irradiating geometry of a source completely in terms of the geometrical conditions. For practical radiometric calculations it is therefore necessary to know the distribution function $L(\theta,\varphi,\lambda)$ with the azimuthal angle φ of a light source exactly. The measurement of this quantity is a relatively elaborate task which nevertheless is necessary in order to produce reasonable simulations of illumination problems [13–15].

The radiance related to the surface must be integrated over the solid angle and is therefore the classical intensity in the sense of the energetic area density (active for emitting sources) or irradiance (for passively illuminated surfaces). It can be written as

$$E = \frac{dP}{dA} = \int L \cdot \cos\theta \, d\Omega \tag{9.26}$$

For a collimated beam the intensity depends on the incident angle. The projection produces an effect as the size of the surface element changes accordingly, see Figure 9.6.

$$A_{\perp} = A \cdot \cos\theta, \quad E(\theta) = E_0 \cdot \cos\theta \tag{9.27}$$

9.3.2 Radiometric Flux Transfer

The differential flux emitted from a small surface element dA_S orthogonal to the surface normal n into the small solid angle $d\Omega$ along the direction $\vec{s}$ is given as

$$\begin{aligned} d^2P &= L \cdot d\Omega \cdot dA_{e\perp} = L \cdot \cos\theta_e \cdot d\Omega \cdot dA_e \\ &= L \cdot d\Omega \cdot (\vec{s} \cdot d\vec{A}_e) \end{aligned} \tag{9.28}$$

This measure gives the transferred differential radiant power in Watts. The geometry is sketched in Figure 9.7. This expression is the basis for all calculations of radiative power transfer. It has to be integrated according to the geometry of the problem.

If two small surface elements exchange radiation without any further interaction, then the radiation flux onto a receiving detector surface is

$$d^2P_r = L \cdot \cos\theta_r \, dA_r \cdot d\Omega_e \tag{9.29}$$

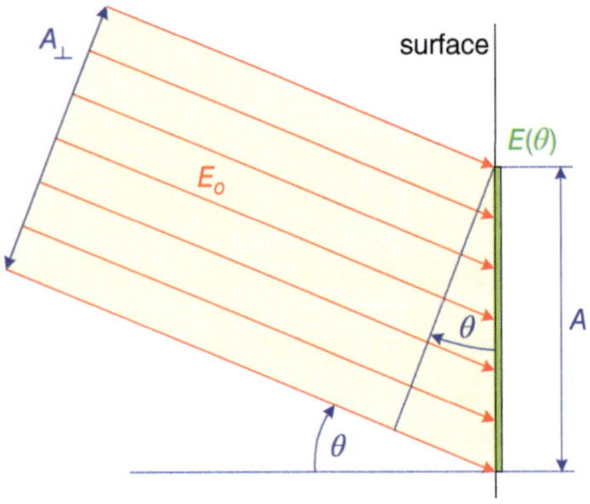

Figure 9.6 Irradiance and incident angle.

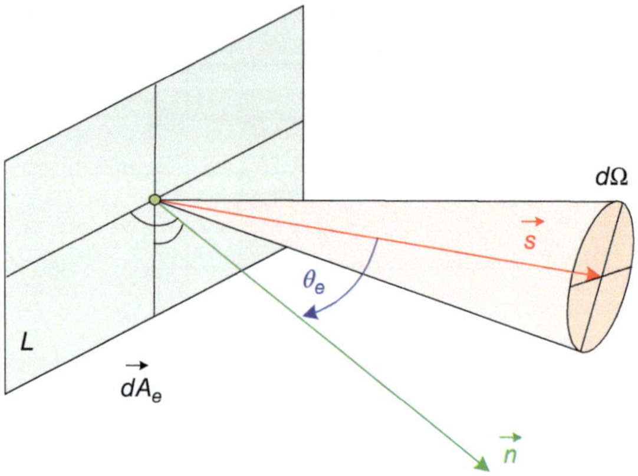

Figure 9.7 Elementary differential flux for the radiance L.

From the detector, the emitting surface is observed under the solid angle

$$d\Omega_e = \frac{\cos\theta_e \cdot dA_e}{r^2} = \frac{dA_{e\perp}}{r^2} \tag{9.30}$$

By inserting this formula into Eq. (9.29), the fundamental law of radiometry is obtained.

$$d^2P = \frac{L}{r^2} \cdot dA_{e\perp} \, dA_{r\perp} = \frac{L}{r^2} \cdot \cos\theta_e \, \cos\theta_r \, dA_e \, dA_r \tag{9.31}$$

The purely geometrical part of the formula, without the radiance L, is called the purely geometrical flux or throughput (etendue, acceptance) of the configuration

$$d^2G = \frac{1}{r^2} \cdot \cos\theta_e \, \cos\theta_r \, dA_e \, dA_r \tag{9.32}$$

Eq. (9.31) is the elementary law for flux transfer; the geometry is sketched in Figure 9.8.

In any practical calculation of radiation transport this geometrical acceptance contains in addition the properties of the transferring optical system, while L characterizes the properties of the light source. Furthermore, the influence of possible interactions between the radiation and the system components like absorption, truncation or scattering has to be taken into account. The radiometric law is a differential description of the radiation transfer between two small surface elements with any orientation. Real configurations can be determined from this formula by integration. For this there are two different cases.

- The radiance L is spatially constant and thus does not depend either on the space coordinates at the emitting surface nor on the direction of the energy flux. Here the quantity L can be written outside the integral as the integration has to cover only the geometrical conditions for surfaces and solid angles.
- The radiance is a function of location and direction according to the indicatrix and the radiance distribution

Figure 9.8 Fundamental law of radiometry.

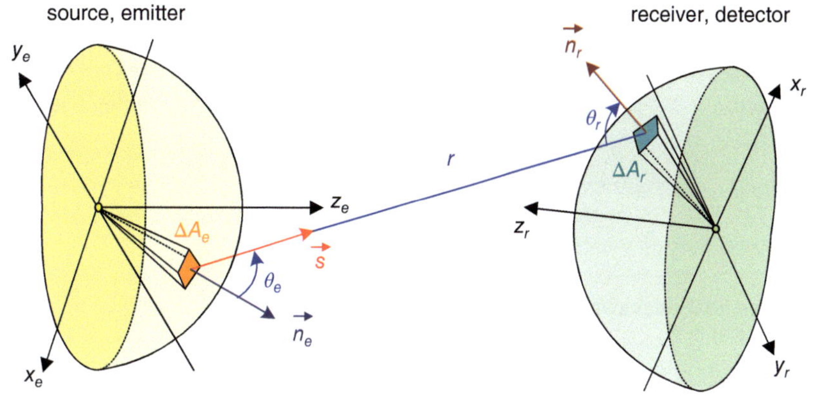

of the light source. Then this function has to be taken into account for the integration over position and solid angle too.

Considering real conditions, the transmission of the optical transfer system as well as the spectral emission of the light bulb will depend on the wavelength. So for the exact calculation of the energy transfer the relevant ranges of wavelengths have to be correctly included in the integration as well [8].

For some special geometries the basic radiometric law can be integrated analytically. This delivers formulas for the calculation of the radiation transport in simple configurations that occur quite often in practical cases. Basically, it is the integration of the solid angle for a certain setup. In addition, to obtain closed formulas, special properties very often have to be assumed for the source distributions; e.g. the Lambert indicatrix [9].

9.3.3 Analytical Solutions for Simple Geometries

9.3.3.1 Surface Element Illuminated by a Point Source

If an idealized point source illuminates a surface element dA_r tilted by θ_r and located at a distance r, then the differential flux is

$$d^2P = \frac{L \cdot dA_{r\perp}}{r^2} = \frac{L \cdot dA_r}{r^2} \cdot \cos\theta_r = \frac{L \cdot dA_r}{z^2} \cdot \cos^3\theta_r \tag{9.33}$$

with the projected distance

$$z = r \cdot \cos\theta_r \tag{9.34}$$

and the effective projected size of the area element (see Figure 9.9).

$$dA_{r\perp} = dA_r \cos\theta_r \tag{9.35}$$

9.3.3.2 Circular Plane Surface Illuminated by a Point Source

If the point source illuminates a circular bounded plane surface with radius a at an axial distance z, for these centred conditions with the solid angle element $d\Omega$ of a ring (see Figure 9.10) the differential flux

$$dP = L \cdot d\Omega = L \cdot 2\pi \cdot \sin\theta_r \cdot d\theta_r \tag{9.36}$$

Considering the relation

$$\cos\theta_r = \frac{1}{\sqrt{1+\tan^2\theta_r}} = \frac{1}{\sqrt{1+\left(\frac{a}{z}\right)^2}} \tag{9.37}$$

Figure 9.9 Illumination of a surface element by a point source of light.

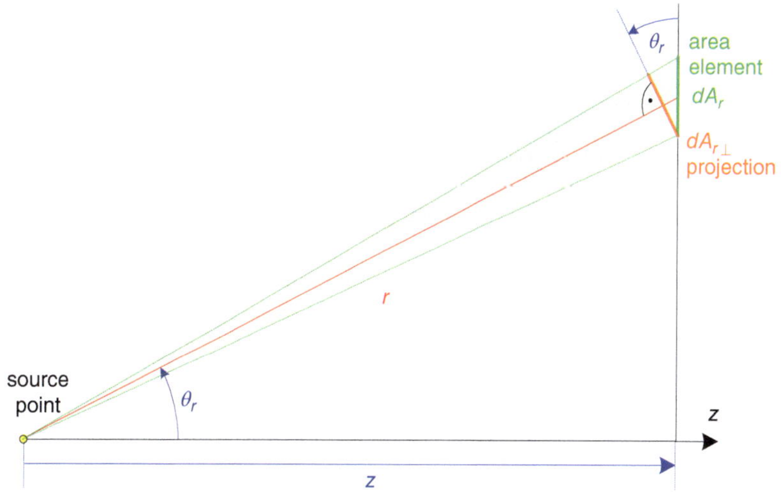

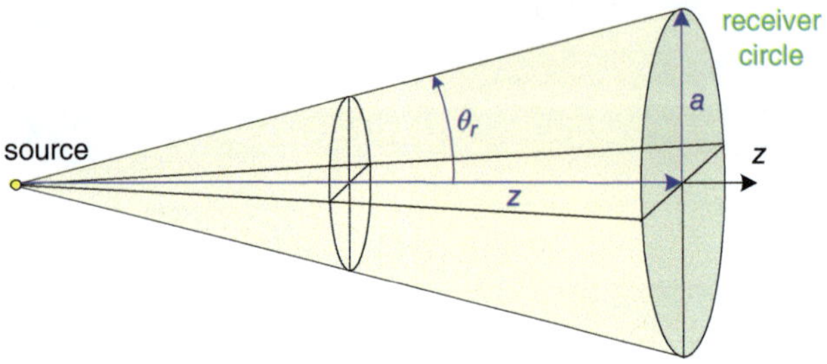

Figure 9.10 Illumination of a circular plane surface by a point source of light.

integration gives the flux

$$P = 2\pi L \cdot \left[1 - \frac{r_\perp}{\sqrt{z^2 + a^2}} \right] \tag{9.38}$$

9.3.3.3 Parallel Surface Elements

If a small element of area dA_e illuminates a second parallel small element of area dA_r at a distance r, with the differential flux as given above and according to the sketch in Figure 9.11

$$\cos\theta_e = \cos\theta_r = \cos\theta \tag{9.39}$$

we obtain the power flux

$$d^2P = \frac{L}{r^2} \cdot \cos^2\theta \cdot dA_e\, dA_r = \frac{L}{z^2} \cdot \cos^4\theta \cdot dA_e\, dA_r \tag{9.40}$$

and the irradiance $E(\theta)$

$$E(\theta) = \frac{d^2P}{dA_r} = \frac{L}{z^2} \cdot \cos^4\theta \cdot dA_e \tag{9.41}$$

From this we have the dependence of the irradiance from the angle θ

$$E(\theta) = E(0) \cdot \cos^4\theta \tag{9.42}$$

This $\cos^4\theta$ dependency describes the natural vignetting found in optical systems where the irradiance decreases accordingly towards the edge of the image field seen from the exit pupil centre.

9.3.3.4 Circular Lambertian Source and Point-Like Receiver

If a circular, extended Lambertian source of radius a illuminates a centred surface element dA_r one obtains

$$\begin{aligned} I &= L \cdot A \cdot \cos\theta_e \\ z &= r \cdot \cos\theta_r \\ dA_{r\perp} &= dA_r \cdot \cos\theta_r \\ \theta_r &= \theta_e = \theta \end{aligned} \tag{9.43}$$

So the differential flux from a ring element of the emitting surface is

$$\begin{aligned} dP &= \frac{L}{r^2} \cdot \cos\theta_e \cdot \cos\theta_r\, dA_r dA_e \\ &= \frac{L}{r^2} \cdot \cos^2\theta_r\, dA_r \left(\frac{r d\theta_e}{\cos\theta_e} \cdot r \sin\theta_e d\varphi_e \right) \\ &= L \cdot \cos\theta_r\, dA_r d\theta_e \sin\theta_e 2\pi \\ &= 2\pi L \cos\theta \sin\theta d\theta\, dA_r \end{aligned} \tag{9.44}$$

and the irradiance is

$$dE = \frac{dP}{dA_r} = 2\pi L \cdot \sin\theta \cos\theta\, d\theta \tag{9.45}$$

After integration, according to Figure 9.12, one gets

$$E = \pi L \cdot \cos^2\theta = \frac{\pi L}{1 + \tan^2\theta} = \frac{\pi L}{1 + \left(\frac{z}{a}\right)^2} \tag{9.46}$$

In the approximation for small luminous surfaces of any shape at a large distance $z \gg a$ and $z^2 \gg A$, respectively, the quadratic law of distance is obtained

$$E = \frac{L\pi a^2}{z^2} = \frac{LA}{z^2} \propto z^{-2} \tag{9.47}$$

9.3.3.5 Circular Lambertian Source and Circular Receiver

If a receiving surface is illuminated by a circular emitting surface with Lambert characteristic in θ_e the differential

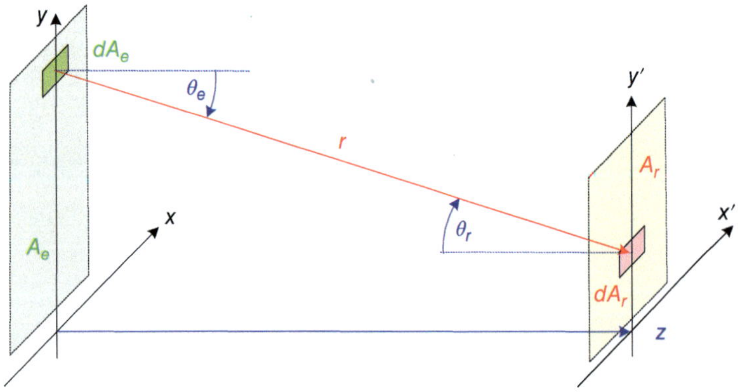

Figure 9.11 Illumination of an area element by another parallel area element.

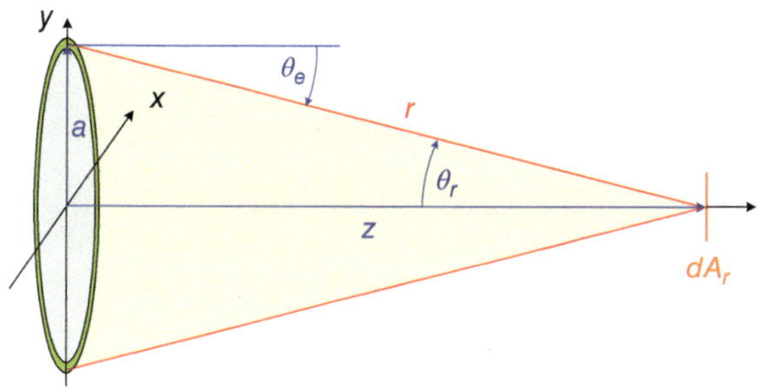

Figure 9.12 Illumination of a surface element by a circular ring.

flux above has to be integrated over the variable angles of the surface. Then for the irradiance we obtain [16]

$$E = \frac{\pi L}{2} \cdot \left[1 - \frac{1 + \tan^2\theta_r - \tan^2\theta_e}{\sqrt{\begin{array}{c}\tan^4\theta_r + 2\tan^2\theta_r \cdot (1-\tan^2\theta_e) \\ +(1+\tan^2\theta_e)^{-2}\end{array}}} \right]$$
$$= \frac{\pi L}{2} \cdot \left[1 - \left(1 + \frac{4\tan^2\theta_e \cos^4\theta_r}{(1-\tan^2\theta_e\cos^2\theta_r)^2} \right)^{-1/2} \right]$$
$$\approx \pi L \cdot [\cos^4\theta_r \tan^2\theta_e + \cos^4\theta_r \tan^4\theta_e \cdot (2\cos^2\theta_r - 3\cos^4\theta_r)] \tag{9.48}$$

Here θ_r is the elevation angle of the receiver element from the centre of the emitting circle (corresponds to the field angle of the chief ray), θ_e is half the cone angle under which the emitting surface appears when observed from the axis point of the receiver surface. This corresponds to the marginal ray aperture angle, see Figure 9.13.

With increasing field angle θ_r the irradiance decreases. For small numerical apertures this decrease starts earlier. This behaviour is shown in Figure 9.14. For comparison, the usual $\cos^4\theta_r$ curve is also given. The last expression in the formula above is an expansion of $\tan\theta_e$. It is seen that the leading term changes with the fourth power of the field and the square of the aperture as it should be. The next expansion terms vary with the aperture to the power of four and higher powers of the field angle cosine.

For application to centred optical systems in the object space, the angle θ_R corresponds to the aperture angle u of the system and θ_e is the field angle w. Accordingly, the emitting surface corresponds to the object plane and the receiver surface to the entrance pupil. In the image space the terms are just reversed. Here it is assumed that the exit pupil illuminates the image plane.

9.3.4 Numerical Radiation Transfer

9.3.4.1 Introduction

There are still some more geometries for which the problem of the solid angle can be solved analytically and there are different methods of approach in approximating the integration of the solid angle. For general cases the calculation of the radiation transport can no longer be carried out analytically. In these cases, the differential radiometric law of distance is integrated numerically by means of raytracing. For raytracing, the sum over many rays is required [17, 18]. As well as the purely geometrical conditions, several physical effects determine the energy transport in real systems can be taken into account. Among these are:

- absorption in media;
- absorption in coating layers;
- remaining residual reflectivities at interfaces due to Fresnel formulae or coatings;
- scattering in media;
- scattering at diffusely transmitting or reflecting surfaces or non-optical surfaces;
- diffraction effects and interferences; in these cases it is necessary to use the phase of the ray paths in addition.

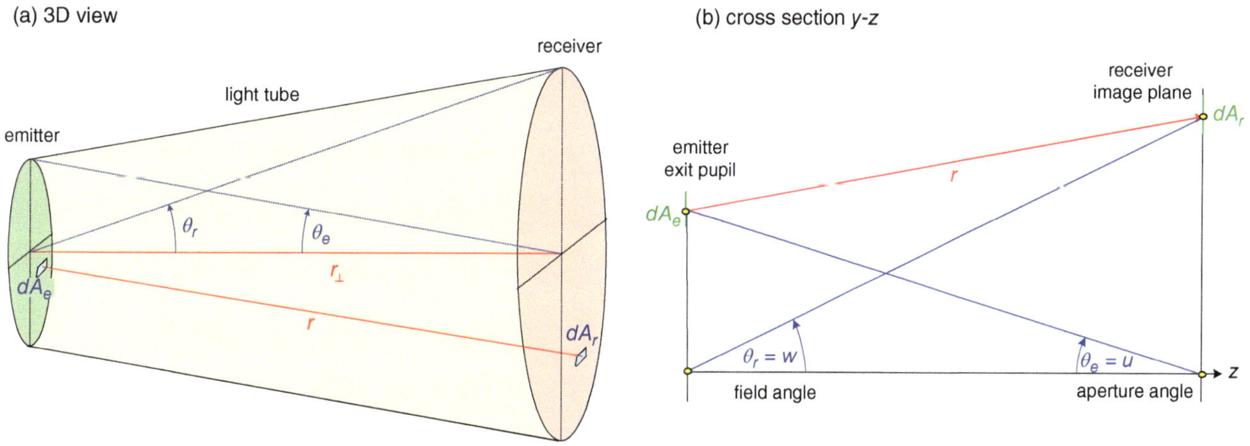

Figure 9.13 Illumination of a circular receiver by a circle for a Lambertian source.

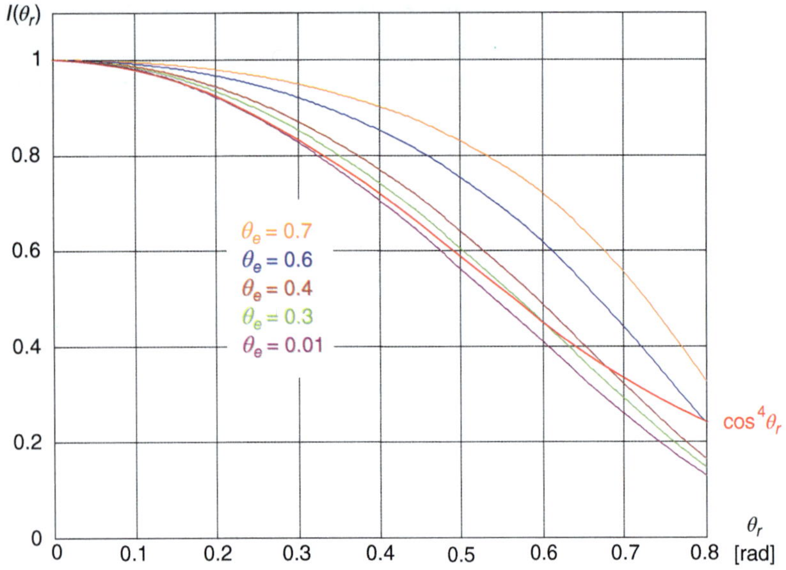

Figure 9.14 Illumination of a circle by a circular Lambertian radiator, the angles are given in radians.

In general, these mechanisms also depend on the wavelength, polarization and the scattered light can be azimuthal dependent and therefore anisotropic.

If the phase as well as the amplitude of a ray are traced by a Monte Carlo calculation method, coherence effects can be simulated too. More or less the Mont-Carlo-approach evaluates the Huygens–Fresnel diffraction integral by corresponding sampled rays or photons, see Figure 9.15. As can be seen in the figure, in case of intermediate stops a resampling of the light is necessary, taking diffraction into account. This is necessary at every plane creating diffraction effects. Critical for the accuracy of the result is the sampling discretization, in particular in the receiving plane. In the case of a Monte Carlo method, the photons are overlaid on one pixel by interference. The small phase differences resulting in the slightly different intersection points on a pixel are neglected.

The sampling of the initial directions and starting points of the light source can be regular or random. Inside one Huygens wavelet, the amplitude is assumed to be uniformly distributed. If it is known that a ray cannot reach the final detector plane, it should not be calculated to save calculation time. This is a kind of bias or important sampling and accelerates the computation.

9.3.4.2 Monte-Carlo Raytracing Approach

The ray-based computation of photometric quantities is quite simple. There are three major steps [17].

- Modelling the light source by a ray set.
- Tracing the rays to the detector.
- Collecting the intensity on the sensor and processing the data, maybe in addition with a direction-sensitivity of the sensor.

The major problem of a ray-based calculation is the discretization of source and detector. This finally can result in errors, which are mostly structured. Therefore, the preferred approach in reality is to make a Monte-Carlo selection of initial rays by random numbers to avoid regular structures and corresponding artefacts. Nevertheless, a remaining noise in the obtained intensity distributions is uncomfortable and a post-processing for smoothing the result is meaningful.

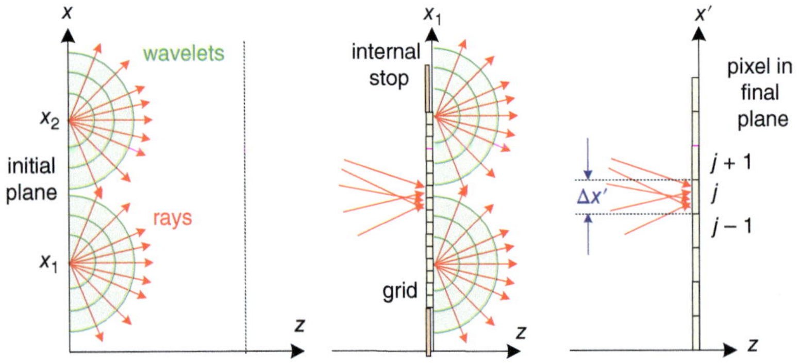

Figure 9.15 Principle of Monte Carlo evaluation of the diffraction integral.

If in total N_e ensemble realizations are used for the averaging and N_r rays are used per realizations, the computational effort is scaled with the product $N_r N_e$. The fidelity or accuracy of the results due to fluctuations of the random numbers measured by a standard deviation follows the classical law

$$\varepsilon_{rms} \sim \frac{1}{\sqrt{N_e \cdot N_r}} \tag{9.49}$$

According to more detailed investigations, the behaviour and prediction accuracy of the final intensity and the degree of coherence is more complicated.

Monte Carlo methods can also be used to simulate the partial coherent light propagation in scattering media. The initial phase distribution can describe the degree of coherence of the light source and during the ray or photon path in the multiple scattering medium, the geometrical phase of the propagation must be recorded. The random interference at a defined position of the exit surface gives rise to a fluctuating intensity.

If statistical Monte Carlo raytrace is used and the selection of rays is governed by random numbers, a statistical noise can never be avoided. Model calculations in the literature to mimic the Huygens–Fresnel diffraction at a single slit has used 20 000 rays, many intermediate planes to resample and still shows a residual error of 3 10^{-4}, independent of the Fresnel number.

If a more complex optical system is modelled by raytrace including diffraction effects, at every location where diffraction effects occur, a new artificial surface must be defined, which acts as a new secondary source. The rays are re-sampled here as is visualized in Figure 9.15 at the middle plane. This can happen at non-smooth surfaces with edges or micro-structure features that change the light propagation direction by diffraction. An adapted selection of discretization point numbers is needed to model the component and the light field appropriately. This is necessary because only a redefinition of spherical Huygens wavelets in new geometrical shadow regions or directions is able to take the diffraction into account. The number of necessary re-sampling planes depends on the inhomogeneities of the system. Similar to stochastic tissue, random diffusor plates can also be simulated by Monte Carlo raytrace. The diffusor plate must be modelled as a statistical random surface in this case with a usual refractive raytrace impact at the dielectric surface.

9.3.4.3 Source Modelling

To model a source accurately, three properties have to be selected per ray:

- the initial starting point, following the spatial distribution;
- the initial starting direction, following the angle characteristic of the source;
- the wavelength, following the spectral distribution, mostly here the sampling can be made more coarse.

Every ray gets a weighting factor corresponding to the relative power from the point in the direction for the current wavelength. In total, the full ray set used must be calibrated to the power of the source under consideration. If z is a uniformly distributed random number between $z = 0\ldots1$ and $\theta_{\min}$, $\theta_{\max}$ are the extreme angles that reach the detector from a starting point x that the angle sampling is given by (in y it is the same)

$$\theta_x = \theta_{\min} + z \cdot (\theta_{\max} - \theta_{\min}) \tag{9.50}$$

This lack of the full hemisphere needs a correction of the amplitude by the factor

$$\eta_x = \frac{\theta_{\max} - \theta_{\min}}{\pi} \tag{9.51}$$

For the statistical simulation of lamp radiation by means of Monte Carlo raytracing calculation, there are two possible ways of generating the data.

- The position and direction of the rays are determined. Then the relative weight according to the emission probability of the lamp characteristic is determined.

  This approach is direct and simple. Its performance, though, is relatively bad because there are a lot of rays with low weight in the peripheral regions of the beam energy density. These do not contribute very much to the total energy.
- The position and angle of the beam are determined directly from the probability of the emission and the beam always begins with the weight 1. This method has a significantly better performance regarding the computing time in connection with the corresponding accuracy of the simulation. For tophat distributions, this gives the same efficiency, but for spatial and angle distributions with long smooth decay ranges it is recommended to use this approach.

With an inclination angle θ of the emitting surface, the power emitted into the solid angle element $d\Omega$ is given by

$$dP = E(\theta) \cdot d\Omega \tag{9.52}$$

with the solid angle element $d\Omega$ for the supported solid angle of the optical system. The density of states as a probability of emission in the normalized form can be written as

$$f(\theta) = \frac{1}{P} \cdot \frac{dP}{d\theta} \tag{9.53}$$

Simple analytical distributions of the luminance function allow the density of states to be given directly, which is necessary for determination of the probabilities in Monte Carlo calculations. These are, for example, the following.

- Lambert's law.

In this case we have an azimuthal independent distribution

$$E(\theta) = E_0 \cdot \cos\theta$$
$$f(\theta) = 2\sin\theta \cdot \cos\theta \qquad (9.54)$$

If the random numbers z_1 and z_2 are calculated normalized to 1, it follows that the ray direction angles are given by the equations

$$z_1 = \int_0^\theta f(\theta)\,d\theta = \frac{1}{2}(1-\cos 2\theta)$$
$$\theta = \frac{1}{2}\cdot \arccos(1-2z_1), \quad \varphi = 2\pi \cdot z_2 \qquad (9.55)$$

- A uniform emitting source inside the cone of $\theta = 0 \ldots \theta_0$

We have in this case

$$E(\theta) = E_0, \quad f(\theta) = \frac{\sin\theta}{1-\cos\theta_0}$$
$$\theta = \arccos[1 - z_1 \cdot (1-\cos\theta_0)], \quad \varphi = 2\pi\cdot z_2 \qquad (9.56)$$

- Gaussian radiator.

For a Gaussian-shaped source with a small far-field divergence angle θ_0, we have

$$E(\theta) = E_0 \cdot e^{-2\left(\frac{\theta}{\theta_0}\right)^2}, \quad f(\theta) = \frac{4\theta}{\theta_0^2}\cdot e^{-2\left(\frac{\theta}{\theta_0}\right)^2}$$
$$\theta = \theta_0 \cdot \sqrt{\frac{1}{2}\cdot \ln(1-z_1)}, \quad \varphi = 2\pi \cdot z_2 \qquad (9.57)$$

If diffraction effects have to be taken into account too, it becomes necessary to monitor the phase in the raytrace and to know about the initial phase. In the case of laser coherent sources, it can be assumed that every point carries the same phase in the source. In the case of incoherent sources, the initial phases are random. A point source with index j in a source plane generates the complex field

$$E_j = A_j \cdot e^{i\varphi_j} \qquad (9.58)$$

with amplitude Aj and initial phase φj.

9.3.4.4 Evaluation of the Detector Irradiance

If a ray intersects the sensor plane at an arbitrary location x, y inside a pixel, the accuracy can be increased if a linear weighting model is used to distribute the ray contribution onto the neighbouring pixels. Let's assume the ray meets the sensor surface between the pixel boundaries x_j and x_{j+1} and y_k and y_{k+1} respectively. Then the signal S carried by the ray can be distributed onto the four neighbouring grid points with the help of the distance-related weights (see Figure 9.16)

$$a_x = x - x_j, \quad a_y = y - y_k$$
$$w_x = 1 - \frac{a_x}{\Delta x}, \quad w_y = 1 - \frac{a_y}{\Delta y} \qquad (9.59)$$

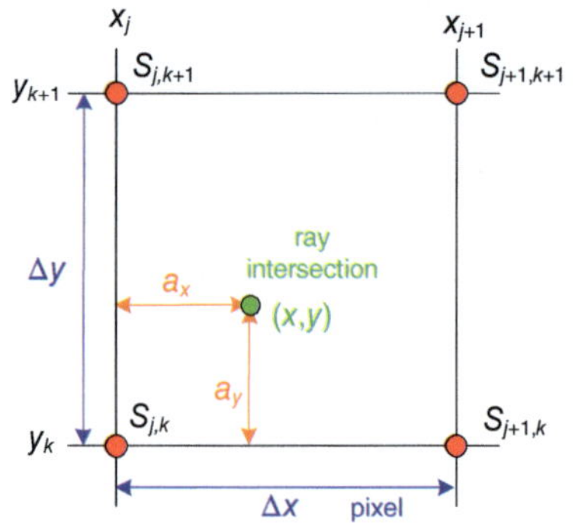

Figure 9.16 Interpolation scheme for one ray intersection point on the detector.

as the signal contributions

$$S_{j,k} = w_x \cdot w_y \cdot S, \quad S_{j+1,k} = (1-w_x)\cdot w_y \cdot S$$
$$S_{j,k+1} = w_x \cdot (1-w_y)\cdot S, \quad S_{j+1,k+1} = (1-w_x)\cdot(1-w_y)\cdot S \qquad (9.60)$$

This interpolation increases the accuracy considerably in comparison to a simple summation inside a finite pixel area. By performing this interpolation, the residual discretization error and the corresponding signal noise is reduced from a linear to a quadratic behaviour.

To detect diffraction and interference effects accurately, the pixel size for a Monte Carlo calculation in case of partial coherence must be in the order of the wavelength. On the other hand, to avoid fluctuations due to the random number process, a large number of rays or photons must meet every pixel. This means that the Monte Carlo method including coherence is quite time consuming. On a detector pixel with index m, all the source points generate the final field

$$E_m = \sum_j A_{jm} \cdot e^{i\varphi_j} \cdot e^{\frac{2\pi i}{\lambda}\cdot \Delta_{jm}} \qquad (9.61)$$

where Δjm indicates the optical path difference from the source point j to the detector pixel m. Here the approximation is made that Ajm is an average value over the source pixel area size. The phase differences over the individual exact intersection points on the extended pixel are neglected; here a mean is taken too. In the practical realization of a ray-based model the pixel size must be small enough to resolve the changes of S and the number of rays N must be large enough to provide smooth results for the ensemble averaging over every pixel.

The incorporation of temporal coherence by selecting different wavelengths and taking the system dispersion into account is straightforward.

9.3.4.5 Examples

The first example shows a system of four confocal parabolic mirrors. The setup is shown in Figure 9.17(a); it

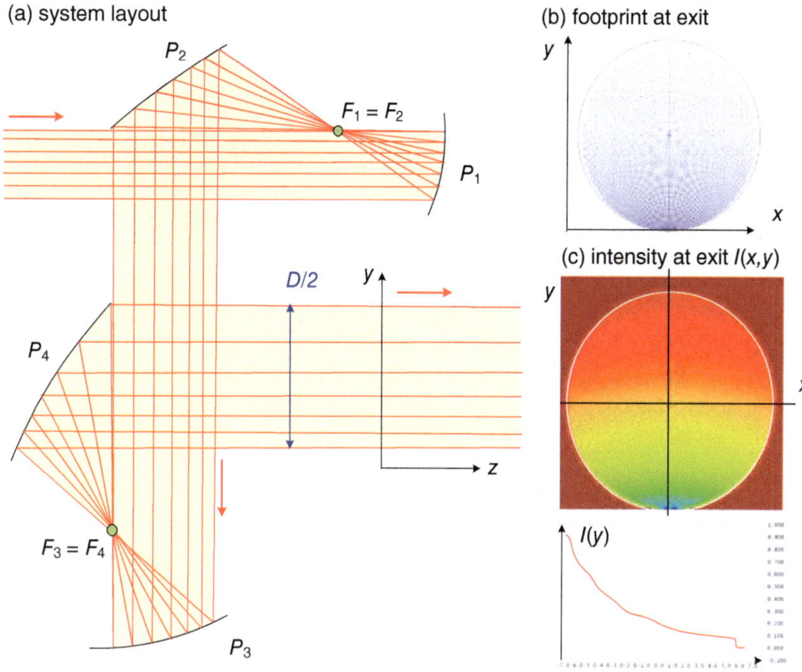

Figure 9.17 Shown are the stigmatic paraboloid system with strong pupil distortion in (a), ray density of pupil distortion in (b) and intensity distribution in (c) with cross section in *y*.

demonstrates the violation of the sine condition in the case of conic sections for the pupil imaging and therefore a strong distortion of the sampling grid between entrance and exit pupil. The points F_j are the focal points of the paraboloids mirrors, $F_1 = F_2$ are coinciding and $F_3 = F_4$. This means that a stigmatic imaging is established without aberrations of the collimated incoming rays. The non-uniformity of the outgoing collimated light can be seen in the layout by the distances of the rays. In (b) a footprint with all ray intersection points is shown and illustrates the change of intensity, which is again shown in (c) as a smooth plot together with a cross section curve $I(y)$.

The second example shows a classical illumination system, a so called RXI collimator. The setup is shown in Figure 9.18(a). It is a system that can be used to collect and collimate a large solid angle of the radiation of a LED light source. The ray path has two channels, an outer one with reflection and an inner one with only two refractive surfaces, which are mostly aspheres. In part (b) of the figure a simple spot diagram shows the two different channels by red and blue ray intersection points respectively. In (c) the intensity calculation by Monte-Carlo raytrace with 5 10^7 rays is shown as a two-dimensional false colour representation and as cross section. Although a huge number of rays are used, the noise is still visible.

9.3.5 Radiation Transport with Interaction

If in the radiation transport other interactions in a system occur, physical effects will influence the energetic and radiometric transport properties of the light. For their characterization some additional quantities need to be defined. The ratio of outgoing and incoming radiant power is called the transmission or transmission factor

$$T = \frac{P_{out}}{P_{in}} \tag{9.62}$$

The logarithm of the inverse transmission is sometimes also used for the description and is called the optical density

$$D = \log_{10}\frac{1}{T} = \log_{10}\frac{P_{in}}{P_{out}} \tag{9.63}$$

If there is a decrease in the radiant power of a medium or a layer, this effect is referred to as absorption. Its characteristic is defined in a similar way, but for one bulk part of the system only

$$\alpha = \frac{P_{out}}{P_{in}} \tag{9.64}$$

A quantity analogous to the optical density is the extinction of a material. It is defined by the logarithm

$$\varepsilon = \log_{10}\frac{1}{\alpha} = \log_{10}\frac{P_{in}}{P_{out}} \tag{9.65}$$

Usually these quantities depend on the wavelength; e.g., $T(\lambda)$. The total transmission is obtained from the spectral distributions of the source power $P(\lambda)$ and the transmission $T(\lambda)$ by integration

$$T = \frac{\int P(\lambda) \cdot T(\lambda)\, d\lambda}{\int P(\lambda)\, d\lambda} \tag{9.66}$$

If a decrease of the radiant power in the path of rays is observed due to scattering, this power will be missing in the

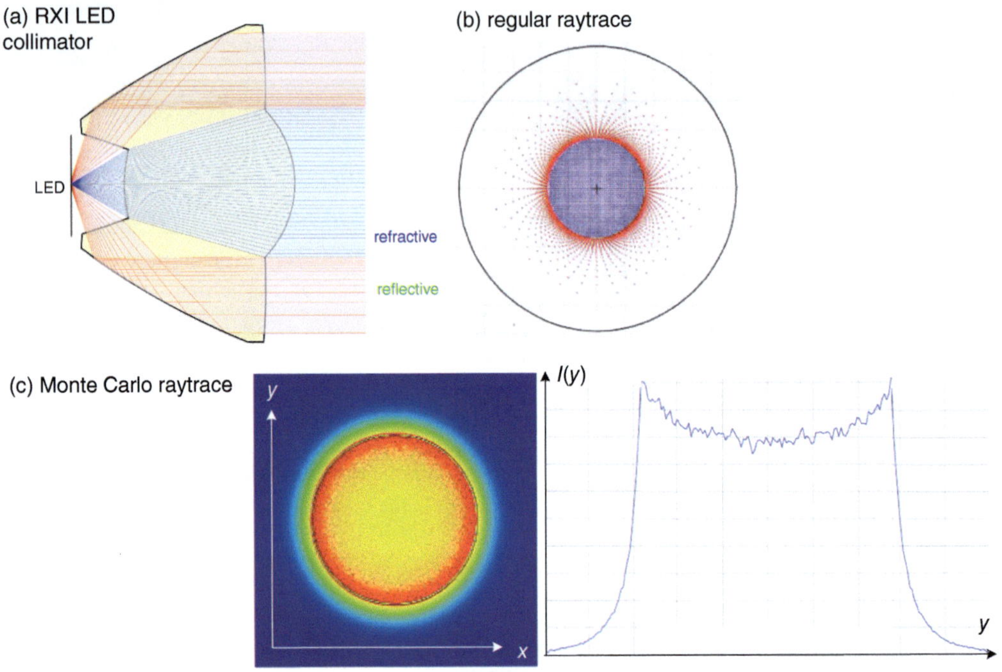

Figure 9.18 Shown are in (a) the setup of a RXI collimator, (b) gives the intersection points after the system with the two channels and in (c) the total intensity is illustrated in two dimensions and as cross section plot.

signal. However, it can be found again by means of scattering or false light in other transport channels. That part of the light being converted into false light is characterized by the scattering power

$$S = \frac{T_{scat}}{T_{total}} = \frac{T_{scat}}{T_{signal} + T_{scat}} \tag{9.67}$$

A corresponding quantity can also be defined for the reflection.

9.3.6 Ray Tube Model

Rays are representatives of local plane waves. According to this geometrical approach they have no transverse extension. On the other hand, radiometric calculations have to describe area-related changes of energy densities. This means that single rays are in principle not truly comfortable for this task, the density can only be obtained by many rays and their transverse distances. Therefore, another approach, the so-called ray tubes, give a much better tool for propagating energy density. In principle they are built analogously to differential or parabasal rays, which are discussed in Section 3.4.9.

For radiometric calculations using ray tubes, narrow beams are considered which, from the raytracing point of view, can be described by three single rays. Within such a ray tube the radiometric law of distance has to be fulfilled in its differential form. The central or chief ray may have the direction cosines (s_x, s_y, s_z) with the radiance $L_{RT}(x,y, s_x, s_y, s_z)$ according to the start position and the start direction. Now a small area element ΔA, perpendicular to the ray, is considered, which changes into $\Delta A'$ by propagation. This area element may be small enough that the radiance can be considered approximately constant over this region. Let's assume the radiation field has the initial radii of curvature R_x and R_y in the local x–y and y–z intersection planes respectively. For this ray tube, the radiometric law is now valid in its differential form

$$\Delta^2 P = \frac{L_{RT}}{r^2} \cdot \Delta A \cdot \Delta A' \tag{9.68}$$

where r is the propagation distance between the area elements and they are both perpendicular on the vector $\vec{s}$. The values of the cosines of the angles θ are equal to 1. Then corresponding to the paraxial circumstances around the reference ray, the area element is transformed according to the equation

$$\Delta A' = \left(1 + \frac{r}{R_x}\right) \cdot \left(1 + \frac{r}{R_y}\right) \cdot \Delta A \tag{9.69}$$

see Figure 9.19. Assuming energy conservation within the ray tube, the radiance at the image position can be obtained from

$$L'_{RT} = \frac{\Delta A}{\Delta A'} \cdot L_{RT} = \frac{L_{RT}}{\left(1 + \frac{r}{R_x}\right) \cdot \left(1 + \frac{r}{R_y}\right)} \tag{9.70}$$

For an improvement of the model, in addition a finite transmission factor $T(x,y)$ can be assumed for the chief

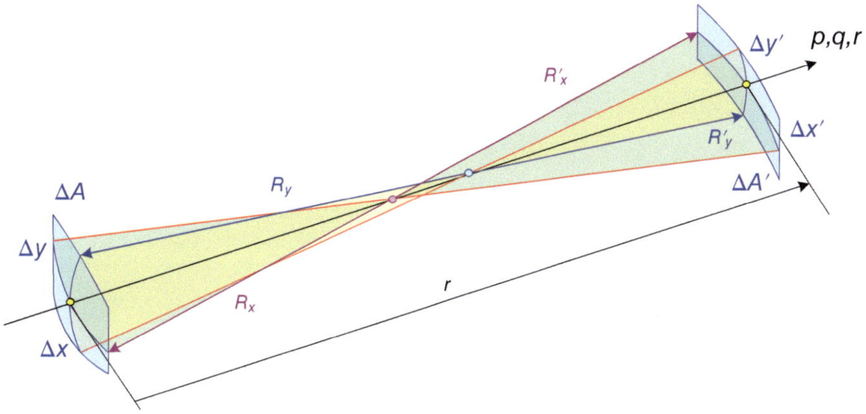

Figure 9.19 Calculation of the radiation transfer by ray tubes.

ray. Also, for the propagation of the narrow beam, the diffraction can be taken into account so that the non-physical singular behaviour for $R_{x/y} = r$ vanishes. Furthermore, the approach can be generalized by allowing the transport of the narrow beam through any parabasal *ABCD*-sub-system. Then the transformation of the surface element is more complicated. It corresponds to the transport of the Jacobi matrix for the transformation of the local coordinate systems along the chief ray. With x, y the coordinates in the starting plane and x', y' those in the target plane, the Jacobi matrix of the transformation $x, y \rightarrow x', y'$ is

$$\underline{J} = \begin{vmatrix} \frac{dx'}{dx} & \frac{dx'}{dy} \\ \frac{dy'}{dx} & \frac{dy'}{dy} \end{vmatrix} = \frac{dx'}{dx} \cdot \frac{dy'}{dy} - \frac{dx'}{dy}\frac{dy'}{dx} \tag{9.71}$$

So, in general, a surface element is transformed as

$$\Delta A' = \underline{J} \cdot \Delta A \tag{9.72}$$

Several simplifications and approximations of this model can be made; e.g. a linear change in the direct vicinity of the ray. In a system with aberrations, the influences on the radiance distribution can be seen from the Jacobi matrix of the coordinate transformation.

For simple geometries, especially single aspherical mirror shapes, the transformation of the Jacobi matrix can be done analytically, thus shortening the computational time for raytracing. Also, in a more general form, the differential raytracing can be applied to the modelling of the transformation in the vicinity of a ray. So changes to the surface elements, and thus the radiometric conditions, can be simulated fast and accurately.

Within an optical system, radiometric modelling can be carried out without statistical means. All effective surfaces are discretized into sufficiently small surface elements. Then the transfer model above can be applied to each conceivable and reasonable combination of the surface elements. Expanding this simplified description, the surface elements are not usually perpendicular to the connecting line. In the same way, the incident angles have also to be taken into account and we have

$$\Delta^2 P = \frac{L_j}{r^2_{j,j+1}} \cdot \cos\theta_j \cdot \cos\theta_{j+1} \cdot \Delta A_j \cdot \Delta A_{j+1} \tag{9.73}$$

Within the narrow ray tube, the radiance L is assumed to remain constant.

Using raytubes is improving radiometric calculations quite a lot. But there are also some problems in real systems with segmentation, non-analytical derivatives in the refraction or reflection and more. This is discussed to some extent in Section 3.4.9. The discussion in Section 8.4 is also directly applicable here for the selection of the parabasal rays and the determination of local 4×4 *ABCD* matrices out of a finite raytrace.

9.4 Radiometry of Optical Systems

9.4.1 Introduction

For the calculation of radiance distribution and radiation transport in rotational-symmetrical optical systems, a small hollow light cone centred around the axis is considered as it propagates through the system. This geometry is illustrated in Figure 9.20. The radiance distribution in the image space results from an analysis of the geometrical change in the aperture angle and the change in the surface elements due to the system's magnification [9, 16]. The law of energy conservation has to be fulfilled locally for every beamlet.

The azimuthal angle around the optical axis is referred to as φ, u and u' are the aperture angles in the object and

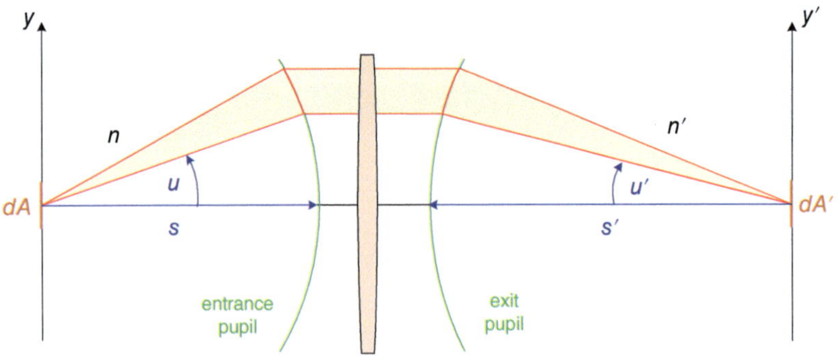

Figure 9.20 Energy conservation of radiation transport in an optical system.

image space, respectively. The transmission is assumed to be equal to 1, so there are no internal losses by absorption.

$$T = 1 \tag{9.74}$$

The solid angle element seen from the object element dA is

$$d\Omega = \sin u \, du \, d\varphi \tag{9.75}$$

The differential local energy conservation in the hollow cone of light

$$\begin{aligned} d^2P &= d^2P' \\ d^2P &= L \cos u \, dA \, d\Omega = L \, \sin u \cos u \, dA \, du \, d\varphi \\ &= \frac{1}{2} L \, dA \, d\varphi \, d(\sin^2 u) \end{aligned} \tag{9.76}$$

If this equation is written without the radiance L of the object surface, an expression for the optical throughput G is obtained. So for compound systems without any absorption and scattering losses, the constancy of the optical flux is a significant condition for a system free from power losses. Combining, e.g., partial systems with different optical throughputs, the system cannot transport the total energy to the receiver or within the image plane. Therefore, for modular compound systems, the correct adjustment of the pupil positions and sizes is also an important requirement to maintain the power flux.

In lossless systems, global energy conservation is of course mandatory. If this is also valid locally for every sub-aperture or sub-solid angle, the sine condition must be fulfilled. This is an additional correction feature and is not forced by physical reasons.

9.4.2 Sine Condition and Photometry

Let's assume an optical system with the refractive indices n and n' in object and image space and the small area elements $dA = dy^2$ and $dA' = dy'^2$ in the object and image plane. See Figure 9.21 for the geometry and the notations. If first a small aperture cone with half angle du_0 is selected centred on the optical axis, the paraxiality is fulfilled and the Lagrange invariant reads

$$dy \cdot n \cdot \sin du_0 = dy' \cdot n' \cdot \sin du_0' \tag{9.77}$$

The light power is given with the radiance L in the distance s as

$$d^2P_o = L \cdot dA d\Omega = L \cdot dA \cdot \frac{\pi (s \cdot du_0)^2}{s^2} = \pi L \cdot dA du_0^2 \tag{9.78}$$

The same is assumed in the image plane and the conservation of energy gives

$$\begin{aligned} d^2P_0 &= d^2P_0' \\ \pi L \cdot dA du_0^2 &= \pi L' \cdot dA' du_0'^2 \end{aligned} \tag{9.79}$$

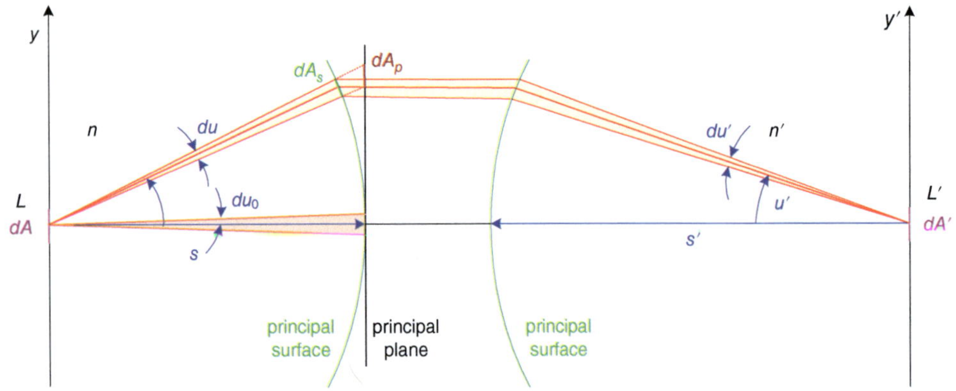

Figure 9.21 Energy conservation of radiation transport in an optical system.

Squaring (9.77) with $\sin u_0 \approx u_0$ gives

$$dy^2 \cdot n^2 \cdot du_0^2 = dy^{2\prime} \cdot n^{2\prime} \cdot du_0^{\prime 2} \tag{9.80}$$

Inserting (9.80) into (9.79) gives the relation between the radiances in case of energy preservation with $dA = dy^2$

$$\frac{\pi L}{n^2} \cdot \left(n^2 \cdot dy^2 du_0^2\right) = \frac{\pi L'}{n'^2} \cdot \left(n'^2 \cdot dy'^2 du_0'^2\right) \tag{9.81}$$

$$\frac{L}{n^2} = \frac{L'}{n'^2} \tag{9.82}$$

This is a very important result and shows that the radiance scaled by the index squared is an invariant in an optical system during propagation. In a second step the infinitesimal solid angle is inclined by the finite angle u according to the figure below by assuming an off-axis field position is considered. The area elements in the principal plane dA_p and a principal sphere dA_s are related by

$$dA_s = dA_p \cdot \cos u \tag{9.83}$$

If the light flux is now considered under the angle u by the projection of the area element dA we get

$$d^2P = L \cdot dA \cdot \cos u \cdot d\Omega = L \cdot dA \cdot \cos u \cdot \sin u \cdot du \cdot d\varphi \tag{9.84}$$

with the azimuthal angle φ. The local conservation of energy requires now again

$$d^2P = d^2P' \tag{9.85}$$

With Eq. (9.84) we get

$$\begin{aligned} &L \cdot dA \cdot \cos u \cdot \sin u \cdot du \cdot d\varphi \\ &= L' \cdot dA' \cdot \cos u' \cdot \sin u' \cdot du' \cdot d\varphi' \end{aligned} \tag{9.86}$$

with invariant azimuth due to circular symmetry $d\varphi = d\varphi'$, $dA = dy^2$ we have

$$\begin{aligned} L \cdot dy^2 \cdot \cos u \cdot \sin u \cdot du &= L\frac{n'^2}{n^2} \cdot dy'^2 \cdot \cos u' \cdot \sin u' \cdot du' \\ n^2 dy^2 \cdot \cos u \cdot \sin u \cdot du &= n'^2 \cdot dy'^2 \cdot \cos u' \cdot \sin u' \cdot du' \\ n^2\, dy^2 \cdot d(\sin^2 u) &= n'^2\, dy'^2 \cdot d(\sin^2 u') \\ d(n^2\, dy^2 \sin^2 u) &= d(n'^2\, dy'^2 \sin^2 u') \\ n^2\, dy^2 \sin^2 u &= n'^2\, dy'^2 \sin^2 u' \\ n \cdot dy \cdot \sin u &= n' \cdot dy' \cdot \sin u' \end{aligned} \tag{9.87}$$

which is the sine condition. Since the solid angle is here used in spherical coordinates

$$d\Omega = \sin u \cdot du \cdot d\varphi = \frac{dA_s}{s^2} \tag{9.88}$$

this means that the principal surface must be a sphere with radius s. This is one alternative interpretation of the sine condition. Due to this relation the local energy conservation cannot be obtained by a plane principal surface. Systems with a spherical principal surface are called aplanatic.

9.4.3 Aplanatic Systems

In the following we assume that the system is aplanatic corrected. The energy conservation is then given by

$$\frac{L}{n^2} = \frac{L'}{n'^2} \tag{9.89}$$

If the two media in the object and image space are the same, one gets

$$L = L' \tag{9.90}$$

If the differential flux is integrated, taking the sine condition into account, the irradiance is

$$E = \pi\, L \cdot \sin^2 u \tag{9.91}$$

From this equation it can be seen that the irradiance in an optical system is proportional to the square of the numerical aperture. If there is additionally an integral transmission factor T, which has been taken into account for the optical system, this equation takes the generalized form

$$E = \pi\, L \cdot T \cdot \sin^2 u \tag{9.92}$$

For a circular aperture and Lambertian-shaped object emission on the axis, the irradiance given as a function of the object radiance and the aperture angle of the image space is

$$E' = \pi\, L' \cdot \sin^2 u' = \left(\frac{n'}{n}\right)^2 \cdot \pi\, L \sin^2 u' \tag{9.93}$$

If in this equation the stop number $F_\#$ is introduced

$$F_\# = \frac{f}{D_{ExP}} = \frac{1}{2 \sin u'} \tag{9.94}$$

with the diameter D_{ExP} of the exit pupil, we get the relation

$$E'_\infty = \left(\frac{n'}{n}\right)^2 \cdot \frac{\pi \cdot L}{4 \cdot F_\#^2} \tag{9.95}$$

In this form the relation is valid for a system with an object located at infinity according to the $F_\#$ definition. The irradiance in the image of, e.g., a photographic system, increases quadratically with the inverse $F_\#$ number. Therefore, the $F_{\#\_}$ numbers are usually discretized with steps in $\sqrt{2}$ so that one step in the $F_\#$ number corresponds to a factor 2 in the irradiance. Table 9.4 give some values of $F\#$ numbers, the squares of the sine of the aperture angle and the corresponding change of the irradiance E.

Considering an image with a finite object position, the magnification m of the system as well as the magnification m_p of the pupil image, have to be taken into account. The pupil magnification m_p determines the behaviour of the diameters from the entrance to the exit pupil during the imaging through the system

$$m_p = \frac{D_{ExP}}{D_{EnP}} \tag{9.96}$$

Table 9.4 Irradiance as a function of the $F_\#$ - number.

$F_\#$ - number	$\text{Sin}^2\, u'$	E - relative [a.u.]
1 : 1	1/4	512
1 : 1.4	1/8	256
1 : 2	1/16	128
1 : 2.8	1/32	64
1 : 4	1/64	32
1 : 5.6	1/128	16
1 : 8	1/256	8
1 : 11	1/512	4
1 : 16	1/1024	2
1 : 22	1/2048	1

In the approximation for relatively short systems, the image position is at the rear focal point F'. The effective exit pupil is situated in front of F' at the distance $m_p f'$. So for the aperture angle on the image side we get

$$\sin u' = \frac{D_{ExP}}{2f' \cdot (m + m_P)} = \frac{1}{2F_\# \cdot \left(1 + \frac{m}{m_P}\right)} \tag{9.97}$$

and the image irradiance is, therefore, by inserting Eq. (9.97) into (9.93)

$$E' = \left(\frac{n'}{n}\right)^2 \cdot \frac{\pi \cdot L}{4 \cdot F_\#^2 \cdot \left(1 + \frac{m}{m_P}\right)^2} \tag{9.98}$$

The pupil magnification depends on the position of the aperture and can typically vary in the range $m_p = 0.5 \ldots 2$. For roughly symmetrical aperture positions, m_p is equal to 1. Then the simplified equation

$$E'(m) = \left(\frac{n'}{n}\right)^2 \cdot \frac{\pi \cdot L}{4 \cdot F_\#^2 \cdot (1+m)^2} = \frac{E'_\infty}{(1+m)^2} \tag{9.99}$$

can be used. So for a 1 : 1 imaging, about four times the light is necessary than is needed for an imaging from infinity in order to obtain the same irradiance.

9.4.4 Natural Vignetting

For an off-axis field point with the field angle w in the object space and w' in the image space, a natural decrease in brightness can be observed towards the edge of the image field relative to the axis point. This is also discussed in Chapter 5.5. Depending on the configuration, these effects can contribute to the result in different ways. For an accurate description, some specific aspects need to be considered. The basic geometry is shown in Figure 9.22 [19, 20].

The analytical calculation equations found in the following sections all assume that no vignetting occurs at other surfaces than the stop. But it should be clear that this is an idealization, which in general must not be true.

9.4.4.1 General Case Free of Vignetting

Each optical system has at least one limiting aperture which determines the transversal extension of the ray bundles. If this limiting effect changes with the azimuthal angle of the ray bundle, or if it varies with the field height, the model representations used here are not valid. So in the following it is always assumed that the systems stop limits the light flux completely, no effects of additional vignetting occur and circular symmetry is fulfilled. Determining the light flux always requires an integration of the radiance over the stop surface. Depending on the setup of the system, the calculation of the total power transfer from the object to the image has to take into account how the radiance changes over the integration range. This affects the variation of the cross-section area as well as the ray directions since both contribute to the solid angle.

In the general case, the limiting aperture is located inside the system, see Figure 9.23. This means that the optical components from the object to the stop space as well as from the stop to the image truncate the ray bundles. Since of course the transfer of the radiance from the object to the image is of interest, the calculation has to be carried out in two steps. We assume w_{stop} is the chief ray angle for a

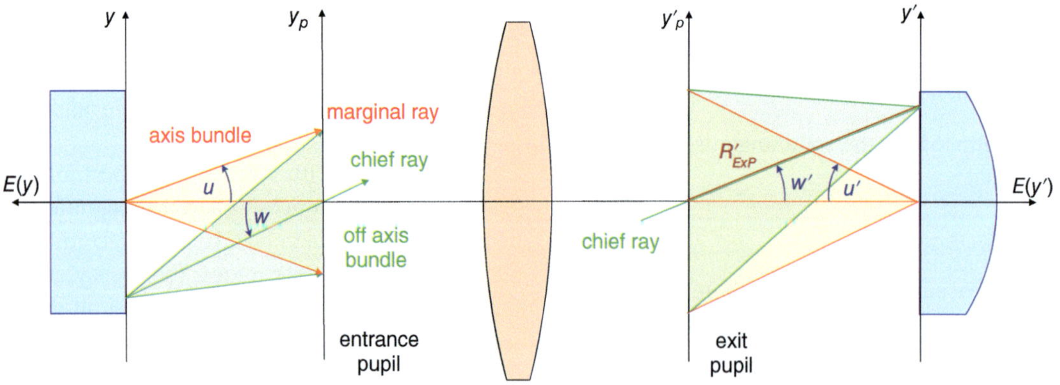

Figure 9.22 Natural decrease in the irradiance field for an off-axis image point with the field angle.

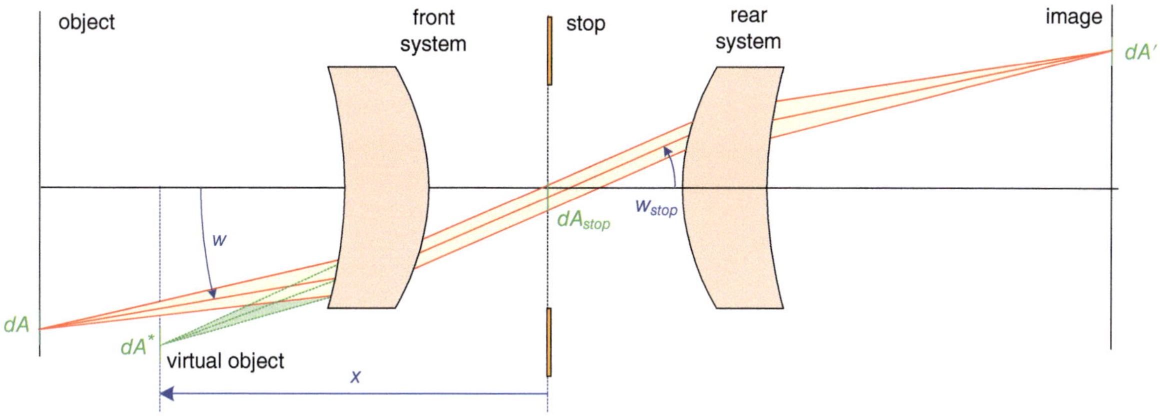

Figure 9.23 Radiation transport in an optical system with internal stop.

narrow ray bundle at the stop and dA_{stop} the cross section area of the ray bundle in the reference plane of the stop space. The area element in the object is dA; it is seen from the stop space with the virtual size dA^*, see Figure 9.23. Then the differential irradiance is given by [16]

$$dE = L \cdot \frac{dA^*}{dA} \cdot \cos^4 w_{stop} \cdot \frac{dA_{stop}}{x^2} \tag{9.100}$$

where x is the distance of the aperture image from the stop along the axis through the front group, see the figure. By integrating over the stop area, the total irradiance is obtained.

$$E = \int L \cdot \frac{dA^*}{dA} \cdot \cos^4 w_{stop} \cdot \frac{dA_{stop}}{x^2} \tag{9.101}$$

with the distortion of the pupil imaging related to the area transform ratio D_p as

$$D_p = 1 - \sqrt{\frac{dA^*}{dA}} \tag{9.102}$$

In the special case of a front system free of pupil distortion, the relation $dA^* = dA$ is valid. If moreover the object has a Lambert characteristic with $L =$ const., we can use the well-known $\cos^4$ law for the decrease in the intensity towards the edge of the field

$$E(w) = E(0) \cdot \cos^4 w_{stop} \tag{9.103}$$

9.4.4.2 Formulation with Entrance and Exit Pupil

If the internal stop is imaged as the entrance pupil into the object space, with a distance p between the object and the entrance pupil, as it is seen in Figure 9.24, then the radiation flux from the object plane to the entrance pupil is given by

$$dP = L \cdot dA \cdot \cos^4 w \cdot \frac{dA_{EnP}}{p^2} \tag{9.104}$$

In complete analogy with the expression using the exit pupil, the expression that is valid for the image space is obtained using the distance p' of the exit pupil

$$dP' = L' \cdot dA' \cdot \cos^4 w' \cdot \frac{dA_{ExP}}{p'^2} \tag{9.105}$$

Energy conservation requires that $dP = dP'$. From the sine condition, resulting in pupil distortion, and assuming the same refractive indices in the object and the image

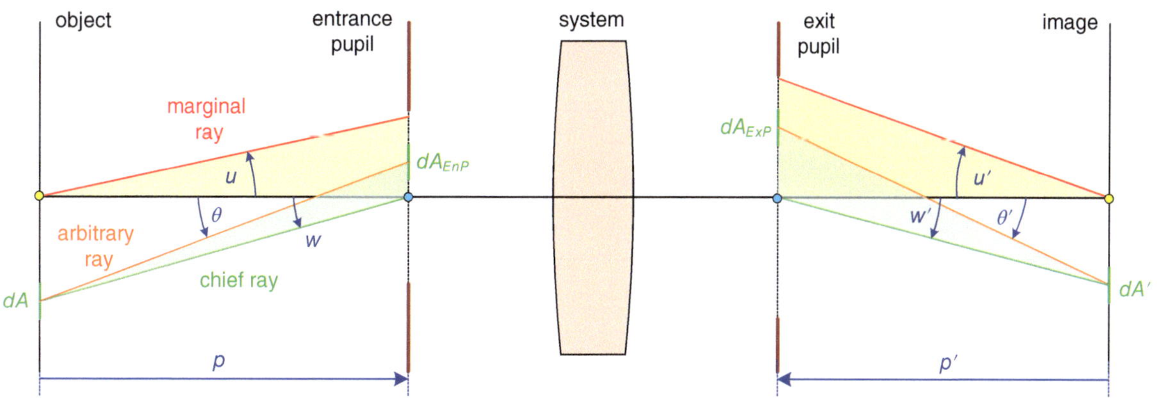

Figure 9.24 Radiation transport in an optical system, formulated with entrance and exit pupil.

domain with $L = L'$ we have

$$\frac{dA_{ExP}}{dA_{EnP}} = \left(\frac{n}{n'}\right)^2 \cdot \frac{dA}{dA'} \cdot \left(\frac{p}{p'}\right)^2 \cdot \frac{\cos^4 w}{\cos^4 w'} \tag{9.106}$$

In order to determine the total energy flux through the system, the incremental flux dP can be integrated over the object space or dP' over the image space.

9.4.4.3 Monocentric and Telecentric 4-f-Systems

Monocentric systems where the object and image surface are spherically curved constitute a very special case. The aperture is at the centre. In this case one obtains a specific field dependency of the form with only one $\cos(w')$ factor

$$E_{con}(w') = \left(\frac{n'}{n}\right)^2 \cdot L \cdot \pi \cdot \sin^2 u' \cdot \cos w' \tag{9.107}$$

This single projection factor results from the skew field angle in the pupil, reducing its area; see Figure 9.25.

In case of a telecentric 4-f-system, the conditions are identical and only a single cos factor for the projection of the internal stop remains, see Figure 9.26. No projection or distance related cos factors occur in the object and image domain in this case.

9.4.4.4 System with Rear Stop

In the particular case where the stop lies behind the system, the conditions are simplified quite significantly; see Figure 9.27.

- The size of the effective area of the exit pupil seen from O' decreases proportionally to $\cos w'$

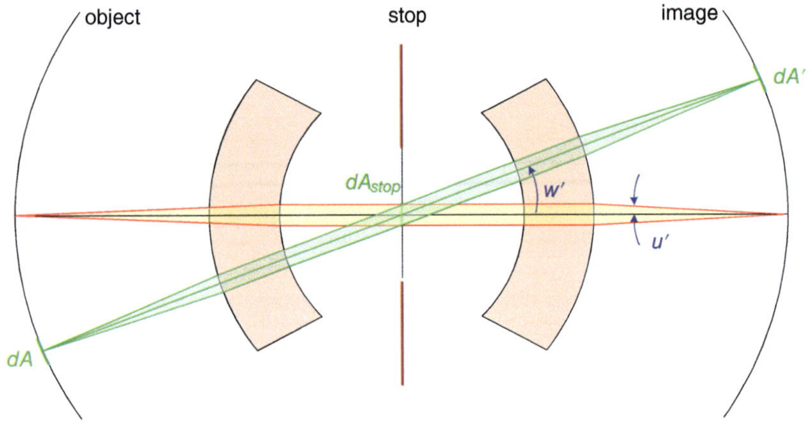

Figure 9.25 Radiation transport in a monocentric optical system.

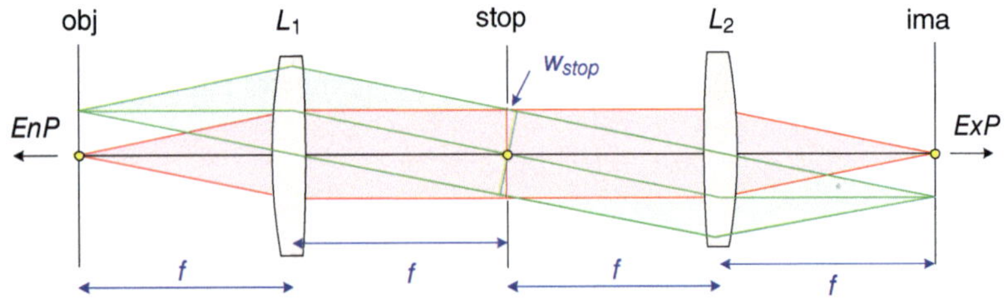

Figure 9.26 Radiation transport in a both-sides telecentric system.

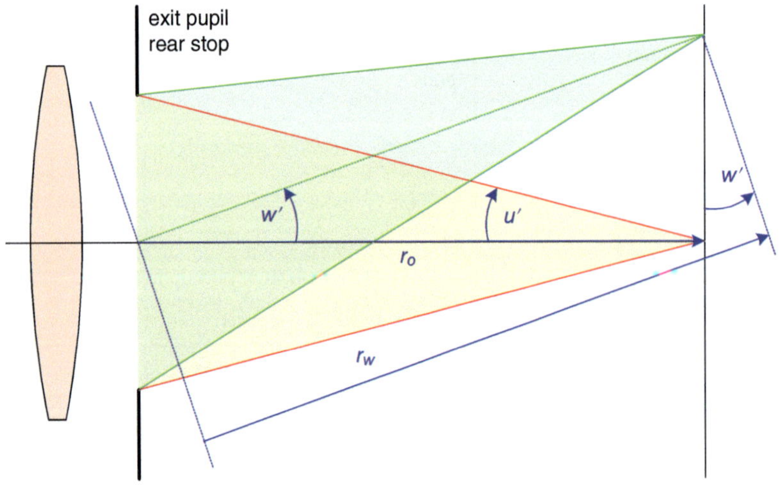

Figure 9.27 Radiance for an off-axis image point.

- The projection of a small surface element in O' perpendicular to the direction of the chief ray also produces a factor $\cos w'$

$$dP = L \cdot \frac{dA_{ExP} \cdot \cos w \cdot dA' \cdot \cos w}{r_w^2} \tag{9.108}$$

- The distance from the centre of the exit pupil to the image point O' grows proportionally to $\cos^2 w'$

$$r_w = \frac{r_0}{\cos^2 w} \tag{9.109}$$

By performing an exact integration with finite angles, one obtains

$$E(w') = \frac{\pi \cdot L}{2} \cdot \left(\frac{n'}{n}\right)^2 \cdot \left[1 - \left(1 + \frac{4\cos^2 w' \cdot \tan^2 u'}{(1-\cos^2 w' \cdot \tan^2 u')^2}\right)^{-1/2}\right] \tag{9.110}$$

On the optical axis one has the expression

$$E'(0) = \pi L' \cdot \sin^2 u' = \left(\frac{n'}{n}\right)^2 \cdot \pi \cdot L \sin^2 u' \tag{9.111}$$

In the approximation of a small image aperture $\sin U' \ll 1$, the term above can be expanded. Then the familiar $\cos^4 w'$ law of natural vignetting can be found in the simplified form:

$$\begin{aligned} E(w') &= E(0) \cdot \cos^4 w' \\ &= \left(\frac{n'}{n}\right)^2 \cdot L \cdot \pi \cdot \sin^2 u' \cdot \cos^4 w' \end{aligned} \tag{9.112}$$

Typical values for this decrease due to natural vignetting are listed in Table 9.5 and are illustrated in Figure 9.28.

Table 9.5 $\cos^4 w$ values.

w	$\cos^4 w$
0	1
10°	0.94
20°	0.78
30°	0.56

The exact dependence of the natural vignetting can also be formulated by introducing a correction factor f_{corr} into the $\cos^4 w'$ law. For small image-side apertures this correction factor converges to 1.

$$E(w') = \pi L \cdot \left(\frac{n'}{n}\right)^2 \cdot \sin^2 u' \cdot \cos^4 w' \cdot f_{corr} \tag{9.113}$$

This geometric correction factor always leads to an increase of the irradiance. It is similar to Eq. (9.48) given by

$$f_{corr}(w', u') = \frac{1 - \frac{1+\tan^2 w' - \tan^2 u'}{\sqrt{\tan^4 w' + 2\tan^2 w' \cdot (1-\tan^2 u') + (1+\tan^2 u')^{-2}}}}{2\sin^2 u' \cdot \cos^4 w'} \geq 1 \tag{9.114}$$

and depends in addition on the aperture angle u' as shown in Figure 9.29. The values of the correction factor lie between 1 and 2.1; for usual cases with $w < 20°$ and $u < 30°$ the correction is in the range of smaller than 10%.

In principle, the $\cos^4 w$ law can also be formulated using the object-side field angle w. In this case, however, an additional correction factor has to be introduced that describes the influence of the distortion inside the optical system, see next section.

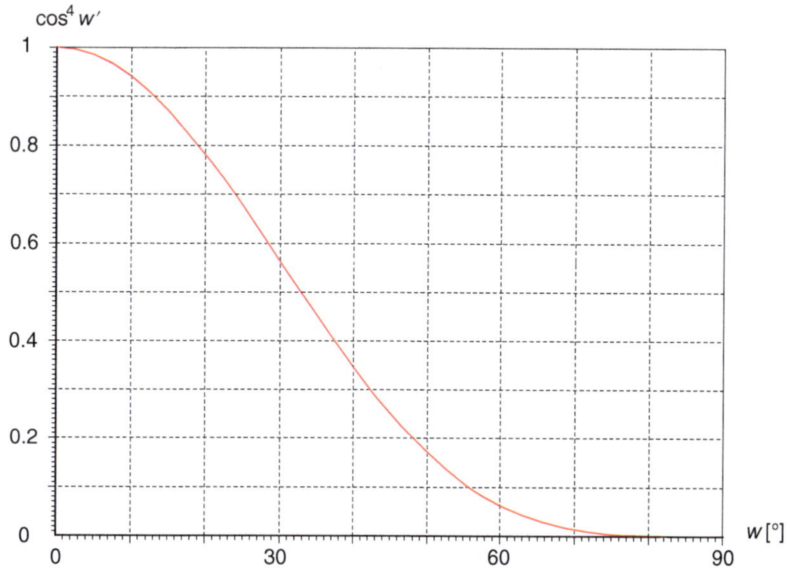

Figure 9.28 $\cos^4$ decrease in the irradiance for an off-axis image point.

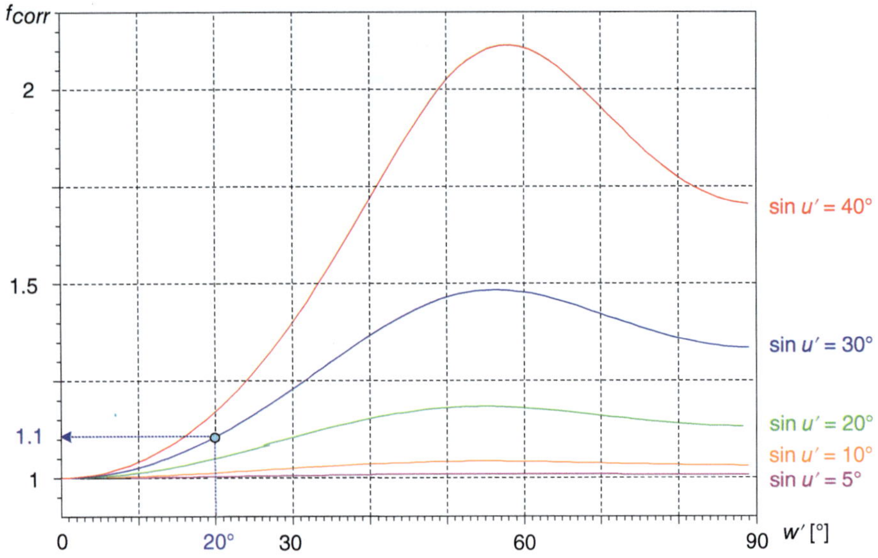

Figure 9.29 Correction factor of the irradiance relative to the natural vignetting as a function of the image-side field angle for different aperture angles u'.

9.4.4.5 System with Front Stop

If the stop lies in front of the system, there are analogous simple conditions. For the determination of the image-side irradiance, as it is shown in Figure 9.30, the magnification of the surface elements is included as an additional factor. Then the calculation is carried out in the image space as a function of the object-side field angle w by

$$E'(w) = \frac{\pi \cdot L}{2} \cdot \frac{dA}{dA'} \cdot \left[1 - \left(1 + \frac{4\cos^2 w \cdot \tan^2 u}{(1-\cos^2 w \cdot \tan^2 u)^2}\right)^{-1/2}\right] \tag{9.115}$$

In the special case of small object apertures this gives approximately

$$E'(0) = \pi L \cdot \frac{dA}{dA'} \cdot \sin^2 u \tag{9.116}$$

$$E'(w) = E'(0) \cdot \cos^4 w \tag{9.117}$$

For a distortion-free system with a constant magnification m it is

$$m = \sqrt{\frac{dA'}{dA}} \tag{9.118}$$

With the pupil magnification

$$m_p = \frac{D_{ExP}}{D_{EnP}} \tag{9.119}$$

it follows that

$$E'(w) = \frac{\pi L n'^2}{4n^2} \cdot \frac{D_{EnP}^2}{f'^2 \cdot \left(1 - \frac{m}{m_p}\right)} \cdot \cos^4 w \tag{9.120}$$

If there are distortions in the system with rotational symmetry, then the formula for the natural vignetting as a function of the object-side field can be calculated in the following way:

$$m_A = \frac{dA'}{dA} = \frac{r'}{r} \cdot \frac{dr}{dr'} \tag{9.121}$$

Finally, one obtains the expression

$$E' = \pi L \cdot \sin^2 u \cdot k \cdot \frac{r}{r'} \cdot \frac{dr'}{dr} \cdot \cos^4 w \cdot f_{corr} \tag{9.122}$$

It should be noticed here that the ratio r/r' describes the classical distortion, the second ratio dr'/dr describes the local slope distortion.

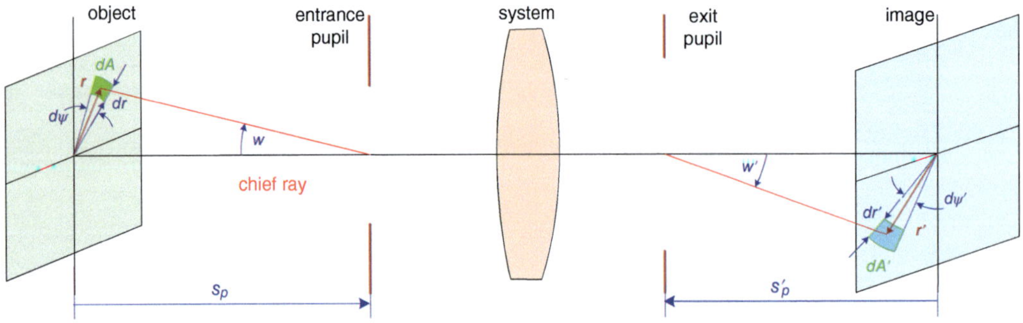

Figure 9.30 Radiation transport in an optical system with rotational symmetry.

If instead of the usual relation

$$y' = f \cdot \tan w \tag{9.123}$$

the equation

$$y' = f \cdot \sin w \tag{9.124}$$

is valid for an imaging system with a corresponding correction of a large distortion, the brightness does not decrease due to natural vignetting towards the edge. In this case the homogenization is achieved by a barrel distortion of the size

$$D = \frac{\sin w}{\tan w} - 1 = \cos w - 1 < 0 \tag{9.125}$$

9.5 Radiometry with Partial Coherent Light*

(*) The content of this section is quite specialized and for advanced reading. Knowledge of this section is not necessary to understand the following chapters. Beginners are recommended to skip this section.

9.5.1 Introduction

In the case of partial coherent light, the radiometric definitions must be generalized, because the simple geometrical behaviour is modified by interference and diffraction effects. A more physical definition is needed; this is found to be the quantity B instead of L [21–32]. The definition of B given in the next section starts from the cross spectral density and therefore is physically related to coherence theory.

The classical incoherent definition of the transfer of energy is given by the formula

$$d^2P = \frac{L}{r^2} \cdot dA_{e\perp}\, dA_{r\perp} = \frac{L}{r^2} \cdot \cos\theta_e \, \cos\theta_r \, dA_e \, dA_r \tag{9.126}$$

where P is the power, r the distance between the emitting source (e) and receiver (r), dA are the area elements and L the radiance. The receiving area divided by r^2 corresponds to the solid angle $d\Omega$ of the receiver element as it is seen by the source.

If a ray represents the connection between the two area elements, from the viewpoint of geometry, ignoring absorption effects, the radiance L is invariant along a ray, despite the ray propagating in free space or through components. If the geometry of Figure 9.31 is inspected, this property can be easily verified.

If the unit direction vector of the initial ray is given by

$$\vec{s}_e = (s_{e\,x}\; s_{e\,y}\; s_{e\,z})^T \tag{9.127}$$

starting from the rectangular source area element dxe, dye, the four rays comprising the area element in the receiver plane with side lengths dxr, dyr, the directions must be changed by $dxsx$, $dxsy$, $dxsz$ and $dysx$, $dysy$, $dysz$ to reach the displacements dxr and dyr. These two infinitesimally changed directions span the area that describes the solid angle. This area can be formulated by the determinant

$$d\Omega_e = \begin{vmatrix} d_x s_{e\,x} & d_x s_{e\,y} & d_x s_{e\,z} \\ d_y s_{e\,x} & d_y s_{e\,y} & d_y s_{e\,z} \\ s_{e\,x} & s_{y\,e} & s_{e\,z} \end{vmatrix} \tag{9.128}$$

the normalization of the unit vector length requires the additional constraint

$$s_{e\,x} ds_{e\,x} + s_{e\,y} ds_{e\,y} + s_{e\,z} ds_{e\,z} = 0 \tag{9.129}$$

this gives the solid angle

$$\begin{aligned} d\Omega_e &= \frac{1}{s_{e\,z}} (ds_{e\,x} ds_{e\,y} - ds_{e\,y} ds_{e\,x}) \\ &= \frac{1}{s_{e\,z}} \left(\frac{\partial s_{e\,x}}{\partial x_r} \frac{\partial s_{e\,y}}{\partial y_r} - \frac{\partial s_{e\,x}}{\partial y_r} \frac{\partial s_{e\,y}}{\partial x_r} \right) dx_r dy_r \end{aligned} \tag{9.130}$$

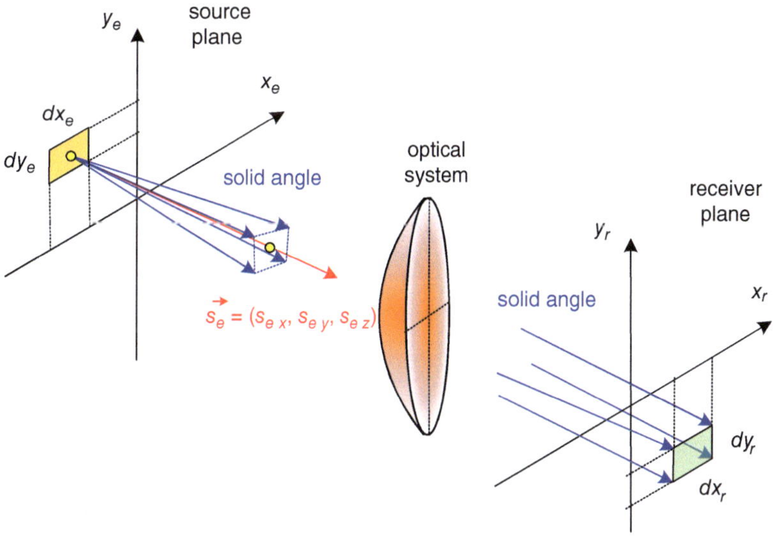

Figure 9.31 Transfer of the solid angle and invariance of the radiance.

This gives the power transfer with the cosine value s_z of the angle between the surface normal and the ray

$$d^2P_e = L_e \cdot \left(\frac{\partial s_{e\,x}}{\partial x_r} \frac{\partial s_{e\,y}}{\partial y_r} - \frac{\partial s_{e\,x}}{\partial y_r} \frac{\partial s_{e\,y}}{\partial x_r} \right) dx_e dy_e dx_r dy_r \tag{9.131}$$

The same consideration delivers the relation from the receiver to the source plane and gives

$$d^2P_r = L_r \cdot \left(\frac{\partial s_{r\,x}}{\partial x_e} \frac{\partial s_{r\,y}}{\partial y_e} - \frac{\partial s_{r\,x}}{\partial y_e} \frac{\partial s_{r\,y}}{\partial x_e} \right) dx_e dy_e dx_r dy_r \tag{9.132}$$

If $W = W_{PP}$ is the point–point eikonal of Bruns describing the optical path difference the Fermat principle requires, the equations of motion (see Chapter 13) for the light propagation follows as

$$\frac{\partial W}{\partial x_e} = -n_e\, s_{e\,x}, \quad \frac{\partial W}{\partial x_r} = -n_r\, s_{r\,x},$$
$$\frac{\partial W}{\partial y_e} = -n_e\, s_{e\,y}, \quad \frac{\partial W}{\partial y_r} = -n_r\, s_{r\,y} \tag{9.133}$$

are fulfilled. By calculating the mixed derivatives here, this allows us to write the Jacobian as

$$n_e \cdot J_e = n_e \cdot \left(\frac{\partial s_{r\,x}}{\partial x_e} \frac{\partial s_{r\,y}}{\partial y_e} - \frac{\partial s_{r\,x}}{\partial y_e} \frac{\partial s_{r\,y}}{\partial x_e} \right)$$
$$= n_e \cdot \left(\frac{\partial^2 W}{\partial x_e \partial x_r} \frac{\partial^2 W}{\partial y_e \partial y_r} - \frac{\partial^2 W}{\partial x_e \partial y_r} \frac{\partial^2 W}{\partial y_e \partial x_r} \right) \tag{9.134}$$

where the mixed derivatives can be exchanged. Due to the preservation of energy, $d2Pr = d2Pe$ is required. Therefore the relation

$$\frac{L_e}{n_e^2} = \frac{L_r}{n_r^2} \tag{9.135}$$

is obtained, which shows that the scaled radiance is invariant along the ray. In addition it is found that

$$n_e^2 J_e = n_r^2 J_r$$
$$n_e^2 dx_e dy_e ds_{e\,x} ds_{e\,y} = n_r^2 dx_r dy_r ds_{r\,x} ds_{r\,y} \tag{9.136}$$

If in a very raw approximation the dispersion of glasses is written as

$$n(\lambda) = \frac{n_0}{\lambda} \tag{9.137}$$

the last equation shows that the phase space volume is scaled in wavelengths as

$$\frac{1}{\lambda_e^2} dx_e dy_e ds_{ex} ds_{ey} = \frac{1}{\lambda_r^2} dx_r dy_r ds_{rx} ds_{ry} \tag{9.138}$$

which describes the invariance of the scaled Jacobian or the constance of the phase space area during propagation, if the energy is preserved. Due to the change of the wavelength with the refractive index, the last version is obtained.

By inserting the equations, it can also be seen that the radiance is associated with a ray; one ray in phase space carries the energy

$$d^2P = L(x, y, s_x, s_y) \cdot dx\, dz\, ds_x\, ds_y \tag{9.139}$$

9.5.2 Partial Coherent Light

If partial coherent light is considered, the complex field changes statistically and a meaningful observable is the correlation of the field at two different locations. If quasi-monochromatic light is considered, no wavelength-related partial coherence occurs and the coherence function can be defined as a time average

$$\Gamma(\Delta\vec{r}) = \Gamma(\vec{r}_1 - \vec{r}_2) = \langle E(\vec{r}_1, t+\tau)\, E^*(\vec{r}_2, t)\rangle_t$$
$$= \frac{1}{T} \int_t^{t+T} E(\vec{r}_1, t+\tau) \cdot E^*(\vec{r}_2, t)\, dt \tag{9.140}$$

and the cross spectral density (CSD) S reads

$$S(\vec{r}_1, \vec{r}_2) = \langle E^*(\vec{r}_1, \nu) \cdot E(\vec{r}_2, \nu)\rangle = \iint \Gamma(\Delta\vec{r}) \cdot e^{2\pi i \vec{\nu} \cdot \Delta\vec{r}} d^2\Delta\vec{r} \tag{9.141}$$

If $\vec{s}$ is a unit vector giving the direction of observation from a source plane, the radiant intensity in this direction can be written with S in the form

$$I(\vec{s}) = \frac{\cos^2\theta}{\lambda^2} \cdot \iint S(\vec{r}_1, \vec{r}_2) \cdot e^{-ik\,\vec{s}_\perp(\vec{r}_2 - \vec{r}_1)}\, d\vec{r}_1\, d\vec{r}_2 \tag{9.142}$$

If the CSD is expanded into a coherent mode Mercer sum in the form

$$S(\vec{r}_1, \vec{r}_2) = \sum_n \lambda_n \varphi_n^*(\vec{r}_1) \varphi_n(\vec{r}_2) \tag{9.143}$$

with the eigenvalues λn, the radiant intensity can be expressed as

$$I(\vec{s}) = \frac{\cos^2\theta}{\lambda^2} \cdot \sum_n \lambda_n |\psi_n(k\vec{s}_\perp)|^2 \tag{9.144}$$

with the Fourier transformed mode functions

$$\psi_n(k\vec{s}_\perp) = \int \varphi_n(\vec{r}) \cdot e^{-ik\vec{r}\vec{s}_\perp} d\vec{r} \tag{9.145}$$

9.5.3 Generalized Radiance

For a partial coherent source, the formulation of the photometry is not as trivial as in the incoherent case. The behaviour is similar and a corresponding generalized radiance B with invariance along a propagation ray can only be defined if the source is quasi-homogeneous and the paraxial approximation is assumed for the propagation.

If $S(r_1,r_2)$ is the cross spectral density, a generalized radiance is given by

$$B(\vec{r},\vec{s}) = \frac{\cos\theta}{\lambda^2}\int_{-\infty}^{+\infty} S\left(\vec{r}+\frac{\vec{r}'}{2},\vec{r}-\frac{\vec{r}'}{2}\right)\cdot e^{-ik\vec{s}_\perp\cdot\vec{r}'}\,d^2\vec{r}' \tag{9.146}$$

and is valid for any type of coherence shape. $\vec{s}_\perp$ is here the lateral component of the direction unit vector, so we have the projection of $\vec{s}$ into the source plane. The dependence of B on r and $\vec{s}$ indicates the interpretation that the radiance is the energy flow locally in the point r into the direction $\vec{s}$. If this last equation is inverted, a generalized van Cittert–Zernike theorem is obtained in the form

$$S(\vec{r},\Delta\vec{r}) = \int B(\vec{r},\vec{s})\cdot e^{ik\vec{s}\cdot\vec{r}}\,d\Omega \tag{9.147}$$

Only if the source is quasi-homogeneous can this general expression be simplified to

$$B_{QH}(\vec{r},\vec{s}) = k^2\cdot\cos\theta\cdot I(\vec{r})\cdot\mu(k\cdot\vec{s}_\perp) \tag{9.148}$$

where the spatial frequency spectrum in the source plane is real and non-negative and is given by

$$\mu(\vec{v}) = \frac{1}{(2\pi)^2}\int_{-\infty}^{+\infty}\mu(\vec{r}')\cdot e^{-i\vec{v}\cdot\vec{r}'}\,d^2\vec{r}' \tag{9.149}$$

This definition can also be written with the help of the complex field E as

$$B(\vec{r},\vec{s}) = \mathrm{Re}[\cos\theta\cdot\langle E^*(\vec{r})\cdot E(\vec{s})\rangle\cdot e^{ik\vec{r}\cdot\vec{s}}] \tag{9.150}$$

where the definition of the CSD as a correlation is used.

9.5.4 Propagation of the Generalized Radiance

9.5.4.1 Introduction

The definition of the generalized radiance can also be defined for an arbitrary value of z as the light propagates outside of the source plane as the Fourier transform

$$B(\vec{r},\vec{s},z) = \frac{\cos\theta}{\lambda^2}\int_{-\infty}^{+\infty} S\left(\vec{r}+\frac{\vec{r}'}{2},\vec{r}-\frac{\vec{r}'}{2},z\right)\cdot e^{-ik\vec{s}_\perp\cdot\vec{r}'}\,d^2\vec{r}' \tag{9.151}$$

It can be shown that the generalized radiance propagates along a ray as

$$B(\vec{r},\vec{s},z) = B\left(\vec{r}-\frac{z}{\cos\theta}\vec{s}_\perp,\vec{s},0\right) \tag{9.152}$$

so the invariance is not perfectly fulfilled. But it can be shown that in the paraxial approximation of small θ and a quasi-homogeneous case, the radiance again follows the invariance law during propagation

$$\frac{dB_{par}(\vec{r},\vec{s},z)}{ds} = 0 \tag{9.153}$$

or alternatively

$$\vec{s}\cdot\nabla_r B_{par}(\vec{r},\vec{s},z) = 0 \tag{9.154}$$

It should be noted here that the generalized radiance B as defined here has some different properties in comparison to the usual radiance in incoherent photometry. In particular it can take negative values, therefore it cannot be simply interpreted as a positive flux density. This situation is quite similar as in the case of the Wigner function. It can be shown that the relation between the generalized radiance and the Wigner function is simply (see Chapter 10)

$$B(\vec{r},\vec{s}) = \frac{1}{\lambda^2}\sqrt{1-s_\perp^2}\cdot W(\vec{r},k\vec{s}_\perp) \tag{9.155}$$

If the generalized radiance is integrated over the spatial coordinate, we obtain the generalized radiant intensity as an angular distribution of the flux density in the far zone as

$$I(\vec{s},\omega) = \cos\theta\cdot\int_{-\infty}^{+\infty} B(\vec{r},\vec{s},\omega)\,d^2\vec{r} \tag{9.156}$$

If the propagation of the complex field

$$E(\vec{r},z) = \int G(\vec{r}-\vec{r}')\cdot E(\vec{r}',0)\,d^2\vec{r}' \tag{9.157}$$

with the Green's function G

$$G(\vec{r}-\vec{r}') = -\frac{1}{2\pi}\frac{\partial}{\partial z}\left(\frac{e^{ik|\vec{r}-\vec{r}'|}}{|\vec{r}-\vec{r}'|}\right) \tag{9.158}$$

is considered, the propagation of the generalized radiance can be written in the exact form

$$B(\vec{r},\vec{s},z) = \int K(\vec{r}-\vec{r}',\vec{s})\cdot B(\vec{r},\vec{s},0)\,d^2\vec{r}' \tag{9.159}$$

with the kernel function

$$K(\vec{r}-\vec{r}',\vec{s}) = \int G^*\left(\vec{r}-\vec{r}'+\frac{\Delta r}{2}\right)\cdot G\left(\vec{r}-\vec{r}'-\frac{\Delta r}{2}\right)\,d^2\Delta r \tag{9.160}$$

If the paraxial approximation is assumed, from the corresponding Fresnel–Green's function in case of an *ABCD*-optical system this representation allows us to write the propagation of the generalized radiance similarly as for the Wigner function in one lateral dimension simply as

$$B'(x,s_\perp) = B(Dx-Bs_\perp,-Cx+As_\perp) \tag{9.161}$$

As a special case from this expression, the free-space paraxial approximation is obtained as

$$B'(x,s_\perp) = B\left(x-\frac{z}{s_z},s_\perp\right) \tag{9.162}$$

9.5.4.2 Example Lambertian Source

If a Lambertian source is considered with a uniform spatial intensity inside the source circular area of radius a, in the

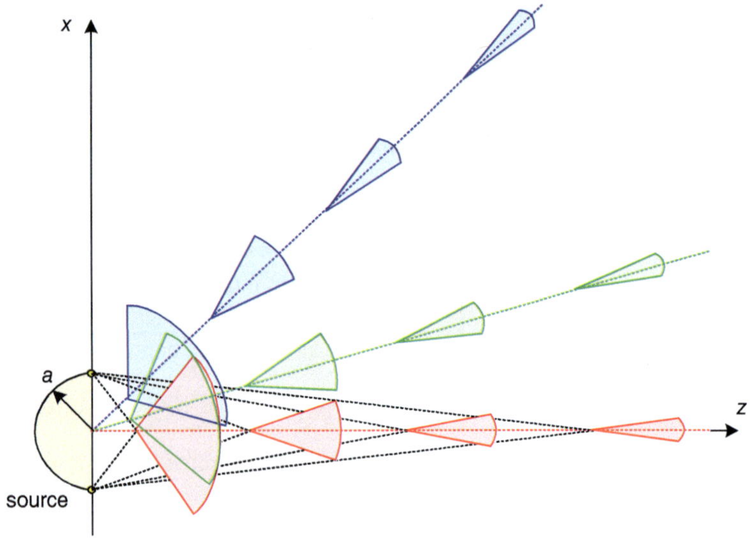

Figure 9.32 Polar diagram of the radiance for a homogeneous Lambertian source.

source plane the spectral density is given by

$$S(\vec{r}) = \begin{cases} S_o & for\ r < a \\ 0 & for\ r > a \end{cases} \tag{9.163}$$

the degree of spectral coherence is

$$\mu(\vec{r}_2 - \vec{r}_1) = \frac{\sin(k \cdot |\vec{r}_2 - \vec{r}_1|)}{k \cdot |\vec{r}_2 - \vec{r}_1|} \tag{9.164}$$

The Fourier transform gives the spectral degree of coherence

$$\mu(k \cdot \vec{s}_\perp) = \frac{1}{2\pi k^2 s_z} \tag{9.165}$$

and the radiance function is

$$B(\vec{r}, \vec{s}, z) = \begin{cases} \frac{1}{2\pi} S_0 & if\ r < a\ and\ \vec{s} \in \Omega_s \\ 0 & else \end{cases} \tag{9.166}$$

where the solid angle ΩP is given by the source area seen by the observation point P. In one transverse dimension the angle range is given by the limiting angles

$$\tan\alpha_{\max} = \frac{x+a}{z}, \quad \tan\alpha_{\min} = \frac{x-a}{z} \tag{9.167}$$

The behaviour of B is sketched in Figure 9.32.

If a Gaussian–Schell model source is considered instead, the behaviour is very similar, only the hard boundaries are substituted by soft Gaussian decay curves.

References

1 Botella, A., Winston, R., and Jiang, L. (2023). *Field Theory of Nonimaging Optics*. CRC Press.

2 Kingslake, R. (1965). *Applied Optics and Optical Engineering*, vol. I. New York: Academic Press.

3 McCluney, R. (1994). *Introduction to Radiometry and Photometry*. Boston: Artech House.

4 Muschaweck, J. and Rehn, H. (2022). *Designing Illumination Optics*. SPIE Press.

5 Lindsay, J.L. (1997). *Applied Illumination*. Lilburn: The Fairmont Press.

6 Winston, R., Minano, J., and Benitez, P. (2005). *Nonimaging optics*. Elsevier.

7 Chaves, J. (2016). *Nonimaging Optics*. CRC Press.

8 DeCusatis, C. (1998). *Handbook of Applied Photometry*. Springer.

9 Koshel, J. (2013). *Illumination Engineering*. Wiley, IEEE Press.

10 Winston, R., Jiang, L., and Oliker, V. (2021). *Nonimaging Optics*. CRC Press.

11 Gitin, A.V. (1998). Radiometry. A comprehensive approach. *Soviet Journal of Optical Technology* 65: 132.

12 Butler, D.J., Koehler, R., and Forbes, G.W. (1996). Diffraction effects in the radiometry of coherent beams. *Applied Optics* 35: 2162.

13 W. J. Smith (1985). Elements of radiometric and photometric calculations. *SPIE Conference Proceedings* 531: 168.

14 Tryka, S. (1997). Angular distribution of the solid angle at a point subtended by a circular disk. *Optics Communication* 137: 317.

15 Mahajan, V.N. (1998). *Optical Imaging and Aberrations, Part I, Ray Geometrical Optics*. Bellingham: SPIE Press.

16 Kingslake, R. (1965). *Applied Optics and Optical Engineering*, vol. II. New York: Academic Press.

17 H.-J. Frasch and G. Steimetz (2002). Monte-Carlo methods in raytracing software. *SPIE Conference Proceedings* 4769: 55.

18 Jannson, T., Aye, T., Tengara, I., and Erwin, D. (1996). Second order radiometric ray tracing. *Journal of the Optical Society of America A* 13: 1448.

19 Reiss, M. (1948). Notes on the cos**4 law of illumination. *Journal of the Optical Society of America* 38: 980.

20 Cassarly, W.J., Fest, E.C., and Jenkins, D.G. (2002). Error estimation and smoothing of 2D illumination and chromaticity distributions. *SPIE Conference Proceedings* 4769: 93.

21 Walther, A. (1995). *The Ray and Wave Theory of Lenses*. Cambridge University Press.

22 Walther, A. (1968). Radiometry and coherence. *Journal of the Optical Society of America* 58: 1256.

23 Walther, A. (1973). Radiometry and coherence. *Journal of the Optical Society of America* 63: 1622.

24 Marchand, E. and Wolf, E. (1974). Radiometry with sources of any state of coherence. *Journal of the Optical Society of America* 64: 1219.

25 Friberg, A. (1979). Effects of coherence in radiometry. *SPIE Conference Proceedings* 194: 55.

26 Friberg, A. (1979). Radiation from partially coherent sources. *SPIE Conference Proceedings* 194: 71.

27 Friberg, A. (1981). On the generalized radiance associated with radiation from a quasihomogeneous planar source. *Optica Acta* 28: 261.

28 Wolf, E. (1994). Radiometric model for propagation of coherence. *Optics Letters* 19: 2024.

29 Wolf, E. (1978). Coherence and radiometry. *Journal of the Optical Society of America* 68: 6.

30 Carter, W. and Wolf, E. (1977). Coherence and radiometry with quasi homogeneous planar sources. *Journal of the Optical Society of America* 67: 785.

31 Walther, A. (1978). Propagation of the generalized radiance through lenses. *Journal of the Optical Society of America* 68: 1606.

32 Petrucelli, J. and Alonso, M. (2009). Propagation of nonparaxial partially coherent fields across interfaces using generalized radiometry. *Journal of the Optical Society of America A* 26: 2012.

10

Phase Space Representation

10.1 General Aspects

10.1.1 Motivation

The concept of the phase space is a very important approach that is well known from quantum physics. In optics, the phase space is spanned by spatial and angle coordinates; alternatively the angle can be expressed as a spatial frequency. In the general case, two spatial coordinates x and y as well as two angle coordinates u and v are used. This means that in total the phase space corresponds to a four-dimensional representation. This is also one of its drawbacks: in the full description the light description needs tremendous computer resources. Therefore, mostly only one lateral dimension is used or a circular symmetry is assumed to get rid of two dimensions.

Due to this property, the question is whether the phase space is a comfortable tool for calculations. In any case, it is quite useful for illustration and visualization. In particular, more complicated setups for illumination can be understood very well within this representation. From the physical point of view, it is quite interesting that in the phase space we have the opportunity to model and to interpret the light by all of the following means.

- Geometrical optics by rays, vignetting, aberrations in space as well as in angle, caustics.
- Illumination optics by special segmented array components or mixing rods, multi-aperture effects and photometry.
- Diffraction effects, like Gaussian beams or interference fringes, gratings.
- Partial coherence, deterministic as well as statistical light properties.
- Resolution and information.
- Fermat principle, Lagrange and Hamiltonian principles of light equations of motion.
- Invariance properties like the Lagrange–Helmholtz invariant, uncertainty relation.
- Propagation of light in non-homogeneous gradient index media.
- Diffraction related beam propagation, Fresnel diffraction and its sampling issues.
- Laser beams and intensity moments for their quality description.
- Optical transfer function, ambiguity function.
- Similarities of the optical models of geometrical and wave optics with the situation in mechanics with classical mechanics and quantum mechanics.

It will be seen that all of these effects and properties can be easily described in the phase space, therefore it's a quite general representation. There are a lot of publications and books on this topic that deal with the mathematical relations of the phase space [1–5]. In this book we don't want to go into deep theoretical details, we only want to demonstrate the usefulness of phase space considerations in many aspects of lens design.

In this section some simpler examples are shown, which illustrate the application of a phase space representation to understand the behaviour of light in the spatial-angle domain. Figure 10.1 shows a synthetic ray path through a system consisting of two thin lenses and a gradient index component in one transverse dimension only. In the free space transitions only x changes and vertical lines are given. The thin lenses and the refractive surfaces change only the angles and horizontal lines are obtained. In total the path of the phase spaces characterizes the z-propagation of the ray. The path is governed by the Fermat principle and follows a Lagrange or Hamilton equation of motion.

If now the wave equation is expanded into a series, asymptotic for small wavelengths according to the description in Section 3.4.14, a bundle of rays in the vicinity of a guide ray is obtained. This aggregate of rays [6, 7] is demonstrated in Figure 10.2. In part (b) of this figure a vicinity according to the uncertainty relation can be imagined, see Section 10.1.3. This corresponds to a parabasal field around the main ray or the sum of adjacent Gaussian beams.

Lens Design for Imaging: Fundamentals of Optical Systems, First Edition. Herbert Gross.

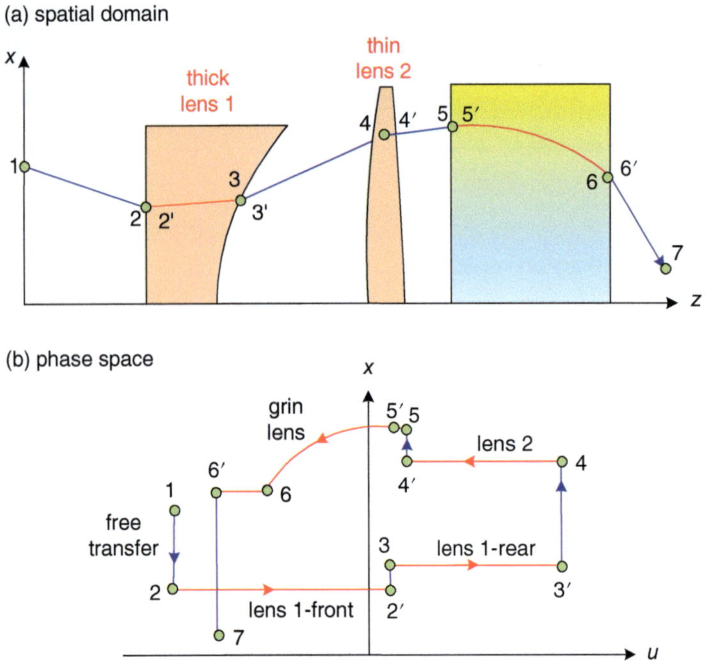

Figure 10.1 Illustration of the propagation of a single ray in an optical system with different components in (a) spatial domain and (b) phase space. The drawing in (b) is schematical and not to scale.

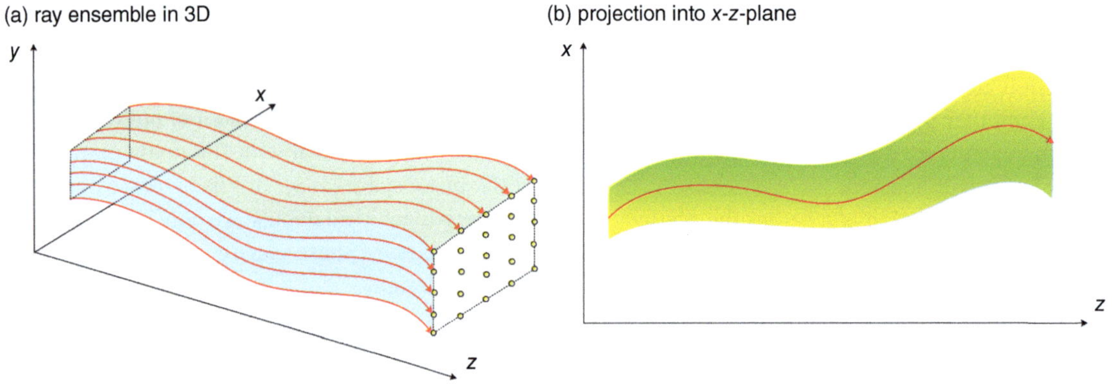

Figure 10.2 Ensemble of rays and continuous vicinity aggregate.

Finally, if the transition to a physical beam with finite extension is carried out, an area-filled picture of the phase space representation is obtained. The Wigner function as a quasi-density function in the phase space of a physical beam can be used to illustrate the development of the energy density in spatial and angle coordinates of a light field. For a propagating beam, this function in one lateral dimension $W(x, u)$ evolves with z. Figure 10.3 illustrates this in a slice representation as $W(x, u, z_j)$ in selected z-planes.

10.1.2 Legendre Transform and Equation of Motion

The usual description of optical systems and rays is done in the spatial domain. If we move from this coordinate picture into the phase space, a Legendre transform has to be performed to switch from the z-coordinate to the angle variable u. This idea is sketched in Figure 10.4; the derivation is here performed for simplicity in one lateral coordinate x only. Let's assume a ray runs from point P to P'. The description of its path can be realized by the axis coordinate z, by its local slope $\tan u(z)$ or by the angle $u(z)$. Both variables are possible [7–10].

The realized path of a ray between two points is governed by Fermat's principle, the ray takes the shortest time of arrival. The path length is the geometrical arc length, but the velocity depends on the local refractive index $n(x, z)$. The momentum variable p can be defined by the slope multiplied with the refractive index.

$$p(z) = n \cdot \dot{x} = n \cdot \frac{dx}{dz} = n \cdot \tan u \tag{10.1}$$

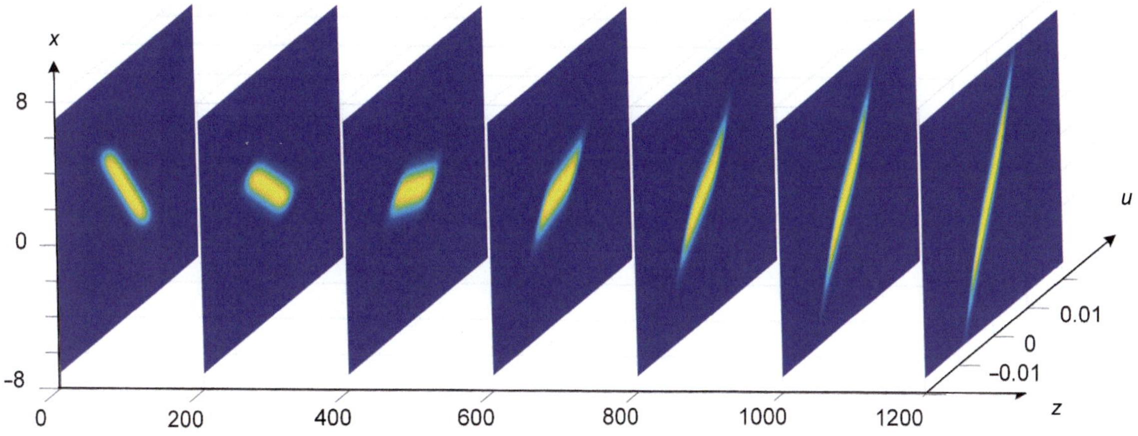

Figure 10.3 Illustration of the propagation of the Wigner function of a beam with z in the phase space. The unities of the axes are arbitrary.

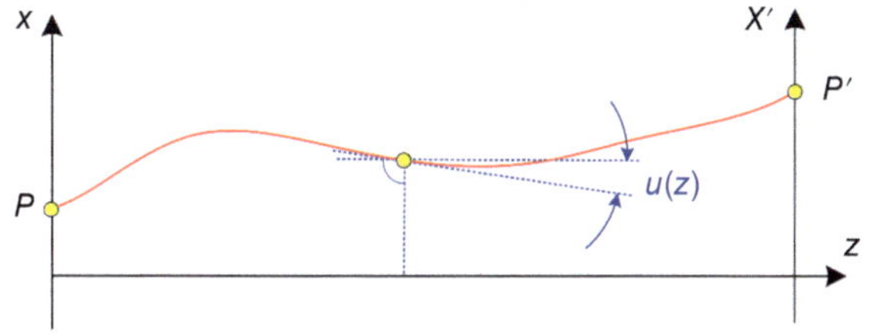

Figure 10.4 Ray path as a function of the distance coordinate z and its slope $\dot{x}=dx/dz = \tan(u)$.

The transit time of the ray or photon for the path from point P to P' is given by the velocity $v = c/n$ as

$$t = \int \frac{ds}{v} = \frac{1}{c}\int n(x,z)\cdot\sqrt{1+\dot{x}^2}dz$$
$$= \frac{1}{c}\int L(x,\dot{x},z)dz = \frac{1}{c}S(x,\dot{x},u) \tag{10.2}$$

with the Lagrangian function

$$L(x,\dot{x},z) = n(x,z)\cdot\sqrt{1+\dot{x}^2} \tag{10.3}$$

and the optical path length S. According to the Fermat principle, the transit time is a minimum and the optical path length variation has to vanish for a change of the parameter u. This means

$$\frac{\partial S(x,\dot{x},\mathrm{u})}{\partial u} = \int\left(\frac{\partial L}{\partial x}\frac{\partial x}{\partial u} + \frac{\partial L}{\partial \dot{x}}\frac{\partial \dot{x}}{\partial u}\right)dz$$
$$= \int \frac{\partial x}{\partial u}\left[\frac{\partial L}{\partial x} - \frac{\partial}{\partial z}\left(\frac{\partial L}{\partial \dot{x}}\right)\right]dz = 0 \tag{10.4}$$

Now we get as a condition the vanishing of the square bracket, which is the equation of motion or Euler equation of the variation problem in geometrical optics and corresponds to the eikonal equation

$$\frac{\partial L}{\partial x} - \frac{\partial}{\partial z}\left(\frac{\partial L}{\partial \dot{x}}\right) = 0 \tag{10.5}$$

The Hamiltonian picture uses not the velocity v, but the momentum p as variable, which is in optics defined as

$$p = n\sin\theta = \frac{n\dot{x}}{\sqrt{1+\dot{x}^2}} = \frac{\partial L}{\partial \dot{x}} \tag{10.6}$$

As $\dot{x}$ it describes the direction of the ray as well, not the angle tangent value, but the sine multiplied by the refractive index. The change of the variable $\dot{x}$ into p can be described mathematically as a Legendre transform $\hat{L}$ of the Lagrange function, which delivers the Hamilton function H. It is given by

$$H(x,p,z) = \hat{L}[L(x,\dot{x},z)] = L(x,\dot{x},z) - \dot{x}\cdot p \tag{10.7}$$

By inserting the Lagrange function we obtain explicitly for the Hamiltonian

$$H(x,p,z) = \hat{L}[L(x,\dot{x},z)] = -\sqrt{n^2(x,z) - p^2}$$
$$= -n\cdot\cos\theta(x,z) \tag{10.8}$$

The Lagrange condition to fulfil the Fermat principle now delivers the equation of motion for the Hamiltonian by the two equations of motion

$$\dot{p} = -\frac{\partial H}{\partial x},\ \dot{x} = \frac{\partial H}{\partial p} \tag{10.9}$$

Now two differential equations are obtained. Figure 10.5 illustrates the change of variables from L to H and the meaning of the derivatives as slopes in a local point Q_j. In the point Q_j at $\dot{x}_j$ in the Lagrange domain, the local slope is given by p_j. According to the Legendre transform, the corresponding point Q_j in the Hamilton domain lies a shift of $p_j\dot{x}_j$ lower. This value corresponds to point Q, where the tangent meets the L-axis for $\dot{x}_j = 0$. The second Hamilton equation gives the slope in Q_j as $\dot{x}_j$ at the coordinate p_j. The tangent in Q_j of Hamilton's diagram intersects the H-axis in the point R, which has the same value as Q_j in

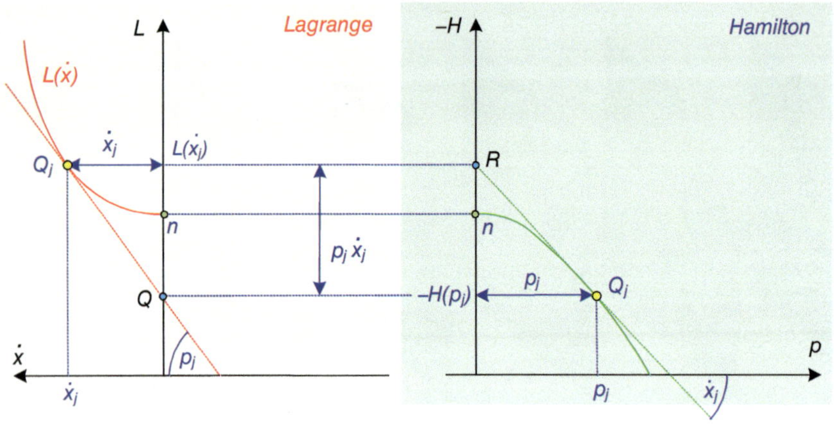

Figure 10.5 Relation between L in the spatial domain (left) and H in the phase space (right) by the Legendre transform.

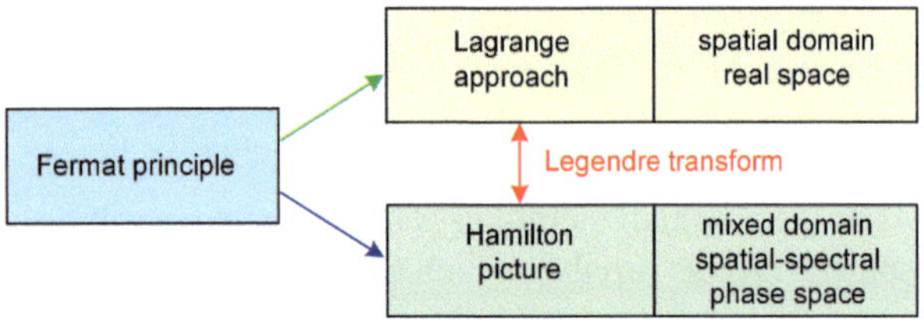

Figure 10.6 Relation between the Fermat principle, and the Lagrange and Hamilton formulation by Legendre transform.

the Lagrange domain. For $p_j = 0$ and $\dot{x}_j = 0$ L and H have the same value n.

If the phase space concept is considered, the Legendre transform describes mathematically the transition from the z-coordinate to the angle coordinate u and therefore conversion into the phase space picture.

Figure 10.6 illustrates again the connection between the two representations and the minima principles of Fermat, resulting in the Lagrange or Hamilton formalism respectively [5].

10.1.3 Uncertainty Relation

As has been already mentioned, there are interesting similarities between mechanics/quantum mechanics and geometrical/wave optics. Therefore it is not surprising that there is a formulation of the well-known uncertainty relation in optics too. If the sketches in Figure 10.7 are considered, the two cases of slit diffraction show that the angle spreading due to diffraction is given by

$$\theta = \frac{\lambda}{D} \tag{10.10}$$

and grows if the slit width D decreases. In the end, according to the slit diffraction the product of both quantities is constant and only fixed by the wavelength. On the other side, a smaller slit means that the spatial position of the light bundle is better known. The cost is poorer information about the angle direction afterwards. This is exactly the interpretation of the uncertainty relation.

The same behaviour is found in the case of Gaussian beam propagation and their basic properties. In Figure 10.8 a Gaussian beam is shown for different values of the waist size w_0. According to the basic relationships of Gaussian beam parameters, the waist size and the far-field divergence angle θ_0 are coupled by the equation

$$w_0\,\theta_0 = \frac{\lambda}{\pi} \tag{10.11}$$

The interpretation of this equation is completely identical to the previous case of the slit. The structure of the equation

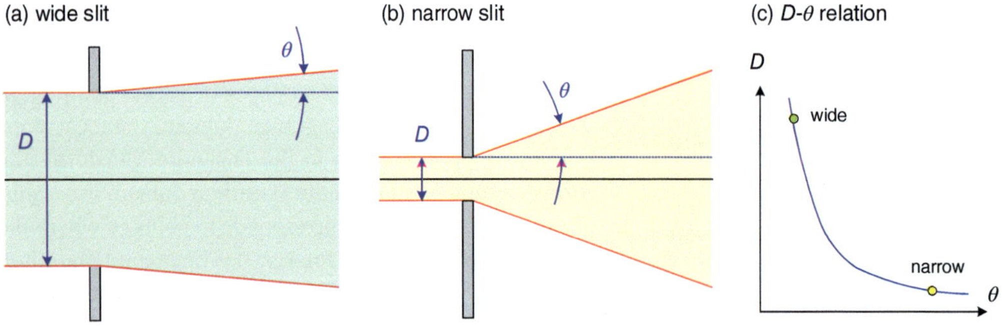

Figure 10.7 Slit diffraction and comparison of the geometry in case of (a) a wide and (b) a narrow slit respectively, (c) shows both cases in a D-θ diagram.

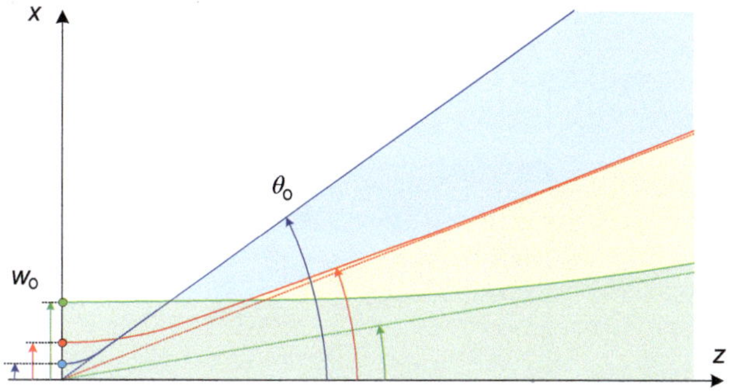

Figure 10.8 Behaviour of a Gaussian beam in case of different waist sizes, the far-field divergence angle grows reciprocal to the waist size w_0.

is also the same, only the numerical pre-factor is changed. Again, we have the reciprocity relation between spatial and angle properties.

Since the spatial extent and the angle spreading are exactly the two coordinates spanning the phase space, the interpretation of the optical uncertainty can also be formulated as follows: there is a minimum area size in the phase space for a light distribution, which is scaled only by the wavelength. This area is the minimum value that can exist, it can be changed in shape, but not in area size. It's the elementary size of a light distribution due to the uncertainty properties, determined by diffraction. A Gaussian beam represents the smallest physical possible area with the value given by Eq. (10.11).

10.1.4 Analogy to Mechanics

There are several aspects in optics that are comparable and similar to mechanics. It is known that in classical mechanics the quantum character is ignored and it is a continuum approach for a huge number of degrees of freedom. Similarly in optics, the geometrical approach corresponds to an approximation ignoring the wave nature of light and does not contain diffraction effects. In Table 10.1 a comparison is listed for the equivalent terms in optics and mechanics.

10.2 Geometrical Ray Model

10.2.1 Introduction

The phase space has been well known in mechanics and quantum mechanics for a long time. In mechanics, it is spanned by the spatial and the momentum coordinate of a body or particle. The same concept can be used in optics to great advantage, in particular for illustrating more complicated relationships of light distributions. In optics, the momentum is substituted by the angle or direction coordinate. The angle coordinate can also be substituted by the spatial frequency. In the simplest case of only one transverse dimension, the x–u diagram can be used to visualize rays and ensembles of discrete rays. Figure 10.9 illustrates a single ray, a ray fan as well as in (c) an ensemble of rays characterizing a planar source. A single ray is represented in the phase space by a point, a ray fan emitted from a point source as a line and the ensemble as an area.

Interestingly this concept can be extended to physical beams and the phase space allows us to take diffraction into account. The uncertainty relation is easily obtained by zooming in and it is seen that the geometrical-optical

Table 10.1 Analogue terms and equations of mechanics and optics.

Term	Mechanics	Optics
Spatial variable	x	x
Momentum variable	Momentum $p = mv$	Angle u, slope $\tan(u)$, momentum, direction cosine p lateral spatial frequency ν_x
Propagation variable	Time t	Distance z
Equation of motion spatial domain	Lagrange equation	Eikonal equation
Equation of motion phase space	Hamiltonian equation	Wigner propagation
Minimum principle	Hamilton principle	Fermat principle
Potential	Mass m	Refractive index n
Liouville invariance	Number of particles	Energy
System response function	Impulse response	Point spread function
wave equation	Schrödinger	Helmholtz
Uncertainty parameter	$h/(2\pi)$	$\lambda/(2\pi)$
Continuum approximation	Classical mechanics	Geometrical optics
Exact description	Quantum mechanics	Wave optics, modes

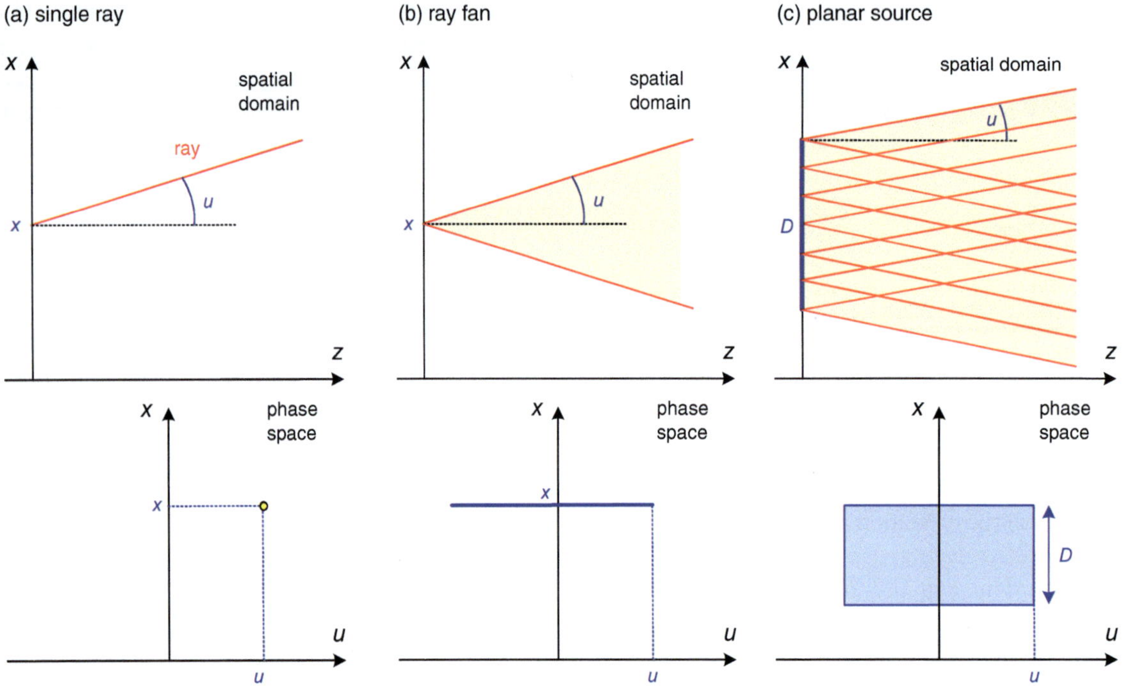

Figure 10.9 Move from the spatial domain into the phase space for (a) a single ray, (b) a ray fan and (c) a planar source in one dimension.

world corresponds to rays in a similar manner to classical mechanics in comparison to quantum mechanics.

In the simple model of an incoherent radiation field, the phase space is homogenously filled. These filled areas do not have any further inner structures; maybe an apodization gives a smooth density distribution.

Figure 10.10 shows the well-known free space propagation of two single rays in a homogeneous medium. The angle is unchanged, the height or lateral coordinate x grows corresponding to the angle u. The phase space diagrams in the lower row have to be drawn for every z-value separately. If the data of the two rays are compared, the shift of the two points 1 and 2 to 1′ and 2′ respectively is emphasized by the yellow and blue dots respectively. It is seen that the propagation of the rays can be understood as a shearing of the grid in the phase space domain. The shearing angle is related to the propagation distance z.

In Figure 10.11 an ensemble of collimated ray bundles is considered. The shearing behaviour as a change during propagation is the same. This kind of ray selection corresponds to a bundle of collimated rays which intersect in the z_1-position. In the interpretation of an optical system, we have an afocal system with the pupil position at z_1. If the density in the phase space, which is here assumed to be constant, is integrated over the angle u, we get the spatial intensity $I(x)$. The corresponding distributions are also shown in the figure for both positions z_1 and z_2. It is seen that the intensity in the pupil has a rectangular shape, in a defocused plane z_2 it becomes trapezoidal.

10.2.2 Phase Space Representation of Ray Bundle Transport

In the field of classical radiometry, a radiation transport in one dimension can be represented very clearly by inspecting the changes in the phase space. The areas with finite light density change their geometrical shapes under propagation. The kind of deformation can be determined very easily by the trajectory of the rays through the system. In Figure 10.12(a) a simple system is sketched and three ray bundles are drawn. The corresponding phase space picture for the object plane but with some more rays is shown in part (b) of the figure. Every green dot represents one ray; the marginal rays start on axis and they are indicated by the blue line. The chief rays meeting the pupil in the centre are marked red. The rays fill a parallelogram, the outermost coma rays are the corners of the shape.

If we consider a simple real imaging through a single lens with telecentricity at the object side as shown in Figure 10.13(a) in green, the green areas in the phase space as in figure (b)–(d) are found similarly to Figure 10.12. In the object plane the filled phase space area is given as a rectangle by the object size $2y_1$ and the numerical aperture $2\sin u$ if the object space is telecentric. If the stop

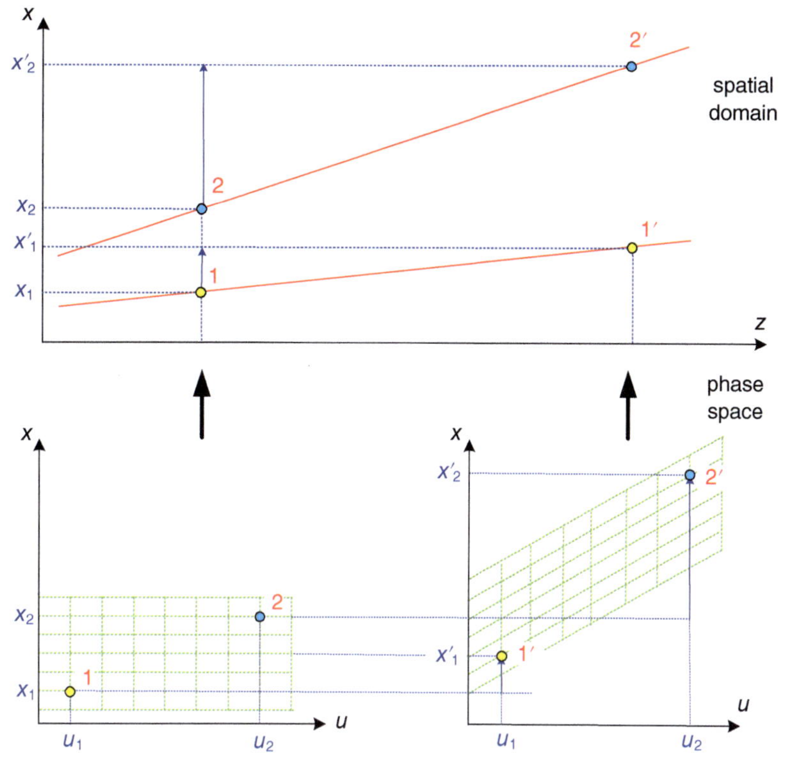

Figure 10.10 Scheme of two single rays, their propagation in the spatial domain as well as in the phase space for two different z-locations.

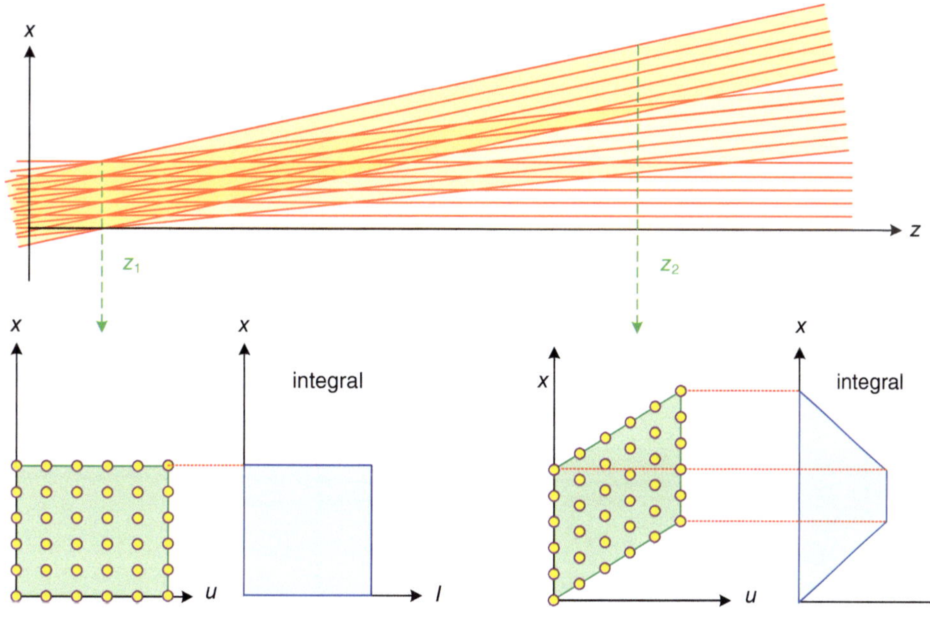

Figure 10.11 Scheme of an ensemble of collimated rays and its propagation, similarly to collimated ray bundles in an afocal system; the lower rows show the phase space dots as well as the intensity $I(x)$ as u-integrated densities.

is at the lens, the off-axis ray cones are inclined and the phase space is represented by a parallelogram in the object space as is illustrated in the figure as the purple case. In the image space, however, there is a finite angle w' of the chief ray. This leads to the phase space area becoming a parallelogram because the aperture angle is superimposed onto the chief ray (CR) angle. However, the areas have the same size in the object and the image space. In comparison it is seen that for the telecentric case, the required diameter of the lens to avoid vignetting is larger.

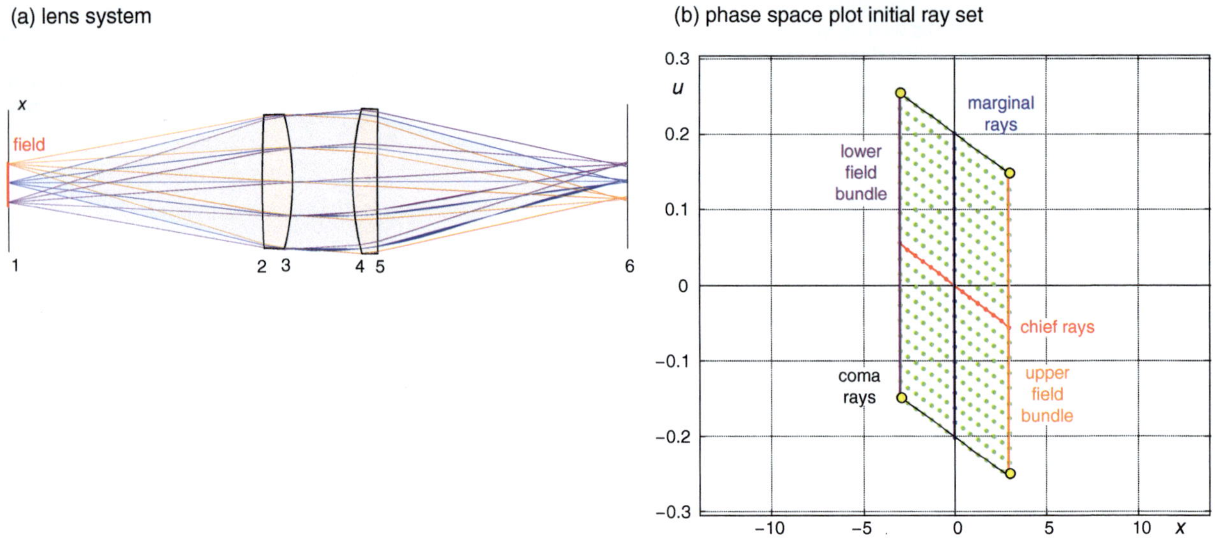

Figure 10.12 Illustration of the telecentric ray bundles (a) in the spatial domain and (b) in the phase space in the object plane for a two-lens system. u is scaled in rad, the x-units are arbitrary.

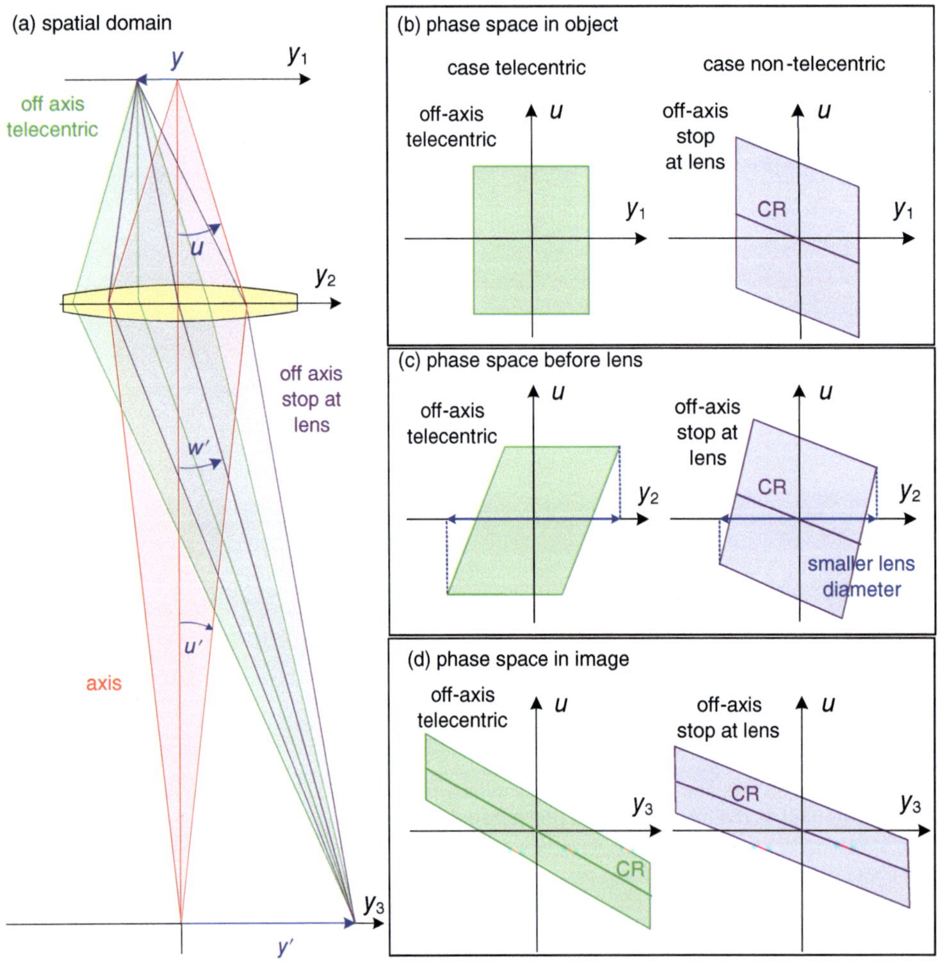

Figure 10.13 Simple imaging system with (green) and without (purple) telecentricity in (a) spatial domain and phase space (b) in object (c) before lens and (d) in image plane.

10.2.3 Ray Aberrations

If paraxial raytracing is used, the ray bundles in the phase space are limited by straight lines, the areas are rectangles or parallelograms due to the underlying linearity. If the rays are calculated in reality, the straight boundaries of the bundle cross section areas are slightly non-linear due to aberrations [11, 12]. This means that the aberrations can be detected as the deviation from linearity. Interestingly in the phase space the deviations in space and in angle can be clearly distinguished. In the spatial domain the lateral aberrations are of interest in the image plane, in the phase space the aberrations are already seen in the pupil as angle deviations. This is sketched in Figure 10.14. In part (b) it is indicated that in the case of additional grid distortion in the pupil, the changes are seen in the angle as well as the coordinate direction. Unfortunately, the scales of the axis are mostly not very convenient and small deviations are hard to observe. Therefore, a visualization of aberrations needs to subtract the paraxial angle and spatial coordinate values to visualize just the differences. Only very large aberrations comparable to the macroscopic values are directly seen.

Figure 10.15 shows the collimation and refocusing behaviour of a divergent ray bundle in a parabolic gradient index lens with a second order radial index variation. The phase space diagrams are shown at three positions: the start, the nearly collimated position in the middle of the lens and near to the end, where the focus point suffers from spherical aberration. The aberrations are already seen in the position z_2 as angle deviations. The transition from z_1 to z_2 corresponds to a Fourier lens or to the move from the

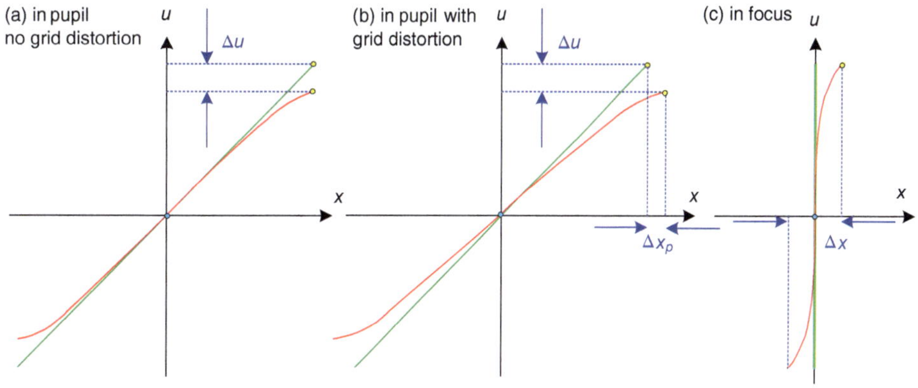

Figure 10.14 Phase space representation of aberrations. The linear straight lines in green correspond to the paraxial reference and are bent in reality due to aberrations shown in red. In the pupil as is shown in (a) the deviation is mostly seen as the angle, in (b) as additional grid distortion and (c) in the image the aberrations are observed as a spatial deviation.

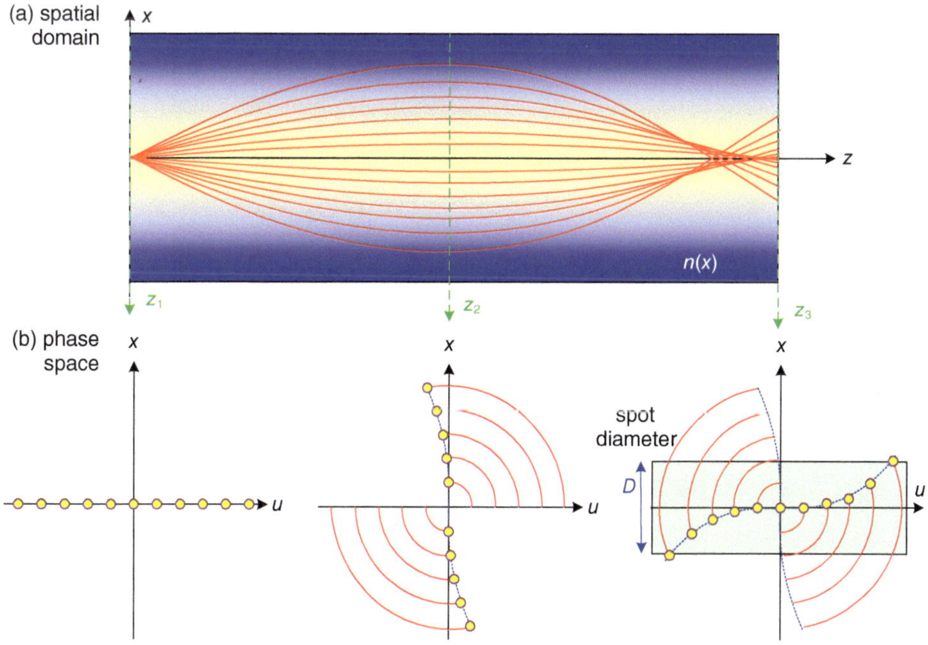

Figure 10.15 Illustration of the ray behaviour in a gradient index lens with collimation and refocusing together with spherical aberration effects in the spatial domain in (a) and in three z-locations in the phase space in (b).

object point into the pupil. This Fourier conjugation corresponds to a switch between angle and spatial coordinates; therefore, in the phase space we see a 90° rotation of the pattern in addition to the small aberration deviations. It is also seen that in the image plane z_3 the pattern again rotates to be 90° and the aberrations are again seen in the spatial domain.

Figure 10.16 shows a case of a simple relay system. The layout is indicated in Figure 10.16(a). In the object space the ray bundles are telecentric; therefore, the area of the bundles forms a rectangle. The phase space pattern is calculated after every surface. The growing non-linearities are seen in the angle as well as the spatial coordinates with the propagation along z. Here the system is not corrected; the deviations are therefore huge. The cubic polynomial boundaries in the image plane correspond to spherical aberration. It can be seen here that the marginal rays in blue suffer from aberrations, while the red chief rays nearly do not. This means that there is spherical aberration present and distortion is quite small.

To make small aberrations in phase space more visible it is also possible to emphasize the transverse aberrations in the ray pattern in the image space by colour. This is seen in Figure 10.17. The ray pattern is represented by the coloured area. In every point the colour indicates the transverse aberration Δx. A small 4 μm distortion is seen for the CR, the spherical aberrations of the marginal ray (MR) create a 0.69 mm lateral aberration in a symmetrical way and the coma spreading (TCU, TCL in the corners) is slightly asymmetric as expected with 0.87 and 0.75 mm for the upper and lower field point.

10.3 Wigner Distribution Function*

(*) The content of this section is quite specialized and for advanced reading. Knowledge of this section is not

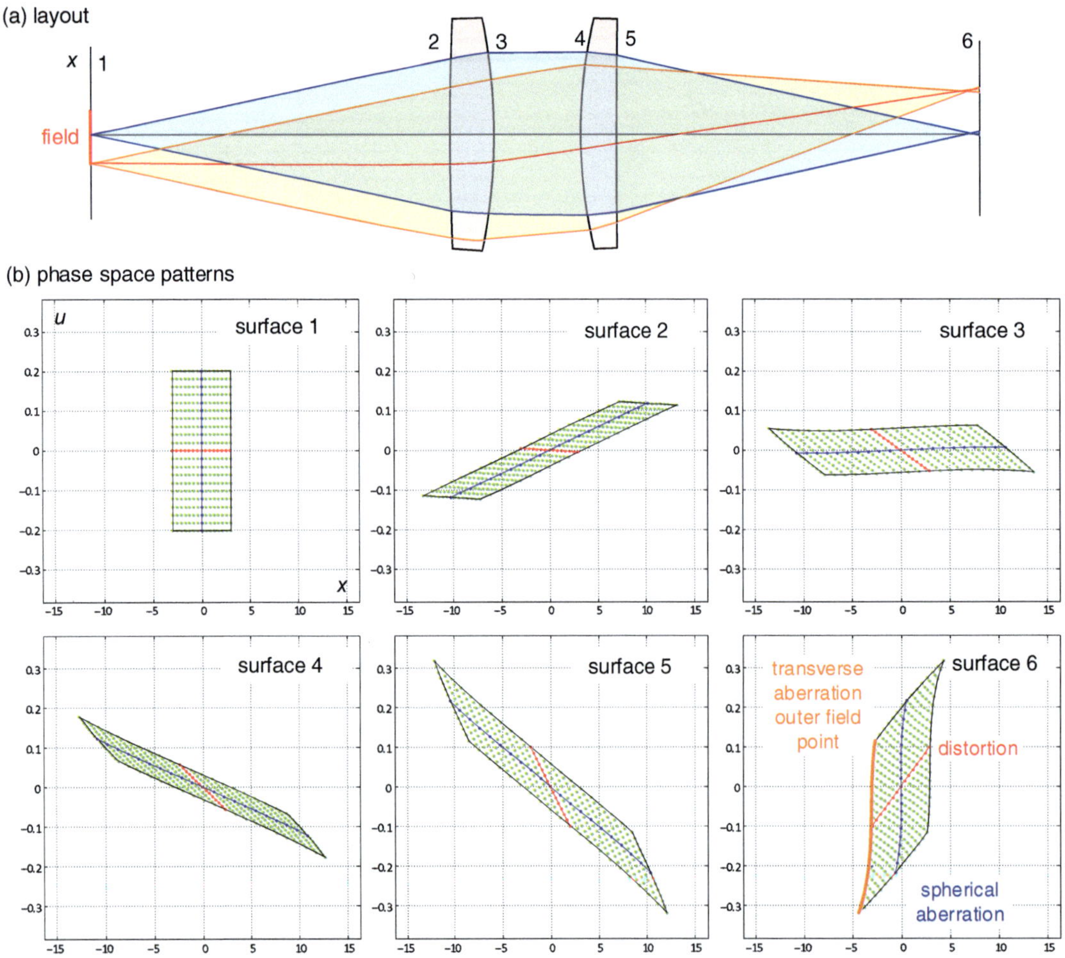

Figure 10.16 Illustration of the ray bundle changes in a two-lens system showing the growing non-linearities during propagation. (a) shows the lens system in the spatial domain, in (b) the phase space picture of the ray bundles is visualized in the six indicated z-positions. u is scaled in rad, the x-units are arbitrary. A very small deviation of the chief rays is obtained due to distortion; the red line is not perfectly straight.

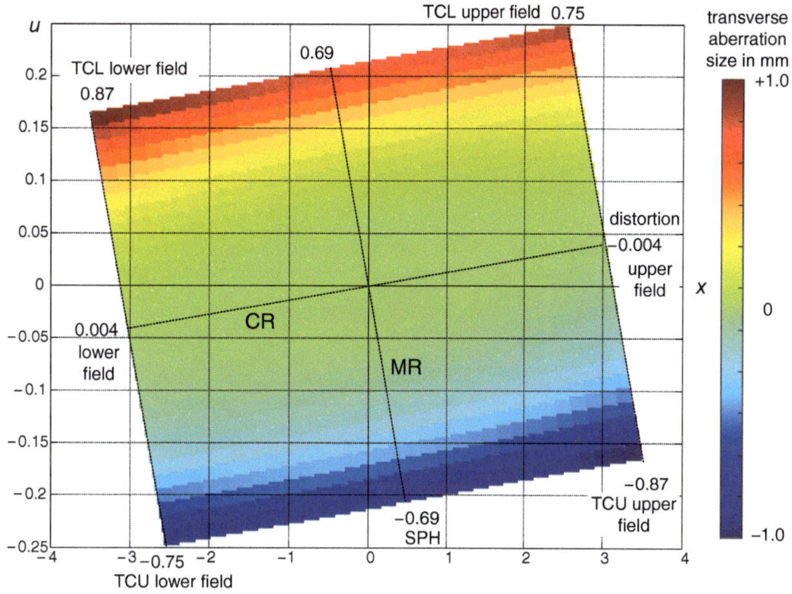

Figure 10.17 Illustration of the aberrations in a corrected two-lens-system by the phase space pattern. To make the small aberrations more visible, here the grid points are drawn as small square symbols and their colour indicates the absolute value of the lateral aberration Δx as given by the numbers. u is scaled in rad, the x-units are arbitrary.

necessary to understand the following chapters. Beginners are recommended to skip this section.

10.3.1 Introduction

The geometrical description of the phase space with rays and paraxial area preservation does not take the physical nature and phenomena of light like diffraction into account. But it is possible to extend this model in the physical sense; this approach is discussed here and is related to the Wigner distribution function (WDF).

In the more general case, the energy density is a function of the spatial as well as the angle coordinate. This means that a density function must be assumed in the phase space and the overall power is obtained by integration over the boundaries of this density function according to the basic photometric law. The density is called the Wigner function and it is a real function. Figure 10.18 illustrates some typical light sources in their phase space shape. A ray is represented by a single dot. A plane wave has a spatial extent, but is infinitesimally thin in the angle coordinate. A point source, on the other hand, has a huge angle extent according to the radiated spherical wave, but is very small in the spatial extent. A typical light emitting diode (LED) is nearly constant in the spatial domain, but has a wide Lambertian angle distribution. A Gaussian beam has a corresponding energy density in the angle as well as in the spatial domain. Correspondingly the colours of the LED and the Gaussian laser beam are shaded, illustrating the varying energy density as the WDF.

A geometrical-optical quasi-continuum of rays can be represented by areas that are densely filled with light. The areas as the product of spatial and angle extent correspond to the Lagrange invariant. In this case, filling of the phase space is smooth; maybe hard boundaries limit the density distribution. To be more specific, if we assume a light source of 10 mm diameter and a numerical aperture of NA = 0.2, we end up with the value $L = 1.0$ mm rad. In the case of a single mode laser beam like a Gaussian beam, the physical relations give for the space-bandwidth product values of $L = w_0\theta_0 = \lambda/\pi = 2 \cdot 10^{-4}$ mm × rad if a wavelength in the red is assumed. Furthermore, the laser beam can exist only in the form of certain modes. This means that if diffraction limited light distributions are considered, the filled areas of the phase space are several orders of magnitude smaller than in the geometrical optical regime. Furthermore, they are discretized into single dots of finite

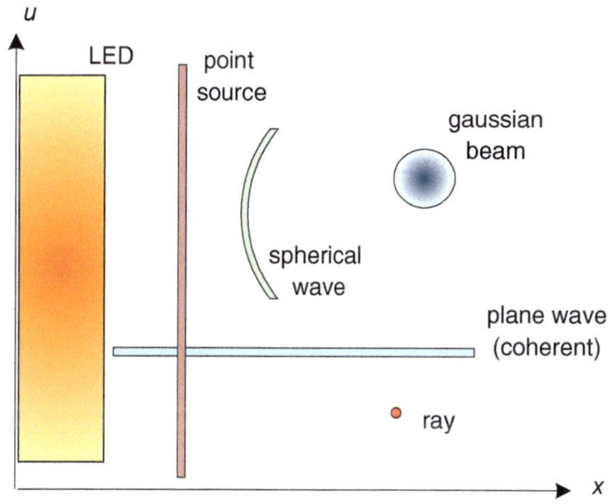

Figure 10.18 Representation of various light distributions in the phase space. The area sizes are not to scale here.

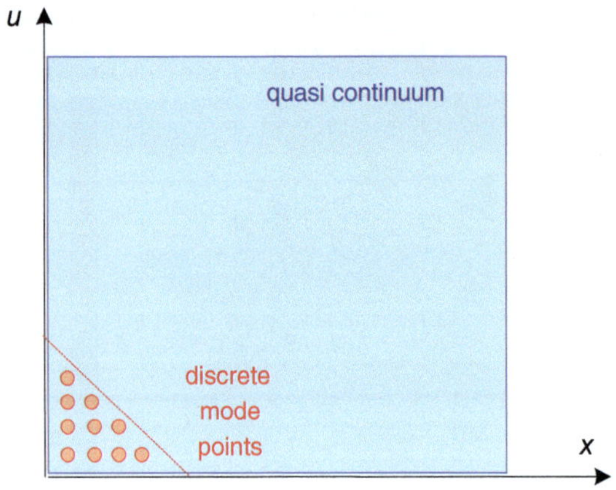

Figure 10.19 Illustration of the quasi-continuous distribution in geometrical optics with the discretized modal structure in the case of diffraction limited beam optics and discrete modes.

sizes corresponding to the modes. This is visualized in Figure 10.19.

The rays in the geometrical approach of the phase space represent plane waves. This singular wave forms simplify the properties of this approach. In general, an arbitrary wave field must be considered. The simple paraxial *ABCD*-transfer propagation of the rays is then modified into an integral form, in the paraxial case into the Collins diffraction integral, also depending on *ABCD*. The probability function in the phase space is the Wigner function, which corresponds to a local angle spectrum of the wave. In contrast to the usual spatial description of a wave, the Wigner function contains a combined spatial-angle behaviour and is bilinear.

When modelling light fields with the Wigner function, a discretized model on a computational grid is of course needed, but this discretization should not be confused with the discrete mode structure of coherent fields. Due to the uncertainty relation in optics, it is impossible to determine or accurately measure spatial and angle values simultaneously. This means that a point in the phase space is an idealization, which is unphysical. In reality a certain minimum area of the size λ/π can never be reduced. This physically minimal area is exactly realized by a Gaussian beam according to the uncertainty relation (Table 10.2).

10.3.2 Theory of the Wigner Function

The WDF is a quasi-density function in the phase space. In the physical interpretation, it is equivalent to the coherence function Γ as a correlation of the light field at two different locations and can be calculated as the Fourier transform of Γ in the form

$$W(\vec{r},\vec{v}) = \int \Gamma\left(\vec{r}+\frac{\Delta\vec{r}}{2},\vec{r}-\frac{\Delta\vec{r}}{2}\right)\cdot e^{-2\pi i\Delta\vec{r}\vec{v}}d\Delta\vec{r} \tag{10.12}$$

with the spatial frequency $\vec{v}$. Here the formulation of Γ is used with the centre location $\vec{r}$ and the difference $\Delta\vec{r}$ between the two correlated points. The Wigner function is real and its interpretation is easier as the coherence function. Its propagation calculation through paraxial systems is quite simple. If a lateral angle with x and y-component instead of the spatial frequency is used, the above equation is equivalent to the Fourier integral

$$W(\vec{r},\vec{u}) = \int \Gamma\left(\vec{r}+\frac{\Delta\vec{r}}{2},\vec{r}-\frac{\Delta\vec{r}}{2}\right)\cdot e^{-i\,k\,\Delta\vec{r}\vec{u}}\,d\Delta\vec{r} \tag{10.13}$$

because the angle is a scaled version of the lateral spatial frequency. The WDF is a quasi-density and can be negative. Negative values occur in the case of diffraction related destructive interference. If an integration over the minimum area of λ/π is performed, W is always positive and behaves as a real density. It can be interpreted as a local Poynting vector, $W(x,y,u,v)$ describes the energy flowing from the point x,y into the direction u, v; see Figure 10.20.

In the case of coherent field E, the Wigner function can be directly calculated as a Fourier transform of the correlation of the complex field E

$$W(x,u) = \int E\left(x+\frac{x'}{2}\right)\cdot E^*\left(x-\frac{x'}{2}\right)\cdot e^{-ikx'\cdot u}\,dx' \tag{10.14}$$

Table 10.2 Different model depths for the light description in the phase space.

Model	Property	Equation of motion paraxial	Equation of motion general
Single rays	Discrete	ABCD law for paraxiality	Real ray trace, eikonal equation, Lagrange and Hamilton formulation of Fermat's minimal principle
Ensemble of rays, energy density, still geometrical	Quasi continuum	Liouville's law for the energy density, preservation of phase space area envelope	Non-linear transform of Jacobian matrix
Wave fields, Wigner function	Local angle spectrum	Collins integral for coherent fields, linear canonical transform	General integral transform of the WDF

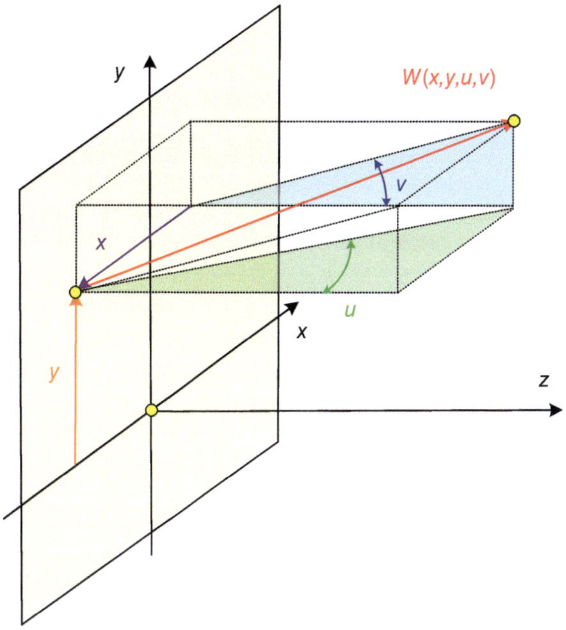

Figure 10.20 Wigner function of the four coordinates *x,y,u,v* as a local Poynting vector.

By integrating the density W, the spatial and angle energy density is obtained respectively

$$I(\vec{r}) = \frac{1}{(2\,\pi)^2} \int W(\vec{r},\vec{u})\, d\vec{u}$$
$$I(\vec{u}) = \int W(\vec{r},\vec{u})\, d\vec{r} \tag{10.15}$$

If the integral is calculated over both variables x and u in a one-dimensional formulation, the power of the light field is obtained

$$P = \iint W(x,u)\, dx\, du \tag{10.16}$$

This again shows that in the geometrical picture the conservation of the phase space area corresponds to the preservation of energy. The property of $W(x,u)$ as a density function is displayed in the calculation of the averaged values of beam properties like the second order moments in space and angle

$$\langle x^2 \rangle = \frac{1}{P} \cdot \iint x^2 \cdot W(x,u)\, dx\, du$$
$$\langle u^2 \rangle = \frac{1}{P} \cdot \iint u^2 \cdot W(x,u)\, dx\, du \tag{10.17}$$

By using the exact propagation algorithms of the Wigner function, it can be shown that the value of the function is constant along the propagation path of a ray following the Fermat principle, which can be written as

$$\frac{dW}{ds} = 0 \tag{10.18}$$

with the arclength s. This is the basis for the argumentation that the WDF can be used to describe single rays in the geometrical Wigner model.

10.3.3 Wigner Function of Gaussian Beams

For a fully coherent Gaussian beam field in one lateral dimension with the beam radius w_0 and the far-field divergence θ_0, the Wigner function is given by

$$W_{GB}(x,u) = N \cdot e^{-2\left(\frac{x^2}{w_0^2}+\frac{u^2}{\theta_0^2}\right)} \tag{10.19}$$

with a normalization constant N. In the case of a partially coherent Gauss–Schell beam without twist, radius of curvature R and lateral coherence lengths L_c it is meaningful to define a normalized degree of coherence β as (see also Section 8.9)

$$\beta = \frac{1}{\sqrt{1+(w/L_c)^2}} \leq 1 \tag{10.20}$$

The smaller the L_c, the smaller the degree of coherence and the smaller is β. $L_c \to \infty$ means full coherence and $\beta = 1$. With β we can define the parameters deviating from the classical coherent Gaussian beams as

$$z_0' = \beta \cdot z_0, \quad \theta = \theta_0/\beta \tag{10.21}$$

and the paraxial free space propagation reads

$$w^2(z) = w_0^2 \left[1 + \left(\frac{z - z_T}{z_0'}\right)^2\right] \tag{10.22}$$

with the waist position z_T. In one lateral dimension the WDF is then written as

$$W_{GSB}(x,u) = \frac{I_0\, w^2(z)}{2\,\pi} \cdot e^{-\frac{2x^2}{w^2(z)}} \cdot e^{-\frac{1}{2}k^2 \cdot \frac{w_0^2 \cdot L_c^2}{w_0^2+L_c^2} \cdot \left(u - \frac{x}{R}\right)^2} \tag{10.23}$$

The modified beam width and divergence due to the partial coherence has the consequence that a Gauss–Schell beam has a larger space-bandwidth product and the area in the phase space is enlarged.

10.3.4 Wigner Function of a Decomposed Field

If a signal is decomposed into several components, due to the bilinear property the Wigner function is not simply the sum of the parts. If two signal fields E_1 and E_2 are considered, the Wigner function as a Fourier transform of the field correlation for the combined signal [6] reads

$$\begin{aligned} W_E(x,u) &= W_{E1}(x,u) + W_{E2}(x,u) + W_{E1-E2}(x,u) \\ &\quad + W_{E2-E1}(x,u) \\ &= W_{E1}(x,u) + W_{E2}(x,u) + 2 \cdot \mathrm{Re}[W_{E1-E2}(x,u)] \end{aligned} \tag{10.24}$$

The first two contributions are the auto or self Wigner parts of the signals

$$W_{Ej}(x,u) = \int E_j\left(x + \frac{\Delta x}{2}\right) \cdot E_j^*\left(x - \frac{\Delta x}{2}\right) \cdot e^{-ik_0 \Delta x \cdot u}\, d\Delta x \tag{10.25}$$

and the last two summands are the cross or mixed Wigner functions, describing the correlation between the two fields

$$W_{E1-E2}(x,u) = \int E_1\left(x + \frac{\Delta x}{2}\right) \cdot E_2^*\left(x - \frac{\Delta x}{2}\right) \cdot e^{-ik_0 \Delta x \cdot u}\, d\Delta x \tag{10.26}$$

The cross terms become Hermitean by interchanging the signal sequence. The cross terms describe the interference ability of the individual signal parts, usually they are located between the individual signal centres and are highly oscillatory. By calculating the integrals over x or u, this superposition adds to zero intensity. This is illustrated in Figure 10.21 for an example of two combined coherent Gaussian beams. The oscillating part of the Wigner distribution around $x = 0$ gives in the u-integration a value of zero. Such a constellation is for example obtained in a shearing interferometer. If both fields E_1 and E_2 are completely independent, they cannot interfere.

If in the general case the signal consists of many contributions, we have

$$E(x) = \sum_j a_j E_j(x) \tag{10.27}$$

and

$$W_E(x,u) = \sum_j |a_j|^2\, W_{Ej}(x,u) + 2 \cdot \sum_{j>m} \sum_m \mathrm{Re}\left[a_j^* a_m W_{Ej-Em}(x,u)\right] \tag{10.28}$$

It can be shown that the mixed terms are related to the self terms by a convolution in spatial and angle space as

$$|W_{Ej-Em}(x,u)|^2 = \iint W_{Ej}\left(x - \frac{x'}{2}, u - \frac{u'}{2}\right) \cdot W_{Ej}\left(x + \frac{x'}{2}, u + \frac{u'}{2}\right) dx'\, du' \tag{10.29}$$

In the special case of an identical replica of the same field, the cross term contains a cos-oscillatory function, describing the interference.

$$E(x) = E_0(x - x_1) \cdot e^{ikxu_1} + E_0(x - x_2) \cdot e^{ikxu_2} \tag{10.30}$$

$$W(x,u) = W_{E_0}(x - x_1, u - u_1) + W_{E_0}(x - x_2, u - u_2) + 2\cos\left[(u_1 - u_2)x - (x_1 - x_2)\left(u - \frac{u_1 + u_2}{2}\right)\right] \cdot W_{E_0}\left(x - \frac{x_1 + x_2}{2}, u - \frac{u_1 + u_2}{2}\right) \tag{10.31}$$

This equation describes the generation of fringes in an interferometer. The spatial separation of the two signals determines the modulation of the mixed term. If as a special example two Gaussian modes are considered, which

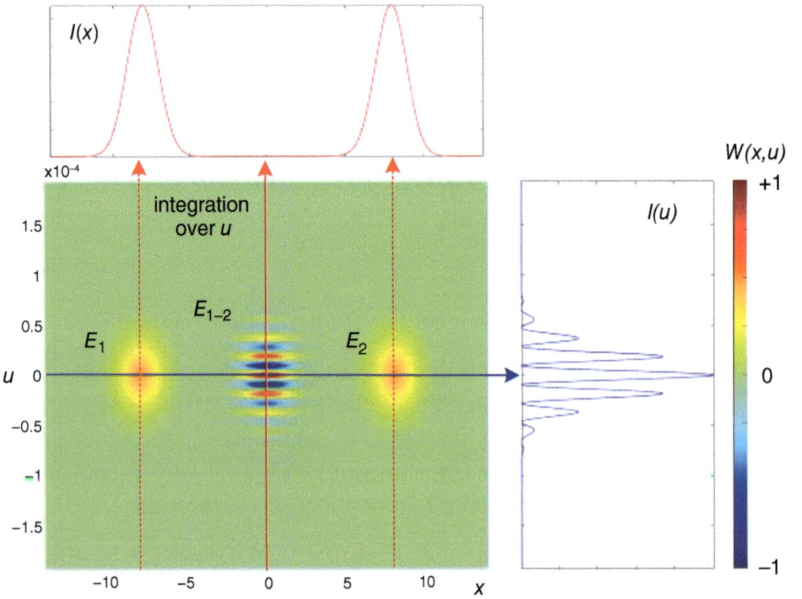

Figure 10.21 Interference and cross term of the WDF for a decomposed signal. The units of x and u are arbitrary. The values of W in the false colour plot are normalized as can be is seen in the colourbar.

have a lateral shift of $2x_o$ and the same parameter as in Figure 10.21, in the waist plane the field reads

$$E(x) = c_1 \cdot e^{-\frac{(x-x_0)^2}{w_0^2}} + c_2 \cdot e^{-\frac{(x+x_0)^2}{w_0^2}} \tag{10.32}$$

where $c_{1/2}$ are arbitrary amplitude coefficients. The WDF is given by

$$\begin{aligned} W(x,u) = &\frac{w_0 \cdot \sqrt{2\pi} \cdot |c_1|^2}{\lambda} \cdot e^{-2\frac{(x-x_0)^2}{w^2} - \frac{2\pi^2 \cdot w_0^2 u^2}{\lambda^2}} \\ &+ \frac{w_o \cdot \sqrt{2\pi} \cdot |c_2|^2}{\lambda} \cdot e^{-2\frac{(x+x_0)^2}{w^2} - \frac{2\pi^2 \cdot w_0^2 u^2}{\lambda^2}} \\ &+ 2\frac{w_0 \cdot \sqrt{2\pi} \cdot |c_1 c_2|}{\lambda} \cdot e^{-2\frac{x^2}{w^2} - \frac{2\pi^2 \cdot w_0^2 u^2}{\lambda^2}} \\ &\cdot \cos\left(\frac{4\pi x_0 \cdot u}{\lambda} - \arg\left(c_1 c_2^*\right)\right) \end{aligned} \tag{10.33}$$

In the case of partially coherent fields, first a coherent mode decomposition is necessary; in a second step the coherent approach of a sum of modes can be used. This discussion of the non-linear behaviour of the Wigner function in the case of superposition is due to the bilinearity, and precisely this property allows us to also describe partially coherent fields in this formalism.

10.3.5 Propagation of the Wigner Function

In the general case, the propagation of the Wigner function is given by an integral transform of the form

$$W_{out}(x,u) = \frac{1}{2\pi} \iint K(x_1, u_1, x, u)\, W_{in}(x_1, u_1)\, dx_1\, du_1 \tag{10.34}$$

with kernel function K, for details see [4]. For a paraxial segment with *ABCD* matrix

$$\vec{r}\,' = \underline{M}\,\vec{r} = \begin{pmatrix} A & B \\ C & D \end{pmatrix} \cdot \vec{r}, \quad \vec{r} = \underline{M}^{-1}\vec{r}\,' = \begin{pmatrix} D & -B \\ -C & A \end{pmatrix} \cdot \vec{r}\,' \tag{10.35}$$

the propagation is given quite simply with the inversion of the equation as

$$W_{out}(\vec{r}\,') = W_{in}(\underline{M}^{-1} \cdot \vec{r}) \tag{10.36}$$

In components this can be written by using Eq. (10.35) as

$$W_{out}(x', u') = W_{in}(D \cdot x - B \cdot u, -C \cdot x + A \cdot u) \tag{10.37}$$

Figure 10.22 illustrates this calculation as the interpolation. It should be emphasized that the interpolation has to be performed on the original grid and the inversion of the usual *ABCD* equation is required to finally end up in a desired grid point.

In particular, a free space segment with distance z gives

$$W_{out}(x, u, z) = W_{in}(x - z \cdot u, u, 0) \tag{10.38}$$

and a single lens with focal length f changes only the angle correspondingly as

$$W_{out}(x, u) = W_{in}(x, u - x/f) \tag{10.39}$$

As in the case of the geometrical paraxial approximated phase space picture, a shear along x is obtained for the free space, and a shear in u for the lens as is shown in Figure 10.23.

In the more general case of non-paraxial angles, the tangent function has to be used and the non-linear propagation in free space

$$\begin{aligned} W_{out}(x, u, z) &= W_{in}\left(x - \frac{u}{\sqrt{1-u^2}} \cdot z, u, 0\right) \\ &= W_{in}(x - \tan u \cdot z, u, 0) \end{aligned} \tag{10.40}$$

is obtained. The shear has then a certain distortion according to the tangent function.

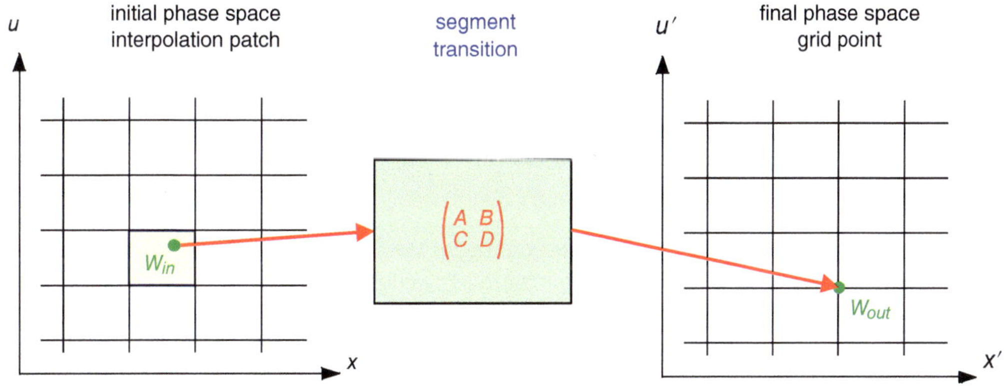

Figure 10.22 Scheme of interpolation for the paraxial propagation of the Wigner function in the initial plane.

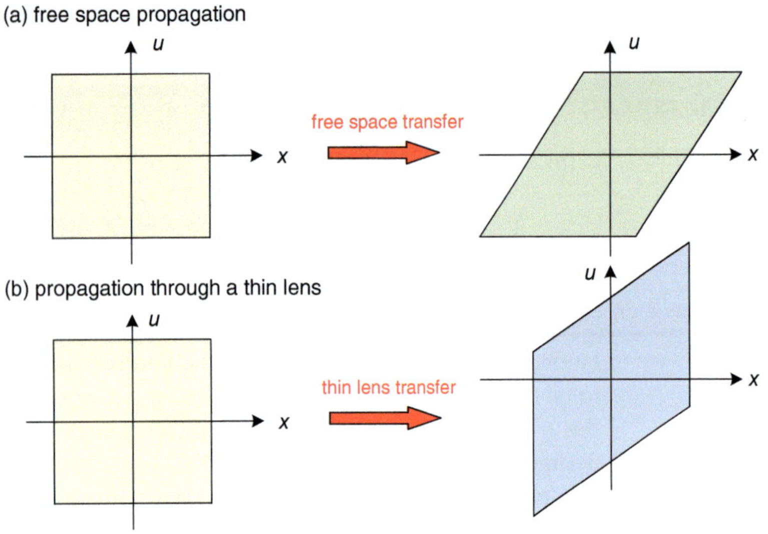

Figure 10.23 Scheme of interpolation for the paraxial propagation of the Wigner function through (a) free space and (b) a thin lens.

10.3.6 Transfer Through Thin Masks

The propagation of the Wigner distribution function through a thin complex mask with amplitude A and phase Φ

$$T(x) = A(x) \cdot e^{i \cdot \Phi(x)} \tag{10.41}$$

changes the complex field by a simple multiplication

$$E_{out}(x) = T(x) \cdot E_{in}(x) \tag{10.42}$$

Now it is well known that the spatial coherence function is given as the time averaged correlation integral and the definition of the Wigner function can be formulated as the Fourier transform of the coherence function. This means that the impact of the filter mask onto the Wigner function can be described by

$$W_{mask}(x, u) = \int T\left(x + \frac{x'}{2}\right) \cdot T^*\left(x - \frac{x'}{2}\right) \cdot e^{-\frac{2\pi i}{\lambda}x'u} dx' \tag{10.43}$$

and the change of the Wigner function by the mask can be formulated therefore as a convolution of the incoming WDF with the mask WDF

$$W_{out}(x, u) = W_{mask}(x, u) * W_{in}(x, u) \tag{10.44}$$

In the special case of a thin phase mask with phase profile $\Phi(x)$, the Wigner function of the element can be written approximately as

$$W_{mask}(x, u) = \delta\left(u - \frac{\lambda}{2\pi} \cdot \frac{d\Phi(x)}{dx}\right) \tag{10.45}$$

A thin lens with focal length f has for example the phase function

$$\Phi_{lens}(x) = \frac{x^2}{2\lambda \cdot f} \tag{10.46}$$

The convolution with the Dirac delta function can now be easily calculated and gives for the lens the new Wigner function

$$W_{out}(x, u) = W_{in}\left(x, u - \frac{x}{2\pi \cdot f}\right) \tag{10.47}$$

which means that the lens only bends the angles linearly with the height x, a well-known result from geometrical optics. If the lens suffers from primary spherical aberration with polynomial coefficients W_4, this can be included by an additional phase mask

$$\Phi(x) = W_4 \cdot x^4 \tag{10.48}$$

The modified Wigner function behind the mask is

$$W_{sph}(x, u) = W_{in}\left(x, u - \frac{2\lambda \cdot W_4 \cdot x^3}{\pi}\right) \tag{10.49}$$

10.3.7 Wigner Function of a Slit

For a slit with width $2a$, the Wigner function reads by inserting the rect-function in Eq. (10.43)

$$W_{slit}(x, u) = \frac{\sin\left[\frac{2\pi \cdot u}{\lambda} \cdot (2a - 2|x|)\right]}{\frac{2\pi \cdot u}{\lambda}} = (2a - 2|x|) \cdot \text{sinc}\left[\frac{2\pi \cdot u}{\lambda} \cdot (2a - 2|x|)\right] \tag{10.50}$$

corresponding to the far-field angle distribution of the slit diffraction, see Section 6.9.7. According to Figure 10.24 the Wigner function is in this case obtained as the Fourier transform of the diamond shape in (a) and looks like the distribution in (b). In part (a) of the figure the conditions for transmission $x + x'/2 > -1$, $x + x'/2 < +1$ gives the green stripe, the additional condition $x - x'/2 > -1$, $x - x'/2 < +1$ the blue band. In total the orange area is

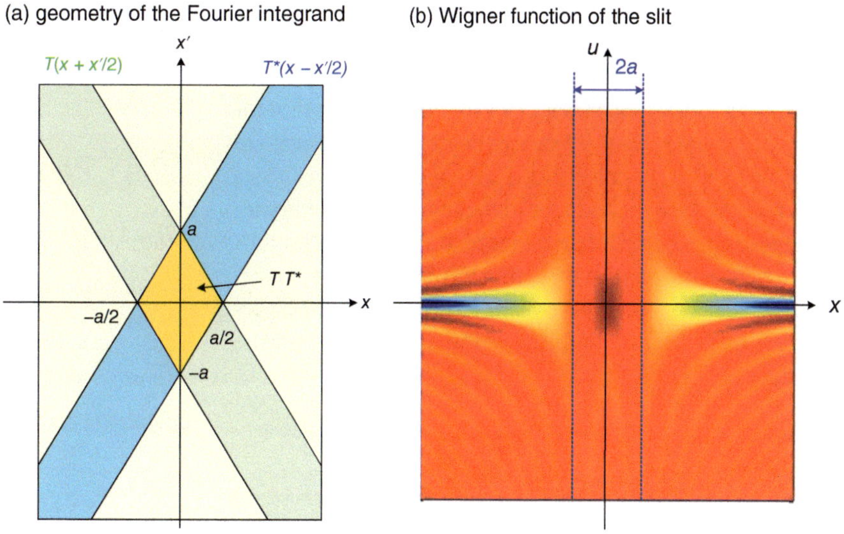

Figure 10.24 Geometry of the Fourier integrand of a slit (a) and Wigner function of the slit (b).

obtained as the range with finite values for W. In part (b) the Wigner function distribution of the slit transmission is visualized. The diffraction ripples and their behaviour with the size of the angle deviation in the geometrical shadow region are seen. This function has to be convolved with the incoming Wigner function. The shown distribution is not the Wigner function of the light behind the slit, this can be seen in Figure 10.25.

10.3.8 Wigner Function for Gauss–Schell Beams

In the case of a partially coherent Gauss–Schell beam, the Wigner function can be written as

$$W_{GSB}(x,u,0) = \frac{I_0\, w_0^2}{2\,\pi} \cdot e^{-2\frac{x^2}{w_0^2}} \cdot e^{-\frac{1}{2}k^2 \cdot \frac{w_0^2 \cdot L_c^2}{w_0^2+L_c^2} \cdot \left(u-\frac{x}{R}\right)^2} \tag{10.51}$$

In generalization to the coherent Gauss beams, here an additional parameter L_c describes the transverse coherence length. The larger the L_c, the more the beam resembles a classical Gaussian beam; for smaller L_c the field becomes more incoherent and the possibility of interference shrinks. It is seen that for a constant waist size w_0 the partial coherence changes the angle width; small values of L_c widen the far-field cone.

If a stop truncates such a partial coherent beam, diffraction effects are observed. These effects grow with the clip level and the degree of coherence. In the case of nearly incoherent fields, the interference ability is strongly reduced. In the Wigner distribution $W(x,u)$, the diffraction effects are seen by the typical fringing effects. They occur spatially dependent at the stop edges. The spatial extent of the fringes grows with the degree of coherence and are in the range of the lateral coherence length L_c.

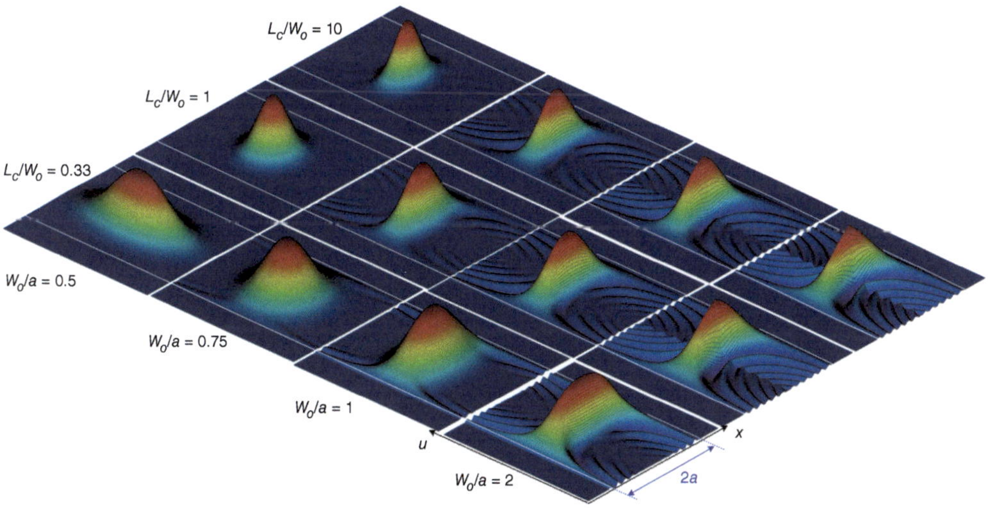

Figure 10.25 Wigner function of a GSB with truncation for different clip levels characterized by w_0/a and different degrees of coherence L_c/w_0.

An example of this behaviour is shown in Figure 10.25. Here we have a one-dimensional Gauss–Schell beam with parameters w, L_c is truncated by a slit with width $D = 2a$. By varying the parameters w_0 and L_c it can be seen that for a low clip level $w_0/a = 0.5$ nearly no diffraction effects are seen, only for large coherence $L_c/w_0 = 10$ do small stripes occur. The effects increase for higher clip levels and higher coherence is expected. Furthermore, it can be seen that for short coherence lengths, the diffraction fringes are spatially located near to the slit edges $x = \text{a}$ and $x = -\text{a}$. For larger coherence, the impact of the edge diffraction extends to the centre of the slit.

10.3.9 Examples

In the first example the propagation of a Gaussian beam is illustrated. For a Gaussian beam, the Wigner function has an elliptical shape. Figure 10.26 shows schematically the change of this ellipse by propagation through the waist region by a shear in x.

In Figure 10.27 the change of the Wigner function of a divergent Gaussian beam, which is then focused by a lens, is illustrated in the four indicated planes 1–4. The perfect elliptical shape is preserved here, because the lens is assumed to be aberration free. The beam ellipse changes due to shearing in x in the free space regions and due to u at the lens. It should be noted that in the various distributions, the scale of the x- as well as the u-axis are adapted and are not the same.

Figure 10.28 shows again a Gaussian beam, which is now impinging onto a real lens off-axis. As a result, an asymmetric non-ideal angle distribution is observed after the lens and in the focus plane a crescent shape indicates coma aberrations also in the spatial domain.

10.4 Photometry in Phase Space

10.4.1 Introduction

We define here x and p as the phase space coordinates in one dimension and x, y and p, q respectively in the two-dimensional case. p and q are the direction cosines of the rays, they correspond to the angle coordinate of the phase space. Then the expression of the differential solid angle can be converted using the projection factor into

$$\cos\theta \cdot d\Omega = dp \cdot dq \tag{10.52}$$

From the unit vector in spherical coordinates in air $n = 1$

$$\vec{p} = \begin{pmatrix} \sin\theta \cdot \cos\varphi \\ \cos\theta \cdot \cos\varphi \\ \sin\varphi \end{pmatrix} \tag{10.53}$$

there follows the differential

$$\begin{aligned} dp \cdot dq &= \left(\frac{\partial p}{\partial \theta} \cdot \frac{\partial q}{\partial \varphi} - \frac{\partial p}{\partial \varphi} \cdot \frac{\partial q}{\partial \theta} \right) d\theta\, d\varphi \\ &= \cos\theta \sin\theta\, d\theta\, d\varphi = \cos\theta\, d\Omega \end{aligned} \tag{10.54}$$

as is illustrated in Figure 10.29.

The radiance can be interpreted as a function of x and p (one-dimensional). The radiant intensity and the irradiance are then projections of the radiance onto the position and the angular coordinate, respectively.

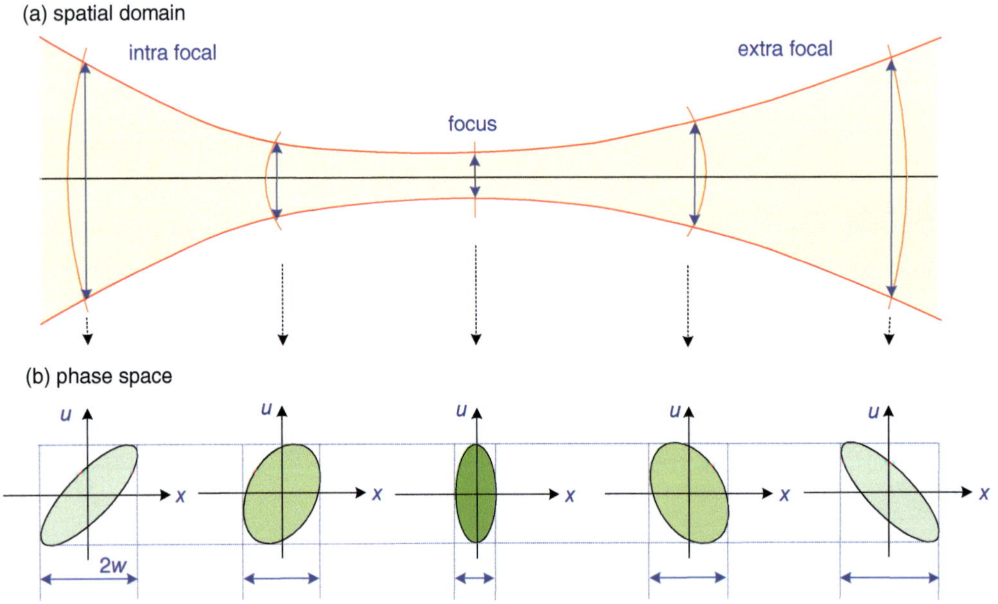

Figure 10.26 Change of the Wigner function distribution during free space propagation in a caustic region of a Gaussian beam in (a) the spatial domain and (b) the phase space.

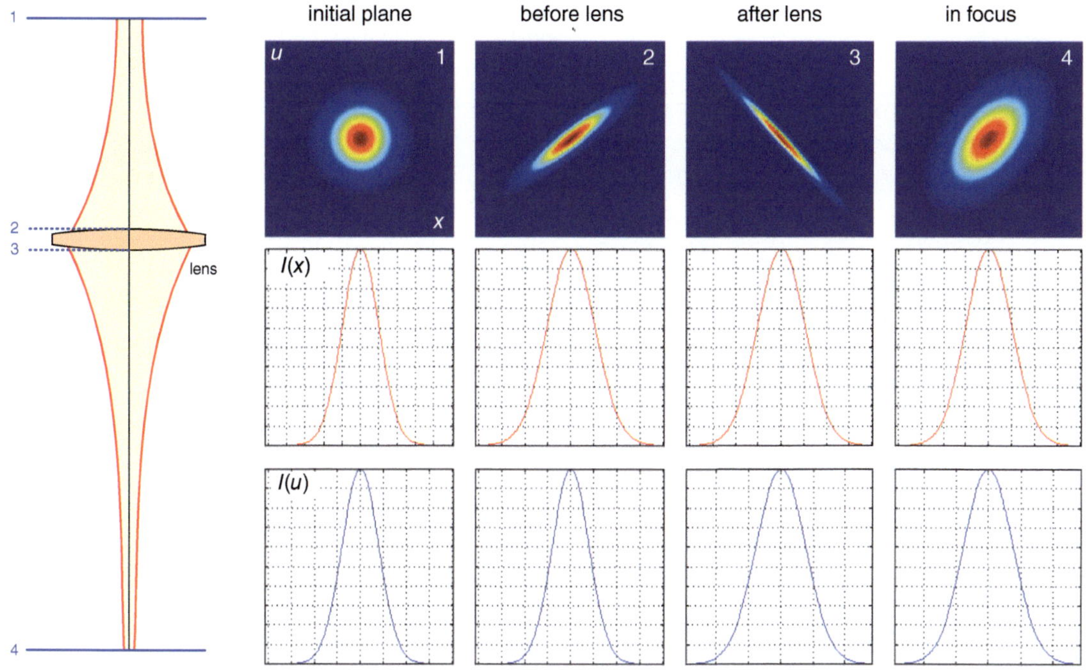

Figure 10.27 Propagation of a Gaussian beam through a paraxial thin lens.

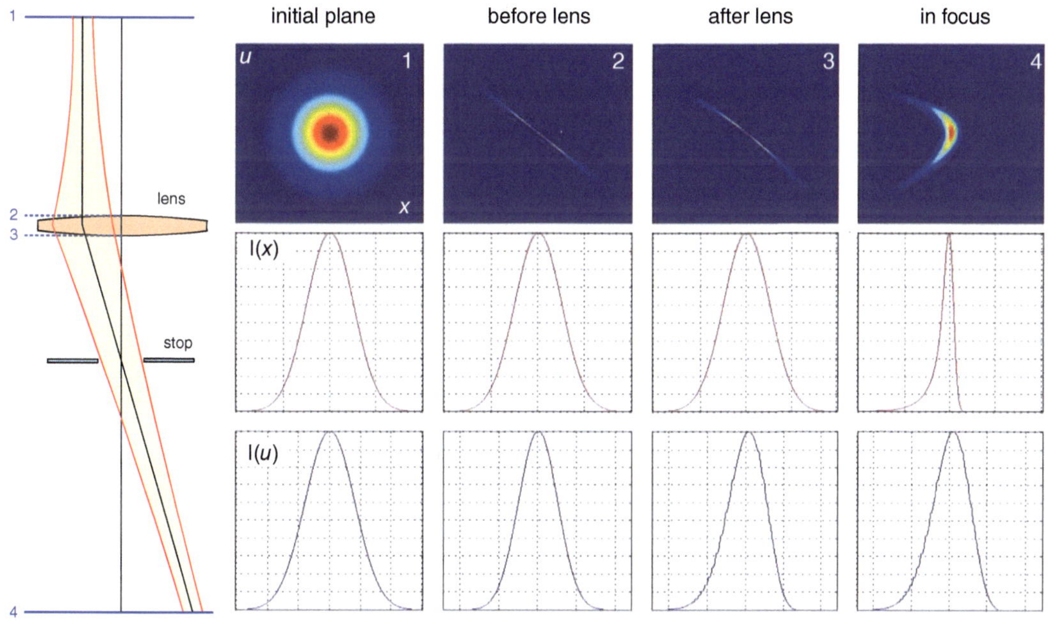

Figure 10.28 Propagation of a Gaussian beam through a lens with coma aberration. The axes are not to scale.

The radiation flux can be integrated in phase space coordinates in the form

$$dP = L(x, y, p, q)\, dx\, dy\, dp\, dq \tag{10.55}$$

In this representation the irradiance can be written as the integral of the radiance over the direction cosine. The intensity function is the integral of the radiance over the position coordinate.

$$E(x) = \int L(x, p)\, dp$$
$$I(p) = \int L(x, p)\, dx \tag{10.56}$$

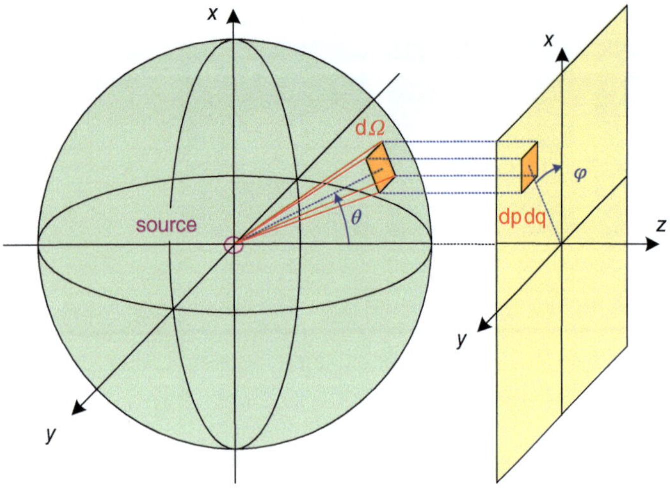

Figure 10.29 Projection of the phase space coordinates p and q, relation with the solid angle.

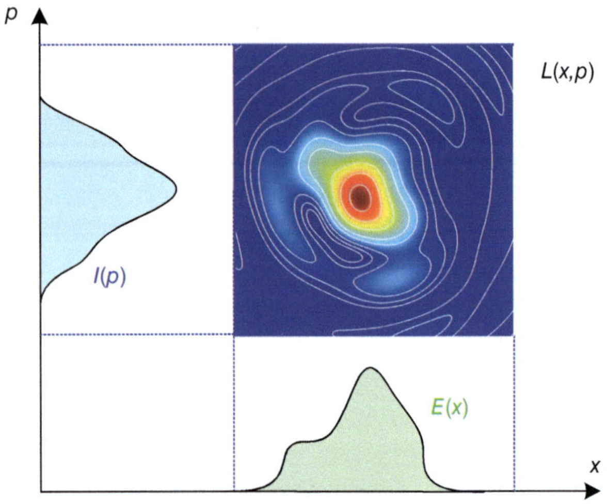

Figure 10.30 Calculation of the irradiance and the radiant intensity as projections in the phase space.

The intensity I determined in this way does not correspond exactly to the radiometric definition where the angle and not the sine of the angle has to be the argument. In the phase space, the integrals above correspond to a projection of the two-dimensional function $L(x,p)$ onto the axes as shown in Figure 10.30.

The application of phase space consideration for understanding concepts in illumination systems are in widespread use, see for example [13, 14].

10.4.2 Conservation of Energy

The conservation of energy is described in optics in the paraxial approximation by the Helmholtz–Lagrange invariant L. Since L is the product of the spatial extent with the angle extent formulated as the sine of the angle u, we can interpret the energy as being proportional in the paraxial domain to the area of a light distribution in the phase space. This means that the preservation of energy guarantees a constant area of the light in the phase space during propagation through paraxial components and free space. Figure 10.31 illustrates this behaviour for an artificial system; the three rectangles are obtained by transforming the light with lenses, mirrors and free space transitions.

This description assumes a constant intensity, which means that the energy density in the pupil is a tophat distribution. This quasi-continuum approach is only valid in geometrical optics, the conservation corresponds to a constant Jacobian matrix J in the case of propagation or changes of the coordinates.

$$\underline{J} = \frac{\partial L}{\partial x} \cdot \frac{\partial u}{\partial y} - \frac{\partial L}{\partial y} \cdot \frac{\partial v}{\partial x} \tag{10.57}$$

Radiometric considerations can be illustrated and discussed very clearly in the phase space. The energy conservation in the optical system then corresponds in four dimensions to a constant volume in the phase space as the general etendue. If a ray is described by the four coordinates x, y (position relative to a reference plane in the object space $z = 0$) and p, q (direction cosines of the ray direction vector), the Abbe sine condition in the x-section plane can be formulated as

$$n \cdot dx \cdot dp = n' \cdot dx' \cdot dp' \tag{10.58}$$

By analogy, in the y-plane it is

$$n \cdot dy \cdot dq = n' \cdot dy' \cdot dq' \tag{10.59}$$

By multiplying both conditions, the equation

$$n^2 \cdot dx \cdot dy \cdot dp \cdot dq = n'^2 \cdot dx' \cdot dy' \cdot dp' \cdot dq' \tag{10.60}$$

is obtained. This formula represents the Liouville theorem. It implies that in the four-dimensional phase space x, y, p, q

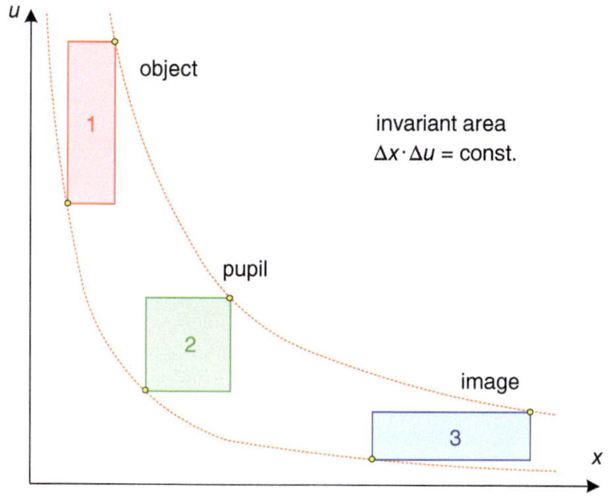

Figure 10.31 Illustration of the conservation of energy in a paraxial optical system by keeping the area in the phase space constant; all three rectangles have the same area size.

of the ray coordinates, the volume of the phase space is constant. An optical system transforms the coordinates into the primed ones, thus fulfilling the condition above. Each ray here represents a point in the phase space. The constancy of the phase space volume corresponds to an invariant optical flux given by the optical system. The Liouville theorem can also be written in the form

$$n^2 \cdot \frac{\partial (p,q)}{\partial (x',y')} = n'^2 \cdot \frac{\partial (p',q')}{\partial (x,y)} \tag{10.61}$$

For a generalization of this purely geometrical discussion, the concept of the phase space can be extended by the introduction of a density function. In this way, the complete radiation field with non-homogenous spatial radiances and angle dependencies can be taken into account. This density function is the Wigner function, which is also particularly used for the description of coherence effects.

10.4.3 Vignetting

In general, vignetting is in the geometrical sense an effect that is governed by the ray heights and the angles. Therefore, it is obvious to describe it in the phase space. In the space–angle domain the phase space defines a window in the sense of a clear area in one transverse dimension that allows rays to pass. This has been already discussed in Section 5.5 from the geometrical point of view assuming a constant homogeneous energy density [15–17]. One more extension of the vignetting model in the phase space is to take photometric distributions and effects into account. Instead of simply calculating the area of clearance, now the integration over the Wigner function as an energy density is performed. This allows us to take apodization effects on the surfaces of the system into account.

In Figure 10.32 an example illustrates the vignetting of a beam with apodization. The Wigner function serves as a weighting function for a Gaussian distribution. Three stops truncate the beam. Although the pure area ratio gives a transmittance of $T = 0.30$, due to the pronounced energy density in the centre the energetic truncation gives the relative power of 0.52 as a result of a numerical integration

The relative power transmission of a given input light field and a system or component, for example a fibre or a light guide, can be easily calculated by the normalized overlap integral of the corresponding Wigner functions [2, 4]

$$\eta = \lambda^2 \cdot \frac{\iint W_{in}(\vec{r},\vec{u}) \cdot W_{sys}(\vec{r},\vec{u})\, d\vec{r}\, d\vec{u}}{\iint W_{in}(\vec{r},\vec{u})\, d\vec{r}\, d\vec{u} \cdot \iint W_{sys}(\vec{r},\vec{u})\, d\vec{r}\, d\vec{u}} \tag{10.62}$$

By using the Wigner function the spatial as well as the angle acceptance are taken into account and the radiation can be coherent or not. This is a more general formulation of the coupling problem in comparison to the

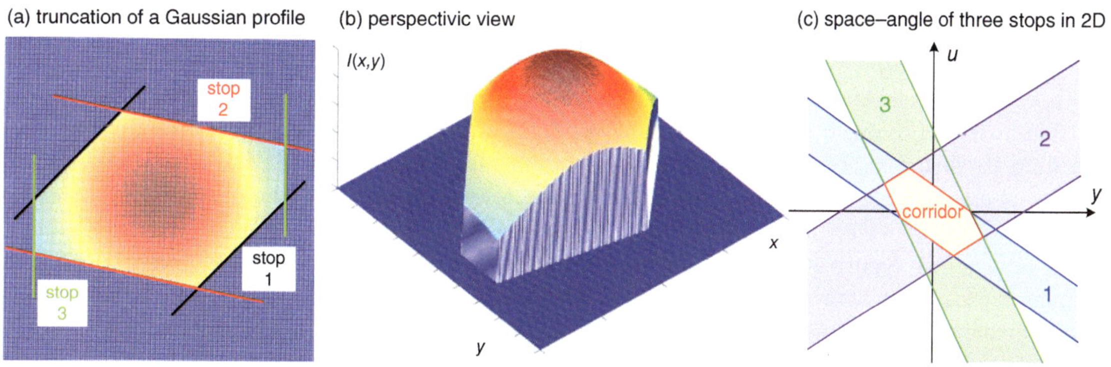

Figure 10.32 Vignetting with Wigner weighting function by three stops shown in (a) and (b) in the spatial domain, (c) illustrates in only two transverse dimensions the phase space corridor in the spatial–angle domain of the phase space.

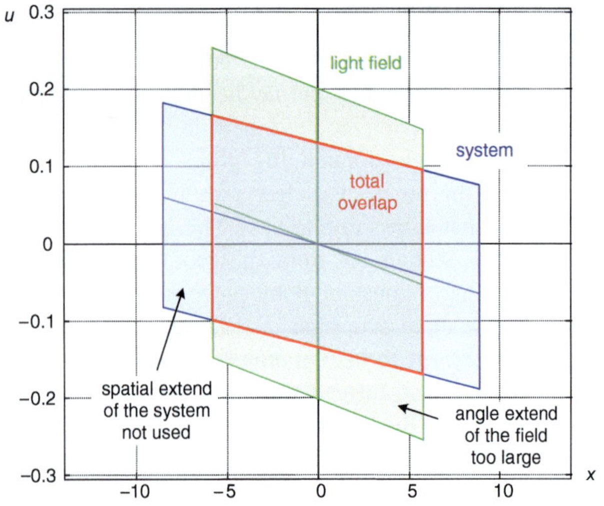

Figure 10.33 Illustration of the ray bundles in the interface plane in case of a mismatch in the spatial and angular domain. u is scaled in rad, the x-units are arbitrary.

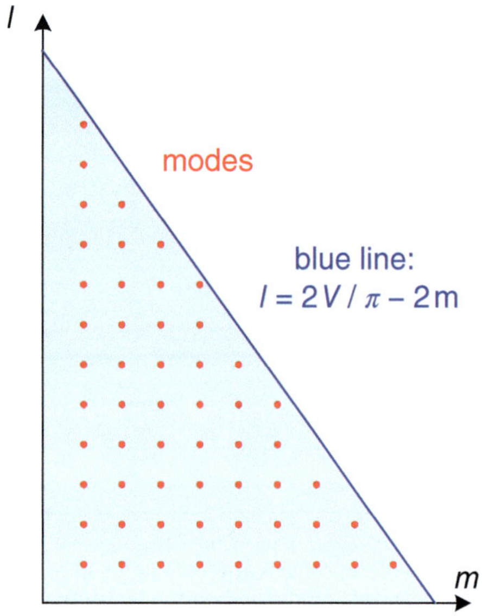

Figure 10.34 Mode spectrum of a fibre with azimuthal mode number l and radial mode index m.

purely coherent field overlap discussed in Section 8.8. In Figure 10.33 this principle is sketched in one transverse dimension.

10.4.4 Fibre Illumination

In a weakly guided fibre, the optical field is formed by discrete modes. There is a radial as well as an azimuthal mode index l and m respectively. If the diameter of the fibre is quite large, many modes can propagate in the fibre and the discrete mode structure can be considered as a quasi-continuum and a geometrical optical approximation is meaningful. If V is the structural parameter of the fibre, the total number of modes is given by

$$M = \frac{4}{\pi} V^2 \tag{10.63}$$

Figure 10.34 illustrates the discrete pattern of allowed mode indices.

In practice, how to couple a given illumination light field into a fibre with optimal efficiency η is often an important question. This corresponds to the question of what the optimal selection of the focal length of the coupling system for given fibre parameter is. The behaviour can be illustrated in the phase space rather comfortably. If step index and gradient index fibres are distinguished, a different behaviour is obtained. Figure 10.35 shows the problem in the phase space; here both coordinates are squared and the acceptable space-bandwidth products are illustrated for both fibre types and two different core radii a and aperture angles θ_c. An incoherent light source is assumed here, so the problem can be considered in the geometrical quasi-continuum approach. In case of the step index fibre, a rectangle is obtained; every overfilling of the core diameter or the acceptance numerical aperture angle θ_c generates loss. In the case of the gradient index fibre, there is a quadratic dependence of the acceptance angle with the radial coordinate r; in the representation here the function becomes linear as indicated in the figure. If now a given light source is considered, for example in the figure a source with phase space covering the area with the green border, it is seen that the eight different selections of the fibre opportunities deliver different coupling efficiencies. The values are listed in Table 10.3.

Figure 10.36 illustrates the optical focal length selection for coupling two different given light sources, represented by the hyperbolic curves in red and purple, into a step index fibre with radius a and acceptance aperture θ_c. It can be seen that the laser with the smallest bandwidth product L as shown in red can be fully coupled into the fibre. The optimal choice of the focal length for focusing is indicated by the condition that the angle as well as the core diameter are underfilled by the same amount. This means that focal lengths between f_1 and f_2 are meaningful. For the laser with the larger etendue drawn in purple, a loss must be accepted in any case; the best result is again obtained along the line through the origin, which violates angle and spatial acceptance by the same value. If the focusing lens suffers from aberrations, mainly in the upper part of the hyperbolic curves a discrepancy between ideal and real conditions occurs. This is indicated by the dotted lines.

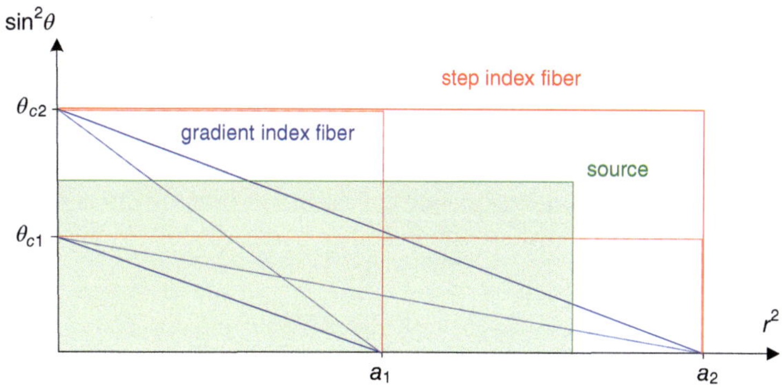

Figure 10.35 Phase space overlap for a fibre coupling problem with squared radial and angle coordinates.

Table 10.3 Fibre coupling efficiencies.

Fibre type	Core size	Aperture	Efficiency
Step index	a_1	θ_1	0.41
		θ_2	0.63
	a_2	θ_1	0.68
		θ_2	1.0
Gradient index	a_1	θ_1	0.20
		θ_2	0.41
	a_2	θ_1	0.38
		θ_2	0.78

creates separated areas in the phase space, so the filling areas are not connected. The focusing again is seen as a shear and in the final plane six, an integration along the angle direction due to the overlap of the isolated sections generates a homogeneous light distribution.

In the second example a system with an axicon and a condenser lens is shown in Figure 10.38 for nearly incoherent light. It is seen that the axicon splits the distribution in the angle direction, by the lens and the free space propagation the two domains are reshaped and overlaid in the spatial direction x. See also Section 4.8.3 for the working principle of an axicon. It should be noted that the very small Wigner contribution between the two large domains is the interference term.

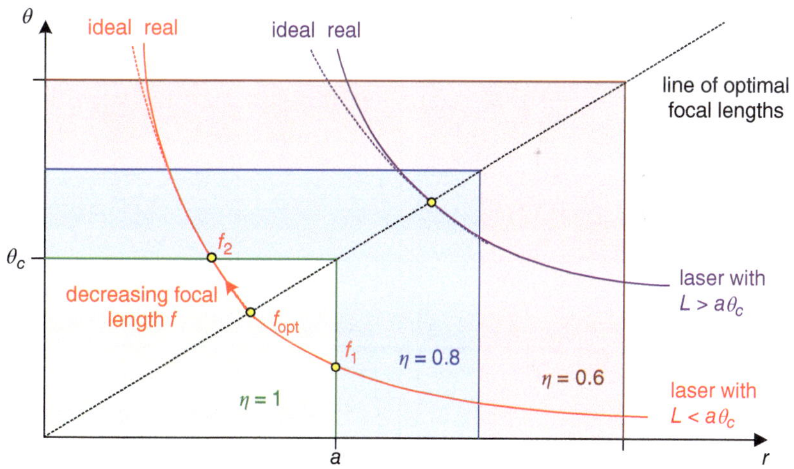

Figure 10.36 Coupling efficiency of a laser illuminating a fibre with two different space-bandwidth products into a fibre with lenses of different focal lengths, which are varied along the curves.

10.4.5 Further Examples

Figure 10.37 shows a more complicated example that illustrates the behaviour of the light shaping in case of a lenslet array in a fly eyes condenser in geometrical optics approximation. For the setup and principle of the system, see Section 4.8.5.4. Six planes are illustrated in the phase space pictures. The sub-aperture impact of the lenslets

Another example to see the oscillating Wigner contributions in the case of interference is illustrated in Figure 10.39. Here in (a) a shearing interferometer is sketched, (b) shows the Wigner representation in the four indicated planes. In plane two the splitting is observed, in plane three this distribution is sheared due to the focal power. Plane four represents the focal plane of the lens;

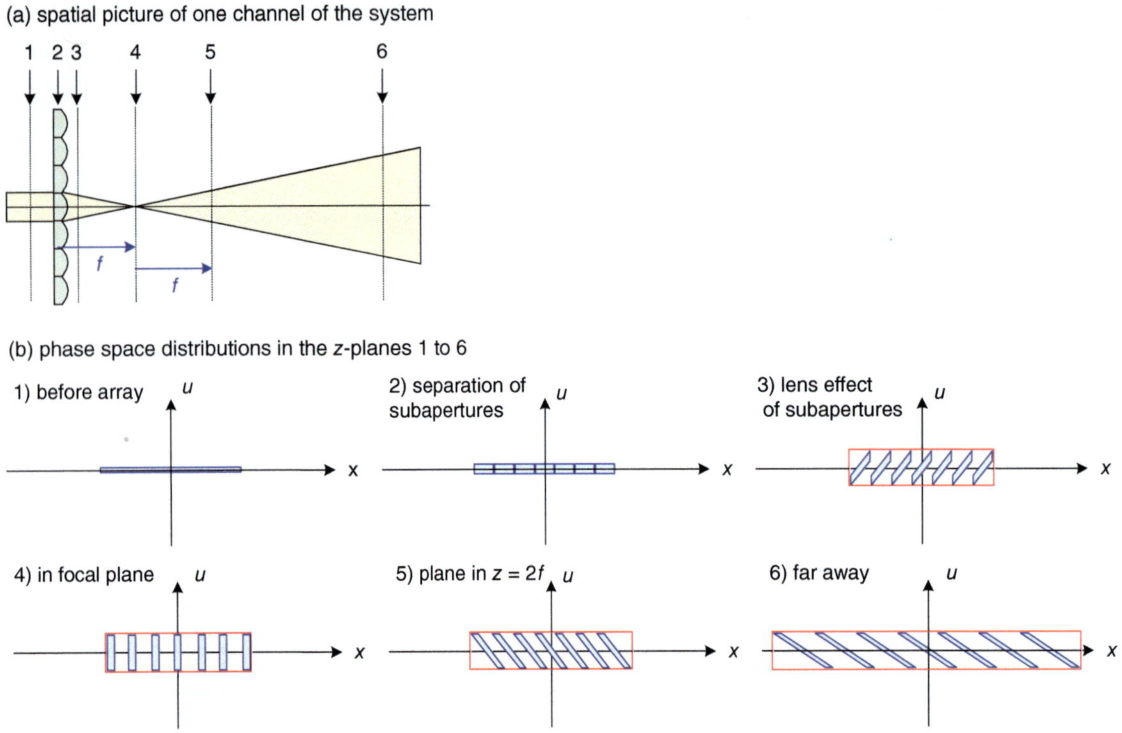

Figure 10.37 Illustration of the change of the geometrical optical light distribution in a fly eyes condenser. (a) illustrates the setup and the selected planes, (b) gives the phase space distributions in these planes.

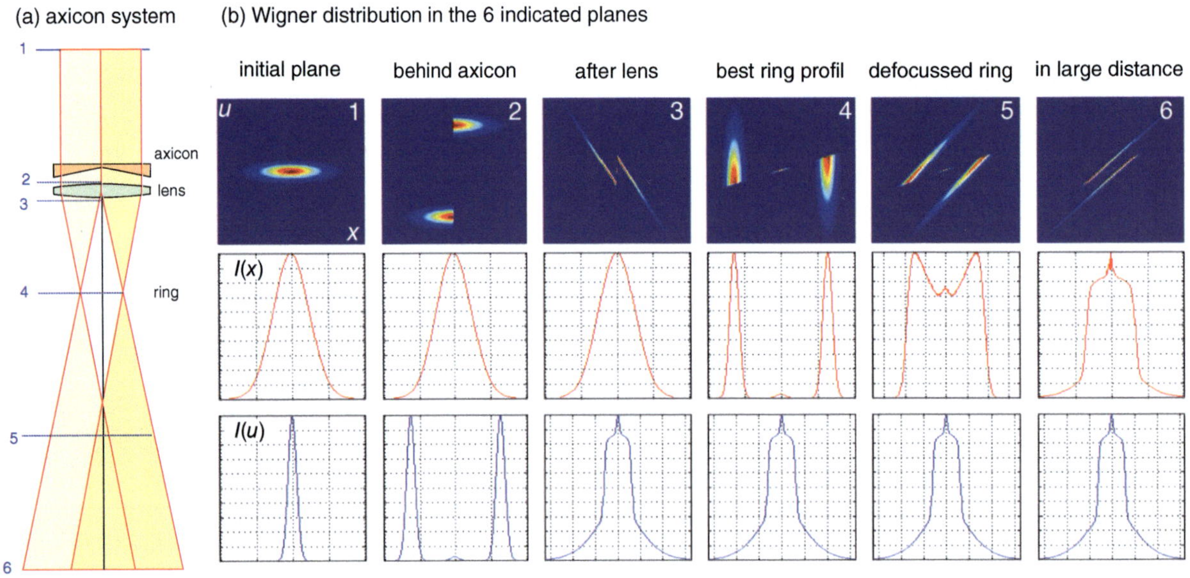

Figure 10.38 Axicon illumination system (a) and Wigner distributions (b).

here the two beam parts are superposed. The system is illuminated with a coherent nearly tophat shaped beam. The splitting by the birefringent plate generates a pronounced oscillatory part in between the bright patches of W. By integrating the Wigner function in u-direction, the oscillating parts vanish; no light contribution is seen between the two beam parts. The lens and the propagation now shear the pattern and as a result in the focal plane the integration gives the interference pattern exactly produced by the oscillatory part.

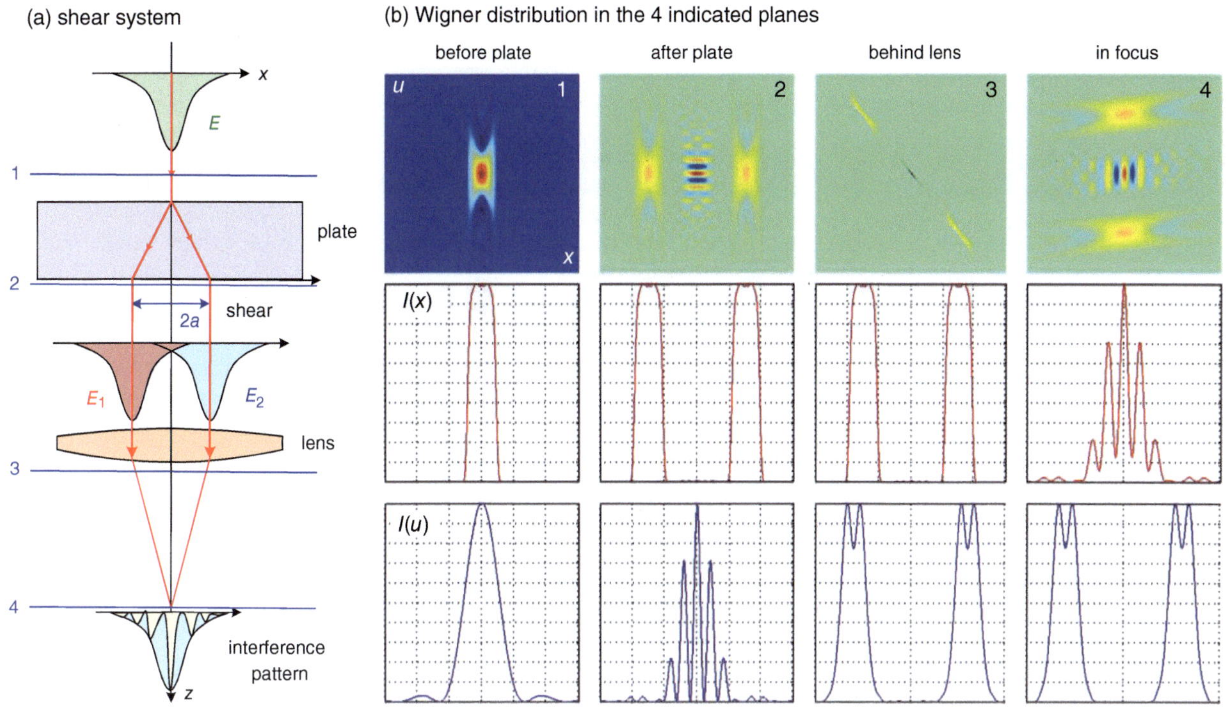

Figure 10.39 Shearing interferometer representation in phase space, illustrating the appearance of interference by the Wigner function.

10.5 Miscellaneous*

(*) The content of this section is quite specialized and for advanced reading. Knowledge of this section is not necessary to understand the following chapters. Beginners are recommended to skip this section.

10.5.1 Caustic in Phase Space

If a ray pattern forms a caustic, the phase space equivalent gets a characteristic property: in a certain range, for one x coordinate, two angle values exist. As can be is seen in Figure 10.40, depending on the z-position the phase space curves are bent and have a point with a vertical tangent. In these cases, in the region indicated as green two angle values belong to one x-coordinate. The vertical extent of the green areas in part (c) indicate the angle spreading for one z-position. If the points with the vertical slope are connected, the envelope of the ray-caustic is obtained [10].

10.5.2 Phase Space Discussion of Sampling

The phase space allows the representation of a signal in the spatial-angle domain. In the discussion of the sampling for diffraction calculations, the spatial coordinate corresponds to points in the exit pupil of an optical system, the angle or spatial frequency coordinate to points in the image plane. The influence of the zooming capability by using the chirp-z-transform is illustrated in Figure 10.41 in one lateral dimension x/u. If a zoom factor of $s = 2$ is applied, the density of points along the u-axis is doubled, and the overall range of the u-window shrinks correspondingly.

Usually the image plane corresponds to the far-field of the field in the exit pupil. This means that the parabolic phase curvature in the exit pupil is automatically removed in the case of the diffraction integral calculation. If now the phase surface of the wave is further changed, for example by defocusing the image plane, an additional phase chirp is generated. This means that the rectangular area of the signal is sheared along the angle direction in the phase space, as illustrated in Figure 10.42(a). It is immediately seen that a larger angle region must be sampled, and the effort of sampling is increasing. If a wave with a finite aperture extent is propagated over a distance z, the spatial coordinate increases and a corresponding shear along the x-direction is observed as is seen in Figure 10.42(b).

The effect of the equivalence transform described in Section 6.2.6.4 can also be easily interpreted in the phase space. According to Figure 10.43, the chirp corresponds to an adaptation of a sheared parallelepiped around the signal area. In the rotated coordinates x'-u' again a narrow angle range $2u'_{max}$ has to be sampled. The value

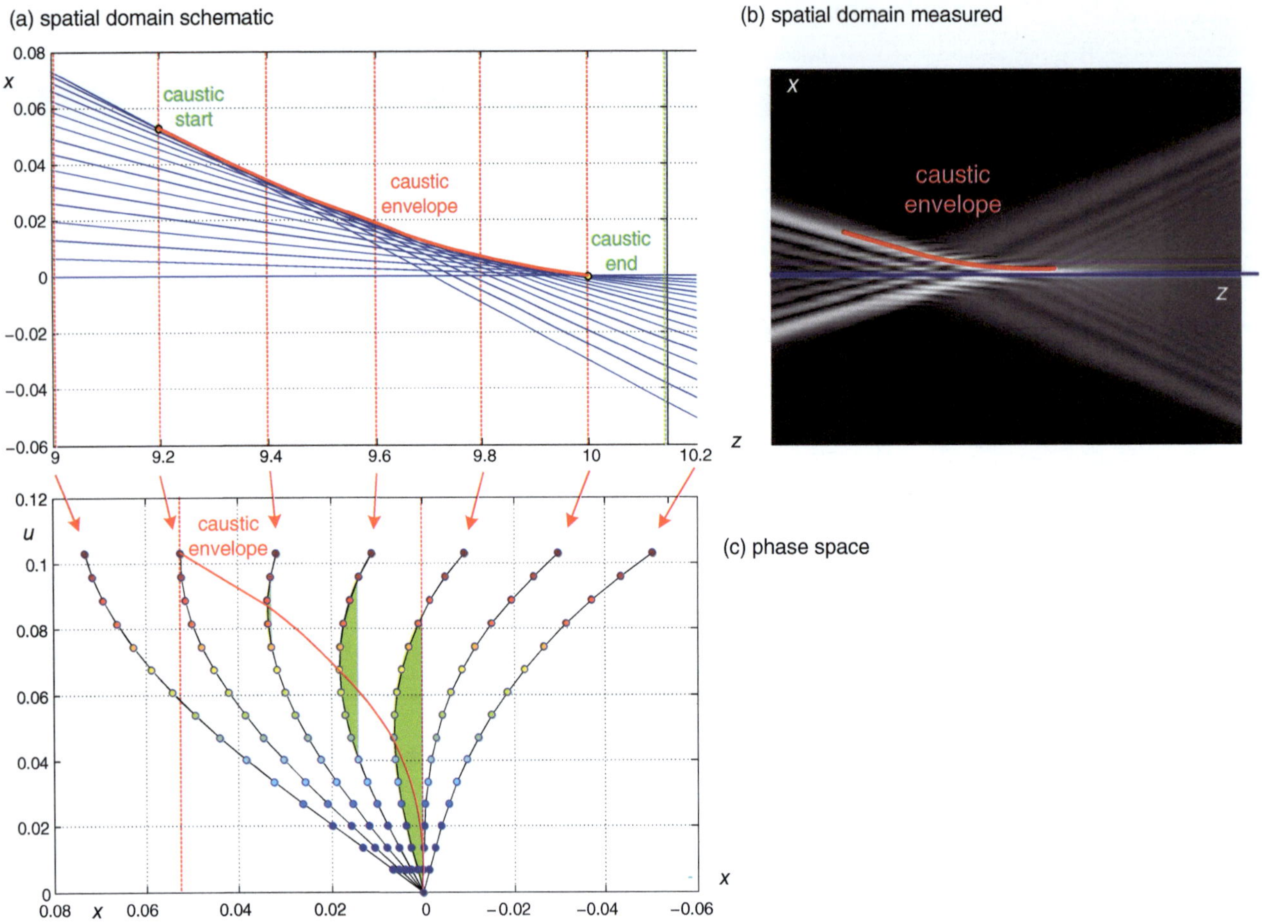

Figure 10.40 Spatial domain with rays (a) and measured as intensity (b) as well as the phase space pattern (c). u is scaled in rad, the x-units are arbitrary.

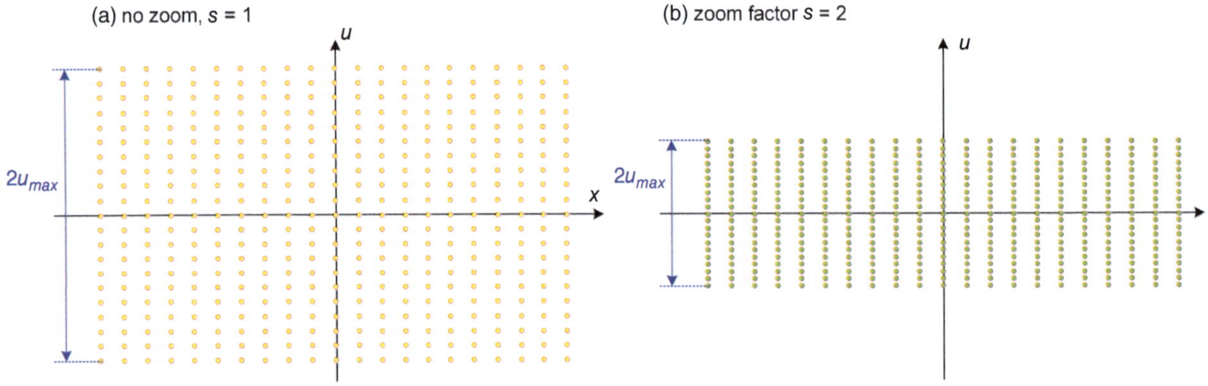

Figure 10.41 Change of the point density and the angle window size in the case of a scaling factor $s = 2$ for the chirp-z-transform; (a) gives the original grid pattern, (b) the transformed grid.

of the rotation angle depends on the radius of curvature, which is needed for the quasi-flattening of the phase. This transform then delivers an optimal compact support area for a more comfortable adapted representation in the paraxial approximation.

10.5.3 Phase Space Analyzer

As discussed above, the Wigner function represents mainly an energy density. It is well known that densities cannot be directly measured. A phase space analyzer is a special optical system with a non-axis symmetrical component,

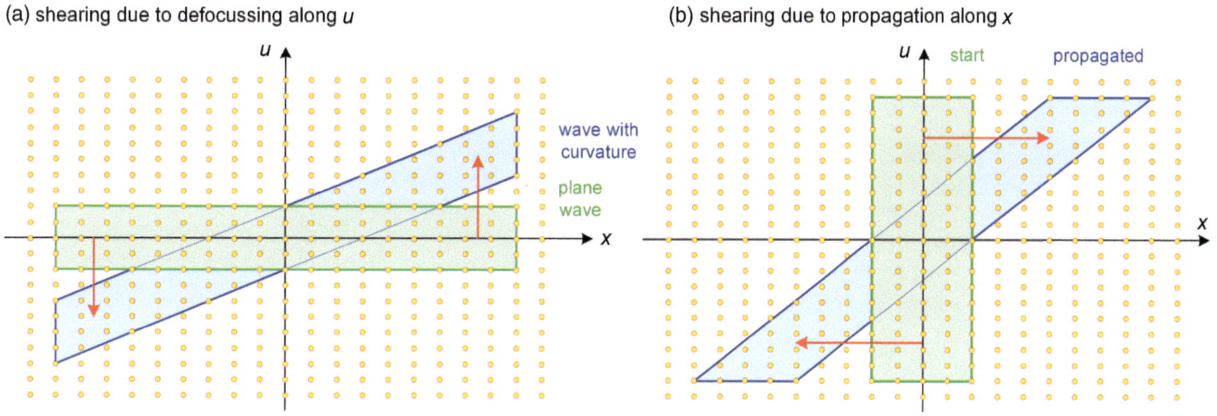

Figure 10.42 Change of the signal pattern in (a) the case of defocusing of the sampled wave and (b) due to propagation in z to corresponding shearing transforms in both directions respectively.

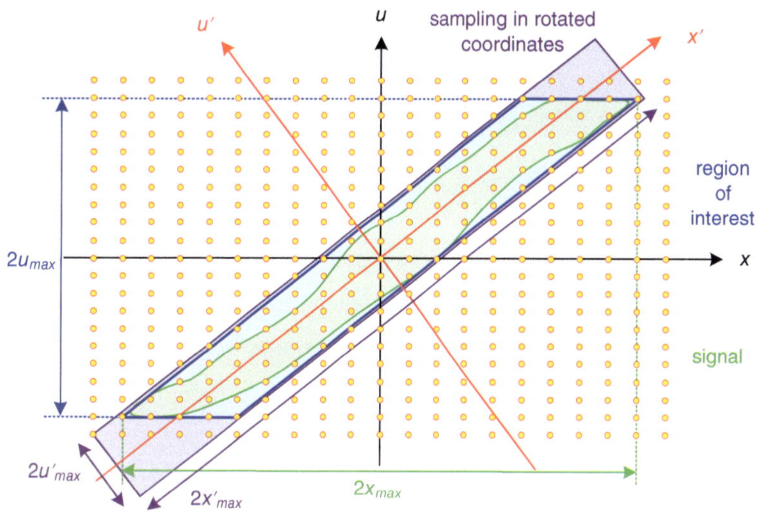

Figure 10.43 Interpretation of the equivalence flattening transform as a rotation of coordinates to calculate the Fourier transform more comfortably.

which allows the analysis of the spatial and angle or wavefront errors of a beam in a simple way [18]. Figure 10.44(a) shows the optical system of the phase space analyzer, where the coherent field of a slit is propagated through a special system. A lens and a toroidal lens, which is rotated through 45° around the optical axis in azimuthal orientation, are located at appropriate distances. The lens is a distance a from the slit, the toroidal lens follows at a distance d. The detector plane is at a distance z from the toroidal lens and a distance $b = d + z$ from the spherical lens. In the image plane of the slit, wave aberrations in the slit plane are transformed by the non-orthogonal system into transverse deviations in the spatial domain. By analyzing the wave front deformation in the slit, the wave aberration can be evaluated by geometrical means.

The most interesting component in the phase space analyzer setup is the toroidal lens with the azimuthal rotation angle of 45°. Figure 10.44(b) shows the effect of this lens on an angle error u in the x-direction. In the lens, this deviation is transformed into an angle v in the y-direction. As a result, an angle deviation u is transformed into an angle v. The mixing effect of the rotated toroidal lens can be used to analyze wave or angle aberrations of the incoming beam of the exit pupil for an optical system along the direction of the slit. Since the illumination is performed in the form of a thin slit, only a radial section of the pupil of the system is observed. The wave aberrations are located along the slit and they are transformed in a perpendicular direction. If a proper imaging of the slit is realized by the measuring setup, the straight slit is deformed and curved if wave aberrations exist. Figure 10.45 shows some examples of defocusing as well as low-order wave aberrations and their impact on the shape of the imaged slit. The aberrations are scaled here as Zernike coefficients, see Chapter 15. The tilt of the slit images indicates a defocus, the s-shaped slit images show a residual primary spherical aberration.

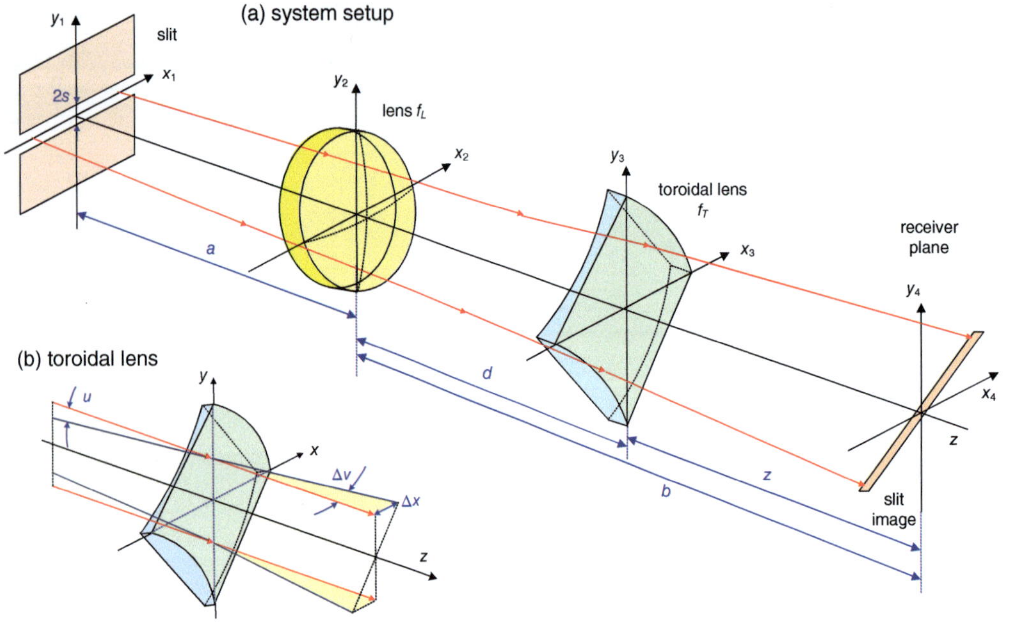

Figure 10.44 Layout of the phase space analyzer in (a), in (b) the special impact of the 45°-rotated toroidal lens is illustrated.

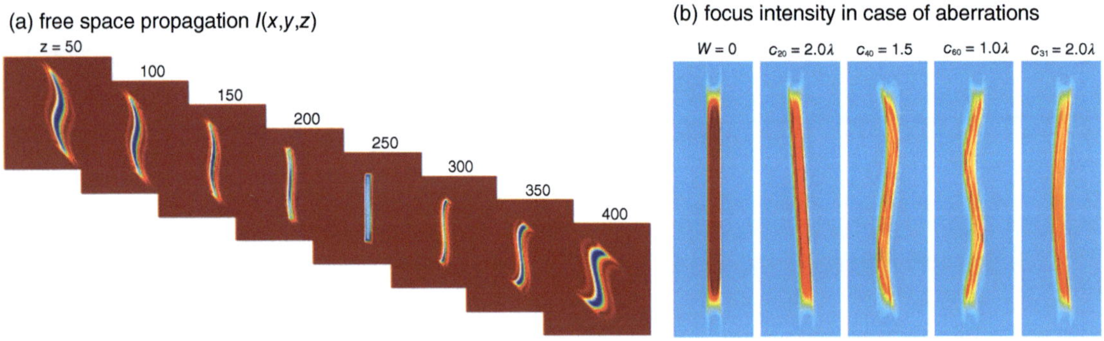

Figure 10.45 Examples of PSA results, shown are in (a) a free space propagation and in (b) slit image for different Zernike aberrations.

The initial slit of the system extracts one direction in the incoming wave. To measure the deviations in the full pupil of a system, the slit must be rotated in azimuthal direction and a kind of tomography can be performed and analyzed. Due to the transfer from an incoming angle u into a special deviation Δx in a certain plane, the system really gives a one-dimensional phase space cross section as a result.

10.5.4 Fractional Fourier Transform

10.5.4.1 Definition

In the framework of the phase space it is obvious that the transition between the spatial and the angle domain can be considered as a 90° rotation and is performed by a Fourier transform. This process is also obtained for a free space propagation between exit pupil and image plane of an optical system. For the transition into a plane in between, a rotation with an arbitrary angle θ occurs and the propagation is described by a Fresnel transform [2].

There is a well-known mathematical formulation of these operators: the so-called fractional Fourier transform (FrFT). In one transverse coordinate the transform reads

$$F_\theta[f(x')] = \frac{e^{\frac{i\pi}{4}}\, e^{-\frac{i\theta}{2}}}{\sqrt{2\pi \cdot \sin\theta}} \cdot e^{-\frac{i}{2}x'^2 \cot\theta} \cdot \int f(x) \cdot e^{-\frac{i}{2}x^2 \cot\theta}\, e^{\frac{i\,xx'}{\sin\theta}}\, dx \qquad (10.64)$$

Corresponding to the classical scaling law of the Fourier transform between pupil and image plane in optics with $x' = \lambda \cdot f \cdot x_p$, the fractional Fourier transform has a scaling parameter f_s too, which serves as a reference scaling length. In two dimensions we get

$$F_\alpha[f(x',y')] = \frac{i\, e^{-i\alpha}}{2\pi \sin\alpha} \cdot e^{-\frac{i}{2}(x'^2+y'^2)\cot\alpha} \cdot \iint f(x,y) \cdot e^{-\frac{i}{2}(x^2+y^2)\cot\alpha}\, e^{\frac{i\,(xx'+yy')}{\sin\alpha}}\, dx\, dy \qquad (10.65)$$

In contrast to a classical Fourier transform, the angle θ or alternatively the order $\alpha = 0\ldots1$ with

$$\theta = \alpha \cdot \frac{\pi}{2} \tag{10.66}$$

describes the transform. The usual Fourier transform is obtained for $\theta = \pi/2 = 90°$ or $\alpha = 1$

$$F_{\frac{\pi}{2}}[f(x',y')] = \frac{1}{2\pi} \cdot \iint e^{i(xx'+yy')} f(x,y)\, dx\, dy \tag{10.67}$$

The additional feature of the fractional Fourier transform in comparison to the classical Fourier transform is a chirp factor in Eq. (10.64) and (10.65), which is scaled by the rotation angle. This shows that the FrFT is a transform with one more parameter α. The relation in the phase space is sketched in Figure 10.46.

If this picture of a phase space rotation is generalized, two FrFT operations should be performed in a cascade and the sum of the angles is the overall angle. To guarantee this property, the scaling must be taken into account appropriately. If two operations are cascaded, the angle-additive property is obtained in the form

$$\widehat{F}_{\alpha 1}[\widehat{F}_{\alpha 2}\{f(x)\}] = \widehat{F}_{\alpha 1+\alpha 2}[f(x)] \tag{10.68}$$

if the standard length f_s is the same for both transforms. If the input and output coordinates are scaled by a factor M in the form

$$\widehat{x}_i = M \cdot x_i \qquad \widehat{x}_o = M \cdot x_o \tag{10.69}$$

the angle of rotation is changed by

$$\tan\widehat{\theta} = M^2 \cdot \tan\theta, \qquad \sin\widehat{\theta} = \frac{\tan\theta}{\sqrt{M^{-4} + \tan^2\theta}} \tag{10.70}$$

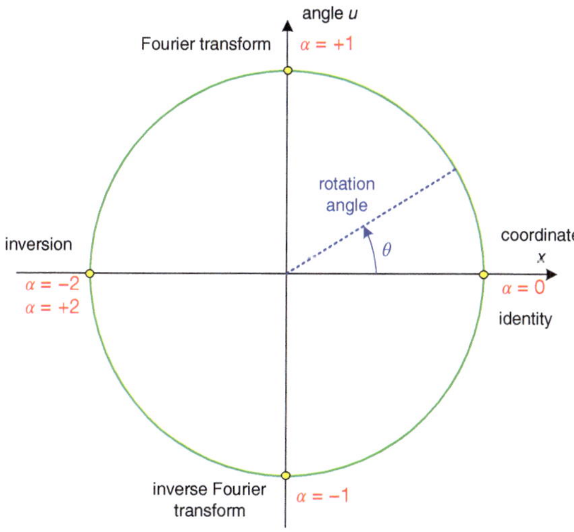

Figure 10.46 Phase space rotation and FrFT for arbitrary angles θ.

and an additional chirp factor $Q(x)$ is obtained

$$Q(x) = e^{i\pi x^2 \cdot \frac{M^4-1}{M^4\tan\theta + \cot\theta}} \tag{10.71}$$

This behaviour describes the spatial scaling for arbitrary rotation angles. The factor M must be the same in case of two transforms to guarantee the cascadability of the FrFT. The reference variable f_s must be the same for both operations.

10.5.4.2 Fresnel Integral and Fractional Fourier Transform

The propagation of a coherent field according to a Fresnel diffraction integral can be described by a FrFT. With the abbreviation and relation between rotational angle α and propagation distance z

$$\tan\theta = \tan\left(\frac{\pi}{2}\alpha\right) = k \cdot z \tag{10.72}$$

the field after propagation can be written as

$$E(\vec{r},z) = \frac{k}{1 - i\tan\theta} \cdot e^{\frac{i}{4}r^2k^2\sin(2\theta)} \cdot F_\alpha(k\vec{r}\cos\theta) \tag{10.73}$$

It is seen that an additional quadratic phase factor is added to the FrFT, so the fractional Fourier transform alone does not describe a Fresnel propagation. In the simple case of a transform between a spherical initial wave with radius R_1 and a final wave with R_2, the relation

$$\tan^2\theta = -\frac{z \cdot (z + R_1 + R_2)}{(z + R_1) \cdot (z + R_2)} \tag{10.74}$$

with the propagation distance z can be derived. In the more general case of a propagation over an $ABCD$ segment and the description of the diffraction by the Collins integral with $A = D$

$$E(x') = \sqrt{\frac{i}{\lambda B}} \cdot e^{-\frac{i\pi D}{\lambda B}\cdot x'^2} \cdot \int E(x) \cdot e^{-\frac{i\pi A}{\lambda B}\cdot x^2} e^{-\frac{2\pi i}{\lambda B}\cdot x \cdot x'}\, dx \tag{10.75}$$

the three parameters M, f_s and R are related to the $ABCD$ matrix elements with the relations

$$A = M \cdot \cos\theta, \qquad B = f_s \cdot M \cdot \sin\theta$$
$$C = -\frac{\sin\theta}{f_s \cdot M} + \frac{M\cos\theta}{\lambda \cdot R}, \qquad D = \frac{\cos\theta}{M} + \frac{f_s \cdot M \sin\theta}{\lambda \cdot R} \tag{10.76}$$

are valid. The FrFT therefore describes the transform between curved waves with parabolic phase and is a sub-operation of the general Fresnel transform. By resolving the equations for the scaling factor M, the two equations

$$\widehat{x}_i = x_i \cdot \sqrt{\frac{2\pi \cdot A}{\lambda \cdot B} \cdot \tan\theta}, \quad \widehat{x}_0 = x_0 \cdot \sqrt{\frac{2\pi \cdot D}{\lambda \cdot B} \cdot \frac{\sin 2\theta}{2AD}} \tag{10.77}$$

or explicit

$$M_i = \sqrt[4]{-\frac{BD}{\lambda^2 \cdot AC}}, \qquad M_o = \sqrt[4]{-\frac{AB}{\lambda^2 \cdot CD}} \tag{10.78}$$

and the relation

$$\cos^2\theta = AD + \frac{AB}{R} \tag{10.79}$$

for the equivalent rotation angle are found.

10.5.4.3 Gradient Index Lenses

If a paraxial gradient index rod lens with length L is considered with a quadratic profile of the refractive index

$$n(r) = n_o \left(1 - \frac{1}{2} n_{r2} r^2\right) \tag{10.80}$$

the $ABCD$ matrix of this component can be written as

$$\mathbf{M} = \begin{pmatrix} \cos(\sqrt{n_{r2}}L) & \frac{1}{\sqrt{n_{r2}}}\sin(\sqrt{n_{r2}}L) \\ -\sqrt{n_{r2}}\sin(\sqrt{n_{r2}}L) & \cos(\sqrt{n_{r2}}L) \end{pmatrix} \tag{10.81}$$

If this matrix is compared with the general FrFT matrix, it can be seen that this kind of component (see Figure 10.47) exactly mimics a fractional Fourier transform with the parameters

$$\theta = L \cdot \sqrt{n_{r2}}, \quad f_s = \frac{1}{\sqrt{n_{r2}}} \tag{10.82}$$

10.5.5 Linear Canonical Transform

The linear canonical transform (LCT) is a generalization of the fractional Fourier transform. It is a general formulation of an integral transform with a kernel of quadratic phase. There is a lot of basic mathematical context found in literature [19], from complex generalization, group theory and canonical operators. Here only a short discussion should be given that is more related to phase space and optical applications. The LCT can be written in one transverse dimension as

$$\hat{C}[f(x)] = \int C_{lct}(x, x') \cdot f(x')\, dx' \tag{10.83}$$

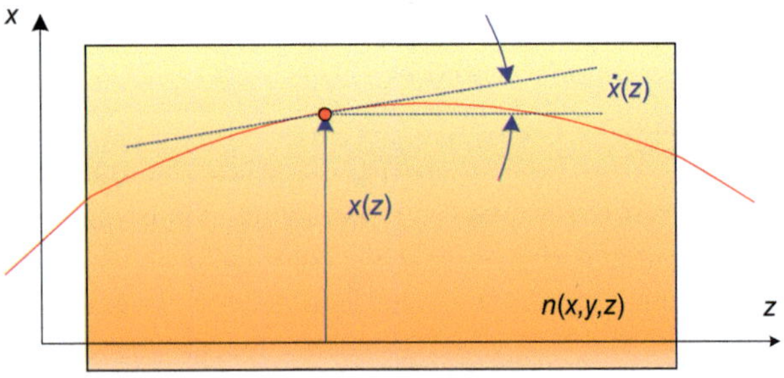

Figure 10.47 Raytrace in a gradient index lens.

with the kernel function

$$C_{lct}(x, x') = \frac{1}{\sqrt{B}} e^{-\frac{i\pi}{4}} \cdot e^{\frac{i\pi}{B}(Dx^2 - 2xx' + Ax'^2)} \tag{10.84}$$

if $B \neq 0$. It is an operator with the three parameters A, B, D. It is also usual in literature to define the parameter as α, β, γ with

$$\alpha = \frac{A}{B}, \quad \beta = \frac{1}{B}, \quad \gamma = \frac{D}{B}, \quad A = \frac{\gamma}{\beta}, \quad B = \frac{1}{\beta}, \quad D = \frac{\alpha}{\beta} \tag{10.85}$$

which gives

$$C_{lct}(x, x') = \sqrt{\beta} e^{-\frac{i\pi}{4}} \cdot e^{i\pi(\gamma x^2 - 2\beta xx' + \alpha x'^2)} \tag{10.86}$$

This transform describes a Fresnel transform over a paraxial $ABCD$ segment with identical refractive indices $n = n'$ or $AD - BC = 1$. The classical Fourier transform, the FFT with a chirp multiplication and a chirp convolution are special cases of the LCT. According to the Iwasawa decomposition, the LCT can always be represented by three operators with FrFT (rotation by θ), scaling by M and chirp q as

$$\underline{M}_{LCT} = \begin{pmatrix} 1 & 0 \\ -q & 1 \end{pmatrix} \begin{pmatrix} M & 0 \\ 0 & 1/M \end{pmatrix} \begin{pmatrix} \cos\theta & \sin\theta \\ -\sin\theta & \cos\theta \end{pmatrix} \tag{10.87}$$

The chirp operator corresponds to the multiplication by

$$Q_q = e^{-i\pi q x^2} \tag{10.88}$$

Therefore, the transform operator can be written as

$$\hat{C}_{lct}(x') = e^{-i\pi q x^2} \cdot \frac{1}{\sqrt{M}} \hat{F}_\alpha \left[\frac{x}{M}\right] \tag{10.89}$$

with the parameters

$$M = \sqrt{A^2 + B^2}, \quad \alpha = \begin{cases} \frac{2}{\pi} arc\tan \frac{B}{A} & if A \geq 0 \\ 2 + \frac{2}{\pi} arc\tan \frac{B}{A} & if A < 0 \end{cases},$$

$$q = \begin{cases} -\frac{D}{B} & if A = 0 \\ -\frac{C}{A} - \frac{B/A}{A^2 + B^2} & else \end{cases} \tag{10.90}$$

If two LCT operations are concatenated, again an LCT is obtained. The three parameters with index 1 and 2 finally deliver a transform with the data

$$\alpha_3 = \alpha_2 - \frac{\beta_2^2}{\alpha_1 + \gamma_2}, \quad \beta_3 = \frac{\beta_1 \beta_2}{\alpha_1 + \gamma_2}, \quad \gamma_3 = \gamma_1 - \frac{\beta_1^2}{\alpha_1 + \gamma_2} \tag{10.91}$$

The LCT is a generalization of the Fresnel transform. Of course, the transform can be extended in a straightforward manner to two lateral dimensions. Therefore, it describes all paraxial wave propagation effects in a more general context. The LCT is not commutative, so the sequence of application matters. This can directly be seen in Eq. (10.91) for the parameters of concatenation.

For more details the reader is referred to the literature.

References

1 Mecklenbräuker, W. and Hlawatsch, F. (1997). *The Wigner Distribution*. Elsevier.
2 Ozaktas, H., Zalevsky, Z., and Kutay, A. (2001). *The Fractional Fourier Transform*. Wiley.
3 Wolf, K.B. (2004). *Geometric Optics on Phase Space*. Springer.
4 Tesdorf, M., Hennely, B., and Ojeda-Castaneda, J. (2010). *Phase-Space Optics*. McGraw Hill.
5 Torre, A. (2005). *Linear Ray and Wave Optics in Phase Space*. Elsevier.
6 Alonso, M. and Forbes, G. (2000). Phase-space distributions for high-frequency fields. *JOSA A* 17: 2288.
7 Alonso, M. (2011). Wigner functions in optics: describing beams as ray bundles and pulses as particle ensembles. *Advances in Optics and Photonics* 3: 272.
8 Gitin, A. (2009). The Legendre transformation as a way of constructing a phase portrait of beams of light rays. *Optics Communication* 282: 757.
9 Gitin, A. (2010). The Legendre transformation in Hamiltonian optics. *Journal of the European Optical Society* 5 (10): 022.
10 Gitin, A. (2008). Legendre transformation: connection between transverse aberration of an optical system and its caustic. *Optics Communication* 281: 3062.
11 Herkommer, A. and Yang, H. (2014). Tackling freeform aberrations by ray-mapping analysis. *SPIE Conference Proceedings*, 9293. https://doi.org/10.1117/12.2075537.
12 Herkommer, A. (2018). Phase-space representations of freeform optical systems. *Encyclopedia of Modern Optics, Second Edition* 4: 205.
13 Rausch, D., Rommel, M., Herkommer, A., and Talpur, T. (2017). Illumination design for extended sources based on phase space mapping. *Optical Engineering* 56: 065103.
14 Rausch, D. and Herkommer, A. (2012). Phase space approach to the use of integrator rods and optical arrays in illumination systems. *Advanced Optical Technologies*, 1: https://doi.org/10.1515/aot-2011-0002.
15 Gitin, A. (2012). Phase space description of vignetting in optical systems. *Optics Communication* 285: 2485.
16 Gitin, A. (2009). Aberrational vignetting and eikonal theory. *Journal of the European Optical Society* 4: 09046.
17 Ozaktas, H. and Oktem, F. (2013). Phase-space window and degrees of freedom of optical systems with multiple apertures. *Journal of the Optical Society of America A* 30: 682.
18 Weber, H. (1992). Wave optical analysis of the phase space analyser. *Journal of Modern Optics* 39: 543.
19 Healy, J., Kutay, M., Ozaktas, H., and Sheridan, J. (2016). *Linear Canonical Transforms*. Springer.

11

Computation and Digital Processing of Images

11.1 Introduction

The major task of optical design is to provide certain performance predictions for an optical system. Mostly, the designer translates the requested system performance into special quality numbers and aberration values. This assumes that it is known how the application works, what quality is needed and how the classical aberration numbers have to be selected and interpreted. This is not trivial and needs usually a lot of experience and discussions with the users and customers.

The use of digital processing of data is of growing importance in lens design. This is a wide field and only some brief selected topics will be discussed in this book. The use of digital methods in the development of imaging instruments has several aspects. Some important ones are the following.

- Simulation of quasi-realistic images as expected performance of the system by computational means.
- Using special imaging modalities, which requires a final computational signal generation.
- Gathering the image by digital sensors like CCD, which generates problems of discretization.
- Postprocessing of digital images to achieve special effects like resolution improvement, contrast enhancement, automatic refocusing.
- Joint optimization and function of analogue optical systems with digital postprocessing.
- Direct digital algorithmic extraction of information out of an image.

Several topics in this context are discussed in this chapter. But for a more complete survey the reader is referred to the general literature about digital imaging [1–28].

11.2 Image Computation

11.2.1 Introduction

One opportunity to check for the correct performance of an imaging system is the computational generation of quasi-realistic images per simulation. In any case this helps to get an impression of the overall appearance of an image produced by the optical system. In reality several approximations are necessary and the assessment is only quantitative if the parametrization is established appropriate. Only then can a check against the specification numbers be expected.

The methods required to compute an image are separated into two groups. The first and simplest approach uses only raytrace. This has the consequence that diffraction blurring is excluded, which restricts the application of this method. The second opportunity is to use Fourier imaging theory; here the impact of diffraction and coherence can be included. Due to the numerical problems in calculating systems with large wave aberrations, this approach is restricted to systems with a correction near to the diffraction limit. In addition, the algorithms work fast and efficiently if the point spread function can be assumed to be spatial invariant. If this so-called isoplanatic condition is not fulfilled and a spatial varying performance must be considered, the calculation schemes become more complicated and time consuming.

In reality it is therefore often the question whether the complete full field of view or only a smaller part should be calculated. Since aberrations are usually quite small in a full field representation, the local degradation quality is hard to resolve. Figure 11.1 illustrates this problem. In particular, problems with colour fringes and resolution can only be investigated locally in small parts of the field of view. On the other hand, issues with contrast, brightness distribution and distortion can be observed more comfortablly in a global inspection of the complete field of view. Physically the image performance usually changes over the size of the field with the best quality in the centre and degradation towards the edges and corners. This means, for a more comprehensive quality inspection, a global as well as a local enlarged view is necessary. In addition, fidelity of colour can be observed best in a natural scene, issues of distortion and uniformity of contrast and brightness are better seen in synthetic pattern images.

Lens Design for Imaging: Fundamentals of Optical Systems, First Edition. Herbert Gross.

Figure 11.1 Full field of view and inspected local selected zoomed sub-fields of 1% and 0.01% of the total area.

If sub-field sizes of 1% are considered, several positions inside the full field must be inspected and the orientation relative to the centre point matters. In circular symmetric systems, the image quality degrades with the radial distance from the centre. Simultaneously the image formats are often rectangular in shape with mostly an aspect ratio of 3 : 4 or 9 : 16 for typical CCD sensors. The selection of the sub-fields and the choice of their size to keep the quality invariant inside the field part is somewhat critical. In Figure 11.1 the original image has 2048×2048 pixels, the first sub-field contains 200×200 pixels and shows the local quality in the image centre here. In the second zoomed part with 20×20 pixels the resolution can be seen best.

As a conclusion it can be seen that it is not possible to assess all questions of performance with one image and one subject only.

11.2.2 Ray Based Image Calculation

One possibility to perform a quasi-realistic image quality is to use raytrace. Usually a Monte-Carlo random selection of rays is used and the image is divided into pixel areas according to the real sensor data. In every pixel the incoming rays are counted separately for the selected colours. To get a smooth image and reduce the noise generated by the statistical ray selection and the pixel discretization, many rays have to be calculated in practice. Typical values are 10^6–10^9, depending on the choice of the size of the field.

The properties of the geometrical raytrace approach are as follows.

- The calculation scheme is easy.
- The diffraction is excluded; therefore, a simplified physical model restricts the range of application.
- The field-dependent change of aberrations and vignetting is automatically obtained; isoplanatic assumptions are not needed.
- It is easy to take colour effects into account.
- Since the rays give only local information, many rays have to be calculated to get a smooth image.

There are several recommended methods to improve this computation from the viewpoint of gained information and computational effort. One aspect has been already mentioned in Section 9.2.4.4. The impact of pixel discretization and noise can be considerably improved if the assignment of single rays is not only collected in the image plane pixel, but interpolated with an appropriate weighting scheme between the grid points in the vicinity of the ray intersection point.

As an example, Figure 11.2 shows a high-quality camera lens of the Distagon type with opening F# = 1.32 and a field of view (FoV) of 27° [29]. As can be is seen from Figure 11.3(a) and (b), the lens suffers from a large vignetting towards the field corners and the peak-valley spot size performance decreases significantly for field angles larger than 15°. The diffraction limited size of the spot is approximately 1 μm, the geometrical spot spreadings are more in a range of 20 μm. This means that the system is far beyond a diffraction limited performance. Figure 11.3(c) shows the geometrical spots as a function of wavelength and field position in more detail. It is seen that the correction delivers a strong field dependence. Furthermore, it is observed that there is a chromatical lateral aberration. This is seen in the spot diagram by a *y*-variation of the centroid reference.

Figure 11.4 gives the results of some ray based image simulation results for different numbers of rays for the camera lens characterized above. From the visual perception, the

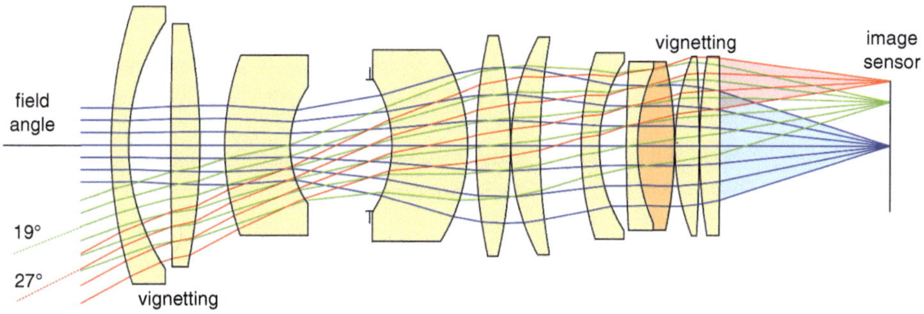

Figure 11.2 Distagon camera lens; the vignetting can be seen in the change of aperture cone cross sections in the image domain, the centre bundle and the outermost field 27° are highlighted by colour.

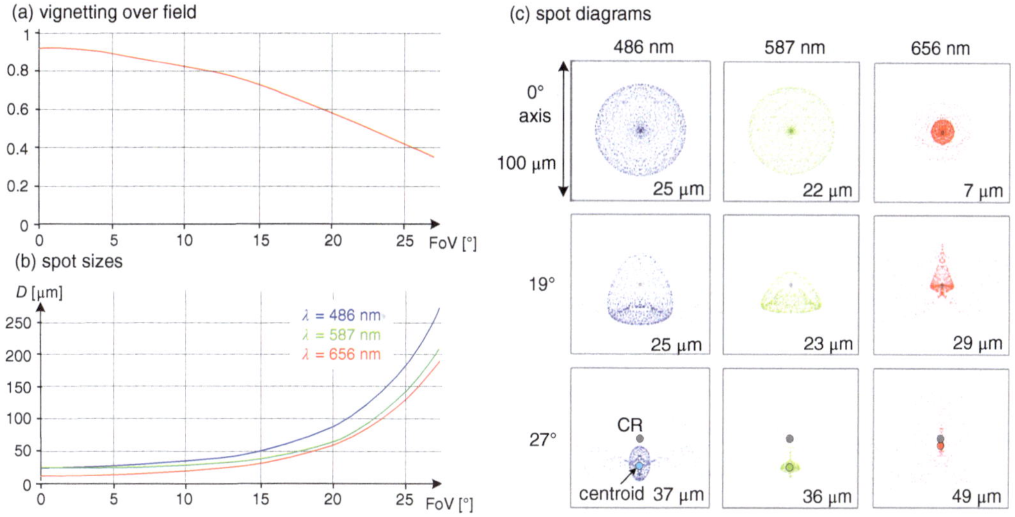

Figure 11.3 Performance of the camera lens with (a) vignetting, (b) spot size over the FoV and (c) spot diagrams for three field points and red, green, blue wavelength. The given spot sizes are defined as rms values.

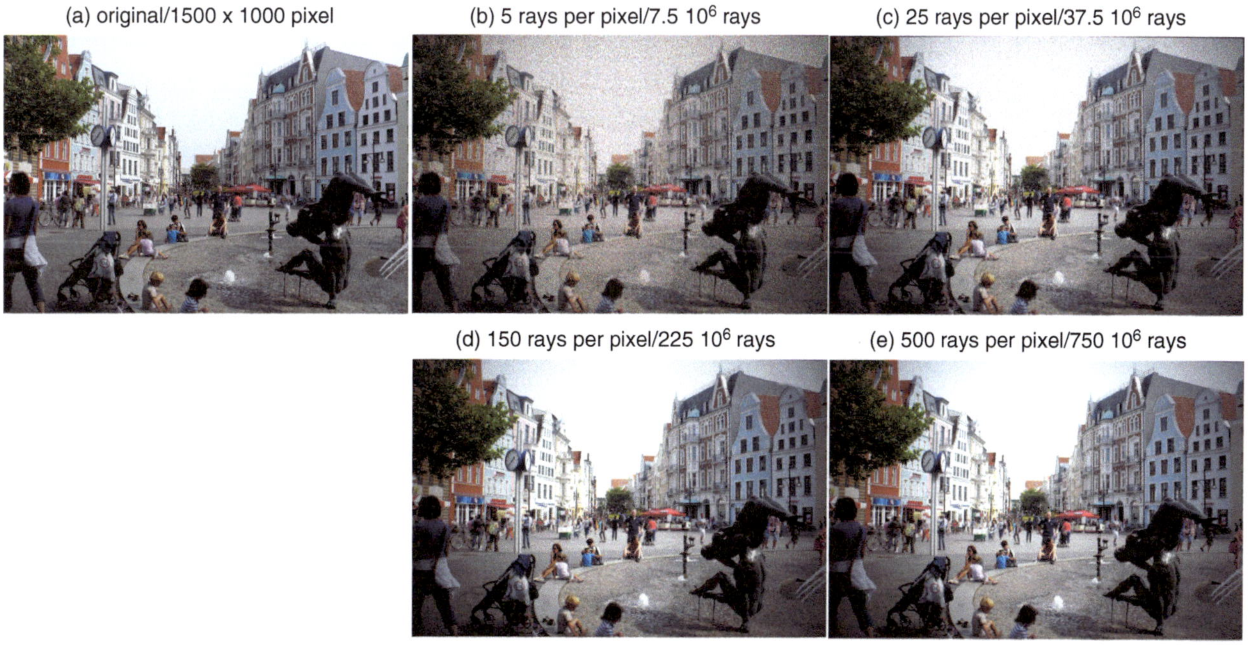

Figure 11.4 Image simulation by raytrace for different number of rays as full-field image. (a) shows the original image, in the parts (b) to (e) a growing number of rays is used according to the denoted numbers.

ray and pixel noise disappears earliest for 10^8 rays corresponding to 60 rays per pixel. For the three colours RGB as above and the pixelization of 1500 × 1000 in the original picture, this corresponds to approximately 20 rays per pixel and colour. Furthermore, the vignetting is clearly seen: the corners are significantly darker than the centre. The strong vignetting has in addition the effect that in this case in total only 62 % of the rays reach the image plane. In the corners only 10 pixels per colour and pixel are obtained. In the series of pictures it is observed that the shot noise of discrete rays is no longer observed if more than approximately 200 rays per pixel are selected. Due to the vignetting, this is achieved in the corner regions for 500 rays per pixel in the last picture (e).

Figure 11.5 shows the results of a calculation with sub-fields of 1° side lengths with a simple geometrical checkerboard object at three different field positions for the same lens. The total field is 54° and measured along the diagonal line, so the sub-field size is approximately 2% of the whole FoV. As a result of the non-uniform spot correction on axis, the squares appear green in the centre of the image; in the corner this colour effect is absent. In the edges the chromatical lateral colour can be observed; due to the special spot shapes in the corner field at 27°, the squares seem to be better in resolution and contrast here in comparison to the edge point at 19°. When observing colour aberrations it is always an advantage to use synthetic objects with black-white edges.

11.2.3 Physical Model of Image Formation

11.2.3.1 Introduction

If a more physical model is used to compute images, the diffraction effects are included and if it is required it is possible to take the coherence of the illumination into account. This gives a much more convincing approach to estimate quality and in particular resolution. If systems are corrected and working in the diffraction limited performance range, like microscopes, lithography projection optics or astronomical telescopes, this approach is mandatory due to the dominant role of diffraction effects of resolution. Historically, this approach was proposed by Fourier and Abbe. According to the size of the Airy diameter of an optical system in diffraction limited quality for a numerical aperture NA

$$D_{airy} = \frac{1.22 \cdot \lambda}{NA} \tag{11.1}$$

the spot sizes for this approach should not be much larger than this value. To get a raw idea about the resolution and the numerical requirements, let's assume a system working in the green with $\lambda = 0.5\ \mu m$, a numerical aperture of NA = 0.25 and a sensor chip with 1 in. size $D_{ima} = 25.4$ mm. Thus, inside the complete image we have with $D_{airy} = 2.4\ \mu m$ in total

$$N = N_x^2 = \left(\frac{D_{ima}}{D_{airy}}\right)^2 = \left(\frac{25.4\ \text{mm}}{0.0024\ \text{mm}}\right)^2 \approx 10^8 \tag{11.2}$$

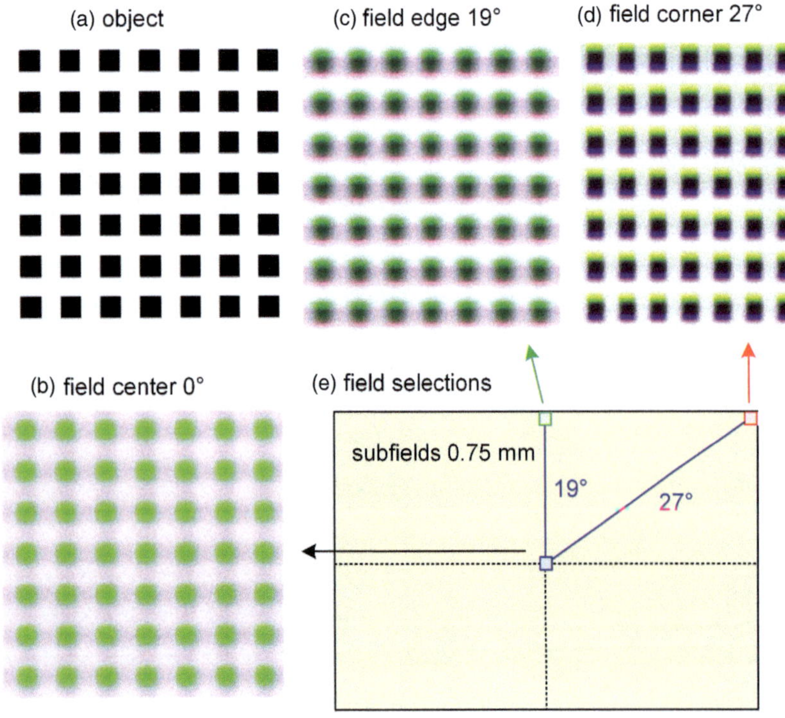

Figure 11.5 Image simulation by raytrace for local field positions for the camera lens. (a) shows the original object, the parts (b) to (d) are the images of the small sub-fields in the centre, edge and corner according to the scheme illustrated in (e).

point spread functions constituting the image. If we assume roughly 10 grid points to sample one point spread function, 10^{10} grid points are needed to compute the full image. This corresponds approximately to a two-dimensional Fourier transform with 81922 points. This is not impossible with today's computers, but is at the limit of the required memory resources. Therefore in reality one tries to simplify the calculation.

In contrast to the raytracing approach, in the physical model some additional numerical problems occur. The calculation of the PSF and the OTF must fulfil the sampling theorem; in the case of optical systems with a performance far from the diffraction limit, this is a challenging task. On the other hand, the convolution with the point spread function delivers smoother images. But if the fast Fourier transform is used for the computations, an equidistant sampling of the exit pupil is required, which may create in the case of pupil distortions the need for ray aiming. The classical approach can be calculated as a convolution of the object intensity with the point spread function intensity. This fast and efficient calculation scheme uses the shift invariance of the PSF. This is realistic within smaller sub-fields, which fulfil the isoplanatism condition to a desired accuracy. One challenging task is to handle the more realistic fact that the point spread function changes with field position; the degradation grows with the radial distance from the optical axis. This is needed if a larger or the complete field of view has to be imaged in one shot.

Another interesting possibility in physical image formation is the incorporation of the degree of coherence into the simulation, which finally influences the image quality. The extreme cases of fully incoherent and fully coherent illumination are comparable from the viewpoint of computational effort. The more realistic case of partial coherent illumination, which is in particular important for microscopic as well as lithographic imaging tasks, is more challenging and of course requires information about the coherence properties of the illuminating source.

An additional important consideration is the simulation in three dimensions including the depth z; in particular the modelling of a defocused image z-stack is straightforward but increases the numerical issues. This problem is easily included in the physical imaging approach and finds several relevant applications in microscopy and human visual seeing with multifocal intra ocular lenses (IOL).

11.2.3.2 Image Formation According to Fourier and Abbe

The physical image formation approach of Abbe is quite similar to that of Fourier's theory. The basics have already been discussed in the Chapters 6 and 7. Abbe considered the imaging of gratings as objects. Only those gratings are resolved whose first diffraction order falls within the numerical aperture of the optical system and is supported by the pupil area of the system, see Figure 11.6.

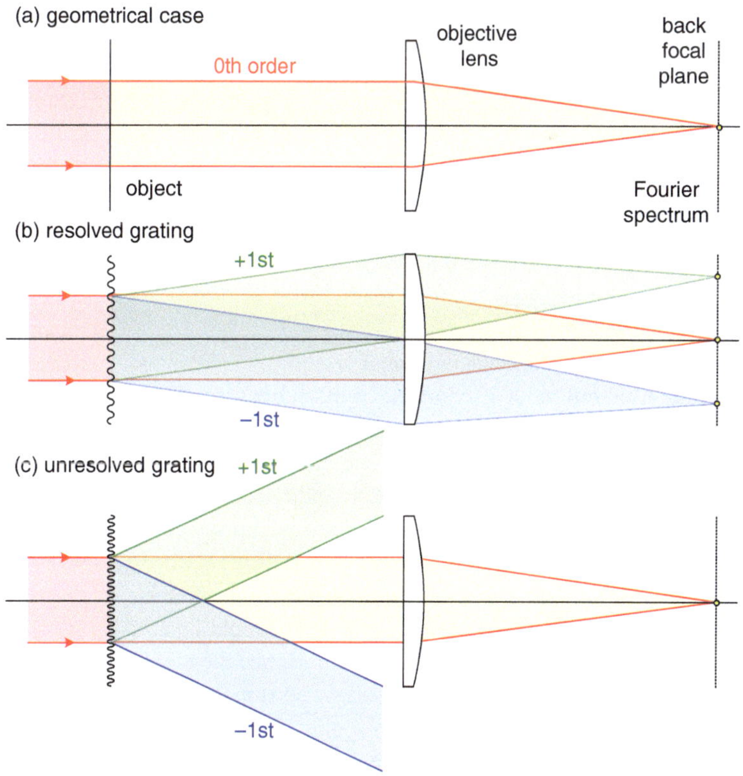

Figure 11.6 Model of image formation after Abbe with a grating object and interference of the diffraction orders, which are accepted inside the aperture cone. (a) shows the geometrical zeroth order on axis, (b) a grating object which is resolved with full contrast and (c) a case where the diffraction orders are outside of the acceptance cone and the grating period is not resolved.

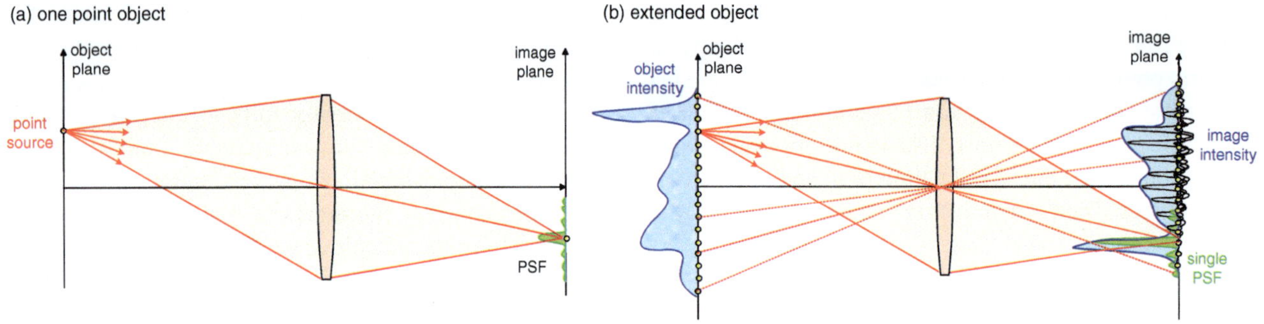

Figure 11.7 Principle of formation of the point-spread function, (a) single light point as object and (b) extended object with incoherent illumination.

From the Fourier model of resolution it follows that the Fourier spectrum of an object is always truncated by a transfer system at a given limiting spatial frequency. Therefore, even a point which, according to Fourier, possesses a continuous and not band-limited Fourier representation, cannot be observed as a point in the image plane. In other words a point is transformed into a field distribution with finite dimensions in the spatial domain, the so-called point-spread function (PSF), see Figure 11.7(a) and Chapter 6.

If now an arbitrary object is considered which consists of many discrete points with different intensities, in the approximation of incoherent imaging the individual point images cannot interfere. The image can be again constructed from the individual weighted point image distributions. The intensities are then superimposed as is sketched in Figure 11.7(b). The spread of each individual point image has the result that:

- very fine details are not resolved in the image;
- fine details are broadened in the image;
- steep intensity gradients are smoothed in the image.

This corresponds to the descriptions of Chapter 7. The imaging system works as a low pass filter for the spatial spectrum of the object.

11.2.3.3 4-*f*-Fourier Imaging Model

As an illustration of a complete optical imaging using the Fourier approach, an arrangement can be considered which consists of two identical sub-systems with focal lengths f' and finite image formation and telecentric stop position in the form of a 4-f-system. The first sub-system is at a distance f from the object plane and therefore produces an image at infinity. The second sub-system is at a distance $2f$ behind the first one and the image plane would lie at a distance f behind the second sub-system. Hence the whole arrangement has a length of $4f$. The system stop is located in the common intermediate focal point plane, which is the pupil plane. The spectrum of the object is found here.

According to the properties of a single lens, an angular coordinate in the object turns into a spatial coordinate in the Fourier-plane and vice versa. Consequently, the plane waves diffracted at the object structures can be directly observed in the mid-plane as a spatial spectrum. In this sense the mid-plane, where the Fourier spectrum of the object appears, is called Fourier-conjugated to the object plane. The rear mirror-symmetric part of the system transforms the spectrum back to a spatial distribution, which then appears in the image plane. This approach is sketched in Figure 11.8. It should be noted that smaller feature details in the object create larger spatial frequencies. Correspondingly the rectangular object in the figure with smaller x-extension has a larger x-extension of the spectrum in the pupil plane.

In general, optical imaging can be modelled by a linear system theory if the non-linearities over the field size are negligible. The object is characterized by its spatial frequency spectrum. It is considered to be decomposed of gratings with different periods. Mathematically this corresponds to a Fourier transform, every Fourier component represents one grating. Fine details and sharp edges in the object structure consequently have large amplitudes for the components at large frequencies. Conversely, the optical system acts as a low-pass filter: large spatial frequencies mean a large inclination angle of the marginal rays relative to the optical axis. Such rays are transferred, however, only by systems with large numerical apertures. The aperture restricts the transmitted spatial frequencies and suppresses frequencies that are too large. The corresponding description in the spatial domain leads to the convolution of the object structure with the point spread function. The sharpness of imaged object details depends in turn on the size of the aperture. Structures smaller than the point spread function itself cannot be resolved. When the first diffraction order is transmitted through the aperture of an optical system the corresponding object structure appears in the image. The image increasingly resembles the object when more diffraction orders are accepted, see Chapter 7.

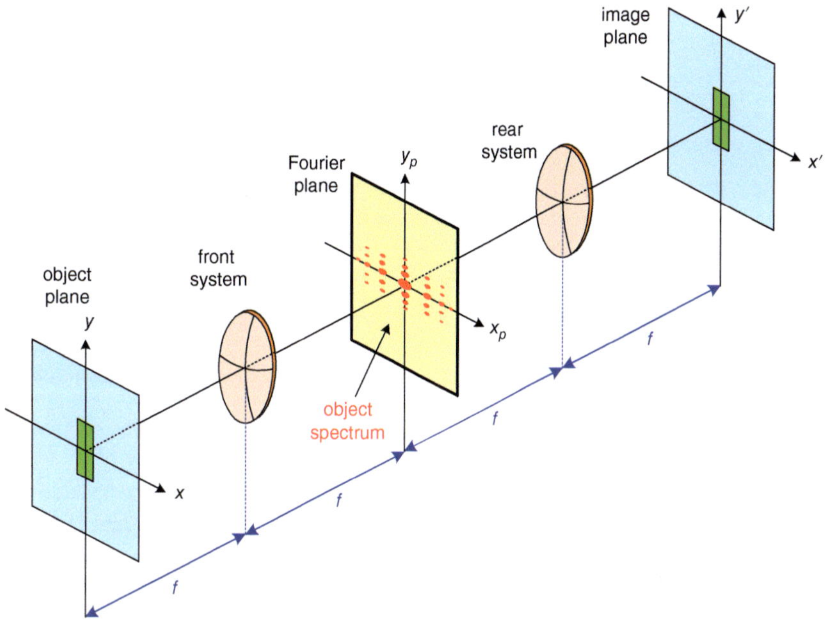

Figure 11.8 Fourier-optical 4-f-arrangement.

To be more accurate, the problem must be discussed in two transverse dimensions and the size of the diffraction order area supported by the pupil area size determines the contrast of the corresponding feature in the image.

In the following, the Fourier-optical simulation of the image formation process will be described step-by-step. For simplification, only one transverse dimension x and a magnification $m = -1$ will be considered. The two-dimensional extension is straightforward.

- The object will be described by a complex transfer function, containing amplitude and phase properties

$$T_{obj}(x_o) = A_{obj}(x_o) \cdot e^{2\pi \cdot i\Phi_{obj}(x_o)} \tag{11.3}$$

with a Fourier spectrum

$$t_{obj}(x_p) = \hat{F}[T_{obj}(x_o)] \tag{11.4}$$

- If a plane wave is incident on the object from the coordinate x_L of the light source with the spatial frequency corresponding to the direction x_L/f_{con} of the illumination condenser

$$\nu_L = \frac{x_L}{\lambda \cdot f_{con}} \tag{11.5}$$

the following approximate modified field distribution appears behind the object

$$E(x_o) = E_L(x_L) \cdot e^{2\pi \cdot ix_o\nu_L} \cdot T_{obj}(x_o) \tag{11.6}$$

The thin element approximation is used here and assumes a negligible thickness of the object. The exact expression for the transfer function including thickness effects depends on the direction of incidence illumination wave $\vec{s}_L$.

- If P is the pupil function and W is the wave aberration of the imaging objective, by using the coherent transfer function H_{CTF}

$$H_{CTF}(x_p) = P(x_p) \cdot e^{2\pi \cdot iW(x_p)} \tag{11.7}$$

or equivalently its Fourier transform in the spatial domain, the PSF amplitude as a function of the image coordinate x_i

$$H_{CTF}(x_i) = \hat{F}\left[P(x_p) \cdot e^{2\pi \cdot iW(x_p)}\right] = E_{PSF}(x_i) \tag{11.8}$$

one obtains the following object spectrum in the pupil plane

$$u(x_p) = t_{obj}(x_p - x_L) \cdot H_{CTF}(x_p) \tag{11.9}$$

which has an off-set x_L determined by the illumination direction.

- After a second Fourier transform one obtains the complex field in the image plane

$$E(x_i, x_L) = \hat{F}[t_{obj}(x_p - x_L) \cdot h_{CTF}(x_p)] \tag{11.10}$$

Hence, applying the convolution theorem and substituting the Fourier transform of the point spread function by the CTF, the amplitude in the image plane can be derived as a convolution

$$E(x_i, x_L) = \int T_{obj}(x_o) \cdot e^{2\pi \cdot i \cdot x_o \nu_L} \cdot H_{CTF}(x_i - x_o)\, dx_o \tag{11.11}$$

In a 4-f-configuration, the pupil with coordinates x_p, lies in the mid-plane of the system. x_p is proportional to the spatial frequency ν_x. The finite pupil extent of D_{pupil} truncates the Fourier spectrum in the Fourier plane as shown in Figure 11.9.

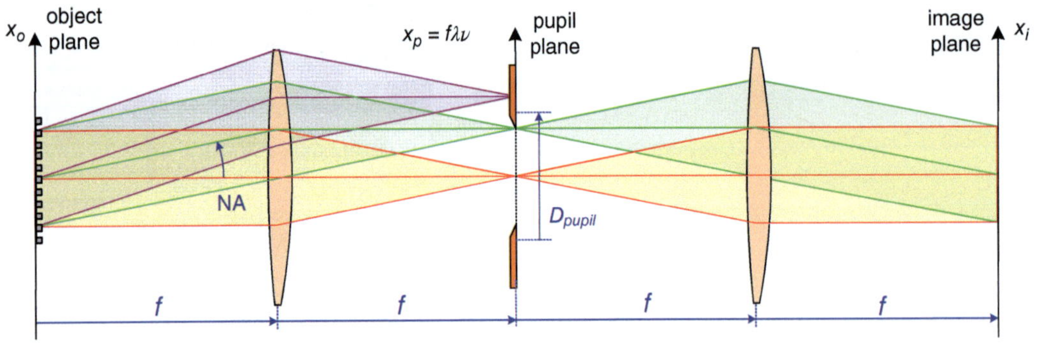

Figure 11.9 Effect of the pupil size on the image formation by filtering the spatial frequencies.

11.2.3.4 Complete 6-*f*-Fourier Model

As has already been seen in Eq. (11.11), for the complete description of the image formation process, the illumination system has to included. The illumination allows us to take the coherence as well as the wavelength spectrum into account. Apart from the boundary cases of modelling purely coherent or purely incoherent illumination, it is necessary to take into account the properties of the illumination system. In the case of the usual Köhler illumination setup, we therefore end up at a 6-*f*-arrangement which is schematically shown in Figure 11.10. The separation of the imaging part into two objective lenses with an accessible pupil plane in between is not necessary in practice. In the case of microscopy it is realized, and the second lens then corresponds to the tube optic.

Figure 11.11 shows this full setup in more detail and explains the terms of use in the Fourier treatment of calculation. A light source of finite size illuminates the object with a finite angular spectrum. The mentioned Köhler illumination means that the illumination source is conjugated to the system pupil and Fourier conjugated to the object. A single discrete illumination point creates in the object domain a plane wave.

In Fourier representation the general partially coherent image formation according to this 6-*f*-arrangement can be expanded into the following single steps. The first steps are

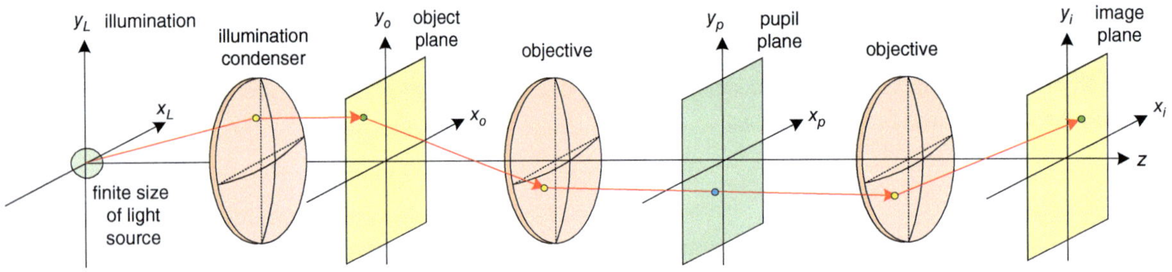

Figure 11.10 Fourier-optical two-stage imaging system with illumination.

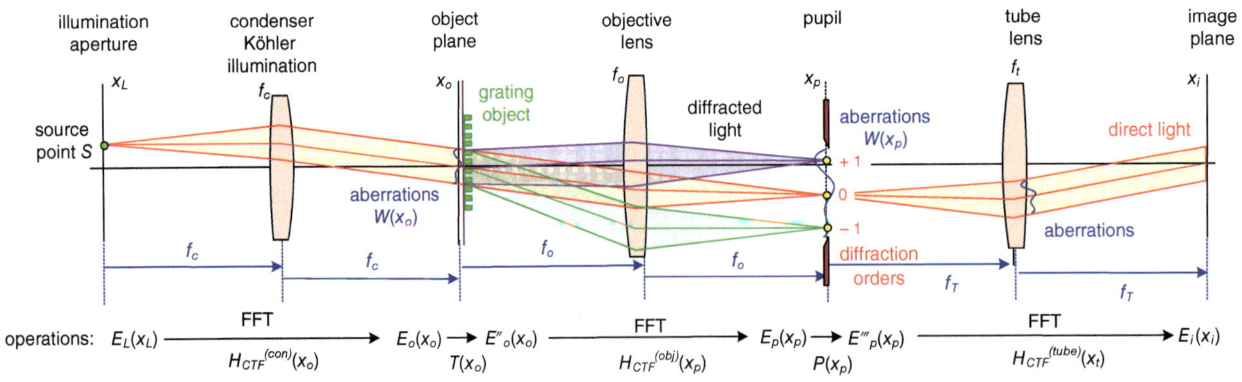

Figure 11.11 Finite diffraction orders for a light source of finite extent with off-axis source point.

thereby analogous to the presentation for the 4-f-system in the previous section.

- The light source has a finite extent and therefore has a position-dependent field strength $S_L(x_L)$. At first only one light source point x_L will be considered.
- The light source point is transformed by the condenser into a plane wave with spatial frequency ν_L in front of the object plane

$$E(\nu_L) = S_L(x_L) \cdot e^{2\pi \cdot i \cdot x \cdot \nu_L} \tag{11.12}$$

- The plane wave illuminates the object in the corresponding direction $\theta_L = x_L / f_C$ and spatial frequency $\nu_L = \theta_L / \lambda$. The modified field behind the object, which is modulated by the transfer function in the thin element approximation, appears as

$$E(x_o) = S_L(x_L) \cdot e^{2\pi \cdot i \cdot x_o \nu_L} \cdot T_{obj}(x_o) \tag{11.13}$$

- With the pupil function P and the wave aberration W of the imaging objective, by using the coherent transmission function H_{CTF}

$$H_{CTF}(x_p) = P(x_p) \cdot e^{2\pi \cdot iW(x_p)} \tag{11.14}$$

one obtains the field in the pupil with the Fourier transformed object spectrum of T as

$$E(x_p) = S_L(x_L) \cdot e^{2\pi i \nu_o x_L} \cdot T_{obj}(x_p - x_L) \cdot H_{CTF}(x_p) \tag{11.15}$$

which is blurred by the CTF and has an off-set due to the illumination direction.

- Finally the field strength in the image plane can be derived by a Fourier transform

$$E(x_i, x_L) = \hat{F}\left[S_L(x_L) \cdot e^{2\pi i \nu_o x_L} T_{obj}(x_p - x_L) \cdot H_{CTF}(x_p)\right] \tag{11.16}$$

Hence, performing the Fourier transform one obtains

$$E(x_i, x_L) = S_L(x_L) \cdot \int T_{obj}(x_o) \cdot e^{2\pi \cdot i \cdot x_o \nu_L} \cdot E_{PSF}(x_i - x_o)\, dx_o \tag{11.17}$$

Here Eq. (11.5) connects the light source coordinate x_L with the corresponding spatial frequency ν_L. The form of Eq. (11.17) is a convolution of the skew illuminated object transmission with the amplitude point spread function.

- Now the finite extent of the light source is considered, this means that the above amplitude has to be integrated over all illumination points x_L of the light source. If $S_L(x_L)$ is the weighting distribution function of the light source amplitude, the integral delivers the field amplitude in the image as

$$E(x_i) = \int S_L(x_L) \int T_{obj}(x_o) \cdot e^{2\pi \cdot i \cdot x_o \nu_L} \cdot E_{PSF}(x_i - x_o)\, dx_o\, dx_L \tag{11.18}$$

- The image intensity can be obtained in the general case by taking the square of the absolute value. At this point of modelling it should be kept in mind that the source itself may possess coherence properties leading to interference and in this way may affect the intensity. This can be written more generally as

$$I(x_i) = \left| \iint E(x_i, x_{L1}) \cdot E^*\left(x_i, x_{L2}\right)\, dx_{L1}\, dx_{L2} \right|^2 \tag{11.19}$$

With a substitution using the coherence function Γ_L of the light source in the object plane

$$\Gamma_L(x_{o1}, x_{o2}) = \iint S_L(\nu_{L1}) \cdot S_L(\nu_{L2}) \cdot e^{2\pi \cdot i(x_{o1}\nu_{L1} + x_{o2}\nu_{L2})}\, ds_{L1}\, ds_{L2} \tag{11.20}$$

one obtains the expression

$$I(x_i) = \left| \iint \Gamma_L(x_{o1}, x_{o2}) \cdot T_{obj}(x_{o1}) \cdot T^*_{obj}(x_{o2}) \cdot E_{PSF}(x_i - x_{o1}) \cdot E^*_{PSF}(x_i - x_{o2})\, dx_{o1}\, dx_{o2} \right|^2 \tag{11.21}$$

Eq. (11.21) is the general one-dimensional expression for the image formation taking arbitrary partial coherent properties of the light source into account. If the second transverse coordinate y is included, an equation with four integrations is obtained.

11.2.3.5 Coherent Image Formation

In the case of coherent image formation only one point of the light source is considered. The integration over the light source disappears, only one plane coherent wave illuminates the object. The resulting image amplitude is calculated directly by the convolution integral of the object wave field and the transfer function

$$E(x_i, y_i) = \int_{-\infty}^{\infty} \int_{-\infty}^{\infty} E_{PSF}(x_i - x_o, y_i - y_o) \cdot E(x_o, y_o)\, dx_o\, dy_o \tag{11.22}$$

Accordingly, the image intensity is given by

$$I_{coh}(x_i, y_i) = \left| \int_{-\infty}^{\infty} \int_{-\infty}^{\infty} E_{PSF}(x_i - x_o, y_i - y_o) \cdot E\left(x_o, y_o\right)\, dx_o\, dy_o \right|^2 \tag{11.23}$$

In frequency space this convolution integral can be written as a product of the corresponding Fourier transformed functions

$$E_i(\nu_x, \nu_y) = H_{CTF}(\nu_x, \nu_y) \cdot E_o(\nu_x, \nu_y) \tag{11.24}$$

The coherent transfer function is equal to the Fourier transform of E_{PSF}

$$H_{CTF}(\nu_x, \nu_y) = \int_{-\infty}^{\infty} \int_{-\infty}^{\infty} E_{PSF}(x, y) \cdot e^{-2\pi\, i \cdot [x\nu_x + y\nu_y]}\, dx\, dy \tag{11.25}$$

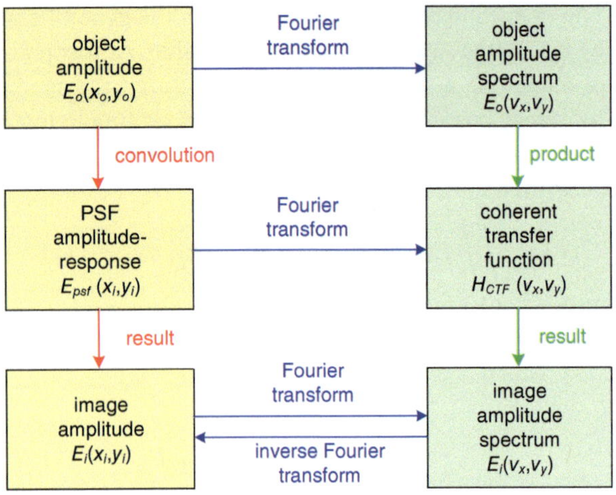

Figure 11.12 Schematic relations of coherent image formation.

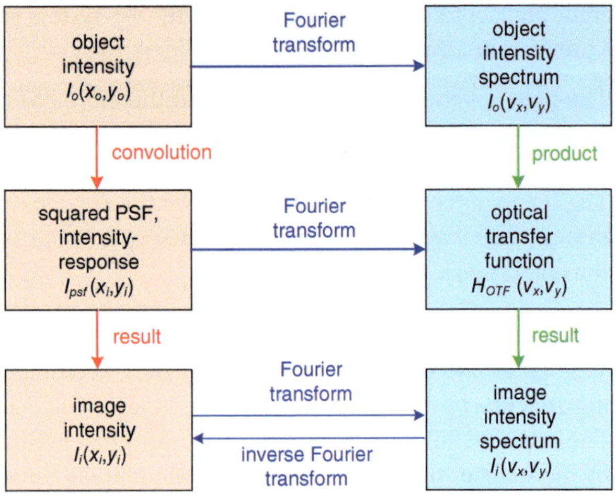

Figure 11.13 Schematic relations of incoherent image formation.

and thus represents the pupil function in a scaled form

$$H_{CTF}(v_x, v_y) = \lambda \cdot z \cdot P(-\lambda z \cdot v_x, -\lambda z \cdot v_y) \quad (11.26)$$

This form describes the well-known fact that the extent of the pupil acts directly as a frequency filter in image formation. The scheme of the image formation is illustrated in Figure 11.12. In the spatial domain as the left column, the operation is a convolution, in the Fourier domain on the right side it is simply a multiplication with the CTF as a filter function.

11.2.3.6 Incoherent Image Formation

The phase correlation between the individual object waves is absent in the case of an incoherent illumination. The coherence function becomes a Dirac δ-function

$$\langle E(x_{01}) \cdot E^*(x_{02}) \rangle = I(x_{01}) \cdot \delta(x_{01} - x_{02}) \quad (11.27)$$

Therefore, the intensity in an image point is determined solely by the intensity of the corresponding object point and the transfer function. There is no coherent interference of other object positions. Formulated in both transverse dimensions this reads

$$I_{inc}(x_i, y_i) = \int_{-\infty}^{\infty} \int_{-\infty}^{\infty} |E_{PSF}(x_i - x_o, y_i - y_o)|^2 \cdot I(x_o, y_o)\, dx_o\, dy_o \quad (11.28)$$

In simplified notation, one can represent the image intensity as a convolution of the object intensity, and the PSF intensity distribution as a point response

$$I_{inc}(x_i, y_i) = I_{PSF}(x_i, y_i) * I_{obj}(x_o, y_o) \quad (11.29)$$

In the spatial frequency domain this convolution is simply equivalent to the product of the spectra

$$I_{inc}(v_x, v_y) = H_{OTF}(v_x, v_y) \cdot I_{obj}(v_x, v_y) \quad (11.30)$$

Therefore it is clear that the point spread function and its spectrum, respectively, act as a filter in the image formation of the object structures. The incoherent image formation is sketched in Figure 11.13 similarly as in the coherent case but now with intensities.

11.2.3.7 Isoplanatic Condition

In the usual approach of image calculation according to Fourier it is preferred that the convolution formulation is applied, because this is fast and efficient. But this mathematical advantage requires a spatially invariant PSF. This principle is sketched in Figure 11.14. But it is well known that in reality the PSF degrades for larger fields of view; this effect is not incorporated into this so-called isoplanatic assumption. The term isoplanatic comes from aberration theory and describes a shift invariant PSF. The field behaviour of related aberrations like astigmatism, field curvature and coma are the worst offenders of the invariance assumption.

A perfect constant PSF is in reality impossible to obtain, this is simply unphysical. But for the practical purpose of the requested accuracy of image calculation it makes sense to define a quantitative criterion of the PSF change. If the real change of the PSF lies below a certain threshold, approximately isoplanatism can be assumed. Two possible options can be recommended to assess the invariance of the PSF: the change in the Strehl number or a correlation of the changed PSF distributions. Figure 11.15 shows for illustration these two criteria for two microscopic lenses, one with and one without a field flattening correction. As criteria for a raw estimation of the isoplanatic area, here around the image centre a yellow range is shown, inside which the Strehl ratio is $D_s > 0.9$. Second for an intermediate field

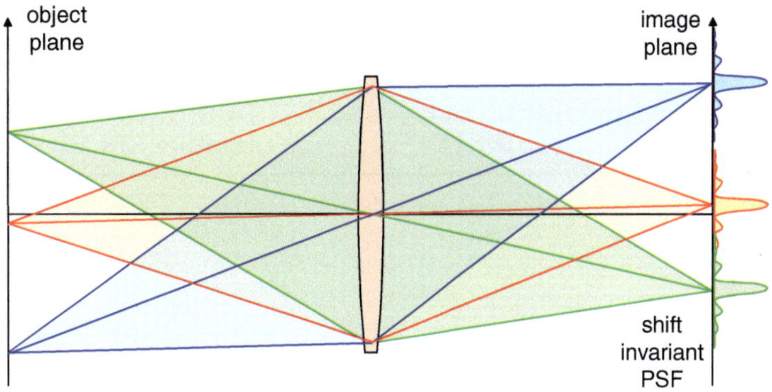

Figure 11.14 Schematic relations of isoplanatic incoherent image formation.

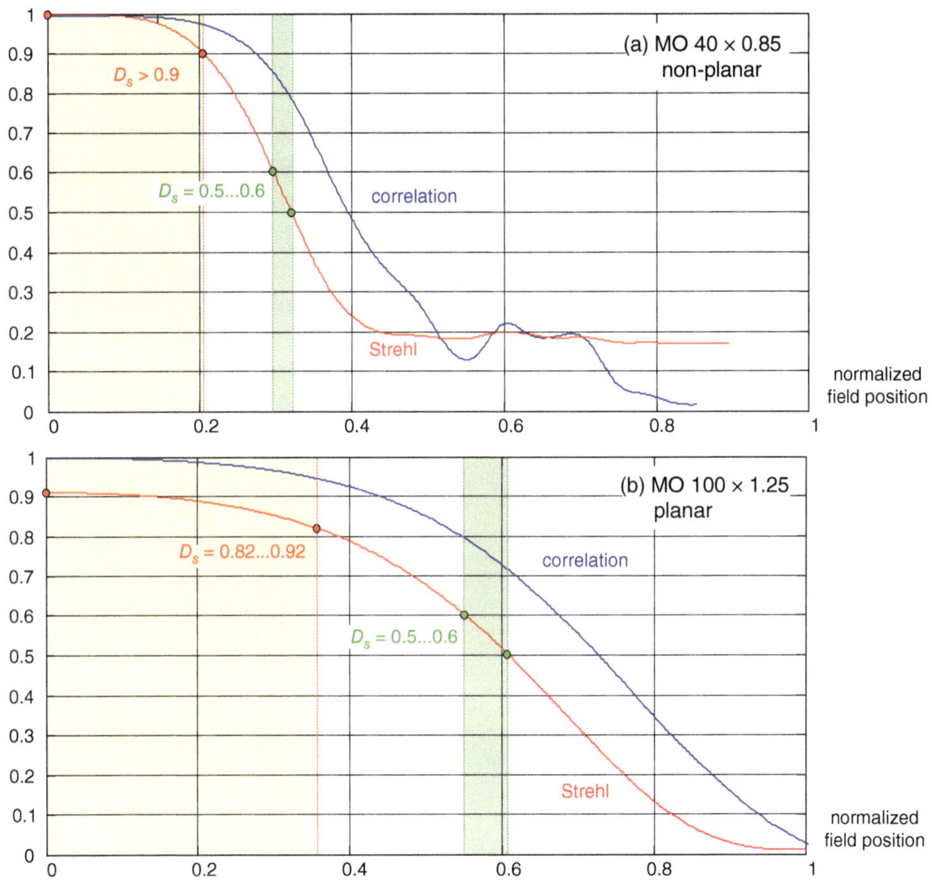

Figure 11.15 Comparison of the isoplanatic patch sizes of (a) a non-planar and (b) a flattened microscopic lens.

position with Strehl $D_s = 0.6$ a range with a 10% drop of the Strehl ratio is marked in green. It is seen that in the image centre, the isoplanatic size is very large and covers more than 30% of the field size for the flattened lens, but only 20% for the simpler system. In any case the decrease of the Strehl for larger field heights delivers only a very narrow isoplanatic ring in the zone, which means that the PSF change is considerably strong here and an image simulation using the convolution approach produces large errors. If the correlation is used, the behaviour is more stable, but similar.

In Table 11.1 some numbers of the examples from Figure 11.15 for both criteria and accuracy requests are listed. In the table, smaller realistic values of 1% Strehl change are considered and compared with a correlation criterion (CC). The isoplanatic field sizes are circular on axis and ring width for medium fields. For comparison the total fields is given in the headline, they are different for both lenses.

To come to a more realistic description of the PSF changes, special interpolation approaches for spatially varying PSF are needed, this is discussed in Section 11.3.

Table 11.1 Quantitative estimations for the size of isoplanatic patch sizes of the lenses shown in Figure 11.15.

System	MO plane 100 × 1.25 field size 250 μm isoplanatic patch size [μm]		MO not plane 40 × 0.85 field size 625 μm isoplanatic patch size [μm]	
	Strehl 1%	**PSF correlation 0.5%**	**Strehl 1%**	**PSF correlation 0.5%**
On axis	70	72	81	100
Half field	3.8	3.8	27	3.1
Field zone	2.5	2.5	29	39
Full field	45	3.8	117	62

From a practical point of view, the quantitative criterion to measure the degree of shift-invariance by the cross correlation of the point spread function $I_{PSF}(x,y)$ between two distant locations

$$C(\Delta, x_1, x_2) = \int I_{PSF}(x + \Delta/2 - x_1) \cdot I^*_{PSF}(x - \Delta/2 - x_2)\, dx \qquad (11.31)$$

is more useful than the Strehl ratio. The reason is an easy evaluation by digital means. In a normalized definition of the relative degree of space invariance, the formula

$$\sigma(x_1, x_2) = \frac{C(x_1 - x_2, x_1, x_2)}{\sqrt{C(0, x_1, x_1) \cdot C(0, x_2, x_1)}} \qquad (11.32)$$

can be used. Values of 1 correspond to a fully invariant PSF. By integrating this measure over the field of view, we get an averaged degree of spatial invariance of 'fulfilment of isoplanatism' as

$$\overline{\sigma} = \frac{\iint \sigma(x_1, x_2)\, dx_1 dx_2}{\iint dx_1 dx_2} \qquad (11.33)$$

In any case the invariance is an idealized assumption, and it is a question of accuracy and compromise how large the field of views are selected for an image calculation using this model.

11.2.3.8 Multifocal Visual Perception*

(*) The content of this section is quite specialized and for advanced reading. Knowledge of this section is not necessary to understand the following chapters. Beginners are recommended to skip this section.

The physical optical image formation simulation is in particular also of interest for cases where the object is located in different depths [30]. There are new developments of IOL which provide a simultaneously sharp imaging of objects in different distances z. Figure 11.16 illustrates the idea, which is a comfortable way to enable the observer to see several object details sharply at the same time without changing eye glasses. The images are overlaid on the human retina and the main point of this approach is how the conditions of brightness and magnification are handled by the human brain. The experience of field tests shows that the human brain can distinguish the objects in different depth rather comfortably after a certain time of adaptation and training.

Figure 11.17 shows the result of some calculations regarding this approach based on the Fourier scheme. Three letters are located at different distances z at slightly different lateral locations. If for a fixed human pupil size the imaging is calculated for a model eye transfer function, the monochromatic view is seen in part (a), the eye is refocused for the distances separately and the three letters are sharp. In part (b) a bifocal IOL is inserted. The distribution on the right side shows the caustic intensities along z on the axis for an eye pupil size of $D = 4$ mm. The image result is shown on the left. It is seen that both letters corresponding to the added power according to the distance difference are clearly imaged. In case (c) the same calculations are performed for three distances and a trifocal lens; the result is similar. The brightness of the images is related to the relative intensity heights of the caustic as well as the changing $1/z^2$ – distance law. The z-values are here scaled in diopters.

The convolution process and the corresponding calculation scheme for the bifocal case is again sketched in

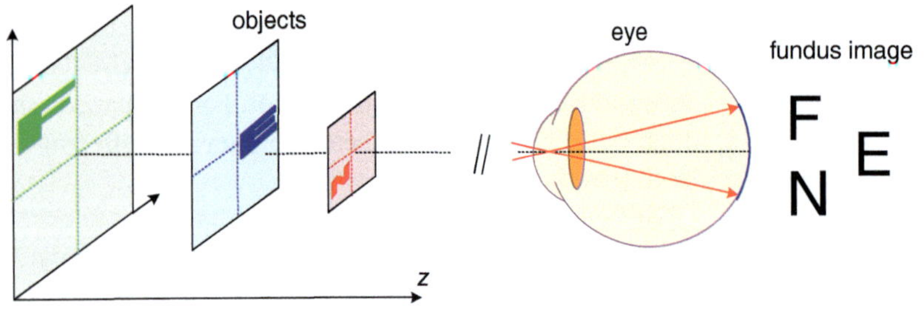

Figure 11.16 Imaging of objects in different distances by the human eye. Here the eye is accommodated to the middle distance and therefore only the letter 'E' is sharp.

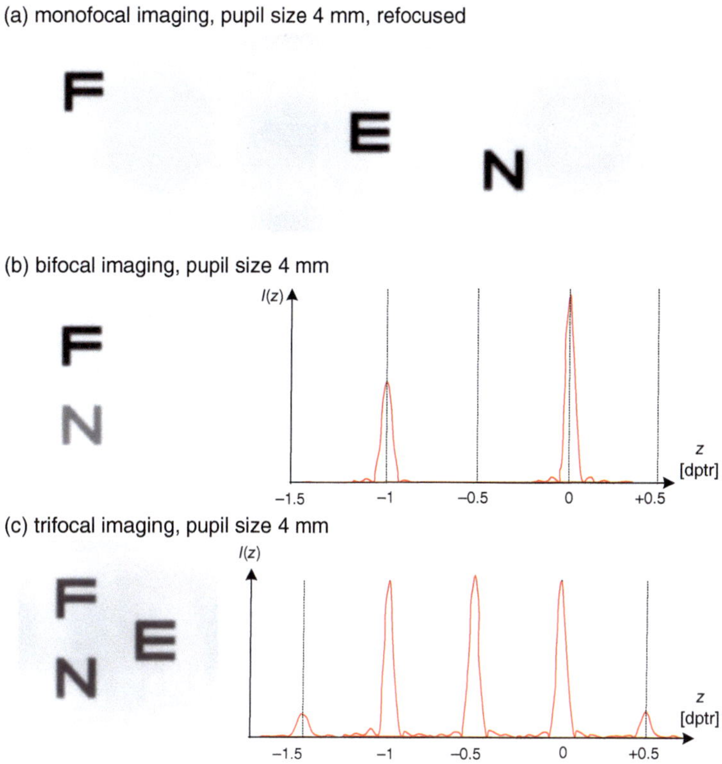

Figure 11.17 Imaging of objects at different depths by the human eye (a) without a special lens, (b) with a bifocal intra ocular lens and (c) with a trifocal intra ocular lens [30].

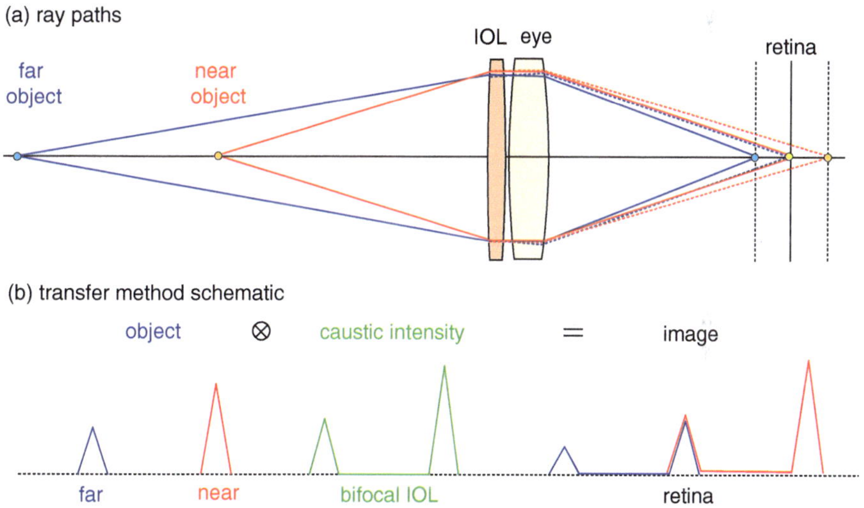

Figure 11.18 Imaging of objects in different depths by a bifocal intra ocular lens schematically.

Figure 11.18. The bifocal intra ocular lens creates two peaks in the focal caustic. In the image processing, the objects in the far and in the near distance have to be convolved with this shape. In the case of the right scaling in distance as well as in energy height, two different distances are sharp imaged onto the retina simultaneously. The two defocused cases are also present, but they have a lower contrast and the brain is able to suppress them in the perception.

11.2.3.9 Image Computation Examples

A comparison of a pattern image formation under coherent and incoherent conditions can be seen in Figure 11.19 for a biological sample of a crowded cell cluster. In the coherent image formation one can see diffraction effects at the edges which lead to a disturbed negative impression of the image. The effect of granulation in the case of coherent illuminated imaging due to interference effects usually perturbs the visual image impression and the expected fidelity of the image with the object. One can claim in general that an incoherent imaging delivers the better quality of images according to human expectation.

Figure 11.20 shows a simple example of two different images to demonstrate the change in resolution and the

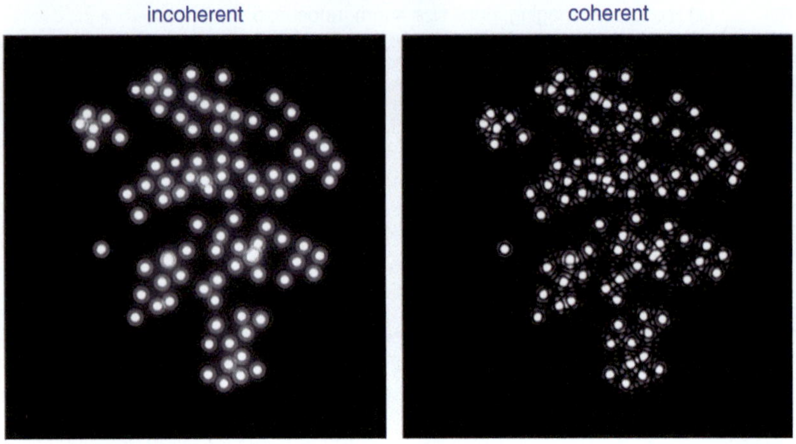

Figure 11.19 Comparison of incoherent and coherent imaging for a microscopic sample.

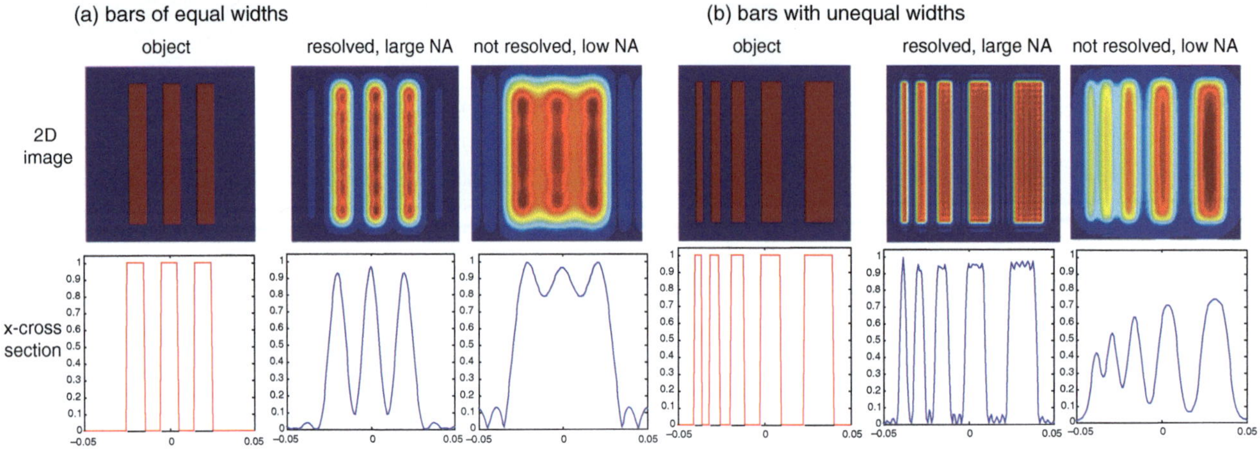

Figure 11.20 Example of image formation for an incoherent illuminated pattern. In (a) a bar pattern with equal widths is imaged with high and low numerical aperture to illustrate the change in contrast and resolution. The same imaging conditions are applied on a chirped bar pattern, showing the change of resolution and contrast in one image. The units are arbitrary here.

reduction of contrast for smaller feature sizes of a synthetic bar pattern, respectively. It is clearly observed that the resolution and contrast decrease with finer object features according to what is expected. In this figure two different numerical apertures are used to visualize the changes in the image. These cases are all incoherently illuminated.

Figure 11.21 shows some practical measurement results for the resolution in a microscope together with the corresponding illumination pattern in the pupil [31]. The truncation of higher diffraction orders of the characteristic regular pattern of the sample is clearly seen in pupil pattern for the case NA = 0.8, correspondingly the features are not well resolved. In case (b) the first diffraction orders are fully supported due to a larger numerical aperture and as a consequence, the hexagonal fine structures in the sample are resolved. It should be noted here that the object

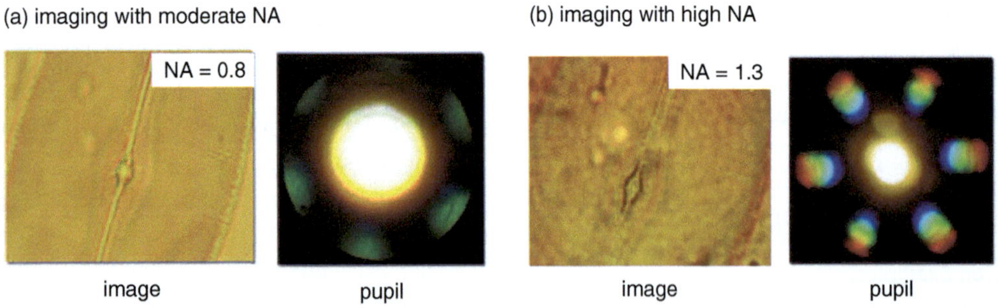

Figure 11.21 Resolution of a sample and pupil illumination for (a) the numerical aperture NA = 0.8 and (b) NA = 1.3. The resolution in the second case is considerably improved.

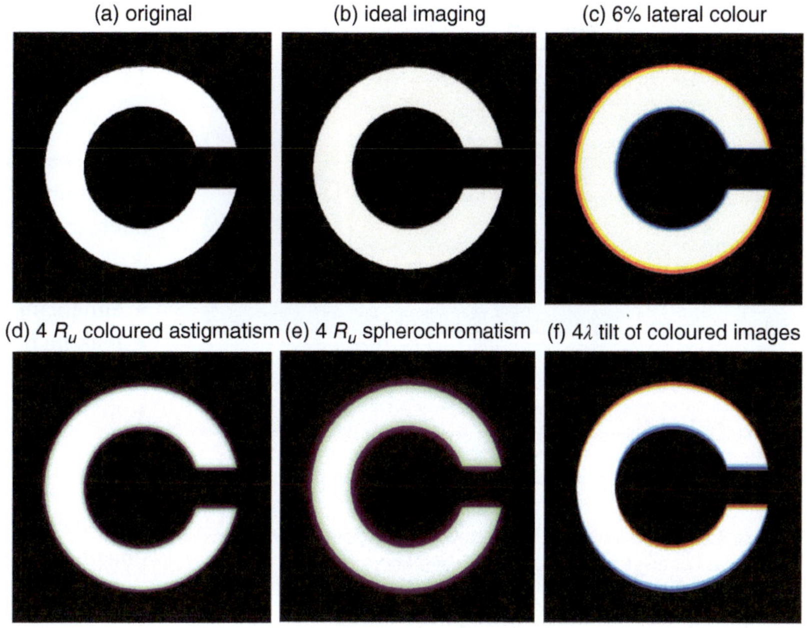

Figure 11.22 Imaging of a black–white Landolt ring with different chromatical aberrations according to the numbers in the captions.

contains a mixture of colours, the diffraction orders are therefore coloured too. Due to the grating equation the longer wavelength for red has a larger radial position in the spectrum.

Figure 11.22 shows the images of a black–white Landolt ring in the case of an ideal imaging as well as for four different colour aberration types. The colour fringes at the edges are seen, they can be detected very sensitively by the human eye. The interpretation of the colour effects and the underlying colour aberrations are discussed in Volume 2, Chapter 14 in more detail.

Figure 11.23 shows the impact of defocusing and primary aberrations on the imaging of a synthetic pattern for incoherent as well as coherent illumination near the

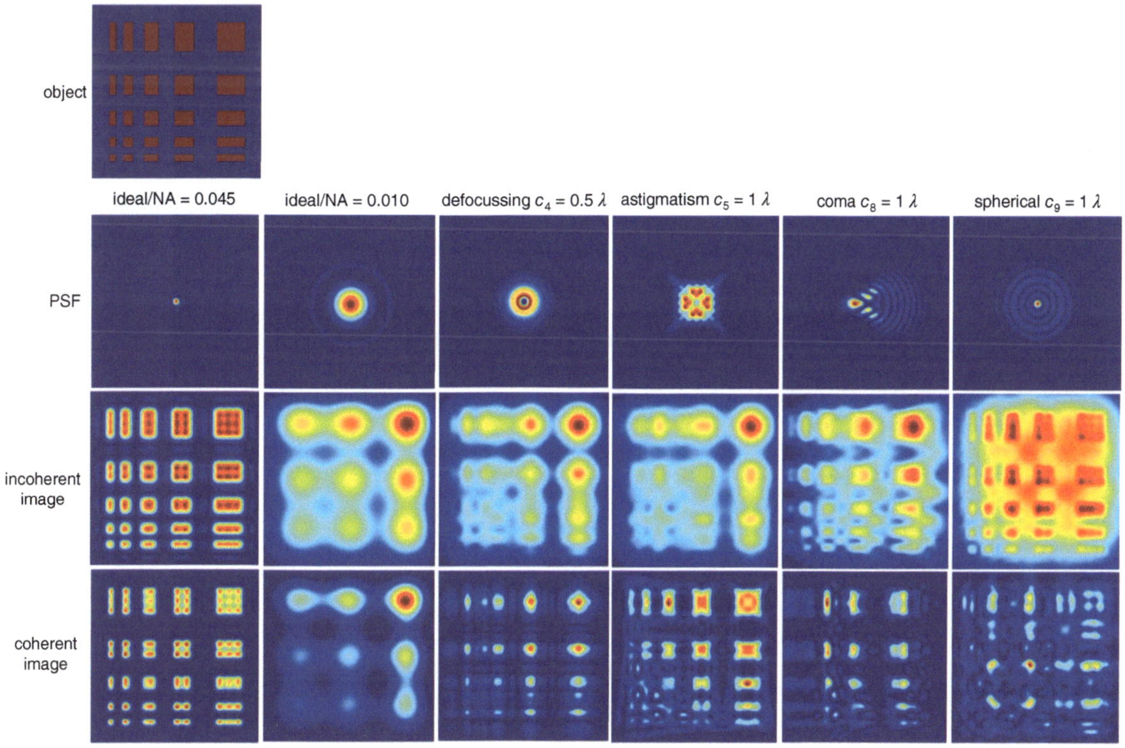

Figure 11.23 Imaging of a bar pattern coherent and incoherent with different aberrations as indicated in the figure as Zernike coefficients. The PSFs as well as the images have the same lateral scale.

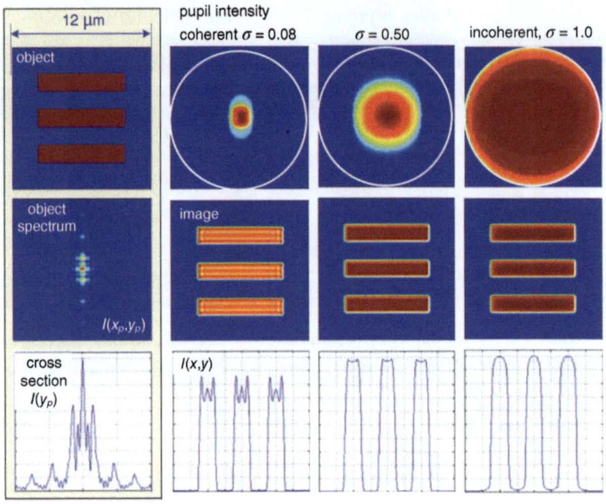

Figure 11.24 Imaging of a bar pattern with different conditions of partial coherence, for explanation see text.

resolution limit. In addition, the PSFs as the convolutions kernels are shown. Their shape influences the orientation of the resolution in various cases and it is seen that the aberrations deteriorate the coherent images more severely.

Figure 11.24 shows the impact of partial coherence on the image formation. It can be seen that the typical interference fringes are reduce, if the coherence goes down for growing σ-values as the parameter of the relative coherence as expected. The case in the second column with $\sigma = 0.08$ is nearly coherent, the pupil intensity is concentrated in the pupil centre, the image shows artificial ringing. In the case $\sigma = 1$ to the right, the illumination is nearly incoherent, the pupil is completely filled, and the image looks more natural as expected.

11.3 Confocal Imaging

11.3.1 Introduction

The classical imaging modes described above for coherent as well as incoherent illumination conditions are not all possible options to generate images. In particular in microscopy, many different modes are used with several goals, some of them are:

- enhancing the contrast;
- making phase objects visible;
- enhancing the resolution;
- making polarizing objects visible;
- allowing for a three-dimensional imaging.

It is not the goal of this introductory text to go into all these methods in detail. But to show at least one alternative imaging mode, in this chapter a short discussion of confocal scanning microscopy imaging is presented [32–34].

In confocal microscopy a point source (preferably a laser is used for illumination) is focused to a spot in the sample and the light emerging from this spot by direct scattering or after fluorescence excitation is collected by a pinhole in a conjugate image plane. Behind the pinhole, a detector gathers all the transmitted light. As it is seen in the setup in Figure 11.25 a beam splitter is needed to separate the illumination beam from the signal light. This arrangement prevents out-of-focus and multiply scattered light being detected since it is mostly rejected at the point detector, see Figure 11.25 for the green ray bundle. Since at any given time only one spot is imaged, this spot has to be scanned across the sample and the image is built up sequentially point-by-point by digital means. The scanning

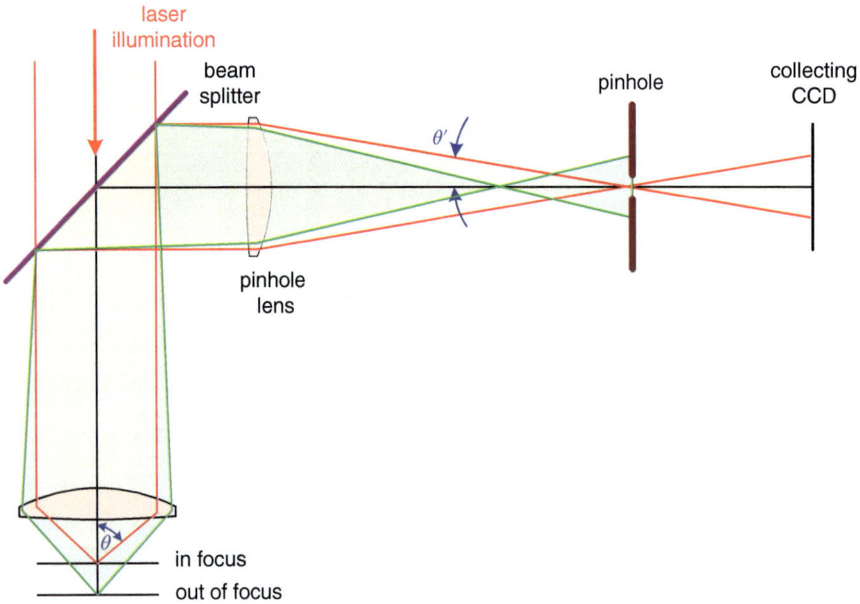

Figure 11.25 Principle of the laser scanning microscope; only the signal light in the conjugated plane with the yellow bundle reaches the detector, the light resulting out of the defocused planes shown in green is blocked by the pinhole.

can be achieved by moving the spot across the stationary sample (beam scanning) or by moving the sample relative to the stationary spot (object scanning). In most confocal microscopes the beam is scanned across the sample for maximum flexibility. However, object scanning has the advantage of better optical quality or simpler optics since the critical optical elements are used on-axis only. Special configurations allow multiple spots (a spot array) to be imaged simultaneously. This enhances the speed of image gathering by parallelized processing. Also in these cases scanning is required. One example is the spinning disk arrangement according to Nipkow where the illumination and detection pinholes are identical and realized by holes in a disk. By rotating the disk, the scanning process is realized.

Since the focal plane that may be located inside the sample contributes to the signal only, the imaging process is called optical z-sectioning. In order to obtain a three-dimensional image of the sample the scanning process can be realized in all three dimensions x, y and z. In most cases a lateral image is acquired first followed by an axial scanning step and the acquisition of another image plane (so-called en-face imaging). If a series of images is obtained this way (called the z stack), whereby all scanning steps are chosen according to the Nyquist theorem (i.e. the step size is about the half width at half maximum of the PSF in the lateral and axial direction, respectively), and light from outside the focal plane defined approximately by the axial extent of the PSF caustic is effectively rejected, this stack represents the three-dimensional image of the sample in a direct way. In order to make this representation viewable on a two-dimensional screen, a number of processing procedures must be applied. One often used approach is the sum projection where for each lateral pixel the sum of all images in the stack is obtained. The result is an image that appears to be sharp throughout the sample scanned. Another representation is the three-dimensional surface rendering and pseudo-three-dimensional representation. In practice the stack is often viewed en-face with options to navigate through the stack.

The gathering of confocal images can be combined with most of the other options to change the image mode, for example 2-photon imaging. Figure 11.26 shows three different confocal imaging modes in comparison to a usual bright field setup. The objects are pairs of small circles in different distances, so the two-point resolution can be investigated. The improved lateral resolution for confocal and in particular for a 2-photon confocal mode is clearly seen. In (c) a case with fluorescence and an additional wavelength shift of the signal light degrades the resolution only slightly.

In the confocal laser scanning microscope, the image is formed by a computed reconstruction from all three-dimensional point signals $I(x,y,z)$ of the sample (so-called voxels) that are covered by the scanning procedure. The size of the pinhole and the quality of the optical train determine the depth discrimination, i.e. the ability of the confocal microscope to reject out-of-focus light. The squaring of the amplitude point spread function for an illumination scanning due to the identical distributions for illumination and imaging changes the three-dimensional imaging

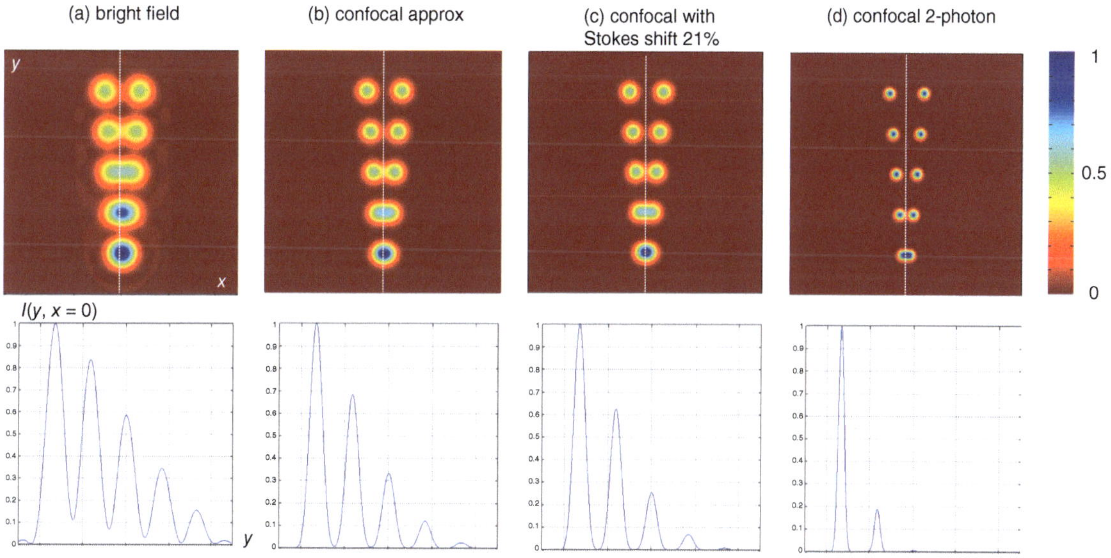

Figure 11.26 Bright field and three different imaging modalities with confocal detection. The object is a pattern of two small circle pairs in five different transverse x-distances. The lower row shows the cross section through the x–y-distributions at $x = 0$ of the upper row along the white dotted lines. (a) shows the classical bright field imaging for comparison, (b) is a confocal mode, (c) a confocal mode with wavelength shift. Finally (d) shows a 2-photon confocal imaging mode, resulting in the best lateral resolution.

essentially. Due to this property, the object is illuminated only in a rather small sheet, therefore a z-sectioning is possible.

11.3.2 Scanning Options*

(*) The content of this section is quite specialized and for advanced reading. Knowledge of this section is not necessary to understand the following chapters. Beginners are recommended to skip this section.

From the fundamental point of view, different principles of scanning can be used for the confocal principle. These are as follows.

- The illumination is performed over the full area of the FoV, a point detector (or pinhole with detector behind) is scanning and gathers the image point-by-point.
- The illumination is performed only by one point, which is scanned over the field of view, a full-field detector collects the image.
- Illumination as well as detector are point like and both are scanned in a synchronized way. So the detector descans the illumination; this is a full confocal mode.

These three basic possibilities are sketched in Figure 11.27 for illustration. In the first case (a), the resolution enhancement due to the squaring of the PSF is not obtained. Furthermore, bleaching effects in the fully illuminated sample may be an issue in critical cases.

The main point in all these imaging modalities is that the image is collected point-by-point and is assembled in the computer in a digital manner. The full confocal imaging case has the special property that both PSFs of the illumination as well as the imaging influences the result. This gives the special advantage of a z-sectioning and an improved lateral resolution.

11.3.3 Image Model

The classical wide field imaging is limited in the transverse resolution by the well-known Airy formula, see Chapter 6. If it is assumed that the illumination and the observation are using the same lens and have the same wavelength, the image intensity is given in the full confocal mode by the square of the ideal caustic. This can be assumed in the lateral as well as in the axial direction and gives the cross sections in the caustic in normalized u- and v-coordinates as

$$I(v) = \left[\frac{2J_1(v)}{v}\right]^2_{ill} \cdot \left[\frac{2J_1(v)}{v}\right]^2_{obs} = \left[\frac{2J_1(v)}{v}\right]^4$$
$$I(u) = \left[\frac{\sin(u/4)}{u/4}\right]^2_{ill} \cdot \left[\frac{\sin(u/4)}{u/4}\right]^2_{obs} = \left[\frac{\sin(u/4)}{u/4}\right]^4 \tag{11.34}$$

This has the positive effect of an improved resolution by a factor of 1.393 for a FWhM definition as indicated in Figure 11.28(a) in transverse direction and by 1.389 in the axial direction, see Figure 11.28(b). Therefore, in a transparent biological sample, all the straylight out of other depths is suppressed and by scanning the depth, a volume imaging is possible. The point like illumination together with the point related observation allows for a three-dimensional assignment. The drawback is obvious: a scan in three dimensions is required which takes a long time to gather a complete three-dimensional image.

If the imaging mode uses fluorescence, a small Stokes shift of the wavelength must be taken into account. For a

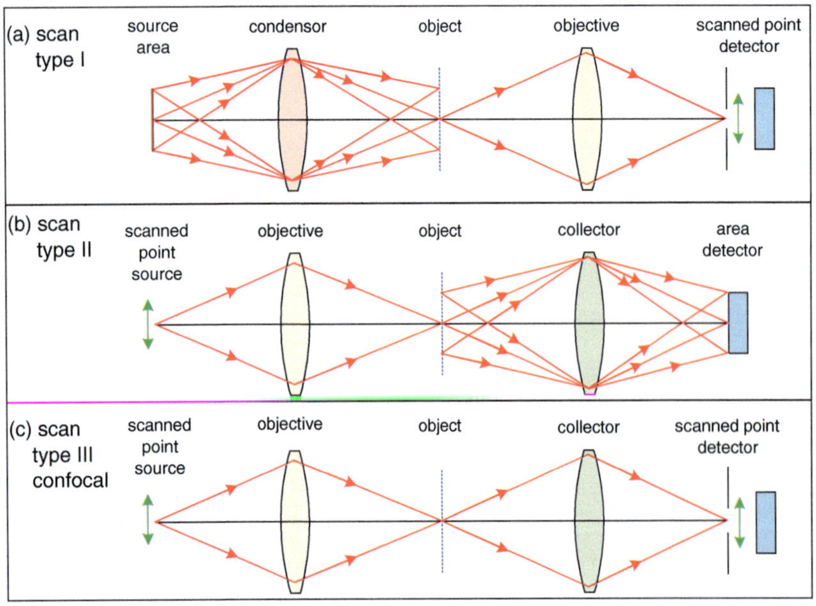

Figure 11.27 Possible imaging modalities for scanning and confocal imaging. In (a) the detector point is scanned, in (b) the illumination and in (c) both.

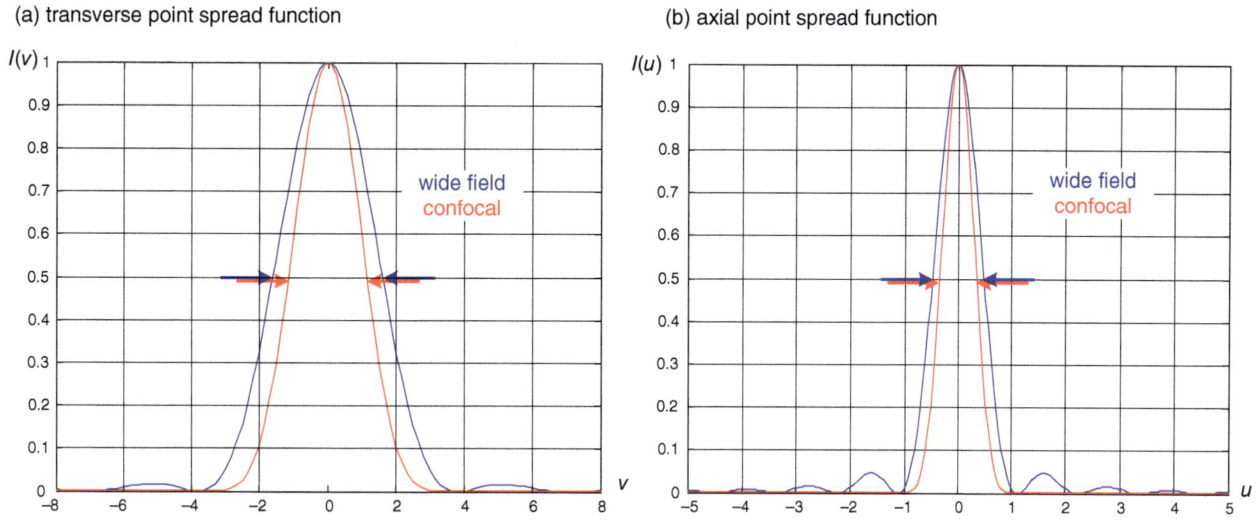

Figure 11.28 PSF of wide field (shown in blue) and confocal imaging (red) in (a) transverse and (b) axial direction with a resolution improvement of approximately 1.4 in both cases.

comfortable number of selected molecules, this is a small effect and very often neglected in reality.

The confocal signal is obtained according to the intensity distribution in Eq. (11.34) by integrating the distribution over the circle area of the confocal pinhole.

$$S(u) = \int_0^{a^*} I_{PSF}^2(u, v)\, 2\pi\, v\, dv \tag{11.35}$$

where a^* is the pinhole size in the scaled lateral coordinate v. According to the Fourier imaging model, the confocal imaging can be described in the following single steps. Figure 11.29 shows the corresponding notations

- Illumination field E_{ill}
- Transfer in front of the object by the illumination amplitude PSF (corresponding to the CTF)

$$E_{obj} = E_{PSF}^{(ill)} * E_{ill} \tag{11.36}$$

- Field modification by the complex object transmission t_{obj}

$$E'_{obj} = t_{obj} \cdot E_{obj} \tag{11.37}$$

- Transfer into the pinhole plane

$$E_{ima} = E_{PSF}^{(ima)} * E'_{obj} \tag{11.38}$$

- Field modification by the pinhole transmission t_{PH}

$$E'_{ima} = t_{PH} \cdot E_{ima} \tag{11.39}$$

- Integration over the pinhole area and detection. It should be noticed that the integration must be extended also over z, so a small part of tissue scattering light out of different depths is included in the confocal signal. The confocal signal is not an intensity, therefore the letter S here.

$$S_{conf} = \iint |E'_{ima}|^2\, dxdydz \tag{11.40}$$

There are several possible idealizations considering the imaging equations. These are as follows.

- Ideal small object point like $t_{obj}(x,y) = \delta(x)\delta(y)$.
- Pinhole considered to be infinitesimal small $D_{PH}(x,y) = \delta(x)\delta(y)$.
- Ideal coherence of the illumination source.
- Ideal incoherence for fluorescence.
- Stokes wavelengths shift neglected in case of fluorescence.
- Illumination source apodization uniform or Gaussian.
- Assumption of perfect z-sectioning and neglecting straylight contributions from other depths.

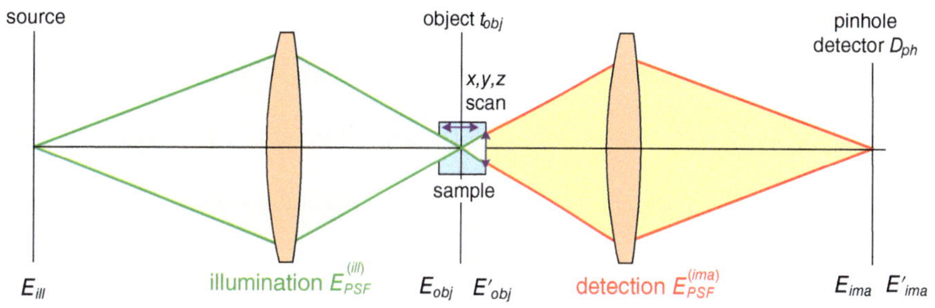

Figure 11.29 Scheme of confocal imaging.

- Objective lenses free of aberrations.

Correspondingly, several different equations should be used to calculate the confocal signal depending on the exact realization and the approximations. If E denotes the coherent point spread function amplitude, we get the following confocal signals S for the various modalities.

- Bright field, incoherent, pinhole point-like

$$S_{ima} = |(E_{ill} \cdot E_{obs}) * t_{obj}|^2 \tag{11.41}$$

- Fluorescence incoherent, pinhole point-like, no Stokes shift

$$S_{ima} = |E_{ill} \cdot E_{obs}|^2 * t_{obj} \tag{11.42}$$

- Object point-like

$$S_{ima} = |E_{ill}|^2 \cdot \left(|E_{obs}|^2 * t_{obj}\right) \tag{11.43}$$

- Object and pinhole point-like

$$S_{ima} = |E_{ill}|^2 \cdot |E_{obs}|^2 \tag{11.44}$$

- Object and pinhole point-like, 2-photon excitation

$$S_{ima} = |E_{ill}(\lambda/2) \cdot E_{obs}(\lambda)|^2 * t_{obj} \tag{11.45}$$

11.3.4 Influence of Pinhole Size and Aberrations*

(*) The content of this section is quite specialized and for advanced reading. Knowledge of this section is not necessary to understand the following chapters. Beginners are recommended to skip this section.

The above considerations assumed a perfect confocal imaging arrangement using a point detector that is much smaller than the image PSF. In this case one essentially detects the maximum of the image PSF only. In most cases an adjustable pinhole at the image plane in front of an extended detector is used. Often the pinhole will be comparable or even larger than the image PSF since the detected signal strength scales with the area of the pinhole in front of the detector. In many applications the signal strength that determines the imaging quality via the signal-to-noise ratio (SNR) cannot be simply increased by increasing the illumination intensity. This is particularly true for biomedical fluorescence imaging – one of the most important applications of confocal microscopes. Here the sample damage by the incident light and bleaching of the fluorescence by photochemical reactions set a limit to the illumination intensity. In these cases, the confocality of the imaging and the signal have to be traded off by using an extended pinhole. The effect of a finite sized pinhole on the imaging is described by integration of the detection PSF over the pinhole area D.

Since the relative size of the pinhole to the detection PSF is of relevance only, a natural scale for the pinhole diameter is the size of the first dark ring of the Airy distribution. The size of the pinhole therefore is comparable to the Airy diameter. The larger the pinhole, the better is the SNR, but the confocality shrinks and lateral as well as axial resolution are reduced. This transition is sketched in Figure 11.30. In principle this means that the confocality is a continuous property and can be changed smoothly by changing the pinhole size. For the limiting case of a very large pinhole, brightfield conditions are obtained.

Since the pinhole has a filtering effect, the confocal signal strongly depends on residual aberrations, in particular for small pinholes. Figure 11.31 shows a case with $c_{40} = 2\ \lambda$

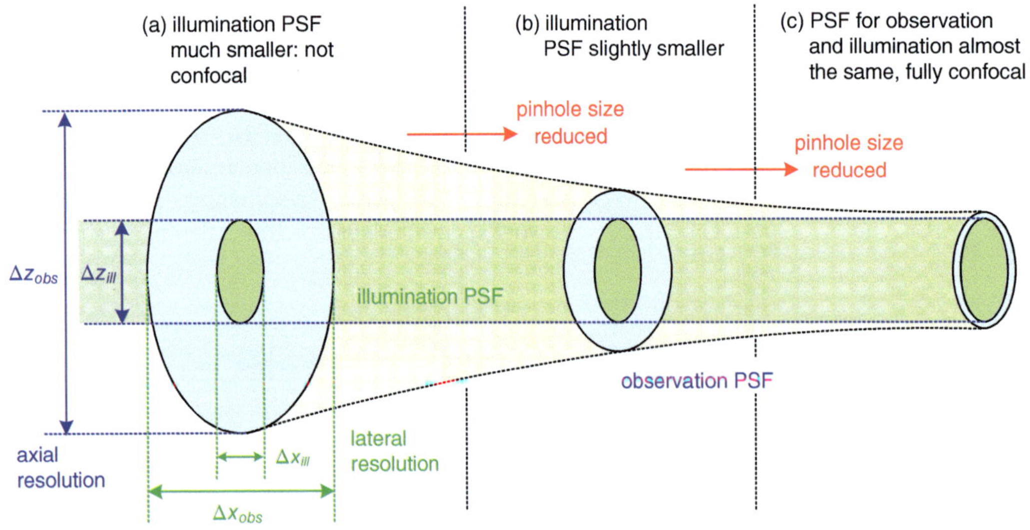

Figure 11.30 Transition of changing pinhole size from very large D_{ph} in (a) to similar size as Airy in (c).

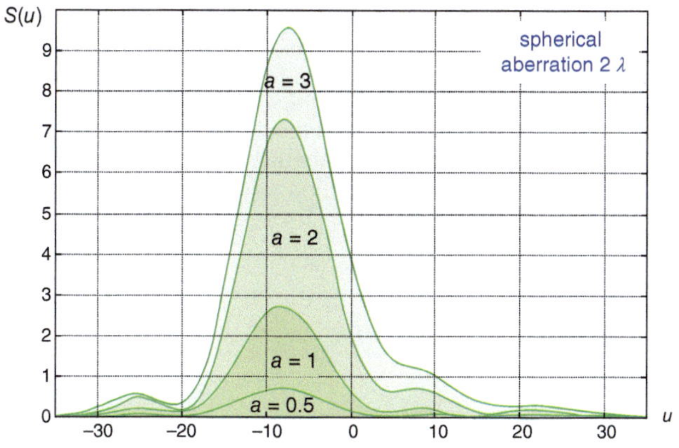

Figure 11.31 Confocal depth signal for different values of the normalized pinhole size a for a spherical aberration of $c_{40} = 2\ \lambda$.

spherical aberration and the normalized pinhole size $a = D/D_{airy} = 0.5\ldots3$. The well-known z-asymmetry of the caustic influences the confocal signal, which is asymmetric too. To determine the exact z-position, usually the centroid of the signal curve $S(u)$ is calculated. This is systematically wrong in the case of spherical aberration. Interestingly very small pinholes extract only the centre part of the peak and have a smaller residual error. A larger pinhole, in the figure for example the case with a = 3, develops a pronounced shoulder, which influences the centroid calculation. But the reduced peak height in case of a small pinhole has the consequence of a lower SNR value, which is a problem in the case of weak microscopic signal strengths.

Figure 11.32 shows some typical confocal signals for different amounts of spherical aberration (scaled as Zernike coefficient c_{40}) and different values of the pinhole size. The curves shown are the power signals on the detector behind the pinhole on-axis, when the axial scan is performed in the object space. The signal peaks are renormalized here. The broadening of the signal and the decrease in the axial resolution with increasing pinhole diameter can be seen.

11.4 Anisoplanatic Imaging

11.4.1 Introduction

It is well known that usually the performance of optical systems decreases towards the edge of the field of view. According to the Seidel theory of primary aberrations, there are types growing with the field height in the first, second and third power. This means that in general the field dependence of the aberrations and the quality data behaves complicatedly. There is a lot of literature found for the modelling and restoration under spatially variant conditions [35–53].

For illustration, Figure 11.33 shows the change in size as well as orientation of the PSF for a single lens in the case of primary coma. Figure 11.34 illustrates the PSF for the full FoV in the case of coma and field curvature aberration. Coma in the third order grows linearly with the field size; field curvature quadratic. Coma in addition as an odd

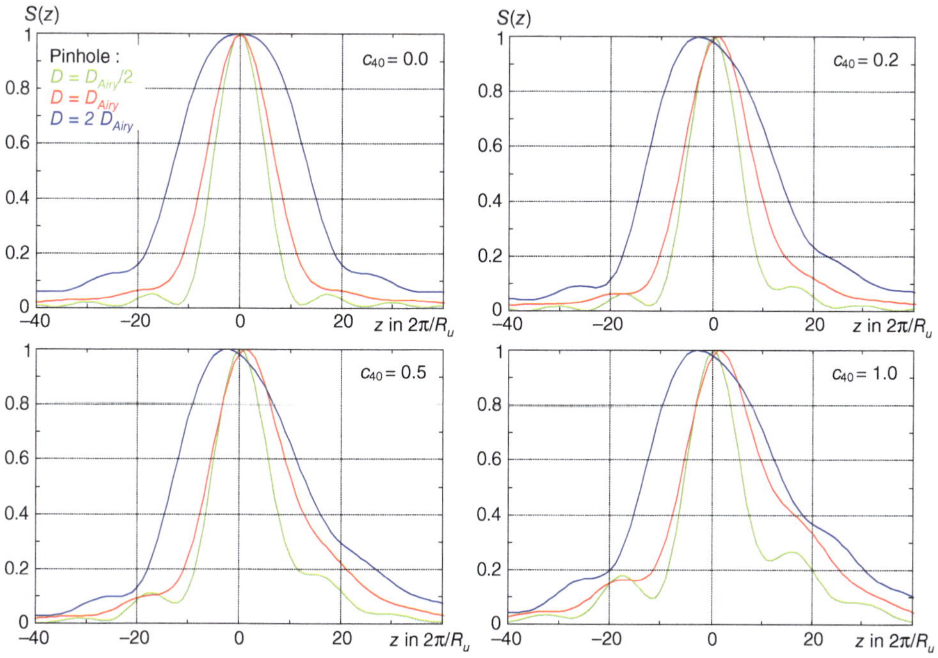

Figure 11.32 Confocal signal for different values of the spherical aberration c_{40} and for different pinhole sizes.

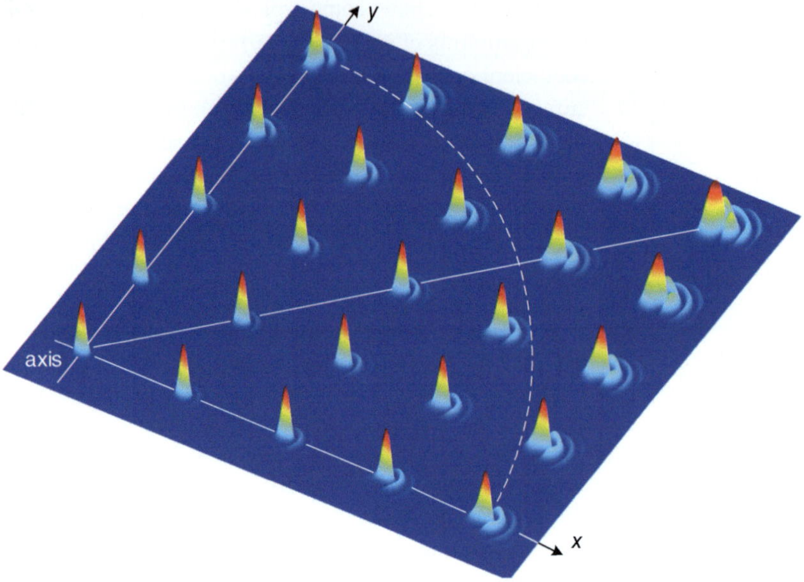

Figure 11.33 Change of the coma aberration in strength and orientation as a function of the field position x/y. The peak heights are renormalized, the coma grows linearly with the distance to the origin. Only the first quadrant with $x, y > 0$ is shown. The individual PSFs are exaggerated inside the field of view.

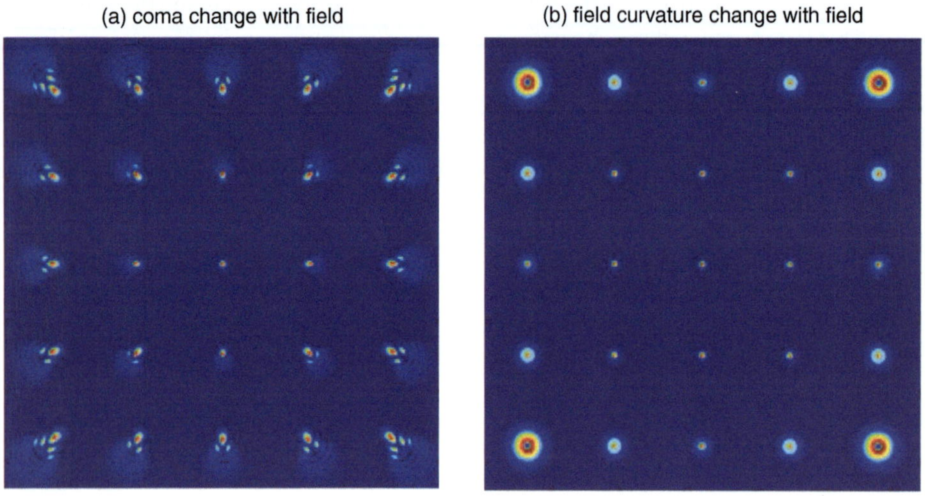

Figure 11.34 Typical change of the PSF for (a) coma or (b) field curvature with the field coordinates [37].

aberration type rotates with the azimuthal angle as seen in Figure 11.33 and 11.34(a).

This change of the PSF with the field position means that the simulation of diffraction models cannot be performed by simple convolution with a constant invariant PSF; the isoplanatic assumption is violated. In Figure 11.35 the result of an imaging calculation of a simple three-lens-camera system is shown for illustration purposes. The field dependence of brightness as well as sharpness, contrast and colour fringes are observed in the exaggerated zoomed parts of one square sub-field. The distortion can be seen from the complete FoV.

If the PSF changes with the field position x_o, y_o, in general the image intensity can be written in the case of incoherent illumination as an integral transform

$$I(x_i, y_i) = \iint O(x_o, y_o) \cdot I_{PSF}(x_i, y_i, x_o, y_o)\, dx_o dy_o \quad (11.46)$$

If a system is space invariant and an object $O(x_o, y_o)$ gives the image $I(x_i, y_i)$, the shifted object gives an identical image at a shifted position:

$$\begin{aligned} O(x_o, y_o) &\rightarrow I(x_i, y_i) \\ O(x_o + \Delta x_o, y_o + \Delta y_o) &\rightarrow I(x_i + m \cdot \Delta x_o, y_i + m \cdot \Delta y_o) \end{aligned} \quad (11.47)$$

with magnification m for every arbitrary shift vector $(\Delta x_o, \Delta y_o)$. In this case the integral transform simplifies to a convolution integral as already discussed. Only if these conditions are valid, can the linear system theory be

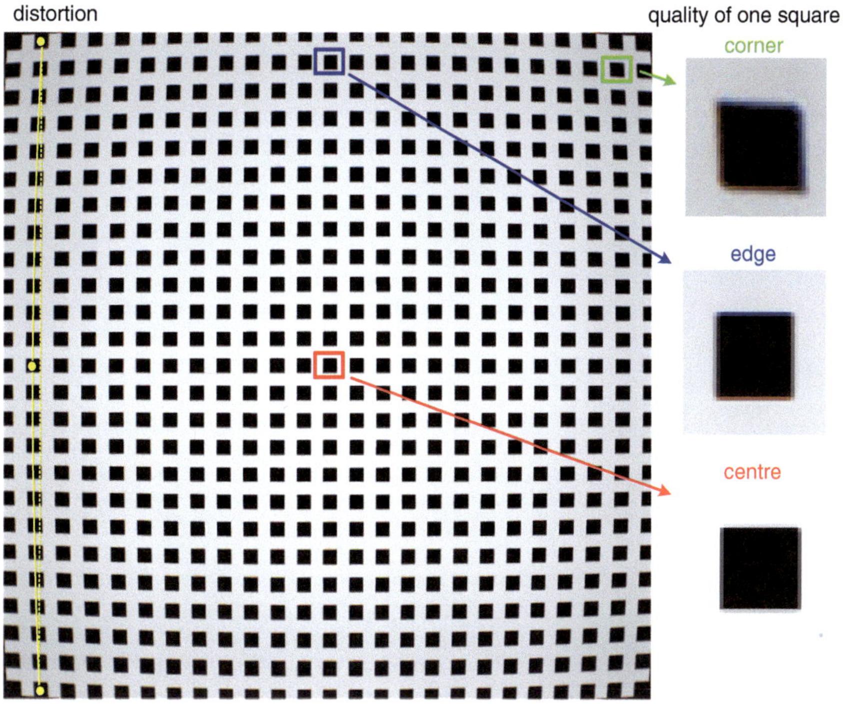

Figure 11.35 Imaging of a checkerboard object by a camera lens with field dependent aberrations for three different field positions and distortion as a global perturbation.

applied to use a convolution for image simulation. This must not be confused with the linear superposition of object details and an additive intensity. From a physical point of view, there are several possible indications for a lack of isoplanatism.

- Photometric inhomogeneity, perhaps due to vignetting.
- Distortion, modification of the object geometry.
- Wave aberrations, blur and change of the width of the PSF.

If the PSF changes with the field coordinates x_o, y_o, one idea is to interpolate the changing PSF function for the simulation in the spatial domain. But the PSF as a complicated intensity distribution cannot be easily interpolated. To solve this problem, there are several opportunities as follows.

- Interpolation of the Zernike coefficients in the pupil.
- Modelling the variant PSF functions by a Taylor expansion with a restricted range of validity inside the FoV for a coarse sampling of selected expansion points.
- Decomposition of a discretized PSF matrix into principal components and interpolation of the expansion coefficients. This corresponds to an interpolation in the spectral domain as in the first option.
- Computing a space variant deconvolution with a non-constant kernel.

Both solutions therefore work with only a few parameters, which change smoothly over the FoV. The interpolation of arbitrary field points on the Zernike level is relatively easy to implement; the drawback is that for every field pixel, the PSF must be calculated again for the local values of the Zernikes coefficients. If a coarse grid of PSFs is used and a principal component decomposition is calculated, this corresponds to significant 'eigen-PSFs'. These functions are independent and allow for interpolation of the expansion coefficients. This procedure is much more efficient if the necessary number of required PSFs is small, which means that the corresponding Karhunen–Loeve expansion is fast converging. Therefore, the decay of the singular values of the SVD governs the speed and this is the faster method, the smaller the changes of the PSF are. If a large distortion is present, a dense PSF-grid is needed for high precision and many eigenfunctions have to be used for convergence.

In reality, due to the optical system correction it can be often assumed that the change of the PSF over the field of view is smooth and weak. This means that a perturbation approach can be used and one image pixel is only influenced by the neighbouring PSF functions. This means that the kernel function of an interpolation has a sparse structure and can be efficiently represented by eigen blur kernels for restoration [43].

Another approach [46] uses a Taylor expansion of the PSF around every supporting grid point. The problem is then the robust calculation of the derivatives, but the interpolation can then be performed locally fast.

It should be mentioned here that beyond the lateral blur issue described here, a spatially variant image simulation

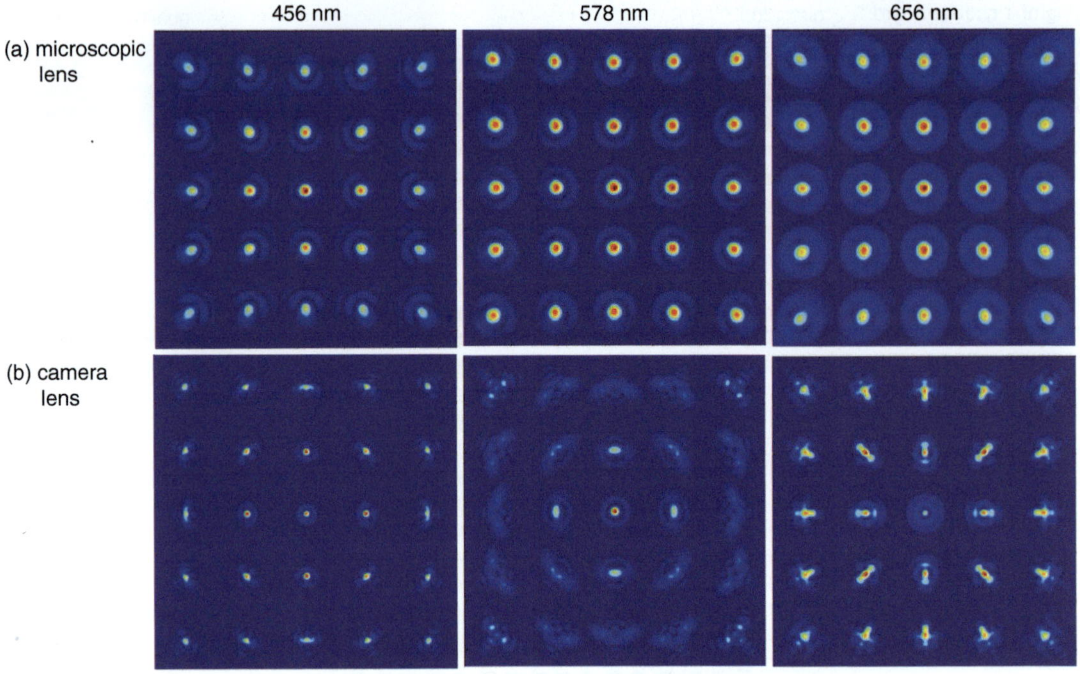

Figure 11.36 Variation of PSF with the field location x and y for (a) a high NA microscopic lens and (b) a camera lens for the three main wavelengths.

or restoration can also be performed in three dimensions in confocal microscopy with one more dimension [48, 49].

Figure 11.36 shows the PSF as a function of the field for a typical microscopic lens in (a) as well as for a classical triplet camera lens in (b) for the three main wavelengths. To get a nearly diffraction limited performance, the camera lens is stopped down. It can be seen that the field variation in case of the camera lens with its significant larger relative field of view is much larger.

11.4.2 Interpolation of Zernikes

It can be easily checked that the variation of the Zernike coefficients with the field position is usually very smooth. Figure 11.37 shows a typical example for a circular symmetric system. In this special case of symmetry only the field height y must be considered as a simple parameter. It can be seen that the spherical aberration is not zero on axis, astigmatism and field curvature are even and coma and

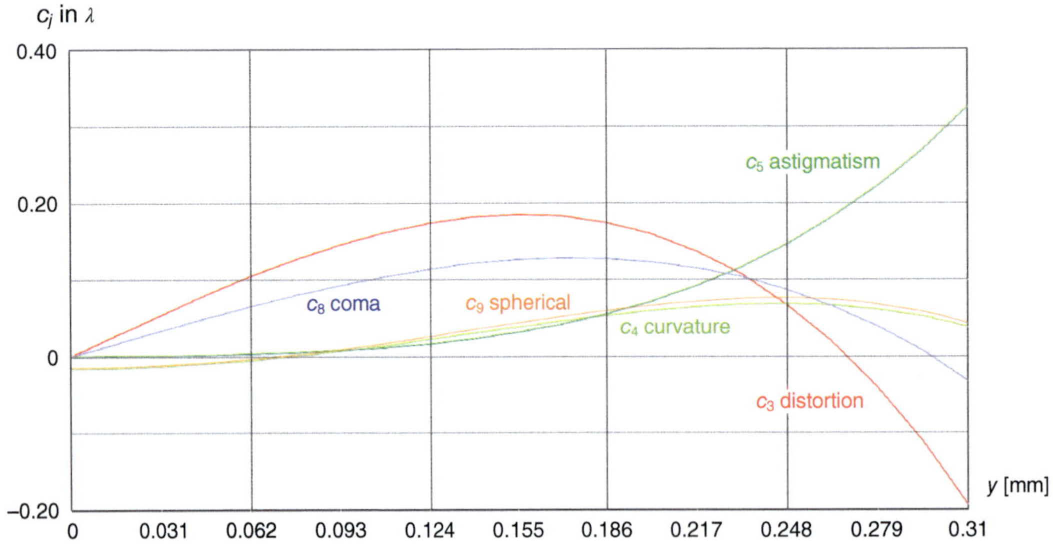

Figure 11.37 Variation of the primary Zernike coefficients with the field height y for a camera system.

distortion are odd aberrations. Therefore, a first approach can be written in this one lateral dimension for only positive $+y$ by taking only primary aberrations into account [37]

$$\begin{aligned} c_9(y) &= c_9^{(0)} + c_9^{(2)} \cdot y^2 + c_9^{(4)} \cdot y^4 + \ldots \\ c_4(y) &= c_4^{(2)} \cdot y^2 + c_4^{(4)} \cdot y^4 + \ldots \\ c_5(y) &= c_5^{(2)} \cdot y^2 + c_5^{(4)} \cdot y^4 + \ldots \\ c_3(y) &= c_3^{(3)} \cdot y^3 + c_3^{(5)} \cdot y^5 + \ldots \\ c_8(y) &= c_8^{(1)} \cdot y + c_8^{(3)} \cdot y^3 + c_8^{(5)} \cdot y^5 + \ldots \end{aligned} \tag{11.48}$$

This simple polynomial expansion is of high accuracy, the necessary number of expansion terms as well as the number of Zernikes to consider corresponds to the overall size of aperture, field and quality of correction.

Figure 11.38 shows the result of the Zernike field maps for the first 25 coefficients of a microscopic lens [37]. Although in the full plane as well as for the higher orders the field variation is more complicated, it remains smooth and is comfortable for interpolation.

11.4.3 Spatially Variant Convolution*

(*) The content of this section is quite specialized and for advanced reading. Knowledge of this section is not necessary to understand the following chapters. Beginners are recommended to skip this section.

The imaging equation can be formulated in general for simplicity in one transverse coordinate as

$$I(x_i) = \int O(x_o) \cdot I_{PSF}(x_i, x_o)\, dx_o \tag{11.49}$$

If now the PSF kernel function changes with position x_o, an expansion into a mode functional system $\varphi_n(x_i)$ in the sense of principal components (see Chapter 12) is performed as

$$I_{PSF}(x_i, x_o) = \sum_{n=1}^{N} a_n(x_o) \cdot \varphi_n(x_i) \tag{11.50}$$

where the spatial dependence is incorporated into the expansion coefficients $a_n(x_o)$. φ_n are the PSF functions on a coarse sampling grid. The interpolation is then done by the smooth coefficient functions $a_n(x_o)$ on the basis of the modes. We assume that at an image location x_o, the convolution operation is valid locally as

$$I(x_i) = \int O(x_o) \cdot I_{PSF}(x_i - x_o)\, dx_o = O(x_i) * I_{PSF}(x_i, x_o) \tag{11.51}$$

By inserting into the image equation, we have

$$\begin{aligned} I(x_i) &= O(x_i) * \sum_{n=1}^{N} a_n(x_o) \cdot \varphi_n(x_i) \\ &= \sum_{n=1}^{N} a_n(x_o) \cdot [O(x_i) * \varphi_n(x_i)] \end{aligned} \tag{11.52}$$

This means that the convolution is done locally with the expansion functions and then the sum over the expansion is calculated. Therefore, an interpolation over the convolutions is obtained which usually behaves smoothly. A convolution is a linear operator and the formula above describes a weighted sum of convolutions. In principle this approach can be used, if the PSFs are known for discrete sampling points, and the weighted sum at the end

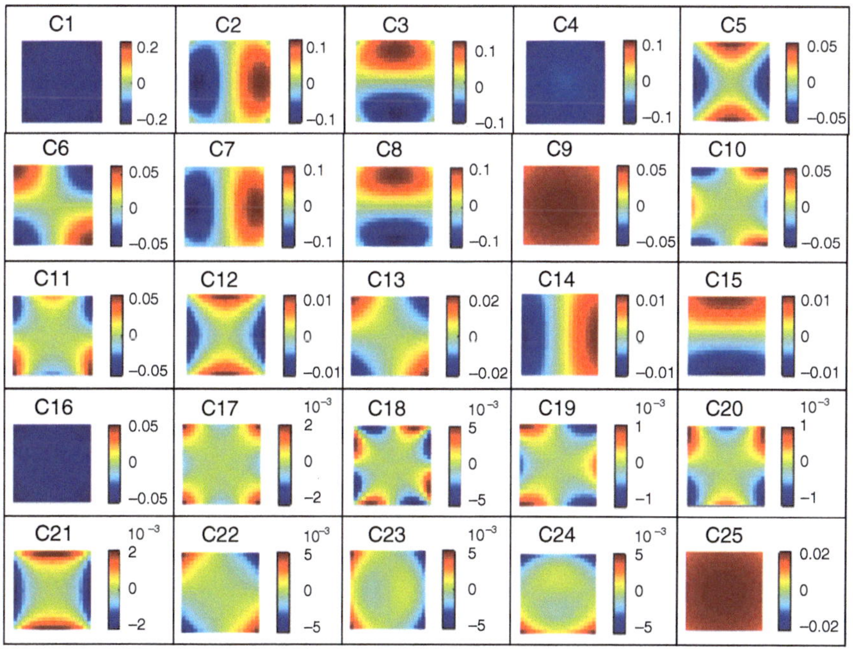

Figure 11.38 Variation of the lower order Zernike coefficients $c_j(x,y)$ with the field location x and y for a high NA microscopic lens.

corresponds to an overlay with a kind of blending function, taking the finite support of every discrete PSF into account. The overlap of the blending functions should smear out some stitching problems at the borders. The finite support gives a chance to calculate this approach with a sparse formulation rather efficiently. In Figure 11.39 the idea of the finite support blending functions is sketched.

The interesting question in this method is how to find and select the expansion functions φ. They should describe the change of the PSF shapes over the field of view in the best way and simultaneously be numerically attractive. This means that ideally they are orthogonal and have a finite support area. If the functional expansion of the PSF is in particular selected to be most efficient and orthogonal, the so-called principal component expansion is obtained as a special case for this approach. This mathematical approach is discussed in Chapter 12 in more detail.

If the mode functions have a finite support, the corresponding matrix representation of the weighted decomposition looks like Figure 11.40 [43]. The diagonal structure of the φ-matrix allows for a sparse calculation scheme, which is numerically comfortable for memory consumption and computational time.

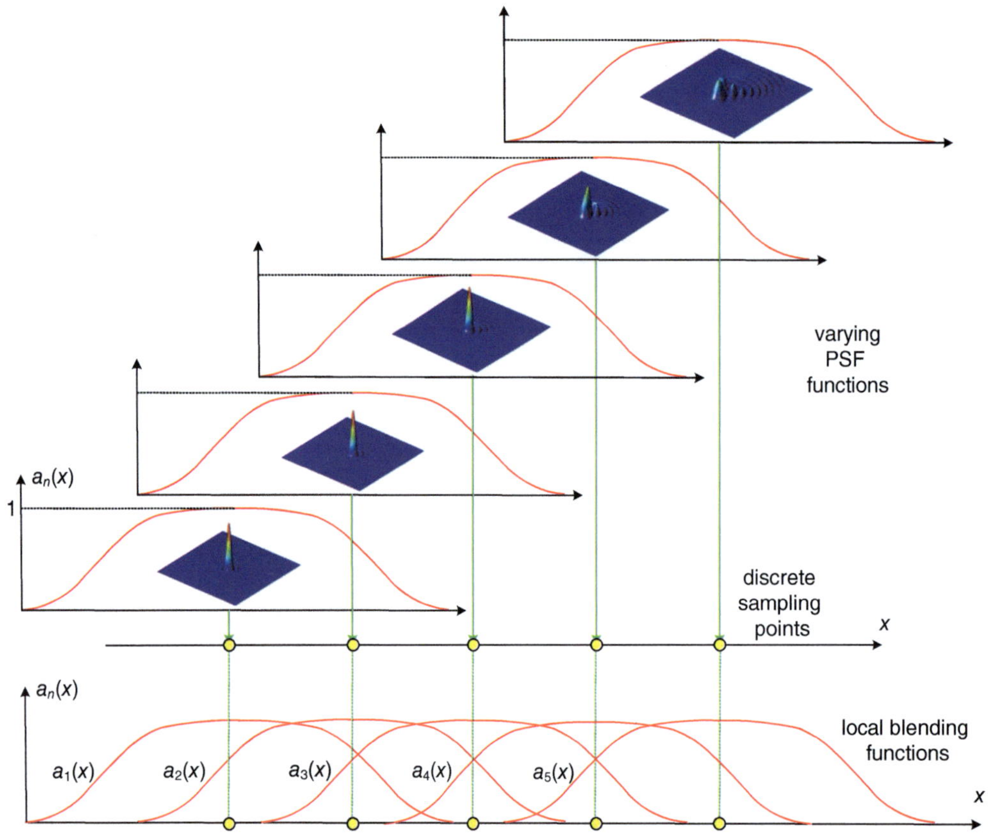

Figure 11.39 Schematic representation of the overlapping of local points spread functions with corresponding blending functions $a_n(x)$.

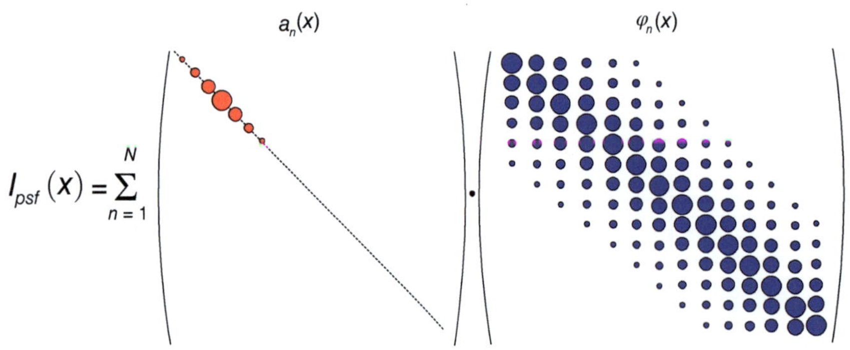

Figure 11.40 Schematic representation of the overlapping of local points spread functions with corresponding blending functions $a_n(x)$.

11.4.4 Interpolation by Principal Components*

(*) The content of this section is quite specialized and for advanced reading. Knowledge of this section is not necessary to understand the following chapters. Beginners are recommended to skip this section.

If in a real imaging setup the PSF is spatially varying, the idea of an image calculation is to decompose the matrix of PSF functions over a finite grid of points inside the field of view into a set of independent eigenfunctions, the so-called orthogonal principal components. More details about principal component analysis (PCA) are found in Chapter 12. The calculation of these eigen-PSFs is performed by a singular value decomposition. The benefit of this approach is that the PSF can be expressed as a sum of eigen-PSFs without coupling terms, so with vanishing correlation

$$I_{PSF}(x_i, y_i) = \sum_{j=1}^{j_{\max}} a_j(x_i, y_i) \cdot I_j^{(PCA)} \tag{11.53}$$

the eigen solutions $I^{(PCA)}$ are independent from the image coordinates x_i, y_i, only the expansion coefficients a_j change with the field position. To find the eigenfunctions, n known PSFs at selected positions with m pixel each are collected in a common matrix A according to the scheme illustrated in Figure 11.41. Due to the sparsity of the local PSFs this helps to reduce the four dimensions to two and to use matrix computations for the decomposition.

The basic idea of the PCA is to find the eigenvectors of A with the covariance matrix of the singular value matrix S. If the calculation is performed by singular value decomposition (SVD), the eigenvalues are ordered in a descending manner and the corresponding matrix contains the eigenfunctions. These eigenfunctions are the principal components. They must be rearranged in a two-dimensional scheme as the eigen-PSFs. If the sequence of singular values is fast converging, for the decomposition in the Eq. (11.53) a truncation can be used to take only the relevant components into account. In particular according to the laws of linear algebra, if the matrix A has the rank r, only r eigenvalues are different from zero. This gives a faster computation and corresponds to a dimensional reduction. The decomposition expansion coefficients a_j can then be two-dimensional interpolated for any arbitrary position in the field of view. This means that the PCS analysis and construction of the eigenvectors gives the opportunity to interpolate a given data set.

In principle, clever combinations of the original PSF functions now give independent new distributions. To make the new decomposition meaningful it must be taken into account that all components have the same mean value of zero. The principal components are eigenfunctions of the covariance matrix $C = AA^T$. One critical point is the selection of the PSF-grid over the field of view. If it is quite dense, a large computational burden is created; if it is too sparse, the error of the representations grows. Therefore,

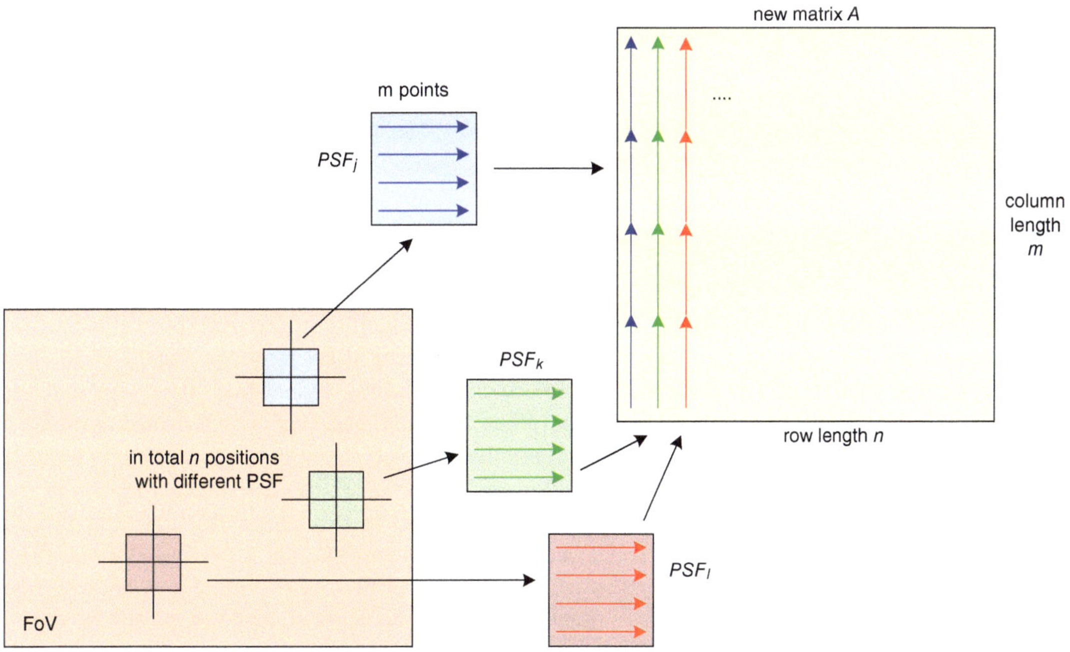

Figure 11.41 Rearrangement of the n local varying PSFs with m pixels each into a common matrix A.

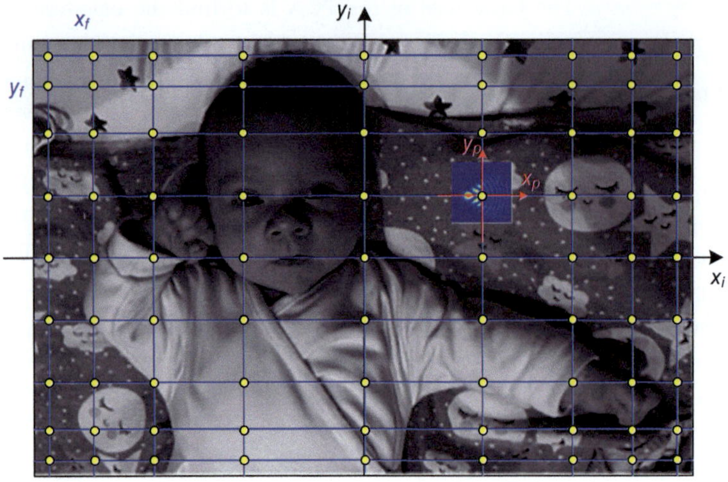

Figure 11.42 FoV, PSF pattern and adapted calculation grid.

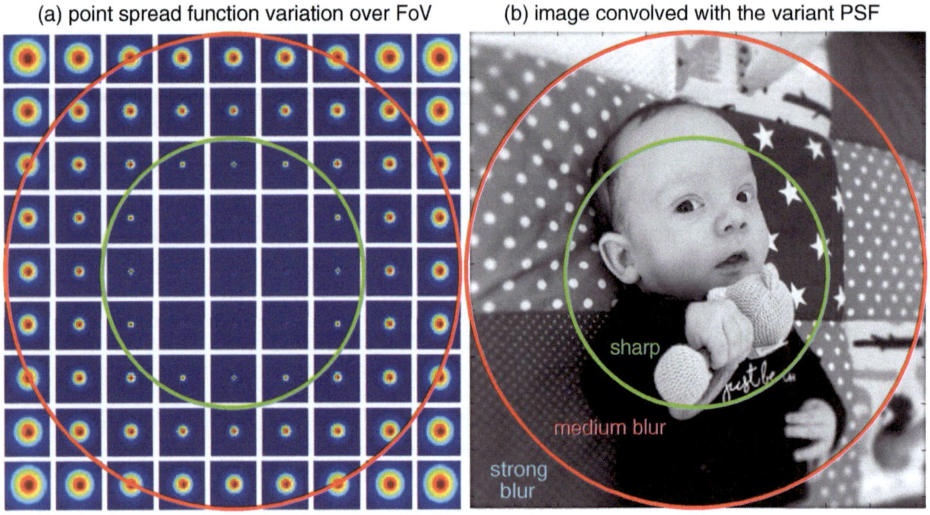

Figure 11.43 Field varying (a) blur with quadratic growth of the PSF (scale enlarged) and (b) image with this 9 × 9 grid of PSFs.

it should be checked whether the eigenvalues of the PSF samples converge appropriately. A typical picture looks like Figure 11.42. In this example 8 × 8 = 64 sampling points are used with locally different PSFs on a non-equidistant grid as indicated by the yellow dots.

To get information on the necessary number of field positions and local PSF distributions, a picture like the one in Figure 11.43 helps. In part (a) the PSFs on a grid of 9 × 9 field positions are shown. An image produced with this data is seen in part (b). The growing blur for growing distances to the image centre is clearly seen in the image. The raw corresponding zones are indicated by the coloured circles for orientation.

Figure 11.44 shows an example of a PSF distribution on a 15 × 15 grid for a system which suffers from astigmatism and field curvature. The strong variance of the spots with the radial distance is observed and deteriorates the image in the corners. Figure 11.45 shows the corresponding first 66 principal components to describe this behaviour on an orthogonal system of basic modes. Figure 11.46 visualizes the corresponding eigenvalues. It is seen that the eigenvalues are fast converging, which means that only a small number of principal functions is needed for a rather accurate description. With the first 12 eigenfunctions, the PSF description has in this case a residual error of 0.27 %, which is mostly good enough. In Figure 11.45 also one of the drawbacks of the PCA method can be seen: the functions are purely mathematical results of the eigenvalue calculation; the individual function cannot be easily interpreted in optical terms.

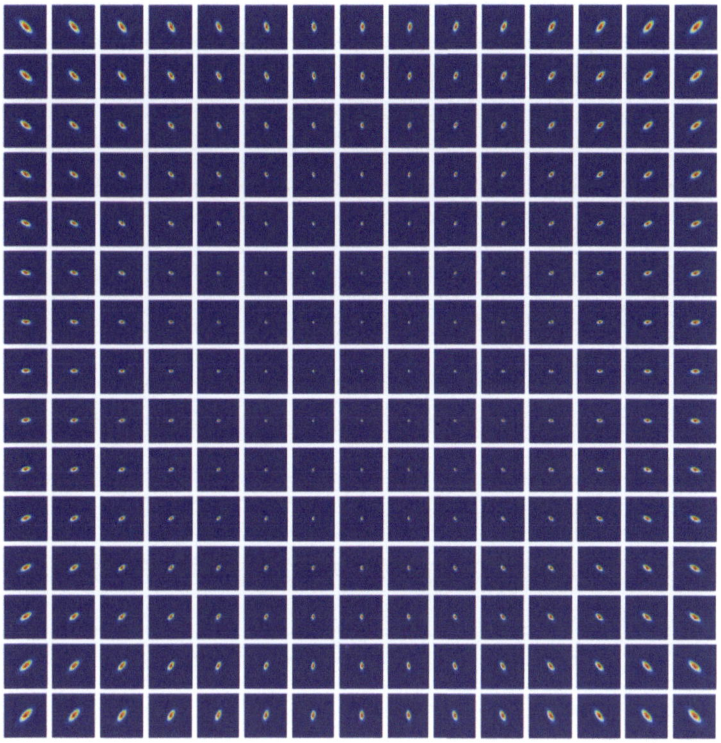

Figure 11.44 PSF grid with 15 × 15 functions over the FoV for a system with significant contributions of astigmatism and field curvature.

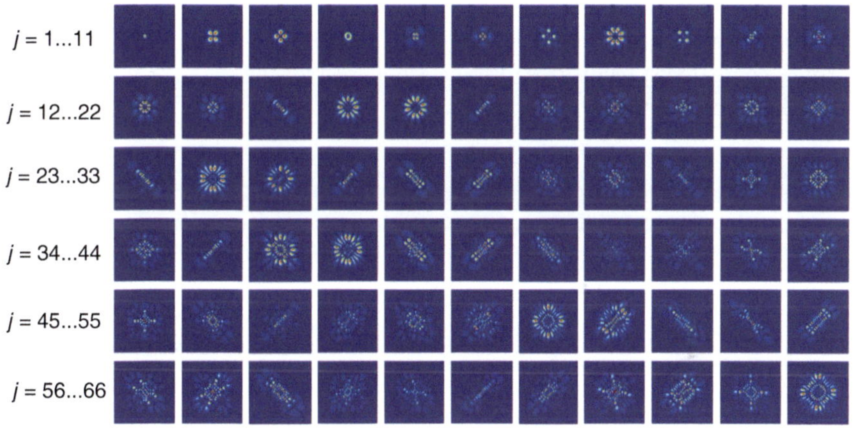

Figure 11.45 Sequence of the lowest 66 principal eigen functions for the example of Figure 11.44.

11.4.5 Taylor Expansion of the PSF*

(*) The content of this section is quite specialized and for advanced reading. Knowledge of this section is not necessary to understand the following chapters. Beginners are recommended to skip this section.

The basic idea of this approach is to assume a weak spatial dependence of the PSF over the field of view [46]. In this case only a nearest neighbour effect must be taken into account. The changes are considered by performing a Taylor expansion of the image function and using only the first order perturbation, which here means the second derivative in the Taylor series. Locally we have a convolution integral of the form

$$\begin{aligned} I(x_i) &= \int O(x_o) \cdot I_{PSF}(x_i - x_o)\, dx_o \\ &= \int O(x_i - x_o) \cdot I_{PSF}(x_o)\, dx_o \end{aligned} \tag{11.54}$$

as discussed before. If now the image is expanded as

$$I(x_i) = \sum_{k=0} a_k \cdot \frac{d^k O(x_i)}{dx_i^k} \tag{11.55}$$

where the expansion coefficients are the moments of the PSF

$$a_k = \frac{(-1)^k}{k!} \cdot \int x^k I_{psf}(x)\, dx \tag{11.56}$$

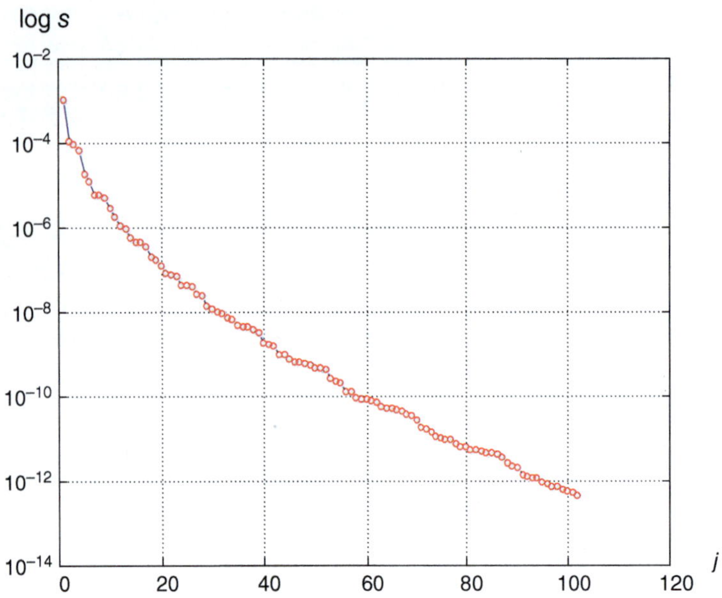

Figure 11.46 Convergence of the PCA eigenvalues s with growing index j on a logarithmic scale.

Now the object function is expanded similarly as

$$O(x_o) = \sum_{m=0} b_m \cdot \frac{d^m I(x_o)}{dx_o^m} \tag{11.57}$$

For simplicity we assume here a magnification of $m = 1$, so the coordinates in object and image are the same. If now both expansions are inserted into each other, equal powers can be compared; a recurrence formula is obtained for the unknown coefficients b_m in the form

$$b_0 = \frac{1}{a_0}, \quad b_m = -b_0 \cdot \sum_{j=0}^{m-1} b_j \cdot a_{m-j} \quad \text{for } m = 1, 2, 3, \ldots \tag{11.58}$$

By selecting the coordinate origin appropriately, it can be achieved that $a_1 = 0$ and then the first perturbation correction term gives

$$O(x) = \frac{1}{a_0} \cdot I(x) - \frac{a_2}{a_0^2} \cdot \frac{d^2 I(x)}{dx^2} \tag{11.59}$$

Now the question of how many terms are necessary to consider depends on the accuracy needed and the amount of change of the PSF over the coordinates in the FoV. In principle the second order approximation of the second derivative means that the change over the width extension of the PSF can be neglected. This approximation is mostly appropriate. In two lateral dimensions x, y this approach works similarly.

11.4.6 Distortion

A special case of a spatially variant change of the image properties is distortion. It is not a degradation with blur, but it changes the right image point position as a function of the field coordinates x and y by aberrations. Figure 11.47 shows some more complicated distortion types together with the classical barrel and pincushion shapes. In this case due to the regular initial pattern, the distortion is clearly observed. For a more natural scene, these geometrical perturbations may be hard to detect as is illustrated in Figure 11.49.

Pure distortion is from the viewpoint of aberration theory not affecting resolution. This means that in a more academic consideration, distortion can be easily removed and restored by digital means. The two grid patterns can be mapped onto each other by interpolation. But in reality this decoupling is not trivial, because distortion is usually not obtained alone. Figure 11.48 shows an example with a combined distortion and blur degradation. In principle, the distortion is related to the energetic centroids of the PSFs. Odd aberrations like coma are also shifting the intensity peaks, so a separation of these effects is in reality cumbersome.

A digital removal of distortion by pixel interpolation in any case also causes a loss in resolution and requires a slightly larger field of view due to boundary effects. Furthermore, in real systems the distortion is dispersive which means that every colour in a RGB image has an individual distortion value.

If a natural scene without straight lines is considered, the distortion is rather less critical. Figure 11.49 shows an example of this type; here the 10% distortion can be accepted without disturbing the image too much.

11.4.7 Special Varying Blur Types*

(*) The content of this section is quite specialized and for advanced reading. Knowledge of this section is not

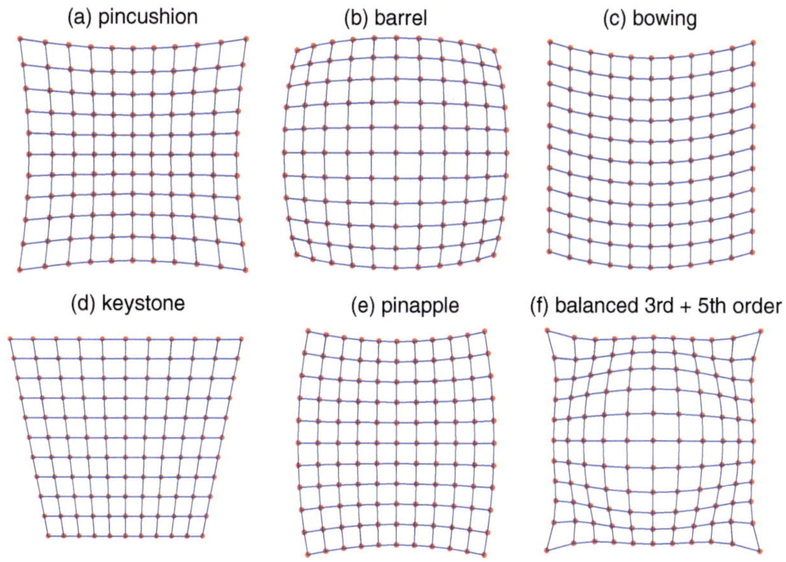

Figure 11.47 Different geometries of distortions. The pincushion type in (a) and the barrel type in (b) are the classical third order distortion shapes of circular symmetric systems. Keystone in (d) are found in Scheimpflug imaging and for off-axis field sections like in a projector system.

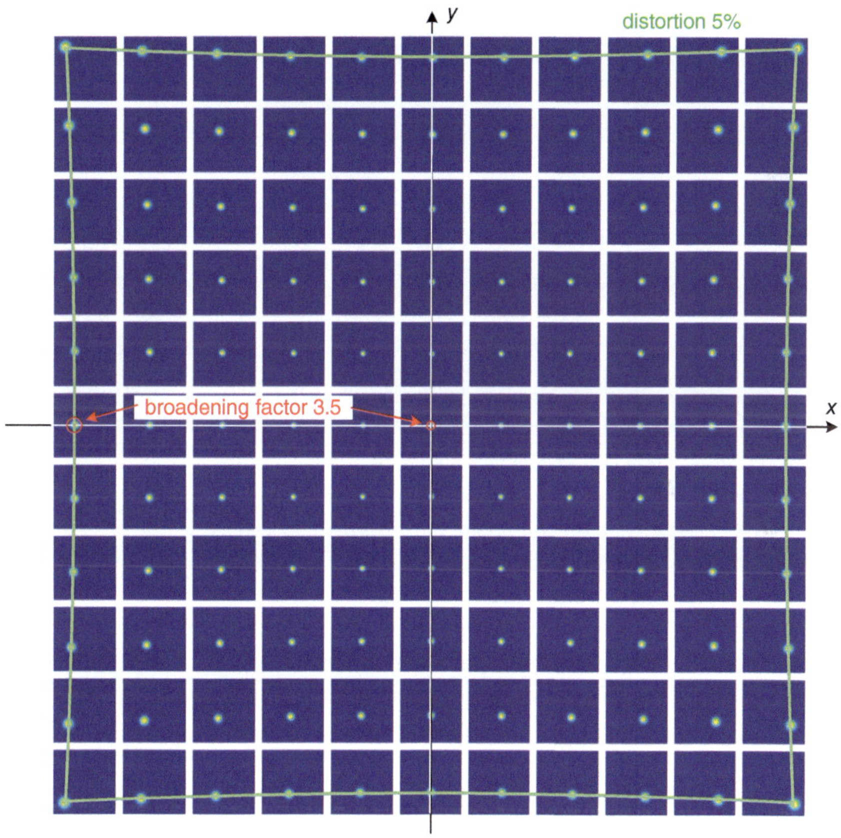

Figure 11.48 Image with a combined broadening and distortion.

necessary to understand the following chapters. Beginners are recommended to skip this section.

In modern microscopy three-dimensional imaging modes are possible, for example by confocal methods. If refractive index mismatch problems are present, the spherical aberration changes with the depth and therefore spatially varying blur effects are also seen in the z-direction. This finally means correspondingly simulating images suffer from degradation in all three dimensions. The easiest method to deal with this problem is to perform a PCA in the third dimension too.

A very special case of image blur with spatial varying properties is the imaging by a multimode fibre [50]. The PCA approach to interpolate the PSF is also applicable here.

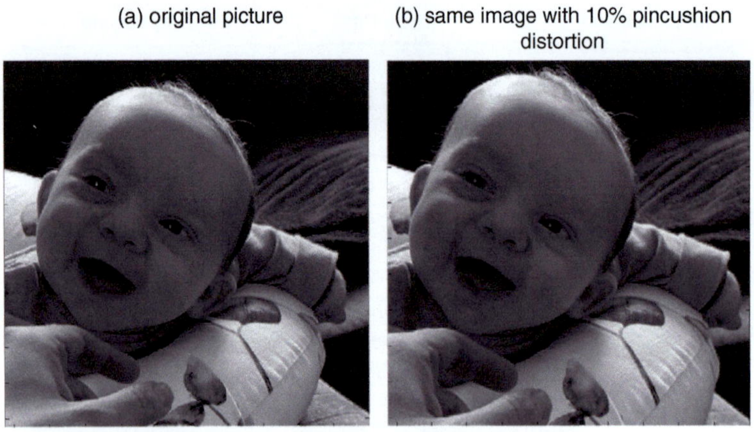

Figure 11.49 Simple pincushion distortion of 10% for a natural scene.

Figure 11.50 Motion blur examples with (a), (b) and (c) linear blur and (d), (e) and (f) circular motion structure. Source: b) Frank John Aleksandrowicz / U.S. National Archives and Records Administration / Public domain, d) rionStar / Wikimedia Commons / Public domain f) Diana Juncher / ESO / CC BY 4.0.

Another interesting case of spatially varying blur in reality is motion blur [53]. This is a very commonly occurring degradation; Figure 11.50 shows some examples of this type. In the easiest case the motion blur of the PSF is linear only, this means that the blur can be described in one dimension only. This is seen in the figure in parts (a), (b) and (c). But also more complicated motion paths are found, for example in long exposure astronomical observations or by rotational movements, see parts (d), (e) and (f). As a more special case of motion blur a vibration blur is in particular characterized by a mostly linear movement, but with varying speed. For more details the interested reader is referred to the literature.

11.5 Digital Imaging Processing

11.5.1 Introduction

Every physical signal exhibits discrete behaviour below a certain limit. Then the signal changes in finite small steps. Consequently, the transfer function becomes a step curve and is no longer continuous. For such a discretization of the signal there can be obvious physical reasons but there may also be a technical necessity. So, for example, for the conversion of an optical signal into an electronic one, every electric light-sensitive detector is characterized by its bit number and is limited as far as its signal resolution is concerned. When a signal is received, a discretization is carried out at three points.

- At the sensor during image detection (spatial discretization in pixel).
- In the data representation during signal processing in the computer (binary representation of the signal strength).
- On the display when the result is reproduced (spatial discretization in pixel).

The schematic procedure for getting a typical signal is shown in Figure 11.51.

The information chain for a transfer process from the object to the received information includes a series of steps as shown in Figure 11.52. The information content is affected by the generation of noise in the different phases of the information transfer, by discretization and quantization as well as by the parameters of spatial, time, energetic and spectral sensitivity and other disturbances. The algorithms of image processing are modifying the signal too.

The purely spatial aspect of sensor technology is briefly discussed first. The conversion from the radiation field $I(x,y)$ in front of the detector for the generation of the signal $S(x,y)$ is described and is directly related to the resolution. For stationary conditions this can be described by the equation

$$S(x,y) = [K \cdot I(x,y) * T(x,y)] \cdot \mathrm{Comb}(x,y) + N_D(x,y) \tag{11.60}$$

In this eq. K is a conversion factor, $T(x,y)$ is the spatial response function of the sensor, in general also as a function of the incidence angle of the detector. The comb-function describes the effect of the discrete pixels and N_D the noise. The symbol * here indicates a convolution.

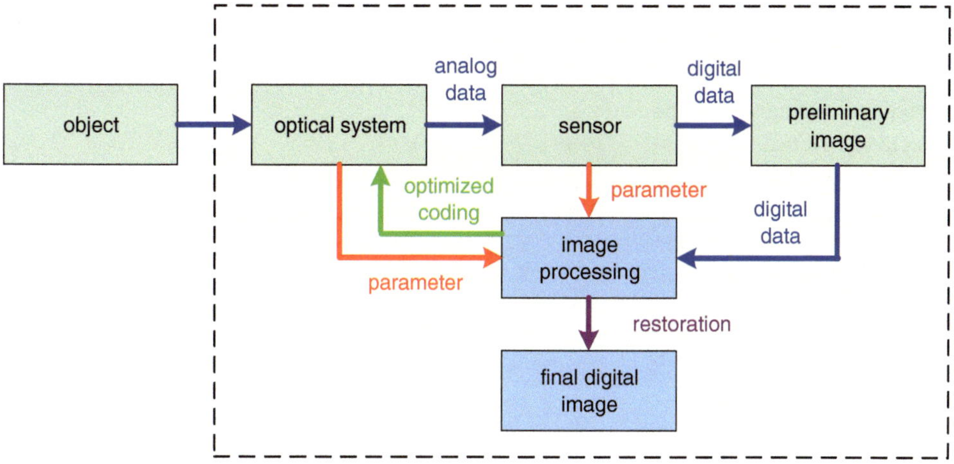

Figure 11.51 Principle of the detection, processing and reproduction of a signal.

Figure 11.52 Process steps for optical signal processing; generally the signal depends on position, time and wavelength.

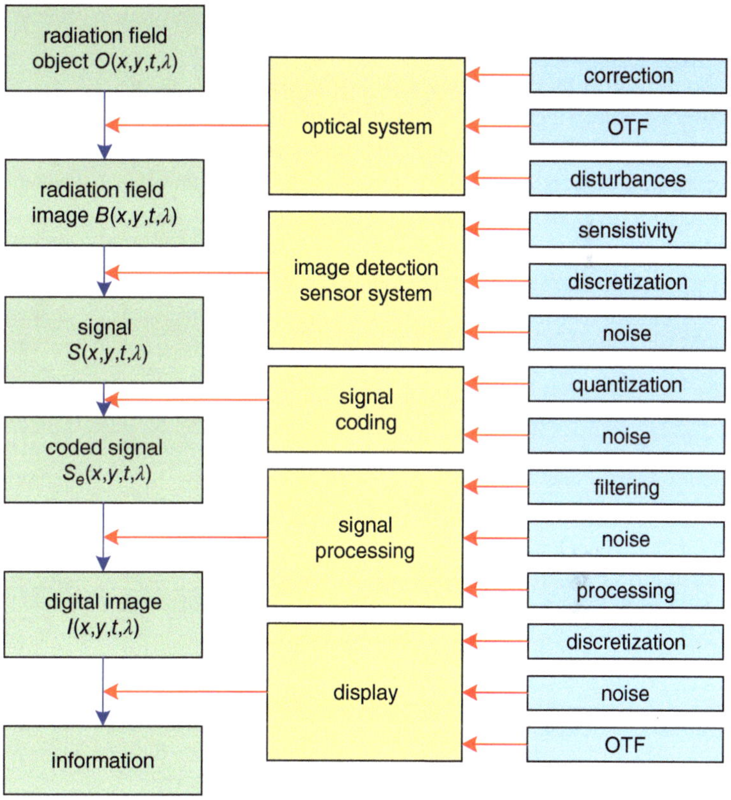

In the spatial transfer function $T(x,y)$ of a camera, its tolerances and error influences are also included. For example, an additional optical system images the radiation field which is to be detected onto the sensor. For an exact spatial resolution, the geometry of the setup has to be calibrated in order to eliminate individual influences and perturbations. Among these are size and position variations of the individual pixels and distortion of the imaging system. It is not always possible to consider all effects using the equation above. In particular, the convolution does not describe the response correctly in every case.

11.5.2 Spatial Discretization and Resolution

The intensity distribution taken by a sensor contains systematic errors in the signal representation due to the spatial discretization into finite sized pixel. For an image the resolution plays an important role. Here the following effects have to be taken into account.

- In the dead area between the active pixel areas some signal energy, and thus essentially some information, is lost; the (SNR) is decreased.
- The registered signal is integrated and averaged over the width or area of a pixel, respectively. Hence a falsified signal evaluation occurs if the variation of the signal distribution over this width cannot be neglected.
- Due to the discretization, only band-limited information about the structure can be extracted. Therefore, the spatial resolution is limited.

If D_{pixel} is the pixel width of the detector, any information about the fine structure above the Nyquist frequency

$$\nu_{Ny} = \frac{1}{2D_{pixel}} \tag{11.61}$$

is lost. The ratio of the spot size D_{spot} to the pixel width D_{pixel}

$$k = \frac{D_{spot}}{D_{pixel}} \tag{11.62}$$

is referred to as the sampling factor of the sensor. In an approximation it can be used as a measure for the structural content of the detected signal. In the range $k = 1....1.6$ noticeable errors can still be observed. In particular the exact relative position of pixel and spot plays a role due to the discretized structure. Only for sufficiently high values, i.e. $k > 6$ do these effects vanish almost completely.

A kind of fidelity-criterion in the form of a rms-deviation between original signal and detected quantity can be used as a measure of quality

$$F = 1 - \frac{\langle [I_{meas}(x) - I_{orig}(x)]^2 \rangle}{\langle I_{meas}^2(x) \rangle} \tag{11.63}$$

For a simple analytical example with a known result (e.g. bar object and Gaussian-shaped point image) it can be shown that despite a large value of k (i.e. $k = 2$) only 96% fidelity can be expected. For $k = 6.5$ we get $F = 99\%$. So, due to pixelation, the registration of a signal by a sensor always has a discretization error. The signal is integrated

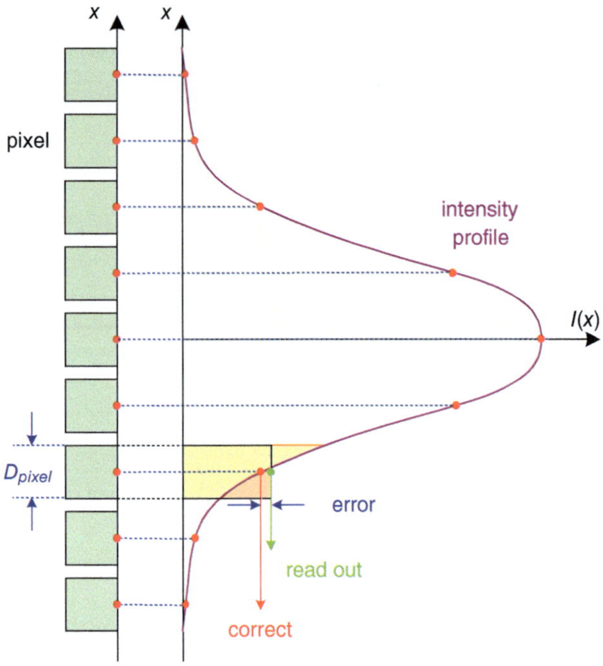

Figure 11.53 Discretized signal registration, determination of averaging over a finite pixel size.

over the finite width of a pixel leading to the determination of an average read out value, see Figure 11.53. As can be is seen in the figure, if the signal curve has a significant curvature over the pixel width, the averaged signal is not exact the same as the exact one assigned to the pixel centre. Figure 11.54 illustrates the Cartesian discretization scheme of a typical CCD sensor with dead zones between the sensitive areas of the pixel. This finally is the reason for a finite fill factor. In addition, in reality a small crosstalk between neighbouring pixel must be assumed; there is no electronic hard border between the pixels.

Considering, e.g. a CCD sensor, there is a difference between the active and the light-sensitive pixel area of width p and scan width $D_{pixel} > p$. Between the pixel areas

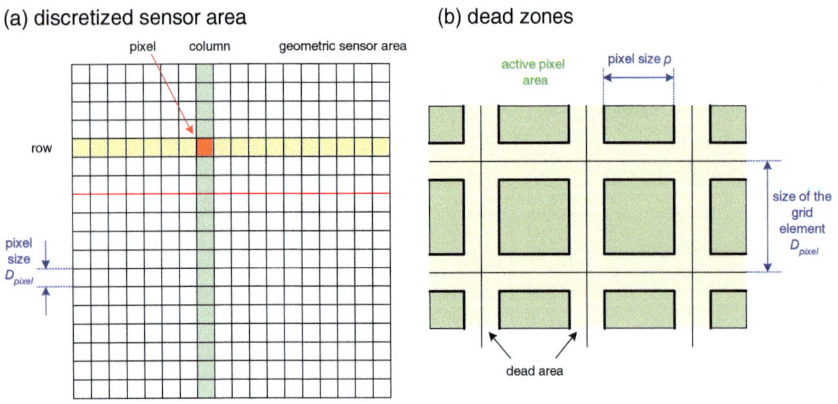

Figure 11.54 Real detectors with (a) discretized sensor area and (b) dead zones in two dimensions.

there is usually a dead area due to electrode stripes. This leads to a filling factor smaller than 1. The filling factor is defined as

$$\eta = \left(\frac{p}{D_{pixel}}\right)^2 \tag{11.64}$$

Real values of η strongly depend on the technology used but are nowadays close to 100%. Sensors can consist of individual light-sensitive small surfaces, like linear arrays for the spatial resolution in just one direction, as well as two-dimensional arrays. Depending on the application there are three outcomes regarding whether the spatial resolution can be directly obtained by means of a sensor: no spatial resolution, in one dimension or in two dimensions.

In order to detect an optical signal, the geometry of the radiation field and the sensor have to be adjusted to each other. Here the size of the beam at the position of the sensor is important. Furthermore, different requirements need to be defined.

- The sensor has to detect only the integrated radiant power.

  In this case the focusing or the beam shaping, respectively, have to be carried out in such a way that the sensor is not swamped, otherwise some power is lost.
- The sensor has to detect the radiant power with spatial resolution.

  In many cases the sensor is positioned in an image plane and simultaneously registers the complete image surface. If this functionality is required, then the beam has to be adjusted to the sensor size in such a way that the size of the pixels yields the desired spatial resolution. The size of the point image of the optical image ought to correspond approximately to the size of the pixels if a resolution loss by the sensor cannot be tolerated.

  If the mid position of a focused beam has to be determined, it is meaningful to make the size of the spot significantly larger than the pixel size. In this way, the radiation's centre-of-mass can be detected with a sub-pixel resolution. If, however, the spot is smaller than the pixel size, the measurement of the position cannot be more accurate than the pixel size. This is because the sensor cannot differentiate between different positions of the spot within the pixel. This condition, which at first appears not plausible, is illustrated in Figure 11.55.

Regarding the interface between optical system and sensor, it should be observed furthermore that the angular acceptance of each sensor is limited. It must not be smaller than the maximum divergence angle of the radiation field. Otherwise, misleading paths of the brightness will occur.

For the colour-sensitive detection of radiation there are several different possibilities.

- Each pixel is equipped with an optical colour filter and therefore detects only a narrow spectral component of the radiation. In the sense of the RGB decomposition of the signal, three different pixel types are detected independently.

  For this kind of colour filter array there are many usual setups. Some examples are shown in Figure 11.56.

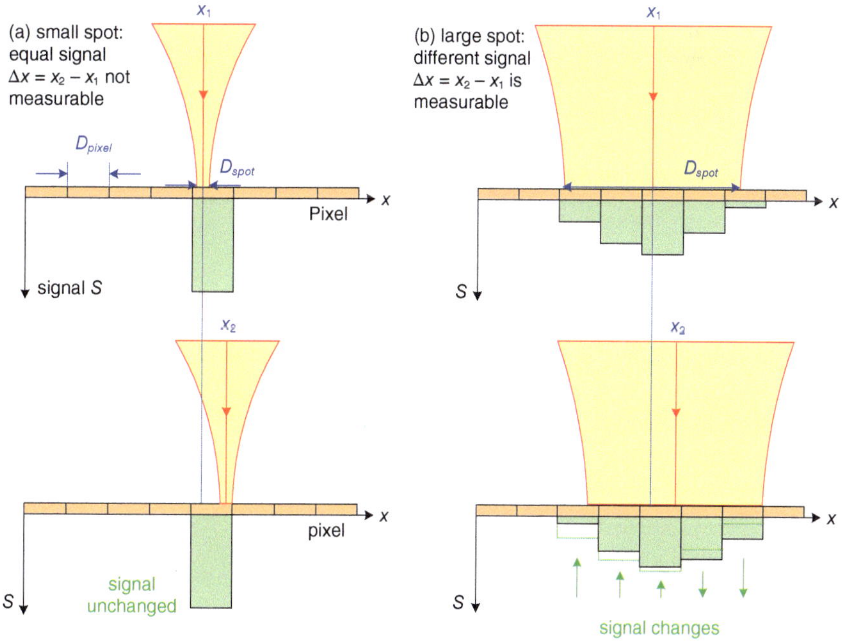

Figure 11.55 Position accuracy of a discretized sensor for a small and large spot size, respectively. The difference in the position Δx of the signal can be measured only if the diameter of the spot is larger than the pixel size. The first column (a) illustrates a spot much smaller than the pixel, in (b) the spot illuminates several pixels. The difference of the obtained signal for a small position difference is sketched in the lower and upper row for comparison.

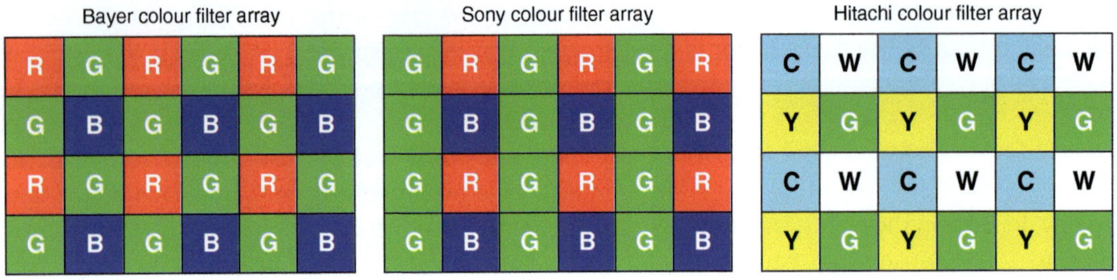

Figure 11.56 Colour filter arrays with different patterns and colour selections as usual found in typical CCD sensors.

Besides additive RGB filters there are also subtractive colour combinations. These variants differ from each other in terms of transmission and speed. It should also be observed that the actual number of colour pixels is different in order to achieve a coarse adaptation to the sensitivities and colour weights, respectively. This kind of colour detection has a direct influence on the spatial discretization of the signals. The spatial resolution is lower than in the monochromatic case and it is not isotropic.

- The incident ray is split into three parts with different wavelengths by means of corresponding colour-selective filter components. Each partial ray is detected by its own sensor. This approach is realized in film cameras by using dichroitic coatings on corresponding prisms for geometric ray division. In systems of this kind, however, images with coloured boundaries can be observed due to the non-telecentric path of the chief ray as well as in the case of defocusing. Another issue is the exact overlap of the sensor arrays for the different colours and a longitudinal colour aberration, which occurs in the optical system.
- By means of an LCD monitor or a mechanical colour wheel, a time sequence of the colour channels is generated which becomes synchronized with the frame transfer and is evaluated separately.

Each of these methods has its strengths and weaknesses. But this gain in spectral information has to be paid for by disadvantages in the spatial or time resolution or the sensitivity. Newer developments have led to image sensors capable of detecting colours which have a z-stack of three photo sensors in the light direction. Thus they detect the colours one after the other in different depths without affecting the spatial resolution. Often protecting windows or micro lenses are used in order to increase the efficiency. These have a chromatic effect. Hence the spectral distribution which arrives at the semiconductor is already filtered and this has to be taken into account by a calibration.

11.5.3 Management of Image Colour

If a RGB-coloured image is considered, the spectral weighting of the illumination source, the impact of the object on the wavelength and the spectral sensitivity of the sensor matters. If the spectral weights of the illumination are given by a function $w(\lambda_j)$ and the spectral sensitivities of the sensor for red, green and blue are $r(\lambda_j)$, $g(\lambda_j)$ and $b(\lambda_j)$ respectively, the coefficients are calculated by

$$\overline{R} = \int w(\lambda) r(\lambda)\, d\lambda,\ \overline{G} = \int w(\lambda) g(\lambda)\, d\lambda,$$
$$\overline{B} = \int w(\lambda) b(\lambda)\, d\lambda \tag{11.65}$$

and are to be normalized according to

$$R = \frac{\overline{R}}{\overline{R}+\overline{G}+\overline{B}},\quad G = \frac{\overline{G}}{\overline{R}+\overline{G}+\overline{B}}\quad B = \frac{\overline{B}}{\overline{R}+\overline{G}+\overline{B}} \tag{11.66}$$

The following Figure 11.57 shows the spectral sensitivity distributions $r(\lambda)$, $b(\lambda)$ and $g(\lambda)$ for some detector types. In particular the low sensitivity of the human eye for blue should be noticed.

The three spectral matching functions have a small overlap, but are not linearly dependent and allow the formation of a colour space with three coordinates. By a weighted addition new colours can be mixed. Then the weights are the colour values referred to the basic system used. There are numerous basic systems with their own advantages and disadvantages but in principle they are all equivalent. The following three systems are distinctive and most important for practical applications. Their colour stimulus functions are shown in Figure 11.58.

- The LMS system, which is equivalent to the natural sensor physiology of the human eye for the cones.
- A normalized XYZ system, which obeys special optimality conditions.
- The RGB system, which is used preferably for technical colour processes.

The conversion of all these colour specification systems can be carried out with matrix equations due to the fact that all colour coordinates can be superimposed linearly. Simple

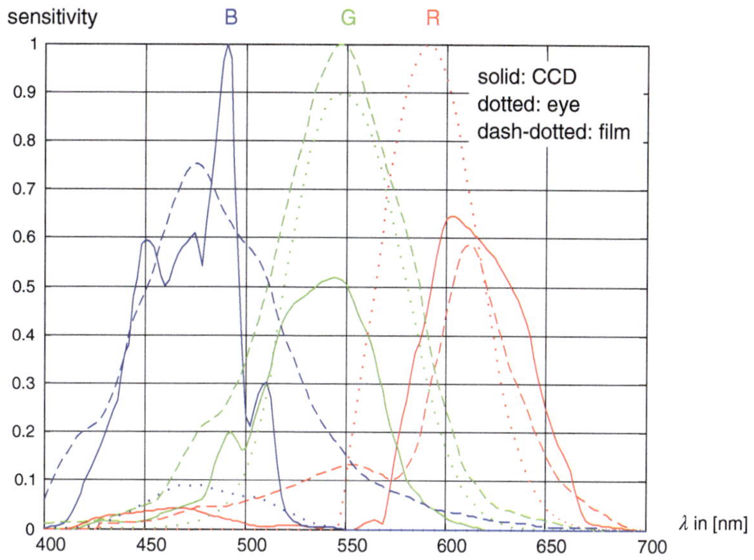

Figure 11.57 Spectral sensitivity curves of a CCD, a film and the human eye.

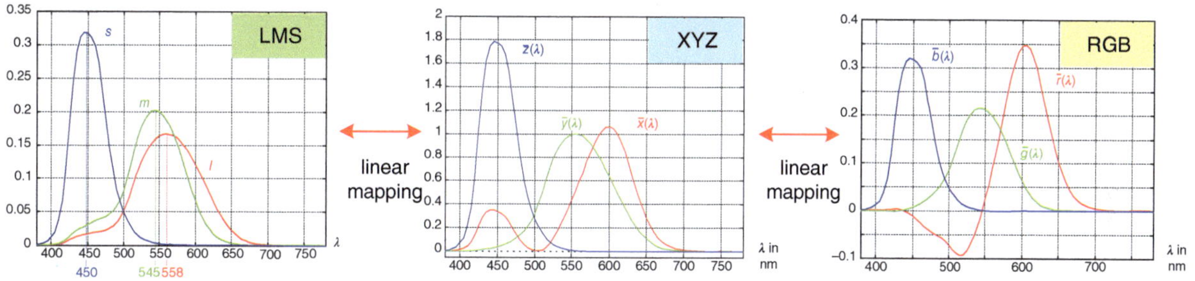

Figure 11.58 Transition between different basic colour valence systems.

3×3 transformation matrices allow the change from one system to the other. An example is given with the equation

$$\begin{pmatrix} X \\ Y \\ Z \end{pmatrix} = \begin{pmatrix} 2.7689 & 1.75175 & 1.13016 \\ 1 & 4.5907 & 0.06010 \\ 0 & 0.05651 & 5.5943 \end{pmatrix} \cdot \begin{pmatrix} R \\ G \\ B \end{pmatrix} \tag{11.67}$$

which converts a RGB-vector into a XYZ-colour vector.

In the image calculation, the individual PSF functions for every colour must first be calculated for one wavelength. Then the polychromatic PSF is calculated by an incoherent superposition as

$$I_{PSF}^{(poly)} = \sum_{R,G,B} w(\lambda) \cdot I_{PSF}(\lambda) \tag{11.68}$$

11.6 Image Quality Metrics

11.6.1 Introduction

There are three different categories of objective image quality metrics [54–60].

- Full reference metric.
  Here only the comparison of an assessed image to an original perfect image is performed. One of the problems with this type of metric is the alignment of the images in case of geometrical transforms like translation, rotation or distortion.
- No-reference metric.
  Here no reference information is needed, it is a kind of absolute quality measurement. The difficulty here is the distinction between regular content and degradation effects. The experience and context of human perception can barely be taken into account.
- Reduced reference metric.
  In this case, special features are extracted and are compared.

If an optical system is imaging an object (O), this first gives a physical image (P). This can be seen by the human observer (V). Alternatively, it is detected by a sensor and processed. A digital image (D) is obtained, which can also be recognized by the visual system of the observer. This scheme is sketched in Figure 11.59. If now an image quality is defined, the question is on what level of this chain the assessment takes place. In particular it makes a difference whether the digital processing and the visual perception with human brain processing are taken into account or not.

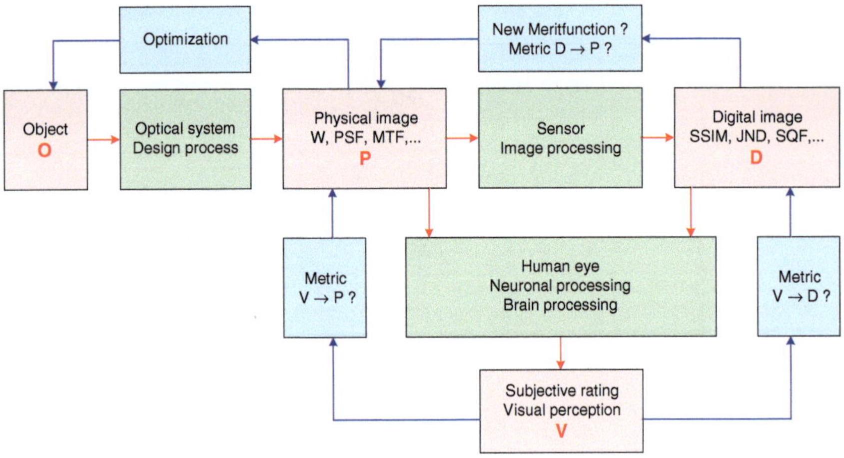

Figure 11.59 Processing of an image physical by the optical system, the digital processing and the human brain.

There are different basic approaches proposed in literature to define image quality metrics.

- Error visibility techniques, which are based on the human visual system (HVS) with its specific image processing capabilities and properties. These approaches are also called feature-based and are mainly heuristic.

  Empirical experience concerning the human visual perception is incorporated into the metric definition, for example orientation and scale decomposition, contrast sensitivity, masking capabilities and pooling strategies of the eye.
- Structural similarity methods.
  In these methods, the quality is quantified by statistical comparison of different features between the test and the reference image. One prominent example is the SSIM criterion with its modifications.
- Techniques that consider information theoretical aspects. Here the principle extractable information content of an image is analysed and measured.

In practice, there are many reasons for the degradation of images. Some prominent types of degradation are:

- blur, resolution decrease;
- noise, different types;
- contrast reduction;
- illuminance, brightness variations;
- geometric changes, shift, rotation, zoom, distortion;
- artefacts, mostly localized, for example stray light;
- changes of colour balance.

In principle, there are three hierarchical levels for image quality assessment to evaluate an optical imaging system.

- Wavefront errors in the system pupil, rms value.
- Quantitative image resolution and contrast measures in the physical image formation. One has to distinguish depending on the geometry:
  - point object: PSF, Strehl, PiB, moments, width;
  - edge and line objects, edge criteria, LSF, ESF;
  - regular grating like structures: MTF, PTF, Hopkins factor, SQF.
- Subjective visual quality, taking the digitalization as well as the psycho-physical properties of the sensor and in particular the human image processing into account:
  - MSE (mean square error);
  - JND (just noticeable difference);
  - SSimM (structural similarity index).

Principal problems, which are not addressed sufficiently by these metrics until now, are:

- supra-thresholding;
- pooling other than Minkowski approaches;
- especially incorporating eye tracking and image resting times due to saccadic movements;
- geometrical limitations.

Therefore, until now, the correlations between these image quality metrics have not been satisfactory in any case. For the assessment of moving images and videos, special approaches are necessary that address the fast changes of the visual impression by a movie.

Here all these special ideas and approaches are not discussed in detail; the reader is referred to the literature on this topic. Only some major terms are briefly explained in the following.

11.6.2 MSE Criterion

The most simple error metric is the MSE. If the image as well as the original object is given in N digital values on a discretized matrix, the MSE is defined as the variance [60]

$$MSE = \frac{1}{N}\sum_{j=1}^{N}\left(I_j - I_j^{(ideal)}\right)^2 \tag{11.69}$$

The MSE is additive for independent error sources and it is independent of the sign of the error. The MSE is independent of the spatial ordering of the pixels. Due to the following reasons, the MSE value is not adequate for describing the human image quality perception.

- Small geometrical deviations like rotation, shift, zooming or distortion cause quite large MSE values.
- MSE is independent of the absolute signal strength.
- There is an equal weighting for all pixels.

If $N_q = 2^M - 1$ is the dynamic range of the pixel intensity quantization values, the signal to noise ratio (SNR) is related to the MSE value by

$$SNR = 10 \cdot \log_{10} \frac{N_q^2}{MSE} \tag{11.70}$$

There are quite simple examples which show a quite different subjective image quality but an identical MSE value. This is demonstrated in Figure 11.60. According to the MSE all the degraded images are of the same quality, but on visual inspection, they are assessed quite differently. Especially a change of contrast seems to be no problem for the HVS as shown in (b), while the blur in part (e) JPEG compression in part (f) of the figure is rated quite low in quality.

11.6.3 SSim Criterion*

(*) The content of this section is quite specialized and for advanced reading. Knowledge of this section is not necessary to understand the following chapters. Beginners are recommended to skip this section.

11.6.3.1 Introduction

Image regions of a strong structured object detail (pattern, texture) are usually not very affected by white noise. On the other hand, smooth regions are disturbed considerably by small values of noise. This is only a simple example for the insufficient properties of simple criteria like the MSE. To assess image quality more realistically, a model must be used that is adapted to the HVS. There are the following steps that must be considered.

- Source and sensor.
  Transmitter of the signal, registration of the image, colour space transform and low pass filtering by the optical system of the eye.
- Transmission part.
  Decomposition into several channels, according to the visual cortex system of the neurons there are channels with different frequencies and orientations, this corresponds to a wavelet transform.
- Normalization and error pooling.
  Pre-processing, comparison with reference signals, normalization of signal strength due to the contrast sensitivity function (CSF) of the eye.
  Masking of details by reduced visibility in the direct neighbourhood in the spatial and frequency domain.
  Adaptive spatial weighting of individual signals and compression.
- Fidelity.
  Comparison with prior information and experience.

One famous approach towards this goal is the structural similarity index measure (SSim) [5]. To validate the usefulness of this more complicated measure, clinical trials are worked out to compare real human perception with the resulting numbers of a criterion.

11.6.3.2 Definition and Computation

The SSim is a quite generic image fidelity criterion. It is quite a good match to human visual perception. Since

Figure 11.60 Images degraded through different reasons with the same value of MSE. Part (a) shows the original picture, the parts (b) to (f) degraded versions according to the data given.
Source: Dwight Hooker / University of Southern California.

the eye tries to extract structural information out of an image, the SSim compares local patches for similarity. Non-structured distortions are for example luminance, shift, contrast. The eye compensates for them; they are often caused by environmental effects and do not contain structural information. Structural distortions contain information, for example, blur, noise, compression. The human eye is quite sensitive to this type of deviation. In the SSim, the following three terms are locally compared:

- luminance/brightness $l(x,y)$;
- contrast $c(x,y)$;
- structures $s(x,y)$.

If μ_j are the values in the pixels, σ_j the standard deviations and cross correlations respectively, the SSim is defined as a function of the image coordinates or alternatively the pixel indices j by

$$S(x,y) = l(x,y) \cdot c(x,y) \cdot s(x,y) = \frac{2\mu_x\mu_y + C_1}{\mu_x^2 + \mu_y^2 + C_1} \cdot \frac{2\sigma_x\sigma_y + C_2}{\sigma_x^2 + \sigma_y^2 + C_2} \cdot \frac{\sigma_{xy} + C_3}{\sigma_x\sigma_y + C_3} \tag{11.71}$$

The C_k are small value constants, they guarantee regularized values for S. S is bounded between −1 and +1. A value of $S = +1$ corresponds to identical images.

The complete information as a result of a comparison is a matrix, which is determined by a sliding window procedure and has values for all pixels. By averaging on ranges or the complete image, the finite extension of structures is taken into account. The size of the sliding window is in principle a parameter that can be selected in certain ranges. It is seen in practical examples that especially structured deviations have a large influence on the S-criterion, while non-structured deviations influence S only weakly. This means that the SSim can be determined locally resolved, clustered or global for an image. This gives a good flexibility to assess also local degradations.

The SSim shows a quite good correlation to human visual perception except for a large sensitivity for zooming, shift and rotation.

11.6.3.3 Examples

Figure 11.61 shows some degraded images with the MSE, SSim criteria as well as a simple image CC. For the MSE, a square root is incorporated here to get a more comfortable normalized value. In addition, for every image the map of local SSim deviations is shown, this indicated the positions where the discrepancies are smaller or larger. A comparison of these cases is summarized in

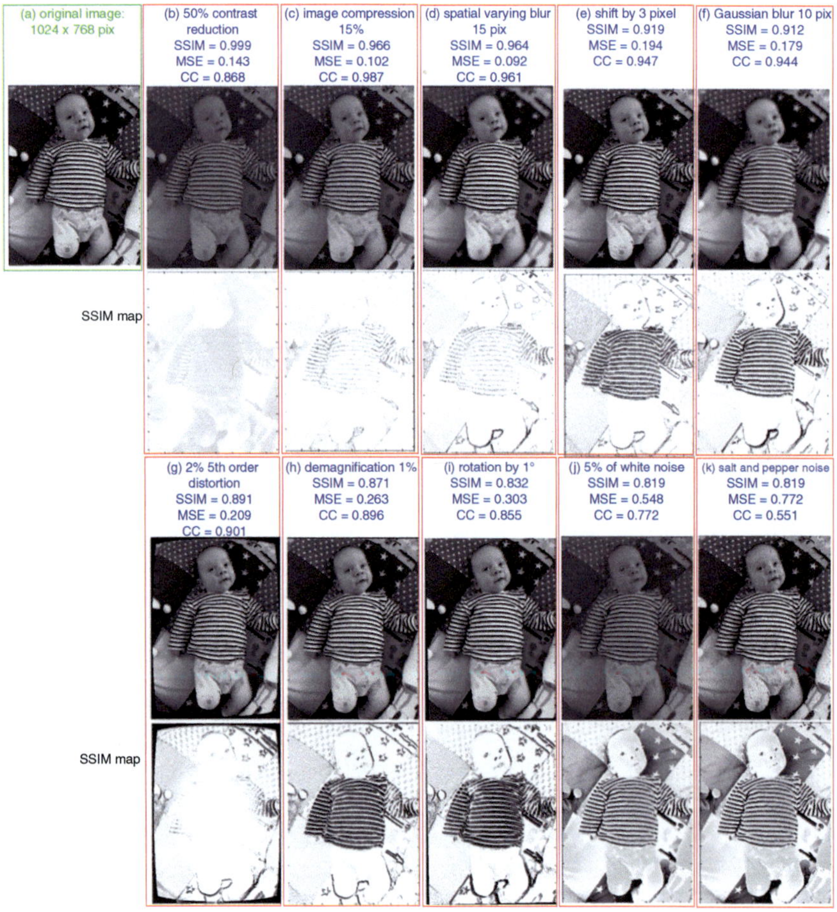

Figure 11.61 Comparison of degraded images of the object shown in (a) for different reasons and values and concrete calculation results of SSim, correlation and the normalized root of MSE. The perturbation types and values are indicated in each headline in parts (b)–(k). In addition to the images the change maps of the SSim are given, which indicate the localizations of the deviations inside the FoV. The cases are ordered according to a decreasing SSim.

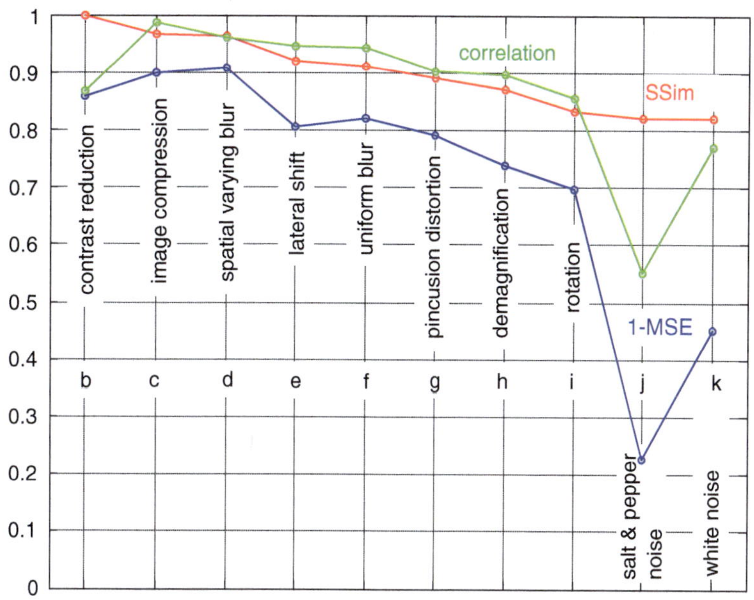

Figure 11.62 Comparison of the criteria values of Figure 11.61. The cases are ordered by the SSim values and follow the sequence in Figure 11.61. It should be noted that instead of the usual MSE the normalized 1-MSE is shown to get an ideal reference of 1.

Figure 11.62. There are several typical insights visualized in this example.

- Image compression and contrast reduction gives a rather uniform degradation as expected.
- In the case of higher order distortion, the changes are mostly located at the border.
- The largest difference between the criteria is seen according to Figure 11.62 for noise, here the SSim is much more tolerant.
- Contrast changes are nearly not penalized by SSim.
- For blur as well as the geometrical changes like magnification, shift, rotation and distortion, SSim and correlation are quite close.

It should be mentioned here that the types of degradations can be separated into:

- structural distortions: noise, blur, compression;
- non-structural distortions: contrast, spatial deformations like distortion.

In this particular example, the structure of the image shows very different regions. Especially the black-white stripes give very different values of the SSim in comparison to the more natural areas of the picture. This illustrates one of the major problems of all criteria: the type of image structure matters.

Very often where the locations of the image errors are found and what kind of information is lost is of importance. In the case of the SSim and the MSE, a spatially resolved map can be calculated by a sliding window technique. Figure 11.63 illustrates an example of a blurred image and a comparison. Both maps are very similar and show in this special case mostly the edges, which are smeared out. The maps therefore look like a gradient card of the image [5]. This is observed in Figure 11.61 as well.

11.7 Digital Image Restoration

11.7.1 Introduction

The improvement of images in digital format is a quite an important topic nowadays. Thanks to the growing computer resources and better resolving sensors, this can

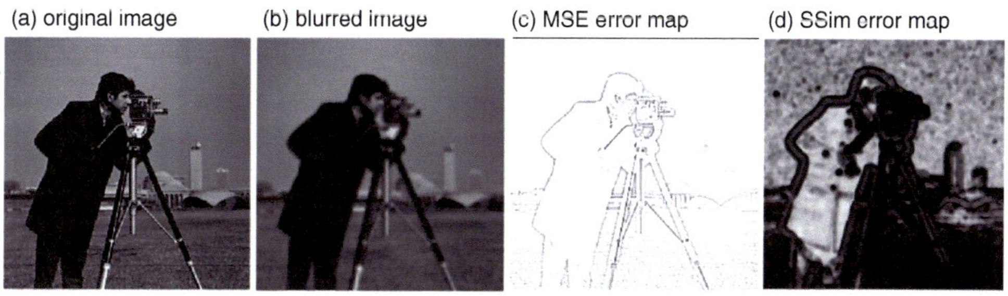

Figure 11.63 Image shown in (a) with spatially resolved maps of the (d) criteria SSim and (c) MSE in case of a blurred image shown in (b). Source: Ali Alzahrani (2022) / John Wiley & Sons / CC BY 4.0.

be done mostly in real time today. The term restoration contains several aspects, which must be distinguished [36, 61, 62].

- Denoising.
  Here only statistical degradations due to noise are removed. This improves the appearance of an image but should not change its information content. The resolution is not improved.
- Deconvolution or deblurring methods describe a special class of restoration, where dominantly the PSF caused blur is removed. Due to the filtering effect of the methods, usually a denoising is performed simultaneously.
- Restoration.
  If a deterministic degradation due to distortion or blur is present, the hidden information or perturbed geometry can be restored to a certain extent. In the best case, the perturbations are well known; for example, the variation of the blur due to the impact of aberrations onto the PSF. This means that additional information about the system properties is used to deconvolve the image and improve the resolution. In reality, noise usually comes on top, so deconvolution without noise is academic.

There are a large number of methods proposed in the literature to perform corresponding restoration calculations. The intention here is only to give a short overview and to illustrate with some examples typical results. For a more mathematically rigorous and detailed representation, the reader is referred to the literature. In sorting the various methods, first iterative and non-iterative methods must be distinguished. In Table 11.2 some important approaches are listed to give an overview without further details and explanations.

In general, the image degradation can be written as an integral equation with an additional noise term. For simplification only one transverse coordinate x is used here.

$$I_{ima}(x) = \int I_{psf}(x,x') \cdot I_{obj}(x,x')\, dx' + I_{noise}(x) \qquad (11.72)$$

If the system can be considered to be isoplanatic with invariant PSF, the equation simplifies to a convolution form

$$I_{ima}(x) = I_{psf}(x) * I_{obj}(x) + I_{noise} \qquad (11.73)$$

in the spatial frequency domain

$$I_{ima}(\nu) = H_{OTF}(\nu) \cdot I_{obj}(\nu) + I_{noise} \qquad (11.74)$$

The easiest approach to deconvolve the effect of the PSF broadening is a linear system theory filter in the frequency domain. This is the classical Wiener method, it is then assumed the PSF or OTF are known. The method is linear and non-iterative. A simple algebraic method uses a singular value decomposition or PCA of the image. This is a non-iterative approach too. By truncating the singular value vector, a sparse representation can be achieved to reduce the numerical burden.

Table 11.2 List of methods for image restoration.

	Method	Properties
1	Direct non-iterative	
1.1	Wiener, inverses Fourier filtering	Bad conditioned, Tikhonov regularization recommended
1.2	SVD-base method	Algebraic methods
2.	Iterative non-linear methods	
2.1	Landweber, error reduction, successive approximation	General equation: $x_{k+1} = x_k + \tau \cdot (b - Ax_k)$
2.2	Steepest descent	Gradient NLSQ method, slow convergence
2.3	POCS method	Projection P in convex space $AP\vec{x} = P\vec{b}$
2.4	Maximum likelihood (ML)	Maximum entropy function
2.5	Lucy-Richardson	Error reduction algorithms for ML-functional
2.5	Blind deconvolution	Restoration without knowledge of PSF
2.6	Simulated annealing	Global optimization, slow

Non-linear optimization-related methods can improve the deterministic as well as the stochastic degradation reasons simultaneously. They are iterative and therefore take a longer computational time. Here a non-least square functional is minimized, taking certain constraints into account. The calculation can be performed in the spatial or in the frequency domain; often a certain regularization and pre-conditioning is comfortable to accelerate the computation. Since only forward calculations are performed, noise is not amplified in these cases. Usual boundary conditions that must be fulfilled are as follows.

- Object intensity positive.
- Smallest derivative for an optimal damping of noise, especially suppressing pixel noise.
- Maximization of the entropy, this guarantees avoiding artefacts like ringing structures.
- Keeping the overall energy constant.
- Allowing image signal only inside a certain outer boundary.

It should be mentioned that it is also possible to restore an image blind.

11.7.2 Deconvolution

11.7.2.1 Wiener Method

The Wiener method is a simple inverse filtering process using a known PSF. In the easiest formulation it follows Eq. (11.74) ignoring the noise term as a linear equation in the Fourier space as

$$I_{obj}(\nu) = \frac{I_{ima}(\nu)}{H_{OTF}(\nu)} \tag{11.75}$$

or explicitly in the spatial domain

$$I_{obj}(x) = \hat{F}^{-1}\left[\frac{\hat{F}[I_{ima}(x)]}{\hat{F}[I_{psf}(x)]}\right] \tag{11.76}$$

Due to this approach, it can be seen that isoplanatism is assumed in this method. The problems with this simple approach are as follows.

- The equation fails in the case of zeros of the transfer function H_{OTF}, which occurs in the denominator.
- Any noise is amplified at locations with small contrast values.
- This linear inverse filtering method in the simple formulation of Eq. (11.75) does not in any case generating positive values for the image intensity.

If the Wiener filtering is represented in the frequency space, it can be seen that due to the typical shapes of behaviour of a PSF, the larger spatial frequencies are amplified. This can be seen in the Figure 11.64. The increase of the filter spectrum towards larger frequencies means that the high frequency noise parts are amplified in comparison to the signal parts, which are usually convergent for larger frequencies. In the figure, a uniform white noise is assumed.

To overcome the ill-conditioning at the OTF zero points, a regularization in the denominator can be used according to Tikhonov. But this approach simultaneously has a filtering effect. This means in addition to the denoising the signal is affected. Therefore, the choice of the regularization constant is critical. If the equation is multiplied by the complex conjugated OTF

$$\begin{aligned} H^*_{OTF}(\nu) \cdot I_{ima}(\nu) &= H^*_{OTF}(\nu) \cdot I_{obj}(\nu) \cdot H_{OTF}(\nu) \\ &= |H_{OTF}(\nu)|^2 \cdot I_{obj}(\nu) \end{aligned} \tag{11.77}$$

the solution reads

$$I_{obj}(\nu) = \frac{H^*_{OTF}(\nu) \cdot I_{ima}(\nu)}{|H_{OTF}(\nu)|^2} \tag{11.78}$$

here the regularization adds a constant value μ in the positive denominator to reduce the filter wings and stabilize the calculation

$$I_{obj}(\nu) = \frac{H^*_{OTF}(\nu) \cdot I_{ima}(\nu)}{|H_{OTF}(\nu)|^2 + \mu} \tag{11.79}$$

It is observed in practice that the choice of the regularization parameter is critical. If μ is too small, a granulation is observed, if μ is too large, a strong filtering blurs the object. Therefore, a more empirical optimized value for μ is mostly used. This is visualized in Figure 11.65 and Figure 11.66. It can be seen that a value of μ which is too small is not able to denoise the image; the resultant picture gets a texture. If μ is too large, the noise is gone but the deblurring cannot be achieved. The optimal value for the regularization depends on the blur as well as the noise in the image and is therefore hard to predict from the image itself in advance.

11.7.2.2 Algebraic Deconvolution*

(*) The content of this section is quite specialized and for advanced reading. Knowledge of this section is not necessary to understand the following chapters. Beginners are recommended to skip this section.

If the image to process is given as a discretized matrix, and due to the fact that in the case of an invariant PSF the

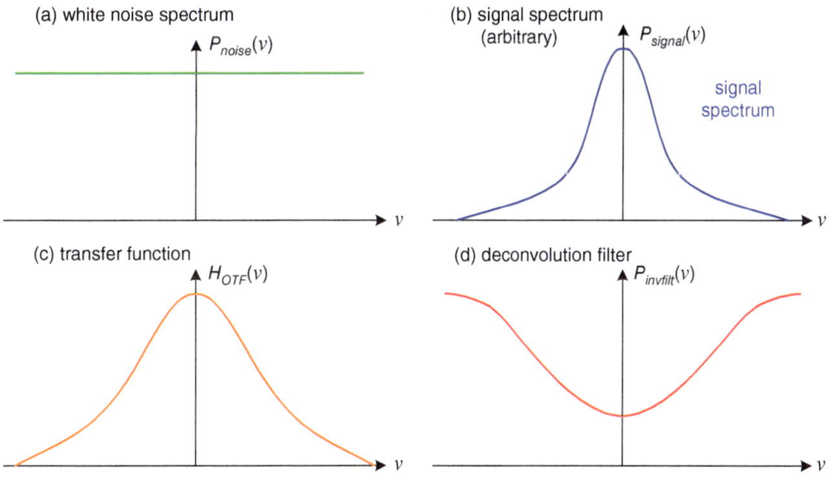

Figure 11.64 Schematic visualization of the Wiener filtering process in the frequency space. (a) shows the spectrum of white noise, (b) the assumed arbitrary signal spectrum, (c) the transfer function and (d) the deconvolution filter function as the inverse value of the OTF.

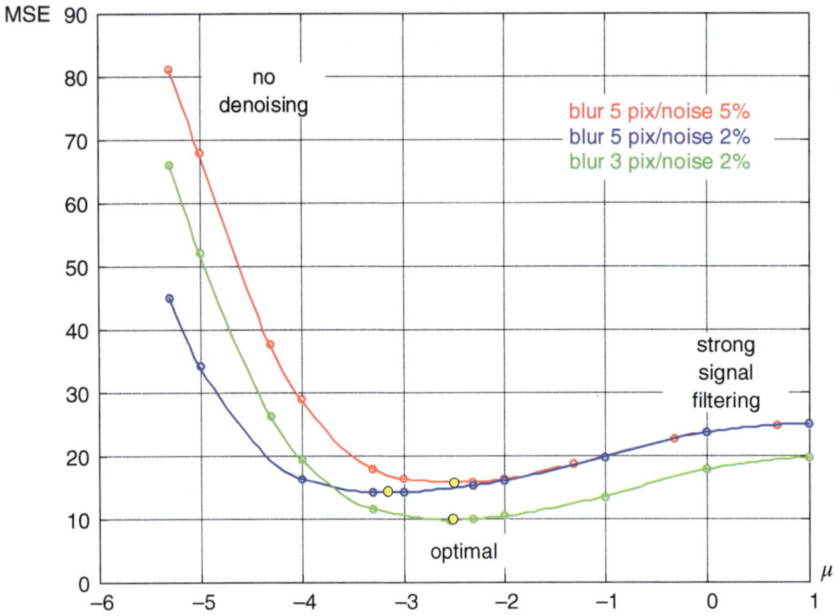

Figure 11.65 Three cases of Wiener deconvolution with varying regularization parameter μ. The corresponding images are seen in Figure 11.68. The trade-off between denoising and deblurring is clearly seen.

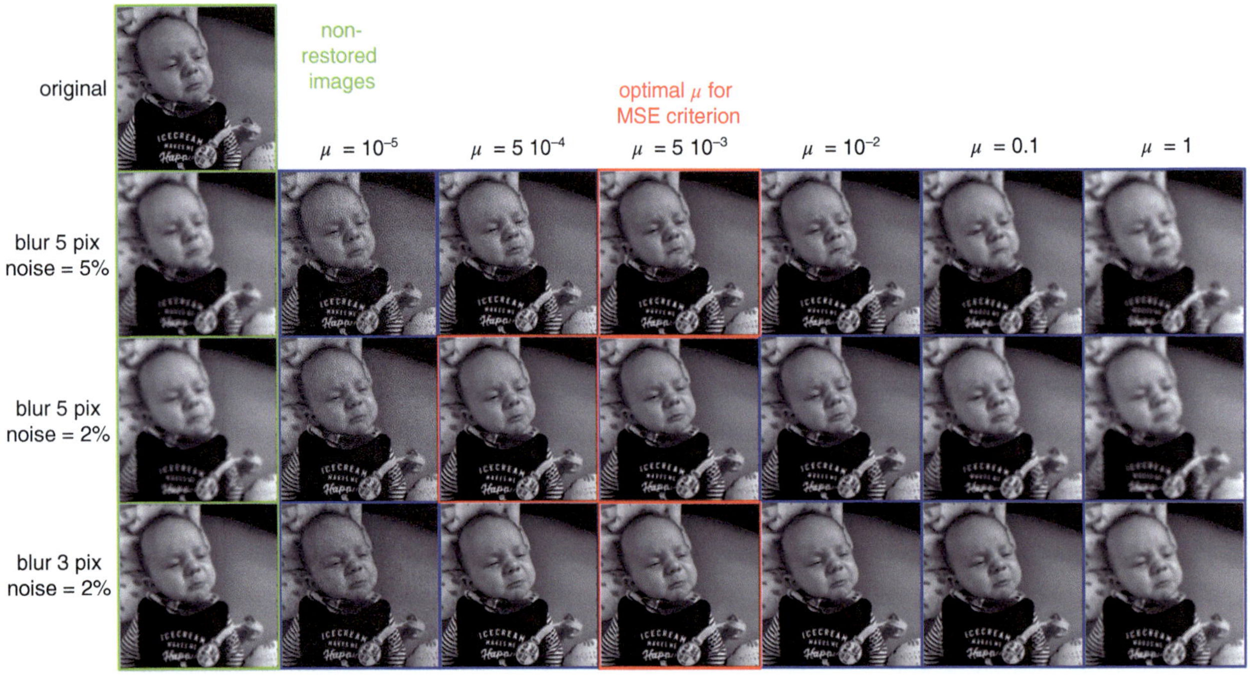

Figure 11.66 Three cases of Wiener deconvolution with varying regularization parameter μ. The trade-off between denoising and deblurring is clearly seen. In the rows two to four different combinations of blur and noise are applied. The images with the green frame are not processed, the images with the red frame are optimal from the inspection of a MSE criterion corresponding to Figure 11.67.

equation in the frequency domain is linear

$$I_{ima}(\nu) = H_{OTF}(\nu) \cdot I_{obj}(\nu) + I_{noise} \tag{11.80}$$

the problem can be formulated as an algebraic linear system as follows. The image is assumed to be a matrix with N^2 pixels. The above equation can then be written in a discretized form as a matrix equation

$$I_{j,k}^{(ima)} = \sum_{j'}\sum_{k'} H_{j-j',k-k'}^{(OTF)} \cdot I_{j',k'}^{(obj)} + I_{j,k}^{(noise)} \tag{11.81}$$

by remapping the two-dimensional matrices onto one-dimensional vectors, the equations reads

$$I_{j}^{(ima)} = \sum_{k} H_{j,k}^{(OTF)} \cdot I_{k}^{(obj)} + I_{j}^{(noise)} \tag{11.82}$$

where the matrix H has the dimension N^4. This equation corresponds to a usual linear system of equations

$$\begin{aligned} \vec{g} &= \underline{A} \cdot \vec{x} + \vec{n} \\ \underline{A} \cdot \vec{x} &= \vec{b} = \vec{g} - \vec{n} \end{aligned} \tag{11.83}$$

with the formal solution

$$\vec{x} = \underline{A}^{-1} \cdot \vec{b} \tag{11.84}$$

which can be calculated by a SVD decomposition. If only the significant singular values are considered, the solution usually is truncated and can be calculated faster. In general, this system is ill conditioned and approximate methods for solving it are used. Due to the noise term, which does not belong to the signal, only an approximate solution ca be obtained

$$|\underline{A} \cdot \vec{x} - \vec{b}\,|^2 = \min \tag{11.85}$$

The additional constraints can be added to this non-least square problem by formulating them as Lagrange multipliers.

11.7.2.3 Further Methods*

(*) The content of this section is quite specialized and for advanced reading. Knowledge of this section is not necessary to understand the following chapters. Beginners are recommended to skip this section.

If the PSF is not invariant, but changes with field position, the restoration is more complicated [36, 51–62]. The general imaging equation reads in one transverse dimension

$$I_{ima}^{(real)}(x) = I_{noise}(x) + \int I_{obj}(x,x') I_{psf}(x',x)\,dx' \tag{11.86}$$

To solve this equation, an iterative method can be used with an appropriate goodness-of-fit criterion. One approach is to expand the image into a Taylor series [46] as explained in Section 11.1.5. Unfortunately, the calculation of the intensity moments of higher order is not very accurate in the case of noise. The solution of the matrix equation is proposed by different authors together with noise reducing filtering steps.

If the usual equation of imaging is written in the spatial domain with a noise term as

$$I_{ima} = H_{OTF} * I_{obj} + I_{noise} \tag{11.87}$$

it can be iteratively solved by an error reduction method according to

$$I_{ima}^{(k+1)} = I_{ima}^{k} + \left[I_{ima}^{k} - H_{OTF} * I_{obj}^{k} \right] \tag{11.88}$$

Since a simple error reduction method has a slow convergence, Landweber has proposed the iteration equation in the object space, which provides better numerical properties

$$I_{obj}^{(k+1)} = I_{obj}^{(k)} + H_{OTF}^{-1} * \left[I_{ima}^{(k)} - H_{OTF} * I_{obj}^{(k)} \right] \tag{11.89}$$

Furthermore, there are also methods proposed in literature which are based on wavelet filtering.

One very famous algorithm to deconvolve an image is the method of Lucy–Richardson. It is non-linear and iterative and follows the equation

$$I_{obj}^{(k+1)}(x,y) = I_{obj}^{(k)}(x,y) \cdot \left[I_{psf}(-x,-y) * \frac{I_{ima}(x,y)}{I_{psf}(x,y) * I_{obj}^{(k)}(x,y)} \right] \tag{11.90}$$

By introducing a reduced step width with an additional parameter τ, the iteration is more robust but slower

$$I_{obj}^{(k+1)}(x,y) = I_{obj}^{(k)}(x,y) \cdot \left[1 + \tau \cdot I_{psf}(-x,-y) * \frac{I_{ima}(x,y)}{I_{psf}(x,y) * I_{obj}^{(k)}(x,y)} \right] \tag{11.91}$$

It should be further mentioned that, in particular in microscopy, very often three-dimensional images are recorded. The depth resolution is typically worse, so the overall blur is anisotropic in the lateral and axial direction. Corresponding methods for an appropriate three-dimensional restoration under these conditions have also been developed; the reader is referred to the literature.

11.7.2.4 Examples

Again, the purpose of this chapter is not to provide the reader with a comprehensive and rigorous presentation of image processing restoration mathematics and theory. But some examples to illustrate the results of a non-expert in restoration and deconvolution gives at least an impression of the large variety of possibilities and shows the complexity of this field.

Figure 11.67 shows the result of a deconvolution with six different algorithms of a technical resolution pattern. Here

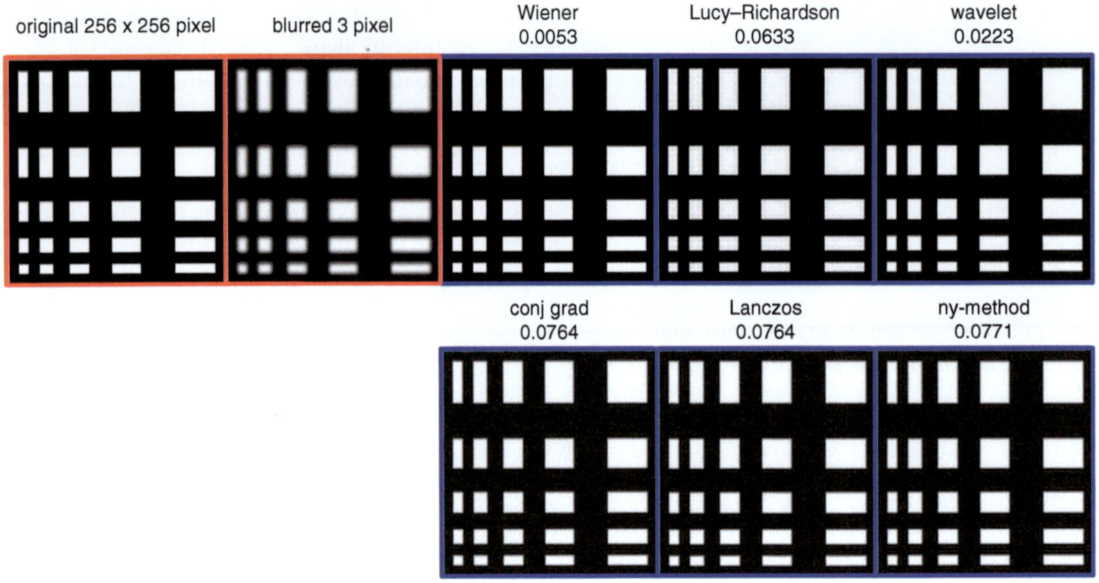

Figure 11.67 Deconvolution of a synthetic image (a) with the blurred version (b) by six different methods, in the lower row algebraic procedures are used. The numbers give the resulting MSE assessment values.

no noise is present to better separate the corresponding effects of deconvolution and deblurring from denoising. The Wiener and Wavelet methods show the best results; in the case of the algebraic approaches, which are also limited by memory consumption of the huge matrix A with dimension N^4 and the Lucy–Richardson method, a residual texture can be observed. Furthermore, several other algorithms are compared here, which are not discussed in detail. These are the Lanczos method, the ν-approach and optimization by conjugate gradients. If noise is included in this example, the rating of the methods will be different.

If a real grey shading image is used as an example as is the case in Figure 11.68, it can be found that the algebraic methods have many more problems achieving reliable results, therefore they are skipped here. Again, the Wiener filtering and the wavelet filtering provide the best results.

11.7.3 Denoising

There are many methods and algorithms to remove or reduce the statistical noise in images. The deblurring and restoration methods usually also filter noise out of the image. But there are some methods with the primary goal to reduce noise level. An improvement of resolution cannot be expected in these cases, Some of them are as follows.

- Frequency filtering by damping of the Fourier spectrum. The image is Fourier transformed, the spectrum is hard truncated or a smooth window function removes high frequency parts. The shortened spectrum is back transformed in the spatial domain.
- Filtering by wavelets.
  The image is expanded into wavelets with a truncated level, this truncated spectral representation is back transformed.

Figure 11.68 Deconvolution of a real image by three different methods with the corresponding MSE values.

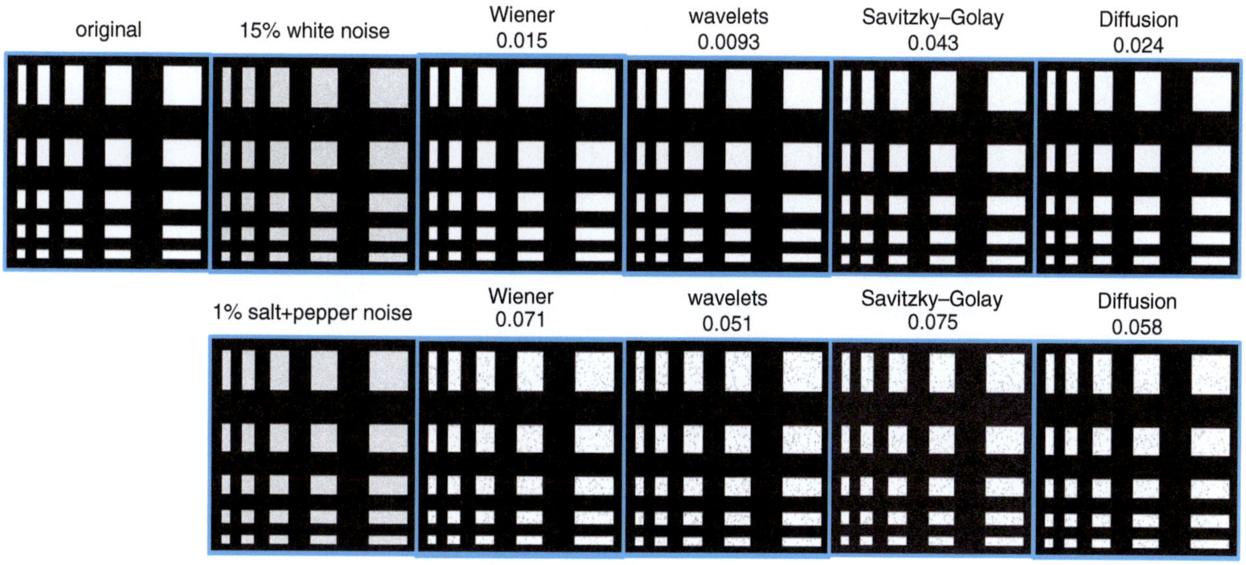

Figure 11.69 Denoising of a geometrical object with white noise (upper row) and salt and pepper noise (lower row) by Wiener filtering, wavelet filtering, polynomial filtering and diffusion.

- Filtering by polynomials.
  A sliding window is scanned over the image. For a rather small number of pixels, a low order polynomial is fitted to the data and the pixel values are substituted by the interpolated values. The size of the window function as well as the selected order of the polynomial are parameters. One prominent example is the Savitzky–Golay approach.
- Anisotropic diffusion.
  The image is smoothed by the diffusion equation. By forcing anisotropic diffusion constants, sharp edges can be preserved.

Due to the different methods for denoising an image, the various parameters to adapt the noise filters and the different types of noise give a huge range of options. It is far beyond the scope of this chapter to go into all these details. A few examples should give an idea of the opportunities and the complexity of establishing a perfect denoising. Figure 11.68 shows a geometrical image with white as well as salt and pepper noise. The noisy images are processed with the four main approaches mentioned above: the Wiener method, a wavelet filter, the Savitzky–Golay polynomial filter and the diffusion method. This is illustrated in Figure 11.69 for a synthetic pattern and several noise types as well as denoising approaches. The indicated numbers give the resultant MSE error in normalized representation. In this case it is seen, for example:

- the wavelet filter shows the best results;
- the salt and pepper noise is generally a larger problem to remove as white noise;
- the salt and pepper noise is only removed by the wavelets in the black regions;
- the Savitzky–Golay filter affects the average values of the black regions.

Figure 11.70 gives a second example for a natural scene, with white as well as Gaussian distributed noise. The denoising algorithms are the same as above. Here it is seen that:

- the Gaussian noise is somewhat harder to remove than the white noise;
- for white noise, all the four algorithms have nearly the same quality of result;
- for the Gaussian noise, the diffusion denoising is the most comfortable.

11.8 Plenoptical Imaging

(*) The content of this section is quite specialized and for advanced reading. Knowledge of this section is not necessary to understand the following chapters. Beginners are recommended to skip this section.

11.8.1 Introduction

The principle of plenoptical imaging is a very special and interesting method of a combined analogue–digital imaging mode, which has the attractive property of a digital refocusing option. The underlying system uses a lenslet array in the image domain; several different versions exist for the best locations of this component [63–77]. The main idea is very old and has been reinvented several times in

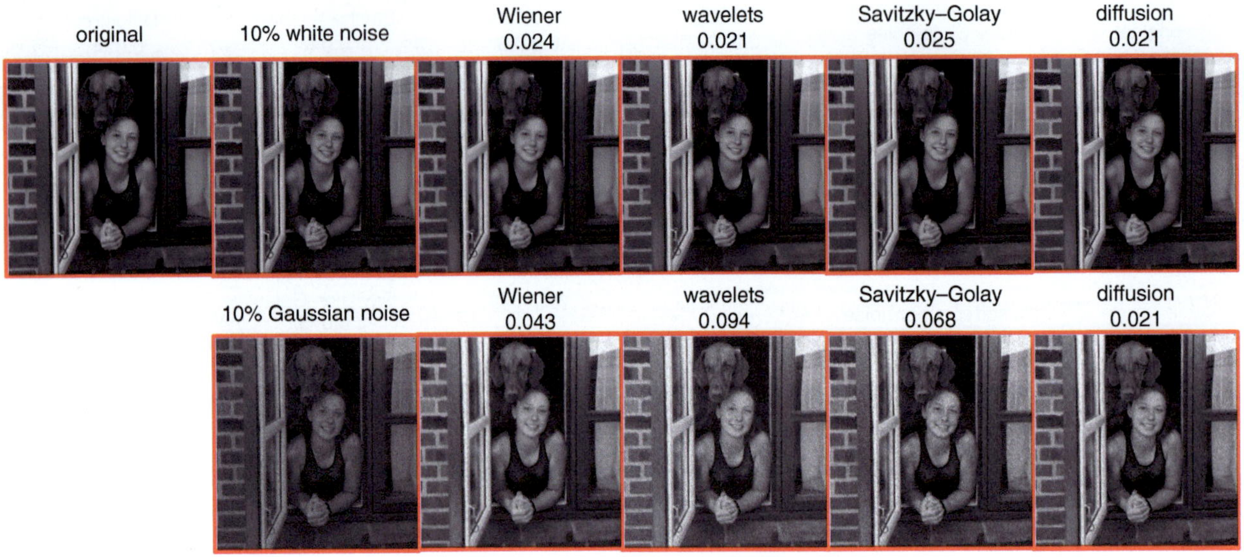

Figure 11.70 Denoising of a real image with white noise (upper row) and Gaussian noise (lower row) by Wiener filtering, wavelet filtering, polynomial filtering and diffusion.

the literature; it is well known as plenoptical, integrated or light-field imaging. Some preliminary ideas and properties have already been presented in Section 4.8.5.4.

The array with a small focal length allows the gathering of the rays in a plane with mixed spatial-angle distribution. This is also the reason that the principle can be well illustrated in the phase space and changes of the Wigner function are used to describe the properties. In a classical imaging mode, every pixel of the sensor captures the rays coming from one image point, but the angle variation inside the numerical aperture does not carry object information. This is different in the integrated imaging concept. Since the spatial-angle balance is changed, if the object distance is altered, the reconstruction of the image out of the raw intensity distribution allows for a digital refocusing into different distances. This can be used for sharp three-dimensional imaging or simply allows us to remove uncomfortable defocusing blur. This means that in integral imaging the full four-dimensional spatial-angular ray field is detected and the angle integration as in classical imaging is modified. By a clever rendering of the light field, the rays are re-sorted. Due to the disparity in angle, different depths can be distinguished and a digital refocusing is possible.

Figure 11.71(a) sketches the basic layout of a plenoptical camera of type 1.0 as the standard model. In the image plane a lenslet array mixes the spatial and the angle properties of the rays. The sensor is then located in the focal plane of the array. As can be seen in part (b) of the figure, the spatial-angle distribution in the sensor plane changes if the distance s of the object point is changed. The combined changes allow us to recover the image sharply for all positions s in a limited range and a limited resolution by digital postprocessing.

As already mentioned, the exact position of the micro lens array allows for different solutions. In particular the

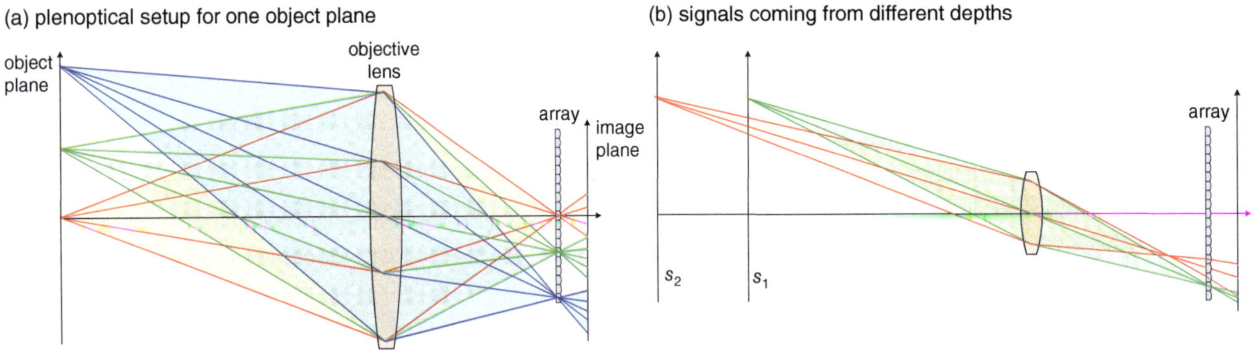

Figure 11.71 Principle of the standard plenoptical camera. (a) shows the main layout with the sensor in the focal plane of the lenslets, (b) illustrates the changes if the object point distance is changed from s_1 to s_2.

following three major setups have been discussed in the literature.

- Version 1.0.
  The array lies in the image plane and the detector is conjugated to the objective pupil.
  The size of the micro-lenses determines the spatial resolution. This case is sketched in Figure 11.71 (a).
- Version 2.0 Keplerian.
  The array serves as a relay with intermediate image plane (Keplerian setup), the sensor is positioned in the image plane of the array.
  The resolution depends on the distances and focal lengths and usually is improved in comparison to the version 1.
- Version 2.0 Galilean.
  The array again is a relay, but in this Galilean setup a virtual intermediate image behind the sensor plane is obtained.

These three options behave differently, in particular the versions 1.0 and 2.0 that are sketched in Figure 11.72. In Table 11.3 some of the major properties of the versions are compared. In particular the behaviour of the lateral resolution and the depth recovery is of interest. For the type 2.0 camera, the magnification of the lenslet array can be varied by position. What changes then is the fill factor of the lenslets and perhaps an overlap.

It should be mentioned that there are some further problems in realizing this ingenious idea in practice. Some of them are as follows.

- The digital reconstruction algorithm should be fast and 'real time'. The direct gathered intensity distribution is meaningless.
- Aberrations of real components in the system.
- Additional effects like vignetting.
- For small aperture angles, additional diffraction spreading effects.
- Edge effects coming from the array boundaries.
- Photometric issues; in the case of reconstruction from different depths, a calibration is needed.

It should be mentioned that there are more setups proposed in literature than these three major arrangements. These are, for example:

- telecentric ray path;
- Fourier approach for microscopy [75, 76];
- negative array lenses.

These special setups are not considered here.

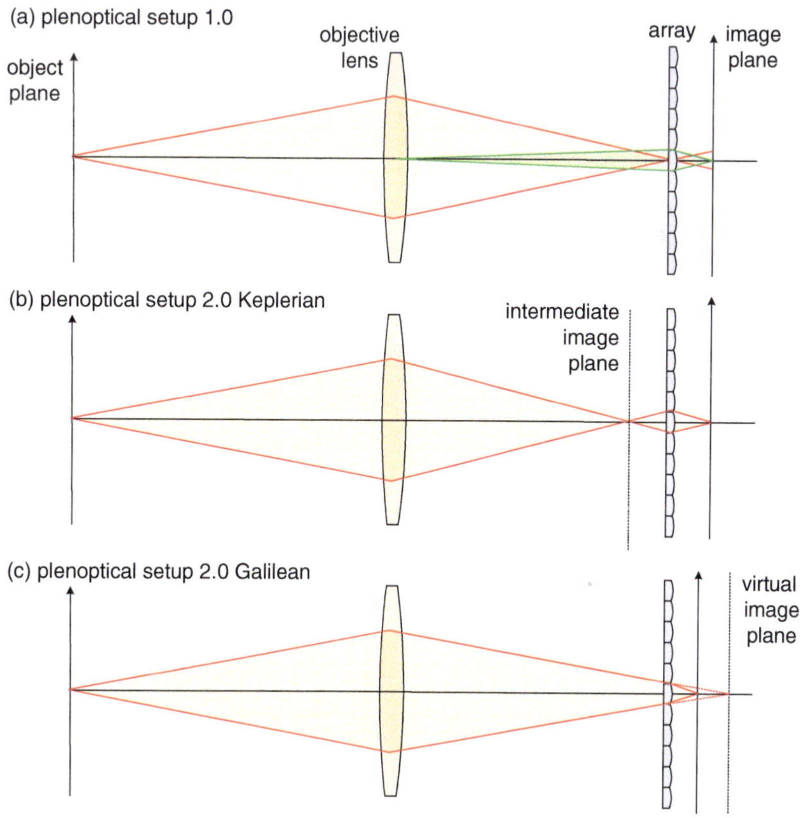

Figure 11.72 Principles of the plenoptical camera in the setups (a) version 1.0 and (b) version 2.0 Keplerian with intermediate image and (c) 2.0 Galilean without intermediate image.

Table 11.3 Comparison of some properties of the plenoptical setups 1.0 and 2.0. Nominal means that this is the usually addressed object distance case, but according to the three-dimensional imaging opportunities, defocused object points can be processed too.

Property	Version 1.0	Version 2.0
Nominal imaging	Defocused (Fourier conjugated)	Focused
Microlens array position	In intermediate image	Conjugated to the systems pupil
Lateral resolution	Small Grows with number of lenslets N_a	Large Depends on aperture and aberrations
Axial recovery range	Large More angular information	Small
Focal length of lenslets f_a	Small: not fully illuminated sensor Large: overlap of micro-images	
Layout versions		Real intermediate image: Keplerian Virtual intermediate image: Galilean

11.8.2 Model of the Plenoptical Camera 1.0

To illustrate how the modelling of a light-field imaging system can be done, here a short derivation is shown for the version of the plenoptical camera 1.0. The basic terms are shown in Figure 11.73.

In the 1.0-version the microlens array with focal length f_a is located in the image plane of the objective lens with focal length f_o and the sensor plane is conjugated to the objective pupil. These relations are illustrated in Figure 11.73(a).

With the distances s, s' of the main objective, for the first imaging stage the equations

$$\frac{1}{s'} - \frac{1}{s} = \frac{1}{f_o}, \quad m_o = \frac{y_i}{y_o} = \frac{s'}{s}, \quad D_p = 2s \cdot \sin u \qquad (11.92)$$

follow. The imaging of the array lenses with $z = s'$ is then described by

$$\frac{1}{z'} - \frac{1}{s'} = \frac{1}{f_a}, \quad m_a = \frac{y_s}{y_i} = \frac{z'}{s'}, \quad D_s = m_a \cdot D_p \qquad (11.93)$$

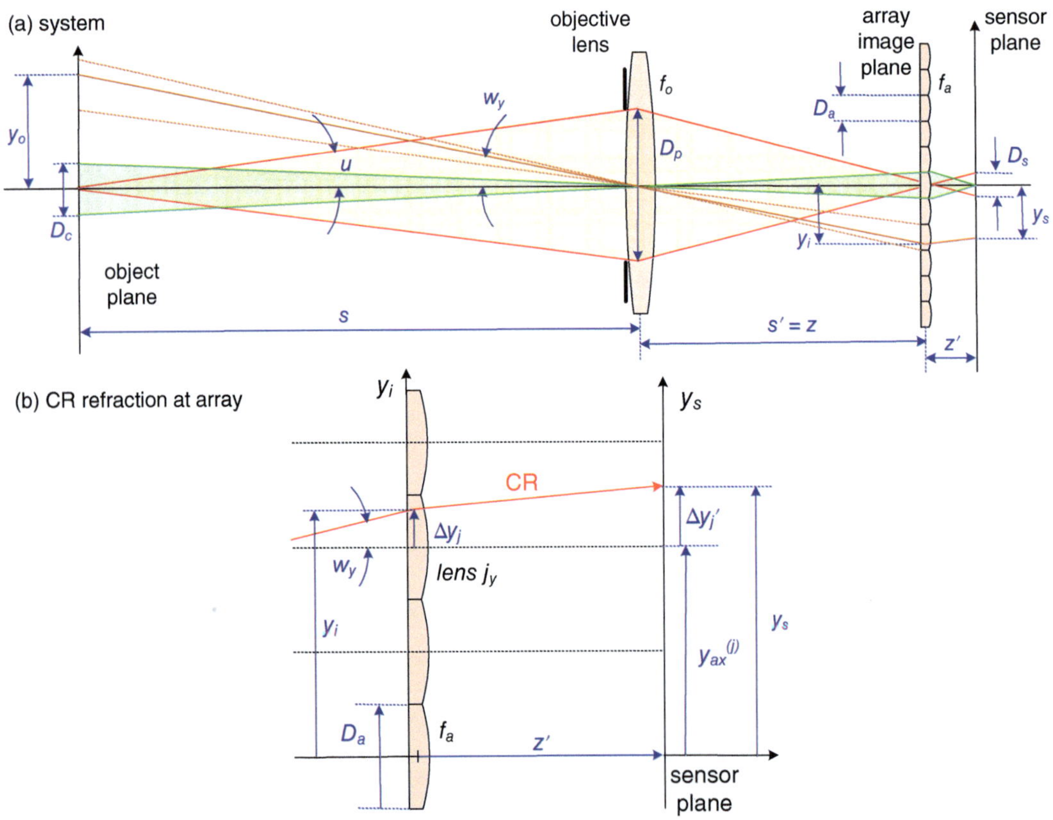

Figure 11.73 Parameter of a plenoptical camera type 1.0 with (a) complete system and (b) chief ray refraction at the lenslet.

and the chief ray with angle w_y meets the final sensor plane at

$$y_s = y_i + z' \cdot \tan w_y = y_i + z' \cdot \frac{y_o}{s} \tag{11.94}$$

As can be seen by the geometrical relations of the system with the lenslet lens diameter D_a, every lenslet captures the size

$$D_c = m_o \cdot D_a \tag{11.95}$$

of the object plane. One object point generates a circle in the final sensor plane of the size

$$D_s = -m_a \cdot D_p = 2s \cdot \sin(u) \cdot \frac{f_a \cdot (s + f_o)}{s \cdot (f_o + f_a) + f_o f_a} \tag{11.96}$$

An arbitrary chief ray has relative coordinates x, y at a lenslet with indices j, k respectively. The axis of this lenslet has the coordinates if the number of lenslets N_a is assumed to be odd

$$\begin{aligned} x_{ax,k} &= x_k + D_a - \mathrm{mod}(x_k, D_a), \\ y_{ax,j} &= y_j + D_a - \mathrm{mod}(y_j, D_a) \end{aligned} \tag{11.97}$$

and the refraction of the ray has to be calculated relative to the local lenslet axis position $x_{ax,k}$, $y_{ax,j}$ with the final coordinates in the sensor plane as the centre of the blur circle as

$$\begin{aligned} x'_k &= x_{ax,k} + (1 - z'/f_a) \cdot (x_k - x_{ax,k}) + w_{x,k} \cdot z' \\ y'_j &= y_{ax,j} + (1 - z'/f_a) \cdot (y_j - y_{ax,j}) + w_{y,j} \cdot z' \end{aligned} \tag{11.98}$$

It is also possible to describe the paraxial functionality by a matrix approach if additional shift or translation operands are included to mimic the shifted sub-apertures of the array. Furthermore, a more physical approach based on Fourier optics has been proposed in the literature. If all the real-world conditions are intended to be included, like diffraction, aberrations, tolerances and vignetting, a more rigorous model is needed.

11.8.3 Phase Space Discussion

Since the light-field is a mixed representation in the spatial-angle domain, a discussion in the phase space is meaningful and shows in particular the effect of the micro-lens array function more clearly. In the special arrangement of version 1.0 the array is located in the image plane. The rays in this plane form lines inside a rectangle as can be seen in Figure 11.74(a). If only the rays through the centres of the lenslet are considered, they have the same angles, but in the focal plane of the array a changed coordinate x. The rectangular pattern was sheared as is illustrated in Figure 11.74(a) due to the finite chief ray angle.

If all the light rays are considered that come from one object point, in the classical case without the array the coloured lines are obtained as shown in Figure 11.74(a). The angle extent due to the numerical aperture corresponds to the lengths of the lines, the inclined dotted red lines as an envelope show the chirp of the angles due to a finite field angle in the image. In part (b) of the figure now the array is included and the sensor plane is represented, and the lines are tilted. This has the consequence that in every finite width of a pixel, the integration corresponds to a summation over the angle θ in the finite width yellow rectangle, the contributions come from several object points. In the sketched case, three object points contribute to the energy in one pixel. This mixing makes the raw image unusual and not simply interpretable. This constellation corresponds to one certain object distance $s = s_1$.

The intensity distribution can now be deconvolved by digital processing to generate the desired image.

In modified setups, the array is placed in an arbitrary plane between camera lens and image plane, in this case every lens of the array capture in principle the full image information. As can be seen in Figure 11.74(c), in case of

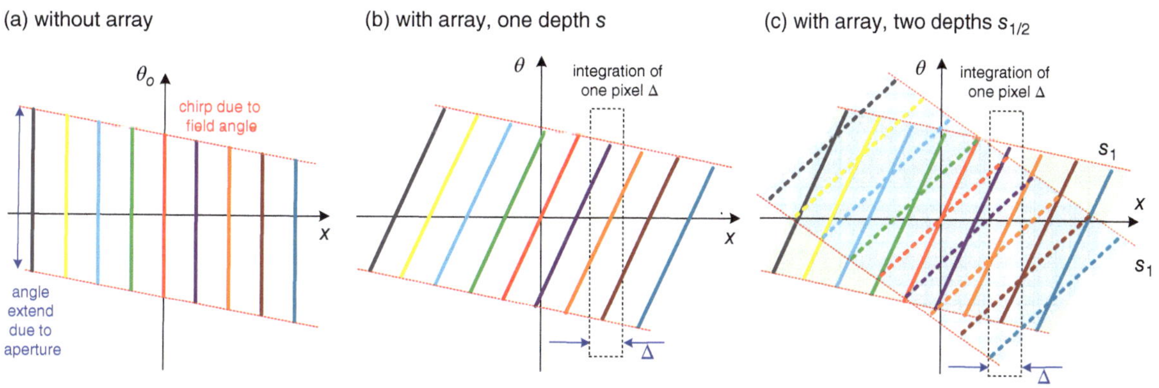

Figure 11.74 Intersection points and angles of the rays from one object point (a) without and (b) with the array and one object depth s and (c) with two different object depths $s_{1/2}$.

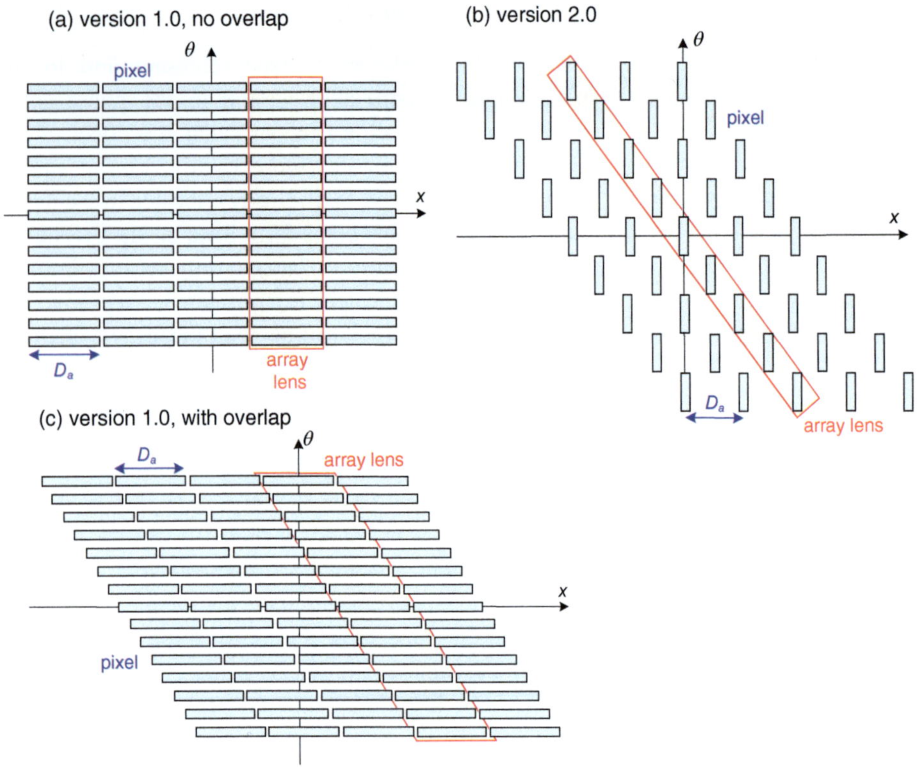

Figure 11.75 Comparison of the plenoptical light-field in the phase space between the versions (a) 1.0 without overlap, (b) version 2.0 and (c) version 1.0 with overlap.

a modified object distance s_2, the pixel collects a signal from three points, in case of s_2 from five points.

If the versions with defocused and focused operation are compared in the phase space, the picture of Figure 11.75 is obtained. In case of the 1.0 version, the case is shown where the array lens sections do not just overlap in the sensor plane. In the 1.0 type, the pixels are in a conjugated pupil plane and they are therefore horizontal; in version 2.0 they are in an image conjugated plane and the pixels are therefore lying vertically. The pixels with underlying yellow colour are those directly behind one lenslet sub-aperture. What is rather nicely illustrated here is that the transverse spatial resolution is governed by the sub-aperture size D_a and pixel width Δx respectively.

The re-ordering of the spatial-angle properties in types 1.0 and 2.0 can be understood by Figure 11.76. The full area information (in one lateral dimension for simplicity) is re-ordered after the first spatial and second angle in the case of type 1.0 or the angle first and spatial second for type 2.0. In reality the four-dimensional phase space is then mapped onto a two-dimensional sensor. This different sorting scheme makes clear why the resolution is different and the reconstruction algorithms are not the same for both setups.

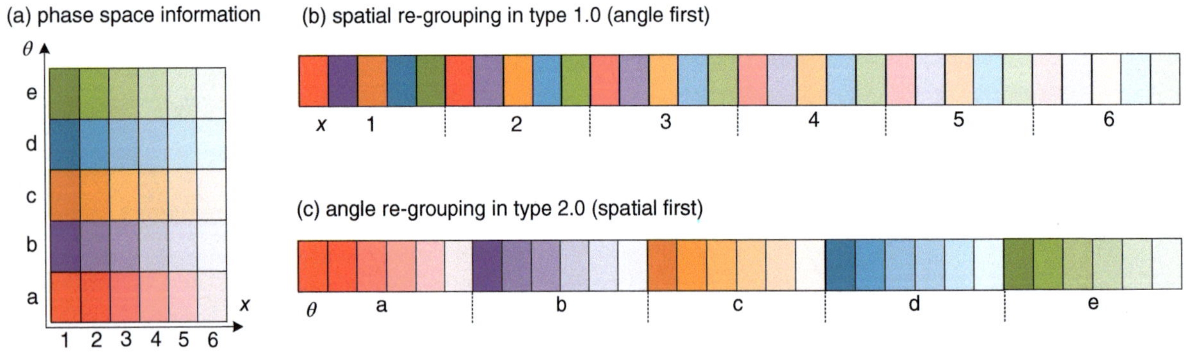

Figure 11.76 Re-ordering the (a) phase space area after (b) spatial information as columns form (a) or (c) angle information as rows from (a).

11.8.4 Digital Refocusing

The clearest view on what happens in plenoptical imaging is illustrated by the rays of the plenoptical distribution of radiance $L(x,\theta)$ in the phase space. Of course in general this is a four-dimensional problem, but without lack of generality the major principles can be discussed in two dimensions. As is well known from the properties of the Wigner function and the phase space, an integration over the angle delivers the spatial intensity distribution $I(x)$ of the field. If the transfer of the light from the object to the sensor is defocused, a tilt or shear of the distribution is obtained, the micro-lens array splits the distribution into several stripes. The angle integration at the sensor location x can be written as

$$I(x) = \int L(x, \theta)\, d\theta \tag{11.99}$$

If a defocused detail of the object is considered, this equation reads

$$I(x) = \int L(x + \Delta s \cdot \theta, \theta)\, d\theta \tag{11.100}$$

This means that the depth information is coded in the modified lateral spatial distribution via the angle change in the form of a shear or rotation in the non-paraxial approach. If the image entropy is optimized, the unknown distance s can be determined. In general, the integration in the direction of a skew angle different from the value θ of the sensor corresponds to a digital refocusing into a modified distance $s + \Delta s$. The plenoptical light field codes the distance over the angle. In the special setup of the array location shown here the new sensor plane is in the focal plane of the arrays and the array is in the previous image plane. The relation between the image and the new intensity distribution is simply a Fourier transform and therefore can be reconstructed by an inverse operation. Mathematically the integration under an arbitrary angle corresponds to the projection slice theorem, which can be calculated very efficiently by Fourier transform and interpolation techniques, see Chapter 12. This is sketched in Figure 11.77. Figure 11.78 illustrates again the plenoptical approach in the spatial domain. To be clearer, different lateral positions are assumed for different depth values. It can be seen that the angle-integration delivers the spatial illumination on the sensor as a combined effect of depth s and lateral coordinate x.

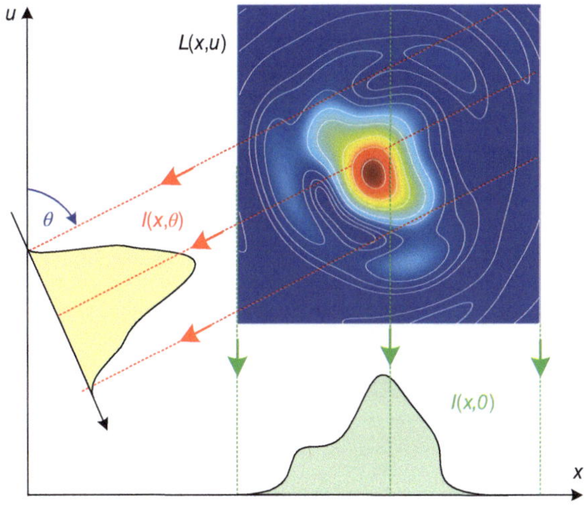

Figure 11.77 Calculation of the irradiance by integration in the phase space under different directions θ.

11.8.5 Depth of Focus

One of the important trade offs and differences between the plenoptical arrangements 1.0 and 2.0 is the lateral resolution and the depth range, which can be digitally recovered. Unfortunately, these two properties are also coupled; a digital refocused image shows a loss in resolution and also in SNR, which means reduced contrast. In total, there are many parameters that influence the physical properties of imaging and there is a lot of specialized literature about this topic, which should not reproduced here [74]. The essential criteria are:

- lateral resolution;
- axial resolution, depth of focus;
- if the object is in the nominal plane;
- if the object lies in a distance Δs defocused to the nominal plane;
- if the object is digitally refocused;
- the Δs_{max}-range, which can be captured for refocusing.

To estimate the geometrical conditions of lateral resolution and depth of focus, it makes sense to inspect the conditions of imaging by a single micro-lens of the array in detail, starting with the intermediate image of the main objective lens. The general geometry and the meaning of the distances and diameters are sketched in Figure 11.79. In particular there are three cases depending on the distance t between array and sensor:

- $t = f_a$ type 1.0 plenoptical camera for the nominal distance with infinity distance of the intermediate image;
- $t > f_a$ type 2.0 with Keplerian setup;
- $t < f_a$ type 2.0 with Galilean setup.

The diameter of the blurred light spot on the sensor is denoted by D_s.

Here only the special case of the type 1.0 plenoptical arrangement is discussed, therefore we have $s'_a = f_a$. If in addition different field points or angles are considered, Figure 11.80 is obtained for one micro-lens no. j. The image

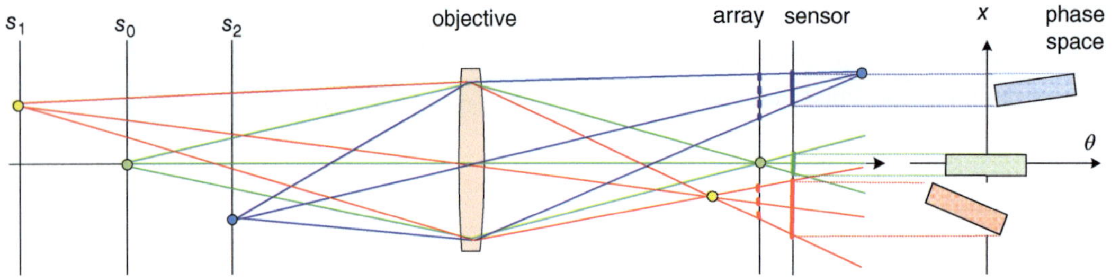

Figure 11.78 Schematic view on three object points at different x-positions and in different depths s_j together with the phase space diagram on the sensor. Furthermore, in the plane of the lenslet array it can be seen that different numbers of lenslets are used.

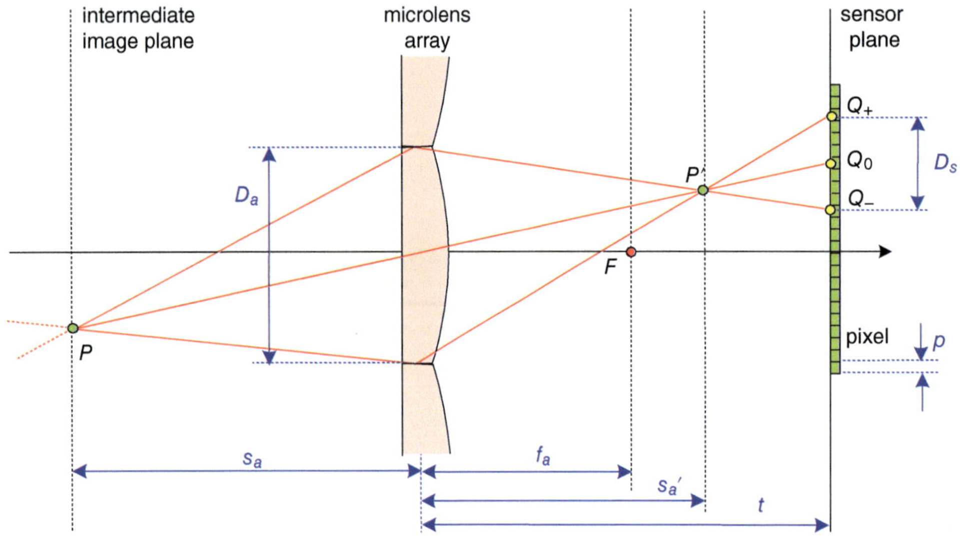

Figure 11.79 General imaging at a single micro-lens.

with field angle w_j spans on the sensor a pixel range from j_k+ to j_k- with a central index k_{jc}. The central point $P_{jc} = F_j$ corresponds to the focal point of the lenslet no. j. The range between the points $P_{jk-} \ldots P_{jk+}$ spans the micro image of the lenslet no. j.

Figure 11.80 illustrates the imaging for a type 1.0 system in more detail. Part (a) shows the conditions with the object in the nominal distance and two chief rays that exactly meet the centres of micro-lenses, a zoomed part illustrates the micro-image on the sensor. In part (b) the nominal object point Q_j as well as a defocused image point Q_{j-1} are sketched to visualize the change of conditions in the case of a defocused object plane. A chief ray belonging to the object height x_j originating from the centre of the objective lens, exits the pupil centre of an object point in the nominal distance that meets the centre of the lenslet with index j, has the angle w_j and intersects the sensor plane in the point P_{cj}; c stands for centre of the micro-image. The corresponding coma rays at the rim of the aperture cone meet the sensor at P_{jk+} and P_{jk-}. k_j+ and k_j- are here the indices of the pixels, so the micro-image covers

$$\Delta k_j = k_{j+} - k_{j-} \tag{11.101}$$

pixels and with the pixel pitch p the diameter of the micro-image has the size

$$D_{sj} = \Delta k_j \cdot p = (k_{j+} - k_{j-}) \cdot p \tag{11.102}$$

The chief ray angle is given by

$$\tan w_j = \frac{y_j}{s} \tag{11.103}$$

The lateral position of the lenslet j is at x_{aj}, the chief ray angle is preserved and the coordinate of the centre P_{cj} is at x'_{jc}. Usually it can be assumed that $f_a \ll s'$, and therefore the distance of the sensor is approximately at the focal plane of the array. Then the lateral shift of the micro-image centre from the axis of the corresponding micro-lens is given as

$$x'_{jc} - x_{aj} = f_a \cdot \tan w_j = \frac{f_a x_j}{s} \tag{11.104}$$

With the preserved numerical aperture angle of the main lens, the diameter of the micro-image and the index range

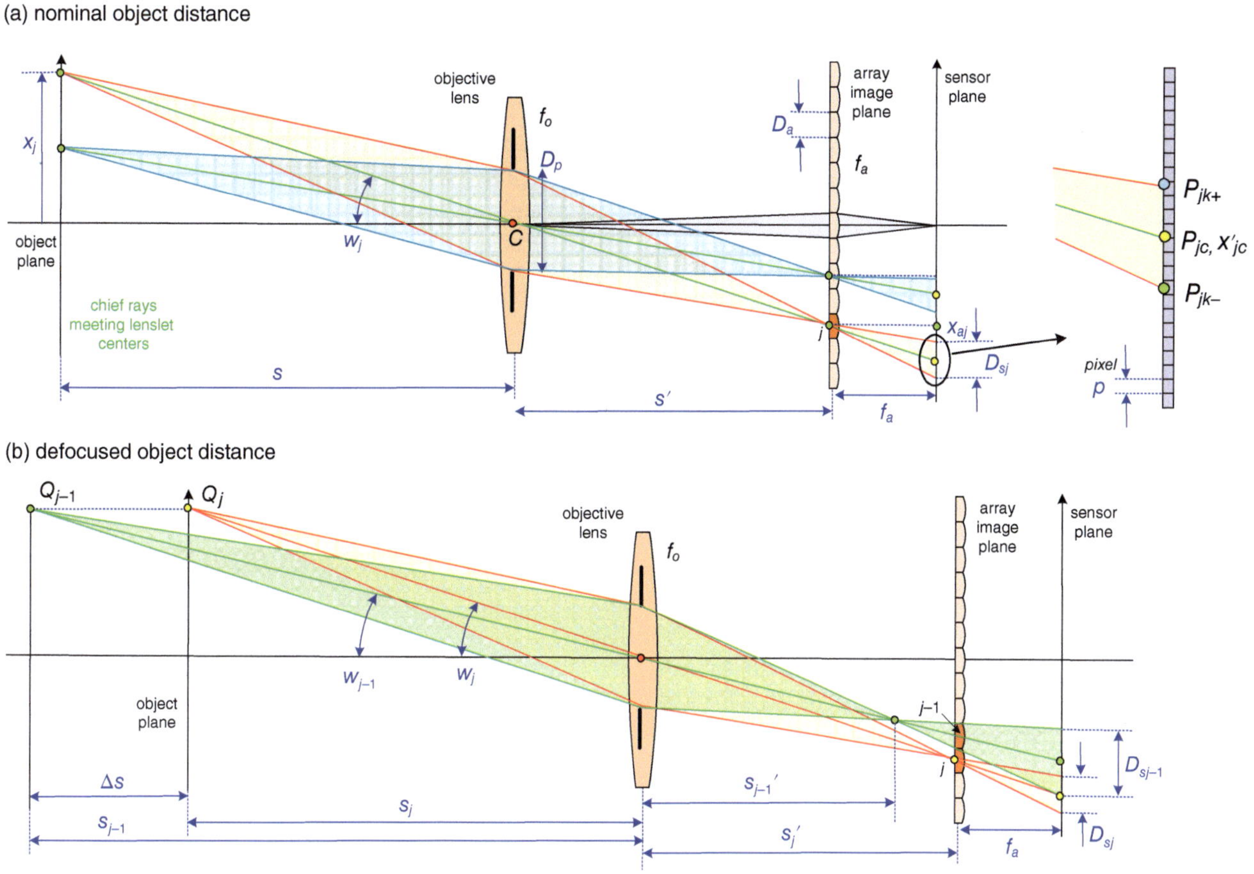

Figure 11.80 Parameter at a plenoptical camera of 1.0 type with (a) complete system in a nominal object distance and (b) considering an object point defocused by Δs.

are

$$D_{sj} = \Delta k_j \cdot p = 2f_a \sin u, \qquad \Delta k_j = \frac{2f_a \sin u}{p} \tag{11.105}$$

Since the object for now is in the nominal plane, the axis of the lenslet no. j is located at

$$x_{aj} = x_j' = m_o \cdot x_j = \frac{s'}{s} \cdot x_j \tag{11.106}$$

Together with the thin lens equation for the main objective lens

$$\frac{1}{s'} - \frac{1}{s} = \frac{1}{f_o} \tag{11.107}$$

we can eliminate all unwanted parameters and find the sensor chief ray position as a function of the object distance as

$$x_{jc}' = x_j \cdot \left(\frac{f_a}{s} + \frac{f_o}{f_o + s}\right) \tag{11.108}$$

If now a second object point is considered that is defocused by Δs, it is convenient to look at the special case where its chief ray exactly meets the lenslet j–1 in the centre. This discretization trick can later be generalized to arbitrary values of Δs. We then have

$$x_{a,j-1} = x_{aj} - D_a \tag{11.109}$$

and with the same approach as before

$$x_{j-1,c}' = x_j \cdot \frac{s(f_o + f_a) - f_o f_a}{(f_o + s)(s + \Delta s)} \tag{11.110}$$

The difference between x_{jc}' and $x_{j-1,c}'$ is the disparity of the two images on the sensor and governs the depth resolution. If the shift of the micro-image due to the defocusing Δs is set to the pixel based lateral resolution limit, the achievable accuracy of the camera for defocusing or the depth of focus is obtained. In the approximation $f_a \ll f_o$ one gets

$$\delta s_j = p \frac{s(f_o + s)}{x_j f_o - p(f_o + s)} \tag{11.111}$$

It should be mentioned that beneath the shift of the micro-image also its size $D_{s,j-1}$ changes.

All the estimations above depend on many assumptions; in the case of type 2.0 the formulas are different but the behaviour is comparable. But in reality it depends in addition on:

- the reconstruction algorithm;
- aberrations of real components in the system;
- additional effects like vignetting;

- the aperture of the main objective lens;
- for small aperture angles on additional diffraction spreading effects.

Therefore, there is simply not one set of formulas which is valid in any case. To give an idea, here only one simple case is derived.

11.8.6 Examples

To gain an impression of the look of plenoptical ray images, a few examples are shown. In Figure 11.81, a type 1.0 plenoptical setup is used to generate a light-field image without overlapping pupils in (b) and with overlapping in (c). As has been mentioned above, in the 1.0 arrangement the lateral resolution is dominated by the size D_a of the micro-lenses. This is clearly observed in (b) with 51 micro-lenses, the micro-images are circular in diameter according to the pupil shape of the main objective lens. If the pupils are overlapped as in (c), the individual micro-lenses create more complicated structured over-folded pupil circles, and the interpretation of the image without reconstruction is not possible.

In Figure 11.82 three plenoptical ray images of the object in (a) are shown. In (b) 101 micro-lenses form a light-field, similar to the figure before. In this case the micro-lenses are exactly conjugated to the pupil. In case (c) an additional defocus corresponding to an object located in another distance shows a certain blur of the micro-images. In (d) a type 2.0 plenoptical setup is used with the same number of micro-lenses in a perfect focused case. Every patch contains a sub-image, these are laterally shifted so most of the image points occur in a neighbouring micro-image under a different angle again. This is exactly the disparity that allows us to refocus the image.

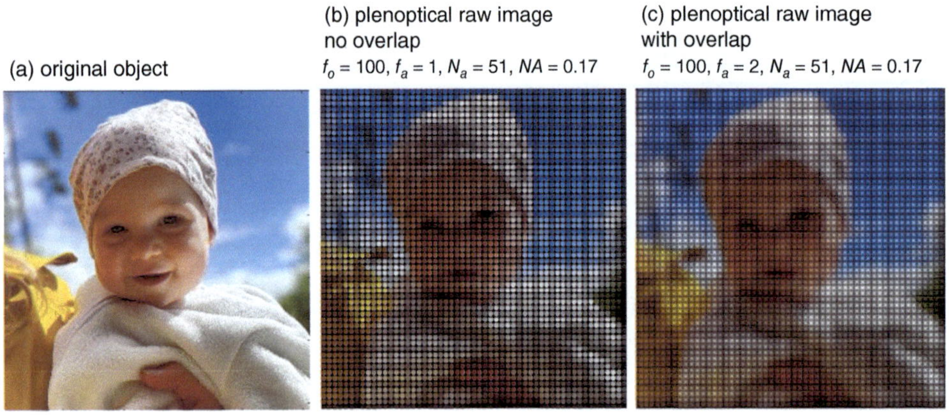

Figure 11.81 Example of plenoptical ray images for version 1.0. (a) is the original, (b) a case without overlap of the circular pupil images and (c) a case with overlap. The system data are listed in the headlines, the focal lengths are given in mm.

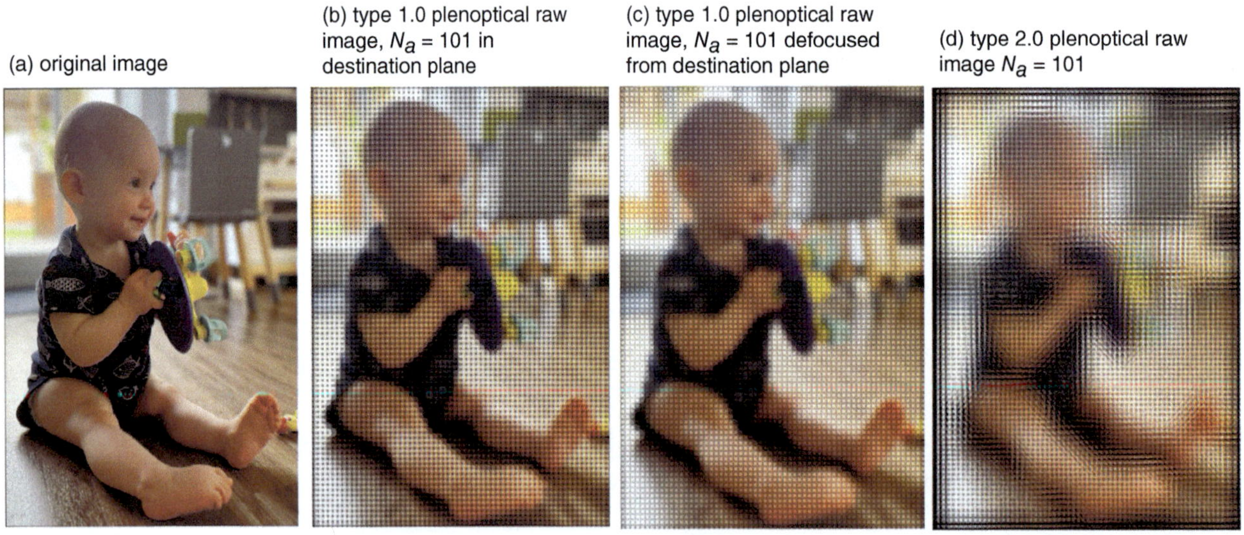

Figure 11.82 Example of plenoptical images of the original (a) for (b) the version 1.0, (c) type 1.0 with defocus from the nomination plane and (d) for type 2.0.

11.9 Digital Phase Imaging*

(*) The content of this section is quite specialized and for advanced reading. Knowledge of this section is not necessary to understand the following chapters. Beginners are recommended to skip this section.

11.9.1 Introduction

It is well known that a classical detector cannot directly measure the phase of a complex field. Therefore, special approaches are needed to make phase structures visible that are often quite important in medicine or biology. In microscopy a very long time ago, phase contrasting was developed by using a corresponding hardware in the form of a Zernike phase contrast, differential interference contrast, dark-field imaging and more. These methods use Fourier optics to modify the balance between amplitude and phase of the diffraction orders. Another direction became increasingly attractive as computer resources became larger and digital algorithms allowed us to extract the phase information from specially gathered intensity measurements. The phase modulates the intensity by changing the propagation direction, therefore a three-dimensional intensity distribution has coded the initial phase and contains all the information about it.

The most important approaches are listed below [78–84]. An issue in all of these methods is the achieved lateral resolution. Furthermore, a question is whether the obtained phase 'image' is qualitative or gives the phase values exactly quantitatively. Further, if digital algorithms are involved, the generation of the phase image needs time and therefore is not obtained immediately which means that the image is not in any case obtained in real time. Usually coherent light is necessary to reconstruct the phase and complicated conditions concerning sampling, coherence and monochromaticity have to be fulfilled to get good results.

It should be noted that the reconstruction of a phase out of only one intensity plane is ill-conditioned. Due to the property that a complex conjugated field gives the same intensity, the so-called twin-image problem exists and often results in a poor convergence of the iterative recovery. Therefore, much more robust conditions are achieved if the intensity gathering is multiplexed under slightly different conditions. The redundant information then helps considerably in being more insensitive against noise, resolving to get the result without ambiguity and overcoming discretization and aberration influences. This means, for example, the image is gathered for different focus planes, different illumination directions, structured changed illumination etc. Some of the most important digital methods to visualize phase structures are as follows.

- Coherent digital imaging (CDI) by phase retrieval.
  Gathering the image in several distances or configurations, digital recovering of the phase by Gerchberg–Saxton-, HIO-, ER- or TIE-algorithms (hybrid input-output, error reduction, transport of intensity).
- Digital holographic imaging.
  Using a coherent superposition of the object wave with a reference wave.
- Ptychography.
  Illumination coherent under different directions and tomographic reconstruction.

Figure 11.83 illustrates the basic setup for one of the approaches. A phase object is Fourier imaged into the spectral domain, in several planes around the focal plane of a system the intensity spectra are recorded.

In these special imaging modes and in particular if digital methods are involved, often the classical physical relations between resolution and contrast are lost. Especially digital methods can enhance the contrast by clever algorithms. In reality there are in addition several issues about getting an

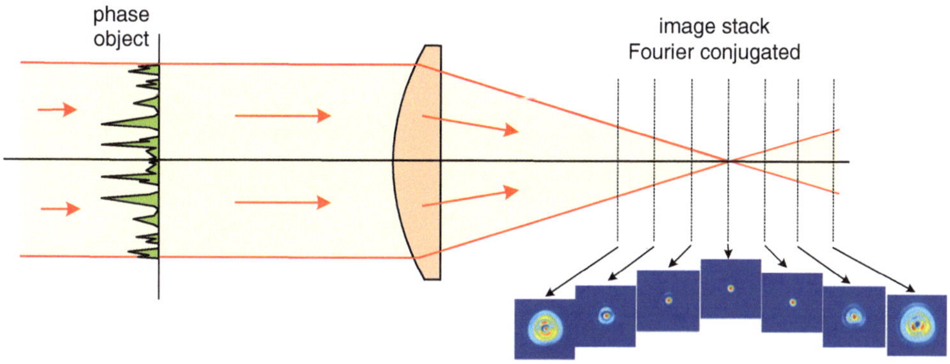

Figure 11.83 Phase imaging by analysing an intensity stack with defocus diversification.

accurate quantitative phase image by digital means. These are. for example:

- discretization of the detected images;
- SNR problems and noise sensitivity of the algorithms;
- alignment of the images in the case of a z-stack;
- truncation of images, violating the condition of energy conservation;
- quantization depth of the camera and intensity discretization;
- remaining partial coherence of the illumination in the spatial (extended source) or temporal (not perfect monochromatic) sense;
- hard edges of the phase structure with a poor Fourier convergence;
- problems with the boundaries and support area.

Figure 11.84 illustrates as an example the change of contrast for a phase object in a microscope. Usually, phase objects don't create any contrast in conventional imaging. But if the Fourier spectrum and the balance between the various diffraction orders is modified by corresponding components or filters, a contrast can be achieved. One well known method is dark-field imaging, here the zeroth diffraction order is suppressed and therefore the interference of the diffraction orders changes. In the figure a phase object shown in (a) with a small transmission contribution is shown in (b) for bright-field with low contrast and in (c) if the dark-field technique is applied. There are many options to increase the contrast of pure phase object especially developed for microscopy. The reader is here referred to the specialized literature, for example [78, 79].

11.9.2 Coherent Diffraction Imaging

One of the earliest and simplest approaches of a phase retrieval out of two measured intensities in object plane and far-field was proposed by Gerchberg and Saxton (GS-algorithm) [81]. The basic scheme is illustrated in part (a) of Figure 11.85. The coherent field is propagated back and forth between image and far-field plane, the phase is always updated and in the far-field the measured intensity is substituted. The far-field of the pupil plane is the image here. If the spectral object intensity in the pupil is known in addition, it can be optionally used in every iteration loop too. Here the case is shown where the phase Φ in the pupil of an optical system is searched for. Despite some

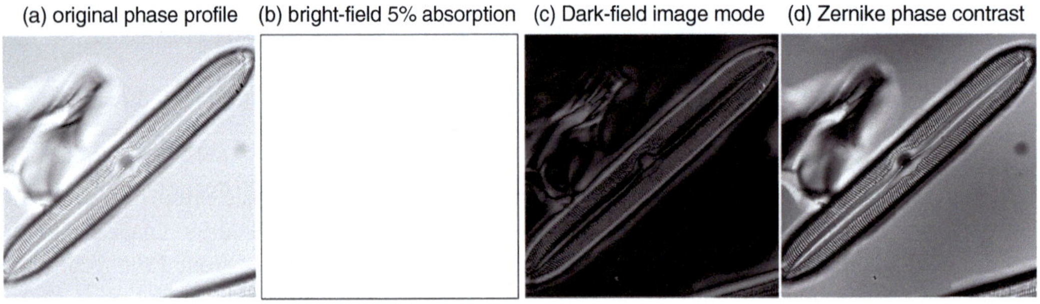

Figure 11.84 Imaging of a diatomee with the original phase shown in (a) by changing the illumination mode from (b) bright-field into (c) dark-field or (d) Zernike phase contrast. Courtesy to Dietzel65, commons.wikimedi.

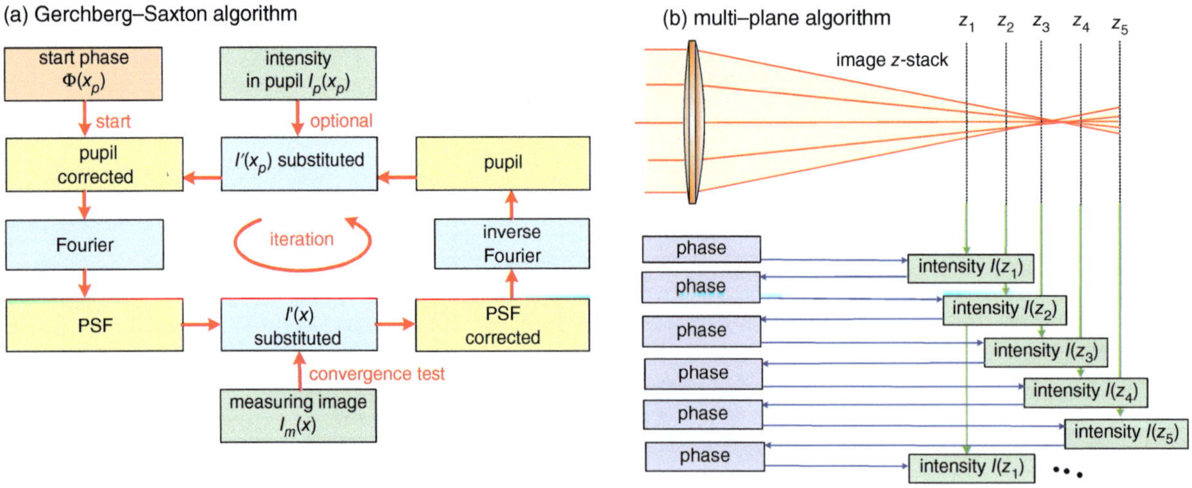

Figure 11.85 Gerchberg–Saxton algorithm principle (a) and case of several z-planes (b), the intensities are here denoted with $I = |A|^2$.

convergence problems due to the twin image issue, for a clever initial guess the algorithm converges towards the correct intensity. But usually the speed of convergence is slow. Mathematically the iteration process can be described by the following set of equations. First an initial guess for the phase is used to calculate the first complex pupil field. $\Phi^{(0)}$. If no a priori information is available, a random first guess can be used. Here the amplitudes are used instead of the intensities as shown in the figure. Furthermore, only one transverse coordinate is formulated here for simplicity, the real algorithm of course works in two transverse dimensions.

$$A^{(0)}(x_p) = |\hat{F}^{-1}[A_m(x)]| \cdot e^{2\pi\, i\Phi^{(0)}(x_p)} \tag{11.112}$$

The iteration with counter n uses the two equations

$$\begin{aligned}\Phi^{(n)}(x) &= \arg\{\hat{F}[A^{(n)}(x_p)]\}\\ A^{(n)}(x) &= \sqrt{I_m(x)} \cdot \Phi^{(n)}(x)\\ A^{(n+1)}(x_p) &= |\hat{F}^{-1}[A^{(n)}(x)]|\end{aligned} \tag{11.113}$$

with the measured intensity $I_m(x)$ and the phase $\Phi(x)$. If the changes between two successive iterations are neglectable, the iteration can be stopped.

Many proposals are found in the literature to overcome the twin image issue, to accelerate the convergence and to achieve an algorithm that is more robust against noise. One option is to use more than one Fourier far-field plane of measured data. If several z-planes around the image plane are available, Fresnel diffraction data can be used to make the process much more stable; this illustrated in part (b) of Figure 11.85. In principle this is a kind of defocus diversification, which describes a tomographic projection in the phase space.

More refined approaches like the error-reduction algorithm or the so-called hybrid input-output method improve the poor convergence properties a lot. The basic ideas are to manage the non-negativity of the intensity and the boundary constraints in a clever way. One of the problems of the original Gerchberg–Saxton algorithm occurs if the field signal is not band limited.

Figure 11.86 illustrates the result of a retrieval for a synthetic phase pattern with seven z-planes for reconstruction. It can be seen that the result is quite convincing in those regions, where the amplitude is large enough. This is one of the well-known problems of nearly all algorithms: that they fail in the case of a phase point with zero energetic weighting. Here it is seen that outside of the rectangular amplitude pattern the phase behaves undefinedly.

Figure 11.87 shows the result of a phase recovery by using the HIO algorithm for a constant amplitude (not shown). It is seen that many iterations are needed, the hard edges in the phase pattern provoke a bad convergence. Finally, a residual texture is seen in the recovered phase distribution.

11.9.3 Transport of Intensity Approach

If the complex field is separated into the real amplitude and the scaled phase in the form

$$E(x,y) = A(x,y) \cdot e^{ik\Phi(x,y)} \tag{11.114}$$

the paraxial wave equation for homogeneous media delivers two coupled partial differential equations for amplitude

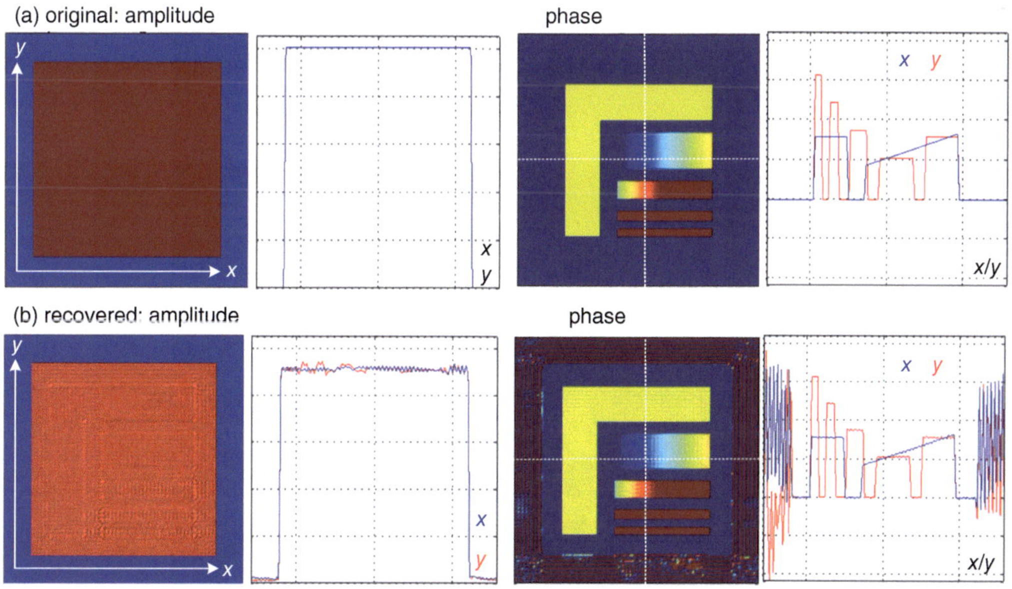

Figure 11.86 Example of a phase retrieval calculation. In the upper row (a) the original amplitude and phase distributions are shown, (b) illustrates the result of a retrieval out of seven defocused planes.

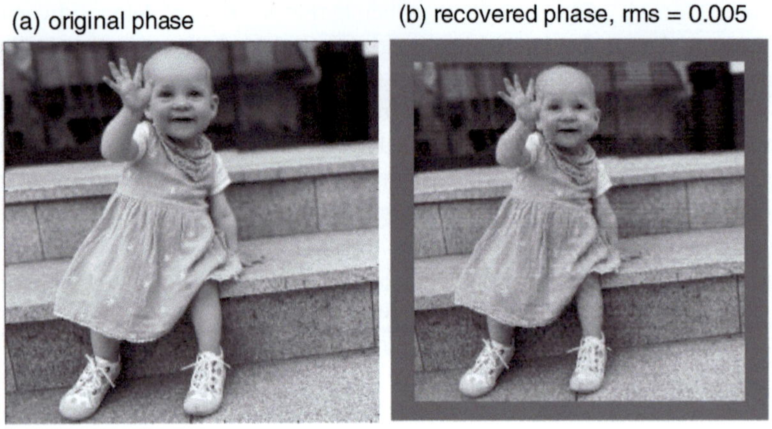

Figure 11.87 Phase image (a) reconstructed by HIO phase retrieval with 2300 iterations.

and phase by separating the real and the imaginary parts in the form

$$\frac{\partial A}{\partial z} = \frac{A}{2}\left(\frac{\partial^2 \Phi}{\partial x^2} + \frac{\partial^2 \Phi}{\partial y^2}\right) + \frac{\partial A}{\partial x}\frac{\partial \Phi}{\partial x} + \frac{\partial A}{\partial y}\frac{\partial \Phi}{\partial y} \tag{11.115}$$

$$\frac{\partial \Phi}{\partial z} = -\frac{1}{2Ak^2}\left(\frac{\partial^2 A}{\partial x^2} + \frac{\partial^2 A}{\partial y^2}\right) + \frac{1}{2}\left[\left(\frac{\partial \Phi}{\partial x}\right)^2 + \left(\frac{\partial \Phi}{\partial y}\right)^2\right] \tag{11.116}$$

If the intensity is calculated from the amplitude in Eq. (11.125) by squaring, the transport of intensity equation (TIE) in paraxial approximation is obtained [85–87]. It can be written as

$$\begin{aligned} k \cdot \frac{\partial I}{\partial z} &= -\nabla_T I \cdot \nabla_T \Phi - I \cdot \nabla_T^2 \Phi \\ &= -\nabla_T [I \cdot \nabla_T \Phi] \end{aligned} \tag{11.117}$$

This equation shows that the transverse gradient of the phase is weighted by the intensity and then dominates the z-propagation of the intensity. Figure 11.88 sketches two simple cases of a curvature or a tilt of the phase and the corresponding impact onto the intensity caustic schematically. Of course, the z-derivative is a function of the lateral coordinate x.

A formal solution of the TIE can be written in the form

$$\Phi = -k \cdot \nabla_\perp^{-2} \left\{ \nabla_\perp \left[\frac{\nabla_\perp (\nabla_\perp^{-2} \partial_z I)}{I} \right] \right\} \tag{11.118}$$

where $\nabla_\perp^{-2}$ is the inverse Laplace operator for the lateral coordinates only. This equation can be solved with Fourier algorithms. In practice, to evaluate the derivative of the intensity $I(x,y)$ after z the intensity has to be determined in at least two planes with a small separation as a finite difference. One of the problems with this equation is again the zero-intensity issue, clearly seen in the denominator. This can be overcome in a certain sense by a regularization procedure avoiding the division by zero.

Figure 11.89 demonstrates an example of a pure phase object and a digital phase retrieval by using the transport of intensity algorithm. In three planes the intensity is detected and the changes code the initial phase. The z-separation to determine the finite difference for the z-derivative is varied in the four rows. It is seen that by selecting the differences to be too large, the lateral resolution of the reconstructed phase pattern is degraded. Similarly as in Figure 11.86 in the exact image plane without defocusing no contrast is observed.

11.9.4 Ptychography

In ptychography (ptychographic iterative engine: PIE) some of the problems in the classical CDI approaches as discussed above are tackled by a diversification with a modified illumination. If the phase probe is illuminated under several modified plane wave illumination conditions, an

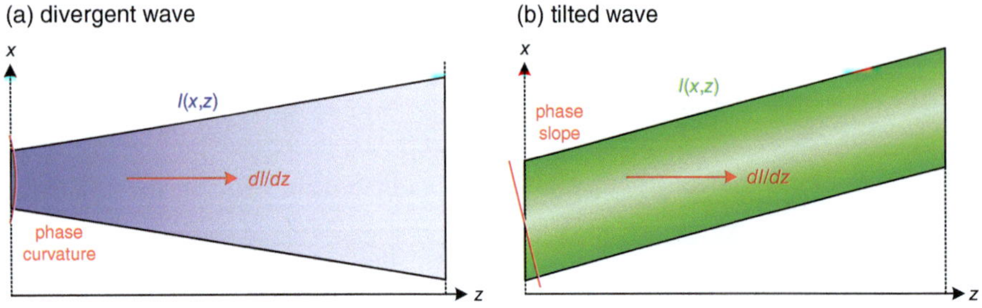

Figure 11.88 Principle of the dependence of the intensity propagation from the phase for (a) a divergent wave and (b) a tilted wave.

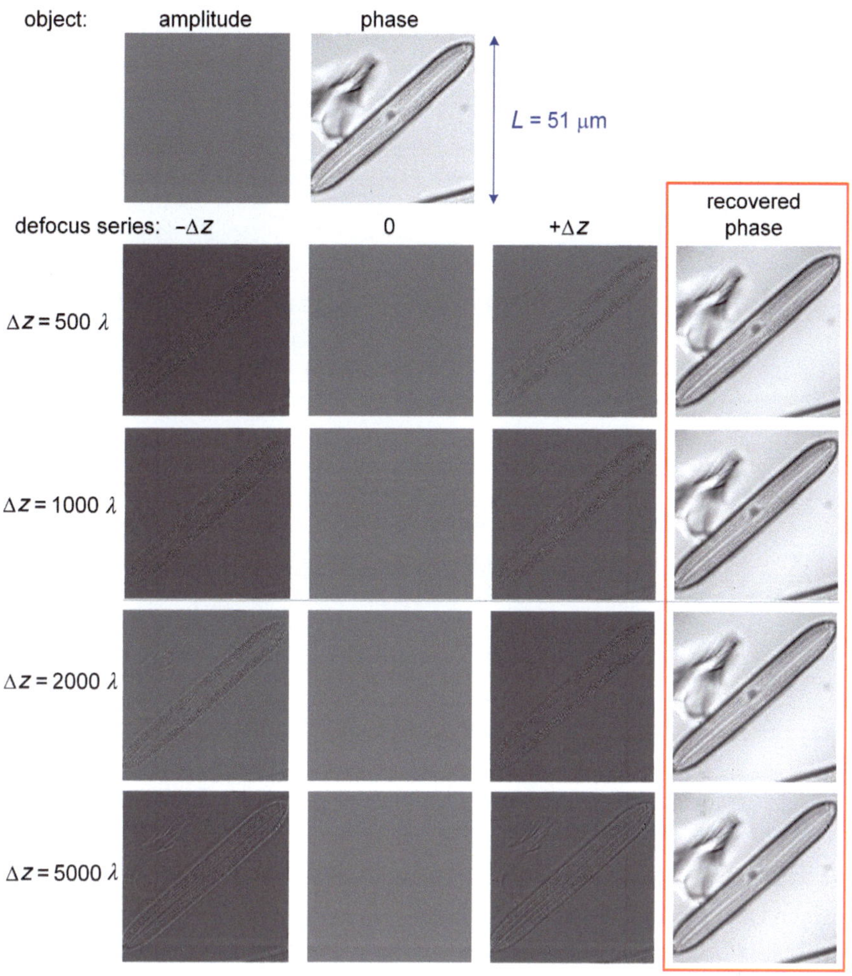

Figure 11.89 Imaging of a pure phase object, recovered by TIE in three planes for four different separation distances Δz in the four rows in arbitrary units. The right column shows the original phase distribution at the top and the results of the four TIE-base reconstructions. Courtesy to Dietzel65, commons.wikimedi.

overlap makes the reconstruction quite robust against noise and other perturbations [88].

In contrast to the defocusing diversification discussed in CDI here a lateral shift or tilt or both of a restricted illumination beam in lateral direction is used. The iterative evaluation with back-and-forth calculation in far-field or optionally in the Fresnel field distance follows a similar scheme. Typically for highly redundant data with large overlap of the illuminations, the convergence of the iterative reconstruction is improved. This can be seen in Figure 11.90, where a phase recovery is obtained after approximately 100 iterations with 11 illumination sets in x and 11 in y direction, see part (c) of the figure. This is somewhat faster than the 1000 and more iterations of the classical CDI computations. One of the drawbacks of this method is also observed: due to the change of the illumination conditions, the boundary of the field of view is not clearly defined. There is a certain outer stripe of the image that has a reduced number of diversifications and cannot be recovered satisfactory.

11.10 Extended Depth of Focus*

(*) The content of this section is quite specialized and for advanced reading. Knowledge of this section is not necessary to understand the following chapters. Beginners are recommended to skip this section.

11.10.1 Introduction

According to the principles of Fourier optical image formation, it is always possible to modify an image by manipulation of the pupil and therefore the frequency response function of a system. In the literature this is often called wavefront coding or PSF engineering. There are many different goals connected to this attempt and a huge amount of literature exists dealing with this topic. Here only one special purpose should be discussed to give the reader an insight into the ideas and problems.

The so-called extended depth of focus (EDF) is an approach which has the aim to shape the focal caustic in a manner to achieve an enlarged depth of focus,

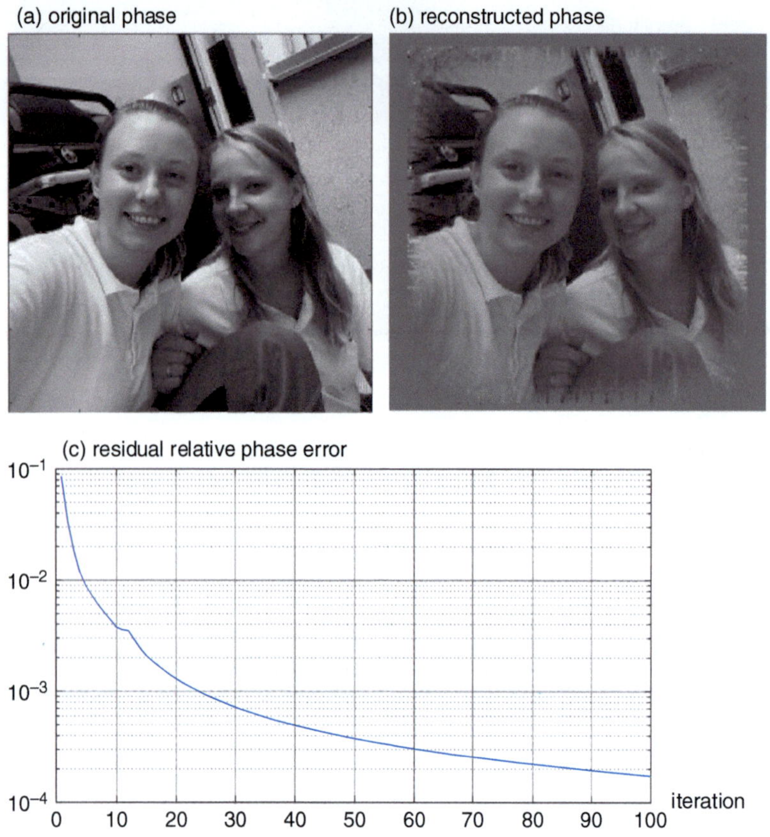

Figure 11.90 Phase image (a) reconstructed by ptychography on a slightly reduced field in (b) and (c) residual relative error of the phase as a function of iteration.

which can be useful in case of three-dimensional objects or non-refocusable systems. In the best case the caustic around the image plane is quite narrow over a large z-depth, keeping the lateral size and the resolution invariant. This means that in observing an object with extension in z a sharp image is obtained for a considerable distance range somewhat larger than the usual depth of focus. The basic idea is sketched in Figure 11.91.

It is possible to construct pupil mask filters in many different ways, but real world experience has shown that there always some drawbacks that come with this approach that can be partly repaired by digital means. Therefore, a setup scheme as shown in Figure 11.92 should be discussed here.

The most prominent results and filter designs proposed in literature for EDF-goals are:

- Toraldo phase rings [89];
- cubic phase plates [90];
- logarithmic shaped phase masks [91]
- masks with a polynomial expansion [92].

In principle there are two different modes of operation. In the first method, the raw image is considered, in the second more highly sophisticated approach, the raw image is repaired by digital means, taking the data of the mask into account. All of these solutions suffer from certain limitations and drawbacks, so the extension of the depth

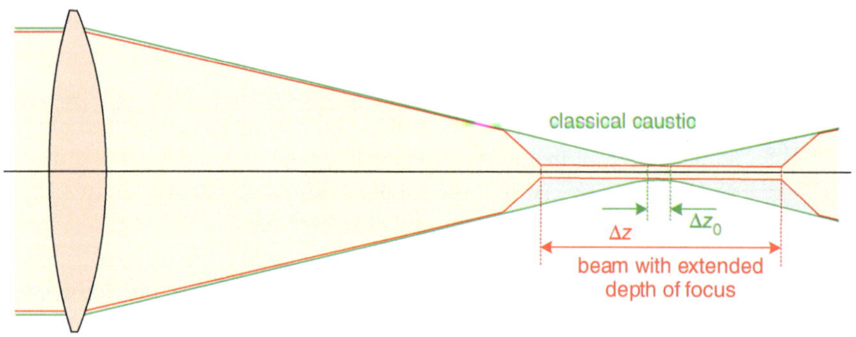

Figure 11.91 Schematic illustration of a classical caustic (green) and a desired profile with extended depth of focus (red).

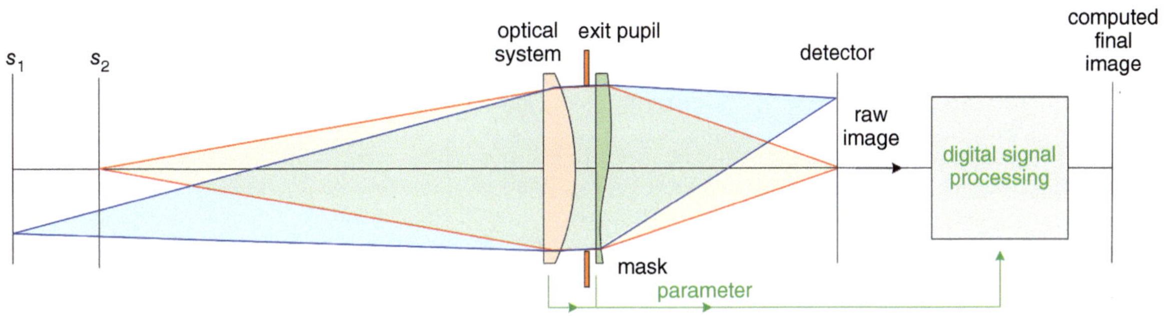

Figure 11.92 Setup of a system for extended depth of focus with digital image restoration.

of focus must be paid for. The most severe problems are as follows.

- Behaviour and function under skew ray bundles for a larger field of view.
- Degradation in case of broadband illumination, mostly the masks are refractive.
- Reduction of lateral resolution.
- Reduction of contrast.
- Reduction of transmission and therefore SNR.
- In the case of diffractive solution, stray-light occurrence.
- Imaging artefacts like lateral shifts, texture structures in the computed image.
- Time delay due to the required computational algorithms.

Many of the solutions reported follow first principles to define and quantitatively evaluate the EDF property, others are more or less numerical results of comprehensive optimizations. Possible goals from first principles are for example:

- stationary point of the PSF caustic on axis $\frac{\partial I(0,0,z)}{\partial z} = 0$;
- stationary point of the MTF for defocusing $\frac{\partial H_{OTF}(0,0,z)}{\partial z} = 0$;
- maximal axial second moment of the axial intensity $I(0,z)$;
- maximum uniformity of the Strehl ratio for defocusing;
- optical Fisher defocus criterion;
- second order invariance of the wavefront rms for defocusing.

Essential for attractive solutions are acceptable degradations of quality. To reach this goal in reality the range of the achieved depth has to be limited and pure phase masks used that do not absorb a part of the signal. To come to a quantitative specification for the EDF functionality, it is essential to define this property in a more general sense. The following definitions to measure the EDF-feature make sense and can be used as a design criterion for EDF-systems [13]:

- Deviation of the field width as a function of z for a single object point on axis

$$\varepsilon(z) = \sqrt{1 - \left| 2\pi \cdot \int_0^\infty E_{ref}(r) \cdot E^*(r,z)\, rdr \right|} \tag{11.119}$$

For this function a certain threshold can be defined as a goal. E_{ref} is the field in the ideal focal point.

- Peak-valley value of the uniformity of the intensity for the caustic height in the EDF-range on axis $I(z)$

$$U = \left(1 - \frac{I_{min}}{I_{max}}\right)_{I>0.8 \cdot I_{max}} \tag{11.120}$$

This number measures the uniformity along z, but doesn't give any information about broadening.

- Variation of the normalized second moment diameter over the range Δz normalized by the Rayleigh range R_u

$$\eta = \frac{\Delta D^{(2nd)}}{\Delta \bar{z}} = \frac{\left(D_{max}^{(2nd)} - D_{min}^{(2nd)}\right) \cdot R_u}{D_{min}^{(2nd)} \cdot \Delta z} \tag{11.121}$$

- Normalized change of the intensity correlation of the caustic along z, the so-called Hilbert space angle

$$\cos\theta_H(z) = \frac{\int I_{psf}(r,z) \cdot I_{psf}(r,0)\, 2\pi r dr}{\sqrt{I_{psf}^2(r,z)\, 2\pi r dr} \cdot \sqrt{I_{psf}^2(r,0)\, 2\pi r dr}} \tag{11.122}$$

- Fisher criterion as the z-integrated deviation from the stationary behaviour of the MTF variation with z

$$J(z) = \int \left| \frac{\partial H_{OTF}(v,z)}{\partial z} \right|^2 dv \tag{11.123}$$

Finally the decision about the most appropriate criterion for an EDF-system is the application and the importance of the desired z-range and the lateral resolution properties. There are always two different approaches: looking for a stationary point in the image plane or optimizing the criterion over the postulated range of EDF as a trade-off.

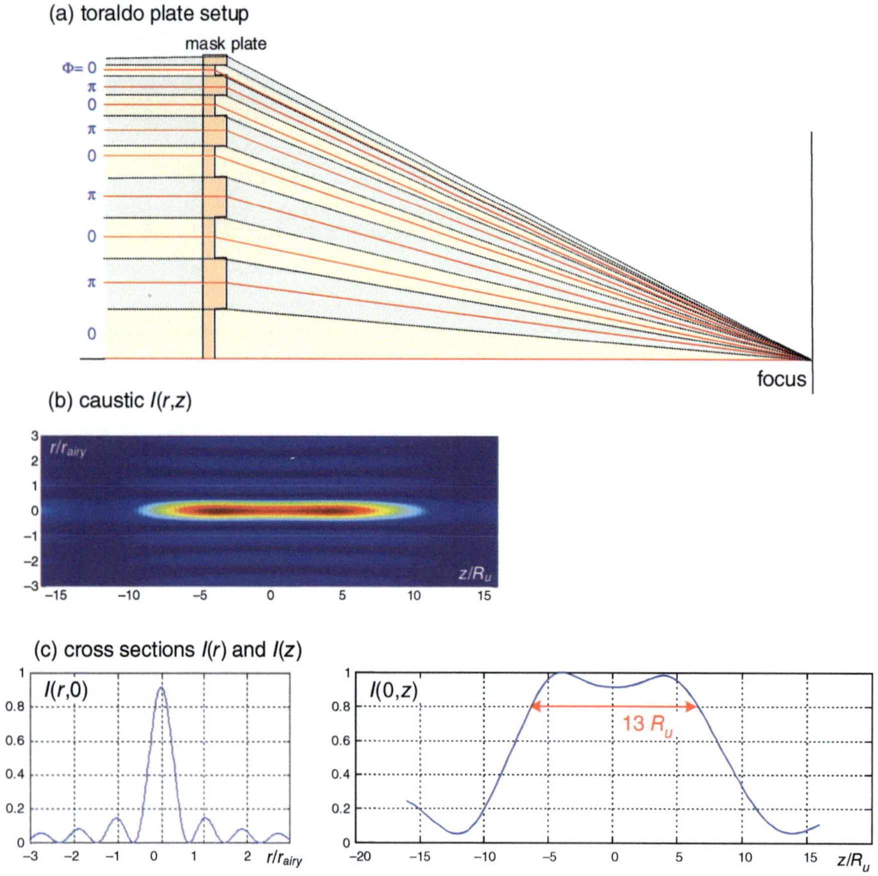

Figure 11.93 Setup (a) and exemplary results of the intensity in the caustic in (b) with lateral and axial PSF cross sections in (c) for a Toraldo phase mask.

11.10.2 Toraldo Phase Mask

One of the oldest approaches to achieve extended depth properties is the Toraldo ring mask. Discrete rings have optimized radii and can be absorbing or only phase shifting by π. The number of rings usually depends on the Fresnel zones, which are involved in the image formation. This means that the number is determined by the focal length and the numerical aperture. If ε_j are the relative radii of the rings, normalized on the pupil radius and Φ_j are the phase values switching between 0 and π, the point spread function field for circular symmetry with n rings and aperture $\sin(u')$ for wavenumber k is given by

$$E(r') = \sum_{j=1}^{n} e^{2\pi \cdot i\Phi_j} \cdot \left[\varepsilon_j^2 \cdot \frac{2J_1(kr' \sin u' \varepsilon_j)}{kr' \sin u' \varepsilon_j} - \varepsilon_{j-1}^2 \cdot \frac{2J_1(kr' \sin u' \varepsilon_{j-1})}{kr' \sin u' \varepsilon_{j-1}} \right] \tag{11.124}$$

Figure 11.93 (a) illustrates such a ring phase mask setup. Figure 11.93 (b) and (c) show as an example the resulting intensity caustic. The mask has 10 rings, the achieved depth at an 80% intensity drop criterion is $\Delta z = 13\ R_u$ instead of the usual value $\Delta z = 1\ R_u$.

11.10.3 Cubic Phase Plate

A very special solution for a EDF mask is given in [77], a phase mask with separable Cartesian third power phase profile with the function

$$P(x_p, y_p) = \begin{cases} e^{i\alpha \cdot \left(x_p^3 + y_p^3\right)} & \text{for } |x_p| \le 1 \text{ and } |y_p| \le 1 \\ 0 & \text{else} \end{cases} \tag{11.125}$$

is proposed, the so-called cubic phase mask. Figure 11.94 shows a sketch of the component and the phase function. Interestingly the phase function is asymmetric. Therefore, the point spread function looks asymmetric as well, the cubic phase corresponds to coma. So depending on the strength parameter α a macroscopic coma spot is obtained, see part (c) of Figure 11.94. This has the consequence that the raw image is of very poor quality.

If now the transfer function of the system with the cubic mask is used to deconvolve the broadened PSF with known system data, a rather well resolved image is obtained over a large range of object distances as desired. This is demonstrated in Figure 11.95 together with a schematic behaviour of the transfer function for several values of Zernike defocus c_4 in part (b). It can be seen that for moderate spatial frequencies the contrast

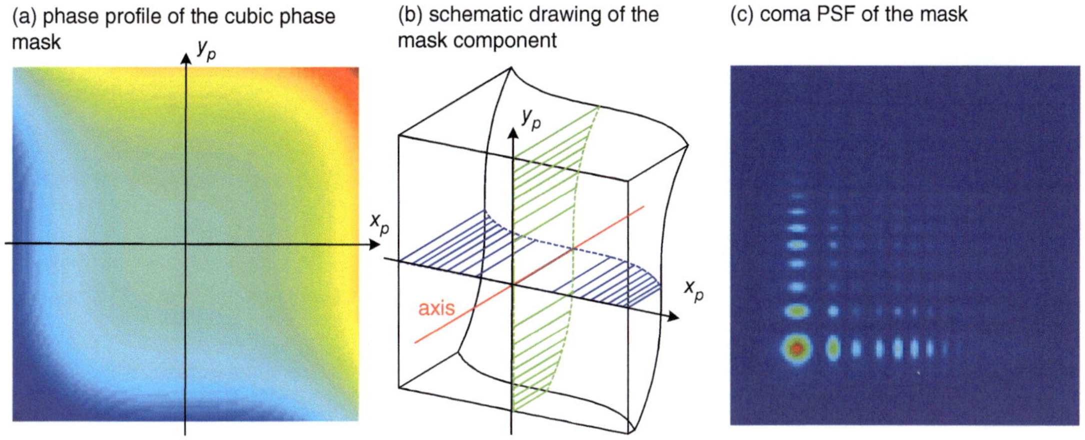

Figure 11.94 Phase profile (a) and mask shape (b) of a cubic cartesian filter and (c) corresponding point spread function.

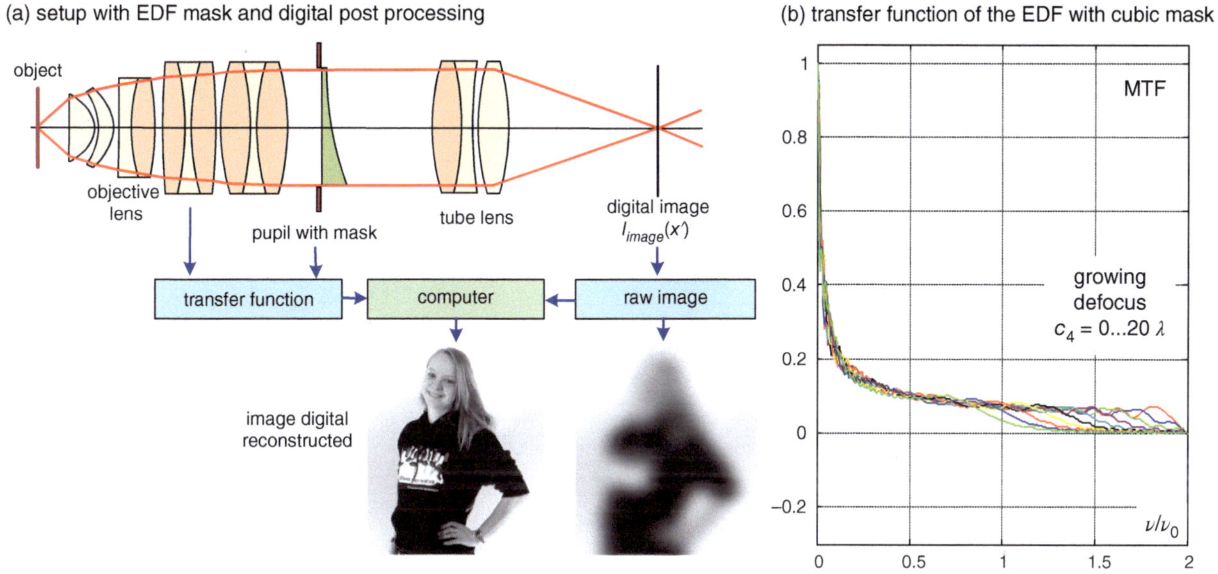

Figure 11.95 Extended depth of focus imaging setup (a) with a pupil mask and a digital post-processing and (b) corresponding transfer function with defocus $c_4 = 0\ldots20\ \lambda$.

and the resolution is nearly invariant during defocusing. The drawbacks of this approach are as follows, they are not all seen in the curves of Figure 11.95(b).

- For large frequencies, the contrast breaks down dependent on the defocus value, this means the resolution shrinks with growing defocus. This is seen in the diagram.
- The value of the contrast is rather small, depending on the parameter α. In this example it is around 10%. By digital contrast stretching to get a good image in the deconvolution, noise is amplified too.
- What is not seen in the MTF diagram: coma shifts the peak value; if the coma is large as here, a considerable lateral shift takes place so an 'image jump' is observed. This lateral shift corresponds to a contribution of the PTF. If the complex OTF is drawn with a separated real and imaginary axis it is seen that the curve looks like a spiral with nearly constant radius. This means the contrast is constant, but the PTF is of macroscopic size. As a consequence, a considerable lateral shift of the image is obtained.

This shows that the price of the EDF typically comes with some shortcomings. It depends on the application whether these can be accepted. Figure 11.96 illustrates an example of a cubic phase mask application in microscopy. It is seen in the last row that the performance in case of defocusing is somewhat better if the reconstruction calculation is performed, but the image changed and was overlaid by a certain texture. The defocus range in this example is 18 times the Rayleigh unit.

Figure 11.96 Image with defocussing for the original shown in (a) by standard setup in (b), with cubic mask as raw image in (c) and after deconvolution in (d). The shown re-scaled point spread function indicates the large coma of the cubic phase mask.

References

1 Rowlands, A. (2017). *Physics of Digital Photography*. IOP Publishing.
2 Andrews, H. and Hunt, B. (1977). *Digital Image Restoration*. Prentice Hall.
3 Mitra, S. (2011). *Digital Signal Processing*. McGraw Hill.
4 Castleman, K. (1996). *Digital Image Processing*. Prentice Hall.
5 Bovik, A. (2009). *The Essential Guide to Image Processing*. Academic Press.
6 Goodman, J. (2017). *Fourier Optics*. Freeman.
7 Bose, T. (2004). *Digital Signal and Image Processing*. Wiley.
8 Pratt, W. (2001). *Digital Image Processing*. Wiley.
9 Bhandari, A., Kadambi, A., and Raskar, R. (2022). *Computational Imaging*. MIT Press.
10 Reynolds, G., deVelis, J., Parrent, G., and Thompson, B. (1989). *Physical Optics Notebook: Tutorials in Fourier Optics*. SPIE Press.
11 Wu, Q., Merchant, F., and Castleman, K. (2008). *Microscope Image Processing*. Academic Press.
12 Katkovnik, V., Egiazarian, K., and Astola, J. (2006). *Local Approximation Techniques in Signal and Image Processing*. SPIE Press.
13 Török, P. and Kao, F.-J. (2007). *Optical Imaging and Microscopy*. Springer.
14 Coelho, R., Nascimento, V., de Queiroz, R. et al. (2016). *Signals and Images*. CRC Press.
15 Lukac, R. (2011). *Computational Photography*. CRC Press.
16 Milanfar, P. (2011). *Super-Resolution Imaging*. CRC Press.
17 Javidi, B., Tajahuerce, E., and Andres, P. (2014). *Multi-dimensional Imaging*. Wiley, IEEE Press.
18 Liu, Z., Zhou, X., and Liu, S. (2024). *Computational Optical Imaging*. Springer.

19 Ma, X. and Arce, G. (2010). *Computational Lithography*. Wiley.

20 Merchant, F. and Casteleman, K. (2023). *Microscope image processing*. Academic Press.

21 Yroslavsky, L. and Eden, M. (1996). *Fundamentals of Digital Optics*. Birkhäuser.

22 Vollmerhausen, R., Reago, D., and Driggers, R. (2010). *Sampled Imaging Systems*. SPIE Press.

23 Liang, J. (2024). *Coded Optical Imaging*. Springer.

24 Fiete, R. (2010). *Imaging Chain of Digital Cameras*. SPIE Press.

25 Francon, M. (1979). *Optical Image Formation and Processing*. Academic Press.

26 A. Lohmann (2006). *Optical Information Processing*. Universitätsverlag Ilmenau.

27 Voelz, D. (2011). *Computational Fourier Optics*. SPIE Press.

28 Christobal, G. and Schelkens, P. (2011). *Optical and Digital Image Processing* (ed. H. Thienpont). Wiley.

29 E. Glatzel, (1977), High speed, wide-angle lens system. USP 4136931.

30 B. Boehme, (2010). Simulation of visual performance for bifocal and monofocal intraocular lenses. Paper presented at *EOS Topical Meeting: Visual and Physiological Optics*, Stockholm, 2010.

31 L. Wenke, (2004), *University of Jena, Personal Communication, with Permission*

32 Mertz, J. (2019). *Optical Microscopy*. Cambridge University Press.

33 Gu, M. (1996). *Principles, of Three-Dimensional Imaging in Confocal Microscopes*. World Scientific.

34 Corle, T. and Kino, G. (1996). *Confocal Scanning Optical Microscopy and Related Imaging Systems*. Academic Press.

35 Lohmann, A. and Paris, D. (1965). Space variant image formation. *Journal of the Optical Society of America* 55: 1007.

36 Bertero, M., Boccacci, P., and De Mol, C. (2022). *Introduction to Inverse Problems in Imaging*. CRC Press.

37 C. Chen (2018). Parametrized PSF for Model Based System Engineering. Master's Thesis, University of Jena.

38 E. Thiebaut, L. Denis, F. Soulez, and R. Morya, (2019). Spatially variant PSF modeling and image deblurring. *SPIE Conference Proceedings*, 9909. http://dx.doi.org/10.1117/12.2233647.

39 Gilad, E. and von Hardenberg, J. (2006). A fast algorithm for convolution integrals with space and time variant kernels. *Journal of Computational Physics* 217: 326.

40 Denis, L., Thiebaut, E., Soulez, F. et al. (2015). Fast approximation of shift-variant blur. *International Journal of Computer Vision* 115: 253.

41 Gerbrands, J. (1981). On the relationships between SVD, KLT and PCA. *Pattern Recognition* 14: 375.

42 Gwak, M. and Yang, S. (2020). Modeling nonstationary lens blur using eigen blur kernels for restoration. *Optics Express* 28: 39501.

43 S. Chan (2010). Constructing a sparse convolution matrix for shift varying image restoration problems. Proceedings of the IEEE 17th International Conference of Image Processing. September 26-29, Hong Kong, China.

44 Rerabek, M. (2008). Space variant PSF-deconvolution of wide-field astronomical images. *Acta Polytechnica* 48: 79.

45 Löfdahl, M. (2007). Multiframe deconvolution with space-variant point-spread functions by use of inverse filtering and fast Fourier transform. *Applied Optics* 46: 4686.

46 Gureyev, T., Nesterets, Y., Stevenson, A., and Wilkins, S. (2003). A method for local deconvolution. *Applied Optics* 42: 6488.

47 J. Nagy, and D. O'Leary, (1997). Fast iterative image restoration with a spatially varying PSF. *SPIE Conference Proceedings*, 3162. https://ui.adsabs.harvard.edu/link_gateway/1997SPIE.3162..388N/doi:10.1117/12.279513.

48 Patwary, N. and Preza, C. (2015). Image restoration for three-dimensional fluorescence microscopy using an orthonormal basis for efficient representation of depth-variant point-spread function. *Biomedical Optics Express* 6: 3826.

49 Yan, T., Richardson, J., Zahng, M., and Gahlmann, A. (2019). Computational correction of spatially variant optical aberrations in 3D single-molecule localization microscopy. *Optics Express* 27: 12582.

50 Turcotte, R., Sutu, E., Schmidt, C. et al. (2020). Deconvolution for multimode fiber imaging: modeling of spatially variant PSF. *Biomedical Optics Express* 11: 4759.

51 Lynch, C., Devaney, N., and Dainty, C. (2019). Locally adaptive super-resolution through spatially variant interpolation. *Applied Optics* 58: 2920.

52 M. Subbarao, Y. Kang, S. Dutta, and X. Tu (2008). Localized and computationally efficient approach to shift-variant image deblurring. *15th IEEE International Conference on Image Processing*, 12–15 October 2008, San Diego, California, p. 657.

53 Yitzhaky, Y. and Stern, A. (2003). Restoration of interlaced images degraded by variable velocity motion. *Optical Engineering* 42: 3557.

54 R.B. Jenkins, S. Triantaphillidou, M.A. and Richardson (2007). Effective pictorial information capacity as an image quality metric. *SPIE Conference Proceedings* 6494. https://doi.org/10.1117/12.703973.

55 M. Chambah, A. Rizzi and C.S. Jean (2007). Image quality and automatic color equalization. *SPIE Conference Proceedings* 6394. https://doi.org/10.1117/12.697778.

56 S. Aja-Fernandez, R.S. Estepar, C.Alberola-Lopwz and C.-F. Westin (2006). Image quality assessment based on local variance. *Conference Proceedings of the IEEE Engineering in Medicine and Biology Society* 1. https://doi.org/10.1109/IEMBS.2006.259516.

57 Moorthy, A., Wang, Z., and Bovik, A. (2011, p419). Visual perception and quality assessment. In: *Optical and Digital Image Processing* (ed. G. Cristobal, P. Schelkens, and H. Thienpont). Wiley.

58 X. Ho, L. Thibos and A. Bradley (2001). Predicting visual performance from scalar metrics of optical quality. Paper presented at the *OSA Annual Meeting*, 16 October 2001, Long Beach, California.

59 Laparra, V., Munoz-Mari, J., and Malo, J. (2010). Divisive normalization image quality metric revisited. *Journal of the Optical Society of America A* 27: 852.

60 Wang, Z. and Bovik, A. (2009). Mean squared error: Love it or leave it. *IEEE Signal Processing Magazine* 26 (1): 98–117.

61 Gunturk, B. and Li, X. (2013). *Image Restoration.* CRC Press.

62 Haykin, S. (2002). *Adaptive Filter Theory.* Prentice Hall.

63 Ng, R. (2006). *Digital Light Field Photography, Thesis.* Stanford University.

64 Kwan, E. and Hua, H. (2022). Calibration of transverse ray and pupil aberrations for light field cameras. *Applied Optics* 61: 6974.

65 Hedayati, E. and Bos, J. (2020). Modeling standard plenoptic camera by an equivalent camera array. *Optical Engineering* 59: 073101.

66 Martinez-Corral, M. and Javidi, B. (2018). Fundamentals of 3D imaging and displays: a tutorial on integral imaging, light field and plenoptic systems. *Advances in Optics and Photonics* 10: 512.

67 Zhu, S., Lai, A., Eaton, K. et al. (2018). On the fundamental comparison between unfocused and focused light field cameras. *Applied Optics* 57: A1.

68 Lam, E. (2015). Computational photography with plenoptic camera and light field capture: tutorial. *Journal of the Optical Society of America A* 32: 2021.

69 Z. Zhang and M. Levoy (2009). Wigner distribution and how they relate to the light field. *Proceedings of the IEEE International Conference on Computational Photography* 48:12.

70 Okano, F., Arai, J., and Kawakita, M. (2007). Wave optical analysis of integral method for three-dimensional images. *Optics Letters* 32: 364.

71 A. Lumsdaine and T. Georgiev (2009). The focused plenoptic camera. *Proceedings of the IEEE International Conference on Computational Photography* 1. https://doi.org/10.1109/ICCPHOT.2009.5559008

72 M. Levoy and P. Hanrahan, (1996). Light field rendering. *SIGGRAPH* Proceedings of the 23rd Annual Conference on Computer Graphics and Interactive Techniques 31. doi.org/10.1145/237170.2371.

73 C. Perwaß and L. Wietzke (2012). Single lens 3D camera with extended depth-of-field. *SPIE Conference Proceedings* 12: 8291.

74 V. Eckstein, T. Schmid-Schirling, D. Carl and U. Wallrabe (2021). Depth-of-field comparison between the plenoptic camera 1.0 and 2.0. *SPIE Conference Proceedings* 11814. https://doi.org/10.1117/12.2594175.

75 Galdon, L., Saavedra, G., Garcia-Sucerquuia, J. et al. (2022). Fourier lightfield microscopy: a practical design guide. *Applied Optics* 61: 2558.

76 Saavedra, G., Martinez-Cuenca, R., Martinez-Corral, M. et al. (2008). Digital slicing of 3D scenes by Fourier filtering of integral images. *Optics Express* 16: 17154.

77 Hahne, C., Aggoun, A., Velisavljevic, V. et al. (2016). Refocusing distance of a standard plenoptic camera. *Optics Express* 24: 21521.

78 Popescu, G. (2011). *Quantitative Phase Imaging of Cells and Tissues.* McGraw Hill.

79 Liu, C., Wang, S., and Veetil, S. (2022). *Computational Optical Phase Imaging*. Springer.

80 Glückstad, J. and Palima, D. (2009). *Generalized Phase Contrast*. Springer.

81 Gerchberg, R. (1972). A practical algorithm for the determination of phase from image and diffraction plane pictures. *Optik* 35: 237.

82 Fienup, J. (2012). Phase retrieval: a personal tour. *Applied Optics* 52: 45.

83 Fienup, J. (1982). Phase retrieval algorithms: a comparison. *Applied Optics* 21: 2758.

84 Konijnenberg, A. (2019). *Computational Methods for Phase Retrieval, Thesis.* Technical University Delft.

85 Teague, M. (1983). Deterministic phase retrieval: a green's function solution. *Journal of the Optical Society of America* 73: 1434.

86 Gureyev, T., Roberts, A., and Nugent, K. (1995). Partially coherent fields, the transport-of-intensity equation and phase uniqueness. *Journal of the Optical Society of America A* 12: 1942.

87 Waller, L., Tian, L., and Barbastathis, G. (2010). Transport of intensity phase-amplitude imaging with higher order intensity derivatives. *Optics Express* 18: 12552.

88 Rodenburg, J. and Faulkner, H. (2004). A phase retrieval algorithm for shifting illumination. *Applied Physics Letters* 85: 4795.

89 S. Prasad, V. Pauca, R. Plemmons, T. Torgersen and J. van der Gracht (2005). Pupil-phase optimization for extended-focus aberration-corrected imaging systems. *SPIE Conference Proceedings* 5559: 335.

90 Dowski, E. and Cathey, T. (1995). Extended depth of field through wavefront coding. *Applied Optics* 34: 1859.

91 George, N. and Chi, W. (2003). Extended depth of field using a logarithmic asphere. *Journal of Optics A: Pure and Applied Optics* 5: S157.

92 Wang, H. and Gan, F. (2001). High focal depth with a pure-phase apodizer. *Applied Optics* 40: 5658.

12

Mathematical Appendix

12.1 Fourier Transform and Related

12.1.1 Fourier Series

If the function $f(x)$ is periodic with a period X

$$f(x+X) = f(x) \tag{12.1}$$

it can be approximately written as a Fourier series as a discrete sum over complex base functions $e^{2\pi\, i\, k \frac{x}{X}}$ as

$$f(x) = \sum_{k=-K}^{K} F(k) \cdot e^{2\pi\, ik\, \frac{x}{X}} \tag{12.2}$$

The base functions fulfil the orthogonality relation

$$\int_0^X e^{2\pi\, ik\frac{x}{X}} \cdot e^{2\pi\, ik'\frac{x}{X}}\, dx = X \cdot \delta_{k\,k\prime} \tag{12.3}$$

The coefficients $F(k)$ of the series can therefore be computed as

$$F(k) = \frac{1}{X}\int_0^X f(x) \cdot e^{-2\pi\, ik\frac{x}{X}}\, dx \tag{12.4}$$

The Fourier series is an optical approximation of $f(x)$, the coefficients describe the frequency amplitudes and are not dependent on the length X. The series is a harmonic analysis of the function, $F(k)$ is also called the spectrum.

12.1.2 Fourier Transform

In optics, and in physics generally, one often deals with wave phenomena. Hence the harmonic functions are widely used as orthogonal basis functions. Therefore, it is important to present here the basic principles and the essential properties of the Fourier series expansion and the Fourier integral representation [1–10].

For the general case of non-periodic functions, the Fourier integral transform has to be used as a generalization of the Fourier series. The transform and the inverse transform are given by:

$$U(x) = \int u(\nu)\ e^{2\pi i x\nu} d\nu$$

$$u(\nu) = \int U(x)\ e^{-2\pi i x\nu} dx \tag{12.5}$$

In computers or experimental measurements one mostly deals with discrete data. This leads to the implementation of the discrete Fourier transform (DFT) which reads in its one-dimensional form as:

$$U(x_k) = \sum_{m=-N/2}^{N/2-1} u(\nu_m) \cdot e^{i2\pi\cdot x_k\cdot \nu_m}$$

$$u(\nu_m) = \frac{1}{N}\sum_{k=-N/2}^{N/2-1} U(x_k) \cdot e^{-i2\pi\cdot x_k\cdot \nu_m} \tag{12.6}$$

The most important properties of the Fourier transform are collected in the following list. In any case the functions $u(x)$ and $U(\nu)$ are Fourier related here

$$u(\nu) = \int U(x) \cdot e^{2\pi\cdot i\nu x} dx \tag{12.7}$$

- Linearity.
  The Fourier transform $F[U]$ of a function U is linear.

$$\hat{F}[a\cdot f(x) + b\cdot g(x)] = a\cdot \hat{F}[f(x)] + b\cdot \hat{F}[g(x)] \tag{12.8}$$

- Scaling (or similarity).
  If a function is 'stretched' by a factor a, its Fourier transform is 'squeezed' by the same factor in the frequency domain:

$$\hat{F}[u(ax)] = \frac{1}{|a|}\cdot U\left(\frac{\nu}{a}\right) \tag{12.9}$$

- Symmetry – real spectrum.
  Symmetric functions can be expanded in cosine functions only. Therefore, their spectrum is real.

$$U(x) = U(-x)\quad \Leftrightarrow\quad Im[u(\nu)] = 0 \tag{12.10}$$

Lens Design for Imaging: Fundamentals of Optical Systems, First Edition. Herbert Gross.

- Anti-symmetry – imaginary spectrum.
  Conversely, anti-symmetric functions exhibit a pure imaginary spectrum.

$$U(x) = -U(-x) \quad \Leftrightarrow \quad Re[u(\nu)] = 0 \tag{12.11}$$

- Conjugation.
  The Fourier transform of the complex conjugate function is reflected around the origin in the frequency domain.

$$U^*(x) \Leftrightarrow u^*(-\nu) \tag{12.12}$$

- Area.
  The area under a function or its integral corresponds to the constant component or DC-term (value at the origin) of its spectrum.

$$u(0) = \int U(x)dx \tag{12.13}$$

- Shifting and translation.
  If a function is shifted, its spectrum is modulated by a phase factor. Modulation of a function by a harmonic function with frequency ν, corresponding to the carrier frequency, shifts the spectrum by an amount ν.

$$\begin{aligned} \widehat{F}[u(x+a)] &= e^{2\pi\ i\nu\ a} \cdot U(\nu) \\ U(\nu - \nu_0) &= e^{2\pi\ i\nu_0\ a} \cdot \widehat{F}[u(x)] \end{aligned} \tag{12.14}$$

- Parity.
  Depending on the symmetry properties of the function $f(x)$ or its real and imaginary part respectively, the Fourier transform shows the symmetry according to the cases in the Table 12.1.
- Differentiation.
  Differentiation of a function corresponds to multiplication of its Fourier transform by a factor proportional to the frequency.

$$\frac{d}{dx}U(x) \Leftrightarrow 2\pi i\nu \cdot u(\nu) \tag{12.15}$$

- Integration.
  Integration of a function corresponds similarly to dividing its Fourier transform by a factor proportional to the frequency.

$$\int U(x)\,dx \Leftrightarrow \frac{u(\nu)}{2\pi i\nu} \tag{12.16}$$

- Double use; in the case of two successive transforms, the arguments switch their signs

$$\widehat{F}\{\widehat{F}[u(x)]\} = u(-x) \tag{12.17}$$

- Moment theorem.
  The nth derivative with respect to ν is given by

$$\frac{d^n u(\nu)}{d\nu^n} = \int (2\pi \cdot ix)^n U(x) \cdot e^{2\pi \cdot i\nu x} dx \tag{12.18}$$

  For $\nu = 0$, one obtains the moment theorem that establishes a relationship between the nth moment of a function and the nth derivative of its Fourier transform at the origin $\nu = 0$

$$M_n = \int x^n \cdot U(x)\,dx = \frac{1}{(2\pi \cdot i)^n} \cdot u^{(n)}(0) \tag{12.19}$$

- Parseval or Plancherel theorem.

Integration of the auto correlation theorem over ν gives the following relation

$$\int_{-\infty}^{\infty} |f(x)|^2\,dx = \int_{-\infty}^{\infty} |F(\nu)|^2\,d\nu \tag{12.20}$$

In principle this relation between spatial and frequency domain can be considered as the conservation of energy.

12.1.3 Discrete Fourier Transform

12.1.3.1 Definition

Often the data are given on a discrete equidistant grid as

$$x_j = j \cdot \Delta x = j \cdot \frac{X}{N} \tag{12.21}$$

with the indices $j = 0, 1, 2, \ldots, N-1$ and the values $f_j = f(x_j)$. In this case a special algorithm according to Cooley–Tukey

Table 12.1 Parity properties.

u(*x*)	*Re*[*u*(*x*)]	*Im*[*u*(*x*)]	*U*(*k*)	*Re*[*U*(*ν*)]	*Im*[*U*(*ν*)]
Real				Even	Odd
Imaginary				Odd	Even
	Even	Odd	Real		
	Odd	Even	Imaginary		
Imaginary even			Imaginary even		
Imaginary odd			Imaginary odd		
Complex even			Complex even		
Complex odd			Complex odd		

allows us to perform the Fourier transform very quickly. The number of operations shrinks from N^2 to $N \cdot \log N$, this is in particular for two dimensions a huge acceleration. There are several special approaches to realize the basic idea; the original one uses values of N, which are powers of 2. But this so-called fast Fourier transform (FFT) has the additional constraint that the grid spacing in the spatial and the frequency domain is coupled. In the frequency domain the same number of N values F_j are created, the spectrum is symmetric around $\nu = 0$ with the spacing

$$\Delta\nu = \frac{1}{X}, \quad \nu_{\max} = \frac{1}{2\,\Delta x} = \left(\frac{N}{2} - 1\right)\Delta\nu,$$
$$\nu_{\min} = -\frac{N}{2}\,\Delta\nu$$
$$\nu_j = \left(j - \frac{N}{2}\right) \cdot \Delta\nu, \; j = 0, 1, 2, \ldots, N-1 \tag{12.22}$$

The Fourier integral can in this case be written as the finite sum

$$F(\nu_m) = \sum_{j=0}^{N-1} f(x_j)\, e^{-2\pi i \cdot \frac{\nu_m x_j}{N}}$$
$$f(x_j) = \sum_{m=-\frac{N}{2}}^{\frac{N}{2}-1} F(\nu_m)\, e^{2\pi i \cdot \frac{\nu_m x_j}{N}} \tag{12.23}$$

The discrete Fourier coefficients depend on the number N of grid points. The Parseval theorem is now formulated by a sum over the coefficients as

$$\sum_n |f_n|^2 = \frac{1}{N} \sum_n |F_n|^2 \tag{12.24}$$

12.1.3.2 Sampling and Aliasing

The discrete nature of data derived from measurements and the computer processing of such data require a detailed consideration of the DFT. The discrete sampling of a continuous function $U(x)$ replaces the integral transform by a discrete summation. The appropriate choice of the sampling interval and the truncation window (number of terms) is essential in order to avoid errors between the computed and the desired transform. The discrete sampling of a function $U(x)$ can be represented as:

$$U_d(x) = U(x) \cdot \mathrm{comb}\left(\frac{x}{\Delta x}\right) \tag{12.25}$$

where the sampling interval is introduced by the comb function

$$\mathrm{comb}\left(\frac{x}{\Delta x}\right) = \sum_n \delta(x - n \cdot \Delta x) \tag{12.26}$$

If the sampling interval is Δx, the maximum frequency that can be resolved (truncation in the frequency domain) is given by the reciprocal value of Δx:

$$\nu_{\max} = \frac{1}{\Delta x} \tag{12.27}$$

The spectrum of the sampled function will be convolved in the frequency domain with a comb function:

$$u_d(\nu) = u(\nu) \otimes \Delta x \cdot \mathrm{comb}(\Delta x \cdot \nu) \tag{12.28}$$

Hence the sampling in space with an interval dx results in periodic repetition of the spectrum in the frequency domain with a period $\nu_{\max} = 1/\Delta x$. The sampling interval should be small enough that the periodically repeating spectrum corresponding to the continuous function does not overlap in the frequency domain (aliasing effect). This is illustrated in Figure 12.1.

In general, an overlap of the spectra creates a distortion of the spectrum in the frequency domain since due to the periodic repetition of the spectrum, high frequencies can interfere with low frequencies. Hence, the discrete sampling is applicable only to bandwidth-limited signals. In this case it is sufficient to select the sampling interval to be small enough in order to avoid such a cross-talk effect:

$$\Delta x < \frac{1}{\nu_{BW}} \tag{12.29}$$

ν_{BW} corresponds to the spectral bandwidth of the continuous function. When a function exhibits higher frequencies, the above-mentioned errors will occur. This is the sampling theorem of Shannon.

Let us take as an example a function $U(x)$ which is sampled over a window x_{max} with a finite number of points N. The use of the series expansion and hence the DFT implies periodicity of the function. Hence the actual signal function $S(x)$ is obtained by truncation with a rect function and convolution with a comb function:

$$S(x) = \left[U(x) \cdot \mathrm{rect}\left(\frac{x}{x_{\max}}\right) \cdot \mathrm{comb}\left(\frac{x}{\Delta x}\right)\right]$$
$$\otimes\, \mathrm{comb}\left(\frac{x}{x_{\max}}\right) \tag{12.30}$$

The period (window) x_{max} is chosen as N times the sampling interval, $x_{max} = N \cdot \Delta x$. For the corresponding spectrum one obtains in the frequency domain:

$$s(\nu) = \{\,[u(\nu) \otimes \mathrm{sinc}(x_{\max} \cdot \nu)] \otimes \mathrm{comb}(\Delta x \cdot \nu)\}$$
$$\cdot\, \mathrm{comb}(x_{\max} \cdot \nu) \tag{12.31}$$

The convolution with the first delta comb results in a periodic repetition of the spectrum. The multiplication with a rect function reduces the spectrum to the first period. This corresponds to a convolution of the sampled function in the spatial domain with a sinc function

$$\mathrm{sinc}(x/\Delta x) = \mathrm{sinc}(\nu_{BW} \cdot x) \tag{12.32}$$

Therefore, the spectrum of a bandwidth-limited signal can be expressed as

$$s(\nu) = [u(\nu) \otimes \mathrm{sinc}(x_{\max} \cdot \nu)] \cdot \mathrm{comb}(x_{\max} \cdot \nu) \tag{12.33}$$

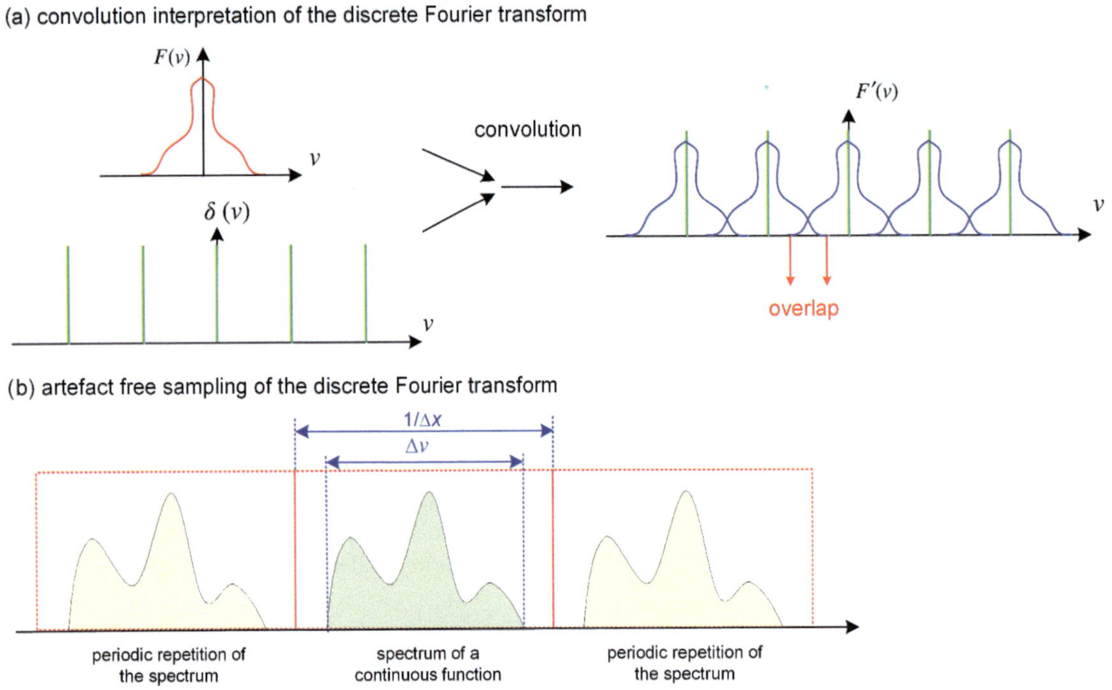

Figure 12.1 Convolution interpretation of the DFT (a) and necessary sampling (b) to avoid aliasing.

where

$$\Delta x = \frac{x_{\max}}{N} = \frac{1}{\nu_{BW}}, \quad \Delta \nu = \frac{\nu_{BW}}{N} = \frac{1}{x_{\max}},$$
$$\Delta x \cdot \Delta \nu = \frac{1}{N} \tag{12.34}$$
$$\nu_{BW} \cdot x_{\max} = N, \quad \nu_{BW} = 2\nu_{\max}, \quad x_{\max} \cdot \nu_{\max} = \frac{N}{2}$$

hold. This coupling between the grid spacings in the spatial and the frequency domain is another formulation of the sampling theorem. It should be noticed here that the spatial domain is asymmetric sampled from $0 \ldots x_{max}$, while the frequency axis is symmetrical around $\nu = 0$. The bandwidth goes thus from $-\nu_{max} \cdots + \nu_{max}$ and we have the factor 2 between the bandwidth and the maximum frequency.

The coupling means that a coarse grid in the spatial domain results in a fine grid in the frequency domain and vice versa. In practical applications the decision of which domain to sample with which pixel size is often not trivial. Zero padding is often a meaningful possibility to adjust the frequency gridding accordingly. If the signal in the spatial domain contains certain fine structures, which can only described by corresponding high frequencies, the spatial sampling is pre-given. If in this case the number of grid points N is too small, the spectra which occur due to the underlying periodicity overlap and the signal is under sampled. This is sketched in Figure 12.2, here in (a) a well sampled signal creates separated spectra, in (b) the case of under sampling shows on overlap of the spectra.

In optical applications related to diffraction calculations it is mostly desirable to have in the spatial and also the spectral domain the zero variables $x = 0$ and $\nu = 0$ included. In this case an even number of sampling points N gives one more pixel as needed. This additional pixel is typically added in the negative range. Therefore, the sampling is selected as

$$\Delta x = \frac{2x_{\max}}{N-2}, \quad x_j = -x_{\max} - \Delta x + (j-1) \cdot \Delta x, \quad \times j = 1, 2, \ldots N \tag{12.35}$$

in the spatial domain and similarly as

$$\Delta \nu = \frac{2\nu_{\max}}{N-2} = \frac{\nu_{BW}}{N-2}, \quad \nu_j = -\nu_{\max} - \Delta \nu + (j-1) \cdot \Delta \nu, \quad j = 1, 2, \ldots N \tag{12.36}$$

in the frequency domain. This scheme is sketched in Figure 12.3 for the spatial domain. The frequency case looks the same.

12.1.3.3 Leakage

Consider the sampling of a sine function at N points with a window x_{max}, a multiple of the period $p = x_{\max}/4$. Since $\Delta \nu = 1/x_{\max}$ one would expect the frequencies of the sine signal to occur at $\pm 1/p = \pm 4/x_{\max} = \pm 4\,\Delta\nu$, i.e. at $m = \pm 4$.

It follows that

$$s(\nu) = [\delta(\pm 4\Delta\nu) \otimes \mathrm{sinc}(x_{\max} \cdot \nu)] \cdot \mathrm{comb}(x_{\max} \cdot \nu) \tag{12.37}$$

i.e. when the sampling window is matched, the DFT of a sine function yields only two frequency components in the spectrum. Consider now the case when the same sine

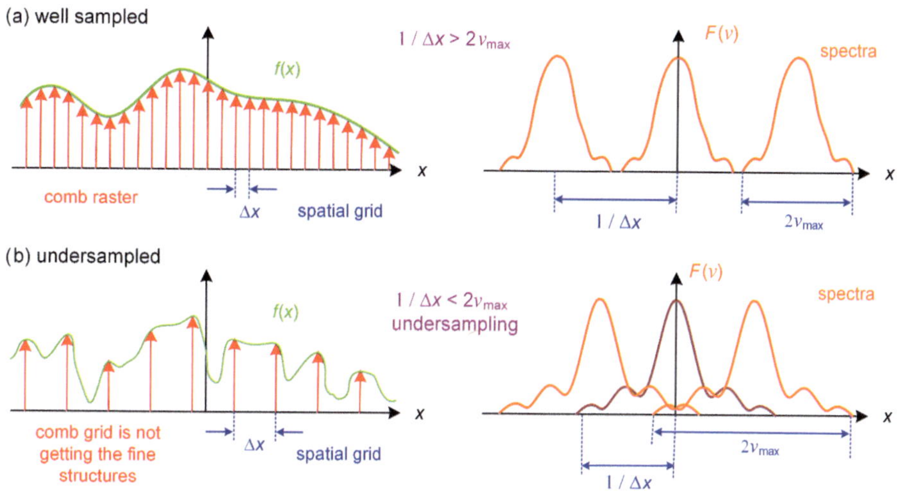

Figure 12.2 The sampling condition is fulfilled in (a) and violated in (b).

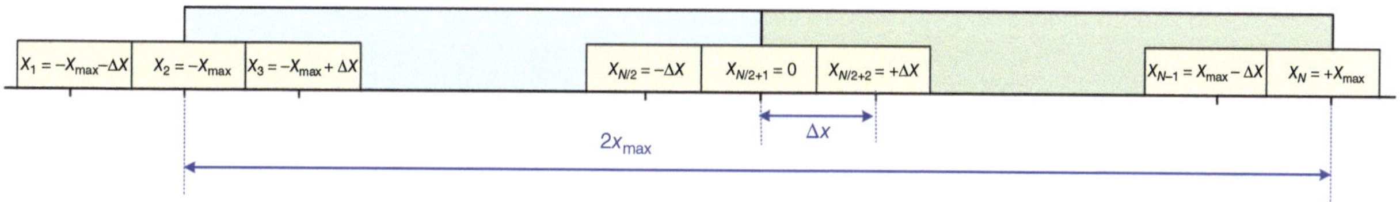

Figure 12.3 Symmetrical sampling including the zero case in the spatial domain.

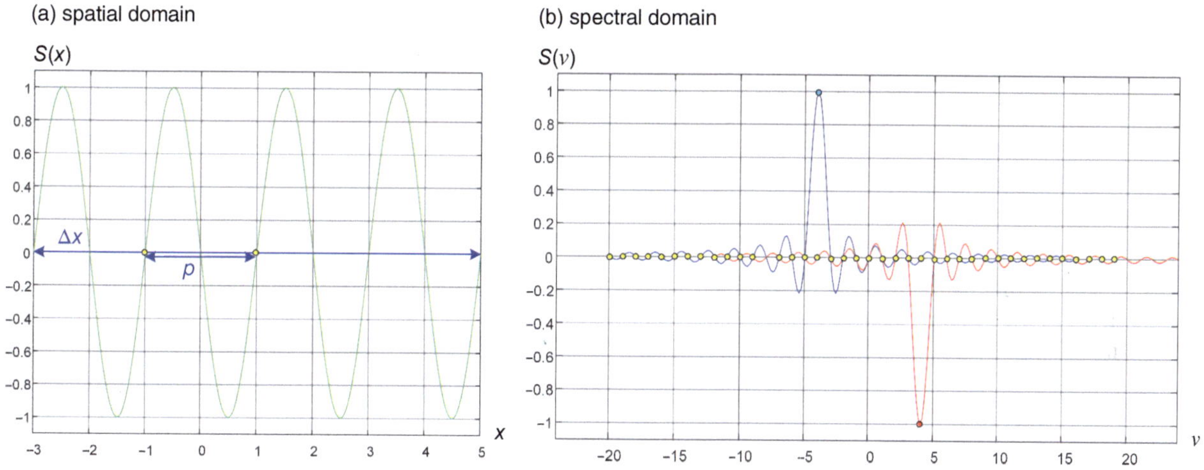

Figure 12.4 Sampling interval equal to an integer multiple of the signal period, (a) shows the spatial and (b) the spectral representation.

signal is sampled with a window x_{max} that is not an integer multiple of the period but equals, e.g. 3.8 periods. From the imaginary part of the spectrum shown in Figure 12.5, it can be seen that the mismatch between the sampling window and the period results in noticeable oscillations accompanying the peaks.

$$s(\nu) = [\delta(\pm 3.8\Delta\nu) \otimes \mathrm{sinc}(x_{\max} \cdot \nu)] \cdot \mathrm{comb}(x_{\max} \cdot \nu) \tag{12.38}$$

The exact frequencies must be evaluated by fitting with a sinc function. This phenomenon is known as leakage. Leakage has to be taken into account, e.g. in the phase-shift interferometry where the wave phase is measured. In this case either a large number of phase measurements or a precisely aligned phase step is required. The origin of leakage is that the rectangular function of the spatial window from $x = 0 \ldots x_{\max}$ creates a oscillatory sinc function in the frequency domain. If this periodic function interferes with the signal, missing information can result.

Due to the description of the DFT, the sinc function is used to interpolate the equidistant points of the signal. For a continuous signal, this kind of interpolation is optimal. If

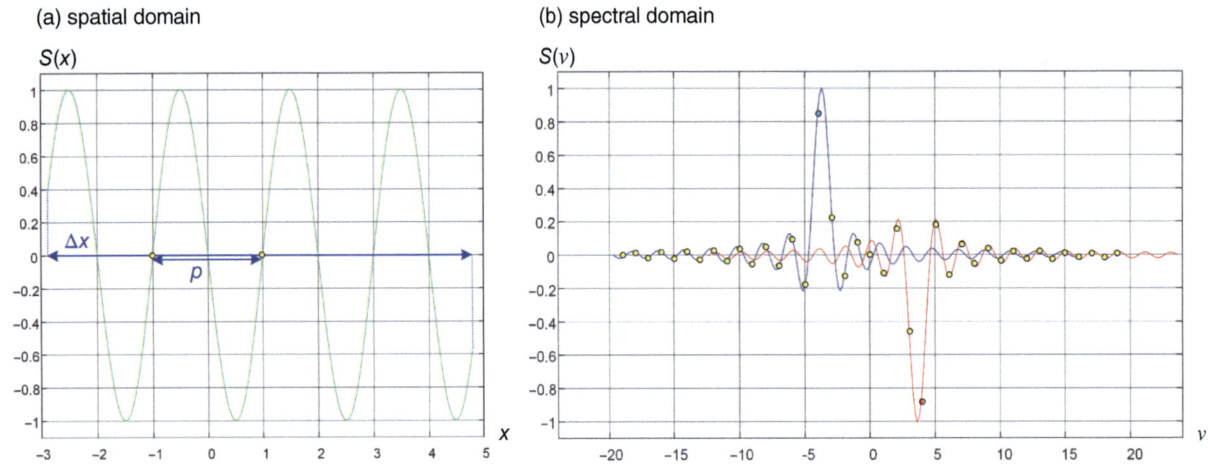

Figure 12.5 Sampling interval not equal to an integer multiple of the signal period, (a) shows the spatial and (b) the spectral representation.

in addition the signal is band-limited, the sinc interpolation is exact. This interpolation can be written as

$$a_{resample}(x) = \sum_{k=0}^{N-1} a_k \cdot \frac{\sin\left[\pi \cdot \left(\frac{x}{x_{\max}} - k\right)\right]}{N \cdot \sin\left[\frac{\pi}{N} \cdot \left(\frac{x}{x_{\max}} - k\right)\right]} \tag{12.39}$$

Therefore, an interpolation can be numerically performed by a Fourier transform, a change of the frequency gridding by zero-padding and a back transform. The same result can be obtained by a chirp-z-transform.

12.1.3.4 Zero Padding and Balanced Sampling

According to the sampling theorem, the spacing in the initial plane and in the final plane of a Fourier based diffraction calculation are coupled. If an optical system is considered this means that the spacing Δx_p in the exit pupil and the spacing $\Delta x'$ in the image plane are connected. If the wave is of high quality, it can be observed that for a dense sampling of the pupil only very few signal points are obtained in the point spread function. If the resolution of $I(x')$ is unsatisfactory and the number of sampling points in the pupil is fixed to obtain a serious sampling of boundary, phase and maybe apodization, the easiest possibility is to increase the number N of points, keeping Δx_p constant. This means that the lateral computational extent in the pupil plane is increased to much larger values as the clear diameter dictates. The points are here filled with zero fields, this is the so-called zero padding. $\Delta x'$ then decreases and the resolution is improved. The disadvantage of this method is the huge memory consumption and a rather ineffective calculation process in dealing mainly with zeros.

To get a better idea about realistic numbers for this method, as an example here a Gaussian paraxial profile is considered. Since the Fourier transform of a Gaussian function is again Gaussian, the estimation can be calculated analytically and compared with numerical results. A Gaussian profile in the spatial domain is given as

$$f(x) = e^{-\frac{x^2}{w_x^2}} \tag{12.40}$$

The Fourier transform of this function is

$$F(\nu) = w_x\sqrt{\pi}\cdot e^{-\pi^2 w_x^2 \nu^2} = w_x\sqrt{\pi}\cdot e^{-\frac{\nu^2}{w_\nu^2}} \tag{12.41}$$

with

$$w_\nu = \frac{1}{\pi \cdot w_x} \tag{12.42}$$

If the FFT is calculated on a spatial domain D with N discrete points, we have

$$D = N \cdot \Delta x, \quad \nu_{\max} = \frac{N}{D} = N \cdot \Delta\nu, \quad \Delta\nu = \frac{1}{D} \tag{12.43}$$

A proper distribution of the resolution in the spatial and in the frequency domain needs a clever choice of the relative filling and the zero padding. The number of points above $1/e^2$ is given by

$$N_x = \frac{2\sqrt{2}w_x}{\Delta x}; \quad N_\nu = \frac{2\sqrt{2}w_\nu}{\Delta\nu} = \frac{2\sqrt{2}D}{\pi w_x} \tag{12.44}$$

to get equal resolution in both spaces $Nx = N\nu$, the condition

$$w_x^{(opt)} = \frac{D}{\sqrt{\pi N}} \tag{12.45}$$

should be fulfilled. The number of points above $1/e2$ are then in the spatial and the frequency domain the same

$$N_x^{(opt)} = N_\nu^{(opt)} = \sqrt{\frac{8N}{\pi}} \tag{12.46}$$

This case corresponds to a balanced sampling in both domains. Here is a short example.

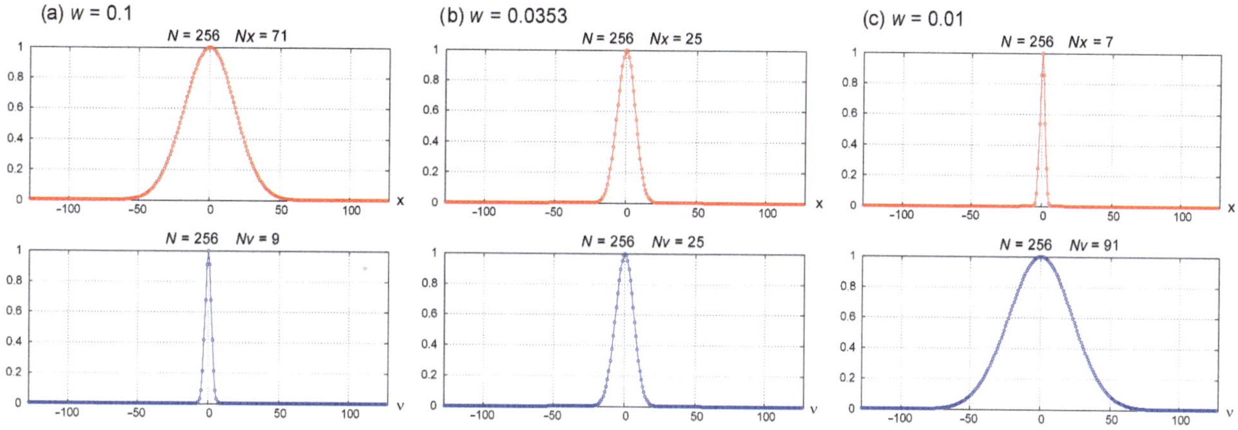

Figure 12.6 Gaussian profiles in the spatial domain (upper row, in red) and in the frequency domain (lower row in blue) for $N = 256$ sampling points and different numbers N_x of points inside the central lobe above $1/e^2$. In (a) the frequency space are badly sampled and in (c) the spatial signal; (b) illustrates a balanced equal resolution in both spaces. The numbers are the indices, symmetrical around zero.

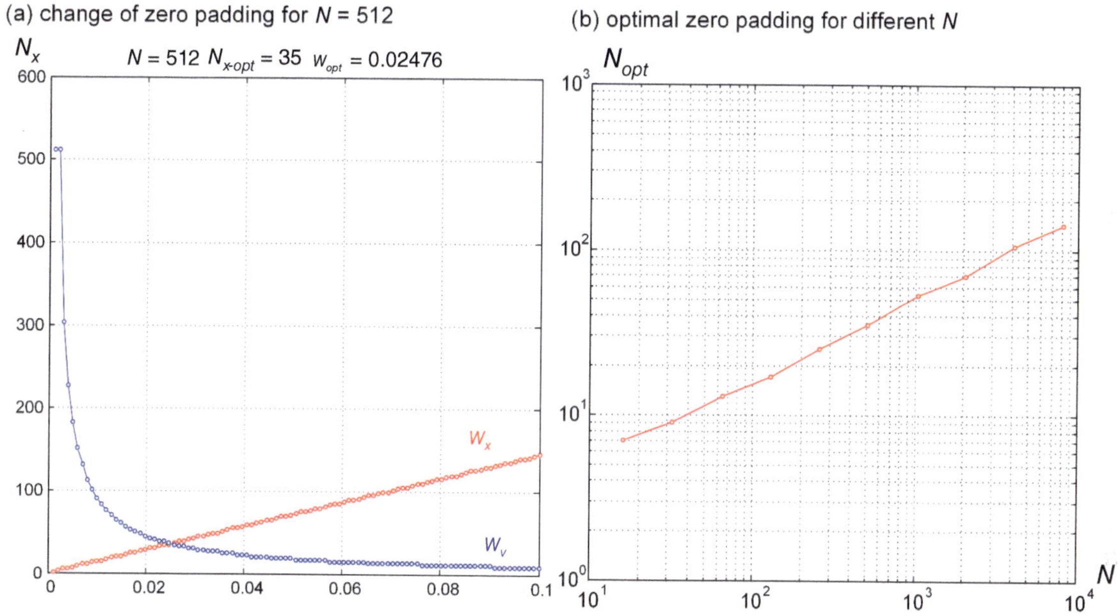

Figure 12.7 Change of (a) the zero padding for $N = 512$ by increasing N_x and w_x simultaneously and (b) optimal zero padding for various values of N.

Sampling for $D = 1$ and $N = 256$ for three different values of w gives the profiles shown in Figure 12.6.

The following diagram in Figure 12.7 shows the relative size changes of w_ν if the point number $N = 512$ is fixed and N_x or w_x are scaled. This corresponds to a change in the zero padding. It can be seen that for $N = 512$ the numbers $N_x = N_\nu = 35$ are identical, which means that here the resolution and number of significant points in the spatial as well as in the frequency domain are the same, which corresponds to an optimal zero padding. This means, 512–35 = 477 or 93% sampling points lie outside of the 13.5% level.

If the optimum value for w_x and the corresponding relevant number of points are shown as a function of the total number of points N, we get Figure 12.7 (b).

12.1.4 Chirp-z-Transform*

(*) The content of this section is quite specialized and for advanced reading. Knowledge of this section is not necessary to understand the following chapters. Beginners are recommended to skip this section.

The problematic fixed coupling of the spatial and spectral resolution in the numerical DFT, dictated by the sampling theorem, can be avoided by applying the z-transform

[11–14] The general discrete z-transform of the function f, for given discrete data $f(x_j)=f_j$ with $j=0,1,2,\ldots,N-1$ and a likewise discrete sequence of numbers z_k with $k=0,1,2,\ldots,M$, reads

$$F_k = F(z_k) = \sum_{j=0}^{N-1} f_j\, z_k^{-j} \tag{12.47}$$

In accordance with its application in signal processing, this algorithm is named the chirp-z-transform (CZT). Special cases are:

- in the particular case when z is a real exponential function, the z-transform is equivalent to the well-known Laplace transform

$$z_k = e^{sT} \tag{12.48}$$

- For a set of points (function z) equally spaced around the unit circle

$$z_k = e^{\frac{2\pi i k}{N}} \tag{12.49}$$

where $N=M$, one obtains the DFT. For a more general form of the z-function

$$z_k = AW^{-k} \tag{12.50}$$

and a spiral contour in the complex z-plane has to be evaluated. It can be efficiently computed by a convolution using an algorithm similar to that of the FFT. Consider

$$A = A_0 \cdot e^{2\pi\, i\theta_0},\quad W = W_0 \cdot e^{2\pi\, i\Phi_0} \tag{12.51}$$

where $2\pi\Phi_0$ is the angular spacing on the contour with respect to the origin. Insertion of the z-series into the expression for the transform yields

$$F_k = \sum_{n=0}^{N-1} f_n\, A^{-n}\, W^{n\,k} \tag{12.52}$$

The classical DFT implies that the function is periodic. In the complex z-plane this corresponds to sampling at equidistantly spaced points around the unit circle with a frequency separation of $2\pi/N$, see Figure 12.8.

In the case of the CZT, this sampling is confined to a narrow frequency interval. In the general case, a spiral contour occurs. If the frequency sampling is applied without filtering, a frequency zoom appears as in the sampling diagram shown in Figure 12.8(b). With the help of the modified binomial formula

$$nk = \frac{1}{2}[n^2 + k^2 - (k-n)^2] \tag{12.53}$$

the evaluation of the z-transform is reduced to an algorithm due to Bluestein

$$\begin{aligned} F_k &= W^{\frac{k^2}{2}} \sum_{n=0}^{N-1} \left[f_n A^{-n} W^{\frac{n^2}{2}} \right] W^{-\frac{(n-k)^2}{2}} \\ &= W^{\frac{k^2}{2}} \left[f_n A^{-n} W^{\frac{n^2}{2}} \right] \otimes W^{-\frac{n^2}{2}} \end{aligned} \tag{12.54}$$

The computation of the above expression can be realized as a convolution and requires the following steps.

- Weighting or multiplication of the given series f_n by $A^{-n}\, W^{\frac{n^2}{2}}$.
- Fourier transform of the obtained chirped function.
- Calculation of a function $W^{-\frac{n^2}{2}}$.
- Fourier transform of this function.
- Multiplication of the Fourier series in order to compute the convolution.
- Inverse transform of the product and multiplication by $W^{\frac{k^2}{2}}$.

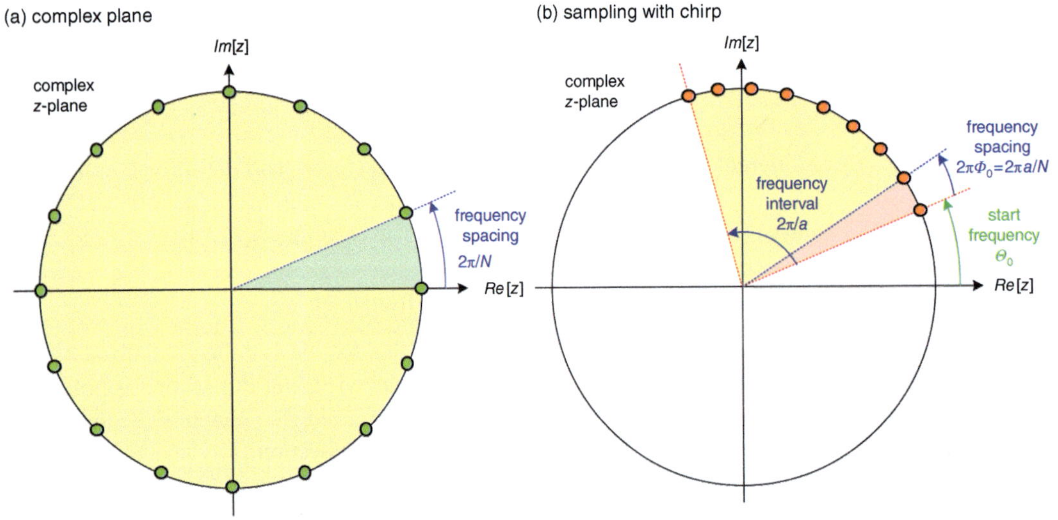

Figure 12.8 Sampling points (a) around the unit circle for the classical Fourier transform and (b) in the case of an additional chirp.

The computation of the z-transform is hence reduced to a convolution, which can again be evaluated by a FFT algorithm. The additional computational steps required result in an increased complexity in comparison to the simple Fourier transform. The computational time is increased about three times, but the exact factor depends on the choice of the number of points N, M and the selected FFT algorithm. The advantages of the z-transform over the Fourier transform are related basically to two properties.

- The desired number of points M does not have to coincide with the given number of points N. Hence one can save computer time if it is necessary to calculate only a few points in the frequency domain.
- Since W itself is a complex number, it can be represented in general in the form

$$W = W_0 e^{2\pi i \Phi_0} \tag{12.55}$$

The case of the Fourier transform corresponds to the special choice

$$\Phi_0 = \frac{1}{N} \tag{12.56}$$

and the points occupy exactly one angular circuit on the unit circle. As a consequence of this property, the frequency values ν of the result of the transform are fixed by the sampling theorem in the form

$$\Delta x \cdot \Delta \nu = 1/N \tag{12.57}$$

If the phase of W in the CZT is reduced to a multiplication by a factor $s > 1$

$$\Phi_0 = \frac{1}{N \cdot s} \tag{12.58}$$

this results in sampling in the frequency domain with a step

$$\Delta \nu = \frac{1}{\Delta x \cdot s \cdot N} \tag{12.59}$$

Therefore, the choice of resolution in the frequency domain is not restricted. In particular, one can compute the spectrum on a specified grid and avoid any interpolation. This is of special importance for optical transfer computations using Fourier algorithms where only fields within free apertures need to be considered. In addition, the advantages of the z-transform are significant when sequential transforms have to be performed. This is the case, e.g. in optical transfer problems related to cavity calculations or beam propagation in inhomogeneous media or for cascaded systems. The change in the number of points of a sampled signal or their spacing can be carried out by interpolation. According to the sampling theorem the best method for interpolating a continuous signal is by the sinc function. However, this method can be applied only to equidistantly sampled functions. The application of the chirp-z transform is more favourable than zero-padding, because an arbitrary scale factor s can be selected.

If the computational effort between the usual fast Fourier transform and the chirp-z-transform is compared, a behaviour like the one illustrated in Figure 12.9 is obtained. The basis of this graph is a straightforward implementation in Matlab. It is seen that in the double logarithmic representation the run time increases nearly linear to the relevant range of point numbers $N = 128\ldots4096$ in one dimension. The FFT is roughly a factor of 20 faster; for the same run time with the FFT roughly 4 times more sampling points can be processed. What is not directly seen in this curve is the growing memory consumption, this changes the linearity for point numbers larger than 4096, depending on the hardware used. A zero padding for the FFT increases the resolution in the image plane by a factor of 4, in the case of the CZT the resolution is improved. This means that the CZT solution for resolution improvement is more efficient for $N > 2\pi = 6.3$.

If in an optical application for propagation of a complex field the wave front is aberrated and the sampling conditions are not ideal, as in the case of the Gaussian profile,

Figure 12.9 Comparison of the computational time for two-dimensional calculations by fast Fourier or chirp-z-transform respectively.

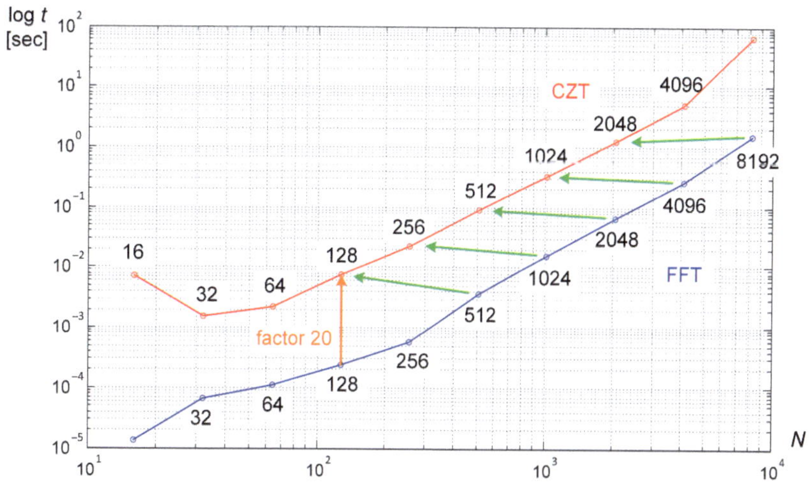

the comparison is less comfortable for the CZT. In reality it makes sense to sample a discrete field in a way that includes the point on axis $x = 0$. This means that for an even number of points ($N = 2^m$) there is one point, which is not needed. This means, the centre $x = 0$ is selected at the index $N/2+1$ and we have

$$\Delta x = \frac{2x_{\max}}{N-2}, \quad x_{\max} = \left(\frac{N}{2} - 1\right) \cdot \Delta x$$
$$x_j = -x_{\max} - \Delta x \ldots x_{\max} = \left[-\frac{N}{2} : 1 : \frac{N}{2} - 1\right] \cdot \Delta x \tag{12.60}$$

The exact relationship between the spatial and frequency (or angle u) domain is then

$$\Delta v = \frac{1}{(N+2) \cdot s \cdot \Delta x}, \quad \Delta u = \frac{\lambda}{(N+2) \cdot s \cdot \Delta x} \tag{12.61}$$

To let the Parseval theorem be valid, a pre-factor of

$$y(v) = \frac{1}{\sqrt{N \cdot s}} \cdot \hat{Z}[y(x)], \quad y(u) = \frac{\lambda}{\sqrt{N \cdot s}} \cdot \hat{Z}[y(x)] \tag{12.62}$$

must be foreseen by application of the z-transform $\hat{Z}$.

12.1.5 Semi-Analytical Fourier Transform*

(*) The content of this section is quite specialized and for advanced reading. Knowledge of this section is not necessary to understand the following chapters. Beginners are recommended to skip this section.

The so-called semi-analytical approach helps to numerically relax the sampling conditions of a complex field that carries a strong linear and quadratic part in the phase function [15–17]. These first and second order contributions can be treated separately in an analytical way. The Fourier transform of the remaining field should then be somewhat better conditioned with small phase gradients of the remaining complex field. The given complex field is written as

$$E(\vec{r}) = A(\vec{r}) \cdot e^{i\,\Phi(\vec{r})} \tag{12.63}$$

For a better numerical condition in calculating the propagation, a quadratic phase is extracted out of the full phase distribution Φ:

$$E(\vec{r}) = U(\vec{r}) \cdot e^{i \cdot (a_x x^2 + a_y y^2 + b_x x + b_y y + cxy + d)} = U(\vec{r}) \cdot e^{i\,\Psi(\vec{r})}$$
$$\Psi(\vec{r}) = a_x x^2 + a_y y^2 + b_x x + b_y y + cxy + d \tag{12.64}$$

with a quasi flattened field $U(r)$. The Fourier transform of $E(r)$ can now be separated as a convolution of the Fourier transform of the quasi-flattened residual field $U(r)$ and the exponential function.

$$E(\vec{k}) = \hat{F}[E(\vec{r})] = \frac{1}{2\pi} \iint E(\vec{r}) \cdot e^{-i\,\vec{k}\vec{r}} d^2\vec{r}$$
$$= \frac{1}{2\pi} \iint U(\vec{r}) e^{i\Psi(\vec{r})} \cdot e^{-i\,\vec{k}\vec{r}} d^2\vec{r} = \hat{F}[U(\vec{r})] \otimes \hat{F}[e^{i\,\Psi(\vec{r})}] \tag{12.65}$$

The first term of this convolution must be treated numerically, the second one can be calculated analytically by the expression

$$\hat{F}[e^{i\,\Psi(\vec{r})}] = -\frac{2\pi \cdot sign(4a_x a_y - c^2)}{\sqrt{-4a_x a_y + c^2}} \cdot e^{i\left[d - \frac{1}{4a_x a_y - c^2}\left(\begin{array}{c} a_x k_x^2 + a_y k_y^2 + (2a_y b_x - b_y c)k_x \\ + (2a_x b_y - b_x c)k_y - c k_x k_y \\ + a_x b_y^2 + a_y b_x^2 - c b_x b_y \end{array}\right)\right]}$$
$$= \alpha \cdot e^{i\,\Phi(\vec{k})} \tag{12.66}$$

with the parameters

$$\gamma = 4a_x a_y - c^2$$
$$\alpha = -\frac{2\pi \cdot sign(\gamma)}{\sqrt{-\gamma}} \Phi(k_x, k_y) = d - \frac{1}{\gamma}\left[\begin{array}{l} a_x k_x^2 + a_y k_y^2 + (2a_y b_x - b_y c)k_x + (2a_x b_y \\ - b_x c)k_y - c k_x k_y + a_x b_y^2 + a_y b_x^2 - c b_x b_y \end{array}\right] \tag{12.67}$$

With this expression, the convolution in Eq. (12.65) reads

$$E(\vec{k}) = \hat{F}[E(\vec{r})] = \alpha \iint \hat{F}_{k\prime}[U(\vec{r})] \cdot e^{i\,\Phi(\vec{k}-\vec{k}\prime)} d^2\vec{k}' \tag{12.68}$$

The same procedure can be used for the inverse Fourier transform, and in the case of a propagation calculation this is needed. It should be further mentioned that there are several singular cases considering the value of the mixed term with coefficient c in two lateral dimensions. If the phase is purely cylindrical, a special treatment is necessary.

In principle, a calculation of the plane wave expansion for propagation using the semi-analytical Fourier transform is nothing other than the quasi-flattening transform in Fresnel approximation. In the real implementation, the choice of a proper quadratic form to reduce the sampling requirements is crucial. Therefore, an unwrapped representation of the phase is required. If this is given, there are furthermore two options to quasi-flatten the field:

- best approximation on axis by using a Taylor expansion of the phase;
- best approximation over the whole domain by fitting the quadratic polynomial onto the field. This guarantees a better uniformity and is more recommended.

If the classical Fourier transform of a complex function

$$f(x) = e^{-\left(\frac{x}{w}\right)^2 - i \cdot a^2 x^2} = e^{-\left(\frac{x}{w}\right)^2 - \frac{i \cdot x^2}{\lambda R}} \tag{12.69}$$

is considered, the performance of the numerical transform decreases with growing values of a. In the optical application, a is related to a curvature of a parabolic wave. If the normalized product $a \cdot w$ is increased, the rms grows and the error of the transforms becomes unacceptable. In Figure 12.10 values between $a \cdot w = 2\ 10{-}4 \ldots 3$ are calculated for different numbers of sampling points N. It can be seen that in a log–log representation, the error behaves nearly linearly. Furthermore, it is observed that a growing number of points N helps, but is not very effective. By switching from $N = 64$ to 2048 less than one order of magnitude is gained in accuracy.

This simulation shows that the additional effort in algorithmic implementation by extracting the quadratic part makes sense; a brute force method to improve the precision is not meaningful.

12.1.6 Non-Uniform Fourier Transform*

(*) The content of this section is quite specialized and for advanced reading. Knowledge of this section is not necessary to understand the following chapters. Beginners are recommended to skip this section.

The sampling of the FFT signal has to be equidistant by the requirement of the algorithm in the space and in the frequency domain. If one of the domains is non-equidistant, the so called non-uniform Fourier transform (NUFT) can be used [18–21]. One possibility to get a NUFT is the elementary direct integration with unequal grid points. If, for example, the frequency axis is non-equidistant sampled, in the discrete case with a non-uniform sequence ν_n we have

$$F(\nu_n) = \sum_m f(x_m) \cdot e^{-2\pi\, i\, x_m \nu_n} \tag{12.70}$$

As an example, a case is considered in Figure 12.11 with dense sampling on the left side and a coarse grid on the

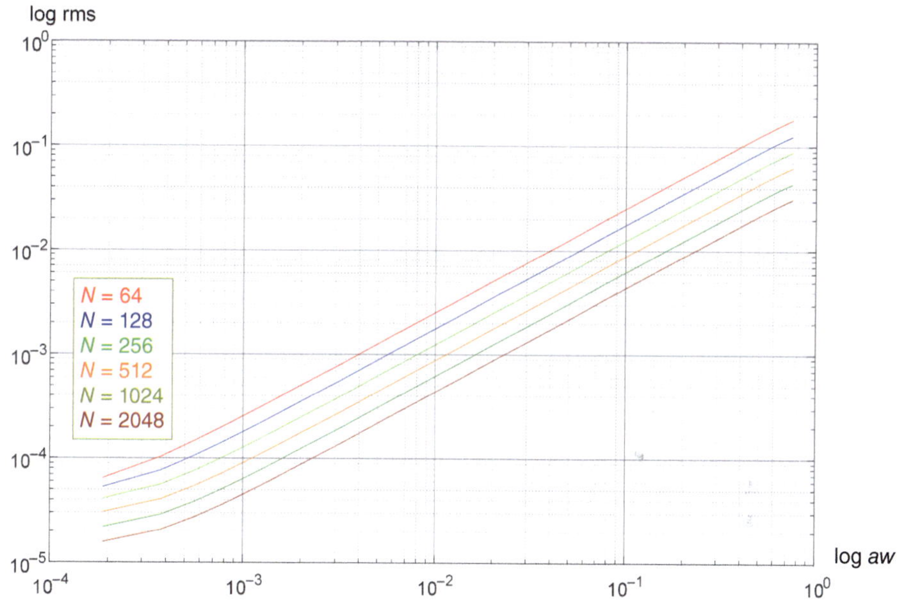

Figure 12.10 rms error of the numerical Fourier transform of a complex oscillatory function.

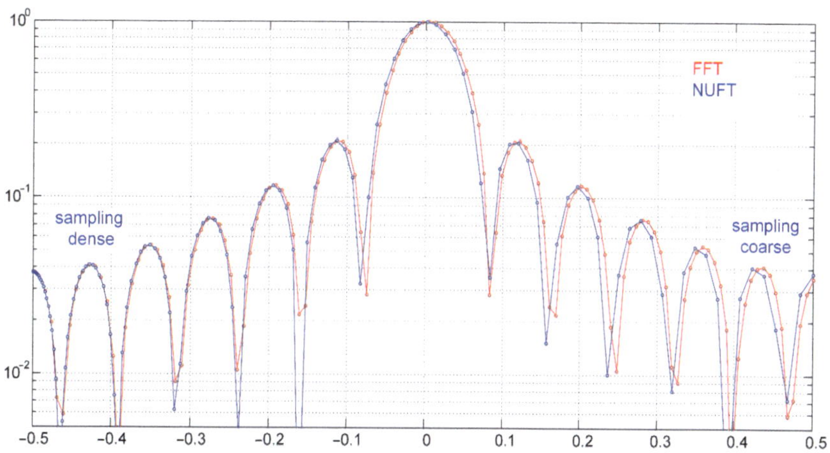

Figure 12.11 Fourier transform and NUFT for a non-equidistant sampling.

Table 12.2 Run time performance comparison of the FFT, direct calculated NUFT and NUQFFT in one dimension.

N	FFT	Direct NUFT	NUQFFT
1024	0.00018	0.070	0.015
2048	0.00022	0.30	0.022
4096	0.00033	1.21	0.042
8192	0.0006	5.5	0.1

Table 12.3 Performance of the NUFT in two dimensions.

N	FFT	Direct FT	NUQFFT
64	0.00035	1.68	0.028
128	0.00070	13.1	0.111
256	0.0020	126.0	0.42
512	0.018	2105.0	1.6
1024	0.044		6.2

right side. In optical applications the need to allow for non-equidistant sampling points occurs during the calculation of a point spread function for a system suffering from pupil aberrations of the rays. The equidistant grid in the entrance pupil is then distorted in the exit pupil.

There are no algorithms for the calculation of NUFT that are as fast as the FFT. But there are iterative approximative calculation schemes that are somewhat faster than the elementary direct integration method, they are called non-uniform quasi-fast Fourier transform (NUQFFT). The accuracy of this approximate interpolation method strongly depends on the degree of non-uniformity. The following Table 12.2 lists the computational times for the one-dimensional case for orientation; Figure 12.12 shows the results of a computation by all the three methods for a simple one-dimensional case with an equidistant grid. As can be is seen here, the accuracy of the results is quite good until relative signal heights of 10^{-4}.

The following Table 12.3 lists the computational times for the two-dimensional case. The last column contains the execution time values for the quasi-fast algorithm of Greengard [18].

In general, there are three different types of NUFT scenarios proposed in literature, they are listed in Table 12.4.

In the case of a type 1 transform, typically a convolution-based interpolation on the basis of Gaussian functions is calculated. First an up-sampling is performed, which defines the speed and accuracy of the algorithm. The choice of the interpolation kernel is adapted to the spacing of the non-uniform grid. This is sketched in Figure 12.13.

Table 12.4 Different NUFT options.

Type	Grid transform
1	Non-uniform → uniform
2	Uniform → non-uniform
3	Non-uniform → other non-uniform

The transform can be written in the form

$$a(z_m) = \frac{1}{M}\sum_{n=0}^{M-1} I(k_n)\cdot e^{-\frac{2\pi\, i\, m\, k_n}{K}} \tag{12.71}$$

with the non-uniform grid values k_n, the complete interval is K and the uniform sampling is given by the values at z_m with M sampling points. The interpolation kernel is given by

$$G_\tau(k) = e^{-\frac{k^2}{4\tau}} \tag{12.72}$$

This kernel must be truncated to M_{sp} points to both sides of the given data point. If R is the up-sampling ratio, the width of the Gaussian kernel is given by

$$\tau = \frac{\pi \cdot M_{sp}}{R(R-0.5)\cdot M^2} \tag{12.73}$$

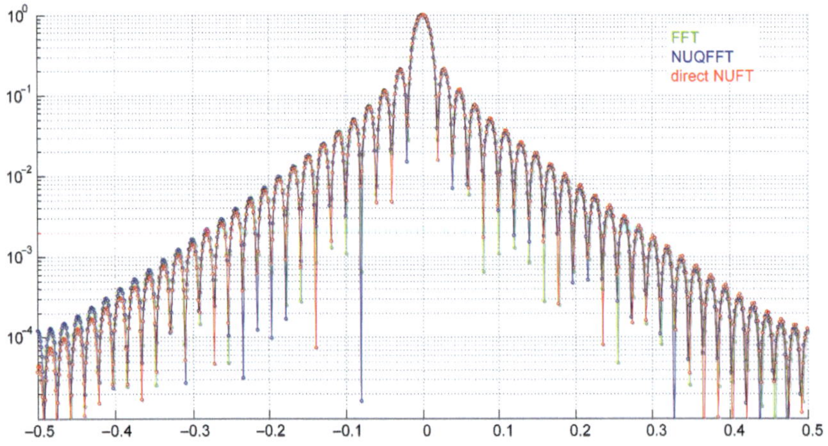

Figure 12.12 Comparison of the precision of a direct calculated NUFT, the NUQFFT and the FFT.

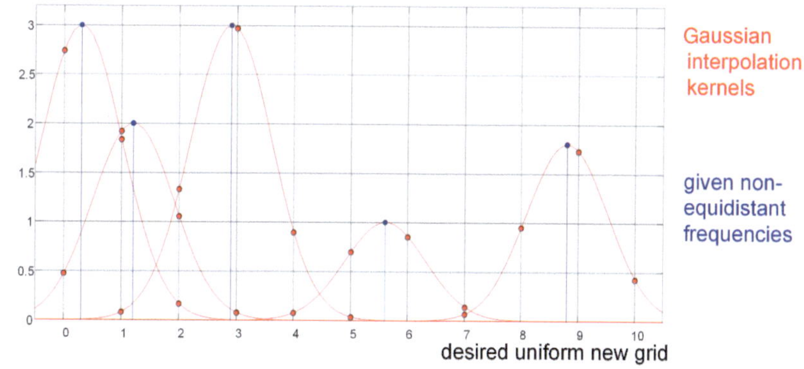

Figure 12.13 Interpolation schemes of NUFT, the given values are shown in blue and carry Gaussian functions, which are interpolated at the equidistant sampled red dots.

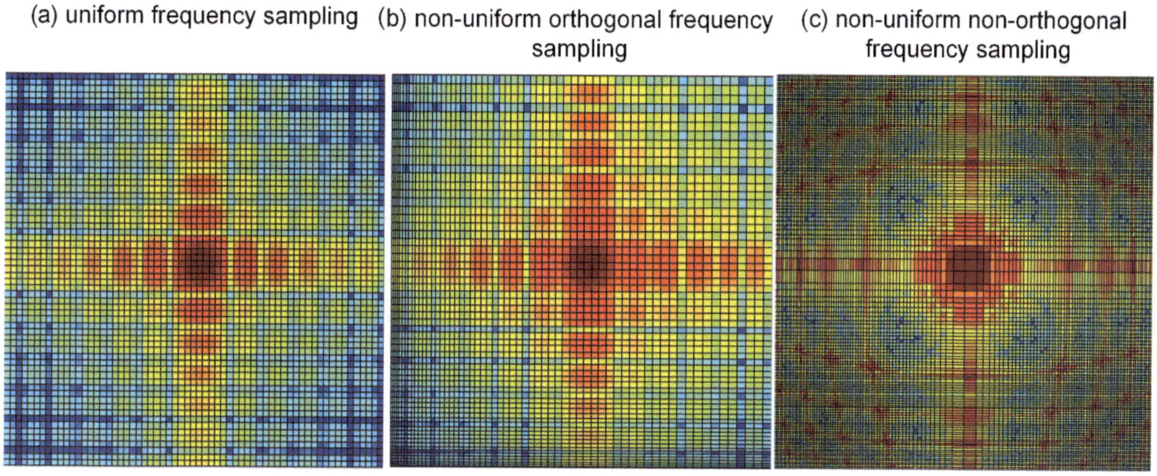

Figure 12.14 Example plots of grids in two dimensions. (a) Uniform frequency sampling, (b) non-uniform but orthogonal frequency sampling and (c) non-uniform non-orthogonal sampling in case of distortion.

Typically to balance the time and the accuracy requirements, the values $M_{sp} = 3$ and $R = 2$ are chosen. An intermediate function $I_\tau(k)$ is defined by the convolution

$$I_\tau(k) = I(k) * G_\tau(k) = \int I(y) G_\tau(k-y)\, dy \qquad (12.74)$$

This function is sampled on an over-sampled uniform grid with M_r points. From this, we get

$$a_\tau(z_m) = \frac{1}{M_r} \sum_{l=0}^{M_r-1} I_\tau\left(\frac{l \cdot K}{M_r}\right) \cdot e^{-\frac{2\pi\, i\, m\, l}{M_r}} \qquad (12.75)$$

Finally, a deconvolution in the k-space gives the desired function. Due to the Gaussian form of g and the transform

$$g(z_m) = \widehat{F}^{-1}[G_\tau(k)] = \sqrt{2\tau} \cdot e^{-\tau \cdot z_m^2} \qquad (12.76)$$

this can be calculated by a simple division as

$$a(z_m) = \frac{1}{\sqrt{2\tau}} \cdot e^{\tau \cdot z_m^2} \cdot a_\tau(z_m) \qquad (12.77)$$

A problem in calculating a NUFT arises in two dimensions, because an arbitrary distorted grid cannot be considered a Cartesian product. Here special algorithms are necessary to compute the transform with reasonable speed on arbitrary distorted two-dimensional-grids. Figure 12.14 shows example plots in two dimensions with uniform frequency sampling, non-uniform but orthogonal stretching of the frequency scale as well as a general case of non-orthogonality.

12.1.7 Special Fourier Related Operations

12.1.7.1 Convolution Integral

The convolution integral between two functions u and v is defined as

$$u(x) * v(x) = \int u(x-t) \cdot v(t)\, dt \qquad (12.78)$$

If the function $u(x)$ is defined between $x = a \ldots b$ and $v(x)$ between $x = c \ldots d$, the convolution is defined in the range $x = a + c \ldots b + d$. If $U(\nu)$ and $V(\nu)$ are the Fourier transform spectra of u and v, the Fourier transform of the convolution is simply the product of the spectra

$$U(\nu) = \widehat{F}[u(x)], \quad V(\nu) = \widehat{F}[v(x)]$$
$$\widehat{F}[u(x) * v(x)] = U(\nu) \cdot V(\nu) \qquad (12.79)$$

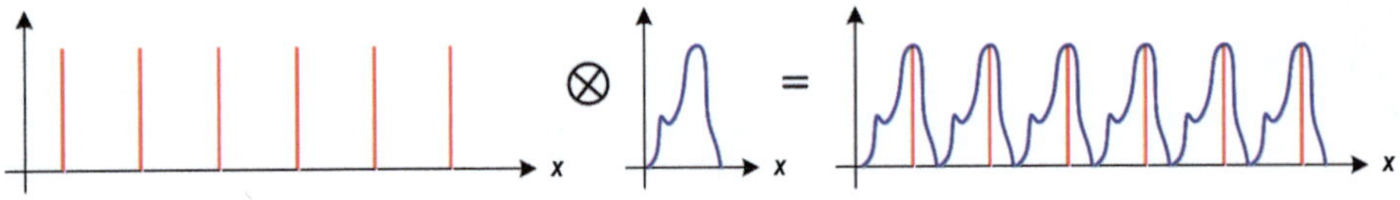

Figure 12.15 Convolution of two functions.

The interpretation of a convolution is sketched in Figure 12.15. In this simple case of a comb function, the second factor is simply reproduced several times.

Convolution obeys the commutative, associative and distributive laws

$$U \otimes V = V \otimes U$$
$$U \otimes (V \otimes W) = (U \otimes V) \otimes W$$
$$U \otimes (V + W) = (U \otimes V) + (U \otimes W) \qquad (12.80)$$

If a function $f(x)$ is only known by its coevolution with a given and known function g(x), the convolution theorem can be used to resolve the unknown function by

$$f(x) = \hat{F}^{-1}\left[\frac{\hat{F}[f(x) * g(x)]}{\hat{F}[g(x)]}\right] \qquad (12.81)$$

as long as the spectrum $G(u)$ has no zeros.

12.1.7.2 Auto-Correlation

A special important case of a convolution is the auto-correlation, a convolution of a function with a shifted version of itself.

$$C_{ACF}(x) = u(x)\circ u^*(x) = \int u(t)\ u^*(t+x)dt \qquad (12.82)$$

With the Fourier transform the relation

$$C_{ACF}[u(x)] = \hat{F}^{-1}[|F[u(x)]|^2] \qquad (12.83)$$

holds and shows that the auto-correlation can be computed by Fourier transform in a comfortable manner. The auto-correlation indicates a similarity of features in a function, in practice mostly the correlation decreases with growing distance of the compared points; as is sketched in Figure 12.16, the similarity is lost.

The correlation length corresponds to the distance $x = L_C$, where the normalized auto-correlation drops to the value $1/e$ (or another threshold value). The Fourier transform of the autocorrelation corresponds to the power spectral density

$$F_{PSD}(\nu) = \hat{F}[u(x)\circ u^*(x)] = |U(\nu)|^2 \qquad (12.84)$$

The value for $x = 0$ (correlation at the same position) gives the rms value squared as a measure of the quality of the self-similarity of the function

$$C_{ACF}[u(0)] = \sigma_{rms}^2 = \int |U(\nu)|^2\, d\nu \qquad (12.85)$$

12.1.7.3 Cross Correlation and Covariance

The cross correlation of two functions is given by a similar integration, but now two different functions are compared

$$C_{CF} = u(x)\circ v(x) = \int u(t)\ v(t+x)dt \qquad (12.86)$$

The correlation function is commonly used for comparative analysis of two signals or images. The correlation function is a direct measure for the similarity of the fluctuations of the two signals u and v, see Figure 12.17 for illustration. If the function $u(x)$ is defined between $x = a\ldots b$ and $v(x)$ between $x = c\ldots d$, the correlation is defined in the range $x = a\text{-}d\ldots b\text{-}c$.

The correlation is not commutative

$$w(x) = u(x)\circ v(x)$$
$$w(-x) = v(x)\circ u(x) \qquad (12.87)$$

The correlation can be represented as a convolution as:

$$u(x)\circ v(x) = u(x) \otimes v(-x) \qquad (12.88)$$

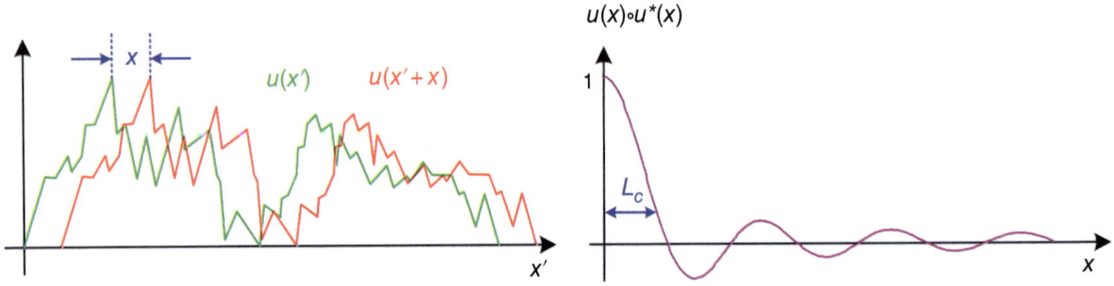

Figure 12.16 Building of the autocorrelation of a function $u(x)$.

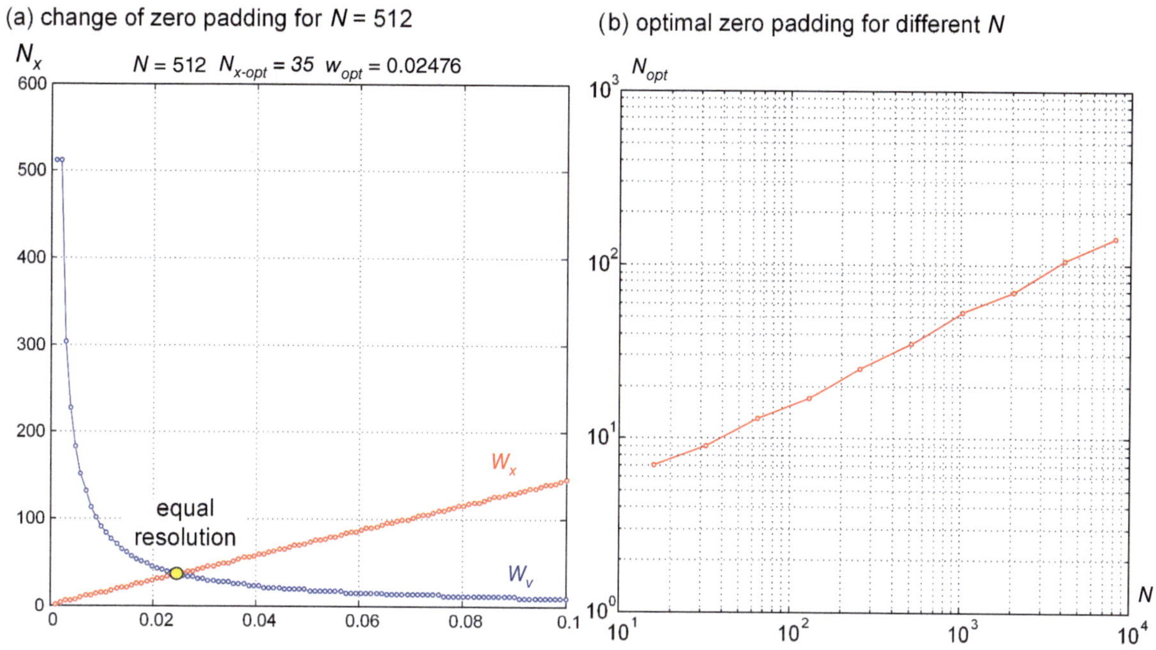

Figure 12.17 Visualization of the correlation and similarity between two functions $u(x)$ and $v(x)$.

The covariance of two signals u and v is defined as

$$\begin{aligned} C_{Cov} &= u(x) \odot v(x) = \int [u(t) - \overline{u}] \cdot [v(t+x) - \overline{v}] dt \\ &= C_{CF} - \overline{u} \cdot \overline{v} \end{aligned} \tag{12.89}$$

and is therefore tightly related to the cross correlation. The only difference is the subtraction of the offset $\overline{u} \cdot \overline{v}$, so the functions are shifted to zero means.

12.1.8 Fourier Slice Theorem and Radon Transform*

(*) The content of this section is quite specialized and for advanced reading. Knowledge of this section is not necessary to understand the following chapters. Beginners are recommended to skip this section.

If a two-dimensional function $f(x,y)$ is projected into several special directions with different angles θ, these slices should often be used to reconstruct the Cartesian two-dimensional function [4–10]. This is the typical task of a tomography reconstruction. If the rotated coordinates are x' and y' with

$$x' = x \cdot \cos\theta + y \cdot \sin\theta, \quad y' = -x \cdot \sin\theta + y \cdot \cos\theta \tag{12.90}$$

The projection into the direction θ is given by

$$\begin{aligned} g_\theta(x') = \hat{P}_\theta[f(x,y)] = \int_{-\infty}^{\infty}\int_{-\infty}^{\infty} f(x,y) \\ \cdot \delta(x' - x \cdot \cos\theta - y \cdot \sin\theta)\, dx\, dy \end{aligned} \tag{12.91}$$

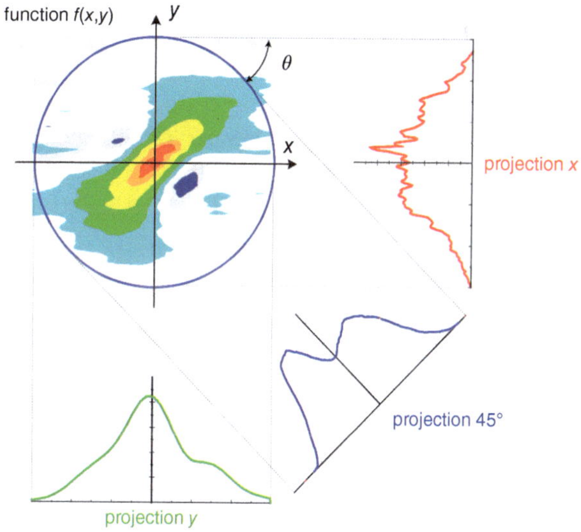

Figure 12.18 Projection and Radon transform.

which is called the Radon transform. P_θ is the projection operator. Figure 12.18 illustrates this operation, where projection means integration along the direction with angle θ.

Now the one-dimensional Fourier transform along θ is obtained if first the projection is calculated as before in Eq. (12.91) along the slice, and then one calculates the one-dimensional Fourier transform of $g_\theta(x')$. This relation is called the projection slice theorem

$$G(v_\theta) = G(v \cdot \cos\theta, v \cdot \sin\theta) \tag{12.92}$$

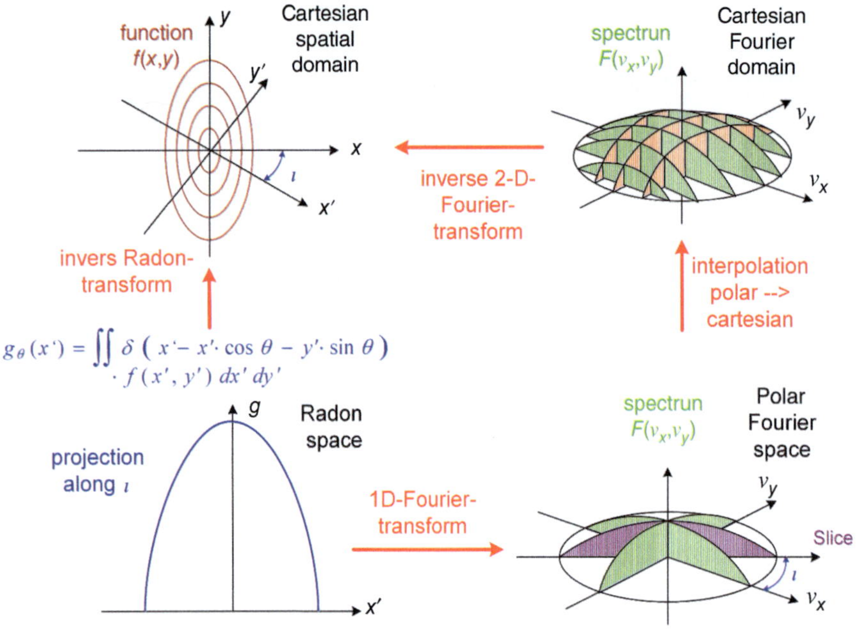

Figure 12.19 Projection slice theorem and relation between Radon transform and Fourier transform.

By using the projection operator, the Fourier slice theorem can be written in the form

$$\hat{F}[\hat{P}_\theta[f(x,y)]] = F(v_x, v_y)\Big|_\theta \tag{12.93}$$

with the Cartesian two-dimensional Fourier transform $F(vx,vy)$ of $f(x,y)$. This relation is again visualized and summarized in Figure 12.19; the left part is the spatial, the right part the spectral domain. This then gives the receipt to compute the inverse Radon transform by a polar-Cartesian interpolation in the Fourier space.

Prominent applications of these mathematical relations are x-ray tomographic reconstruction, the imaging in plenoptical imaging in the phase space and the measurement of Wigner distributions in the phase space.

12.2 Miscellaneous

12.2.1 Singular Value Decomposition

Every matrix A with dimension $m \cdot n$ (m columns, n rows) can be decomposed into a product of three matrices in the form [22]

$$\underline{A} = \underline{U} \cdot \underline{S} \cdot \underline{V}^T = \begin{pmatrix} \cdots & \cdots & \cdots & \cdots \\ u_1 & u_2 & \cdots & u_m \\ \cdots & \cdots & \cdots & \cdots \\ \cdots & \cdots & \cdots & \cdots \end{pmatrix}$$
$$\times \begin{pmatrix} s_1 & \cdots & \cdots & \cdots \\ \cdots & s_2 & \cdots & \cdots \\ \cdots & \cdots & s_r & \cdots \\ \cdots & \cdots & \cdots & 0 \end{pmatrix} \begin{pmatrix} \cdots & \cdots & v_1^T & \cdots \\ \cdots & \cdots & v_2^T & \cdots \\ \cdots & \cdots & \cdots & \cdots \\ \cdots & \cdots & v_n^T & \cdots \end{pmatrix}$$
$$= u_1 s_1 v_1^T + u_2 s_2 v_2^T + u_3 s_3 v_3^T \ldots u_r s_r v_r^T \tag{12.94}$$

Here S is the diagonal matrix of the singular values

$$\underline{S} = \begin{pmatrix} s_1 & 0 & 0 & \cdots \\ 0 & s_2 & 0 & \cdots \\ 0 & 0 & s_3 & \cdots \\ \cdots & \cdots & \cdots & \cdots \end{pmatrix} \tag{12.95}$$

Usually, the singular values are sorted in descending order and this eigenfunction expansion shows a comfortable convergence if the first few terms in the sum carry the most important information. If the matrix A is symmetric with $n = m$, the row vector v and the column vector u of the matrices V and U form the Karhunen–Loeve basis. They build a functional system for expansion. If not, the dimensions of the three matrices have the dimensions $U(m,r)$, $S(r,r)$ and $V^T(r,n)$, which means that U and V are rectangular. The columns of U are the eigenvectors u_j. This is illustrated in Figure 12.20.

The $m \cdot m$ matrix U and the $n \cdot n$ matrix V are defined by

$$\underline{C}_{ACF} = \underline{A} \cdot \underline{A}^T = \underline{U} \cdot \underline{S}^2 \cdot \underline{U}^T$$
$$\underline{A}^T \cdot \underline{A} = \underline{V} \cdot \underline{S}^2 \cdot \underline{V}^T \tag{12.96}$$

with the auto-correlation matrix CACF. This means that the squared singular values s_{j2} are the eigenvalues of the auto-correlation matrix and the vectors u_j are the eigenvectors of CACF. The second row of Eq. (12.96) can be written as the outer product of the vectors u and v with the singular values as weights

$$\underline{A} = \underline{S} \cdot \vec{u} \otimes \vec{v} \tag{12.97}$$

The interpretation of this equation is: the matrix A with rank r is represented as an expansion into two rank r–1 matrices. If the singular values are sorted in descending

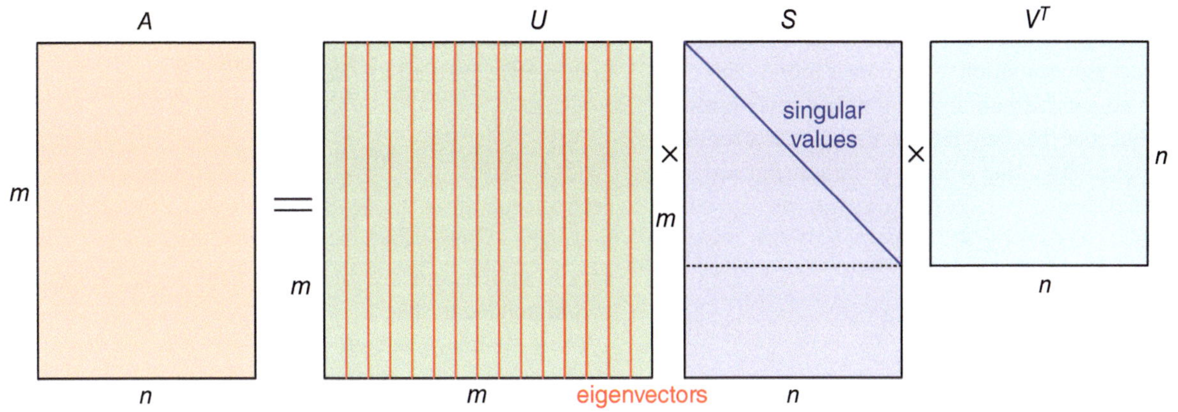

Figure 12.20 Schematic of the *SVD* matrix decomposition for general dimensions *m*, *n*.

order, this expansion can be truncated and the small neglectable contributions not used. This then means that the matrix A is approximately represented by the main contributions only. Due to the orthogonality of the eigenvectors u and v this decomposition is also orthogonal and represents the matrix A best in a non-least-square sense. If only a few singular values are essential, a large reduction in complexity is achieved. In this sense, the eigenvector representation is an optimal sparse representation of the information given by the matrix A.

The matrices U and V are orthogonal. If the Eq. (12.94) is multiplied by V from the right side, we get

$$\underline{A} \cdot \underline{V} = \underline{U} \cdot \underline{S} \tag{12.98}$$

A more descriptive explanation of the *SVD* decomposition is as follows: the matrix V describes a unit sphere, the diagonal matrix S stretches the main axis by the singular values into a hyper ellipse and finally the matrix U rotates the orthogonal ellipsoid into an arbitrary direction.

The *SVD* decomposes the arbitrary matrix A into singular values and two different bases with singular vector V from the right and matrix U from the left side. Any matrix can be decomposed into this type of product. The eigenvalue decomposition on the other side uses identical bases from left and right in the form

$$\underline{A} = \underline{X} \cdot \underline{\Lambda} \cdot \underline{X}^T \tag{12.99}$$

The eigenvalue matrix Λ is not the same as the singular value matrix S and not every matrix can be decomposed into eigenvalues and vectors. It should be mentioned that the number of non-zero singular values is identical to the rank of a matrix and that all singular values are positive.

12.2.2 Principal Component Analysis and Karhunen–Loeve Transform

The *SVD* decomposition, the principal component analysis (PCA) and the Karhunen–Loeve transform (KLT) are strongly related, but are not identical [22]. Mostly the *SVD* as a well-established and available tool to calculate the singular values is used to numerically compute the eigenvectors of a matrix A.

The principal component analysis is a method to express a stochastic or unknown functionality into an orthogonal eigenvector basis. This transform has the property to express the stochastic data with a minimized least square error. The functions are obtained out of the covariance matrix and are of statistical nature; they cannot be easily interpreted. The functions are uncorrelated and in this sense they are orthogonal. This decomposition is called the principal component analysis.

If x is a random vector with dimension n and mean $\overline{x}$ and a matrix A performs a linear transform into a vector y in a way such that the rows of A are the eigenvectors of the covariance matrix $C_x = C_{ACF}$ and the mean vanishes we get

$$\vec{y} = \underline{A} \cdot (\vec{x} - \overline{x}) \tag{12.100}$$

The covariance C_y of the new vector y is now given by

$$\underline{C}_y = \underline{A} \cdot \underline{C}_x \cdot \underline{A}^T = \begin{pmatrix} \lambda_1 & 0 & \dots & 0 \\ 0 & \lambda_2 & \dots & 0 \\ \dots & \dots & \dots & \dots \\ 0 & 0 & \dots & \lambda_n \end{pmatrix} \tag{12.101}$$

This new matrix due to the eigenvector property is diagonal and the eigenvectors are uncorrelated with vanishing off-diagonal elements. Each eigenvalue λ_j is the variance of y_j. If now the matrix and the eigenvalues are ordered in a descending manner, an approximate reduced representation can be obtained by discarding the higher order small value eigenvalues. The usefulness of the PCA representation is that due to the orthogonality of A, it can easily be inverted to calculate the original data back as

$$\vec{x} = \overline{x} + \underline{A}^{-1} \cdot \vec{y} = \overline{x} + \underline{A}^T \cdot \vec{y} \tag{12.102}$$

In particular it is of interest to simplify and reduce this calculation by shortening the dimensionality of A. Then a more compact representation of the data is found. This corresponds to an approximation with a rather small error. It can be shown that this new representation has a residual error in representation that is equal to the sum of the disregarded eigenvalues

$$\vec{x} \approx \vec{x}_{approx} = \underline{A}_{red}^T \cdot \vec{y} \tag{12.103}$$

$$\varepsilon = \mathrm{rms}(\vec{x} - \vec{x}_{approx}) = \sum_{j=j_{\max}+1}^{n} \lambda_j \tag{12.104}$$

Eq. (12.101) gives the computational receipt of the principal component analysis: the covariance matrix C of the observables x with zero mean must be constructed and the eigenvectors are constructed by the usual algorithms, for example *SVD*. They build the principal components functions.

The only difference between the PCA and the KLT is that the subtraction of the mean in the constitutive Eq. (12.100) is omitted in case of the KLT. The KLT is defined as the optimal decomposition of a matrix A into orthogonal components. This means that in the case of the PCA the coordinate system is shifted to the origin and the function has a resultant mean value of zero. The transform matrix of both methods is identical and the transforms can be written as

$$\begin{aligned} \vec{y}_{PCA} &= \underline{A} \cdot (\vec{x} - \overline{x}) \\ \vec{y}_{KLT} &= \underline{A} \cdot \vec{x} \end{aligned} \tag{12.105}$$

A simple method to calculate the eigenvectors or KL-basis function is using the covariance. If n measurements with m variables are considered, the matrix of all measurements can be formulated as matrix in the form

$$C_{\text{var}} = A_{mn} \cdot A_{mn}^T \tag{12.106}$$

where the matrix A contains the measurement vectors with normalized unit lengths. Then the diagonal elements are 1 and the off-diagonal elements describe the relative correlation of the measurements. The eigenvectors of the covariance matrix are then the KL functions.

12.2.3 Quasi-Random Sampling*

(*) The content of this section is quite specialized and for advanced reading. Knowledge of this section is not necessary to understand the following chapters. Beginners are recommended to skip this section.

There are many different algorithms proposed in mathematics to generate sequences of random numbers. In general two types of numbers must be distinguished [23–25]:

- pseudo random numbers that appear to be random, but are generated by a deterministic process; and pseudo random numbers that pass most of the randomness criteria, but not all;
- quasi random numbers that build so-called low discrepancy quasi random sequences. They are of interest in practice because they have a somewhat better uniform distribution, although they are not perfectly random. This gives better results and convergence in the case of integration and Monte Carlo calculations or numerical integrations.

This difference between both types is seen in Figure 12.21 in two dimensions and in Figure 12.22 as a result of a one-dimensional integration result very clearly. The quasi random point sets in two dimensions have no gaps or clusters. In one dimension they can be projected onto a uniform sequence with the same size and a nearly one-to-one correspondence is observed. The same can be done in two dimensions, approximately for every regular square one point of the quasi distribution is found [25]. The errors of pseudo random distributions reduce with the number N of samples by

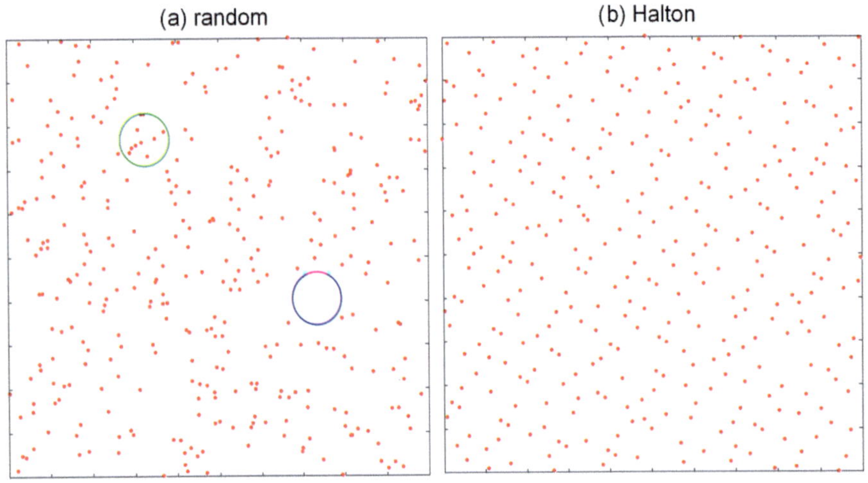

Figure 12.21 Comparison of random uniform and quasi random Halton distributions of points in two dimensions.

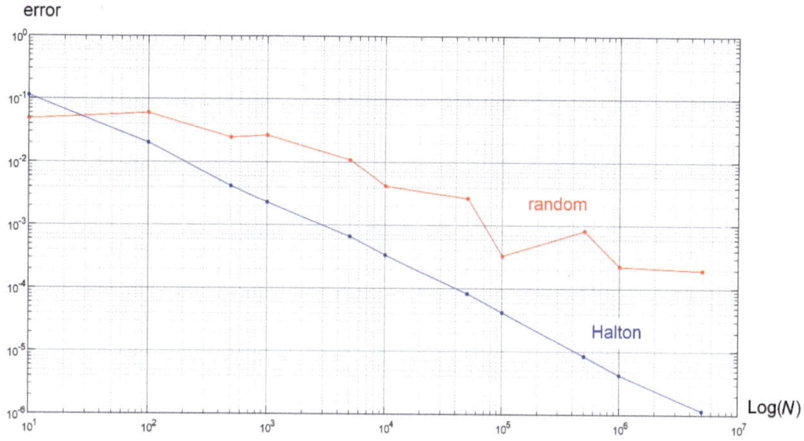

Figure 12.22 Error convergence of random and quasi random Halton distribution of points as a function of the point number N in one dimension.

$$\varepsilon_{rnd} = c \cdot \frac{1}{\sqrt{N}} \tag{12.107}$$

in contrast the quasi random Halton sequence error decays as

$$\varepsilon_{qrd}^{(Halt)} = c \cdot \frac{\log N}{N} \tag{12.108}$$

which is faster, see Figure 12.22. More usually the so-called discrepancy as a deviation from the uniformity or equi-distributed sequence is used to qualify a quasi random sequence. It describes and quantifies the deviation from a constant point density. In the case of a symmetrical distributed area between $-1 \ldots +1$ for x and y, the sum of all values should vanish. A deviation from the zero value measures the error of uniformity. This error is drawn in Figure 12.22.

There are several approaches to quasi random distributions after Halton, Sobol, Faure, Niederreiter, Fibonacci, Weyl, von Neumann, Hammersley and more. All of the quasi random sequences are in their original version deterministic and if the algorithms are called several times, as in the case of an ensemble averaging approach, the results are identical. Mostly a generalized recurrence formula is used to find a number from its previous number. Nearly all algorithms allow generation of sequences in several dimensions simultaneously. But it is found that for very high dimensions, all procedures create regular structures and the discrepancy grows.

Therefore, very recently, several proposals have been given in the literature to use some initial conditions in the algorithms to randomize their results for subsequent calls. There are different approaches for randomizing the initial or seed conditions of a quasi sequence. These are, for example, a nested scrambling, randomization of the generator matrix, shuffling, digital random permutation or random shifting. Shifting is the easiest solution. If $x_{j,qrd}$ is a sequence of one-dimensional quasi-random numbers between 0 and 1 und z a random number, the new sequence

$$y_{j,qrd} = \mathrm{mod}(x_{j,qrd} + z, 1) \tag{12.109}$$

is again quasi random.

12.2.4 Window Functions*

(*) The content of this section is quite specialized and for advanced reading. Knowledge of this section is not necessary to understand the following chapters. Beginners are recommended to skip this section.

An important step in many calculations involving the fast Fourier transform for diffraction modelling is the use of window functions. The window function $w(x)$ has the task of keeping the modified function unchanged as much as possible and forcing it to zero at the border. This avoids leakage errors in the discrete Fourier transform calculation. But every multiplication with a window function also changes the signals in the inner region and therefore acts as a filter in the frequency domain, which is convolved with the spectrum of the function. Therefore, in the best case the spectrum $S(\nu)$ of the window function looks similar to a Dirac delta function, not changing the spectrum of the product.

In the literature, many different window functions are proposed. Most are constructed according to the goals above. Some of the most common proposals are listed in thc following equation. The parameters a, α, m allow us to adapt the function for a given signal.

$$w(x) = \frac{1-\alpha}{2} + \frac{1}{2}\cos(\pi x) - \frac{\alpha}{2}\cos(2\pi x) \quad \textit{Blackman}$$

$$\begin{aligned} w(x) &= 0.3635819 + 0.4891775 \cdot \cos(\pi x) \\ &\quad - 0.1365995 \cdot \cos(2\pi x) \\ &\quad + 0.0106411 \cdot \cos(3\pi x) \quad \textit{Blackman–Nutall} \end{aligned}$$

$$\begin{aligned} w(x) &= 0.35875 + 0.48829 \cdot \cos(\pi x) \\ &\quad - 0.14128 \cdot \cos(2\pi x) \end{aligned}$$

$$
\begin{aligned}
&+ 0.01168 \cdot \cos(3\pi x) && \textit{Blackman–Harris}\\
w(x) &= 0.355768 + 0.487396 \cdot \cos(\pi x)\\
&- 0.144232 \cdot \cos(2\pi x)\\
&+ 0.012604 \cdot \cos(3\pi x) && \textit{Nutall}\\
w(x) &= 0.53836 + 0.46164 \cdot \cos(\pi x) && \textit{Hamming}\\
w(x) &= \frac{1}{2}(1 + \cos(\pi x)) && \textit{Hann}\\
w(x) &= \frac{I_0(\pi\alpha\sqrt{1-x^2})}{I_0(\pi\alpha)} && \textit{Kaiser}\\
w(x) &= 0.21557895 + 0.41663158 \cdot \cos(\pi x)\\
&- 0.277263158 \cdot \cos(2\pi x)\\
&+ 0.083578947 \cdot \cos(3\pi x)\\
&+ 0.006947368 \cdot \cos(4\pi x) && \textit{Flat–top}\\
w(x) &= e^{-\frac{1}{2}\left(\frac{x}{a}\right)^m} && \textit{SuperGauss}\\
w(x) &= \begin{cases} 1 & \textit{for } |x| \le a \\ \frac{1}{2}\left[1 + \cos\left(\pi \frac{|x|-a}{1-a}\right)\right] & \textit{for } |x| > a \end{cases} && \textit{Tukey}\\
w(x) &= e^{-\frac{1}{2}\left(\frac{x}{a}\right)^2} && \textit{Gauss}\\
w(x) &= 0.62 - 0.24 \cdot |x| + 0.38 \cdot \cos(\pi x) && \textit{Bartlett}\\
w(x) &= \cos\left(\frac{\pi}{2}x\right) && \text{Cos}\\
w(x) &= 1 - x^2 && \textit{Welch}\\
w(x) &= 1 - |x| && \textit{triangle}
\end{aligned}
\tag{12.110}
$$

Figure 12.23 shows these 16 window functions $w(x)$ in a normalized way in exactly the same order as in the formula. Figure 12.24 shows the spectra $S(\nu)$ in a similar representation, but a logarithmic ordinate.

Figure 12.25 shows all these functions ordered after the width of the spectrum for a relative decay of 10^{-4}. The widths at 10^{-5} and 10^{-6} are also indicated. The widths are normalized for a data window of diameter $L = 2x = 2$. It is seen that the decays are not the same. The fastest decay to 10^{-4} is obtained for the Blackman window; the fastest decay on 10^{-6} is seen for the Nutall window. What is not so directly seen in the spectral representation is the change of the original signal in the spatial domain, this is better observed in Figure 12.23. From this point of view, the Tukey and the super-Gaussian windows have the best performance.

In any case it can be seen that the very simple approaches, like triangle, cos or Tukey (parabolic) functions, have significantly worse behaviour in the spectral domain.

12.2.5 Fresnel Integrals*

(*) The content of this section is quite specialized and for advanced reading. Knowledge of this section is not necessary to understand the following chapters. Beginners are recommended to skip this section.

The Fresnel integral in its complex formulation is defined as [26]

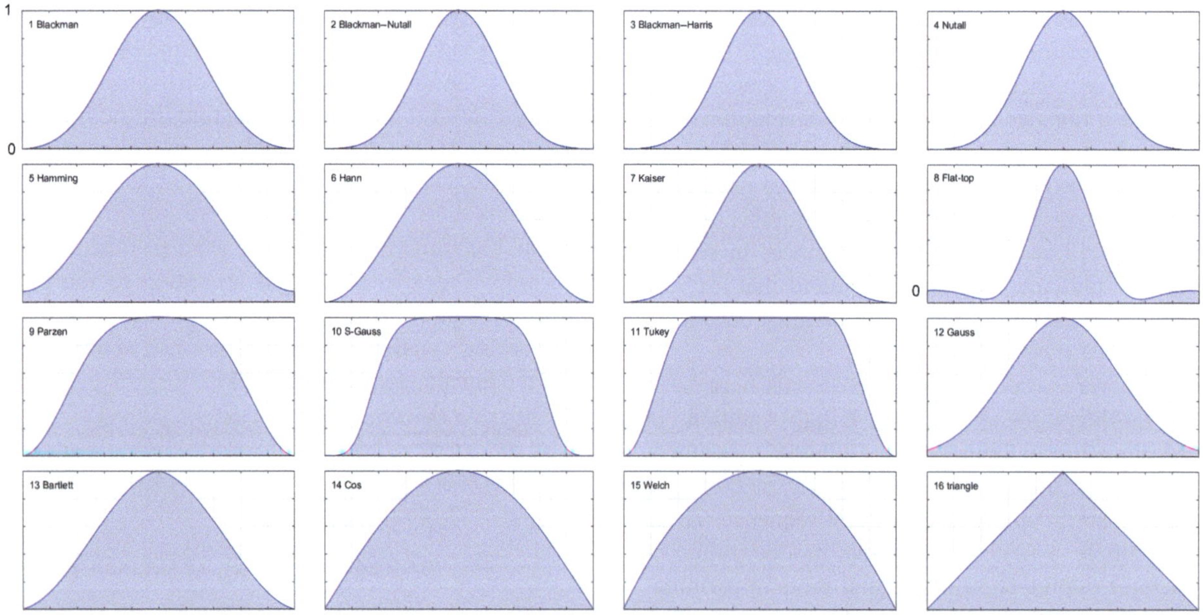

Figure 12.23 Window functions.

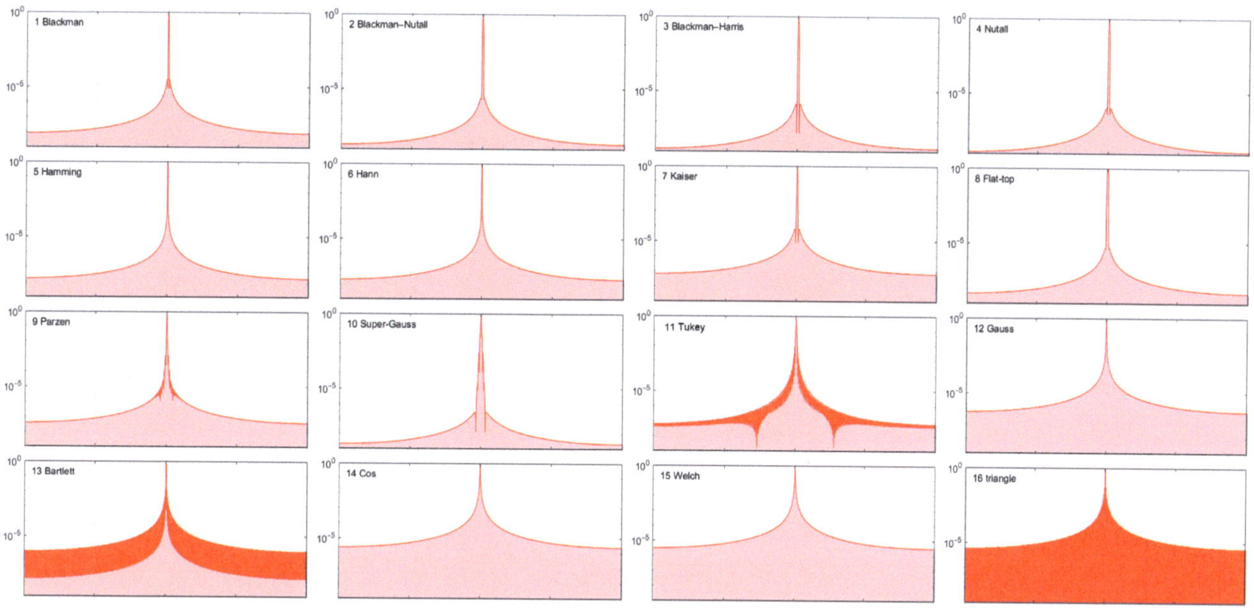

Figure 12.24 Spectrum of the window functions. The dark red regions have a strong and fast oscillating behaviour, therefore they look different because of dense lines.

Figure 12.25 Spectral widths of various window functions for three different relative thresholds.

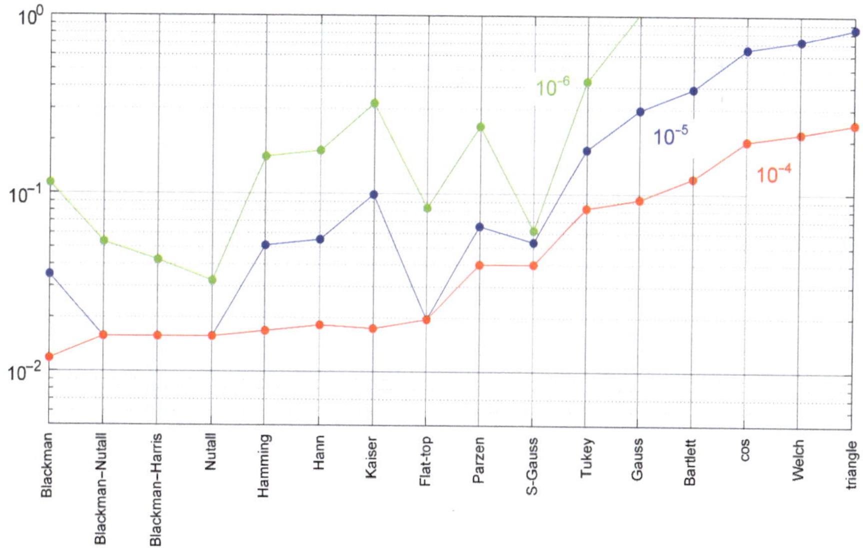

$$F(x) = \int_0^x e^{\frac{1}{2}i\pi t^2} dt \tag{12.111}$$

It is related to the complex Gaussian error function $erf(z)$ by

$$F(x) = C(x) + i \cdot S(x) = \frac{1+i}{2} \cdot erf\left[\frac{\sqrt{\pi}}{2} \cdot (1-i) \cdot x\right] \tag{12.112}$$

There are asymptotic approximations of the Fresnel integral valid in the limit of very small and large values of the argument, respectively:

- Small values of the argument $x \ll 1$

$$F(x) = x \cdot e^{\frac{1}{2}i\pi x^2} \tag{12.113}$$

- Large values of the argument $x \gg 1$

$$F(x) = \frac{i+1}{2} - \frac{i}{\pi x} \cdot e^{\frac{1}{2}i\pi x^2} \tag{12.114}$$

If the real and imaginary parts of the integral are separated

$$F(x) = C(x) + i \cdot S(x) \tag{12.115}$$

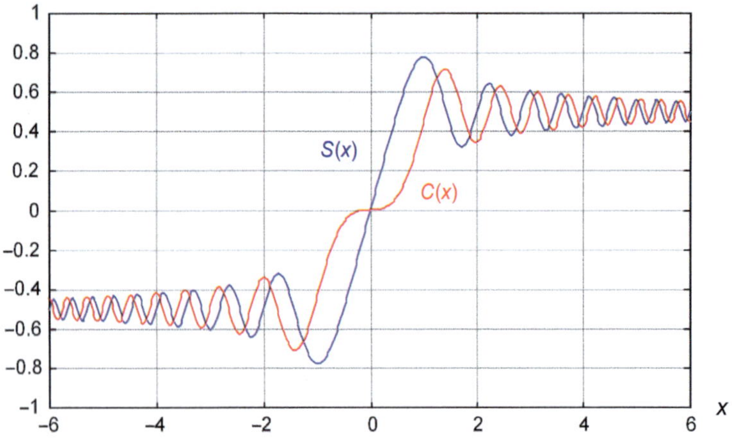

Figure 12.26 Real part $C(x)$ and imaginary part $S(x)$ of the Fresnel integral.

one obtains the functions $C(x)$ and $S(x)$

$$C(x) = \int_0^x \cos\left(\frac{1}{2}\pi\, t^2\right) dt, \quad S(x) = \int_0^x \sin\left(\frac{1}{2}\pi\, t^2\right) dt \tag{12.116}$$

The behaviour of the real and imaginary parts as functions of the real argument x are illustrated in Figure 12.26.

For coherent Fresnel diffraction of an infinite plane wave incident on a long edge, the intensity can be presented in terms of Fresnel integrals as

$$I(t) = \frac{1}{2} \cdot \left[\left(\frac{1}{2} - C(t)\right)^2 + \left(\frac{1}{2} - S(t)\right)^2\right] \tag{12.117}$$

where t is a scaled dimensionless argument related to the Fresnel number N_F:

$$t = \sqrt{\frac{k}{z \cdot \pi}} \cdot x = \sqrt{\frac{2}{\lambda \cdot z}} \cdot x = \sqrt{2N_F} \tag{12.118}$$

The Fresnel diffraction at an edge can be illustrated by means of the so-called Cornu spiral [27] shown in Figure 12.27. The amplitude of the diffracted field is decomposed into real and imaginary parts which equal the Fresnel integrals $C(t)$ and $S(t)$, respectively. The Fresnel integrals are plotted on the horizontal and vertical axes of a diagram. The resulting curve has a spiral shape with asymptotic points at (1/2, 1/2) and (−1/2, −1/2) for very large and small values of the argument t. These correspond to large separations from the geometrical shadow boundary in the light and dark regions, respectively. The amplitude of the field is given by the distance from a point on the curve to the starting point (−1/2, −1/2). The convergence towards the asymptotic points reflects the decay of the diffraction ripples for large separations from the edge.

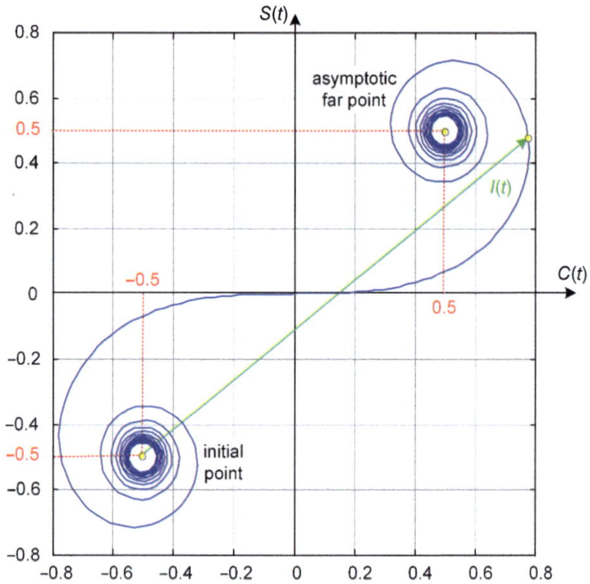

Figure 12.27 The Cornu spiral as a visualization of the Fresnel integrals.

12.2.6 Method of Stationary Phase*

(*) The content of this section is quite specialized and for advanced reading. Knowledge of this section is not necessary to understand the following chapters. Beginners are recommended to skip this section.

12.2.6.1 Introduction

If a one-dimensional integral of the form

$$I(x) = \int_a^b f(t) \cdot e^{ix \cdot g(t)} dt \tag{12.119}$$

is considered with a slowly varying factor $f(t)$ and a highly oscillating second factor with a large value of the parameter x, the method of the stationary phase claims that in the asymptotic approximation for large x the integral value is mainly dominated by the points where the first derivative

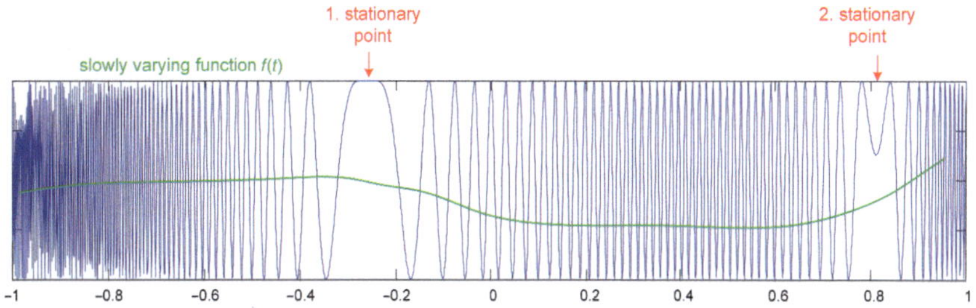

Figure 12.28 Principle of the method of stationary phase, the blue curve is the phase of the fast oscillating integrand.

of the function $g(t)$ is zero. Only here does the exponent not change quickly and the value of $f(t)$ contributes to the integral. This is sketched in Figure 12.28.

This heuristic explanation is derived in the next section and a quantitative formula gives the integral value under these conditions. Beneath the internal locations x with vanishing derivative the end points of the integral may contribute too. All the locations with a finite contribution are called critical points. The internal points are called type 1, the boundary points type 2. In two dimensions, the situation is more complicated, in particular the partial derivatives on the boundary curve of the integration area are of concern.

12.2.6.2 One-Dimensional Case

If an integral has a highly oscillatory integrand, it can be written in the form

$$I(k) = \int_a^b f(x) \cdot e^{ik \cdot g(x)}\, dx \tag{12.120}$$

By partial integration, the expression

$$I(k) = \left. \frac{f(x) \cdot e^{ik \cdot g(x)}}{ik \cdot g'(x)} \right|_{x=a}^{x=b} - \frac{1}{ik} \int_a^b \left[\frac{f(x)}{g'(x)} \right]' \cdot e^{ik \cdot g(x)}\, dx \tag{12.121}$$

is obtained. If inside the interval between $x = a \ldots b$ a stationary point is located at xs, the derivative of $g(x)$ vanishes here $g'(xs) = 0$ and a Taylor expansion can be therefore written as

$$g(x) = g(x_s) + \frac{1}{2} \cdot (x - x_s)^2 \cdot g''(x_s) \tag{12.122}$$

If this expression is inserted, the result for the type 1 critical points is obtained as

$$I^{(1)}(x) = f(x_s) \cdot \sqrt{\frac{2\pi}{x \cdot |g''(x_s)|}} \cdot e^{ix \cdot g(x_s) + \frac{i\pi}{4} \cdot sign[g''(x_s)]} \tag{12.123}$$

In this approximation the dominant contribution to the original integral therefore comes from the stationary point of a locally weakly changed phase. The contributions of all other regions of x are cancelled out due to the fast oscillations with positive and negative values. The assumption for large k indicates for the internal points that the order of approximation is proportional to $1/\sqrt{k}$.

If the end points $x = a, b$ of the integral are considered, a simple mathematical estimation of their contribution to the integral gives

$$I^{(2)}(k) = \frac{1}{ik} \left[\frac{f(b)}{g'(b)} \cdot e^{ikg(b)} - \frac{f(a)}{g'(a)} \cdot e^{ikg(a)} \right] \tag{12.124}$$

for these type 2 critical points. The approximation scaling goes here with $1/k$, therefore the type 2 points have a smaller contribution to the integral value.

12.2.6.3 Two-Dimensional Case

In two dimensions x and y the double integral

$$I(x,y) = \iint_A f(x,y) \cdot e^{ik \cdot g(x,y)} dxdy \tag{12.125}$$

over area A is considered [27]. If the partial derivatives of second order are used, parameter D is defined as

$$D = g_{xx} g_{yy} - g_{xy}^2 \tag{12.126}$$

The internal critical points (x_s, y_s) of type 1 are given by the condition similar to that in one dimension by the vanishing first derivatives

$$g_x(x_s, y_s) = g_y(x_s, y_s) = 0 \tag{12.127}$$

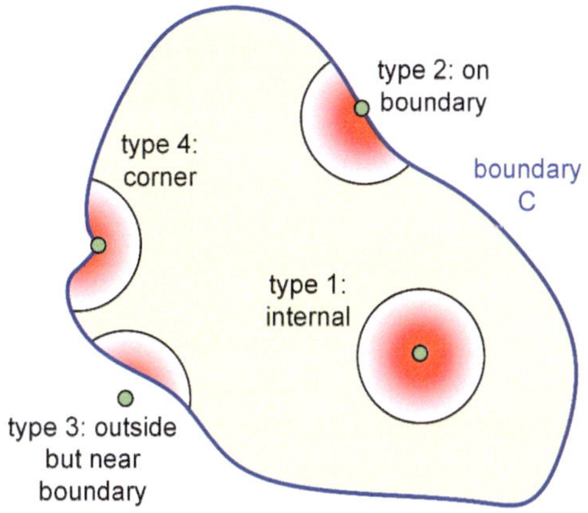

Figure 12.29 Types of critical points in stationary phase in two lateral dimensions.

The value of the integral contribution is then

$$I^{(1)}(k) = \frac{2\pi i\sigma}{k\sqrt{|D|}} f(x_s, y_s) \cdot e^{ik \cdot g(x_s,y_s)} \tag{12.128}$$

with $\sigma = \pm 1$ depending on the sign of D and the shape of the extreme point.

The determination of the other types of stationary points in two dimensions is more complicated. The easiest case is a type 2 point, where on the boundary the direction derivative vanishes along the boundary curve C and all derivatives are smooth and continuous.

Through a detailed analysis it is further found that there are more possibilities for contributions. So-called type 3 critical points are obtained at a boundary corner with a non-continuous tangent direction. Figure 12.29 sketches these different options. Similar to the one-dimensional case, the relative approximation orders are different, type 1 behave as $1/k$, type 2 have $1/k^{3/2}$, type 3 have $1/k^2$.

References

1 Papoulis, A. and Pillai, S. (2002). *Probability, Random Variables and Stochastic Processes*. McGraw Hill.

2 Owen, A. (2017). A randomized Halton algorithm in R. arxiv, 1706.02808. doi:10.48550/arXiv.1706.02808.

3 Strang, G. (2007). *Computational Science and Engineering*. Wellesley-Cambridge Press.

4 Bracewell, R. (1995). *Two-Dimensional Fourier Transform*. Prentice Hall.

5 Goodman, J. (2017). *Fourier Optics*. Freeman and Company.

6 Castleman, K. (1996). *Digital Image Processing*. Prentice Hall.

7 Bose, T. (2004). *Digital Signal and Image Processing*. Wiley.

8 Bracewell, R. (1986). *The Fourier Transform and its Applications*. McGraw-Hill.

9 Nussbaumer, H. (1982). *Fast Fourier Transform and Convolution Algorithms*. Springer.

10 Barret, H. and Myers, K. (2004). *Foundations of Image Science*. Wiley.

11 Rabiner, L.R., Schafer, R.W., and Rader, C.M. (1969). The chirp z-transform algorithm. *IEEE Transactions on Audio* 17: 86.

12 Committee, D.S.P. (ed.) (1979). *Programs for Digital Signal Processing*. New York: IEEE Press.

13 Roose, S., Brichau, B., and Stijns, E.W. (1993). An efficient interpolation algorithm for Fourier and diffractive optics. *Optics Communication* 97: 312.

14 Bakx, J.L. (2002). Efficient computation of optical disk readout by use of the chirp z transform. *Applied Optics* 41: 4897.

15 Asoubar, D., Zhang, S., Wyrowski, F., and Kuhn, M. (2012). Parabasal field decomposition and its application to non-paraxial propagation. *Optics Express* 20: 23502.

16 Asoubar, D., Zhang, S., Wyrowski, F., and Kuhn, M. (2014). Efficient semi-analytical propagation techniques for electromagnetic fields. *Journal of the Optical Society of America A* 31: 591.

17 Wang, Z. (2020). *Fourier Transform Techniques for Fast Physical Modeling*. Doctoral thesis, University of Jena.

18 Greengard, L. and Lee, J.-Y. (2004). Accelerating the nonuniform fast Fourier transform. *Society for Industrial and Applied Mathematics Review* 46: 443.

19 Fessler, J. and Sutton, B. (2003). Nonuniform fast Fourier transforms using min-max interpolation. *IEEE Transactions on Signal Processing* 51: 560.

20 Kuo, J.-T. and Lee, H.-Y. (2009). Improved accuracy factors for the nonuniform fast Fourier transform (NUFFT) algorithm. *IEEE Microwave and Wireless Components Letters* 19: 3.

21 Shimobaba, T., Kakue, T., Oikawa, M. et al. (2013). Nonuniform sampled scalar diffraction calculation using nonuniform fast Fourier transform. *Optics Letters* 38: 5130.

22 Gerbrands, J. (1981). On the relationships between SVD, KLT and PCA. *Pattern Recognition* 14: 375.

23 Bhat, C. (2003). Simulation of mixed discrete choice models using randomized and scrambled Halton sequences. *Transportation Research Part B* 37: 837.

24 Wang, X. and Hickernell, F. (2000). Randomized Halton sequences. *Mathematical and Computer Modelling* 32: 887.

25 Cohen, L. (1995). *Time-Frequency Analysis*. Prentice Hall.

26 Abramowitz, M. and Stegun, I. (1972). *Handbook of Mathematical Functions*. Dover.

27 Born, M. and Wolf, E. (1999). *Principles of Optics*. Cambridge: Cambridge University Press.

Index

Lens Design for Imaging: Fundamentals of Optical Systems, First Edition. Herbert Gross.